Family
Encyclopedia

Family Encyclopedia

Contents

Astronomy   7

Geography and Geology   21

Natural History   45

Science and Technology   63

Computer Terms   147

Sport   203

Art, Music and Literature   223

Famous People   287

British Kings and Queens   371

World History   445

World Facts   517

Space and Astronomy

For many centuries observing space and questioning the origins and evolution of the universe has provided mankind with one of its great challenges. Observational records exist which date from the ancient civilizations of China, Babylonia and Egypt, yet even with the modern development of deep-space probes and powerful radio-telescopes, there are still many questions which remain unanswered.

This section provides an up-to-date dictionary of terms commonly used in the study of space and astronomy. The entries are arranged A to Z with cross-references given in SMALL CAPITALS.

*

antimatter is matter that is made up of antiparticles, that is particles that have the same mass as ordinary matter but have opposite values for other properties (e.g. charge). So if matter is made up of electrons, neutrons and protons, antimatter is composed of *positrons*, anti-neutrons and anti-protons. The coming together of matter and antimatter results in the destruction of both particles and the production of energy.

asteroid alternatively called a *meteorite*, is actually a *planetoid*; a rocky or metallic body that is one of many orbiting the Sun between the orbits of Mars and Jupiter. Most of these tiny planets are very small, some reaching a few hundred kilometres across but the largest, *Ceres*, has a diameter of about 1000 kilometres. Collisions between asteroids sometimes produce smaller pieces that occasionally fall to Earth and these fragments are also called meteorites. The many craters seen on the surfaces of some planets are probably due to asteroid impact.

Meteors, or *shooting stars*, are the remnants of comet debris, which should they fall into the Earth's atmosphere, burn up as they are destroyed by friction (*see also* COMET).

astronaut is someone who has undergone very special and rigorous training to prepare for space travel.

The first astronaut was Yuri Gagarin who orbited the Earth in 1961 and his spacecraft *Vostok I* took 98 minutes for one orbit. Valentina Tereshkova was the first woman in space. Both these astronauts were from the former Soviet Union now the CIS.

July 1969 saw an American team actually landing on the Moon and since then there have been numerous missions by the CIS to an orbiting SPACE STATION, and America has launched several SPACE SHUTTLES, the first reusable vehicle capable of space flight.

When a rocket is launched from Earth there are tremendous forces pushing down on the astronaut, due to the enormously high acceleration required to reach orbit. An astronaut must therefore be very fit, and highly skilled in science and technology so that he or she can operate the equipment in the spacecraft and carry out numerous experiments in space. The early astronauts were test pilots from the armed forces but today more scientists and

doctors are involved. Over 400 people have now flown in space although only a small proportion are women. To help train astronauts for the weightlessness of space, special aeroplanes are used. These dive at high speed from a great height to produce a weightless effect for a short time. When actually in space it is necessary to wear a space suit which is specially designed to cater for the astronaut's safety. The suit is made up of layers of special material with tubular networks: it provides protection from radiation, the cold of space and supplies air from a portable pack (or life support system). The helmet has a visor to cut out harmful rays from the Sun and a microphone with headphones to allow communication with other astronauts on the spacecraft.

astronomy is the scientific study of the various bodies in the 'heavens' and has been in existence for thousands of years. From the earliest times, people have plotted the positions of the stars and planets, devising a yearly calendar governed by the movements of the Earth and Moon relative to the Sun. When making a journey, people in former times always used the positions of the various heavenly bodies to guide the direction which they took. Hence some knowledge of astronomy was very important to early travellers and explorers. Early astronomers gained a great deal of knowledge from their studies and were able to predict the appearance of COMETS and ECLIPSES. It became possible to make many more discoveries in astronomy following the invention of the lens telescope, in 1608. This was devised by a Dutchman, Hans Lippershey, who was a maker of spectacles by profession rather than an astronomer!

Up until the Middle Ages, it was thought that the Earth was at the centre of the universe and that other heavenly bodies revolved around it. Nicolaus Copernicus, during the 1540s, put forward the idea (which had first been taught, in 170 BC, by Aristarchus) that the Earth travelled around the Sun. The idea of the Sun being the centre of the SOLAR SYSTEM, gradually gained acceptance.

In the latter half of the 20th century, SATELLITES and space probes have given us much new and exciting information about the farthest regions of space. For example, modern astronomers have spent (and continue to spend) many years studying the wealth of data sent back to Earth by such space probes as *Viking* 1 and 2 and *Voyager* 1 and 2.

aurora this is the luminous and often highly coloured streaks or sheets of light in the sky, formed by high-speed, electrically charged solar particles entering the upper atmosphere of the Earth. Here, the particles give up electrons and molecules are created associated with the release of light. These effects are related to SUNSPOT activity and are called the *Northern Lights* (aurora borealis) in the northern hemisphere, and the *Southern Lights* (aurora australis) in the southern hemisphere.

big bang is a theory which has been put forward by some scientists to explain the origins of the universe. It proposes that about 15,000 million years ago there was a vast explosion which sent all the material spinning rapidly outwards as a mass of atoms. Physicists think that in a matter of minutes, a huge cloud of hydrogen atoms had formed and continued to spread outwards, dispersing into separate clouds as it did so. These hydrogen atoms make up about 90% of the universe, and the clouds formed GALAXIES of stars including the MILKY WAY. The outward expansion is still continuing but it is thought that this may one day cease. All the material may eventually collapse and collide together (the BIG CRUNCH) but it is thought that before this occurs, the Sun will burn hotter and the Earth will be consumed in about 5000 million years from the present.

There is much speculation about the origin of the dust and gases which were scattered at the time of the big bang, and also the possible cause of the huge explosion.

big crunch gravity attracts the galaxies in the Universe together, and the force it exerts depends on their mass, but it is over-ridden by the expansion and moving apart which is still continuing, possibly as a result of the BIG BANG. The Big Crunch theory proposes that there may be hidden material (consisting of single atomic particles) between the galaxies which, if present in sufficient quantities, could add to the gravitational pull. This extra gravity might be enough to slow down and halt the expansion and indeed put it into reverse. If this occurred the galaxies would be pulled together, collide and be destroyed in the Big Crunch.

binary star a pair of stars held together by the force of gravity which orbit around each other. These double stars may be so close to one another that each goes round the other in a few hours. In this instance they are so close as to be almost touching and travel very fast. Others are vast distances apart and travel around each other in thousands or even millions of years. Many stars in the MILKY WAY are binary stars and each member of the pair may differ in shape, size and brightness.

black hole it is thought that there are black holes in space which are collapsed stars, and the proposal for these was first put forward by Karl Schwarzschild, a German astronomer, in 1916. It is believed that when a star (which may be many times larger than our Sun), comes to the end of its 'life,' it collapses so quickly that no material, not even light, can be emitted. This type of event is known as an *implosion* and because of the enormous gravitational force of the collapsed material, it is seen as a hole from which no light escapes. The space around the black hole is curved into a complete circle. It is thought that many apparent single stars are in fact BINARY STARS with a black hole as a partner. Much recent work on black holes has been carried out by Stephen Hawking, a British physicist (*see also* NEUTRON STAR).

calendar calendars were devised by Man to mark the passage of time, and were devised by early astronomers according to the phases of the Moon or on the orbit of the Earth around the Sun. The 365 day year was based on the *Julian* calendar introduced by the Roman Emperor, Julius Caesar, in 46BC. In fact, the Earth takes $365\frac{1}{4}$ days to go once around the Sun (a solar year) so on this system, every fourth or *leap* year, an extra day has to be added. In fact, the calendar year is still inaccurate and several proposals have been put forward for its reform.

comet is composed of rock, ice, dust and frozen gases and usually cannot be seen. A comet becomes visible when it approaches closer to the Sun because it appears to flare up and look bright. It has a *head*, consisting of a solid core or nucleus of rock surrounded by a layer of gas, the *corona*. The nucleus is irregular in shape and just a few kilometres in diameter (In fact Halley's comet is elongated and is roughly 15 x 4km). A comet moves in a long, eccentric orbit around the Sun and one complete cycle is called a *period*. The orbit may take the comet a long distance from the Sun when it cannot be seen. Near to the Sun, particles of dust and gas are knocked away from the head of the comet and trail out to form a brightly glowing *tail*. When the orbit takes the comet away from the Sun it leaves tail first. Comets are believed to be left over remnants of the BIG BANG explosion and each time they approach the Sun, some of the ice melts. Eventually, it is thought that a comet loses its ice, dust and gas and becomes a piece of orbiting rock. The shortest period belongs to Ericke's comet and is only 3½ years. Others have much longer periods lasting hundreds or even thousands of years (e.g. Ikeya-Seki, 880 years).

The most familiar comet, which has excited interest for a long time, is Halley's comet which appears every 76 years and is next due in 2062. In 1986 the spacecraft, *Giotto*, belonging to the European Space Agency, flew to within 539 km of the comet, gaining valuable information and photographs, until its cameras were obscured by dust.

constellation a group of stars which are placed together and given a name, although there is no scientific basis for the grouping. Many were named by astronomers of the Babylonian empire some 2000 years before the birth of Christ and the most ancient form the twelve signs of the zodiac. Thirty nine constellations are visible in the northern hemisphere and forty six in the southern. Familiar examples in the northern hemisphere include the Great Bear (*Ursa Major*), the Bull (*Taurus*), the Hunter (*Orion*) and the Pole or North Star.

corona the coloured rings (seen through thin cloud) around the Sun or Moon, caused by DIFFRACTION of light by water droplets.

cosmic rays these are 'rays' composed of ionizing radiation from space and consisting mainly of protons (*see* ATOM) and alpha particles (helium nuclei) (*see* RADIOACTIVITY), and a small proportion of heavier atomic nuclei. When these particles interact with the Earth's atmosphere, secondary radiation is created, including mesons, neutrons (*see* ELEMENTARY PARTICLES), positrons (*see* ANTIMATTER) and electrons (*see* ATOM). Cosmic rays seem to come from three sources: galactic rays, possibly formed from SUPERNOVA explosions; solar cosmic rays and the SOLAR WIND.

daylength the length of daylight changes throughout the year except on the equator where there are approximately 12 hours of light and 12 hours of darkness. The changes occur because the Earth is tilted on its axis, and it alters on a daily basis as the Earth proceeds on its orbit around the Sun. At any point in the year, while countries in the northern hemisphere are tilted more towards the Sun, those in the south are further away and vice versa. This governs the onset of the four different seasons of the year. Latitude also affects daylength, and those countries at latitudes greater than 67° north or south experience the midnight sun during midsummer when it does not become dark at all. The lengths of day and night are the same throughout the world on the two equinoxes (21th March and 23rd September).

Earth the Earth is one of a group of nine planets which orbit a star that we call the SUN. It occurs in one small part of the GALAXY (also called the MILKY WAY) which is referred to as the SOLAR SYSTEM. The Earth is the fifth largest of the planets and the third nearest to the Sun. Most of the Earth's surface is covered by the oceans – about two-thirds compared to one third which is land. A layer of air, the ATMOSPHERE, surrounds the Earth which is sub-divided into a number of different regions according to height above the surface. The Earth, as viewed from space, is like a blue ball or sphere which is flattened at both the poles. At the equator, the radius of the Earth is 6378 km or 3963 miles.

The Earth itself is composed of four main layers; an outer *crust* overlying a *mantle*, followed by an *outer core* and *inner core*. The rocky crust varies in thickness, being about 40 km thick under the continents and 8 km under the oceans. It 'floats' on the mantle which is made up of extremely hot rock and is about 2900 km thick (*see* PLATE TECTONICS). The core is so hot that it is part liquid, with a temperature in the region of 6500 K. The outer core is believed to be about 2250 km thick and is thought to be more liquid than the inner core which is solid and composed mainly of iron and nickel. The Earth is the only one of the nine planets able to support life which exists entirely in a narrow band on or near the crust. The Earth's weight increases every day by about 25 tonnes because fine space dust finds its way down to the surface.

eclipse the name given to a total or partial disappearance of a planet or moon by passing into the shadow of another. A *solar eclipse* occurs when a new moon passes between the Earth and the Sun.

A solar eclipse

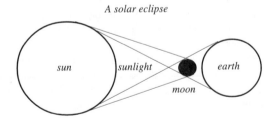

A *lunar eclipse* happens when the Sun, Earth and Moon are in a line and the Earth is situated between the Moon and the Sun and its shadow falls on the Moon. Different types of eclipse happen each year but a solar eclipse usually appears as partial when viewed from Earth. Even during a complete solar eclipse, the Moon is only just large enough to cover the Sun, and the shadow which it throws onto one part of the Earth is only a few kilometres wide so few people are able to see it. It is generally the case that the Moon is not completely in line and so the eclipse is only partial.

escape velocity this is the velocity necessary for a rocket or space probe to escape from the pull exerted by the gravity of a planet. The velocity varies according to the mass and diameter of the planet. For the Earth it is 25,000 mph and for the Moon, 5350 mph. No light escapes from a black hole because the escape velocity is greater than the speed of light.

expansion theory the theory that the galaxies of space are still rapidly moving apart which may have been caused by the BIG BANG explosion. The speed at which the galaxies are travelling is calculated by studying a property of light known as the *red shift*. For every extra million light years of distance (one light year is around 9.5 million million kilometres), the speed the galaxies are travelling increases by about 15 km per second. The relationship between speed and distance is called *Hubble's law*.

galaxy the name given to a huge group or band of stars. Our galaxy, which includes the Sun, is also called the MILKY WAY and it contains about 100,000 million stars. It is thought that there are many million more galaxies as well as the Milky Way, and the nearest large one to us is called *Andromeda*. Galaxies have different shapes, the Milky Way and Andromeda being spiral-shaped, others appearing as saucers. The Milky Way is approximately 10^5 light years across.

heat shield a spacecraft encounters extreme and intense heat especially when it re-enters the Earth's atmosphere from space.

The heat is caused by friction because the space ship is travelling so fast. A heat shield is a special heat resistant layer which prevents the spacecraft from burning up. In earlier space ships the shields were constructed from materials that could not be re-used, as they burned but only very slowly. The space shuttle has special heat-resistant tiles composed of silicon and carbon which can be used again.

implosion is the opposite of an explosion. It is a rapid inrushing or collapse of materials inwards instead of outwards. Stars may *implode*, the gravitational force pulling all materials inwards. These collapsed stars are very dense in the centre and form a BLACK HOLE or NEUTRON STAR.

infrared astronomy the study of infrared radiation (*see* ELECTROMAGNETIC WAVES) given off from heavenly bodies such as the stars and planets. Infrared radiation has a wavelength of 0.8 to 1000µm which is shorter than that of radio waves and longer than that of visible light. Infrared radiation passes through parts of the galaxy obscured by dust, highlighting otherwise invisible structures. More infrared radiation is emitted by hotter structures than by colder ones, and it penetrates the Earth's atmosphere more readily than visible light. The radiation tends to be obscured by water vapour in the Earth's atmosphere but this problem can be partially overcome by siting observatories at high altitudes. Hence infrared astronomers are able to gain valuable information about distant space bodies and have discovered 'young' stars which have not begun to glow brightly.

interstellar medium the matter found throughout space and comprising gas and dust – mainly hydrogen, helium and interstellar molecules. In all it amounts to about 10% of the mass of the galaxy.

Jupiter is the giant planet of our SOLAR SYSTEM, more than one thousand times larger than Earth. It is one of the 'gas giants' being composed mainly of hydrogen. Its atmosphere is made up of hydrogen with approximately 15% helium and traces of water, ammonia and methane. This forms a liquid 'shell' surrounding a zone of metallic hydrogen (that is the hydrogen is compressed so much that it behaves like metal), which itself surrounds a core made partly of rock and ice. This core has a mass ten times greater than that of the Earth. Violent storms and winds rage around Jupiter whipping up bands of frozen chemicals such as ammonia. One such storm is the *Great Red Spot,* visible on the surface as an enormous cyclone that has probably lasted for hundreds of years. Jupiter spins very rapidly so that one of its days lasts for only nine hours and fifty minutes. This rapid spin drags the whirling gases into bands which appear dark and light. A year on Jupiter lasts for nearly 12 Earth years because the planet is farther from the Sun and has a greater orbit. Jupiter is the fifth planet from the Sun and because of its rapid rate of spin, it bulges outwards at its equator. Hence the diameter at the equator is 142,800 km compared to 134,000 km at the poles. The outermost layers of Jupiter are very cold, in the region of -150°C, but the very centre of the planet is extremely hot probably exceeding the temperature of the Sun. Jupiter's great mass means that it exerts a strong gravitational pull and is able to hold down the molecules of gas which swirl around its bulk. A person on Jupiter would be twice as heavy compared to his or her weight on Earth.

Jupiter has its own satellites or moons orbiting around it and some of these are as large as Earth's moon. The *Voyager* spacecraft passed close to Jupiter in 1979 and sent fascinating information about the planet and its moons back to Earth. This revealed that two of the moons, called *Ganymede* and *Callisto*, have craters pitting their surface like Earth's moon. Another moon, *Europa*, was shown to be a ball of yellow ice. The closest moon to Jupiter, *Io*, has several erupting volcanoes and a surface of yellow sulphur. Enormous electrical energy exists between Io and Jupiter estimated to be equivalent, in any one second, to all the electrical power generated in the United States.

light year a measure of the distance travelled by light in one year, which is in the region of 9.467 x 10^{12} kilometres. Light travels at 299,792 kilometres per second and it takes 8 minutes for it to reach Earth from the Sun. The great distances between the various galaxies and heavenly bodies can be measured in light years. The nearest star to Earth (after the Sun), is more than four light years distant and is called *Proxima Centauri* (*see also* PARSEC).

magellanic clouds are two separate GALAXIES, detached from the MILKY WAY which appear, when viewed from the southern hemisphere, as diffuse patches of light. The largest is approximately 180,000 LIGHT YEARS away and the smallest is 230,000 light years distant. Both contain several thousand million stars. They are the nearest galaxies and part of a larger group which includes the Milky Way. They were first discovered by a Portuguese sailor called Magellan in 1520 and can only be viewed from near the equator or in the southern hemisphere. NEBULAE are present in both clouds indicating the formation of new stars.

Mars is the fourth planet in the SOLAR SYSTEM and the one nearest to the Earth. Its orbit lies between that of the EARTH and JUPITER, and it is about half the size of Earth. It has a thin atmosphere, exerting a pressure less than one hundredth of that of the Earth. It also has a small mass about one tenth of that of the Earth so that a person on Mars would weigh about 60% less than Earth weight. Mars is often called the *Red Planet* as it has a dusty, reddish surface strewn with rocks. It is much colder than the Earth with an atmosphere mainly of carbon dioxide which is frozen at the two poles. The polar ice caps melt and re-form as the seasons change. Minimum surface temperatures are in the region of -100°C with the maximum only about -30°C. A year on Mars lasts for 687 Earth days and the length of one day is almost the same, 24 hours and 37 minutes. The diameter of Mars is 6794 km and the crust in the northern part of the planet is composed of basalt (volcanic rock, *see* IGNEOUS ROCKS). There are many extinct volcanoes, canyons and impact craters and evidence of water erosion at some stage in the planet's history. The mountains are much higher, and the valleys deeper than those which exist on Earth, hence there must have been violent movements of the crust during the past. The deepest valley called *Valles Marineris*, is 4000 km long, 75 km wide and up to 7 km deep. The highest mountain, *Olympus Mons*, rises 23 km from the surface of Mars, and is three times taller than Mount Everest.

There are two small, irregularly shaped SATELLITES or moons orbiting Mars, called *Phobos* and *Deimos*. Phobos, the larger, is only 27 km from one end to the other and just 6000 km above the planet's surface. Its orbit is a gradually descending spiral and it is estimated that in 40 million years' time it will collide with Mars. The *Viking* space probe landed on Mars in 1976 and many valuable photographs were taken and soil samples obtained. There had been speculation for many years about the possibility of the existence of life on Mars but the space probe failed to discover any evidence for this.

matter any substance that occupies space and has mass, the material of which the UNIVERSE is made.

megaparsec a unit for defining distance of objects outside the galaxy, equal to 10^6 PARSECS (or 3.26 x 10^6 light years).

Mercury the first planet of the SOLAR SYSTEM and nearest to the SUN. It has no atmosphere, and so during the day the surface temperature reaches 425°C (enough to melt lead), but at night the heat all escapes and it becomes intensely cold, -170°C. A day on Mercury lasts for 59 Earth days but the planet travels its orbit so fast that a year is only 88 days. Mercury is a very dense planet for, although it is only slightly bigger than the Moon, it has an enormous mass which is almost the same as that of the Earth. It is thought that this is accounted for by Mercury having a huge metallic core. Very little was known about the surface of the planet until it was visited by the *Mariner 10* spacecraft which passed to within 800 km of it in 1974. It revealed that Mercury has a wrinkled surface with thousands of craters which have been caused by the impact of meteors and other larger space bodies. The largest crater, 1300 km across and known as the *Caloris Basin*, must have been caused by the collision of an enormous space body. Mercury has very little gravity and an elliptical orbit, which takes it to within 46 million kilometres of the Sun at its nearest point and 70 million kilometres when farthest away.

Milky Way the galaxy, which includes the SOLAR SYSTEM, and consists of millions of STARS and NEBULAE. It stretches for about 100,000 light years and our solar system is in the region of 30,000 light years from the centre. It has a spiral shape with trailing *arms* which slowly revolve around the centre. The solar system is one tiny part occurring a long way out on one of the trailing spiral arms. There are probably millions of galaxies, including others with the same spiral shape. The Milky Way appears as a glowing cloud of light thrown out by the many millions of stars it contains.

Moon the Earth's one satellite, which orbits the Earth at an average distance of 384,000 kilometres (238,600 miles). It has no atmosphere, water or magnetic field, and surface temperatures reach extremes of 127°C and -173°C. It takes nearly 28 days to complete its orbit around the Earth, and always presents the same face towards Earth. As it orbits around the Earth each 28 days, the Moon passes through *phases* from new to first quarter to full to last quarter and back to new again. One half of the moon is always in sunlight and the phases depend upon the amount of the lit half which can be seen from Earth. A *new moon* occurs when the Earth, Moon and Sun are approximately in line with one another, and none of the lit half can be seen at all; it appears that there is no Moon. About a week later a small *sliver* of the lit half can be seen from Earth and this grows throughout the month until the whole of the lit half is visible at *full moon*. In the second half of the cycle, the amount of the lit half of the Moon which can be seen from Earth gradually declines once more. When the Moon is apparently growing in size, it is called *waxing* and when it is declining it is called *waning*.

The diameter of the Moon is 3476 kilometres and its mass is 0.0123 of that of the Earth. Its density is 0.61 of that of the Earth and it has a thick crust (up to 125 kilometres or 75 miles) made up of volcanic rocks. There is probably a small core of iron, with a radius of approximately 300 kilometres (186 miles).

The surface of the Moon is heavily cratered, probably due to meteorite impact, and the largest (viewed from Earth), is 300 kilometres (186 miles) across and surrounded by gigantic cliffs, (up to 4250 metres or 14,000 feet high). The dark side of the Moon was a mystery until 1959 when *Luna 3*, a Russian space rocket took photographs of it as it flew round. The surface is dry and rocky and in 1969, the American astronauts Neil Armstrong and Edwin Aldrin made the first human footprints on its dusty surface. Other spacecraft have since landed and brought back samples of lunar rock and soil, and analysis of these has revealed that the Moon must be at least 4000 million years old. A distinctive feature of the Moon is its *maria (*sing. *mare)* which were once thought to be seas. The name was coined before modern study found them to be dry, but their origin is not yet established although it is thought that they date from 3300 million years ago. The tendency now is to dispense with the Latin, e.g. *Sea of Showers* instead of *Mare Imbrium.*

nebula (pl. *nebulae)* a cloud of interstellar matter, consisting of gases and dust, in which stars originate and also die. When stars die, their gases and dust are poured back out into space forming *planetary nebulae.* Dark patches in the Milky Way galaxy are nebulae which obliterate the light from the stars behind them. Sometimes surrounding stars throw their light onto a nebula making it glow brightly. The *Orion Nebula* has a cluster of stars being formed within its cloud. It is 1300 LIGHT YEARS away and about 15 light years from one side to the other.

Neptune is normally the eighth planet of the SOLAR SYSTEM with its orbit between that of URANUS and PLUTO. However, for about twenty years in every 248 years, Pluto's orbit approaches closer to the Sun than Neptune's. At the time of writing, the two planets are within this period and Neptune will be the outermost one until 1999. Neptune is a vast planet, one of the *gas giants*, and is around 4497 million kilometres from the Sun. It is extremely cold with surface temperatures of approximately -200°C, and the atmosphere consists mainly of methane, hydrogen and helium. The diameter at the poles is 48,700 kilometres and at the equator it is 48,400 km which is about four times that of the Earth. The mass of the planet is 17 times greater than that of Earth and it takes 165 years to circle once around the Sun. Neptune is such a long way from the Earth that it can only be viewed using the most powerful telescopes, and even then it appears to be minute. Two astronomers, John Couch Adams in 1845 in Britain, and Urbain Leverrier in France in 1846, worked out the existence and position of Neptune before it could actually be seen. They noticed that the path of the orbit of nearby URANUS was not as expected, and worked out that it was being affected by the gravitational pull of another planet. A year later, Neptune was viewed for the first time. Neptune spins once on its axis every 18 to 20 hours and probably has a core of frozen rock and ice.

Most information about Neptune has been obtained from the *Voyager 2* space probe in August, 1989. It took 9000 photographs which shows that Neptune is surrounded by a faint series of rings. There is a large dark cloud, the size of the Earth, called the *Dark Spot* which has a spinning oval shape. Winds whip through the atmosphere at velocities of 2000 kilometres per hour and there are white methane clouds which constantly change shape. Neptune has 8 SATELLITES or moons, two of which are large and are called *Triton* and *Nereid*. Triton is the largest moon and is about the same size as MERCURY; its orbit is in the opposite direction to that of the other moons. It has a frozen surface with icy volcanic mountains and appears to have lakes of frozen gas and methane. The thin atmosphere of Neptune is composed mainly of nitrogen gas.

neutron star a small body with a seemingly impossibly high density. A star that has exhausted its fuel supply collapses under gravitational forces so intense that its electrons and protons are crushed together and form neutrons. This produces a star ten million times more dense than a WHITE DWARF – equivalent to a cupful of matter weighing many million tons on Earth. Although no neutron stars have definitely been identified, it is thought that PULSARS may belong to this group.

nova (plural *novae*) in the literal sense this is a new star, (nova being Latin for *new*), but it may also be a star that suddenly burns brighter by a factor of five to ten thousand. It seems that a nova is one partner in a BINARY STAR. The smaller star burns much hotter than the sun while the other partner is a vast expanse of hot red mist called a RED GIANT. An explosion results if cooler gas from the red cloud reaches the hot star, causing it to burn up more brightly – a nova. A red giant may eventually become a WHITE DWARF.

orbit the path of one heavenly body moving around another which results from the gravitational force attracting them together. The lighter body moves around the heavier one which is itself also in motion. The speed at which a heavenly body travels depends upon the size of its orbit. This is itself determined by the distance between the two heavenly bodies.

parallax the apparent movement in the position of a heavenly body due in fact to a change in the position of the observer. It is therefore caused, in reality, by the Earth moving through space on its orbit. The distance of a heavenly body from Earth can be calculated by astronomers using parallax. The direction of the body from Earth is measured at two six month intervals when the Earth is at either side of its orbit. From the apparent change in position, the distance of the body from Earth can be deduced.

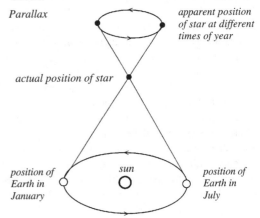

Parallax — *apparent position of star at different times of year* — *actual position of star* — *position of Earth in January* — *sun* — *position of Earth in July*

parsec an astronomical unit of distance which is used for measurements beyond the SOLAR SYSTEM, and which corresponds to a PARALLAX of one second of arc. (There are 3600 arc seconds in one degree). A parsec is 30,857,000,000,000 kilometres or 3.2616 LIGHT YEARS. The nearest star to Earth has been calculated as being 1.3 parsecs distant.

planet the name given originally to seven heavenly bodies that were thought to move among the stars, which were themselves thought to be stationary. It is derived from the Greek word for *wanderer*. The term is now used to describe those heavenly bodies moving in definite orbits around the SUN which, in order of distance are: MERCURY, VENUS, EARTH, MARS, JUPITER, SATURN, URANUS, NEPTUNE and PLUTO. Mercury and Venus are termed the *inferior planets* and Mars to Pluto the *superior planets*, the latter because they revolve outside the Earth's orbit. The planets travel their own particular orbit but are all moving in the same direction.

Pluto the ninth and smallest planet in the SOLAR SYSTEM and the one that lies farthest away from the SUN. The existence of Pluto was predicted by an American astronomer called Percival Lowell from the behaviour of the orbits of its closest neighbours, NEPTUNE and URANUS. Pluto was finally spotted in 1930, fourteen years after Lowell died. Pluto appears as a tiny speck when viewed from Earth and little is known about it, but it probably has an iron core and a rocky surface with a covering of methane ice. Since it is so far from the Sun (a maximum of 7,338 million kilometres), it must be extremely cold, in the region of -230°C. A day on Pluto lasts for almost seven Earth days and a year is 248.4 Earth years. At the equator, the planet has a diameter in the region of 3,500 kilometres. Pluto has a wide elliptical orbit which sometimes brings it closer to the Sun. In 1989, it was at its closest point to the Sun (called its *perihelion*), but this only occurs every 248 years. During this phase of its orbit, Pluto apparently has a thin atmosphere composed of methane gas. However, when it moves away from the Sun again it is possible that this becomes frozen once again. In 1979 it was discovered that Pluto has one small satellite (which was named *Charon*), and is about a quarter the size of Pluto. The two bodies effectively form a double planet system.

Proxima Centauri a small, faint red star which is the closest one to the SOLAR SYSTEM (some 4.3 light years away) yet to be discovered. It occurs in the constellation of *Centaurus* and is the third and smallest member of a series of three stars called *Alpha Centauri*. It takes about a million years for it to orbit the other two stars which are larger and brighter. Proxima Centauri occasionally burns more brightly for short periods of time but is usually quite dim.

pulsar pulsars are thought to be collapsed, rotating NEUTRON STARS which are left after a SUPERNOVA explosion. A pulsar is a source of radio frequency radiation which is given out in regular short bursts. The radiation is in the form of a beam which sweeps through space as the pulsar rotates, and is detected on Earth if Earth happens to lie in its path. The first pulsar was found to emanate from the *Crab Nebula* which is a supernova.

quasar any of the quasi-stellar objects (i.e. like stars) which are extremely compact, give out light and yet are enormously distant bodies – up to 10^{10} LIGHT YEARS away.

radio astronomy the detection of a large range of radio waves (especially radiation given off by hydrogen atoms in space) which are emitted by numerous sources. These include the SUN, PULSARS, remnants of SUPERNOVAe and QUASARS.

radiotelescope the instrument used to detect and analyse extra-terrestrial electromagnetic radiation (light and radio waves). There are two types – the *parabolic reflector*, which focuses the radiation on to an aerial, and the *interferometer*, where an interference pattern (*see* DIFFRACTION) of 'fringes' is formed and precise wave length measurements can be made. The latter is more accurate while the former is easier to move. The parabolic reflector is in the form of a large metal dish which can be pointed in different directions, used in conjunction with others and mounted on a rail so that it can be moved. The largest system in the world is built

across a valley at Arecibo in Puerto Rico, and is 305 metres across.

red giant an ageing star that is extremely hot and has used up about 10% of its hydrogen. The outer layers are cooler than the intensely hot centre. As the name implies, these are very large stars. The red giant, *Aldebaran*, is 35 times larger than the SUN with a diameter of 50,100,000 kilometres. Some red giants go on to become WHITE DWARFS, using up their hydrogen at an increased rate. The Sun will eventually become a red giant in about 5000 million years time.

satellite any body, whether natural or manmade, that orbits a much larger body under the force of gravitation. Hence the MOON is a *natural satellite* of the Earth. All the planets, except for MERCURY and VENUS, have at least one natural satellite. *Artificial satellites* are manmade spacecraft launched into orbit from Earth. The first satellite to be launched was the Russian *Sputnik I* in 1957, but many hundreds have followed since then. Some satellites, especially those used for communications, are placed in a special *geostationary* orbit. This orbit is about 36,000 km above the Earth's surface. The satellite orbits the Earth in the same period of time as Earth rotates on its axis (24 hours). Hence the satellite maintains the same position relative to the Earth and appears to be stationary. Equipment on Earth therefore does not need to be adjusted to follow the satellite.

Many other satellites are used for other purposes such as meteorological recordings and weather forecasting. Each satellite has a dish aerial facing towards Earth and *thruster* motors to help maintain its position. When the fuel supply for the thrusters is exhausted, the satellite drifts out of its orbit and can no longer be used. The equipment on board a satellite or spaceship requires electricity which is usually derived from solar powered cells (*see* PHOTOELECTRIC CELL). (However, spacecraft travelling long distances away from the Sun have electricity generated by small nuclear reactors).

Saturn one of the four *gas giants*, the second largest planet and sixth in the SOLAR SYSTEM, with an orbit between that of JUPITER and URANUS. Saturn has a diameter at the equator of about 120,800 kilometres and is a maximum distance of 1,507,000,000 kilometres from the SUN. Saturn rotates very fast and this causes it to flatten at its poles and bulge at the equator. A day on Saturn lasts for $10\frac{1}{4}$ hours and a year, (or one complete orbit of the Sun), for 29.45 Earth years. Saturn is a cold planet of frozen gases and ice and has a surface temperature in the region of -170°C. It is mainly gaseous with an outer zone of hydrogen and helium over a metallic hydrogen layer and a core of ice silicate. The atmosphere is rich in methane and ethane. Saturn is well-known for its *rings* which are, in fact, ice particles and other debris thought to be the remains of a SATELLITE which broke up close to the planet. The rings are wide, in the region of 267,876 kilometres across, but they are extremely thin (only a few kilometres). The *Voyager* space probes approached close to Saturn in 1980 and 1981 and photographs taken revealed that there were many more rings than had previously been detected. They are brighter than those of any other planet.

Saturn has 24 satellites or moons, some of which were discovered by the *Voyager* spacecraft, including *Atlas, Prometheus* and *Calypso. Titan* is the largest moon and, at 5200 kilometres diameter, is larger than Mercury. It is the only moon known to have a detectable atmosphere, a layer of gases above its surface.

Sirius also known as the *Dog Star* because it occurs in the constellation called the *Greater Dog, Canus Major*, this is the brightest star in the night sky. It is a BINARY STAR with a *white dwarf* as its partner and shines brightly because it lies quite close to Earth, only 0.5 light years distant.

solar energy is energy which reaches Earth from the sun in the form of heat and light. 85% of solar energy is reflected back into space and only 15% reaches the Earth's surface, but it is this energy which sustains all life.

solar system the system comprising the sun (a star of spectral type G), around which revolve the nine planets in elliptical orbits. Nearest the SUN is MERCURY, then VENUS, EARTH, MARS, JUPITER, SATURN, URANUS, NEPTUNE AND PLUTO. In addition, there are numerous SATELLITES, a few thousand ASTEROIDS (so far discovered) and millions of COMETS. The age of

The solar system

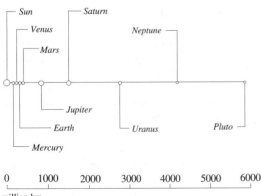

0   1000   2000   3000   4000   5000   6000

million km

the solar system is put at 4.5 to 4.6 billion years, a figure determined by the *radiometric dating* (uranium-lead content) of iron meteorites (*see* DATING METHODS). They are thought to be fragments or cores from early planets and thus representative of the early stages of the solar system.

solar wind the term for the stream of charged, high-energy particles (mainly electrons, protons and alpha particles) given out by the SUN. The particles travel at hundreds of kilometres per second, and the wind is greatest during flare and SUNSPOT activity. Around the Earth, the particles have velocities of 300–500 kilometres per second, and some become trapped in the magnetic field to form the VAN ALLAN RADIATION BELT. However, some reach the upper atmosphere and move to the poles, producing AURORA.

space probe is an unmanned spacecraft, such as *Voyager* 1 and 2, launched to travel through the SOLAR SYSTEM and beyond and programmed to send back data and photographs. Space probes eventually move so far away that contact with them is lost forever but they continue to travel for thousands of years. Vehicles from space probes have landed on Venus and Mars, obtaining valuable information.

space station a manned space research centre, including living accommodation, in which astronauts live and work for several months at a time. The manned space station, *Mir*, of the former Soviet Union was launched in 1986. At the end of the 1990s a new space station designed for eight people will be launched. It is called *Freedom* and is a joint American, Canadian, European and Japanese venture and all the materials for its construction are being carried out by SPACE SHUTTLE and assembled in space. It will be 500km (310 miles) above the surface of the Earth.

space telescope is a telescope launched into orbit in order to obtain images of stars and other heavenly bodies which are much clearer than if viewed from Earth. Images viewed through Earth-based telescopes are subject to distortion due to atmospheric factors, but such problems do not affect telescopes that are sited in space. Space telescopes have a special mirror and the images are radioed back to Earth where they can be studied on a screen and photographs obtained. In April 1990, the SPACE SHUTTLE *Discovery* launched the *Hubble* space telescope to undertake observations of the stars. After the launch it was found that the telescope's huge mirror was not working properly and it was repaired during a later shuttle flight.

space time *see* **relativity**

star a body of fiery gas, similar to the SUN, which is contained by its own gravitational field. Stars are glowing masses that produce energy by thermonuclear reactions (NUCLEAR FUSION). The core acts as a natural NUCLEAR REACTOR, where hydrogen is consumed and forms helium with the production of electromagnetic radiation. A classification system for stars, based upon the spectrum of light they emit, groups stars as various *spectral types*. The sequence is, in order of descending temperature: O – hottest blue stars; B – hot blue stars; A – blue-white stars; F – white stars; G – yellow stars; K – orange stars; M – coolest red stars. Stellar evolution is the various stages in the life of a star, beginning with its creation from the condensation of gas, primarily hydrogen. The growth of the gas cloud pulls in more gas, and the increase in gravity compresses the molecules together, attracting more material and creating a denser mass. The heat normally produced by molecules due to their vibratory motion is increased greatly and the temperature is raised to millions of degrees, which enables nuclear fusion to take place. The supply of hydrogen continues to be consumed (and the star occupies the main sequence of the Hertzsprung-Russell diagram) until about 10% has gone. (The Hertzsprung-Russell diagram is a graph of star types from hot to cooler, depending upon the stage each has reached in its evolution.) When 10% of the hydrogen has been consumed the rate of combustion increases. This is accompanied by collapses in the core and an expansion of the hydrogen-burning surface layers, forming a RED GIANT. Progressive gravitational collapses and burning of the helium (which is generated by the consumption of the hydrogen) results in a WHITE DWARF, which is a sphere of enormously dense gas. The white dwarf cools over many millions of years and forms a *black dwarf* – an invisible ball of gases in space. Other sequences of events may occur, depending upon the size of the star formed. BLACK HOLES and NEUTRON STARS may form red (super) giants via a SUPERNOVA stage.

Sun is the star nearest to Earth around which Earth and the other planets move in elliptical orbits. It is one of millions of stars in the MILKY WAY but is the centre of the SOLAR SYSTEM. It was formed around five thousand million years ago and is about half way through its life cycle. It has a diameter of 1,392,000 kilometres and a mass of 2×10^{30} kilograms. The interior reaches a temperature of 13 million degrees centigrade while the visible surface is about 6000°C. The internal temperature is such that thermonuclear reactions occur, con-verting hydrogen to helium with the release of vast quantities of energy. The Sun is approximately 90% hydrogen and 8% helium and will one day become a RED GIANT.

sunspots the appearance of dark areas on the surface of the SUN. The occurrence reaches a maximum approximately every eleven years in a phase known as the sunspot cycle. They are usually short-lived (less than one month) and are caused by magnetism drawing away heat to leave a cooler area which is the sunspot. The black appearance is due to a lowering of the temperature to about 4000K. Sunspots have intense magnetic fields and are associated with magnetic storms and effects such as the *aurora borealis* (*see* AURORA). They may send out solar flares which are explosions occurring in the vicinity of the sunspots.

supernova (*plural* **supernovae**) a large star that explodes, it is thought, because of the exhaustion of its hydrogen (*see* SUN), whereupon it collapses, generating high temperatures and triggering thermonuclear reactions. A large part of its matter is thrown out into space, leaving a residue that is termed a WHITE DWARF star. Such events are very rare, but at the time of the explosion the stars become one hundred million times brighter than the Sun. A supernova was sighted in 1987 in the large MAGELLANIC CLOUD, 170,000 light years distant.

telescope is an instrument for magnifying an image of a distant object, the main types of astronomical telescopes being *refractors* and *reflectors*. The refracting type have lenses to produce an enlarged, upside-down image. In the reflecting type there are large mirrors with a curved profile which collect the light and direct it onto a second mirror and into the eyepiece (*see also* RADIOTELESCOPE).

universe all matter, energy and space that exists which is continuing to expand since its formation (*see* BIG BANG).

Uranus is the seventh planet in the SOLAR SYSTEM and one of the four *gas giants* with an orbit between those of Saturn and Neptune. Uranus has a diameter at the equator of 50,080 kilometres and lies an average distance of 2,869,600,000 kilometres from the Sun. The surface temperatures are in the region of -240°C. It is composed mainly of gases with a thick atmosphere of methane, helium and hydrogen. Uranus was the first planet to be observed with a telescope and was discovered by William Herschel, a German astronomer in 1781. Uranus remained a mystery until quite recently but in 1986, *Voyager 2* approached close to the planet and obtained valuable information and photographs. The planet appeared blue, due to its thick atmosphere of gases and a faint ring system consisting of 13 main rings. Uranus was known to have five moons but a further ten, some less than 50 kilometres in diameter, were discovered by Voyager 2. A day on Uranus lasts for about $17^1/_2$ hours and a year is equivalent to 84 Earth years. Its largest moon is *Titania* with a diameter of 1600 km. All five moons are very cold and icy with a surface covered in craters and cracks. *Ariel* has deep wide valleys and *Miranda*, the smallest moon, is a mass of canyons and cracks with cliffs reaching up to 20km. Uranus has a greatly tilted axis so that some parts of the planet's surface are exposed to the Sun for half of the planet's orbit (about 40 years) and are then in continuous darkness for the rest of the time. Due to the tilt of the axis, the Sun is sometimes shining almost directly onto each of Uranus' poles during parts of its orbit.

Van Allen belts are two belts of radiation consisting of charged particles (electrons and protons) trapped in the Earth's magnetic field. The result is the formation of two belts around the Earth. They were discovered in 1958 by an American physicist called James Van Allen. The lower belt occurs between 2000 and 5000km above the equator and its particles are derived from the Earth's atmosphere. The particles in the upper belt, at around 20,000km, are derived from the solar wind. The Van Allen belts are part of the Earth's magnetosphere, an area of space in which charged particles are affected by the Earth's magnetic field rather than that of the Sun.

Venus is the second planet in the solar system with its orbit between those of Mercury and Earth and it is also the brightest. It is known as the Morning or Evening star. Venus is about 108.2 million km (67 million miles) from the Sun and is extremely hot with a surface temperature in the region of 470°C. It has a thick atmosphere of mainly carbon diox-

ide, sulphuric acid and other poisonous substances which obscure its surface. The size of Venus is similar to Earth with a diameter at the equator of 12,300 km. The atmosphere of carbon dioxide traps heat from the Sun (the greenhouse effect) allowing none to escape. Hence the surface rocks are boiling hot and winds whip through the atmosphere at speeds in excess of 320 km/hr. Venus is unusual in being the only planet to spin on its axis in the opposite direction to the path of its orbit. Also it spins very slowly so that a 'day' on Venus is very long, equivalent to 243 Earth days. A year is 225 days. Venus has no satellites and because its surface is hidden, much of the known information about the planet has been obtained from SPACE PROBES. The *Magellan* space probe launched by the USA in 1989 visited Venus, sending back valuable photographs. *Venera 13*, a Russian probe landed on Venus in 1982, and obtained a rock sample and other information. The surface of the palnet has been shown to be mountainous with peaks 12 kilometres high. It is covered with craters and also a rift vally. It is possible that there are active volcanoes.

white dwarf a type of star that is very dense with a low luminosity. White dwarfs result from the explosion of stars that have used up their available hydrogen (*see* STAR). Due to their small size, their surface temperatures are high, and they appear white (*see also* SUPERNOVA).

X-ray astronomy stars emit a variety of electromagnetic radiation including X-rays (*see* ELECTROMAGNETIC WAVES), but this is unable to penetrate the Earth's atmosphere to be detected. Very hot stars send out large amounts of X-rays which can be detected by equipment contained on space satellites. The satellite ROSAT launched in 1990 has equipment to undertake X-ray astronomy and obtain information about distant bodies in space (e.g. WHITE DWARFS and SUPERNOVAE) that are emitting X-rays.

GEOGRAPHY AND GEOLOGY

The study of the Earth's surface, climate and natural resources is of great importance to mankind because only by learning more about the planet's past is it possible to prepare for the future.

This section explains the terms commonly used in the study of the Earth's geography and geology. The entries are in A to Z order and cross-references are given in SMALL CAPITALS.

alluvium deposits produced as a result of the action of streams or RIVERS. Moving water carries sediment, particles of sand, mud and silt, and the faster it moves the greater the load it can carry. When the velocity of the water is checked due to it meeting a stationary or slower moving body of water, then much or all of the sediment is dropped forming alluvium. In mountainous regions where streams descend to lowlands, a fan or cone of sediment may build up. Rivers in areas of less extreme geography deposit alluvium on the flood plain where the water changes velocity along the curving course of the river, or when rivers overflow their banks during floods.

aquifer a layer of rock, sand or gravel that is porous and therefore allows the passage and collection of water. If the layer has sufficient porosity and permeability it may provide enough GROUNDWATER to produce springs or wells. If the layers above and below the aquifer are impermeable, the water is under pressure (*hydrostatic pressure*) and can be extracted in an *Artesian Well* whilst the level of the well is lower than that of the WATER TABLE. A significant proportion of London's water supply comes from the London Basin (an Artesian basin), although the supply is diminishing because the water table and pressure have fallen due to prolonged extraction of water. In many cases pumps may be required to raise the water to the surface.

An artesian basin

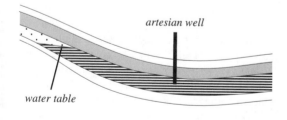

artesian well

water table

> *porosity*– the spaces within a rock or sediment referring to the pores that are able to release their contained water.
> *permeability*– the ability of a material to allow the passage of water, air, etc.
> *impermeable*– any material that does not allow passage of fluids.

atmosphere the layer of gases and dust surrounding the Earth which can be divided into shells, the lowermost being the TROPOSPHERE which is overlain by the STRATOSPHERE. The density falls with height and because it is thinner, breathing is more difficult at high altitudes. Almost 75% of the total mass is contained within the troposphere. The main gases found in the atmosphere are nitrogen (78%), oxygen (21%), the inert gas argon (0.9%), carbon dioxide (0.03%) and then very small amounts of other inert gases, with methane, hydrogen, water vapour and ozone. The ozone exists mainly in a layer (*see* OZONE LAYER) at a height of about 25 to 30 km although it is found elsewhere within the stratosphere.

Atmospheric pressure is the downward force of the atmosphere on the surface of the Earth and is roughly equivalent to a force of $1kg/cm^2$ and originally defined as the pressure that will support a column of mercury 760mm high at 0°C, sea level and a latitude of 45° (*see* **S** - BAROMETER).

Beaufort scale a system for indicating wind strength, developed in 1805 by Admiral F.B. Beaufort. Measurements are taken 10m above the ground and each of 12 levels is characterized by particular features on the landscape or certain effects upon people or objects. Wind speed is measured in metres per second (*see* table on page 24).

boulder clay a deposit that is glacial in origin and made up of boulders of varying sizes in finer-grained material, mainly clay. It is laid down beneath a glacier or ice sheet and shows little or no structure. Large blocks plucked from the terrain over which the ice has moved may be found in a matrix of finer material which has been ground down by the glacier. An alternative and more frequently used term is *till*, of which several types have been defined depending upon their specific mode of formation and position within the ice body. *Moraine* is an associated term referring to ridges of rock debris carried and deposited by ice sheets or glaciers of the various types; lateral moraine accumulates at the edge of a glacier and terminal moraine at the leading edge. Ground moraine is the same as boulder clay.

chalk a soft, fine-grained SEDIMENTARY ROCK made of calcium carbonate formed mainly from the skeletons and shells of very small marine organisms. It is formed primarily as deposits of calcare-

The Beaufort scale

Scale	Wind speed	Weather condition	Identifying features on land
0	< 0.3	calm	smoke rises vertically
1	0.3 – 1.5	light air	smoke indicates wind direction
2	1.6 – 3.3	light breeze	wind vane moves; wind felt on face
3	3.4 – 5.4	gentle breeze	twigs move; light flag extended
4	5.5 – 7.9	moderate breeze	small branches moved; dust raised
5	8.0 – 10.7	fresh breeze	small trees sway; wavelets on inland water
6	10.8 – 13.8	strong breeze	large branches move
7	13.9 – 17.1	near gale	whole trees in motion; some resistance when walking
8	17.2 – 20.7	gale/fresh gale	twigs broken off trees; resistance slows walking
9	20.8 – 24.4	strong gale	chimney pots and slates may be removed
10	24.5 – 28.4	storm	considerable structural damage; trees uprooted
11	28.5 – 32.7	violent storm	widespread damage; lives endangered
12	> 32.7	hurricane	catastrophic damage; loss of life

ous mud in shallow seas, but it can also be formed by chemical precipitation. Chalk deposited in the Upper Cretaceous approximately 100 million years ago now covers much of northwest Europe, and can be seen in the White Cliffs of Dover and the cliffs near Calais. Chalk produces a particular scenery, with low rolling hills, dry valleys and steep scarps (an escarpment or steep slope which ends a flat or gently sloping area).

Chalk is an important AQUIFER and is used in the manufacture of CEMENT and FERTILIZERS.

climate characteristic weather conditions produced by a combination of factors, such as rainfall and temperature. Whether taken singly or jointly these factors, together with influencing features such as altitude and latitude, produce a distinctive arrangement of zones around the Earth each with a generally consistent climate when studied over a period of time. The major climatic zones are, from the equator:

humid tropical—hot and wet

subtropical, arid and semi-arid—desert conditions, extremes of daily temperature

humid temperate—warm and moist with mild winters

boreal (N. hemisphere)—long, cold, winters and short summers

sub-arctic (or sub-antarctic)—generally cold with low rainfall

polar—always cold

Climatic patterns are very dependent upon heat received from the Sun and at the poles the rays have had to travel further and in so doing have lost much of their heat, producing the coldest regions on the Earth. Conversely, air near the equator is very warm and can therefore hold a great deal of water vapour resulting in hot and wet conditions - humid tropical. There are more complex systems of climate classification, e.g. Köppen and Thornthwaite which are based upon precipitation and evaporation, characteristic vegetation and temperature. In each system, the Earth can be split into numerous provinces and smaller areas producing quite a detailed overall picture.

cloud clouds are droplets of water or ice formed by the condensation of moisture in a mass of rising air. Water vapour formed by evaporation from seas, lakes and rivers is ordinarily contained in air and only becomes visible when it condenses to form water droplets. Warm air can hold more water vapour than cold air thus when air rises, becoming cooler, it becomes saturated ('full') of water and eventually droplets form, as cloud, each droplet forming around a central nucleus such as dust, pollen or a smoke particle.

Clouds are classified firstly upon their shape, and also by their height. There are three major groups: *cumulus* ('heap' clouds), *stratus* (sheetlike) and *cirrus* (resembling fibres) which are divided further as cloud forms show a mix of shape, e.g. *cirrocumulus*.

cirrus

cirro-cumulus

stratus   *cumulus*

coal a mineral deposit that contains a very high carbon content and occurs in banded layers (seams) resembling rock. It is formed from ancient forest vegetation that has accumulated and been transformed into PEAT. With time, increasing pressure from sediments above and some heating, the peat becomes coal of various types, each type (or *rank*) being marked by the carbon content which increases with pressure and temperature, and the volatiles content, which decreases at the same time. Volatiles are substances that readily become gas, specifically in this case, a mixture of hydrogen, methane and carbon dioxide which are released when coal is heated.

The coal of highest rank is anthracite which is 95% carbon with a little moisture and volatile matter. Bituminous coals contain from about 45% to 75% carbon while lignite has 70% moisture and volatiles.

Coal beds occur in sequences of sandstone and shale called, collectively, *coal measures*. Most of the world's anthracite and bituminous coals accumulated in the Carboniferous period (*see* GEOLOGICAL TIMETABLE) and the coal measures reach thousands of metres in thickness in many countries. The countries with the greatest reserves of coal include USA, CIS, China, Australia and Germany.

Although coal is used less than it was as a primary energy source, its byproducts are still used in the chemicals industry, for the manufacture of dyes, drugs, pesticides and other compounds.

continent one of several large landmasses covering 29% of the Earth's surface of which 65% is in the northern hemisphere. Where the continent edge meets the sea is a continental shelf and slope which is often cut by submarine canyons. Further ocean-ward lies the deepest parts of the ocean, the abyssal plain. There are seven continents making up the Earth:

	highest point (metres)	area (sq km)
Asia	8848	43,608,000
Africa	5895	30,335,000
North and Central America	6194	25,349,000
South America	6960	17,611,000
Antarctica	5140	14,000,000
Europe	5642	10,498,000
Oceania	4205	8,900,000

crystal a solid that has a particular, ordered, structure and shape with usually flat faces and a chemical composition that varies little. The arrangement of the atoms in the crystal in a specific framework give it a characteristic shape and if conditions permit when the crystal is growing, the crystal faces are arranged in a constant geometric relationship. Many solids have a crystalline structure, from household items such as sugar and salt to chemicals, metals and the minerals within rocks. If a solution of a chemical or a magma (molten rock) is allowed to cool slowly, crystals will develop. Similarly, if a crystal is placed in a saturated solution of the same compound *crystallization* will occur. If, however, the solution or magma is cooled suddenly, either the crystals formed are microscopic or the resulting solid exhibits no crystalline form (is *amorphous*).

The regular formation of crystal faces is part of a crystal's *symmetry* which is described for each crystal system, by three properties: planes, axes and a centre of symmetry. These properties together create six crystal systems. These are:

Cubic	common salt
Tetragonal	zircon
Hexagonal	calcite; quartz
Orthorhombic	topaz; barytes; sulphur
Monoclinic	gypsum
Triclinic	feldspars

Crystallography is the scientific study of crystals, their form and structure.

cyclone an area of low atmospheric pressure with closely packed isobars producing a steep pressure gradient and therefore very strong winds. Due to the Earth's rotation, the winds circulate clockwise in the southern hemisphere and anticlockwise in the northern hemisphere. In tropical regions, cyclones (tropical hurricanes) combine very high rainfall with destructive winds which results in widespread damage and possible loss of life.

Outside the tropics, in more temperate climates, the term cyclone is being replaced by *depression* (or *low*) in which the pressure may fall to 940 or 950 mb (millibars). Compare this with the average pressure at sea-level which is 1013 mb.

An *anticyclone* is a high pressure area (around 1030 - 1050 mb) and usually results in stable weather.

dating methods are used in a number of scientific subjects, whether geography, geology or archaeology, where it is often necessary to date a sample and there are several ways in which this can be achieved. In many instances, a relative age can be determined quite readily, for example in geology one rock cutting across another can provide clues as to the order of formation of rock sequences, or the order in which rocks were folded or deformed (*see* FOLDS AND FAULTS). However, to obtain an *absolute* or true age (bearing in mind any experimental errors), the radioactive decay of ISOTOPES can be employed -*radiometric dating*.

Radiocarbon dating is one such method. This relies upon the uptake by plants and animals of small quantities of naturally-occurring radioactive ^{14}C in the atmosphere. When the organism dies, the uptake stops and the ^{14}C decays with a half-life (see S-RADIOACTIVITY) of 5730 years, and the age of a sample can be calculated from the remaining radioactivity. This method is useful for dating organic material within the last 70,000 years but the accuracy of the methods falls between 40,000 and 70,000 years.

Rocks can be dated by one of several methods each of which relies upon the radioactive decay of one element to another. In these cases the half-life is very long enabling the dating of very old rocks. Uranium-lead was an early method in which uranium-238 (^{238}U) decays to lead-206 with a half-life of 4.5 billion years.

Other systems are:

Rb/Sr	rubidium-87 decays to strontium-87 half-life of 50 billion years.
K/Ar	potassium-40 decays to argon-40 half-life of 1.5 billion years.
Th/Pb	thorium-232 decays to lead-208 half-life of 14 billion years.
Sm/Nd	samarium-147 decays to neodymium-143 half-life of 2.5 billion years.

The long half-lives make these systems useful in dating Lower Palaeozoic and Precambrian rocks (*see* GEOLOGICAL TIMESCALE). Sm/Nd is resistant to metamorphism and can be used on extraterrestrial materials: Th/Pb is often used with other methods because it is not an accurate method alone; Rb/Sr is useful for dating metamorphic rocks and in general the best results for dating are obtained from igneous or metamorphic rocks.

desert an arid or semi-arid (that is, dry and parched with under 25cm (10 in.) of rainfall annually) region in which there is little or no vegetation. The term was always applied to hot tropical and subtropical deserts, but is equally applicable to areas within continents where there is low rainfall and perennial ice-cold deserts.

The vegetation is controlled by the rainfall and varies from sparse, drought-resistant shrubs and cacti to sudden blooms of annual plants in response to a short period of torrential rain. If the groundwater conditions permit, e.g. if the water table is near to the surface creating a spring, or the geology is such that an artesian well (*see* AQUIFER) is created, then an oasis may develop within a hot desert, providing an island of green.

Hot deserts are found in Africa, Australia, United States, Chile and cold deserts in the Arctic, eastern Argentina and mountainous regions.

Hot desert extremes:

extreme of shade temperature: Death Valley, California
maximum 28°C
maximum daily range 41°C
max. ground surface temperature: Sahara 78°C
extreme of rainfall: Chicama, Peru, 4mm per year

The process whereby desert conditions and processes extend to new areas adjacent to existing deserts is called *desertification*.

drainage is the movement of water derived from rain, snowfall and melting of ice and snow, over the land (and through it in subterranean waterways) which results eventually in its discharge into the sea. The flow of streams and rivers is influenced by the underlying rocks, how they are arranged and whether there are any structural features that the water may follow. Further factors affecting drainage include soil type, climate and the influence of man.

There are a number of recognizable patterns which can be related to the geology:

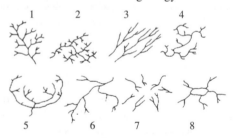

1. Dendritic	a random branching unaffected by surface rocks
2. Trellis	streams aligned with the trend of underlying rocks.
3. Parallel	streams running parallel to each other due to folded rocks, or steep slopes with little vegetation.
4. Rectangular	controlled by faults and joints, the latter often in igneous rocks.
5. Annular	formation of streams in circular patterns around a structure of the same shape (e.g. an igneous intrusion).

6. Barbed   a drainage pattern where the tributaries imply a direction of flow contrary to what actually happens.

7. Radial   streams flowing outwards from a higher area.

8. Centripetal   the flow of streams into a central depression where there may be a lake or river.

When a drainage pattern is a direct result of the underlying geology, it is said to be *accordant* (the opposite case being *discordant*).

earthquake movement of the earth, which is often violent, and caused by the sudden release of stress that may have accumulated over a long period. Waves of disturbance -*seismic* waves - spread out from the origin or *focus*, of the earthquake which is most likely to be movement along a FAULT, although some are associated with volcanic activity.
Earthquakes are classified by their depth of focus: shallow (less than 70km); intermediate (70 to 300km); deep (more than 300km).

Areas of earthquake activity

Over three quarters of earthquake energy is concentrated in a belt around the Pacific and this is because most seismic activity occurs at the margins of tectonic plates (*see* PLATE TECTONICS). This means that certain regions of the world are more likely to suffer earthquakes, for example the west coast of N. & S. America, Japan, the Philippines, S.E. Asia and New Zealand.
The effects of earthquakes are naturally very alarming and can be quite catastrophic. Near the focus, ground waves actually throw about the land surface. Surface effects may include opening of fissures (large cracks), the breaking of roads and pipes, buckling and twisting of railway lines and the collapse of bridges and buildings. Secondary effects can be equally destructive if the ground vibrations initiate landslides, avalanches and TSUNAMI or cause fires.
There are several systems of measuring the intensity of earthquakes, the one in common use being that devised by Richter, an American seismologist. The Richter Scale is:

Rating	Identifying features
1 Instrumental	detected only by seismographs
2 Feeble	noticed by sensitive people
3 Slight	similar to a passing lorry
4 Moderate	loose objects are rocked
5 Rather strong	felt generally
6 Strong	trees sway; loose objects fall
7 Very strong	walls crack
8 Destructive	chimneys fall; masonry cracks
9 Ruinous	houses collapse where ground starts to crack
10 Disastrous	ground badly cracked; buildings destroyed
11 Very disastrous	bridges and most buildings destroyed; landslides
12 Catastrophic	ground moves in waves; total destruction

Because this is a logarithmic scale, the magnitude of one level is very much more than the previous level. Over the last ninety years not many earthquakes have registered over 8 on the Richter scale. Earlier ones include San Francisco (1906), Kansu in China (1920), Japanese trench (1933), Chile (1960) and offshore Japan (1968). *See also* SEISMOGRAPH.

erosion the destructive breakdown of rock and soil by a variety of *agents* that, together with weathering, forms *denudation*, or a wearing away of the land surface. Erosion occurs because of the mechanical action of material carried by agents: water (rivers, currents, waves), ice (glaciers) and wind. Wind laden with sand can scour rock, rocks embedded in glaciers grind down the rocks over which they pass and gravel and pebbles in streams and rivers excavate their own course and may create pot holes, undercut banks, etc. Water can also carry material in solution. There are six different kinds of erosion processes, each with a different effect:

process   *effect*

abrasion   wearing away through grinding, rubbing and polishing.

attrition   the reduction in size of particles by friction and impact.

cavitation   characteristic of high energy river waters (e.g. waterfalls, cataracts) where air bubbles collapse sending out shock waves that impact on the walls of the river bed (a very localized occurrence).

corrasion   the use of boulders, pebbles, sand etc. carried by a river, to wear away the floor and sides of river bed.

corrosion   all erosion achieved through solution and chemical reaction with materials encountered in the water.

deflation   the removal of loose sand and silt by the wind.

The effect of these processes can be very marked, with time, and can combine to deepen gorges, create and enlarge waterfalls and result in the movement of a waterfall upstream. The latter occurs through undercutting of lower, softer rocks which then cause the collapse of overhanging ledges.

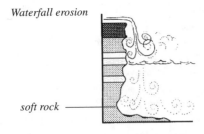

Waterfall erosion

soft rock

evaporite a sedimentary rock formed by the evaporation of water containing various salts, resulting in their deposition. The particular sediment created depends upon the concentration of ions (*see* C-ION) in solution, and the solubility (*see* SOLUTIONS) of each salt formed. The least soluble - calcium carbonate and magnesium carbonate - precipitate first, followed by sodium sulphate, then sodium chloride, potassium chloride and magnesium sulphate. A typical sequence of evaporite deposits could show calcite, possibly dolomite, gypsum (or anhydrite), rock salt, and finally potassium and magnesium salts (e.g. carnallite, $KCl.MgCl_2.6H_2O$).

There are several major deposits of evaporites in the world, e.g. the Stassfurt deposits in Germany, the Wieliczka salt mines in Poland, Texas and New Mexico, Chile and Cheshire in Britain. Some of these deposits are over 3500m thick, and since about 5m of evaporites would require the evaporation of 300m of seawater, it is clear that very special conditions existed in the past. To accumulate such vast thicknesses, there must be a shallow area of water from which evaporation and precipitation proceeds; which is periodically refilled with salt water, *and* there must be a gradual subsidence of the land to permit continued build up of the salts.

Modern evaporites are being formed in places such as the Caspian Sea, the Arabian Gulf and deposits have been formed recently in Eritrea.

faults *see* FOLDS.

floods a flood is where land not normally covered by water is temporarily underwater and can occur for several reasons and the scale of the flood may vary enormously. An increase in rainfall, particularly if prolonged may lead to flooding as may a sudden thaw of lying snow. Flooding of coastal areas by the sea may be caused by the combination of a very high tide and stormy conditions generating high waves. Perhaps most devastating of all are the TSUNAMI caused by earthquakes. Large areas of, for example, S.E. Asia, continually face the threat of floods.

When a river is in flood it carries a far greater load of rocks, boulders, sand and silt and its destructive power is therefore increased. In some cases, flood waters have been so powerful as to carve new channels for the river and in so doing demolish houses, roads, bridges, wash away trees and soil and transport vast piles of boulders and masonry seemingly impossible distances.

floodplain is the area in a river valley which may be covered by water when a river is in flood, and which is built up of ALLUVIUM. The floodplain is developed by a river's movement with time and the deposition and transport of sediment. Whether it is due to existing features along a river's course, or simply due to turbulence of flow and subsequent deposition of material where a river slows, the course of a river will include numerous bends. Subsequent erosion and deposition of sediment results in a meandering (*see* RIVERS) of the river with deposition of alluvium and a progressive movement along of the river bends, forming a floodplain.

Development of a flood plain

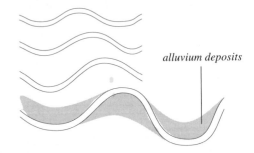

alluvium deposits

folds and *faults* are both geological features developed through tectonic activity. Folds are produced when rock layers undergo compression resulting in a buckling and folding of the rocks - a *ductile* or flowing deformation. There is an almost infinite variety in the shape, size and orientation of folds and an earlier generation of folds may be refolded by subsequent periods of deformation. Their size may range from microscopic to folds occupying hillsides or even whole mountains.

The way folds are formed will depend upon the amount of compression, the rock types involved and the thicknesses of the layers, because different rocks respond in different ways, but each fold possesses certain common features. The zone of greatest curvature is the *hinge* and the *limbs* lie between hinges or on either side of the hinge. The *axial plane* is an imaginary feature bisecting the angle between limbs and the *fold axis* is where the axial plane meets the hinge zone.

Faults are generally planar features, and are caused by *brittle* deformation. Rocks are moved, or *displaced*, across faults by as little as a few millimetres or as much as several kilometres (although possibly not all in the same event). There are several types of faults, depending upon the movement across the fault, and the orientation of the fault plane. The measurement of movement uses a horizontal and vertical component, the heave and throw respectively. If the movement is up or down the fault plane it is a dip-slip fault; if it is sideways, the term is strike-slip.

Two examples of vertical dip-slip fault movement

Two examples of horizontal strike-slip movement

Many faults produce associated features such as zones of crushed rock, or striations (grooves) known as *slickensides*, on adjacent rock surfaces, where minerals such as quartz or calcite may grow.

Major faults today account for many surface features and are an important part of the *plate tectonic* structure and development of the Earth.

fossils are the remains of once-living plants and animals, or evidence of their existence, that have been preserved (usually) in the rock layers of the Earth. The term may include the preservation in ice of woolly mammoths that lived 20,000 years, although anything younger than about 10,000 years is not generally considered a fossil.

Fossils may be bones, shells, borings, trails, casts and the fossilized remains of something associated with or caused by an organism, e.g. tracks or a worm burrow, are called *trace fossils*. Although many rock sequences contain a large number of fossils, conditions prevailing at the time have to be just right to ensure preservation. Vital to the process is that the organism be covered quickly with sediment, something that will happen more readily in marine rather than terrestrial conditions. Even then, significant changes may occur - soft parts usually decay and shells may be replaced chemically, or dissolved away leaving a mould to be filled by a different material. Occasionally, a whole insect may be preserved in amber.

The scientific study of fossils is called *palaeontology.*

fossil fuels include PETROLEUM (oil), COAL and NATURAL GAS - fuels created by the fossilization of plant and animal remains. Oil and gas are often found together with the gas lying over the oil and beneath the impermeable layer which seals the reservoir.

geochemistry a part of GEOLOGY that deals with the chemical make-up of the Earth, including the distribution of elements (and ISOTOPES) and their movement within the various natural systems (ATMOSPHERE, LITHOSPHERE, etc.). Recently, it has also been taken to include other planets and moons within the solar system. The geochemical cycle, in broad terms, illustrates the way in which elements from MAGMA (the 'starting point') move through different processes and geochemical environments.

geochronology the study of time on a geological scale using *absolute* or *relative* dating methods. Absolute methods (*see* DATING METHODS) provide an age for a rock and involve the use of radioactive elements with known rates of decay. Relative ages are determined by putting rock sequences in order through study of their sequence of deposition or folding. Fossils can also be used in relative age dating.

geography the study of the Earth's surface, including all the land forms (*see* GEOMORPHOLOGY), their formation and associated processes, which comprise *physical* geography. Such aspects as climate, topography and oceanography are covered. *Human* geography deals with the social and political perspectives of the subject including populations and their distribution. In addition, geography may cover the distribution and exploitation of natural resources, map-making and REMOTE SENSING.

geological timescale a division of time since the formation of the Earth (4600 million years ago) into units, during which rock sequences were deposited, deformed and eroded and life of diverse types emerged, flourished and, often, ceased.

The following table shows the various subdivisions, and many of the names owe their derivation to particular locations, rock sequences and so on.

Cenozoic	Recent	Modern Man
	Pleistocene	Stone Age Man
	Pliocene	many mammals, elephants
	Oligocene	pig and ape ancestors
	Eocene	
	Palaeocene	horse and cattle ancestors
Mezozoic	Cretateous	end of the dinosaus and the ammonites
	Jurassic	appearance of birds and mammals
	Triassic	dinosaurs appear; corals of modern type
Palaeozoic	Permian	amphibians and reptiles more common; conifer trees appear
	Carboniferous	coal forests; reptiles appear, winged insects
	Devonian	amphibians appear; fishes more common
		ammonites appear; early trees; spiders
	Silurian	first coral reefs; appearance of spore-bearing land plants
	Ordovician	first fish-like vertebrates: trilobites and graptolites common
	Cambrian	first fossils period; trilobites, grapolites, molluscs, crinoids, radiolaria, etc.
	Precambrian	sponges, worms, algae, bacteria; all primitive forms

Derivation of some geological names:

Cenozoic—recent life   *Palaeonzoic*—ancient life
Mesozoic—mediaeval life   *Archaen*—primaeval

The Roman name for Wales (*Cambria*) led to Cambrian, while the names of two Celtic tribes, the *Silures* and *Ordovices* provided the remaining lower Palaeonzoic names. Carboniferous is related to the proliferation of coal (i.e. carbon) and Cretaceous comes from *creta*, meaning chalk. The Triassic is a threefold division in Germany, while the *Jura* mountains lent their name to the Jurassic

geology the scientific study of the Earth, including its origins, structure, processes and composition. It includes a number of topics which have developed into subjects in their own right: GEOCHEMISTRY, MINERALOGY, petrology (study of rocks), structural geology, GEOPHYSICS, PALAEONTOLOGY, STRATIGRAPHY, economic and physical geology. Charles Lyell (1797 - 1875) was an influen-

tial figure in the early years of geological study and wrote *Principles of Geology*.

geomagnetism the study of the Earth's magnetic (*geomagnetic*) field which has varied with time. At mid-oceanic ridges (*see* PLATE TECTONICS) where new crust is created, measurement of the geomagnetic field shows stripes relating to reversals of the Earth's magnetic field which is taken up in the newly-formed rocks. This provides a tool in determining the age of much of the oceanic crust and is a vital piece of evidence in supporting the theory of plate tectonics.

Geometric fields of the mid-Atlantic ridge

geomorphology a subject that grew out of geology around the middle of last century, and is the study of *landforms*, their origin and change, i.e. the study of the Earth's surface. Landforms are composed of various rock types and formed from the surface materials of the Earth by geomorphological processes which originate from tectonic movements and the climate. Landforms can be arranged into certain categories based upon factors such as the underlying structural geology, the nature of the topography, i.e. the surface features, and the terrain (soil, vegetation, etc), and the type of geomorphological processes dominant.

geophysics is the study of all processes *within* the Earth (i.e. the crust and the interior) and is concerned with the physical properties of the Earth. Included are seismology (*see* SEISMOGRAPH), GEOMAGNETISM, gravity, HYDROLOGY, oceanography (the study of the oceans, currents, tides, sea-floor and so on), heat flow within the Earth and related topics. As with many other subject areas, the component topics often develop to the point of becoming a subject in their own right, and boundaries between subjects merge.

geothermal energy the temperature within the Earth increases with depth (this is called the *geothermal gradient*), although not uniformly, and the average gradient is in the range 20 to 40°C per kilometre. At the edges of some tectonic plates (*see* PLATE TECTONICS) the gradient increases dramatically and it is sometimes possible to harness this heat as geothermal energy.

The high heat flow may be caused by magmatic activity (*see* MAGMA) or by the radioactive decay of certain elements. When water or brine (water containing dissolved salts) circulates through these rocks, it becomes heated and may appear at the surface as a warm spring, or if it is temporarily contained and heated further, it may force out a body of steam and water as a *geyser* (derived from geysir, the Icelandic for gusher or roarer).

Geothermal energy is 'tapped' all over the world, including Iceland, New Zealand, California and Italy. Iceland, New Zealand and Kenya have geothermal power stations.

glacier an enormous mass of ice, on land, that moves very slowly. About 10% of the Earth's land is covered by glaciers although during the last glaciation this was nearer 30%. Glaciers that cover vast areas of land, e.g. Greenland, Antarctic are called ice sheets (or ice caps if smaller) and these hide the underlying land features. The more typical glaciers are either those that flow in valleys or those filling hollows in mountains. Glaciers can be classified further into polar (e.g. Greenland), subpolar (e.g. Spitzbergen) and temperate (e.g. the Alps) depending upon the temperature of the ice.

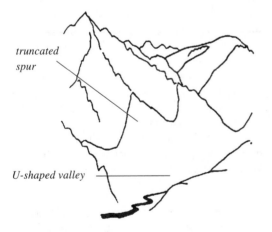

truncated spur

U-shaped valley

Although glaciers move slowly, they act as powerful agents of erosion on the underlying rocks. Large blocks may be dragged off the underlying rock, become embedded in the ice and then scratch and scour the surface as the glacier moves. This produces smaller particles of rock debris and the blocks themselves may be broken. Debris is also gathered from valley sides and carried along.

Ridges or piles of this rock debris are called moraine and depending upon position relative to the glacier, may mark present or former edges of the ice. In addition to the formation of moraine and associated, characteristic formations, glaciers produce some typical large scale features such as U-shaped valleys and *truncated spurs.*

Glaciation is the term meaning an ice age or a part of an ice age when glaciers and ice sheets are enlarged significantly.

grasslands form one of the four major types of vegetation, the others being forest, savanna and DESERT. Grasslands are characterized by seasonal drought, limited precipitation and occasional fires and these all, with grazing by animals, restrict the growth of trees and shrubs. Typical grasslands include the pampas of Argentina, the veldts of South Africa, the Steppes in the CIS and the prairies of central N. America.

The coverage of grasslands expanded after the last glaciation when climates became generally hotter and drier and there was an increase in the number of large grazing mammals.

Savanna is similar to grassland, but with scattered trees and is found extensively in S. America, southern Africa and parts of Australia. There are usually well-defined seasons: cool and dry, hot and dry followed by warm and wet and during the latter there is a rich growth of grasses and small plants. Although savanna soils may be fertile they are highly porous and water therefore drains away rapidly.

groundwater water that is contained in the voids within rocks, i.e. in pores, cracks and other cavities and spaces. It often excludes *vadose* water which occurs between the water-table and the surface. Most groundwater originates from the surface, percolating through the soil (*meteoric* water). Other sources are *juvenile* water, generated during and coming from deep magmatic processes, and *connate* water which is water trapped in a sedimentary rock since its formation.

Groundwater is a necessary component of most weathering processes and of course its relationship to the geology, water-table and surface may lead to the occurrence of AQUIFERS and artesian wells.

hardness of minerals minerals differ in their physical hardness and a test introduced in 1822 is still in use today to aid mineral classification. Moh's scale lists 10 minerals ranked by hardness and each mineral will scratch those lower on the scale where talc is the softest and diamond the hardest. In addition it is common to use the finger nail (equivalent to 2.5) or a pen-knife (5.5) to assist in the determination.

Moh's scale of hardness:

1	talc	6	orthoclase (feldspar)
2	gypsum	7	quartz
3	calcite	8	topaz
4	fluorite	9	corundum
5	apatite	10	diamond

humidity the amount of moisture in the Earth's atmosphere. *Absolute* humidity is the actual mass of water vapour in each cubic metre of air while *relative* humidity is

$$\frac{\text{water vapour content of air at a given temperature}}{\text{water vapour content required for saturation at that temperature}} \times 100\%$$

The relative humidity of air therefore varies with temperature, cold air holds little moisture, warm air much more. A *hygrometer* is the instrument used to measure relative humidity.

hurricane a wind which on the BEAUFORT SCALE exceeds 75 mph (121 km/h). Also an intense, tropical cyclonic storm that has a central calm area – the *eye* – around which move winds of very high velocity (over 160 km/h). There is usually very heavy rain, with thunderstorms and the whole system may be several hundred kilometres across.

Such storms occur mainly in the Caribbean and the Gulf of Mexico and often affect the southern states of the USA, creating considerable destruction.

hydrology the study of water and its cycle, which covers bodies of water and how they change. All physical forms of water - rain, snow, surface water - are included as are such aspects as distribution and use. The way in which water circulates between bodies of water such as seas, the atmosphere and the Earth forms the *hydrological cycle*.

The cycle consists of various stages: water falls as rain or snow of which some runs off into streams and then into lakes or rivers, while some percolates into the ground. Plants and trees take up water and lose it by transpiration to the atmosphere, while evaporation occurs from bodies of water. The water vapour in the air then condenses to cloud which eventually repeats the cycle.

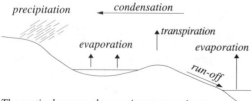

The vertical arrows show moisture returning to the atmosphere where it will condense as cloud

Of the 1.5 billion cubic kilometres of water on earth, the oceans hold 93.9%, groundwater 4.4%, polar ice 1.0%, with rivers, lakes, the atmosphere and soil holding the remainder.

ice age a period in the history of the Earth when ice sheets expanded over areas that were normally ice-free. The term is usually applied to the most recent episode in the Pleistocene (*see* GEOLOGICAL TIMESCALE), but the rock record indicates that there have been ice ages as far back as the Precambrian. Within ice ages there are fluctuations in temperature producing interglacial stages when the temperatures increase.

igneous rocks one of the three main rock types. Igneous rocks crystallize from MAGMA and are formed at the surface as LAVA flows (*extrusive*) or beneath the surface as *intrusions*, pushing their way into existing rocks. There are numerous ways of classifying igneous rocks, from mineral content to crystal size and mode of origin and emplacement. Typical rocks are basalt, granite and dolerite.

Rocks erupted at the surface as lava are called *volcanic*, while *plutonic* rocks are those large bodies solidifying at some depth; *hypabyssal* rocks are smaller and form at shallow depths. When a body of magma has time to crystallize slowly, large mineral crystals can develop while extrusion at the surface leads to a rapid cooling and formation of very small crystals (or none, if molten rock contacts water, as in a glass). A further division into acid, intermediate, basic and ultrabasic rocks is based upon silica (SiO_2) content, this being greatest in acid rocks.

Igneous intrusions may occur in several forms either parallel to or cutting through the existing rocks, and the commonest are *sills* (concordant) and *dykes* (discordant). *Plugs* commonly represent the neck of a former volcano, while *batholiths* are massive, elongate bodies that may be hundreds of kilometres long.

Examples of igneous intrusions

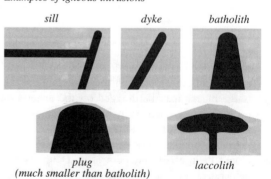

sill   dyke   batholith

plug
(much smaller than batholith)   laccolith

irrigation is when water is taken to dry land to encourage or facilitate plant growth. The water may be applied by means of canals, ditches, sprinklers or the flooding of the whole area. Flood irrigation is not always a good idea because the water evaporates from the flooded field leaving behind any salts that were dissolved in the water. If this is done repeatedly, the build up of salts may harm the soil and make it infertile. Water may be conserved and used more effectively by means of a pipe with small holes being laid around a plant, and the water drips from the holes onto the soil. Irrigation can transform arid regions but rivers from which water is diverted will inevitably be much reduced.

isobar a line on a weather map that joins points of equal pressure corrected for the varying heights of recording. Over a large area, isobars produce a map of pressure lines, similar to contours on a topographic map, which identify the high and low centres, i.e. the anticyclones and depressions.

jet stream westerly winds at high altitudes (above 12 km) found mainly between the poles and the TROPICS, that form narrow jet-like streams. The air streams move north and south of their general trend, in surges, which are probably the cause of depressions and anticyclones. There are a number of separate jet streams but the most constant is that of the subtropics. Jet stream speed and location is of importance to high-flying aircraft.

latitude and **longitude** latitude is the angular distance, north or south from the equator, of a point on the Earth's surface. The equator is 0° and points can therefore be measured in degrees south or north of this line. The imaginary lines drawn on a map or globe are the lines of latitude.

Longitude is a similar concept. It is the angular distance of a point measured on the Earth's surface to the east or west of a 'central' reference point. The referent point in this case is the plane created by a *meridian* (an imaginary circle that cuts the poles and goes over the Earth's surface and the point in question) going through Greenwich in England. A point may be 0° longitude if it sits on this line or a number of degrees east or west. The Greenwich Meridian, based upon the Greenwich Observatory, was established by an international agreement in 1884. There is a time difference of one hour when travelling 15° of longitude at the equator (*see also* TROPICS).

lava molten rock at about 1100 or 1200°C erupted from a VOLCANO or a similar fissure. It may flow onto the ground (subaerial) or onto the sea floor (submarine). Due to rapid cooling in air or water, most lavas show a fine-grained or glassy structure. The way lava is erupted, flows, and its shape as a flow is determined by its viscosity. *Pahoehoe* lava is a fluid basaltic type which forms rope-like flows due to molten lava in the centre of the flow dragging the solidifying crust into folds. A more viscous lava, *Aa* (both terms are Hawaiian words) flows more slowly and forms jagged, pointed blocks. Often a stream of lava will be contained by craggy sides of partly solidified lava. *Pillow lavas* are formed on the sea floor and as their name suggests, consists of pillow-like shapes built up and out from the source with one pillow rupturing at some point to allow more lava out to form another pillow. *Pumice* is lava filled with small air bubbles, creating a light, rough stone.

There are roughly 10,000 active volcanoes occurring in belts coinciding with the margins between tectonic plates. Lavas may cover vast areas, e.g. 250,000 square kilometres of the Deccan Plain of India, and 130,000 square kilometres in the Columbia River plateau region of the western USA. Such eruptions have occurred throughout geological time.

lightning and **thunder** lightning is the discharge of high voltage electricity between a CLOUD and its base, and between the base of the cloud and the Earth. (It has been shown that in a cumulonim-bus cloud, positive charge collects at the top and negative at the base). Lightning occurs when the increasing charge (of electricity) in the cloud overcomes the resistance of the air, leading to a *discharge*, seen as a flash. The discharge to ground is actually followed by a return discharge up to the cloud, and this is the visible sign of lightning. There are various forms of lightning, including sheet, fork and ball, and it may carry a charge of around 10,000 amps.

Thunder is the rumbling noise that accompanies lightning, and it is caused by the sudden heating and expansion of the air by the discharge, causing sound waves. The sound often continues for some time because sound is generated at various points along the discharge – the latter can be several kilometres long. The thunder comes after the lightning because sound travels more slowly than light, and this allows an approximate measure of distance from the flash to be made. For every 5 seconds between the flash and the thunder, the lightning will be roughly one mile away.

limestone a sedimentary rock that is made up mainly of calcite (calcium carbonate, $CaCO_3$) with dolomite ($CaMg(CO_3)_2$). There are essentially three groups of limestone: chemical, organic and detrital. These groups reflect the enormous variety of limestones, which may contain the broken remains of marine organisms – shells, coral, etc (detrital); minute organic remains of, for example, algae, foraminifera as in chalk, make an *organic* limestone; and grains formed as concentrically layered pellets (ooliths and pisoliths) in shallow marine waters, which with EVAPORITES are examples of chemical limestones.

Limestones are important economically and have many uses. They form AQUIFERS and reservoirs for oil and gas and are used in the manufacture of cement, in agriculture and as roadstone, and when cut and finished slabs of it are used to face buildings. Where limestone outcrops appears at the surface, the process of water flow and dissolution of the rock may lead to a *karst* topography, where there is a virtual absence of surface drainage and groundwater moves along joints into holes, enlarging them over time to form potholes and caves that, in well-developed and extensive cases, may result in collapses of rock over caves and voids to produce towering rock pinnacles (as in China).

magma the fluid, molten rock beneath the surface of the Earth. Magma may undergo many stages of change and movement before being extruded at the surface as *lava*, or intruded at some depth as an intrusion (*see* IGNEOUS ROCKS). The composition varies because in moving upwards through the crust, volatiles (gases and liquids) may be lost and some minerals may crystallize out, thus changing the nature of the remaining melt. Magma reaches the surface through pipes into volcanoes, or through fissures, but at depth it may form bodies many kilometres across.

Magmas are formed by the partial melting of mantle (the layer between the Earth's crust and core) in areas of *subduction*, where tectonic plates are destroyed (*see* PLATE TECTONICS) as one plate descends beneath another. Magmas usually undergo many changes in composition as they move towards the surface.

maps are flat, two-dimensional representations of three-dimensional subjects, e.g. an area of land, which contain a variety of data that will differ depending upon the type of map. Differing scales, that is the ratio of a distance on the map to the actual distance on the ground, enable smaller or larger areas of land to be represented.

A standard topographical map indicates the shape of the land by means of *contours* (lines joining points of equal height), and on it the road, rivers, railway lines, towns, forests and parkland may be marked. Maps can be drawn up to illustrate land usage, relief (i.e. the shape of the land surface),

superficial deposits (river and glacial deposits on top of the underlying rocks) or solid geology (the rocks shown with all things above stripped away).

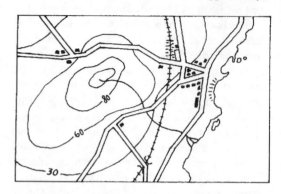

A *map projection* is the representation of the complete surface of the Earth on a plane. There is a large number of such projections and each presents the globe in a different way and thus finds different uses.

Projection	Types	used for
Mercator	cylindrical	navigation
Conical	conic	maps of a small continent
Gnomonic	azimuthal	seismic survey; navigational
Peter's	modified	depicts the Earth's densely
	cylindrical	populated areas in proportion
Stereographic	azimuthal	used widely in structural geology and crystallography

One of the commonest is the Mercator projection, named after the Flemish geographer who used it for his world map of 1569.

Projections are often used in the analysis of directional data. A grid, without the world map, can be used in the study of geological data, e.g. the orientation of folds. This is a sample representation of the polar equal-area net, which is used for plotting a large number of linear elements:

Polar equal area net

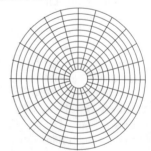

metamorphic rocks are rocks formed by the alteration or recrystallization of existing rocks by the application of heat, pressure, change in volatiles (gases and liquids), or a combination of these. There are several categories of *metamorphism* based upon the conditions of origin: *regional* – high pressure and temperature as found in *orogenic* (mountain-building) areas; *contact* – where the rocks are adjacent to an igneous body and have been altered by the heat (with little or no pressure); *dynamic* – very high, confined pressure with some heat, as generated in an area of faulting or thrusting, i.e. where rock masses slide against each other; *burial* – which involves high pressure and low temperature, e.g. as found at great depth in sequences of sedimentary rocks.

The key feature of all metamorphic rocks is that the existing *assemblage* (group) of minerals is changed by the pressure and/or heat and the presence of volatiles. New minerals grow that are characteristic of the new conditions. Typical metamorphic rocks are schist, slate, gneiss, marble, quartzite and hornfels. Depending on the type of metamorphism, there are systems of classification into *zones* or *grades* where specific minerals appear in response to increasing pressure and/or temperature.

meteorology is the scientific study of the conditions and processes within the Earth's atmosphere. This includes the pressure, temperature, wind speed, cloud formations, etc. which, over a period of time, enable meteorologists to predict likely future WEATHER patterns. Information is generated by weather stations, and also by satellites in orbit around the Earth.

mineralogy the scientific study of MINERALS, i.e. any chemical element or compound extracted from the Earth. It involves the following properties: colour, CRYSTAL form and cleavage, HARDNESS, specific gravity, lustre (how the mineral reflects light) and streak (the colour created by scratching the mineral on a special porcelain plate). Together, these properties help identify and classify minerals and one of the most important features is the *cleavage*.

Cleavage is the tendency for minerals to split along particular, characteristic planes which reflect and are controlled by the internal structure of the crystal. The plane of splitting is that which is weakest, due to the atomic structure of the crystal.

minerals are naturally-occurring inorganic substances that have definite and characteristic chemical compositions and CRYSTAL structures. Minerals have particular features and properties (*see, for example* HARDNESS *and also* MINERALOGY). Some elements occur as minerals, such as gold,

diamond and copper but most minerals are made up of several elements (*see* ST-ELEMENT). Rocks are composed of minerals. There are about 2000 minerals, although the commonest rocks contain combinations from about 30 minerals, which in addition to quartz and calcite come from five or six main groups, all silicates. Silicate minerals are the most abundant rock-forming minerals and comprise about 95% of the Earth's crust, reflecting the fact that oxygen and silicon are the two commonest elements in rocks. The average make up of rocks in the crust is as follows:

silicia	SiO_2	59.3%
alumina	Al_2O_3	15.3%
iron	FeO	3.7%
oxides	Fe_2O_3	3.1%
lime	CaO	5.1%
soda	Na_2O	3.8%
potash	K_2O	3.1%
magnesia	MgO	3.5%
titania	TiO_2	0.7%
water	H_2O	1.3%
phosporus pentoxide	P_2O_5	0.3%

Silicate minerals are based upon a tetrahedral arrangement of four oxygen ions with one silicon ion, and these tetrahedral units are joined in various ways to each other and differing metal ions.

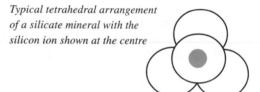

Typical tetrahedral arrangement of a silicate mineral with the silicon ion shown at the centre

Some of the mineral groups:

group	example	structure
Feldspars	orthoclase	framework (tectosilicate)
Pyroxenes	augite	single-chain (inosilicate)
Amphiboles	horneblende	double-chain (inosilicate)
Micas	biotite	flat sheets (phyllosilicate)
Clays	kaolinite	flat sheets (phyllosilicate)
Garnet	pyrope	linked tetrahedra (neosilicate)

Minerals are mined and extracted for diverse purposes, e.g. ornamental stone (in the form of rocks), to generate metals for further use (gold, copper, iron, etc.) or as valuable items in their own right, e.g. diamonds.

monsoon generally refers to winds that blow in opposite directions during different seasons. Monsoons are related to temperature changes in the subtropics and pressure alterations associated with changing JET STREAMS. The word is derived from the Arabic *mausim* which means season, and its meaning has been extended to include the rains that accompany the wind. The Indian subcontinent has a rainy season in its southwesterly monsoon and other areas where monsoons are seen to strongest effect are S.E. Asia, China and Pakistan. However, monsoons also occur in N. Australia, E. & W. Africa.

mountains the formation of mountain chains clearly involves phenomenal movements of the Earth's crust and unimaginable forces, even though the process takes place over many millions of years. The process of mountain building (OROGENY or orogenesis) involves the accumulation of enormous thicknesses of sediments which are subsequently folded, faulted and thrusted, with igneous intrusions at depth (plutons of granite), producing rock complexes involving sedimentary, igneous and metamorphic rocks. A massive linear area that has been compressed in this way is called an orogenic belt and the formation of such belts is interpreted by means of PLATE TECTONICS. Different mechanisms are postulated for the formation of mountain chains, e.g the Andes by subduction of oceanic lithosphere; the collision of continents for the Himalayas and the addition of vast basins of sedimentary rocks and *island arcs* onto an existing plate in the case of the North American Cordillera (*see* PLATE TECTONICS *for definition of terms*).

The result of these global crustal movements is the mountain ranges as we see them today where the higher peaks belong to the younger ranges. The highest points on the seven continents are as follows:

peak	height	continent	country
Everest	8848m	Asia	Nepal/China
Acongagura	6960m	South America	Argentina
McKinley	6194m	North America	Alaska
Kilimanjaro	5895m	Africa	Tanzania
Elbrus	5642m	Europe	CIS
Vinson Massif	5140m	Antarctica	Antarctica
Mauna Kea	4205m	Oceania	Hawaii

Mountains have a considerable effect on local weather conditions and south-facing slopes in the northern hemisphere are warmer and drier than north-facing slopes because they receive more sun. When warm, moist air reaches a mountain, it cools as it rises, releasing moisture so on the leeward side the air descends and absorbs moisture. In many instances, deserts occur on the leeward side of a mountain range, e.g. the Gobi Desert (Asia), and the Mojave Desert of western North

America. It is well-known that the temperature falls with height on a mountain – approximately 6°C for each 100m. Mountains thus show a variety of plants that vary with altitude.

natural gas hydrocarbons in a gaseous state which when found are often associated with liquid petroleum. The gas is a mixture of methane and ethane, with propane and small quantities of butane, nitrogen, carbon dioxide and sulphur compounds.

As with petroleum, gas owes its origin to the deposition of sediments that contain a lot of organic matter. After deposition and burial and through the action of heat, with time, oil and gas are produced. These migrate to a suitable reservoir rock where they reside until extracted by drilling. World gas reserves:

gas-producing country	%world share	reserves	%world share
CIS	37	CIS	38
USA	25	Iran	14
Canada	5	Adu Dhabi	4.25
Netherlands	3	Saudi Arabia	4.25
Algeria	2.5	USA	4
UK	2.3	Quatar	4
Indonesia	2.3	Algeria	2.7
Norway	1.4	Venezuela	2.5

The massive gas reserves of western Canada were first discovered towards the end of the last century when drilling for water for construction of the Canadian Pacific Railway.

oceans technically, those bodies of water that occupy the ocean basins, the latter beginning at the edge of the continental shelf. Marginal seas such as the Mediterranean, Caribbean and Baltic are not classed as oceans. A more general definition is all the water on the Earth's surface, excluding lakes and inland seas. The oceans are the North and South Atlantic; North and South Pacific; Indian and Arctic. Together with all the seas the salt water covers almost 71% of the Earth's surface.

From the shore the land dips away gently in most cases - the continental shelf - after which the gradient increases on the continental slope leading to the deep sea platform (at about 4 km depth). There are many areas of shallow seas on the continental shelf (*epicontinental seas*) e.g. North Sea, Baltic and Hudson Bay. In the ice age, much of the shelf would have been land and conversely should much ice melt, the continents would be submerged further. The floors of the oceans display both mountains, in the form of the mid-oceanic ridges, and deep trenches. The ridges rise 2-3 km from the floor and extend for thousands of kilometres while the trenches reach over 11 km below sea level, at their deepest (Mariana Trench, south-east of Japan).

The six ocean zones are:

littoral zone	between low and high water spring tides
pelagic zone (0–180m)	floating *plankton* and swimming *nekton*
neretic zone (low tide –180m)	*benthic* organisms
bathyal zone (180–1800m)	beyond light penetration, but much benthic life (crawling, burrowing or fixed plants and animals)
abyssal zone (>1800m)	
abyssal plain (~4000 m)	ooze of calcareous and siliceous skeletal remains; red clay only below 5000 m

The oceans contain *currents*, i.e. faster-moving large-scale flows (the slower movements are called *drifts*). Several factors contribute to the formation of currents, namely the rotation of the Earth, prevailing winds, differences in temperature and sea water densities. Major currents move clockwise in the northern hemisphere and anticlockwise in the southern hemisphere. Well-known currents include the Gulf Stream and the Humboldt current.

The major currents

Oceanography is the study of all aspects of the oceans from their structure and composition to the life within and the movements of the water.

ore a naturally-occurring substance that contains metals or other compounds that are commercially useful and which it is economically feasible to mine for profit. *Native ores*, such as gold and copper, occur as the metallic element itself and not in a compound but most metals have to be extracted from compounds, commonly the oxide or sulphide. Minerals containing metallic elements are called ore minerals and ore deposits are aggregates of these minerals.

Extraction of an ore commonly involves removal from a pit (open cast mine) or shaft (in a deep mine). The ore rock is then crushed and possibly washed and sorted before being treated with heat or mixed with chemicals. A furnace may then be used to smelt the ore to produce the metal which separates off from the waste or slag.

Ores are produced geologically in several ways. Some mineral concentrations are associated with magmas, e.g. nickel, cobalt, chromium, platinum or tungsten compounds. Ore minerals are usually found in veins with no obvious connection to igneous activity and in this case may be formed by percolation of hot water laden with metals. Weathering and erosion can also lead to the concentration of certain ore minerals (bauxite or gold and zircon respectively).

orogeny *see* MOUNTAINS.

ozone layer a part of the Earth's atmosphere, at approximately 15–30 km height, that contains ozone. Ozone is present in very small amounts (one to ten parts per million) but it fulfils a very important role by absorbing much of the Sun's ultraviolet radiation, which has harmful effects in excess, causing skin cancer and cataracts and unpredictable consequences to crops, and plankton.

Recent scientific studies have shown a thinning of the ozone layer over the last 20 years, with the appearance of a hole over Antarctica in 1985. This depletion has been caused mainly by the build-up of CFCs (chlorofluorocarbons) from aerosol can propellants, refrigerants and chemicals used in some manufacturing processes. The chlorine in CFCs reacts with ozone to form ordinary oxygen, lessening the effectiveness of the layer. CFCs are now being phased out but the effects of their past use will affect the ozone layer for some time to come.

peat an organic deposit formed from plant debris which is laid down with little or no alteration or decomposition (break down) in a water-logged environment. Bogs, fens, swamps, moors and wetlands are all sites of peat production with some variation in peat structure depending upon the acidity of the conditions.

The conditions are vital so that oxygen is not available for the decay of plant material. As peat accumulates over the years, water is squeezed out and the lower peat layers shrink. However, air drying is essential when peat is cut for fuel. Acid conditions are commonest and *sphagnum moss* is the dominant peat vegetation. However, peat may form in shallow lakes through the gradual takeover by marsh vegetation, or it may develop in shallow lagoons, flood plains or deltas. Conditions for peat formation are currently found all over the world, and in addition to Ireland and Scotland, also on the coastal plains of Virginia and North Carolina, in the Everglades of Florida, Indonesia, India and Malaysia. After the last ice age there were lengthy periods of peat formation. The preserving properties of peat bogs have been seen to great effect over recent years when whole baby woolly mammoths and humans have been found.

permafrost is ground that is permanently frozen, save for surface melting in the summer. It is technically defined as being when the temperature is below 0°C for two consecutive years, and it can extend to depths of several hundred metres. The top layer that thaws in the summer is called the *active layer* and there may be unfrozen ground between this and the permafrost, a zone which is called *talik*. Depths of 1500 metres in Siberia and 650 metres in North America have been recorded and today permafrost underlies 20 – 25% of the Earth's land area – a figure that was much greater during parts of the Pleistocene.

The ground in areas of permafrost shows distinctive features including patterns of circles, polygonal cracks; mounds and pingos. Polygonal cracks are due to the contraction caused by cooling in winter and in Spitzbergen the polygons may reach 200 metres across. Mounds are caused simply by the increase in volume that accompanies freezing of water which pushes up surface layers of soil. Large mounds, up to 40 to 50 metres high are called *pingos*.

petroleum (or *crude oil*) is a mixture of naturally-occurring hydrocarbons formed by the decay of organic matter which, under pressure and increased temperatures, form oil. The often mentioned '*reservoir*' is the rock in which oil (and gas) is found and common types of reservoir rock are sandstone, limestone or dolomite. The oil migrates, after formation from the source rocks to the reservoir (because such vast quantities could not have been formed in place) where it must be contained by a *trap*. A trap is a particular geological configuration where the oil is confined by impermeable rocks.

Most of the world's petroleum reserves are in the Middle East although the CIS and USA currently produce a significant proportion of the world's oil. The modern oil industry began over a century ago when a well was bored for water in Pennsylvania, and oil appeared. Petroleum also occurs in the form of *asphalt* or *bitumen*, syrupy liquids or near solid in form, and there are significant deposits to-

day. The Pitch Lake of Trinidad, over 500 metres across and about 40 metres deep, is fed from beneath as the asphalt is removed. There are similar occurrences in Venezuela and California and in Alberta, Canada are the famous Athabasca Oil Sands, where the sandstone is full of tar, an oil of asphalt.

Typical geological traps

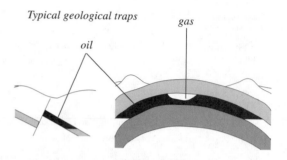

Petroleum consists of many hydrocarbons (*see* S-HYDROCARBONS) of differing composition with small amounts of sulphur, oxygen and nitrogen. The components are separated and treated chemically to provide the basic building blocks and products for the vast petrochemicals industry (*see* S-PETROCHEMICALS).

plate tectonics a concept that brings together the variety of features and processes of the Earth's crust and accounts for continental drift, sea-floor spreading, volcanic and earthquake activity, and crustal structure.

It has long been noticed how coastlines on opposite sides of oceans, e.g. the Atlantic, seemed to fit together. Other geological features led to the theory that continents were joined together millions of years ago. This theory was supported by a reconstruction of fossil magnetic poles and in 1962 by the idea of sea-floor spreading where ocean ridges were the site of new crust formation, with slabs of crust moving away from these central sites. All this was brought together with the idea that the *lithosphere* (the crust and uppermost part of the mantle) is made up of seven large and twelve smaller plates composed of oceanic or continental crust. The plates move relative to each other with linear regions of creation and destruction of the lithosphere.

There are three types of plate boundary; *ocean ridges* where plates are moving apart (constructive); *ocean trenches* where plates are moving together (also for young mountain ranges) (destructive); *transform faults* where plates move sideways past each other (conservative). At destructive plate boundaries one tectonic plate dips beneath the other at an oceanic trench in a process called *subduction* and in so doing old lithosphere is returned to the mantle. *Island arcs* are an example of volcanic activity associated with subduction at an ocean trench, where there is very often also earthquakes.Where two continental plates converge, the continents collide to produce mountains as seen in the Alps and Himalayas today. The transform faults of conservative plate boundaries are generated by the relative motion of two plates alongside each other and the best known example is the San Andreas fault in California, a region which suffers earthquakes along this major fracture.

plate boundaries

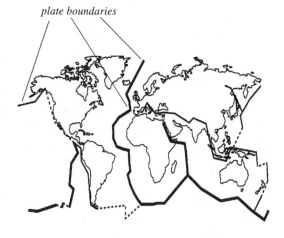

rain one form of precipitation in which drops of water condense from the water vapour in the atmosphere to form rain drops. Other types of precipitation, all water in some liquid or solid form, include snow, hail, sleet, drizzle and also dew. Snow forms below 0°C and depending upon the temperature, occurs in different shapes. When the temperature is well below freezing, it forms ice *spicules* which are small and needle-like. Nearer to 0°C, the characteristic snowflakes grow, but at extremely low temperatures snow becomes powdery. Because snow can vary in its form and accumulation, accurate measurement of falls is difficult, but 25mm of water will be produced by about 300 mm of newly-fallen snow.

Hail is a small pellet of frozen water that forms by rain drops being taken higher into colder parts of the atmosphere. As it then falls, the hailstone grows by adding layers of ice, due to condensation of moisture upon the cold nucleus. *Dew* is the condensation of water vapour in the air caused by a cooling of the air.

remote sensing is the collection of a variety of information without contact with the object of study. This includes aerial photography from both air-

craft and satellites and the use of infrared, ultra-violet and microwave radiation emitted from the object, e.g. an individual site, part of a town, crop and forest patterns. Another type of remote sensing involves the production of an impulse of light, or radar, which is reflected by the object and the image is then captured on film or tape.

Using these various techniques, large areas of the ground can be studied and surprisingly sharp pictures obtained which can be used in many ways. Remote sensing is used in agriculture and forestry, civil engineering, geology, geography and archaeology, amongst others. In addition, it is possible to create pictures with a remarkable amount of detail which would otherwise take a very long time to collect.

rivers streams of water that flow into the sea or in some cases into lakes or swamps. Rivers form part of the cyclical nature of water, comprising water falling from the atmosphere as some form of precipitation (*see* RAIN) and being partly fed by groundwaters or run-off from the melting of glaciers (both of which in any case are derived from atmospheric water).

Rivers develop their own immediate scenery and a river valley will owe its form to the original slope of the land, the underlying rocks and the climate. A river with its tributaries is called a *river system*, and the area from which its water is derived is the *drainage basin* (*see also* DRAINAGE). As rivers grow in size and velocity, rock and soil debris washed into them is carried downstream, eroding the river bed and sides as it goes. As a river continues to flow and carry debris, depositing much material in times of FLOOD, it widens its valley floor, forming a FLOODPLAIN. As it does so, the river swings from side to side forming wide loops, called *meanders*. Eventually as meanders develop into ever more contorted loops, a narrow neck of land may be left which is eventually breached. Thus the river alters and shortens its course, leaving a horseshoe-shaped remnant, or *ox-bow* lake.

Development of river meander leading to the formation of an ox-bow lake

rocks are aggregates of MINERALS or organic matter, and can be divided into three types, based upon the way they are formed: IGNEOUS, SEDIMENTARY and METAMORPHIC.

sedimentary rocks are rocks formed from existing rock sources through the processes of erosion, weathering and include rocks of organic or chemical origin. They can be divided into *clastic* rocks, that is made of fragments, *organic* or *chemical*.

The clastic rocks are further divided on grain size into coarse (or *rudaceous*, grains of 1-2 mm), medium (or *arenaceous*, eg sandstone) and fine (or *argillaceous* up to 0.06 mm). When the grains comprising clastic rocks are deposited (usually in water) compaction of the soft sediment and subsequent *lithification* (that is, turning into rock) produces the layered effect, or *bedding*, that is often visible in cliffs and outcrops in rivers. It is also common for original features to be preserved, for example ripples, small or large dune structures, which in an exposed rock face appear as inclined beds called *current bedding. Graded bedding* shows a gradual change in grain size from the base, where it is coarse, to the top of a bed, where it is fine and this is due to the settling of material onto the sea floor from a current caused by some earth movement.

Many sedimentary rocks, particularly shale, limestone and finer sandstone contain FOSSILS of animals and plants from millions of years ago, and with the original features mentioned above, these are useful in working out the sequence of events in an area where the rocks have been strongly folded.

seismograph in the study of earthquakes (*seismology*), seismographs are used to record the shock waves (*seismic* waves) as they spread out from the source. The seismograph has some means of conducting the ground vibrations through a device that turns movement into a signal that can be recorded. There are numerous seismic stations around the world that record ground movements, each containing several seismographs with numerous *seismome-ters* (the actual detector linked to a seismograph).

soil is the thin layer of uncompacted material comprising organic matter and minute mineral grains that overlies rock and provides the means by which plants can grow. Soil is formed by the breakdown of rock, in a number of ways. Rock is initially fractured and broken up by weathering: the action of water, ice and wind, and any acids dissolved in water moving over, or percolating through the rock. This allows in various organisms that speed up the breakdown process and mosses,

lichens, fungi then take hold and after a while there forms a mixture of organisms including bacteria, decayed organic material, weathered rock and *humus* which is called *topsoil.* Humus is decomposing (breaking down) organic material produced from dead organisms, leaves and other organic material by the action of bacteria and fungi.

The texture of the soil affects its ability to support plants and the most fertile soils are *loams* which contain mixtures of sand, silt and clay with organic material.This ensures there is sufficient water and minerals (which 'stick' to the finer particles) while the coarser sand grains provide air spaces, vital to roots. In addition to plant roots, soil contains an enormous number of organisms including fungi, algae, insects, earthworms, nematodes (roundworms) and several billion bacteria. Earthworms are useful in that they aerate the soil, and the bacteria alter the mineral composition of the soil.

The parent rock is the primary factor in determining the nature of a soil. While sand, silt and clay produce a loam, sand alone is too porous and clay too compacted and impervious (doesn't allow water through). A clay soil can be improved by adding lime, hence a *marl* (a lime rich clay) forms a good soil. Limestone itself does not produce a soil. The rate of breakdown of the rock is also important. Granites decompose slowly but basaltic rocks are the opposite and therefore yield their soil components quickly. This is seen particularly in volcanic areas where lava flows and volcanic ash quickly lead to very productive soils.It can take hundreds of years for soils to become fertile, but to be productive agriculturally, the soil has to be cared for with irrigation, fertilization and prevention of erosion all being important factors. This is apparent when you consider that the soil provides approximately 18 kg of nitrogen, 4 kg of potassium and 3 kg of phosphorus to grow one ton of wheat grain.

stalactites and stalagmites in areas of LIMESTONE where caves form and streams trickle through the rocks and caves, calcium-rich waters tend to drop from cave roofs. As there is a little evaporation from these drops of water, some of the dissolved calcium is deposited, as calcite (calcium carbonate, $CaCO_3$). This deposit builds up very slowly into a stalactite projecting down from the roof. If water continues to drop to the floor, a complementary upward growth develops into a stalagmite - and often the two meet to form a column, or pillar.

Many limestone caves exhibit spectacular developments e.g. White Scar at Ingleton in Yorkshire;

La Cave in the Dordogne; Wookey Hole in the Mendips and many other places.

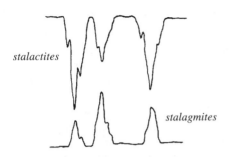

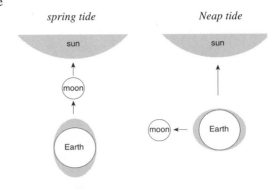

stratigraphy is the branch of geology concerned with the study of stratified rocks, i.e. rocks that were originally laid down in layers. It deals with the position of rocks in geological time and space, their classification and correlation between different areas. Rock units can be identified and differentiated by several means.

Stratigraphy began about two hundred years ago when it was first realized that in a normal sequence of rocks, relatively unaffected by any tectonic movements, younger rocks will lie above older rocks. This was known as the 'law' of superposition, but it soon became apparent that all sorts of geological events could upset this simplistic theory.

stratosphere one of the layers of the atmosphere, lying above the TROPOSPHERE. It lies at a height of between 10 and 50 km and shows an increase in temperature from bottom to top where it is 0°C. (The average temperature above is -60°C). A very large part of the ozone (*see* OZONE LAYER) in the atmosphere is found in the stratosphere and the absorption of ultraviolet radiation contributes to the higher temperature in the upper reaches. This inversion of temperature creates a stability which limits the vertical extent of cloud and produces the sideways extension of a cumulonimbus cloud into the characteristic shape.

thunder *see* LIGHTNING.

tides the regular rise and fall of the water levels in the world's oceans and seas which is due to the gravitational effect of the Moon and Sun. The Moon exerts a stronger pull than the Sun (roughly twice the effect) and variation in tides is caused by the relative positions of the three bodies and the distribution of water on the Earth. When the Sun, Moon and Earth are aligned, the effects are combined and result in a maximum, the high *spring tide* (when the Moon is new or full), conversely when the Sun is at right angles to the Moon, the effect is minimised resulting in a low *Neap tide*.

The effect of tides in the open oceans is negligible, perhaps one metre, and enclosed areas of water such as the Black Sea exhibit differences in the order of centimetres. However, in shallow seas where the tide may be channelled by the shores, tides of six to nine metres may be created.

The tides around Britain's coasts vary markedly due in part to the effect of the *Coriolis force*. This is when air or water is pushed to the side because of the Earth's rotation. Hence in the northern hemisphere, water moving across the surface is pushed to the right (and conversely in the southern hemisphere). Hence the tidal wave which passes northwards up the Irish Sea creates higher tides on the Welsh and English coasts than on the Irish side, and as the tidal wave moves into the North Sea, the Coriolis force pushes the water to the right giving higher tides on the British coastline than on the coasts of Norway and Denmark. The potential for generating energy from tides has long been realized and the first tidal power station was built in France and made operational in 1966.

time zones zones which run north-south, with some variations across the Earth, and which represent different times. Each zone is one hour earlier or later than the adjacent zone, and is 15° of *longitude*. The zones were devised for convenience, but to compensate for the accumulated time change, the *International Date-Line* was introduced. The line runs roughly on the 180° meridian, although it does detour around land areas in the Pacific Ocean, and to cross it going east means repeating a day, while in the opposite direction losing a day.

tornado a narrow column of air that rotates rapidly and leaves total devastation in its path. It develops around a centre of very low pressure with high velocity winds (well over 300 km/hour) blowing anticlockwise and with a violent downdraught. The typical appearance is of a funnel or snake-like column filled with cloud and usually no more than 150 metres across.

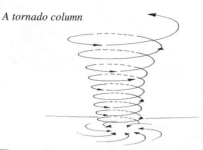

A tornado column

The precise way in which tornadoes form is not known although it involves the interface of warm moist air with dry cooler air and an inversion of temperatures with some event acting as a trigger, possibly an intense cold front. These conditions are found in many countries at mid to low latitudes but particularly so in the mid-west of the USA. The destruction created by tornadoes is due partly to the violent winds and partly to the very low pressure. This has the effect of causing buildings to explode outwards because the pressure outside exceeds that inside and although a tornado may affect an area only 100-150 metres across, the destruction is total. Tornadoes are often unpredictable in their behaviour and can lose contact with the ground or retrace their routes. When a tornado moves out over the sea, and once the tunnel has joined with the waves, a *waterspout* is formed.

tropics are two lines of LATITUDE that lie 23.5° north and south of the equator. The northern line is the *Tropic of Cancer* and the southern one the *Tropic of Capricorn*, and the region between them is called the tropics.

The term 'tropical' is often used to describe climate, vegetation, etc, but it is not an accurate usage of the word. In general a tropical climate does not have a cool season, and the mean temperature never falls below 20°C. Rainfall can be very high indeed, and in many countries these conditions produce a dense, lush vegetation, e.g. tropical rain forests.

troposphere the part of the Earth's atmosphere between the surface and the *tropopause* (the boundary with the STRATOSPHERE). The tropopause is the point at which the change in temperature with height (the lapse-rate) stops, and the temperature remains constant for several kilometres. Within the troposphere itself, the temperature decreases approximately 6.5°C for each kilometre of height. The troposphere is also the layer that contains most of the water vapour and about 75% of the weight of gas in the atmosphere, and it is the zone where turbulence is greatest and most weather features occur.

The level of the tropopause, and therefore the top of the troposphere, varies from about 17 km at the equator, falling to 9 km or lower at the poles. The height variation relates to temperature and pressure at sea-level.

tsunami (plural is also tsunami) an enormous sea wave caused by the sudden large-scale movement of the sea floor resulting in the displacement of large volumes of water. The cause may be an EARTHQUAKE, volcanic eruption, a submarine slide or slump of sediment, which may itself have been started by an earthquake or tremor. The slipping of thousands of tonnes of rock from the sides of fjords may also cause tsunami.

The effect of this sea floor movement in the open ocean may not be seen at all, as the resulting wave may only be one metre or less in height. However, because the whole depth of water is affected, there is a vast amount of energy involved, so when the waves reach shallow water or small bays, the effects can be catastrophic. The waves may travel at several hundred kilometres per hour (600-900) and reach heights of 15-30 metres. The devastation caused is clearly going to be terrible and there are many such instances in the record.

The word originates from the Japanese (*tsu*: harbour and *nami*: waves) where there have been many instances of destructive tsunami. In 1933, an earthquake triggered tsunami with waves up to 27 metres high and thousands of people were drowned along the Japanese coast. The waves were actually recorded about 10 hours later in San Francisco, having crossed the Pacific. It seems that tsunami are generated by submarine earthquakes registering 8 or over on the Richter scale.

tundra is the treeless region between the snow and ice of the arctic and the northern extent of tree growth. Large treeless plains can be found in northern Canada, Alaska, northern Siberia and northern Scandinavia. The ground is subject to PERMAFROST but the surface layer melts in the summer, so soil conditions are very poor, being waterlogged and marshy. The surface therefore can support little plant life. Cold temperatures and high winds also limit the diversity of plants, restricting the *flora* to grasses, mosses, lichens, sedges and dwarf shrubs. Some areas of tundra receive the same low level of precipitation as deserts yet the soil remains saturated due to the partial thaw of the permafrost. Most growth occurs in rapid bursts during the almost continuous daylight of the very short summers.

Due to the inhospitable conditions, animal life is also limited, although more numerous in summer. In addition to insects (midges, mosquitoes, etc.) and migratory birds, there are wolves, arctic foxes, lemmings, hares, snowy owls and the herbivorous reindeer in Europe and caribou in northern America. Polar bears occur at the coast.

In addition to this arctic tundra, there is also *alpine tundra* which is found on the highest mountain tops and is therefore widely spread. However, conditions differ because of daylight throughout the year and plant growth in the tropical alpine tundra also occurs all year round.

unconformity is a break in the deposition of SEDIMENTARY ROCKS representing a gap in the geological record. The unconformity is the junction between younger rocks (above) and older and is formed by a succession of rocks being pushed and possibly folded, eroded and then submerged again so that new sediment is deposited on the older rocks. Usually the rocks above and beneath the unconformity lie at different angles.

It is possible, however, for an unconformity to be represented by differing beds with the same orientation *(disconformity),* or simply by a surface that indicates non-deposition but shows no other apparent breaks, or by sediments being deposited on an igneous intrusion that has been eroded.

volcano a natural vent or opening in the Earth's crust that is connected by a pipe, or *conduit,* to a chamber at a depth that contains MAGMA. Through this pipe (usually called a *vent*) may be ejected LAVA, volcanic gases, steam and ash, and it is the amount of gas held in the lava, and the way in which it is released on reaching the surface, that determines the type of *eruption.*

Volcanoes may be *active,* i.e. actually erupting whether just clouds of ash and steam or lava; *extinct,* i.e. the activity ceased a long time ago; or *dormant.* Dormant volcanoes have often in the past been thought to be extinct, only to erupt again with startling ferocity.

Volcanoes can be described by the type of eruption that is named after a particular volcano that exhibits a specific eruption pattern. These are:

Hawaiian	violent eruptions with viscous lava and *nuées ardentes**
Peléean	moderate eruptions, small explosions and lava of average viscosity
Strombolian	very explosive after a dormant period with ash/gas clouds and gas filled lava
Vesuvian	very explosive with *pyroclastics*** ejected in a column up to 50 kg high producing thick airfall deposits
Plinian	outpouring of fluid lava and little explosive activity

*nuées ardentes: an old term meaning an incandescent ash flow that moves rapidly.

**pyroclastic: volcanic rocks formed from broken fragments, e.g. bombs, pumice, ash, cinders

Volcanoes of the Hawaiian type are also called *shield* volcanoes. The sides of the volcanoes are almost flat because of the rapid flow of the lava. *Composite* volcanoes show greater angles of slope because of a build-up of lava and pyroclastic material. Both shield and composite types are also called *central type* because the supply comes from a central vent, as opposed to *fissure* volcanoes, which erupt through splits where the crust is under tension.

Composite volcano with a central vent

Active volcanoes occur in belts associated with the tectonic plates *(see* PLATE TECTONICS) with about 80% of the active sub-aerial volcanoes at destructive plate margins, 15% at constructive plate margins and the remainder within plates. Most submarine volcanism is at constructive plate margins.

The environmental effect of volcanoes can be very significant, whether it be the enormous amounts of ash ejected into the atmosphere or the consequences of lava flows consuming the countryside. At the time of eruption, volcanic materials are often over 1000ºC, hence flows either burn, push over or cover whatever they meet.

Over 500 volcanoes have been active in historic times but only about 50 erupt each year, often on a very small scale.

water table the level below which water saturates the spaces in the ground; the top of the zone where groundwater saturates permeable rocks. It is where atmospheric pressure is equalled by the pressure in the groundwater. The position (*elevation*) of the water table varies with the amount of rainfall, etc, loss through evaporation and transpiration from vegetation, and percolations through the soil. A spring or seepage occurs when, because of geological conditions, the water table rises above ground level (*see also* AQUIFER).

weather the combined effect of atmospheric pressure, temperature, sunshine, cloud, humidity, wind and the amount of precipitation which together make up the weather for a certain place over a particular (usually short) time period. The weather varies enormously around the Earth but some countries have a stable weather pattern while Britain has a changeable one because it is at a location where many different air masses meet. The surface at which two air masses with different meteorological properties meet is called a *front*.

A *warm front* occurs in a depression, between warm air moving over cold air, and it heralds drizzle followed by heavy rain that then gives way to rising temperatures. A *cold front* is the leading edge of a cold air mass, which moves under warm air, forcing the latter to rise. The result is a fall in temperature, with rainfall passing behind the front.

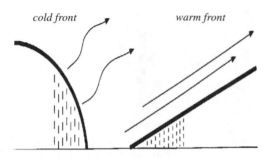

weathering is a combination of chemical and physical processes on the surface of the Earth, or very near to it, that breaks down rocks and minerals. Weathering takes various forms and can be divided into mechanical, chemical and organic:

mechanical	*freeze-thaw action* – alternate freezing and thawing of water in cracks, producing widening or break-up.
	exfoliation – peeling off in thin rock layers (like onion skin).
	disintegration – into grains.
chemical	*carbonation* – the reaction of weak carbonic acid (H_2CO_3) with the rock.
	hydrolysis – combination of water with minerals to form insoluble residues (e.g. clay, minerals).
	oxidation and *reduction*.
organic	*breakdown* by flora and fauna, e.g. burrowing animals, tree roots, and the release of *organic acids* from decomposed plants that react with minerals.

These weathering processes together produce a layer of material that may then be moved by processes of EROSION.

wind a generally horizontal or near-horizontal movement of air caused by changes in atmospheric pressure in which air normally moves from areas of high to low pressure. Wind speed is greater when the ISOBARS (lines joining points at the same pressure) are closely packed on weather maps, and the *Beaufort scale* provides a systematic guide to windspeed. Because of the Earth's rotation and the effect of the Coriolis force (*see* TIDES), air in the northern hemisphere flows clockwise around a high pressure area and anticlockwise around a low.

The *trade winds* play an important part in the atmospheric circulation of the Earth, and they are mainly easterly winds that blow from the subtropics to the equator. The *westerlies* flow from the high pressure of the subtropics to the low pressure of the temperate zone. The westerlies form one of the strongest wind flows, and their strength increases with height (*see* JET STREAM). Depressions are most common in this wind system. The *doldrums* is a zone of calms or light winds around the equator, applied particularly to the oceans, with obvious links to the time when sailing ships were becalmed. Also linked to sailing are the *Roaring Forties*, which are westerlies in the southern hemisphere where they tend to be stronger. However, the supposed link of trade winds with early travel on the sea is incorrect – their origin is from the Latin word meaning 'constant'.

The trade winds

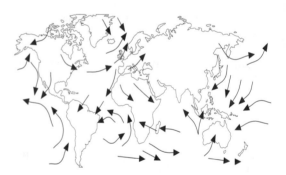

In addition to its destructive power, wind provides an additional hazard when combined with cold. *Wind chill* is the effect wind has in lowering apparent temperatures through increasing heat loss from the body. For example, in calm conditions at 10°F there is little danger for someone properly clothed, but if the wind speed is 25 mph, then the wind chill creates an equivalent temperature of -29°F, which is potentially harmful.

NATURAL HISTORY

The animal and plant kingdoms contain so many millions of species that it would be impossible to cover them all in one volume. All living things can be classified, however, according to their stucture and anatomy. This section of the *Family Encylopedia* lists the different classes and groups found in nature and explains their common characteristics. There are also explanations of more general terms commonly used to describe features of the natural world. The entries are in A to Z order, and there are cross-references, indicated in SMALL CAPITALS, to enable you to relate specific animals or plants to their parent group.

acid rain contains a high concentration of dissolved chemical pollutants such as sulphur and nitrogen oxides. The pollutants arise as gases, given off mainly by the burning of fossil fuels (coal and oil) by industries, vehicles and power stations. The wind may carry the acid gases a long way from their source, but eventually they dissolve in water and fall elsewhere as acid rain. Both the gases and the acid rain cause damage to plant and animal life and lead to an increase in the acidity of water sources such as lakes and rivers. This may cause long-term deterioration in the natural environment affecting numerous plants, animals and micro-organisms.

Actinozoa or **Anthozoa** includes the sea pens, sea pansies, sea feathers, sea fans, sea anemones and corals. It is a class of small marine animals belonging to the Phylum *Coelenterata* (CNIDARIA) which also includes jellyfish and hydras. They are very simple animals with a circular body plan (known as *radial symmetry*) in the form of a structure called a *polyp*. A polyp is a small cylinder attached to an underlying surface such as a rock, the free end usually having a ring of tentacles for feeding. The polyps may exist alone or in colonies, as with many of the corals. Corals have hard external skeletons (exoskeletons) made of calcium carbonate, which are often delicately shaped and coloured. Other animals in this group have structures forming a simple internal skeleton.

A sea anenome

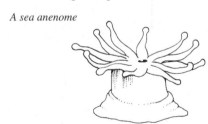

adaptation a feature or characteristic of an organism that has evolved under the processes of natural selection. It enables the organism to exploit a particular aspect of its environment more efficiently. All organisms have adaptations enabling them to survive in their habitat.

 Familiar adaptations include the shape and size of birds' beaks and animals' teeth (according to what they eat); the powerful, digging front legs and claws in such burrowing animals as moles and

flippers in swimming animals such as seals. However, if they become too specialized, they may not be able to survive a sudden change (e.g. in climate) and may become extinct. There are fears for the survival of the panda, which is highly specialized and adapted in terms of its habitat and diet (because this consists solely of bamboo shoots). The illustration below shows how the human foot has adapted for an upright stance whereas the chimpanzee foot has adapted for gripping and climbing.

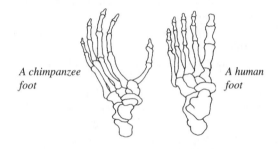

A chimpanzee foot

A human foot

adaptive radiation the separation of species through evolutionary process and natural selection (*see* DARWIN and NATURAL SELECTION) into many descendant species which are able to exploit a variety of different habitats. An example is the vast numbers of different amphibians which have evolved since their early ancestors moved on to land.

Agnatha lampreys and hagfishes (order Cyclostomata) belong to the vertebrate class called Agnatha. These are aquatic, eel-like animals belonging to a very ancient group. Their characteristic feature is a lack of jaws and they have a sucking mouth with horny teeth. They are scavengers or parasites on larger fish.

alga the common name for a simple water plant which lacks roots, stems and leaves but is able to PHOTOSYNTHESIZE. *Algae* range in size from single cells to plants many metres in length.

amoeba *see* PROTOZOAN.

Amphibia salamanders, newts, frogs and toads, are members of this vertebrate class. Amphibians were the first vertebrates to colonize the land about 370 million years ago. The modern ones tend to be highly specialized and not typical of their fossil ancestors, and many of their characteristics are adaptations for life on land. However, most have to return to water to breed and the young (larvae – *see* **ST** – LARVA and METAMORPHOSIS) are aquatic and breathe through gills. In order to become adults they undergo METAMORPHOSIS. Adult amphibians respire through nostrils that are linked by a passage to the roof of the mouth, and also through their skin which is kept moist.

anaconda *see* SQUAMATA.

Angiospermae a class of flowering plants. These are the most complex and highly developed plants and this enables them to live in a great many different habitats as they have evolved various specializations. The female reproductive cell or gamete is formed within a structure called an *ovule*, itself protected by a closed sheath known as the *carpel*. After fertilization, the ovule develops into a seed and seeds may be contained by fruits, (*see* also FLOWERS and PLANTS).

Annelida ragworms (class Polychaeta), earthworms (class Oligochaeta) and leeches (class Hirudinea) belong to the invertebrate phylum Annelida. These familiar animals have an outer layer or cuticle usually composed of *collagen*-like proteins. The body is cylindrical in shape and divided into segments each bearing stiff bristles (*chaetae*) made of chitin (a hydrocarbon with nitrogen). The body cavity is called the *coelom* and is fluid filled. It provides a firm base (*hydrostatic skeleton*), against which muscles present in the body wall can contract to cause movement. A simple nervous system (*see* **ST**) consisting of a pair of *nerve cords* with *ganglia* (swellings) is present. Some worms live in the soil and leaf litter whereas others inhabit marine or freshwater habitats. Earthworms are a vital part of the cycle of decay, eating soil which passes through the gut, food substances being absorbed along the way. Waste material passes out of the body as *castings*. The activity of earthworms helps to till and aerate the soil and the castings improve the texture. Hence they are of great value to farmers, gardeners and horticulturists.

An earthworm

apes *see* **primates**

Arachnida spiders, scorpions, mites, ticks, king crabs, harvestmen, etc. all belong to this class of INVERTEBRATES called Arachnida.

Some animals in this group, especially the spiders, are very familiar to, and hold a special fasci-

nation for man. Most arachnids are terrestrial animals and, unlike insects, they have a head and thorax (middle region of the body) which are not divided from one another. These are formed from eight segments and make up a structure called the *prosoma*. The rest of the body is made up of 13 segments and is called the *opithosoma*. In the head region there are two pairs of projections or *appendages*. The first are called *chelicerae* and are adapted for piercing and grasping prey as most arachnids are carnivorous. The second pair are called the *pedipalps* and these can be specialized to perform a variety of different functions as sense organs, for copulation or noise production. Behind these are four pairs of walking legs and on the *opithosoma*, there may be other specialized appendages, e.g. for silk spinning (spinnerets in spiders) or for injecting poison (as with scorpions). Ticks and mites are parasites on other animals but most arachnids are free-living.

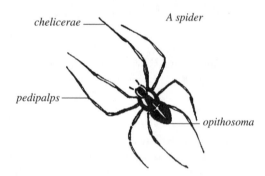

A spider

chelicerae

pedipalps

opithosoma

Arthropoda the largest phylum of INVERTEBRATE animals containing over one million known species which occupy many different habitats. They are often highly specialized and include such classes as the Crustacea (lobsters, crabs, shrimps), Insecta (insects), Arachnida (spiders, mites, scorpions) and Myriapoda (millipedes and centipedes).

Artiodactyla is a mammalian order that includes cattle, camels, hippopotamuses, pigs, deer, antelope, goats, sheep and llamas. These animals are even-toed with the third and fourth toes equally developed to support the whole weight of the body. They are one order of mammals which make up the group commonly called the Ungulates, the other being the Perissodactyla (horses, tapirs, rhinoceros and zebra).

Aves (birds) a *class* of VERTEBRATES which evolved from flying reptiles and still show some features of their ancestry. The most notable of these is the production of reptilian-like eggs and the presence of scales on the legs. Typically, the body of a bird is highly modified for flight with a lighter skeleton than that of other vertebrates. Some organs may even be absent to reduce weight, e.g. females possess only one ovary. The front limbs are developed as wings which are aerodynamically specialized for flight. Birds are covered with feathers and those of the wings are modified for flight while others provide insulation. The breastbone of the skeleton is well developed and has a structure called a *keel* to which the large flight muscles are attached. Birds lack teeth but grind up food in a special part of the gut called the *gizzard*. The jaws are developed to form a beak or bill and a great many different sizes and shapes occur enabling birds to exploit numerous different habitats and types of food.

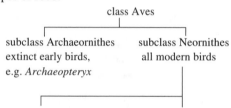

class Aves

subclass Archaeornithes extinct early birds, e.g. *Archaeopteryx*	subclass Neornithes all modern birds

super order Palaeognathae (the walking birds or ratites) usually flightless and large, e.g. ostrich, emu, cassowary, kiwi, elephant bird, moa, rhea.	super order Neognathae (the flying birds) contains numerous orders, e.g. perching, (Passeriformes) including all the song birds, parrots, (Psittaciformes) and eagles, hawks, vultures, etc. (Falconiformes)

Bacillariophyta is the division to which *diatoms*, a type of ALGAE, belong. They are simple, single-celled plants, which sometimes form chains or colonies, and live in both marine and fresh water environments. They are vital ORGANISMS in the PLANKTON and, since they PHOTOSYNTHESIZE, form the basis of FOOD CHAINS. They have beautiful and unique glassy cell walls made of silica and are important fossils. Enormous accumulations of their cell walls are mined as *diatomaceous earth* (or kieselguhr) which has various uses in industry. Planktonic diatoms store food as oil and this has contributed to the petroleum reserves that are extracted and used today.

barrier reef a reef of coral built up in a line running parallel to the shore but at some distance from it, so as to create a lagoon. A good example is the Great Barrier Reef, Australia which is almost 2000 km (1250 miles) long. A barrier reef usually provides a rich and varied habitat and is home to many different kinds of plants and animals.

beaver *see* RODENTIA.

benthos the benthos consists of all the plants and animals living in or on the bottom of the sea, lake or a river. The organisms are described as *benthic* and they may burrow, crawl or remain attached.

biome a geographical area that is characterized by particular ECOSYSTEMS of plants, animals and micro-organisms. It is created by and related to climate and is usually mainly distinguished by the form of vegetation it supports. Examples of biomes are desert, savanna, tropical forest, chaparral (scrubland), temperate grassland, temperate deciduous forest, taiga (coniferous or boreal forest) and tundra.

biosphere the region of the Earth's surface (both land and water) and its immediate atmosphere, which can support any living organism.

birds *see* AVES.

Bivalvia many species of shellfish, e.g. mussels, oysters, scallops, clams and razorshells belong to this class and are called bivalves (class Bivalvia) of the phylum Mollusca.

These animals are characterized by having a body enclosed within a shell divided into two halves (bivalve shell) which is flattened and hinged along the mid-line. They have a smaller head than that of most other molluscs and usually do not move around, living on the bottom of lakes or the sea. However, more mobile kinds such as scallops occur and also burrowing ones such as the shipworm and razorshell.

boa *see* SQUAMATA.

bony fishes most fish have a bony skeleton containing calcium phosphate and a body covered with overlapping scales. These are the bony fishes or class Osteichthyes and include many familiar freshwater and marine species. They have jaws (compare AGNATHA), gills covered by a protective flap called the *operculum* and often possess an air sac called a *swim bladder* for buoyancy. They usually lay large numbers of eggs which are fertilized externally and examples include cod, haddock, salmon, trout and sticklebacks.

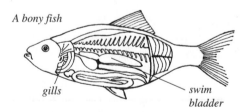

A bony fish

gills

swim bladder

Brachiopoda lampshells are an ancient group of marine animals belonging to the phylum Brachiopoda. Along with two other phyla, they possess a special structure called a lophophore which is used for feeding and RESPIRATION. They have a bivalve shell and a stalk, known as a pedun-cle, which attaches the animal to the surface on which it lives. Brachiopods are very important IN-DEX FOSSILS and one kind, *Lingula* has survived to the present day little changed from the Ordovician geological period, 500 – 440 million years ago.

Lingula

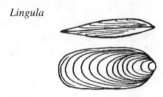

brackish water occurs in estuaries (*see* ESTUARY) and is intermediate in saltiness between sea and fresh water.

Bryophyra is the division of the plant kingdom to which mosses belong. These simple plants commonly thrive in shady, wet conditions and have a waxy, outer layer (*cuticle*) which helps to prevent them from drying out. Mosses lack true roots but may be attached to a surface by hairs (*rhizoids*) or use other plants for support. The hornworts and liverworts are other simple plants which are commonly included in the Bryophyta, although this is now considered to be scientifically incorrect. The three types, although sharing several common features, are probably not closely related.

butterflies *see* LEPIDOPTERA.

camels *see* ARTIODACTYLA.

camouflage a kind of colour pattern or body shape which helps to disguise an animal and enable it to blend into its background. This is useful for both predators and prey animals and can be seen in many different species from invertebrates to large MAMMALS. Patterns of camouflage have evolved through the processes of natural selection (*see* DARWIN) and may be quite intricate and complex. Examples include moths patterned to look like bark or possessing false eye spots, preying mantis, many fishes and animals such as tiger and zebra.

capybara *see* RODENT.

Carnivora an order of mammals which are largely flesh-eaters, including both predators and carrion eaters and those which combine the two. Many of these animals show adaptations of the jaws and teeth. They possess a powerful bite and large piercing and shearing teeth. They often have re-tractile claws and two sub-orders are recognized, the toe-footed carnivores (*Fissipedia*) such as wolves, dogs, cats, raccoons, weasels and badgers, and the fin-footed carnivores (*Pinnepedia*), e.g. walruses, sea lions and seals.

carnivorous plants there are about 400 different species of carnivorous plants that are usually in-sectivorous. They often grow in poor conditions which would not normally supply enough minerals and nourishment. Hence they are specially adapted to attract, trap and digest mainly insect prey in ingenious and highly effective ways. For example, *Sundews* secrete a sticky substance on the leaves which holds insects fast. *Pitcher plants* have modified leaves forming a water-filled trap with smooth sides into which an insect falls and drowns. *Venus fly traps* have hinged leaves with spines at the edges which snap shut over an insect. Once an insect (or other small prey) is caught, these carnivorous plants release digestive ENZYMES and eventually dissolved food substances are ab-sorbed.

The pitcher plant

cartilaginous fishes a class of fishes called the Chondrichthyes which are highly specialized, having a skeleton composed entirely of CARTILAGE. The most familiar examples are the *rays*, *sharks* and *skates* (belonging to the sub-class Elasmobranchii) and also the ratfish, sub-class Holocephali (Chimeras).

cats *see* CARNIVORA.

caterpillars *see* LEPIDOPTERA.

cattle *see* ARTIODACTYLA.

centipede *see* MYRIAPODA.

Cephalopoda nautilus, squids, cuttlefish, octo-puses and the (extinct) fossil ammonites belong to this class. The most advanced group in the phylum Mollusca.

The ammonites had well-developed spiral shells with the animal living inside the last chamber, but *Nautilus* is the only living species that has this structure. In the modern species the shell is much reduced as in the squids, or is inside the body as in cuttlefish or absent altogether as in octopuses. Cephalopods are mainly marine predators, with a well-developed head surrounded by a ring of ten-tacles used to seize their prey. They swim by jet propulsion – expelling water from the body using muscular action. They have complicated eyes and excellent vision and a well-developed nervous system which makes them the most intelligent in-vertebrate animals. Cuttlefish are able to show remarkable colour changes possibly as a means of communication.

A squid

Cestoda a class of the phylum Platyhelminthes. These are the parasitic tapeworms which live inside the gut of a vertebrate. They have a head with suckers and hooks which attache the animal to the intestinal lining of its host. Behind the head there is a series of sacs (called *proglottids*) which give the appearance of segments. Each of these has male and female sex organs and these animals are *hermaphrodite*. The outer surface is protected from the digestive juices of the host's gut by a tough, outer CUTICLE. Tapeworms do not have a gut but dissolved food substances are absorbed over the whole body surface. The life cycle requires both a primary and secondary host. In one tapeworm that affects humans *Taenia saginata*, the beef tapeworm, the primary host containing the adult animal is man and the secondary hosts are cattle. Human beings are infected by eating poorly cooked meat containing cysts (immature worms with a protective shell) which develop into adults once they have been eaten.

*The sucker head of   A cross-section
the tapeworm   of the body*

Cetacea an order of marine mammals comprising the dolphins and whales. They are characterized by having forelimbs developed as *flippers*, and a powerful, flattened tail with two *flukes*, and a dorsal fin. These animals are excellent swimmers and have a thick layer of blubber (fat) beneath a nearly hairless skin serving as insulation and a food store. Air is drawn into the lungs through a blowhole on the upper surface and can be shut off when the animal dives. Dolphins, killer whales and toothed whales are predators belonging to the sub-order *Odontocetti*. Many of the great whales, e.g. the blue whale, are plankton-feeders belonging to the sub-order *Mysticetti*. They filter vast quantities of sea water through whalebone plates in order to ex-

tract their food. The blue whale is the largest known living animal, in excess of 150 tonnes in weight and 30 metres (100 feet) in length.

chameleon *see* SQUAMATA.

Chelonia an order of the class Reptilia (see REPTILE) comprising the turtles, terrapins and tortoises. The body of these animals is encased in a bony shell covered by horny plates. Both marine and land-dwelling species occur and they have jaws developed as horny beaks without teeth. These are a very ancient group of reptiles apparently little changed throughout their evolutionary history.

chimera *see* CARTILAGINOUS FISHES.

Chiroptera bats belong to the mammalian order Chiroptera, the main characteristic of which is flight. They have membrane-like wings spread between elongated forelimbs and fingers and the hind limbs and, occasionally, the tail. Most bats are nocturnal insect-eaters, catching their prey on the wing. They have large ears adapted for *echolocation* by means of which they locate prey and sense their surroundings. Some are nectar-feeders, others eat fruit and *vampire* bats suck blood.

Chlorophyta green algae, belonging to the division Chlorophyta are the largest and most varied group of algae, possessing chlorophyll and able to photosynthesize. They are mainly aquatic in both marine and freshwater but are able to live on land in damp conditions. A well-known type is *Spirogyra*.

Cnidaria hydras, jellyfish and sea anemones and corals belong to the phylum Cnidaria (or *Coelenterata*). These simple aquatic invertebrates have two body forms, an attached stalk called a *polyp* and a free-swimming bell called a *medusa*. Depending upon the species, one or other body form may predominate or one may succeed another in what is known as an 'alternation of generations.' These animals are carnivorous, with stinging cells which immobilize their prey, and tentacles around the 'mouth.' (*see* ACTINOZOA and HYDROZOA).

cobra *see* SQUAMATA.

cockroach *see* ORTHOPTERA.

coelacanth a type of bony fish, belonging to the order Coelacanthiformes, sub-order Crossypterygii, the flesh-finned fishes. All were thought to be extinct but a living representative was discovered in 1938. Living coelacanths are members of the genus *Latimeria* and have a characteristic three-lobed tail fin. They are large fish living at great depths on the floor of the Indian Ocean and 'walk' on their crutch-like pectoral fins. Coelacanths give birth to live young and do not possess lungs. They are a remnant of numerous kinds of lobe-finned fishes that were present in the *Devonian* geological period (408 – 360 million years ago). However, these extinct forms were freshwater animals with lungs and some were the ancestors of the amphibians.

Coleoptera beetles and weevils belong to this vast order (the largest in the animal kingdom) of the class Insecta, and there are about 500,000 known species. They are found in a great number of different habitats, some feeding on vegetation, others on decaying material and the rest are predatory and carnivorous. They have two pairs of wings but the first are protective sheaths (called *elytra*) which are hard and, when folded, meet precisely in a line along the middle of the back. They protect the delicate, membranous hindwings which are used for flying (but are reduced or not present in other species). The mouth parts are often enlarged for biting, and both young (larvae) and adult beetles of certain species can be serious economic pests of crops, timber and stored food. Many beetles are beautifully coloured and patterned.

colony a group of individuals of the same species, living together and to some extent dependent on one another. In some cases the individuals may actually be joined, as in corals, and function as one unit. In others, e.g. social insets such as bees and ants, they are separate individuals but have a complex organization with specialized functions.

commensalism *see* SYMBIOSIS.

conservation means managing the environment in such a way as to preserve the natural resources of plants, animals and minerals etc. and to minimize the impact of man. The natural environment is in a constant state of change, and species become extinct and landscapes and climate alters, but this is usually over a long period of time. Man's activities are often devastating and whole ECOSYSTEMS can be lost very rapidly (e.g. clearance of the tropical rainforest). The environment which is left after such activities is usually significantly poorer and man's gains tend to be short-lived. Hence conservation involves planning ahead, preservation, restoration and reconstruction all the while minimizing damage. This may be on a small scale or global, and conservation is now recognized to be of vital importance to the survival and quality of life of human beings as well as other plants and animals.

Copepoda tiny marine and freshwater invertebrate animals which belong to this large sub-class of the class Crustacea. They are all between 0.5 – 2mm long and do not have a shell (carapace) or com-

pound eyes. They have five or six pairs of swimming legs and are very important species in the PLANKTON. Some may be free-living filter feeders while others (e.g. the fish louse) are parasitic. A well-known freshwater type has one simple eye (ocellus) in the mid-line and is called *Cyclops*.

corals *see* ACTINOZOA.

cricket *see* ORTHOPTERA.

Crustacea lobsters, crayfish, crabs, shrimps, pill bugs, prawns, barnacles, water fleas and COPEPODS belong to this class of the phylum Arthropoda. They have a well-defined head and the body (thorax and abdomen) is divided into segments although these may not be obvious. Various projections called appendages are used for a variety of different purposes, e.g. swimming, feeding and as gills for RESPIRATION.

cyanobacteria the commonly-named 'blue-green algae' which are now regarded as bacteria. They are *procaryotic* organisms and may be single-celled or occur in colonies or as filaments. They are found in all aquatic environments and are often very specialized, growing, for example in hot springs. They possess the ability to photosynthesize and some are able to trap nitrogen and are important in soil fertility. Some kinds occur in the PLANKTON, others form *symbiotic* (SYMBIOSIS) relationships with other organisms (e.g. lichens). Blooms of some species in freshwater lakes which have been enriched with agricultural fertilizers can be a problem, as toxins are produced that are lethal to fish and other animals.

cyclostomata *see* AGNATHA.

Decapoda prawns, shrimps, lobsters and crabs belong to this order of the class Crustacea (*see* CRUSTACEA). These animals possess five pairs of walking legs, the first pair of which (in the crawling varieties) are modified to form powerful pincers. The others are used for feeding or swimming.

decomposers are organisms that break down dead organic matter such as plants, animals and animal waste in order to obtain energy, leaving behind simple organic or inorganic waste. Carbon dioxide is produced and heat is released during decomposition and the main organisms involved are bacteria and fungi but also earthworms (*see* ANNELIDA) and other INVERTEBRATES.

deforestation the felling and removal of trees from an area of natural forest which usually has a devastating effect on the environment. It is carried out to provide wood for fuel (for local people) or for the timber trade, or to clear land for agriculture. Often the soil is very low in minerals and its fertility is soon exhausted. On slopes the soil can be washed away by water gushing down, leading to erosion and landslides. The end result of deforestation can be the creation of near-desert conditions.

deer *see* ARTIODACTYLA.

diatoms *see* BACILLARIOPHYTA.

dinosaurs were a large group of reptiles that flourished some 190 to 65 million years ago during the *Jurassic* and *Cretaceous* geological periods (*see* GEOLOGICAL TIMESCALE). They showed a tremendous variety of body shape and size and included flying species. It was once thought that they were all slow, cold-blooded creatures but now scientists believe that many were fast-moving and able to keep their bodies warm by metabolic activities.

The climate in the Cretaceous period changed considerably. For reasons that are still not absolutely clear, the vast majority of the dinosaurs disappeared in a mass extinction occurring over 5 to 10 million years at the end of the Cretaceous.

A dinosaur

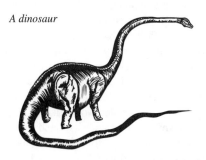

Diptera flies belong to this large insect order which containins about 80,000 species. These are called the true or two-winged flies but the back pair are modified to form knobs (*halteres*) used for balance. The mouth of a fly is adapted to pierce, lap or suck and they feed on sap, nectar and rotting organic matter. Some, e.g. mosquitoes, feed on blood. The LARVA (maggot) eventually forms a pupa and becomes an adult through 'metamorphosis.

A common house fly

dog *see* CARNIVORA.

dolphin *see* CETACEA.

earthworms *see* ANNELIDA.

Echinodermata starfish, brittle stars, sea urchins, sea cucumbers, sea lilies and sea daisies belong to the phylum Echinodermata. These marine invertebrates have existed for a very long time as fossil forms have been found in rocks 500 million years old. They commonly have five arms radiating from a central disc which is the body. The arms bear many tiny tube feet which act like suction pads and are used for movement, feeding and respiration. They have an external shell of plates containing calcium often with bumps or spines and some are able to regenerate lost arms.

echo location the use by animals of high-pitched pulses of sound which bounce off surrounding objects allowing the location of these to be determined. Nocturnal animals such as bats, and aquatic mammals like dolphins produce these sounds and echo location enables them to make sense of their environment and to locate prey.

Some bats have large ears adapted for echo location

ecology this is the study of the relationship between plants and animals and their environment. Ecology (also known as *bionomics*) is concerned with, for example, predator-prey relationships, population changes (in the species present) and competition between species.

ecosystem all the biological life and non-biological components (e.g. minerals in the soil) within an area and the inter-reactions and relationships between them all. Thus an ecosystem includes all the organisms in a community and the geology, chemistry and climate and it may be a lake, forest, region of tundra, etc. Various cycles operate within an ecosystem. Energy, in the form of sunlight, is converted to chemical energy by green plants and is then consumed by animals and released as heat. Food substances, or nutrients, are returned to the cycle as wastes.

Ectoprocta *see* MOSS ANIMAL.

Elasmobranchii *see* CARTILAGINOUS FISHES.

elephant *see* PROBOSCIDEA.

environment the area which surrounds and supports an organism, including all the other living creatures which share the area as well as all the inorganic elements (e.g. minerals contained in the soil).

epiphyte a plant that grows up on another plant purely for support and which is not rooted in the ground. The commonest examples are *mosses, lichens* and some tropical orchids.

estuary a coastal inlet that is affected by marine tides and freshwater in the form of a river draining from the land. An estuary is usually a drowned valley created by a rise in sea level after the last ICE AGE. Estuaries usually contain a lot of sediment (sand, silt, etc.) washed from the land and the tidal currents may produce channels, sandbanks and sand waves. Organisms which inhabit an estuary are adapted to cope with the changes in the *salinity* (saltiness) of the water. They often provide rich feeding grounds especially for many birds, e.g. waders, ducks and geese.

eutherian mammals which have a placenta with which to nourish their embryos within the womb of the mother, belong to the sub-class Eutheria.

eutrophic a body of water, usually a lake or slow-moving river, which is over-rich in nitrogen and phosphorus due to contamination by agricultural fertilizers or sewage. This leads to rapid growth of algae (*algal blooms*) which uses up the oxygen available to other species due to bacteria decomposing dead algae. The final outcome is the death of fish and other organisms. The process is known as *eutrophication*.

fanworm *see* POLYCHAETA.

Foraminifer a phylum of small marine animals (Protozoans) which have hard shells made of calcium carbonate or silica and are important in the fossil record. Sedimentary rocks such as chalk may contain many foraminiferan fossils and modern species are often found in deep sea oozes.

fossil *see* GG.

frog *see* AMPHIBIA.

fungus any one of a number of kinds of simple plants without the photosynthetic pigment CHLOROPHYLL. Fungi cause decay in fabrics, timber and food and diseases in some plants and animals. They may be single-celled or form strands called filaments and are used in BIOTECHNOLOGY in baking, brewing and to make antibiotics.

Gastropoda invertebrate animals, typically (but not always) with a shell such as the limpets, whelks, conches, snails and slugs belonging to this class of the phylum Mollusca. They have a large muscular foot used for movement (hence *gastropod*), a distinct head with eyes and a coiled or twisted shell. Most are marine but some live in freshwater and others are adapted for life on land.

A snail

gila monster *see* SQUAMATA.

goats *see* ARTIODACTYLA.

grasshopper *see* ORTHOPTERA.

greenhouse effect and global warming normally the solar radiation from the sun is absorbed by the Earth's surface and re-emitted as INFRARED radiation. However, this radiation can become trapped in the Earth's atmosphere by carbon dioxide, water vapour, clouds and ozone causing a heating up and consequent rise in the global temperature. Due to the activities of man, the amount of CO_2 in the atmosphere is rising caused, for example, by the clearance of forests (trees trap CO_2) and the burning of fossil fuels. It has been estimated that the global temperature could rise by between 1.5°C and 4.5°C over the next 50 years. This would cause significant melting of the polar ice caps and a consequent rise in sea level causing widespread flooding of coastal areas throughout the world.

growth rings the rings seen if you cut through the woody stem of a plant, for example, the trunk of a tree. Each ring represents one year's growth and by counting the number of rings, the age of the tree can be discovered. Also, each ring is different and comparison of dead and living trees provides a dating method (and information about past climates) for periods within the past eight thousand years. This is known as DENDROCHRONOLOGY.

Gymnospermae conifers, cycads and ginkgos are seed plants belonging to this class (or subdivision) of the division Spermatophyta. They have an ancient history as fossils and are found in rocks dating back to the Devonian geological period (408 – 360 million years ago). They produce *naked seeds* (hence *gymnosperm*) which are not enclosed by a protective sheath (*carpel*). *Compare* ANGIOSPERMAE.

habitat the place where a plant or animal normally lives, e.g. river, pond, seashore.

hagfish *see* AGNATHA.

halophyte a plant that can tolerate a high level of salt in the soil. Such conditions occur in salt marshes, tidal river estuaries (and on motorway verges and central reservations) and a typical species is rice grass (*Spartina*).

hedgehogs *see* INSECTIVORA.

Hemiptera bugs belong to this order of the class Insecta. This is a very large group and includes leaf hoppers, scale insects, bed bugs, aphids, cicadas and water boatmen. The typical body shape is a flattened oval with two pairs of wings folded flat down on the back of the resting insect. Their mouths are tubes adapted for piercing and sucking; some are herbivorous, others carnivorous and yet others parasitic. Some are serious agricultural pests transmitting plant diseases through feeding on sap.

A land bug

herbaceous (*herb-like*) describing a plant with little or no woody stem in which all the green parts die back at the end of each season. Either the whole plant dies (having produced and dispersed seeds) as in *annuals*, or parts of the plant (roots) survive beneath the ground and grow up again the following year, as in *perennials*.

hibernation an adaptation seen in some animals in which a sleep-like state occurs that enables them to survive the cold winter months when food is scarce. Some marked changes occur, notably a drop in body temperature to about 1°C above the surrounding air, and a slowing of the pulse and metabolic rate to about 1% of normal. In this state the animal's energy requirements are reduced and it uses up stored body fat. Bears, bats, hedgehogs, some fish, amphibians and reptiles are examples of animals that hibernate.

hippopotamus *see* ARTIODACTYLA.

holocephali *see* CARTILAGINOUS FISHES.

homeothermy (homiothermy) means warm-bloodedness and is the ability of an animal to keep up a constant body temperature, independent of the heat or cold of its environment. It is a typical feature of birds and mammals and various metabolic processes are involved (*see also* POIKILOTHERMY).

homing the ability of animals to find their way home, especially referring to birds. Many bird species fly thousands of miles over land and sea during annual migrations to return to the same area year by year. It is thought that they navigate by means of the sun and stars and by using magnetic fields.

hominid a member of the primate family Hominidae, which includes modern man as well as fossil hominids that may have evolved around 3.5 million years ago. All belong to the genus Homo. Their main physical characteristic is the ability to walk upright (known as *bipedalism*), and there are also changes in the shape of the skull and teeth compared to those of other primates. They have (or had) complex social and cultural behaviour patterns.

horse *see* PERISSODACTYLA.

human *see* HOMINID.

humus material that makes up the organic part of soil being formed from decayed plant and animal remains. It has a characteristic dark colour and its composition varies according to the amount and type of material present. It holds water (and it forms a physical state known as a COLLOID), which can then be used by growing plants and helps to prevent the loss of minerals from the soil by leaching. Hence it is very important in determining soil fertility. Humus may be more ACIDIC (*mor*), as in the soil of coniferous forests, or ALKALINE (*mull*), as found in the soil of deciduous woodlands and grasslands. Humus contains numerous MICRO-OR-GANISMS and INVERTEBRATE animals, and its presence is of obvious economic importance in the cultivation of food crops.

Hydrozoa corals known as millepore corals, hydras and such animals as the Portuguese man-of-war (*Physalia*) and *Vellela* belong to this class of the phylum Cnidaria. These are mostly marine invertebrates although freshwater types (e.g. *Hydra*) do occur. Many are colonial so that what appears to be a single animal is, in reality, made up of numerous small individuals as with *Physalia* and *Vellela*. *Physalia* is a colony of dozens or perhaps hundreds of individuals. Some form the gas-filled float, others are for feeding having tentacles and a mouth, yet others are protective, possessing stinging cells and some are for reproduction.

Hymenoptera bees (*see below*), wasps, ants, sawflies and ichneumon flies belong to this order of the class Insecta.

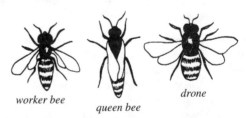

worker bee *queen bee* *drone*

These insects have a narrow 'wasp waist' and usually possess wings. The mouthparts are adapted in a variety of ways, i.e. for biting, sucking or lapping. A structure called an *ovipositor* at the hind end of the body is used for egg laying. It may be adapted for piercing, sawing or stinging and is often long and looped forward.

Some are social insects living in colonies, notably the ants and some bees and wasps. Honey bees have always been an important species to man.

ichneumon fly *see* HYMENOPTERA.

iguana *see* SQUAMATA.

insecta a very important class of invertebrate animals belonging to the phylum Arthropoda, with over three quarters of a million species known to Man. They inhabit a wide variety of habitats, both on land and in water, and there are 26 orders including beetles (Coleoptera), flies (Diptera), bees and ants (Hymenoptera), butterflies (Lepidoptera), grasshoppers (Orthoptera) and dragonflies (Odonata). The body of an insect is divided into three regions known as the head, thorax and abdomen. The head often bears projections called *antennae* which are sense organs, and eyes and mouthparts adapted to feed on various food substances. The thorax bears three pairs of legs and also possibly wings and is composed of three segments. The abdomen or rear part of the body is made up of 11 segments.

insectivora moles, shrews and hedgehogs are small mammals belonging to the order Insectivora. They are mainly nocturnal, with long snouts bearing sensitive hairs which respond to touch, and tweezer-like (incisor) teeth. Their diet is mainly insects although other foods are taken and they are quite a primitive group which seem to have changed very little since their emergence in the *Cretaceous* geological period about 130 million years ago.

The mole has well developed feet which help it to dig and shovel earth

isopod woodlice, fishlice and pill bugs belong to this order Isopoda of the class Crustacea. These familiar invertebrates inhabit marine, freshwater and terrestrial environments and many are parasites. Woodlice are important DECOMPOSERS living in damp places under stones and near the soil surface.

kangaroo *see* METATHERIA.

killer whale *see* CETACEA.

king crab *see* ARACHNIDA.

koala bear *see* METATHERIA.

lamprey *see* AGNATHA.

leaf hopper *see* HEMIPTERA.

leech *see* ANNELIDA.

Lepidoptera moths and butterflies belong to the order Lepidoptera of the class Insecta. The body, wings and legs of the adult insects are covered with numerous, minute scales and the mouth parts are in the form of a long tube (called a *proboscis*)

which is often held coiled beneath the head. This is uncoiled and extended to enable the insect to feed on nectar. Two large pairs of fine wings, often brightly coloured, are characteristic of this group and are held vertically in resting butterflies whereas moths hold their wings in various positions. The immature insects are called caterpillars and they have a well-defined head with chewing mouthparts and a segmented body, usually with each segment bearing a pair of legs. Many caterpillars are serious economic pests of food crops and trees and may transmit plant diseases. The caterpillar becomes an adult by undergoing METAMORPHOSIS via a *pupa* (or *chrysalis*). The pupa may be surrounded by a silk cocoon which the caterpillar produces from special silk glands. Others use leaves or similar material to construct a cocoon.

lichen a plant-like growth formed from two organisms which live together in a symbiotic relationship (*see* SYMBIOSIS). The two organisms involved are a FUNGUS and an ALGA. A lichen forms a distinct structure which is not similar to either partner on its own. Usually most of the plant body is made up of the fungus with the algal cells distributed within. The fungus protects the algal cells and the alga provides the fungus with food through PHOTOSYNTHESIS. A lichen is typically very slow growing and varies in size from a few millimetres to several metres across. It may form a thin flat crust, be leaf-like or upright and branching. Often, lichens are found in conditions that are too cold or exposed for other plants such as in the Arctic and mountainous regions. Reindeer moss and Iceland moss are lichens that provide an important food source in Arctic regions. Other lichens contain substances that are used in dyes, perfumes, cosmetics and medicines, as well as poisons.

limpet *see* GASTROPODA.

littoral a term for the shallow water environment of a lake or the sea that lies close to the shore. In the sea, the littoral zone includes the tidal area between the high and low water marks. Sunlight is able to penetrate so that rooted aquatic plants can grow and usually there is a wide variety and abundance of organisms.

llama *see* ARTIODACTYLA.

lobster *see* DECAPODA.

locust *see* ORTHOPTERA.

lugworm *see* POLYCHAETA.

Malacostraca prawns, and other similar marine and freshwater animals belong to the subclass Malacostraca of the class Crustacea. They typically have compound eyes on stalks and numerous legs or appendages used for a number of different functions including swimming, and a hard shell or carapace.

mamba *see* SQUAMATA.

mammal belonging to the vertebrate class Mammalia which contains approximately 4500 species. Mammals have a number of distinguishing characteristics. They have hair; four different types of teeth (*heterodont*); small bones in the middle ear which conduct sound waves; they are warm-blooded and able to regulate their body temperature and produce sweat as a cooling mechanism. The heart is four-chambered and keeps oxygenated and deoxygenated blood separate and a layer of muscle called the diaphragm is used to fill the lungs with air. The young of mammals develop within the body of the mother and fertilization of the egg is internal. Mammary glands produce milk with which to feed the young once they are born. Mammals possess very complex sense organs and larger brains than other vertebrates of a similar size. They are adaptable and are found in almost all habitats.

mammoth *see* PROBOSCIDEA.

Metatheria a marsupial mammal belonging to the subclass Metatheria of the class Mammalia. The distinctive feature is that the female has a pouch on the lower part of the body (abdomen) which is called the *marsupium*. Marsupials are born at a very early stage of development and crawl into the pouch and fix on to a teat. Here they complete their development until they are able to leave their mother's pouch. Familiar marsupials include kangaroos, koala bears and opossums, and these mammals are only found in Australia and America. The group evolved about 80 million years ago during the late *Cretaceous* geological period. The break up of early land masses meant that the spread of marsupials was restricted. As a result, and especially in Australia, the marsupials have adapted to occupy habits and lifestyles which might otherwise have been filled by placental mammals.

An opossum

mice *see* RODENT.

migration the seasonal movement of animals especially birds, fish and some mammals (e.g. porpoises). Climatic conditions usually trigger off mi-gration where perhaps lower temperatures result in less food being available. Some animals, particularly birds, travel vast distances, e.g. golden plovers, fly 8000 miles from the Arctic to South America. Migrating animals seem to use three mechanisms for finding their way. Over short distances an animal moves to successive familiar landmarks (*piloting*). In *orientation*, a straight line path is taken, based upon the animal adopting a particular compass direction. *Navigation* is the most complex process because the animal must first determine its present position before taking a direction relative to that. It seems that some birds use the sun, stars (often the North star which moves very little), and an 'internal clock' which makes allowances for the relative positions of these heavenly bodies. Even when the sun is hidden by cloud, many birds are able to continue their migration quite accurately by plotting their direction with respect to the Earth's magnetic field.

millipede *see* MYRIAPODA.

millipore coral *see* HYDROZOA.

mimicry a resemblance in which one species has evolved to look like another species. Mimicry occurs in both the plant and animal kingdoms but is mainly found in insects. There are two main types of mimicry. The first is called *Batesian mimicry*, (named after the British naturalist H.W. Bates, 1825—1892). In this, one harmless species mimics the appearance of another, usually poisonous, species. A good example of this is the non-poisonous Viceroy butterfly mimicking the orange and black colour of the poisonous Monarch butterfly. Predators learn to avoid the harmful butterfly and leave the mimic alone as well because of the close resemblance. The second type is *Mullerian mimicry* (named after the German zoologist J.F.T. Müller, 1821—1897). In this, different species which are either poisonous or just distasteful to the predator, have evolved to resemble each other. The resemblance ensures that the predator avoids all the similar looking species so all achieve protection.

mites *see* ARACHNIDA.

mole *see* INSECTIVORA.

Mollusca invertebrate animals, many with shells, such as slugs and snails (class Gastropoda), mussels and oysters (class Bivalvia) and cuttlefish, squids and octopuses (class Cephalopoda) all belong to the phylum Mollusca. Most molluscs are marine but there are freshwater and land-living varieties and it is a large group of animals including over 50,000 species.

monitor lizard *see* SQUAMATA.

monkey *see* PRIMATE.

mosquito *see* DIPTERA.

moss animals and sea mats (phylum Ectoprocta or Bryozoa). These are a group of marine invertebrate animals, many of which are important reef builders. Individual animals are tiny, (only about 1mm across) and resemble the polyps of the phylum Cnidaria. However, they form colonies which may extend for over half a metre. There are around 5000 species and many have a hard external skeleton containing calcium into which the body can be withdrawn. They have a mouth surrounded by tentacles which waft in the water and trap particles of food.

moths *see* LEPIDOPTERA.

Myriapoda centipedes (subclass Chilopoda) and millipedes (subclass Diplopoda) belong to this class of the phylum Arthropoda. These familiar invertebrate animals have a distinct head with antennae and a long segmented body. Centipedes have one pair of legs per segment and are carnivorous. Millipedes have two pairs of legs on each segment and are herbivorous.

nautilus *see* CEPHALOPODA.

nekton animals which swim in the PELAGIC (*see also* OCEANS) zone of the sea or a lake such as jellyfish, fish, turtles and whales.

nematoda roundworms are invertebrate animals belonging to the phylum Nematoda. They have a characteristically rounded body which tapers at either end and some are only about 1mm in length while others may reach 1 metre. There are numerous species and they may inhabit water, leaf litter and damp soil and also occur in plant tissues and as PARASITES in animals. They only have muscles running lengthways (longitudinally) down the body and this is unique to the roundworms. Some free living species are of vital importance as DECOMPOSERS in the soil but others are serious economic pests attacking the roots of crop plants. Yet others are serious parasites of man and animals. Humans can be affected by over 50 nematode species, e.g. threadworms.

neritic zone the shallow water marine zone near the shore, which extends from low tide to a depth of approximately 200 metres. Most *benthic* organisms (*see* BENTHOS) live in this zone because sunlight can penetrate to these depths. Sediments deposited here are sand and clays, and those laid down in ancient seas (now sandstones and mudstones), preserve features such as ripple marks and fossils of marine organisms.

newt *see* AMPHIBIA.

niche all the environmental factors that affect an organism within its community. These factors include living space, available food, climate and all the conditions necessary for the survival and reproduction of the species. Although many species co-exist side by side, each occupies a specific niche within a community with circumstances which apply uniquely to them.

octopus *see* CEPHALOPODA.

Odonata dragonflies and damselflies belong to the order Odonata of the class Insecta. These familiar large insects are carnivorous both as adults and *nymphs* (LARVAE). Eggs are laid in or near water and the aquatic nymphs which hatch out have gills for breathing. They are skilful and voracious hunters with biting mouthparts that are on a plate or *mask* held beneath the head. This is shot out suddenly to seize the prey, which includes invertebrates and even small fish. The nymph crawls up a stalk of a reed, or other plant, right out of the water just before its final moult. The adult which emerges has a pair of large compound eyes and two pairs of large wings with conspicuous veins. In dragonflies the wings are always held out horizontally at right angles to the body in the resting insect. In damselflies, the resting insect holds its wings folded over its back. These insects have very good eyesight and are excellent hunters catching insect prey in flight with their legs. They are often brightly and beautifully coloured and are a very ancient group. Some fossil species found in rocks from the Carboniferous era had wingspans of over half a metre.

A damselfly

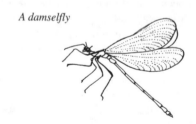

Oligochaeta earthworms and some other worm species belong to the class Oligochaeta of the phylum Annelida. They are distinguished by having very few bristles (chaetae) which are not borne on special structures called *parapodia* (*see* ANNELID).

opossum *see* METATHERIA.

orthoptera grasshoppers, crickets, locusts, preying mantis and cockroaches belong to the order Orthoptera of the class Insecta. This is a large order of insects containing many species familiar to Man. They are usually large and mainly herbivorous with biting mouthparts. Two pairs of wings may be present but some species are wingless. The hind pair of legs are often large and modified for jumping. Many produce sounds by rubbing one part of the body against another (often wings and hind legs), using special *stridulatory* organs to produce noise which is called *stridulation*. Some species, especially locusts and cockroaches are pests of crops or buildings.

A cricket

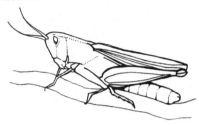

osteichthyes *see* BONY FISHES.

paramecium *see* PROTOZOA.

pelagic any organisms swimming and living in the sea between the surface and middle depths including many fish, marine mammals and plankton. Pelagic sediments (called *ooze*) are deep water deposits made up of the shells of minute organisms and small quantities of fine debris.

Perissodactyla herbivorous grazing mammals such as horses, zebra, rhinoceros and tapir belong to this order of the class Mammalia. These animals characteristically have hoofed feet and an odd number of toes. The teeth are specialized for grinding and the gut adapted for the digestion of cellulose. This was a distinct mammalian line 60 million years ago in the EOCENE geological period (see MAMMAL).

photic zone the uppermost, surface layer of a lake or sea where sufficient light penetrates to allow PHOTOSYNTHESIS to take place. The extent varies according to the quality of the water, (i.e. whether it is cloudy due to suspended particles), but it can extend to a depth of 200 metres.

physalia *see* HYDROZOA.

pig *see* ARTIODACTYLA.

pitcher plant *see* CARNIVOROUS PLANTS.

plankton *see* ST.

Plasmodium *see* SPOROZOA.

Platyhelminthes flatworms, of which there are about 20,000 species, belong to the phylum Platyhelminthes. There are four classes in this group; the Turbellaria (free-living worms), Cestoda (tapeworms), Trematoda (parasitic flukes) and Monogenea (flukes).

platypus *see* PROTOTHERIA.

poikilothermy means cold-bloodedness and is a feature of all animals except mammals and birds. The body temperature of these animals varies ac-

cording to that of the surrounding environment, but they adjust by seeking sun or shade as required. Also, the blood flow to some body tissues is adjusted according to whether heat needs to be lost or conserved (*see also* HOMEOTHERMY).

pollution is contamination of the natural environment with harmful substances which have usually been released due to the activities of Man. The substances which cause the pollution are called *pollutants* and include chemicals from industry and agriculture, gases from the burning of fossil fuels, industrial and domestic waste in landfill sites and sewage. Pollution is regarded as a serious problem because of the threat it poses to the health, not just of humans, but of all animals and plants. Many countries have now introduced stringent regulations in order to control pollution levels.

polychaeta worms such as the ragworm (*Nereis*), lugworm (*Arenicola*) and fan worm (*Sabella*) belong to the class Polychaeta of the phylum Annelida. These characteristically possess a protruding pair of bumps or lobes on each segment, each having numerous bristles or chaetae. Most of these worms are marine animals and live in tubes constructed from pieces of shell and sand. Lugworms and ragworms make burrows in the mud or sand. Most species have a distinct head often with jaws and eyes.

population a group of individuals which belong to the same species. Populations may be genetically isolated, rarely breeding with other members of the same species outwith their particular area. Or, one dense population may merge with another. Usually, a population is affected by such factors as birth and death rates, density (numbers) and immigration and emigration.

population dynamics the study of the changes that occur in population numbers, whether plant or animal, and the factors that control or influence these changes. Distinction is made between factors that are dependent or independent of population density. For example, a natural occurrence such as a flood is an independent factor while food supply is dependent.

porifera sponges, of which there are about 9000 species, belong to this phylum of invertebrate animals. Most sponges are marine and live attached to rocks or other surfaces. They have a hollow, sac-like body punctured with holes and an internal skeleton made up of tiny spines (spicules) of calcium carbonate, silica or protein. Sponges are filter feeders taking particles of food from the water and digesting them. They are very simple animals with no muscle or nerves and are able to regenerate lost parts of their body structure.

A sponge

Portuguese man-of-war *see* HYDROZOA.

prawn *see* MALACOSTRACA.

Praying mantis *see* ORTHOPTERA.

Primates the mammalian order which includes monkeys, lemurs, apes and humans. Evidence suggests that primates evolved from tree-dwelling INSECTIVORES late in the Cretaceous geological period. Characteristic features are manual dexterity (permitted by thumbs and big toes that can touch and grasp), good binocular vision and eye-to-hand co-ordination. The brain is large and the primates are highly intelligent. There are two suborders, the Prosimii (lemurs) and Anthropoidea (monkeys, apes and humans).

Proboscidea elephants belong to this order and are placental mammals characterized by having a trunk (proboscis) and tusks which are modified incisor teeth. The order evolved during the Eocene geological period and was much larger in numbers of species containing, for example, the extinct mammoths. There are just two modern species, the African and Indian elephant.

Prototheria two groups of mammals, the platypuses and spiny anteaters (Echidnas) belong to the subclass Prototheria, the *monotremes*. These are the only living mammals that lay eggs and today they are only found in Australia and New Guinea. The monotremes resemble reptiles in the structure of their skeleton and in the laying of eggs. They are warm-blooded although the body temperature is variable and lower than that of most other mammals. It is thought that the monotremes originated about 150 million years ago early in the development of the mammals.

protozoa used to be the term for a phylum of small, single-celled (mainly) micro-organisms, but now a general word for the PROTISTA. Included are such organisms as *Amoeba* and *Paramecium* and they are very widely distributed in aquatic and damp terrestrial habitats. Some feed on decaying organic material (*saprophytes*) while others are parasites,

e.g. *Trypsanosoma* which causes sleeping sickness.

pterophyta ferns belong to a division of the plant kingdom called the Pterophyta. There are about 12,000 species of ferns which emerged as far back as the Carboniferous geological period.

A fern

Ferns are just about all terrestrial and the majority are found in tropical regions. The leaves are usually called *fronds* and each is made up of numerous smaller leaflets. They are considered to be a fairly primitive group of plants and many of the extinct ferns grew to an enormous size and are well preserved as fossils.

quadrat a small sampling plot (usually one square metre), which is chosen at random and within which organisms found are counted and studied. This is like a 'window' into the area as a whole and the types of animals and plants to be found there. This may be done in relation to a particular species or several different types of organisms.

race a group of individuals which is different in one or more ways from other members of the same species. They may be different because they occupy another geographical area, perhaps showing a variation in colour, or they may exhibit *behavioural*, *physiological* or even genetic differences. The term is sometimes used in the same way as sub-species. Usually, animals from different races are able to interbreed.

racoon *see* CARNIVORA.

Radiolaria tiny, marine protozoan animals, present in PLANKTON, belong to the order Radiolaria. These organisms have a spherical body shape and characteristically possess a beautiful intricate skeleton often composed of silica. This is perforated with holes to make a large variety of patterns, and may be in the form of a lattice of spheres, one inside the other. Often there are many spines or hooks projecting outwards. When Radiolarians die their skeletons, which are highly resistant to decay, settle on the ocean floor to accumulate as *Radiolarian ooze*. When compressed into rock it forms *flint* and *chert*. Radiolarian fossils are very important to geologists and are among the few found in the most ancient rocks from the Precambrian geological period (more than 590 million years old).

Various forms of radiolarian

ragworm *see* POLYCHAETA.

rat *see* RODENTIA.

rattlesnake *see* SQUAMATA.

ray *see* CARTILAGINOUS FISHES.

reindeer moss *see* LICHEN.

reptilia reptiles, a class of VERTEBRATES which were the first animals to truly colonize dry land and to exist entirely independently of water, often live in extremely hot conditions such as semi-desert. They have a dry skin covered with horny scales which protects them from drying out. Reptiles usually lay eggs in holes which they excavate in the sand or mud. The eggs are covered by a protective shell which is porous and allows for the passage of air. Some retain the eggs inside the body until they hatch. Reptiles are cold-blooded (*see* POIKILOTHERMY), but are able to maintain an even body temperature by behavioural means such as basking in the sun. There is evidence that some extinct reptiles (e.g. some of the DINOSAURS), were warm-blooded. Modern reptiles include snakes, lizards, turtles, tortoises and crocodiles.

rhinoceros *see* PERISSODACTYLA.

rice grass *see* HALOPHYTE.

Rodentia a widespread and successful order of mammals containing rats, mice, squirrels, capybaras and beavers. The upper and lower jaws contain a single pair of long incisor teeth which grow continuously throughout the animal's life. These are adapted for gnawing, and the absence of enamel on the back means that they wear to a chisel-like cutting edge. Rodents are herbivorous or omnivorous and tend to breed rapidly. Some, notably rats, are serious pests, spoiling food and are responsible for the spread of diseases.

roundworm *see* NEMATODA.

salamander *see* AMPHIBIA.

sawfly *see* HYMENOPTERA.

sea anemone *see* ACTINOZOA.

sea cucumber, **sea daisy**, **sea lily**, **sea urchin** *see* ECHINODERM.

sea lion and **seal** *see* CARNIVORA.

sea snake *see* SQUAMATA.

scale insect *see* HEMIPTERA.

scorpion *see* ARACHNIDA.

shark *see* CARTILAGINOUS FISHES.

sheep *see* ARTIODACTYLA.

shrimp *see* MALACOSTRACA.

skate *see* CARTILAGINOUS FISHES.

slug and **snail** *see* GASTROPODA.

snake *see* SQUAMATA.

spiny anteater *see* PROTOTHERIA.

spirogyra *see* CHLOROPHYTA.

Sporozoa a class of protozoan organisms which are parasites of higher animals and some cause serious diseases in Man. An example is *Plasmodium* which causes malaria and is transmitted by the mosquito.

Squamata lizards and snakes belong to the order Squamata of the class Reptilia. They are covered by horny scales and male animals are unique in possessing a paired penis. Lizards possess movable eyelids and most have four limbs, e.g. iguanas, monitor lizards, chameleons and gila monsters. Snakes lack movable eyelids and are able to dislocate their lower jaw in order to swallow large prey. Some are constrictor types, e.g. boas, pythons and anacondas, while others may be venomous, e.g. cobras, vipers, mambas, rattlesnakes and sea snakes.

A sea snake

squid *see* CEPHALOPODA.

squirrel *see* RODENTIA.

starfish *see* ECHINODERMATA.

sundew *see* CARNIVOROUS PLANTS.

tapeworm *see* CESTODA.

tapir *see* PERISSODACTYLA.

terrapin *see* CHELONIA.

territory a specific area that animals, whether singly or in groups, defend to exclude other members of their species. A territory may be used for feeding, mating and breeding or all these activities and the size varies greatly from the small nesting territory of sea birds to the large areas used by red squirrels. Many mammals mark their territories with scent while birds use song. Territory differs from home range, the latter being the area in which an animal roams but it is not defended.

tick *see* ARACHNIDA.

toad *see* AMPHIBIA.

tortoise *see* CHELONIA.

trypanosome *see* PROTOZOA.

Turbellaria flatworms such as *Planaria*, belong to the class Turbellaria of the phylum Platyhelminthes. These are simple, small animals, mostly marine although Planarians are common in fresh water streams and ponds. They are carnivorous and move by undulating the body in a wave-like way, or by means of small hairs called *cilia* on their underside.

A flatworm

turtle *see* CHELONIA.

ungulate *see* ARTIODACTYLA and PERISSODACTYLA.

variation the differences between individuals of the same population or species.

venus fly trap *see* CARNIVOROUS PLANTS.

Vertebrata a major subphylum which includes all the animals with backbones: fishes, amphibians, reptiles, birds and mammals (*see* individual entries).

viper *see* SQUAMATA.

walrus *see* CARNIVORA.

water boatman *see* HEMIPTERA.

weevil *see* COLEOPTERA.

whelk *see* GASTROPODA.

zebra *see* PERISSODACTYLA.

SCIENCE AND TECHNOLOGY

The term *Science and Technology* covers many areas, including biology, anatomy, chemistry, mathematics, engineering and physics. This section explains concepts and theories that are essential in the study of all these subjects. The entries are arranged in A to Z order, and cross-references are given in SMALL CAPITALS.

*

acceleration *see* VELOCITY.

acids and bases *acids* are chemical compounds in a liquid form which, depending upon their strength, may occur naturally in foods as very weak acids (e.g. citric acid in lemons) or may form corrosive liquids (e.g. sulphuric acid) which has many industrial uses. Most acids have a sour taste, although it is not advisable to taste them as many are poisonous and can burn skin. Acids turn litmus paper from blue to red (*see* INDICATOR) and an acidic solution has a pH lower than 7. To aid digestion, our bodies produce acids to break down food.

Some examples of acids		
citric	$C_3H_5O(COOH)_3$	found in lemons and fizzy drinks
hydrochloric	HCl	occurs in the stomach, used a lot in the chemical, food and oil industries
sulphuric	H_2SO_4	a strong acid used to make fertilizers, dyes, explosives, iron and steel; also used in car batteries
nitric	HNO_3	used to make explosives, fertilizers, and dyes
formic	HCOOH	an organic acid found in stinging nettles

A mixture of concentrated nitric and hydrochloric acids (in the ratio 1:4) is called *aqua regia* ('royal water') because it will dissolve the 'noble' metals such as gold.

Bases are the opposite to acids and if a base is reacted with an acid in the correct proportions, the two are cancelled out (*neutralized*), producing water and a salt. Nevertheless, bases can be highly reactive and corrosive. Bases turn red litmus paper blue.

Some examples of bases		
lime/calcium hydroxide	$Ca(OH)_2$	used industrially and in cement production
caustic soda/sodium	$Na(OH)$	used to make other chemicals; also hydroxide in paper, aluminium, soap, petrochemicals manufacture
magnesium oxide	MgO	used in antacid preparations

If a compound (usually the oxide -O, or hydroxide -OH of a metallic element) can function as an acid *or* as an alkali (a soluble base) it is said to be *amphoteric*, that is it can combine with an acid or a base to form a salt.

acoustics *see* SOUND.

adhesives are materials that will bond (stick) solids together, through initially wetting the surfaces to be joined and then solidifying to form a joint. Some adhesives, or glues can be made from natural materials such as bones, horn and hides by boiling and simple chemical treatment. Glue made in this way are a mixture of proteins.

Adhesives made from chemicals are called synthetic, and these tend to be stronger than the natural forms. There are several types of synthetic adhesive including thermosetting resins, thermoplastic resins and epoxy resins, the last being among the strongest. Synthetic adhesives have numerous uses including in the manufacture of furniture, textiles, car windscreens, electrical products and also in construction.

In recent years, 'superglues' have been introduced which, it is claimed, are capable of sticking together most surfaces. Great care has to be taken when working with these compounds as they can also stick fingers together and can be dangerous if they come into contact with eyes, etc.

aerodynamics the study of the movement and control of solid objects such as aircraft, rockets, etc. in air. As objects move through the air there is a resistance or *drag* which works against the direction of movement, and the drag increases with speed. All machines used in transport: aircraft, cars, trains, lorries, are designed with streamlined bodies to reduce drag to a minimum. This involves using smooth curved shapes rather than angular shapes.

The aerodynamics of an aircraft wing

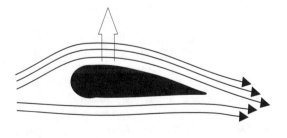

Another important aspect of aerodynamics is *lift*, which is an upward force vital in the take off and flight of aircraft. All aircraft have wings with a curved top surface, and a flatter lower surface. This feature is based upon *Bernoulli's Theorem* (which applies to both liquid and air) which proves that when the flow of air speeds up, the air pressure is reduced. The characteristic *aerofoil* shape of a wing means that air travelling over the top, curved surface has to travel further than air beneath the wing. It therefore moves faster, causing a reduction in air pressure and the wing is lifted, or pushed upwards.

The aerofoil shape is not only used on aircraft—the same principle is applied to rudders, the vanes on a windmill, and even on racing cars, although in the latter case it is to keep the car on the racetrack!

aerosol a fine mist or fog consisting of liquid or solid particles in a gas. It also refers to a pressurized container with a spray mechanism for dispersal of a fine mist of fluid droplets. This system is used for deodorants, paints, insecticides and other chemicals. Within the can is the liquid to be sprayed, and a propellant gas which is under pressure. When the valve is opened by pressing the button, the gas forces (propels) the contents out through the nozzle. Until very recently, the propellants used were commonly CFCs or chlorofluorocarbons, but because it has been proven that these chemicals damage the ozone layer (*see* **GG**: OZONE LAYER.), alternative compounds have been sought.

An *atomizer* is another way of creating an aerosol, but does not require a propellant gas.

ageing includes all the changes which take place as an animal grows older. Animals age at different rates, from some adult insects which only live for a day (mayflies), to elephants (up to 70 years) and tortoises and parrots which may have a life span of well over 100 years. Ageing is perhaps best understood and most studied in human beings. In people, the cells which make up the different parts of the body age at different rates, and while most are renewed some are not (e.g. brain cells), so the whole process is very complicated. As ageing occurs, noticeable physical changes take place which include wrinkling of the skin, greying hair, deterioration in sight, hearing, taste, smell, memory and stiffening of limbs and joints. However, there is a great difference between individual people in the ways in which they are affected by these processes depending upon their inherited (genetic) make up and their lifestyle and environment.

alcohols organic compounds containing carbon, oxygen and hydrogen and with a particular structure where a hydroxyl group, (oxygen and hydrogen, OH), is linked to a carbon. Alcohols are mainly liquids and are used in a variety of ways:

alcohol	formula	uses
methanol	CH_3OH	solvent; manufacture of paints, varnishes, etc.
ethanol	CH_3CH_2OH	alcoholic drinks; solvent; food industry; perfumes
propanol	$CH_3CH_2CH_2OH$	solvent
butanol	$C_4H_{10}O$ (4 different forms)	solvents; artificial flavourings

The alcohols are produced by various chemical reactions, but ethanol occurs naturally by fermentation in which yeast converts sugar to alcohol with carbon dioxide being the side product.

Antifreeze is a liquid added to the water in the cooling systems of car engines. The antifreeze lowers the freezing point of the water, avoiding freezing in the winter. The main component of antifreezes is ethylene glycol, although alcohols may be used.

aldehydes (also known as *alkanals*) another group of organic compounds containing carbon, hydrogen and oxygen which have a particular group, CHO, joined to another carbon atom. The oxygen is joined to the carbon by a double bond, the hydrogen by a single bond. Aldehydes are mainly colourless liquids or solids with typical odours and wide uses:

aldehyde	formula	uses
formaldehyde	HCHO (methanol)	as a solution in water for its germicidal action; also for sterilizing instruments and storing specimens in pathology
acetaldehyde	CH_3CHO (ethanal)	as a stepping stone in the manufacture of other chemicals
propanal	CH_3CH_2CHO	a chemical intermediate

algebra is the use of symbols, usually letters, to help solve mathematical problems where the object is to find the value of an unknown quantity or to study complex theories. In its simplest form it may involve the determination of an unknown in an equation such as: $3x + 1 = 16$, where x is obviously equal to 5. Albert Einstein used a very advanced form of algebra in deriving the equations for his general theory of relativity. The formal rules of algebra were developed by al Khuwãrizmi, an Arab mathematician in the ninth century.

algorithm a set of rules that together form a mathematical procedure that allows a problem to be solved in a certain number of stages. Each rule is very carefully defined so that, theoretically, a machine can undertake the process.

An algorithm is a way of reaching the right answer to a difficult problem by breaking the problem down into simple stages. The stages allow the action to be successfully achieved without necessarily understanding fully the whole process. To be an algorithm, a list of instructions must have an end and be short enough to allow each to be completed, and the process must end at some point. Computer programs contain algorithms, but simpler more everyday examples include making a telephone call and following a recipe. Long division is another example.

alimentary canal *see* DIGESTION.

alkali a base which when dissolved in water produces a solution with a pH over 7, e.g. Na_2CO_3, $Ca(OH)_2$. These are alkaline solutions. Hydroxides of the metallic elements sodium and potassium (NaOH and KOH respectively) produce strong alkalis, as does ammonia in solution (as ammonium hydroxide, NH_4OH).

alloys a mixture or compound containing two (or more) metals, or a metal with a non-metal, e.g. iron-carbon where the metal is the larger component. Alloys tend to have properties which are described as metallic, but alloying is a means of increasing a particularly useful property such as strength, resistance to corrosion or high temperatures.

Bronze was probably the first man-made alloy, made of copper and tin, and it was used for weapons, pans, ornaments. Other common alloys are *brass*, which is copper and zinc, and *steel* which is iron and carbon. There are hundreds of alloys, each created for a particular use, from coins to aircraft parts. This is shown by the different types of brass and bronze now available: brass may also contain aluminium, iron, nickel, tin, lead or manganese while the term bronze is now applied to alloys of copper not containing tin, e.g. aluminium bronze (corrosion resistant), beryllium bronze (high hardness) and manganese bronze (high strength).

A special type of alloy is an *amalgam*—an alloy of a metal with mercury. Mercury is an unusual metal because it is a heavy, silver liquid at room temperature and most metals will mix with it to form an amalgam. For many years dentists have used such an amalgam, with silver, for fillings in teeth, although modern materials are now being used more often.

aluminium is a silvery-white metallic element that in air forms a protective coating of aluminium oxide which prevents further corrosion. It is ductile and malleable (*see* DUCTILITY) and a good conductor of electricity and is used for the manufacture of a large number of things, from drinks cans and foil to pans, cars and aircraft. Aluminium is the commonest metallic element in the Earth's crust and the third most common overall (at about 8%). It is extracted from the mineral *bauxite* (aluminium hydroxides formed by weathering of aluminium-bearing rocks, in tropical conditions) by electrolysis (*see* ELECTROLYSIS). Salts of aluminium are important in the purification of waters and as catalysts (*see* CATALYSIS). The oxide is important in the production of cement, abrasives (*corundum*), refractories and ceramics.

Aluminium occurs in a number of drugs and medications but there is a body of opinion that an excess of aluminium in some form may damage the brain of old people or the very young.

ammonia is a colourless gas made up of nitrogen and hydrogen (NH_3) and it is a very useful chemical compound. It has a sharp, pungent smell, is lighter than air and very soluble in water. It is used in the manufacture of nitric acid, nylon, plastics, explosives and fertilizers. Ammonia is also used in refrigeration equipment as a coolant. Ammonia gas can be turned into a liquid by cooling to -33.5°C, and when it becomes gas again, it takes a lot of heat from the surroundings, thus producing the cooling effect.

Ammonia occurs naturally in some gases given off at hot springs, and it results from the breakdown of proteins and urea by bacteria. It is manufactured industrially by the *Haber process* which involves the reaction of nitrogen with hydrogen under very high temperature (500°C) and pressure (300 atmospheres) and in the presence of a catalyst (*see* CATALYSIS).

ampere the unit used for measuring electric current, usually abbreviated to amps, which was named after the French physicist André Ampère who did much important work in the early 19th century. Machines and appliances use differing currents and with a mains supply of 240 volts, an electric fire will use between 4 and 8 amps. The ampere has a very specific definition, created in 1948, relating to the force produced between two wires in a vacuum, when current is passed. Prior to that the unit related to the amount of silver deposited per second from a silver nitrate solution when one ampere was passed!

analogue *see* DIGITAL AND ANALOGUE.

analysis (in chemistry) involves the use of one, or a

combination of several analytical techniques and pieces of equipment to identify a substance or a mixture of substances. *Qualitative* analysis is when the components are identified, and *quantitative* analysis involves the determination of the relative amount of constituents, usually elements, making up a sample.

Qualitative analysis of inorganic (*see* CHEMISTRY) substances can be done using a variety of *reagents* (a known substance or solution that produces a characteristic reaction) and/or one of several physical methods of analysis, such as X-ray analysis or spectroscopy. Spectroscopy (infra red, ultra violet and also nuclear magnetic resonance) and mass spectrometry are used in organic qualitative analysis and in addition the compound type can be identified by specific reactions which produce known products.

Further methods are adopted for quantitative analysis including the use of certain reactions to precipitate a compound which is then weighed accurately; physical techniques such as spectroscopy and others which relate to a particular property, e.g. electrical and optical, and *volumetric analysis* which involves *titrations*. Titration is when measured amounts of a solution are added, progressively, to a known volume of a second solution until the chemical reaction between them is complete (shown by an indicator) thus enabling the unknown strength of one solution to be calculated. Chromatography is very useful in separating the components of a mixture prior to analysis.

Chemical analysis is used widely in areas such as chemistry itself, medicine, food technology, geology, environmental control, drug testing in sport, forensic science and many more.

anatomy is the scientific study of the structure of living things, which enables an understanding to be built up of how different parts of a plant or animal work. Anatomical study often involves dissection of dead specimens.

animals—*kingdom animalia* modern biologists usually group animals together in the biological kingdom called *Animalia* which contains many different *phyla* (*see* CLASSIFICATION). The simplest animals are the sponges (phylum *Porifera* and the most complicated, including man and other mammals, belong to the phylum *Chordata*. These animals are constructed of many cells (multicellular) performing a variety of functions which, unlike those of plants, do not have outer cell walls. Animals cannot manufacture their own food but must take it in from the outside in some way. Also. animals possess nerves and muscles which vary

greatly in complexity between the different groups. Most animals reproduce by means of sexual reproduction and store carbohydrates as an energy reserve in the form of glycogen (a compound made up of glucose units). In plants, carbohydrate is stored as starch.

anodizing is a process for depositing a hard, very thin layer of oxide on a metal. The metal is usually aluminium or an alloy and the oxide layer provides a non-corrosive coating of aluminium oxide (Al_2O_3) which is resistant to scratches and which protects the underlying metal. Anodizing is achieved by electrolysis and the metal made is the anode, often in a bath of sulphuric, chromic or oxalic acid. Such metals are used in many ways: for window frames, and in trains and aircraft, for example. The oxide coating may be rendered decorative by the use of a dye during electrolysis. Anodized items can be made to retain their natural brightness through the use of very high purity aluminium or an aluminium-magnesium alloy, because in these cases the oxide layer formed is transparent.

Archimedes' principle a law of physics, named after the Greek mathematician Archimedes (287–212 BC). It states that when a body is partly or totally immersed in a liquid, the apparent loss in weight equals the weight of the displaced liquid. Archimedes was a prolific inventor and investigator. He did much scientific work with levers and pulleys and invented the Archimedean screw which is essentially a hollow screw which, with its base in water, when turned, lifts water to a higher level. The device is still used for irrigation if pumps are not available.

An Archimedean screw

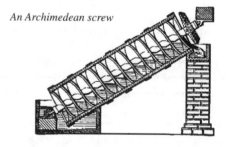

arithmetic the science of numbers which includes addition, subtraction, division, multiplication, fractions, decimals, roots and so on. Arithmetic was first used by the Babylonians about 4000 years ago but the system currently used is that of Arabic numerals. The Romans used a different system in which letters represent quantities:

Roman	Arabic	Roman	Arabic
I	1	X	10
II	2	L	50
III	3	C	100
IV	4	D	500
V	5	M	1000

For example, XIV = 14 and MDCCCCVC represents 1995.

asexual reproduction any form of reproduction in which a new individual is produced from a single parent without the production or fusion of sex cells (gametes). It occurs in plants, micro-organisms and some of the simpler animals, often alternating with a sexual phase. Different types of asexual reproduction occur, one of which is by division of a single cell into two, known as *fission,* as seen in *Amoeba*, a member of the kingdom *Protista*. Another method is *fragmentation* in which parts of the plant or animal separate off, or the whole breaks up with new individuals being formed from the portions. This is seen in some invertebrate animals, particularly earthworms.

Budding is another form of asexual reproduction which occurs in such animals as hydras, sea anemones and corals. A small bud is produced from the parent which may break off and form a new animal or remain attached. Asexual reproduction is also seen in plants, e.g. in the production of bulbs, corms and tubers known as *vegetative propagation*. Such plants as ferns and mosses may have an asexual stage, producing spores, and this is known as the *sporophyte generation.*

atom the smallest particle that makes up all matter, and yet is still chemically representative of the element. Atoms are less than microscopic in size and yet they are made up of even smaller particles. At the centre of an atom is a relatively heavy *nucleus* containing *protons* and *neutrons*. The protons are positively charged and they are offset by an equal number of electrons with negative charge that occupy *orbitals* around the nucleus. The atom is overall electrically neutral and the various elements in the periodic table each have a unique number of protons in their nucleus. The electron weighs a little more than one two thousandth part of a proton!

Orbitals provide a means of simplifying and expressing *atomic structure* and related processes, such as bonding. The theory put forward by Niels Bohr, the famous Danish physicist, proposed that electrons encircled the nucleus in definite orbits. This was discovered to be too simple and the new theory visualized a charge cloud or orbital which represents the likely distribution of the electron in space. There are various types of orbital, each with a different cloud shape around the nucleus (dependent upon atomic structure) and these orbitals interact, merge and change when chemical bonds are formed. The precursor to this picture of atomic structure began with Democritus several hundred years BC. In the 19th century Dalton put forward theories to explain the structure of matter and was the first to calculate the weights of some atoms.

Elements that have more than 83 protons are unstable and decay radioactively, while those with over 92 protons occur only in the laboratory, as a result of high energy experiments (*see also* BONDS). Atoms of an element that have differing numbers of neutrons are called isotopes.

atomic number (*symbol* Z) is the number of protons in the nucleus of an atom. All atoms of the same element have the same number of protons but may have differing numbers of neutrons (producing isotopes). *Atomic weight* or *relative atomic mass* is the ratio of the average mass of an element (i.e. the average for the mixture of isotopes in a natural element) to one twelfth of the mass of carbon -12 atom. Chemical notation places the atomic number as a subscript before the symbol for the element, e.g. chlorine has 17 protons -$_{17}$Cl.

autotrophism means 'self-feeding' and describes the situation in which an organism manufactures the materials it requires for life from inorganic (*see* CHEMISTRY) substances. Plants are *photoautotrophs*, using light as an energy source and carbon dioxide, minerals and water taken from the air and soil. Some bacteria are *chemoautotrophs* obtaining energy from chemical reactions using such substances as ammonia and sulphur.

averages in statistics, average is a general term for *mean, median* and *mode*. The arithmetic mean is the value obtained from a set of figures by adding all the values in the *data set* (collection of numbers) and dividing by the number of readings. The median is the middle reading in a set, so if there are nine readings for temperature, height, weight or a similar quantity, arranged in order, the median is the middle, or in this case, the fifth value. The mode is the value that occurs most frequently in a set of readings, hence in a range of temperatures such as 17°, 18°, 12°, 17°, 21°, 16°, 19°, 17°, the mode is 17°.

bacteria are single celled organisms belonging to the kingdom *Monera* and have a unique type of cell called procaryotic (procaryote). They were the first organisms to exist on Earth but their activities have always been extremely significant as they are important in all life processes. Bacteria may have a protective, slimy outer layer called a capsule and in some, 'hairs' known as filaments are present which cause movement. Two types of bacteria, which each have a different cell wall structure (Gram positive or Gram negative) are recognized by a test known as Gram's stain. Bacteria show a number of different shapes and forms which help to distinguish between the various types. These are *spiral* (spirilli, spirillus), *spherical* (cocci, coccus), *rod-like* (bacilli, bacillus), *comma-shaped* (vibrio) and *corkscrew-shaped* (spirochaetae). They may require an environment containing oxygen, in which case they are described as *aerobic*, or exist in conditions without oxygen, when they are known as *anaerobic*. Bacteria are vitally important in the decomposition of all organic (*see* CHEMISTRY) material and are the main agents in the chemical cycles of carbon, oxygen, nitrogen and sulphur. Most bacteria obtain their food from decaying or dead organic material (*saprotrophs*). Others are parasites and some are *autotrophs* (*see* AUTOTROPHISM), obtaining energy from chemical reactions using inorganic materials such as sulphur and ammonia. Bacteria usually reproduce asexually (*see* ASEXUAL REPRODUCTION) by simple fission but some have a form of sexual reproduction. Some are responsible for very serious diseases in animals, plants and man, e.g. typhoid, syphilis, tuberculosis, diphtheria and cholera.

ballistics is the study of the *flight path* of *projectiles* (i.e. an object that has been projected into the air) under the influence of gravity. The trajectory is the path taken by an object and is usually a particular length, distinguishing it from an orbit. A *ballistic missile* is a missile fired from the ground which follows a flight path shaped like a parabola and which is powered and guided only during the first phase of flight.

Ballistics is a term applied to bullets, rockets, missiles and similar objects.

bar code is a series of black bars printed on a white background that represent codes that can be scanned and interpreted by a reader and computer respectively. Bar codes are found on most products in supermarkets, on books and other items and they contain a product number and maker's identification number.

As the product is dragged past an optical scanner at the checkout, or a light pen is moved over the bar code, the computer matches the product with the price list in its memory. In addition to speeding up the process of adding up items, the use of bar codes helps an accurate analysis of stock to be made, allowing restocking to be undertaken before a product line is exhausted. It also facilitates the output of a detailed, itemized receipt.

barometer is an instrument used to measure atmospheric pressure, i.e. the pressure that the atmosphere exerts on the Earth. The principle of the barometer was discovered by Torricelli, a 17th century Italian physicist. It is used to help predict forthcoming weather changes which result from changes in atmospheric pressure. A mercury barometer is a thin tube, closed at one end which is almost filled with mercury and the open end is stood in a mercury-filled container or reservoir. The vertical height of the mercury column that can be maintained by the atmospheric pressure pushing down on the reservoir is taken as the value of atmospheric pressure at the time of reading. The column is usually about 760 mm high. The mercury barometer is most accurate but a more compact device is the *aneroid barometer* (meaning 'without liquid'). This consists of a small metal box-like chamber (with a thin corrugated lid) which is evacuated (to create a vacuum). Subsequent changes of pressure move the lid and this movement is magnified and translated onto a dial.

bases *see* ACIDS

battery a device, usually portable, in which a chemical change is harnessed to produce electricity. Many batteries are to all intents and purposes non-rechargeable and these are called primary cells and the most important is the Leclanché type. One of the commonest arrangements is a positive electrode made up of a rod of carbon surrounded by a mixture of manganese dioxide and powdered carbon in a case and the whole thing stands in an *electrolyte* (*see* ELECTROLYSIS) of ammonium chloride in the form of a paste. The outer case of zinc acts as the negative electrode. When the battery is connected, a current flows due to reactions and the movement of ions between the various components, thus creating a voltage.

The other important type of battery is the secondary cell, which can be recharged, e.g. the lead *accumulator*, or car battery. In this cell, plates of lead and lead oxide are suspended in sulphuric acid and on charging there are chemical changes (because of *electrolysis*) which are complete when it is charged. When the terminals are wired up, the chemical changes are reversed and current flows until the battery is discharged. The sulphuric acid used has to be of a certain relative density and discharge should not be continued beyond a certain point because insoluble salts build up on the plates. Other types of accumulator include nickel-iron, in which the nickel oxide plate is positive and the iron is negative, in a solution of potassium hydroxide; zinc-air, sodium/sulphur and lithium/chlorine. Although the last two are higher energy cells than most accumulation, they require high operating temperatures.

behaviour (in animals) this term includes all the activities that an animal (or person) carries out during its life and is a major area of scientific study. Those who study behaviour wish to find out why an animal acts in a particular way and what factors might affect this. Behaviour can be divided into three categories. These are known as *reflex*, *instinctive* and *learned*. Reflex behaviour is totally spontaneous and does not involve any conscious act of the will. For example, if you accidentally touch something hot, your hand is immediately and rapidly withdrawn from the source of heat. Instinctive behaviour is inborn and does not need to be learned, being present in all members of a species. The classic example is the male three-spined stickleback fish protecting his territory. Male sticklebacks have a red belly, and this acts as a trigger for the defensive behaviour. A male stickleback will attack models that only vaguely resemble a rival fish provided that the red coloration is present. More realistic stickleback models without a red underside are ignored. Much animal behaviour is *learned*, i.e. it is changed by experience, and this can be seen especially in the young of a species as they play and experiment in the environment that surrounds them. Various types of learned behaviour are recognized. It is generally agreed that all behaviour is a mix, instinct combined with learning and modification by environmental conditions.

benzene a hydrocarbon that is toxic and carcinogenic (may cause cancer). At room temperature it is a colourless liquid with a characteristic smell and it has the formula C_6H_6. The six atoms of carbon form a ring with bond angles of 120°. The ring structure is very stable because although the convention is to show benzene with three double bonds, the bonding electrons are not localized and are spread around the ring.

ring structure for benzene

Benzene was isolated by Michael Faraday, the British physicist, in 1825. He distilled coal tar, and the lightest fraction obtained was condensed to form benzene liquid. Benzene is currently manufactured from petroleum. The benzene ring forms the basis for a large branch of chemistry that deals with *aromatic* hydrocarbons, and benzene is used principally as the starting point for production of other chemicals, e.g. phenol, styrene and nylon.

bimetal strip comprises a bar with strips of two different metals welded together. Because of the different rates of expansion, each expands (or contracts) by a different amount when subjected to heat (or cold), and the bar bends (the metal on the outside being the one that expands more). The two metals used are often brass and invar (an iron-nickel alloy), the brass expanding about twenty times more than invar. This property enables bi-metallic strips to be used in thermostats, thermometers and thermal switches. Also the indicator bulbs in a car flash because of a small bimetal strip.

binary numbers is a system that uses a combination of two digits, 0 and 1, expressed to the base 2, thus beginning with the value 1 in the right-hand column, each move to the left sees an increase to the power of 2:

1×2^5	1×2^4	1×2^3	1×2^2	1×2^1	1×2^0	origin of binary numbers
32	16	8	4	2	1	binary numbers
1	0	1	1	1	0	example

Hence in the example, the figure is $32 + 8 + 4 + 2 = 46$.

The binary number system is the basis of code information in computers, as the digits 0 and 1 are used to represent the two states of on or off in an electronic switch in a circuit.

biochemistry the scientific study of the chemistry of biological processes occurring in the cells and tissues of living organisms. Chemical compounds are the building blocks of all plants and animals, and studying the ways in which these react with one another and how they contribute to life processes is the realm of biochemistry. Biochemists are involved in a wide variety of scientific studies including the search for cures of serious diseases such as cancer and AIDS.

biological control a means of controlling pests by using their natural enemies rather than chemicals which might prove harmful to the environment. The control is artificially brought about in that large numbers of the 'predators' are introduced by human beings in order to control the pests. The method can be very successful, an example being the spread of the prickly pear cactus in Australia which was successfully controlled by the introduction of the cactus moth, whose caterpillars fed on and destroyed the young plant shoots. However, all the biological implications have to be thoroughly researched in advance as sometimes the attempted control has gone very wrong. An example of this was the introduction of large Hawaiian toads into Australian sugar cane plantations with the belief that these would eat the sugar cane beetles that damaged the crop. In fact, the toads did not eat the cane beetles but proved seriously damaging to other animals and are now a great problem in themselves. Another method of biological control used with insect pests (especially some which carry disease, such as the tsetse fly) is to use natural chemicals called pheromones which attract them so that they can be caught. Sometimes the males are caught and radiation is used to make them sterile and then they are released to mate with females. No offspring are produced from these matings and hence the number of insects decreases.

biology the scientific examination of all living things, including both plants, animals and micro-organisms which is further subdivided into specialized areas of study.

bioluminescence the production of light by living organisms which is seen in many sea-dwelling animals, e.g. deep sea fish (angler fish), hydrozoans such as *Obelia*, molluscs, e.g. squids, and some crustaceans. It is also seen in some land-dwelling insects, e.g. fireflies and glow-worms and in some bacteria and fungi. Some animals have bacteria living with them (symbiosis) and it is these which actually produce the light. The light may be given out continuously as in bacteria or 'switched' on and off as in fireflies. The light may be used to attract prey or a mate, or as a warning. A special protein (called a photoprotein) known as luciferin combines with oxygen and during this chemical reaction light is produced.

biophysics a combination of biology and physics in which the physical properties of biological mechanisms are the subject of study. The laws and techniques of physics can be used, for example, in the study of how the heart and muscles work, or in the mechanisms of respiration and blood circulation.

biotechnology the use by man of living organisms to manufacture useful products or to change materials. Examples include the fermentation of yeast to produce alcohol and in bread making, and the growth of special fungi to make antibiotics. Biotechnology mainly uses micro-organisms especially bacteria and fungi. In recent times, genetic engineering, which alters the genes (*see* CHROMOSOMES AND GENES), or 'building blocks', that determine the nature of an organism, is increasingly being used in the field of biotechnology.

biotic *see* ORGANISM.

blast furnace is a furnace comprising a large vertical tower used mainly for the smelting of iron from iron oxide ores. The tower is made of steel plates lined with refractory bricks (firebricks) to withstand the very high operating temperatures (in excess of 1600°C). After some time, about three years, the lining will wear out and the furnace is then shut down to allow the bricks to be replaced.

The furnace is fed from the top with ore, coke and limestone ($CaCO_3$). From near the base of the tower is introduced a high pressure blast of very hot air which ignites the coke, producing carbon monoxide (CO) which then reacts with the iron oxide to produce iron (molten) and at the same time the limestone breaks down to form lime (CaO) and carbon dioxide (CO_2). The lime subsequently combines with impurities from the ore to form a molten slag. As the iron is formed it collects at the base of the furnace, and the slag floats on top. Each is run off periodically, the iron to form bars of 'pig iron' (or cast iron) which contains about 4.5% carbon. The blast furnace can also be used for smelting copper.

The slag also finds a use as *ground granulated blast-furnace slag* (ggbfs) in which it is granulated, ground and then mixed with Portland cement for use in construction. Cements made with ggbfs improve workability and resistance to deleterious reactions and although it is slower to harden, it heats up less than most cements.

blood *see* CIRCULATION.

boiling point is the temperature at which a substance changes from the liquid to the gaseous state. This occurs when the vapour pressure of a liquid is equal to atmospheric pressure. This explains why at high altitudes, water boils at a lower temperature because air pressure decreases. Boiling points of liquids are therefore quoted as the value at standard atmospheric pressure (one atmosphere).

When a substance is dissolved in a solvent such as water, the boiling point of the solvent is elevated. This is called the *molecular elevation of boiling point*, and is caused by a lowering of vapour pressure (because fewer solvent molecules escape from the liquid), hence the temperature has to be higher.

bonds are forces that hold atoms together to form a molecule; or molecules and ions to form crystals or lattices. Bonds are created by the movement of electrons and the type of bond depends upon the atoms involved and their *electronegativity* (a measure of an atom's attraction for electrons, within a molecule). Three basic bond types are recognized: *ionic* (or electrovalent), where one atom loses an electron to another atom, as in NaCl which becomes Na^+ (positive ion or *cation*) and Cl (negative ion or *anion*). Such compounds dissociate (split) into ions in solution and therefore are able to conduct electricity; *covalent* bonds where electrons are shared, often one from each participating atom, although the sharing is equal only when the atoms are identical. Covalent bonding is common in organic compounds. *Metallic* bonding is where electrons are shared between many nuclei thus facilitating conduction of electricity.

bone the hard material that forms the skeleton of most vertebrate animals and contains living cells, called *oestoblasts* and *oestocysts*, responsible for the formation and repair of bones. Bone is made up of hard connective tissue, itself consisting of a tough protein called collagen and bone salts (crystals of calcium phosphate–hydroxyapatite) which both make it strong. The skeleton of an adult human being has 206 bones, some of which are fused together and others are very small.

bone marrow a soft and spongy tissue which is found in the centre of the long bones (e.g. those of the leg and arm) and also within the spaces of other bones. The bone marrow produces blood cells especially in young animals. In older animals only some bone marrow, known as red marrow and mainly found in the vertebrae, pelvis, ribs and breastbone produces blood cells. It contains special (*myeloid*) tissue in which are found cells known as erythroblasts which develop into red blood cells or *erythrocytes*. The cells of the myeloid tissue also develop into leucocytes or white blood cells which are an essential part of the body's immune system.

botany the scientific study of all aspects of the life of plants, compare ZOOLOGY.

brain *see* NERVOUS SYSTEM.

breathing *see* RESPIRATION.

breeding and **hybrid** breeding is the production of offspring and usually refers to the interference by man to produce animals or plants with particular desirable characteristics. The results of breeding are particularly noticeable in domestic farm animals such as cattle or dogs, which look very different to their wild ancestors. Selective breeding has given us animals which provide more milk, wool or meat, and crops which produce more grain

or larger, juicier fruits and succulent leaves.

A hybrid animal is the offspring of parents that are of two different species (*see* CLASSIFICATION), the most familiar example of which is the mule. A mule is the offspring of a female horse and a male donkey. Hybrid animals are usually sterile. N.B. hybrid has a different meaning in genetics where the differences in the parents are at the gene (*see* CHROMOSOMES AND GENES), rather than the species level.

bud (plant) a compact and immature plant shoot made up of many folded and unexpanded leaves or flower petals. These are tightly folded or wrapped around one another and may be enclosed in a protective, sometimes sticky sheath. When the bud opens, the petals or leaves grow and expand, a process which marks the onset of spring moving into summer. *See also* ASEXUAL REPRODUCTION.

bridges a structure used in civil engineering to cover a gap, whether over road, water or railway. For short *spans* (i.e. the distance between supports) of 300 to 600 metres, steel arches, suspension or cantilever bridges are suitable. Above 600 metres, steel suspension bridges are often used. In such bridges, the bridge deck (i.e. the floor that carries the load to the beams and supports) is hung from vertical rods or wire ropes connected to cables that go over the towers and are anchored in the ground. In the UK, the Severn bridge is such a case.

A *cantilever bridge* is usually symmetrical, comprising three spans. The two outer spans are anchored at each shore and then extend beyond the supporting pier to cover one third of the remaining span. This gap is then bridged by another element which rests on the arms of the two cantilevers.

There are other types of bridge, several with particular functions, e.g. swing bridges that rotate on a pivot to allow vessels to pass (often used for canals).

bromine is a typical member of the halogen group (*see also* PERIODIC TABLE). It is a dark red liquid with noxious vapours. It occurs naturally as bromide compounds in sea water and salt deposits, and also in marine plants. When sea water is evaporated, and after sodium chloride ('salt') crystals have precipitated out, the remaining fluid (or *bittern*) contains magnesium bromide.

Bromine is very reactive and has numerous industrial uses including the manufacture of disinfectants, fumigants, anti-knock agents for petrol and in photography.

Brownian Movement is the erratic and random movement of microscopic particles when suspended in a liquid or gas. The movement is due to the continuous movement of molecules in the surrounding gas or liquid which then impact with the particles. This phenomenon accounts for the gradual dispersal of smells in the air and the mixing of fluids which will happen with time, and which will be accelerated if the gas or fluid is hot. As the particles become larger, there is a greater chance of impacts from all sides cancelling each other out and so the movement can no longer be observed. This occurs when the particles are 3-4μ (3-4 microns, or micrometres μm; equal to 3-4 millionths of a metre).

Brownian movement is named after its discoverer, Robert Brown, a Scot (1773–1858). He was a botanist and first observed this movement when studying pollen. It is taken as evidence that kinetic energy applies to all matter and exists even when it cannot be seen.

bubbles may be of a gas within a liquid, e.g. air in water, or air surrounded by a thin soap film as with soap bubbles. Detergent or soap in water creates good bubbles due to the detergent molecules forming a stable alignment. The shape of a soap bubble is due to a balance between the pressure inside trying to blow it apart and the surface tension of the liquid trying to contract the bubble. The pressure inside a bubble decreases as the size of the bubble increases, so smaller bubbles have greater pressure inside than larger ones.

building and construction is central to the development of countries and their economies, and to the operation of almost all aspects of everyone's daily life. There are many professional people; planners, surveyors, civil and structural engineers, architects, builders and more involved, and the major parts of the construction process are: planning, design, construction for new projects, maintenance and refurbishment and restoration for existing buildings.

There is a great variety of projects from railways, houses, hospitals, offices, schools, roads, harbours to dams, airports, bridges and tunnels. Each requires numerous specialists to take a project from the drawing board in a design office to completion. Large projects in particular, which may cost many millions of pounds have to be planned and organized as near to perfection as possible to ensure smooth progress and to make best use of time, people and resources. The basic constructional materials used most often are concrete and steel, but of course brick, timber, stone etc. are also important.

Bunsen burner *see* LABORATORY.

caffeine and tannin *caffeine* is an *alkaloid* (basic organic substance found in plants) and *purine* that occurs in tea leaves, coffee beans and other plants. It acts as a *diuretic* (a substance that increases urine formation) and as a weak stimulant to the central nervous system. It is also found in cola drinks–up to 10% in some cases.

Tannins are a large group of plant substances of use in the treatment of hides. However, the term is misused in connection with tea because tannic acid is not present and tannin is simply a general collective term. Tea does contain *polyphenols* which in some mixtures may resemble tannin in their chemical structure.

calculators are not a recent invention–only the development and inclusion of integrated circuits on silicon chips has been a recent and revolutionary step. Calculators began as mechanical devices in the 17th century and were, inevitably, laboriously slow. The first electronic calculator was produced in 1963, but its scope was very limited and calculators became generally available in the early 1970s. The first calculator that could be programmed to solve mathematical problems was made by Hewlett-Packard in 1974. Now there is an enormous variety of calculators available, some as small as credit cards.

calculus a large branch of mathematics that enables the manipulation of and working with quantities that vary continuously. The techniques of calculus were developed by two scientists at the same time, the German Gottfried Leibniz and Isaac Newton in Britain. It had its origin in the study of falling objects. *Differential calculus* concerns the rate of change of a dependent variable and involves the gradient, maxima and minima of the curve of a given function. *Integral calculus* deals with areas and volumes, and methods of summation and many of its applications developed from the study of the gradients of tangents to curves.

calcium a soft silver-white metal at room temperature which tarnishes very quickly in air, and reacts violently with water. Calcium compounds are very widely distributed; as calcium carbonate ($CaCO_3$) in limestone and chalk; as calcium sulphate ($CaSO_4$) in gypsum; within silicates occurring in minerals and there are many other occurrences. In biological systems, Calcium ions (Ca^{2+}) are vital to life as constituents of bone and teeth.

Calcium is also important industrially as lime (CaO), as $CaCO_3$ in the production of iron from its ores, and in the preparation of certain transition element metals, (*see* PERIODIC TABLE).

camera *see* PHOTOGRAPHY.

camouflage *see* **NH** CAMOUFLAGE.

cancer a disease of animal cells in which there is a

rapid and uncontrolled rate of growth and reproduction. A collection of cancer cells is known as a
tumour which is termed *benign* if it remains localized in one place. A *malignant* tumour spreads, invades and destroys surrounding healthy tissue. It
spreads via the bloodstream and produces secondary cancers in other parts of the body by a process
known as *metastasis*. These malignant cancers can
often prove fatal as they disrupt the normal functioning of cells in the affected tissues. The search
for the cure and causes of various kinds of cancer
is a vast area of scientific and medical research.
The treatment and life expectancy of people affected by most kinds of cancer has greatly improved in recent years. Some substances and environmental factors (known as *carcinogens*) are
known to increase the risk of cancer developing
and the best known of these is cigarette smoke.
Cancer treatment involves surgery, radiotherapy
(the use of radiation to destroy tumours), chemotherapy (the administration of drugs to destroy tumours) and the giving of more specialized drugs
(cytotoxic) which attack the malignant cells in a
special way.

capacitor a device for storing electric charge comprising two plates separated, in its simplest form,
by air but usually by an insulating material called
a *dielectric*. Commercial capacitors have foil
plates separated by a dielectric such as Mylar,
forming a sandwich which is rolled up into a cylindrical shape. This gives a large area of plate in a
small device. Other dielectrics include mica, certain plastics and aluminium oxide. Capacitors are
used a great deal in electrical circuits in televisions, radios and other electronic equipment. The
capacitor works by a charge on one plate inducing
an equal charge on the opposite plate and a change
in the charge is mirrored between plates.

capillary *see* CIRCULATION.

capillary action a phenomenon related to surface
tension in liquids in which there is an attraction
between the sides of a glass tube and water molecules. The result is that water is pulled up a very
thin glass tube (a *capillary tube*), and it will move
further the thinner the tube. This action happens
with other liquids including alcohol, but mercury
moves in the opposite way. Many occurrences involving the movement of water involve capillary
action, e.g. tissues soaking up water, walls taking
in water producing damp.

capsule a widely used term in biology which describes a structure found in flowering plants,
mosses and liverworts, bacteria and animals.

1. In flowering plants, a capsule is a dry fruit

containing seeds which splits when ripe. The splitting and releasing of the seeds occurs in different
ways, e.g. splitting into different parts, known as
valves, in the iris, through holes or pores as in the
snapdragon, or through a lid as in the scarlet pim-
pernel.

2. In plants called bryophotes (*see* **NH**:
BRYOPHOTES), which includes liverworts and
mosses, a capsule contains spores and is produced
on the end of a thick stalk (seta). It is known as a
sporophyte.

3. Some bacteria produce an outer sheath of ge-
latinous material surrounding the cell wall which
is protective and is called a capsule.

4. In vertebrate animals a connective tissue sheath
known as a capsule surrounds some of the joints in
the skeleton. Also, a membrane or fibrous connective tissue sheath called a capsule encloses and
protects some vertebrate organs such as the kidney.

carbides are compounds of metals with carbon
which in many instances result in hard, refractory
materials with very high melting points. Tungsten
or tantalum carbide are used for tools as are mixtures with cobalt and nickel. Other uses include:

metal	property of resulting carbide	use
titanium	resists molten metals; high melting point, hardness heat resistance	parts in fast reac- tors; gas turbine blades; coatings on rocket components
chromium	inert chemically	electrodes, parts for use with cor- rosive chemicals
tantalum	high melting point; metallic conduction	elements in elec- tric furnaces

carbohydrates are chemical compounds made up
of carbon, hydrogen and oxygen forming a very
large and important group of substances. The most
familiar ones are sugars, starch and cellulose and
carbohydrates are manufactured by green plants
during the process of photosynthesis. Carbohydrates are a vital food source for animals as they
are broken down within the body to provide the
energy required for all the various life processes.

carbon a non-metallic element that occurs as the element in two forms, graphite and diamond, and in
numerous compounds. Carbon is unique in the
enormous number of compounds it can form both
in inorganic form, e.g. carbon dioxide, carbonates,
and in being the basis of organic chemistry. Carbon is *allotropic*, i.e. it exists in several elemental

forms–carbon black, in addition to graphite and diamonds, each differing in their crystalline structure and thus their properties.

Carbon is used extensively; in steelmaking, for motor brushes, cathode ray tubes and *active carbon* is used as a purifying agent in removal of air pollutants, water treatment and purification of chemicals, and as carbon fibres. Carbon is the standard for measuring atomic masses (*see* ATOMIC NUMBER) and the radioactive carbon -14 is used in radio carbon dating (*see* **GG**: DATING METHODS).

carbon dioxide is a colourless gas at room temperature that *sublimes* at -78.5°C. It occurs naturally in the atmosphere and is the source of carbon for plants, playing a vital role in plant and animal respiration and metabolism (since it is exhaled by animals and absorbed by plants during photosynthesis). Carbon dioxide forms when coal, wood etc. are burned and it is a byproduct of fermentation. It is produced industrially from *synthesis gas* (a mixture of hydrogen and carbon monoxide) used in the production of ammonia. Its physical properties make it a useful refrigerant and it is dissolved in mineral waters and fizzy drinks, creating bubbles. Also, since it is heavier than air and does not support combustion it is used in fire extinguishers.

Sublimation is when a solid substance changes to the vapour phase *without* first becoming liquid.

carbon fibres fibres consisting of oriented carbon chains which produce a material that is very strong for its weight. Carbon (or graphite) fibres are manufactured by heating polyacrylonitrile $(CH_2.CHCN)_n$ fibres initially at low temperatures to stabilize the fibres and then at high temperatures (2500°C) and under tension to orientate the graphite along the fibres. The resulting fibres are usually 8–11μm in diameter and are not generally totally crystalline, with graphite crystals in amorphous carbon. The fibres are used as a reinforcing material in resins and ceramics producing strong materials able to withstand high temperatures and with a higher strength to weight ratio than metals.

Comparative strength of carbon fibres and selected materials			
Material	Density	Tensile strength	Temp. of use (°C)
aluminium	2.7	0.06	600
steel	7.9	2.0	1200
aluminium oxide fibre	4.0	2.0	800
carbon fibre	1.9	2.3	2500
glass	2.5	0.07	700
silicon fibre	2.2	5.8	750
magnesium	1.74	0.04	200

carbon monoxide a colourless gas with no smell that is formed when coke and similar fuels are burnt in a limited supply of air, i.e. incomplete combustion. It also occurs in the exhaust emissions of car engines. It is used industrially because of its reducing properties (*see* OXIDATION AND REDUCTION), for example in metallurgy.

One of its most important properties is its toxicity. It is poisonous when breathed in because it combines with haemoglobin in the blood, which reduces the capacity of the blood to carry oxygen (*see* CIRCULATION).

carnivore any animal that eats the flesh of other animals, often referring to mammals belonging to the order *Carnivora* (*see* **NH**). Some highly specialized plants are carnivorous, and these usually trap insects or small frogs, etc, and secrete enzymes that break them down, enabling the products of digestion to be absorbed.

cartilage a special type of connective tissue that, together with bone, forms the skeleton of vertebrate animals. It is composed of proteins and carbohydrates, especially collagen, which is a fibrous substance that makes cartilage tough and elastic. Cartilage is found in various parts of the skeleton, e.g. between the vertebrae of the backbone and around joints. It is also present in rings around the windpipe (trachea) and in the ears and nose.

cast iron or *pig iron* is the impure form of iron produced in a blast furnace. It contains 2–4.5% carbon in a form called *cementite* (Fe_3C) and other elements such as silicon, phosphorus, sulphur and manganese. Cast irons are often too brittle to use, but additions of other metals can improve its workability. Wrought iron is made by melting cast iron and scrap iron in a special furnace, removing most of the impurities. Cast iron melts at 1200°C, whereas wrought iron softens at 1000°C and melts at 1500°C (*see also* STEEL).

catalysis the use of a *catalyst* to alter the rate of a chemical reaction. A catalyst usually increases the speed at which a chemical reaction proceeds, but is itself unchanged at the end of the reaction. In practice the catalyst may be changed physically if not chemically. Catalysts are used a great deal in the chemical industry, iron and platinum being common examples. Metal oxides are also used (copper, zinc, aluminium, titanium), and *zeolites* have become more important over recent years. In biological systems, enzymes are the natural organic equivalent of catalysts.

A *catalytic converter* is fitted to new cars to reduce the volume of exhaust gases produced. The gases pass over a honeycomb structure (which greatly increases the surface area) that is coated with a catalyst such as platinum or palladium, and reactions produce water vapour and carbon dioxide.

Zeolites are natural or synthetic alumina silicates of sodium, potassium calcium and barium that contain cavities with water. The water can be removed by heating and then occupied by other molecules. With small amounts of platinum or palladium they are used in the catalytic cracking of hydrocarbons.

cathode ray tube is an evacuated tube in which *cathode rays* are produced and strike a fluorescent screen; it forms the basis of a television and a *cathode ray oscilloscope* which is used in laboratories and radar systems. At one end of the tube is an assembly (*electron gun*) that produces a beam of electrons which can be focused. The beam passes between two sets of plates, each of which can deflect the electron beam when a voltage is applied to the plates. The resulting beam then impinges upon the screen which is coated with zinc sulphide and fluoresces where the electrons strike.

A cathode ray tube

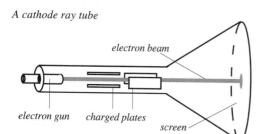

electron beam

electron gun   charged plates

screen

caustic soda chemically is sodium hydroxide (NaOH), a whitish substance that gives a strongly alkaline (*see* ALKALIS) solution in water and will burn skin. It is *deliquescent*, that is it picks up moisture from the air and may eventually liquefy. It is used a great deal in the laboratory and is a very important industrial chemical, being used in the manufacture of pulp and paper, soap and detergents, petrochemicals, textiles and other chemicals. Sodium hydroxide is itself manufactured by the electrolysis of sodium chloride solution, chlorine being the other product, or by the addition of hot sodium carbonate solution to quicklime (calcium oxide, CaO).

cell the smallest and most basic unit of all living organisms. A cell is able to perform the main functions of life which are metabolism, growth and reproduction and there are two main types, procaryote and eucaryote. Bacteria and blue-green algae are procaryotes and all other animals and plants are eucaryotes. The simple organisms consist of a single cell and are termed unicellular, e.g bacteria. However, most living creatures are formed of vast numbers of cells which congregate together to form highly specialized tissues and organs. A typical cell has an outer layer or membrane surrounding a jelly-like substance called *cytoplasm*. Various structures called organelles (*see* ORGAN) are contained within the cytoplasm which are responsible for a number of different activities. The largest structure present is the nucleus surrounded by a (nuclear) membrane which contains the chromosomes and genes. Plant cells differ from animal ones in having an outer wall composed of a carbohydrate substance called cellulose. These walls may be thick giving strength and shape to the structure of the plant. Substances pass into and out of cells across the cell membranes e.g. food materials and oxygen and carbon dioxide.

A typical cell   nucleus   cytoplasm

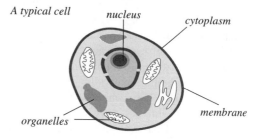

organelles   membrane

cell division the formation of two daughter cells from one parent cell, with the nucleus dividing first by a process known as *mitosis*. The end result is that the nucleus of each daughter cell contains the same number of chromosomes (carrying the genes) as the original parent cell. The chromosomes are in the form of strands of material which come together in pairs. These eventually arrange themselves across the centre of the nucleus and the nuclear membrane disappears. Each member of a pair is now composed of two strands, known as *chromatids,* which are joined at a single point called a centromere. As mitosis progresses, the centromeres split and one chromatid from each of the pairs moves to opposite ends. A nuclear membrane forms around each group of chromatids, now the new chromosomes and then the cytoplasm itself divides to form two daughter cells each with its own nucleus. Another type of cell division occurs in the formation of sex cells (sperm and eggs), and this is known as *meiosis*. In meiosis, the stages of mitosis are repeated again after a resting phase. The end result is that four daughter cells are produced, each with half the number of chromosomes of the parent cell, a condition known as *haploid*. At fertilization, two haploid cells (a sperm and an egg) fuse together so that the new individual has the full number of chromosomes, half from each parent, and this condition is known as *diploid*. Most cells are diploid with only the sex cells (gametes) being haploid. *See also* FERTILIZATION, REPRODUCTION (diagram of mitosis) CHROMOSOME AND GENE.

cellular communications are radios, but particularly telephones that can be used almost anywhere (car phones, portable telephone) and linked to exchanges and standard telephone lines by means of an almost countrywide network. The network is divided into cells and when a user with a portable or car phone makes a call, it is transmitted to a base station and then to a telephone exchange. From the exchange the signal is then directed to the recipient either via a landline or through a similar route of cells.

Each cell has a relay station to deal with all calls in that cell and when a user moves to another cell another station takes over.

cellulose a type of carbohydrate (known as a *polysaccharide*) which is composed of chains of glucose (sugar) units. It makes up the cells walls of many plants and has a fibrous structure which gives strength. Cellulose is not easily digested and some herbivores have evolved a strategy for dealing with this in that they possess a four-chambered stomach containing bacteria which are able to break down cellulose. These are known as *ruminant* animals. In humans, cellulose is important as dietary fibre.

cement is any compound or mix that binds materials together, but generally refers to Portland cement used in building and construction. Cement is manufactured from chalk or limestone, with clay which provides the necessary lime, silica and alumina. The mix is crushed and heated in a kiln to form a *clinker* (a partly melted, dense ash) which contains calcium silicates and aluminates. The clinker is finely ground and gypsum (calcium sulphate, $CaSO_4$) added. The cement product then goes through complex *hydration* (combination with water) reactions which eventually result in a hard stone-like material.

There are now many different cements which have particular properties depending upon their composition. These include: ordinary Portland cement, rapid-hardening cement, Portland blast-furnace cement, sulphate-resistant cement and micro-silica cement.

Enormous quantities of cement are used in construction for mortars or in concrete.

centre of gravity of a body is the point through which the total effect of gravity acting upon all parts of the body are effectively concentrated. This usually coincides with the *centre of mass* and is the point at which the weight or downward force due to gravity acts. When applied to everyday objects, this means that some will fall over more easily than others. The stability of an object can be increased by lowering its centre of gravity, perhaps by enlarging its base. A good example is in transport where large tall loads are unstable (due to a high centre of gravity) whereas racing cars are designed to have a low centre of gravity to allow them to move at high speeds.

centrifuge a machine for separating out particles suspended in a liquid. It comprises two arms about a central pivot, each arm holding a tube. When the arms are rotated rapidly about the central point the suspended particles are forced outwards to the bottom of the tubes and different particles will separate at different speeds, depending upon their size and mass. An *ultracentrifuge* rotates at very high speeds (up to 60,000 rpm) and is used for determining the masses of polymers and proteins. In industry, centrifuges are used for separating mixtures and in the production of cream and sugar.

centripetal force is the force necessary to move an object in a circular path, such as a ball on the end of a rope. The ball follows a circular route because there is a force in towards the centre of the circle– the centripetal force. If the rope is released, the ball continues in the direction at the time and moves off at a tangent. The force can be felt easily when some weighty object is swung around and the person holding the rope stays in the same place, and an object will only move in a circle if the centripetal force is provided.

Centripetal force applies when a car corners, the friction between tyres and road providing the force and in some fairground rides. The same applies to a satellite orbiting the Earth where gravitational attraction between the bodies supplies the centripetal force.

ceramics is the term generally applied to items manufactured from clays and includes pottery and *porcelain* where a moist clay is moulded to shape and fired in a kiln until it hardens. Decorative glazes can be added and many domestic and commercial items are produced in this way. These are *traditional ceramics* made of 50–60% clay to which animal bone is added (to control the porosity) producing a non-porous *china*. The mineral *feldspar* can be added in place of bone, producing porcelain goods, e.g. insulators. Clay minerals with no additions are used to produce bricks, tiles, pipes and similar construction materials.

Considerable advances have been made in recent years in the field of materials technology and many ceramics or *composites* (a mix of materials, often as fibres of one material in a different matrix) have been developed.

Light ceramics	carbon fibres; carbides and nitrides; silicon dioxide; aluminium dioxide
Glasses (see GLASS)	there are three major commercial systems: a) soda-lime-silica (Na_2O–CaO–SiO_2) b) lead crystal (PbO–SiO_2) c) low expansion borosilicate (B_2O_3-SiO_2 -Na_2O-CaO)
Glass ceramics	glass containing very small crystals
Refractory oxides	able to withstand the action of corrosive solids, liquids and gases at high temperatures e.g. alumina (Al_2O_3) in lasers, zirconia (ZrO_2) in developing ceramic steels, magnesia (MgO) in pollution control, titania (TiO_2) in enamelling and catalysis
Refractory metals	metal carbides, nitrides, borides and silicides, for use in cutting tools, electrodes, heating elements and protective coatings

cereals are important food crops derived from wild grasses which have been grown by man for many thousands of years. Cereals are grown for their seeds, which are usually ground to produce flour and contain important nutrients such as carbohydrates, proteins, vitamins and calcium. Over the centuries farmers have selectively cultivated cereal crops only sowing the seeds from the best and biggest plants and those that produced the most grain. This is an example of plant breeding and the end result is that cultivated cereals now look very different from the wild grasses that are their 'ancestors.'

chemical reaction the interaction between substances (reactants) producing different or altered substances (products) in which the bond elements are broken and reformed. Reactions may proceed at normal pressure and temperature, or special conditions may be necessary, e.g. high pressure, or the presence of a catalyst (*see* CATALYSIS). Some terms refer to typical reactions; for example, *hydrolysis* is the breakdown of a substance by water. The *heat or reaction* is the heat given out or required when certain amounts of substances react under constant pressure. Other heat quantities of a similar nature include:

heat of combustion	the heat evolved when one mole (*see* MOLECULE) of a substance is burned in oxygen.
heat of dissociation	the heat required to dissociate (splitting a molecule into simpler fragments) one mole of a compound.
heat of formation	the heat given out or required when one molecule of a substance is formed from its constituent elements at standard pressure and temperature.

chemistry is the study of the composition of substances, their reactions and effects upon one another and the resulting changes. There are three main branches: physical, inorganic and organic. *Physical chemistry* deals with the link between chemical composition and physical properties and the physical changes due to reactions. *Inorganic chemistry* is the study of the elements and their compounds (excluding the organic compounds of carbon), and the group and period characteristics as arranged in the periodic table. *Organic chemistry* is the study of carbon compounds, of which there are an enormous number and many are compounds with hydrogen, the hydrocarbons. Organic chemistry also deals with compounds of significant benefit to man–drugs, vitamins, antibiotics, polymers and plastics, and many more.

The study of chemistry really began in the 17th century when Boyle, an Irish scientist introduced the idea of elements. Since then many scientists have become famous for diverse discoveries–from the discovery of hydrogen by Cavendish (1766) and the synthesis of an organic compound from inorganic substances by Wöhler (1828) to the determination of the structure of DNA by Watson and Crick in 1953.

chlorine the second element in the halogens group which is a yellow-green choking gas which is harmful if inhaled. It occurs as the chloride, mainly sodium chloride ($NaCl$), but also magnesium chloride and others. It is manufactured by the electrolysis of brine (which also produces sodium hydroxide, $NaOH$) and stored as a liquid. It is a very important material in the chemicals industry and its powerful oxidizing properties are utilized in bleaches and disinfectants. It is also used in the production of hydrochloric acid and numerous organic chemicals. Compounds derived from chlorine are used in water sterilization, paper manufacture, solvents, PVC and other polymers, refrigerants and more. Hydrocarbons containing chlorine account for a very large proportion of chlorine derivatives.

chlorofluorocarbons or CFCs are organic compounds that contain chlorine and fluorine (both halogens). Although CFCs have in the past had numerous industrial uses, as aerosol propellants, in refrigerants and are produced when manufacturing foam plastics, their use is now being curbed and much reduced because of their harmful environmental effect.

Although CFCs are stable for a long time, they eventually break down and release their chlorine which reacts with ozone in the atmospheric ozone layer (*see* **GG**: OZONE LAYER) producing oxygen and chlorine oxide. This reduction in ozone then has direct detrimental effects upon plants and animals on Earth.

chloroplast and chlorophyll chloroplasts are the organelles (*see* CELL) present within the cells of green plants and algae, e.g. *Spirogyra*, which are the site of photosynthesis. They contain chlorophyll which is the pigment that gives the plant its green colour. Chlorophyll is essential in photosynthesis as it traps energy from sunlight and helps the plant to manufacture its food from carbon dioxide and water. The chloroplasts (and chlorophyll) are mainly found in the leaves but also in the stems of green plants.

cholesterol a type of fatty substance which occurs naturally and is produced within animal bodies. In mammals such as man it is made mainly in the liver, carried in the bloodstream and stored within cell membranes. Cholesterol is used to make some hormones and also a substance known as bile which is important in digestion.

Cholesterol can cause a problem in some people when too much is present in the blood. It may build up on the inner walls of arteries (a condition known as *arteriosclerosis*), causing them to become narrow and more easily blocked by a blood clot which can lead to a heart attack (*see* CIRCULATION).

chromatography is a technique used in chemical and biological analysis in which the constituents of a mixture can be separated. There are numerous closely-related techniques but each depends upon one principle–the differing rate at which components of a mixture move through a stationary phase due to the presence of a mobile phase. The simplest case is that of a piece of blotting paper dipped into water. A drop of ink placed on the blotting paper will move up as the water ascends the paper, and in so doing will be separated into its component pigments. In this case the water is the mobile phase and the paper is the stationary phase, and this is essentially a form of paper chromatography. Other chromatographic techniques include:

gas chromatography where a gas is the mobile phase. Used for complex mixtures of organic materials.

high performance liquid chromatography passage of the sample through a column by means of a liquid mobile phase under pressure.

thin layer chromatography separation by solvent movement across a flat surface of special paper, powdered cellulose, silica gel, alumina (aluminium oxide) or other stationary phase. It is used widely in qualitative analysis.

ion-exchange chromatography separation of ionic materials by a solution passing over the surface of a resin which contains ions that can be exchanged. It is used in inorganic chemistry to separate metal mixtures or to separate amino acids.

gel permeation chromatography is where materials are separated by molecular size by a solution passing through a porous polymer gel, in which small molecules are trapped, while larger molecules move.

chromosomes and **genes** chromosomes are threads of material that are found within the nucleus of each living cell. They consist mainly of chains of DNA (composed of a substance called nucleic acid), and parts of these are called genes, which are the 'blueprints' or plans that determine everything about the organism. Each living organism has a particular number of chromosomes, which contain very many genes.

Human beings have 23 pairs of chromosomes, 22 of which look the same in both males and females. The 23rd pair are the sex chromosomes, which in males look like an XY and in females an XX. Although the chromosomes look similar in all individuals belonging to a particular species, the genes that they carry are all slightly different, hence each one is totally unique. *See also* HEREDITY and MUTATION.

circuit an electrically-conducting path that when complete allows a current to flow through it. A circuit may be very simple consisting merely of a battery connected by copper wire to a bulb and then back to the opposite terminal of the battery. When cells and batteries were first made (in the early 1800s by the Italian scientist Volta), how charge moved around a circuit remained a mystery. The convention became that current flowed as a positive charge from the positive terminal around to the negative terminal. This is contrary to what actually happens since electrons flow in the opposite direction, but the convention remains (*see also* ELECTRICITY).

A simple circuit diagram

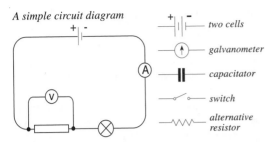

— two cells

— galvanometer

— capacitator

— switch

— alternative resistor

circulation (of the blood) is the process by which blood is moved around an animal's body by the pumping action of the heart.

The blood carries oxygen and food to all the cells of the body and also takes carbon dioxide from them to the lungs where it is eliminated. Blood which contains oxygen (oxygenated blood), is pumped through blood vessels called *arteries* by the left side of the heart. As the arteries reach the tissues and organs they become very tiny (arterioles and capillaries). Here blood releases its oxygen (becoming deoxygenated), and this is picked up and used by cells. Cells release carbon dioxide into the deoxygenated blood as it passes through more capillaries. The blood is now transported through tiny vessels (venules) which become larger *veins* and then it passes back to the right side of the heart. From here it passes to the lungs where carbon dioxide is released and oxygen is picked up before it returns to the left side of the heart once again.

This type of blood circulation is described as 'double'. The heart is divided into two sides which each act as an independent pump with no communication between them. Each side is further divided into two chambers, the upper one is called the *atrium* and collects incoming blood passing it to the lower *ventricle*. This is a strong, muscular pump which contracts and pumps blood out either to the body or to the lungs. Arteries always carry oxygenated blood except for the pulmonary artery which takes blood from the heart to the lungs. Veins always transport deoxygenated blood except for the pulmonary vein which takes oxygenated blood from the lungs to the heart.

The human heart

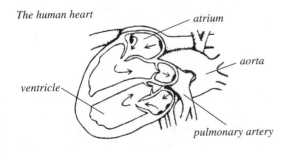

classification a means of grouping living organisms according to how similar they are to each other. For the early biologists, the physical similarities between organisms were the most important feature. However, increasingly in modern times the similarities between organisms at a genetic level have become more important in understanding their relationships.

One main method of classification is most commonly recognized and this was originally devised by a Swedish naturalist, Carolus Linnaeus (1707–88) and is known as the *Linnaean system*. In this, all living things belong to a particular species and very similar species are grouped together in a *genus* (*pl.* genera), (*see* CLASSIFICATION). Hence in biology, each living organism has a double-barrelled name, first its generic (genus) and then its specific (species) 'label' and these are given in Latin. Examples are *Lumbricus terrestris* (the earthworm), *Rana temporaria* (the common frog), *Homo sapiens* (human beings) and, among plants, *Quercus robur* (the common oak), *Primula vulgaris* (primrose) and *Taraxacum officinale* (the dandelion). Genera which show similar features are grouped into families and these in turn into orders. Numbers of orders are put into classes which are grouped into a phylum (*pl.* phyla).

Finally, phyla are placed together in a kingdom. In plants, the grouping division usually replaces phylum but otherwise the classification is the same.

clone a living organism which is an exact genetic copy of another individual and is produced from a single cell by an asexual process (*see* ASEXUAL REPRODUCTION).

cloning is an artificial process engineered by man and much used in the breeding of plants (e.g. cuttings). However, some animal clones have also been produced in the laboratory under strictly controlled conditions.

colloids were defined originally as substances in solution that could not get through a dialysis (semipermeable) membrane. The definition of a colloid now is one of particle size – when the size range is 10^{-4} to 10^{-6}mm, and it therefore falls between a coarse suspension and a true solution. Most substances can now be prepared to exist in this state, and many exist naturally as colloids. Common examples of colloidal solutions include starch and albumen, butter and cell cytoplasm.

An *emulsion* is a colloidal solution of one liquid in another; generally one is water or a solution in water and the other is an oil or a similar liquid that does not mix with water. Emulsions are used widely in industry in food and pharmaceuticals, cosmetics, paints and lubricants.

combustion or burning is actually a chemical reaction that occurs quickly and involves the high temperature oxidation of a substance, i.e. it combines with oxygen, to produce heat, light and flame, and of course ashes (oxides). Combustion of fossil fuels is the source of most energy used in homes, factories and offices. When coal or a similar fuel is burned, the carbon is converted to carbon dioxide or carbon monoxide and the hydrogen to water vapour. Combustion is a vital industrial process and although it generally refers to burning in oxygen, it is used on other occasions when, for example, a substance is burnt in hydrogen.

compact disc is a disc used to hold music, graphics or data for replaying. The disc has a layer of aluminium, and music is recorded as very small pits etched into the surface. The pits are tiny – about half a micrometre (μm) wide and one to three wide. Each track is about 1.5μm from the next. The music is replayed by focusing a laser beam on the disc. The beam is partially reflected, depending upon whether or not it strikes a pit, and the reflected beam is detected as a series of pulses that are changed back into a copy of the original recording. Because there is no physical contact between the disc and the playing medium, compact discs should last much longer than their earlier counterparts: magnetic tapes and vinyl records.

compound is when two or more elements combine chemically in a substance and in definite proportions to produce molecules held together by chemical bonds. The formation of a compound necessitates a chemical reaction, and the elements cannot be separated physically. Also, the same compound is formed irrespective of its origin so one molecule of water is two atoms of hydrogen and one of oxygen no matter how it is made.

The above can be regarded as the specific definition of compound but there are instances when the boundaries are blurred. For example, many silicate minerals, e.g. the feldspars, have varying compositions, as do many polymers although all are chemically combined. Indeed materials such as glass and steel are not mixtures and yet they do not fall readily into the definition of compound.

concentration *see* SOLUTIONS.

concrete a building material comprising a mixture of sand, cement, stone and water, which after it is set becomes very hard. Concrete is used in vast quantities for all types of building and construction and often steel rods or meshes are set into the concrete to increase its strength–*reinforced* concrete. *Prestressed concrete* is when the concrete is under compression, achieved by stretching the reinforcing rods and keeping them in tension after the concrete has set. Prestressed concrete is useful for large spans or where beams have to be as light as possible.

There is a whole technology surrounding concrete, with a variety of types and numerous compounds and materials that can be added to a mix to confer particular properties. *Aerated concrete* made up solely of cement, water and gas bubbles is used for its insulating properties; *lightweight concrete* utilizes lightweight aggregates (synthetic or natural) to form an insulating and lighter concrete; *high strength concrete* has obvious benefits while a denser concrete for protective shields can be made by using barytes, iron or lead shot in place of stone. Additives to concrete (known as *admixtures*) are used to alter the properties of the mix or the hardened material and include water reducers, accelerators, plasticizers, corrosion inhibitors and many others. This variety of concrete type renders it one of, if not the most versatile building material.

condensation is the process by which a substance changes from the gaseous state to the liquid state. In so doing energy is lost, i.e. it is cooled. There are many everyday examples of condensation–water forming on a pan lid when vegetables are cooking; water droplets on a cold tap in the bathroom; the steam from a kettle which is actually the hot water vapour condensing as tiny water droplets in the cooler air. *Cloud* (*see* **G**: CLOUD) is another example which is due to warm air saturated with water vapour being cooled.

The physical process of condensation is used in the chemicals industry when substances are purified by distillation. In chemistry a *condensation reaction* is the reaction of one molecule with another and the elimination of a simple molecule such as water.

conduction and convection *conduction* of heat is a process of heat moving through a material and is due to molecular vibrations. When a material is heated, the molecules vibrate rapidly and knock into neighbouring molecules which transfers the heat (or thermal energy) along the material. Metals e.g. copper and aluminium are the best conductors of heat while liquids and gases are progressively poor conductors.

In a gas or liquid that is free to move, *convection* is the process that moves heat from one part to another. As water in a tank is heated, the hot water rises and cooler water sinks. This establishes a convection current. Convection is used to great effect in the hot water systems in houses. A similar effect is seen with gases and when air is warmed, this sets up currents which on a large scale create onshore and offshore winds at the coast. On a smaller scale, radiators warm the air in a room by convection.

conductor and insulator electric conductors are materials that allow a flow of electrons, producing an electric current. As with conduction of heat, metals form the best electrical conductors because electrons around the outside of the atoms are loosely held and can move freely. Poor conductors include water, glass and air, indeed most non-metals.

A material that does not conduct electric charge is called an *insulator*. Plastics, rubber, glass and air form insulators because their electrons are not usually free to move.

connective tissue a type of tissue which is commonly found in the bodies of animals. It is further divided into a variety of different sorts depending upon the materials from which it is composed. It is usually composed of a non-living core containing various fibres in which are spread a number of cells.

construction *see* BUILDING AND CONSTRUCTION.

contraction and expansion when a solid is heated and its molecules vibrate more due to the input of thermal energy, the result is that the molecules move apart a little and the solid *expands*, almost imperceptibly. When the reverse happens, the solid *contracts*. Although the expansion in a solid may be negligible, the resulting force can be very large. In construction particularly, account has to be taken for the expansion of steel and concrete and all bridges have expansion joints to avoid damage which would otherwise be caused. Railways lines have a similar feature but in this case, line ends have overlapping joints.

The property of expansion also has its useful aspects and it can be applied to numerous devices which contain a bimetal strip.

Liquids generally expand more than solids, producing an increase in volume. Water is a notable exception to this statement and its behaviour is quite complex. As water cools from boiling it contracts a very small amount until it reaches 4°C at which point it expands a little. As we all know, at 0°C it forms ice and expands a great deal but on further cooling it contracts more. At 4°C water has its least volume and therefore its greatest density and it will sink beneath colder or warmer water. This is the reason for ponds freezing on the surface while fish can survive in the slightly warmer water at depth.

convection *see* CONDUCTION AND CONVECTION.

copper is a red-brown metal which is very malleable and ductile (*see* DUCTILITY) and has numerous uses. It occurs as native copper (as the metal itself, often with silver, lead and other metals) and in a variety of mineral forms, e.g. malachite ($CuCO_3.Cu(OH)_2$), bornite ($CuFeS_3$) and chalcopyrite ($CuFeS_2$). The ores are concentrated and copper is extracted by smelting and refining by electrolysis. Copper has been an important metal for thousands of years in its alloys, brass and bronze, and it is now used in coins as an alloy (with nickel).

Pure copper is an excellent electrical conductor and is used in wiring and a significant proportion of copper production is taken in electrical applications. It is also used for pipes in plumbing although plastics are being used increasingly in this context. Copper is also employed in fungicides, paints, pigments and printing.

corrosion is the process of metals and alloys being attacked chemically by moisture, air, acids or alkalis. If left to continue, the metal will be gradually worn away. Corrosion may occur uniformly or it may be concentrated at weak points or joints. It may also produce an oxide layer, as on aluminium which protects against further attack. The more serious effects are seen with corrosion where some moisture is present because this sets up an electrolytic process and with underground corrosion the soil acts as the electrolyte (*see* ELECTROLYSIS).

Corrosion can be prevented or slowed by applying protective layers (paint, anodizing, or plating with zinc, nickel, chromium etc.). *Cathodic protection* is also used in the protection of underground structures (e.g. pipelines) by making the object/structure in question the cathode in a circuit which has a voltage higher than the estimated voltage of corrosion. The anode in this circuit may dissolve away and be replaced when necessary. Steel reinforcement (*see* CONCRETE) can be cathodically protected providing it is continuous electrically.

coulomb is the unit of electrical charge and is defined as the charge passing a point in a circuit when a current of one ampère flows for one second, thus a charge of 8 coulombs passes if a current of 2 ampères flows for 4 seconds. One coulomb is equal to the charge on approximately 6.25 x 10^{18} electrons!

cracking *see* PETROCHEMICALS.

cytology the branch of biology devoted to the scientific study of cells, including both their structure and function, that depends very much on the use of the light and electron microscope.

Darwin and natural selection. Charles Darwin (1809–1882) was one of the most famous naturalists ever to have lived and he devised the theory of evolution (known as *Darwinism*) to explain the great variety of plants and animals which he saw around him. He arrived at his theories during a five year voyage around the world (the voyage of the Beagle) and when he returned to England in 1859 he published a scientific paper (with the title *Origin of Species*). He proposed that some individuals in a species are more successful than others (they have a greater degree of 'fitness"). In the competition for food or for a mate they are more likely to be successful and these characteristics are inherited by their offspring which means that eventually these features become more widespread. He called this 'survival of the fittest.' Following from this, plants and animals were gradually able to change and adapt to new conditions and environments. Hence new species eventually evolved from an original, ancestral stock. Darwin's theories were not accepted at the time and caused outrage because they questioned the Biblical version of God's Creation as a one-off event. However, now they are largely accepted and have been expanded by the modern study of genes and inheritance. Darwin's theory can be summarized as follows: from organisms which have the ability to change, new species can emerge that will adapt to new environments. Old species which are no longer suited to the surrounding environment will eventually die out.

decibel *see* SOUND.

deciduous and evergreen deciduous plants shed their leaves at the end of the growing season, which is the Autumn in temperate regions such as Britain. Examples include familiar trees such as the oak and sycamore. Evergreen plants on the other hand, keep their leaves all through the year and these include the cone-bearing coniferous trees (conifers) such as Sitka spruce.

decimal numbers is the most commonly used number system based on powers of ten. The *decimal point* is the dot that divides the number's whole part from the fractional part (i.e. that which is less than one). However, numbers need not contain a decimal point : 789 is also a decimal number. A decimal is itself less than one and is written after the decimal point e.g. 0.789 and 0.00987. Decimal numbers are written within the place value system, that is, the value given to a digit depends upon its position in the number and with the decimal system each column has ten times the value of the column to the right thus:

The number 7891 is really:

seven 1000s	eight 100s	nine 10s	and one 1
(10^3)	(10^2)	(10^1)	(10^0)
7	8	9	1

A common fraction such as $1/4$ can be changed into decimal form by dividing the 1 by 4 to give 0.25.

dehydration in a chemical reaction or process is the removal by heat of water held in a molecule or compound. Sometimes a catalyst is used, or a dehydrating agent such as sulphuric acid (H_2SO_4). Dehydration is used a great deal in the food industry in the production of coffee, soups, sauces, mashed potato, milk and so on. It arrests the processes of natural decay because there is no moisture available for micro-organisms to survive and chemical reactions are slowed or stopped. Of course, dehydration produces a reduction in volume and particularly weight which is useful for storage and transport.

In medicine, dehydration is the excessive, often dangerous, loss of water from body tissues, accompanied by loss of vital salts. The average daily intake of water is about 2 litres but lack of water for just a few days can be dangerous because the heart can be affected.

dendrochronology a technique of dating past events using the growth rings of trees. Each year a new ring of wood is added to the trunk just beneath the bark and this is called an annual ring. It is possible to date the rings in living trees by working back year by year. Then this pattern of rings can be used to date fossil trees or specimens of wood found at archaeological, or other sites. The longest living trees are the most useful and the standard one is the bristle cone pine, which can live for up to 5000 years and be used to date specimens older than this.

density is the mass of a substance, per unit volume, given by the following equation:

$$\text{density (d)} = \frac{\text{mass (M)}}{\text{volume (V)}}$$

It is measured in kilograms per cubic metre (kg/m³), although it may be more convenient on occasion to use grams per cubic centimetre (g/cm³). Density varies with temperature; only a little in the case of solids and liquids which usually expand and therefore become less dense, with gases the density varies a great deal depending upon its container and the surrounding pressure.

Relative density is the density of a material compared to that of water, given by the following equation:

$$\text{relative density} = \frac{\text{density of a substance}}{\text{density of water}}$$

Relative density used to be called *specific gravity*. It has no units, but is the same value as the density when measured in g/cm^3. Some typical densities:

	kg/m^3	g/cm^3
air	1.3	0.0013
soft wood	450	0.45
hard wood	800	0.80
petrol	800	0.80
water	1000	1.00
hardened cement	2200	2.2
granite	2600	2.6
aluminium	2700	2.7
diamond	3500	3.5
steel	7700	7.7
lead	11400	11.4
mercury	13600	13.6
gold	19300	19.3

detergents and soaps are both cleaning agents that remove grease and dirt and hold it in suspension for washing away. Soap acts as a detergent but there are now many synthetic detergents derived from petroleum. The detergent molecules contain two distinct groups; one which gives the molecule solubility in water, e.g. a sulphate, and long hydrocarbon chains which enable it to dissolve oily materials. When detergent molecules come into contact with grease, the hydrocarbon chains which are *hydrophobic* (water-hating) attach to the grease, and the other end of the molecule which is *hydrophilic* (water-loving) is in the water. The grease is then enclosed and can be removed from the garment. Detergents are made in many forms: washing-up liquids, powders, shampoo, etc.

Soaps are sodium and potassium salts of *fatty acids* (a type of organic acid of animal or vegetable origin) which are heated in large vats with dilute sodium hydroxide (caustic soda) to effect *hydrolysis* (*see* CHEMICAL REACTION). Sodium chloride is then added to precipitate the soap from the solution. The soap may then be treated with perfumes before being made into bars or flakes.

Metallic soaps are a very different group of compounds, being insoluble in water. They are metal salts (metals such as lithium, aluminium, calcium, and zinc) of long carboxylic acid chains (organic acids with one or more carboxyl, -COOH, groups) and are used in cosmetics, pharmaceuticals, fungicides and lubricating oils.

differentiation a mathematical operation used in calculus for finding the derivative of a function. Depending upon the complexity of the function there are different methods of differentiation. The simplest relates to the common function;

$$f(x) \text{ or } y = x^n$$

This has the derivative (or differential coefficient);

$$f^1(x) \text{ or } \frac{dy}{dx} = nx^{n-1}$$

Thus, for example, if $y = 4x^3$, $\frac{dy}{dx} = 12x^2$

and for $y = 3x^3 + 4x^2$, $\frac{dy}{dx} = 9x^2 + 8x$

diffraction is the bending of waves around an obstacle and as they pass through a narrow gap. This applies to all waves; water, sound, light and electromagnetic, and can be detected by a change in the shape of the wavefront and by *interference* patterns. When a beam of light passes through a narrow slit it is diffracted but the slit has to be very narrow indeed (less than 0.01mm) to have any effect. If *monochromatic* (one wavelength) light is used, and the diffracted light is passed through two further slits, then an interference pattern of light and dark fringes is created. In 1801, Thomas Young the physicist used an experimental procedure such as this to measure the wavelength of light. The fact that sound can be diffracted is easily shown because it is possible to hear round corners.

diffusion is the process that occurs in gases and liquids whereby one liquid is spread throughout the body of another, e.g. an ink drop in water, due to the molecular motion of the water (*see* BROWNIAN MOTION), producing a more uniform concentration. The same process occurs with gases and accounts, for example, for the smell of a gas leak filling a room. The molecular movement of gases is more vigorous than liquids and the molecules distribute themselves equally within the volume in which they are enclosed.

Diffusion also occurs across cell membranes and a similar mechanism is used in dialysis as a means of separating certain molecules.

digestion the process by which organisms break down solid food into small particles which can be used by the body.

In human beings, the digestive process starts in the mouth where food is cut up into smaller particles by the teeth. The food is mixed with saliva containing an enzyme which breaks down starch into sugar. The food is then swallowed and passes via a tube called the *oesophagus* into the stomach.

In the stomach a fluid, the gastric juice, is released which contains hydrochloric acid and enzymes. The stomach has muscular walls and is able to expand and contract and further manipulate the food. Proteins are broken down and eventually a semi-solid acidic mass (known as *chyme*), is passed into the small intestine. Alkaline fluid from an organ called the *pancreas* (pancreatic juice) is added here and this contains more enzymes which further break down the food. Also, *bile*, a thick fluid produced by the liver and stored in the *gall bladder* located nearby, is added to the food in the intestine. This contains bile salts, bile pigments and cholesterol and aids in the digestion of fatty substances. As the food passes along the highly coiled length of the small intestine it continues to be broken down into minute particles (molecules) which can be absorbed into fine blood vessels present in the intestinal wall. The blood circulation carries the food to all parts of the body where it is used by cells to perform all the functions of life and to provide energy.

Any food substances that cannot be digested, such as fibre, are passed to the large intestine which is the final part of the alimentary canal or digestive system. Here water is removed and reabsorbed into the body, and the final waste (*faeces*) is passed to the outside through the anus.

The human digestive system

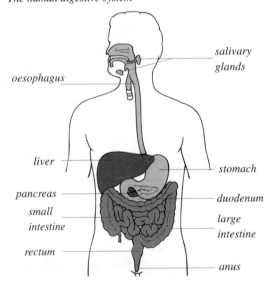

oesophagus

salivary glands

liver

stomach

pancreas

small intestine

duodenum

large intestine

rectum

anus

digital and analogue are two different ways of measuring a value – a digital system uses distinct units, e.g. electrical pulses, while analogue is a quantity that is similar to another quantity (e.g. a thermometer where the column of liquid represents a temperature).

The digital system uses binary numbers, which renders it admirably suited to use in computers. Numbers and letters are coded by groups of the digits 1 and 0, and each digit is represented in an electronic circuit by a component being on or off, e.g. passing or not passing current. In a magnetic system it could be magnetized or not magnetized, and that produces the on/off effect.

diode a device with two terminals that allows current to flow in only one direction. Modern varieties are *solid state* (electronic devices made up of solids, with no moving parts, filaments, etc.) and usually are made up of a special *silicon chip*. Silicon is a semiconductor used to make diodes. A crystal of silicon with phosphorus and boron added to opposite halves (a process called *doping*) has a poorly-conducting junction called the *depletion layer* between the two halves. The phosphorus increases the number of electrons available to move through the material and the boron makes holes into which electrons can move. Depending upon the current flow in the circuit, the diode passes current easily when the depletion layer is thin but when the current is reversed, the depletion layer thickens and no current is passed.

This property of diodes means they can be used to turn alternating current into a direct current, a process called rectification. Then the diode is called a *rectifier*. The diode may be used for *half-wave rectification* in which it simply blocks the backward half of the current, or with a more complex arrangement of diodes, *full-wave rectification* can be achieved in which the blocked half of the alternating current is reversed and flows through as direct current. This is the principle employed in radios and tape recorders that can use a mains adaptor, or batteries.

diploid *see* CELL DIVISION.

disease any illness that affects an organism which, in man, is caused in two main ways. Often diseases are caused by infectious micro-organisms e.g. bacteria, viruses and fungi (*see* FUNGUS). Bacteria can be killed with antibiotic drugs such as penicillin and examples of bacterial diseases are cholera, tuberculosis and typhoid. Viruses are not susceptible to antibiotics and the illnesses they cause have to be fought off by the body's immune system, although protection can be given against some of these by means of vaccination. The common cold is a viral disease, and those which can be prevented by vaccination include mumps and measles. The second group of diseases are caused by the failure of a body system to work properly. This may be an inherited disorder, e.g. cystic fibrosis, or occur at some stage in a person's life for an unknown reason. A common example is *Diabetes mellitus* which occurs when an organ called the pancreas fails to produce enough of the hormone insulin, itself responsible for the breakdown of sugars in food.

distillation is the separation of a liquid mixture into its components. The liquid is first heated to vapour which is then cooled so that it condenses and can be collected as a liquid. The mixture can then be separated if the components liquids have different boiling points, each one vaporizing at a different temperature.

Distillation is one of the most common separation processes in industry and it permits a high degree of separation. It is performed in a large column within which there are a number of trays (or plates) at different levels. The liquid is fed in through a heater and a constant flow is maintained between the ascending vapour and descending liquid. The vapour is then taken out at the top of the column through a condenser. *Fractional distillation* is the term applied to this process and the column is the fractionating column. Whisky is distilled in this manner and crude oil is refined by the same means. *Flash distillation* involves the rapid removal of solvent by moving the liquid from high to low pressure so that part turns immediately to vapour and is condensed. This technique is used in the desalination of sea water, i.e. the removal of salts from sea water to provide fresh water.

DNA stands for deoxyribonucleic acid and this is the material of which the chromosomes and genes are composed. It occurs as strands in the nucleus of each cell and contains all the instructions which determine the structure and function of that cell. One of the most exciting discoveries of this century occurred in 1953 when two scientists, James Watson and Francis Crick, worked out and demonstrated the structure of DNA. They found that it occurs as two spiral threads coiling round each other (a double helix) with 'bridges' across at intervals connecting the two, as in a ladder. Four types of molecules (known as bases) occur on the 'rungs' of the ladder and these pair up in particular ways (*base pairing*). These four molecules store all the genetic information by being built up in different combinations. In cell division, the two DNA threads split apart and each reproduces the missing half to rebuild the double helix. Sometimes the copying does not occur properly and can result in a mutation. Many mutations are lethal but minute changes can occur in this way and this is the genetic basis for evolution and species change.

DNA structure

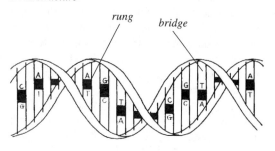

Doppler effect is the change in the observed frequency or wavelength of a wave when the source producing the wave is moving. If the source, e.g. a vehicle, is moving towards the listener, the frequency increases (and the wavelength decreases) and the opposite happens as it moves away. The result is the pitch of the sound changes as the vehicle approaches, passes and moves away. The principle applies to all electromagnetic radiation including light. The light from a moving object appears more red when it is moving away from the observer. This is taken to mean that stars in distant galaxies are moving away from us when they exhibit the *red shift*.

drugs and pharmacology *drugs* are substances which, when introduced into an animal's body, produce some kind of effect or response. Most drugs are given to help cure illnesses or other medical conditions and there are a great number of different kinds. **Pharmacology** is the name of the scientific study of all aspects of drugs and medicines including their preparation, uses, properties and effects. Drugs may be given in a variety of different ways: 1. by mouth (orally) as capsules, tablets or in liquid form, by injection; 2. as a cream or ointment applied to the skin; 3. by inhalation (sometimes using an inhaler); 4. by means of skin patches or insertion just beneath the skin and 5. suppositories which are placed via the anus into the end of the digestive tract (the rectum). New drugs are rigorously tested before their use is allowed but many occur naturally in plants. caffeine (in coffee and tea) and alcohol are examples of drugs which are in widespread use.

ductility and malleability ductility is the property of metals or alloys that allows them to be drawn out into a thin wire and although their shape is permanently changed they retain their strength and do not crack. *Malleability* is similar but is the property that allows metals and alloys to have their shape altered by hammering, rolling or similar process, into thin sheets.

Most metals are ductile, notable examples being

copper and gold. The same metals are very malleable and gold can be produced as very thin gold leaf which is 2μm (micrometres) thick. The high ductility and malleability of metals is due to their molecular structure, in which the lattice can be greatly altered before the atoms are torn apart and the metal breaks.

dyes (or dyestuffs) are substances with a strong colour that can be fixed to material to be dyed, such as fabrics and fibres, plastics, etc. The synthetic dye manufacturing industry began in 1857 when Perkin & Sons set up a factory to produce mauveine. At that time the wool and cotton trade was undergoing vast expansion and demand for dyes was heavy. In the past dyes had been made mainly from plants but mauveine was soon followed by the production of other synthetic dyes. Alizarin appeared in 1869, Congo Red in 1884 and indigo in 1897. Now dye production is a vast industry undertaken in complex chemical plants.

In some cases when a dye does not fix to the material, a *mordant* may be necessary. This is a compound (aluminium, chromium and iron hydroxides) that impregnates the fabric and then the dye reacts with the mordant to ensure stability. There are many types of dye, which are categorized according to application.

ear *see* HEARING.

ecdysis and moulting *ecdysis* is the process of shedding skin. It is undergone by a group of invertebrate animals and it enables them to grow. It is a characteristic of arthropods such as insects, spiders and crabs which possess a tough, outer shell. At the start of ecdysis a hormone is released under the influence of which the new shell (or *exoskeleton*) begins to form beneath the old one. In addition some materials are reabsorbed from the old exoskeleton and these can then be used again. Eventually, the old exoskeleton splits and the animal extracts itself from its former covering. To begin with the new exoskeleton is very soft and the animal is often rather vulnerable at this stage. The new exoskeleton expands, often involving the animal taking in air or water to increase its body size and it hardens by the incorporation of a special substance called *chitin* (or *calcium* salts in crabs). The animal is thus larger than it was before and may undergo this process several times during the course of its life (*see also* METAMORPHOSIS).

moulting is a term which is sometimes used to describe ecdysis. However, more familiarly, it is applied to the loss of fur or hair or feathers which often takes place in vertebrate animals, especially in the spring.

echo is when a reflected sound is heard a short time after the original sound was made. There is a delay between making a sound and hearing its echo because it takes the sound waves a little time to travel the distance (light travels so quickly that it does not exhibit this phenomenon). The time delay can be used in echo-sounding machines to locate the sea bed or submerged objects. SONAR (Sound Navigation Ranging) is such a device that operates with high frequency sound, collecting the returning waves that have been reflected from submerged objects. In all such cases, the distance can be calculated from the time taken for the sound to return. Echo-sounding equipment is used in ships and boats and *radar* is a similar principle but uses microwaves.

In small spaces, particularly rooms and halls, the echo time may be very short and the echo is not heard but becomes mixed up with the original sound, and the whole sound seems to be extended. This is called *reverberation* and may be a problem in concert halls where walls and ceilings may have to be modified to reduce the effect.

egg an egg, also known as an *ovum* (*pl.* ova), is the female reproductive cell. An egg may have a hard outer shell, as in birds and turtles, and is usually surrounded by one or more protective layers called membranes. After the egg is fertilized by a male sex cell (*sperm*), it divides into several cells which become an embryo. The embryo is nourished by a supply of food from the yolk within the egg. Some eggs contain very little yolk whereas others, such as those of birds, have a large amount which enables a chick to grow quite large before it hatches. Eggs may be extremely small, as in female mammals, which keep the developing embryos inside their bodies. The young of mammals are supplied with all that they need from the body of the mother and grow until they are ready to be born. There are two exceptions to this, the duck-billed platypus and spiny anteater, which are unique among mammals in that they lay eggs (*see* REPRODUCTION).

elasticity is the property of any material such that it stretches when forces are applied and recovers its original form when the forces are removed. To stretch a spring or other elastic material a stretching force must be applied to both ends. As the stretching force is increased, the extension becomes greater and up to a certain point the extension is proportional to the force. However, there is a limit to this proportional response and beyond a material's *elastic limit*, it is permanently stretched and will not return to its original length. *Hooke's Law* records this physical relationship but applies only to materials if the elastic limit is not exceeded. All materials are elastic to some extent but in many cases the extension is a very small amount.

electricity is a general term to cover the energy associated with electric charges, whether static or dynamic. In a simple circuit the *potential difference* between the battery terminals (i.e. the difference in potential) causes a current to flow. Current is measured in ampères and potential difference in volts. Batteries supply *direct current* but the electricity supply to homes and offices is *alternating current*. Power stations generate electricity by means of massive alternators (*see* GENERATOR) driven by turbines which are themselves driven by steam from coal, gas or oil-fired boilers (or heat from a nuclear reactor). The alternators generate a current of 20, 000 amps at a voltage of 25, 000 volts, which passes through a transformer and then into overhead power cables at a reduced current but a higher voltage–up to 400, 000 or 400 kilovolts. The current is reduced to minimize the power losses due to heating in the long cables. The cables feed the nationwide network (the *Grid*), and power from the Grid is distributed by substations where transformers reduce the voltage to 240 volts for ordinary consumption. Heavy industry uses supply at 33, 000 volts and light industry 11, 000 volts. *See also* STATIC ELECTRICITY.

electrocardiogram (ECG) and **electroencephalogram (EEG)** are both records of the electrical activity in the heart and brain respectively. The ECG is recorded on an electrocardiograph connected through leads to pads on the chest and legs or arms. It often indicates abnormal heart activity and is therefore a useful diagnostic tool.

The EEG records the brain's electrical activity on an electroencephalograph. Electrodes placed on the scalp record activity, or brain waves, of which there are four types associated with particular phases of activity or rest.

electrolysis is the chemical decomposition (breaking down) of a substance in solution or molten state when an electric current is passed. The solution is called an *electrolyte* and it permits the passage of a current because it forms ions in solution. A strong electrolyte, e.g. sulphuric acid, undergoes complete ionization. When the current passes through the solutions the ions move to the electrodes of opposite charge; the *cathode* being negatively charged and attracting positively charged ions (*cation*); the *anode* being positively charged and attracting *anions*, which are negatively charged ions. At the electrodes the ions give up their charges to form atoms or groups. Gases are liberated, and solids deposited.

Electrolysis is used a great deal in industry to extract and purify metals and also to electroplate objects. *Electroplating* consists of metal salts dissolved in a solution into which are put the electrodes. The cathode has the metal plated onto it, and the anode dissolves into the solution, replacing the metal ions. This technique is used to plate with chrome, nickel, gold, silver and other metals, and also to anodize (*see* ANODIZING).

electromagnet is a magnet created by passing a current through a coil of wire that is wound round a soft iron core; the core becomes a magnet while the current is on. The coil of wire is called a *solenoid*. This magnetic effect of electric current was discovered by Oersted, a Danish scientist. There is a cumulative magnetic effect because in producing a magnetic field, the coil magnetizes the core, which produces a magnetic field that can be about a thousand times stronger than the field from the core alone.

Electromagnets are clearly useful because the magnetic field generated can be controlled easily. They are used in televisions to control the electron beams in the cathode ray tube, and also in devices such as switches, electric bells, loudspeakers and in the earpiece of the telephone.

An associated phenomenon is that of *electromagnetic induction* in which an electric current is produced in a conductor when it is moved through a magnetic field (*see also* GENERATOR).

electromagnetic waves come from a number of sources and are the effect of oscillating electric and magnetic fields. The wavelength of these waves varies but all travel through free space (a vacuum) at approximately $3 \times 10^8 ms^{-1}$ (300, 000 kilometres per second), which is the speed of light. The *electromagnetic spectrum* contains waves from low frequency/long wavelength radio waves (long wave) through microwave, infrared and the visible spectrum to ultraviolet x-rays and the short wavelength/high frequency gamma rays.

Approximate wavelengths and frequencies for electromagnetic waves		
	wavelength (m)	*frequency* (Hz)
gamma rays	10^{-13}–10^{-10}	10^{19} -10^{21}
X-rays	10^{-10}-10^{-8}	10^{17}–10^{19}
ultraviolet radiation	10^{-8}–10^{-7}	10^{15}–10^{17}
visible light	10^{-7}–10^{-6}	10^{14}–10^{15}
infrared radiation	10^{-6}–10^{-2}	10^{11}–10^{14}
microwaves	10^{-2}–10^{-1}	10^{10}–10^{11}
radio waves	10^{-1}–10^{4}	10^{5}–10^{10}

Electromagnetic waves are generated when particles with an electrical charge change their energy, e.g. when an electron changes orbit around a nucleus. It also happens when electrons or nuclei oscillate and their kinetic energy changes. A large change in energy produces high frequency/short wavelength radiation.

Radio waves are the longest in the spectrum and are used to transmit sound and pictures. Microwaves have wavelengths of a few centimetres and have numerous uses. *Infrared* (IR) waves are generated by the continuous motion of molecules in materials, and hot objects give out most. When an electric fire is switched on the infrared radiation is felt in the heat. As objects become hotter and hotter, their molecules vibrate more rapidly and the wavelength of the radiation becomes shorter. Eventually it impinges on the visible spectrum and the object appears 'red-hot'. *Ultraviolet* (UV) radiation occurs beyond the violet end of the visible light spectrum, is a component of sunlight and is emitted by white-hot objects. Ultraviolet light from the sun converts steroids in the skin to essential vitamin D but an excess of UV light can be harmful. However, much of the sun's ultraviolet radiation is stopped by the Earth's ozone layer (*see* **GG**: OZONE LAYER). *Gamma rays* are very short wavelength radiation released during radioactive decay and are the most penetrating of all radiations.

electronics is an important area of science and technology that deals with electrical circuits using semiconductors, diodes, transistors, etc. and other devices in which the movement of electrons is controlled to create switches and other components. Technology has advanced so much in recent years that electronic circuits can fit onto a single *silicon chip* and highly complex circuits are constructed on *printed circuit boards* where individual components are linked by metal traces printed on the board and through which the current flows.

The use of such boards and microelectronic components is widespread and they are found in computers, cars, watches, televisions, space craft and many other machines and pieces of equipment.

electroscope is an instrument used in physics for the detection of small electrical charges. It consists of a metal cap joined to a rod which projects down into a case. At the bottom of the rod is a gold leaf. When a charged object touches the cap, some charge is transferred to the rod and gold leaf and because like charges repel each other, the gold leaf rises. When it is charged, the electroscope can be used to determine whether the charge on an object is positive or negative.

element a pure substance that comprises atoms of the same kind and which cannot be broken down into simpler substances in ordinary chemical reactions (nuclear reactions can, however, alter elements). There are 103 elements known to us, of which 92 occur naturally and the rest have been created in the laboratory. Indeed scientists continue to experiment and occasionally claim the existence of another element.

Elements combine together to form compounds and under normal conditions all but two elements (bromine and mercury) are either a solid or a gas. The elements are classified by their atomic number into the periodic table (*see* PERIODIC TABLE), which comprises groups and periods with similar properties and behaviour.

elementary particles (or *fundamental particles* or *subatomic particles*) are the basic particles and building blocks of which all matter is made. The three key particles in all atoms – electrons, neutron and protons – have now been supplemented by new particles. Essentially two types are thought to exist, *leptons* and *hadrons*, and these are identified by the different ways in which they interact with other particles. Leptons include the electron and the *neutrino*, the latter having no charge and virtually no mass. The neutrino was originally proposed on the basis of theory, to preserve the physical laws of mass, energy and momentum, and it has since been established experimentally. The proton and neutron are called hadrons, although they are not truly elementary particles, and it is now thought that these are composed of real elementary particles called *quarks*. Quarks have become part of a highly elaborate theory of hadron structure in which hadrons occur in two forms, *baryons* and *mesons*, the first comprising three quarks and the latter two plus a quark and its antiquark (*see* **SA**: ANTIMATTER). In addition, quarks have properties termed 'flavour' and 'colour charge', producing a highly complex character for each particle. Although this theory seems to be generally accepted by physicists, quarks have yet to be confirmed experimentally.

embryo the stage in the development of a new plant or animal that follows on from the fertilization of an egg by a sperm. It is most often used to describe the young of a mammal before birth while they are developing within the mother, and, in birds, to the growing chick while it is still inside the egg. Doctors define an 'embryo' as the stage in the development of a human being, from two weeks after fer-

Elements and their symbols

Element	Symbol	Atomic Number	Element	Symbol	Atomic Number
Actinium	Ac	89	Mercury	Hg	80
Aluminium	Al	13	Molybdenum	Mo	42
Americium	Am	95	Neodymium	Nd	60
Antimony	Sb	51	Neon	Ne	10
Argon	Ar	18	Neptunium	Np	93
Arsenic	As	33	Nickel	Ni	28
Astatine	At	85	Niobium	Nb	41
Barium	Ba	56	Nitrogen	N	7
Berkelium	Bk	97	Nobelium	No	102
Beryllium	Be	4	Osmium	Os	76
Bismuth	Bi	83	Oxygen	O	8
Boron	B	5	Palladium	Pd	46
Bromine	Br	35	Phosphorus	P	15
Cadmium	Cd	48	Platinum	Pt	78
Caesium	Cs	55	Plutonium	Pu	94
Calcium	Ca	20	Polonium	Po	84
Californium	Cf	98	Potassium	K	19
Carbon	C	6	Praseodymium	Pr	59
Cerium	Ce	58	Promethium	Pm	61
Chlorine	cl	17	Protactinium	Pa	91
Chromium	Cr	24	Radium	Ra	88
Cobalt	Co	27	Radon	Rn	86
Copper	Cu	29	Rhenium	Re	75
Curium	Cm	96	Rhodium	Rh	45
Dysprosium	Dy	66	Rubidium	Rb	37
Einsteinium	Es	99	Ruthenium	Ru	44
Erbium	Er	68	Samarium	Sm	62
Europium	Eu	63	Scandium	Sc	21
Fermium	Fm	100	Selenium	Se	34
Fluorine	F	9	Silicon	Si	14
Francium	Fr	87	Silver	Ag	47
Gadolinium	Gd	64	Sodium	Na	11
Gallium	Ga	31	Strontium	Sr	38
Germanium	Ge	32	Sulphur	S	16
Gold	Au	79	Tantalum	Ta	73
Hafnium	Hf	72	Technetium	Tc	43
Helium	He	2	Tellurium	Te	52
Holmium	Ho	67	Terbium	Tb	65
Hydrogen	H	1	Thallium	Tl	81
Indium	In	49	Thorium	Th	90
Iodine	I	53	Thulium	Tm	69
Iridium	Ir	77	Tin	Sn	50
Iron	Fe	26	Titanium	Ti	22
Krypton	Kr	36	Tungsten	W	74
Lanthanum	La	57	Uranium	U	92
Lawrencium	Lr	103	Vanadium	V	23
Lead	Pb	82	Xenon	Xe	54
Lithium	Li	3	Ytterbium	Yb	70
Lutetium	Lu	71	Yttrium	Y	39
Magnesium	Mg	12	Zinc	Zn	30
Manganese	Mn	25	Zirconium	Zr	40
Mendelevium	Md	101			

tilization until two months, and after this the word 'foetus' is used (*see* REPRODUCTION).

embryology the branch of biological or medical study that is concerned with all aspects of the growth and development of embryos.

emulsion *see* COLLOID.

endocrine system this is the name given to a network of small organs (known as *glands*) within the body of an animal, which are responsible for the production of chemical signalling substances called hormones. A hormone is released into the bloodstream and travels in the body until it reaches its target cells or organ somewhere else, where it causes a response to occur. There are several endocrine glands in the body of a human being including the *pituitary* gland (at the base of the brain), the *thyroid* gland (in the neck), and the paired *adrenal* glands (one above each kidney). The male and female sex organs (the *testes* and *ovaries*), are also endocrine glands which produce hormones that are responsible for the changes which occur at puberty and control fertility. (*See* HORMONE, REPRODUCTION.)

energy is the capacity to do work, and there are many different forms of energy: light, heat, sound, electrical, kinetic, potential and more, and all are measured in joules (J). *Kinetic energy* is possessed by moving objects and for a mass m, with a constant speed v, the kinetic energy is $\frac{1}{2}mv^2$. So a ball kicked by a footballer has kinetic energy which it loses when it hits the net of the goal, pushing the net outwards. Objects have *potential energy* by virtue of their position, that is, they have been moved and when released can do work. Hence a stretched spring, a car at the top of a hill, or a weight on a shelf all have potential energy. The energy is defined as *mgh* where m is the mass which is raised through a height h and g is the acceleration of free fall. *Thermal energy* (sometimes called heat energy) is that kinetic and potential energy possessed by an object's molecules and it rises with an increase in temperature. *Electrical energy* is that stored in batteries, and *electromagnetic waves* and sound waves also possess energy.

The law of conservation of energy states that energy cannot be created, nor can it be destroyed, however it can be changed from one form to another. This means that in any action all energy can be accounted for. This may be a simple procedure such as throwing a ball where chemical energy in the arm launches the ball which then has kinetic energy. Depending upon the throw it may have potential energy if the ball stops momentarily before falling, again with kinetic energy. Then when it hits the ground the kinetic energy becomes sound and thermal energy.

Society today requires vast amounts of energy to survive and this is supplied as electricity, oil, gas, coal, etc. However, the primary source of energy is the sun which is stored in plants which in turn provide energy either as food or as fuels. In addition to burning fuels for energy, alternative and renewable sources are being exploited such as tidal, wind, solar and hydroelectric.

engines are machines that convert energy into work, and fuel undergoes combustion to supply the energy. Petrol and diesel engines use the chemical energy from their fuels and electric motors use electrical energy stored in a battery or from a generator. The human body is also an engine, and food is the fuel. There are essentially two types of engine: those in which the combustion is internal as in the internal combustion engine, and those where the fuel is burned outside the engine itself–external combustion.

Engines are used to power all sorts of vehicles such as boats, aeroplanes and cars and although the internal combustion engine is not particularly efficient, it does provide a means of turning fuel into mechanical work very rapidly indeed. In the 1940s a British engineer, Frank Whittle, invented the gas turbine or *jet engine*. Air enters the front of the engine, is compressed and then enters a combustion chamber where liquid fuel is burnt. The energy produced expands the gas and shoots it out where it provides both thrust for motion and energy to drive the turbine which operates the compressor.

The *efficiency* of an engine is a ratio of the work provided for the energy put in. In general, most systems that burn fuels are very inefficient because so much energy is lost as heat. As a percentage, the efficiency of petrol and diesel engines is 25% and 35% respectively, while power stations are only around 30% efficient in producing electrical energy. Although electric motors are in themselves about 75% efficient, the process supplying the electrical input energy is only around 30% efficient.

entomology the specialized branch of biology which is concerned with the study of insects. A person who studies insects is known as an entomologist.

entropy is a measure of the disorder or randomness of a system which tends always to increase. The increase is because at every stage of energy transfer, some energy is wasted and the greater the disorder, the higher is the entropy. One result of entropy is that heat always flows from a hot to a cold body and this is the basis of the second law of thermodynamics which can be rewritten as: any system will always undergo change so as to increase the entropy.

enzyme enzymes are naturally-occurring protein molecules that are found in all living things, and which act as catalysts (i.e. they speed up and activate chemical reactions, *see* CATALYSIS) within cells. Enzymes are very specialized and each only acts on a certain substance. Also, conditions of temperature and acidity or alkalinity have to be just right or the reaction will not take place. This is one of the reasons why a stable body temperature is maintained in mammals.

Some enzymes are involved in breaking down processes, e.g. digestive enzymes such as *ptyalin* which is present in saliva and breaks down starch to sugar. Others are involved in reactions to build up more complex molecules from simpler ones such as in tissue growth.

equilibrium is when a system, whether chemical or physical, remains the same over time. In physics an object is in equilibrium if all the forces acting on it are equal and opposite. However, there are three equilibrium states. If a system returns to equilibrium position after being moved slightly, then it is in stable equilibrium, e.g. tipping slightly a box with a wide base. Unstable equilibrium is when the system moves from equilibrium when moved slightly, e.g. a pencil 'stood' on its point, and neutral equilibrium is when a movement results in a new equilibrium position, as with a ball. In chemistry equilibrium is reached in a chemical reaction when the proportion of reactants and products is constant, as the rate of the forward and reverse reactions is the same. Equilibrium is affected by changes in temperature, pressure or concentration of the reactants.

eucaryote the type of cell found in all plants and animals (but not bacteria or blue-green algae (cyanobacteria) in which the nucleus is bound within a membrane. *See also* PROCARYOTE.

evaporation is the process that occurs when a liquid turns into a vapour. Heat accelerates the process, which happens because some molecules near the surface of the liquid gain sufficient kinetic energy (*see* ENERGY) to overcome the attractive forces of the liquid's molecules and escape into the surrounding atmosphere. During the process of evaporation from a container, the temperature of the liquid falls until heat is replaced from heat in the surroundings. This is the reason why swimmers feel cold when leaving the water–because

heat energy is taken from the body, converted into kinetic energy enabling some water molecules to escape.

Evaporation is occurring all around us as rainwater puddles dry in the sun or as moisture evaporates from lakes, rivers, etc. eventually to form cloud and then rain. The principle is also used in industry where solutions are made more concentrated by evaporating off the solvent, and also in cooling and refrigeration systems. In a refrigerator, a volatile liquid evaporates and its vapour is pumped away. As it evaporates, it draws heat from the stored food. The pump then compresses the vapour back to liquid, and heat is given off via the cooling fins at the back of the refrigerator.

evolution the gradual change, over a long period of time, of one species of animal or plant. The organisms eventually acquire characteristics that are different from those of the ancestral species. This is able to occur, firstly, if there has been genetic mutation that allows for different information to be passed on from a parent to its offspring. Secondly, if the offspring (one or several) that received the different characteristic proves better suited to its environment than other members of the species then it is more likely to survive and reproduce. In this way the new characteristic tends to be preserved while those individuals not possessing it are more likely to die out. These changes are very small and take place slowly over many thousands of years, but it is thought that all living organisms have evolved from different ancestors in this way. The study of fossils (*see* G FOSSILS) has helped in the understanding of how this may have taken place in particular species. *See also* DARWIN, CHROMOSOME AND DNA.

Homonoid evolution

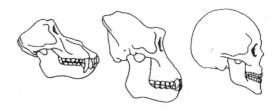

excretion the name given to the process by which an organism gets rid of the waste products of metabolism and eliminates them from its body. It differs from the process of getting rid of food waste, which is called egestion and is concerned with eliminating material that has been taken in. Excretion gets rid of waste products manufactured within the organism itself. The main waste products are carbon dioxide, water and nitrogen-containing substances from the breakdown of protein.

One way that these are disposed of is by diffusion or leakage to the outside (in plants and simple animals) either from a single cell or through the body. Higher animals have developed specialized organs for excretion of waste products, and these include gills, lungs and kidneys.

expansion *see* CONTRACTION AND EXPANSION.

explosives are substances or mixtures that, when heated or subjected to a shock or a blow, release a very large amount of energy very violently. The chemicals are actually undergoing rapid decomposition, producing large volumes of gas and quantities of heat. Explosives have numerous uses, both military and civil. In construction they are used for clearing land and blasting new cuttings through rock, in mining, quarrying and tunnelling, and also to demolish large structures, e.g. old cooling towers or multi-storey buildings.

There are basically three groups of explosives. *Propellants* are compounds that burn at a steady speed and can be detonated (set off) only under very specific conditions, e.g. as used in rockets. *Initiators* (or primary explosives) are extremely sensitive to heat and shock and are used in very small quantities to initiate explosions in large masses of less sensitive explosive. Compounds such as mercury fulminate ($Hg(ONC)_2$) and some metallic azides, e.g. lead azide, $Pb(N3)_2$, are initiators. *High explosives* are very powerful yet more stable and are used in bombs and shells.

Dynamite is a mixture of nitroglycerine with other compounds, while TNT (trinitrotoluene) is a very violent explosive made from toluene with nitric and sulphuric acids. Plastic explosives can be moulded by hand and are made of a high explosive such as cyclonite mixed with an oil binder.

extinction when a plant or animal species dies out completely this is called extinction, and it has occurred many times to thousands of plant and animal species in the course of the Earth's history. There have been times of mass extinctions, e.g. at the end of the Palaeozoic geological era (*see* **GG**: GEOLOGICAL TIMESCALE), about 248 million years ago, when it is thought about 90% of species ceased to exist. A further similar event took place some 65 million years ago, in the early Tertiary period, when the dinosaurs became extinct. Until the appearance of modern man, which was recent in terms of geological time (about 100,000 years ago), extinctions could be described as 'natural', occurring because of the processes of evolution.

Mass extinctions were probably brought about by climatic (especially temperature) changes, volcanic activity and possibly collision of asteroids with the Earth. Man's impact upon the Earth has been enormous as whole environments have been changed by tree felling, development of modern agriculture and industry and by pollution. Sadly, the extinction of many species which used to inhabit the Earth has been brought about by the destructive activities of human beings.

extrusion is a manufacturing process used in the production of shaped metal but particularly plastic goods. It is the most economical of plastic shaping methods and the products are simple in shape and have features in just two dimensions so that they can be extruded in the third, continuous, dimension. Pipes and gutters, strips, tubes, fibres can all be produced in this way. The essence of the process is a large screw which receives grains or pellets of plastic which have been heated and compressed and the melt is then forced out through a die which gives the section its shape. PVC foam can also be produced by extrusion.

In the extrusion of metal a block is forced by a ram out through a die. Some metals are extruded while cold but most are heated to increase the malleability (*see* DUCTILITY and MALLEABILITY).

eye *see* SIGHT.

fats are a group of naturally-existing compounds known as *lipids*. They are composed of combinations of one molecule of a substance called glycerol and three of fatty acids. They are found widely in plants and animals and are very important as long term energy stores having twice the number of calories as carbohydrates. In mammals there is a layer of fat deposited beneath the skin which provides insulation, preventing heat loss from the body. This is an absolutely vital provision for many animals, enabling some to inhabit the coldest regions of the Earth, e.g. seals, polar bears and penguins. These fat reserves enable some animals, e.g. bears and hedgehogs, to hibernate through the cold winter months and when they emerge in the spring they must immediately replenish their fat stores. The layer of fat beneath the skin helps to cushion the body against injury and at deeper levels it is stored as fatty (adipose) tissue.

fermentation is a process carried out by certain micro-organisms, e.g. yeast, bacteria and moulds which break down organic substances (those containing carbon, hydrogen and oxygen) into simpler molecules, producing energy. Alcoholic fermentation is a process which has been harnessed by man for centuries, and in this yeast converts sugar to alcohol and carbon dioxide, and it is used to produce such drinks as wine, beer and cider. Fermentation is one of the processes now used in biotechnology and is important in the manufacture of cheese, yoghurt and bread. Also, and most importantly, it is used in the production of drugs such as antibiotics and in genetic engineering.

fertilization is the fusion or joining together of the male (sperm) and female (egg or ovum) sex cells which is the essential part of sexual reproduction. Fertilization describes the process in which the two cells come together to become one and it sets in motion a chain of events, (involving further cell division and growth) which eventually gives rise to a new individual. It is a common event in both plants and animals and enables genetic 'mixing' to occur as the new organism receives its characteristics from each parent. In many animals, e.g. most fish, fertilization is described as *external* as the eggs are laid outside the body and sperm are shed over them. In many other animals, e.g. mammals, fertilization is *internal* as the male sex cells are released inside the body of the female.

fertilizers are chemicals added to the soil to improve crops and their yield and the growth of plants and flowers. Fertilizers replace the nutrients in the soil that are extracted by growing plants. Modern farming is very intensive and natural processes are unable to provide all the necessary nutrients required. In addition to carbon, hydrogen, oxygen there are other *essential elements* such as nitrogen, phosphorus, potassium, calcium, magnesium and sulphur; plants require *trace elements* (perhaps in parts per million quantities) such as iron, boron, manganese, zinc, copper, molybdenum and chlorine. Artificial fertilizers make up a lot of the deficits in these elements.

Ammonium sulphate is an important nitrogenous fertilizer (i.e. nitrogen supplying) and other chemicals used include sodium nitrate, urea and ammonia. *Superphosphates* contain phosphorus, the chemical used being calcium hydrogen phosphate, $Ca(H_2PO_4)_2$.

It is essential that fertilizers are used correctly and that they are not overused as the excess nutrients can have detrimental effects on the land, and particularly on streams and rivers. As the nutrients drain into water they encourage the growth of algae and surface plants which choke the stream and results in a lack of oxygen in the water which eventually kills animal and plant life beneath the surface.

filters and filtration a filter is a device for separating solids or particles suspended in solution from the liquid, and filtration is the separation process. It may also involve removing particles from a gas. There are numerous materials which are used as filters; filter paper–a pure cellulose paper used in laboratories; cloth and paper filters are used in engines to clean oil and air; crushed charcoal and sand are used in industry and in the medical field dialysis machines use membranes as filters to cleanse the blood of patients with defective kidneys.

The *filtrate* is the clear liquid that results from filtration while the solid particles left are called the *residue*.

flash point is the lowest temperature at which certain liquids give off sufficient flammable vapour to produce a brief flash when a small flame is applied. The term is used particularly for products such as petrol which vaporize very easily, because it is for all practical purposes the temperature at which petrol burns. It is important to be aware of this when petrol is used, transported or stored.

The same applies to industrial solvents such as toluene, benzene, ethanol, etc. which have flash points up to 13°C. Benzene has a melting point of about 5°C but will generate an explosive vapour while solid.

flight a few specialized groups of animals possess particular features which enable them to fly and these include insects such as beetles, flies, dragonflies and butterflies and birds and bats. In beetles, the forewings are hardened and form protective covers for the rear pair of flying wings which are moved in co-ordination and these are the most ancient group known to have possessed flight. Bees and wasps also have two pairs of flying wings which are hooked together and move as one. Similarly, in butterflies, the two pairs of wings are overlapped and move as a single pair and are covered with numerous minute scales. Insect wings are extensions of the outer covering (*cuticle*) of the middle part of the body (*thorax*) behind the head. They are membranes with veins running through them, and are moved by large flight muscles which bend the thorax out of shape. As they move up and down the angle of the wings in relation to the body alters which allows for lift on both the up and down strokes of the beat.

In birds and bats the wings are modified forelimbs or arms. In bats the membranous wings are a layer of skin spread between the long forelimbs and fingers and the body and hindlimbs. Birds possess several adaptations for flight including a lighter skeleton than other vertebrate animals, fewer organs to reduce weight and aerodynamic wings with modified flight feathers.

flotation and buoyancy buoyancy is the upward thrust felt by an object in a fluid and is equal to the weight of the fluid displaced (*see* ARCHIMEDES' PRINCIPLE). An object in water experiences this up-thrust because although the object is under pressure on all sides from the liquid, the pressure is greatest where the water is deepest, i.e. underneath the object. Hence an object will float if the up-thrust is more than its weight and the *law of flotation* states that a floating object will displace its own weight of the fluid in which it floats. Thus if a block is floated in water and then in a less dense fluid, it will float lower in the less dense fluid to displace a greater volume.

This principle applies to ships, and because salt water is denser than fresh water a ship floats lower in fresh water. Water temperature also affects density and therefore affects flotation. All ships have a line marked on their side (the *Plimsoll line*) which indicates the point beyond which the ship cannot be loaded. This is particularly important for a ship sailing from cold salt water to warm, less salty water.

flowers are the reproductive organs of a group of plants called *Angiospermae* (or *Anthophyta*) which are the flowering plants. They vary greatly in size from the very small and insignificant to the large and magnificent, showing an enormous range of colours and patterns of petals. Before the flower bud opens it is usually tightly folded and enclosed by green, leaf-like structures called *sepals*. These together make up an outer supporting structure for the flower called the *calyx*.When the flower bud opens the coloured petals expand and the sepals may wither and fall off. Petals may attract insects to the flower for pollination, but the actual reproductive parts are contained inside them. These are the *stamens* which are the male organs and the *carpels* which are female and contain egg cells or *ovules*. These structures are supported at the base by a portion of the flower called the *receptacle*. Following fertilization, the ovules eventually form seeds.

A flower

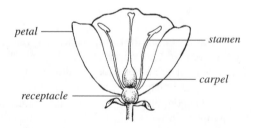

petal — — stamen

— carpel

receptacle —

food chain in simple terms, a food chain is the route by which energy is transferred through a number of organisms by one eating another from a lower level (called a *trophic* level). At the base of the chain are the primary producers which are the green plants. These are able to use energy from the sun to manufacture food substances from carbon dioxide and water. These food substances (glucose, cellulose and starch) are made use of by animals at the next level in the chain. The herbivores which eat the green plants are known as primary consumers. These in turn are eaten by carnivores (flesh-eating animals) which are called secondary consumers. There may be more than one level of secondary consumer (a flesh-eater may itself be eaten by a larger carnivore) ending up with animals at the end of the chain, e.g. lions, which are not preyed upon and are called the top predators. However, when these and all organisms die, they are eaten by scavenging animals and the remains eventually broken down by micro-organisms so none of the energy is lost but is used again. Food chains are often highly complicated and all of those that exist in a given environment are interlinked and form a food web. A food chain should be perfectly balanced with many more organisms at the lower levels than at the higher ones. Sometimes the natural balances are upset and this may be due to human interference. For example, in Britain there are no large carnivores such as wolves because they were hunted and killed off during the Middle Ages. In Scotland red deer numbers are too high and (although the situation is complicated) these would, at one time, have been hunted by wolves.

fluids are substances that flow easily and readily alter their shape in response to outside forces. Liquids and gases are both fluids. Liquids have freely-moving particles which tend to be restricted to the one mass, but gases expand to fill their containing space and do not maintain the same volume. These properties of fluids are very useful in many ways, particularly in industry where machines utilize fluid or gas-filled chambers or cylinders to operate mechanisms, e.g. the braking system on a car.

Hydrostatic is the term applied to a machine using fluid pressure, and *pneumatic* is a system that uses compressed air, e.g. air-brakes, pneumatic drills and other tools.

food additives are chemicals added to foods by manufacturers to improve a particular property whether it is colour, shelf-life, taste or appearance. In addition to colourings and preservatives there are anti-oxidants to stop reaction of the food with oxygen, sweeteners, flavour enhancers (e.g. sodium glutamate), emulsifiers, pH adjusters and many more.

There is now a regulatory system in Europe whereby each additive that can be used in food is given a number and it is then listed on the packaging. Although the vast proportion of these compounds cause no problems, some do create side effects in some people. *Tartrazine* is a well-known example. It is a yellow colouring agent but can cause hyperactivity in children and also skin complaints and breathing problems. Other *E factors* may cause dizziness, vomiting or muscular weakness.

food preservation is the prevention of food spoilage through chemical decomposition and action of micro-organisms. It is generally achieved by the sterilization of the food which destroys any bacteria by heating in sealed containers (canning), or by pickling, drying, freezing, smoking, etc. Pickling, drying and salting are methods that were used over history and these kept foods through the use of agents (acid, i.e. vinegar, and smoke) in which bacteria would not survive, or by the removal of water (essential to the growth of bacteria) as in drying.

These established methods have been supplemented by more modern techniques such as canning and freezing and also *freeze-drying*. Freeze-drying is used for numerous foods, notably coffee (and has been used for some time in the medical, veterinary and pharmaceutical fields) and perishable foods. The process involves freezing, producing ice from the liquid content of the material and then *sublimation*, i.e. the ice is extracted as vapour at low pressure and temperature. Foods preserved in this way can be kept for very long periods.

Another recent innovation is that of *irradiation* in which food is subjected to ionizing radiations (radioactivity) to kill micro-organisms. The technique is still under scrutiny and not everyone is fully convinced that it is suitable.

force is the push or pull upon a body which may cause it to move, stop moving or alter direction of motion. Force is defined as the mass of a body multiplied by its acceleration. If the mass is in kilograms and the acceleration in metres per second (m/s^2) then the force is in *newtons*. An object will continue to move at a constant speed and in a straight line unless another force acts upon it. For example, a craft in space under the influence of no forces will maintain the same speed in the same direction and would need a force to change its di-

rection. This is the basis of *Newton's first law of motion* which states that an object will continue in a state of rest or uniform motion in a straight line unless an external force acts upon it. This second law of motion relates to momentum and the third law can be given concisely as–to every action there is an equal, opposite reaction.

On Earth, we have to contend with friction and gravity but objects can move at a constant velocity or be at rest if forces acting on them are balanced. So an aeroplane can maintain a constant air speed because lift and weight are balanced, as are thrust and air resistance. However, should any of these constituent forces change then the velocity of the aeroplane would change.

formula (*plural* **formulae**) in mathematics or physics is a law or relationship denoted by symbols and figures and possibly expressed in algebraic (*see* ALGEBRA) form.

In chemistry it is a type of shorthand notation that enables a substance to be written in terms of elements and molecules using letters to represent the elements (*see* SYMBOLS). There are three types of chemical formula: an *empirical formula* shows the simplest ratio of atoms present in a compound, e.g. butane has the empirical formula C_2H_5, although it is really C_4H_{10}. The number and type of atoms present is shown in the *molecular formula*, in the case of butane, C_4H_{10}, which means there are four carbon and ten hydrogen atoms in every molecule. The *structural formula* indicates the structure of a molecule and shows the bonds between atoms.

fossils *see* **GG**: FOSSILS

freezing is the change in a material's state from liquid to solid brought about by reducing its temperature. The temperature at which this change occurs is called the *freezing point*. For pure substances the freezing point is the same as the *melting point*. The freezing point varies enormously between materials as the following table shows:

hydrogen	-259°C	water	0°C
oxygen	-218°C	sodium	98°C
nitrogen	-210°C	silicon	1410°C
argon	-189°C	carbon	3550°C
mercury	-39°C		

Impurities reduce the freezing point, e.g. salt reduces the freezing point of water and this property is exploited in car engines by adding *antifreeze* to the coolant to avoid freezing in winter.

frequency represented by the symbol f is the number of complete wavelengths (*see* WAVE) of a wave motion, per second. Frequency is measured in hertz (Hz) and is calculated from the formula; $c = f\,l$, where c is the speed of the wave, and l (lambda) is the wavelength. In the *electromagnetic* spectrum there is a large range of frequencies from low frequency radio waves to very high frequency gamma rays.

Sound waves are very different from light waves, and the human ear can hear sounds with frequencies between 20 Hz and 20, 000 Hz (or 20 KHz). A different frequency is heard as a different sound with high frequencies being sounds of high pitch. The scientific pitch of middle C on the piano has a frequency of 256 Hz and the Cs below and above are 128 Hz and 512 Hz, although when a piano is tuned these frequencies are changed slightly.

friction is the force that acts against motion, trying to stop materials and objects sliding across each other. A moving object will tend to slow down due to friction, and a force has to be exerted to keep it moving. The force required will differ depending upon the surface and the nature of the material moving across it. Friction is higher between solids (or a solid and liquid) than between solids and air.

The reason for friction is that rough surfaces have minute projections which restrict movement, and also there is a tendency for molecules to stick together under pressure. Friction in solids can be divided into *static friction* and *dynamic friction*. If an object is being pushed across a surface, the static friction is the maximum force, applied just before the object moves while the dynamic friction is that in action when the object is moving and it is much less than the static friction. There is also *fluid friction* when an object moves through a liquid or a gas. The effects of friction can be seen all around us the difficulty of pushing a heavy box across the floor; being able to walk and run because of the friction between our shoes and the ground; a car's tyres gripping the road and a train's wheels gripping the rails. In all these cases, friction causes the loss of energy as heat, something that applies to all machines.

fruit a fruit develops from the ovary of a flower and is given the name when it is mature and ripe. It contains the seeds and there are two main kinds, dry, e.g. an acorn and succulent or juicy, e.g. a tomato. The fruit is the means by which the seeds, which will become new plants, are protected until they are ready to be dispersed. Most juicy fruits are eaten by animals and the seeds pass through the digestive system without harm and are scattered in the droppings to grow elsewhere. Dry fruits often split open to release the seeds which may be shot out explosively, be carried away by wind or water, or cling to the fur of animals to drop off elsewhere. There are several different kinds of juicy fruit including a berry (e.g. blackberry) which is an *aggregate* fruit formed from one flower, but with lots of seeds) a pineapple which is a *multiple fruit* formed from a cluster of flowers and a cherry, a *simple fruit* also known as a drupe containing a stone surrounding the seed. Many of the food plants that we think of as vegetables are actually fruits, e.g. peas, beans and marrows. Pears and apples are known as *pomes* and they and some other kinds of fruit such as strawberries are also called *false fruits*. This is because the fruit does not just develop from the ovary of the flower but also from the receptacle (*see* FLOWERS).

fuel is a material that stores energy and upon combustion will release that energy. Fossil fuels (*see* **GG** FOSSIL FUELS) are the most widely used and account for most of the world's energy supply, with petroleum being the largest contributor. When these fuels are burnt (oil, gas, coal) energy is released and the other products are carbon dioxide, water and a variety of other gases and solids that depend upon the original composition or purity of the fuel. Nuclear fuels such as plutonium and uranium are unstable and release large amount of energy in nuclear reactions.

fuel cell is a cell that generates electricity directly by the conversion through electrochemical reactions, of fuels (gas or liquid) fed into the cell. The two components required are a fuel and an oxidant which are supplied to the electrodes and invariably a catalyst is used (*see* CATALYSIS). Fuels used include hydrogen (H_2), hydrazine (N_2H_4), ammonia (NH_3), and methanol (CH_3OH) and the oxidant is usually oxygen (O_2) or air. The electrolyte (*see* ELECTROLYSIS) in the cell can be a solution or solid, or special ion-exchange resins which, as the name suggests, contain ions that can be replaced by other ions.

Fuel cells have been used on spacecraft. In these, hydrogen and oxygen are combined to produce electricity and water is formed as a very useful byproduct. However, the efficient fuel cells have failed to gain a foothold for use in industry.

fungus (Kingdom Fungi) all fungi are simple organisms which may be one cell or exist as threads (or *filaments*) of many cells. They were once classified as simple plants but as they contain no chlorophyll and cannot photosynthesize, they are now placed in their own kingdom–fungi. Fungi absorb their food from other organic material and are vital in the breakdown and recycling of organic substances. They are essential in that they make minerals available to the roots of growing plants. Fungi are vital organisms for man, being used in the processes of biotechnology and fermentation. Some are harmful causing diseases in plants and animals, others are parasites and a few are edible, e.g. mushrooms. The scientific study of fungi is called *mycology*.

fuse is a very useful, protective device for electrical circuits. Most electrical appliances have their own fuse in the plug and in addition circuits, whether domestic or industrial, have fuses.

In all circuits there is the possibility of a fault developing and too much current flowing, which could damage the circuit or cause a fire. The fuse is placed in the circuit to avoid this possibility. The commonest form of fuse is a short piece of thin wire encased in a small glass tube with metal ends which overheats, melts and breaks if too high a current flows through it. It is placed in the live wire of the circuit so that if a fault develops, the current is switched off. The fuse value is greater than, but as close as possible to the current that usually goes through the appliance; 3 amp and 13 amp fuses are commonest.

More recently, *circuit breakers* have replaced fuses. These are switches which automatically break the circuit in the event of an overload and they can be reset when the fault has been eliminated.

galvanizing is a process whereby one metal is coated with a thin layer of another, more reactive metal. It is performed to offer protection to the coated metal and iron and steel are often treated in this way. Galvanizing is done in two ways, by dipping into molten zinc or by electrodeposition, i.e. electrolysis. When iron or steel is dipped into zinc, a little aluminium or magnesium is added to prevent a zinc iron alloy forming, because this is very brittle.

With electrodeposition, the object to be coated is connected to the cathode (*see* ELECTROLYSIS) and zinc ions from the electrolyte coat the object while current flows. The layer of zinc then protects the underlying metal because corrosion affects it before the iron or steel beneath.

galvanometer is an instrument used to measure small currents and is often called a *milliammeter* if its scale is calibrated accordingly. It uses the physical property that a wire in a magnetic field experiences a force when a current passes through the wire. The current to be measured passes through a coil in a magnetic field and as a result the coil turns, and in turn it moves a pointer across the scale. Not surprisingly, this is called the *moving coil galvanometer*. The movement of the coil is resisted by springs and it comes to rest when the force generated by the coil in the magnetic field is balanced by the springs. The higher the current, the greater the force generated and the further the pointer moves across the scale. The sensitivity of the meter (i.e. giving more pointer movement for a particular current) can be increased in several ways and some galvanometers use a light indicator in place of the pointer. A beam of light is shone onto a mirror positioned on the coil and any movement deflects the beam along the scale.

Although galvanometers are used a lot and can be converted for use as an ammeter or voltmeter, many modern versions are digital instruments.

gas is the fluid state of matter (solid and liquid being the others). Gases are capable of continuing expansion in every direction because the molecules are held together only very loosely. A gas will therefore fill whatever contains it and because the molecules move around rapidly and at random, they bump into each other and the walls of the container which results in a PRESSURE being exerted on the walls. If a certain amount of gas in a container is put into another container half the size, the pressure doubles (if the temperature is constant). Also, heating a gas in a container increases the pressure. It can be seen therefore that the temperature, pressure and volume of a *fixed mass* of gas are all related and many years ago, early experimentation with gases resulted in three *gas laws*, which when combined can be stated as:

For a fixed mass of gas, $\dfrac{pv}{T}$ is constant

Where p is pressure, v is volume and T is temperature. The three laws individually are

$\dfrac{p}{T}$ = constant if v is unchanged   the Pressure law

$\dfrac{v}{T}$ = constant if p is unchanged   Charles' law

pv = constant if T is unchanged   Boyle's law

Only an *ideal gas* obeys these laws exactly and no gas can be considered ideal in this sense, although many approach this point at medium pressures and temperatures.

gas turbine is a type of internal combustion engine used in aircraft and ships, which is often called the *jet engine* because it involves the production of a jet of hot gas which provides the propulsion. The engine was invented by Frank Whittle, a British engineer, in the 1940s (*see* ENGINES).

The jet engine produces a forward force by thrusting out gas behind and it takes in large quantities of air for this purpose. The air also supplies the oxygen needed for combustion of the fuel. The air intake is therefore at the front of the engine, and behind is the compressor which consists of a number of blade-like fans. The compressor forces air under high pressure into the combustion chamber where the fuel (kerosene) burns to produce a hot gas which expands and is thrust out of the rear of the engine, creating a forward thrust on the engine. This gas passes through the turbine before being expelled providing the rotational force to turn the compressor.

Geiger counter (**Geiger tube** *or* **Geiger-Müller tube**) is an instrument that can detect and measure ionizing radiations, mainly alpha, beta and gamma rays. It was named after the German physicist Hans Geiger. It is made of a sealed and enclosed tube with a fine wire down the centre of the tube. The tube is filled with argon gas at low pressure and the wire is the anode and the tube forms the cathode. The end of the tube is covered by a mica window through which the radiations pass. When a particle with a charge (or gamma rays) enters the tube, the argon is ionized into electrons and positive *ions* which move to their respective and oppositely charged electrodes and for a moment the gas conducts and a small current flows in the circuit. This is registered by a *ratemeter* (or *scaler*) and can be converted to a series of clicks.

generator is a machine that produces an electric current from mechanical motion. It is based upon the principle of electromagnetic induction (*see also* ELECTROMAGNET). If a coil in a magnetic field is rotated, a current is generated (induced) as the coil cuts through the magnetic field. As the coil completes its rotation, it cuts the field in the opposite direction and an induced current flows in the opposite direction. This is the basis of an alternating current generator, or *alternator*. To enable the coil to continue turning, carbon brushes form the connection between the coil and the outside circuit by rubbing against slip rings fixed to the ends of the coil. To generate direct current from a generator, the device is fitted with a *commutator* and instead of having two slip rings the generator has one ring which is split (i.e. the split ring, or commutator) and the coil ends connect to each half. Then every time the coil passes the split the connections are reversed and current flows in a constant direction to the outside circuit.

Cars have alternators for charging the battery (after the supply has been turned into direct current) and enormous versions are used in power stations for generating mains electricity.

genes *see* CHROMOSOMES AND GENES.

genetics and genetic engineering genetics is the name given to the branch of science which deals with the study of *genes* and *chromosomes* and the way in which characteristics are passed on from parents to offspring (called heredity). It is one of the most important areas of modern scientific study especially in helping us to gain an understanding of how certain hereditary diseases and disorders are passed on. Also, it enables new strains of organisms with useful characteristics to be bred more easily. Genetic engineering is the term given to the modification by human beings of an organism's genetic make-up and this is done in two main ways. Firstly, DNA from one organism might be transferred to another where it would not normally occur. This has been carried out in the fight against various diseases and is a technique within *biotechnology*. An example is the gene that codes for the *hormone* insulin in human beings which has been inserted into the cells of certain bacteria. The bacteria have been harnessed to produce insulin which is then used to treat diabetes. Secondly, DNA from two different organisms has been combined to produce an entirely new species. It is this second area of research which has caused a great deal of concern and fears that harmful organisms might be produced. Hence it is subject to extremely strict regulation and controls so that all the organisms involved remain securely within the laboratory.

geometry is a branch of mathematics that deals with the properties of lines, curves and surfaces. It includes the study of planar (flat) figures such as the circle and triangle and also three-dimensional figures such as the sphere and cube. Geometry (meaning measurement of the Earth) was first used in the measurement of land areas and today it is the basis of much calculation undertaken by engineers, builders and architects.

Co-ordinate geometry is where points, lines and shapes are represented by algebraic (*see* ALGEBRA) expressions. In two dimensions the plane containing a point is represented by x and y axes at right angles to each other which meet at the origin, O. The position of a point can then be defined by two distances, one along the x and one along the y axis, which intersect at the point in question. Lines are represented by equations, whether straight or curved. The values given to the particular position of a point are its *Cartesian co-ordinates*.

A well-known *theorem* (a rule proven by reasoning) in geometry is *Pythagoras' theorem* which states that in a right-angled triangle, the (area of the) square on the hypotenuse (the longest side opposite the right angle) equals the sum (of the areas) of the squares on the other two sides.

gestation and birth is the period of time in a mammal between fertilization of the egg and birth of the young which, in human beings, is also called *pregnancy*. The gestation period is characteristic of the species concerned and varies from 18 days in the mouse to 9 months in human beings and 18 to 23 months in the Indian elephant. Usually, larger mammals have longer gestation periods and their offspring require care for a greater length of time before they can live independently. Following fertilization of the egg in a mammal such as man, cells divide rapidly to become an embryo which becomes attached, by means of a special organ called the *placenta*, to the wall of the *womb* (*uterus*). The placenta allows oxygen and food to pass to the baby through a connecting cord called the *umbilical cord.*

When the gestation period is completed and the baby is ready to be born, the process of birth takes place. This is triggered off by hormones which cause the womb to contract by means of the powerful muscles which are present in its wall. The baby is gradually forced through the birth canal (*vagina*) to the outside and afterwards the placenta is also shed. Most mammals eat through the umbilical cord (but it is cut after a human birth) and the portion which is left attached to the baby soon dries, shrivels and drops off leaving a mark called the *umbilicus* or navel.

Human foetal development

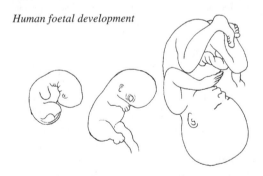

Animals which give birth to live young as described above are called *viviparous*, whereas those which lay eggs that develop and hatch outside the mother's body (e.g. birds and some reptiles) are called *oviparous*. Some animals keep their eggs inside the body for protection. The young are nourished within the egg and not by the body of the mother and these animals (e.g. some fish, reptiles and insects) are called *ovoviviparous*.

gills *see* LUNGS.

gland *see* ENDOCRINE SYSTEM.

glass is the hard transparent material from which windows, bottles, jars and glasses, lenses and laboratory ware is made. There are several types of glass but the essential ingredient in all is silica (SiO_2).

Man-made glasses appeared about 4000 BC in Egypt and by 1500 BC it had developed into both an art and a technology. Glass science was then not developed until the work of numerous scientists including Faraday, then later Zeiss, Abbé and Schott. Over 30 elements had been used in experimental glasses by the turn of this century and now 70 have been tried but just three commercial systems account for nearly all glass production (*see* CERAMICS).

The raw materials are heated in a furnace forming a red-hot liquid. A blob of this molten glass on the end of a 'blow-pipe' can be blown into intricate and beautiful shapes by glass-blowers. However, most glassware is produced from moulds (for bottles, etc.) and sheet glass is made by floating molten glass on a bed of molten tin. Glass is used in many ways and its properties can be altered by the processing or by addition of other materials. *Toughened glass* is manufactured by cooling glass rapidly under cold air; laminated glass has a plastic sandwiched by two layers of glass, increasing toughness and glass fibres in a resin form a useful *composite* (i.e. a mix or combination of materials) for making strong but light structures e.g. boat and car bodies. *Bullet-proof* glass is a very effective composite which consists of several sections with different properties.

glycerol (or glycerin) is a sweet-tasting, viscous (thick, syrupy) liquid with no colour or smell. It belongs to the alcohol group but has a structure similar to a sugar. Its formula is $HOCH_2CH(OH)CH_2OH$ and it is derived synthetically from propane (C_3H_6), or as a byproduct in the production of soap. It occurs naturally in plants and animals as a component of stored fats. It is an extremely useful compound and is used in the manufacture of ice cream, sweets and other foodstuffs, toilet preparations, resins and explosives. It also has the useful property that it can absorb up to 50% of its weight of water vapour which means it can be used as a moisturizing agent.

gold and the noble metals gold is a bright yellow metal and a good conductor of electricity and heat. It is soft and malleable (*see* DUCTILITY) and can be produced as very thin (even see-through) leaf or drawn out into wire and it is very resistant to corrosion. Of course its primary use for thousands of years has been in the making of jewellery and coins, and today its value is fundamentally important to the financial stability of the world's currency markets.

Gold occurs as nuggets or veins or smaller particles in quartz or in streams after it has been weathered out of its original site. It is also found in the residue after copper has been purified by *electrolysis*. It is extracted from ore by dissolving the gold in potassium cyanide (KCN) solution (the cyanide process) or by the amalgamation process which involves treatment with mercury to form an amalgam which is processed further.

In addition to the applications mentioned above, gold is used in dentistry, photography, medicine, in electrical contacts (in microchips) and conductors. The purity or fineness of gold is measured in *carats*, which are parts of gold in 24 parts of the alloy. Thus pure gold is 24 carat. Jewellery is often made of 9 carat gold, in which the remaining 15 parts are copper.

Gold is also one of the *noble metals*, with silver, platinum etc. and all have similar properties.

graph is a drawing or picture that represents data and numerical values, or shows the mathematical relationship between two or more variables. Often this takes the form of plotting (positioning) points at a certain place, relative to two values measured along axes at right angles to each other (*see cartesian co-ordinates* in GEOMETRY). A *histogram* is a type of graph consisting of a number of blocks drawn with reference to two axes such that the area of each block is directly proportional to the value of the frequency. A *bar graph* looks similar to a histogram but the height of the block is then the relevant factor. A *pie chart* is another graphical method of representing data in which a circle is divided into different sized sectors where the angle of the sector is proportional to the size of the sample, expressed as a percentage.

gravity or gravitational force is the downward pull exerted by the material making up the Earth. A gravitational attraction exists between all objects and this will increase with their mass and with a lessening of the distance between them. Although this attraction is always present, in most instances it is so very small as to be unnoticeable and the attraction which dominates is that of the Earth, which affects objects on its surface (*see* WEIGHT). If gravity is thought of as a gravitational field exerting a force on a mass, the value of that field strength, g, can be calculated as almost 10 m/s^2 (metres per second, per second). This means that a falling object near the Earth's surface accelerates

at 10 m/s^2; alternatively a mass of one kilogram near the Earth's surface has a force acting on it due to gravity, of 10 N (newtons).

Gravity on the Moon is much less than on Earth–about one sixth, hence although an astronaut's *mass* remains the same, his or her weight is much less.

growth the process by which living organisms increase in size and often in weight. It is brought about by cells dividing and often becoming more specialized to form the bulk of a particular tissue or *organ*. All organisms have a maximum size which is determined by their genetic make-up and the growth process is controlled by HORMONES. Once the individual has reached its full size there is no more growth but cell division occurs to repair and replace worn out or damaged tissues. Growth often occurs in spurts as can be seen in the development of a human child. A baby grows fastest during the first six months of life and there is a further growth spurt at the time of adolescence.

habitat *see* NH.

hair a hair is any fine outgrowth from the surface of a plant or animal. Hairs are found on many living organisms and may be used for a variety of different purposes. The hair or fur of mammals is composed of dead cells in which a substance called *keratin* has been laid down. It is used for insulation to keep the animal warm and may occur in a variety of colours. The colour is determined by the presence and amount of a certain dye or *pigment* known as *melanin*. The base of each hair is embedded in the skin and has a tiny muscle attached to it. In cold conditions the hair is raised by these muscles to stand up straight and this traps a layer of air which has a warming effect. People notice this happening when they have 'goose pimples' in response to the skin feeling cold.

halogens are the elements fluorine, CHLORINE, BROMINE, iodine and astatine which form a group in the PERIODIC TABLE. They are the extreme form of the non-metals and show typical characteristics forming covalent diatomic molecules, i.e. form molecules of two atoms, e.g. Br$_2$. At room temperature fluorine and chlorine are gases, bromine is a volatile liquid and iodine a volatile solid. Astatine exists only as short-lived, radioactive isotopes. The reactivity increases from iodine to fluorine and the latter reacts with all elements save helium, neon and argon to form compounds. Halogens occur naturally in salt deposits and as IONS in sea water.

haploid *see* CELL DIVISION.

hard water is water that does not readily produce a

lather with soap (*see* DETERGENTS AND SOAP) because of dissolved compounds (carbonates, sulphates and chlorides) of calcium, magnesium, sodium and iron. The use of soap results in the formation of a scum due to a chemical reaction between the metal IONS and fatty acids in the soap, producing salts. Water that does produce a good lather is called soft water.

Temporary hardness is one of two types and is due to water passing over carbonate-rich rocks such as chalk or limestone. This produces metal hydrogen carbonates which dissolve in the water and upon boiling these salts form insoluble carbonates which in a kettle results in kettle fur. This leaves the water soft. *Permanent hardness* is caused by metal sulphates (calcium or magnesium sulphates or chlorides) and cannot be removed by boiling. Special ion-exchange water softeners must be used in this case.

hearing the means by which animals are able to receive and decode sound waves which is usually closely related to balance, especially in vertebrates. Specialized small organs called RECEPTORS, which are often hair-like, vibrate in response to sound waves. This triggers off an electrical impulse in a sensory nerve (one which travels to the brain) which is in contact with the receptor (or receptors). The information is transmitted to the brain where it is decoded and interpreted as sound. In many invertebrates sound receptors may be relatively simple structures. However, in other animals such as mammals, complex and specialized organs, the ears, are used to detect sound.

The human ear

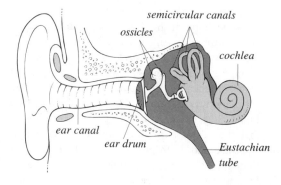

Fish have simpler ears than mammals and also a *lateral line system* which has sensory cells along the body, both of which are used for hearing. Amphibians (frogs and toads), reptiles and birds possess ears with a simpler structure than those found in mammals.

Lateral line system

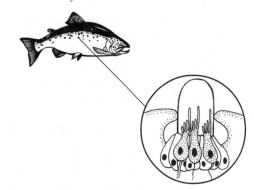

heart *see* CIRCULATION.

heat is ENERGY, and is measured in JOULES. It is stored in materials as vibrations of the molecules and the vibration increases with temperature. Heat travels by CONDUCTION, CONVECTION and by electromagnetic RADIATION and material states, i.e. whether something is liquid, solid or gas, depends upon the heat available. In a solid the molecules are attracted together and this overcomes the kinetic energy possessed by virtue of the molecules vibrating. If energy is put in and the molecules can be separated enough to become free, then the solid becomes a liquid. The work done in separating the molecules is called the *latent heat of fusion.* Similarly, the *latent heat of vaporization* is the heat required to turn a liquid into gas. When these changes are reversed, the material gives out heat.

The human body generates heat from food and the liver is a major heat-producing organ. Muscles also generate heat as a byproduct during activity.

heat exchanger is used to transfer heat from one fluid (i.e. a liquid or gas) to another without the fluids coming into contact. Heat exchangers are used a great deal to cool machines, e.g. the radiator in a car engine, or to conserve heat in an industrial (often chemical) process so that it can be used elsewhere in the process.

helium is one of the INERT GASES (or noble gases), so called because it has a stable electronic configuration and no chemical reactivity as such. It occurs naturally in very small quantities and being non-inflammable is used as an inert atmosphere for arc welding, for airships and balloons, in gas lasers and with oxygen as the atmosphere for deep sea divers. Helium liquefies below 4K (-269°C) and is used extensively in *cryogenics*–the study of materials at very low temperatures.

Helium has no colour, taste or smell and was named after *helios*, the greek for Sun. It is formed in stars such as the Sun as hydrogen nuclei are pressed together in the processes of nuclear fusion (*see* NUCLEAR FISSION AND FUSION).

herbivore any animal that feeds on plants, e.g. the familiar grazing mammals such as cows, sheep and horses.

heredity the passing on of characteristics from parents to offspring which is accomplished through the transfer of genes (*see* CHROMOSOMES AND GENES). Some of the basic laws of heredity were studied and worked out by an Austrian monk, Gregor Mendel (1822–1884) who carried out experiments with pea plants. He noticed that some characteristics were dominant over others, e.g. tallness is dominant to shortness in these plants. The study of genetics during this century has established that there are *dominant* and *recessive* genes for many characteristics and this is particularly important in some inherited diseases. A dominant gene is one which will always be seen in the offspring. A recessive gene will only be seen if it is present as a 'double dose', i.e. one from each parent. *See* also DNA, GENETICS.

Mendel's experiment with pea plants

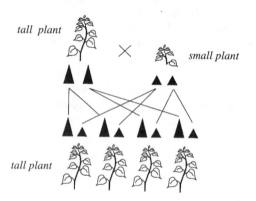

tall plant

small plant

tall plant

hertz *see* FREQUENCY.

holography is a method of recording and then projecting a three-dimensional image using light from a LASER but without the need of a camera. A single laser beam is split into two by a special mirror, with half the beam continuing straight through the mirror onto photographic film. The other part of the beam is reflected by the mirror onto the object in question and then onto the film. The two beams produce an interference pattern, or *hologram* on the film which when illuminated with laser light recreates a three-dimensional image of the object. A screen is not required and the light forms an image in mid-air.

There are now holograms that do not need laser light to produce an image, for example, some credit cards have holograms that work with reflected daylight.

homing *see* **NH**.

hominid *see* **NH**.

hormone an organic substance produced in minute amounts by special cells within plants and animals, which acts as a 'chemical messenger' causing a powerful response somewhere else in the body. In mammals hormones are produced and released (secreted) by special small organs called glands (the *endocrine* glands), and travel in the bloodstream. The site at which they produce a response is called the 'target,' (*see* ENDOCRINE SYSTEM).

horsepower is a measure of power which originates from the time when steam engines began to replace horses. The name was coined by James Watt the British engineer and one horsepower is equivalent to about $3/4$ kW, or more accurately, 746 W. It has become the standard practice to measure engine power in *brake horsepower*. This is horsepower measured by means of the resistance offered by a brake and shows the useful horsepower that can be produced by an engine. A typical car develops around 90–100 brake horsepower (bhp) while an articulated lorry develops 400 or 500 bhp.

horticulture this refers to the cultivation of all kinds of garden and greenhouse plants including flowers, vegetables and fruit. Horticulture is concerned with the breeding of new varieties of plants by propagation and seed production. Many of our food plants, especially fruit and vegetables, result from the techniques of horticulture.

hybrid *see* **breeding**

hydraulics is the study of fluid flow, and the practical application of the dynamics of liquids in science and engineering. Hydraulics is thus a very important subject as it is relevant to the design of harbours, canals, dams or the study of the flow of water in pipes. *Hydraulic machinery* is operated, and the power transmitted using the pressure of a liquid. Such machines use the properties of liquids that they cannot, for all practical purposes, be squashed and when pressure is applied to a confined liquid, the pressure applies to all parts of the liquid. The transference of a force through the liquid is arranged in such a way as to produce a greater force in the machine, e.g. in a hydraulic car jack.

Most car braking systems use this hydraulic principle. When the brake pedal is pushed, a piston in a cylinder forces brake fluid through narrow pipes which go to the wheels. At the wheels, the fluid pressure pushes pistons in cylinders which then move the brakes which are in the form of pads or shoes which move against the drum or disc attached to the wheel.

hydrocarbons are *organic* compounds (*see* CHEMISTRY) that contain only CARBON and HYDROGEN. There are many different hydrocarbon types and they form one of the most important groups of organic compounds. There is a fundamental division into two groups based on structure: *aliphatic* hydrocarbons comprise carbon atoms in open chains and in addition to the main groups such as alkanes (the paraffins), this term includes all compounds made by the replacement of atoms in a molecule. *Aromatic* hydrocarbons are characterized by a ring structure made up of six carbon atoms as found in BENZENE (C_6H_6). These are also called closed-chain or cyclic compounds.

Aliphatic hydrocarbon groups include the *alkanes*, with a general formula C_nH_{2n+2}, e.g. ethane C_2H_6. The first four compounds in the series are gases, then liquids and the higher members (above $C_{16}H_{34}$) are waxy solids. Alkanes are the main components of petroleum and in general they are quite resistant to chemical action. The *alkenes* have the formula C_nH_{2n} and resemble the alkanes, but are more reactive. They are used as fuels and in the manufacture of other substances, e.g. alcohols and glycerols. The *alkynes* (or acetylenes) have triple bonds between two carbons, and a general formula of C_nH_n. Acetylene is a well-known member of this series.

hydroelectricity is ELECTRICITY that is generated from water moving from a high to a low level. If a river or lake is dammed to form a large reservoir then the water is held back, high above the original river level, and the body of water has enormous *potential* ENERGY. To turn this into electrical energy the vast concrete dam contains a tunnel down which water can be directed and at the foot of the tunnel, before the water rejoins the river, the force of the flow is used to turn the blades of a large turbine linked to a generator, thus creating electricity.

Some power plants of this nature are called *pumped storage schemes* in which water flows through the day from a high level to a low level reservoir to generate electricity and at night when demand for electricity is low, water is pumped back up to the higher level ready for supply the next day.

Hydroelectric schemes are an alternative, or 'green' energy source because they do not cause pollution. However, they can change the local environment very considerably because they alter the balance of plant and animal life around the dam, and particularly when the valley is flooded to create the reservoir.

hydrogen and the hydrides hydrogen is the lightest element and forms molecules comprising two atoms, H_2. It occurs free and is widely distributed in water (H_2O), organic matter (HYDROCARBONS and CARBOHYDRATES) and minerals. The ordinary hydrogen atom has a nucleus of one proton, with one electron. There are other ISOTOPES and in addition to *protium*, there is *deuterium* which contains one neutron and tritium, containing two neutrons.

Hydrogen is manufactured by the ELECTROLYSIS of water and is produced in the treatment of petroleum with catalysts (*see* CATALYSIS). It is explosive over a wide range of mixtures with oxygen and because of this is no longer used in ballooning (where it has been replaced by HELIUM). It is used in numerous industrial processes including the production of methanol and ammonia and in metallurgy.

Hydrogen reacts with most elements to form **hydrides** of which there are several different types. Some are ionic and salt-like while most of the non-metals form covalent compounds (*see* BONDS), e.g. methane, CH_4. There are then many more complex compounds. Hydrides are often used as catalysts in hydrogenation reactions, i.e. the addition of hydrogen to a substance, frequently used in the refining of petroleum.

hydrometer is an instrument used for measuring the relative DENSITY of a liquid. It is a quick and convenient tool, although not as accurate as some other methods. The hydrometer consists of a tubular stem with a weighted bulbous base. It floats in the liquid to be measured and the density can be read from the graduated scale on the stem. The narrow stem ensures that small changes in density create readily visible changes in the reading. Hydrometers are used in checking the quality of beer and also in testing the level of charge in a car battery, because the relative density of the acid in a battery varies with the available charge.

hydroxides are compounds that contain the OH group which is present as the hydroxyl ion OH^-. ALKALIS are the hydroxides of metals and are strongly basic. Hydroxides such as caustic soda (NaOH), potassium hydroxide (KOH) and ammonium hydroxide (NH_4OH) are commonly used in industry in the production of soap, detergents, bleaches, paper and many other items.

hypertonic and hypotonic a hypertonic solution is one which has a higher *osmotic* pressure (*see* OSMOSIS) than another, or a standard, with which it is being compared. Hypotonic is the opposite, i.e. it has a lower osmotic pressure than the solution to which it is compared. *Isotonic* solutions are those with the same concentration.

These concentration differences are important in cells (*see* OSMOREGULATION).

immune system the natural defence system which operates within a vertebrate animal to protect it from infections, caused mainly by such micro-organisms as BACTERIA and VIRUSES. The cells which operate the immune system are the white blood cells or *leucocytes* of which there are a number of different types. The immune system is provoked into action by the presence of 'foreign' PROTEIN substances which are called *antigens*. Many substances are antigens but they are commonly found on the surface of bacteria and viruses. When the white blood cells encounter antigens in the body, they produce special proteins called *antibodies* which trap the foreign material. The antibody and antigen are locked together like two pieces of a jigsaw puzzle and are eventually 'eaten up' by other cells of the immune system called *macrophages*. More antibodies are rapidly produced by the cells and will always recognize, and be ready to bind to, the antigens. In this way, the animal becomes *resistant* to a particular disease and this is the basis of vaccination.

Vaccination A preparation of a virus or bacteria which causes a particular disease is treated to kill it or make it harmless, but it still provokes the production of antibodies (an immune response) when it is injected into the bloodstream. These antibodies are then always available to attack any live micro-organisms of that type which the animal might encounter in the future.

Particular antibodies are produced to each kind of antigen and the immune system operates in two ways. Firstly, there is the inborn or *natural immunity* present from birth which is not specialized and operates against almost any substance that threatens the body. Secondly, there is *acquired immunity* which is the situation described above and brought about by an encounter with a particular foreign substance.

indicators are substances used to detect the presence of other chemicals. Most often an indicator is used in the laboratory to show the pH of a liquid, i.e. its acidity or alkalinity. Indicators are used in *titrimetry* which is the fast reaction of two solutions to an end point determined by a visual indicator and the two solutions are commonly an acid and a base.

The simplest indicator is *litmus paper* which

shows red in acid and blue in alkali. *Universal indicator* is a mix which shows a gradual series of colour changes over a range of pH. However, most such indicators operate over a narrower pH range, e.g. methyl orange is red below pH 3.1 and changes to orange and then yellow as the pH reaches 4.4 Other indicators in acid-base reactions include:

Indicator	Low pH colour	High pH colour	Range
thymol blue	red	yellow	1.2–2.8
methyl red	red	yellow	4.2–6.3
bromothymol blue	yellow	blue	6.0–7.6
phenol red	yellow	red	6.8–8.4
phenolphthalin	colourless	red	8.3–10.0
alizarin yellow	yellow	orange	10.1–12.0

Indicators tend to be complex organic molecules and, for example, phenolphthalin comprises three benzene rings arranged thus:

Another group, called *redox indicators*, show different colours depending upon whether they are oxidized or reduced (*see* OXIDATION AND REDUCTION). The change occurs over a narrow range of electric potential within the system under study.

inductance is when a circuit carrying an electric current is characterized by the formation of a magnetic field. Similarly a current can be made to flow when the position of the magnetic field moves relative to the circuit (*see* ELECTROMAGNET, GENERATOR). The unit of inductance is the henry (H), named after the American physicist Joseph Henry who was a pioneer in this field. Due to inductance a change in the current within one circuit can cause a current to flow in a nearby circuit. This is because if current flows through a circuit containing a coil, a magnetic field is created and an adjacent coil, in a separate circuit but within the magnetic field, experiences an *induced current*.

inert gases are the gases in group 0 of the PERIODIC TABLE, known also as the *noble gases*. These are helium, neon, argon, krypton, xenon and radon and they make up approximately 1% of air, by volume, with argon the most abundant. As their name suggests, these elements are chemically unreactive, due to their stable electronic configuration. They can be extracted from liquid air by fractional DISTILLATION and helium occurs in natural gas deposits. Their stability makes them useful in many applications, e.g. helium and argon are used as inert atmospheres for welding, in light bulbs and fluorescent lamps. Helium in its liquid form is important in low temperature research.

inertia the property of a body that causes it to oppose any change in its velocity, even if the velocity is zero. An object at rest requires a FORCE to make it move, and a moving object requires a force to make it slow down or accelerate or change direction. Newton called this resistance to a change of velocity inertia. The greater the mass of a body, the higher is its inertia.

infection and toxin an infection is a disease or illness which is caused by the invasion of a micro-organism which, because it is able to do this, is called a PATHOGEN. In animals and man most pathogens are bacteria or viruses but in plants a wide range of fungi are pathogenic. A high standard of hygiene is the best means of preventing infections from occurring, and sometimes people who are sick with an infectious disease require special isolation, nursing and care.

There are a number of ways in which a pathogen can enter an animal's body. 1. It might be breathed in; 2, it may enter the bloodstream through a cut or wound; 3, it might be carried by a blood-sucking insect such as a mosquito which passes the pathogen on when it bites or stings; 4, it may be present on food (called contaminated) which is then eaten; 5, in drinking water, particularly that which is contaminated by sewage and 6, through sexual intercourse–a sexually transmitted disease.

Often the illness is caused not by the micro-organism itself but by a poison, known as a *toxin*, which it produces and this is especially true of bacteria.

infinity is the term used to describe a number or quantity which has a value that is too large to be measured. For example, outer space is regarded as infinite since it has no limits. By convention, infinity (designated by the symbol ∞) is the result of dividing any number by zero. If a value is so small as to be incalculable, it can be written as negative infinity ($-\infty$) and it is called infinitesimal.

infrared and **ultraviolet** *see* ELECTROMAGNETIC WAVES.

injection moulding is an industrial process, similar to EXTRUSION, used for the production of plastic mouldings. Granular or powdered plastic is fed into a hopper which feeds into a screw mecha-nism. The heat generated by the screw and outside heaters produces a plastic melt by the time it reaches the end of the screw. The melt is then injected or forced through a nozzle into a cool mould and pressure is maintained while the entrance to the mould is blocked by solidified melt. The moulding is then released and the process is repeated. Most injection moulding machines run automatically and large quantities can be produced at low cost. There is a large range of products which can be manufactured in this way including small boats, buckets and bowls, crates, telephone hand-sets and small gears.

instinct *see* BEHAVIOUR.

insulation is the means of preventing the passage of heat or heat loss by conduction, convection or radiation (it also applies when an electric current is prevented from passing). There are many ways of providing thermal insulation, the commonest example being the insulating materials used in houses to reduce heat loss. Cavity walls can be filled with foam, roof spaces lined with fibreglass or polystyrene beads and windows can be double glazed.

A vacuum prevents conduction and convection and this property is utilized in a vacuum flask to retain heat in hot drinks. There are several examples of natural insulation, notably the fur and feathers of mammals and birds respectively.

insulator *see* CONDUCTOR.

intelligence the ability of animals to learn and understand and so to be able to live effectively in their surroundings. Human beings are considered to be the most intelligent of animals and this ability depends upon GENETIC make-up and environment. The potential for developing a person's intelligence is considered to be at its greatest during childhood, especially in the early years. Also, it is thought to be important to use intellectual abilities throughout life in order to exercise our intelligence to its full power.

invertebrate animals are those 'without a backbone' or rather which lack an internal skeleton. They make up 95% of animal species and show many different kinds of specializations which allow them to adopt all sorts of lifestyles axnd habitats.

ion is an atom or molecule that is charged due to the loss or gain of an electron or electrons. A *cation* is positively charged and an *anion* negatively charged. *Ionization* is the process that results in ions and it can occur in a number of ways including a molecule breaking down into ions in solution, or the production of ions due to the bombardment of atoms by radiation. During electrolysis, ions are attracted to the electrode with an opposite charge.

iron is a metallic element in group 8 of the PERIODIC TABLE, which in its pure form is silver-white. It is a common component of minerals in clays, granites and sandstones and is a dominant element in meteorites. It occurs naturally as several minerals: magnetite (Fe_3O_4), haematite (Fe_2O_3), limonite ($FeO(OH)_nH_2O$), siderite ($FeCO_3$) and pyrite (FeS_2). In combination with copper it occurs as chalcopyrite, $CuFeS_2$. Pure iron melts at 1535°C and is extracted from its ores by the BLAST FURNACE process.

Iron is a widely used metal, in combination with other metals. Cast and wrought iron (*see* CAST IRON) are both used extensively but most iron goes into the production of STEEL.

Iron is also of biological significance because it is essential to the red blood cells. These cells contain the pigment *haemoglobin* which is made up of the iron-containing pigment haem and the protein globin. Haemoglobin is responsible for the transport of oxygen around the body, hence if there is insufficient iron in the diet anaemia may result.

irradiation is the exposure of an object to radiation which may be electromagnetic radiation such as X-rays or gamma rays. The radiation may also come from a radioactive source. Above a certain level radiation harms organisms and this principle is adopted when using irradiation as a technique for food preservation. Very specific irradiation is used in medicine when obtaining an X-ray or in the use of ionizing radiations to treat cancer.

isomers are chemical compounds with the same molecular formula, i.e. same composition and molecular weight, but which differ in their chemical structure. This property is called *isomerism*, of which there are two types. *Structural isomers* differ in the way that their atoms are joined together whether it be changes in the arrangement of carbon atoms in the chain or the position of a group or atom on the chain or ring. *Tautomerism* is a special case of structural isomerism which is termed dynamic. This is because a compound exists as a mixture of two isomers in equilibrium and removal of one isomer results in conversion of the other to restore the equilibrium. The other type of isomerism is *stereoisomerism* in which isomeric compounds have atoms bonded in the same way but arranged differently in space. One type of stereoisomerism is *optical isomerism* due to asymmetry of the molecules and in this case the isomers differ in the direction in which they rotate a plane of polarized light (*see* POLARIZATION OF LIGHT).

isotope is one of several atoms of the same element that have the same ATOMIC NUMBER (i.e. same number of protons) but differing numbers of neutrons in the nucleus (affecting their atomic mass). The chemical properties of isotopes are therefore the same, only physical properties affected by mass differ. Most elements exist naturally as a mixture of isotopes but can be separated upon their slightly different physical properties. For laboratory purposes MASS SPECTROMETRY is often used. *Radioisotopes* are isotopes which emit RADIOACTIVITY and decay at a particular rate, e.g. carbon -14.

Chemical symbols can be used to show the different configurations of isotopes, thus for hydrogen there are three isotopes:

	Protium	Deuterium	Tritium
	ordinary hydrogen		
protons	1	1	1
neutrons	0	1	2
electrons	1	1	1
symbol	$_1^1H$	$_1^2H$	$_1^3H$

When deuterium replaces hydrogen in a water molecule, the resulting compound is called *heavy water*.

joule (J) is the unit for measurements of ENERGY and WORK. It is equal to a force of one newton moving one metre (one newton is the force required to give a mass of one kilogram an acceleration of one metre per second, per second). The unit is named after the British physicist James Prescott Joule who investigated the link between mechanical, electrical and heat energy.

One thousand joules is called a kilojoule. Energy used to be measured in *calories*, one calorie being 4.1868 joules.

kidney *see* OSMOREGULATION.

kiln is a furnace or large oven that has many uses. The most obvious is its use for drying, baking and hardening clay objects such as plates, cups and similar domestic items, but also bricks and other building components. Kilns vary greatly in size depending upon their use and modern versions are heated by gas or electricity.

The minerals extraction industry also uses kilns for drying ore, driving off carbon dioxide from limestone or roasting sulphide ores to remove sulphur as sulphur dioxide, prior to further processing.

laboratory any room or building that is especially built or equipped for undertaking scientific experiments, research or chemicals manufacture. The study of physics, chemistry, biology, medicine, geology and other subjects usually involves some work in a laboratory and each is equipped with special instruments.

A chemical laboratory typically contains balances for weighing samples, bottles of chemicals including dangerous acids, a vast array of glassware (test tubes, flasks etc.) and bunsen burners. The *bunsen burner* is a gas burner consisting of a small vertical tube with an adjustable air inlet at the base to control the flame. The flame produced by burning the hydrocarbon gas/air mix has an inner cone where carbon monoxide is formed and an outer fringe where it is burnt. When the gas is burnt completely the flame temperature is very high, around 1450°C. The burner was invented by Robert Wilhelm Bunsen, a German scientist.

laminates are materials produced by bonding together under pressure two or more different materials. Plastics are commonly used in this way, bonded to chipboard or another building board, to make kitchen units and worktops and other items of furniture. Wood, such as *plywood*, is made of veneers bonded together which is in effect a laminate. Laminated plastics are composites made up of plastics and reinforcing or strengthening materials, e.g. carbon fibres, glass fibres. Glass sheet can also be made in laminated form with layers of glass joined by tough plastic, and these are used for security glass, sound insulation, fire resistance, etc. The overriding feature of laminates is that the combination of materials, albeit in thin sheets, produces a composite material that is much stronger than the individual components alone.

larva and metamorphosis a larva is a young or immature form of an animal that looks and often behaves differently from the adult. There may be several larval stages, and this is a common feature of invertebrate animals. Familiar larvae are those of butterflies and moths (caterpillars), those of flies (maggots) and those of many marine animals, huge numbers of which make up PLANKTON. Larvae can feed and lead an independent life but (with one or two exceptions) are not able to reproduce. Usually an animal passes through several larval stages before undergoing a process known as *metamorphosis* to become an adult. In insects, the final larval stage is a *pupa* or *chrysalis*, during which all activities such as feeding and walking cease. The body is enclosed within an outer case and, under the influence of HORMONES, it gradually changes to become an adult. Usually, there is a breakdown of some of the larval tissues, which are then re-used, and some of the cells that were inactive in the larva divide to form the body of the adult. When the process is complete, the pupal case splits and the adult emerges.

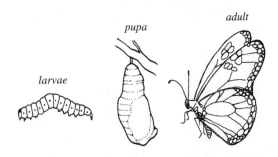

larvae   pupa   adult

laser is an acronym for *light amplification by stimulated emission of radiation* and is a device that produces an intense beam of light of one wavelength (*monochromatic*) in which the waves are all in step with each other (*coherent* light). In its simplest form a ruby crystal shaped like a cylinder is subjected to flashes of white light from an external source. The chromium atoms in the ruby become excited through absorbing photons of LIGHT and when struck by more photons, light energy is released. One end of the cylinder is mirrored to reflect light back into the crystal and the other end is partially reflecting, allowing the escape of the coherent light. The ruby laser produces pulses of laser light and is called a *pulse laser*.

Lasers have also been constructed using INERT GASES (helium and neon mixed; argon alone) and carbon dioxide. These are called *gas lasers*, and produce a continuous beam of laser light. Lasers are being used for an increasing number of tasks including printing, communications, compact disc players (*see* CD TECHNOLOGY), cutting metals, HOLOGRAPHY and for an ever-widening range of surgical techniques in medicine. Lasers are also used in shops at the checkout, to read BAR CODES.

learning *see* BEHAVIOUR.

lens is a device that makes a beam of rays passing through it either converge (meet at a point) or diverge. Optical lenses are made of a uniform transparent medium such as glass or a plastic and they refract the light (*see* REFRACTION). They are either convex (thickest in the middle) or concave (thickest at the edges) in shape and lenses can be made with combinations of these profiles or one side may be flat (plane). Light rays that pass through a convex lens are bent towards the *principal axis* (or optical axis–the line joining the centre of curvature of the two lens surfaces), and away from the axis with a concave lens. The *focus* is where the light rays are brought together at a point and the focal length is the distance of the focus from the centre of the lens.

A convex lens forms a small image of the object which is inverted (upside down) and on the opposite side of the lens. This is called a *real image* and can be seen on a screen. As an object in the distance is brought nearer to a convex lens the image moves away from the lens and becomes larger. The image formed by a concave lens is always upright, smaller than the object and it is a *virtual image*, that is it cannot be projected because although the rays of light appear to come to the observer from the image, they do not actually do so.

Lenses can be found in many optical instruments including the camera and the telescope and also in the human eye (*see* SIGHT).

lever a simple machine which at its simplest consists of a rigid beam which pivots at a point called the *fulcrum*. A load applied at one end can be balanced by an *effort* (a force) applied at the opposite end. There are three classes of lever depending upon the position of fulcrum, load and effort. The fulcrum may be between the effort and load (sometimes called a first-class lever); if the load is between the fulcrum and effort it is second class; and a third class lever has the effort between the fulcrum and load. Examples of these are pliers, a wheelbarrow and the shovel on a mechanical digger.

Although the work done by both effort and load must be equal, it is possible by moving a small effort through a large distance to move a very large load, albeit through a small distance.

light is ELECTROMAGNETIC WAVES of a particular wavelength which are visible to the human eye. Objects can only be seen if light reflected from them or given off by them reaches the eye. Light is given out by hot objects and the hotter the object the nearer to the blue end of the spectrum is the light emitted. The Sun is our primary source of light and the light travels at almost 3×10^5 (300, 000) km/s through space. The light waves are made up of packets or quanta (singular, quantum) of energy, called *photons*. Every photon can be considered to be a particle of light energy, the energy increasing as the wavelength shortens.

White light is actually made up of a range of colours. This can be shown by passing a very narrow beam of light through a prism. Because the individual colours have different wavelengths they are refracted by differing amounts and the resulting *dispersion* produces the characteristic *spectrum*.

A beam of light can often be seen particularly if a torch is shone on a misty night or in a dark room where there is smoke, or if sunlight highlights dust particles. In each case the edge of the beam indicates that light travels in straight lines and because of this a shadow forms when an object is placed between a light source and a flat surface.

light bulbs consist of a glass bulb within which is a filament of tungsten. When a current is passed, the tungsten glows white hot, giving out light. Tungsten is used because it has a very high melting point (3410°C) and because the thin wire of the filament would burn up quickly in air, the bulb contains argon and/or nitrogen to provide an inert atmosphere. Only a small proportion of the energy of a light bulb is given out in the form of visible light. Fluorescent lights are often used because to achieve the same illumination, they use only a third of the power. Fluorescent lamps, or strips, contain a vapour, e.g. mercury, under low pressure. Electrons from an electrode bombard the mercury atoms making electrons move into a higher orbit around the mercury nuclei. When the electrons move back into their normal orbit, ultraviolet light is given off which causes a coating on the tube to glow. Neon and other INERT GASES are used for this purpose, as is sodium.

lightning conductor was invented by the American Benjamin Franklin. It is made of a metal rod (often copper) attached to the top of a building and which is then connected by cable to a metal plate buried in the ground. If and when lightning strikes, the lightning conductor provides a suitable path for electrons to move to the ground, without causing damage to the building. The ground, or as it is known, the *earth*, has an unlimited capacity to accept electrons. A lightning conductor may also help reduce the possibility of lightning striking. This is because there is a flow of IONS from the point of the conductor which lowers the charge induced on the roof by the thundercloud. The result is that some of the charge on the cloud is cancelled out, lowering the risk of a strike.

linear equation is an equation in which the variables in the equation are not raised to any power (squared, etc.) but may have a coefficient. Thus $y = mx + c$ is a linear equation and the slope of the line plotted has a gradient m and c is the value where the line crosses the y-axis. This is known as the *intercept*. Sets of linear equations are used in engineering to describe the behaviour of structures.

linear motor otherwise known as a *linear induction motor*, relies essentially upon the principle of ELECTROMAGNETIC induction, that is when current flows through a coil, magnetic forces cause the coil to spin. In the linear motor, electromagnets are laid flat and when a current is passed, a metal bar skims across them, i.e. it becomes a method of propulsion. Linear motors are now used in trains that do not run on conventional rails, but 'float' over a guiding rail due to the magnetic fields generated.

liquid is a fluid state of matter that has no definite shape and will take on the shape of its container. A liquid has more kinetic energy than a solid but less than a gas, and although it flows freely the molecules stick together to form drops, unlike gases which continue to spread out whenever possible. Liquids can form solids or gases if cooled or heated but many substances such as oxygen, nitrogen and other gases will only form liquids if cooled a great deal or put under pressure.

liquid crystals are substances that are liquids which on heating become cloudy and in this state they show alignment of molecules in an ordered structure, as in a crystal. At higher temperatures still there is a transition to a clear liquid. The application of a current disrupts the molecules, causing realignment and optical effects, darkening the liquid and this property has been exploited for use in displays. *Liquid crystal displays* are used in calculators and watches and also in thin thermometers which work by laying a strip containing the liquid crystals on a child's forehead and the temperature is shown by the segments of crystal changing colour.

liver the liver is a large and very important organ present in the body of vertebrate animals, just below the ribs in the region called the ABDOMEN. It is composed of many groups of liver cells (called lobules) which are richly supplied with blood vessels. The liver plays a critical role in the regulation of many of the processes of METABOLISM. A vein called the hepatic portal vein carries the products of digestion to the liver. Here any extra glucose (sugar) which is not immediately needed is converted to a form in which it can be stored known as glycogen. This is then available as a source of energy for MUSCLES. PROTEINS are broken down in the liver and the excess building blocks, (amino acids), of which they are composed, are changed to ammonia and then to urea, a waste product which is excreted by the kidneys. Lipids, which are the products of the digestion of fat, are broken down in the liver and CHOLESTEROL, an essential part of cell membranes, some hormones and the nervous system, is produced. Bile, which is stored in the gall bladder and then passed to the intestine, is produced in the liver. In addition, poisonous substances (toxins) such as alcohol are broken down (detoxified) by liver cells. Important blood proteins are produced and also substances which are essential in blood clotting. Vitamin A is both produced and stored in the liver and it is a storage site for vitamins D, E and K. Iron is also stored and some hormones and damaged red blood cells are processed and removed.

logarithm abbreviated to log, is a mathematical function first introduced to render multiplication and division with large numbers more simple. However, the advent of calculators and computers has reduced the former dependency on logs. The basic definition is that if a number x is expressed as a *power* (i.e. the number of times a quantity is multiplied by itself) of another number, y, that is $x = y^n$, then n is the logarithm of x to the base y, written as $\log_y x$.

There are two types of log in use; *common* (or Briggs') and *Napierian* or *natural*. Common logs have base 10, $\log_{10} x$. Addition of the logs of two numbers gives their *product*, while subtraction of two numbers' logs is the means of division. Natural logs are to the base e where e is a constant with the value 2.71828. The two logs can be related by the function:

$$\log_e x = 2.303 \log_{10} x$$

loudspeaker is a device for turning an electric current into sound and is found in radios, televisions, and many other pieces of equipment that output sound. The commonest design is the *moving-coil loudspeaker*. This consists of a cylindrical magnet with a central south pole surrounded by a circular north pole producing a strong radial magnetic field. There is also a coil sandwiched between the two poles of the magnet which is free to move forwards and backwards and a stiff paper cone that is fastened to the coil. Because the wire of the coil is positioned at right angles to the magnetic field, when current flows through the coil, it moves.

When an alternating current passes through the coil it moves forwards and backwards and the paper cone vibrates resulting in sound waves.

lubricants are a vital group of materials used in modern industry to make surfaces slide more easily over each other, by reducing FRICTION. Without *lubrication* surfaces grind against each other, producing wear and this can shorten the working life of a machine or engine. In the main, lubricants are liquid in form and are made from oil. However, there are other liquid lubricants (vegetable and mineral oils), some plastic lubricants (fatty acids, soaps) and also some solids, such as graphite and talc. Vehicle engines use oil as a lubricant and turbines for aircraft use synthetic fluids. The main synthetic lubricants are silicones, polyglycols, esters and halogenated (*see* HALOGENS) HYDROCARBONS.

Greases made from a liquid lubricant with a thickening agent are also used to combat wear where temperature and shearing forces apply. A petroleum oil forms the base, and a soap mixture the thickener, often with additives such as graphite, glycerine or fatty acids.

luminescence is when a body gives out light due to a cause other than a high temperature. It is due to a temporary change in the electronic structure of an atom, and involves an electron taking in energy and moving to a higher orbit in the atom which is then re-emitted as light when the electron falls back to its original orbit. The energy required to promote the electron to a higher orbit may come from light (*photoluminescence*) or from collisions of the atoms with fast particles (*fluorescence*). When materials continue to give out light after the primary energy source has been removed, this is called *phosphorescence*.

This phenomenon of luminescence is put to use in the CATHODE RAY TUBE of TELEVISIONS. It also occurs in nature and is called BIOLUMINESCENCE.

lungs and gills lungs are the sac-like organs which are used for RESPIRATION in air-breathing vertebrate animals. In mammals a pair of lungs are situated within the rib cage in the region of the body behind the head known as the thorax. Each lung is made of a thin, moist membrane which is highly folded, and it is here that oxygen is taken in to the body and carbon dioxide is given up. The lungs do not have muscles of their own but are filled and emptied by the muscular movement of a sheet-like layer dividing the thorax from the abdomen, known as the DIAPHRAGM. The diaphragm flattens, which reduces the pressure in the thoracic cavity enabling the lungs to expand and fill with air (inhalation). When the diaphragm muscles relax it arches upwards forcing air out of the lungs (exhalation). This is accompanied (and the effect made greater) by outward and inward movement of the ribs which are controlled by other (intercostal) muscles. Gills, which are present in many invertebrate animals and also in fish, are organs which fulfil the same function as lungs.

In mammals such as man, air enters through the nose (and mouth) and passes into the windpipe or TRACHEA which itself branches into two smaller tubes called bronchi (*singular* bronchus). Each bronchus goes to one lung and further divides into

smaller, finer tubes known as bronchioles. Each bronchiole is surrounded by a tiny sac called an alveolus (*plural* alveoli) formed from one minute fold of the lung membrane. On one side of the membrane there is air and on the other there are numerous tiny blood vessels called capillaries. Deoxygenated blood, that is, blood which contains little oxygen, is brought via a branch of the pulmonary artery to each lung. The artery divides many times eventually forming capillaries surrounding the alveoli. In the same way, other capillaries unite and become larger eventually forming the pulmonary veins which take oxygenated blood back to the heart. The heart pumps it around the body. Carbon dioxide passes out from the capillaries across the alveoli into the lungs and oxygen passes into the blood in the opposite direction, this process being known as gaseous exchange. The numerous folds of the membrane forming the alveoli increase the surface area over which this is able to take place.

The human lung and respiratory system

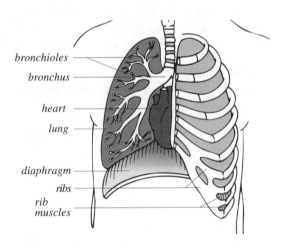

lymph system is a network of fine tubules occurring throughout the body of a vertebrate animal. In places the tubules are swollen to form lumps of tissue known as *lymph nodes* or glands.

This network transports the lymph–a colourless, watery fluid which is mainly water but also contains white blood cells (called *leucocytes* and *lymphocytes*) PROTEIN and digested fats. The lymphocytes are an essential part of the body's IMMUNE SYSTEM and collect in the lymph glands and nodes. In the event of an infection these cells multiply rapidly and this often causes a swelling which can be felt through the skin. (An example of this is the adenoids and tonsils which often become enlarged in the case of an infection of the nose or throat.) Lymph, which contains waste matter from cells, drains into the tubules from all the tissues of the body. It is filtered when it reaches the lymph nodes, where bacteria and other foreign and waste material are destroyed. The lymph tubules finally drain into two major lymphatic vessels that empty into veins at the base of the neck. The circulation of lymph is by muscular action rather than by the beating of the heart and it is essential in the transport of digested fats as well as in the immune system.

The lymph system

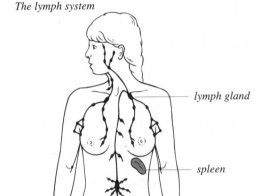

— lymph gland

— spleen

— lymph vessels

In children, white blood cells are produced in the thymus gland but this shrinks as adulthood is reached. In adults, white blood cells are produced in the bone marrow, spleen and lymph nodes.

machine is basically a means of overcoming resistance at a point by applying a FORCE at another point. Although a machine does not reduce the amount of work to be done to achieve a task, it does allow the work to be done more conveniently.

There are six *simple machines* in the study of physics. These are the wheel and axle, wedge, LEVER, PULLEY, screw and the inclined plane. Each in its own way can be used for a particular task–the lever or pulley to raise a load, the wheel to transport a load and so on. More complex machines usually involve the input of energy either for modification or for driving a mechanism to achieve a task.

The *mechanical advantage* of a machine is the ratio of the load moved to the effort put in to achieve the movement. The *velocity ratio* is the distance moved by the effort divided by the distance moved

by the load. An inclined plane, or *ramp*, doesn't appear very much like a machine but it is because it enables a load to be taken gradually to a height to which the load could not have been lifted vertically. In this case the velocity ratio is essentially the length of the ramp over the height of the ramp.

Mach number is the speed of a body expressed as a ratio with the speed of sound. It was named after an Austrian physicist Ernst Mach. If the Mach number is below 1 then the speed is *subsonic*; above 1 is *supersonic* and as an aircraft increases its speed to go over Mach 1 it is said to *break the sound barrier*. *Hypersonic* is when speeds are in excess of Mach 5. Most passenger airliners travel subsonically but Concorde flies at Mach 2. Certain military planes reach over Mach 3.

The noise associated with breaking the sound barrier is a *sonic boom*. A subsonic aircraft produces pressure waves in front of itself and these waves travel at the speed of sound. A supersonic aircraft overtakes the pressure waves creating a shock wave like a cone with the point of the cone at the nose of the aircraft. The shock wave creates a typical double bang which can be strong enough to damage buildings.

magnetism is the effective force which originates within the Earth and which behaves as if there were a powerful magnet at the centre of the Earth, producing a magnetic field. The magnetic field has its north and south poles pointing approximately to the geographic north and south poles and a compass needle or freely swinging magnet will align itself along the line of the magnetic field. With the correct instrument it can also be seen that the magnetic field dips into the Earth, increasing towards the poles.

A *bar magnet* has a north and south pole, so named because the pole at that end pointing to the north is called a north-seeking pole, and similarly with the south pole. When dipped in a material that can be magnetized, such as iron filings, the metal grains align themselves along the magnetic field between the poles of the magnet. Some materials can be magnetized in the presence of a magnet, e.g. iron and steel. Iron does not retain its magnetism, but steel does. These are called *temporary* and *permanent magnets*. A more effective way to produce a magnet is to slide a steel bar into a solenoid (coil) through which current is passed and magnetism is *induced* in the steel (*see* ELECTROMAGNET).

In addition to iron, cobalt and nickel can also be magnetized strongly and these materials are called ferro-magnetic. Non-metals and other metals such as copper, seem to be unaffected by magnetism, but very strong magnets do show some effect.

The origin of magnetism is actually unknown, although it is attributed to the flow of electric current. On the electronic scale within magnetic materials it is thought that electrons act as minute magnets (because electrons carry a charge) as they spin around their nuclei in atoms. In some elements, this electron spin is cancelled out but in others it is not and each atom or molecule acts as a magnet contributing to the overall magnetic nature of the material.

A bar magnet

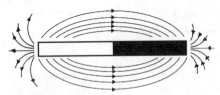

magnetometer is an instrument for measuring the strength of a magnetic field. The *deflection magnetometer* consists of a long magnetic pointer pivoted on a short magnet. The pointer swings along a scale, allowing small deflections of the magnet to be measured. Other versions have a small coil which generates a voltage on moving through a magnetic field. More complex and sensitive magnetometers are used for special purposes, for example, it is possible to tow a magnetometer behind an aircraft to detect changes in the Earth's magnetic field that may be due to mineral deposits, including oil.

magnification is the ratio of the size of the image to the object in an optical magnifying system. Magnification of an object can be achieved by simple means, e.g. a hand lens or *magnifying glass* which is merely a convex LENS. A far greater magnification can be achieved by using a *compound microscope* which consists essentially of two convex lenses with a short focus, called the objective and the eyepiece. The two lenses are at opposite ends of a tube beneath which is a stage upon which the sample (object) is placed. Magnifications of several hundred can be obtained with this microscope. For greater magnification still (above 1500) an electron microscope is used.

The *electron microscope* uses a beam of electrons striking the object. In the *transmission electron microscope* (*tem*) the electron beam passes through a thin slice of the sample and the resulting image is formed by the scattering of the beam which is enlarged and focused on a fluorescent screen. The *scanning electron microscope* (*sem*) actually scans the surface of the sample and the image is created by secondary electrons which are emitted from the sample. Scanning does not magnify the object as much as transmission electron microscopy, but the image is three-dimensional. Magnifications up to 200, 000 can be achieved with *tem*.

Microscopes are used throughout science and engineering, often with modifications to fulfil a particular role. They are used for studying samples of plant and animal tissue, rock samples to determine the minerals and structures present, metals and non-metals to determine structure, and much more.

malleability *see* DUCTILITY.

mammals *see* NH.

maser stands for *m*icrowave *a*mplification by *s*timulated *e*mission of *r*adiation and is a microwave amplifier/oscillator that works in a similar way to the LASER and, after it was discovered, prompted the research that resulted in the laser. An atom already in an excited state, because of absorption of energy, gives out a photon because of absorption of further energy. The 'active' material of the maser is therefore built up to an excited state and enclosed to generate a wave of just one frequency. The microwaves produced can also be used in a clock because of their very precise frequency.

mass *see* WEIGHT.

mass number is the total number of *protons* and *neutrons* in the nucleus of an ATOM. Atoms of a particular element may have differing mass numbers because of different numbers of neutrons in the nucleus, i.e. they are ISOTOPES.

mass spectrometry is a technique to analyse samples to determine the elements in a compound or the various isotopes in an element. The *mass spectrometer* bombards the sample with high energy electrons, producing charged IONS, and fragments that are neutral. The ions are then deflected in a magnetic field that separates them according to their mass, and then pass through a slit to a collector. The output is a printed chart (a *mass spectrum*) containing a series of lines or peaks, where each peak corresponds to a particular ion and its mass. If isotopes are being identified, the mass spectrometer provides the mass and relative amount of each isotope.

Mass spectrometry is used for the identification and structural analysis of organic compounds and the determination of elemental traces in inorganic materials.

mathematical symbols in addition to the well-known symbols of arithmetic, such as = (equals),– (minus), + (plus), x (multiplied by) and ÷ or / (divided by), there are many other useful symbols that permit the shorthand writing of scientific formulae, equations and statements. These are used in mathematics, physics, chemistry, astronomy, geology and others scientific and engineering disciplines. A sample is shown below:

± plus or minus	Σ the sum of
> greater than	a proportional to
< less than	≈ similar or equal to
≥ greater than or equal to	∫ equivalent to
≤ less than or equal to	∴ therefore
= equal by definition	∵ because

mathematics is the science of relationships that involve numbers and shapes and has been around for thousands of years since the ancient Egyptians used geometry in construction. It is divided into two main categories–*pure* and *applied mathematics*. Pure mathematics includes algebra, arithmetic, geometry, trigonometry and calculus (*see individual entries*). Applied mathematics verges onto other subject or parts of subjects and includes statistics, mechanics, computing and mathematical aspects of topics such as thermodynamics, astronomy and optics.

matrix (*plural* **matrices**) is an array of elements, that is, numbers or algebraic symbols, set out in rows and columns. It may be a square or a rectangle of elements. Matrices are very useful for condensing information and are used in many ways, for example in solving simultaneous LINEAR EQUATIONS (simultaneous equations are two or more equations with two or more unknowns which may have a unique solution). The order of a matrix refers to the number of rows and columns, thus;

$A = \{2 \ 1 \ 3\}$ has one row and three columns while

$B = \begin{cases} 2 \ 6 \\ 1 \ 4 \end{cases}$ has two rows and two columns

Matrices with one row are called *row vectors* and similarly *column vectors* are so named because they have just one column.

Only matrices of the same order can be added or subtracted and to be multiplied the second matrix must have the same number of columns as the first has rows. The multiplication is carried out by combining the rows and columns of each matrix to form a new matrix where the various elements are the products of the rows of the first and the columns of the second matrices. Thus if

$$A = \begin{cases} a \ b \\ c \ d \end{cases} \text{ and } B = \begin{cases} w \ x \\ y \ z \end{cases}$$

$$\text{then } AB = \begin{cases} aw + by \quad ax + bz \\ cw + dy \quad cx + dz \end{cases}$$

matter is any substance that occupies space and has mass and is the material of which the universe is made. Matter exists normally in three states: gas, liquid and solid. Most substances can be made to exist in all three states at different temperatures, thus by cooling a gas it eventually becomes liquid and then a solid. *Plasma* is considered a fourth state and consists of a high temperature gas of charged particles (electrons and ions). Although it contains charged particles, a plasma is neutral overall, but it can support an electric current. Stars are made of plasma, but it finds applications in the study of controlled nuclear fusion (*see* NUCLEAR FISSION and NUCLEAR FUSION).

mechanics is the part of physics that deals with the way matter behaves under the influence of forces. It involves:

dynamics the study of objects that are subjected to forces that result in changes of motion.

statics the study of objects subjected to forces but where no motion is produced.

kinematics the study of motion without reference to mass or force but which deals with velocity and acceleration of parts of a moving system.

Newton's law of motion (*see* FORCE) forms the basics of mechanics except at the atomic level, when behaviour is explained by QUANTUM MECHANICS.

meiosis and mitosis *see* CELL DIVISION.

melting point *see* FREEZING.

membrane a membrane is a thin sheet of tissue widely found in living organisms. It covers, lines or joins cells, organelles (small organs), organs and tissues, and consists of a double layer of lipids (fats) in which protein molecules are suspended. Water and fat-soluble substances are able to pass across a membrane but sugars cannot. Other substances or ions are actively carried across a membrane by a complex system known as *active transport*.

memory a function of the brain in animals that enables information to be stored and brought back for use later. Several areas of the brain (such as the temporal lobes) are involved in memory, and although, in humans, this function has been widely studied it is not entirely understood. Placing information in the memory involves three stages – registration, storage and recall. The information is committed either to the *short term memory* or the *long term memory*. It tends to fade quickly from the short term memory if not needed but, if used often, is transferred to the long term memory. Information is usually lost (forgetfulness) during the process of retrieval or recollection. Sometimes this can be helped if the circumstances in which the information was registered can be recreated. This is a technique which is used by the police when they are trying to gain information (which might have been only fleetingly seen) from witnesses at the scene of a crime.

metabolism is the name given to all the chemical and physical processes which occur in living organisms. These are of two kinds, *catabolic* (or breaking down) as in the digestion of food, and *anabolic* (or building up) as in the production of more complicated MOLECULES from simple ones. All these processes require energy and ENZYMES in order to take place. Plants trap energy from the sun during PHOTOSYNTHESIS and animals gain energy from the consumption of food. The metabolic rate is the speed at which food is used or broken down to produce energy, and this varies greatly between different species of animals. In people, children have a higher metabolic rate than adults and more energy is required by someone during hard work than by someone who is at rest.

metal fatigue is the structural failure of metals due to repeated application of stress, which results in a change to the crystalline nature of the metal. *Stress* is force per unit area and when it is applied to a material, a *strain* is developed, that is a distortion. Minor defects in the surface of a metal may provide points where stress can build up and a crack may form. This is particularly important in the design and construction of aircraft, bridges, machinery and buildings to ensure that there is not a sudden, catastrophic failure.

metallurgy is the scientific study of metals and their alloys. It includes the extraction of metals from their ores and their processing for use. *Extractive metallurgy* deals with production of the metals from their ores and *physical metallurgy* is the study of the structure of metals and their properties.

Metals are studied in the laboratory using a microscope to allow small flaws and cracks to be seen. Such studies are important in designing to avoid or minimize METAL FATIGUE.

metals are materials that are generally ductile, malleable (*see* DUCTILITY), dense with a metallic lustre (or sheen) and usually good conductors of heat and electricity. There are about 80 metals in all, some of which occur as the pure metal, but most are found as compounds in rocks and deposits.

Pure metals are used for certain applications (e.g. GOLD in jewellery and copper in electrical goods) but it is usually the case that mixtures of metals (ALLOYS) provide the properties required for specific applications.

Metallic elements are generally electropositive, that is they give up electrons to produce a cation (e.g. Na^+) in reactions. When reacting with water, bases are produced. Elements that show some features of both metals and non-metals are called METALLOIDS or semimetals, e.g. arsenic, bismuth and antimony.

metamorphosis *see* LARVA.

methane is the first member of the alkane (*see* HYDROCARBONS) series, and has the formula CH_4. It is a colourless, odourless gas and the main component of coal gas and a byproduct of decaying vegetable matter. It is the primary component of natural gas and occurs in coal mines where mixed with air it forms highly explosive *firedamp*. It is highly flammable and is used in the manufacture of hydrogen, ammonia, carbon monoxide and other chemicals.

metric system is a system of measurement of weights and measures based upon the principle that levels of units are related by the factor 10, i.e. it is a decimal system that began with the metre. There are now seven basic units of measurement: metre, second, kilogram, ampere, kelvin, candela (light) and mole and each has a standard value and is defined very precisely. *See also* SI UNITS.

microbiology and micro-organism microbiology is the scientific study of all aspects of the life of micro-organisms or microbes. These organisms are so called because they can only be seen with a microscope. They include bacteria, viruses, yeasts and moulds, many of the (largely single-celled) organisms called protozoa (*see* PROTISTA) and some algae and fungi. Many of these organisms are highly significant for man because of their use in BIOTECHNOLOGY and GENETIC ENGINEERING and also because some cause diseases of various kinds. The study of microbiology has advanced significantly along with the technological development of microscopes, particularly the electron microscope and also with more advanced laboratory facilities.

microphone is a device for changing sound into electric current. It is essentially the reverse process to that which happens in a LOUDSPEAKER. In a moving coil microphone, sound waves cause vibration in a diaphragm made of paper or plastic and this moves a small coil which rests in the field of a cylindrical magnet. A small current is *induced* in the coil due to it moving through the magnetic field and this can be amplified and output through a loudspeaker. A variety of microphones are now in use: *carbon microphone* where the diaphragm changes the resistance in a carbon contact; *condenser microphone* and a crystal microphone which relies upon the PIEZOELECTRIC EFFECT in crystals.

microwaves are electromagnetic waves with wavelengths of a few centimetres or less. Microwaves are used for communications, e.g. via satellites, and intense beams are produced in a MASER. Microwaves are easily deflected and as they have a shorter wavelength than radiowaves, they are suitable for use in radar systems because they can detect small objects. Heat is produced when microwaves are absorbed and this effect is utilized in microwave ovens. A unit called the MAGNETRON generates the microwaves and these impart their energy to the water in the food, producing heat and the cooking effect.

mirror an object or surface that reflects light. Mirrors are made from glass coated with a thin metallic layer but polished metal surfaces have the same effect. When light strikes a mirrored surface, it is reflected in a particular way, according to the laws of REFLECTION. A plane mirror has a flat surface which bounces back light, creating an image 'behind' the mirror. Curved mirrors are convex or concave. Concave mirrors produce a magnifying effect if the object is sufficiently near to the mirror, but distant objects are smaller and inverted. Convex mirrors produce upright images that are smaller than the object. Concave mirrors are used in car headlights while convex mirrors are often used for driving mirrors because they produce a wide angle of view.

mitochondrion (*pl.* mitochondria) a type of rod-shaped organelle found in the cytoplasm (cell contents except for the nucleus) of EUCARYOTIC cells which is surrounded by a double membrane. Mitochondria have been called the 'power houses' of the cell as they are very important in the generation of energy in a form called ATP. ATP production is the end result of cellular respiration and provides energy for all metabolic processes. As mitochondria are the sites where this takes place, they are especially abundant in cells which require lots of energy such as those of muscles. Mitochondria contain a form of DNA, structures called ribosomes in which proteins are manufactured with numerous enzymes, each specific to a particular metabolic process.

molecule and mole a *molecule* is the smallest chemical unit of an element or compound that can exist independently. It consists of atoms bonded together in a particular combination, e.g. oxygen, O_2, is two oxygen atoms and carbon dioxide, CO_2, is on carbon and two oxygen atoms. Molecules may contain thousands of atoms. Gases, organic compounds, liquid and many solids consists of molecules but some materials, e.g. metals and ionic substances are different and comprise changed atoms or IONS.

The *mole* is the unit of substance that contains the same number of elementary particles as there are in 12 grams of carbon. One mole of a substance contains 6.023×10^{23} molecular atoms, or ions, or electrons.

momentum is the property of an object defined as the product of its velocity and mass and it is measured in kgm/s. Momentum is related to force as follows:

force = the rate of change of momentum.

Changes in momentum occur mainly due to the interaction between two bodies. During any interaction the total momentum of the bodies involved remains the same, providing no external force, such as FRICTION, is acting. Newton's second law of motion states that the rate of change of momentum of a body is directly proportional to the force acting and occurs in the direction in which the force ends.

If a body is rotating around an axis then it has *angular momentum* which is the product of its momentum and its perpendicular distance from the fixed axis.

Momentum can be seen around us all the time, but one of the most obvious examples is the collision of balls on the snooker table.

monera the biological kingdom in which all the organisms have PROCARYOTIC cells and are the bacteria and cyanobacteria (formerly called blue-green algae). *See* PROCARYOTE.

motors (**electric**) the basic principle behind most common electric motors is that when current is passed through a coil in a magnetic field, a turning effect is produced. In a d.c. (direct current) motor a coil rotates between the poles of a permanent magnet and is connected to a battery via carbon brushes which contact a split ring or *commutator*. The commutator reverses the current direction every half-turn of the coil ensuring that the coil continues to rotate (*see also* GENERATOR). Motors are used in many different devices and pieces of equipment, such as hair-dryers, mixers and food processors, drills and so on. Such motors operate from an a.c. (alternating current) supply and in this case an electromagnet is used because it can match the change in current and yet maintain a constant turning effect. In addition, the smooth and reliable running of these motors is ensured by: the incorporation of several coils at different angles each connected to a set of commutator pieces; several hundred turns of wire in each coil, on a soft iron core (as *armature*) which becomes magnetized, increasing the strength of the magnetic field; the poles of the magnet are shaped into a curve creating a magnetic field that is almost radial, giving a constant turning force (*see also* INDUCTANCE).

muscle a special type of tissue which is responsible for movements in the body of an animal and which often works by pulling against the hard SKELETON. Muscle cells are elongated and arranged as bundles (called fibres) and in mammals there are three types controlled by different parts of the nervous system. *Voluntary* or striated muscle operates limbs and joints and is under the conscious control of the will. Each muscle fibre consists of smaller elongated *fibrils* (myofibrils) and these either lengthen and become thinner, or shorten and become fatter, as they slide over one another, depending upon whether the muscle is contracting or relaxing. *Involuntary* or smooth muscle is not under conscious control but is regulated by a special part of the nervous system (called the *autonomic nervous system*). These muscles work automatically and are responsible for the contractions of the gut, and of the womb at the time of birth. This type of muscle also occurs at various other sites within the body. *Cardiac muscle* is the third specialized kind and is responsible for the beating of the heart which continues throughout life. This muscle is involuntary but the rate of beat is affected by activity of an important nerve called the vagus nerve. The *pacemaker* is a collection of specialized cardiac muscle cells situated in the wall of the right atrium (one of the upper chambers of the heart). These cells are under the control of the autonomic nervous system and produce the contractions which stimulate the rest of the heart to contract.

mutation any change which occurs in the DNA in the chromosomes of cells. Mutation is one way in which genetic variation occurs and allows natural selection to operate (*see* GENETICS). This is because any alteration in the sex cells may produce an inherited change (mutation) in the characteristics of later generations of the organism. Mutations occur naturally during the time when DNA is copying itself, as in cell division although they are relatively rare and most are harmful or lethal. They can occur much more frequently following exposure to certain chemicals and radiation (X-rays), and these agents are known as *mutagens*. A *mutant* is an organism which shows the effects of a mutation.

mycology the name given to the scientific study of all aspects of the life of FUNGI.

natural selection *see* DARWIN.

nervous system a network of specialized cells and tissues which is present in all multicellular animals to a greater or lesser degree (with the exception of sponges). The activity of the nervous system consists of electrical impulses which are caused by the movement of chemical (sodium and potassium) IONS. The nervous system includes RECEPTORS, which receive information from the surrounding environment and are called *sensory*. These are concerned with the senses such as sight, sound, touch and pressure and they transmit the information which they detect along nerves (called sensory nerves) and these travel to the central nervous system. In the simpler animals such as invertebrates, this often consists of a paired nerve cord with swellings along its length called *ganglia* (*sing*. ganglion). In vertebrate animals, the central nervous system is highly complex and consists of the brain and spinal cord. Within the central nervous system all the information is decoded and, if appropriate, a response is initiated. This often consists of a signal being sent outwards along a nerve (called a motor nerve) which travels to a muscle causing a contraction to occur. In vertebrates, a part of the nervous system is concerned with the control of the involuntary or smooth MUSCLE of the body. This is called the autonomic nervous system and consists of two divisions, the *sympathetic* and *parasympathetic* which act in opposite ways (antagonistic). It is sometimes called *involuntary* because its activity regulates the internal environment of the body and it supplies the smooth muscle (heart, gut, etc.) and glands with their motor nerve supply.

A nerve is made up of numerous nerve cells or 'neurons'. some nerves are sensory, others are motor and yet others are mixed carrying both types of neurons. Each neuron has a cell body (containing the neuclus) and many fine projections called 'dendrites'. Dendrites from surrounding neurons are able to communicate with one another across a gap called a 'synapse'. A long, fine projection runs out from the neuron cell body and this is called an 'axon'. It may be surrounded by a fatty sheath (called a 'myelin sheath') which is restricted at intervals at sites which are known as 'nodes of Ranvier'.

nitrides are very similar compounds to CARBIDES and form hard materials with high melting points. Metals such as titanium, chromium, vanadium, zirconium and hafnium are used and these result in nitrides with high hardness, resistance to molten metals and a high corrosion resistance. As such they are used in crucibles for melting metal, protective coatings for moving surfaces and cutting tools.

nitrogen is a colourless gas existing as a diatomic molecule, N_2. It occurs in air (75% by weight), and as nitrates, ammonia and in PROTEINS. It is relatively unreactive at room temperature but reacts with some elements on heating. It is used extensively in the production of ammonia in liquid form as a refrigerant (its melting point is -210°C) and as an inert atmosphere. It is obtained industrially by the fractionation of liquid air.

Nitrogen is a vital element in the life cycle of both plants and animals and there is a regular circulation, called the *nitrogen cycle*. Bacteria take nitrogen from the atmosphere (*nitrogen fixation*) and plants use nitrate ions (NO_3^-) from the soil. The nitrogen is incorporated into plant tissue, which is then eaten by animals. The nitrogen is returned to the soil by the decomposition of dead plants and animals and by excretion.

noise is a random mixture of changing frequencies of sound as produced, for example, by an electric drill. It is also the term describing the background disturbance registered when a signal of some description is being measured. The effect occurs in electrical circuits and is due to interference arising from other sources, e.g. lightning, electric motors, random motion of electrons making up the current. This often happens with a radio receiver and electronic filtering is often undertaken to improve the signal and reduce the noise.

nose *see* SMELL.

nuclear energy is the energy produced by the controlled decay of radioactive elements. Upon decay, an element such as uranium releases energy as heat which can be harnessed. During radioactive decay the energy given off per atom is thousands and thousands times more than during burning. However, the decay often has to be accelerated by bombarding the material with neutrons.

NUCLEAR FISSION is the splitting of such atoms and is the way in which electricity is generated for nuclear power. In a *nuclear reactor* heat from the nuclear reactions heats water into steam, which drives the turbines. The core of a reactor contains the nuclear fuel, which may be uranium dioxide with uranium-235. Neutrons produced by the fission reactions are slowed down by a graphite core to ensure the *chain reaction* continues. The graphite core is called the *moderator*. *Control rods* of boron steel are lowered into or taken out of the reactor to control the rate of fission. Boron absorbs neutrons, and so if rods are lowered there are fewer neutrons available for the nuclear fission, and the reactor core temperature will fall. This is a *thermal reactor*. In a *fast breeder reactor*, low grade uranium surrounds the core, and impact from neutrons creates some uranium-239, which forms plutonium, which itself can be used as a reactor fuel. *Nuclear fusion* (*see* NUCLEAR FUSION) has not yet been harnessed for commercial power production.

Nuclear energy has the benefit of producing a lot of energy from a small amount of fuel. It doesn't produce gases that contribute to the *greenhouse effect* (*see* **NH**: GREENHOUSE EFFECT), but the waste produced is very dangerous and must be stored or treated very carefully. The waste products are from the reprocessing of used fuel in which unused uranium is separated from the waste and small quantities of plutonium-239 (as produced and used in fast breeder reactors). Plutonium-239 is very hazardous indeed and is used in the production of nuclear weapons.

nuclear fission nuclear fission is the splitting process that results when a neutron strikes a nucleus of, for example, uranium-235. The nucleus splits into two and releases more neutrons and a lot of energy. A *chain reaction* develops when the neutrons go on to split further nuclei, and the energy released becomes enormous. The splitting of the uranium creates two other nuclei, both radioactive:

$$\text{uranium-235} + \text{neutron} \rightarrow \text{krypton-90} + \text{barium-144} + \text{neutrons} + \text{energy}$$

This chain reaction occurs when there is a certain mass of uranium-235, known as the *critical mass*. This was the principle employed in the first *atom bomb*, when two pieces of uranium-235 were brought together to exceed the critical mass and produced a destructive force of unimaginable proportions.

nuclear fusion is where two nuclei are combined to form a single nucleus with an accompanying release of energy. Ordinarily nuclei would repel each other because of the like electrical charge, and so very high collision speeds have to be used, which in practice means the use of incredibly high temperatures. A fusion reaction may be:

deuterium + tritium → helium + neutron + energy

$$_1^2H \ + \ _1^3H \ \rightarrow \ _2^4He \ + \ _0^1n$$

It is necessary to raise the temperature of the hydrogen gas to around 100 million K. Because thermal energy has to be supplied before the nuclear reactions occur, fusion is often called *thermonuclear fusion*. Fusion occurs in the Sun and, in an uncontrolled way, in the *hydrogen bomb*, but it is technically very difficult to control in the way that nuclear fission is managed. Research into this subject involves the use of a *tokamak*, which employs a magnetic field shaped like a doughnut to trap the hot gas. Thermonuclear reactors could be a solution to energy supply problems because the fuel is readily available and the waste product is not radioactive and it is also inert (unreactive).

nuclear reactor *see* NUCLEAR ENERGY.

numbers our numbers are based on Arabic numerals, which in turn were based on a Hindu system, and the system is constructed upon powers of ten, i.e. the decimal system (*see* DECIMAL NUMBERS). BINARY NUMBERS use the base two and are used extensively in computer programs. *Real numbers* include all *rational numbers* (whole numbers, or integers, and fractions) and *irrational numbers*, that is those that cannot be expressed as an exact fraction, e.g. some square roots such as 2, 5 and values such as the constant π. *Complex numbers* are those written in the form $a + ib$ where a and b are real numbers and i is the square root of -1.

A *prime number* is a number that can be divided only by itself and 1, such as 2, 3, 5, 7, 11, 13, 17, 19, 23. There is an infinite number of prime numbers, and over the years mathematicians have spent much time trying to find a general method for calculating primes. A *surd* is an expression that contains the root of an irrational number and that can never be expressed exactly (e.g. 3 = 1.7320508 . . .).

oestrous cycle *see* REPRODUCTION.

oils are greasy liquid substances obtained from animal or vegetable matter or mineral sources, and they are complex organic compounds. There are also many synthetic oils. There are basically three groups of oils: the *fatty oils* from animal and vegetable sources; *mineral oils* from petroleum and coal; and *essential oils* derived from certain plants. Typical vegetable oils are extracted from soya beans, olives, nuts and maize. The essential oils are volatile (evaporate quickly) and are used in *aromatherapy* and in making perfumes and flavourings. Examples are peppermint oil, clove oil, oil of wintergreen and rose oil.

Mineral oils are actually fossil fuels and come under the general term of petroleum.

omnivore any animal that eats both plant and animal material, e.g. human beings (*compare* CARNIVORE and HERBIVORE).

optical fibres are made of very thin glass rods, which are flexible. When light rays enter one end of the fibre it is reflected completely within the fibres, being reflected from side to side until it emerges at the other end. Bundles of fibres are called *light pipes*, and they have several very important uses. Surgeons use them to see inside the body of a patient, often without the need for an incision. Also the pipes can carry an enormous number of telephone calls, where the calls are coded and sent along the fibres as pulses of LASER light. Another remarkable feature of optical fibres is that the light exits almost as strongly as it entered, even if the path is several kilometres, and this eliminates the need for numerous stations to boost the signal as required with ordinary telephone cables.

organ and tissue a tissue is a collection of cells within an ORGANISM, which are specialized to perform a particular function. An *organ* is a distinct and recognizable site or unit within the body of an organism that consists of two or more types of tissue. It is specialized in terms of its structure and function, and, in animals, examples include the kidneys, liver, skin and eyes. In plants, flowers, stems, roots, leaves and flowers are all organs. An *organelle* is a structure within a cell that performs a particular function and is surrounded by a membrane to separate it from other cell contents, e.g. the nucleus and MITOCHONDRIA.

organism any living creature including micro-organisms, plants and animals. There are very many different kinds of organism with new species being discovered all the time. At the other end of the scale, over the course of the Earth's history, numerous types of organism have become extinct. *Biotic* is an adjective relating to life or living things (hence *biota*, the plant and animal life of a region). Thus, for any organism, the other organisms around it make up the biotic environment.

osmoregulation this describes any process or mechanism in animals that regulates the concentration of salts (e.g. sodium chloride) and water in the body. Depending upon the environment it inhabits, there is a tendency for water to pass into or out of an animal's body by *osmosis*. Water tends to pass into the body of an animal which inhabits fresh water as the concentration of salts within its body is higher than that outside. Animals have a variety of structures to rid the body of excess water. In simple organisms (e.g. single-celled ones) an organelle called a *contractile vacuole* fills with water and expels it to the outside through a pore in the cell membrane. In marine animals there is a tendency to lose water from the body to the surrounding environment where the concentration of salts is higher. In vertebrate animals, whether the problem is one of water gain or loss, the kidneys are the main osmoregulatory organs. In land-dwelling animals, the outer covering of skin or cuticle forms a barrier to excess water gain or loss but many other mechanisms are also at work (including kidneys, sweating, panting, behavioural responses) so that a correct water/salt balance is maintained.

In humans, there is a pair of kidneys situated at the back of the abdomen and these are responsible for cleaning the blood and removing waste products which are then excreted. The kidney contains numerous tubules called *nephrons*, each with an expanded cup-shaped portion at one end called the Bowman's capsule. Behind this there is a folded length of tubule, known as the proximal convoluted tubule then a straight hairpin-shaped loop, the loop of Henle, and finally another looped portion called the distal convoluted tubule. Blood enters the Bowman's capsules from tiny capillaries which form a knot, called the glomerulus, inside the cup. This blood is brought to the kidney by the renal artery. Water and waste substances, such as urea, a breakdown product of protein digestion, pass along the length of the nephrons and this is known as filtrate. Useful substances, including water and salts are reabsorbed. Many capillaries surround the nephrons and cleaned or filtered blood eventually leaves the kidney in the renal vein. The distal convoluted tubules, containing the waste products which have not been re-absorbed, empty into a collecting duct and final processing takes place. The liquid which is left, known as *urine* enters the ureter which is a narrow tube leading to the *bladder.*

A human kidney

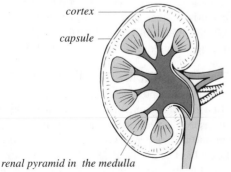

cortex

capsule

renal pyramid in the medulla

From the bladder another tube called the urethra leads to the outside. The two kidneys daily receive between 1000 and 2000 litres of blood and process about 180 litres of filtrate. 1.5 litres of urine is produced and the rest of the filtrate is reabsorbed. If the kidneys fail, their function has to be carried out by a machine and this is known as dialysis.

osmosis is the process whereby molecules of solvent (usually water) move through a semi-permeable membrane to the more concentrated solution. This is due to the size of the molecules compared to the holes in the membrane. The holes permit the small water molecules through but not the larger solvent molecules so there is a tendency for the molecular concentrations to approach equality. *Osmotic pressure* is the pressure that must be applied to prevent osmotic flow.

Osmosis is an important mechanism in living organisms in the movement of water across cell membranes and particularly in the uptake of water by plant roots. Certain mechanisms have also evolved to prevent too much water entering cells causing rupture, or leaving cells, causing shrinkage.

oxidation and reduction are processes that occur during chemical reactions. *Oxidation* is the gain of oxygen or loss of electrons from the reactant. Oxidation can occur in the absence of oxygen as it can also be represented by the loss of hydrogen. Similarly, *reduction* is the loss of oxygen or gain of electrons from one of the reactants, and also the gain of hydrogen.

oxygen is a colourless, odourless gas vital for the respiration of most life forms. It is the most abundant of all the elements forming 21% by volume in air, almost 90% by weight of water and nearly 50% by weight of rocks in the crust. It is manufactured by the fractional DISTILLATION of liquid air and on heating reacts with most elements to form oxides. It is used extensively in industry for steel making, welding, rocket fuels and in chemical synthesis (forming other chemicals).

paint is a liquid made up of the pigment (the coloured material) in SUSPENSION within a non-volatile and a volatile part. The non-volatile oil or resin holds the pigment in place, while the volatile part enables the paint to be applied easily and it eventually evaporates (and may be water or a HYDROCARBON solvent). When applied to a surface the liquid evaporates leaving the pigment as an adhesive skin. Many paints are oil-based and are waterproof when dry and may contain additives to speed drying or improve coverage. Linseed oil was used with a thinner and a drier but synthetic compounds are now commonly used. Paints based on water are emulsions (*see* COLLOID), hence emulsion paint. These are often acrylic resins or polyvinyl acetate in water.

Of course, the earliest examples of paint were obtained from natural materials. Now there are special paints for particular purposes, e.g. anti-fouling paint to stop the growth of barnacles on a ship's hull.

pancreas an important gland present in VERTEBRATE animals situated behind and just below the stomach. It has two functions as it produces both digestive ENZYMES and the hormones insulin and glucagon which regulate the amount of sugar present in the blood. The digestive enzymes produced include trypsin, which breaks down protein; amylase which digests starch and lipase which aids in the digestion of fats. These are produced in an alkaline fluid which counteracts the acid effect of the stomach's gastric juice. The pancreatic juice, with its enzymes, passes through a tube or duct into the first part of the small intestine below the stomach which is called the *duodenum*.

paper consists of sheets of hydrated cellulose fibres derived from wood pulp. Pulp is made from timber mechanically or chemically. The mechanical method is simpler and more economical but produces weaker, poorer quality papers (e.g. newsprint). Pulp from chemical processes is stronger and brighter but more expensive. The pulp is washed, bleached and broken down and delivered to the paper mill where any fillers, size (a glue) and dyes are added. *China clay* is a common filler, providing bulk and opacity. The pulp is then watered down to make a slurry, treated further and then output onto a moving wire mesh and through presses which remove most of the water. Heated rollers then dry the paper which is then smoothed and finished (which involves pressing the surface fibres–a process known as *calendering*). The paper is then taken onto rolls for use or for cutting into sheets.

The whole process from pulp to paper can be a continuous process requiring machinery occupying hundreds of feet but it can produce sheets of paper in minutes.

parasite and symbiosis a *parasite* is an organism which obtains its food by living in or on the body of another living organism, without giving anything in return. The organism on which the parasite feeds is known as the *host*. Usually, the parasite does not kill the host as its future depends upon their mutual survival. However, sometimes a host species can become so seriously ill or weakened by the presence of the parasite that it dies. Parasites are of two kinds: *ectoparasites* attach themselves to the host's surface or skin and examples include the blood-sucking headlice, ticks and fleas. *Endoparasites* live inside the host's body, often within the gut or muscle, and examples are tapeworms, roundworms and liver flukes. Parasitic plants usually twine themselves around their host. They send projections, called haustoria (*sing.* haustorium) into the sap of the host to draw up water and food substances. A parasitic plant of this type is the dodder which lacks leaves and CHLOROPHYLL and cannot PHOTOSYNTHESIZE. Others, such as the mistletoe, are partial parasites, able to photosynthesize but obtaining minerals and water from the host.

Some other organisms live together in a way which brings mutual benefit rather than harm, and this type of relationship is called *mutualism*. Sometimes one organism benefits while the other remains unharmed and this is termed *commensalism*. These three types of relationship between organisms–parasitism, mutualism and commensalism usually of two different species– are grouped together and called *symbiosis*.

particle accelerator is a machine for increasing the speed (and therefore the kinetic energy) of charged particles such as protons, electrons and helium nuclei by accelerating them in an electric field.

Accelerators are used in the study of subatomic particles. To split an atom, particles travelling close to the speed of light are required. There are two types of accelerator *linear* and *cyclic*. Linear acceleration, or *linacs*, have to be very long, up to several kilometres, and consist of rows of electrodes separated by gaps. The ions in the beam accelerate across each gap due to a high frequency potential between alternate electrodes. *Cyclic accelerators* use a magnetic field to bend the path of the charged particles. The *cyclotron* is an example in which energies of several million electron volts are imparted to the particles as they travel along a spiral path between D-shaped electrodes. The *synchrotron* is another example. In the cyclic accelerators the stream of charged particles is accelerated to the required level and then deflected out of the ring.

The first accelerator was built by Cockcroft and Walton in 1932, and with it they split the atom for the first time. The energies of modern machines are measured in GeV (gigaelectron-volts, i.e. billions) and at the Fermi Laboratory in the USA, 800 GeV has been reached. Higher energies can be achieved using a *storage ring* which is a toroidal (doughnut-shaped) component in some accelerators. Particles enter the ring and can stay there for many weeks or months. A ring 300 metres in diameter in Geneva produces energies up to 1700 GeV.

pasteurization is the process named after the French chemist and biologist, Louis Pasteur. It involves the partial sterilization of food and kills potentially harmful bacteria. Milk is an obvious example. If it is heated to 62°C for 30 minutes, it kills bacteria that could cause tuberculosis and it increases the shelf life by delaying fermentation because other bacteria have also been killed or damaged. An alternative treatment involves heating milk to 72°C for 15 seconds. Higher temperatures still are used to produce 'long-life' milk. Pasteurization is also used with beer and wine, to eliminate any yeast which would create cloudiness in the drink.

pathogen and pathology a *pathogen* is any organism that causes disease in another organism. Most pathogens that affect humans and other animals are BACTERIA or VIRUSES, but in plants a wide range of fungi also act as pathogens. *Pathology* is the medical science and speciality in which the area of study is the causes of diseases and the ways in which these affect the body. Pathology relies on the use of powerful MICROSCOPES to study samples, and also on the techniques of MICROBIOLOGY.

periodic table is an ordered table of all the ELEMENTS arranged by their ATOMIC NUMBERS, i.e. the number of protons and electrons in an atom. The arrangement means that elements with similar properties are grouped near to each other. The horizontal rows are called *periods* and the vertical rows are *groups*. Elements with the same number of electrons in their outer shell behave in a similar way and this is the basis of the vertical group. Moving from left to right along the periods corresponds to the gradual filling of successive electron shells and an increase in the size of the atom.

The elements and their symbols are listed on page 91. The periodic table is illustrated on page 124.

There are various sections within the periodic table as follows:

alkali metals	Li, Na, K, Rb, Cs, Fr
alkaline earth metals	Ca, Sr, Ba, Ra
chalcogens	O, S, Se, Te, Po
halogens	F, Cl, Br, I, At
inert gases	He, Ne, Ar, Kr, Xe, Rn
rare earth elements	Sc, Y, La to Lu
lanthanides	Ce to Lu inclusive
actinium series	Ac onwards
transuranium elements	elements after U
platinum metals	Ru, Os, Rh, Ir, Pd, Pt

Mendeléef was a Russian chemist who constructed the first periodic table, but based upon atomic weights. This basic principle, modified to use atomic numbers rather than atomic weights formed the basis of the modern table and even then, in 1869, allowed Mendeléef to predict the existence of undiscovered elements.

pesticides are chemical poisons designed to kill insects (*insecticides*), weeds (*herbicides*), fungi (*fungicides*) and other pests. Until the advent of organic pesticides in the 1930s the compounds used were inorganic mixtures such as Bordeaux mixture (copper sulphate and lime) and calcium arsenate. Some naturally-occurring organic insecticides were also used, such as pyrethrum produced from the flowers of the chrysanthemum. Now most are organic, save for some well-known examples such as DDT.

Fungicides traditionally contained sulphur, or copper or mercury although the toxicity of mercury has led to a decline in its use. Now there are a number of organic compounds and many are *systemic* i.e. they actually enter the plant. *Herbicides* are the single most important group of pesticides, as they kill weeds which compete with crops for light and nutrients and which may be sources of other pests and diseases. Herbicides may be selective (killing only certain plants) or total (non-selective) and in the main are complex organic chemicals. *Insecticides* are very important in controlling insects that consume or destroy crops, and also in limiting diseases spread by insects, such as malaria and sleeping sickness. There are naturally-occurring insecticides, for example, nicotine, derris and pyrethrum, and synthetic compounds which form the greater proportion. Synthetic varieties include organochlorines, organophosphates and carbamates.

Pesticides do not achieve 100% success and some individuals, e.g. insects, survive to reproduce. It is possible that successive generations have a greater resistance to a particular pesticide which is overcome only by using higher concentrations of the pesticide. However, the pest may eventually become resistant to the pesticide, as happened with DDT.

petrochemicals are chemicals derived from crude oil (petroleum) and natural gas which are used to manufacture an enormous range of compounds and materials including plastics, drugs, fertilizers, solvents and detergents. Over 90% of synthetic organic materials come from these sources. A major factor in the development of petrochemicals was the dramatic rise of the motor car in the early part

The Periodic Table

Group

1A	2A	3B	4B	5B	6B	7B	8	8	8	1B	2B	3A	4A	5A	6A	7A	0
H 1																	He 2
Li 3	Be 4											B 5	C 6	N 7	O 8	F 9	Ne 10
Na 11	Mg 12											Al 13	Si 14	P 15	S 16	Cl 17	Ar 18
K 19	Ca 20	Sc 21	Ti 22	V 23	Cr 24	Mn 25	Fe 26	Co 27	Ni 28	Cu 29	Zn 30	Ga 31	Ge 32	As 33	Se 34	Br 35	Kr 36
Rb 37	Sr 38	Y 39	Zr 40	Nb 41	Mo 42	Tc 43	Ru 44	Rh 45	Pd 46	Ag 47	Cd 48	In 49	Sn 50	Sb 51	Te 52	I 53	Xe 54
Cs 55	Ba 56	La* 57	Hf 72	Ta 73	W 74	Re 75	Os 76	Ir 77	Pt 78	Au 79	Hg 80	Tl 81	Pb 82	Bi 83	Po 84	At 85	Rn 86
Fr 87	Ra 88	Ac° 89															

← —————— Transition Elements —————— →

* Lanthanides	La 57	Ce 58	Pr 59	Nd 60	Pm 61	Sm 62	Eu 63	Gd 64	Tb 65	Dy 66	Ho 67	Er 68	Tm 69	Yb 70	Lu 71
° Actinides	Ac 89	Th 90	Pa 91	U 92	Np 93	Pu 94	Am 95	Cm 96	Bk 97	Cf 98	Es 99	Fm 100	Md 101	No 102	Lr 103

of this century, and the discovery of oil in large quantities. The hydrocarbons left from petroleum after the removal of the gasoline meant that producers and government sought ways of using these chemicals and consequently production of organic chemicals from these sources grew dramatically.

Petroleum is a complex mixture and *distillation* produces a number of *fractions* (see table). The operations undertaken in a *refinery*, to produce lighter fractions, are *cracking* and *reforming*. Catalytic cracking, which has replaced the old thermal process is an accelerated decomposition of middle to higher fractions over a solid catalyst usually consisting of zeolites. Catalytic reforming, which uses a platinum or platinum/rhenium catalyst, is undertaken at about 500°C and 7 to 30 atmospheres pressure. Various reactions occur simultaneously with straight-chain alkanes being converted to isomers and gasoline of a higher octane number being produced. There are other processes, including hydrocracking, steam cracking and steam reforming, which are implemented to act on a particular feedstock to produce certain chemicals.

Petroleum fractions	
methane and ethane	natural gas
propane and butanes	liquefied petroleum gases
light naphtha	
naphtha	motor spirit (gasoline)
kerosine	jet fuel
gas oil	diesel fuel
heavy distillates	feedstocks for lubricants, waxes etc.
bitumen/asphalt	

pH of a solution is a measure of that solution's acidity or alkalinity. It shows the concentration of hydrogen IONS (H^+) in an aqueous solution and is the negative logarithm (to base 10) of H^+ concentration calculated with the formula:

$$pH = \log_{10}(1/H^+)$$

The scale ranges from 1 (very acidic, e.g. concentrated hydrochloric acid) through the neutral point of 7 (pure water) to 14 (very alkaline, e.g. CAUSTIC SODA). Since pH is a logarithmic value, one unit of change is equivalent to a tenfold change in the H^+ ion concentration. The pH of solutions is checked by means of INDICATORS.

pharmacology *see* DRUGS AND PHARMACOLOGY.

pheromones a pheromone is a chemical substance that acts as a communication signal between individuals of the same species. Pheromones are found widely throughout the animal kingdom and have a number of different functions, e.g. sexual attraction (common in insects) and marking of territory (used by many mammals either by urine spraying as in dogs and members of the cat family, or by means of special scent glands often on the head or bottom). Pheromones act as *external hormones* and have been shown to be effective at very low concentrations. They are often organic acids or alcohols which are usually termed *volatile* because their effect is short-lived. Pheromones are important in techniques of biological control.

Pheromones are much rarer in plants but one of the most important economically and environmentally is produced by a plant called the 'Scary Hairy Wild Potato' (*Solanum berthaultii*). The leaves produce a pheromone which is chemically identical to the warning signal produced by aphids (small insect pests of many garden and crop plants). Breeding this aphid-repellent character into cultivated crops will hopefully lead to a reduction in aphid damage and less need for chemical insecticides.

phosphorus is a non-metal which occurs naturally as compounds and mainly as calcium phosphate ($Ca_3(PO_4)_2$), but also the mineral apatite. Phosphorus has several forms: red, white and black (a property called *allotropy*) and the white form is the most reactive. It is obtained by heating calcium phosphate with sand and carbon in an electric furnace.

Phosphorus is essential to life because calcium phosphate is a vital component of animal bones. It is also important in the compounds that it forms, e.g. phosphates are widely used in fertilizers. Phosphorus compounds are also used in the manufacture of glass and china ware, matches, detergents, special steels and foods and drinks.

photochemistry is the study of chemical reactions brought about by light. Only light that is actually absorbed will produce any effects and it is necessary to determine which parts of the spectrum are appropriate. The essential step in a photochemical reaction is the raising of an atom or molecule to an excited state by the absorbed light. Ultraviolet light is often the vehicle for such reactions and radiation from the far ultraviolet can break chemical bonds. However, the light may not actually produce a reaction directly. The excited molecule may emit the energy absorbed, affecting a neighbouring molecule which then undergoes a reaction.

Absorbed light may act as a catalyst or supply energy which renders a reaction possible. PHOTOSYNTHESIS is an example of a photochemical reaction.

photoelectric cell (or *photocell*) is a device used for the detection of light and other radiations. One type of cell, the *photoemissive cell*, makes use of the *photoelectric effect*. This is when light energy striking a substance causes energy to be transferred to electrons in the substance. When light above a certain (threshold) frequency is used, photoelectrons are generated and can create a current in a circuit. The photoemissive cell is in effect a light-powered electric cell and comprises a metal base and a transparently thin metal layer coated with selenium. Light entering the cell causes electrons to be released from the selenium and they move across a barrer to the metal layer, setting up a potential difference which can be used to drive a current. Cells of this nature can be used on solar cells and in camera light meters.

The other types of photoelectric cell are *photovoltaic* and *photoconductive* and are used to detect ultraviolet and infrared radiation, respectively.

photography is the process of capturing an image on photographic film (or plates) by means of LENSES in a *camera*. The earliest cameras produced images on metal or glass plates. Cameras consist essentially of a box with a variable aperture and a timed shutter through which light enters after which it is focused by lenses onto the light-sensitive film. The film is coated with an emulsion containing a silver halide (chloride or bromide) and on exposure to light the silver becomes easily reduced and when the film is developed a black deposit of fine silver gives a negative image. By further exposure of the negative and an underlying sensitive paper to light, a positive image is produced which is fixed and washed, producing a photograph.

A *polaroid camera* is different in that the photograph develops immediately after exposure to light. The technique of photography is based upon the incidental discovery by a German doctor, Schulze, who noticed that a silver nitrate solution on chalk turned black in sunlight. Daguerre then took the first photographs of a living person over one hundred years later, in 1839. The modern face of photography owes much to William Fox-Talbot, a British scientist who invented the negative to positive process and the American George Eastman who developed the rolls of film for cameras and who in 1892 founded Kodak.

The first moving photographs were made in 1893 by Edison (US) and Dickson (UK) in America, although only one person could view at a time. The Lumière brothers made the first practical projector–the start of the movies.

photon *see* LIGHT.

photosynthesis and transpiration *photosynthesis* is the complicated process by which green plants use the energy from the sun to make carbohydrates from carbon dioxide and water, releasing oxygen as a result. There are two stages known as the *Calvin Cycle* and the *Light Reactions* of photosynthesis which are very complex but result in the production of sugars and starch. Photosynthesis can only occur if light trapping pigments are present, and the main one of these is CHLOROPHYLL which is green-coloured and occurs in stems and leaves. Chlorophyll captures light energy and this initiates a series of energy transfer reactions which enable simple organic compounds to be made from the splitting of carbon dioxide and water. Photosynthesis is the basis of all life on Earth and regulates the atmosphere as it increases oxygen concentration while reducing the carbon dioxide concentration. *Transpiration* is the loss of water in the form of vapour through pores, known as *stomata*, in the leaves of plants. As much as one sixth of the water taken up by the roots can be lost in this way. The transpiration rate is affected by many environmental factors such as temperature, light and carbon dioxide levels (i.e. whether photosynthesis is taking place), humidity, air currents and water uptake from the roots. Transpiration is greatest when a plant is photosynthesizing in warm, dry, windy conditions.

phototropism the response by plants in the form of growth movement to the presence of light. Plant shoots show *positive phototropism* as they grow towards the light but roots tend to display *negative phototropism* because they grow away from the light source. Phototropism is caused by *auxin* (a plant growth HORMONE). The hormone is more abundant on the dark side of the plant and this side is induced to grow more by elongation of cells resulting in a curving towards the light.

physics is the study of matter and energy, and changes in energy without chemical alteration. Physics includes a number of topics such as magnetism, electricity, heat, light and sound (*see individual entries*). The study of modern physics also encompasses QUANTUM THEORY, atomic and nuclear physics (i.e. subatomic particles and their behaviour (*see* ELEMENTARY PARTICLES) and physics of NUCLEAR FISSION AND FUSION). As the research into topics has expanded over recent years, so new subjects begin to develop often on the boundaries of two major disciplines. This has happened in geophysics (geology and physics), biophysics (biology and physics) and astrophysics, which combines astronomy with physics.

physiology the study of all the METABOLIC functions of animals and plants including the processes of

RESPIRATION, REPRODUCTION, EXCRETION, working of the NERVOUS SYSTEM, PHOTOSYNTHESIS, etc. It covers all aspects of the life of organisms and may be one specialized area of research or broadly based.

piezoelectric effect is the effect within certain crystals whereby positive and negative charges are generated on opposite faces when the crystal is subjected to pressure. The charges are reversed if the crystal is put under tension and the whole effect is reversible, i.e. the application of an electric potential produces an alteration in size of the crystal. Quartz is the commonest piezoelectric crystal and very pure crystals grown in the laboratory can be cut to vibrate at one frequency when a voltage is applied. The vibrations are used in watches and enable near perfect time to be kept. The piezoelectric effect is also used in crystal microphones and pickups. Other crystals that show the effect are Rochelle salt (sodium potassium tartrate) and barium titanate.

pig iron *see* CAST IRON.

pigments are compounds that produce colour and occur naturally in plants and animals. In plants, *chlorophyll* imparts a green colour, and animals contain *melanin*, which produces the black or brown colour of hair or skin. *Carotenoids* are plant pigments, orange, red and yellow, that occur in carrots and tomatoes.

Synthetic pigments are used to colour plastics, textiles, inks, etc. Pigments are different from dyes and tend to be insoluble and occur as particles and many are inorganic:

Colour	Pigment
white	titanium dioxide
	lead carbonate and sulphate
	zinc oxide and sulphide
red/brown	iron oxides
	red lead
	cadmium red, orange, scarlet
yellow	iron oxides
	lead, zinc and cadmium chromates
blue	ultramarine (an aluminosilicate with sulphur)
black	carbon

pitchblende is an important ore of uranium, a mineral called Uraninite, made up primarily of uranium oxide. It is also the principal source of radium, an element discovered by Marie and Pierre Curie in 1898 (they also discovered polonium from the same source).

Pitchblende occurs as a black mass, resembling tar and the uranium in it decays to form radioactive radium and radon gas. Deposits occur in Canada, East Africa, Saxony and Colorado.

plankton very small organisms, often microscopic and including both plants and animals, which drift in the currents of oceans and lakes. The plants (or *phytoplankton*) consist mainly of single-celled algae (*Bacillariophyta*) called *diatoms* which photosynthesize and form the basis of food chains. The animals (or *zooplankton*) include the *larval* stages of larger organisms, some protozoans (Kingdom PROTISTA) and small creatures called copepods which are related to crabs (*see* **NH:** CRUSTACEA). Plankton provide food for many larger animals, e.g. whales, and are of vital importance in the FOOD CHAIN.

plants Kingdom Plantae forms one of the major kingdoms of life. Plants are distinguished from animals by their ability to manufacture food by PHOTOSYNTHESIS. This type of nutrition is termed *autotrophic* whereas that of animals, relying on taking in food from outside, is called *heterotrophic*. The photosynthetic cells of plants contain organelles (*see* ORGAN) called *chloroplasts* which contain the pigment CHLOROPHYLL and this traps light energy from the sun. Plant cells have walls (absent in animal cells), and the main substance of which these are composed is a carbodyhydrate called CELLULOSE. Plant cells store carbohydrate in the form of *starch* whereas animal cells store it as glycogen.

There are twelve divisions of the plant kingdom (the same as phyla in animals). In three of these divisions, there is no true system of structures or tubes to transport water and food. These are called *non-vascular* plants and include mosses, liverworts and hornworts. In the other divisions this *vascular tissue* is present but they are further divided into those which produce *seeds* and those which do not. The *seedless plants* are the horsetails, ferns, club mosses and whiskferns. The seed plants are divided into two groups, the *Gymnosperms* which produce 'naked' unprotected seeds, e.g. the conifers, and the *Angiosperms* (or *Anthophyta*), the flowering plants. These are the ones most familiar to us and are the most complex among plants, producing seeds within special protective coverings.

Most plants have sexual REPRODUCTION and many are also able to reproduce asexually. In the plant life cycle an *alternation of generations* occurs with a *sporophyte* generation which is *diploid* (i.e. has the full number of CHROMOSOMES) producing the *gametophyte generation* which is *haploid* (half the number of chromosomes). The sporophyte generation is nearly always the most prominent form, and in most plants the gametophyte is represented by very small structures.

Flowering plants are either *monocotyledons*, with one seed leaf or *dicotyledons*, with two. Some flowering plants are called *annuals* as they grow from seed, produce flowers and die. Others are *perennials* which grow up from the same root stock year after year even though the leafy parts die back during the winter.

plastics is a group name for mainly synthetic organic compounds which are mostly POLYMERS, formed by polymerization that can be moulded when subjected to heat and pressure. There are two types: *thermoplastics* (e.g. PVC or polyvinyl chloride) which become plastic when heated and can be heated repeatedly without changing their properties; *thermosetting plastics* such as phenol/formaldehyde resins lose their plasticity after being subjected to heat and/or pressure. Plastics are moulded and shaped while in their softened state and then cured by further heat (thermosetting e.g. epoxy resins, silicones) or cooling (thermoplastics e.g. perspex and polythene).

The first synthetic plastic was *Bakelite*, invented in 1908. Since then the plastics industry has become vast and an enormous range of domestic, leisure, industrial and commercial items are now produced. Plastics can be shaped by blow moulding, vacuum forming, EXTRUSION and INJECTION MOULDING and they are used extensively in composite materials and LAMINATES.

plutonium *see* NUCLEAR ENERGY.

pneumatics refers to the use of compressed air, usually to power machines. A supply of air is piped to the pneumatic motor which commonly features a piston within a cylinder. Compressed air pushes the piston one way and a spring or air pressure pushes it back, producing a hammer action used in drills for mixing or construction.

The greatest benefit of pneumatic tools over electrically-powered tools is that there is no chance of electric shocks or sparks which could cause a fire or explosion in certain circumstances.

polarization of light is when light is made to vibrate in one particular plane. Light normally consists of waves vibrating in many directions and because it is electromagnetic radiation there is an electric and a magnetic field vibrating at right angles to each other. If light is polarized, the electric field vibrations are confined to one plane (*plane polarized light*) called the plane of vibration and the magnetic vibrations are in one plane at right angles, the plane of polarization.

Polarization occurs when light passes through certain crystals (quartz, calcite) or is reflected from some surfaces (e.g. the sea). A polarizer produces polarized light and a polaroid filter is one such material used in sunglasses. Polarized light also has uses in mechanical engineering to reveal stress patterns in materials. Only transverse waves such as light can be polarized; longitudinal waves such as sound cannot.

pollen pollen grains are the male sex cells (*gametes*) of flowering PLANTS. They occur in small pollen sacs contained within a structure called the *anther*, which is part of the male reproductive organ of the flower. The anther occurs on the end of a thin stalk called the stamen. Pollen grains and their female equivalents, the *embryo sacs*, are the *gametophyte generation* (*see* PLANTS) in flowering plants.

pollination is the transfer of pollen from an anther to a stigma. This is part of the female reproductive organ of the flower, the other portions being the *style* and *ovary* which together make up a *carpel*. The ovary contains one or more *ovules* which develop into seeds. Pollen may be transferred by means of the wind, insects, birds, water, etc. and the grains vary in shape according to the method of pollination used by the plant. Wind-pollinated plants have light, smooth grains while insect-pollinated ones have grains which are rough or spiny. This is called *cross-pollination* as the pollen is transferred from one flower to another of the same species. *Self-pollination* is where pollen is transferred from an anther to a stigma in the same flower.

polygon is a term used in GEOMETRY to describe a closed plane (i.e. two-dimensional) figure with three or more straight line sides. Common polygons are figures such as the triangle, quadrilateral and hexagon. A square is a regular polygon where the sides and all angles are equal.

For a polygon with n sides, the sum of the interior angles = $180°$ $(n-2)$. Except for the triangle and square, polygons are named after their number of sides so a ten-sided figure is called a *decagon*. Polygons are common around us, whether in a natural form as in honeycomb, or the faces of crystals or man-made as with tiles and nuts and bolts.

polymer a large, usually linear molecule that is formed from many simple molecules called *monomers*. Synthetic polymers include PVC, Teflon, polythene and nylon while naturally-occurring polymers include starch, cellulose (found in the cell walls of plants) and rubber (*see individual entries*). Early versions of polymers were modified natural compounds, e.g. the vulcanization of rubber by heating with sulphur. The first fully synthetic polymer to be developed was

Bakelite (*see* PLASTICS) followed by urea-formaldehyde and alkyd resins in the late 1920s. Polyethylene (that is polythene) was first produced by ICI on a commercial basis in 1938 and the Dupont company in America produced the first nylon in 1941. Many synthetic polymers are produced from alkenes (*see* HYDROCARBONS) in reactions called addition polymerizations which are rapid and require only relatively low temperatures. Condensation polymerization is another means of producing polymers e.g. the nylons (polyamides) and the silicones (polysiloxanes) in which some molecule (often water) is removed at each successive reaction stage.

potassium is an alkali metal that is silver white and highly reactive. In fact, it reacts violently with water and therefore in nature occurs only as compounds. It occurs widely in silicate rocks as alkali feldspar, in blood and milk and in salt beds and as potassium chloride in sea water. Potassium is an essential element for plants and it is added to the soil by farmers, as fertilizers (such as potash). Fertilizers are the primary use of potassium but it is also used in making batteries, ceramics and glass.

power is the rate at which WORK is done. It is also regarded as the rate at which energy is converted. The unit of power is the *watt* (W) which is measured in joules per second. Thus if a machine does 10 joules of work every second, it has a power output of 10W. Electrical power can be calculated from the product of the voltage and the current.

Power is very important because many routine processes and functions require energy to do work, whether it concerns an engine, light bulbs or our bodies. In most cases, when energy is expended it is not all turned into the useful work required but some escapes or is lost, e.g. engines lose energy in vibrations and heat.

pressure is the force exerted per unit area of a surface. The pressure of a gas equals the force that its molecules exert on the walls of its container, divided by the surface area of the vessel. The pressure of a gas varies with temperature and volume (*see* GAS) and the pressure in a liquid or in air equals the weight of liquid or air above the area in question and therefore as the depth increases, the pressure also increases. This explains why air pressure decreases with height (*see* ATMOSPHERE). If force is measured in N (newtons) and area in m^2 (square metres), pressure is N/m^2 or *pascals* (Pa).

primates an order of mammals which includes monkeys, lemurs, apes and humans. Primates have highly mobile hands and feet due to the presence of thumbs and big toes which can grasp. They also have a large brain (especially the part called the *cerebrum*) and are very intelligent.

prism is a solid figure that is essentially triangular in shape (resembling a wedge) and made of a transparent material. It is used in physics to deviate or disperse a ray in optical instruments or laboratory experiments. If a narrow beam of white light passes through a prism it is split into a range of colours–the *spectrum*. The light is split because each of the colours is refracted by a different amount, because each is light of a different *wavelength* (*see* WAVE).

probability is the chance of something happening, or the likelihood that an event will occur, expressed as a fraction or decimal between 0 and 1 (or as odds). If something is absolutely certain, the probability is 1; if it is impossible, then the probability is 0, and all probabilities lie within these two figures. Thus if a coin is flipped, there is an equal chance of gaining 'heads' and 'tails', i.e. one chance out of two for heads or tails giving a probability of _ or 0.5. The probability of a particular number coming up on rolling a dice is one in six, i.e. 1/6 or 0.1666, and it is 0.5 for rolling an even number (3 possibilities out of the 6 numbers).

procaryote any *organism* which has a nucleus that is not surrounded by a true membrane. These organisms all belong to the Kingdom MONERA and are the BACTERIA and blue-green *algae* (*cyanobacteria*). They have a single chromosome and do not undergo meiosis or mitosis (*see* CELL DIVISION) but reproduce ASEXUALLY by a method called *binary fission*. In this, the whole cell divides and each 'daughter' cell receives a copy of the single parental chromosome. Procaryotes are very important organisms, vitally involved in all life-sustaining processes. A few are disease-causing organisms, but others are used in BIOTECHNOLOGY. They were the first organisms to evolve being the only life forms on Earth for about 2 billion years.

protein and amino acid a protein is a type of organic (*see* CHEMISTRY) compound of which there are many different kinds, and it usually contains nitrogen and sulphur. The individual MOLECULES are made up of building blocks called *amino acids* arranged in long chains (known as *polypeptide* chains). There are 20 different amino acids but a huge number of possible arrangements in a polypeptide chain or protein. Most proteins consist of more than one polypeptide.

protista a biological kingdom containing numerous different kinds of simple organisms, most of which are single cells. They are the simplest eucaryotic (*see* EUCARYOTE) organisms because they are usually unicellular (single-celled). However, since each protist is able to carry out lots of different functions within its one cell, and many are highly complex, some have the most elaborate of all cells. Most protists have extensions or projections out from the *cytoplasm* of the cell, called cilia, (*sing.* cilium) or flagella (*sing.* flagellum), which are used for movement. These may not be present at all stages of the life cycle but usually occur at some time or other. All protists can reproduce asexually and some also have sexual reproduction. They are important organisms in the PLANKTON and hence significant in food chains. Most require oxygen for RESPIRATION and so are called *aerobic*. Some are PHOTOSYNTHETIC (i.e. they are *autotrophic* and make their own food), others are *heterotrophic* (i.e. they absorb or ingest food) and yet others combine the two, a condition known as *mixotrophic*. Protists inhabit a wide range of environments. Many are free-living but others live inside cells and tissues of other organisms as PARASITES or in other relationships (known as *symbiosis*).

pulley is one of the six varieties of simple MACHINE. It is a wheel with a groove around the edge around which a rope can be passed. An individual pulley, suspended on an axle may make lifting a load a little easier but in general several pulleys are combined in a *block-and-tackle* system to enable the lifting of larger loads that would not normally be feasible. A rope wound around the pulleys in such a system can be pulled a long way to raise a heavy load a short distance.

pumps are machines that move gases and liquids. A basic pump is the force, or reciprocating pump in which a piston moves in and out of a cylinder. Fluid or gas is taken into the cylinder on one stroke and pushed out through a valve when the piston moves in the opposite direction. A bicycle pump or a water pump are examples of this type. Another type is the rotary pump which uses some form of spinning blade assembly to push liquid along a pipe.

Pythagoras' Theorem *see* GEOMETRY.

quantum mechanics is the system of mechanics that facilitates an explanation of the structure of the atom and the behaviour of small particles within atoms (*see* ATOM). The electronic structure of the atom comprises a nucleus around which electrons orbit at various levels. Putting energy into an atom causes electrons to move temporarily into a higher orbit, through absorption of the en-

ergy. When the electron returns to its original orbit, light energy is given off as a *quantum* or PHOTON. This principle was discovered by Max Planck (and called *quantum theory*) and it was realized that all electromagnetic radiations can be thought of in this way, i.e. a photon is a quantum of electromagnetic radiation.

Further developments were made in the subject by Heisenberg who used quantum mechanics to explain atomic structure and behaviour of subatomic particles. Essentially he stated that it is not possible to know both the position and momentum of a particle such as the electron, *at the same time* (this is called *Heisenberg's uncertainty principle*). Quantum mechanics was gradually superseded by *wave mechanics* and in the mid-1920s Schrödinger evolved his wave equation which describes the behaviour of a particle in a force field. It permitted the description of the electrons in an atom in terms of waves.

quantum numbers a set of four numbers used to describe atomic structures (*see* ATOM and QUANTUM MECHANICS). The principal quantum number (n) defines the shells (or orbits) which are visualized as orbitals (that is, a charge cloud which represents the probability distribution of an electron). The orbit nearest the nucleus has 2 electrons and $n = 1$. The second shell contains 8 electrons (and $n = 2$) and so on, the maximum number of electrons in each shell being defined by the formula $2n^2$. The orbital quantum number (l) defines the shape of the orbits and these are designated s, p, d and f, the letters arising purely for historical reasons. The magnetic orbital quantum number (m) sets the position of the orbit within a magnetic field and s is the spin quantum number. The latter is based upon the assumption that no two electrons may be exactly alike and pairs of electrons are considered to have opposite spin.

The quantum numbers n, l and m are related and allow the electronic structure of any atom to be determined.

quantum theory *see* QUANTUM MECHANICS.

radiation (thermal) is a process by which heat is transferred from one place to another. Objects give out infrared waves and hot objects may also emit light into the ultraviolet region. Some surfaces are better emitters and absorbers of thermal radiation. Dull black surfaces are better at absorbing and emitting thermal radiation than white or silver surfaces. The nearer to a silver, mirror-like finish is a surface, the greater will be the reflection and lower the absorption of radiation.

Vacuum flasks, silver body blankets and greenhouses all make use of this property. In the case of the greenhouse, the glass permits light and short wavelength infrared radiation through from the Sun, which warms the air and the plants. The plants re-emit radiation, but of a longer wavelength and this does not pass through the glass, but is reflected back into the greenhouse, helping to maintain the higher temperature.

radio uses electromagnetic waves to send and receive information without wires (hence the former term wireless, for the radio). It includes, in its widest sense, radio, television and radar. Radiowaves are created in an antenna or aerial by making electrons oscillate. Long and medium waves are sent around the world by bouncing them off the *ionosphere* (a layer of charged particles in the upper atmosphere) but VHF (very high frequency) and UHF (ultra-high frequency waves for television) waves require a straighter path between the receiver and the transmitting aerial because these waves are not reflected by the ionosphere. Similarly, the longer wavelengths can diffract around hills and other obstacles, unlike the short wavelength VHF and UHF waves.

When a receiver is tuned to the frequency of the appropriate wave sent from the transmitter, it can amplify and rectify the signal to produce a varying current which matches the frequency of the sound wave at the microphone. This current is then used to work a loudspeaker, thus reproducing the original sound.

radioactivity means that a material naturally gives out radiations while undergoing spontaneous disintegration. The (nuclear) radiation given out by a radioactive material is one of three types: a–particles (alpha), ß–particles (beta) or gamma (g) rays and each is a different entity with varying effects.

Radioactivity was discovered in 1896 by Henri Becquerel who found that uranium salts emitted some form of radiation capable of ionizing the air and also capable of affecting a photographic plate, through its wrapping. Further work determined the properties of the three types of radiation.

An *alpha particle* is a helium nucleus (2 protons and 2 neutrons, He^{2+}), and it has a strong ionizing effect but little penetration, having a range in air of a few centimetres. *Beta particles* are electrons moving at different speeds, in some cases almost at the speed of light. They have a negative charge, and although their ionizing effect is very weak, they are more penetrating, with a range in air of about one metre. Beta particles can be stopped by a few millimetres of aluminium. *Gamma rays* are short wavelength electromagnetic waves with lit-

tle ionizing effect but greatest penetration. They are never completely absorbed, but lead 25mm thick reduces their intensity to half. Radioactive materials can give one or a combination of radiation types.

Radioactive decay is the emission of particles by an unstable nucleus, which in so doing becomes another atom. The decaying nucleus is the parent, the resulting nucleus is the daughter, which together with particles emitted are called the decay products. During decay, nuclei disintegrate randomly but at a different set rate for different atoms. The *half-life* of a radioactive *nuclide* (i.e. the radioactive form of an element) is the time taken for half the atom to decay, i.e. for the activity to fall by half. For example, uranium-238 has a half-life of 4.51×10^9 years; radium-226, 1620 years; and sodium-24, 15 hours; and radon-220 just 52 seconds.

Artificial nuclides are used in many ways, e.g. in medicine cobalt-60 for the treatment of cancer. On industrial plants radio- isotopes are used to check the flow of fluids and gases, leaks, residence times in chemical processes, and a variety of isotopes are used. (*See also* DATING METHODS).

rainbow is the characteristic display of the colours of the spectrum, which may form a large arc of a circle. The Sun must be behind the viewer for the rainbow to be seen. Light enters droplets of water in the sky and is refracted and internally reflected. Because the raindrop acts like a prism, the light is split up into the constituent colours.

receptor a special type of animal cell, which is called *excitable* because it is sensitive to a particular type of stimulus. When the receptor cell is *excited*, electrical impulses are sent along a (sensory) nerve to the central nervous system. Some receptors respond to factors or stimuli outside the animal's body, and others are present internally. Receptors are sensitive to a variety of different stimuli and are often called after the *sense* that they detect. Examples are *chemoreceptor* (chemicals), *photoreceptor* (light), *mechanoreceptor* (touch) and *proprioreceptor* (pressure, movement or stretching within the body). Receptors may be grouped together within a special organ, such as the ear, which detects sound waves and also controls balance and posture.

recording media there are many ways in which sounds and pictures can be recorded. One medium that is fast being replaced by new technologies is the vinyl record. These are plastic discs upon which the sound is recorded in a groove cut in the disc. The sound to be recorded is fed via microphones to cutters operated by electromagnets. Thus, when a stylus is placed on the record, it vibrates along the groove, and the vibrations are turned into electric currents to operate loudspeakers.

Magnetic tape can be used for recording sound (audio tape) or pictures (video tape) and consists of a plastic tape coated with magnetic particles, often iron oxide or chromium dioxide for higher quality. Recordings are made by magnetizing the tape in a specific way. It is a flexible medium as it can be re-used and edited.

More recent media include the compact disc, which is a form of laser disc and has a very large capacity, often used in the storage of computer data.

rectifier *see* DIODE.

reflection light striking a surface may be reflected but in most cases the reflections are in all directions. When the surface is smooth, e.g. a mirror, an image is produced. When light is reflected from a surface it follows certain laws. The incoming (*incident*) ray makes an angle (angle of incidence) with the line drawn at 90° to the surface (the normal) equal to the angle of reflection which is the angle between the normal and the *reflected* ray. This encompasses the *laws of reflection*. A *reflector* is any surface that reflects electromagnetic radiation.

refraction is the bending of, most commonly, a light ray when it travels from one medium to another, e.g. air to water. The refraction occurs at the point where the light passes from one material to another and is caused by the light travelling at different velocities in the different media. The incident (*see* REFLECTION) ray passing into a material becomes the refracted ray and in an optically more dense medium is bent towards the normal to the interface. The two angles – of incidence and refraction – are related by *Snell's Law* which states that the ratio of the sines (*see* TRIGONOMETRY) of the two angles is constant for light passing from one given medium to another. The value of the ratio of the sines of the angles is called the *refractive index*, measured when light is refracted from a vacuum into the medium.

refrigeration *See* EVAPORATION.

relativity is the theory developed by Einstein and which is made up of two parts. The *special theory* states that the speed of light is the same for all observers, whatever their speed, that is light from an object travels at the same velocity whether the object is moving or stationary. Nothing may move faster than the speed of light. Further important implications of this theory are that the mass of a body is a function of its speed, and Einstein derived the *mass-energy equation*, $E = mc^2$ where c is the speed of light. As a result of this theory the concept of *time dilation* was proposed which essentially means that for someone travelling at very great speed, time passes much more slowly for them that it does for a stationary observer.

The *general theory of relativity* relates to gravity. Matter in space is said to cause space to curve so as to set up a gravitational field and gravitation becomes a property of space. The validity of Einstein's theories has been tested with experiments in modern atomic physics.

reproduction the production of new individuals of the same species either by *asexual* or *sexual* means. The term usually refers to sexual reproduction which involves the joining together (called *fusion*) of special sex cells, one of which is female (the egg or *ovum*) and the other is male (e.g. *sperm* in animals and pollen in flowering plants). The sex cells of any organism are known as *gametes*. Many organisms produce gametes within special reproductive organs. In the flower of a seed plant, the male sex organs are the stamens which produce pollen. The *carpels* are female and produce *ovules* which later develop into seeds after *fertilization*. Gametes are special cells which contain half the number of chromosomes (called haploid, *see* CELL DIVISION) of the parent. When a male and female gamete join together (*fusion*), fertilization takes place and a single cell (now called a zygote) is produced which contains the full number of chromosomes (diploid). This goes on to produce a new individual, usually by a process of many cell divisions. In mammals (and many other animals), the male sex organs are the *testes* and the female ones are the *ovaries*. These reproductive organs (also called *gonads*), are specialized structures, the cells of which produce the sperm and ova under the influence of hormones. Most adult female mammals have a reproductive cycle, called an *oestrous cycle*, during which time the eggs develop and the animal becomes ready to mate (a period known as *heat* or *oestrus*). Mating occurs at the time when the animal is most likely to become pregnant. Some mammals have a definite *breeding season* and only one oestrous cycle in a year (*monoestrous*). Others have several cycles and are *polyoestrous*. In female humans, the *menstrual cycle* replaces the oestrous cycle. Mating usually results in the fertilization of one or more egg. The cells rapidly divide and become embryos which grow and develop inside a muscular, bag-like organ called the *uterus* or womb. The animal is now said to be *pregnant*. A special organ of pregnancy, called the *placenta*, develops and this attaches the embryos to the uterus by a cord called the *umbilical cord*. This provides for the passage of food, oxygen and some other substances (e.g. vitamins and antibodies) to the embryos while waste products, (mainly carbon dioxide and urea), pass in the opposite direction and are removed by the mother's blood circulation. When the young are fully developed they are *born*, which means that they are pushed to the outside through a passage leading from the womb called the *vagina*. This is brought about by contractions of the muscles in the wall of the uterus under the influence of hormones. Immediately after birth, the placenta is also pushed out.

resistance (R) is measured in *ohms* and is the potential difference between the ends of a conductor divided by the current flowing. Superconductors apart, materials resist the flow of current to varying degrees, and some of the electrical energy is thereby converted to heat. The resistance of a wire depends on the ability of the material to conduct electricity, and the dimensions of the wire. In general, a short wire has less resistance than a long one and a thick wire less resistance than a thin one. *Resistors* are devices made to produce resistance and they control the current in a circuit. They are found in radios, televisions and similar pieces of equipment. A variable resistor is called a *rheostat* and by means of a sliding contact can alter the current flowing. Wires used in circuits have a low resistance to minimize heat loss, but in some cases thermal energy is required, as in heating elements for electric kettles, fires, immersion heaters and cookers. Nichrome wire is commonly used for such applications.

resonance is the creation of vibrations in a system, such that it is vibrating at its natural frequency, due to vibrations of the same frequency being received from another source. Resonance occurs in many instances; the strings of an instrument, columns of air in a wind instrument and even an engine causing vibrations in a bus or van.

respiration and ventilation *respiration* is often used to mean *breathing* but more correctly it is a metabolic (*see* METABOLISM) process which occurs in the cells of an organism. It is the process by which living cells release energy by breaking down complicated organic substances (food molecules) into simpler ones using enzymes. In most organisms, respiration occurs in the presence of oxygen and is called *aerobic*, with water and car-

bon dioxide produced as waste products and energy released. Some organisms (e.g. a number of species of bacteria and yeasts) do not require oxygen but use alternative chemical reactions. This is known as *anaerobic respiration* and it is possible for it to occur for a short time in the muscles of mammals when these are being very hard worked. However, it is short-lived and results in the build up of a substance called *lactic acid* which causes muscular *cramps*. *Ventilation* is the actual process of breathing or drawing air in and out of an animal's body. Various mechanisms and structures are involved in different animals, using muscles to pump air in and out (*see also* LUNGS AND GILLS).

rubber is a high molecular weight polymer which occurs naturally, being obtained from the tree *Hevea braziliensis*. It is an elastic solid from the latex (a colloid of rubber particles in a watery base) of the tree which contains about 35% rubber. After straining, the latex is coagulated and the raw rubber is 'compounded' and other substances are added to increase strength and it is then *vulcanized* (heated with sulphur) to increase the cross-linking in its structure. This provides more elasticity and the rubber becomes less sticky. Natural rubber is a form of polyisoprene -$(CH_2.CH : C(CH_3) : (H_2)_n$.

Synthetic rubbers are polymers of simple molecules and include butyl rubber, neoprene, styrene-butadiene rubber and silicone rubber. Butyl rubber is used for the inner tubes of tyres; styrene-butadiene for car tyres; and silicone rubbers for applications requiring temperature stability, water repellence and chemical resistance.

salt is the common name for sodium chloride, NaCl. It occurs in sea water and in hot climates shallow ponds of sea water are left to evaporate to dryness, providing deposits of salt. Significant salt deposits were formed in the geological past and are now mined (*see* evaporite). In chemistry, a salt is produced by the reaction of equivalent amounts of an acid and a base, with the production of water. Salt molecules contain metal atoms with one or more non-metal atoms and are named from the acid and base from which the salt is formed, e.g. hydrochloric acid and sodium hydroxide produce sodium chloride.

saturated solution is a solution of a substance (solute) that exists in equilibrium when there is excess solute present (*see* SOLUTIONS). Heating a saturated solution allows more to dissolve producing a *supersaturated solution*. Cooling or loss of solvent will cause some of the solute to come out of solution, i.e. it will crystallize.

screw is one of the six varieties of simple *machine*.

It is an extremely effective device and is used extensively for fastening things together. Screws or screw threads are also used in instruments to enable controlled movement, e.g. on a microscope stage, or in a micrometer when making accurate measurements. A screw jack is one type of jack for lifting a car to allow a wheel to be changed. This means that a heavy object like a car can be raised relatively easily.

seed *see* FLOWERS *and* PLANT.

semiconductors materials, such as silicon, that can act as a conductor of electricity or an insulator. Pure semiconductors are insulators when cold and allow current to pass when heated although at room temperature they conduct only poorly. However, by *doping* them, with small amounts of other substances, semiconductors can be made to conduct. Depending upon the material used for doping, a p- or n-type semiconductor is created (*see* DIODE).

Semiconductors may be compounds or elements. In addition to silicon, germanium selenium and lead telluride are used. Diodes, rectifiers and transistors all utilise semiconductors, as do silicon chips used in microprocessors.

sense organ *see* RECEPTOR.

sight the sense of sight is possessed by many animals. Some simple, single-celled (unicellular) organisms have cells which are sensitive to light, but cannot be said to 'see' in the way we understand it. They have *eyespots* which contain pigments (called *c̄arotenoids*) and these are sensitive to light. They cause the cell or organism to move in a particular way in response to light. Other invertebrate animals have more complicated eyes, ranging from a fairly simple type called ocelli (*sing.* ocellus) to a more advanced kind known as a *compound* eye. Compound eyes are a feature of many Arthropod invertebrates (spiders, insects, beetles, etc.) and give very good vision especially in some insects, e.g. dragonflies and flies. Squids and octopuses (which are molluscs), have eyes that are similar to those of vertebrate animals and have excellent powers of vision.

The eyes of vertebrates are almost spherical balls filled with fluid, and contained within bony sockets in the skull. Light enters the eye through the transparent *cornea* and is bent (or refracted) as it passes through a small space filled with a fluid, the *aqueous humour*. The light passes through the *pupil* which is a hole in the centre of the coloured *iris*. The light rays pass through the lens and on through a larger cavity filled with fluid called the *vitreous humour*. The lens bends the light rays and

focuses them so that they form an *image* on the layer at the back of the eye which is called the *retina*. The retina contains the actual light receptors (photoreceptors). These are pigment-containing cells of two types, *rods* and *cones*. The pigments undergo chemical changes (*bleaching*) in light of different wavelengths and this generates electrical impulses which travel to the brain along a special sensory *optic nerve*. Cones contain pigments which allow for colours to be detected and there are about 6 million in a human retina. Rods are sensitive in very dim light and allow for night vision but do not detect colour. The human retina contains about 125 million of these. Most mammals are nocturnal and have poor colour vision, but keen night sight is allowed for by the large number of rods that are present.

The human eye

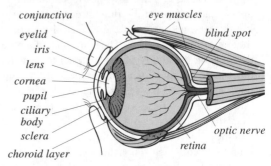

silicon is a non-metallic element and the second most abundant in the Earth's crust. It does not occur as free silicon, but is found in abundance as numerous silicate minerals including quartz (SiO_2). Silicon is manufactured by reducing SiO_2 in an electric furnace but further processing is necessary to obtain pure silicon. When doped with boron or phosphorus it is used in semiconductors, (*see also* DIODE). Quartz and some silicates are used industrially to produce glass and building materials. Silicon melts at 1410°C. *Silicones* are polymers built on SiR_2O groups (where R is a hydrocarbon). Simpler substances are lower melting point oils, which are used as lubricants. More complex varieties are solid and very stable and are used as electrical insulators.

siphon is a bent tube, shaped like an upside-down U, and it can be used for transferring liquid from a higher to a lower level. One end of the tube is placed in the higher container, and the other end is primed by sucking. When the liquid flows through the tube, the end is placed in the lower container and the flow continues even though the suction is no longer there. The liquid flows due to a pressure difference between the two ends of the tube.

SI Units (Système International d'Unités) is a system of units agreed internationally. It comprises seven basic units and some supplementary units with a larger number of derived units. It also established the prefixes used for decimal multiples where the practice is to raise to 10 by a power that is a multiple of three.

basic units		derived units		
m	metre	Bq	becquerel	(radioactivity)
kg	kilogram	C	coulomb	(electric charge)
s	second	F	farad	(capacitance)
A	ampere	Gy	gray	(ionizing radiation)
K	kelvin	H	henry	(inductance)
mol	mole	Hz	hertz	(frequency)
cd	candela (light)	J	joule	(work or energy)
supplementary		lm	lumen	(luminous flux)
rad	radian (angular measurement)	lx	lux	(illuminance–one lumen/m²)
sr	steradian (solid angle)	N	newton	(force)
		Ω	ohm	(resistance)
		Pa	pascal	(pressure)
		S	siemens	(conductance)
		Sv	sievert	(dose equivalent)
		T	tesla	(magnetic flux density)
		V	volt	(electric potential)
		W	watt	(power)
		Wb	weber	(magnetic flux)

decimal multiples					
10^{-18}	atto	a	10	deca	da
10^{-15}	femto	f	10^2	hecto	h
10^{-12}	pico	p	10^3	kilo	k
10^{-9}	nano	n	10^6	mega	M
10^{-6}	micro	μ	10^9	giga	G
10^{-3}	milli	m	10^{12}	tera	T
10^{-2}	centi	c	10^{15}	peta	P
10^{-1}	deci	d	10^{18}	exa	E

skeleton the whole structure that provides a framework and protection for an animal's body and within which organs are protected and muscles are attached. In many invertebrates there is an *exoskeleton* which may have to be shed to allow for growth (*see* ECDYSIS AND MOULTING).

Other invertebrates add to the outer edge of their exoskeleton to allow for growth, e.g. molluscs such as snails. Other animals have an *endoskeleton* which lies inside the body. Vertebrate animals have a skeleton made up of numerous bones and cartilage. In mammals over 200 bones are present, some of which are joined by *ligaments* while oth-

ers are fused together. Both types of skeleton are flexible due to the presence of joints which allow for the movement of limbs, etc.

The posterior view of the human skeleton

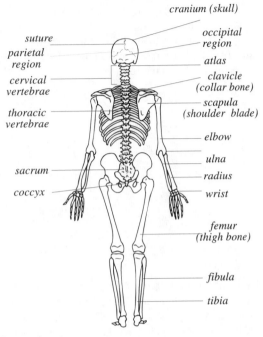

cranium (skull)
suture
parietal region
cervical vertebrae
thoracic vertebrae
sacrum
coccyx
occipital region
atlas
clavicle (collar bone)
scapula (shoulder blade)
elbow
ulna
radius
wrist
femur (thigh bone)
fibula
tibia

Snails have a shell hardened by calcium (*calcareous*), in arthropods the skeleton is hardened by the presence of a substance called *chitin* and in crabs and lobsters the shell (carapace) is hardened with chitin and calcium salts.

skin an important organ which forms the outer covering of a vertebrate animal. There may be a variety of structures protruding from the surface of the skin (i.e. hair, fur, feathers or scales) depending upon the type of animal. The skin provides a protective layer for the body, helping to cushion it in the event of accidental knocks and preventing drying out. Also, skin helps to maintain the correct body temperature. When the body is hot the many blood *capillaries* in the skin widen allowing more blood to flow through them, and heat is lost to the outside by radiation. Sweat *glands* present in the skin secrete a salty fluid which evaporates from the surface and forms another cooling mechanism. If the body is cold, the capillaries contract to decrease blood flow and conserve heat. The layer of fat in the skin has a warming effect and small erector muscles attached to hair roots contract to raise the hair. The hairs trap a layer of air which helps to warm the body. Skin is a physical barrier to harmful substances or organisms which might otherwise enter more easily. It contains many receptors sensitive to pain, touch, pressure and temperature and these connect with sensory nerves (*see* NERVOUS SYSTEM) relaying messages to the brain. Hence skin is very important in enabling an animal to live within its environment. The outer layer is dead and is continually being shed and replaced by cells from underneath. In amphibians (frogs, toads, etc.) the skin has to be kept moist and there is some exchange of gases (oxygen in and carbon dioxide out) through the surface.

A cross section of human skin

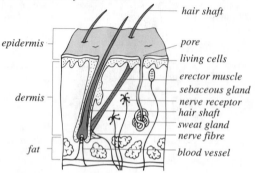

epidermis — hair shaft
pore
living cells
erector muscle
sebaceous gland
dermis — nerve receptor
hair shaft
sweat gland
nerve fibre
fat — blood vessel

sleep a period of deep rest shown by many animals during which the metabolic rate (*see* METABOLISM) is lowered, awareness is reduced and the body is relaxed. Most animals sleep but some need more than others. Adult humans require about eight hours sleep each night. Other animals, (e.g. some grazing mammals), need only a few minutes sleep at any one time. Many animals are easily and rapidly awakened from sleep and this is more likely in those which are hunted as prey. There is a change in the electrical activity of the brain of a person entering into sleep and this can be recorded, producing a trace called an *electroencephalogram* (EEG). *Alpha* brainwaves are produced when an adult person is awake and *delta* waves during sleep. At the onset of sleep, low frequency waves of high amplitude are produced (known as *slow wave sleep*). These are interrupted by short periods of high frequency, low amplitude waves, during which time the person may be restless and show rapid eye movements behind closed eyelids. This is known as Rapid Eye Movement (REM) sleep and it is the time when dreaming occurs. A person who wakens during REM sleep remembers dreams but otherwise these are usually forgotten. The part of the brain especially involved in the control of sleep is called the *reticular formation*.

smell a keen sense of smell or *olfaction* is characteristic of many animals. This is achieved by special receptors (*chemoreceptors*) which are able to detect chemicals carried in the air or dissolved in water. They may be grouped together within *olfactory organs* such as the *nose* of vertebrates. The lining or *epithelium* of the *nasal cavity* contains *olfactory cells* which respond to chemical molecules dissolved in the moisture of the surface mucous membranes. These cells are neurones (*see* NERVOUS SYSTEM) which connect with branches of the *olfactory nerve* within a special area of the brain called the *olfactory bulb*. The olfactory nerve transmits the information to the *cerebral cortex* of the *brain* where it is decoded and detected as a particular smell.

sodium is an alkali metal which occurs widely—its principal source being sodium chloride (salt) in sea water and salt deposits. It is obtained by electrolysis of fused sodium chloride, but does not exist in its elemental form because it is highly reactive. When prepared, the metal is sufficiently soft to be cut with a knife and it is a silvery white colour. However, it reacts violently with water and rapidly with oxygen and the halogens. It is essential to life, particularly in the biological mechanism involved in the transmission of nerve impulses.

It forms numerous compounds with diverse uses and is itself used as a heat-transfer fluid in reactors. Compounds and their uses include: hydroxide, numerous uses; benzoate, antiseptic; carbonate, in glass, soap and other manufacturing processes; chlorate, herbicide; citrate, medicinal; hypochlorite, bleaches; nitrite, in dyes; and many more.

The other alkali metals are lithium (Li), potassium (K), rubidium (Rb), caesium (Cs) and francium (Fr). Francium is radioactive, and caesium is extremely reactive.

sodium hydroxide *see* CAUSTIC SODA.

solar cell is an electric cell that produces electrical energy from light. It is based upon a semiconductor device, called a photoelectric cell, which creates a small current because of the movement of electrons. Solar cells are ideal for use in spacecraft to power electronic equipment, but to produce an appreciable output cells have to be put into panels. In this way they are used to complement domestic heating systems.

solid a state of matter in which the component atoms, ions or molecules maintain a constant position relative to each other. Some solids are *crystalline*, with a regular and repetitive arrangement of atoms, while others are totally disordered (amorphous–no shape). When solids are heated, their atoms absorb the energy and vibrate more. Eventually the energy intake is sufficient to break down the structure, and it melts to form a liquid. A few solids go directly from solid to gas, they *sublime,* e.g. solid carbon dioxide.

solution a mixture of two or more components producing a single, homogeneous phase from which there is no settling out. The term usually applies to a solution of a solid in a liquid, and often the liquid is water (producing an *aqueous* solution). However, it is possible to have solutions of gases in liquids, gases in solids, liquids in liquids, and solids in solids (a solid solution).

The *solute* is the substance that dissolves to make a solution, and the *solvent* is the substance in which the solute dissolves. The solvent is usually a liquid, and water is the commonest. *Solubility* is the concentration of a saturated solution, i.e. the maximum amount of one substance dissolved in another.

A *solid solution* is when two or more elements share a common crystalline framework. The composition can vary, although within limits, and two types of solid solution are found. A substitutional solid solution is when atoms of one element are replaced by another, e.g. nickel and copper; an interstitial solid solution is when small atoms rest in lattice spaces of the structure, as with carbon in metals. Minerals commonly exhibit solid solutions.

sonar *see* ECHO.

sound the effect upon the ear created by air vibrations with a frequency between 20 Hz (hertz) and 20 kKz (20,000 Hz). More generally, sound waves are caused by vibrations through a medium (whether gas, liquid or solid). One of the commonest sources of sound is a loudspeaker. When it produces sound, the cone vibrates, producing a series of compressions in the air. These are called longitudinal progressive waves, that is the oscillations occur in the same direction that the wave is travelling. The sound waves so produced enter the ear causing pressure changes on the ear drum, causing the brain to register the sound. Most items produce sound when they vibrate or are moved or banged together, but sound can only be transmitted through a medium and it cannot travel through a vacuum.

The speed of sound varies with the material it is travelling through, moving most quickly through solids, then liquids and gases. In air, sound travels at approximately 350 m/s. The speed increases with temperature, but is unaffected by pressure. Frequency of sound waves relates to *pitch*; high frequencies produce a sound of high pitch, e.g. a whistle at 10 kHz, while low pitch is caused by low frequencies, e.g. a bass voice at 100 Hz.

Sound intensity is measured in *decibels* (db) and it is a logarithmic scale. Ordinary conversation might register 40–50 decibels, traffic 80 and thunder 100 while jet aircraft can exceed 125 dB.

specific gravity *see* DENSITY.

specific heat or specific heat capacity is the heat required by unit mass to raise its temperature by one degree, that is the gain in thermal energy divided by mass and temperature. The units are joules per kgK. The specific heat capacity varies with the material and metals have a much lower value than, for example, water. This means that much more energy is required to create a 1K (1°C) rise in 1kg of water than in 1kg of copper:

water	4200 joules required per kgK
ice	2100 " " "
aluminium	900 " " "
glass	700 " " "
copper	400 " " "

If the specific heat capacity of a substance is known, and its mass, it is possible to work out the thermal energy (heat) required to give a certain temperature rise.

spectroscopy the study of spectra using *spectroscopes* (this includes spectrometers and spectrographs, etc.). Light emitted from a hot object, or given out by a substance upon excitation can be analysed. The light or radiation passes through an analyser or *monochromator* which usually incorporates a prism to split the light into its components. This enables monochromatic light (light of a specific wavelength) to be studied. An alternative is to use a filter which absorbs unwanted frequencies. The light then passes into a detector, and there are numerous types, some employing a photoelectric cell.

Spectrometry is used to study light from the stars and also extensively in chemistry. There are numerous techniques, some of which are listed below:

type of spectrometry	applications
X-ray fluorescence	determination of most elements in e.g. rock sample
atomic absorption	determination of metals, whether in trace or minor amounts
infrared	identification of organic compounds
mass spectrometry	organic compound identification and structural analysis

Others include spectrometry, ultraviolet, plasma emission and flame photometry.

spore and sporophyte a small reproductive structure, usually consisting of one cell only, which detaches from the parent and is dispersed. If environmental conditions are favourable it grows into a new individual. Spores are commonly produced by fungi and bacteria, but also occur in all groups of green land plants especially *ferns, horsetails* and *mosses*. The sporophyte is the phase in the life cycle of a plant that produces spores. The sporophyte is *diploid* but it produces haploid spores (*see* CELL DIVISION). It may be the dominant stage in the life cycle of a plant (as in the seed plants) or be mainly dependent upon the gametophyte structure for water and nourishment as in mosses (*Bryophyta*). (*See also* GAMETE AND GAMETOPHYTE).

starch is a *polysaccharide* found in all green plants. Poly-saccharides are a large group of natural carbohydrates in which the molecules are made from simple sugars (*monosaccharides*) of the form $C_6H_{12}O_6$ (hexoses) or $C_5H_{10}O_5$ (pentoses). Starch is built up from chains of glucose ($C_6H_{12}O_6$) units arranged in two ways, as amylose (long unbranched chains) and amylopectin (long cross-linked chains). Potato and some cereal starches contain 20–30% amylose and 70–80% amylopectin. Amylose contains 200–1000 glucose units while amylopectin numbers about 20. Starch is formed and broken down in plant cells and is stored as granules and it occurs in seeds. It is insoluble in cold water and is obtained from corn, wheat, potatoes, rice and other cereals by various physical processes. It is used as an adhesive for sizing paper and has many uses in the food industry.

static electricity is electricity or electric charges, at rest. The structure of the atom is visualized as having electrons 'orbiting' a central nucleus and this means that in some materials, electrons can be removed by rubbing with a cloth. When, for example, a Perspex rod is rubbed with a woollen cloth, the cloth pulls electrons away from the rod, becoming negatively charged and leaving the rod positively charged. The reverse happens when the cloth is rubbed on a polythene rod, the rod becoming negatively charged. No charge is created, it is just that charges are separated. Charges on materials can be registered using the gold-leaf electroscope and, as with other phenomena involving charges, unlike charges attract while like charges repel (*see also* VAN DER GRAAFF GENERATOR).

statistics is the part of mathematics that deals with the collection, analysis, interpretation and presentation of quantitative data. It involves processing data with a view to predicting future outcomes, based upon the information available. Probability plays an important role in statistics. The figures can be presented in numerous graphical ways but it is important to select relevant features to illustrate a point, i.e. statistics can be very misleading, depending upon how they are presented. Statistics are used extensively by manufacturers, in medical research, insurance and, of course, in politics.

steel is iron that contains up to 1.5% carbon in the form of *cementite* (Fe_3C). The properties of steel vary with iron content and also depend upon the presence of other metals and the production method. *Alloy steels* contain alloying elements while *austenitic steel* is a solid solution of carbon in a form of iron and is normally stable only at high temperatures but can be produced by rapid cooling. *Stainless steel* is a group of chromium/nickel steels which have a high resistance to corrosion and chemical attack. A high proportion of chromium is necessary (12–25%) to provide the resistance and a low carbon content, typically 0.1%. Stainless steel has many uses: cutlery, equipment in chemical plants; ball bearings and many other items of machinery.

stroboscope is an instrument that is used to view rapidly moving objects. By shining a flashing light onto a revolving or vibrating object, the object can be made to appear stationary providing the frequency of the flashes of light match the revolutions or vibrations of the object in question. This technique is used in engineering to examine, for example, the blades of a propeller, or an engine part. It is also used to set the ignition timing in a car engine.

sulphur is a yellow, non-metallic element that exhibits *allotropy*, i.e. exhibits several physical forms. It is widely distributed in both the free state and in compounds (*see* SULPHUR COMPOUNDS), forming sulphates, sulphides and oxides, amongst others. It is manufactured by heating pyrite or purification of the naturally-occurring material. The primary use of sulphur is in the manufacture of sulphuric acid (H_2SO_4), but it is also used in the preparation of matches, dyes, fireworks, fertilizers, fungicides and the photographic industry.

sulphur compounds are very common, and include sulphates such as gypsum ($CaSO_4.2H_2O$) and anhydrite ($CaSO_4$). Metal sulphides often form minerals e.g. FeS_2 iron pyrite (also chalcopyrite, $CuFeS_2$) from which the elements can be separated. Sulphur forms several oxides including sulphur dioxide (SO_2), one of the primary gases causing acid rain. Sulphuric acid is the commonest product of sulphur and is a very strong and corrosive acid that is used very widely in the manufacture of dyestuffs, explosives and many other products.

superconductor is a material which shows practically no resistance to electric current when maintained at temperatures approaching absolute zero (which is -273°C). Each material has a critical temperature above which its behaviour is normal, resistance decreasing with falling temperature. At the critical temperature, the resistance disappears almost to nothing. If a current is induced in superconducting material by a changing magnetic field, then the current will continue to flow long after its source has been removed. Metals such as aluminium, lead, and tin become superconducting as do some ceramics, and the phenomenon has been applied to electromagnets because large currents can flow without the supply of large amounts of energy. Other potential uses of superconductors include larger and faster computers, and the transmission of electricity without heat loss.

surface tension is the 'tension' created by forces of attraction between molecules in a liquid resulting in an apparent elastic membrane over the surface of the liquid. This attraction between molecules of the same substance is called *cohesion* and the result is that it tries to pull liquids into the smallest possible shapes. This can plainly be seen in water which forms round droplets and also supports the feet of insects on ponds and puddles. The same phenomenon is demonstrated by a needle on a piece of blotting paper which is then placed gently on water. When the paper absorbs sufficient water to sink, the needle remains afloat, because of the surface tension of the water. Droplets of mercury show the same effect, forming compact globules on a surface.

Adhesion is when molecules of two different substances are attracted to each other, as shown by water wetting glass. This attraction is also responsible for the *meniscus* formed where water meets glass. The meniscus is the upward-curving surface of the water upon meeting the glass. Mercury forms a meniscus curving down because its cohesion is greater than its adhesion with glass (*see also* CAPILLARY ACTION).

suspension is a two-phase system with denser particles distributed in a less dense liquid or gas. Settling of the particles is prevented or slowed down by the viscosity of the fluid or impacts between the particles and the molecules of the fluid. Fog is a suspension of liquid particles and smoke a suspension of solid particles.

symbiosis describes various relationships between organisms, usually two different species, which co-exist to the benefit of at least one of the parties involved. In *mutualism* both parties benefit and neither is harmed. In *commensalism* one party benefits and the other is unharmed but in *parasitism*, one organism thrives at the expense of the other (*see also* PARASITE).

symbols in chemistry are used to represent elements, atoms, molecules, etc. Each element has its own symbol of one or two letters (*see* PERIODIC TABLE), thus fluorine is F, and chlorine is Cl. Symbols are used further in formulae, i.e. the shorthand representation of a compound, for example NaCl is sodium chloride. The formulae can then be used in equations to represent chemical reactions and processes, e.g.

$$NaOH + HCl \rightarrow NaCl + H_2O$$

This formulae states that caustic soda and hydrochloric acid, when combined, will react to form sodium chloride and water.

Symbols are used elsewhere, particularly in the sciences, to provide a convenient shorthand, e.g. prefixes in decimal numbers (*see* SI UNITS), concepts in physics such as m for mass and I for electric current and in mathematics where a letter or figure represents a word or sentence.

symmetry is the property of a geometrical figure whose points have corresponding points reflected in a given line (axis of symmetry), point (centre of symmetry) or plane. Symmetry is closely related to balance in nature and many forms exhibit bilateral symmetry, humans included. Symmetry is very evident in crystals that have grown in ideal conditions because then the crystals faces are apparent and most crystals exhibit several symmetrical features (*see also* CRYSTALLOGRAPHY).

synthetic fibres are used widely in producing cloth and for reinforcements in composite materials. They are manufactured from polymers or modified natural materials and the first synthetic fibres, e.g. *rayon*, were made from cellulose. Rayon is produced from wood fibre treated with alkali and carbon disulphide and it is then extruded into a bath of sulphuric acid to harden the fibres. Completely synthetic fibres include: *nylons* which are polyamides, and are used in textiles, insulation and cables; *polyesters*, used in textiles and film and also for reinforcement in boats etc; *acrylics* and *glass fibres* which are incorporated in resins to increase strength.

There are inorganic fibres such as alumina (Al_2O_3) and glass wool many of which are used in insulation and packing. Asbestos was used for the production of fire resistant textiles but it is being replaced because it is hazardous to health.

tannin *see* CAFFEINE AND TANNIN.

taste and taste bud *taste* is one of the senses and is enhanced by that of smell. Hence if a person has a cold, he or she is not able to taste things properly.

The organs of taste are called *taste buds* and are situated on the tongue and sides of the mouth. They consist of groups of cells which are able to detect four different tastes—salt, sweet, bitter and sour. All the flavours which we detect are combinations of these in different proportions and the taste buds for each are grouped together in various parts of the tongue. Substances can only be tasted if they start to dissolve in the saliva of the mouth, hence certain hard materials have no taste at all.

teeth are hard structures used by animals for biting and chewing food and also for attacking, grooming and other activities such as behavioural displays. In the more advanced vertebrate animals the teeth are collected in the *jaws* but fish and amphibians (frogs and toads) have teeth all over the *palate* (roof of the mouth). Teeth evolved from the scales of cartilaginous fish and are adapted according to the lifestyle of the animal. In mammals a tooth consists of a central *pulp cavity* supplied with nerves and blood vessels, surrounded by a layer of *dentine* and an outer thickness of enamel. Enamel is extremely hard to resist wear and decay. The *root* of the tooth is embedded in a socket in the jaw bone. Four different types of teeth are present—molar, premolar, canine and incisor— but the numbers and arrangement differ between animals. This is known as *dentition*. Also, *jaw of carnivore*— carnivores have well-developed canine teeth for biting and the last premolars (upper jaw) and first molars (lower jaw) have cusps with sharp, cutting edges for shearing through flesh. These are called *carnassial teeth*.

A cross section of the human molar tooth

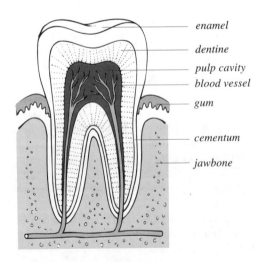

enamel

dentine

pulp cavity

blood vessel

gum

cementum

jawbone

Jaw of herbivore–herbivores have large, ridged premolars and molars for grinding vegetation and incisors for cutting through stems.

Rodent jaw–in rodents the upper and lower jaws contain a single pair of long incisors which grow continuously throughout the animal's life. These are adapted for gnawing and there is an absence of enamel on the back which means that they wear to a chisel-like cutting edge.

telegraph an early device for transmitting messages first developed in the late 1700s. By the mid 19th century the technique had been refined so that information was sent along a wire as electrical pulses. Morse introduced his code in which each letter of the alphabet was represented by a different set of pulses. Telegraph messages were also used to control train movements and eventually the Post Office formed a network. The telegraph gradually faded into history when it became possible to use the telephone on an international basis.

telephone is an instrument that enables speech to be transmitted by means of electric currents or radio waves. It was invented by Alexander Graham Bell in 1876 and a public service was begun three years later after Bell brought his invention to the UK. The modern telephone consists of a mouthpiece containing a thin diaphragm of aluminium which moves with the sound of speech. This movement presses carbon granules which produces a surge of current and in the earpiece of the receiving set, these surges are changed back into sound. An electromagnet reacts to the current charges and vibrates a diaphragm, thus reproducing the voice of the speaker.

Advances in technology have improved the transmission of telephone calls and satellites are used for international calls and recently optical fibres have been introduced as a new medium of transmission.

television is the transmission of moving image by electrical means and the television set is essentially a complex cathode ray tube. The tube consists of an electron gun and a screen that fluoresces and the beam of electrons is deflected by magnetic coils. On a black and white screen, the picture is composed of lines as the spot moves across the screen and the spot is also moved down the screen, although at a slower rate. The result is that 25 images per second are shown on the screen, which to the human eye appears to be a moving picture.

Producing colour pictures is much more complex. There are three electron guns for red, green and blue and the screen is coated with thousands of tiny strips of the same colours which, when struck by electrons, glow in combination to produce a colour picture. Accurate targetting of the strips is achieved by a *shadow mask* through which the electron beams are fired.

The pictures and sounds are converted by cameras and microphones into electrical signals which are sent to transmitting aerials. Receiving aerials then pick up the signal, pass them to the television set where they form pictures and sound.

temperature is a measure of an object's overall kinetic energy. When an object is cold, its molecules, atoms or ions have less kinetic energy–its temperature is lower. Although molecules do not stop moving altogether, at–273°C they possess the minimum possible energy. This is *absolute zero* and this temperature cannot be exceeded. The temperature scale which has -273°C (actually -273.15°C) as its zero is called the *Kelvin* scale. The degrees are the same size as centigrade degrees, but the temperatures are stated as figures without the degree. In addition to the centigrade (or *Celsius*) scale which has water freezing at 0° and boiling at 100°, there is the *Fahrenheit* scale where the respective temperatures are 32°F and 212°F. Fahrenheit can be converted to Celsius using the equation:

$$F = 1.8C + 32 \quad \text{and the reverse is} \quad C = \frac{5(F - 32)}{9}$$

tempering is the process whereby (usually) steel is heated to a particular temperature and then cooled quickly in oil or water. The heating permits stresses in the metal to be relieved and results in a toughened, less brittle material.

tensile strength of a material is the force that is required to stretch it until it breaks. It is measured using a special machine upon which the 'breaking' force can be read from a dial. Since tensile strength is the force per unit area, the breaking force is divided by the material's cross-sectional area. The units are newtons per square metre.

thermocouple is a device for measuring temperatures and consists of two different types of metallic wire joined at both ends. The temperature is measured at one join and the other join is kept at a fixed temperature. A temperature difference between the two joins causes the metals to produce a small electric current which can be metered. Thermocouples are used in furnaces because they have a range up to about 1600°C. The two metals used are often copper and *constantan*, the latter being a copper/nickel alloy which has a constant resistance irrespective of the temperature.

thermodynamics is the study of laws affecting processes that involve heat changes and energy transfer. Heat transfer from one body to another, the link between heat and work and changes of state in a fluid all come within the field of thermodynamics, it is the prerequisite to analysis of work by machinery. There are essentially three *laws of thermodynamics*. The first law says that heat is a form of energy and is conserved and any work energy produced in a closed system must arise from the conversion of existing energy, i.e. energy cannot be created or destroyed. The second law states that the entropy of any closed system cannot decrease and if the system undergoes a reversible process it remains constant, otherwise it increases. The result of this is that heat always flows from a hot body to a cooler one. The third law states that absolute zero (*see* TEMPERATURE) can never be attained.

thermometer an instrument used to measure temperature. The basis of a thermometer is a property of a substance that varies reliably with temperature, e.g. expansion. Thermometers that utilise a liquid in glass are based upon the property that liquids expand slightly when they are heated. Both mercury and alcohol are used and when the bulb at the base of the thermometer's stem is heated, the liquid expands up the stem to create a reading. More sensitivity is gained by using a narrower tube. This is the case with *clinical thermometers* where the scale covers just a few degrees on either side of the normal body temperature of 37°C.

In industry, thermocouples are used for temperature measurement in furnaces and other instruments that provide an electrical measurement are used in preference to liquid in glass thermometers which have a limited range. *Resistance thermometers* are based on the property that the electrical resistance of a conductor normally increases with heat and so it becomes more difficult to pass an electric current. A spiral of platinum wires is used in this case. A *thermistor thermometer* works on the same principle, but consists of a semiconductor in which the resistance decreases with a temperature increase, e.g. 100,000 ohms at 20°C and just 10 ohms at 100°C.

thermostat *see* BIMETAL STRIP.

tissue *see* ORGAN.

titanium is a malleable and ductile silvery-white metal that melts at 1660°C. It occurs as the minerals rutile (TiO_2) and ilmenite ($FeTiO_3$) and is obtained by reducing titanium chloride ($TiCl_4$) with magnesium. It is characterized by its lightness, strength and high resistance to corrosion and is used in alloy form (with aluminium, manganese, chromium and iron) in the aircraft industry. It is also used in missile manufacture, engines and chemical plant.

Titanium compounds have many important uses. Titanium carbide (TiC) is very resistant to chemical attack and is used in tool tips. Titanium dioxide (TiO_2) is used as a white pigment in the production of paints and printing inks and as a filler in paper, rubber, fabrics and plastics.

touch the sense of touch is made possible by specialized sets of receptors which are located in the skin and also in muscles and other internal areas of an animal's body. Touch has different elements including pressure, pain and temperature. The receptor cells which detect pain and pressure tend to be concentrated in certain areas (e.g. fingertips) and distributed less thickly elsewhere. Sensory receptor cells are associated with the hairs covering an animal's body so that the slightest movement of air can be detected. The sense organs involved in touch are specially adapted to respond to a particular sensation. Different nerve pathways are used to transmit the information to the brain where it is decoded and detected.

trace elements is a term used both in biology and geology. In biology it refers to an element that an organism requires in very small quantities. These elements may be necessary for the formation or action of vitamins, enzymes and hormones.

In geology, trace elements occur in small quantities in rocks but can be detected by geochemical analysis. Most elements except the few commonly occurring ones are present as trace elements, if at all, and in quantities of 1% or less. Often the concentrations are just a few parts per million.

transformer is a device for changing the voltage of an alternating current. It is based upon the principle of mutual induction whereby an alternating current passing through a coil (the primary) on a soft iron core induces current flow in another (secondary) coil on the core. The primary and secondary coils form a transformer. Transformers can be made to step-up or step-down a voltage by varying the number of turns in the coils. For example, if the primary coil has ten times the number of turns compared to the secondary, then the voltage in the secondary is one tenth that of the supply, and the associated current increases tenfold.

Transformers have many uses but practical versions are not 100% efficient and some energy is lost as heat. Huge transformers are used in the mains power supply between the power station and the domestic supply. Current generated at a power station goes through a step-up transformer, creating voltages of up to 400,000 volts, at much reduced currents (thus minimizing heat loss) for transmission through the power lines of the grid. Power from the *grid* then goes to substations where transformers step the voltage down by a series of transformers to 132,000 volts, then 33,000v (for heavy industry), then 11,000v for light industry and finally 240v for offices and homes.

transistor is a semiconductor that can be used for three functions: as a switch, a rectifier and as an amplifier. A transistor consists essentially of a semi-conductor chip of silicon which is doped (*see* DIODE *and* SEMICONDUCTOR) to form two p-n junction diodes back to back (p-type diode is silicon doped with boron; n-type is doped with phosphorus). Current cannot flow through a transistor unless a small current is applied to the p-type region of the semiconductor called the *base* circuit, but when this current is applied an enlarged current flows in the output or *collector* circuit. A transistor can be used to amplify current changes and practical amplifiers contain several transistors, as used in radio, to increase currents to output a signal through the loudspeaker.

Because a current must flow in the base circuit to allow current to flow in the collector circuit, a transistor can be used as a switch and is turned on and off by a change in the base current.

trigonometry is the branch of mathematics that involves the study of right-angled triangles including problem-solving involving the calculation of unknown sides and angles from known values. It involves the use of the *trigonometrical ratios*, sine, cosine and tangent. In the right-angled triangle below the ratios are:

$$\sin \theta = \frac{\text{opposite}}{\text{hypotenuse}} = \frac{yz}{xy}$$

$$\tan \theta = \frac{\text{opposite}}{\text{adjacent}} = \frac{yz}{xz}$$

$$\cos \theta = \frac{\text{adjacent}}{\text{hypotenuse}} = \frac{xz}{xy}$$

Trigonometry is used in surveying and navigation.

ultrasound (or ultrasonic waves) is sound with a frequency beyond the range of human hearing, i.e. around 20,000 Hz. Ultrasound is used extensively in industry and medicine. It is used to detect faults and cracks in metals and to test pipes; it can clean surfaces due to the rapid, small vibrations; in ultrasonic welding, soldering and machining; and in medicine ultrasound is used to scan a growing foetus and also to destroy kidney stones or gall stones. A recent development is the use of ultrasound in chemical processes, to initiate reactions in the production of food, plastics and antibiotics. Ultrasound can make chemical processes safer and cheaper as it eliminates the need for high operating temperatures and expensive catalysts.

vaccination *see* IMMUNE SYSTEM.

vacuum is defined as a space in which there is no matter. In practice a perfect vacuum cannot be achieved although interstellar space comes very close indeed. Also, special equipment in the laboratory can reach very low pressures, but in general vacuum is taken to be air or gas at very low pressure, 10^{-4} mm Hg or lower.

The *vacuum flask* uses a vacuum to help keep liquids or gases cold or hot. It was invented by James Dewar at the end of the 19th century and it consists of a double-walled glass bottle. A vacuum is created between the glass walls, and the surfaces are silvered so that together, transfer of heat by convection and radiation is reduced to a minimum.

valency is the bonding potential or combining power of an atom or group, measured by the number of hydrogen ions (H^+, or equivalent) that the atom could combine with or replace. In an ionic compound, the charge on each ion represents the valency e.g. in NaCl, both Na^+ and Cl^- have a valency of one. In covalent compounds (*see* BONDS), the valency is represented by the number of bonds formed. In carbon dioxide, CO_2, carbon has a valency of 4 and oxygen 2. The *valency electrons* of an atom are those in the outermost shell that are involved in forming bonds and are shared, lost or gained when a compound or ion is formed, i.e. they determine the chemical reactivity.

The *electronic theory of valency* explains bonds through the assumption that specific arrangements of outer electrons in atoms (outer shells of eight electrons) confers stability (as with the inert gases, which have such a structure) through the transfer or sharing of electrons. Thus with the combination of sodium with chlorine, sodium has one electron in the outer shell, which it loses (to form the Neon stable structure) to chlorine, giving it also a stable structure, in this case, that of Argon.

Van der Graaff generator is a machine that provides a continuous supply of electrostatic charge (*see* STATIC ELECTRICITY) and which can build up a very high voltage. It consists essentially of a hollow metal sphere supported on an insulating tube. A motor-driven belt of rubber or silk carries charge (positive or negative) from the driving roller up the moving belt to the sphere, where the charge collects. Very high voltages (up to 13 million volts) can be produced and in conjunction with high voltage X-ray tubes and other equipment, these voltages are used in research to split atoms. The apparatus was named after the American physicist R.J. Van der Graaff.

velocity is the rate of change of an object's position, i.e. the speed at which an object travels. Velocity provides both the magnitude (size) and direction of travel (per unit time) and and such is called a *vector quantity*. The units are metres per second (ms^{-1} or m/s). On plotting a graph of distance moved, against time, a straight line would represent an object moving at constant velocity. However, it is usually the case that velocity changes with time, in which case an object is said to be accelerating (or decelerating). *Acceleration* is the gain in velocity of an object divided by the time taken to achieve the gain and the units are metres per second, per second. As with velocity, it is a vector quantity. Graphical plots of velocity against time are useful because the gradient of the line plotted is equal numerically to the acceleration and the area under the graph represents the distance moved.

The *terminal velocity* of an object falling through air, for example, is the constant velocity reached when the object's weight is matched by the air resistance.

vertebrate any animal with a backbone including fish, amphibians, reptiles, birds and mammals.

virus the smallest kind of micro-organism which is completely parasitic (*see* PARASITE) and exists and reproduces within the cells of a host organism. Most, but not all viruses cause diseases in plants, animals and even bacterial cells. Viruses operate by invading and taking over the metabolism of the cells which they inhabit and using the metabolic processes to reproduce. Diseases caused by viruses include influenza, herpes, AIDS, mumps, chicken pox and polio, and possibly some cancers.

viscosity is a property of fluids that indicates their resistance to flow. Oil is more viscous than water and an object will take much longer when falling through oil than when it falls through water. All fluids show this resistance to shear forces but a perfect fluid would be non-viscous. Viscosity measurements can be used to find the molecular weight of polymers.

vitamins a group of organic substances that are required in very small amounts in the human diet to maintain good health. A lack of a particular vitamin results in a *deficiency* disease. There are two groups of vitamins; those which are fat-soluble including A,D, E and K and those which are water-soluble, C (ascorbic acid) and B (Thiamine).

The six vitamin groups are as follows;

A or Retinol

Source; green vegetables, dairy produce, liver, fish oils

Needed for; the manufacture of rhodopsin–a pigment needed by the rod cells of the eye for night vision. Also for the maintenance of the skin and tissues

Deficiency; night blindness and possible total blindness

B complex including Thiamine, Riboflavin, Nicotinic acid, Pantothenic acid, Biotin, Folic acid, B_6, Pyroxidine, B_{12} (cyanocobalamin), Lipoic acid

Source; green vegetables, dairy produce, cereal, grains, eggs, liver, meat, nuts, potatoes, fish

Needed for; the manufacture of red blood cells, for enzyme activity and for amino acid production. Also for maintaining the fatty sheath (myelin) around nerves

Deficiency; Beri beri (B_1 deficiency) anaemia and deterioration of the nervous system (B_{12} deficiency)

C (ascorbic acid)

Source; citrus fruit, green vegetables.

Needed for; maintaining cell walls and connective tissue. Aiding the absorption of iron by the body

Deficiency; scurvy–affects skin, blood vessels and tendons

D

Source; fish oils, eggs, dairy produce

Needed for; controls calcium levels required for bone growth and repair.

Deficiency; Rickets in children–deformation of the bones. Osteomalacia in adults–softening of bones

E

Source; cereal grains, eggs, green vegetables

Needed for; maintenance of cell membranes

Deficiency; unusual as common in the diet

K

Source; leafy green vegetables, especially spinach, liver

Needed for; clotting of blood

Deficiency; rare as it is also manufactured by bacteria in the gut

volt (V) is the SI unit of potential difference. It is defined formally as the difference of potential between two points on a conducting wire that is carrying a current of one ampere when the power given out between these points is one watt. The voltage across battery terminals is a measure of the potential energy given to each coulomb of charge and a potential difference of one volt exists if each coulomb has one joule of potential energy, so there is a potential difference of 6 volts if each coulomb is given 6 joules of potential energy.

water is the normal oxide of hydrogen, H_2O, and is found in one form or another in most places on Earth. It can occur in solid, liquid or gas phases, forms a large part of the Earth's surface, and is vital to life. Natural water is never absolutely pure, but contains dissolved salts, organic material, etc. Pure water freezes at 0°C and boils at 100°C and has its maximum density of $1gm/cm^3$ at 4°C. It occurs in all living organisms and has a remarkable combination of properties in its solvent capacity (i.e. its ability to dissolve so many substances), chemical stability, thermal properties and abundance. Water has an almost unlimited range of uses and can provide power through hydroelectric schemes.

watt *see* POWER.

wave is a periodic displacement that repeats itself and a mechanism of transferring energy through a medium. In the simple waveform shown the *amplitude* is the maximum distance moved by a point from its rest position as the wave passes. The *wavelength* is the length of one complete wave.

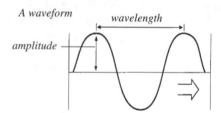

A waveform

The origin of a wave is vibrating particles which store and release energy while their average position remains constant, as it is only the wave that travels. Waves are either *longitudinal* with oscillations occurring in the direction of wave travel (e.g. sound), or *transverse* because the oscillations are from side to side, at 90_ to the direction of travel (e.g. light).

The *wave equation* relates frequency (f) and wavelength (λ):

$$c = f \lambda \quad \text{where c is the speed of the wave}$$

All forms of wave show the properties of reflection, refraction, interference and diffraction. Electromagnetic waves have these properties but have one additional feature in that they can travel through a vacuum.

weight is the gravitational force of attraction exerted on an object by the Earth and although such an attraction exists between all objects, it is so infinitesimally small as to be non-existent and the Earth is so massive that its attraction dominates. Weight is a force, measured in newtons (N), and is the force that makes an object accelerate downwards when falling to the Earth's surface. The weight of an object is therefore dependent upon its distance from the Earth, and the further away it is, the smaller the pull of the Earth and the less is the object's weight.

Mass is a measure of the quantity of matter that a substance possesses and it remains constant, but the weight of an object varies. A constant mass will therefore have different weights depending upon the gravitational effect of nearby bodies. The weight of a constant mass on Earth will be approximately six times greater than its weight on the Moon and in space the weight of that same mass would be zero. Because weight on Earth is directly proportional to mass; any instrument that measures one property also measures the other.

work is done when something is moved by a force. It is also a transfer of energy, i.e. energy is changed into a different form when work is done. Work is measured in the unit of energy, the joule and is calculated as the force, multiplied by the distance moved in the direction of the force. One joule of work is done when a force of one newton is moved through one metre, *in the direction of the force*.

X-rays are electromagnetic waves with a short wavelength (approximately 10^{-10}–10^{-8}m) and are produced when electrons moving at high speed strike a target, are stopped very quickly and X-rays are emitted. They were discovered in 1895 by Röntgen. He found that wrapped photographic plates left near a working cathode ray tube became fogged as if they had been exposed to light and he called this unknown radiation X-rays. In fact X-rays were being emitted as electrons hit the anode and walls of the cathode ray tube.

Atoms of all elements give out characteristic X-rays when hit by electrons. The stream of electrons colliding with the atom displaces electrons from inner orbitals and vacant places are then filled by electrons from the outer orbital, which give out energy as they move down. X-rays have the properties of electromagnetic radiation and also penetrate solid matter, cause ionization (by removal of electrons from atoms), make some materials fluoresce and, as mentioned, they affect photographic film. These properties render X-rays both useful and hazardous. Their ionization effect damages living tissue, but by using very small doses, they can be used in medicine to take X-ray photographs of the body. The extent to which the rays are absorbed depends upon the density and the atomic weight (or relative atomic mass, *see* ATOMIC NUMBER) of the material; the lower these factors, the more easily will the rays penetrate. The greater density of bone means it is possible to take an X-ray photograph because the flesh appears transparent while the bones are opaque.

X-rays are also used in industry for checking joints in metal and examining flaws. They are also used in *X-ray diffraction* (or X-ray crystallography) which is an analytical tool in geology, crystallography, and biophysics. X-rays directed at the sample are diffracted off the planes of atoms in the crystal. By repeating the procedure and then calculating the spacing between atomic planes, a representation of the crystal's structure can be determined. In this way the structure of some proteins and nucleic acids has been analysed.

yeast *see* FERMENTATION.

zoology the scientific study of animals.

zygote *see* REPRODUCTION.

COMPUTING TERMS

At school or in the home many of us use computers on a daily basis, and they play an increasingly important part in all our lives. This section provides explanations of current terms used in computing and information technology. The entries are in A to Z order, and cross references are given in SMALL CAPITALS.

✳

abort to cancel or terminate a process or procedure while it is in progress.

accelerator board an adapter for the computer containing a more advanced microprocessor than the one already in the computer. Adding an accelerator can speed up the computer generally or can speed up a particular function such as the graphics display.

access to locate and retrieve the information, whether in the form of data or PROGRAM instructions, stored on a disk or in a computer. Nowadays the term usually refers to the amount of time it takes to transfer information from one source to another, and is called the **access time**. It is measured in nanoseconds (ns) for memory chips and in milliseconds (ms) for data transfer from the hard disk. Typical access times for hard disks on personal computers range between the fast (9ms) and the slow at around 100ms. The access time is determined by the time required for the disk heads to move to the correct track (the seek time), and then to settle down (the settle time) and the time needed for the sector to move under the head (latency).

acoustic coupler cradles that hold a telephone handset and allow MODEMS to communicate through the telephone.

active cell the cell in a SPREADSHEET in which the cursor is currently positioned, allowing a number or formula to be entered.

active matrix display a liquid crystal display (LCD) in which each of the display's electrodes is under the control of it own transistor. Active matrix displays are more expensive than the lower resolution PASSIVE MATRIX DISPLAYS.

active window in an APPLICATION PROGRAM or OPERATING SYSTEM that can display multiple windows, the active window is the one that accepts commands or typing and is indicated by a coloured bar at the top of the window.

The active window has a shaded title bar

Ada a programming language designed for US Government operations. It is a high-level structured language developed around 1980 that is designed to be readable for ease of maintenance.

adapter card a card complete with electronic components that plugs into the computer's main circuit board with the aim of providing enhanced capabilities. Adapter cards can be used to provide high quality graphics, modems, etc.

ADb *see* APPLE DESKTOP BUS.

add in program a small program that is designed to complement an application and add to the capabilities of the host application.

address a location in a computer system that is identified by a name or code. The address is specified by the program or the user.

address bus an electronic channel linking the microprocessor to the random access memory (*see* RAM) along which the addresses of memory locations are transmitted.

Adobe Type Manager (ATM) a program that smoothes out the edges of type for presentation on the screen. This is especially useful where the printer prints exactly what is on the screen. If it looks good on the screen it will look good in print.

AI *see* ARTIFICIAL INTELLIGENCE.

AIX (Advanced Interactive eXecutive) a variant of the UNIX operating system developed by IBM primarily for their workstations.

alert box a WINDOW appearing on a screen providing a warning that an error has occurred or the command chosen may result in lost work.

Alert box associated with an erase disk command

algorithm a set of straightforward logical mathematical steps that, if followed, provide a solution to a problem.

alias a representation of an original document or file that can be used as if it was the original. An alias's name appears in italics. An alias is not a copy of the file but a small file that directs the computer to the original file. Using an alias can assist in file storage and retrieval.

aliasing *see* JAGGIES.

alignment *see* READ/WRITE HEAD ALIGNMENT; TEXT ALIGNMENT.

alphanumeric character any keyboard character that can be typed, such as A to Z in either uppercase or lowercase, numbers 1 to 10 and punctuation marks.

Amstrad a computer company founded by Alan Sugar. It made its name with a word processing computer that sold well, mainly because of its low price. However, it was rapidly superseded by generations of PERSONAL COMPUTERS.

analogue the opposite of DIGITAL. An analogue signal varies continuously to reflect the changes in the state of the quantity being measured, e.g. sound waves vary continually. A thermometer is an analogue measurement device, and as the temperature varies so does the height of the mercury. On the other hand, digital signals are either on or off.

AND *see* FORMULA; LOGICAL OPERATOR.

animation a graphic creation that gives the impression of movement by showing a series of slides on the computer screen. Each slide is slightly different from the previous slide and when the images are played back fast enough, the movement appears smooth.

ANSI (American National Standards Institute) a non-governmental organization founded in 1981 to approve the specification of data processing standards in the USA. It is also responsible for the definition of HIGH-LEVEL PROGRAMMING LANGUAGES.

anti-virus program a program that checks for the existence of a VIRUS and informs the user if there is one present on the computer's secondary storage. The virus is identified either by the existence of the virus codes or by identifying the effect of the virus (e.g. corrupted files).

APL (A Program Language) a HIGH-LEVEL PROGRAMMING LANGUAGE designed for handling engineering and mathematical functions and as a notation for communication between mathematicians. It is quick, efficient and well suited to WHAT IF applications.

Apple Computer, Inc. a pioneering and innovative company in the information industry. It creates solutions based on easy to use PERSONAL COMPUTERS, SERVERS, PERIPHERALS, SOFTWARE and PERSONAL DIGITAL ASSISTANTS. Based in California, Apple develops, manufactures, licenses and markets products, technologies and services for education, business, consumers and for industrial scientific, and government organizations in over 140 countries. It is one of the largest manufacturers of personal computers in the world. The brand name of its range of computers is MACINTOSH.

Apple desktop bus (ADb) a standard INTERFACE for Apple computers that allows connection of input devices such as keyboards, mice, graphics tablets and trackballs to all Macintosh computers. Up to 16 ADb devices can be connected to the computer, which can receive data at approximately 4.5 kilobits (i.e. 4500 bits) per second.

AppleShare a NETWORK OPERATING SYSTEM that converts an Apple Macintosh computer into a file SERVER for a network. The server can be accessed by the network as if it were an added HARD DISK.

AppleTalk a program that connects computers and other hardware such as printers together in a LOCAL AREA NETWORK. Every Apple Computer has Appletalk network facilities built in.

application program a computer program that performs specific tasks such as letter writing, statistical analysis, design, etc. Application programs work with the OPERATING SYSTEM to control peripheral devices such as printers.

archive a store of files (either PROGRAM or DATA) that is kept as a backup in case the original files are corrupted or damaged. An archived file is often stored in a compressed format (*see* COMPRESS) in order to use up less disk space.

arithmetic operator a symbol that indicates which arithmetical operation to perform. In a SPREADSHEET, arithmetic operators are used to compile formulae for adding (+), subtracting (−), multiplying (*) or dividing (/) the contents of cells to obtain the desired result.

array a form in which data is stored by computer programs. An array is a table of data. The data is accessed by naming the array and then by the X and Y coordinates.

artificial intelligence (AI) the ability of an artificial mechanism to exhibit intelligent behaviour by modifying its actions through reasoning and learning from experience. Artificial intelligence is also the name of the research discipline in which artificial mechanisms that exhibit intelligence are developed and studied. It was first discussed by the British mathematician Alan Turing, who is regarded as one of the founders of the subject. In recent years artificial intelligence has been used in the development of expert systems that use knowledge-based information to make seemingly rational decisions. However, this rationale is limited to a specific knowledge area. It is generally accepted that AI has not met its objectives and the necessary solutions are probably many, many years away.

ascending order a method of sorting data in a list with the result that the data is arranged from 1 to

10 or A to Z. Descending order reverses the sort order.

1	.	A	*Each column is*
2	/	a	*sorted in*
3	r	B	*ascending order*
6	t	b	
8	u	C	
15	w	c	
22	y	D	
44	^	d	
66	2	E	
75	5	e	

ASCII (American Standard Code for Information Interchange) one of several standard sets of codes devised in 1968 that define the way information is transferred from one computer to another. It is one method of representing BINARY code, whereby specific binary patterns are represented by alphanumeric codes. The standard ASCII code contains 96 upper and lower case characters and 32 control characters that are not displayed.

assembly language a LOW-LEVEL PROGRAMMING LANGUAGE that is based on instructions that relate directly to the processing chip. (*See also* HIGH-LEVEL PROGRAMMING LANGUAGE.)

assistant a series of steps in an application program, similar to a WIZARD, that assists the user to create a particular document style such as mailing lists, advertising flyers, newsletters, envelopes, etc.

asynchronous communication a commonly used mode of transmitting data over telephone lines. The two communicating systems are not synchronized and characters are sent or received one at a time. The start and ending of a stream of data are denoted by a START BIT and a stop bit.

ATM *see* ADOBE TYPE MANAGER.

Auto CAD a widely used but very expensive CAD program.

auto dial a feature of most communications programs in which the program automatically dials the appropriate phone number and makes a connection with the answering computer.

auto dial/auto answer modem a MODEM that is able to generate tones to dial the receiving computer and also to answer the telephone and establish a link when a call is received.

auxiliary storage another term for SECONDARY STORAGE.

auto save a utility that regularly saves the work being done on a computer onto the hard disk. It is important to save work regularly otherwise there is a risk of loss of work because of, for example, a power or computer failure. The auto save utility sets the computer to save at specific intervals, allowing the user to progress with productive work.

BABT (British Approvals Board for Telecommunications) an organization that approves all devices for use with the public telephone network, e.g. MODEMS, FAX cards and SERVERS (not ACOUSTIC COUPLERS).

background task some computer operating systems allow more than one task to be completed at a time. The main or high priority procedure is carried out in the FOREGROUND and lower priority procedures are carried out in the background.

backlit screen a type of display mostly used in NOTEBOOK or LAPTOP COMPUTERS. The display uses LCD technology to light the display behind the text. The text is contrasted against the background.

backspace a keyboard key that moves the CURSOR to the left, deleting any previously typed characters.

backup a copy of a program or data file that is kept for ARCHIVE purposes usually on a removable floppy or hard disk.

backup utility an easy to use program that automatically makes a copy of the main storage disk of a computer. The copy or backup is usually made onto tape, optical disk or floppy disks. The better backup utility programs copy only the files that have been updated since the previous backup, thus saving time and disk space.

backward compatible an application program that works not only with the most recent version of an operating system but also with previous versions.

bad sector a part of a disk (HARD DISK or FLOPPY DISK) that cannot be used to record data. Bad sectors can be generated in manufacturing, caused by damage in handling the disk or by dust. When a bad sector is identified, the disk should be discarded if it is a floppy or reformatted if it is a hard disk.

bandwidth a measure of the range of frequencies that can pass through an INTEGRATED CIRCUIT. The greater the bandwidth the more information that can pass through the circuit.

bar code a series of printed vertical bars of differing widths that represent numbers. These codes are printed on virtually every supermarket product

A representation of a bar code

and are then used by appropriate software to identify the product for stock control and sales pricing.
Bar codes are also found on books and other publications and many other items.

BASIC (Beginner's All-purpose Symbolic Instruction Code) one of the most popular computer programming languages. It was developed at
Dartmouth College in the USA in 1964 by John
Kemeny and Thomas Kurtz. BASIC is available
on a wide range of computer platforms and is one
of the easiest languages in which to program. It is
normally an interpretative language, with each
statement being interpreted and executed as it is
encountered. It can also be a compiled language,
with all the statements being compiled into machine code before execution. Early versions of
BASIC were criticized for not encouraging structured programming. New versions of BASIC, i.e.
Visual Basic, are very powerful indeed.

BAT the common extension used at the end of a
BATCH FILE name.

batch file a file containing DOS commands that is
accessed by typing the file name at the DOS
prompt. This file type has the extension .BAT.

batch processing a type of computer operation that
processes a series or batch of commands at one
time without user intervention. A BATCH FILE, for
example, processes a group of DOS commands on
start-up to assist the user.

baud a measure of telecommunications transmission speed denoting the number of discrete signal
elements that can be transmitted per second. Devised by a 19th-century French telecommunications pioneer, J.M. Baudot, the **baud rate** is the
standard way of representing information in telex
and telegraph communications. It commonly refers to the changes in signal frequency per second,
and not BITS per second unless at low baud rates
(300) where it is equal to bits per second. At
higher rates, the number of bits per second transmitted is higher than the baud rate, and one change
in the electrical state of the circuit represents more
than one bit of data. This means that 1200 bits per
second can be sent at 600 baud.

bells and whistles the advanced features that an application program contains. Most users find that
they regularly use only a small percentage of the
features of a full-featured program, whether it is a
word processor, a SPREADSHEET or a graphics program. This is an argument for buying a low-fea-
tured program, which is usually substantially
cheaper in price and yet meets most requirements.

benchmark a measurement standard used to compare the performance of different computers and
equipment. Standard measures of processor
speeds do not take account of the speed of other
devices such as disk drives or communications,
whereas modern benchmark tests take all factors
of the computer system into account.

Bernoulli a type of DISK DRIVE, usually holding over
20 megabytes of data, that uses removable cartridges.

beta testing the final stage of testing of a computer
program before it is released for general sale. The
testing is usually performed by a number of selected users who have knowledge of, and skill in,
using the particular type of application.

Bezier curve a style of curve that depends on vector
forces of power and angle to determine its shape.
In computer graphics these curves are manipulated by control handles at the midpoints of a
curve.

bidirectional a term to describe a device, usually a
printer, that is capable of printing from right to left
as well as from left to right. There are also bidirectional printer ports, capable of sending and receiving data. SERIAL cables are bidirectional, as they
are capable of transmitting and receiving information.

Big Blue the colloquial name given to IBM (International Business Machines), one of the largest
computer companies in the world.

binary the language of all computers in which all
numbers, letters and special characters are represented by 1 and 0. It is called base two notation.
(*See* BINARY NUMBERS).

binary digit *see* BIT.

binary numbers the use of a base notation of two
compared with a base notation of 10 in normal
decimal numbers and 16 in a HEXADECIMAL numbering system. In base two, there are only two
states, 0 or 1 ('on' or 'off').

BIOS (Basic Input Output System) a code that re-
sides on a ROM chip and controls basic hardware
operations such as the PROCESSORS dealings with
disk drives, displays, and keyboard.

bit (Binary digIT) the smallest unit of information
in a digital computer. It has a value of 0 or 1 that
represents yes/no and either/or choices. A collection of eight bits is called a BYTE.

bitmap a method of storing graphics information in
memory in which a BIT is devoted to each PIXEL
(picture element) on screen. A bitmap contains a
bit for each point or dot on a video display, and
can allow for fine RESOLUTION because each point
on the display can be addressed. Unfortunately,
bit-mapped graphics require a great deal of
memory, often in excess of one megabyte. A BIT-

MAPPED font is a font for screen and/or printer in which every character is made up of a pattern of dots. To allow display or printing, a full record must be kept in the computer's memory. Again, this is a memory-intensive process.

bits per second (bps) a measure of speed of data transmission, especially in connection with the performance of MODEMS.

board *see* CARD.

block a selection of information that can be dealt with by one series of commands. For example, in SPREADSHEET applications a block of information can be selected and copied to another part of the sheet.

boilerplate a standard passage of text used in memos, reports or letters. The passage is often stored in a SCRAPBOOK and pasted into a document as required. Letterheads are examples of a boilerplate that can easily be retrieved when required. Contracts of employment can also be saved in a standard format and recalled as required. Only minor amendments then need to be made to the document before printing.

Fred's Dictionary Corner	Fax Transmission
	❏ Please call to confirm receipt
Date: Mar 23, 1996 To: Fax number:	❏ Please respond by return fax
From: Fred Smith Our phone: (1555) 666-0002	❏ Call only if transmission is incomplete
Our fax: (1555) 666-0001 No. of pages including cover page: 1	

An example of a boilerplate

bold a style of text that adds emphasis to a character by making it darker and heavier than normal type. The headwords in this book are printed in **bold type**.

bomb an unexpected termination of an application, similar to a CRASH. This can indicate a serious problem with the hardware, but most often it is caused by a software conflict.

bootstrapping a program routine designed to make a computer ready for use. To 'boot the system' or INITIALIZE it is to make it ready for normal use. In modern computers booting the system loads its OPERATING SYSTEM into memory and prepares it to run application programs. Within the ROM of a computer is a program that is started when the power is switched on, and this tells the computer to search for the computer's operating system. This is called a **cold boot**. A WARM BOOT is achieved either by pressing the reset button on newer IBM computers and compatibles or by pressing the Ctrl/Alt/Del keys simultaneously. This is usually done to unlock the system.

bps *see* BITS PER SECOND.

branch a section of a program that causes the program to divert to a subroutine when certain criteria are met. On completion of the subroutine the program returns to the main trunk. The term can also apply to a section of a DIRECTORY.

broad band an ANALOGUE communications method using high BANDWIDTH. Broad band communications operate at high speeds and can be sent over long distances.

browse to display records in a format suitable for quick on-screen review and editing.

bubblejet *see* INKJET.

buffer an electronic memory storage device that is used for temporary storage of data passing in or out of the computer. A common use for a buffer is as a temporary holding area between a slow-moving device and a fast-moving device, i.e. between a computer and a printer. A buffer therefore allows different parts of the computer system to operate at optimum speeds.

bug a mistake—a hardware or software error. The term was first coined in the early days of computing when a butterfly was found to have been the cause of a malfunction in an early Mark I computer. A programming error may be very serious and can cause the computer to perform incorrectly or even crash.

bulletin board a computer service set up by organizations or clubs with the aim of providing or exchanging information. It is accessed through a MODEM and telephone lines and can provide entertainment services as well as information.

bundled software software that is provided with a computer as part of the overall purchase price. For example, some computer manufacturers would like their products to be seen as easy to use and so they try to cut out the difficulties that a new user associates with setting up programs. To achieve this, the manufacturer bundles software that is already set up in the computer system. The new user can then take the machine from the box, set it up and be productive very quickly.

burn-in a period of time during which computer components are tested. CHIPS and other components have a tendency to fail early or late in their lives. Early testing of the computer, say for 48 hours, is recommended before delivery to a customer.

bus a channel through which data passes. It refers to either the **data bus** (the route data takes to the processor) or the **expansion bus** (the route by which data moves to expansion cards).

button an on-screen area in a GRAPHICAL USER INTERFACE computer that is used to select a particular command. Pointing the mouse at the button and clicking will select the command.

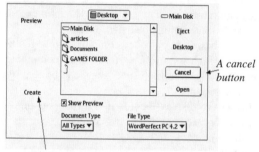

A button to create a new file

byte a combination of BITS used by the computer to represent an alphabetical character, a numerical digit or a special character such as an accent. For example, the letter A is represented by 01000001. A page of text will require slightly more bytes than there are characters to store it, to include spaces, control characters, etc. A large document will therefore need very many bytes, and many computers have millions of bytes (MEGABYTES) of memory. Units that are used most often are the kilobyte and megabyte, representing 1 thousand and 1 million bytes. These are, however, inaccurate because a kilobyte is actually 1,024 bytes (2 to the power 2) and a megabyte 1,048,576 bytes.

C or **C language** a computer language developed in 1972. It is regarded as a medium-level language combining attributes from both LOW-LEVEL and HIGH-LEVEL PROGRAMMING LANGUAGES. Programs in C are easy to read, they can run very quickly, and they can run on most computers.

cache an area of the random access memory (*see* RAM) in a computer that is used as a temporary storage for frequently used data. It allows faster access to the data than if the data were held on a physical disk, thus speeding up processing.

CAD (Computer-Aided Design) special software that will create and manipulate graphics shapes in the same manner in which an architect or designer might operate. The draughtsperson can change, edit, save and reprint drawings without the problem of redrawing everything over again. Until quite recently CAD was within the realm of DEDICATED computer systems, but the increased speed, memory and processing power of modern systems

means that it can be undertaken on more ordinary machines. It is used in many disciplines, including architecture, interior design, civil engineering, mechanical engineering, and so on.

CAI (Computer-Assisted Instruction) a form of teaching by computer. The CAI program leads a student through a series of tutorials, question and answer sessions, or other tests. The student can use CAI techniques to learn a wide variety of subjects, from computer programming to chess.

calculated field in a DATABASE MANAGEMENT PROGRAM a FIELD that contains the result of a formula that may be based on results in other fields. It may also contain dates or logical statements; e.g., a field could be based on the formula:
=if(a<31,Date,'late'),
which means that if the value of a is less than 31 then put the value in Date into the field, otherwise put the word late in the field.

calculator an on-screen utility that can be used in a fashion similar to a hand-held calculator.

A calculator

CAM (Computer-Aided Manufacture) the use of computers in manufacturing, which usually refers to the control of robots.

cancel if the user has made an error the command can be cancelled by pressing the ESCAPE KEY in a DOS operating system or by clicking a 'cancel' button in a GRAPHICAL USER INTERFACE.

capacity the amount of information that can be held in a storage device. A 3¹/₂-inch FLOPPY DISK will typically contain 1.4 MEGABYTES while a HARD DISK can contain up to several GIGABYTES.

card a circuit board that is made up of plastic backing with circuits etched onto the plastic. CHIPS are then attached to this base. The card is fitted into a slot in the main board of the computer. Different cards provide different functions such as communications cards, graphic accelerator cards and video capture cards.

cartridge a removable unit used as a secondary or backup storage, e.g. magnetic tape or optical disks. In a printer, it is a removable unit that contains ink that is fed to the print heads and onto the paper.

cascading windows in a GRAPHICAL USER INTERFACE environment, several windows can be open at any one time. The cascading effect is attained by overlapping the windows so that the title bar of each window is visible.

These windows cascade down the computer screen

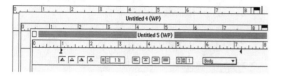

case sensitive the ability of a program to differentiate between UPPERCASE and lowercase characters. In DOS, it does not matter whether lowercase or uppercase is used as the operating system is not case sensitive. However, in a case-sensitive search of a database, a search for 'Bill' will result in a different found set from a search for 'bill'. Some word processing programs have the capacity to convert text from lowercase to uppercase characters, and vice versa.

cathode ray tube the standard type of computer VDU that uses an electron gun to fire a beam of electrons at the phosphor screen.

CDEV (Control panel DEVice) a UTILITY PROGRAM that is designed to make a computer easier to use. Examples of CONTROL PANEL devices are for mouse settings, keyboard function keys, date and time settings, network settings, etc.

CDI (Compact Disc Interactive) a standard that refers to the design of systems for viewing audio-visual compact disks using a TV monitor and a CDI player.

CD ROM (Compact Disk Read Only Memory) a system invented by Phillips in 1983. CD ROM can store much larger amounts of data than conventional storage. Although slower than hard disks, a CD ROM will store in excess of 600 megabytes of data. They are very useful for the storage of archival data. Video and sound can also be stored on a CD disk, making possible MULTIMEDIA applications.

cell an element or block of a SPREADSHEET into which data, numbers or formulae are placed. A cell is created at the intersection of a column and a row.

central processing unit (CPU) the core of a computer system, which contains the INTEGRATED CIRCUITS needed to interpret and execute instructions and perform the basic computer functions. At one time it was used to describe the box that housed

the electronics of the computer system. In modern computer systems. It is the integrated circuit that makes use of VLSI (Very Large Scale Integration—up to 100,000 transistors on one chip) to house the control transistors for the computer system.

character a single letter, number, space, special character or symbol that can be made to appear on screen by using the keyboard.

character set the full set of numbers, punctuation marks, alphabetic characters and symbols that a particular computer system uses and that a printer is capable of producing.

characters per inch (cpi) the number of characters that occupy one inch of text when printed. A normal size would be 10 cpi.

characters per second (cps) a measurement of the speed at which a printer can produce type. A DOT MATRIX PRINTER will have a speed of between 120 and 240 cps depending on the quality of print desired.

chat forum a conference area provided by an online service provider, such as COMPUSERVE, which allows two or more users to type messages and converse in real time. It is possible to find chat forums on virtually any subject.

check box a small box that is used to TOGGLE between different options in a DIALOG BOX. When the box has a cross or X in it the option is selected; when empty the option is deselected.

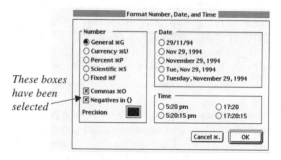

These boxes have been selected

check sum an error detection technique commonly used in data communications. The sending computer adds the number of bits in a piece of data, sends it along with the data to allow the receiving computer to check that the data is complete. If the check sum is not correct the data is incomplete and an error message is sent.

child *see* GRANDPARENT; PARENT.

chip a tiny chip or wafer of silicon that contains minute electronic circuitry and forms the core of a MICROPROCESSOR or computer. After the initial discovery of semiconductors, technological advance was rapid. Early INTEGRATED CIRCUITS duplicated the functions of a number of electronic components, but now it is possible to create chips that contain unimaginable numbers of components. In fact, it is possible to have 16 million components on a chip smaller than the tip of a finger. Chips can be mass-produced, and after their design, which is undertaken on an enlarged circuit diagram, the circuitry is transferred to plates called photomasks. Using a succession of photomasks, the chip is coated with materials that result in several layers of doped SILICON, and it then forms the equivalent of a highly complex electronic circuit.

chooser a UTILITY PROGRAM for the APPLE MACINTOSH that controls the selection of printers, fax cards and file servers. Once a device is selected, every application can use the device without the necessity for further selections to be set up.

circuit board a plastic board onto which circuits are etched and to which components such as CHIPS and other electronics are connected. The main circuit board for a computer is called the MOTHERBOARD.

circular reference a situation occurring in a SPREADSHEET that is the result of a cell containing a formula that depends on the result of the formula. For example, cell A7 could contain the formula 'sum (a1.a5)' in which the cell a1 contains the formula '=a7'. The addition of a1 to a5 cannot therefore be completed until a7 has a total, and this total will change each time a summation occurs. This process could go on forever in a circular fashion. The user can stop the process by limiting the number of ITERATIONS that can occur.

CISC (Complex Instruction Set Computer) a type of processor CHIP in which an instruction may take several operations and cycles to execute. (*See also* RISC).

clear a command that is used to remove a part of a document. This may be an unwanted paragraph or sentence from a document in a word processing program or a selection of cells from a spreadsheet. Clear commands can usually only be undone until another command is selected. It is important to ensure that you really wanted to clear the selection before you move on to the next command.

click to press and release the MOUSE button. This procedure is done in order to select an item such as an option in a CHECK BOX. The mouse pointer is positioned over the check box, the mouse button is clicked and the check box is selected. (*See also* DOUBLE CLICK).

client a personal computer or workstation in a LOCAL AREA NETWORK that is used to request information from a network's FILE SERVER.

client document a document that is connected to another document (primary) in another computer on a network. When the primary document is updated the client document is updated immediately.

clip art a collection of ready-drawn pieces of art that are available to copy and paste into any document. The purpose is to enhance the look of the document, whether it is a newsletter or promotional flyer. Many graphics packages provide thousands of pieces of clip art.

The signs of the zodiac are available as clip art. This is the sign of Taurus.

clipboard a temporary storage area into which is placed any selection of a document resulting from a cut and paste or copy and paste command. The clipboard holds the text or graphics after the copy command is issued and until the paste command is issued, at which point the text or graphic is placed into the active document at the selected place.

clock speed a description of the speed of a MICROPROCESSOR usually described in megahertz (one million cycles per second). The system clock emits a stream of electrical pulses or clicks that synchronize all the processor's activities.

clone the original name given to computers that succeeded in replicating the features of an IBM personal computer. Many companies now produce computers that are compatible with the IBM PC but have more features, better components and are built to higher specifications.

close the command to finish working with a computer file. In a GRAPHICAL USER INTERFACE environment, several windows can be open at any one time. Each window has a **close box** that, when checked (*see* CHECK BOX), closes the window.

CMOS (Complementary Metal-Oxide Semiconductor) a chip that features low power consumption.

coaxial cable a cable of the type used for television aerials. It is constructed of an insulated central wire with surrounding mesh enclosed in a plastic cover. The cable type is used in network systems such as ETHERNET.

COBOL (COmmon Business Oriented Language) one of the most commonly used computer programming languages for large mainframe business applications. It has never achieved the popularity of BASIC on smaller computers such as PCs. For large businesses, however, it became the choice for invoicing, salary records and stock control because its programs are easy to read and amend. Its function is to store, retrieve and process such data, and it therefore became useful in automating such processes.

code a list of instructions written to solve a particular problem.

cold boot *see* BOOTSTRAPPING.

cold link *see* LINK.

colour graphics adapter a VDU display standard offering RESOLUTIONS of 640 x 200 pixels in monochrome and 320 x 200 in four colours.

colour monitor a VDU that displays images in multicolours as opposed to a monochrome screen, which is only black and white (or black with green or another colour).

colour separation to print colour commercially, an image has to be separated into its component colours by scanning. The colour scanner directs laser or high intensity light at the image and through filters and a computer converts it into individual pieces of film for each colour used in the printing process. Cyan (blue), magenta (red), yellow and black dots then combine in the four-colour printing process to recreate the chosen image.

column in a SPREADSHEET program a column is a vertical block of CELLS that extends from the top to the bottom of the spreadsheet. The column is usually identified by an alphanumeric character. In a word processing program a column is normally a newspaper style column in which the text flows from the bottom of one column to the top of the next column on the same or succeeding page.

COMDEX the largest computer show in the world. It takes place in the USA twice per year.

comma delimited file a method of saving a file in which each data field is separated by a comma. The comma is a DELIMITER that indicates the end of one field of data and the beginning of the next (or the end of the file). Saving a file in this way makes it easier to transfer files between different programs or computers.

command an instruction or set of instructions that will start or stop an operation in a computer program, e.g. RUN, PRINT, EXIT.

command button a button appearing in dialog boxes that initiates a command such as continue with the operation, cancel the operation or help.

command.com an essential command file that is required for DOS to run. The file controls the on-screen prompts, interprets the typed commands and executes the required operations.

command key on a Macintosh keyboard a special key that is used in conjunction with other keys to provide shortcuts for commands. All Macintosh programs use the same command shortcuts.

command line a line of instructions or commands input to the computer through a keyboard. For example, 'format a:' could be typed to tell the computer to format a disk on a: drive.

communications port a PORT at the rear of a computer into which a serial device such as a printer (*see* SERIAL PRINTER) or a MODEM can be plugged.

communications program a program that allows a computer to connect with another computer. This is achieved through a MODEM. Communications programs include telephone directories, facilities that automate the dial-up process, log-on procedures, etc.

communications protocol a list of standards that control the transfer of information or exchange of data between computers connected via the telephone network.

communications settings when you access an on-line service you must set your computer to the same set of standards as the main computer. The main standards are BAUD RATE, PARITY BIT, DATA BITS, STOP BITS duplex, and HANDSHAKING. Settings would be:

no parity; 8 data bits; 1 stop bit; full duplex; handshaking would normally be XON/XOFF.

community the population of an ON-LINE INFORMATION SERVICE or BULLETIN BOARD expressing itself in conferences, discussion boards and ELECTRONIC MAIL.

compact disc *see* CD ROM.

compact disc interactive *see* CDI.

Compaq a major manufacturer of IBM-compatible computers. In fact, the computers designed and manufactured by Compaq are of a generally higher quality than most IBM-compatible machines.

compatible a characteristic of word or data processing equipment that permits one machine to accept and process data prepared by another machine, without conversion. This commonly refers to data but also can refer to hardware such as printers, monitors, i.e. IBM-compatible. To be really compatible, it should be possible for a program or PERIPHERAL to run on a system with no modification and with everything running as intended.

compiler a program that translates code that has been written in a HIGH-LEVEL PROGRAMMING LANGUAGE into an executable program.

compress to reduce the space taken up by files in order to reduce the disk space that is used to store the compressed files. (*See also* FILE COMPRESSION.)

Compuserve an ON-LINE INFORMATION SERVICE provider that provides a wide variety of services such as ELECTRONIC MAIL, news services, sports results and information, encyclopedia, financial information, computer information, many on-line forums and files to download. The INTERFACE is generally good for GRAPHICAL USER INTERFACE computers, and there is access to the INTERNET.

Compuserve's main options screen

computer an electronic data processing device, capable of accepting data, applying a prescribed set of instructions to the data, and displaying the result in some manner or form. Also any CONFIGURATION of the devices that are interconnected and programmed to operate as a computer system. Typically, this includes a CENTRAL PROCESSING UNIT with keyboard, VDU, printer and some form of disk drive. It also refers to the setting up of a computer system or program to ensure that it matches the needs of the user.

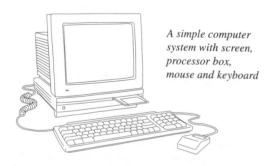

A simple computer system with screen, processor box, mouse and keyboard

computer system *see* SYSTEM.

concatenation the adding of two or more fields or pieces of text together to form one item, commonly used in SPREAD-SHEET or DATABASE MANAGEMENT PROGRAMS to manipulate data.

concordance file a file that contains a list of words that are to appear in an index. The word processing program uses the concordance file to create the index along with the page numbers that relate to the index words.

condensed type a type style that reduces the width of characters so that more characters are printed per inch of space. Dot matrix printers will print 17 characters per inch in condensed mode.

conditional statement a statement used in computer programming to determine the next operation. Conditional statements are also used in SPREADSHEET and database programming. For example, a salesman's commission can be calculated with a conditional statement such as:

=if(sales>=5000, sales*.10, 400),

which means that if the salesman generates sales greater than £5,000 then a 10 per cent commission is payable, otherwise a flat rate commission of £400 is due. In other words, the result is conditional on the level of sales.

config.sys a DOS file that contains commands that set up the computer's operating system. DOS requires that peripherals and applications have specific start commands that are held in the config.sys file. It is therefore important that the file is not deleted or the various programs or peripherals will not function properly.

configuration the machines that are interconnected and programmed to operate as a computer system. Typically, this includes a CENTRAL PROCESS UNIT with KEYBOARD, VDU, printer and some form of DISK DRIVE. It also refers to the setting up of a computer system or program to ensure that it matches the needs of the user. Configuration has to be performed at the outset, and while modern applications software has automated the procedure to some extent there are certain elements that have to be done manually. Once established, the set-up is saved in a configuration file that should not be erased or altered.

console the terminal that is used to control the computer system. It is applied as a descriptive term to a control device on large MAINFRAME systems. It is also used to describe the KEYBOARD and VDU in PERSONAL COMPUTER systems.

constant a fixed value used in a SPREADSHEET. When using a spreadsheet program, the use of constants has to be carefully monitored. Where possible, constants in repeated formulae should be avoided. One constant in a primary CELL should be used, and all the repeat formulae should be based on the primary cell. If the constant in the primary cell is changed all repeat formulae will automatically change.

context sensitive help an information system incorporated into application programs that automatically finds the relevant pages to assist with a command or operation with which the user is having difficulty. Context sensitive help systems reduce the time spent searching through HELP FILES for the appropriate section, thus making the program more user-friendly and allowing the user to be more productive.

control panel a utility program designed to allow the user to alter the look and feel of the computer environment. These utilities control such aspects as screen colours, monitor settings, date and time display, sound and speech settings, mouse controls, etc. (*See also* CDEV.)

Copeland the code name for the next major version of the MACINTOSH operating system.

coprocessor a secondary or support CHIP that is used alongside the main chip to provide added power for specific operations such as graphics display or mathematical calculations.

copy to create a duplicate of a file, graphic or program without changing the original version. Some copy protected programs, however, do not allow an exact copy of an original or key disk to be made.

copy protection a method of preventing, or at least reducing, the user's ability to copy a program illegally. A special code is written into the program that requires that the user type in a PASSWORD or inserts a special disk in order to use the program.

corrupted file a file or part of a file that has become unreadable. Causes of file corruption could include improper handling of a disk, a power surge, flaws on the disk surface or damaged READ/WRITE HEADS.

CP/M (Control Program Monitor) the operating system that dominated the desktop computer world before DOS was introduced in 1981. Control Program Monitor was the trade name used for the operating system for microcomputers based on the Z80 microprocessor chip.

cpi *see* CHARACTERS PER INCH.

cps *see* CHARACTERS PER SECOND.

CPU *see* CENTRAL PROCESSING UNIT.

crash an unexpected termination of an application, usually resulting in the freezing of the computer, i.e. it becomes completely unresponsive. The computer often has to be rebooted to recover (*see* BOOTSTRAPPING).

cropping a feature of graphics programs that allows electronic trimming of a picture either to get rid of unwanted parts of the image or to fit the image into a predefined space.

cursor an indicator on the screen of a VDU, used by a computer to direct a user to the starting position—the point at which data is entered. It can be a small line, a square of light on the screen or an arrow symbol, and can be controlled by use of the mouse or the arrow keys on the numeric pad on the right-hand side of the keyboard.

custom software a computer program that is written specifically for a client to match the systems that the client operates in his or her business. The program is useful only to one client and will probably not be usable by others.

cybernetics a branch of science that is concerned with computer control systems and the relationship between these artificial systems and biological systems.

cyberspace in modern computer communications a user connects with cyberspace when he or she logs on to an on-line service or connects with another computer. ELECTRONIC MAIL and forum messages move around cyberspace.

daisywheel the print wheel for a **daisywheel printer**, which produces LETTER QUALITY printing. It does this by rotating a print element resembling a wheel with spokes. Each spoke contains two characters of the alphabet. Daisywheel printers were once the first choice but have now been overtaken by the technology of printing that has made INKJET, bubble-jet and LASER PRINTERS more readily available.

data jargon for information. Data can be groups of facts, concepts, symbols, numbers, letters, or instructions that can be used to communicate, make decisions, etc.

data bits the elements of a character sent during ASYNCHRONOUS COMMUNICATIONS that contain the actual data.

data bus *see* BUS.

data file a computer file containing data as opposed to an application or program.

data processing the preparing and storing, handling or processing of data through a computer.

database a file of information (data) that is stored on a computer in a structured manner and used by a computer program such as a DATABASE MANAGEMENT SYSTEM. Information is usually subdivided into particular data FIELDS, i.e. a space for a specific item of information.

database management system (or DBMS) a software system for managing the storage, access, updating and maintenance of a database. Users can use it to edit the database, save the data, and extract reports from the database.

daughter board a printed circuit board that plugs

into the main board, or MOTHERBOARD, in a computer with the purpose of adding processing power or other facilities.

DBMS *see* DATABASE MANAGEMENT SYSTEM.

debug to locate and correct errors (BUGS) occurring in a application, e.g. when writing the code for a program there will undoubtedly be mistakes made. In order to make the program work correctly the errors must be eliminated. The program must be 'debugged'.

decollator a machine that separates the sheets of a multipart form or continuous paper, i.e. it separates the top sheet from the second sheet.

decryption the process of decoding or deciphering data from an encrypted form (*see* ENCRYPTION) in order that the data can be read and used.

dedicated a term that describes a computer or hardware device that is used solely for one purpose, e.g. when a computer is dedicated to act as a SERVER for a NETWORK.

dedicated line a communications cable line that is dedicated exclusively to a particular communication function. For example, a dedicated line may be used in a building to connect up a number of computers.

default a pre-set preference that is used by a program, the 'fallback' position. For example, in a word processor the defaults for style of type and font may have the default of SANS SERIF 12 BOLD.

delete to erase a character, word, command or program. Once the item is deleted, it may not be possible to recover it, so it is important to be careful about which files, etc, are deleted.

delimiter a character that is used to show the end of a command or the end of a field of data in a data record. Characters commonly used as delimiters are the comma (,), semi-colon (;) or tab.

demo (short for **demonstration**) a program that is restricted in some way but still shows a potential user the main features of the program. Usually the features to be disabled are the SAVE and print features. Inevitably, **demo disks** are used to promote and sell software.

demodulation *see* MODEM.

density a measure of the amount of information (in BITS) that can be stored on magnetic media such as a FLOPPY DISK. Single density allows for a measured quantity, but there is also double density. Quad density, or high density, uses very fine-grained magnetic particles, and although they are more expensive to produce than double density, they can store one MEGABYTE or more on a single disk.

descending order *see* ASCENDING ORDER.

desk accessory a small UTILITY program that can help in a computer user's productivity. Desk accessories include items such as notebooks, address books, on-screen calculators and scrapbooks.

This example of a desk accessory utility program is useful for finding files stored on disk

desktop in an operating system environment that uses a GRAPHICAL USER INTERFACE the desktop is the computer representation of a physical desk top on to which files and folders can be placed.

desktop publishing (DTP) the software and hardware that makes possible the composition of text and graphics as normally done by a printer or in a newspaper office. Desktop publishing requires the use of a computer, LASER printer, and various software programs to prepare and print documents. It is possible to produce anything from a single page of text to advertisements, pamphlets, books and magazines. Computer-aided publishing has been possible since the early 1970s for organizations willing to invest large sums of money, for example, traditional printers or publishing houses. Desktop publishing as a function of PERSONAL COMPUTERS became possible on a broad scale only in 1985, with the introduction of the first relatively inexpensive laser printer producing LETTER QUALITY for type and visuals.

A basic desktop publishing system allows its printer to produce print by employing a variety of FONTS and type sizes, type JUSTIFICATION, hyphenation, and other typesetting capabilities provided by publishing software programs. Page layouts, based on a template, can be set up on the monitor and transferred, as seen on the monitor, to the printer. Many types of GRAPHICS can be created, and the system may also incorporate art and photographs from sources inside the computer. The command codes for producing text and graphics are comparatively simple: some computers use symbols and a pointer controlled by the mouse; others use word and letter commands. A basic desktop publishing system includes a MICROCOMPUTER, a laser printer that is able to print at 300 DOTS PER INCH or more, word processing software, a PAGE DESCRIPTION LANGUAGE, and a software PAGE LAYOUT PROGRAM that enables its user to position, size and manipulate blocks of type and pictures.

In contrast to professionally printed matter, 300 dpi provides relatively LOW RESOLUTION, although greater resolution is now available. More complex laser printers or the use of an added photo-typesetting unit produces finer quality print and illustrations. The addition of a computer-connected SCANNER allows the use of text and visual material from other sources.

device driver in DOS a UTILITY PROGRAM that extends the capabilities of the operating system to allow hardware devices such as a mouse, CD ROM drive, printer or hard disk to work with the computer.

diagnostic program a UTILITY PROGRAM designed to test computer hardware and operating systems for errors that may cause or be causing system ERROR MESSAGES.

dialog box a WINDOW that is an integral part of a program and is used to convey information or request information from the user about the operations of the program. A dialog box could have: OPTION buttons, which are 'either/or' buttons; CHECK BOXES, which allow several options from a menu to be selected; LIST BOXES, which present a list of options, one of which can be selected; and COMMAND BUTTONS, which allow the user to continue the operation with the selection or to cancel the operation.

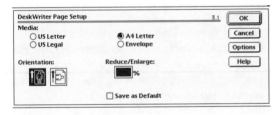

A dialog box concerned with page set-up options

dial-up the process of locating and retrieving information over telephone lines.

digital a term used to describe the use of two states, on or off, in order to represent all data. A computer is digital since it represents all data in a series of 1s and 0s. Using these 1s and 0s, calculations and other operations can be performed in an exact manner.

Digital Research a major manufacturer of computers, including IBM-compatible machines. It was formed in 1957. It also created its own operating system, DR-DOS, similar to MS-DOS.

digitize to convert text, images or sounds into a series of dots that can be read by a computer. It is also the term used to describe the process of scanning (*see* SCANNER).

DIMM (Double Inline Memory Module) a small

CIRCUIT BOARD containing RAM chips that increase the amount of memory available to a computer. (*See also* SIMM).

dip switch (Dual In-line Package switch) one of a collection of small 'on' and 'off' switches used to select options on a CIRCUIT BOARD without having to modify the hardware. They are frequently found inside printers to control vertical spacing and other variable functions and in computers and other electronic devices. A dip switch is the complete unit of plastic that contains the circuit and leads for fitting into the device.

directory the table of contents of a computer file system that allows convenient access to specific files. A directory is an area of the disk that stores files. It is common practice to store the files from one particular application in a specific directory so that they do not get mixed up with other files. Files can then be recognized by their names. When a directory is called up on screen it usually provides several items of information. A name can be given when a disk is formatted and this is the VOLUME LABEL.

In addition, the size of files in kilobytes is stated and the directory/file structure is shown.

disk *see* FLOPPY DISK; HARD DISK.

disk cache an area of computer RAM that is used as a temporary storage for frequently used data. It allows faster access to the data than if the data were held on a physical disk. It therefore speeds up processing.

disk drive the piece of hardware and electronics that enables information to be read from, and written to, a disk. The recording and erasing is performed by the READ/WRITE HEAD. The circuitry controlling the drive is called the **disk drive controller**.

dither to combine small dots of different colours or shades to produce the effect of a new colour or shade. For example, the combination of blue dots and yellow dots produces a green image. If the dots of blue are slightly larger than the yellow dots the shade of green becomes darker and moves towards purple. Use of dithering and a palette of 256 colours can produce a continuously variable colour range.

docking station a hardware device into which a NOTEBOOK COMPUTER can be connected to provide added facilities such as DISK DRIVE, CD ROM, colour VDU, PRINTER access, etc. The notebook computer can easily be inserted and extracted from the docking station.

document traditionally, a piece of work created in a word processing program such as a letter, memo or report. Recently the term has been expanded to include work created in a DATABASE MANAGEMENT PROGRAM or SPREADSHEET.

documentation books that provide information and instruction in the use of a piece of hardware or software. Since the books are bulky and expensive to print some manufacturers provide the information on disk. The information can be accessed as a TUTORIAL file or often as part of an ON-LINE HELP system.

document reader a hardware device that scans printed text, converting the text into digital signals. Software can be obtained to convert the digitized files into readable text that can be edited as any word processing document. (*See also* SCANNER.)

DOS (Disk Operating System) the program responsible for communications between a computer and its PERIPHERAL DEVICES such as the DISK DRIVE, PRINTER or the VDU. It controls and manages all the peripheral devices connected to the computer system. It therefore must be the first program to be loaded when the computer is switched on. The commonest OPERATING SYSTEM is MS-DOS (produced by Microsoft Corporation in the USA), which was introduced in 1981.

dot matrix printer a piece of equipment for printing characters. It is an IMPACT PRINTER, and as such is comparatively noisy when compared to the non-impact printers such as INKJET or LASER. A dot matrix image is created by a number of pins striking a ribbon and forming the image on the paper. Printers with just 9 pins produce poor quality output, and although there are versions with more pins, which give better quality, they come second in effect and quality to the newer technologies. They are fast, however, and are still much used for large volumes of repetitive work.

dots per inch (dpi) a measure of the RESOLUTION of a screen or printer. The more dots per inch that the computer can display or print the higher the resolution. (*See also* DOT PITCH).

dot pitch a measure of the RESOLUTION of computer screens or printers. The smaller the dot pitch the sharper the image that is displayed. A dot pitch of 0.28mm is HIGH RESOLUTION while 0.4mm is LOW RESOLUTION.

double-click to click the mouse button twice in quick succession. A control panel utility can be used to set the time delay between clicks of the mouse. A double click can extend the use of the single click. For example, a single click positions the cursor in a word while double clicking selects the whole word. Single clicking on a program icon selects the icon while double clicking will select the icon and open the program. A file can similarly be selected with a double click.

double density disk a FLOPPY DISK that can store approximately 720 kilobytes of data.

double-sided disk a type of FLOPPY DISK with both surfaces available for storage of data. Two READ/WRITE HEADS are required for double-sided disks.

download to copy a file from an ON-LINE INFORMATION SERVICE or from another computer to your computer. It is the opposite of UPLOAD.

down time the time when computer equipment is not available for use because of hardware or software malfunction. This is a very frustrating time since the investment in equipment is not producing results. The selection of reliable equipment is, therefore, important.

dpi *see* DOTS PER INCH.

draft mode the quickest, LOW RESOLUTION output from a DOT MATRIX or INKJET PRINTER. It is used to produce a document used for initial review and editing prior to producing the final full quality output.

drag to hold down the mouse button and move the mouse pointer across the screen. The drag technique is used to select an area of text in a word processor document or to select a group of cells in a spreadsheet or to select a group of document icons in a desk top window.

drag and drop having selected an area of text (for example, in a word processing document) with the drag command, the selection can be dragged from one part of the document to another and then dropped into its new place in the document.

DRAM (Dynamic RAM) a type of computer memory chip that cannot retain memory and so has to be continually refreshed (*see* REFRESH). This type of chip is used to transfer data within the computer.

draw program *see* OBJECT-ORIENTED PROGRAM.

drop down menu a list of command options that appears only when the main command is selected. Use of drop down menus allows programmers to provide many options to the user without cluttering the screen.

drum scanner *see* SCANNER.

DTP *see* DESKTOP PUBLISHING.

dumb terminal a computer terminal that lacks its own CENTRAL PROCESSING UNIT and DISK DRIVES.

dump the process of transferring the contents of memory in one storage device to another storage device or item of hardware. For example, it may be a dump from disk to printer, disk to tape or screen to printer. Dumps are often performed when programmers are debugging programs (*see* DEBUG).

duplex *see* FULL DUPLEX.

DVI (Digital Video Interface) a set of standards or specifications for combining conventional computer techniques with those of video.

Dvorak keyboard an alternative KEYBOARD from the normal QWERTY keyboard. Some 70 per cent of the keystrokes are made on the home row compared with around 30 per cent with the QWERTY layout.

dynamic data exchange an interprocess channel through which correctly prepared programs can exchange data and control other programs.

dynamic link a method of linking data shared by two separate programs. When data is changed by one program it is changed immediately for use by the other. This type of link is required in MULTIUSER networks.

echo to show on screen the commands being executed by a computer as they are being performed.

edit to change or alter text, graphics or values that appear in a file. The edit process is required to correct mistakes previously made in a file and is a core function of all word processing software.

edutainment the term given to a growing selection of computer software that educates the user while being entertaining. An example of such a program is Sim City, in which the user becomes the mayor of a town and has to make a variety of decisions that affect its survival and growth.

EGA (Enhanced Graphic Adapter) a colour bit-mapped (*see* BITMAP) VDU display adapter for IBM-compatible PERSONAL COMPUTERS. It displays up to 16 colours simultaneously with a RESOLUTION of 640 x 350 PIXELS.

electronic mail or **email** the use of a NETWORK of computers to send and receive messages. Growth has been restricted in the past because of the variety of incompatible email systems. However, this is being corrected with the INTERNET's standard platform for worldwide electronic mail communications. The use of electronic mail has the advantage over conventional communication of cutting out unnecessary chat and can connect groups of people on a worldwide basis for collaboration on projects.

electronic marketplace every day more businesses are offering their goods and services for sale over the INTERNET. Payment for goods or services in this electronic marketplace is made by credit card and goods are shipped by courier as in normal mail order.

electronic publishing the use of the INTERNET to publish and distribute work. At no time need the work be printed on paper. The type of work that can be the subject of electronic publishing includes on-line news services, on-line encyclopedia or computer-based training manuals.

email *see* ELECTRONIC MAIL.

emulate to duplicate the function of a program, operating system or hardware device in another computer system.

encryption the method of encoding data so that unauthorized users cannot read or otherwise use the data. Data characters can be jumbled by a computer program, communicated to another computer and, as long as the receiving computer has the same encryption program, recompiled into meaningful information.

environment the style or setting in which the user enters commands into or performs tasks with the computer. The GRAPHICAL USER INTERFACES provide an environment or setting that looks similar to a desktop while a DOS system provides a command line environment.

EPROM (Erasable Programmable Read Only Memory) a memory CHIP that can be programmed, erased and reprogrammed.

EPS graphic (Encapsulated PostScript graphic) an object-oriented graphics file format developed by Apple. The format uses separate graphic objects such as lines, rectangles, arcs, ovals, each of which can be independently moved or sized. A file saved in this format can be read by many programs.

erasable storage a READ/WRITE secondary storage device in which data can be written and erased repeatedly. A HARD DISK is such a device whereas a CD ROM is not since it cannot be erased once data is written to it.

erase to rub out or delete from a STORAGE device.

error message a message displayed on a screen that indicates that the computer has detected an error or malfunction. **Error trapping** is the ability of a program to recognize and almost anticipate an error and then carry out a pre-set course of action in response to the error.

escape key (esc) a nonprinting character or keyboard control key that causes an interruption in the normal program sequence. Within a software program it is usually pressed to cancel a command or operation.

Ethernet a LOCAL AREA NETWORK hardware standard capable of linking up to 1,024 computers in a network. Ethernet can transfer up to 10 megabits per second.

EtherTalk an implementation of the ETHERNET local

area network developed by APPLE and the 3com corporation, designed to work with the APPLESHARE network system.

event-driven program a program that is constructed to react to the computer user who initiates events such as clicking a mouse rather than a COMMAND-driven program, which requires specific commands to be typed into the computer to obtain results.

execute to carry out the individual steps called for by the program in a computer.

expansion bus *see* BUS.

expansion card a printed CIRCUIT BOARD that is fitted into the main computer board. Expansion boards are fitted to enhance the power of the computer, providing facilities such as MODEMS, added memory and high speed graphics.

expansion slot a PORT in the main computer system that allows the fitting of an expansion card. There are several slots available for fitting expansion cards in computers.

expert system a program that uses the accumulated expertise in a specific area of many people in order to assist nonexperts who wish to solve problems.

export to create a data file in one program that can be transferred to another computer and be read by another program. Exported files can usually be transferred in a particular format to ensure that they can be read in the new system.

extended memory specification *see* XMS.

FAT *see* FILE ALLOCATION TABLE.

fax or **fax machine** or **facsimile** a device capable of transmitting or receiving an exact copy of a page of printed or pictorial matter over telephone lines in, usually, less than 60 seconds. It is currently the preferred method for the rapid transmission of printed material. Facsimile transmission in some form has been available since the end of the last century. The fax was invented by a Scot, Alexander Bain, from Caithness. From the 1920s, newspapers used slow facsimile devices, equipped with photoelectric cells that scanned material placed on rotating drums, to transmit photographs. Police departments, the military and some businesses sent printed matter via matched facsimile machines. However, fax remained a relatively specialized communications device until the development of sophisticated scanning and digitizing techniques in computer and communications technologies and the establishment of standards that made it possible for all fax machines to communicate with one another over ordinary telephone lines.

A standard fax machine

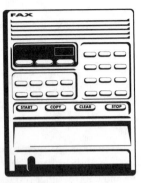

Most contemporary fax machines conform to a set of standards, known as Group III, that were implemented in 1980 and that require digital image scanning and data compression. Machines built to conform to Group III standards can transmit data at a maximum 9,600 bits per second. To transmit, the original document is fed into the machine, where it is scanned by a mirror-and-lens-aided device, or, in some faxes, by a series of light-emitting diodes (LEDs). Light and dark picture elements—PIXELS—are described digitally, and the message is shortened by compressing much of the white space. The receiving machine, which is addressed through its telephone number, translates the code it receives back into a pattern of greys, black and white. The reconstituted message is printed out on heat-sensitive paper, using techniques similar to those of photocopying machines. Some fax machines can double as copiers, and modern machines use ordinary paper, which eliminates the use of heat-sensitive paper that browns over a period of time.

feed the process of supplying paper to a printer. Paper can be fed into the printer either by a friction system or a TRACTOR FEED device, which gives **line feed** when the printer moves the paper forward one line at a time. Laser printers use **page feed**, which ejects one page at a time. (*See also* SHEET FEEDER.)

fibre optics a method of carrying information along cables using light. This method of transmitting data is faster and more reliable than conventional wires. Recent developments have allowed scientists to pass an amount of information equivalent to 1,000 bibles per second along a fibre optic cable.

field a defined group of characters or numbers, e.g. a customer number, a product description, a telephone number or address within a specific space in a DATABASE program.

fifth generation computer the computer of the future. It is the next stage of computer development, which will incorporate technologies such as PARALLEL PROCESSING, SPEECH RECOGNITION, integrated communications, and much more.

file a collection of data that is given a distinct name and is stored on the computer's SECONDARY STORAGE. Files are stored within directories, analogous to the old system of filing cabinet, drawers and folders.

file allocation table (FAT) a table held on a computer disk that keeps a record of the location on a disk of all the files. Files can be distributed in many locations on a disk, and the FAT keeps a record of the locations so that the file appears contiguous, or in one piece, to the user.

file association a link between a document file and the program that created it so that when a document file is selected by double clicking it opens the program and hence the file. Without file association, double clicking the document would have no effect.

file attribute information held in a file directory that contains details about the file and how the computer can access it.

file compression the process of condensing a file with the result that it takes up around half the normal space on a disk. Usually files that are not frequently used, are being archived or are being prepared for electronic transmission are compressed in this way.

file compression utility a program that is designed to COMPRESS files. There are various programs available that will compress files, e.g. Stuffit, Disk Doubler and JPEG for graphics. (*See also* FILE COMPRESSION*).

file conversion utility a program that is designed to convert files created in one program for use by another program. For example, files created in Word can be converted and used by Word Perfect.

file extension *see* FILE NAME.

file format the method that an OPERATING SYSTEM or program uses to store data on a disk. Different software companies have different methods of storing data, with the result that it is difficult for one program to read a file created in another program. (*See also* FILE CONVERSION UTILITY*).

File Maker Pro a popular DATABASE MANAGEMENT PROGRAM created by the Claris Corporation. It has versions for both the Macintosh operating system and Microsoft Windows operating system.

file manager a utility program that allows the user to copy, delete, add or move files around without reverting to the DOS commands and to create directories. Microsoft Windows uses a file manager program to assist its users.

file name a name given to a file by the computer user so that the operating system recognizes the file. Every file on a disk directory must have a distinct name. Some operating systems restrict the length of a file name to eight characters that are separated by a full stop from a three character **file extension**. The file extension is often added by default by the program and identifies the file with the software with which it was produced.

file recovery the process of retrieving or restoring a file that has been previously erased.

file server a PERSONAL COMPUTER in a network that provides access to the storage media for workstations or other computers in the network. The operation of the network operating system ensures a seamless view of the server's files from each workstation.
The server is usually a high-powered computer with a very large storage capacity that is set aside as the controller for the clients on the network. (*See also* SERVER.)

file transfer protocol a standard that controls ASYNCHRONOUS COMMUNICATIONS by telephone to ensure error-free transmission of files.

fill an operation that is used in a SPREADSHEET program to enter values in a range of CELLS. For example, a range of dates can be 'filled' into cells to act as headings for a monthly cash flow report.

filter to select certain files from a DATABASE by setting up a set of criteria. For example, to find those records that are dated between 1 and 31 January *and* contain a reference to a particular salesman. An alternative filter would be to find those records containing a reference to a particular salesman *or* referring to a particular customer.

finder a UTILITY PROGRAM that manages memory and files in conjunction with the Macintosh operating system.

firmware the part of the system software that is stored permanently in the computer's read only memory (*see* ROM). Firmware cannot be altered or modified.

fixed disk *see* HARD DISK.

flash memory a type of memory device that can be programmed, erased and reprogrammed. It is retained when the power is turned off. Flash memory cards, similar in size to credit cards, are used to store programs and files for PERSONAL DIGITAL ASSISTANTS, LAPTOP COMPUTERS and NOTEBOOK COMPUTERS where size and space-saving are crucial.

flatbed scanner a hardware device that is used to

transfer text and graphics from paper into a digitized format that can then be edited in a computer program. (*See also* OPTICAL CHARACTER RECOGNITION).

A flatbed scanner is similar in looks to a photocopier

size	density	drive	capacity
$3^1/_2$ inch	double	standard 720K	
$3^1/_2$ inch	double	standard 800K (on a Mac)	
$3^1/_2$ inch	high	high density	1 . 4 4 M (megabytes)
$3^1/_2$ inch	high	high density	2.88M
$5^1/_4$ inch	double	standard 360K	
$5^1/_4$ inch	high	high density	1.2M

A $3^1/_2$ inch floppy disk

flat file database a DATABASE MANAGEMENT PROGRAM that can access only one record or file at a time. This restricts the usefulness of the program compared with a RELATIONAL DATABASE management program.

flicker a distortion that occurs on a VDU, caused by a low rate of refreshment of the screen, i.e. the electron beam does not progress over the screen fast enough to reflect changes in the display when the display is constantly changing.

floating point calculation a form of calculation that the computer employs for calculating numbers. The decimal point in a number is not fixed but floats, allowing a high level of accuracy in calculations. The floating point calculation in some programs can handle numbers up to 10^{20} accurately. Other programs, however, may effectively limit the accuracy of a calculation by reducing the size of the numbers to, say, 10^{15}.

floppy disk a removable secondary medium of storage for computers. The disks are made of a plastic that is coated with a magnetic material (of which the main component is ferric oxide), and the whole thing is protected by a rigid plastic cover (in the case of $3^1/_2$ inch disks). The disk rotates within its cover, and an access hole allows the READ/WRITE HEAD of the disk drive to record and retrieve information. There is a WRITE-PROTECT notch on the disk cover that can be set so that the disk drive cannot change the disk but can only read the data stored there. The $5^1/_4$ inch disks are more susceptible to damage than the $3^1/_2$ inch disks because they do not have a rigid outer cover and are gradually being superseded. However, floppy disks are an essential part of any computing system because they are the means of installing software, backing up files and transferring data between users if no others means exists. The storage capacity of a disk varies and depends upon the size, the density of the magnetic particles coating the disk's surface and the drive used. Some examples are:

floptical disk a FLOPPY DISK that, because of its construction, allows the disk drive's READ/WRITE HEADS to align very accurately with the disk. This allows a far greater amount of information to be stored on a disk.

flow when data is being imported into a word processing document or page layout document, the imported text will flow into the available columns and around any graphic images. When one column is filled the text will flow into the next column.

folder in GRAPHICAL USER INTERFACE systems a folder is the DIRECTORY in which files are located or stored. The folder is represented on screen by an ICON styled like a physical folder in a filing cabinet.

font a complete set of letters, numbers, special characters and punctuation marks of a particular size and for one identifiable typeface whether roman or bold (the WEIGHT), italic or upright (the posture). The term is often used to refer to a family of fonts or TYPEFACES, although this is technically incorrect. Fonts come as bit-mapped (*see* BITMAP) or OUTLINE FONTS.

font family a set of FONTS sharing the same TYPEFACE but differing in the size and the boldness of the type.

footer text positioned at the foot of a page by a word processing program, for example. The type of text could vary from the file name, date, time, page number, originator or other relevant text.

footnote a note at the bottom of a page in a word processing or page layout document that is used to explain a word or phrase or concept. The word being footnoted is identified by the placement of a superscripted number after the word. This number corresponds to the footnote number.

footprint a physical measure of the amount of desk space that a computer and its peripheral devices take up when sitting on the user's desk. If the computer footprint takes up the whole desktop space there is no room left to work.

forecasting a method of using past results to project results into the future. For example, future sales can be forecast by analysing the results of past months and years. It is important to look at past years since there may be seasonal trends, such as increasing sales at Christmas, that will have to be taken into account in creating the forecast.

foreground task the priority job that the computer is undertaking in precedence to the other tasks that are being processed in the BACKGROUND. The foreground task is the task that you are monitoring on the computer screen. To support foreground and background processing the computer must be capable of MULTITASKING.

format the preparation of a HARD DISK or FLOPPY DISK for use by laying down clearly defined recording areas (*see also* INITIALIZE). The format is the way in which the magnetic pattern is laid down on the disk.

In particular programs, e.g. a SPREADSHEET, the format is the overall arrangement of labels and values in the separate cells of the spreadsheet. This may relate to the layout of decimal numbers or the alignment of entries within columns. A similar concept applies to DATABASE MANAGEMENT and word processing programs. In the latter the format will encompass all aspects of the typeface, the page layout (numbering, headers and footers) and the paragraph styles.

formatting the process of instruction that produces the desired format of text in a document for on-screen display or printing.

formula a calculation in a program, such as a SPREADSHEET, that defines a relationship between values that can be directly input or are already present in the spreadsheet. For example, a simple task is to add a range of cells with the formula 'sum(a1.a6)'. Or a cell d3 with the value of 66 can be added to the value 24 with a formula 'd3+24', which gives the result 90. There are many functions, such as 'sum', that can be used in a formula, and relatively complex calculations can be carried out automatically. Others include average, cosine, lookup. LOGICAL OPERATORS such as and, or and not can also be incorporated into formulae.

FORTRAN (FORmula TRANslation) a HIGH-LEVEL PROGRAMMING LANGUAGE designed for mathematical, engineering and scientific work. It was developed in the 1950s by IBM.

forum a designated group in which discussion takes place via the electronic network. ON-LINE INFORMATION SERVICE providers set up resources to allow people to choose from a wide range of subjects or to choose a general area where any subject can be discussed (subject to the rules of the service provider). (*See also* CHAT FORUM).

fourth generation computer the current generation of computers that use CHIPS. The use of microchip technology allows computers to be small and lightweight. However, peripheral devices such as power supplies and hard disks restrict how small a computer can be constructed.

fractals groups of shapes that are alike but not identical, such as leaves or snowflakes. No two snowflakes are identical but they have generally similar patterns. Computer programs can create fractals to provide artists and designers with a huge variety of graphic images.

fragmentation the storage of files on a disk that uses non-contiguous sectors to store the file. On a newly formatted disk, a file will be stored in its entirety in one location. The next file to be stored will take up the next sectors, and so on until the disk is full. When several files are deleted the relevant sectors are free for new storage of files, but one sector may not be large enough for a file. The file is therefore fragmented, or split, between two or more available sectors. The process of saving and deleting files will result in a particular file being located in many sectors on the disk. The disk READ/WRITE HEADS will take longer to retrieve the file in such circumstances. Disks can be defragmented using a utility program and thus as much as 50 per cent of the time taken to retrieve a file can be saved.

free form database a form of DATABASE that has no preset structure of information on each record. The information that can be held on one record can be completely different from the information on another record. Such free form databases are useful for storing general notes that are accumulated on a desk notepad, for example.

freeware copyrighted programs that are provided by the author free of charge. These programs, although free, are often troublesome as they may not have been fully tested, resulting in errors or system crashes. More importantly, they may carry a computer VIRUS that could cause complete loss of all files. It is therefore important to use virus detection utilities before running freeware programs.

frequency a measure of the speed at which a computer processor operates. It is measured in MEGAHERTZ.

FST (Flatter Squarer Tube) the technology for producing a VDU or TV screen that is flat rather than having the traditional convex surface.

full backup utility a utility program that creates a full backup of the files on a disk. It is different from an INCREMENTAL BACKUP, which backs up only the files that have been altered since the previous backup.

full duplex a protocol for ASYNCHRONOUS COMMUNICATIONS, which allows the sending and receiving of signals at the same time. Asynchronous communication requires the correct standard of cabling.

function any single operation of a computer or word processor, e.g. editing. Also, within certain programs such as SPREADSHEETS, a procedure that is stored in the program and that will perform a particular sequence of operations or calculations to produce an end result.

function key a special purpose key on the keyboard of a word processor or computer system that enables the user to perform a particular task or execute a command that might otherwise take several keystrokes. Functions differ depending on the program. However, there is a general convention to program the keys for the same function, e.g. F1 is used generally to display the help screen.

These are the function keys on a keyboard

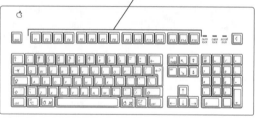

fuzzy logic a description of the development away from strict logical arguments to take account of human or non-logical behaviour. The development of ARTIFICIAL INTELLIGENCE requires a degree of fuzzy logic since human decisions are rarely based on strict rules of logic.

gateway a device that converts communications from one PROTOCOL or BANDWIDTH to another. This function allows two different types of NETWORK to communicate with each other. For example a LOCAL AREA NETWORK (LAN) can communicate with a WIDE AREA NETWORK (WAN), and a LAN can be connected with the INTERNET through an appropriate gateway.

GIF (Graphics Interchange Format) an efficient file compression system for graphic images (pictures). Because of its efficiency, GIF files are used widely for downloading from ON-LINE services.

giga a prefix meaning one billion (a thousand million), and abbreviated g.

gigabyte one billion BYTES or 1,000 MEGABYTES, although, strictly speaking, a gigabyte is 1,073,741,824 bytes. Hard disk capacities of one gigabyte on PERSONAL COMPUTERS are increasingly common.

GIGO (Garbage In, Garbage Out) a common situation where poor or distorted results from a program are caused by incorrect input or mistakes in the input. Thus the quality of the data output is only as good as the quality of the input.

glare the reflection of light from the computer screen. This can be very distracting and can cause stress if not corrected by using a glare filter or by moving the screen.

glitch a malfunction caused by a hardware fault. The malfunction is most often caused by a power surge or interruption.

global a style or format that is applied throughout a document or program. For example, all ruler settings in a word processing document can be set to the same tabs, etc. Similarly, all cells in a SPREADSHEET program can be set to the same numeric format.

glossary in word processing documents a glossary can be used to store phrases or styles that are commonly used. This saves time when keying in text as an abbreviation for the common phrase can be used and then automatically substituted.

goto a programming phrase that directs the program logic to a part of the program in order to accomplish a specific function. Also, a command feature that allows the user to select a page to move to in a word processing document or allows the user to select a cell to go to in a SPREADSHEET program.

grabber a representation of the mouse pointer with which images, text or cells are selected by moving the pointer across the selection required.

grandparent the oldest file in a grandparent, PARENT, son BACKUP system. The grandparent file should not need to be used unless both the parent and child backups have been used and corrupted.

graph a pictorial representation of a set of values, for example, numbers or quantities, that is used to show the relationships between those values. This may be in the form of a line graph, whereby a series of dots are plotted in relation to axes and then joined to one another by a line, or a bar graph, whereby the values are represented as vertical or horizontal bars.

Graphs are an invaluable way of getting a message across. The old adage is 'a picture paints a thousand words'.

A simple bar chart graph

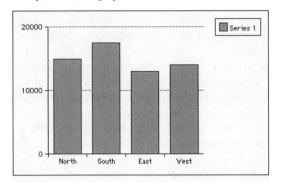

graphical user interface (GUI) the part of the software program that communicates and interacts with the user by means of pull-down MENUS, DIALOG BOXES and ICONS. The GUI makes computers easier to use because people recognize and respond to graphic representations of concepts, etc, much more readily than if they have to read words. Microsoft Windows utilizes this system.

graphics a generic term used to describe anything to do with pictures as opposed to text. There are two types of graphic images used by computer programs: OBJECT-ORIENTED PROGRAMS or draw programs and bit-mapped (*see* BITMAP) or PAINT PROGRAMS. Object-oriented programs are used where precision graphics are required, e.g. in CAD or architecture programs. Bit-mapped graphics are useful in artistic applications where shading and patterns are more important than precision.

The hand shows a bit-mapped graphic while the line shape is a vector graphic

graphics tablet an input device that uses a touch sensitive pad and a stylus. The movement of the stylus over the tablet generates an electrical pulse that is recorded by the computer and translated into a digital form as a screen PIXEL. The drawing on the pad is thus transferred to the screen.

grey scale the shades of grey from white to black that a computer can display. The more grey scales that are used the more realistic the picture will look. However, memory and storage required increase with the number of grey scales used.

groupware software that is created to increase the productivity of a group of workers in a team using a LOCAL AREA NETWORK.

guest an access privilege in a LOCAL AREA NETWORK that allows an infrequent user to examine certain files on the network without having a PASSWORD.

GUI *see* GRAPHICAL USER INTERFACE.

gutter an additional margin added to a word processing document or page layout document that allows space for a binder without obscuring the text.

hacker an individual who is obsessed with finding out more about computers. The term has evolved to refer to people who access other people's computers, usually with the aid of communications technology and without permission. There have been some notable instances of hackers gaining access to top security government systems.

half height drive a DISK DRIVE that occupies approximately 1.6 inches height in a computer drive bay. Originally the drive bay was 3$^1/_2$ inches high to accommodate the original size of a disk drive. Disk drives have shrunk in size since those early days.

halftone the shading in an image created by use of dots of various sizes and densities. Light areas are represented by small dots spaced apart while dark areas are created by larger dots placed close together. A similar technique can be used to produce flowing colours. Images scanned into a computer can be created in halftones to create usable images for reproduction.

hand-held scanner a scanning device (*see* SCANNER) that can be held in the hand. The scanning head is moved over the text or image to be copied. The image or text is digitized and can be stored in the computer. The image is held as a bit-mapped file (*see* BITMAP) and as such any text in it cannot be edited. However, OPTICAL CHARACTER RECOGNITION software can be used to translate the text into a usable file that can be manipulated like any text file.

handle a small black square that surrounds a GRAPHICS image in an OBJECT ORIENTED PROGRAM, allowing the user to change the size of the image or to reshape the image or to move the image around the screen. By choosing the appropriate handle, the image can be changed in one or two dimensions.

handshaking a greeting between two devices, such as a modem to modem or computer to printer, that signals that data transmission between the devices can proceed. The two types are **hardware handshaking** and **software hand-shaking**. In hardware handshaking a control wire is used to signal when transmission can proceed. In software handshaking, such as XON/XOFF, a control code is sent to control the flow of data.

hanging indent in a word processing program this is the format of a paragraph that has the first line starting on the left margin and the subsequent lines starting further to the right, as in this text.

hard a term used to describe a hyphen or PAGE BREAK inserted by the user in a word processing, page layout or spread-sheet program, as opposed to a SOFT command inserted by the program.

hard card a printed CIRCUIT BOARD (PCB) that plugs into the EXPANSION SLOT of a computer. The PCB contains a hard disk drive and controller circuitry. This is an easy way of adding extra storage capacity to a computer but it is an expensive option.

hard copy a document or file that is printed as opposed to one that is stored in a computer's memory or stored on disk.

hard disk a fixed disk that forms a storage medium within the computer. It was developed in 1973 by IBM and initially called the Winchester disk, but early versions were very expensive. Today they are a standard component of just about every computer and their mass production has reduced the price enormously. At the same time, the capacity has risen dramatically, from the early sizes of 10, 20 and 40 MEGABYTES to several hundred megabytes and even one GIGABYTE (i.e. one billion or 1,073,741,824 bytes or 1000 megabytes). The hard disk includes the storage medium, the READ/ WRITE HEAD and the electronics to connect it to the computer. There are several disks, or platters, that revolve at 3600 rpm, and the head floats just above the disk surface to eliminate wear. The large capacity of hard disks means that several software programs can be installed on one machine plus innumerable data files. Hard disks may fail, however, and it is important to back up data regularly.

hardware interrupt *see* INTERRUPT.

hardware platform the physical equipment of a computer system such as the CENTRAL PROCESSING UNIT, DISK DRIVE(S), VDU and PRINTER. In fact, anything that can be connected to a computer.

Hayes command set a standard set of instructions that have been developed to control communications through MODEMS. Commands include:

AT	attention command prefix	Hn	hang up
A	answer	Ln	speaker volume
D	dial	P	pulse dialling
Fn	select line modulation		

Hayes compatible modem a MODEM that recognizes the HAYES COMMAND SET, which is the *de facto* standard for ASYNCHRONOUS COMMUNICATIONS between modems.

HDLC (High-level Data Link Control) a PROTOCOL for synchronous communications.

head the device used by a DISK DRIVE to read a disk. As each side of a disk can be read by a separate head the number of heads is often used as shorthand for the number of sides (e.g. double-sided disk).

head crash the physical impact of a disk head on the disk, resulting in damage to its surface and a serious equipment malfunction that usually destroys data stored on the disk. It was relatively common on older systems, but modern high-tolerance engineering ensures that head crashes rarely occur.

header text that is placed at the top of every page in a document. The header normally contains the date, page number and document title. Different word processing programs have a variety of controls over the headers in a document.

heap a part of the computer's memory that is set aside for specific instructions that control such aspects as user input, menus and icons.

help file a file built into a software program that provides assistance and further information about selected topics. The file can be opened while the program continues to run.

hertz (Hz) a measure of the frequency at which electrical waves repeat each second. It is a measurement used to show the speed of a computer CHIP. Generally, the higher the speed of a chip the better is the performance. However, two different chips can operate at the same speed but have different performance levels. For example, an Intel 386 operating at 33 megahertz has lower performance than an Intel 486 operating at the same speed. A computer with a fast PROCESSOR is not necessarily ideal since it depends on the quality of DATA BUS, display (*see* VDU) and DISK DRIVES to determine the performance of a system.

heuristic a method used by experts to solve problems using a rule of thumb rather than strict logic. This is important in developing ARTIFICIAL INTELLIGENCE and knowledge systems.

hexadecimal a numbering system that uses a base of 16. Decimal uses a base of 10 and BINARY uses a base of 2.

Decimal	*Hexadecimal*	*Binary*
0	0	0000
1	1	0001
2	2	0010
3	3	0011
4	4	0100
5	5	0101
6	6	0110
7	7	0111
8	8	1000

9	9	1001
10	A	1010
11	B	1011
12	C	1100
13	D	1101
14	E	1110
15	F	1111
16	G	10000

HFS (Hierarchical File System) a disk storage system developed by APPLE to organize files on a HARD DISK. The system allows storage of files in a series of FOLDERS. A folder can be stored in a folder within a folder, etc. A major drawback of this type of filing system is that the user cannot define a path that an application can follow to find a file. Files can be 'lost' within many nested folders and so a FINDER utility is a useful addition to a Macintosh system.

hidden code an invisible code or instruction in a document that controls the appearance of the document when printed. Different codes control styles such as bold type or paragraph indents.

hidden file a file that is rendered invisible because of the way its file attributes have been set. The file cannot be seen in directory listings because it is judged so important that the file should not be altered or deleted.

high density a storage technique for FLOPPY DISKS that can store over 1 megabyte of data on the disk (normally 1.44 megabytes). The disk media must use relatively expensive fine-grained magnetic particles to be capable of storage in high density format.

high-level programming language a set of commands for computers that people can understand. Once the programmer has completed the program, all the commands in the high-level programming language are compiled into their equivalent machine code. The use of high-level languages such as BASIC, C or PASCAL allows the programmer to concentrate on solving the problem rather than on how to tell the computer to perform calculations.

highlight to select an area of a document in order to apply a command to that area or otherwise work with the selection. The selected area is often displayed in REVERSE VIDEO. It is most commonly defined using the mouse by clicking and holding until the desired area is selected.

> This is a highlighted portion of text to which a command, such as style in italics, could be applied

high resolution the extra sharpness of the RESOLUTION of high quality PRINTERS and VDUS that produces output with smooth curves and well-defined

fonts with no jagged edges. The resolution of a printer or screen is measured in DOTS PER INCH or in the number of PIXELS that can be displayed.

HMA (High Memory Area) an area of 64 KILOBYTES of memory in a DOS system above the first MEGABYTE of memory.

holographic storage a storage technology that uses three-dimensional images created by light patterns projected and stored on photosensitive material. When it becomes available it will store a greater amount of information than CD ROMs.

home computer a computer that is designed or marketed for home use as opposed to office or work use. It is generally perceived as being of lower power or capability than a business computer. However, this distinction is becoming less obvious as the technology advances.

home key a key on a keyboard that has various uses depending on the program being utilized. Normally the home key will move the cursor to the beginning of the current line, current paragraph or current document.

host the computer in a computer NETWORK that provides information, files or programs to other computers or WORK-STATIONS on the network. The host computer can provide information to a LOCAL AREA NETWORK, a WIDE AREA NETWORK or over the INTERNET.

hot key a keyboard key combination shortcut that gives access to a menu command or direct access to a program.

hot link a connection between two distinct documents that automatically copies information from one document (the source) to the other document (the target). Changing the information in the source document will result in a similar change in the target document.

housekeeping activities that are performed to reduce clutter on the computer desktop and disks and generally make for efficient use of the computer. Housekeeping includes deleting unwanted files and programs, reorganizing files into the most appropriate directories, defragmenting disks (*see* FRAGMENTATION), etc.

hypercard an accessory program authored by APPLE. Originally this program was shipped with every Macintosh but it is now supplied commercially.

hypermedia a term used to describe how hypertext concepts can be applied to multimedia.

hypertalk a computer scripting language that is used to create instructions for HYPERCARD programs. Hypertalk is an event-oriented language (*see* EVENT-DRIVEN PROGRAM).

hypertext the ability to pick up on one word in a document as a route to another area of a document. For example, in a hypertext dictionary a link would exist between a head word and the same word when used in a definition. By clicking on 'document' in this definition the computer would be directed to the definition of 'document'. Such a system is used in the worldwide web to connect pages of related information.

Hz *see* HERTZ.

IBM personal computer (IBM PC) a PERSONAL COMPUTER developed by IBM that was released in 1981. Since then the computer technology has developed significantly.

icon a symbol on screen that represents something or some process or function in the computer. Icons are used in a GRAPHICAL USER INTERFACE, and the image of an icon resembles the result of choosing that particular option or command. Programs resident within WINDOWS use numerous icons for tasks such as opening and closing files (which use small pictures of files), printing (a printer), discarding files (a dustbin), and so on. Icons can also represent software programs and enable rapid access to the appropriate program.

Examples of two icons representing a folder and a graphic document

IDE interface a type of disk controller that is built into the hard disk drive, cutting out the need for a separate controller or ADAPTER CARD. The drive that connects directly to the MOTHERBOARD is relatively fast and inexpensive.

idle time the time during which the computer is turned on but is not processing any instructions. The computer is waiting for a COMMAND.

if a LOGICAL OPERATOR that tests a CONDITIONAL STATEMENT and, if it is true, performs one task; if it is false, it performs another task.

illegal character a character that is not recognized by a command-driven operating system (*see* EVENT-DRIVEN PROGRAM) in a particular situation. For example, in DOS you cannot use an asterisk (*) or a space when naming a file.

image enhancement the improvement of a GRAPHICS image by smoothing out the jagged edges, changing the colours, adjusting the contrast or removing unwanted details.

image processing any process that relates to manipulation of images from the initial digitizing to manipulating the image (embellishing and refining), saving the image and printing the image.

image setter a high quality, professional grade typesetting machine that creates images at RESOLUTIONS of 1200 DOTS PER INCH or more.

impact printer a printer that relies on contact with the paper and an ink ribbon to imprint the character. It is noisy but has the advantage of being able to producing multiple copies of documents

import to open a file that has been created in one application in another application. The file must be in a form that the new application can read it or the new application must have conversion codes available to it.

incremental backup a backup procedure that takes a copy of only the files on a disk that have been updated since the previous backup was taken. FULL BACKUP takes a copy of all files irrespective of when they were last backed up.

index a list of key words created at the end of a document. The index contains the word and the page references where that word can be found. Some word processor and desktop publishing programs create indexes automatically.

index file a file in a DATABASE MANAGEMENT PROGRAM that keeps a list of the location of records using a pointer system. This allows the sorting and searching of a database to be much faster if the whole record is used.

infection the state of having a VIRUS in a computer system. The virus may not be immediately obvious as it can be present on a system for many months before it is activated. This can be caused by a particular series of keystrokes or it may happen on a special date. The activated virus can cause severe problems or can simply display a rude message.

information DATA that has been compiled into a meaningful form. Information is often used interchangeably with data but this is incorrect.

information superhighway the global network of computers connected by satellites and telephone lines. (*See also* INTERNET).

information technology (IT) a jargon term used to describe all computer, telecommunications and related technology that is concerned with the handling or transfer of information. IT is a vast field incorporating the collection, handling, storage and communication of INFORMATION.

init in the Macintosh operating system a UTILITY file that is executed at start-up. It is similar to 'terminate' and 'stay' resident programs in the DOS operating environment. Inits can conflict with each other and cause a system CRASH. Inits include disk drive drivers, fax card drivers, etc.

initialize to start up or set up the basic conditions. When disks or diskettes are initialized, they are formatted to accept data that will be stored later. (*See also* FORMAT.)

inkjet printer a printer type that forms an image by spraying ink on to a page from a matrix of tiny spray jets. Print RESOLUTIONS of 300 DOTS PER INCH are not uncommon with inkjet printers.

input the INFORMATION to be entered into a computer system for subsequent processing.

input device any peripheral device that provides a means of getting data into the computer. The term thus includes the keyboard, mouse, modem, scanner, graphics tablet.

input/output (I/O) the general term for the equipment and system that is used to communicate with a computer. It ensures that program instructions and data readily go into and come out of the CENTRAL PROCESSING UNIT.

insert mode the input mode that allows input to be typed into a document at the CURSOR point. Text already in the document will be moved to allow for the new entries. Overtype mode, on the other hand, deletes previous type as new material is inserted.

insertion point the point at which text can be entered into a document when typing. It is analogous to the CURSOR in old DOS systems.

installation program a UTILITY PROGRAM that is commonly supplied with application software with the purpose of assisting the user to install the software correctly on a hard disk. The utility takes a step by step approach to the installation to ensure that the correct system files are located on the hard disk and to ensure that the various files are located in the correct DIRECTORY or FOLDER.

integrated circuit a module of electronic circuitry that consists of transistors and other electronic components, usually contained on a rigid board. A variety of boards are plugged into a computer to enable it to perform its various tasks.

integrated program a group of software packages each with a logical relationship to the other components. For example, a typical integrated package may include a word processor, a spreadsheet, a database, a graphics application and perhaps a communications application. The common link is that all these applications operate in a similar manner and it is possible to transfer data between them. Major software manufacturers produce many such packages.

integrity the quality associated with a file that is complete and uncorrupted. For various reasons a file can be corrupted. In this case the file is said to have 'lost its integrity'.

Intel a major manufacturing company that makes integrated CHIPS. The range of chips started with the popular 80286 processor and has progressed to the PENTIUM chip.

interactive processing a system in which the user can monitor the computer's processing directly on the computer screen and make any corrections to the process that are required. In the early days of computers, processing relied on BATCH PROCESSING, when the user had to wait hours to obtain the results of the program.

interface the term for the PORTS and the correct electronic CONFIGURATION between two or more devices that help them exchange data. (*See also* USER INTERFACE.)

interlaced a VDU display technology that produces HIGH RESOLUTION pictures but rapidly moving pictures may appear to flicker or streak. Only half the screen is refreshed (*see* REFRESH) on the first pass with the second half of the screen refreshed on the second pass.

internal command a DOS command that is always available at the DOS PROMPT. The COMMAND.COM program is loaded on start-up and contains codes for common internal commands such as copy, dir, prompt and CD (change directory). External commands run separate program files.

internal hard disk a HARD DISK that is located inside the PERSONAL COMPUTER's case. It uses the main computer's power supply and is consequently cheaper than an external hard drive.

internal memory another name given to RAM and ROM, which is where the computer stores information being used by a program or file.

internal modem a MODEM that is located inside the PERSONAL COMPUTER's case and connected directly to the EXPANSION SLOT. It uses the main computer's power supply.

internet a worldwide system of linked computer NETWORKS. The system can link computers that have different operating systems and storage techniques. There is no main source of information or commands as the system was designed to operate even if one network were destroyed.

interpreter a routine that translates a program written in a HIGH-LEVEL PROGRAMMING LANGUAGE into MACHINE LANGUAGE. The interpreter translates each command at a time and then, once the computer has executed the command, it moves to the next line. If an error has been made in the program the interpreter stops and reports an ERROR MESSAGE. This process allows a novice programmer to learn programming from his or her mistakes as he or she writes the program. (*See also* COMPILER.)

interrupt a signal from the microprocessor that temporarily halts or interrupts processing to allow another operation such as receipt of input to take place. As soon as the operation has been completed the original process continues. The computer is constantly faced with such situations. These are called **hardware interrupts** as opposed to **software interrupts**, which are interrupt signals generated by a computer program.

I/O *see* INPUT/OUPUT.

iteration a COMMAND or program statement that is continually repeated until a particular condition is met. A simple iteration is:

add one to a number until the number is equal to 10.

jaggies or **aliasing** the ragged edges that appear on computer GRAPHICS. They are caused by the square edges of PIXELS, which show up when a curve is drawn.

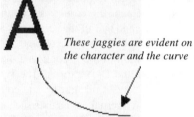

These jaggies are evident on the character and the curve

job an item of work that is performed by a computer, such as BACKGROUND printing of documents.

job queue a series of jobs that a computer is to perform in sequence.

join in a RELATIONAL DATABASE, information from two separate data tables is combined or joined to create another data table that contains summary information.

joystick an INPUT DEVICE controlling the cursor of a computer. The joystick is normally used for controlling computer games.

JPEG (Joint Photographic Experts Group) a FILE COMPRESSION technique that is used to reduce the size of GRAPHICS files by close to 100 per cent. The technique results in some loss of detail but this is minor in comparison to the size of the reduced file.

justification the alignment of lines of text in a paragraph along the margins. Text can be aligned with the left margin, right margin or both.

Line one is left justified

Line two is centre justified

Line three is right justified

Each line of text is shown in a different justification mode

K the abbreviation for kilo as in kilometres. It actually means 1000, but in the computer world it is used rather more loosely because 1 KILOBYTE is actually 1024 bytes. It is commonly used to refer to the relative size of a computer's main memory. 64K is equal to approximately 64,000 characters of information.

KB, kbyte *see* KILOBYTE.

kermit an ASYNCHRONOUS COMMUNICATIONS protocol that is used for telephone communications.

kern to reduce or increase the space between two characters in a display font with the result of placing the characters in a pleasing style.

key a button on a keyboard.

keyboard a set of alphabetic, numeric, symbol and control keys that relays the character or command to a computer, which then displays the character on the screen. The keyboard is the most frequently used INPUT DEVICE.

key field the FIELD that is used as the one for sorting data. For example, a SORT of records in a database of customers using the surname of a customer as the key field will provide an alphabetic list of customers.

keypad the same as the numeric keypad, which is the group of numbers at the right-hand side of a keyboard.

keystroke the action of pressing a key on the keyboard resulting in a character being entered or a command being initiated.

keyword a word in a programming language that describes an action or operation that the computer recognizes.

kilobyte (K, KB, kbyte) the basic unit of measurement for computer memory equal to 1,024 BYTES.

knowledge engineering the process of extracting information from experts and expressing this knowledge in a form that an EXPERT SYSTEM can use.

label text in a SPREADSHEET program as opposed to a number or formula. A label is used for descriptive purposes such as a heading for a row or column.

LAN *see* LOCAL AREA NETWORK.

language a method of communicating. Humans use languages such as English, Spanish, French, etc, while computers use languages such as C, FORTRAN, BASIC, etc.

landscape orientation an optional way of printing a page of text where the page is turned on its side so that it is wider than it is long. (*See also* PORTRAIT ORIENTATION.)

laptop computer a small portable computer that can operate from its own power supply and can be used almost anywhere. It consists of an integrated

LCD screen, keyboard and TRACK-BALL. It is constructed in such a way that it can be carried and operated away from an office base.

Laser-jet a LASER PRINTER manufactured by Hewlett-Packard. Because of its quality and price it has come to be regarded as an industry standard.

laser printer a HIGH RESOLUTION printer that uses a technology similar to photocopiers to fuse the text or graphic images to the paper. Output varies from 300 DOTS PER INCH and greater, although 300 and 600 dpi are the commonest resolutions.

A laser printer

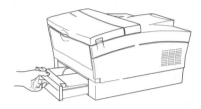

launch to start an application or program.

layer an on-screen sheet on which text or graphic images are placed. These images are independent from text or graphics on another sheet or layer. Such layers are used in page layout programs, graphics programs or CAD programs.

layout the process of arranging text or graphics on a page in programs such as word processing or database management systems

LCD (Liquid Crystal Display) a low power display system that uses crystal molecules to display or not, depending on the connection of an electric current. The displays are difficult to read for long periods and are therefore of limited use for computer screens. The use of backlighting makes the screen easier to use but at the expense of using more power.

LED (Light Emitting Diode) a small light used by various computer devices to communicate information about the status of the device.

legend the key on a GRAPH that shows the meaning of the different colours or shades.

letter quality a style of print that matches the quality of impact printing on a typewriter. The LASER PRINTER has replaced the DAISYWHEEL PRINTER as the standard for letter quality printers.

libraries stores of prewritten programming routines for use in generating applications.

light-emitting diode *see* LED.

light pen a stylus used for INPUT, pointed at a computer display that is sensitive to the light from the display.

line art a computer drawing that consists of only

black and white areas. There are no shades of grey or halftones. Thus line art can be printed on LOW RESOLUTION printers.

Line art in black and white

line feed *see* FEED.

line graph a style of graph using lines to show the relationship between the variables being plotted.

A line graph

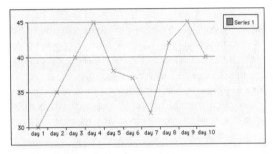

line spacing the space between lines of text in a word processing document or page layout program. Most programs allow at least single spacing or double spacing.

link to establish a connection between two computers (as in a NETWORK) or two programs or two files. Where two files are connected the purpose of the connection is to allow the changes in one file to be reflected in the other file. With a cold link the user must initiate a command to update the target file whereas with a HOT LINK the computer performs the task automatically.

liquid crystal display *see* LCD.

LISP (LISt Processing) a HIGH-LEVEL PROGRAMMING LANGUAGE used to a great extent in the development of ARTIFICIAL INTELLIGENCE.

list box a box that appears as part of a DIALOG BOX and lists various options from which the user can make a choice.

load to transfer a program from a computer's secondary storage to the primary memory (RAM) so that it can be activated.

local bus a high speed EXPANSION SLOT that allows high speed transmission of information to travel between the computer processor and a PERIPHERAL DEVICE such as a monitor. The alternative would be to use an expansion BUS, which is slower

local area network (LAN) a grouping of personal computers that are linked by cables within a restricted area. This enables the users to share peripheral devices and information stored either on the individual machines or on a FILE SERVER. The flow of information around the network is controlled by programs using PROTOCOLS or rules. ETHERNET and APPLETALK are examples of protocols.

A simple local area network

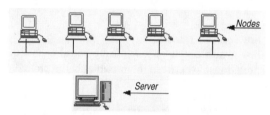

local drive in a network of computers the WORKSTATION might have a built-in DISK DRIVE, which is referred to as the local drive, as opposed to the server drive or remote drive.

lock to protect a file being altered or changed either accidentally or deliberately. Files can be locked by a software utility or by physically locking a FLOPPY DISK with a WRITE/PROTECT tab.

log off to end a session working at a computer terminal or system.

log on to begin working at a computer terminal or system. In MS-DOS (Microsoft disk operating system) to log on means to activate a drive. In networks, a PASSWORD may be necessary to log on to the system.

logical drive *see* PHYSICAL DRIVE.

logical operator a special word (e.g. AND, OR, NOT) used in a programming statement that expands or limits a search. For example, when searching a database the query may be to find all occurrences of customers living in London. This could be restricted with the revised query find all customers living in London AND who have purchased goods in the last month.

logo a HIGH-LEVEL PROGRAMMING LANGUAGE that is commonly used in education to teach programming concepts.

look-up function in programming, a procedure in which the program consults a pre-defined data list (**look-up table**) to obtain information or for comparison purposes.

loop in programs and MACROS a loop is a set of instructions that tells the computer to continue performing a task until a certain condition is met or the loop has been repeated a certain number of times.

Lotus a major software design company that made its first major impact with the SPREADSHEET package Lotus 123.

lowercase *see* UPPERCASE.

low-level programming language a style of computer language that uses codes or expressions that are similar to the MACHINE CODE instructions understood by the processor chip. (*See also* HIGH-LEVEL PROGRAMMING LANGUAGE.)

low resolution screen or printer output that is of low quality. The fewer DOTS PER INCH that a printer can produce, the lower the quality. The fewer PIXELS on a screen the lower the quality of output. Lower resolution produces more JAGGIES on an image.

luggable a PERSONAL COMPUTER that is too big to be described as portable but is small enough to be transported easily from place to place.

machine code the basic 1s and 0s a computer processor uses as its instructions.

machine language a BINARY language that all computers must use. Machine code uses the lowest form of coding, binary, to instruct the machine to change the numbers in memory locations. All other computer languages must be compiled from their high-level code into machine code before the programs can be executed.

Macintosh a line of computers designed and manufactured by Apple Computer. First released in 1984, they introduced GRAPHICAL USER INTERFACE to the PERSONAL COMPUTER world.

Mac OS the version of the disk operating system written by APPLE that is packaged with their MACINTOSH computers.

macro a record of commands used regularly in an application that can be activated by a keystroke. The macro could be a list of commands used to print a report. Without the macro, the report will require several commands to be executed while if they are recorded in a macro, one command or keystroke can be initiated to print the report.

magnetic disk a secondary storage device that consists of a plastic disk coated with magnetically sensitive material. Magnetic disks are usually described as floppy disks or hard disks depending on their construction. Hard disks generally have a higher storage capacity.

magnetic field a force surrounding electrical devices that can have an adverse effect on data stored on MAGNETIC MEDIA.

magnetic media any of a wide variety of disks or tapes, coated or impregnated with magnetic material, on which information can be recorded and stored. The magnetic coating is repositioned when influenced by a MAGNETIC FIELD, and the READ/ WRITE HEAD emits a magnetic field when writing to the disk or tape, which produces a positive or negative charge corresponding to that item of data. When reading, the head senses the charges and decodes them. Disks are used universally but for very high capacity storage, magnetic TAPE is ideal.

magnetic tape *see* TAPE.

mail gateway an electronic path that allows ELECTRONIC MAIL to be sent between different mail services or direct to a computer on the INTERNET.

mailbox within the ELECTRONIC MAIL system, a disk file or memory area in which messages for a particular destination (or person) are placed. Modern BULLETIN BOARD communications systems use a mailbox metaphor to store messages for electronic mail users. The bulletin board system is a telecommunications utility that facilitates informal communication between computer users.

mail merge the process of merging two files for the purpose of creating a mail shot. One file consists of a letter while the second file consists of a database of names and addresses. Each name and address in the database is merged with the letter, creating a letter addressed to each name in the database.

mainframe any large computer such as an IBM or a Cray. They do not use the same architecture as small desktop computers and are intended for use by many people, usually within a large organization. To begin with, it referred to the large cabinet that held the CENTRAL PROCESSING UNIT and then to the large computers, developed in the 1960s, that could accommodate hundreds of DUMB TERMINALS. Now the word mainframe applies to a computing system that serves the needs of a whole organization.

main memory *see* RAM.

margin the space between the edge of a page and the start of the text.

math coprocessor a chip used for performing FLOATING POINT CALCULATIONS.

MB, mbyte *see* MEGABYTE.

megabyte (MB, mbyte) one million bytes (characters) of information. The common storage measurement for memory and hard disks, e.g. 4 megabytes of RAM, with a 210-megabyte HARD DISK drive.

megahertz (MHz) a measurement of one million HERTZ.

megastream a name used by British Telecom for its high speed digital communication lines.

membrane keyboard a style of keyboard covered by a touch sensitive material to prevent liquid or dirt entering the keyboard circuits.

memory the circuitry and devices that are capable of storing data as well as programs. Memory must be installed in all modern computer systems. It is the computer's primary storage area, e.g. RAM as distinguished from the SECONDARY STORAGE of disks. Typical memory devices are SIMMS, which are plugged into the MOTHERBOARD of the computer. SIMMS are plug-in modules that contain all the necessary chips to add more RAM to a computer. The motherboard is the large circuit board that contains the CENTRAL PROCESSING UNIT, RAM, EXPANSION SLOTS and other microprocessors.

memory address a code or name that refers to a specific location where data is stored in a computer's RAM.

memory cache *see* CACHE.

memory management the process of efficiently using a computer's memory. Most OPERATING SYSTEMS have built-in memory management systems to control the use of memory and its allocation between conflicting programs.

memory map a map that shows how the OPERATING SYSTEM utilizes the RAM.

memory resident program a program that remains in memory ready for use at any time. The program occupies a proportion of the RAM.

menu a list of commands or options that are available to the computer user on a monitor or VDU. A user is presented with a menu that will give a choice of commands or applications. Menus make the computer system easier to use. A PULL DOWN MENU is a selection of commands that appears after a command on the MENU BAR of a program has been selected. A command or action is selected and often another menu will appear. The term originated from an idea by MACINTOSH. Menu-driven software contains programs that proceed to the next step only when the user responds to a menu prompt.

menu bar the area of a screen that is given over to the listing of menu items.

A menu bar

menu-driven program a program that proceeds to the next step only when the user responds to a menu prompt.

merge to draw two pieces of information or records together to create a new file or for a particular purpose, such as merging a letter file in a word processor with a data record in a database to create a mailshot.

microchip *see* CHIP; MICROPROCESSOR.

microcomputer a small computer, traditionally the smallest size of computer, which is desktop size. The modern use of the term includes any small computer system. In many ways the term has lost its relevance, and its meaning has been somewhat blurred by technological developments. Initially it referred to any computer that had certain key units on one INTEGRATED CIRCUIT called the MICROPROCESSOR. The first PERSONAL COMPUTERS, designed for single users, were called microcomputers because their CENTRAL PROCESSING UNITS were microprocessors. However the distinction between a microcomputer and a minicomputer has all but disappeared, and many microcomputers are now more powerful than the MAINFRAMES of a few years ago.

microfloppy a 3½ inch FLOPPY DISK. It is encased in a plastic shell to protect it from superficial damage.

microprocessor or **microchip** an electronic device (INTEGRATED CIRCUIT) that has been programmed to follow a set of logic-driven rules. It is essentially the heart of any computer system. Also, a processor that is contained on one chip.

microsecond (μs) one millionth of a second.

millisecond (ms) one thousandth of a second.

Microsoft the world's largest software company. Microsoft has developed numerous items of software including operating systems (DOS, Windows, and OS/2) and applications programs (Excel, Word).

MIDI (Musical Instrument Digital Interface) a set of standards that can be used to connect musical instruments, such as digital pianos, to computers.

migrate to move from using one computer platform to another or from one software application to another. A user can migrate from Windows to OS/2 operating systems.

minicomputer a computer system, usually smaller than a mainframe but larger than a microcomputer, designed for many users.

mini tower a small tower style computer system designed to sit on a desk rather than the floor where a normal tower system would sit.

This mini tower sits beside the screen

MIS (Management Information Systems) the current name given to the subject of data processing.

MISC (Minimum Instruction Set Chip) the basis of the next generation of computer chips. They take the concept of RISC chips one stage further.

MNP (Microcom Network Protocol) a STANDARD developed by the communications company Microcom. It is primarily aimed at error detection and correction between communications devices.

mode the state of operation of a computer. In COMMAND mode, the computer will accept commands, in INSERT MODE, text can be inserted into an existing sentence, in EDIT mode text can be amended. A computer responds in different ways depending on the mode.

mode indicator a message displayed on screen that indicates the MODE of operation in which the computer is set, such as EDIT mode, INSERT MODE, sleep mode, wait mode, etc).

modem (MOdulator/DEModulator) a device for converting a computer's DIGITAL signals into ANALOGUE signals that can be transmitted down a telephone line. The modem is an extremely important device, enabling communication and transfer of data all over the world. In order that modems facilitate communication between computers, the modems at each end of the line must conform to the same PROTOCOL.

module part of a program or set of programs capable of functioning on its own.

moiré a type of graphic distortion seen as flickering on the screen caused by placing several high contrast line patterns too close to one another.

monitor another name for display, screen or VDU.

monochrome a type of monitor that displays only black and white pixels (or black with green or another colour).

monospace a font type that uses an equal amount of space for each character in the font family:
'Courier is a monospaced font.'
'Courier is a monospaced font.'
The difference between a monospaced font and a proportionally spaced font can be seen in the above text.

morphing a technique that appears to melt one image into another image to create a special effect, such as creating the impression that a person changes into a panther. The effect is created by filling in the blanks between the figures so that the change from one figure to another is gradual.

motherboard the main printed circuit board in a computer. It contains the main processor chips, the display controllers, sound chips, etc.

Motorola a major manufacturer of processor chips, such as the Power PC range, which rival Intel's PENTIUM processor range.

mouse an input device that controls the on-screen CURSOR. Movement of the mouse on the desktop causes a similar movement of the cursor around the computer screen.

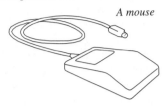

A mouse

MS-DOS Microsoft's disk operating system.

MTBF (Mean Time Between Failures) a measure of the reliability of a computer or, more particularly, of the reliability of a component used in the manufacture of the computer.

multimedia the process of combining computer data, sound and video images to create an environment similar to television. The market for multimedia on compact disks is now expanding rapidly.

multisync monitor a COLOUR MONITOR that automatically adjusts to the input frequency of the adapter card that is used by the computer (VGA, super VGA, etc).

multitasking where a computer processor can undertake more than one task or operation at a time. For example, a print job can be processed at the same time as a spreadsheet is calculating. Compare with BACKGROUND operations.

multithreading the procedure used to describe when a program splits itself into separate tasks or threads. Each thread can operate concurrently with the others.

multiuser system a system that allows more than one user to operate the system at any one time.

multiplexing a technique that is used in LOCAL AREA NETWORKS to allow several signals to pass along the cables at one time. In this way several computers can access the network simultaneously. Special multiplexing devices must be incorporated, which mix the frequency of the signals being sent along the network. The presence of these devices increases the cost of the network.

nano a prefix representing one billionth.

nanosecond (ns) one billionth of a second. These units of measure are used to indicate the speed of operation of a computer CHIP.

nanotechnology the study of how to make computers smaller and more efficient. The term also covers the science associated with the effect of making materials and components smaller.

native file format the format in which a particular program saves a file. The file is saved with certain key characters or codes that tell the program about the various display options associated with the file. The native file format refers to the coding style used by the particular program. Different programs have different formats but can often use the formats for another program to assist FILE TRANSFER.

natural language a language such as English, French, etc, as opposed to an artificial language such as BASIC or COBOL.

near letter quality a mode of operation for DOT MATRIX PRINTERS that produces characters at typewriter quality. This mode has become outdated because of the advent of LASER PRINTERS and INKJET PRINTERS, which produce high quality output faster and quieter.

Netware the Novell company's operating system for LOCAL AREA NETWORKS.

network the interconnection of a number of terminals or computer systems by data communication lines. It may consist of two or more computers that can communicate between each other. Networks for PERSONAL COMPUTERS differ according to scope and size. LOCAL AREA NETWORKS (LANs) usually connect just a few computers (although it may be more than 50), perhaps in order that they can share the use of an expensive PERIPHERAL DEVICE. Large systems are called WIDE AREA NETWORKS and use telephone lines or similar media to link computers together. In general LANs cover distances of a few miles, and some of the largest versions are found in universities and large companies. Each user has a WORKSTATION capable of processing data, unlike the DUMB TERMINALS of a MULTIUSER system.

network administrator the individual who is in charge of a LOCAL AREA NETWORK assisting users and ensuring the correct software is used.

network interface card an adapter card that allows networking cable to be connected directly to the computer. MACINTOSH computers have a basic networking system built into the computer.

network operating system the operating system that is used as a controller for all network components. The network operating system controls FILE SERVER software, the individual workstation software and the network hardware.

network server *see* FILE SERVER; SERVER.

news group a group of people who use an on-line service to discuss interactively topics of mutual interest.

neXT a computer workstation designed by Steve Jobs who was a founding member of APPLE COMPUTER. The computer uses the UNIX operating system with a GRAPHIC USER INTERFACE.

nickel cadmium battery or **NiCad battery** a type

of rechargeable battery used in NOTEBOOK and LAPTOP COMPUTERS.

nickel metal hydride battery a rechargeable battery that is more powerful than the NICKEL CADMIUM BATTERY and so is more suitable for NOTEBOOK and LAPTOP COMPUTERS.

node a connection point that joins two devices, such as the joining of a WORKSTATION to a NETWORK. The workstation is commonly referred to as a node of the network.

noise static that is caused by electrical interference and that can reduce the effectiveness of data communications. DIGITAL communication lines do not suffer interference in the way that ANALOGUE lines do.

non-impact printer a PRINTER that produces text output on plain or special paper without contact between the printing mechanism and the paper. Typical examples are INKJET, bubblejet or LASER PRINTERS. Most are capable of high quality print.

Norton utilities a suite of UTILITY programs from Symantec Corporation, which include undelete options, performance testing programs, and so on.

NOT *see* LOGICAL OPERATOR.

notebook computer a small computer that is generally more compact than a LAPTOP. These are useful for mobile users but are not very satisfactory for sustained usage.

Novell a corporation that specializes in software solutions for networks.

null modem cable a cable used to connect two computers without using a MODEM. These cables are generally used to transfer files from a mobile computer, such as a notebook, to a desktop machine at an office to ensure that files are consistent between the computers.

numeric format in a SPREADSHEET, the way a number can be displayed is controlled by use of the numeric format command. The number can be displayed in a variety of ways, including with no decimals; with two decimals; with currency prefix; as a date or time.

numeric keypad a section of the keyboard that allows numbers to be entered in an easy format (*see* KEYPAD).

num lock key a keyboard key that when pressed fixes the keypad to NUMERIC FORMAT rather than the optional controls or characters.

object linking and embedding (OLE) a set of STANDARDS designed to allow links to be created between documents and applications, thereby enabling information in one document to be automatically updated when the information in the other is changed.

object oriented graphics graphics that are created by a program that creates an image by a mixture of lines, rectangles, ovals, etc, which can be moved independently. Object oriented graphics can be resized without distorting the image.

object oriented programming system a programming environment that consists of a range of objects that have their own programming code. The objects are incorporated into a program by combining them in the sequences required. A drawback with object oriented programming is that it operates slowly and uses a large amount of memory. It is, however, an easy way to be introduced to programming.

OCR *see* OPTICAL CHARACTER RECOGNITION.

OEM (Original Equipment Manufacturer) a business that makes a piece of hardware as opposed to the company that buys the hardware, reconfigures it, relabels it and sells it to the end user. There may be only a few OEMs in the industry that make laser printer drivers but there may be many companies selling laser printers, many of which will have different features.

off-line equipment that is not under the direct control of the CENTRAL PROCESSING UNIT. A printer may be in an off-line state when it is switched on but not capable of receiving data from the computer.

offset similar to GUTTER, it is the space added to a left margin to allow for the document binding.

off the shelf software a software application that is mass-marketed and serves a general purpose, rather than CUSTOM SOFTWARE, which is developed for a specific customer. (*See also* PACKAGED SOFTWARE.)

OLE *see* OBJECT LINKING AND EMBEDDING.

on-line the operation of terminals and peripherals in direct communication with (and under the control of) the CENTRAL PROCESSING UNIT of a computer, i.e. switched on and ready to receive data.

on-line help a utility associated with a particular application that provides a help system for reference while the application is being operated.

on-line information service a profit-making organization, such as COMPUSERVE or America On-line or , that makes information available to its members or subscribers via telephone services. The on-line providers also provide CHAT FORUMS and libraries of information on a vast range of subjects.

open to access a file with its associated application in order to edit the file or print a hard copy of it.

open bus system a design of the MOTHERBOARD where the expansion BUS has slots into which EXPANSION BOARDS can be fixed.

operating environment *see* ENVIRONMENT.

operating system a suite of computer programs (systems software) that control the overall operation of a computer. Operating systems perform the housekeeping tasks such as controlling the input and output between the computer and the PERIPHERAL DEVICES, and accepting information from the keyboard or other INPUT DEVICES. Common operating systems are MS-DOS (*see* DOS), OS/2, UNIX, XENIX. OS/2 (Operating System/2) was developed ultimately by IBM although initially it was a joint development with Microsoft. However, when Microsoft devoted its energies to the improvement of WINDOWS, IBM took OS/2 and it has now gained more acceptance although it is nowhere near rivalling the market share of Microsoft Windows. UNIX is written in language C and was developed by AT & T Bell Laboratories in the 1970s. It can be used on computers from PERSONAL COMPUTERS to MAINFRAMES, and it is suited to MULTIUSER systems. XENIX was produced by Microsoft for use on IBM-compatible computers.

optical character recognition (OCR) an information processing technology that can convert readable text into computer data. A SCANNER is used to import the image into the computer, and the OCR software converts the image into text. No single software package provides a foolproof conversion of text, but the more sophisticated ones have various means of highlighting queries and even recognizing certain unclear characters once the user has responded to the first case. It is a useful tool for inputting large amounts of typewritten or already printed text for editing and changing.

optical disk a type of disk that uses light to write data to the disk and read the data from the disk. It can hold large amounts of information but has slower access times than other SECONDARY STORAGE such as hard disks. CD ROMS use optical disk technology

optical fibre a glass filament that is used to transmit data. Optical fibres can carry huge amounts of information over long distances and do not suffer from electrical interference as do conventional copper wires.

optical mouse an input device that is connected to a computer by light beams as opposed to wires. This allows the mouse much more freedom on the user's desk.

option a choice that the user faces when operating a computer. The simplest options appear in DIALOG BOXES where the choice of command is large but may be print, cancel, options or print preview.

OR *see* FORMULA; LOGICAL OPERATOR.

orientation type can be laid down on a page in a vertical format (PORTRAIT ORIENTATION) or in a horizontal format (LANDSCAPE ORIENTATION). Portrait orientation shows the page taller than it is wide and landscape orientation shows the page wider than it is tall.

OS/2 an operating system created by IBM, which brought a GRAPHICAL USER INTERFACE environment to the the PERSONAL COMPUTER along with MULTITASKING and other advanced features.

outline font a font for printer and screen in which each character is generated from a mathematical formula. This produces a very much smoother outline to a character than can be obtained from a bitmapped font (*see* BITMAP). The use of mathematical formulae means that characters can be changed in size quite easily and only one font need reside in the memory. (*See also* TRUETYPE).

outline utility a UTILITY PROGRAM that is often incorporated in word processing programs and that allows the user to organize thoughts and concepts before creating a report.

output most computer output comes in the form of printed reports, letters and other printed data, or information sent to a mass storage device such as a hard disk drive or a tape drive. The most common output form, however, is the image on the screen. Output devices are peripherals, such as a printer, a magnetic tape drive or floppy disk, that will accept information from the computer.

output device any device that produces a usable form of output from the computer. Printers produce a hard copy of files, documents, etc. A fax machine produces output to another fax machine. A screen (VDU) lets the user view output and a sound card allows the user to monitor sound output.

overtype mode *see* INSERT MODE.

overwrite to save a file on to a disk under a file name that already exists. The original file is deleted—overwritten by the new version.

pack to COMPRESS a file. It means saving files in a way that utilizes a minimum of disk space.

packaged software software that is mass-produced, marketed and sold. The software is the same for all users, unlike CUSTOM SOFTWARE, which is written specifically for a client. (*See also* OFF THE SHELF SOFTWARE).

page break a mark in a document that indicates the end of one page and the start of the next. A page break can be generated in a document by use of a menu or embedded command. The page break is extensively used in word processing or spreadsheet programs to provide the presentation required.

It is possible to differentiate between page breaks. **Soft page breaks** are placed automatically by a program, and as text is inserted, the page breaks move automatically. **Hard page breaks** are inserted into a document by the user and retain their position relative to the existing text.

page description language (PDL) a programming language that tells a printer how to print out a page of text or graphics. Any application that can generate output in a PDL can drive any printer, thus making it device-independent. In order that the printer can interpret the PDL it requires its own processor and memory.

page feed *see* FEED.

page layout program an application that allows the user to mix text and graphics in a document of virtually any page size and almost unlimited extent. Both text and graphics can be inserted from other software programs. DESKTOP PUBLISHING is made possible by page layout programs such as PAGEMAKER and Quark XPress.

PageMaker a leading market PAGE LAYOUT PROGRAM published by the Aldus Corporation (now Adobe). One of the pioneers of DESKTOP PUBLISHING, it is highly flexible and permits the incorporation of text and graphics into a document, achieving electronically what hitherto was accomplished by the typesetter and designer.

page preview a feature of many programs that shows the user the way a full page will appear in print. Page preview can show exactly the format to be printed and will also allow checking of margins, headers and footers along with the placement of graphics.

pages per minute (ppm) a measurement of the number of pages that a printer can output per minute.

pagination the process of dividing a document into pages and numbering the pages ready for printing.

paint program an application that allows the user to create pictures or drawings on the computer by selecting the individual PIXELS that make up the screen display. Paint programs have progressed since the first versions of programs, like Macpaint for the Apple MACINTOSH and Paintbrush for DOS.

palette the menu of colours, brush styles or patterns that can be chosen to create an image in GRAPHICS programs.

pane a term used for each section of a SPLIT SCREEN window. In SPREADSHEET programs a window can be split into four separate panes to make moving around it easier.

Pantone a system that allocates numbers to a range of colours in order that the exact colour match is

A palette of shading options

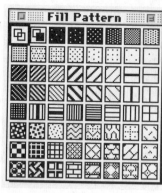

made. The system is used by graphics studios and print shops to ensure that the colour used in the printing is the colour that the artist or designer requires.

paperless office an office in which paper is no longer used or generated. The term was coined some years ago when computers became commonplace, but in practical terms the computer revolution has tended to produce more paper rather than less. However, as technology advances with electronic mail, scanners, fax cards, etc, the era of the truly paperless office may be approaching.

parallel columns a feature of word processing or page layout programs that sets two or more columns side by side.

parallel communication *see* SERIAL COMMUNICATION.

parallel port a port or slot on the back of a computer that is used to transmit high speed SYNCHRONOUS data streams. It is an extension of a computer's internal DATA BUS and is used primarily for connection to printers.

parallel printer a printer designed to be connected to a PARALLEL PORT. The printer cable connection should not be more than three metres long as the risk of interference increases with the length of the cable.

parallel processing the use of two processors combined to undertake one task. The technique is used where there is a requirement for a massive number of calculations, such as in predicting the weather or high quality graphic processing. Parallel processing should be compared with MULTITASKING, which is the use of one processor to undertake two or more tasks simultaneously.

parameter a step in any program sequence that will cause the program to take a specific course of action. It is a value that is added to a command to ensure the task is undertaken in the desired manner. For example, to format the disk in drive A one would type FORMAT A: and the A: is the parameter of the FORMAT command.

parent file in a series of three BACKUPS of a file the parent file is the second oldest file. A conventional backup procedure is to keep three copies of important files. The first is the child, the next the parent and the oldest is the GRANDPARENT.

parity bit an extra BIT added to transmitted data that allows checking for communications errors. The parity bit is attached to each BYTE of data and indicates whether the sum of the bits is odd or even. When the receiving MODEM receives the data, a check is performed to ensure that the sum of the bits (odd or even) is the same as the parity bit. If not, an ERROR MESSAGE is reported.

park to remove the READ/WRITE HEAD from a hard disk to an area of the disk that contains no data in order to protect files during transportation of the disk. If during transport the read/write head touches the surface of the disk there is the risk of data loss. The procedure to park the read/write heads can be achieved automatically or by using a **park utility** program each time the computer is to be moved.

partition a section of a HARD DISK that is created for a particular purpose. A hard disk can be divided, or partitioned, into several parts. For example, one partition could hold the main OPERATING SYSTEM, a second could hold an alternative operating system such as UNIX. A third and fourth could contain data files relating to each operating system. The hard disk can also be partitioned for security purposes, with a PASSWORD required to access the different partitions.

Pascal a HIGH-LEVEL COMPUTER PROGRAMMING LANGUAGE. It has waned in popularity with the advent of C. It was originally designed by Niklaus Wirth for the teaching of structured programming. It is very popular in colleges and schools, and is used as a teaching language and in the development of applications. It is similar to BASIC in that the computer is told what to do by statements in the program. It has some disadvantages in that it is too slow for large-scale development, and commercial versions are changed sufficiently to make them individually isolated.

passive matrix display a form of LCD screen display used for LAPTOP and NOTEBOOK COMPUTERS. A transistor is used to control an entire row or column of the display's electrodes. This type of display is cheap to produce and uses lower battery power than the more expensive but higher quality ACTIVE MATRIX DISPLAY, which uses a single transistor for each display electrode.

password a key word that is selected by a user to protect files from unauthorized access.

password protection a means of allowing files to be protected from unauthorized use. The password may be set up in such a way as to protect the file at different levels. For example, entry of the password could allow the user to read the file but not to edit it, or to have full access to edit and copy the file.

paste a part of a 'cut and paste' procedure in text editing by which a selection of text from a document is moved to or copied to a CLIPBOARD. Once the text is on the clipboard it can be pasted or moved back to a selected area of the document.

path the means of pointing to the exact location of a file on a computer disk. It uses a hierarchical structure of directories (*see* DIRECTORY) in which files are stored. It is the name given to the location of the file, using the directory structure. For example, a file called THISYEAR stored on the C drive in a directory called ACCOUNTS, has the path C:\ACCOUNTS \THISYEAR.

PC (Personal Computer) a MICROCOMPUTER that can be programmed to perform a variety of tasks for home and office. PCs can be equipped with all the necessary software and other devices to perform any task. They can be part of a large DATABASE or a small NETWORK. The boundary between the PC and former specific computing set-ups such as WORKSTATIONS and MINICOMPUTERS is now blurred. Workstations are powerful computers for use by designers, architects and engineers and they feature computer-aided design (*see* CAD) software. PCs are now evolving to this level of sophistication, and similarly are in many instances performing the functions asked of a minicomputer.

In colloquial use, PC is used to refer to IBM-compatible computers, i.e. those running DOS or WINDOWS operating systems as opposed to UNIX or MAC OS.

PCB *see* HARD CARD.

PC card *see* PCMCIA.

PC DOS the version of the disk operating system written by Microsoft Corporation that is packaged by IBM with their personal computers.

PCI (Peripheral Component Interconnect) a STANDARD created by INTEL dealing with the process of communicating directly between PERIPHERAL DEVICES and the computer processor. This standard supports PLUG AND PLAY, whereby the peripheral device can be connected to the computer and immediately used without further setting of DIP SWITCHES.

PCMCIA (Personal Computer Memory Card International Association) a group of manufacturers that has set a STANDARD for credit card-sized PERIPHERAL DEVICES such as memory cards, fax modems, sound cards, etc. The devices are designed primarily for NOTEBOOK COMPUTERS. The credit card-sized expansion cards originally referred to as PCMCIA cards are now more commonly referred to as PC cards.

PDA *see* PERSONAL DIGITAL ASSISTANT.

PDL *see* PAGE DESCRIPTION LANGUAGE.

peer to peer a style of LOCAL AREA NETWORK where all the computers are connected to one another and have access to all the information in the network rather than having a central computer that is used solely as a FILE SERVER containing the files accessible by the network users.

pen computer a style of computer that can recognize handwriting as a method of input. The computer has a touch-sensitive screen onto which the user writes with a pen-like device called a STYLUS. The computer interprets the writing and converts it to a digital form as if the input had been typed. Handwriting recognition software is constantly improving but still requires that the input writing is in the form of print rather than cursive writing.

pentium the name of the latest microprocessor chip from INTEL. The pentium is a RISC that contains over three million transistors. It can operate at speeds of more than double that of it predecessor, the 486DX2 chip. (*See also* POWER PC.)

peripheral device a generic term for equipment that is connected to the computer. These are external (to the CENTRAL PROCESSING UNIT) and include such devices as external DISK DRIVES, PRINTERS, MODEMS, CD ROMS, SCANNERS and VDUS.

personal computer *see* PC.

personal digital assistant (PDA) a portable battery powered computer, slightly larger than the palm of a hand, which is generally used for a small range of specific purposes such as note taking, address book, agenda calendar and to do lists. Some PDAs, such as the Newton, have handwriting recognition systems, and most have the facility to use a fax, modem and PC card.

personal information manager a DATABASE MANAGEMENT PROGRAM designed specifically to emulate a diary, address book and notebook. These are best used with a NOTEBOOK COMPUTER or PERSONAL DIGITAL ASSISTANT, since a diary or notebook held on a desktop computer is not very useful while the user is away from the desk.

physical drive the hardware that is used as the storage device for a computer. A computer can have more than one physical drive. This can be compared with a logical drive, which could be a partition of a physical drive or a section of the RAM set aside to act as a storage area.

pica a measure of FONT size equal to 12 POINTS ($^1/_6$ of an inch).

PICT (abbreviation for picture) an object oriented graphics file format developed by Apple for the MacDraw program. The format uses separate GRAPHICS objects such as lines, rectangles, arcs and ovals, each of which can be independently moved or sized. A file saved in this format can be read by many programs.

pie graph a form of graph used to present data in a visually attractive fashion. Pie charts are generally shaped like a circle and may have offset or exploded sections to highlight a particular figure.

PIF (Program Information File) a file containing information about DOS applications that assists Microsoft WINDOWS in running the application. The file contains data such as the filename, how to display the file and the amount of memory to use. Windows can run a DOS application even if there is no PIF available.

pin feed similar to TRACTOR FEED.

piracy the unauthorized copying of software, the rights of which belong to someone else. This is theft. (*See also* SOFTWARE PIRACY).

pitch a measurement of the number of characters that a printer prints in a linear inch. Pica pitch uses 10 characters per inch while elite pitch uses 12 characters per inch.

pixel a picture element that is the smallest dot that can be displayed on a screen. The pixels make up the picture to be displayed on screen.

plasma display a type of display screen that uses charged gas particles to illuminates the screen.

platen part of the friction device that pulls paper through a printer. The platen also acts a solid surface on to which the write head can impress the image onto the paper.

platform independent the description of a NETWORK that allows computers using different operating systems to be present. For example, a network of Macintosh computers can have one or more PCs on the network, and vice versa.

plot to create an image using lines rather than a series of dots.

plotter a hardware device that creates drawings by moving a series of pens, usually of different colours, across a page. Plotters are commonly used in computer-aided design (*see* CAD) applications and other detailed graphic presentations.

plug and play a technology that allows a PERIPHERAL DEVICE to be connected to a computer and then used without further setting of DIP SWITCHES. The computer will automatically create the necessary connections. This technology was advertised in the WINDOWS 95 operating system and has been used in MACINTOSH computers since 1985.

point the size of a FONT generated by a printer. There are 72 points in an inch. Normal fonts are printed at around 10 points. It also refers to the use of the mouse or other pointing device, such as a TRACKBALL, to position the CURSOR at a specific place on the screen. Additionally, it is to point to commands or data located in a separate record in a DATABASE MANAGEMENT PROGRAM.

point and click a description of the process of using a device that POINTS, such as a MOUSE or TRACKBALL, to select a command. The user points to a menu command or file and clicks the mouse or trackball button once or twice and the command is selected or the file is opened.

pop up menu a menu of command options that appears on the screen when the user points and clicks a particular part of the screen or uses a second mouse button. (*See also* PULL DOWN MENU).

port a plug or socket through which data may be passed into and out of a computer. Typically, each input and output device requires its own separate port. Most computers have a PARALLEL and a SERIAL PORT. The parallel port is a high-speed connection to printers while the serial port enables communication between the computer and serial printers, modems and other computers. In addition to receiving and transmitting data, the serial port also guards against data loss.

portable computer a computer that can be packed up and moved to a different location. Some so-called portable computers would be better described as LUGGABLE. Truly portable machines weigh less than 10 pounds, can be contained in one case and have their own battery power. (*See also* NOTEBOOK COMPUTER).

portrait orientation a description of the normal way of printing a page of text. The page is longer than it is wide. (An A4 page is 297 x 210mm ($11^1/_2$ x $8^1/_4$ inches).

POST (Power On Self Test) a test that a computer carries out on start-up to ensure that the main components are working correctly.

post to add data to a record in a DATABASE MANAGEMENT PROGRAM or similar program, e.g. for keeping accounting records.

Postscript an example of a PAGE DESCRIPTION LANGUAGE for printing, created by Adobe. It is the most popular PDL in use because it is regarded as an industry standard.

power down to turn off a computer. It is important to power down correctly as most programs must be shut down in a proper sequence otherwise data may be lost or corrupted.

power PC a MICROPROCESSOR chip manufactured by Motorola and used by IBM and APPLE. It uses RISC technology and has the advantages of being relatively cheap to manufacture and consuming less power than other chips. (*See also* PENTIUM).

power supply a device in a computer that converts the AC mains supply to DC current used by a computer. The power supply tends to be heavy and contributes greatly to the weight of a computer.

power up to switch on a computer and load the OPERATING SYSTEM ready for use.

power user a computer user who is able to use all the advanced features of a program or series of programs. A power user would be generally regarded as an expert.

ppm *see* PAGES PER MINUTE.

precedence the order in which arithmetic operations are performed. The order is important when creating formulae, e.g. in a computer program or spreadsheet. The rules of precedence are (1) exponential equations; (2) multiplication and division; (3) addition and subtraction. For example, the formula 6+5+9+8/4 does not give the average of the four numbers since the division is carried out before the additions. The result of the above is 22. In order to obtain the average of the four numbers, the formula must force the additions to be performed first. This is done with brackets as follows (6+5+9+8)/4, giving the correct average of 7.

primary document *see* CLIENT DOCUMENT.

primary storage a computer's main RAM or ROM, unlike SECONDARY STORAGE such as hard disks, compact disks and optical disks.

printer the device that produces hard copy. Two main technologies are used: IMPACT PRINTERS, which operate by striking an ink ribbon onto paper to produce an image; and NON-IMPACT PRINTERS, which make no contact with the paper. Printers vary enormously in quality of output, speed, fonts available, and so on. In addition to the DAISYWHEEL and DOT MATRIX PRINTER, which are impact printers and have distinct disadvantages in terms of flexibility and quality, there are the non-impact printers. The INKJET PRINTER creates text and graphics by spraying ink on to the page, and there is always some risk of smudging the image. Although their RESOLUTION and quality does not approach that of the LASER PRINTER, however, they do tend to be less expensive.

Laser printers use a process similar to that of photocopiers whereby ink powder is fused on to the paper. To begin with, the major disadvantage of laser printers was their high cost, but with the advent of reasonably priced machines, the benefit of high quality print, quiet running and built-in fonts is more widely available. There are other types of printer available, e.g. THERMAL, but they do not form a significant part of the market.

printer font a FONT that a printer keeps in memory and uses to produce output on a page. The printer fonts sometimes differ from the display fonts used on screen, but with TRUE-TYPE and POSTSCRIPT, the display fonts are the same as the printer fonts.

printer port a port or slot on the computer to which a printer is connected. The port may be a SERIAL or a PARALLEL PORT.

print queue a list of files to be printed that are temporarily held by a PRINT SPOOLER. The print spooler operating in the BACKGROUND sends each file in turn to the printer while the computer can operate normally in the FOREGROUND.

print spooler a UTILITY PROGRAM that maintains a queue of files waiting to be printed. The print spooler sends a file to the printer whenever the printer is ready to receive another document.

processor a device in computing that can perform arithmetical and logical operations.

processing the normal operation of the computer acting upon the input data according to the instructions of the program in use.

Prodigy an ON-LINE INFORMATION SERVICE providing on-line shopping, news reports, CHAT FORUMS, ELECTRONIC MAIL, etc. It was developed by IBM and Sears.

program a set of instructions arranged for directing a digital computer to perform a desired operation or operations. Programmers use a variety of HIGH-LEVEL PROGRAMMING LANGUAGES such as BASIC, C and FORTRAN to create programs (*see* PROGRAM LANGUAGE). At some stage the program is converted to MACHINE LANGUAGE so that the computer can carry out the instructions. Computer programs fall into one of three categories:

(1) system programs, which are the programs required for the computer to function. They include the OPERATING SYSTEM (e.g. Microsoft disk operating system, MS-DOS).

(2) UTILITY PROGRAMS, which includes those programs that help keep the computer system functioning properly, providing facilities for checking disks, etc.

(3) APPLICATION PROGRAMS, the programs that most people use on their computers, such as word processing, database management, financial pack-

ages, desktop publishing, graphics and many more.

programmable something is programmable if it is capable of receiving instructions to perform a specific task. A computer is programmable since you can instruct, or program it, to perform a variety of tasks.

programming the procedure involved in writing instructions that the computer will follow to perform a specific task. Programming is part art and part science. The process can be summarized as follows:

(1) decide on the purpose of the application.

(2) collect and write down all the important factors and variables, using flow charts to depict the decision processes.

(3) translate the ideas into a programming language.

(4) compile the program to convert it into machine language that is understood by the computer.

(5) test and eliminate errors from the program.

(6) over a period of time make adjustments to enhance the program.

programming language a language that is used by computer programmers to write computer routines, e.g. COBOL, BASIC, FORTRAN, PASCAL, C, Visual Basic. The HIGH-LEVEL PROGRAMMING LANGUAGES, such as BASIC, C, Pascal, are so called because the programmer can use words and arrangements of words that resemble human language, and this leaves the programmer free to concentrate on the program without having to think of how the computer will actually carry out the instructions. The language's COMPILER or INTERPRETER (both programs the former of which is the faster) then turns the programmer's instructions into MACHINE LANGUAGE that the computer can follow.

prompt a symbol or message that informs the user that the computer is ready to accept data or input of some form. A prompt could be 'Are you sure you want to quit the program?' Yes or No. A more common prompt in DOS is C:>, which tells you that the computer is waiting for input.

proportional sizing the process by which the user changes the size of a GRAPHICS object without altering the relative dimensions of the image.

proportional spacing a font in which each letter takes up space relative to the size of that letter. For example, the letter m takes up more space than the letter i. This compares with a MONOSPACE font, which allocates the same space to each letter.

proprietary a term for technology that is developed and owned by a person or company who restricts the use of the technology. If anyone wants to use that technology a fee or licence has to be negotiated. Some developers, on the other hand, make their technology open or available to anyone who has a use for the technology. APPLE uses proprietary systems for its MACINTOSH, and IBM restricts the use of the code in its ROM.

protocol the conventions or rules that govern how and when messages are exchanged in a communications network or between two or more devices. There are many different protocols, but communicating devices must use the same protocols in order to exchange data.

public domain software a computer program that the author has decided to distribute free to users. Others can then redistribute the software without any permission being required. The author may put restrictions on the use of the program, but most SHAREWARE is not advanced enough to restrict code use.

pull down menu a selection of sub-options related to a command name on the main menu bar. For example, a style command on the main menu bar could have the sub-options of bold, italic, underline, double underline, superscript, subscript, text colour.

quad spin CD ROM a CD ROM that spins at a rate four times faster than the original CD ROM drives. Information can be accessed the more quickly the faster the CD ROM spins.

query in a DATABASE MANAGEMENT PROGRAM, a query is when the user asks the program to find a particular reference or type of data that is in the records of the database.

queue when two or more files are waiting for an action to take place. For example, in a LOCAL AREA NETWORK several users may send files to the printer for output. Each print job is added to a queue and is processed in turn.

Quickdraw the OBJECT-ORIENTED graphics and text display technology that is coded into the ROM of MACINTOSH computers. It is used for drawing GRAPHICS on screens and printers.

quit to exit from a program. It is important to quit from a program in a proper fashion, i.e. through the correct menu command, as failure to do this could result in loss of data or program preferences.

QWERTY the standard typewriter/computer keyboard, denoted by the letters on the top line of characters. It was originally developed to slow down typists to stop the manual typewriters jamming the keys. There are now other alternatives to QWERTY keyboards that enable faster use and easier learning. One such system is the DVORAK KEYBOARD, which places the most frequently used letters together.

Key layout on a QWERTY keyboard

radio button a round button that allows the user to choose one of a range of OPTIONS in a DIALOG BOX. Option buttons differ from CHECK BOXES in that several options can be checked in check boxes but only one radio button option can be chosen.

RAM (Random Access Memory) or **main memory** or **internal memory** the memory that can be, and is, altered in normal computer operations. The RAM stores program instructions and data to make them available to the CENTRAL PROCESSING UNIT (CPU), and the CPU can write and read data. Application programs use a part of the RAM as a temporary place of storage, allowing modification of the file in use until it is ready for storage permanently on disk. Any work in the RAM that has not been saved will be lost if the power fails.

random access the retrieval of information randomly from any part of the computer memory or from magnetic media, which means that the computer can reach the information straight away without having to go through a series of locations to reach the desired point.

random access memory *see* RAM.

RAM cache a part of the RAM that is set aside to store data and programs in order that the computer can operate more speedily. The computer processor can transfer data and program code to and from the RAM cache many times faster than it can transfer data and program code to and from a hard disk. If the CENTRAL PROCESSING UNIT had to wait on the hard disk to complete an operation there would be no point in having faster processors.

RAM disk an area of RAM set aside by a UTILITY PROGRAM that is formatted to act like a disk drive. The RAM disk is volatile, and the contents will be lost when the computer is shut down. There is a significant advantage in speed to be gained from using a RAM disk, but it is important to ensure that data is stored onto a floppy or hard disk on a regular basis to avoid the risk of data loss.

range a CELL or a group of contiguous cells in a SPREADSHEET program. A range of cells could include a ROW, part of a row, a COLUMN, a part of a column or a group of cells spanning several rows and columns. Ranges are usually given names that relate to the information contained in the range. For example, the range D5.. D9 might contain sales for a division of a business and could be given the name 'salesdiv1'. Any future reference in a formula to the range D5.. D9 could use 'salesdiv1', e.g. Sum(salesdivl).

raster display the type of display found in TV sets. An electron beam scans the screen many times a second, moving in a zigzag pattern down the screen. Each horizontal line is made up of dots that are lit up individually to create a pattern

reboot to restart the computer without turning off the power. The computer's primary RAM memory is initialized and the operating system is reloaded. (*See also* BOOTSTRAPPING.)

read to retrieve information stored on magnetic media and transfer it to the memory of the computer.

read.me file a text file that is often included with program disks and contains up-to-date information about the program, provides updates to the program's instruction manual or gives technical hints or tips about the operation of the program.

read only attribute information stored in a file's directory that tells the computer whether or not the file can be modified or deleted. If the file is read only, it cannot be changed in any way.

read only memory *see* ROM.

read/write head the electromechanical means whereby information stored on magnetic media can be retrieved and transferred to the memory of the computer.

read/write head alignment a DISK DRIVE must be correctly set up otherwise it may not be able to read a disk correctly. Incorrect alignment may be caused by jolting the computer during transport.

real time the near-instantaneous processing of data and feedback so that the user can respond immediately to the computer program. Programs that use real time processing include flight simulator games, on-line chat forums and point-of-sale recording.

recalculation method the selected method chosen to recalculate a SPREADSHEET after the values in a CELL or number of cells have been changed. The spreadsheet can be automatically recalculated or it can be manually recalculated when the user commands.

recalculation order the sequence that a SPREADSHEET program uses to calculate a spreadsheet. When spreadsheets first appeared, the method of recalculation was restricted to COLUMN-wise or ROW-wise, i.e. the calculation proceeded down one column before moving to the next column. This produced some errors if the spreadsheet logic was not well thought out. More recent spreadsheets use a natural recalculation method that scans the sheet to find the logical re-

calculation order and calculates a CELL only once all dependent cells have been calculated.

record to store data on a disk. Also, all the information related to a topic in a database of information. For example, in a database of customers of a business a record will contain all the relevant information about one customer

recover to restore lost or damaged files. The files can be recovered by restoring from a BACKUP copy or undeleting the file. The backup may not have the most recent changes recorded, and so the full, up-to-date information may not be recovered. The undelete method depends on the information on the disk not having been overwritten. UTILITY PROGRAMS that undelete files are commercially available.

redlining the process of marking changes or additions to the text of a document when comparing different versions.

> the circuitry and devices that are capable of storing data as well as programs. Memory must be installed in all modepn computer systems. It is the computer's primary storage area, e.g. RAM (random access memory) as distinguished from the secondary storage of disks. Tipical memory devices are SIMMs, Single In-line Memory Modules, which are plugged into the motherboard of the computer.

Errors are highlighted for redlining

reformat to repeat a FORMATTING operation on a disk or to proceed with a formatting operation on a disk that has already been formatted. In word processing, page layout or spreadsheet program, to reformat means to change the arrangement or style of the text.

refresh to update the image on the computer screen. The screen is being constantly updated by an electron gun firing electrons at the phosphorous screen. When the computer processor sends a change to the video output it must be reflected on the screen. Each time the screen is refreshed it will reflect the changes.

refresh rate the speed at which the monitor updates its display.

relational database a type of DATABASE MANAGEMENT PROGRAM in which data is stored in two-dimensional tables that are indexed for cross-reference. Reports can be created using data from two files that are related in a particular way, such as by customer name. For example, one file could contain information on a customer's name and address and another could contain details about stocks of goods. A report can be created showing the amount of one type of stock bought by a particular customer.

relational operator a sign that is used to specify the relationship between values. Relational operators are used in queries on databases. For example, to obtain a report on transactions between two dates an expression such as

=>'1/12/95' & =<'31/12/95'

will extract a list of transactions between 1 and 31 December 1995. Other relational operators are:

=	equal to
<	less than
>	greater than
<=	less than or equal to
>=	greater than or equal to
<>	not equal to

relative cell reference a reference to a CELL in a SPREADSHEET that refers to its position with regard to another cell rather than an absolute reference.

release number the decimal number that is used to identify an improvement in a version of a software program.

removable storage any SECONDARY STORAGE system where the storage medium can be extracted and taken away from the computer. Magnetic tapes, floppy disks and disk cartridges are examples of removable storage.

rename to change the name of a file, directory or disk.

repaginate documents are divided into pages for reporting purposes and when text is added to a page the division between the pages is changed. For the user to see where PAGE BREAKS occur, the document must be repaginated. Some programs repaginate the document automatically.

repeat rate the rate at which a character will be typed on the screen when a particular key is kept depressed.

report a presentation of information in print. The report will be complete with page numbers, footers and headers. It will be formatted in such a way as to make it attractive and easy to read.

reset button a button on a computer (often on the front panel) that allows the user to perform a WARM BOOT or restart of the computer. It does not switch the computer off. (*See also* BOOTSTRAPPING.)

resolution a measurement of the sharpness of an image generated by a printer or VDU. Printer resolution is measured in DOTS PER INCH. The more dots per inch, the greater the sharpness of the image on the paper. VDU resolution is measured in the number of PIXELS and their size. The smaller the pixel and the more there are on screen, the greater the resolution of the image. (*See also* DESKTOP PUBLISHING, DOT PITCH, HIGH RESOLUTION, LOW RESOLUTION).

restore to recreate the conditions or state of a disk, file or program before an error or event occurred to destroy or corrupt the data. Restoration would normally involve use of a BACKUP file made previously.

retrieve to obtain data previously stored on file in order that work can be done on the data.

return a COMMAND KEY on the keyboard that is used to initiate a chosen command.

reverse video a state in monochrome monitors in which instead of black text on a white background displays white text on a black background.

rich text format a STANDARD relating to the creation of a TEXT FILE in a way that the FORMATTING details are available to other programs. Use of the standard allows files to be transported between applications, possibly by telephonic communication, without loss of the format of the text.

right justification the alignment of text in a word processing document along the right margin of the document.

Right justified text, such as this, often looks odd.

ring back systems a security system for communications over telephone lines. When the system answers a call from a remote computer system, it confirms the identity of the remote computer and returns the call using a pre-stored telephone number. Unauthorized access to the main computer is therefore restricted and only authorized users can communicate in this way.

RISC (Reduced Instruction Set Chip) a chip that has a limited number of instructions that the processor can execute, thus increasing the speed of the PROCESSOR. The processor is designed to emphasize the most common instructions and to allow these instructions to work as fast as possible.

robot an electromechanical device that may perform programmed tasks. Robots are commonly used in automated factories to perform repetitive functions. The first industrial robots were developed in the early 1960s, although they consisted of little more than an automated hand. Japan subsequently invested heavily in robots and other countries have now followed suit. Primary uses include welding, painting and component assembly.

A common idea of a robot

ROM (Read Only Memory) the part of a computer's internal memory that can be read but not altered. It contains the essential programs (system programs) that neither the computer nor the user can alter or erase. The instructions that enable the computer to start up come from the ROM, although the tendency now is to put more of the operating system on ROM chips, rather than putting it on disk.

root directory in DOS, the top level or first directory on a disk in which subdirectories are created.

row in a SPREADSHEET program a row is a horizontal block of CELLS that extends from the left to the right of the spread-sheet. The row is usually identified by an alphanumeric character.

ruler in a word processing environment, a bar at the top of the page to assist the user in setting margins and tab stops.

run to initiate or execute a program. The computer reads the code from the disk and stores all or part of the code in the RAM. The computer can then perform tasks.

run time version a special version of an INTERPRETER that allows one application only to be run. For example, a program may have been written in PASCAL and will require a Pascal interpreter. However, the user may not have a Pascal interpreter in his or her computer system. The solution is that the program has its own limited or special interpreter that will only work only with that program.

sans serif a plain type style or FONT, such as Helvetica, that has no detail at the end of the main character stroke.

This is sans serif text

This is serif text

save to transfer the contents of a computer's RAM to a less volatile memory such as a HARD DISK or FLOPPY DISK. It is recommended that work is saved at regular intervals otherwise there is the risk of losing work if the RAM memory is deleted. This can happen if there is a power supply failure or a system CRASH.

scalable font a FONT that can be reduced or enlarged to any size required without distorting the font shape. Postscript and TRUETYPE fonts are examples of scalable fonts.

scaling in presentation graphs, the *y*-axis can be scaled to produce a GRAPH that displays the data in a more visually pleasing form or in a way that emphasizes the results being depicted on the graph.

scanner a piece of hardware that copies an image or page into a computer by creating a DIGITAL image. A scanner works by bouncing a beam of light off the paper and recording the reflected light as a se-

ries of dots similar to the original image. If the dots are created as a variation of 16 grey values, the scanner is using a tagged image file format (*see* TIFF), otherwise a dithered image is created (*see* DITHERING). This is a simulation of a HALFTONE image, which is created by varying the size of, and the space between, the dots to create the image. Digital halftones become distorted when they are sized. Scanners are available in three basic types: a drum scanner, a FLATBED SCANNER or a HAND-HELD SCANNER.

scrapbook a UTILITY PROGRAM that can be used to retain frequently used images, pictures or text. Company letterheads, logos or party invitations are among the items that can be stored permanently in the scrapbook file.

A scrapbook

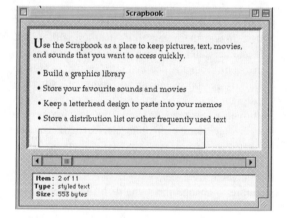

screen another name for the VDU, display or monitor.

screen capture to take a 'snapshot' of the screen at a particular moment. The snapshot can then be used to explain an entry in a technical manual or dictionary of terms.

screen dump a printed output of a snapshot of the screen.

screen font a FONT that is used by a program to display text. ADOBE TYPE MANAGER and TRUETYPE fonts produce screen fonts that are SCALABLE and produce exactly the same design in printed output.

screen saver a UTILITY PROGRAM that is designed to prolong the life of a screen. Over time, images burn into the screen, causing a ghost image or reduced sharpness. The screen saver utility prolongs the life of the screen by switching off the image after a period of non-use. Alternative screen savers create moving images on the screen, such as stars, bouncing balls, fish, flying toasters, etc.

script a list of instructions that automatically perform a task within an application program. The script is a program within a program but can easily be written by automatically recording a series of keystrokes. For example, a user manually enters the commands, telephone numbers and passwords to allow entry to an on-line information service. Once the user is logged on, the keystrokes to retrieve certain information are recorded. The final sequence to log off is also recorded. The next time that the information is to be retrieved from the on-line service the script can be selected to control the process automatically.

scroll to move the ACTIVE WINDOW over a document so that a different part of the document is visible in the window. Scrolling can be vertical or horizontal. In a word processing document scrolling vertically allows sight of the various pages in the document. Horizontal scrolling allows sight of a horizontally oriented spreadsheet.

SCSI (pronounced 'scuzzi', Small Computer System Interface) an INTERFACE connection that allows high speed transfer of information between a computer and one or more peripherals such as hard disks, scanners, or printers. The devices can be connected in a chain so that one SCSI port can support several devices at a time, although data can only be transferred from one device at a time.

second generation the era in computing technology that is represented by use of transistors rather than vacuum tubes in computing devices. This generation was evident in the early 1960s and was superseded by the THIRD GENERATION, when INTEGRATED CIRCUITS replaced the transistor.

secondary storage or **auxiliary storage** a form of permanent storage of data on disk drives. The drives can be HARD DISKS, magnetic tapes, FLOPPY DISKS, CD ROMs or OPTICAL DISKS. The main feature of secondary storage is that it is not deleted when the computer's power is turned off.

sector a storage area on a disk. When a disk is formatted it is organized into TRACKS and sectors. A sector is a pie-shaped division of the magnetic surface on the disk that separates information into individual sections or zones. Data can be stored in these sectors and read by the disk drive. A unit of storage of one or more sectors is called a cluster.

security the method of protecting files or programs so that unauthorized people cannot copy or access them. There are several methods of securing files, including the use of a PASSWORD, physically locking the computer, the use of data ENCRYPTION or by downloading files to removable disks for safe storage. Security is a major concern for computer users since a great deal of damage can be done to files or systems if someone has unauthorized access to a computer.

seek to locate a file on a disk. A FILE ALLOCATION TABLE indicates the part of the disk on which the file or program is stored. The READ/WRITE HEADS are directed to the correct location, and the information is read to the processor. The time that it takes for the read/write heads to reach the correct sector is called the **seek time**.

select to chose a portion of a document or database in which to perform a particular task or to review the selected records. The **selection** in the document may be made in order to change the text formatting or to move it from one part of the document to another. The selected records in a database of sales may be used to analyse results for a particular salesperson.

serial communication a method of transferring data over a single wire, one BIT at a time. Data can be transferred over a relatively long distance using serial communications compared with parallel communication, which is restricted to a distance of around three metres because of the problems of interference.

serial mouse an input pointing device that is connected to a serial port on the computer as opposed to the bus mouse, which is connected to the main processor board.

serial port a PORT on the back of a computer that is set up to allow SERIAL COMMUNICATIONS between the computer and another device.

serial printer a printer that connects to a serial port of the computer. The serial printer is slower than a parallel printer but can be sited at the other end of the office whereas the parallel printer must be within about three metres of the computer.

serif the detailed strokes at the end of the main strokes of a character. Times is a FONT with serifs. (*See also* SANS SERIF.)

server a computer used in a LOCAL AREA NETWORK that is the main source of programs or shared data. The server also controls the use of peripherals such as printers or modems by creating queues of requests that are answered in a sequential order.

service provider a company that provides a connection to the INTERNET in return for a monthly subscription and a charge per hour of use. The service provider may facilitate a simple connection or may make available added services such as news and sports services, on-line shopping, etc.

set up to install a piece of hardware or software into a computer system so that it works with the system. For example, when you buy a new program it has to be set up by creating the correct system files and by placing the various parts of the program in the correct directories on the storage disk. A scanner has to be set up by providing the correct connections to the computer and also by providing the correct software to operate the scanner.

shareware software that can be obtained on a trial basis but to continue to use the program a fee must be paid to the author. Shareware programs are copyrighted. Shareware can be obtained from a variety of sources but most commonly can be downloaded from an ON-LINE INFORMATION SERVICE such as Prodigy or America Online

sheet feeder a device that FEEDS individual pages of paper into a printer or a scanner. The paper is drawn through by a series of friction rollers. For printing, the sheet feeder is predominantly used on LASER PRINTERS and inkjet printers.

shell a UTILITY PROGRAM that operates as an INTERFACE between the user and an operating system that is regarded as difficult to use. For example, WINDOWS is a shell utility for DOS.

shift key a key on a keyboard that allows the user to select an alternative range of characters such as uppercase letters, brackets, pound signs, etc. The key is identified by a white, upward facing arrow or the word 'shift'.

shortcut key a key or more commonly a key combination that allows the user to bypass the normal menu selection process by pressing a key or keys simultaneously. The shortcut is used for commonly used commands such as cut and paste (Ctrl x and Ctrl v), open file (Ctrl o) or create new file (Ctrl n).

silicon the material from which computer chips are made. It is a naturally occurring semiconductor material found in sand and clay.

Silicon Valley an industrial area in the Santa Clara Valley in California that has a high concentration of information technology industries. It can also refer to an area with a concentration of information technology businesses. **Silicon Glen** is an area in central Scotland that has a similar cluster of information technology companies.

Sim City *see* EDUTAINMENT.

SIMM (Single Inline Memory Modules) a small plug-in circuit board that contains MEMORY chips required to add more RAM to a computer. (*See also* DIMM).

simulation an application based on assumptions about behaviour that can be used to produce a model of real life effects. For example, a business simulation can be produced on a SPREADSHEET to show a manager the effect that differing sales levels would have on profit or how advertising may affect sales. In the motor industry, a simulation of a wind tunnel can tell the designers the best shape for a new car design. The basic assumptions are important as they determine the accuracy of the simulation. The more assumptions, the more complex the model but the more accurate the results.

single density a disk storage technology that has been superseded because of the small amount of data that can be stored on the disk. The magnetic particles on the disk are relatively large compared with the much finer particles used in modern high density drives, which can typically hold about four times the amount of information.

single in-line package (SIP) a small circuit board similar to a SIMM. Because it is more difficult to install, it is usually installed by the computer manufacturer.

single sided disk an old type of storage disk that allows only one side to be used to read or write data. Most modern disks allow data to be written on both sides of the disk.

site licence an agreement between the authors of a computer program and a user of the program that allows the user to run the program on an agreed number of computers at one time.

slide show a preset list of graphic presentations that are displayed on a screen in a predefined order. The purpose of the slide show can vary from a sales presentation to a presentation for a directors' meeting.

slot *see* EXPANSION SLOT.

small cap or **small capital** a capital letter that is smaller than a normal capital. The cross-references in this book are in SMALL CAPS, while this is the NORMAL CAPITAL.

small computer system interface *see* SCSI.

smart terminal in a NETWORK of computers, a smart terminal is one that has its own processor and secondary storage. In a network, terminals do not need to be smart, they can be DUMB TERMINALS, in which case they have no processing power or storage other than that of the network server.

snail mail the term used by some to describe the postal service, which can take days to deliver a letter as opposed to the ELECTRONIC MAIL service which can deliver the mail instantly. Electronic mail is much faster as long as the recipient opens his or her computer mailbox.

soft a term used to describe a hyphen or PAGE BREAK inserted by a word processing, page layout or spreadsheet program, as opposed to a HARD command inserted by the user.

software any program that is loaded into a computer's internal memory and that tells the computer what function to perform. Software is also known as programs or applications.

software licence an agreement between the user of a program and its author that gives the user the right to use one copy of it. The user pays a fee for this facility. When payment has been made for the program, the user often considers that he or she has paid for it outright, but this is not the case. The user has paid for a licence to use the software and has not bought it.

software piracy the illegal, unauthorized copying of software. Since software can be copied very quickly, piracy is common and costs software publishers vast amounts of money. COPY PROTECTION can be incorporated in programs, but this often penalizes legitimate users, who require a program copy because of a disk failure, so it has become less common. Some analysts argue that as the cost of software reduces, piracy will decline as more people would rather pay a reasonable price for the official package with all the manuals and technical backup than have a pirated copy.

software publisher the company that writes, markets, sells and distributes a computer program. Major software publishers are Microsoft, Lotus, Word Perfect.

sort a command that organizes data into a particular order. The order can be alphabetical or numeric and can be ascending (from A to Z or 1 to 10) or descending (from Z to A or 10 to 1). A sort command is available in a wide range of applications.

sort key in a DATABASE MANAGEMENT PROGRAM the sort key is the FIELD name by which a SORT is to be conducted. For example, in a database of stock items the sort key could be by stock name or by stock code.

sort order the way in which a sort places data. The most common sort order is to use the ASCII character set. This is a set of 128 characters of the alphabet (upper and lower case) and numbers plus other characters such as ',. } {( )'.

sound board an add-on board for PERSONAL COMPUTERS that gives digital sound capabilities to a computer. MACINTOSH computers have stereo sound facilities built into the system so no added board is required.

source the file or disk from which information or data is taken by the processor. Once the processor has performed a task the data or information is sent to its destination disk or file.

source code the instructions that a programmer creates when writing a PROGRAM. It has to be translated or compiled into MACHINE CODE before the computer can run the program.

SPARC (Scaler Processor ARChitecture) a processor chip design by Sun Microsystems for computers that are used for CAD design work or other high power requirements.

special interest group a group of like-minded users who regularly get together to discuss their chosen topic on-line on an INTERNET connection.

speech recognition an information processing technology in which the spoken word is converted into signals that can be recognized by speech recognition software and converted into commands for the computer to follow.

speech synthesis the process by which the computer translates text into computer generated output that simulates human speech. This type of technology may assist visually impaired people to use computers.

spell checker a UTILITY PROGRAM or part of a larger application that individually checks each word in a document against a dictionary file. The spell checker can show the correct spelling of a word and a list of alternatives if it cannot recognize a word in the text. The chosen version is then inserted into the document.

spike a surge of electricity that at best causes a system CRASH or at worst can burn out components inside the computer. A device to protect against power surges can be bought and placed between the mains supply and the computer.

split screen a facility offered by some wordprocessing and spreadsheet programs that allows the screen to be divided in to two or four panels so that different parts of a document can be viewed at one time. This facilitates editing processes such as copying and pasting from one part of a document to another.

spooler a UTILITY PROGRAM that is used to facilitate printing. When the processor is sending information direct to the printer it must wait for the printer to deal with the output. Since the printer is significantly slower, the processor is not being used effectively. The processor can alternatively send the output to a spooler, which saves all the printing commands. The spooler saves the commands faster than the printer can process them so the processor is free to perform other tasks while the spooler communicates with the printer at the speed of the printer.

spreadsheet a program that creates an on-screen worksheet, which is a series of ROWS and COLUMNS of CELLS into which values, text and formulae can be placed. The spreadsheet is recalculated whenever a change to a cell's value or formula is made. The original spreadsheet (Visicalc) was created in

1978 for use on an Apple II computer. Since then spread-sheets have become more and more sophisticated with database management, analytical graphics, statistical analysis and many more functions available. One of the main uses of spreadsheets is WHAT IF ANALYSIS. By creating a model of a business or economy, many variables can be input in turn to assess the effect of changes in values on the economy or business. Spreadsheets are a very effective tool for businesses of all sizes, for example, in the production of cash flow analysis.

	A	B	C	D	E
1					
2					
3					
4					
5					
6					
7					
8					
9					
10					
11					
12					
13					

A spreadsheet window

SQL *see* STRUCTURED QUERY LANGUAGE.

SRAM (Static Random Access Memory) a fast but relatively expensive MEMORY that is used to create CACHES. Information put into SRAM is held there as long as the computer power is switched on. Information from SRAM can be accessed very quickly by the processor.

stand-alone a term applied to a computer system that is self-contained and has only the hardware and software required by the user.

standard a predefined set of guidelines that is set by the industry's manufacturers to determine the type of INTERFACE between a PERIPHERAL DEVICE and the computer and the way the device communicates with the processor.

standard mode an operating mode for MICROSOFT WINDOWS that speeds up the operation of Windows applications.

star dot star the colloquial name given to the WILDCARD search in DOS and WINDOWS. The use of *.* finds all files because the asterisk can represent any set of characters.

star network a representation of a NETWORK where the network SERVER is located in a central position and the user stations are connected around this central point. The terminals have no connection to one another, only to the server.

start bit the initial BIT sent in SERIAL COMMUNICATIONS that indicates to the receiving computer that the BYTE of data is about to be sent.

start-up disk a disk that contains the operating sys-

tem code required to start the computer. The start-up disk is usually a hard disk but can be a floppy. Having a floppy disk as a start-up disk is very useful, especially if something happens to the hard disk that makes it unreadable.

start-up screen a Macintosh graphics file that, when placed in the START-UP DISK, will be displayed on start-up. Any picture can be used, from one of a car to one of the user.

stationery document *see* STYLESHEET.

stop bit the final BIT sent in SERIAL COMMUNICATIONS, which indicates to the receiving computer that the BYTE of data has been sent.

storage the retention of programs and data in a computer in such a way as to allow the computer processor to access the information when required. The primary storage is the RAM and ROM, and the SECONDARY STORAGE is a device such as an optical or magnetic drive.

street price the price at which a computer or other hardware can be bought as compared with the official retail price set by the manufacturer. The street price is usually below the recommended price since the resellers discount the goods in order to boost or maintain their sales volumes.

string a series of characters that can be used as a basis for a search. For example, the search string 'help' will bring up all occurrences of the word or part of a word 'help'.

structured programming language a computer programming language that encourages the programmer to think logically about the purpose of the program and to avoid the use of the GOTO statement (which can result in a messy unreadable program). The use of named procedures and branch control structures is encouraged. Examples of structured programming languages are C, PASCAL and ADA.

structured query language (SQL) a set of commands used to assist users in obtaining information from a database. SQL has 30 commands and is thus relatively easy to learn and use. SQL can be used for querying databases in both personal computers and mainframe computers.

stylesheet a file that has been saved with all the FORMATTING required for a particular task. The stylesheet has all the required fonts, paragraph indents, font sizes and margins. Stylesheets or stationery documents can be created for many documents, such as standard letters, news sheets, promotional fliers, monthly reports, etc.

stylus a device resembling a pen (with no ink) that is used as an INPUT DEVICE on a GRAPHICS TABLET, screen or PERSONAL DIGITAL ASSISTANT.

subdirectory a DIRECTORY within a directory. For example, a main directory may be used to list correspondence that may be grouped into three different areas—letters, memos and faxes. The three subdirectories of the main directory would therefore be letters, memos and faxes. Letters may then have its own subdirectories for different categories.

submenu a secondary MENU that appears as a set of OPTIONS associated with an option chosen in the main menu. In a GRAPHICAL USER INTERFACE environment using drop down menus, the submenu appears to the side of the main menu choices.

subscript text that is printed slightly smaller and below the main body of text. The numbers in this chemical formula are in subscript: H_2SO_4

suite a collection of several programs that fit together to provide a comprehensive set of tools that a business person may use. It should include a word processing program, a spreadsheet, an organizer, a database, a communications program and a presentation program.

super computer a computer designed to execute very complex calculations at very high speeds. The type of problems that super computers are good at solving include weather forecasting, global warming analysis and economic analysis.

superscript text that is printed slightly smaller and above the main body of text. The numbers in the following are in superscript: $Y= X^2+Z^3$

super VGA a GRAPHICS display STANDARD that can display from 800 pixels by 600 vertical lines to 1024 pixels by 768 lines with 256 colours.

surge protector a device placed between a computer and the mains power supply to protect the computer from momentary increases in the voltage of the power supply.

swap file a file used when a computer's RAM is not large enough to store the full program. A swap file is held on the hard disk, and the computer processor moves information between the RAM and the swap file as required.

synchronous a term used to describe a method of communication that is synchronized with electronic signals produced by a computer. A data BIT is sent with every tick of the computer. (*See also* PARALLEL PORT.)

syntax the set of rules that govern the way in which a command or statement is given to a computer so that it recognizes the command and proceeds accordingly.

syntax error an error resulting from the incorrect spelling of a command or in the way the commands are entered. For example, DOS commands must be entered in strict order of command first, then parameters and then switches.

SYSOP (SYStem OPerator) the person who is in charge of a BULLETIN BOARD or an area in an ON-LINE INFORMATION SERVICE. The SYSOP is responsible for helping users and ensuring that the rules of the bulletin board are maintained.

system or **computer system** all the necessary hardware and software required in an installation, all of which is interconnected and set up to work together (central processing unit, disk drives, monitor, printer, keyboard, and so on).

system 7 a version of the MAC OS, an operating system for the Macintosh range of computers.

system date the date that is held in a computer's internal memory. This memory is not subject to loss when the computer's power is switched off as it is protected by a battery backup.

system disk a disk containing the operating system and all related files. The system disk can be a floppy or a hard disk.

system error an error that occurs at the SYSTEM level of operation of a computer, as opposed to a user-generated error. System errors may be caused by bad programming.

system program *see* ROM.

system prompt an indicator to show that the OPERATING SYSTEM is ready to accept a command. The system prompt is seen in command line operating systems as opposed to GRAPHICAL USER INTERFACE systems. In DOS the system prompt is C> where **c** represents the current disk drive.

system software the group of codes that the computer requires to start up. It includes the OPERATING SYSTEM, which controls all the major functions of the computer.

tab key in a text editing program a key that is used to move text to the right by a fixed or preset number of spaces. The tab key can be used to indent the first line in a paragraph or to create a table of columns. The number of spaces that the text moves to the right is determined by the TAB STOPS. These can be set at any value. The tab key is also used to move the CURSOR between on-screen command OPTIONS.

table text that is arranged in rows and columns in order to display information. A table is also the basic structure for storage of data in DATABASE MANAGEMENT PROGRAMS. Rows correspond to records, and columns correspond to fields of data.

tab stop the point where the cursor stops when the tab key is pressed. Normally the tabs are preset at half an inch but can be altered to any value. There are normally four types of tab stops: left tab stop aligns the text to the left side of the tab stop; right tab stop aligns itself to the right of the tab stop; centre tab stop aligns the text centrally under the tab stop; decimal tab stop aligns the decimal point of a number under the tab stop.

tagged image file format *see* TIFF.

Tandy a major corporation that manufactures and sells a range of personal computers and electronic goods. Originally Tandy used its own OPERATING SYSTEM but since the mid-1980s has incorporated IBM compatibility into its computer systems (*see* COMPATIBLE).

tape a thin strip of plastic, coated with a magnetic recording material, that is normally held in a plastic case. It is used principally as a backup storage medium. It is not used for secondary storage because of the relatively long access time to retrieve data that is a long way into the tape. Data is held sequentially on the tape and since the tape is held on reels the last data to be saved to the tape is at the end of the reel. If the user wants to access the data at the end of the reel the tape drive must spin the reels from the start to the end of the tape. This could take several minutes thus precluding the tape drive as anything but backup storage.

target the destination for a copied file. A file is copied from a SOURCE to a target.

TB, tbyte *see* TERABYTE.

technical support the provision of information, usually by a telephone hotline, for registered users of hardware or software by the manufacturer if they have a problem. The support for the initial period after purchase is normally free of charge. After the initial period (often three months) the technical support is charged at a set rate per hour.

telecommunications the use of the telephone systems, either land lines or satellite, to transmit information (voice, video or computer data).

telecommute to work from a home base rather than commute to the office. Telecommuting has been made possible by the efficiency of the telephone systems. An employee can perform tasks at home and send results to the office computer by using a MODEM. In addition, messages can be transmitted to the employee at home through an ELECTRONIC MAIL system. It has been relatively slow to catch on since employees miss the social contact of work at the office and managers do not like the reduced supervision over workers who are not in the office. Where home working has grown is in subcontracted work as supervision is a lower requirement. People who work at home in this way are often called **teleworkers**. This is becoming more popular, and groups of teleworkers can work to-

gether on a large contract. Much work is undertaken for foreign companies, when advantage can be taken of time differences, e.g. to update records and have them ready for the next working day.

Telenet a commercial ON-LINE INFORMATION SERVICE that allows connection to other computers worldwide. Telenet is usually used as access to the INTERNET.

teleworker *see* TELECOMMUTE.

template a document that is prewritten and formatted and ready for final editing or adjustment before printing. For example, an invoice can be set up with all the headings, automatic date entry, formulae to calculate VAT, etc. The only data required before printing is the customer name, description of the goods or service and the amount.

tera (T) a prefix representing one trillion (10^{12}).

terabyte (TB, tbyte) a measurement of MEMORY capacity that is approximately equal to one trillion BYTES. The actual number is 1,099,511,627,776 bytes.

terminal an INPUT/OUTPUT device consisting of a monitor, keyboard and connection to a central server. A terminal can be a DUMB TERMINAL, which has no processor or secondary storage, or a SMART TERMINAL, which has these facilities and can thus operate on its own as well as operate as part of the NETWORK.

terminal emulation a procedure whereby a TERMINAL or PERSONAL COMPUTER acts like another in order that communications can take place between computers. Since there are many different types of personal computer, STANDARDS have been devised that allow the computers to communicate. These are incorporated into communication programs so that, for example, a home computer acts like the terminal of a large computer, which is the centre of an on-line service. Terminal emulations include TTY, VT42 and VT52.

text alignment the JUSTIFICATION of text in a word processing or page layout document. Text is aligned with reference to the right and left margins of the page. Text can be lined up with the right margin (unusual), the left margin (most common) the centre of the page (for headings) or can be fully justified (spread out so that the text reaches both the left and right margins).

text chart in business presentations a slide presentation that contains no graphics. The slide consists only of text, e.g. showing a menu list of items.

text editor a basic word processing program that is used mainly for writing computer programs and batch files. It has very limited facilities for formatting and printing.

text file a file that contains only ASCII characters. This file type is used mainly for transfer of information between different programs or computers. The file does not contain any formatting codes that could indicate bold or italic text or differing fonts or font sizes. The use of text files is diminishing as many programs have automated FILE TRANSFER procedures to allow one program to read a file created in another program.

text wrap *see* WRAP AROUND TYPE.

thermal printer a printer that uses heat to form an image on special paper. The thermal heads heat the paper, which has a wax-type coating causing a discoloration that results in print. Fax machines use this type of printing device. Despite its speed and quietness it is not popular, mainly because of the waxy feel of the paper.

thesaurus a book or file that lists alternative meanings for words, i.e. synonyms. Most word processing programs now include an electronic thesaurus that displays a list of synonyms for each word selected. Once the new word has been chosen, the program will replace the old word with the new.

third generation an era in computing around the mid 1960s when the transistor was replaced by INTEGRATED CIRCUITS and disk storage and ON-LINE terminals were introduced.

third party vendor a business that buys and sells computer equipment and accessories. The company buys from a computer manufacturer and sells to the end user (a business or the general public). The third party vendor makes life easier for the manufacturer, who does not need to worry about the selling or marketing function, and provides a point of contact for buyers, who can view and assess a range of equipment before purchase.

three-dimensional spreadsheet a SPREADSHEET program that consists of several layers of related pages or worksheets. For example, if a company has four shops, a worksheet can be created for each shop to show its income and expenditure. In order to produce a consolidation of the results of the four shops a fifth worksheet is created that calculates the sum of the figures for each shop. The formulae in this sheet use three dimensional statements such as:

'= Shop1! C9 + Shop2! C9 + Shop3! C9 + Shop4! C9'

throughput a measure of a computer's overall speed of performance as opposed to the speed of a particular element of the computer or system. The slowest component will effectively determine the performance of the computer. For example, a slow disk drive will counter the effect of a fast processor chip. Similarly fast RAM and a RAM cache will be useless if the chip is slow.

tick a single beat of the MICROCHIP that determines the number of instructions that a chip can process. Normally a chip can process one instruction per tick.

TIFF (Tagged Image File Format) a STANDARD relating to graphic images. A TIFF file contains a series of dots (*see* BITMAP) that makes up the image. The dots can be printed, stored on a disk or displayed on a monitor.

tile to set windows in a side by side fashion on the desktop. Tiling windows shrinks the size of the windows so that more can be seen on the screen. Windows can be tiled or shown as CASCADING WINDOWS.

time sharing a technique for sharing resources in a MULTI-USER system. Users do not notice that they are sharing the system resources. If the system does become overloaded with users, they will notice a decline in the operating speed.

title bar a shaded bar containing the name of the file that is found at the top of an on screen window in GRAPHICAL USER INTERFACE systems. The title bar is shaded when the WINDOW is active.

A title bar

toggle a key OPTION that allows switching back and forth between states of operation, i.e. the FUNCTION is switched on and off. The caps lock key is a toggle key that alternatively switches to uppercase mode and back to lowercase mode each time the key is pressed.

token passing a PROTOCOL in which tokens move around a NETWORK. When a NODE wants to send a message over the network it has to obtain a free token. The node that controls the token controls the network until the message has been passed and acknowledged.

token ring network a LOCAL AREA NETWORK that uses token passing technology as the basis for communications.

toner powdered ink that is electrically charged and is used in laser printers and photocopying machines. The toner is applied to a charged drum and fused to the paper with a heating element.

toolbar a strip of buttons that appears at the top of the screen that are used to select commands without using menus. The tool bars can be edited so that the user can choose the buttons that are most appropriate to the tasks that are being undertaken.

Tool bars are commonly used in word processing and other software packages.

A summon button toolbar

toolbox a set of prewritten programs or routines used by programmers for incorporation into larger programs. This saves time and ensures that there is consistent implementation of the toolbox routines, such as printing.

top down programming a method of designing programs. It starts with a basic statement of the program's main objectives, which is then divided into sub-objectives, and so on. The sub-objectives or subcategories are of a type that can be programmed easily. C and PASCAL are STRUCTURED PROGRAMMING LANGUAGES that lend themselves to top down programming.

topology a LOCAL AREA NETWORK layout. Topologies can be centralized or decentralized. A centralized network is like a star and a decentralized network is like a ring.

touch sensitive display a type of screen with a pressure sensitive panel in front of the screen. The panel is effectively lined up with the display options, which are selected by pressing the panel at the correct place. This type of screen technology is used for public information access rather than in business.

tower system of style of computer system in which the electronics, disk drives, expansions cards, etc, are contained in a box resembling a tower that usually sits on the floor beside the user's desk. The VDU and keyboard sit on the desk. (*See also* MINI TOWER.)

tpi *see* TRACKS PER INCH.

track one of a number of concentric circles on a floppy or hard disk. The track is encoded on the disk during FORMATTING and is a particular area on the disk for data storage.

trackball an INPUT DEVICE that is similar to a mouse that is turned upside down. Instead of moving the mouse over the desk to move the ball, the ball is moved within a static unit. The trackball takes up less space on a desktop than a mouse, and in some NOTEBOOK COMPUTERS the trackball is embedded in the case of the computer.

trackpad an INPUT DEVICE that is a development of the trackball. It consists of a square pad embedded in the case of a LAPTOP COMPUTER, and movement of the finger over the pad moves the cursor on the screen.

tracks per inch (tpi) a measure of the density of data storage on floppy disks, higher figures representing a greater capacity for data.

tractor feed a mechanism, similar to PIN FEED, that is used in dot matrix printers to push the paper past the print head. The mechanism has sprockets that engage in prepunched perforations at the edges of the paper.

transfer to move information from disk to memory and vice versa.

translate to change a file that has been saved in one file format to another so that the file can be opened in a different program.

tree structure a way of organizing directories on a disk that shows the core or main directory at the top with the various subdirectories and sub-subdirectories shown like the branches of a tree extending downwards and outwards.

triple spin CD ROM a CD ROM drive that operates three times faster than a standard CD ROM drive. CD ROMs are relatively slow, and so manufacturers have attempted to speed up processing by spinning the drives faster.

Trojan horse a program that appears to perform a valid function but, in fact, contains hidden codes that can cause damage to the system that runs the program. It does not replicate itself or infect other files as a VIRUS does.

troubleshoot to investigate the reason for a particular occurrence or malfunction in a computer system. Very often a failure is caused, not by a major problem, but a small error that can be rectified quite easily.

TrueType a font technology that rivals Postscript and displays on-screen fonts exactly as the printer prints them out. The fonts are SCALABLE and so, no matter which font size is chosen, the screen display and the printer will be the same. TrueType fonts do not require any special printer processors, unlike Postscript fonts, to enable them to be printed, and so the TrueType document is portable between systems.

truncate to cut off part of an entry (a number or character string) either to ensure that it fits a predefined space or to reduce the number of characters for easier processing.

tutorial a process of instructions that guides the user through a series of steps designed to show the features of a program such as a word processor, database or spreadsheet.

twain a STANDARD connected with scanning that allows a document to be scanned without leaving the application into which the image is to be inserted.

type ahead buffer a memory BUFFER that stores the characters being typed on the keyboard so that they can be processed by the RAM when the processor is free.

typeface a set of characters sharing a unique design, such as Courier or Times. Typefaces can be SERIF or SANS SERIF.

type style refers to the WEIGHT of type or the slope of the type as opposed to the size of the type or the typeface. The weight of type refers to how **bold** the type appears and the slope of type refers to whether *italic style* is chosen.

undelete program a UTILITY PROGRAM that is used to restore files that have been deleted from a disk, possibly by accident. The recovery (*see* RECOVER) of the file depends on there having been no other data written onto the disk since the file was deleted as this would affect the storage areas.

underline a command used to highlight text by placing a line under the selected word or portion of text. Double underline can also be used in some word processing packages.

undo a command available in programs that reverses the effect of the previous command given. It allows the user to cancel the effect of a command that has had an unforeseen or disastrous effect.

undocumented features of programs that have not been documented in the program manual.

unformatted a term indicating that an item MAGNETIC MEDIA requires FORMATTING prior to being put into operation.

uninterruptable power supply (UPS) a power supply that switches to an alternative power source, such as a battery, in the event of the main supply crashing. The alternative will probably have a short life but will be long enough to allow proper shutdown to take place, thus preserving data.

UNIX an OPERATING SYSTEM suitable for a wide range of computers from MAINFRAMES to PERSONAL COMPUTERS. It is suited to MULTIUSER situations and can handle MULTITASKING.

unprotected software software that can be copied from the original program disks on to other floppy disks or onto a hard disk. Retaining a copy of a program disk is sensible in case of damage to the original. Some software is protected so that copies cannot be taken, which can create problems if disk damage occurs (*see* COPY PROTECTION).

update to revise the contents of a file, usually a database file, so that the contents reflect the correct and current state.

upgrade to purchase the most recent version of software released by the author or to purchase a new hardware update, such as a new computer system.

upload to copy a file from your computer to another computer connected through the telecommunication system. (*See also* DOWNLOAD.)

uppercase type in capital letters as opposed to lower case.

THESE ARE UPPERCASE CHARACTERS
these are lowercase characters

upward compatibility the ability of an application program to run under a more advanced computer or operating system than it was originally designed for.

user the person who is operating the computer.

user-defined a selection of preferences chosen by the user of a computer.

user-friendly jargon for a system that is easy to learn and operate. A user-friendly program often involves the use of MENUS rather than commands that have to be remembered; help that is available on screen at the touch of a key; ERROR MESSAGES with some explanation of the fault and solution; PROMPTS when performing potentially damaging procedures, e.g. questions such as 'Do you really want to . . .' when deleting a file. Applications software also comes with numerous manuals including one presenting the basics of a program in a way that is easily understood.

Computers using GRAPHICAL USER INTERFACE are generally thought to be more user-friendly than those systems that rely on typed commands, e.g. DOS.

user group a gathering of people with similar objectives who communicate through computers.

user interface a means by which the user communicates with the computer. The different user INTERFACES include a GRAPHICS USER INTERFACE, in which the user communicates through MENUS and ICONS, and the command line interface in which the user has to type the appropriate command at a PROMPT.

utility program a program that helps the user to obtain the most benefit from a computer system by performing routine tasks, e.g. copying data from one file to another. (*See also* DESK ACCESSORY.)

V42 bis a data compression and error correction STANDARD used for communications between MODEMS.

vaccine a UTILITY PROGRAM designed to prevent a computer VIRUS from attacking a system. The vaccine has three things to do: prevent the virus attacking the system; detect the presence of a virus once it is in the system; remove the virus. Some vaccines do all three while others manage only one.

value a numeric CELL entry in a SPREADSHEET program. A value can be a constant that is entered directly into a cell or it can be the product of a formula.

vapourware software under development that is marketed in advance of its release. Often the release is delayed because of technical problems but the marketing goes on, creating the impression that the software does not exist.

VDT (Video Display Terminal) *see* VDU.

VDU (visual display unit) or **monitor** or **screen** or **video display terminal** a device that incorporates a cathode-ray tube and produces a picture of computer input or output. The display is created by firing a series of rays at a phosphorus-coated screen. The screen colours are created by mixing red, green and blue. There is some concern that the electromagnetic radiation given out by VDUs may be hazardous. They emit X-rays, ultraviolet radiation, electrostatic discharge and electromagnetic fields, but it is not proven that these emissions are dangerous to users. However, some studies have indicated that at a distance of 18 inches the radiation is less than that of the background, and this, and the use of low-radiation VDUs is being included in regulations for workers using computers.

vector graphics *see* OBJECT ORIENTED GRAPHICS.

verify a computer procedure that ensures that an operation was completed correctly.

version number the number assigned to a version or release of a program. Each time the author revises the program the version number is amended. Minor changes are indicated with a change in the decimal point while major changes are reflected in the main number.

VESA (Video Electronics Standards Association) a grouping of manufacturers who have devised STANDARDS to ensure that computers and VDUs are compatible.

video adapter the electronic card that generates the graphic output that is displayed on a VDU.

video card *see* **video adapter**.

video disk an optical storage device used for pictures, movies and sound. The disk has a high storage capacity holding up to two hours of TV pictures. Video disks can be used for interactive video when the disk is placed under the control of a computer.

video RAM *see* VRAM.

view an on-screen display of the contents of a file or part of a file. In DATABASE MANAGEMENT PROGRAMS a view can be generated to look at a selection of records. The selection depends on the criteria specified in a QUERY command.

virtual drive part of a computer's internal memory that is defined to act like a DISK DRIVE. Data can be stored temporarily in the virtual drive and accessed very quickly by the main processor.

virtual machine a computerized version of a computer that acts as if the computer was real and can actually run applications.

virtual memory (VM) the use of disk drive storage to extend the RAM of a computer. Because the processor treats virtual memory like RAM, the use of VM slows down the computer processing speed.

virtual reality a computer-generated environment that allows the user to experience various aspects of life without the travel or the danger that may be associated with the activities. A head-mounted display and sensor glove are used to create the effect.

virus a program that is designed to cause damage to systems that the virus infects. A virus program can copy itself from file to file and disk to disk and can therefore spread quickly through a computer system. It can also move between computers through the use of infected disks and also through telecommunication systems. A virus can be detected and removed by a VACCINE.

visual display unit *see* VDU.

voice/data switch a switch that identifies the type of call being received over a telephone line and that routes the call to the appropriate device. Voice calls go to a telephone and fax calls to a fax machine.

voice mail a system that stores voice communication on disk and can replay the message on command. Voice mail systems can be combined with telephone systems to provide computerized answer machines.

voice recognition the ability of a computer to recognize a voice, translate it into a digital pattern and reproduce the pattern as text or as computer generated speech. Voice recognition has improved in recent years, but voice recognition programs are still limited to several hundred words and require considerable training to recognize properly a user's speech.

voice synthesis *see* SPEECH SYNTHESIS.

volatile storage storage of which the contents are lost when power is removed, e.g. the RAM.

volume label a name that identifies the disk. The name is given when the disk is formatted (*see* FORMAT).

VRAM (Video RAM) RAM used in conjunction with a video card together to enhance the performance of video displays.

WAIS (Wide Area Information Server) an application that is used to search the thousands of databases connected to the INTERNET for a selection of keywords. It would be totally impractical to search the Internet manually for a particular reference so an automated program is used to perform the search.

wait state the interval programmed into a computer during which the MICROPROCESSOR waits for the RAM to catch up. As processors are built to run at faster speeds, the memory must be designed to keep up. If this does not happen then the faster processors will be worthless.

wallpaper an on-screen design that acts as a backdrop to the ICONS, WINDOWS, etc. GRAPHICAL USER INTERFACE computer operating systems have facilities to change the patterns through CONTROL PANELS.

WAN *see* WIDE AREA NETWORK.

warm boot a system restart that is initiated usually because a system error has occurred during operations. A warm boot will reset the memory and reload the OPERATING SYSTEM but may not reset PERIPHERAL DEVICES such as modems. A complete power shutdown and restart may be required for such operations. (*See also* BOOTSTRAPPING).

weight the relative thickness and thinness of type. Different TYPE STYLES can have different weights and within a particular style the weight can be graduated between extra light and extra bold.

what if analysis a procedure using a SPREADSHEET to explore the effect of changes in one input into a calculation. For example, what will happen to profits if sales volumesare increased by reducing selling price?

wide area network (WAN) a computer NETWORK that connects computers over long distances using telephone lines or satellite communications. A wide area network can span the world whereas a LOCAL AREA NETWORK can cover only a few kilometres.

wild card a special character that is substituted for another character or range of characters in a search of FILENAMES. In DOS, an asterisk is a substitute for a number of characters while a question mark is a substitute for one character. For example, a search for *ook will find book, look, shook, and so on, whereas a search for ?ook will find book and look, but not shook.

Winchester a type of HARD DISK used for data storage.

window an on-screen frame, usually rectangular in shape, that contains the display of a file. Several windows can be open on a desktop at a time. For example, two different files in a word processor may be open along with a window in a database management program.

Windows (Microsoft Windows) a comprehensive software facility that utilizes the GRAPHICAL USER INTERFACE features that were once the domain of the MACINTOSH. These include PULL-DOWN MENUS, a variety of accessories and the powerful facility of moving text and graphics from one program to another via the CLIPBOARD (an area for temporarily storing items that have been copied from one area, and that are to be pasted elsewhere). It is possible to run several programs at once, each within a separate window, and to move from one to the other very quickly. All applications that run within the Windows system have a common way of working with windows, dialog boxes, etc.

wizard a computerized EXPERT SYSTEM that leads the user through the sometimes complex process of creating a document such as an advertising flyer or a newsletter.

word count a feature of many programs that provides the user with a total number of words contained in a document. It is an invaluable feature of word processing programs.

Word Perfect a popular word processing program written by the Word Perfect Corporation.

word processing a method of document preparation, storage and editing, using a microcomputer/personal computer. Word processing is the most widespread computing application, primarily because of the ease with which documents can be amended before printing. It is used to create, edit, format and print documents. This type of application is one of the most popular to be used on a desktop computer. The main software programs include Word Perfect, Word, AMIPro and MacWrite.

word wrap a feature of word processing programs that automatically moves words down to the next line if they go beyond the right-hand margin.

Word wrap takes the text automatically to the next line when the text goes beyond the right-hand margin.

Word wrap takes the text automatically to the next line when the text goes beyond the right-hand margin.

The right-hand margins are different, thus the text wraps at a different place

workbook a three-dimensional SPREADSHEET.

work group a small group of employees assigned to work together on a specific project. The work group can become more productive if personal computer technology is used to its best effect, i.e. if LOCAL AREA NETWORKS, ELECTRONIC MAIL, shared databases, and so on, are utilized.

worksheet a matrix of rows and columns in a SPREADSHEET program into which are entered headings, numbers and formulae.

workstation a desktop computer in a LOCAL AREA NETWORK that serves as an access point to the network. Programs can be run from the workstation, and all network resources can be accessed.

worldwide web a HYPERTEXT-based document retrieval system linked to the INTERNET. Each page is indexed and can be linked to a related document. The worldwide web is allowing easier access to information available on the Internet.

WORM (Write Once Read Many) an OPTICAL DISK drive that can store up to one TERABYTE of data. Once the data is written to the disk it cannot be altered or erased.

wrap around type type that is contoured so that it surrounds a graphic item in a document. This is a feature of desktop publishing programs such as PageMaker, where it is called text wrap.

write an operation of the CENTRAL PROCESSING UNIT that records information on to a computer's RAM. It more commonly refers to the recording of information on to SECONDARY STORAGE media such as disk drives.

write/protect to protect a file or disk so that a user cannot modify or erase its data. A write/protect tab can be seen in the bottom right-hand corner of the back of a floppy disk, which, when slid down, protects the disk in this fashion.

WYSIWYG (What You See Is What You Get) the feature that what is seen on the screen is exactly what is replicated when the information is printed.

x-axis the horizontal axis of a GRAPH. The horizontal axis usually contains the categories of values being plotted. For example, creating a graph of monthly sales will require that time be plotted on the horizontal or x-axis and sales be plotted on the vertical or y-axis.

x-y plotter a printer that creates a drawing by plotting x and y coordinates provided by the application program. This type of drawing is high precision and is commonly used for CAD drawings, architectural drawings and blueprints.

xmodem an asynchronous file transfer protocol that facilitates error-free transmission of computer files through the telephone system.

XMS (e*X*tended *M*emory *S*pecification) a set of guidelines that standardize the method a programmer can use to access memory above 1024 KILOBYTES, which is the memory limit associated with Intel 8088 and 8086 microchips.

XON/XOFF a method of communicating between MODEMS, whereby software controls the flow of data. In order that the two modems do not send data at the same time, one indicates that it wishes to send data by sending an XON code, and when it is finished it sends an XOFF code. *See also* COMMUNICATIONS SETTINGS.

YMCK an abbreviation for yellow, magenta (red), cyan (blue) and black, which are the basic colours used in printing.

ymodem an asynchronous file transfer protocol that is similar to xmodem but sends files in batches of 1024k as opposed to 128k.

z-axis represents the third dimension in a three-dimensional graphic image. The third dimension represents depth.

zap to delete or get rid of a program or file from a computer memory.

Zapf Dingbats a set of decorative symbols developed by Herman Zapf, a German TYPEFACE designer.

zip to COMPRESS files so that they utilize less space on a disk.

zone a subgroup of networked computers in a LOCAL AREA NETWORK. Messages or ELECTRONIC MAIL can be addressed to everyone in the subgroup.

zoom either to enlarge a window in order that it fills the screen or to enlarge or reduce the size of a page so that the full page can be seen on screen (**zoom out**) or the enlarged detail of the information in a smaller area can be seen (**zoom in**).

zoom box a small box positioned at the edge of a screen window that is used to ZOOM a window.

This is a zoom box on the window title bar

SPORT

A brief explanation of its origin and development is given for each sport listed, together with brief histories of its major events. The entries are in A to Z order, and cross-references are given in SMALL CAPITALS to enable you to relate major competitions and events to their sport.

Admiral's Cup *see* YACHTING.

American football has its origins in the undisciplined ball games introduced by the early colonists to America. The basic rules of the modern game were established by college teams, and the first professional game was played in Latrobe, Pennsylvania, on 31 August 1895. Today it is America's most popular spectator sport.

The most important game of the American football season is the *Super Bowl*, contested by the winners of the two divisional leagues, the NFC (NPL before 1970) and the APC (APL). Inaugurated in 1966-7, it is now America's greatest sporting event. The winners have been:

1967 Green Bay Packers	1968 Green Bay Packers
1969 New York Jets	1970 Kansas City Chiefs
1971 Baltimore Colts	1973 Miami Dolphins
1974 Miami Dolphins	1975 Pittsburgh Steelers
1976 Pittsburgh Steelers	1977 Oakland Raiders
1978 Dallas Cowboys	1979 Pittsburgh Steelers
1980 Pittsburgh Steelers	1981 Oakland Raiders
1982 San Francisco 49ers	1983 Washington Redskins
1984 Los Angeles Raiders	1985 San Francisco 49ers
1986 Chicago Bears	1987 New York Giants
1988 Washington Redskins	1989 San Francisco 49ers
1990 San Francisco 49ers	1991 New York Giants
1992 Washington Redskins	1993 Dallas Cowboys
1994 Dallas Cowboys	1995 San Francisco 49ers
1996 Dallas Cowboys	

archery is an ancient skill that was first established as a competitive sport in the mid-19th century. The first national meeting was held in York in 1844, and the Grand National Archery Society, which still governs the sport, was founded in 1861.

The sport was included in the OLYMPIC GAMES from 1900 to 1908 and again in 1920, and was reintroduced in 1972 with team events added in 1988. The *World Championship* takes place every two years and is contested on a knockout basis.

association football has its origins in ancient ball games played in various parts of the world, but much of the game's development came in England. Three British monarchs banned the sport at various times for its disruptive influence until it began to be organized in the 19th century. The first attempts at framing a set of rules for football were made in the mid-19th century by a group of representatives of the English public schools and universities. Out of their meetings the Football Association (FA) was founded in 1863, and the 11 players per side game became standard soon afterwards.

The international governing body, the Fédération Internatio-nale de Football Association (FIFA), was founded in Paris in 1904. By 1992 it had 179 members. International games have been played since 1872 when amateur teams representing Scotland and England first met for a friendly match in Glasgow. The first international championship was contested by the four home nations and lasted from 1883 to 1984.

The first *World Cup* competition was held in Uruguay in 1930, and it has been held every four years since (except 1942 and 1946). The *European Championship*, instituted in 1958, is also held every four years, and the African Nations Cup and the South American Championship every two years.

The first *World Cup* for the Jules Rimet Trophy was contested by 13 nations in Uruguay in 1930. Brazil kept the original trophy after their third victory in 1970, and teams now compete for the FIFA World Cup. Host nations qualify automatically for the final stages of the tournament. The results of the finals have been as follows:

	Host	*Winner*	*Runner-up*
1930	Uruguay	Uruguay	Argentina
1934	Italy	Italy	Czechoslovakia
1938	France	Italy	Hungary
1950	Brazil	Uruguay*	Brazil
1954	Switzerland	West Germany	Hungary
1958	Sweden	Brazil	Sweden
1962	Chile	Brazil	Czechoslovakia
1966	England	England	West Germany
1970	Mexico	Brazil	Italy
1974	W. Germany	W. Germany	Holland
1978	Argentina	Argentina	Holland
1982	Spain	Italy	W. Germany
1986	Mexico	Argentina	W. Germany
1990	Italy	W. Germany	Argentina
1994	USA	Brazil**	Italy

* after extra time ** after extra time and penalties

France will host the finals of the competition in 1998.

The *European Championship*, originally called the European Nations Cup, it took its present name in 1968. The teams compete for the Henri Delaunay Cup, named after the former General Secretary of UEFA. The results of the finals have been as follows:

Host	Winner	Runner-up
1960 France	USSR*	Yugoslavia
1964 Spain	Spain	USSR
1968 Italy	Italy	Yugoslavia
1972 Belgium	W. Germany	USSR
1976 Yugoslavia	Czechoslovakia*	W. Germany
1980 Italy	West Germany	Belgium
1984 France	France	Spain
1988 W. Germany	Holland	USSR
1992 Sweden	Denmark	Germany
1996 England	Germany	Czech Republic

*after extra time ** after extra time and penalties*

The *European Champion Clubs' Cup* is popularly known as the European Cup. It is an annual knock-out competition first held in 1955/6 for the best league champions of all UEFA affiliated countries. Since 1991/2 a Champions' League has been included, together with a number of qualifying stages, and the structure of the competition is now more flexible. The competition winners have been:

1956 Real Madrid	1957 Real Madrid
1957 Real Madrid	1958 Real Madrid
1959 Real Madrid	1960 Real Madrid
1961 Benfica	1962 Benfica
1963 AC Milan	1964 Inter Milan
1965 Inter Milan	1966 Real Madrid
1967 Celtic	1968 Manchester United
1969 AC Milan	1970 Feyenoord
1971 Ajax	1972 Ajax*
1973 Ajax*	1974 Bayern Munich*
1975 Bayern Munich	1976 Bayern Munich
1977 Liverpool*	1978 Liverpool
1979 Nottingham Forest*	1980 Nottingham Forest
1981 Liverpool	1982 Aston Villa*
1983 SV Hamburg	1984 Liverpool
1985 Juventus	1986 Steaua
1987 FC Porto	1988 PSV Eindhoven
1989 AC Milan*	1990 AC Milan*
1991 Red Star Belgrade	1992 Barcelona
1993 Marseilles	1994 AC Milan
1995 Ajax	1996 Juventus

The *European Cup-Winners Cup* is open to winners of domestic senior cup competitions in UEFA-affiliated countries. Originally the final play-off was decided over two games, but this changed in 1961, and it is now one game. The winners of the competition have been:

1961 AC Fiorentina	1962 Athletico Madrid
1963 Tottenham Hotspur	1964 Sporting Club, Lisbon
1965 West Ham United	1966 Borussia Dortmund
1967 Bayern Munich	1968 AC Milan
1969 Slovan Bratislava	1970 Manchester City
1971 Chelsea	1972 Rangers
1973 AC Milan	1974 FC Magdeburg
1975 Dynamo Kiev	1976 Anderlecht
1977 SV Hamburg	1978 Anderlecht*
1979 Barcelona	1980 Valencia*
1981 Dynamo Tblisi	1982 Barcelona
1983 Aberdeen*	1984 Juventus*
1985 Everton	1986 Dynamo Kiev
1986 Dynamo Kiev	1987 Ajax
1988 Mechelen*	1989 Barcelona
1990 Sampdoria	1991 Manchester United*
1992 Werder Bremen	1993 Parma
1994 Arsenal	1995 Real Zaragoza
1996 Paris St Germain	

* Winner of the European Super Cup, a competition first contested in 1974 by the winners of the European Cup and the Cup-Winners Cup on a play-off basis.

The *UEFA Cup* is competed for by the teams not eligible for the other two competitions, the final is decided over two legs on a home and away basis. Originally called the International Industries Fairs Inter-Cities Cup, or Fairs Cup, it was established in 1955, and the first competition took three years to complete. It became the UEFA Cup in 1971. The winners since 1958 have been:

1958 Barcelona	1960 Barcelona
1961 Rome	1962 Valencia
1963 Valencia	1964 Real Zaragoza
1965 Ferencvaros	1966 Barcelona
1967 Dynamo Zagreb	1968 Leeds United
1969 Newcastle United	1970 Arsenal
1971 Leeds United	1972 Tottenham Hotspur
1973 Liverpool	1974 Feyenoord
1975 Borussia M'bach	1976 Liverpool
1977 Juventus	1978 PSV Eindhoven
1979 Borussia M'bach	1980 Eintracht
1981 Ipswich Town	1982 Gothenburg
1983 Anderlecht	1984 Tottenham Hotspur
1985 Real Madrid	1986 Real Madrid
1987 Gothenburg	1988 Beyer Leverkusen
1989 Napoli	1990 Juventus
1991 Inter Milan	1992 Ajax
1993 Juventus	1994 Inter Milan
1995 Parma	1996 Bayern Munich

The *English Football League* was established in 1988/9 by twelve clubs from the Midlands and

North of England. By 1958 there were four divisions and 92 clubs, and this set-up was retained until 1993 when the top 22 clubs formed the FA Premier League and the three divisions of the Football League succeeded the original Divisions 2 to 4. The First Division/FA Premiership League champions have been:

1889 Preston North End	1890 Preston North End
1891 Everton	1892 Sunderland
1893 Sunderland	1894 Aston Villa
1895 Sunderland	1896 Aston Villa
1897 Aston Villa	1898 Sheffield United
1899 Aston Villa	1900 Aston Villa
1901 Liverpool	1902 Sunderland
1903 Sheffield Wednesday	1904 Sheffield Wednesday
1905 Newcastle United	1906 Liverpool
1907 Newcastle United	1908 Manchester United
1909 Aston Villa	1910 Aston Villa
1911 Manchester United	1912 Blackburn Rovers
1913 Sunderland	1914 Blackburn Rovers
1915 Everton	1920 West Bromwich Albion
1921 Burnley	1922 Liverpool
1923 Liverpool	1924 Huddersfield Townl
1925 Huddersfield Town	1926 Huddersfield Town
1927 Newcastle United	1928 Everton
1929 Sheffield Wednesday	1930 Sheffield Wednesday
1931 Arsenal	1932 Everton
1933 Arsenal	1934 Arsenal
1935 Arsenal	1936 Sunderland
1937 Manchester City	1938 Arsenal
1939 Everton	1947 Liverpool
1948 Arsenal	1949 Portsmouth
1950 Portsmouth	1951 Tottenham Hotspur
1952 Manchester United	1953 Arsenal
1954 Wolverhampton	1955 Chelsea
1956 Manchester United	1957 Manchester United
1958 Wolverhampton	1959 Wolverhampton
1960 Burnley	1961 Tottenham Hotspur
1962 Ipswich Town	1963 Everton
1964 Liverpool	1965 Manchester United
1966 Liverpool	1967 Manchester United
1968 Manchester United	1969 Leeds United
1970 Everton	1971 Arsenal
1972 Derby County	1973 Liverpool
1974 Leeds United	1975 Derby County
1976 Liverpool	1977 Liverpool
1978 Nottingham Forest	1979 Liverpool
1980 Liverpool	1981 Aston Villa
1989 Arsenal	1990 Liverpool
1991 Arsenal	1992 Leeds
1993 Manchester United	1994 Manchester United
1995 Blackburn Rovers	1996 Manchester United

The *English FA Cup* competition was established as the Football Association Challenge Cup with 15 teams from across Britain entering first competition. It is contested annually on a knockout basisn and since 1923 the final has been played at Wembley Stadium. The winners have been:

1872 Wanderers	1873 Wanderers
1874 Oxford University	1875 Royal Engineers
1876 Wanderers	1878 Wanderers
1879 Old Etonians	1880 Clapham Rovers
1881 Old Carthusians	1882 Old Etonians
1883 Blackburn Olympic	1884 Blackburn Rovers
1885 Blackburn Rovers	1886 Blackburn Rovers
1887 Aston Villa	1888 West Bromwich Albion
1889 Preston North End	1890 Blackburn Rovers
1891 Blackburn Rovers	1892 West Bromwich Albion
1893 Wolverhampton	1894 Notts County
1895 Aston Villa	1896 Sheffield Wednesday
1897 Aston Villa	1898 Nottingham Forest
1899 Sheffield United	1900 Bury
1901 Tottenham Hotspur	1902 Sheffield United
1903 Bury	1904 Manchester City
1905 Aston Villa	1906 Everton
1907 Sheffield Wednesday	1908 Wolverhampton
1909 Manchester United	1910 Newcastle United
1911 Bradford City	1912 Barnsley
1913 Aston Villa	1914 Burnley
1920 Aston Villa	1921 Tottenham Hotspur
1922 Huddersfield Town	1923 Bolton Wanderers
1924 Newcastle United	1925 Sheffield United
1926 Bolton Wanderers	1927 Cardiff City
1928 Blackburn Rovers	1929 Bolton Wanderers
1930 Arsenal	1931 West Bromwich Albion
1932 Newcastle United	1933 Everton
1934 Manchester City	1935 Sheffield Wednesday
1936 Arsenal	1937 Sunderland
1938 Preston North End	1939 Portsmouth
1946 Derby County	1947 Charlton Athletic
1948 Manchester United	1949 Wolverhampton
1950 Arsenal	1951 Newcastle United
1952 Newcastle United	1953 Blackpool
1954 W. Bromwich Albion	1955 Newcastle United
1956 Manchester City	1957 Aston Villa
1958 Bolton Wanderers	1959 Nottingham Forest
1960 Wolverhampton	1961 Tottenham Hotspur
1962 Tottenham Hotspur	1963 Manchester United
1964 West Ham	1965 Liverpool
1966 Everton	1967 Tottenham Hotspur
1968 W. Bromwich Albion	1969 Manchester United
1970 Chelsea	1971 Arsenal
1972 Leeds United	1973 Sunderland
1974 Liverpool	1975 West Ham
1976 Southampton	1977 Manchester United
1978 Ipswich	1979 Arsenal
1980 West Ham	1981 Tottenham Hotspur

1982 Tottenham Hotspur	1983 Manchester United
1984 Everton	1985 Manchester United
1986 Liverpool	1987 Coventry City
1988 Wimbledon	1989 Liverpool
1990 Manchester United	1991 Tottenham Hotspur
1992 Liverpool	1993 Arsenal
1994 Manchester United	1995 Everton
1996 Manchester United	

The *English League Cup* was first contested in 1960, it was not until 1970 that all 92 league teams competed. The Cup was first sponsored in 1982 by the Milk Marketing Board. In 1992 Coca-Cola became the sponsor. The three most successful English League Cup teams have been Liverpool, Nottingham Forest and Aston Villa, each of which has won the tournament four times.

The *Scottish Football League* was founded in 1890, two years after the English Football League. It was restructured in 1975/6 when the Premier Division, with the top ten clubs, was established with the remaining clubs divided into Divisions 1 and 2. A further restructuring in 1994 introduced a four-league system with ten teams in each. The First/Premier Division Champions have been:

1891 Dumbarton and Rangers	1892 Dumbarton
1893 Celtic	1894 Celtic
1895 Hearts	1896 Celtic
1897 Hearts	1898 Celtic
1899 Rangers	1900 Rangers
1901 Rangers	1902 Rangers
1903 Hibernian	1904 Third Lanark
1905 Celtic	1906 Celtic
1907 Celtic	1908 Celtic
1909 Celtic	1910 Celtic
1911 Rangers	1912 Rangers
1913 Rangers	1914 Celtic
1915 Celtic	1916 Celtic
1917 Celtic	1918 Rangers
1919 Celtic	1920 Rangers
1921 Rangers	1922 Celtic
1924 Rangers	1964 Rangers
1923 Rangers	1925 Rangers
1926 Celtic	1927 Rangers
1928 Rangers	1929 Rangers
1930 Rangers	1931 Rangers
1932 Motherwell	1933 Rangers
1934 Rangers	1935 Rangers
1936 Celtic	1937 Rangers
1938 Celtic	1939 Rangers
1947 Rangers	1948 Hibernian
1949 Rangers	1950 Rangers
1951 Hibernian	1952 Hibernian
1953 Rangers	1954 Celtic

1955 Aberdeen	1956 Rangers
1957 Rangers	1958 Hearts
1959 Rangers	1960 Hearts
1961 Rangers	1962 Dundee
1963 Rangers	1965 Kilmarnock
1966 Celtic	1967 Celtic
1968 Celtic	1969 Celtic
1970 Celtic	1971 Celtic
1972 Celtic	1973 Celtic
1974 Celtic	1975 Rangers
1976 Rangers	1977 Celtic
1978 Rangers	1979 Celtic
1980 Aberdeen	1981 Celtic
1982 Celtic	1983 Dundee Utd
1984 Aberdeen	1985 Aberdeen
1986 Celtic	1987 Rangers
1988 Celtic	1989 Rangers
1990 Rangers	1991 Rangers
1992 Rangers	1993 Rangers
1994 Rangers	1995 Rangers
1996 Rangers	

The *Scottish FA Cup* was initiated at a meeting of eight clubs hosted by Queen's Park in 1873, in order to establish a competition in Scotland similar to the English FA Cup. The winners have been:

1874 Queen's Park	1875 Queen's Park
1876 Queen's Park	1877 Vale of Leven
1878 Vale of Leven	1879 Vale of Leven
1880 Queen's Park	1881 Queen's Park
1882 Queen's Park	1883 Dumbarton
1884 Queen's Park	1885 Renton
1886 Queen's Park	1887 Hibernian
1888 Renton	1889 Third Lanark
1890 Queen's Park	1891 Hearts
1892 Celtic	1893 Queen's Park
1894 Rangers	1895 St Bernard's
1896 Hearts	1897 Rangers
1898 Rangers	1899 Celtic
1900 Celtic	1901 Hearts
1902 Hibernian	1903 Rangers
1904 Celtic	1905 Third Lanark
1906 Hearts	1907 Celtic
1908 Celtic	1909 Celtic
1910 Dundee	1911 Celtic
1912 Celtic	1913 Falkirk
1914 Celtic	1920 Kilmarnock
1921 Partick	1922 Morton
1923 Celtic	1924 Airdrie
1925 Celtic	1926 St Mirren
1927 Celtic	1928 Rangers
1929 Kilmarnock	1930 Rangers
1931 Celtic	1932 Rangers
1933 Celtic	1934 Rangers

1935	Rangers	1936	Rangers
1937	Celtic	1938	East Fife
1939	Clyde	1947	Aberdeen
1948	Rangers	1949	Rangers
1950	Rangers	1951	Celtic
1952	Motherwell	1953	Rangers
1954	Celtic	1955	Clyde
1956	Hearts	1957	Falkirk
1958	Clyde	1959	St Mirren
1960	Rangers	1961	Dunfermline
1962	Rangers	1963	Rangers
1964	Rangers	1965	Celtic
1966	Rangers	1967	Celtic
1968	Dunfermline	1969	Celtic
1970	Aberdeen	1971	Celtic
1972	Celtic	1973	Rangers
1974	Celtic	1975	Celtic
1976	Rangers	1977	Celtic
1978	Rangers	1979	Rangers
1980	Celtic	1981	Rangers
1982	Aberdeen	1983	Aberdeen
1984	Aberdeen	1985	Celtic
1986	Aberdeen	1987	St Mirren
1988	Celtic	1989	Celtic
1990	Aberdeen	1991	Motherwell
1992	Rangers	1993	Rangers
1994	Dundee Utd	1995	Celtic
1996	Rangers		

athletics are thought to have originated in the ancient Greek Olympic Games around 1370 BC. Organized events in England began in the mid-19th century, with the first regular competition taking place at Exeter College, Oxford, in 1850.

The most important competitions in international athletics are the OLYMPIC GAMES and the *World Championships*. There are also a number of events, such as the IAAF (International Amateur Athletic Federation) World Cup, the European Cup and the World Indoor Championships, which attract top-class competition.

The first *World Championships* for athletics alone were held in 1983 in Helsinki, Finland. They are now held biennially. The winners of the men's events in Gothenburg, Sweden, in 1996 were:

Event	*Winner (country)*
100m	D. Bailey (Canada)
200m	M. Johnson (USA)
400m	M. Johnson (USA)
800m	W. Kipketer (Denmark)
1,500m	N. Morceli (Algeria)
5,000m	I. Kifui (Kenya)
10,000m	H. Gebrselassie (Ethiopia)
Marathon	M. Fiz (Spain)
100m hurdles	A. Johnson (USA)
400m hurdles	D. Adkins (USA)
3,000m steeplechase	M. Kiptanui (Kenya)
20 km walk	M. Didoni (Italy)
50 km walk	V. Kononen (Finland)
4 x 100m relay	Canada
4 x 400m	USA
High jump	T. Kemp (Bahrain)
Long jump	I. Pedroso (Cuba)
Triple jump (*)	J. Edwards (UK)
Pole vault	S. Bubka (Ukraine)
Shot	J. Godina (USA)
Discus	L. Riedel (Germany)
Hammer	A. Abduvaliyev (Tadjikstan)
Javelin	J. Zelezny (Czech Republic)
Decathlon	D. O'Brien (USA)

(*) indicates a world record

The winners of the women's events at the World Championships in Gothenburg in 1995 were:

Event	*Winner (country)*
100m	G. Torrence (USA)
200m	M. Ottey (Jamaica)
400m	M.-J. Perec (France)
800m	A.-F. Quirot (Cuba)
1,500m	H. Boulmerka (Algeria)
5,000m	S. O'Sullivan (Ireland)
10,000m	F. Ribeiro (Portugal)
Marathon (not official)	M. Machado (Portugal)
100m hurdles	G. Devers (USA)
400m hurdles (*)	K. Batten (USA)
10 km walk	S. Stankina (Russia)
4 x 100m relay	USA
4 x 400m relay	USA
High jump	S. Kostadinova (Bulgaria)
Long jump	F. May (Italy)
Triple jump (*)	I. Kravets (Ukraine)
Shot	A. Kumbernuss (Germany)
Discus	E. Svereva (Belarus)
Javelin	N. Shikolenko (Belarus)
Heptathlon	G. Shouaa (Syria)

World Mile Record (since 1954)

Year	Runner	Time
1954	Roger Bannister (UK)	3 min 59.40 sec
1954	John Landy (Australia)	3 min 57.90 sec
1957	Doug Ibbotson (UK)	3 min 54.20 sec
1958	Herb Elliot (Australia)	3 min 54.50 sec
1962	Peter Snell (NZ)	3 min 54.40 sec
1964	Peter Snell (NZ)	3 min 54.10 sec
1965	Michael Jazy (France)	3 min 53.60 sec
1966	Jim Ryun (USA)	3 min 51.30 sec
1967	Jim Ryun (USA)	3 min 51.10 sec
1975	Filbert Bayi (Tanzania)	3 min 51.00 sec

1975	John Walker (NZ)	3 min 49.40 sec
1979	Sebastian Coe (UK)	3 min 49.00 sec
1980	Steve Ovett (UK)	3 min 48.80 sec
1981	Sebastian Coe (UK)	3 min 48.53 sec
1981	Steve Ovett (UK)	3 min 48.40 sec
1981	Sebastian Coe (UK)	3 min 47.43 sec
1985	Steve Cram (UK)	3 min 46.35 sec
1993	Noureddine Morceli (Algeria)	3 min 44.39 sec

British Open *see* GOLF.

badminton originated either from Badminton Hall in Avon, *c.*1870, where it was played by the family and guests of the Duke of Beaufort, or from a game played in India around the same time. The modern rules of the sport were established in India, however, where it was very popular for a time among British Army officers. The Badminton Association was founded in England in 1893 and the International Badminton Federation in 1934.

The most important competitions are the Thomas Cup (men), the Uber Cup (women) and the *World Championships*, all of which are held biennially. In recent years teams from Indonesia, China and Japan have dominated international competition.

baseball according to American folklore the game was invented by a West Point cadet, Abner Doubleday, in 1839. A game of the same name, however, is known to have been played in England prior to 1700, and it is more likely that baseball evolved from such popular bat and ball games as cricket and rounders.

The first rules of the modern game were established in 1845 by Alexander Cartwright Jr, and the professional game developed almost entirely in North America where there are two 'major' leagues, the American (AL) and the National (NL), founded in 1901 and 1876 respectively.

The *World Series* is a best-of-seven contest held annually between the winners of the two professional baseball leagues. It was first played in 1903 and is now one of America's most popular sporting events. In 1992 the Toronto Blue Jays became the first non-American team to win the series when they defeated the Atlanta Braves. The winners since 1980 have been:

1980 Philadelphia Phillies	1981 Los Angeles Dodgers
1982 St Louis Cardinals	1983 Baltimore Orioles
1984 Detroit Tigers	1985 Kansas City Royals
1986 New York Mets	1987 Minnesota Twins
1988 Los Angeles Dodgers	1989 Oakland Athletics
1990 Cincinnati Reds	1991 Minnesota Twins
1992 Toronto Blue Jays	1993 Toronto Blue Jays
1994 *not contested*	1995 Atlanta Braves
1996 New York Yankees	

At international level South American teams dominate. The most successful country is Cuba, which has won the World Amateur Championships 20 times since it was established in 1938. Baseball has also been a demonstration sport at six OLYMPIC GAMES and became a medal sport for the first time in 1992.

basketball in its present form was invented in 1891 by Dr James Naismith, a sports instructor from Springfield, Massachussets, although games bearing a strong resemblance to basketball have been played for thousands of years. Early games in America featured large numbers of players but the 5-a-side standard was agreed in 1895 and the first league established three years later.

In the USA there are now 23 professional basketball teams, and it is one of America's most popular spectator sports, with televised games attracting large audiences.

The *NBA Championship* is contested annually by the 23 professional basketball teams in the USA. The teams are divided into two conferences, Eastern and Western, which are subdivided into the Western and Atlantic Divisions and the Midwest and Pacific Divisions respectively. A series of play-offs decide the champions. Recent NBA champions have been:

1980 Los Angeles Lakers	1981 Boston Celtics
1982 Los Angeles Lakers	1983 Philadelphia 76ers
1984 Boston Celtics	1985 Los Angeles Lakers
1986 Boston Celtics	1987 Los Angeles Lakers
1988 Los Angeles Lakers	1989 Detroit Pistons
1990 Detroit Pistons	1991 Chicago Bulls
1992 Chicago Bulls	1993 Chicago Bulls
1994 Houston Rockets	1995 Houston Rockets
1996 Chicago Bulls	

At amateur level it is popular around the world and has been included in the OLYMPIC GAMES since 1936 (the women's game was included in 1976). *World Championships*, first held in 1950 (men) and 1953 (women), are staged every four years and have been dominated in recent years by teams from the USA and the former Yugoslavia.

biathlon is one of the most physically demanding combination sports. Competitors ski over a prepared course carrying a small-bore rifle and stop for either two or four shooting competitions, standing and prone. Men compete over 10km and 20km, and women over 7.5km and 15km. There is also a relay event, which is 4 x 7.5km with every member shooting once.

Men's biathlon has been included in the OLYMPIC GAMES since 1960 and the women's event since

1992. There is also a World Cup and a World Championship, both contested annually.

billiards is believed to have evolved from the game of *paillemalle*, played on grass. It was popular in the 15th century, and Louis XI, king of France (1461-83), is known to have had a table. Today there are World professional and amateur championships, played annually and biennially respectively.

bobsleigh and *toboggan* racing was pioneered by a group of Swiss enthusiasts who organized the first known bobsleigh races in the 1880s. The first two luge runs were built in 1879 at Davos in Switzerland.

There are now bobsleigh events in the OLYMPIC GAMES for two and four man teams and for single man luge toboggans.

bowls the rules of the modern game were framed in 1848-9 by a Scottish solicitor, William Mitchell, although similar games were played as early as the 13th century. There are now two types of greens, the 'crown' and the 'level', the former is almost exclusive to the north of England, while the latter played mostly in the UK and Commonwealth countries.

World Outdoor Championships were established in 1966 for men and 1969 for women. They are held every four years, with singles, pairs, triples and fours events. The Leonard Trophy is presented to the overall best men's team. The results at the 1992 champions were:

Men
Singles	T. Allcock (England)
Pairs	R. Corsie and A. Marshall (Scotland)
Triples	Israel
Fours	Scotland
Overall	Scotland

Women
Singles	M. Johnston (Ireland)
Pairs	M. Johnston and P. Nolan (Ireland)
Triples	Scotland
Fours	Scotland
Overall	Scotland

boxing is one of man's oldest physical contests. The first organized boxing match is thought to have taken place in the late-17th century when the Duke of Albemarle organized a fight between his butler and his butcher. The modern rules, which introduced gloves, were set out in 1867 by the 8th Marquis of Queensbury (who gave his name to them). The first world championship fight took place in New York in 1884 with Jack Dempsey beating George Fulljames in New York for the middleweight title.

In boxing today there are four different governing authorities, all with their own weight classes, titles and fight regulations. This naturally leads to confusion, especially as all four governing bodies recognize world champions. The four are: the World Boxing Association (founded 1960), the World Boxing Council (1963), the World Boxing Organization (1988) and the International Boxing Federation (1983).

The weight categories accepted by the WBC are as follows:

Weight	Limit	Also known as
Heavy	>190 lbs.	
Cruiser	190 lbs.	Junior Heavy (WBO)
Light Heavy	175 lbs.	
Super Middle	168 lbs.	
Middle	160 lbs.	
Super Welter	154 lbs.	Junior Middle (WBO, IBF, WBA)
Welter	147 lbs.	
Super Light	140 lbs.	Junior Welter (WBO, IBF, WBA)
Light	135 lbs.	
Super Feather	130 lbs.	Junior Light (WBO, IBF, WBA)
Feather	126 lbs.	
Super Bantam	122 lbs.	Junior Feather (WBO, IBF, WBA)
Bantam	118 lbs.	
Super Fly	115 lbs.	Junior Bantam (WBO, IBF, WBA)
Flyweight	112 lbs.	
Light Fly	108 lbs.	Junior Fly (WBO, IBF, WBA)
Straw	105 lbs.	Mini Fly (WBO, IBF, WBA)

Calcutta Cup *see* RUGBY UNION.

canoeing as a sport can be attributed to an English barrister, John MacGregor, founder of the Royal Canoe Club in 1866. In the sport today there is flat and wild water competition featuring kayaks (double blade) and Canadian canoe (single blade). It has been included in the OLYMPIC GAMES, with several different categories, since 1936.

Commonwealth Games are held every four years and are contested by representatives of the British Commonwealth. The original idea for the games was suggested by the Rev. J Astley Cooper in the magazine *Greater Britain* in 1891. The first British Empire Games were held in 1930 in Canada, and four sports were contested by representatives from Great Britain, Canada, Australia and New Zealand.

Known as the Commonwealth Games since 1970, recent Games have included ten sports. Athletics and swimming are obligatory, and the other eight are selected from 15 recognized sports, with two demonstration sports also chosen. The venues for the Games have been:

1930 Hamilton, Canada   1934 London, England
1938 Sydney, Australia   1950 Auckland, N. Zealand
1954 Vancouver, Canada   1958 Cardiff, Wales
1962 Perth, Australia   1966 Kingston, Jamaica
1970 Edinburgh, Scotland 1974 Christchurch, N. Zealand
1978 Edmonton, Canada   1982 Brisbane, Australia
1986 Edinburgh, Scotland 1990 Auckland, N. Zealand
1994 Victoria, Canada

The 1998 Commonwealth Games will be held in Kuala Lumpur, Malaysia and the 2002 Games in Manchester, England.

County Championship *see* CRICKET.

cricket the precise origins of cricket are unclear although games with a bat and ball were played as long ago as the 13th century. The first match in which scores were kept was contested in 1744 by England and Kent. The Marylebone Cricket Club was founded in 1787 and remained the governing body for the sport until the formation of the Cricket Council in 1968.

The Imperial (International from 1965) Cricket Conference (ICC) was formed in 1909 and now has nine Test-playing members, 19 associate members and five affiliated members. The nine full, test-playing members are England, Australia, West Indies, India, Pakistan, New Zealand, Sri Lanka, South Africa (ceased to be a member in 1961, readmitted in 1992) and Zimbabwe (from 1992).

The first Test match was played in Melbourne in 1877 between Australia and England, and the first in England was played by the same nations three years later, although in both these contests neither side was truly representative of its country.

The summarized results of Test match series completed by 1 June 1995 are:

		Won	Drawn	Lost
England v	Australia	90	84	111
	India	31	36	14
	New Zealand	34	37	4
	Pakistan	14	31	7
	South Africa	46	38	18
	Sri Lanka	3	1	1
	West Indies	25	38	46
Australia v	England	111	84	90
	India	24	17	8
	New Zealand	13	12	7
	Pakistan	12	17	10
	South Africa	31	15	13
	Sri Lanka	4	3	0
	West Indies	32	21	27
India v	Australia	8	17	24
	England	14	36	31
	New Zealand	12	14	6
	Pakistan	4	33	7
	South Africa	0	3	1
	Sri Lanka	7	4	1
	West Indies	7	31	27
	Zimbabwe	1	1	0
Pakistan v	Australia	10	17	12
	England	7	31	14
	India	7	33	4
	New Zealand	16	16	4
	Sri Lanka	10	5	1
	West Indies	7	11	12
	Zimbabwe	4	1	1
South Africa v	Australia	13	15	31
	England	19	39	47
	India	1	3	0
	New Zealand	11	6	3
	Sri Lanka	1	2	0
	West Indies	0	0	1
Sri Lanka v	Australia	0	3	4
	England	1	1	3
	India	1	6	7
	New Zealand	2	7	4
	Pakistan	1	5	10
	South Africa	0	2	1
	West Indies	0	1	0
West Indies v	Australia	27	21	32
	England	46	38	25
	India	27	31	7
	New Zealand	9	13	4
	Pakistan	12	11	7
	South Africa	1	0	0
	Sri Lanka	1	0	0
Zimbabwe v	India	0	1	1
	New Zealand	0	1	1
	Pakistan	1	1	4
New Zealand v	Australia	7	12	13
	England	4	37	34
	India	6	14	12
	Pakistan	4	16	16
	South Africa	3	6	11
	Sri Lanka	4	7	2
	West Indies	4	13	9

English *County Championship* cricket began with the first recorded match between counties, contested by Kent and Surrey, in 1709. The first county to be declared county champion was Surrey in 1827. From 1827 to 1889 the champions were decided on the basis of fewest matches lost, and Kent, Surrey and Sussex dominated. Since 1890 the championship has been contested on a points system. The postwar winners have been:

1946 Yorkshire	1947 Middlesex
1948 Glamorgan	1949 Middlesex
	and Yorkshire
1950 Lancaster and Surrey	1951 Warwickshire
1952 Surrey	1953 Surrey
1954 Surrey	1955 Surrey
1956 Surrey	1957 Surrey
1958 Surrey	1959 Yorkshire
1960 Yorkshire	1961 Hampshire
1962 Yorkshire	1963 Yorkshire
1964 Worcestershire	1965 Worcestershire
1966 Yorkshire	1967 Yorkshire
1968 Yorkshire	1969 Glamorgan
1970 Kent	1971 Surrey
1972 Warwickshire	1973 Hampshire
1974 Worcestershire	1975 Leicestershire
1976 Middlesex	1977 Middlesex and Kent
1978 Kent	1979 Essex
1980 Middlesex	1981 Nottinghamshire
1982 Middlesex	1983 Essex
1984 Essex	1985 Middlesex
1986 Essex	1987 Nottinghamshire
1988 Worcestershire	1989 Worcestershire
1990 Middlesex	1991 Essex
1992 Essex	1993 Middlesex
1994 Warwickshire	1995 Warwickshire

cross-country running was first staged with an international field in 1903 at Hamilton Park Racecourse in Scotland. The World Cross-Country Championships are now held annually with the senior events being run over 5km and 12km courses for women and men respectively.

cycling the first treadle-propelled bicycle was invented by Scottish blacksmith, Kirkpatrick Macmillan, in 1839, but it was not until the more refined *velocipede* was built by Pierre Michaux of Paris, in 1861, that there was any popular interest in bicycle racing. The first club was the Liverpool Velocipede Club, which was formed in 1867, and the first race (over 1200 metres) took place in Paris a year later.

It has been included in the OLYMPIC GAMES since 1896, and there are currently nine events in the programme, including sprint, pursuit, time trials, and individual and team road-racing events.

The *Tour De France* is the most famous cycle race in the world. First held in 1903, it is now one of the world's most popular spectator events. The course changes every year but always includes a number of mountain stages as well as time trials. It is held over a three week period. The longest race has been over 5745 km (3569 miles). Recent overall winners have been:

1979 B. Hinault (France)	1980 J. Zoetemelk (Holland)
1981 B. Hinault (France)	1982 B. Hinault (France)
1983 L. Fignon (France)	1984 L. Fignon (France)
1985 B. Hinault (France)	1986 G. LeMond (USA)
1987 S. Roche (Ireland)	1988 P. Delgado (Spain)
1989 G. LeMond (USA)	1990 G. LeMond (USA)
1991 M. Indurain (Spain)	1992 M. Indurain (Spain)
1993 M. Indurain (Spain)	1994 M. Indurain (Spain)
1995 M. Indurain (Spain)	

Davis Cup *see* TENNIS.

English FA Cup *see* ASSOCIATION FOOTBALL.

English Football League *see* ASSOCIATION FOOTBALL.

English League Cup *see* ASSOCIATION FOOTBALL.

European Championship *see* ARCHERY, ATHLETICS, BADMINTON, BOWLS.

European Cup *see* ASSOCIATION FOOTBALL.

European Cup-Winners Cup *see* ASSOCIATION FOOTBALL.

European Super Cup *see* ASSOCIATION FOOTBALL.

equestrianism although horse-riding is many thousands of years old, competition jumping became popular only in the late 19th century with the first 'horse show' held in Dublin in 1867. Dressage competition evolved from French and Italian horsemanship academies in the 16th century, while the three-day event has its origins in cavalry endurance tests.

Equestrian events have been included in the OLYMPIC GAMES since 1912, but the first competition to combine all six equestrian disciplines—show jumping, three-day eventing, dressage, carriage driving, endurance riding, and vaulting—was the *World Equestrian Games*, first held in Stockholm in 1990. The second Games were held at the Hague in 1994.

fencing has its origins in the duelling of the Middle Ages, although fighting with swords had been practised as a sport in older civilizations.

There are three types of sword used today: the *foil*, the *epée* and the *sabre*. Each is a separate discipline with a different set of rules applicable to it in competition:

weapon	weight	target
epée	770 grams	whole body
foil	500 grams (maximum)	upper body
sabre	500 grams	upper body and head

Fencing has been included in all OLYMPIC GAMES, and these tournaments count as World Championships in Olympic years. Women first competed in the 1924 Games.

football *see* AMERICAN FOOTBALL; ASSOCIATION FOOTBALL.

golf although the precise origins are unclear, it was in Scotland that the popular pastime was first properly organized and the first club established. The earliest rules of the game were drawn up in 1754 by enthusiasts in St Andrews, who went on to form the Royal and Ancient Golf Club there. In 1764 18 holes was declared as the standard round, but it was not until 1919 that the Royal and Ancient, who had decided the number, was formally recognized as the supreme authority in golfing matters.

In professional golf the most important, and financially rewarding, competitions are the British Open, the US Open, the US Masters and the US PGA. These are played annually at different courses except the US Masters, which is always played at the Augusta National course in Georgia.

Although not offering the same high levels of prize money, there is also a well-established masters' series in women's golf. These are the US Women's Open, the LPGA Championships, the Nabisco Dinah Shore and the Du Maurier Classic.

The *British Open* was first played over three 12-hole rounds at Prestwick, Scotland, in 1860. Prestwick hosted the first 12 Opens, and all subsequent championships have been played over various seaside links courses around Britain. Prize money (totalling £10) was introduced in 1863. Since 1960 the winners of the tournament have been:

1960 K. Nagle (Australia)	1961 A. Palmer (USA)
1962 A. Palmer (USA)	1963 B. Charles (N. Zealand)
1964 T. Lema (USA)	1965 P. Thomson (Australia)
1966 J. Nicklaus (USA)	1967 R. Vicenzo (Argentina)
1968 G. Player (S. Africa)	1969 T. Jacklin (UK)
1970 J. Nicklaus (USA)	1971 L. Trevino (USA)
1972 L. Trevino (USA)	1973 T. Weiskopf (USA)
1974 G. Player (S. Africa)	1975 T. Watson (USA)
1976 J. Miller (USA)	1977 T. Watson (USA)
1978 J. Nicklaus (USA)	1979 S. Ballesteros (Spain)
1980 T. Watson (USA)	1981 B. Rogers (USA)
1982 T. Watson (USA)	1983 T. Watson (USA)
1984 S. Ballesteros (Spain)	1985 S. Lyle (UK)
1986 G. Norman (Australia)	1987 N. Faldo (UK)
1988 S. Ballesteros (Spain)	1989 M. Calcavecchia (USA)
1990 N. Faldo (UK)	1991 I. Baker-Finch (Aus.)
1992 N. Faldo (UK)	1993 G. Norman (Aus.)
1994 N. Price (Zimbabwe)	1995 J. Daly (USA)
1996 Tom Lehman (USA)	

The *Ryder Cup* was launched in 1927, following a successful match between the USA and Great Britain. It is now held biennially with the teams taking it in turn to play host. The Great Britain team was succeeded by a Great Britain and Ireland team in 1973, which in turn was succeeded by a European team in 1979. The winners of the tournament have been:

1927	USA	1929	GB	1931	USA
1933	GB	1935	USA	1937	USA
1947	USA	1949	USA	1951	USA
1953	USA	1955	USA	1957	GB
1959	USA	1961	USA	1963	USA
1965	USA	1967	USA	1969	Drawn
1971	USA	1973	USA	1975	USA
1977	USA	1979	USA	1981	USA
1983	USA	1985	Europe	1987	Europe
1989	Drawn	1991	USA	1993	USA
1995	Europe				

Grand National *see* HORSE RACING.

Grand Prix *see* MOTOR RACING.

Grand Slam *see* RUGBY UNION.

gymnastics although the origins of gymnastics can be traced back to the ancient civilizations of Greece, Persia, China and India, as a modern competitive sport it has been developed primarily in Eastern European countries. The first national federation was formed in Germany in 1860, and gymnastics were included in the first modern OLYMPIC GAMES in 1896, although women's events were not introduced until 1928.

The apparatus used in modern gymnastics differs for men and women. Men use the parallel bars, horizontal bar, pommel horse, rings, and horse vault. Women use asymmetrical bars, the beam, and the horse vault.

The *World Championships* were established in 1903 for men and 1934 for women. Both are now held biennially.

Modern rhythmic gymnastics for women use ropes, hoops, ribbons and balls in musically accompanied floor exercises, and have been included in the OLYMPIC GAMES since 1984.

handball is similar to football, only played with the hands rather than the feet. It was first played at the end of the 19th century in Germany, and it has remained popular in that country.

At the 1936 OLYMPIC GAMES it was played by 11-a-side teams outdoors, but when it was reintroduced in 1972 it was as a 7-a-side indoor event. World Championships take place every four years, and a European Cup for national champions is contested annually.

hang gliding as a modern sport was pioneered by Otto Lilienthal in Germany in the late 19th century. The distances travelled and the duration of flights have increased dramatically in recent years with developments in technology and design. *World Championships* were first held in 1975 and are now held biennially.

hockey was played in primitive form in ancient civilizations, and tomb drawings in the Nile Valley suggest it could be as much as 4000 years old. The modern game became popular in the mid 19th century when many public schools took up the sport.

The most successful teams at international level have traditionally been from India and Pakistan, although recently European teams have dominated the major competitions. The IHF World Cup and the Champion's Trophy are the premier competitions, and the sport has been included in the OLYMPIC GAMES since 1908.

horse racing has its origins in warfare, hunting and chariot racing with the earliest reference to prize money being the purse of gold offered by Richard I in 1195. The oldest known racecourse in Britain is on the Roodee at Chester, which staged its first meeting in 1540. The Jockey Club was formed in 1750, and the first steeplechase was run in 1752 in Cork, Ireland.

There are two kinds of modern racing, flat racing and jumping, with the latter divided into steeplechasing and hurdle racing. The most important flat racing events in Britain are the five English Classics: the Derby, the Oaks, the St Leger and the 1000 and 2000 Guineas. Other famous races held annually include the series of Irish and French Classics, the Coronation Cup at Epsom and the Ascot Gold Cup at Royal Ascot.

In National Hunt Racing, the *Grand National* and the Cheltenham Gold Cup are the two most important events in the British calendar. The Grand National is the oldest and most famous steeplechase in the world. It has been run at Aintree, Liverpool, every year since 1847 with the exception of the war years. The current course takes in 30 fences over two circuits, and is four miles and four furlongs long. The only horse to have ever won the Grand National three times is Red Rum (1973, 1974, 1977). Recent winning horses have been:

1980 Ben Nevis	1981 Aldaniti
1982 Grittar	1983 Corbiere
1984 Hallo Dandy	1985 Last Suspect
1986 West Tip	1987 Maori Venture
1988 Rhyme 'n' Reason	1989 Little Polveir
1990 Mr Frisk	1991 Seagram
1992 Party Politics	1993 *race abandoned*
1994 Minnehoma	1995 Royal Athlete

hurling is one of Ireland's oldest competitive sports but standardized only since the foundation of the Gaelic Athletic Association in 1884. The hurl, or stick, is similar to the hockey stick only flat on both sides. The All-Ireland Championship has been held annually since 1887.

ice hockey was developed in Canada in the late 19th century with the first club founded in Montreal in 1880. World Championships have been held annually since 1930 with teams from North America, Scandanavia and Eastern Europe dominating in recent years. It has been included in the Olympic Games, since 1920.

ice skating is over 2000 years old and first became popular in the Netherlands, although the first known skating club was formed in Edinburgh in the mid-18th century. The first artificial ice rink was opened in London in 1876.

The modern sport can be divided into two: figure skating and speed skating. *World Championships* are contested in both disciplines. Figure skating has been included in the OLYMPIC GAMES since 1908, and speed skating was introduced into the 1924 Games (1960 for the women's event).

International Championships *see* RUGBY UNION.

judo was developed from several different schools of Japanese martial arts. The first training school was founded in Shitaya in 1882 by Dr Jigoro Kano. 'Ju' means 'soft', and the sport relies on speed and skill as opposed to brute force. Classes in judo are divided into *kyu* (pupil) and *dan* (master) grades, differentiated by different coloured belts.

Kyu grades		*Dan grades*	
1st	brown	1st–5th	black
2nd	blue	6th–8th	red and white
3rd	green	9th–11th	red
4th	orange	12th	white
5th	yellow		

It has been included for men in the OLYMPIC GAMES since 1964 (except 1968) and for women since 1992, and there are currently twelve weight divisions.

lacrosse was originally played by North American Indians and was taken up by French settlers, who called it *baggataway*. The modern crosse, or playing stick, evolved from the early implement used by the Indians, which was a curved stick with a net fitted at one end. The women's game evolved separately from that played by men, and they both now have *World Championships*.

lawn tennis *see* TENNIS.

modern pentathlon, as it was introduced at the 1912 OLYMPIC GAMES, consists of horse riding, fencing, shooting, swimming and cross-country running.

For many years the sport was dominated by members of the military forces, and military lore explains the origins of the modem format thus: a messenger has to ride on horseback, fight with sword and pistol, swim a river and run on foot to complete his journey. The format of the sport has been amended by the Olympic Committee and instead of being competed for over three or more days it is staged over one day.

motor cycling the earliest motorcycle race was held in 1897 at Sheen House in Richmond over an oval course of one mile. *World Championship* races began in 1949 with classes for 50cc, 125cc, 250cc, 350cc, 500cc bikes and sidecars. In road-racing the most important series is the Isle of Man TT races, first held in 1907. Motocross, or *scrambling*, is a specialized branch of racing with competitors negotiating undulating dirt tracks on specially adapted machines. A *World Championship* is held annually.

motor racing was inevitable with the invention, and rapid growth in popularity, of the motor car. The first known race took place in Paris in 1887 and was won by Count Jules Felix Philippe Albert de Dion de Malfiance, riding a steam quadricycle.

The most important class of racing, that of Formula One, held its first *World Championship* in 1950, although the first Grand Prix was run in France in 1906. Other notable competitions are the Le Mans circuit (for touring cars), the Indy Car World Series (16 races in North America and Australia), and the many long-distance 'rally' events held around the world.

The *World Championship Grand Prix* was inaugurated in 1949. The first Grand Prix was held at Silverstone, England, and won by the Italian Giuseppe Farina. Grand Prix races are now staged all over the world, and points are awarded for finishing places, which count towards the World Championship. Recent World Champion drivers have been:

1980 A. Jones (Australia)	1988 A. Senna (Brazil)
1981 N. Piquet (Brazil)	1989 A. Prost (France)
1982 K. Roseburg (Finland)	1990 A. Senna (Brazil)
1983 N. Piquet (Brazil)	1991 A. Senna (Brazil)
1984 N. Lauda (Austria)	1992 N. Mansell (UK)
1985 A. Prost (France)	1993 A. Prost (France)
1986 A. Prost (France)	1994 M. Schumacher (Germany)
1987 N. Piquet (Brazil)	1995 M. Schumacher (Germany)
	1996 D. Hill (UK)

NBA Championship *see* BASKETBALL.

netball is a 7-a-side game played primarily by women. It was developed in the USA at the end of the 19th century. A *World Championship* has been held every four years since 1963. In recent years these have been dominated by teams from Australia and New Zealand.

Olympic Games were first staged every four years at Olympia, 120 miles west of Athens in Greece, with the earliest recorded games being held in 776 BC. The games grew in size and importance and reached the peak of their fame in the 5th and 4th centuries BC, when events included running, jumping, boxing, wrestling, discus and chariot racing.

The final games of the ancient era were held in AD 393, before the emperor of Rome, Theodosius I, decreed their prohibition because they were disapproved of by the early Christians.

The first modern Games were instigated by a Frenchman, Pierre de Fredi, Baron de Coubertin. His efforts to bring back the principle of the ancient Games and re-establish the high sporting standards of the early Greeks led to the formation of the International Olympic Committee in 1895 and thence to the staging of the 1896 Games in Athens. The venues of the summer Games have been:

I	1896	Athens, Greece
II	1900	Paris, France
III	1904	St Louis, USA
IV	1908	London, England
V	1912	Stockholm, Sweden
VI	1916	Berlin, Germany
VII	1920	Antwerp, Belgium
VIII	1924	Paris, France
IX	1928	Amsterdam, Netherlands
X	1932	Los Angeles, USA
XI	1936	Berlin, Germany
XII	1940	Not held
XIII	1944	Not held
XIV	1948	London, UK
XV	1952	Helsinki, Finland
XVI	1956	Melbourne, Australia
XVII	1960	Rome, Italy
XVIII	1964	Tokyo, Japan
XIX	1968	Mexico City, Mexico
XX	1972	Munich, FR Germany
XXI	1976	Montreal, Canada
XXII	1980	Moscow, USSR
XXIII	1984	Los Angeles, USA
XXIV	1988	Seoul, South Korea
XXV	1992	Barcelona, Spain
XXVI	1996	Atlanta, USA

The XXVII Games in 2000 will be held in Sydney, Australia. *See also* WINTER OLYMPIC GAMES.

rowing the earliest established sculling race, the *Dogget's Coat and Badge*, was instituted in 1715 by the Irish humorist Thomas Dogget and was rowed from London Bridge to Chelsea. The first regatta was held on the Thames at Putney in 1775.

Rowing has been a part of the OLYMPIC GAMES since 1900, and *World Championships* have been held every year since 1962.

The famous University Boat Race, contested annually between teams from Oxford and Cambridge universities, was first held on the River Thames in 1829. The Putney to Mortlake course, over a distance of 6779 metres (4 miles 374 yards), which is currently used, has not changed since 1863. At present Cambridge leads as the winning team. The only dead heat in the history of the race was recorded in 1877.

rugby league originated as a breakaway from RUGBY UNION in 1895 when the governing body refused to let northern clubs pay players who were losing their Saturday wage in order to play. Today it is most popular in the north of England, Australia, New Zealand and France. The most important tournaments in England are for the Premiership Trophy and the Challenge Cup. Both are contested annually.

A rugby league *World Cup* was played every three years from 1954 to 1975, when the tournament was renamed the International Championship. The World Cup was revived in 1985 and won in 1988, 1992 and 1995 by Australia.

rugby union is thought to have originated at Rugby School, England, as the modern game can supposedly be traced back to the moment when one William Webb Ellis picked up the ball and ran with it during a school football match in November 1823. The first club was formed at Guy's Hospital in London in 1843 and the Rugby Football Union (RFU) was established in 1871. The Fédération Internationale de Rugby Amateur (FIRA) was founded in 1934. Membership reached 49 nations in 1992. The international Rugby Football Board (IRFB) was founded in 1886. The members are: Wales, England, Scotland, Ireland, France, New Zealand, Australia and South Africa.

At international level the annual *International Championship* and the recently established *World Cup* are the most important events. The first Rugby Union World Cup tournament was played in Australia and New Zealand in 1987 by 16 national teams and won by New Zealand. The second was played in Britain and France in 1991 and won by Australia. The third World Cup was held in South Africa in 1995 and won by the host nation.

A home nations select team, the British Lions, has toured Australia, New Zealand and South Africa 22 times since 1888.

The *International Championship* was first contested by the home nations in 1884, with France joining to make it the 'Five Nations' tournament in 1910. Each team plays each other once during the season.

The *Grand Slam* is the beating of all four other nations in one season's games, and the *Triple Crown* is awarded to the home nation team that beats the other three. The *Calcutta Cup* is played for by England and Scotland. The winners of the Championship since 1920 have been:

1920 Wales /Scotland tie	1921 England
1922 Wales	1923 England
1924 England	1925 Scotland
1926 Ireland/ Scotland tie	1927 Ireland/France tie
1928 England	1929 Scotland
1930 England	1931 Wales
1932 England/Wales/ Scotland tie	1933 Scotland
1934 England	1935 Ireland
1936 Wales	1937 England
1938 Scotland	1939 England/Wales/ Ireland tie
1947 England/Wales tie	1948 Ireland
1949 Ireland	1950 Wales
1951 Ireland	1952 Wales
1953 England	1954 England/France/ Ireland tie
1955 Wales/ France tie	1956 Wales
1957 England	1958 England
1959 France	1960 England/France tie
1961 France	1962 France
1963 England	1964 Scotland/Wales tie
1965 Wales	1966 Wales
1967 France	1968 France
1969 Wales	1970 France /Wales tie
1971 Wales	1972 incomplete
1973 all tie	1974 Ireland
1975 Wales	1976 Wales
1977 France	1978 Wales
1979 Wales	1980 England
1981 France	1982 Ireland
1983 France /Ireland tie	1984 Scotland
1985 Ireland	1986 France/Scotland tie
1987 France	1988 Wales/France tie
1989 France	1990 Scotland
1991 England	1992 England
1993 France	1994 Wales
1995 England	1996 France

Ryder Cup *see* GOLF.

Scottish FA Cup *see* ASSOCIATION FOOTBALL.

Scottish Football League *see* ASSOCIATION FOOTBALL.

shooting the first shooting club, the Lucerne Shooting Guild was formed around 1466 and the first recorded match held in Zurich in 1472. It has been a feature of the OLYMPIC GAMES since the first modern games in 1896.

skiing the home of the sport is undoubtedly Scandinavia, where the first competitive slalom was held in Murren in 1922.

There are two kinds of competitive skiing; *Nordic*, which is cross-country or jumping, and *Alpine*, which is racing on prepared slopes. Nordic skiing has been included in the OLYMPIC GAMES since 1924, and *World Championships* have been held biennially since 1939. Alpine skiing has been an Olympic sport since 1936 and *World Championships* have been held biennially since 1939.

snooker it is generally thought that Colonel Sir Neville Chamberlain concocted the game of snooker in 1875 at Madras in India. The name *snooker* is taken from the nickname given to first year cadets at the Royal Military Academy, Woolwich. The first rules were established by The Billiards Association in 1900.

Snooker *World Championships* were first held in 1926. The tournament was discontinued in 1952 and not restored until 1964. In 1969 it became a knockout contest. Since 1979 all finals have been held at the Crucible Theatre in Sheffield. Recent World Champions have been:

1981 S. Davis (England)	1982 A. Higgins (N. Ireland)
1983 S. Davis (England)	1984 S. Davis (England)
1985 D.Taylor (N. Ireland)	1986 J. Johnson (England)
1987 S. Davis (England)	1988 S. Davis (England)
1989 S. Davis (England)	1990 S. Hendry (Scotland)
1991 J. Parrot (England)	1992 S. Hendry (Scotland)
1993 S. Hendry (Scotland)	1994 S. Hendry (Scotland)
1995 S. Hendry (Scotland)	

squash was developed at Harrow School in England in the early 19th century. It was originally a game for practising the game of rackets but using a softer, squashy ball. The Squash Rackets Association was formed in 1928, and this became the World Squash Federation in 1992.

The *World Open Championship* was first held in 1976. Since 1979 they have been contested annually by men (annually since 1990 for women). Recent winners of the singles events have been:

	Men	*Women*
1990	G. Hunt (Australia)	S. Devoy (New Zealand)
1991	R. Martin (Australia)	S. Devoy (New Zealand)
1992	J. Khan (Pakistan)	S. Devoy (New Zealand)
1993	J. Khan (Pakistan)	M. Martin (Australia)
1994	J. Khan (Pakistan)	M. Martin (Australia)
1995	J. Khan (Pakistan)	M. Martin (Australia)

Super Bowl *see* AMERICAN FOOTBALL.

swimming competitively dates back to 36 BC in Japan when the emperor introduced it into schools. In Britain, organized competitive swimming was introduced only in 1837 when the National Swimming Society was formed.

It has been an integral part of the OLYMPIC GAMES since 1896, and water sports now include diving, water polo, synchronized swimming and swimming using breaststroke, butterfly, front crawl and backstroke over different distances.

World Championships, separate from the Olympics, were first held in 1973 and are now staged every four years. The seventh World Championships were held in Rome in 1994.

table tennis was originally known as *ping-pong* and was widely played in England in the early part of the 20th century. The modern game grew in popularity when textured rubber mats were attached to the faces of the wooden bats, allowing the players to put spin on the ball. The International Table Tennis Federation was founded in 1926, and the *World Championships* have been held since 1927. In recent years Chinese players have dominated. The Swaythling Cup is awarded to the winning men's team and the Corbillon Cup to the winning women's team.

tennis evolved from the indoor game of *real tennis* and is thought to have first been played at the end of the 18th century. It was patented by Major Wingfield, regarded as the 'father' of the game, who claimed to have invented lawn tennis at his country house in Nantcwlyd, Wales, in 1874. The Marylebone Cricket Club later revised the major's rules, and in 1877 the All-England Croquet Club added Lawn Tennis to its title.

The All-England Championships at Wimbledon have been held since 1877 and are regarded as the most prestigious tennis championships played on grass in the world. Since 1922 they have been contested on a knockout basis. The Men's Singles Champions have been:

1922 G. Patterson (Aus)	1923 W. Johnston (USA)
1924 J. Borotra (France)	1925 R. Lacoste (France)
1926 J. Borotra (France)	1927 H. Cochet (France)
1928 R. Lacoste (France)	1929 H. Cochet (France)
1930 B. Tilden (USA)	1931 S. Wood (USA)
1932 E. Vines (USA)	1933 J. Crawford (Aus)
1934 F. Perry (UK)	1935 F. Perry (UK)

1936 F. Perry (UK)	1937 D. Budge (USA)
1938 D. Budge (USA)	1939 B. Riggs (USA)
1946 Y. Petra (France)	1947 J. Kramer (USA)
1948 B. Falkenburg (USA)	1949 T. Schroeder (USA)
1950 B. Patty (USA)	1951 D. Savitt (USA)
1952 F. Sedgeman (Aus)	1953 V. Seixas (USA)
1954 J. Drobny (Egypt)	1955 T. Trabert (USA)
1956 L. Hoad (Aus)	1957 L. Hoad (Aus)
1958 A. Cooper (Aus)	1959 A. Olmedo (USA)
1960 N. Fraser (Aus)	1961 R. Laver (Aus)
1962 R. Laver (Aus)	1963 C. McKinley (USA)
1964 R. Emerson (Aus)	1965 R. Emerson (Aus)
1966 M. Santana (Spain)	1967 J. Newcombe (Aus)
1968 R. Laver (Australia)	1969 R. Laver (Australia)
1970 J. Newcombe (Aus)	1971 J. Newcombe (Aus)
1972 S. Smith (USA)	1973 J. Kodes (Czech.)
1974 J. Conners (USA)	1975 A. Ashe (USA)
1976 B. Borg (Sweden)	1977 B. Borg (Sweden)
1978 B. Borg (Sweden)	1979 B. Borg (Sweden)
1980 B. Borg (Sweden)	1981 J. McEnroe (USA)
1982 J. Conners (USA)	1983 J. McEnroe (USA)
1984 J. McEnroe (USA)	1985 B. Becker (Germany)
1986 B. Becker (Germany)	1987 P. Cash (Aus)
1988 S. Edberg (Sweden)	1989 B. Becker (Germany)
1990 S. Edberg (Sweden)	1991 M. Stich (Germany)
1992 A. Agassi (USA)	1993 P. Sampras (USA)
1994 P. Sampras (USA)	1995 P. Sampras (USA)
1962 R. Laver (Aus)	1996 R. Krajicek (Netherlands)

The Women's Singles Champions have been:

1922 S. Lenglen (France)	1923 S. Lenglen (France)
1924 K. McKane (UK)	1925 S. Lenglen (France)
1926 K. Godfree (UK)	1927 H. Wills (USA)
1928 H. Wills (USA)	1929 H. Wills (USA)
1930 H. Moody (USA)	1931 C. Aussem (Germany)
1932 H. Moody (USA)	1933 H. Moody (USA)
1934 D. Round (UK)	1935 H. Moody (USA)
1936 H. Jacobs (USA)	1937 D. Round (UK)
1938 H. Moody (USA)	1939 A. Marble (USA)
1946 P. Betz (USA)	1947 M. Osborne (USA)
1948 L. Brough (USA)	1949 L. Brough (USA)
1950 L. Brough (USA)	1951 D. Hart (USA)
1952 M. Connolly (USA)	1953 M. Connolly (USA)
1954 M. Connolly (USA)	1955 L. Brough (USA)
1956 S. Fry (USA)	1957 A. Gibson (USA)
1958 A. Gibson (USA)	1959 M. Bueno (Brazil)
1960 M. Bueno (Brazil)	1961 A. Mortimer (UK)
1962 K. Susman (USA)	1963 M. Smith (Aus)
1964 M. Bueno (Brazil)	1965 M. Smith (Aus)
1966 B. J. King (USA)	1967 B. J. King (USA)
1968 B. J. King (USA)	1969 A. Hayden-Jones (UK)
1970 M. Court (Aus)	1971 E. Goolagong (Aus)
1972 B. J. King (USA)	1973 B. J. King (USA)

1974 C. Evert (USA)	1975 B. J. King (USA)
1976 C. Evert (USA)	1977 V. Wade (UK)
1978 M. Navratilova (Czech.)	1979 M. Navratilova (Czech.)
1980 E. Cawley (Aus)	1981 C. Evert Lloyd (USA)
1982 M. Navratilova (USA)	1983 M. Navratilova (USA)
1984 M. Navratilova (USA)	1985 M. Navratilova (USA)
1986 M. Navratilova (USA)	1987 M. Navratilova (USA)
1988 S. Graf (Germany)	1989 S. Graf (Germany)
1990 M. Navratilova (USA)	1991 S. Graf (Germany)
1992 S. Graf (Germany)	1993 S. Graf (Germany)
1994 C. Martinez (Spain)	1995 S. Graf (Germany)
1996 S. Graf (Germany)	

To achieve the *Grand Slam* in international tennis is to hold the titles to all four major tournaments simultaneously. These tournaments are: Wimbledon, the French Open, the Australian Open and the US Open.

The *Davis Cup* was established in 1900 by American player Dwight F. Davis as an international challenge match. Since 1972 a World Group of 16 nations have played-off for the Cup. Each match is decided over two pairs of singles and a doubles. Recent winners of this event have been:

1981 USA	1982 USA
1989 FR Germany	1990 USA
1983 Australia	1984 Sweden
1985 Sweden	1986 Australia
1987 Sweden	1988 FR Germany
1991 France	1992 USA
1993 Germany	1994 Sweden
1995 USA	1996 USA

test cricket *see* CRICKET.

toboggan *see* BOBSLEIGH.

Tour De France *see* CYCLING.

UEFA Cup *see* ASSOCIATION FOOTBALL.

University Boat Race *see* ROWING.

volleyball was invented by sports instructor William Morgan in 1895 and was originally intended for those who found basketball too strenuous. The game quickly spread from North America around the world, reaching Britain in 1914. An International Federation was founded in 1948, and the sport has been included in the OLYMPIC GAMES since 1964.

A *World Championship* was first staged in 1949, and the event has been held every four years since 1952. In recent years teams from the USA, Russia and China have dominated.

walking has been included in the OLYMPIC GAMES since 1906. It is defined as 'progression by steps so that unbroken contact with the ground is maintained'. Road walking is now more common than track walking, and the current Olympic distances for men 20km and 50km.

water polo originally known as 'football in the water', the sport was developed in Britain in the late 19th century. It was given official recognition in 1885 by the Amateur Swimming Association Great Britain and has been included in the OLYMPIC GAMES since 1900. A *World Championship* was first held separately from the World Swimming Championships (*see* SWIMMING) in 1991.

water skiing was pioneered by Ralph Samuelson on Lake Pepin, Minnesota, in 1922. Competition today is divided into the events of trick skiing, slalom and jumping. The *World Championship* has been held biennially since 1949.

weightlifting competitions go back to the Chou Dynasty in China and were an integral part of the ancient OLYMPIC GAMES.

The sport has been included in the modern Games since 1896. Contests are decided by the aggregate of two forms of lifting: the snatch, and the clean and jerk. In 1993 it was decided by the International Weightlifting Federation to introduce a new set of 10 weight categories. These are:

Men:	54kg, 59kg, 64kg, 70kg, 76kg, 83kg, 91kg, 99kg, 108kg, over 108kg
Women:	46kg, 50kg, 54kg, 59kg, 64kg, 70kg, 76kg, 83kg, over 83kg

Wimbledon *see* TENNIS.

Winter Olympic Games although ICE SKATING was included in the 1908 Olympic Games, the first international competition to feature a variety of winter sports was not held until 1924. This event, called the International Winter Sports Week, was later ratified by the International Olympic Committee as being the first Winter Olympic Games. The venues of the games have been:

1924 Chamonix, France	1928 St Moritz
1932 Lake Placid	1936 Garnisch-Partenkirchen
1948 St Moritz	1952 Oslo, Norway
1956 Cortina d'Ampezzo	1960 Squaw Valley
1964 Innsbruck, Austria	1968 Grenoble
1972 Sapporo, Japan	1976 Innsbruck, Austria
1980 Lake Placid	1984 Sarajevo, Yugoslavia
1988 Calgary, Canada	1992 Albertville
1994 Lillehammer, Norway	

From 1994 the Winter Games are held in the middle of the four-year Olympic cycle. The 1998 Winter Olympic Games will be held in Salt Lake City, USA.

World Cup *see* ASSOCIATION FOOTBALL, RUGBY LEAGUE, RUGBY UNION.

World Championship *see individual sports.*
World Equestrian Games *see* EQUESTRAINISM.
World Mile Record *see* ATHLETICS.
World Open Championship *see* SQUASH.
World Outdoor Championship *see* BOWLS.
World Series *see* BASEBALL.

wrestling was one of the most important events in the ancient OLYMPIC GAMES and has been included in the modern Games since 1896. There are two styles: Greco-Roman and freestyle, with the main difference being that in Greco-Roman the use of the legs is prohibited. A *World Championship* was first staged in 1904, and it is now an annual event in non-Olympic years.

yachting as a sport is thought to have begun in the 17th century in the Netherlands, then the world's greatest maritime power. The first recorded yacht race took place in 1661 when Charles II challenged the Duke of York to a race on the River Thames.

Yachting has been included in the OLYMPIC GAMES since 1900, and there have been variations in the number of different classes through the years because of changes in technology and design.

The *Admiral's Cup* was first held in 1957 and is now held biennially. National three-boat teams compete over a six-race series combining inshore and outshore racing. The races take place in the English Channel, at Cowes and in the Solent, and culminate in the Fastnet race from Cowes to Fastnet Rock, off the Irish coast, and back. The winners since 1957 have been:

1957 Great Britain	1959 Great Britain
1961 USA	1963 Great Britain
1965 Great Britain	1967 Australia
1969 USA	1971 Great Britain
1973 West Germany	1975 Great Britain
1977 Great Britain	1979 Australia
1981 Great Britain	1983 FR Germany
1985 FR Germany	1987 New Zealand
1989 Great Britain	1991 France
1993 Germany	1995 Italy

The *America's Cup* is an international challenge competition for large yachts. It gained its name in 1851 when the Royal Yacht Squadron presented a trophy to be awarded to the winner of a race around the Isle of Wight. The race was won by the schooner *America*. The Cup was later offered as a challenge trophy by the New York Yacht Club. Yachts from Britain, Canada and Australia competed for it unsuccessfully 24 times before it was won by an Australian yacht, *Australia II*, in 1983. The catamaran *Stars and Stripes* regained it for the United States in 1987, and it has remained there despite two further challenges and a legal battle concerning the use of a catamaran against a monohulled vessel.

The *Whitbread Round the World Race*, inaugurated in 1973, is held every four years and is the longest race in the world. Starting and finishing at Portsmouth in England, the competitors have to sail via the Cape of Good Hope and Cape Horn, a distance of some 32,000 nautical miles (increased in 1990 from the original distance of 26,180 nautical miles). The race is conducted as a handicap, with different classes of yacht competing.

ART, MUSIC AND LITERATURE

This section of the *Family Encyclopedia* provides essential information on terms used in the worlds of art, music and literature to enable a better understanding of these important subjects. The entries are in A to Z order, and there are cross-references, indicated in SMALL CAPITALS, to enable you to relate one term to another to improve understanding. When the name of a person is preceded by the mark ◊, you will find information on that person in the FAMOUS PEOPLE section.

A the sixth NOTE of the SCALE of C; it is the note to which instruments of an orchestra are usually tuned (*see* TUNING).

abbandono (*Italian*) 'passionately'.

abbreviations are employed in music for terms of expression, as *dim.* for DIMINUENDO, *f.* for FORTE; as arbitrary signs, such as two DOTS on either side of an oblique line for repetition of a group of NOTES; or as numerals, which serve as shorthand symbols for various CHORDS in FIGURED BASS.

absolute music *or* **abstract music** instrumental music that exists purely as music and does not attempt to relate to a story or image. It is the opposite of PROGRAMME MUSIC.

absolute pitch *or* **perfect pitch** the sense by which some people can exactly identify, or sing without an accompanying instrument, any NOTE they hear.

abstract art art that intentionally avoids representation of the observed world. Abstraction has long been a feature of the decorative arts and to a large degree continues to dominate 20th-century art. There are two distinct trends: one towards an ordered, hard-edged CONSTRUCTIVISM, the other leaning to a freer, more expressionistic reduction of forms.

abstract expressionism art that is based on freedom of expression, spontaneity and random composition and is characterized by loose, unrestrained brushwork and often indistinct forms, usually on large canvases. The works may or may not be figurative. The term mainly applies to an art movement of the 1940s in New York, although it was first used in 1919 with reference to the early abstract work of ◊Kandinsky. Inspired by SURREALISM, the movement represented a breakaway from the REALISM hitherto dominant in American art and went on to influence European art in the 1950s.

abstract music *see* ABSOLUTE MUSIC.

Absurd, Theatre of the a form of theatre, developed in the 1960s, that characterizes the human condition as one of helplessness in the face of an irrational, 'absurd' universe. ◊Beckett, ◊Pinter and ◊Ionesco, whose characters communicate with one another in disjointed, inconsequential language, are among the best-known practitioners of the form, whose roots go back to DADAism. Absurdist drama reacts strongly against the conventions of naturalism. However, the only absurdist dramatist to become president of a country, Vaclav ◊Havel, has argued that the form is in fact a very efficient vehicle for describing life realistically in a totalitarian society.

a cappella *or* **alla capella** (*Italian*) literally 'in the chapel style'; it is a term that has come to mean unaccompanied choral singing.

accel. the abbreviation for ACCELERANDO.

accelerando (*Italian*) 'quickening'; a term used to indicate a gradual speeding up of pace (abbreviation ACCEL.).

accent the emphasis given to specific notes to indicate the RHYTHM of a piece of music.

acciaccatura (*Italian*) an ornamental or auxiliary NOTE, normally the SEMITONE below, played just before, or at the same time as, a regular note. An *acciaccatura* is written in small type before the regular note and has a stroke through its tail. From the Italian *acciacciare*, to crush.

accidental a NOTE in a piece of music that departs by one or two SEMITONES from the KEY SIGNATURE. It is indicated by a SHARP, FLAT or NATURAL sign before it. An accidental holds good throughout a BAR unless contradicted.

accompaniment music supporting a soloist or CHOIR. An accompaniment may be provided by an orchestra, organ or, most usually, a piano.

accordion a portable REED ORGAN, which was invented in Germany in the early 19th century. Air is forced through the reeds by means of bellows that are operated by the player's arms, and notes and chords are played by pressing buttons. The *piano accordion* has a keyboard (operated by the right hand) for playing MELODY NOTES, and buttons (operated by the left hand) for simple chords. The accordion is associated with informal or FOLK music, but it has been used by serious composers.

Acmeist group a school of Russian poetry which first emerged as a coherent group in Leningrad in 1911. In part a reaction to the more obscure aspects of SYMBOLISM, the group proclaimed the virtue of clarity of expression. It was founded by the poet Anna Akhmatova (1889–1966) whose ex-husband was executed by the Bolsheviks in 1921 and whose son was arrested and imprisoned several times during Stalin's terror.

acoustic guitar *see* GUITAR.

acoustics (1) a branch of physics that is concerned with SOUND. The main characteristics of a sound are its PITCH, intensity, RESONANCE and quality. (2) the characteristics of a hall or auditorium that enable speech and music to be heard without the sounds being distorted. In a concert hall with good acoustics, sounds from the stage can be heard clearly in all quarters.

acrostic a type of poem, in which the initial letters of each line form a word reading downwards. The form was used by Roman poets, and the Elizabethan poet and statesman Sir John Davies (1569–1626) devised in *Hymns of Astraea* (1599) 26 verses, each spelling out the name of Elizabeth I ('Elisabetha Regina').

acrylic paint a versatile synthetic paint that is quick-drying and can be used in thick, heavy layers or thin washes on almost any surface. A range of matt or gloss finishes can be achieved by the use of additives.

action (1) the mechanism of a keyboard instrument that links the keyboard to the strings or, in the case of an ORGAN, to the PIPES and STOPS. (2) the gap between the strings and FINGERBOARD of a stringed instrument as dictated by the height of the BRIDGE.

action painting a form of ABSTRACT EXPRESSIONISM in which the paint is applied to the canvas in the course of a series of actions or movements by the artist.This may involve dancing, cycling or rolling about on the canvas to spread and mix the wet paint. In a less random technique the artist might paint the silhouette of a model in various poses against the canvas.

act tune *or* **curtain music** an instrumental piece of music that is played between the acts of a play while the curtain is down. It is usually associated with 17th- and 18th-century music. *See also* ENTR'ACTE; INTERLUDE; INTERMEZZO.

ad lib. abbreviation for AD LIBITUM.

ad libitum (*Latin*) literally 'at pleasure' and usually abbreviated to AD LIB. In music, the term is used to indicate that, when playing a piece, a performer can: (a) alter the TEMPO or RHYTHM; (b) choose an alternative PASSAGE by the composer; (c) improvise a CADENZA; and (d) include or omit a passage if he or she so chooses.

adagietto (*Italian*) (1) 'slow' but not as slow as ADAGIO. (2) a short composition in an adagio TEMPO.

adagio (*Italian*) (1) literally 'at ease', i.e. at a slow TEMPO. (2) a slow movement.

adagissimo (*Italian*) 'very slow'.

added sixth a frequently used CHORD created by adding the sixth note from the ROOT to a MAJOR or MINOR TRIAD; for example, in the KEY of C major, A is added above the triad of C-E-G.

additional accompaniments new or revised PARTS for extra instruments written by later composers and added to 17th- and 18th-century works in order to increase fullness. In many cases, the additions did not match the quality of the original music, but ◊Mozart once wrote additional music for ◊Handel's MESSIAH when an organ was not available.

a due corde (*Italian*), **à deux cordes** (*French*) 'on two strings'; when applied to music for stringed instruments, the term means that a piece should be played on two strings, not just on one.

Aeolian harp a type of ZITHER which has strings of similar length but of different thickness. The instrument is not actually played but left outside to catch the wind; different CHORDS are sounded according to the speed of the wind, which makes the strings vibrate faster or slower (*see* VIBRATION). The name is derived from Aeolus, who was the Greek god of the wind.

Aeolian mode a MODE that, on the PIANO, uses the white NOTES from A to A.

Aesthetic movement a cultural movement which developed in England in the late 19th century, characterized by a very affected and mannered approach to life and the arts, a fondness for orientalism and decadence, archaic language and pseudo-medievalism. The character Bunthorne in ◊Gilbert and Sillivan's comic opera *Patience* (1881) is a conflation of all the features found ridiculous in the movement by contemporaries: 'Though the Philistines may jostle, you will rank as an apostle in the high aesthetic band, / If you walk down Piccadilly with a poppy or a lily in your medieval hand.' Oscar ◊Wilde is the most prominent literary figure of merit associated with the movement.

aesthetics an area of philosophy concerning the ideals of taste and beauty and providing criteria for critical study of the arts. The term was coined in the mid-18th century by the German philosopher Alexander Baumgarten, and in the 20th century came to include a wider theory of natural beauty.

affettuoso (*Italian*) 'tender' or 'affectionate', i.e. an indication that a piece of music should be played with tender feeling.

African art a term generally used to describe African tribal art in the countries south of the Sahara Desert. Much of this art is of a group nature, in that it has cultural and religious significance at its heart rather than individual ambition. Examples of typical art forms include richly carved wooden masks and figures. Body art is also important in tribal ritual and may involve scarring, tattooing or disfigurement of parts of the body, although it can also make use of paint, beads and feathers. Brightly coloured batiks and printed fabrics are a more recent feature of this rich heritage. African tribal art had a marked influence on 20th-century art styles, as in the works of ◊Picasso and ◊Cézanne.

agitato, agitatamente (*Italian*) literally 'agitated', 'agitatedly', i.e. an indication that a piece should be played restlessly or wildly.

Agnus Dei (*Latin*) 'O Lamb of God'; the concluding part of the Latin MASS. Numerous musical settings have been made for it.

air (1) a simple TUNE or SONG. (2) a melodious BAROQUE composition.

air brush an atomizer, powered by compressed air, that is used to spray paint. It is shaped like a large fountain pen and produces a fine mist of colour, giving delicate tonal gradations and a smooth finish. Its principal use is in the fields of advertising and graphic design.

alabaster a fine-grained type of gypsum that can be translucent, white or streaked with colour. It is soft and easy to carve and is therefore a popular medium for decorative artefacts and statues. It is not as strong or weather-resistant as marble, and is not often used for outdoor works.

Alberti bass a simple accompaniment to a MELODY consisting of 'broken' or spread CHORDS arranged in a rhythmic pattern. It is so called because the Italian singer, composer and harpsichordist Domenico Alberti (1710–40) used it in his keyboard SONATAS.

alborada (*Spanish*) literally 'morning song'; a form of popular Spanish music for BAGPIPES and SIDE DRUM.

Albumblatt (*German*) literally 'album leaf'; a popular title given by 19th-century composers to short, instrumental compositions (often for the PIANO) and of a personal nature.

Aldeburgh Festival an annual music FESTIVAL held in June, which was founded by ◊Britten and the tenor Peter ◊Pears at Aldeburgh, Suffolk, in 1948. It maintains its strong association with Britten, many of whose works were first performed there.

aleatory music music which contains unpredictable or chance elements so that no two performances of a piece are ever similar. It is a form explored since 1945 by composers such as ◊Cage in his *Music of Changes*, and ◊Stockhausen.

alienation effect an effect which is supposed to occur upon an audience when the audience is reminded by action, dialogue or song that it is in fact an audience watching a play, and not, for example, waiting for a bus. The intention is to confront the audience with the artificiality of dramatic representation, through devices such as interrupting the action to address the audience, or by bursting into song, or by stylizing the stage set or action. The theory is especially associated with ◊Brecht, who claimed that his plays created an 'alienating', distancing effect, enabling the audience to resist identifying emotionally with the characters on stage and therefore more capable of absorbing the message of plays, e.g. that traditional notions of morality simply represent 'bourgeois' morality. 'Alienation effects' of the kind described by Brecht have existed in drama since before ◊Aristophanes, and form an important element in the Theatre of the ABSURD.

alla (*Italian*) 'in the style of'.

alla breve (*Italian*) (1) an instruction that a piece of music should be performed twice as fast as the NOTATION would suggest. (2) 2/2 TIME.

allargando (*Italian*) literally 'getting broader', i.e. an indication that a piece should be played grandly whilst at the same time getting slower.

alla tedesca (*Italian*) an abbreviation of *alla danza tedesca,* meaning 'in the style of a German dance'. *See* ALLEMANDE.

allegory a form of narrative in which the characters and events symbolize an underlying moral or spiritual quality, or represent a hidden meaning beneath the literal one expressed. ◊Bunyan's *Pilgrim's Progress* is the greatest English-language example of a sustained allegory.

allegretto (*Italian*) a term indicating light and moderately quick movement, but not as fast as ALLEGRO.

allegro (*Italian*) literally 'lively', i.e. in a quick TEMPO. The term is often used as the title of a bright composition or MOVEMENT.

allemande (*French*) an abbreviation of *danse allemande* or 'German dance', of which there are two forms. (1) a moderately slow dance used by 17th- and 18th-century composers as the first MOVEMENT of a SUITE of four contrasting dances. (2) a brisk dance of the 18th and 19th centuries, similar to the WALTZ.

allentando (*Italian*) 'slowing down', i.e. a term used to indicate that the TEMPO of a piece of music should be slowed down.

all' ongarese (*Italian*) 'in the style of Hungarian [gypsy] music'.

alphabet the letters used in music as they occur in the natural SCALE are C, D, E, F, G, A, B. The oldest HARPS and shepherd PIPES are believed to have had seven TONES, to which the Greeks gave the names of letters, A being the lowest. Greek NOTATION became highly complicated with the development of the MODES, and Pope Gregory the Great (540–604) changed church notation, again employing the first seven letters, indicating the lower OCTAVE by capitals, and the upper by small letters. NOTES were gradually added to the lower A, and when the modern scale was adopted in the 16th century, the lowest tone had become C instead of A. In addition, Germans use H for B natural, B for B flat.

Alphorn (*German*) a primitive type of HORN with no valves, traditionally played by Swiss herdsmen to call in their cattle in the evening. Made from wood and bark, alphorns usually have an upturned bell which rests on the ground, and they can be up to 3 metres (10 feet) long.

al segno (*Italian*) 'to the sign' (i.e. to a standard symbol used in musical NOTATION). The term is used in two ways: it can instruct the player either to go *back* to the sign and start again, or to *continue* until the sign is reached.

alt an abbreviation of the Latin phrase *in alto,* which means 'high'. It is used for the NOTES in the OCTAVE rising from G above the TREBLE CLEF; the notes in the octave above that are said to be *in altissimo.*

Altamira the site in northern Spain of prehistoric rock paintings dating from about 13000 BC, in 1879 the first ever to be discovered. Originally dismissed as forgeries, their age and authenticity were accepted as genuine only in the early 20th century. A variety of animals painted in a lively, naturalistic manner are depicted, including bison, aurochs and wild horses.

altarpiece a decorated wall, screen or sectional painting set behind the altar of a Christian church, a feature of church decor from the 11th century. There are two forms: a *retable* can be fairly large and complex, rising from floor level; a *reredos* is often smaller and may stand on the altar itself or on a pedestal behind it.

alto (*Italian*) 'high' (1) the highest adult male VOICE, which is now used only in male-voice CHOIRS. (2) an addition to the name of an instrument to indicate that it is one size larger than the SOPRANO member of the family; for example alto CLARINET. (3) a low female voice that has a greater compass than the male alto voice (usually, and more properly, called CONTRALTO).

Amati the name of a famous family of violin-makers who worked in Cremona, Italy, in the 16th and 17th centuries.

Ambrosian chants a collection of CHANTS or PLAINSONG, used in Milan Cathedral, which are named after St Ambrose (*c.* 340–97), Bishop of Milan, who greatly influenced church singing and may have introduced the antiphonal singing of the Syrian church. Despite bearing his name, the earliest surviving chants were composed long after his death.

amen (*Hebrew*) 'so be it'.

American organ *or* **cabinet organ** a REED ORGAN, similar to a HARMONIUM except that air is sucked through the reeds instead of being blown through them.

amoroso (*Italian*) 'lovingly', indicating that a piece should be played with warm affection.

amplifier any device, particularly an electric one, which renders a sound louder.

anabasis (*Greek*) a succession of ascending tones.

anacrusis (*Greek anakrousis*, literally 'a prelude') an unstressed NOTE or grouping of notes at the beginning of a musical PHRASE; it can also mean an unstressed syllable at the beginning of a SONG.

Ancients, The a group of Romantic artists working in England between 1824 and the early 1830s. Samuel ◊Palmer was a leading member of the group. Their work was mainly pastoral in theme, much inspired by ◊Blake's illustrations of ◊Virgil.

ancora (*Italian*) 'again', 'yet' or 'still', as in *ancora forte* meaning 'still loud', and *ancora più forte* meaning 'yet louder'.

andante (*Italian*) 'going' or 'moving'; it is usually used to indicate a moderate TEMPO or a walking pace. *Più andante* means 'moving more' or slightly faster. Andante is sometimes used as a title for a moderately slow piece of music.

andantino (*Italian*) 'less slow' (i.e. slightly faster) than ANDANTE.

Anfang (*German*) 'beginning'; *Anfangs* means 'from the beginning'.

anglaise (*French*) short for *danse anglaise* or 'English dance', i.e. a lively dance in quick time, such as a HORNPIPE.

Anglican chant a characteristically English way of setting to music prose, psalms and canticles, in which the number of syllables per line can vary. To accommodate this irregularity, the first NOTE of each musical PHRASE is a RECITING NOTE which is timeless and is used to sing as many syllables as necessary before moving on to notes which are sung in time and which normally carry one syllable each. It is basically a simple form of GREGORIAN CHANT.

Anglo-Saxon art a term for works of art produced in England between AD 5 and 1066. The major source of surviving artefacts is the 7th-century excavation site at Sutton Hoo, and much of the Anglo-Saxon jewellery collection at the British Museum comes from there. The abstract plant and animal designs show the influences of Celtic art typical of Anglo-Saxon craft. The late 7th and early 8th centuries saw the production of the Lindisfarne Gospels in the kingdom of Northumbria. These are famous for their delicate interwoven designs, reminiscent of Irish illuminations, as in the Book of Kells. The other centre for manuscript production was the Winchester school of the 10th century.

Angry Young Men a rather imprecise term used in mid-1950s Britain to denote a group of English writers who had little in common apart from vaguely leftish sympathies and a hatred of English provincialism and intellectual pretentiousness ('madrigalphobia'). Other characteristics of Angry Young Men included a liking for jazz, polo-neck sweaters, coffee bars, and compliant women – like their American counterparts, the ◊Beat Generation writers, Angry Young Men regarded women as appendages.

anima (*Italian*) 'soul' or 'spirit', as in *con anima,* which means that a piece should be played 'with soul' or 'with emotion'.

animato (*Italian*) 'animated'.

animo (*Italian*) 'spirit', so *con animo* indicates that a piece should be performed 'with spirit'.

animoso (*Italian*) 'spirited'.

answer the second entry of the main SUBJECT (theme) of a FUGUE which is played a FIFTH higher or lower than the first entry. In a *real answer,* the subject and answer are identical; in a *tonal answer,* the intervals in the answer are changed.

anthem the Anglican equivalent to the Roman Catholic MOTET. An anthem is usually an elaborate musical setting of non-liturgical words sung by a church CHOIR without the congregation; SOLO parts are common and accompaniment by an ORGAN is usual.

anticipation the sounding of a NOTE (or notes) of a CHORD before the rest of the chord is played.

antinovel *see* novel.

antiphon the sacred words, sung in PLAINSONG by two CHOIRS, before and after a PSALM or CANTICLE in a Roman Catholic service. *Antiphonal* is an adjective applied to the musical effect achieved by two choirs (or groups of instruments) which are positioned in different parts of a hall and sing (or play) alternately, one 'answering' the other.

Antique, The remains of ancient art, in particular Greek and Roman statues, which were taken as a standard of classical order and beauty in the representation of the human form by RENAISSANCE and NEOCLASSICAL artists.

anvil a PERCUSSION INSTRUMENT consisting of steel bars that are struck with a wooden or metal mallet. It is meant to sound like a blacksmith's anvil being struck with a hammer and was used by both ◊Verdi (in *Il Trovatore*) and ◊Wagner (in *Das Rheingold*).

a piacere (*Italian*) 'at pleasure', meaning that the performer of a piece of music is permitted to take a certain amount of liberty, particularly with TEMPO, while playing it. *See also* AD LIBITUM.

appassionato, appassionata (*Italian*) 'impassioned' or 'with passion or feeling', hence the title *Sonata appassionata* which was given to ◊Beethoven's Piano Sonata in F Minor (Op. 57).

applied art art that serves a useful purpose or that ornaments functional objects; often a synonym for *design*. Subjects included under this term are architecture, interior design, ceramics, furniture, graphics, etc. These are usually contrasted with the *fine arts* of painting, drawing, sculpture, printmaking, etc, and the division became more distinct at the time of the Industrial Revolution and the emergence of AESTHETICS. This division is still a matter of important debate.

appoggiando (*Italian*) 'leaning'; when applied to NOTES, this implies that they should pass very smoothly from one to the next.

appoggiatura (*Italian*) a term for a 'leaning' NOTE (*see also* APPOGGIANDO), indicated in the SCORE. (1) a *long appoggiatura* is a note of varying length that is different from the HARMONY note. (2) a *short appoggiatura* is a very short note of indefinite length, sometimes accented, sometimes not. (3) a *passing appoggiatura,* as used in the 18th century, normally occurs when the principal notes of a MELODY form a sequence of thirds and it is played before the beat.

aquarelle the French term for watercolour painting, where a water-based paint is applied to dampened paper in thin glazes that are gradually built up into areas of varying tone.

aquatint an etching technique where a resin-coated metal plate is placed in a bath of acid that bites into the resin, producing a pitted surface. The depth of tone intensifies the longer the plate remains in the acid, and areas required to be lighter in tone are 'stopped out', using washes of varnish. The finished print resembles a watercolour wash, and the technique of overlaying separate plates of different colours can be used to build up a range of depth and colour. The process is often combined with linear etching.

arabesque (1) a florid treatment of thematic music. (2) a lyrical piece of music that employs an exaggerated and elaborate style, as used by ◊Schumann and ◊Debussy.

Arabian Nights Entertainments *or* **The Thousand and One Nights** a compilation of stories, the earliest of which probably originated in Persia, that were translated into Arabic in the mid-9th century AD, and have become classics of Arab and world literature. The stories are told by Scheherazade to her husband, a king who has killed each previous bride following the consummation of the marriage. Scheherazade saves her life by keeping the king in suspense, postponing the conclusion of each story until the following night.

Arcadia a region of ancient Greece which became the archetypal setting for rural bliss and innocence in the arts. ◊Virgil's *Eclogues* in the 1st century BC established the use of Arcadia as a literary device for this purpose, but the ironic undertone in Virgil's work (Arcadia is largely barren) is usually missing from later writers' use of the myth. Virgil's dark undertone returns to Western culture in the work of several 16th and 17th-century painters, e.g. ◊Poussin, with reference to the phrase *Et in Arcadia Ego* ('And I [i.e. death] also in Arcadia').

archet (*French*) a BOW, such as is used to play a stringed instrument.

archi (*Italian*) 'bows', a term that refers to all stringed instruments played with a BOW.

arco (*Italian*) the singular of ARCHI. It is the usual instruction to play with the BOW after playing PIZZICATO.

ardito (*Italian*) 'bold', 'energetic'.

aria (*Italian*) a SONG or AIR. Originally the term was used for any song for one or more voices but it has come to be used exclusively for a long, SOLO song as found in ORATORIO and OPERA.

arioso (*Italian*) 'like an ARIA'. (1) a melodious and song-like RECITATIVE. (2) a short AIR in an OPERA or ORATORIO. (3) an instrumental piece that follows the style of a vocal arioso.

Armory Show the international exhibition of modern art held at the 69th Regimental Armory in New York in 1913, one of the most influential exhibitions ever shown in the US. Effectively two exhibitions in one, it represented not only a fine cross-section of contemporary American art but also a massive selection of modern European art, a total of around 1,600 works. It toured the US, arousing great controversy and excitement among the 250,000 people who paid to see it, but it served the function of restoring the life and vitality of contemporary art and critical debate in the US.

arpeggiare (*Italian*) 'to play the HARP', i.e. to play CHORDS 'spread out' as they are on the harp. *See also* ARPEGGIO.

arpeggio (*Italian*) 'HARP-wise', i.e. an indication that the NOTES of a CHORD should be played in rapid succession, as they are on a harp, and not simultaneously.

arrangement an adaptation of a piece of music for a medium different from that for which it was originally composed.

ars antiqua (*Latin*) the 'old art', i.e. music of the 12th and 13th centuries as opposed to **ars nova**, the new style of music that evolved in the 14th century.

Art Autre *or* **Art Informel** a name coined by art critic Michel Tapie in *Un Art Autre* (1952); he used it to describe nongeometric ABSTRACT EXPRESSIONISM.

Art Brut the work of anyone not linked to the art world either as professional or amateur, for example psychiatric patients or prisoners, etc. The term can also include graffiti and the work of young children. It refers to any work uninfluenced by the art world and its fashions.

Art Deco the decorative art of the 1920s and 1930s in Europe and North America, originally called *Jazz Modern*. It was classical in style, with slender, symmetrical, geometric or rectilinear forms. Major influences were ART NOUVEAU architecture and ideas from the ARTS AND CRAFTS MOVEMENT and the BAUHAUS. The simplicity of style was easily adaptable to modern industrial production methods and contemporary materials, especially plastics. This resulted in a proliferation of utility items, jewellery and furniture in an elegant streamlined form, as well as simplification and streamlining of interior decor and architecture.

Art Nouveau a style of decorative art influential and popular between 1890 and World War I in Europe and North America. Art Nouveau was primarily a design style with its main effects being seen in applied art, graphics, furniture and fabric design, and in architecture. In the fine arts it represented a move away from historical realism, but was not as vigorous or dominant as IMPRESSIONISM or CUBISM. Art Nouveau design is characterized by flowing organic forms and asymmetrical linear structures, although architectural and calligraphic forms were more austere and reserved. Its principal exponents were the Scottish architect and designer Charles Rennie Mackintosh (1868–1928) and the American designer Louis Comfort Tiffany (1848–1933).

Arts and Crafts Movement an English movement in the decorative arts towards the end of the 19th century. It was based on the ideas of the art critic John ◊Ruskin and the architect Augustus Pugin

(1812–52), with reference to the medieval guilds system, and took its name from the Arts and Crafts Exhibition Society formed in 1888. The motive was to re-establish the value of handcrafted objects at a time of increasing mass-production and industrialization. Designers in the movement, with a variety of styles, attempted to produce functional objects of an aesthetically pleasing nature. The most active and important leader of the movement was William ◊Morris.

ascription *see* ATTRIBUTION.

Ashcan School a group of American painters of urban realism between 1908 and 1918. Their aim was to declare themselves primarily American painters, and they painted what they saw as American life, rejecting subject matter of academic approval. Among their influences were the works of ◊Daumier and ◊Goya.

assai (*Italian*) 'very', as in *allegro assai,* 'very fast'.

assemblage any sculptural type of construction using found objects, from pieces of painted wood to old shoes. *See also* COLLAGE.

assez (*French*) 'moderately', as in *assez vite,* 'moderately quick'.

atelier the French term for an artist's studio. In 19th-century France, an *atelier libre* was a studio where artists could go to paint a model. No formal tuition was provided, and a small fee was charged.

a tempo (*Italian*) 'in time', a term that indicates that a piece should revert to its normal TEMPO after a change of speed.

athematic music music that does not have any THEMES or TUNES as such; it is concerned with exploring the unconventional possibilities of sounds.

atonal music music that is not in any KEY. Atonal music is particularly associated with the works of ◊Schoenberg, although he preferred to use the word *pantonality,* meaning a synthesis of all the keys.

attacca (*Italian*) 'attack', i.e. start the next MOVEMENT without a pause.

attribution *or* **ascription** the assigning of an unsigned picture to a painter, using similarity of style or subject as the basis.

aubade (*French*) 'morning music', as opposed to a SERENADE.

augmentation the lengthening of the time values of NOTES in melodic parts with the result that they sound more impressive and grand. The opposite of augmentation is DIMINUTION.

augmented interval the INTERVAL formed by increasing any perfect or MAJOR interval by a SEMITONE.

augmented sixth a CHORD based on the flattened SUBMEDIANT that contains the augmented sixth INTERVAL.

augmented triad a TRIAD of which the FIFTH is augmented.

a una corda (*Italian*) 'on one string' (*compare* A DUE CORDE), meaning left-hand PEDAL (of a piano), i.e. reducing the volume.

Authorized Version *or* **King James Bible** a translation of the Bible published in 1611 in the reign of James I of England. The powerful and poetic language of the Authorized Version has made it perhaps the most influential book on writers in English from the early 17th century on. Although it is often said to be the greatest book ever written by a committee, the work is based for the most part on the great translation of 1525 by William Tyndale (*c.*1495–1536).

autograph in art, a term used to denote a painting by one artist only, and not assisted by pupils or assistants.

autoharp a type of ZITHER in which CHORDS are produced by pressing down KEYS (1) that dampen some of the strings but let others vibrate freely. The instrument was invented in the late 19th century and was popular with American folk musicians.

Automatistes a Canadian group of painters formed in the 1950s whose ideas were based on the spontaneity of creativity. One of its main proponents was Paul Emile Borduas (1905–60).

avant-garde (*French*) 'vanguard', a term applied to music (or any other art) that is considered to break new ground in style, structure, or technique.

Ave Maria (*Latin*) 'Hail Mary', a prayer to the Virgin Mary used in the Roman Catholic Church. It has been set to music by numerous composers.

B (1) the seventh note of the scale of C major. (2) abbreviation for bass or for Bachelor (as in B. Mus., Bachelor of Music).

baby grand the smallest size of grand PIANO.

Bach trumpet a 19th-century valved TRUMPET which was designed to make it easier to play the high-pitched parts that were originally composed by BACH and his contemporaries for a natural (unvalved) trumpet.

back the lower part of the sounding box of string instruments, connected in VIOLS to the sounding board or belly by a sound post set beneath the bridge. Its construction and material vitally affect the quality of the TONE (3) produced.

bagatelle (*French*) 'trifle', a short, light piece of music, usually for piano, for example, ◊Beethoven's FÜR ELISE.

bagpipes a reed instrument in which air is supplied to the PIPE or pipes from an inflated bag. Bagpipes are known to have existed for 3000 years or more and hundreds of different types are found today. The best known form of bagpipe is played in Scotland and consists of a *bag* which is inflated through a pipe and is held under the arm; a CHANTER (a REED pipe with finger holes) on which the MELODY is played; and several DRONE pipes, each of which is tuned to a different note. Air is rhythmically squeezed from the bag by the arm (and is then replenished with more breath) and is forced out through the chanter and drone pipes.

balalaika (*Russian*) a FOLK instrument of the GUITAR family, with a triangular body. It is of Tartar origin and usually has just three strings and a fretted FINGERBOARD. Balalaikas are made in several sizes and are often played in concert with one another.

ballabile (*Italian*) 'in a dancing style'.

ballad a narrative poem or song in brief stanzas, often with a repeated refrain and frequently featuring a dramatic incident. The songs sold by the vagabond Autolycus in Shakespeare's *As You Like It*, describing battles, public executions and the like, are typical examples of the sort of *broadside ballad* popular with the common people up to (and a bit beyond) Victorian times. In 1765, the antiquary Thomas Percy (1729–1811) published his *Reliques of Ancient English Poetry*, a collection of ancient and not-so-ancient traditional English poems, songs and folk ballads that was to prove immensely influential.

Examples of what has been termed the *literary ballad* soon began to appear. Notable examples include ◊Coleridge's 'Rime of the Ancient Mariner' and ◊Keats' 'La Belle Dame sans Merci'.

Among the greatest examples of the form are the *Border ballads*, which stem from the violent world of the English/Scottish borders from the late Middle Ages to the 17th century. Their merit had been recognized by Sir Philip ◊Sidney in his *Defence of Poesie* (1595), but his opinion was to remain a minority one until the late 18th century, when the rediscovery of the great ballad heritage coincided with the birth pangs of the Romantic movement, and with enthusiasm for the 'noble' and 'simple' truths of the world of traditional songs throughout Europe. Representative examples of the Border ballads are the laments 'Sir Patrick Spens' and 'The Bonny Earl of Murray', the haunting song of fairyland 'Thomas the Rhymer', and the macabre 'Twa Corbies'. The most important collection of Border ballads is that issued by Sir Walter Scott in 1802–3, *Minstrelsy of the Scottish Borders*. Many of the ballads here were written down at the dictation of the singers of the old songs themselves.

In music, a ballad is also a sentimental 'drawing-room' song of the late 19th century (sometimes referred to as a *shop ballad* to differentiate it from a broadside ballad) or a narrative song or operatic ARIA.

ballade a type of medieval French poetry, often set to music by TROUBADOURS. In music, it is a 19th-century term, coined by CHOPIN, for a long, romantic instrumental piece. Chopin wrote four outstanding ballades.

ballad opera popular OPERA composed of dialogue and SONGS with tunes borrowed from FOLK music, popular songs, and sometimes opera. They first appeared in the 18th century in England, probably the best-known being *The Beggar's Opera*.

ballet a dramatic entertainment in which dancers in costume perform to a musical ACCOMPANIMENT. Mime is often used in ballet to express emotions or to tell a story. Ballet has a long history that dates back to before the Middle Ages. In the 16th and 17th centuries, ballets often included singing and consequently were closely linked to OPERA. By the end of the 18th century, however, ballet had evolved more gymnastic qualities and, although it was still included as an integral part in many operas, it also kept a separate existence. In the 19th century, ballet achieved new heights of popularity in France and spread to Italy and Russia where several schools of ballet were established that incorporated traditional dancing into their teaching. ◊Tchaikovsky's ballet scores (e.g. *Swan Lake* and *The Sleeping Beauty*) had a massive influence on Russian ballet and greatly added to its international appeal. The Russian choreographer and entrepreneur Sergei Diaghilev (1872–1929) encouraged young composers such as ◊Stravinsky and ◊Ravel to write ballet scores. At the start of the 20th century, ballet became immensely popular in England. ◊Vaughan Williams, ◊Bliss and ◊Britten all composed notable pieces for ballet, and the outstanding British choreographers Sir Frederick Ashton (1904–88) and Sir Kenneth Macmillan (1929–92) helped to maintain the interest. In the USA, the choreographer George Balanchine (1904–83) had an equally powerful influence, and many American composers, such as ◊Copland and ◊Bernstein, have since written ballet music. Meanwhile, ◊Prokofiev's masterpieces (e.g. *Romeo and Juliet*) maintained Russia's tradition in dancing.

Today, ballet is witnessing a new revival, with SCORES being produced by young composers and with energetic companies establishing new dancing techniques.

band a term used to describe virtually any group of instrumentalists except a concert ORCHESTRA, for example dance band, JAZZ band, POP band, MILITARY BAND.

banjo a GUITAR-like, stringed instrument of Black American origin. It comprises a shallow metal (sometimes wood) drum with parchment stretched over the top while the bottom is (usually) left open. Banjos can have between four and nine strings, which are played by plucking either with the fingers or a PLECTRUM.

bar (1) a vertical line (bar line) drawn down one or more STAVES of music. (2) the space between two bar lines.

barber-shop quartet a quartet of amateur male singers who perform CLOSE-HARMONY arrangements. The tradition originated in barber shops in New York in the late 19th century.

Barbizon School a group of French landscape painters in the 1840s who based their art on direct study from nature. Their initial influences included ◊Constable as well as some of the Dutch landscape painters. Their advanced ideas represented a move away from academic conventions, and their interest in daylight effects and their bold use of colour helped prepare the way for IMPRESSIONISM.

barcarolle (*French*) a boating song with a rhythm imitating that of songs sung by gondoliers.

bard a Celtic minstrel, part of whose job it was to compose SONGS for his master. Bards traditionally held annual meetings (*Eisteddfods*) in Wales and these have been revived in recent times as competition festivals.

baritone (1) a male voice, midway between BASS and TENOR with a range of approximately two OCTAVES. (2) a BRASS INSTRUMENT of the SAXHORN family.

Baroque a cultural movement in art, music and science in the 17th century. In terms of art history, the area of reference is slightly broader and takes in the late 16th and early 18th centuries. It specifically indicates the stage between the MANNERISM of the late High RENAISSANCE and ROCOCO, into which Baroque developed. As a style it is characterized by movement, rhetoric and emotion, stemming from the achievements of the High Renaissance, and it represented a reaction away from Mannerist attitudes and techniques. ◊Caravaggio was among its leading figures when it first began in Rome, and ◊Rembrandt's work reflected Baroque trends for part of his career. Adjectivally, *baroque* can also be used to describe art from any

age that displays the richness and dynamism associated with the movement.

barrel organ a mechanical ORGAN of the 18th and 19th centuries in which air was admitted into PIPES by means of pins on a hand-rotated barrel. It was restricted to playing a limited number of tunes but was nonetheless frequently used in church services. *See also* MECHANICAL INSTRUMENTS.

bass (1) the lowest adult male VOICE. (2) an abbreviation for DOUBLE BASS. (3) an addition to the instrument name to indicate the largest member of a family of instruments (except where CONTRABASS instruments are built).

bassa (*Italian*) 'low'.

bass-bar a strip of wood glued as reinforcement under the BRIDGE (1) inside the BELLY of instruments of the VIOLIN family.

bass clarinet a single-REED instrument built an OCTAVE lower than the CLARINET, with a crook and upturned bell.

bass clef *see* CLEF.

bass drum a large PERCUSSION INSTRUMENT consisting of a cylindrical wooden hoop which is usually covered on both sides with vellum. It is common in MILITARY BANDS in which it is suspended vertically from the shoulders and beaten with two sticks.

basset horn an ALTO CLARINET.

bass flute an ALTO FLUTE with a PITCH a FOURTH lower than a 'concert' or normal flute.

basso continuo *see* FIGURED BASS.

bassoon a double-reed instrument of the 16th century, consisting of a wooden tube doubled back on itself. It has a compass from B flat below BASS CLEF to E on the fourth line of the TREBLE CLEF.

basso ostinato (*Italian*) 'obstinate bass', a GROUND BASS, i.e. a bass FIGURE that is repeated many times throughout a composition (or part of a composition) while the upper parts vary.

baton the stick used by a CONDUCTOR to give commands to performers.

batterie (*French*) (*also* **battery**) (1) a 17th- and 18th-century term for ARPEGGIO. (2) the PERCUSSION section of an ORCHESTRA.

battuta (*Italian*) *see* BEAT.

Bauhaus a German school of architecture and applied arts founded by the architect Walter Gropius (1883–1969) at Weimar in 1919. One of its aims, as with the ARTS AND CRAFTS MOVEMENT, was to narrow the gap between fine and APPLIED ARTS; another was to focus on architecture as the environment of art. Each student took a six-month foundation course in practical craft skills such as weaving, glass painting and metalwork. Among the first masters were the EXPRESSIONIST painters ◊Klee and

◊Kandinsky. A more CONSTRUCTIVIST influence came when the Bauhaus moved to Dessau in 1925. Later came a shift in emphasis from craftsmanship towards industrialized mass-production. Gropius resigned from the Bauhaus in 1928; it was moved to Berlin in 1932 and was closed by the Nazis in 1933. A number of Bauhaus masters emigrated to the US, where their ideas were influential.

Bayreuth a town in Germany where WAGNER arranged for the building of a festival theatre which has subsequently become internationally famous for staging his operas. Built to resemble a Greek amphitheatre, the theatre holds 1,800 people, and the special feature of a hood surrounding the orchestra pit gives the auditorium an acoustical excellence ideal for Wagner's music.

beat (1) a unit of rhythmic measure in music, indicated to a CHOIR (2) or ORCHESTRA by the movement of a conductor's BATON. The number of beats in a BAR depends on the TIME-SIGNATURE. (2) a form of 20th-century POPULAR MUSIC with a steady and powerful RHYTHM.

Beat Generation a term invented by Jack ◊Kerouac to describe a group of American writers, artists and musicians in the 1950s. Notable Beat writers included Kerouac himself, whose novel *On the Road* (1952) became the 'Beat bible', the poets Lawrence Ferlinghetti (1919–) and Allen Ginsberg (1926–), and the novelists Neal Cassady (1926–68) and William ◊Burroughs. The Beat writers were anti-Western in their values; they dabbled in communalism, loved modern jazz and took drugs.

The term 'Beat' was said to denote (a) the weariness of struggling against materialist society, (b) jazz rhythm, which they tried to capture in their prose, (c) beautitude. The last quality is somewhat dubious.

bebop a JAZZ development of the 1940s in which complex RHYTHMS and harmonic sequences were carried out against rapidly played melodic IMPROVISATION. It is particularly associated with the jazz saxophonist, Charlie ◊Parker.

Bebung (*German*) 'trembling', i.e. a VIBRATO effect caused by shaking a finger holding down a KEY (1) of a CLAVICHORD.

Bechstein a company of German piano manufacturers established in Berlin in 1856 by Friedrick Wilhelm Karl Bechstein (1826–1900). Branches of his firm were subsequently formed in France, England and Russia.

bel a unit used to measure the intensity of SOUND, named after Alexander Graham ◊Bell. *See also* DECIBEL.

bel canto (*Italian*) 'beautiful singing', a style of singing characterized by elaborate technique, associated with 18th-century Italian OPERA.

bells (*orchestral*) cylindrical metal tubes (*tubular bells*) of different lengths which are suspended from a frame and struck with a wooden mallet.

belly the upper part of the body or soundbox of a stringed instrument.

ben, bene (*Italian*) 'well', as in *ben marcato* meaning 'well marked', 'well accented'.

Benedicite (*Latin*) a CANTICLE known as the *Song of the Three Holy Children;* it is used during Lent as an alternative to the TE DEUM in the Anglican service of Morning Prayer.

Benedictus (*Latin*) (1) the second part of the SANCTUS of a Roman Catholic MASS. (2) the CANTICLE 'Benedictus Dominus Israel' or 'Blessed be the Lord God of Israel'.

berceuse (*French*) a cradle SONG.

bergamasca (*Italian*), **bergomask** (*English*) *or* **bergamasque** (*French*) (1) a popular 16th- and 17th-century dance from Ber-gamo in Italy. (2) a 19th-century dance in quick 6/8 TIME.

Bible, The *see* Authorized Version.

Biedermeier a style in art and architecture in Austria and Germany between 1815 and 1848. It took its name from a fictional character of the time, Gottlieb Biedermeier, who personified the philistine artistic taste of the middle classes. Architecture associated with the style is solid and utilitarian, paintings are meticulous and devoid of imagination.

big band a large BAND, most commonly associated with the SWING era. Such bands were famed for the strong dance RHYTHMS they produced.

Bildungsroman (*German*) 'education novel', a novel that describes the growth of a character (usually based on the author) from youthful naivety to a well-rounded maturity. The term derives from ◊Goethe's *Wilhelm Meister's Apprenticeship.* Two notable examples in English are ◊Dickens' *David Copperfield* and ◊Joyce's *Portrait of the Artist as a Young Man.*

binary form a structure, common in BAROQUE music, consisting of two related sections which were repeated. SONATA FORM evolved from it.

bis (*French*) 'twice', 'again'.

bitonality the use of two keys simultaneously.

blanche (*French*) 'white'; the French word for a MINIM.

blank verse a term sometimes used to denote any form of unrhymed verse but normally applied to unrhymed verse in iambic pentameters, i.e. a line of verse with five short-long 'feet', e.g. the actress

Mrs Siddons's reputed remark: 'You brought me water, boy; I asked for beer' (she was known to speak unconsciously in blank verse). The form was developed in English by the Earl of Surrey (*c*.1517–47), and reached its highest peak in ◊Shakespeare's plays. Other notable works written in blank verse include ◊Milton's *Paradise Lost* and ◊Wordsworth's *Prelude*.

Blaue Reiter, Der (*German*) 'The Blue Rider', the name, taken from a painting by Wassily ◊Kandinsky, of a group of German EXPRESSIONISTS formed in Munich in 1911. Leading members were ◊Kandinsky and Paul ◊Klee, who, although their working styles were diverse, were united by a philosophy of the creative spirit in European art. They organized two touring exhibitions in Germany in 1911 and 1912, and produced an *Almanac* (1912), which included major European avant-garde artists as well as tribal, folk and children's art. The idea of the *Almanac* was to unite music, art and literature in a single creative venture. It was intended to be the first in a series, but the group disbanded in 1914.

block chords a harmonic procedure in which the NOTES of CHORDS are moved simultaneously in 'blocks'.

Bloomsbury Group a group of artists, writers and intellectuals who lived or worked in the Bloomsbury area of London in the early decades of the 20th century and whose members included Virginia ◊Woolf and John Maynard ◊Keynes.

bluegrass a type of FOLK music originally from Kentucky, USA, (where Kentucky bluegrass grows); *see* COUNTRY AND WESTERN.

blues a 20th-century Black American SONG or lamentation following an essentially simple form of twelve BARS to each verse. Blues music formed the basis for JAZZ; musicians favoured such instruments as the GUITAR and HARMONICA.

Bluestocking an originally disparaging term denoting members of small, mostly female groups in English 18th-century social life, who held informal discussion groups on literary and scholarly matters. The term derives from the blue stockings worn by a male member of the groups, the botanist Benjamin Stillingfleet.

bocca chiusa (*Italian*) literally 'closed mouth', i.e. humming.

body paint *see* gouache.

Boehm system an improved system of KEYS (2) and levers for the FLUTE (1) which is named after its German inventor, Theobald Boehm (1794–1881). The system is also applied to other instruments, for example, the clarinet.

bolero (*Spanish*) a moderately fast Spanish dance in triple TIME. ◊Ravel's *Bolero* (1928), a spiralling crescendo based on a repeated theme, was the music for a ballet choreographed by ◊Nijinsky.

bones a pair of small sticks (originally bones) that are held in the hands and clicked together rhythmically.

bongos pairs of small, upright Cuban DRUMS that are often found in dance BANDS. They are played with the hands.

boogie-woogie a JAZZ and BLUES style of piano playing in which the left hand plays a persistent bass RHYTHM while the right hand plays a MELODY.

Book of Common Prayer the once-official book of services for the Church of England, first published in 1549. The most loved version is that of 1662, the language of which, like that of the AUTHORIZED VERSION, has been very influential on English prose and poetry.

bop short for BEBOP.

Border ballad *see* BALLAD.

bouche fermée (*French*) *see* BOCCA CHIUSA.

bourrée (*French*) a lively French dance of the 17th century.

bouzouki a Greek stringed instrument with a long, fretted neck. Its six strings are plucked, often as an ACCOMPANIMENT to songs. The sound of the bouzouki reached a worldwide audience with the film music of the Greek composer Mikis Theodorakis (1925–).

bow a wooden stick which is strung with horse-hair and used to play instruments of the VIOLIN and VIOL families.

bowed harp a primitive VIOLIN, dating back to at least the twelfth century. It was held on the knee and played vertically.

bowing the technique of using a BOW to play an instrument.

brace the vertical line, usually with a bracket, which joins two STAVES of music to indicate that they are played together.

branle (*French*) a French FOLK DANCE from the 15th century, which had a swaying movement.

brass band a type of BAND, particularly associated with the north of England, which consists of BRASS INSTRUMENTS and DRUMS only. Brass bands have been popular in England since the beginning of the 19th century.

brass instruments a family of WIND INSTRUMENTS which are made of metal but not always brass. Instruments with REEDS and those which used to be made from wood (such as the FLUTE) are excluded. A characteristic of the family is that sound is produced by the vibration of the lips which are pressed into a funnel-shaped MOUTHPIECE. A selection of NOTES can be produced by effectively lengthening the tubing, either with a slide (as in the TROMBONE) or with valves (as in the TRUMPET). Brass instruments include the trombone, CORNET, BUGLE, trumpet, French HORN, TUBA, and EUPHONIUM.

bravura (*Italian*) literally 'bravery', as in a 'bravura passage', a passage that demands a VIRTUOSO display by the performer.

break (1) in JAZZ, a short, improvised, SOLO passage. (2) the point in a vocal or instrumental range where the REGISTER changes.

breit (*German*) 'broadly' or 'grandly', a term used to describe the manner in which a piece should be played.

breve originally the short NOTE of music (*c*.13th century), but as other notes have been introduced, it is now the longest note and is only occasionally used.

bridge (1) a piece of wood that stands on the BELLY of stringed instruments and supports the strings. (2) a passage in a COMPOSITION (1) that links two important THEMES together.

brindisi (*Italian*) 'a toast', a drinking song in an opera during which toasts are often given.

brio (*Italian*) 'vigour', so CON BRIO means 'with vigour'.

brisé (*French*) 'broken'; a term which indicates that a CHORD should be played in ARPEGGIO fashion, or that music for stringed instruments should be played with short movements of the BOW.

broadside ballad *see* ballad.

broken octaves a term used to describe a passage of NOTES that are played alternately an OCTAVE apart; they frequently occur in piano music.

bronze a metal alloy of bronze mixed with tin and occasionally lead and zinc, which has been used as a medium for sculpture since ancient times, when it was cast solid using wooden models. Modern techniques use hollow casting methods of sand casting or *cire perdue* ('lost wax').

Brücke, Die (*German*) 'The Bridge', an association of German artists founded in 1905 in Dresden. The name derives from ◊Nietszche's idea that a man can be seen as a bridge towards a better future, and in this the artists saw themselves as a link with the art of the future, in a move away from REALISM and IMPRESSIONISM. They also wanted to integrate art and life, and so lived together in community in the tradition of the medieval guilds system. Their influences included AFRICAN tribal art and the works of Van ◊Gogh and FAUVISM. Their painting was mainly EXPRESSIONIST, although comprehending a

variety of styles and techniques. They concentrated initially on figures in landscape and portrait, and made great use of texture, clashing colour and aggressive distortion to powerful effect. They helped found the Neue SEZESSION in 1910, then split away and exhibited as a group with Der BLAUE REITER.

In 1913 the group disbanded because of conflicts over aims and policies in relation to the development of CUBISM.

brushwork the 'handwriting' of a painter, i.e. the distinctive way in which he or she applies paint, either smoothly or roughly, thinly or thickly, in long strokes or short. Like handwriting, brushwork is individual to a painter.

bugle a simple BRASS INSTRUMENT with a conical tube and a cup-shaped MOUTHPIECE which was widely used for giving military signals. *See also* LAST POST.

bull roarer *see* THUNDER STICK.

burla, burlesca (*Italian*) a short and jolly piece of music.

Byzantine music music of the Christian Church of the Eastern Roman Empire which was established in AD 330 and lasted until 1435. It influenced Western church music.

C (1) the key-note or TONIC of the SCALE of C major. (2) an abbreviation for CONTRALTO; *con* (with); *col, colla* (with the).

c.a. abbreviation for *coll' arco*, meaning 'with the BOW', as opposed to PIZZICATO (with reference to a stringed instrument).

cabaletta (*Italian*) a term for a simple ARIA with an insistent RHYTHM, or an emphatically rhythmical ending to an ARIA or DUET.

cabinet organ *see* AMERICAN ORGAN.

caccia (*Italian*) a 'hunt', as in *corno da caccia*, hunting horn. Also a 14th-century hunting poem about country life set to music.

cachucha a Spanish solo dance in 3/4 time, resembling the BOLERO.

cacophony a discordant muddle of SOUND or DISSONANCE.

cadence literally a 'falling', a term used to describe the concluding PHRASE at the end of a section of music.

cadenza (*Italian*) literally 'cadence', but it has come to have two specific meanings: (1) an elaborate ending to an operatic ARIA. (2) a flourish at the end of a PASSAGE of SOLO music in a CONCERTO.

calando (*Italian*) 'diminishing', i.e. in both volume and speed.

calcando (*Italian*) 'pressing forward', i.e. a term used to indicate an increase in speed.

calypso a kind of song with SYNCOPATED RHYTHMS from the West Indies, notably Trinidad; calypso LYRICS are usually witty and topical, and are often vehicles for political satire.

cambiata (*Italian*) an abbreviation for *nota cambiata*, CHANGING NOTE.

camera (*Italian*) literally a 'room', but in musical terms it refers to a type of music that can be performed in a place other than a church, music hall, or opera house, etc; *see* CHAMBER MUSIC.

campanelli (*Italian*) *see* GLOCKENSPIEL.

campanology the art of bell-ringing or the study of bells.

campus novel a NOVEL with a university setting. They are invariably satirical and are equally invariably written by former or practising academics. An early example is Kingsley ◊Amis's *Lucky Jim* (1954). Some novelists, such as David ◊Lodge, whose *Small World* is perhaps the definitive example, have made a speciality of the genre. Another notable example is *The History Man* (1975) by Malcolm Bradbury (1932–). American examples tend towards anguish and 'relationships'; the British are more concerned with casting a cold eye on the harsh world of academic politics.

cancan a Parisian music-hall dance of the late 19th century in quick 2/4 time.

cancel (*US*) *see* NATURAL.

cancrizans a type of music that makes sense if it is played backwards; RETROGRADE MOTION.

canon a COUNTERPOINT composition in which one part is imitated and overlapped by one or more other PARTS, for example, a SOPRANO lead with a TENOR follow-up. In a 'strict' canon, the imitation is exact in every way.

cantabile (*Italian*) 'song-like', a term applied to instrumental pieces indicating that they should be played in a singing style.

cantata originally a piece of music of the BAROQUE period that is sung (as opposed to a SONATA, a piece which is played). It has come to be a term used to describe a vocal or CHORAL piece with an instrumental accompaniment. German cantatas were generally religious works. In many ways, the form is similar to OPERA and ORATORIO, but it tends not to be so elaborate.

canticle a HYMN that has words from the Bible, other than a PSALM.

cantor (*Latin*) a 'singer'; nowadays the term refers to the chief singer in a CHOIR or the lead singer of liturgical music in a synagogue.

cantus firmus (*Latin*) 'fixed song', i.e. a MELODY in POLYPHONIC music, often taken from PLAINSONG, with long NOTES (1) against which COUNTERPOINT tunes are sung.

canzone, canzona, canzon literally 'song'; a vocal work, or an instrumental piece that is modelled on music for the VOICE.

canzonet (1) a short kind of CANZONE. (2) a type of MADRIGAL or a simple SOLO SONG.

capotasto (*Italian*) the 'head of the fingerboard', i.e. the raised part or 'nut' at the top of the FINGERBOARD of a stringed instrument that defines the lengths of the strings. A moveable capotasto, comprising a wood or metal bar that can be clamped to the fingerboard, is occasionally used on fretted instruments to shorten all the strings at the same time, thus raising the PITCH.

cappella, a *see* A CAPPELLA.

capriccio (*Italian*) 'caprice'; a short, lively piece.

caricature a drawing of a person in which his or her most prominent features are exaggerated or distorted in order to produce a recognizable but ridiculous portrait, possibly suggesting a likeness to another object. The technique was pioneered in the late 16th century and flourished in the 18th and 19th centuries.

carillon (1) a set of bells, usually in a bell tower, which can be played by electrical or mechanical means to produce a tune. (2) an ORGAN STOP which produces a bell-like sound.

carol originally, any medieval English song with a refrain, but now generally a song associated with Christmas.

cartoon (1) a drawing, or series of drawings, intended to convey humour, satire or wit. Cartoons were commonly used from the 18th century onwards in newspapers and periodicals as a vehicle for social and political comment, and in comic magazines for children and adults in the 20th century. (2) a full-size preparatory drawing for a painting, mural or fresco. The drawing was fully worked out on paper and then mapped out on to the surface to be painted.

castanets a Spanish PERCUSSION INSTRUMENT comprising two shell-like pieces of wood which are clicked together by the fingers. In orchestras, they are occasionally shaken on the end of sticks.

castrato (*Italian*) an adult male singer with a SOPRANO or CONTRALTO voice produced by castration before puberty. Castrati were popular singers in the 17th and 18th centuries. The practice was abandoned during the 19th century.

catch a ROUND for three or more VOICES. The words are often humorous and frequently contain puns, for example, *Ah, how Sophia* which, when sung, sounds like 'Our house afire'.

Cavalier Poets a loose grouping of lyric poets associated with the court or cause of Charles I during his clashes with the English Parliament and the ensuing Civil War. (Poets such as Marvell and Milton who supported Parliament, however, are not known as Roundhead poets.) The most notable Cavalier poets are ◊Herrick, Sir John Suckling (1609–41), Thomas Carew (c.1595–1640) and Richard Lovelace (1618–c.1658). The latter's witty and graceful 'To Althea' is perhaps the most loved of the Cavalier poems.

cavatina (*Italian*) a short and often slow SONG or instrumental piece.

Cave Paintings see Altamira; Lascaux.

CB an abbreviation for *contrabasso* (DOUBLE BASS).

cebell a 17th-century English dance similar to the GAVOTTE.

Cecilia, St the patron saint of music who was martyred in the second or third century. Since the 16th century music festivals to commemorate her have been annual events. Her feast day is 22 November.

ceilidh (*Gaelic*) a gathering at which SONGS, FOLK music and dances are performed; ceilidhs are particularly associated with Scotland and Ireland.

celesta a small keyboard instrument in which HAMMERS are made to strike metal bars suspended over wooden resonators; the sound produced has an ethereal, bell-like quality. ◊Tchaikovsky included a celesta in the' Dance of the Sugar-Plum Fairy' in his ballet *The Nutcracker*.

cello *see* VIOLONCELLO.

cembalo (*Italian*) (1) a DULCIMER. (2) an abbreviation of *clavicembalo*, which is the Italian for HARPSICHORD.

CF abbreviation for CANTUS FIRMUS.

chaconne (*French*) a slow dance in triple time that is thought to have originated in Mexico.

chalk a soft stone, similar to a very soft limestone, used for drawing. *Crayon* is powdered chalk mixed with oil or wax.

chalumeau (*French*)) (1) a generic term for a type of REED-PIPE. (2) a term now used for the lower REGISTER of the CLARINET.

chamber music originally, chamber music was a term used to describe any type of music that was suitable for playing in a room of a house as opposed to a church or concert hall. However, it has come to mean music for a small number of instruments (for example, flute and piano) or group of performers (for example, STRING QUARTET, SEXTET, etc.), with one instrument to each part.

chamber orchestra a small orchestra, sometimes solely of stringed instruments, for performing CHAMBER MUSIC.

change-ringing an English method of ringing a peal of church bells; the bells are rung in an established order, which then passes through a series of changes.

changing note *or* **cambiatta** a dissonant PASSING NOTE which is a third away from the preceding note, before being resolved.

chanson (*French*) a song for either a SOLO VOICE or a CHOIR. In some contexts it can also mean an instrumental piece of song-like quality.

chant a general term for a type of music which is sung as part of a ritual or ceremony. It is a term which is used particularly for unaccompanied singing in religious services.

chanter the pipe of a BAGPIPE, on which the MELODY is played.

chanterelle (*French*) literally the 'singing one', i.e the highest string on a bowed, stringed instrument (for example, the E string on a VIOLIN).

character piece a term used by composers for a short instrumental piece, such as may attempt to describe a specific mood. Examples are ◊Schumann's *Fantasiestücke*, *Nachtstücke*, and *Albumblätter*.

charcoal the carbon residue from wood that has been partially burned. Charcoal will make easily erasable black marks and is used mainly to make preliminary drawings, e.g. on walls. When used on paper it has to be coated with a fixative to make the drawing permanent.

charleston a ballroom dance, similar to the FOXTROT, which was evolved by black Americans of the Southern USA.

chest voice the lower REGISTER (2) of VOICE (1), so called because NOTES seem to emanate from the chest. *Compare* HEAD VOICE.

chevalet (*French*) the BRIDGE of a stringed instrument.

chiaro, chiara (*Italian*) 'clear', 'distinct'.

chiaroscuro (*Italian*) literally 'light-dark', used to describe the treatment of light and shade in a painting, drawing or engraving to convey depth and shape. It is particularly used of works by painters like ◊Caravaggio or ◊Rembrandt.

chinoiserie In the 16th and 17th centuries, trade with the Far East created a European market for Chinese art and influenced the development of a vogue for things Chinese. Pagodas and stylized scenes, plants and animals conceived to be in the Chinese style began to decorate pottery, furniture, fabrics and ornaments. These were finally mass-produced, both in Europe and the Far East, specifically for this market. A familiar product in this style is the famous Willow Pattern pottery range.

chitarrone (*Italian*) a large LUTE.

choir (1) the place, defined by special seats or 'choir stalls', in a large church or cathedral where singers are positioned. (2) a body of singers, such as a male-voice choir, church choir. (3) (US) a section of the orchestra, for example 'brass choir'.

choirbook a large medieval volume that (usually) included both words and music and was designed to be read by various members of a CHOIR (2) while it was stationed on a centrally placed lectern.

choir organ the section of an ORGAN that is played from the lowest MANUAL and is soft enough to accompany a church CHOIR (2).

choral an adjective used to describe music that involves a CHORUS, for example CHORAL SYMPHONY.

chorale a HYMN-tune of the Lutheran Church, but dating back to the 15th century.

chorale cantata a CANTATA that was written to be performed in a (Lutheran) church.

choral symphony (1) a SYMPHONY in which a CHORUS is used at some point (or, indeed, a symphony written entirely for voices). (2) the popular name of ◊Beethoven's symphony no. 9 in D Minor, which ends with 'An die Freude' ('Ode to Joy') for chorus and soloists.

chord a combination of NOTES played simultaneously, usually not less than three.

chorus (1) a body of singers. *See* CHOIR. (2) music written for a body of singers (usually to follow an introductory piece). (3) a REFRAIN that follows a SOLO verse.

chromatic (from Greek *chromatikos,* 'coloured') a term used to describe NOTES which do not belong to a prevailing SCALE, for example, in C major all SHARPS and FLATS (1) are chromatic notes. The 'chromatic scale' is a scale of twelve ascending or descending SEMITONES; and a 'chromatic CHORD' is a chord that contains chromatic notes. *See also* TWELVE-NOTE MUSIC.

ciaccona (*Italian*) *see* CHACONNE.

cimbal, cimbalom *see* DULCIMER.

cinquecento the Italian term for the 16th century.

circular breathing the technique of sustaining a NOTE when playing a WIND INSTRUMENT by breathing in through the nose while sounding the note.

cittern a pear-shaped stringed instrument of the GUITAR family popular from the 16th century to the 18th century It was similar to the LUTE except that it had a flat back and wire strings, and was easier to play.

clappers virtually any kind of PERCUSSION INSTRUMENT comprising two similar pieces that can be struck together, for example BONES, spoons and sticks.

clarinet a single-REED WOODWIND instrument dating back to the 17th century. It has a cylindrical tube and nowadays comes in two common sizes: B flat and A. It is an instrument common to both CLASSICAL music and JAZZ.

clàrsach (*Gaelic*) a small harp of the Scottish Highlands and Ireland.

classical (1) a term used to describe a certain form of music which adheres to basic conventions and forms that are more concerned with carefully controlled expression rather than unrestrained emotion. (2) a term used to describe 'serious' music as opposed to popular music.

Classicism a style of art based on order, serenity and emotional control, with reference to the classical art of the ancient Greeks and Romans. It eschews the impulsive creativity and sponteneity of Romanticism in favour of peace, harmony and strict ideals of beauty. Figures drawn in the classsical style were usually symmetrical and devoid of the normal irregularities of nature. *See also* NEOCLASSICISM.

clavecin (*French*) *see* HARPSICHORD.

claves short sticks that are held in the hand and clicked together to emphasize a BEAT (1) or RHYTHM. They originated in Cuba.

clavicembalo (*Italian*) *see* HARPSICHORD.

clavichord a keyboard instrument dating from the 15th century in which the strings are struck by a brass 'tangent' and can be made to sound a note of variable PITCH until the KEY (1) is released. However, the sound is soft and the keyboard is limited; the instrument fell out of favour with the introduction of the PIANO.

clavier (1) a practice keyboard which makes no sound save clicks. (2) any keyboard instrument that has strings, for example, the CLAVICHORD, the HARPSICHORD, the PIANO.

clef a symbol positioned on a line of a STAVE which indicates the PITCH of the line and consequently all the NOTES on the stave. Three clefs are commonly used: ALTO (TENOR), TREBLE and BASS.

Clerihew a verse form invented by Edmund Clerihew ◊Bentley with two rhymed couplets of variable length and often encapsulating an unreliable biographical anecdote: 'Mr Michael Foot/ Had lots of loot;/ He loved to gloat/ While petting his stoat.'

close harmony HARMONY in which the NOTES of the CHORDS are close together. In singing, this means that each VOICE remains fairly close to the MELODY.

coda (*Italian*) literally a 'tail', meaning a PASSAGE at the end of a piece of music which rounds it off.

col, coll', colla, colle (*Italian*) literally 'with the', so *col basso* means 'with bass'; *colle voce*, 'with voice'.

collage a piece of art created by adhering pieces of paper, fabric, wood, etc, on to a flat surface. The technique was popular with the Cubists, Braque and Picasso, and is a precursor of the more sculptural methods of assemblage.

coloratura (*Italian*) the florid ornamentation of a melodic line, especially in opera.

colour (1) in art, an effect induced in the eye by light of various wavelengths, the colour perceived depending on the specific wavelength of light reflected by an object. Most objects contain pigments that absorb certain light frequencies and reflect others, e.g. the plant pigment chlorophyll usually absorbs orange or red light and reflects green or blue, therefore the majority of plants appear to be green in colour. A white surface is one where all light frequencies are reflected and a black surface absorbs all frequencies. Artists' colours are made by combining pigments of vegetable or mineral extraction with an appropriate medium, e.g. linseed oil. The rarity of some mineral pigments has a direct effect on the prices of particular colours. (2) in music, the TONE quality of instruments and voices.

colour field painting a movement begun by abstract expressionists towards a more intellectual abstraction. Their paintings were large areas of pure, flat colour, the mood and atmosphere being created by the shape of the canvas and by sheer scale.

colourist a term in art criticism referring to an artist who places emphasis on colour over line or form. For example, ◊Titian and ◊Giorgione have been called the 'Venetian Colourists'. The term is, however, too vague to be applied consistently.

combination tone a faint (third) NOTE that is heard when two notes are sounded simultaneously; also called a 'resultant tone'.

combo an abbreviation of 'combination', especially a collection of musicians that make up a JAZZ BAND.

comedy a form of drama, usually of a light and humorous kind and frequently involving misunderstandings that are resolved in a happy ending. The first major examples known, those by ◊Aristophanes, are still among the greatest, and became known as *old comedy*. The roots of Aristophanes' work are ancient, based on fertility rituals, and often involve ferocious attacks on named individuals, e.g. ◊Euripedes and ◊Socrates. The outrage caused by Aristophanes' plays led to the Athenians banning the ridicule of named individuals on stage. The so-called *new comedy* that then developed in Greece is known to us largely through adaptations of lost Greek originals by Roman dramatists. It is this new comedy, based upon Aristotle's opinion that the business of comedy is with people of no significance (who can be safely ridiculed), that was to prevail on the stage until the advent of ◊Chekhov, ◊Ibsen and ◊Strindberg, with their downbeat comedies of middle-class life. These latter comedies may loosely be described as *tragicomedies*, although the term is normally reserved to denote those plays of the Jacobean period, such as Beaumont and ◊Fletcher's *Philaster* (1609) and *A King and No King* (1611), and ◊Shakespeare's 'dark comedies', e.g. *Measure for Measure* (1604), in all of which the action seems to be leading inexorably towards a tragic ending, but resolves itself (more or less) happily at the end, after trials, tests and tribulations. Other notable forms of comedy include the comedy of HUMOURS, the COMEDY OF MANNERS, SENTIMENTAL COMEDY and COMMEDIA DELL'ARTE. *See also* MELODRAMA, *compare* TRAGEDY.

comedy of humours see HUMOURS, THEORY OF.

comedy of manners a form of comedy that features intrigues, invariably involving sex and/or money, among an upper region of society. The central characters are usually witty sophisticates, and there is often much mockery of characters from inferior stations who try to imitate the behaviour of their betters, e.g. a merchant trying to live the lifestyle of a gallant. The form is associated particularly with RESTORATION dramatists such as William Congreve (1670–1729), Sir George Etherege (*c.*1634–91) and ◊Wycherley. The finest of these Restoration comedies is Congreve's *The Way of the World* (1700).

comic opera an OPERA that has an amusing plot, or (sometimes) an opera that includes some spoken dialogue.

commedia dell'arte a form of Italian comedy, popular from the RENAISSANCE until the 18th century, that used stock farcical characters (such as Harlequin and Columbine) and plots as a basis for improvization. Actors wore masks representing their particular characters, and were often skilled acrobats. The COMMEDIA DELL'ARTE form was given new life by ◊Goldoni.

common chord a MAJOR or MINOR CHORD, usually consisting of a keynote and its third and FIFTH.

comodo (*Italian*) 'convenient', as in *tempo comodo,* meaning at a 'convenient speed'.

compass the musical RANGE of a VOICE or instrument.

composition in art, the arrangement of elements in a drawing, painting or sculpture in proper proportion and relation to each other and to the whole. In music, (1) a work of music. (2) the putting together of sounds in a creative manner. (3) the art of writing music.

compound interval an INTERVAL which is greater than an OCTAVE.

compound time musical TIME in which each BEAT (1) in a BAR is divisible by three, for example 6/8, 9/8 and 12/8 time.

computer-generated music music that is created by feeding a formula or program into a computer, which then translates the program into sounds.

con (*Italian*) 'with', so *con amore* means 'with love', i.e. lovingly, tenderly.

concert a public performance of secular music other than an OPERA or BALLET.

concert grand a large GRAND PIANO that is used in concert halls.

concertina a type of ACCORDION, hexagonal in shape, with small studs at each end which are used as keys.

concertmaster (*US*) the first violinist, or leader of an orchestra.

concerto (*Italian*) (1) originally, a work for one or several voices with instrumental ACCOMPANIMENT. (2) a work for several contrasted instruments. (3) an orchestral work in several movements, containing passages for groups of SOLO instruments (*concerto grosso*). (4) a piece for a solo instrument and an accompanying orchestra.

concert overture an orchestral piece of one MOVEMENT, similar to an opera OVERTURE, but written solely for performance in a concert hall. It originated in the 19th century. Examples are ◊Mendelssohn's *Hebrides* and ◊Tchaikovsky's *Romeo and Juliet*.

concert pitch the internationally agreed PITCH, according to which A above middle C (in the middle of the TREBLE CLEF) is fixed at 440 hertz (cycles per second).

Concertstück, Konzertstück (*German*) a short CONCERTO.

concitato (*Italian*) 'agitated'.

concord a combination of sounds (such as a CHORD) that are satisfactory and sound agreeable. The opposite of DISSONANCE.

concrete art a term used to describe severely geometrical abstract art.

concrete music *see* MUSIQUE CONCRÈTE.

conducting the art of directing and controlling an ORCHESTRA or CHOIR (or operatic performance) by means of gestures. As well as indicating the speed of a piece, a conductor, who often uses a BATON to exaggerate his or her arm movements, is also responsible for interpreting the music.

conga (1) a tall, narrow DRUM which is played with the hands. (2) an entertaining dance in which the participants form a long, moving line one behind the other.

conjunct a succession of NOTES of different PITCH.

conservatory a school that specializes in musical training. The term originates from the kind of charitable institutions for orphans, called *conservatorii* in 16th- and 17th-century Italy, where music was taught to a high standard.

consort an old spelling of the word 'concert', meaning an ENSEMBLE of instruments, for example, a consort of VIOLS.

Constructivism a movement in abstract expressionism concerned with forms and movement in sculpture and the aesthetics of the industrial age. It began in post-World War I Russia with the sculptors Antoine Pevsner (1886–1962), his brother Naum Gabo (1890–1979) and Vladimir Tatlin (1885–1953). Their work, which made use of modern plastics, glass and wood, was intentionally nonrepresentational. Their ideas were published in Gabo's *Realistic Manifesto* in 1920. Gabo and Pevsner left Russia in 1922 and 1923 respectively and went on to exert great influence on Western art. Tatlin remained to pursue his own ideals of the social and aesthetic usefulness of his art.

Constructivist theatre a form of theatre devised by Vsevolod Meyerhold (1874–1940), Russian theatrical producer and director. By the time the Bolsheviks came to power in 1917, Meyerhold was recognized as the most prominent exponent of avant-garde ideas and productions on the Russian stage. Among the plays he produced at his Meyerhold Theatre were those of Mayakovsky (*see* FUTURISM). His productions (which were strongly influential on ◊Brecht) cheerfully abandoned traditional conventions, such as the proscenium arch, and broke plays up into episodic segments. Stage sets were determinedly Constructivist, i.e. they used technological artefacts derived from industrial processes to emphasize the kinetic, active nature of the stage (in large part, a development from Russian Futurism). Actors were encouraged to think of themselves as 'biomechanisms', and to study circus acrobatics and the conventions of COMMEDIA DELL' ARTE. Meyerhold's ideas inevitably proved intolerable to the Soviet cultural commissars of Socialist realism and he was executed.

continuo (*Italian*) an abbreviation of *basso continuo*. *See* FIGURED BASS.

contrabass (1) (adjective) an instrument that is an OCTAVE lower than the normal BASS of the family, for example, contrabass TUBA. (2) (noun) a DOUBLE BASS.

contralto the lowest female VOICE, which usually has a range of about two OCTAVES.

contrapuntal relating to COUNTERPOINT.

contratenor the 14th- and 15th-century word for a VOICE with approximately the same range as a TENOR.

contredanse (*French*) a French corruption of 'country dance', i.e. a lively dance.

cor (*French*) *see* French horn *under* HORN.

cor anglais (*French*) literally 'English horn', but it is neither English nor a HORN. It is in fact an ALTO OBOE pitched a FIFTH below the standard oboe.

coranto (*Italian*) *see* COURANTE.

corda (*Italian*) a 'string', as in 'piano string'; the term *una corda* literally means 'one string', an indication to use the 'soft' PEDAL on the PIANO.

cornet (1) a BRASS INSTRUMENT with three valves that has a quality of TONE (3) lying between a HORN and a TRUMPET; it has great flexibility and is often used in MILITARY and JAZZ BANDS. (2) an ORGAN STOP used for playing flourishes.

cotillion a popular ballroom dance of the early 19th century.

cottage piano a small upright PIANO.

counterpoint the combination of two or more independent melodic lines that fit together to create a coherent SOUND texture. The CLASSICAL (1) conventions of HARMONY are based on counterpoint.

counter-subject a MELODY, found in a FUGUE, that is CONTRAPUNTAL to the main THEME (SUBJECT), i.e. after singing the subject, a VOICE (2) carries on to sing the counter-subject while the answer is sung.

counter-tenor the highest natural male VOICE (not to be confused with FALSETTO).

country and western a generic term for a form of 20th-century American folk music, originating from the southeast of the USA, with Nashville, Tennessee, as its traditional home. It is usually played by small BANDS using FIDDLES, GUITARS, BANJOS and DRUMS etc. The songs are typically of a sentimental, sometimes tragic, nature. Lively BLUEGRASS music is a form of country and western.

couplet (1) the same as DUPLET. (2) a two-note SLUR. (3) a SONG in which the same music is repeated for every STANZA.

courante (*French*) short for *danse courante* or 'running dance', a lively Baroque dance in triple time.

cow bell as used as a PERCUSSION INSTRUMENT in an orchestra, it is an ordinary square cow bell with the clapper taken out. It is played with a drumstick.

crayon *see* CHALK.

credo (*Latin*) 'I believe', the first word in the Roman Catholic Creed.

crescendo (*Italian*) 'increasing', i.e. getting gradually louder.

croche (*French*) a QUAVER.

crook (1) a detachable section of tubing that was inserted into a BRASS or WOODWIND INSTRUMENT between the MOUTHPIECE and the body of the instrument to give it a different KEY (by increasing the length of the air-column). Performers often had as many as twelve crooks but the introduction of valved instruments in the 1850s virtually dispensed with their necessity. (2) a curved metal tube between the mouthpiece and the body of large wind instruments such as the BASSOON and BASS CLARINET.

crooning a soft, sentimental style of singing, often to DANCE MUSIC. Bing ◊Crosby was a noted 'crooner'.

cross rhythms rhythms that appear to have conflicting patterns and are performed at the same time as one another.

crotchet (*US* 'quarter note') a NOTE with a quarter of the time value of a whole note (SEMIBREVE).

Cruelty, Theatre of a form of theatre devised by the French actor and theatre director Antonin Artaud (1896–1948) which uses non-verbal means of communication such as pantomine, light effects and irrational language, to project the pain and loss fostered by the modern world. His aim was to use drama to subvert the idea of art as a set of concepts separate from real life and he has had a lasting effect on Western drama.

crumhorn *see* KRUMHORN.

Cubism an art movement started by ◊Picasso and ◊Braque, and influenced by AFRICAN tribal masks and carvings, and by the work of ◊Cézanne. They moved away from REALIST and IMPRESSIONIST trends towards a more intellectual representation of objects. Hitherto, painters had observed subjects from a fixed viewpoint, but the Cubists also wanted to represent a more cerebral understanding of their subject. The result was an explosion of multi-viewpoint images, often broken up into geometric shapes and realigned to suggest faces full on and in profile together, to explain the three-dimensional variety of an object or to imply movement. Such fragmented images could be highly complicated. Cubism had an enormous and continuing influence on 20th-century art.

cuckoo a short PIPE with a single finger hole; it gives two notes that imitate the sound of the bird.

cue a catchword or note on a score, used to indicate the entrance of a voice or instrument.

curtain music *see* ACT TUNE.

curtall a small BASSOON of the 16th and 17th centuries.

cycle a series or sequence of pieces of music, by a single composer, which have a common THEME or idea.

cymbalo *see* DULCIMER.

cymbals PERCUSSION INSTRUMENTS comprising two metal plates which are held in the hands and clashed together. They are mounted on stands for JAZZ and POPULAR MUSIC DRUM KITS, where they are operated by pedals or struck with sticks.

czárdás a Hungarian dance with two parts (a slow section and a fast section) that alternate.

D (1) the second NOTE of the SCALE of C major. (2) abbreviation for DOMINANT, and for doctor (as in D.Mus., Doctor of Music).

da capo (*Italian*) 'from the head'; it is an instruction to repeat the beginning of the piece (abbreviation: DC).

Dada an art movement that began in Zurich in 1915, its name randomly chosen from a lexicon. Dada represented a reaction to postwar disillusion with established art. Leading figures included Jean ◊Arp and the poet Tristan Tzara (1896–1963), and, when the movement spread to New York, Francis Picabia (1879–1953) and ◊Marcel Duchamp. Its aim was to reject accepted aesthetic and cultural values and to promote an irrational form of non-art, or anti-art. The random juxtapositions of collage and the use of ready-made objects suited their purpose best. A notable example is Duchamp's *Fountain* (1917), which was an unadorned urinal. Dada gave way to NEUE SACHLICHKEIT around 1924 as the artists associated with the movement diversified. It led, however, to the beginnings of SURREALISM and is the source of other movements in ABSTRACT ART, such as ACTION PAINTING.

dal segno (*Italian*) 'from the sign', i.e. go back to the point in the music marked by the relevant symbol and repeat the music which follows it (abbreviation: DS).

damp to stop the vibrations of an instrument by touching it, or part of it, for example the strings of a HARP, the skin of a DRUM.

Dämpfer (*German*) a MUTE.

Danube School a school of German painters who developed the art of landscape painting as a genre in its own right. Its most prominent member was Albrecht Altdorfer (1480–1538) whose work is characterized by a fantastic inventiveness, distortion of figures and brilliant effects of colour and light.

DC *see* DA CAPO.

début (*French*) 'beginning', i.e. a first appearance.

decibel one tenth of a BEL, a unit for measuring SOUND. A decibel represents the smallest change in loudness that can be detected by the average human ear.

deciso (*Italian*) literally 'decided', i.e. with decision or 'play firmly'.

Deconstruction *see* STRUCTURALISM.

decrescendo (*Italian*) 'decreasing', i.e. getting gradually softer.

degree a step of a SCALE; the position of each NOTE on a scale is identified by its degree.

delicatamente (*Italian*) 'delicately'.

delicato (*Italian*) 'delicate'.

descant (1) a soprano part, sometimes improvised, sung above a HYMN TUNE while the tune itself is sung by the rest of the congregation or CHOIR. (2) (spelt **discant** by music scholars) a general term for all forms of POLYPHONY used from the 12th century.

De Stijl *see* STIJL, DE.

deus ex machina a Latin term meaning 'god from the machine', i.e. a god introduced into the action of a play to resolve some intractable situation in the plot. The device was used commonly in both Greek and Roman drama. In Greek drama, the intervention took the form of the god being lowered onto the stage via some kind of stage machinery. ◊Euripides seems to have been particularly fond of the device, but it was rarely used by ◊Aeschylus and ◊Sophocles. The term has come to denote any twist in a plot which resolves or develops the action in an unexpected way.

The device became an extremely common feature of Victorian literature and drama, in the form of legacies from long-lost relatives enabling otherwise impossible marriages to take place and foiling innumerable wicked schemes. The device has been rarely used in modern serious literature, except for ironic effect, as at the end of David ◊Lodge's novel *Nice Work*.

deux temps (*French*) in 2/2 TIME. *Valse à deux temps* is a WALTZ which has only two dance steps to every three BEATS of the BAR.

development the expansion or changing in some way of parts of a THEME of music that have already been heard, e.g. by varying the RHYTHM or elaborating the PHRASE to give it new impetus.

diapason (1) the term given to a family of ORGAN

STOPS which are largely responsible for the TONE of the instrument. (2) (*French*) a TUNING FORK; *diapason normal* means the same as CONCERT PITCH.

diatonic belonging to a SCALE. The diatonic NOTES of a major scale consist of five TONES (T) and two SEMITONES (S), arranged TTSTTTS. *Compare* CHROMATIC.

dièse (*French*) sharp.

Dies Irae (*Latin*) 'Day of Wrath', a part of the REQUIEM MASS, with a PLAINSONG MELODY which has often been used by ROMANTIC composers such as ◊Berlioz, ◊Liszt, and ◊Rachmaninov.

digital (1) one of the KEYS on the keyboard of a PIANO or ORGAN. (2) in sound recording, a method of converting audio or analogue signals into a series of pulses according to their voltage, for the purposes of storage or manipulation.

diminished interval a PERFECT or MAJOR INTERVAL reduced by one SEMITONE by flattening the upper NOTE or sharpening the lower one.

diminished seventh chord a CHORD which covers a MINOR SEVENTH diminished by one SEMITONE, i.e. C-B flat diminished to C-A. (This is in fact equivalent to a major sixth, but the term 'diminished seventh' is often used.) It is frequently employed as a means of TRANSITION into another KEY.

diminished triad a MINOR TRIAD in which the FIFTH is flattened (diminished), for example in the KEY of C major, C-E-G flat.

diminuendo (*Italian*) 'diminishing', i.e. getting gradually quieter.

diminution the shortening of NOTE TIME-values, so that a MELODY is played more quickly, usually at double speed.

diptych a pair of paintings or carvings on two panels hinged together so that they can be opened or closed.

direct a sign placed at the end of a line or page of old music that indicates the PITCH of the following NOTE or notes.

discant *see* DESCANT.

discord a CHORD or combination of NOTES which creates an unpleasant or jarring sound that needs to be resolved.

dissonance the creation of an unpleasant sound or DISCORD.

distemper an impermanent paint made by mixing colours with eggs or glue instead of oil.

divertimento (*Italian*) an 18th-century term for a piece of music that was intended to be a light entertainment, i.e. a diversion. ◊Mozart wrote many divertimenti.

divertissement (*French*) (1) a short ballet incorporated into an opera or play. (2) a short piece that

includes well-known tunes taken from another source. (3) a DIVERTIMENTO.

divisé (*French*) *see* DIVISI.

divisi (*Italian*) 'divided'; a term used to indicate that, where a PART is written in double NOTES, performers should not attempt to play all the notes but should divide themselves into groups to play them. It is particularly used in music for STRINGS.

division (1) a 17th-century type of VARIATION in which the long NOTES of a MELODY were split up into shorter ones. (2) an obsolete term for long vocal RUNS used by composers such as ◊Bach and ◊Handel.

Divisionism *see* POSTIMPRESSIONISM.

'Dixieland' a simple form of traditional JAZZ, originated in New Orleans at the start of the 20th century.

do (*Italian*) *see* DOH.

dodecaphonic relating to dodecaphony, the TWELVE-NOTE SYSTEM of composition.

doh the spoken name for the first NOTE of a MAJOR SCALE in TONIC SOL-FA.

dolce (*Italian*) 'sweet' or 'gentle'.

dolcissimo (*Italian*) 'very sweet'.

dolente (*Italian*) 'sorrowful'.

doloroso (*Italian*) 'sorrowfully'.

domestic tragedy a form of tragedy that appeared in the Elizabethan period, which focuses on the crises of middle-class domestic life in an unpatronizing and sympathetic manner. An early example is the anonymous *Arden of Faversham* (1592). *The London Merchant* (1731) by George Lillo (1693–1739), which was much admired and very influential throughout Europe, established the genre as a highly popular (and profitable) one. The form is closely related to sentimental comedy, and was also influential on the development of melodrama.

dominant (1) the fifth NOTE above the TONIC of a MAJOR or MINOR SCALE. (2) the name given to the RECITING NOTE of GREGORIAN CHANTS.

doppio (*Italian*) 'double', as in *doppio movimento*, meaning 'twice as fast'.

Dorian mode a term applied to the ascending SCALE which is played on the white keys of a PIANO beginning at D.

dot a MARK used in musical NOTATION. When it is placed after a NOTE, it makes the note half as long again; when it is placed above a note it indicates STACCATO.

dotted note *see* DOT.

double (1) a word used to describe certain instruments that are built an OCTAVE lower than normal, for example a double BASSOON (also called a *contrabassoon*) is built an octave lower than a standard bassoon. (2) a term used to describe a type of VARIATION found in 17th-century French instrumental music in which MELODY NOTES are embellished with ornamentation.

double bass the largest and lowest-pitched of the bowed string instruments. It used to have three strings but now it has four (sometimes five).

double counterpoint COUNTERPOINT in which the two PARTS can change places, i.e. the higher can become the lower and vice versa.

double fugue (1) a FUGUE with two SUBJECTS. In one type of double fugue both subjects are introduced at the start; in another type the second subject appears after the first and the two are eventually combined.

down-beat the downward movement of a conductor's BATON or hand which usually indicates the first BEAT of a bar.

Doxologia Magna (*Latin*) the 'Gloria in Excelsis Deo'. *See* GLORIA.

D'Oyly Carte Company a company founded by the English impresario Richard D'Oyly-Carte (1844-1901) to perform the operas of ◊Gilbert and Sullivan (whom he had brought together) and for which he built the Savoy Theatre in London.

drama (1) a PLAY for the stage, radio or television. (2) dramatic literature as a genre.

dramatic irony a situation in which a character in a PLAY, NOVEL, etc, says or does something that has a meaning for the audience or reader, other than the obvious meaning, that he or she does not understand. Its use is common in both COMEDY and TRAGEDY.

drone a PIPE that sounds a continuous NOTE of fixed PITCH as a permanent BASS. The BAGPIPES, for example, have several drone pipes. Also, a similar effect produced by stringed instruments fitted with 'drone strings'.

drum a PERCUSSION INSTRUMENT of which there are numerous types, including BASS DRUM, SIDE DRUM, TABOR, TENOR DRUM and TIMPANI. Most drums consist of a hollow metal or wood cylinder over which is stretched a skin. Sound is produced by beating the skin with drumsticks or with the hands.

drum kit a set of DRUMS and CYMBALS arranged in such a way that they can all be played by one person sitting on a stool. Some of the instruments (such as the BASS DRUM) are played with a foot PEDAL, but most are struck with sticks or wire brushes. They are used by JAZZ and POP drummers and can vary enormously in size.

DS abbreviation for DAL SEGNO.

due corde (*Italian*) literally 'two strings'; a term used in VIOLIN music indicating that a PASSAGE that could theoretically be played on one string should nevertheless be played on two to produce the desired effect.

duet a combination of two performers or a composition for such a pair, for example, piano duet.

dulcimer *or* **cymbalo** an ancient instrument which was introduced to Europe from the East in the Middle Ages. It consists of a shallow box over which strings are stretched. The instrument is placed on the knees and the strings are struck with small HAMMERS. In the USA, an instrument similar to the ZITHER is sometimes called a dulcimer.

dulcitone a keyboard instrument containing TUNING FORKS which are struck with HAMMERS, as in a PIANO.

duplet a group of two NOTES of equal value which are played in the time normally taken by three.

duple time a form of musical TIME in which the number of BEATS in a BAR is a multiple of two, for example 2/4 (2 CROTCHETS) and 6/8 (6 QUAVERS in two groups of three).

dur (*German*) 'major', as in MAJOR KEY.

dynamic accents ACCENTS which correspond to the regular RHYTHM of a piece of music, as indicated by the TIME SIGNATURE.

E the third note (MEDIANT) of the SCALE of C major.

écossaise abbreviation of *danse écossaise*, i.e. 'Scottish dance', although in fact the term has little to do with Scottish dancing and merely refers to a quick dance in 2/4 TIME.

Eight, The a group of American painters, for the most part REALIST painters who campaigned vigorously on the development of progressive art away from the strictures of academic tradition. *See also* ASHCAN SCHOOL.

eighth-note (US) a QUAVER.

electronic instruments a generic term for instruments that convert electrical energy into sound, such as the SYNTHESIZER.

embouchure (1) the mouthpiece of a BRASS or WIND INSTRUMENT. (2) the correct tensioning of the lips and facial muscles when playing woodwind and brass instruments to create good TONE.

encore (*French*) 'again', the call from an English audience (the French equivalent is BIS) for more music. If the performance does continue, the additional music is also known as an 'encore'.

end pin *see* TAIL PIN.

engraving a technique of cutting an image into a metal or wood plate using special tools. When ink is applied to the plate, the raised parts will print black and the engraved parts white. The term is also used for a print produced in this way.

enharmonic intervals INTERVALS that are so small

that they do not exist on keyboard instruments; an example is the interval from A sharp to B flat.

ensemble (*French*) literally 'together'; a term meaning a group of players or singers, a MOVEMENT in OPERA for several singers, or the precision with which such a group performs together.

entr'acte (*French*) the music played between the acts of a play or OPERA. *See also* ACT TUNE, INTERLUDE, INTERMEZZO.

epic (1) a very long narrative poem dealing with heroic deeds and adventures on a grand scale, as ◊Homer's *Iliad*. (2) a novel or film with some of these qualities.

episode (1) in A FUGUE, a PASSAGE that connects entries of the SUBJECT. (2) in A RONDO, a contrasting section that separates entries of the PRINCIPAL THEME.

epistolary novel a novel in the form of a series of letters written to and from the main characters, sometimes presented by the author in the anonymous role of 'editor'. The form flourished in the 18th century, an example being Samuel ◊Richardson's *Pamela* (1741). The best parody is Jane ◊Austen's *Love and Freindship* (written when she was 14), closely followed by Fielding's (much larger) *Shamela Andrews* (1741). Jane Austen toyed for a while with the form: *Lady Susan* (written 1793–4), a minor work by her own high standards, ends with a lovely send-up of the form's conventions in the 'Conclusion', and the first draft of *Sense and Sensibility* (1797–8), entitled *Elinor and Marianne*, was originally in epistolary form, but the form was too weak for what she planned to write.

equal temperament a convenient, but technically incorrect, way of tuning a keyboard in which all SEMITONES are considered equal, for example F sharp and G flat are taken to be identical NOTES when theoretically they are not. Such a system makes complex MODULATIONS practicable.

escapement the mechanism in a PIANO which releases the HAMMER, allowing a string to vibrate freely after it has been struck.

espressivo (*Italian*) 'expressively'.

esquisse (*French*) a 'sketch', a title sometimes given to short instrumental pieces.

estinto (*Latin*) literally 'extinct', i.e. as soft as possible.

etching a technique of making an engraving in a metal plate, using acid to bite out the image rather than tools. Tones of black or grey can be produced, depending on the extent the acid is allowed to bite. The term is also used for a print produced in this way.

étude (*French*) a 'study' or piece of music evolved from a single PHRASE or idea. Studies are also written purely as exercises to improve technique or FINGERING.

euphonium a large BRASS INSTRUMENT, a tenor TUBA, which is mainly used in BRASS and MILITARY BANDS.

eurhythmics a system of teaching musical RHYTHM by graceful physical movements. It was invented in 1905 by the Swiss composer and teacher Émile Jaques-Dalcroze (1865–1950).

evensong *see* NUNC DIMITTIS.

Existentialism a philosophical position based on a perception of life in which man is an actor forced to make choices in an essentially meaningless universe that functions as a colossal and cruel Theatre of the ABSURD. The main writers associated with Existentialism are French, notably ◊Sartre and ◊Camus, and the bilingual ◊Beckett. The roots of Existentialism are complex; one important source is NIHILISM, the influence of which is apparent in Camus' novel *The Outsider* (1942), described by its author as 'the study of an absurd man in an absurd world'. Another important source is the Danish theologian Søren Kierkegaard (1813–55), who observed that life must be lived forwards but can only be understood backwards. Life is therefore a continual act of faith in which choices are made for a series of actions, the consequences of which must remain unclear. Camus' quest for meaning in life and art led him to an accommodation with Christianity, detailed in his essay *The Rebel* (1951), whereas Sartre came to an uneasy alliance with Marxism; Beckett, however, kept on ploughing his unique and lonely furrow, free of political or religious comfort, until his death.

exposition (1) in the SONATA FORM, the first section of a piece in which the main THEMES are introduced before they are developed. (2) in FUGUE, the initial statement of the SUBJECT by each of the PARTS.

Expressionism a term, derived from the character of some 20th-century Northern European art, which was coined in a description of an exhibition of FAUVIST and CUBIST paintings at the Berlin SEZESSION in 1911 but quickly came to be applied to the works of Die BRÜCKE and Der BLAUE REITER. Expressionist works represented a move away from the observational detachment of REALISM and, to an extent, IMPRESSIONIST trends, and were concerned with conveying the artist's feelings and emotions as aroused by his subject. Any painting technique that helped to express these feelings was considered a valid medium and included bold, free brushwork, distorted or stylized forms, and vibrant, often violently clashing, colours. The term expressionist also refers to an expressive quality of distortion or heightened colour in art from any period or place. *See also* ABSTRACT EXPRESSIONISM. In music it is used to imply the expression of inner emotions.

extemporization *see* IMPROVISATION.

F (1) the fourth note (or SUBDOMINANT) of the scale of C major. (2) in abbreviations, *f* means *forte* (loud); *ff, fortissimo* (very loud) and *fp, forte piano* (loud and then soft).

fa In the TONIC SOL-FA, the fourth degree in any major scale.

fado (*Portuguese*) a type of melancholy song with a guitar accompaniment.

false relation in harmony, the occurrence of a note bearing an ACCIDENTAL, which is immediately followed, in another part, by the same note which does not bear an accidental, or vice versa.

falsetto (*Italian*) an adult male voice, used in the register above its normal range. It has often been used to comic effect in operas.

fancy *see* FANTASIA.

fandango a lively Spanish dance, thought to be South American in origin, in triple time. It is usually accompanied by guitar and castanets. Composers who have included or adapted the fandango form include ◊Gluck in his *Don Juan* ballet, ◊Mozart in *The Marriage of Figaro*, and ◊Rimsky-Korsakov in his *Capriccio Espagnol*.

fanfare a flourish of trumpets, or other instruments (for example, the organ) that imitate the sound of trumpets.

fantasia (*Italian*) a piece in which the composer follows his imagination in free association rather than composing within a particular conventional form; when such a piece is played, it can sound as if it is being improvised. It is also a composer's adaptation or use of another's theme or of a known song, for example. ◊Vaughan Williams's *Fantasia on a theme by Thomas Tallis*.

Fantasie (*German*), **fantaisie** (*French*) a FANTASIA.

Fauvists a group of French painters, including ◊Matisse and ◊Derain, who painted in a particularly vivid and colourful style. The term *fauve* ('wild animal') was coined as a form of derogatory criticism of an exhibition held at the SALON d'Automne of 1905. Their use of strong, bright colours to express their response to the fierce light of the Mediterranean coast owes something to the influence of ◊Gauguin and Van ◊Gogh, but they were less interested in representing what they saw and more concerned to express their own feelings in the boldness and freedom of their compositions. Although Matisse continued to explore Fauvist

techniques, the other artists soon diverged, and the movement as such was fairly short-lived. It was, however, influential in CUBIST and EXPRESSIONIST art.

fauxbourdon (*French*) a 15th-century continental technique of improvising a bass part for a PLAINSONG melody.

feminine cadence an ending in which the final chord occurs on a weak beat of the bar and not the more usual strong beat.

fermata (*Italian*) *see* PAUSE.

ff *see* F (2).

fiddle (1) a generic term for a range of primitive stringed instruments played with a bow, as used in parts of Asia, Africa and Eastern Europe. (2) a colloquial term for a violin, especially in folk music.

fife a small flute still used in 'drum and fife' bands.

fifth an INTERVAL of five notes (the first and last notes are counted) or seven semitones, for example, from C to G.

figurative art *or* **representational art** art that recognizably represents figures, objects or animals from real life, as opposed to ABSTRACT ART.

figure a short musical phrase that is repeated in the course of a composition.

figured bass the bass part of a composition which has numerical figures written below the notes to indicate how the harmony above should be played. It is, in effect, a type of musical shorthand in which the bass line and melody are written down while the numbers indicate which chords should be played. The system was used during the 17th and early 18th centuries.

finale (*Italian*) (1) the last movement of a work. (2) the concluding section of an opera act.

fine arts *see* APPLIED ARTS.

fingering a type of notation that indicates which fingers should be used to play a piece of music.

fino (*Italian*) 'as far as', so *fino al segno* means 'as far as the sign'.

fioritura (*Italian*) 'flowering', i.e. an embellishment.

fipple flute *see* FLAGEOLET; RECORDER.

Five, The the name given to a group of nationalistic 19th-century Russian composers who were known in Russia as *moguchaya kuchka* (The Mighty Handful). The five were ◊Rimsky-Korsakov, Mily Alexeyevich Balakirev (1837–1910), Aleksandr Borodin (1833–87), César Antonovich Cui (1835–1918) and ◊Mussorgsky.

flageolet a small, end-blown FLUTE with six holes, four in front and two at the back, popular in the 17th century.

flamenco a generic term for a type of Spanish song

from Andalusia, usually sad and often accompanied by guitar and dancing. Flamenco guitar playing relies heavily on the strumming of powerful, dynamic rhythms.

flat (1) a note which is lowered by one semitone as indicated by the flat sign (*see* Appendix). (2) a note (or notes) produced at too low a pitch and hence 'out of tune'.

flirts and straights a distinction between authors who specialize in showing off to their readers (flirts) and those whose main concern is getting on with telling the story (straights). A flirt will play with conventions and expectations, and disrupt the narrative with disquistions on the meaning of life or on what the author had for breakfast. A straight will, for the most part, maintain the agreed conventions between author and reader. A writer of detective stories, in which genre the conventions are fairly rigorous, is almost always a straight, though inferior crime writers will make their heroes secret readers of ◊Proust or ◊Nietzsche. Many writers combine both elements: George ◊Eliot is a straight in character development, but a flirt in digression.

flue pipes all ORGAN pipes that have narrow openings, or flues, into which air passes; the other pipes are REED PIPES.

Flügelhorn (*German*) a soprano brass instrument similar to a bugle in shape, but with three pistons.

flute (1) the tranverse or German flute is a member of the WOODWIND family of instruments, although these days it is normally made of silver or other metal. One end of the instrument is stopped and sound is produced by blowing across the mouthpiece formed around an aperture cut into the side of the instrument at the stopped end. The pitch is controlled by means of a lever system. (2) the English flute is a beaked, end-blown, wind instrument with finger holes, now more usually called the RECORDER.

folia (*Portuguese*) 'the folly', a wild and noisy Portuguese dance.

folk dance any dance, performed by ordinary people, in a pre-industrial society, that has evolved over the years and gained a traditional form. Folk dances differ widely in character and some have symbolic significance, such as war dances, etc.

folk song properly, any song that has been preserved by oral tradition. Many composers and pop musicians have written new compositions that imitate old folk songs.

form the structure of a composition. The basic elements of musical composition which define a given piece's form are repetition, variation and contrast. Examples of recognized forms include FUGUE, RONDO, SONATA FORM, etc.

forte (*Italian*) 'loud' (abbreviation *f*).

fortepiano (*Italian*) an early word for PIANOFORTE. Not to be confused with *forte piano* (loud then soft).

fortissimo (*Italian*) 'very loud' (abbreviation *ff*).

forza (*Italian*) 'force', so *con forza* means 'with force'.

forzato (*Italian*) 'forced'.

found object *or* **objet trouvé** a form of art that began with Dada and continued with Surrealism, where an object, either natural or manufactured, is displayed as a piece of art in its own right.

fourth an INTERVAL of four notes (including the first and last) or five semitones, for example C to F.

foxtrot a dance, originating in the USA, in duple time. It first became popular in the ballroom from about 1912 and was at the height of its popularity in the 1930s and 1940s, by which time it had acquired two variations, the quick and the slow foxtrot.

fp abbreviation for FORTE PIANO (*Italian*) 'loud then soft'.

free reed a type of REED found in such instruments as the ACCORDION and HARMONICA. It consists of a small metal tongue that vibrates freely in a metal slot when air is blown over it. The pitch of the reed is determined by its thickness and length.

found object *or* **objet trouvé** a form of art that began with DADA and continued with SURREALISM, where an object, either natural or manufactured, is displayed as a piece of art in its own right.

French horn *see* HORN.

French sixth *see* AUGMENTED SIXTH.

fresco a painting directly painted on to a wall that has previously been covered with a damp, freshly laid layer of lime plaster, the paint and plaster reacting chemically to become stable and permanent. Fresco painting worked particularly well in the warm, dry climate of Italy, where it reached its peak in the 16th century.

fret (*French*) one of a series of thin pieces of metal fitted into the wooden fingerboard of a stringed instrument to make the stopping of strings easier and more accurate. Each fret represents the position of a specific note.

front man the person who stands at the front of the stage during a performance of JAZZ and who is therefore the focus of the audience's attention. He or she is often, but not always, the leader or singer of the band.

fugue a contrapuntal composition for two or more parts (commonly called 'voices') which enter successively in imitation of each other. The first entry is called the 'SUBJECT' and the second entry (a fifth higher or lower than the subject) is called the 'ANSWER'. When all the voices have entered, the EXPOSITION is complete and is usually followed by an EPISODE which connects to the next series of subject entries. A COUNTER-SUBJECT is a melodic accompaniment to the subject and answer, and is often in DOUBLE COUNTERPOINT. A fugue may be written for voices, instruments or both. The form dates back to the 17th century. Possibly the greatest exponent of the fugue was Johann Sebastian ◊Bach, and his *Art of Fugue* is the fullest statement of the form.

full close *see* CADENCE.

full organ a term used in ORGAN music to indicate that all the loud stops are to be used together.

funk a form of heavily syncopated, rhythmic black dance music, originating in the United States. The adjective, often used in JAZZ terminology, is *funky*.

fuoco (*Italian*) 'fire', so *con fuoco* means 'with fire'.

furioso (*Italian*) 'furious'.

Futurism a movement of writers and artists, originating in early 20th-century Italy, that extolled the virtues of the new, dynamic machine age, which was reckoned to have rendered the aesthetic standards of the past redundant. The founding document of the movement is the poet Filipo Marinetti's (1876–1944) *Futurist Manifesto* of 1909, which, in literature, called for the destruction of traditional sentence construction and the establishment of a 'free verse in free words' (*parole in libertà*) owing nothing to the literary standards of the past; the new relationship between words was to be in terms of analogy alone. The English variant of Futurism is VORTICISM, which included writers such as Wyndham ◊Lewis and Ezra ◊Pound. The Russian variant, *Russian Futurism*, was led by the poet Vladimir Mayakovsky (1893–1930), whose curious love affair with the Soviet dictatorship ended with his suicide (Lenin described him as 'incomprehensible'; once he was safely dead, Stalin lauded him as a great Bolshevik poet).

Futurism became a spent force by the 1930s, largely because of its close association with Fascism and the establishment of the Futurist hero Mussolini's dictatorship in Italy (Marinetti's vision of war as 'the hygiene of the world' is also Mussolini's).

In art, Boccioni was among the painters in the group, whose aim was to convey a sense of movement and dynamism, as in Boccioni's *The City Rises* (1910). As a group, the Futurists published manifestos on various aspects of the arts, and exhibitions toured Europe during 1911–12. The original group of painters had broken up by the end of World War I, but their work and ideas had a resounding influence on subsequent art movements.

fz abbreviation for FORZATO.

G the fifth note (or DOMINANT) of the scale of C major.

galant (*French*) 'polite'; a term applied to certain graceful styles of court music, especially of the 18th century.

galanterie an 18th-century German term for a keyboard piece in the GALANT style.

galliard (*French*) a lively court dance, usually in triple time, which dates back to the 16th century.

galop a lively dance in duple time that originated in Germany and was popular in the 19th century.

gamelan a type of traditional orchestra found principally in Indonesia and South-East Asia. Although such an orchestra includes strings and woodwind instruments, it is the array of gongs, drums, chimes, xylophones and marimbas that produces the unique and highly complex rhythms of gamelan music. ◊Debussy was especially influenced by gamelan.

gamut (1) the note G on the bottom line of the bass clef. (2) an alternative (now obsolete) word for the key of G. (3) the whole range of musical sounds, from the lowest to the highest.

gavotte an old French dance, originally of the upper Alps, in 4/4 time, which usually starts on the third beat of the bar. It was favoured by ◊Lully in his ballets and operas and was revived in the 20th century by such composers as ◊Prokofiev and ◊Schoenberg.

gedämpft (*German*) 'muted'.

Generalpause (*German*) a rest of one or more bars for all the members of an orchestra.

genre a distinctive type or category of artistic or literary composition.

genre painting a painting that has as its subject a scene from everyday life, as opposed to a historical event, mythological scene, etc. Genre paintings appear in the backgrounds of medieval paintings, but it was the Dutch painters, e.g. ◊Bruegel, ◊Bosch and ◊Vermeer, who were the first to specialize in them and to continue the tradition. In France, Jean-Baptiste Chardin (1699–1779) used genre scenes to great effect, but they did not become really popular with painters until the REALISTS, e.g. ◊Courbet and ◊Millet. British painters, e.g. ◊Hogarth, ◊Gainsborough and ◊Wilkie, painted genre scenes, but gradually the distinction between such scenes and other genres of painting has blurred.

German sixth *see* AUGMENTED SIXTH.

gigue (*French*) a lively dance or jig.

giocoso (*Italian*) 'merry'.

gioioso (*Italian*) 'joyful'.

giusto (*Italian*) 'exact', as in *tempo giusto*, which can mean either 'strict time' or 'appropriate speed'.

Glasgow Boys *or* **Glasgow School** a group of painters centred in Glasgow in the 1880s and 90s. They represented a move away from academic strictures and were inspired by the plein air BARBIZON SCHOOL. They established an outpost of the European vogue for NATURALISM and ROMANTIC lyricism in landscape painting, and their influence extended into the 20th century.

glass harmonica at its simplest, a set of goblets that are played by rubbing a moistened finger around the rims. This idea was taken further by the American scientist and statesman Benjamin ◊Franklin, who invented a glass harmonica in which a gradated series of glass bowls is fixed to a rotating spindle and played with the fingers. Both ◊Beethoven and ◊Mozart wrote music for the instrument, which produces a high-pitched humming sound.

glee a simple, unaccompanied composition for male voices in several sections.

glissando (*Italian*) 'sliding'; a rapid sliding movement up or down a scale.

Glockenspiel (*German*), **Campanelli** (*Italian*) literally 'a play of bells', an instrument, produced in a variety of sizes, comprising steel bars of different lengths that are arranged like a keyboard; each bar sounds a different note. Played with hammers, it produces sounds that have a bell-like quality. It is used in orchestras and military bands (in which it is held vertically).

Gloria (*Latin*) the first word of *Gloria in excelsis Deo* ('Glory to God in the highest'), the hymn used in both Roman Catholic Masses and in Anglican services. Many composers have set it to music. It is also the first word of the doxology *Gloria Patri* ('Glory be to the Father'), sung after a psalm.

gong a PERCUSSION INSTRUMENT that originated in the Far East. Gongs are made in many sizes and shapes, but an orchestral gong consists of nothing more than a large sheet of metal with a pronounced rim. Sound is produced by striking it with a hammer.

gospel song a type of popular religious song originated by black American slaves who sang hymns to pulsating BLUES rhythms. Such songs, which are still sung fervently today in religious services, were one of the originating forces of JAZZ.

Gothic a style of architecture that lasted from the 12th to the 16th centuries in Northern Europe and Spain. Its effect on art was to produce the INTERNATIONAL GOTHIC style.

Gothic novel a type of NOVEL that was enormously popular in the late 18th century, combining elements of the supernatural, macabre or fantastic, often in wildly ROMANTIC settings, e.g. ruined abbeys or ancient castles. The heroes and/or heroines, whether medieval or modern, for the most part speak in a formal, stilted language curiously at odds (for the modern reader) with the appalling situations they find themselves in. Horace ◊Walpole's *The Castle of Otranto* (1764) and *The Mysteries of Udolpho* (1794) by Ann Radcliffe (1764–1823), are among the best-known examples of the genre. Jane ◊Austen's *Northanger Abbey* (1818) remains the definitive parody.

gouache *also called* **poster paint** *or* **body paint** an opaque mixture of watercolour paint and white pigment.

GP an abbreviation of GENERALPAUSE.

grace note an ornamental, extra note, usually written in small type, used to embellish a melody.

grandioso (*Italian*) 'in an imposing manner'.

grand opera a term originally used to distinguish serious opera, sung throughout, from opera that contained some spoken dialogue. The term is now also used to describe a lavish production.

grand orchestre (*French*) a full orchestra.

grand orgue (*French*) a great ORGAN (as opposed to a swell organ, etc).

grand piano *see* PIANOFORTE.

grave (*Italian*) 'slow' or 'solemn'.

grazia (*Italian*) 'grace'.

grazioso (*Italian*) 'gracefully'.

great stave a STAVE created by pushing the stave with the treble clef and the stave with the bass clef closer together so that both clefs can be located on one exaggerated stave.

Gregorian chant a term that refers to the large collection of ancient solo and chorus PLAINSONG melodies preserved by the Roman Catholic Church. They are named after Pope Gregory I (*c.* 540-604) but date from about 800. Until recently they were sung at specific ceremonies, such as baptism, Mass, etc.

grisaille a monochrome painting made using only shades of grey, often used as a sketch for oil paintings.

grotesque a term for a style of ornamentation that began in Roman times and reached its height with

ROCOCO. It consisted of a series of figurative or floral ornaments in decorative frames that are linked by festoons.

ground bass a BASS line that is constantly repeated throughout a composition, as a foundation for variation in the upper parts.

grupetto (*Italian*) a 'little group'; a general term used to describe various ORNAMENTS of one or more decorative notes.

Guarneri *or* **Guarnerius** a violin made by one of several members of a famous Italian family of violin-makers who were based in Cremona in the 17th and 18th centuries.

guitar a plucked STRING instrument, which may have been introduced to Spain from North Africa. Unlike the LUTE, it has a flat back and usually carries six strings suspended over a fretted fingerboard (twelve-string guitars were favoured by certain BLUES musicians). The acoustic ('soundbox ') guitar has been played by classical, flamenco and folk musicians for generations, but the electric guitar is a comparatively new development. The body may be hollow (or 'semi-acoustic') or solid, with 'pick-ups' (electrically motivated resonators that respond to the vibration of the strings) mounted under the BRIDGE. The vibrations received by the pick-ups have to be electrically amplified or else they are virtually featureless. Electric guitars have a huge COMPASS.

gusto (*Italian*) 'taste', so *con gusto* means 'with taste'.

H (*German*) B natural.

habanera (*Spanish*) a dance of Cuban origin with a powerful, SYNCOPATED RHYTHM, most usually associated, however, with Spain.

Hague School a school of Dutch landscape painters, one of whose prominent leaders was Anton Mauve (1838–88).

haiku a 17-syllable sequential verse form devised by the Japanese poet Basho (1644–94). A Basho haiku gave Ian ◊Fleming the title for one of his novels, *You Only Live Twice*.

half note (US) a MINIM.

half step (US) a SEMITONE.

Hallé Orchestra an internationally famous orchestra founded in Manchester in 1848 by the German-born conductor and pianist Sir Charles Hallé (1819–95).

halling a lively Norwegian dance, in 2/4 TIME, during which men leap high into the air.

hammer that part of the PIANO mechanism which strikes the strings; a mallet for playing the DULCIMER; the clapper of a BELL.

Hammond organ the brand name of an electric OR-GAN first produced by the Hammond Organ Company, Chicago, in 1935. The sound it produces is electronically manufactured and attempts to reproduce the sound of the PIPE ORGAN. It cannot be said to succeed in this, but its unique temperament has been exploited by JAZZ, POP and music-hall musicians the world over.

handbells bells, of various pitch, that are held in the hands of a group of performers and rung in sequence to create a tune.

hardanger fiddle a Norwegian VIOLIN used in FOLK music. It is somewhat smaller than an ordinary violin and has four SYMPATHETIC STRINGS.

harmonica a small, FREE-REED instrument commonly called the *mouth organ*. Although it is a small and apparently inconsequential instrument, many BLUES and FOLK musicians have illustrated its potential by exploiting its emotive power.

harmonic minor MINOR SCALE containing the minor sixth with the MAJOR SEVENTH, in which ascent and descent are without alteration.

harmonics the sounds that can be produced on stringed instruments by lightly touching a string at one of its harmonic nodes, i.e. at a half-length of a string, quarter-length and so on.

harmonium a small, portable REED-ORGAN developed in the 19th century. Air is pumped to the REEDS (which are controlled by STOPS and KEYS) by PEDALS worked continuously by the feet.

harmony (1) the simultaneous sounding of two or more NOTES, i.e CHORDS. A harmonious SOUND is an agreeable or pleasant sound (CONCORD); but harmonization may also produce sounds which, to some ears at least, are unpleasant (*see* DISCORD). (2) the structure and relationship of chords.

harp an instrument, of ancient origin, consisting of strings stretched across an open frame. It is played by plucking the strings, each of which is tuned (*see* TUNING) to a separate NOTE.

harpsichord a keyboard instrument, developed in the 14th and 15th centuries, in which the strings are plucked (not struck) by quills or tongues (PLECTRA). The tongues are connected to the KEYS by a simple lever mechanism. The harpsichord went out of favour during the late 18th century due to the introduction of the PIANO. However, it has seen a revival in the 20th century, and new compositions exploiting its 'twangy' sound have been written for it.

hautbois (*French*) 'high' or 'loud wood'. *See* OBOE. From the Elizabethan period to the 18th century, the English equivalent was *hautboy*.

Hawaiian guitar a style of GUITAR playing in which a steel bar is moved up and down the strings (as opposed to the more usual STOPPING of strings with the fingers) to produce a distinctive slurred sound. The guitar is usually played horizontally.

head voice the upper register of a VOICE, so called because the sound seems to vibrate in the head of the singer. *Compare* CHEST VOICE.

Heckelphone a double-REED instrument which is effectively a baritone OBOE. It was used by Richard ◊Strauss in his opera *Salomé*.

Heldentenor (*German*) a 'heroic tenor ', i.e. a tenor with a strong voice suitable for ◊Wagner's heroic roles.

hemiola a RHYTHM in which two BARS in triple TIME are played as though they were three bars in DUPLE TIME.

hemisemidemiquaver the sixty-fourth NOTE, i.e. a note with a value of a quarter of a SEMIQUAVER or $^{1}/$ 64th of a SEMIBREVE.

heroic tragedy a form of tragedy that became very popular during the RESTORATION. Such tragedies were usually written in bombastic rhymed couplets and featured the adventures in love and war of noble characters in exotic locations, past and present. The characteristic conflict at the heart of such plays was the clash between the equally imperious commands of love and duty.

heterophony (*Greek*) literally 'difference of sounds', i.e. two or more performers playing different versions of the same MELODY simultaneously.

hexachord a SCALE of six NOTES which was used in medieval times.

High Mass MASS that is sung throughout, as distinguished from Low Mass, which is said.

history painting a genre of painting that takes as its subject a scene from history, religious or mythological legend, or from great works of literature, e.g. by ◊Dante or ◊Shakespeare.

hocket (*French*) the breaking-up of a MELODY into very short PHRASES or single NOTES, with RESTS in between them.

homophony a term applied to music in which the PARTS move 'in step' and do not have independent RHYTHMS. *Compare* POLYPHONY.

hook the black line attached to the stem of all NOTES of less value than a CROTCHET.

hootenanny a US term for a small festival of FOLK music.

horn a BRASS INSTRUMENT consisting of a conical tube coiled into a spiral and ending in a bell. The lips are pushed into a funnel-shaped MOUTHPIECE. The modern orchestral horn is called the *French horn* (because that is where it was developed) and is fitted with three (sometimes four, sometimes

seven) valves which open and close various lengths of tubing so that the PITCH of the NOTES can be changed. There are two common 'horns', which are in fact WOODWIND instruments, the BASSET HORN (ALTO CLARINET) and the English horn or COR ANGLAIS (alto OBOE).

hornpipe (1) a single-REED WIND INSTRUMENT played in Celtic countries. (2) a 16th-century dance in triple TIME, originally accompanied by the hornpipe and later erroneously associated with sailors.

Hosanna (*Hebrew*)'Save now', part of the SANCTUS in the MASS.

Hudson River School a group of American landscape painters active in the mid-19th century whose work was concerned with the beauty and mysticism of nature, expressed in romantic terms on a grand and noble scale. They were influenced by the writers Fenimore ◊Cooper and Washington ◊Irving as well as by ◊Turner.

humoresque (*French*), **Humoreske** (*German*) a word used by, e.g. ◊Schumann, as the title for a short, lively piece of music.

humours, theory of (in medieval medical theory) any of the four body fluids: blood, phlegm, yellow (choleric) bile, and black (melancholic) bile. This theory is the basis for the *comedy of humours*, notably Ben ◊Jonson's *Every Man in his Humour* (1598), in which the characters have names and behaviour appropriate to their eccentrically dominant 'humour' or personality trait. For humour in the modern sense *see* wit.

hurdy-gurdy a medieval stringed instrument, shaped like a VIOL. A wooden wheel, coated in RESIN, is cranked at one end to make all the strings resonate. The strings are stopped by rods operated by KEYS. The hurdy-gurdy was often used to provide dance music.

hymn in the Christian Church, a poem sung to music in praise of God.

icon *or* **ikon** a religious image, usually painted on a wooden panel, regarded as sacred in the Byzantine Church and subsequently by the Orthodox Churches of Russia and Greece, where they survive. The word comes from the Greek *eikon*, meaning 'likeness', and strict rules were devised as to the subject, generally a saint, and to the form of the painting and its use, so although icon painting flourished in the 6th century it is extremely difficult to date icons painted then or later. A reaction to what was considered idolatry took place in the 8th century, resulting in *iconoclasm*, the destruction of such images.

iconography, iconology the study and interpretation of representations in figurative art and their symbolic meanings. It is particularly important in the understanding of Christian art, especially of the medieval and RENAISSANCE periods, e.g. the dove signifying the Holy Spirit, or the fish symbolizing Christ.

idée fixe (*French*) 'fixed idea', i.e. a recurring theme.

idiophone any instrument in which SOUND is produced by the VIBRATION of the instrument itself, for example, CYMBALS, BELLS, CASTANETS etc.

ikon *see* ICON.

Imagism a poetry movement of the early 20th century which advocated using everyday language and precise representation of the image of the subject discussed. Imagist poems were short and to the point, anti-ROMANTIC and anti-Victorian in tone, and could be on any subject under the sun. Prominent Imagists included ◊Pound and Amy Lowell (1874–1925), whose *Tendencies in Modern American Poetry* (1917) has the best introduction to the movement.

imitation a device in COUNTERPOINT whereby a PHRASE is sung successively by different VOICES.

impasto an Italian word used to describe the thickness and textures that can be achieved with acrylic or oil paint.

imperioso(*Italian*) 'imperiously'.

impetuoso (*Italian*) 'impetuously'.

Impressionism an art movement originating in France in the 1860s, centred on a fairly diverse group of artists who held eight exhibitions together between 1874 and 1886. The main artists were ◊Cézanne, ◊Degas, ◊Manet, ◊Monet, Morisot, ◊Pissarro, ◊Renoir and ◊Sisley, although they did not all show paintings at all eight exhibitions. The name of the movement was coined by critics from a painting by Monet in the 1874 exhibition entitled *Impression: Soleil Levant*. Members of the group were variously influenced by the BARBIZON SCHOOL, the works of ◊Turner and ◊Constable, and the REALISM of ◊Courbet. The advent of photography and scientific theories about colour also had their impact on the painters' approach to their work. The Impressionists were concerned with representing day-to-day existence in an objective and realistic manner, and they rejected the ROMANTIC idea that a painting should convey strong emotions. They wanted to record the fleeting effects of light and movement, and so their usual subjects were landscapes or social scenes like streets and cafés. They chose unusual viewpoints and painted 'close-ups', probably influenced by photography. They were on the whole much freer in their use of unusual colours and a lighter palette; their subject matter was also less

weighty, and they came in for some criticism over the lack of intellectual content of their painting. Impressionism has had an enormous influence on almost every subsequent major art movement: on CUBISM via ◊Cézanne; on the synthetic art of ◊Gauguin through ◊Seurat and the NEOIMPRESSIONISTS, and on EXPRESSIONISM through the works of Van ◊Gogh. This influence has continued in a large proportion of 20th-century art.

By analogy, Impressionism is also used to identify certain types of atmospheric music, such as the music of ◊Debussy and ◊Ravel.

impromptu a type of PIANO music that sounds as if it has been improvised, i.e. written in a free and easy style.

improvisation the art of playing or 'inventing' music that has not already been composed, i.e. spontaneous composition. Some forms of music (especially JAZZ) often rely heavily on the ability of performers to improvise certain sections. It has the same meaning as *extemporization*.

in alt *see* ALT.

incalzando (*Italian*) 'pressing forward', i.e. working up speed and force.

incidental music music written to accompany the action in a play, but the term is also commonly applied to OVERTURES and INTERLUDES.

inciso (*Italian*) 'incisive', hence an instruction that a strong RHYTHM is required.

indeciso (*Italian*) 'undecided', i.e. the pace of a piece of music can be varied according to the performer's feelings.

indeterminacy a term used by John ◊Cage to describe music that does not follow a rigid NOTATION but leaves certain events to chance or allows performers to make their own decisions when performing it.

inflected note a NOTE with an ACCIDENTAL placed before it, i.e. it is sharpened or flattened.

inner parts the PARTS of a piece of music excluding the highest and lowest; for example, in a work for SOPRANO, ALTO, TENOR and BASS, the alto and tenor roles are inner parts.

In nomine (*Latin*) ('In the name of the Lord') a type of CANTUS FIRMUS used by English composers of the 16th century. It was first used by ◊Taverner in his setting of *In nomine Domini* for one of his Masses.

instrument in music, a device on which or with which music can be played. There are five traditional categories of instrument: WOODWIND, BRASS, PERCUSSION, KEYBOARD, and STRING. However, ELECTRONIC and MECHANICAL INSTRUMENTS also exist.

instrumentation *see* ORCHESTRATION.

intaglio the cutting into a stone or other material or the etching or engraving on a metal plate of an image; the opposite of RELIEF. Intaglio printing techniques include ENGRAVING and ETCHING.

interior monologue a form of STREAM OF CONSCIOUSNESS narrative technique employed by the English novelist Dorothy Richardson (1873–1957) which anticipated ◊Joyce's use of the technique in *Ulysses* (the device was not really new; similar effects can be found in, e.g. ◊Dickens).

interlude a title sometimes used for a short PART of a complete composition, e.g. a piece of music performed between the acts of an opera. *See also* ACT TUNE, ENTR'ACTE, INTERMEZZO.

intermezzo (*Italian*) (1) a short piece of piano music. (2) a short comic opera performed between the acts of a serious opera, especially in the 16th and 17th centuries. *See also* ACT TUNE, ENTR'ACTE, INTERMEZZO.

International Gothic a predominant style in European art covering the period between the end of the Byzantine era and the beginning of the RENAISSANCE, i.e. *c.*1375–*c.*1425. Some variations in styles occurred regionally, but the most influential centres were Italy, France and the Netherlands. Ideas spread widely due to an increase in the art trade, to travelling artists, and to a certain amount of rivalry over royal commissions. The Dukes of Berry and Burgundy were among the major patrons of the time. International Gothic style was characterized by decorative detail and refined, flowing lines; figures were often elongated or distorted to increase an appearance of elegant charm and the use of gilts and rich colours figured strongly. Scale and perspective were more symbolic than naturalistic, although naturalism began to take hold in the later works of the period.

interpretation the way in which a performer plays a piece of composed music. No composer can possibly indicate exactly how a piece should be played and, to some degree, it is up to the performer to play it as he or she thinks fit.

interval the gap or 'sound distance', expressed numerically, between any two NOTES, i.e. the difference in PITCH between two notes. For example, the interval between C and G is called a FIFTH because G is the fifth note from C. *Perfect intervals* are intervals that remain the same in MAJOR and MINOR KEYS (i.e. FOURTHS, fifths, OCTAVES.)

intonation a term used to describe the judgement of PITCH by a performer.

intone to sing on one NOTE. A priest may intone during a Roman Catholic or Anglican service.

introduction a section, often slow, found at the start of certain pieces of music, notably SYMPHONIES and SUITES.

Introit an ANTIPHON, usually sung in conjunction with a PSALM verse, in the Roman Catholic and Anglican LITURGIES.

invention a title used by ◊Bach for his two-part keyboard pieces in CONTRAPUNTAL form.

inversion a term which literally means turning upside-down. It can refer to a CHORD, INTERVAL, THEME, MELODY or COUNTERPOINT. For example, an *inverted interval* is an interval in which one NOTE changes by an OCTAVE to the other side, as it were, of the other note.

Ionian mode a MODE which, on the PIANO, uses the white NOTES from C to C.

irlandais (*French*) 'in Irish style'.

isorhythm a term used to describe a short RHYTHM pattern that is repeatedly applied to an existing MELODY which already has an distinct rhythm.

Italian sixth *see* AUGMENTED SIXTH.

italiano (*Italian*), **italienne** (*French*) 'in Italian style'.

Jacobean tragedy a development of revenge tragedy in the Jacobean period. The distinction between Jacobean and REVENGE TRAGEDY is a disputed one, hinging on the supposed wave of cynicism and pessimism that is alleged by some historians to have accompanied the accession of James VI of Scotland to the English throne, and the end of the Elizabethan era. The debate is a complex one, but it is undoubtedly the case that tragedies such as ◊Shakespeare's *Hamlet*, ◊Middleton's *The Changeling*, and ◊Webster's *The White Devil* and *The Duchess of Malfi*, display an obsession with political and sexual corruption. The language of these plays is highly sophisticated, ironic, and coldly brilliant.

jam session a 20th-century slang expression for an occasion when a group of musicians join forces to improvise (*see* IMPROVISATION) music. It is usually only appropriate to JAZZ, BLUES and ROCK music.

Janissary music the music of Turkish MILITARY BANDS which influenced European composers during the 18th century. It is particularly associated with CYMBALS, DRUMS and TAMBOURINES.

jazz a term used to describe a style of music that evolved in the Southern States of the USA at the turn of the century. It owes a great deal to the RHYTHMS and idioms of BLUES and SPIRITUALS, but many of the favoured instruments (for example, SAXOPHONE, TRUMPET and TROMBONE) were European in origin. Jazz traditionally relies upon a strong rhythm 'section', comprising BASS and DRUMS, which provides a springboard for other instruments. Jazz developed from being a form of music played in the back streets of New Orleans to a sophisticated art form performed by small dedicated groups as well as 'BIG BANDS' or 'jazz orchestras'.

Self-expression, and therefore IMPROVISATION, has always been a crucial aspect of jazz and this has allowed many individuals (such as Louis ◊Armstrong and Benny ◊Goodman) to blossom and further its cause. *See also* BEBOP.

Jazz Modern *see* ART DECO.

Jeune France ('Young France') the name adopted by a group of French composers, including ◊Messiaen, who identified their common aims in 1936.

Jew's harp a simple instrument consisting of a small, heart-shaped metal frame to which a thin strip of hardened steel is attached. The open-ended neck of the frame is held against the teeth and the strip is twanged to produce sound, which is modified by using the cavity of the mouth as a soundbox.

jig a generic term for a lively dance. *See also* GIGUE.

jingle a short, catchy piece of music with equally catchy LYRICS, often used to enliven the commentary of radio stations broadcasting popular music.

jingles an instrument consisting of a number of small bells or rattling objects on a strap, which are shaken to produce sound.

Jugendstil the German form of ART NOUVEAU.

juke box an automatic, coin-operated, machine that plays records.

K when followed by a number, a reference to either a catalogue of ◊Mozart's works compiled by an Austrian scientist Ludwig von Köchel (1800–77), or a catalogue of ◊Scarlatti's works compiled by Ralph Kirkpatrick.

Kapellmeister (*German*) literally 'master of the chapel', i.e. director of music to a noble court or bishop.

kazoo a simple instrument consisting of a short tube with a small hole in the side which is covered with a thin membrane. When a player hums down the tube, the membrane makes a buzzing sound. It is usually considered a children's instrument although it is frequently used by FOLK and JAZZ musicians.

kettledrum *see* TIMPANI.

key (1) on a piano, harpsichord, organ etc., one of the finger-operated levers by which the instrument is played. (2) On woodwind instruments, one of the metal, finger-operated levers that opens or closes one or more of the soundholes. (3) a note that is considered to be the most important in a piece of music and to which all the other notes relate. Most pieces of Western music are 'written in a key', i.e. all the chords in the piece are built around a particular note, say F minor. The concept of a key is alien to certain types of music, such as Indian and Chinese.

key note *see* TONIC.

key signature the sign (or signs) placed at the beginning of a composition to define its KEY. A key signature indicates all the notes that are to be sharpened or flattened in the piece; should a piece move temporarily into another key, the relevant notes can be identified with ACCIDENTALS.

kinetic art an art form in which light or balance are used to create a work that moves or appears to move. More complicated kinetic art objects are made to move by electric motors.

King James Bible *see* AUTHORIZED VERSION.

kit a miniature violin which was particularly popular with dancing masters of the 17th and 18th centuries, who could carry one in the pocket and thereby provide music for lessons.

Klangfarbenmelodie (*German*) literally, 'melody of tone colours'; a term used by ◊Schoenberg to describe a form of composition in which the pitch does not change; colour is achieved by adding or taking away instruments.

klavier *see* CLAVIER.

koto a Japanese zither which has 13 silk strings stretched over a long box. The strings pass over moveable bridges and are played with plectra worn on the fingers. The instrument is placed on the ground and produces a distinctive, somewhat harsh, sound.

Krumhorn *or* **Krummhorn** (*German*) *or* **crumhorn** a double-reed instrument, common in the 16th and early 17th centuries. The tube was curved at the lower end and the reed was enclosed in a cap into which the player blew. It was made in several sizes: treble, tenor and bass.

Kyrie Eleison (*Greek*) 'Lord have Mercy', the formal invocation at the start of the Mass and communion service.

la *or* **lah** (1) the note A. (2) In the TONIC SOL-FA, the sixth note (or SUBMEDIANT) of the major scale.

lacrimoso (*Italian*) 'tearful'.

lai (*French*) a 13th- and 14th-century French song usually consisting of 12 irregular stanzas sung to different musical phrases.

lambeg drum a large, double-headed bass drum from Northern Ireland.

lament a Scottish or Irish folk tune played at a death or some disaster, usually on the bagpipes.

lamentoso (*Italian*) 'mournfully'.

lampon (*French*) a drinking song.

Lancers, The a type of QUADRILLE in which 8 or 16 couples take part.

Ländler a country dance in slow 3/4 time from Austria and Bavaria, from which the WALTZ was probably derived.

langsam (*German*) 'slow'.

languido (*Italian*) 'languid'.

largamente (*Italian*) 'broadly', meaning slowly and in a dignified manner.

larghetto (*Italian*) 'slow' or 'broad', but not as slow as LARGO.

largo (*Italian*) literally 'broad', meaning slow and in a dignified manner.

Lascaux the site, in Dordogne, France, of some outstanding Paleolithic cave paintings and rock engravings. Dating from *c.*15000 BC, they have survived in remarkably good condition and depict local fauna, etc, on a large scale, in a bold, direct style.

Last Post, The a bugle call of the British Army to signal the end of the day at 10 p.m. It is also played at military funerals.

lay a song or ballad.

lay clerk an adult male member of an Anglican cathedral choir.

lead (1) the announcement of a subject or theme that later appears in other parts. (2) a sign giving the cue or entry of the various parts.

leader (1) in Britain, the title of the principal first violin of an orchestra or the first violin of a string quartet or similar ensemble. (2) the leader of a section of an orchestra. (3) In the USA, an alternative term for conductor.

leading motif *see* LEITMOTIF.

leading note the seventh note of the scale; it is so called because it 'leads to' the TONIC, a semitone above.

ledger lines *see* LEGER LINES.

legato (*Italian*) 'smooth'.

leger lines *or* **ledger lines** short lines added above or below a STAVE to indicate the pitch of notes that are too high or low to be written on the stave itself.

leggiero (*Italian*) 'light'.

legno (*Italian*) 'wood'; *col legno* is a direction to a violinist to turn the bow over and to tap the strings with the wood.

leise (*German*) 'soft' or 'gentle'.

leitmotif *or* **Leitmotiv** (*German*) literally a 'leading theme', i.e. a recurring theme of music, commonly used in opera, that is associated with a character or idea, thus enabling the composer to tell a story in terms of music. It has been used by many composers, including ◊Mozart and ◊Berlioz, but it is particularly associated with ◊Wagner. In the last act of *Götterdämmerung*, for example, every leitmotif associated with Siegfried is woven into the death march.

lentamente (*Italian*) 'slowly'.

lento (*Italian*) 'slow'.

LH an abbreviation for 'left hand', found in piano music.

liberamente (*Italian*) 'freely', i.e. as the performer wishes.

libretto (*Italian*) literally 'little book'. It is a term used for the text of an opera or oratorio.

licenza (*Italian*) 'licence' or 'freedom'; *con alcuna licenza* means 'with some freedom'.

Lied, Lieder (*German*) 'song, songs'. The term is now used for songs by the German romantic composers, for example ◊Brahms, ◊Schubert, ◊Strauss, etc.

ligature (1) a 12th-century form of notation for a group of notes. (2) a slur indicating that a group of notes must be sung to one syllable. (3) the tie used to link two notes over a bar line. (4) the metal band used to fix the reed to the mouthpiece of a clarinet, etc.

limerick a humorous five-lined piece of light verse, with the first two lines rhyming with each other, the third and fourth lines ryming with each other, and the fifth line rhyming with the first line. Usually there are three stressed beats in the first, second and fifth lines and two stressed beats on the third and forth lines. Traditionally the name of a place is mentioned in the first line and may be repeated in the last line. The English humorist and painter Edward Lear (1812–88) made the form popular in the 19th century. Limerick is a town in Ireland, but the name of the verse is probably derived from a Victorian custom of singing nonsense songs at parties.

lira da braccio, lira da gamba Italian stringed instruments of the 15th and 16th centuries. The *lira da braccio* had seven strings and was played like a violin; the *lira da gamba* was a bass instrument played between the knees, and had up to sixteen strings.

lira organizzata a type of HURDY-GURDY that included a miniature organ.

literary ballad *see* ballad.

literary criticism the formal study, discussion and evaluation of a literary work.

liturgy a term for any official, and written down, form of religious service.

loco (*Italian*) 'place'. It is used in music to indicate that a passage is to be played at normal pitch, after a previous, contrary instruction, i.e. the music reverts to its original place on the stave.

lontano (*Italian*) 'distant'.

loure a type of bagpipe played in northern France, especially Normandy.

lunga pausa (*Italian*) a 'long pause'.

lusingando (*Italian*) 'flattering', i.e. in a cajoling manner.

lustig (*German*) 'merry'.

lute a plucked stringed instrument with a body resembling that of a half-pear. It is thought to have a history dating back some three thousand years and was particularly popular during the 16th and 17th centuries; it has since been revived by 20th-century instrument makers. It has a fretted fingerboard with a characteristic 'pegbox' (a string harness) bent back at an angle to the finger-board. A lute can have up to 18 strings. It was traditionally used as an instrument for accompanying dances, but many solo works have also been written for it.

Lydian mode (1) a scale used in ancient Greek music, the equivalent of the white notes on a piano from C to C. (2) From the Middle Ages onwards, the equivalent of a scale on the white notes on a piano from F to F.

lyre an instrument familiar to the ancient Greeks, Assyrians and Hebrews. It comprised a small, hollow box from which extended two horns that supported a cross bar and anything up to 12 strings, which could be plucked or strummed. It is traditionally taken to represent a token of love (Orpheus played the lyre).

lyric a short poem, or sequence of words, for a song. The term has a particular application to 20th-century musicals and pop songs. A 'lyricist' is the person who writes the words to a popular tune.

m *abbreviation for* MAIN, MANO, MANUAL.

m (**me**) in TONIC SOL-FA. the third note (or MEDIANT) of the major scale.

ma (*Italian*) 'but', as *andante ma non troppo*, 'slow, but not too slow'.

machete a small Portuguese guitar.

madrigal a musical setting of a secular poem for two or more voices in COUNTERPOINT, usually unaccompanied. The first madrigals date back to the 14th century, and the first publications were made in Italy about 1501. The art of madrigal spread to every part of Europe, with the result that a wealth of polyphonic vocal music was created and reached a high art before the development of instrumental music. In the 17th century madrigals were superseded by CANTATAS.

maestà the Italian word for 'majesty', used in art to denote a depiction of the Virgin and Child enthroned in majesty and surrounded by angels or saints.

maestoso (*Italian*) 'majestic' or 'dignified'.

maestro (*Italian*) literally 'master', a term used for a master musician, particularly a conductor.

maestro sostitutto (*Italian*) 'substitute conductor', an assistant conductor responsible in opera for an offstage chorus or band. Also, a RÉPÉTITEUR.

maggiore (*Italian*) 'major mode'.

magic realism a term devised in the 1920s to describe the work of a group of German painters, part of the NEUE SACHLICHKEIT, whose work exhibited a disquieting blend of surreal fantasy with matter-of-fact representationalism. In literature, the term is often applied with particular reference to the work of certain South American novelists, notably the Peruvian novelist Mario Vargas Llosa (1936–) and the Colombian Gabriel Garcia Marquez (1928–), whose work combines deadpan description of the everyday world with (often equally deadpan) excursions into fantasy. Some European novelists have been characterized as 'magic realists', e.g. ◊Calvino and ◊Rushdie. The techniques of magic realism have a long ancestry; ◊Chesterton's novel *The Man Who Was Thursday* is a notable early 20th-century example, and there are clear links with the GOTHIC NOVEL.

Magnificat (*Latin*) short for *Magnificat anima mea Dominum* ('My soul magnifies the Lord') the canticle of the Virgin Mary sung at Roman Catholic Vespers and Anglican Evensong. It is usually chanted, but many composers, such as ◊Bach, have set it to their own music.

main (*French*) 'hand', so *main droite* means 'right hand' (particularly in piano music).

major (*Latin*) 'greater', as opposed to MINOR or 'lesser'. Major scales are those in which a major third (INTERVAL of four semitones) occurs in ascending from the tonic; while the minor scales involve a minor third (three semitones). A major tone has the ratio 8 : 9 while a minor tone has the ratio 9 : 10.

malapropism the incorrect use of a word, often through confusion with a similar-sounding word. It is called after Mrs Malaprop, a character in ◊Sheridan's comedy *The Rivals* (1775) whose name is derived from the French mal a propos, 'not apposite'. Some of her malapropisms include 'She's as headstrong as an allegory on the banks of the Nile' and 'Illiterate him quite from your mind'.

malinconia (*Italian*) 'melancholy'.

mancando (*Italian*) 'decreasing' or 'fading away'.

Manchester School the name applied to a group of British composers who studied music at the Royal Manchester College during the 1950s. They include Harrison ◊Birtwhistle and Peter Maxwell ◊Davies, among others.

mandolin, mandoline a stringed instrument, similar to the lute, but smaller, usually played with a plectrum. It has four pairs of strings and is occasionally used as an orchestral instrument.

mano (*Italian*) 'hand'.

Mannerism an exaggerated and often artificial sense of style found in Italian art between *c*.1520 and 1600, i.e. between the High RENAISSANCE and BAROQUE periods. It represents a reaction against the balanced forms and perspectives of Renaissance art and is characterized by uncomfortably posed, elongated figures and contorted facial expressions. Harsh colours and unusual modes of perspective were also used to striking effect. The major artists of the period were able to create emotional responses of greater power and sophistication, and they paved the way for the development of Baroque art.

manual a keyboard on an organ or harpsichord; organs may have four manuals, named *solo*, *swell*, *great* and *choir*.

maracas a pair of Latin-American percussion instruments made from gourds filled with seeds, pebbles or shells. Sound is made by shaking the gourds.

march a piece of music with a strict rhythm, usually 4/4 time but sometimes in 2/4, 3/4 or 6/8 time, to which soldiers can march. The pace varies with the purpose of the piece, from the extremely slow *funeral* or *dead march* to the *quickstep* (with about 108 steps a minute) and the *Sturm Marsch* or *pas de charge* with 120 steps a minute.

marcia (*Italian*) 'march', so *alla marcia*, 'in a marching style'.

mariachi (*Spanish*) a Mexican folk group of variable size; it normally includes violins and guitars.

marimba a Latin American instrument which may have originated in Africa. It is similar to a large XYLOPHONE and can be played by up to four people at the same time.

mark a sign or word used in NOTATION to indicate the time, tone, accent or quality of a composition, or the pace at which it should be performed.

martelé (*French*) *see* MARTELLATO.

martellato (*Italian*) literally 'hammered'; a term used mainly in music for strings to indicate that notes should be played with short, sharp strokes of the bow. The term is also occasionally used in guitar and piano music.

marziale (*Italian*) 'warlike'.

masque a spectacular court entertainment that was especially popular during the 17th century. It combined poetry and dancing with vocal and instrumental music to tell a simple story that invariably flattered its aristocratic audience.

Mass (*Latin* **Missa**, *Italian* **Messa**, *French* and *German* **Messe**) in musical terms, the setting to music of the Latin Ordinary of Mass (those parts of the Mass that do not vary). The five parts are the KYRIE ELEISON, GLORIA, CREDO, SANCTUS with BENEDICTUS, and AGNUS DEI.

Master of the Queen's Musick an honorary position (in Britain) awarded to a prominent musician of the time; it is his (or her) duty to compose anthems, etc, for royal occasions.

mastersinger *see* MEISTERSINGER.

Matins the name given to the first of the canonical hours of the Roman Catholic Church. The term also refers to morning prayer in the Anglican Church.

mazurka a Polish folk dance of the 17th century for up to 12 people. The music can vary in speed and is often played on bagpipes. ◊Chopin, amongst other composers, was influenced by the music and wrote some 55 'mazurkas' for piano.

me in the TONIC SOL-FA, the third note (or MEDIANT) of the major scale.

measure (1) a unit of rhythm or notes and rests included between two bars. (2) a stately dance of the minuet or pavanne type. (3) (US) a BAR (of music).

mechanical instruments instruments that can play complex music through the programming of their mechanism (e.g. by punched paper or pins on a spindle) when supplied with power (through foot pedals, clockwork, steam power, electricity, etc).

mediant the third note in a major or minor scale above the TONIC (lowest note), e.g. E in the scale of C major.

medium a material used in art, e.g. OIL in painting, PENCIL in drawing, or BRONZE in sculpture. The term is also used to denote a method, e.g. painting as opposed to sculpture.

Meistersinger (*German*) 'mastersinger', the title of highest rank in the song schools or guilds that flourished in German cities from the 14th century until the 19th century. Where the MINNESINGERS drew their members from the aristocracy, mastersingers were usually craftsmen or tradesmen who composed poems and music and who formed themselves into powerful guilds. They were doubtless of great value as a means of extending musical culture, but in their latter days their original purpose tended to be defeated by pedantic restrictions, as satirized in ◊Wagner's great opera, *Die Meistersinger von Nürnberg*.

melodica (*Italian*) a free-reed instrument which was developed from the harmonica. It is box-shaped and has a small keyboard; the player blows down a tube and plays by pressing the keys.

melodic minor scale *see* SCALE.

melodic sequence *see* SEQUENCE.

melodrama a form of drama (from the Greek for 'song' plus 'drama') that seems to have arisen in 18th-century France and that contained elements of music, spectacle, sensational incidents and sentimentalism (*see* SENTIMENTAL COMEDY). The form reached its peak in the popular theatre of 19th-century England, when quite spectacular stage effects often accompanied the action, and villains became blacker than black in their persecution of pure heroes and heroines, e.g. in *The Miller and his Men* (1813) by Isaac Pocock (1782–1835), a Gothic melodrama with splendid bandits, virtuous and faithful young lovers, battles, a wild ravine, and an exploding bridge at the climax. *See also* DOMESTIC TRAGEDY.

melody a succession of notes, of varying pitch, that create a distinct and identifiable musical form. Melody, HARMONY and RHYTHM are the three essential ingredients of music. The criteria of what constitutes a melody change over time.

membranophone the generic term for all instruments in which sound is produced by the vibration of a skin or membrane, for example, DRUM, KAZOO.

meno (*Italian*) 'less', so *meno mosso* means 'slower' (less speed).

menuet (*French*) MINUET.

Messa, Messe *see* MASS.

mesto, mestoso (*Italian*) 'sad'.

metallophone an instrument that is similar to a XYLOPHONE but has metal bars (usually bronze).

metaphysical painting an art movement begun in Italy in 1917 by Carlo Carré (1881–1966) and de ◊Chirico. They sought to portray the world of the subconscious by presenting real objects in incongruous juxtaposition, as in their *Metaphysical Interiors* and *Muses* series (1917). Carré soon abandoned the movement, and by the early 1920s both artists had developed other interests. Although short-lived, the movement did have some influence.

meter (1) in verse, the measured arrangement of syllables according to stress in a rhythmic pattern. (2) in music a rhythmic pattern. *See* RHYTHM.

metaphysical poetry a poetry movement of the 17th century, noted for intense feeling, extended metaphor and striking, elaborate imagery, often with a mystical element. ◊Donne is regarded as the first important metaphysical poet.

method acting a style of acting, devised by Konstantin ◊Stanislavsky, which involves an actor immersing himself in the 'inner life' of the character he is playing, and, using the insights gained in this study, conveying to the audience the hidden reality behind the words. Stanislavsky's theories were adopted and adapted by the American director Lee Strasberg (1901–82), whose method style of acting achieved world fame (or notoriety) through pupils such as James ◊Dean and Marlon ◊Brando.

metronome an instrument that produces regular beats and can therefore be used to indicate the pace at which a piece of music should be played. The first clockwork metronome was invented patented in 1816 and had a metal rod that swung backwards and forwards on a stand. The speed of ticking could be altered by sliding a weight up or down the rod. Electronic metronomes are also manufactured today.

mezzo (*Italian*) literally 'half', so *mezzo-soprano* means a voice between soprano and contralto.

mf abbreviation for *mezzo forte*, (*Italian*) 'moderately loud'.

microtones INTERVALS that are smaller than a SEMITONE in length, for example, the quarter-tone.

middle C the note C which occupies the first ledger line below the treble staff, the first ledger line above the bass staff, and is indicated by the C clef.

military band a band in the armed forces that plays military music, usually for marching. There are many different types of military band, and the number of players varies. Most bands comprise a mixture of brass, woodwind and percussion instruments.

minim a note, formerly the shortest in time-value, with half the value of a SEMIBREVE; the equivalent of a half-note in US terminology.

Minnesingers the poet-musicians of Germany in the 12th and 13th centuries, who were of noble birth, like the TROUBADOURS of France, and who produced *minnelieder*, or love songs. ◊Wagner's Tannhäuser is a minnesinger. They were succeeded by the MEIS-TERSINGERS.

minor (*Latin*) 'less' or 'smaller'. Minor intervals contain one semitone less than MAJOR. The minor third is characteristic of scales in the minor mode.

minstrel a professional entertainer or musician of medieval times. Such people were often employed by a royal court or aristocratic family.

minuet a French rural dance in 3/4 time that was popular during the 17th and 18th centuries. It was incorporated into classical sonatas and symphonies as a regular movement.

miracle play *see* MYSTERY PLAY.

mirliton (*French*) any wind instrument in which a thin membrane is made to vibrate and make a noise when the player blows, hums or sings into it. It is now known as the KAZOO.

mirror music any piece of music that sounds the same when played backwards.

Miserere (*Latin*) short for *Miserere mei Deus* ('Have mercy upon me, O God'), the first line of the 51st Psalm. It has been set to music by several composers, including ◊Verdi.

Missa (*Latin*) 'Mass', so *Missa brevis* means 'short Mass'; *Missa cantata* means 'sung Mass'; *Missa pro defunctis* is Mass for the dead, or Requiem; and *Missa solemnis* is solemn or high Mass.

misterioso (*Italian*) 'mysteriously'.

misura (*Italian*) 'measure'; equivalent to a BAR.

Mixolydian mode (1) the set of notes, in ancient Greek music, which are the equivalent of the white notes on a piano from B to B. (2) in church music of the Middle Ages onwards, the equivalent of the white notes on a piano from G to G.

mixture an organ stop that brings into play a number of pipes that produce HARMONICS above the pitch corresponding to the actual key which is played.

mobile a type of moving sculpture devised by the American sculptor Alexander Calder (1898–1976) who originally trained as an engineer.

moderato (*Italian*) 'moderate' (in terms of speed).

modes the various sets of notes or SCALES, which were used by musicians until the concept of the KEY was accepted (*c*.1650). Modes were originally used by the ancient Greeks and were adapted by medieval composers, especially for church music. Modes were based on what are now the white notes of the piano.

modulation the gradual changing of key during the course of a part of a composition by means of a series of harmonic progressions. Modulation is *diatonic* when it is accomplished by the use of chords from relative keys; *chromatic* when by means of non-relative keys; *enharmonic* when effected by the alteration of notation; *final*, or *complete*, when a new tonality is established; and *partial*, or *passing*, when the change of key is only transient.

moll (*German*) 'minor' (as opposed to 'major', *dur*).

molto (*Italian*) 'very', so *allegro molto* means 'very fast'.

monochrome a drawing or painting executed in one colour only. *See also* grisaille.

monodrama a dramatic work for a single performer.

monody a type of accompanied solo song which was developed during the late 16th and early 17th centuries. It contained dramatic and expressive embellishments and devices, and consequently had an influence on OPERA.

monothematic a piece of music that is developed from a single musical idea.

monotone declamation of words on a single tone.

montage an art technique similar to COLLAGE, where the images used are photographic.

morbido (*Italian*) 'soft' or 'gentle'.

morceau (*French*) a 'piece' (of music).

mordent a musical ornament whereby one note rapidly alternates with another one degree below it; this is indicated by a sign over the note.

morendo (*Italian*) 'dying', i.e. decreasing in volume.

moresca (*Italian*) a sword dance dating from the 15th and 16th centuries, which represents battles between the Moors and the Christians. It was the origin of the English MORRIS DANCE. It has been included in OPERAS, often to a marching rhythm.

morisco (*Italian*) 'in Moorish style'.

morris dance a style of English dance, the music for which is provided by pipe and tabor. It was orginally a costume dance, the characters often being those from the Robin Hood ballads. Of Moorish or Spanish origin, the dance later became associated with many tunes, some in 4/4, others in 3/4 time.

mosso (*Italian*) 'moved', so *piu mosso* means 'more moved', i.e. quicker, and *meno mosso* 'less speed'.

motet a musical setting of sacred words for solo voices or choir, with or without accompaniment. The first motets were composed in the 13th century.

motif *or* **motive** a small group of notes which create a melody or rhythm, e.g. the first four notes of ◊Beethoven's 5th symphony.

motion the upward or downward progress of a melody. It is said to be *conjunct* when the degrees of the scale succeed each other; *disjunct* where the melody proceeds in skips; *contrary* where two parts move in opposite directions; *oblique* when one part moves while the other remains stationary; and *similar*, or *direct*, when the parts move in the same direction.

moto (*Italian*) 'motion', so *con moto* means 'with motion' or 'quickly'.

motto theme a short theme that recurs during the course of a composition. In this way, it dominates the piece.

mouth organ *see* HARMONICA.

movement a self-contained section of a larger instrumental composition, such as a symphony or sonata.

mp abbreviation for *mezzo piano* (*Italian*), meaning 'half-soft'.

muffled drum a drum with a piece of cloth or towelling draped over the vibrating surfaces. It produces a sombre tone when struck and is usually associated with funeral music.

musette (*French*) (1) a type of small BAGPIPE popular at the French court in the 17th and 18th centuries. (2) an air in 2/4, 3/4 or 6/8 time that imitates the drone of the bagpipe. (3) a dance tune suitable for a bagpipe. (4) an organ reed stop.

musica (*Italian*) 'music'.

musica ficta *or* **cantus fictus** (*Latin*) 'feigned music' or 'feigned song'; it is a term for ACCIDENTALS used in MODE music.

musical a type of play or film in which music plays an important part and the actors occasionally sing, for example, *My Fair Lady*, *West Side Story*.

musical box a clockwork MECHANICAL INSTRUMENT in which a drum studded with small pins plays a tune by plucking the teeth of a metal comb.

musical comedy a term used between 1890 and 1930 to describe a humorous play with light music and singing in it.

music drama a term first used to describe the operas of ◊Wagner, where the action and music are completely interlocked, with, for example, no pauses for applause after an aria, or repetition within a piece.

musicology the scientific study of music.

musique concrète a term coined by the French composer Pierre Schaffer in 1948 to describe a type of music in which taped sounds are distorted or manipulated by the composer. The term ELECTRONIC MUSIC is now more generally used.

muta (*Italian*) 'change', a musical direction: (1) that the key be changed in horn or drum music; (2) that the MUTE be used.

mutation stops organ stops that produce sound, usually a HARMONIC – which is different from the normal or octave pitch corresponding to the key that is depressed.

mute any device used to soften or reduce the normal volume, or alter the tone, of an instrument. With bowed instruments, a small clamp is slotted onto the bridge; in brass instruments a hand or bung is pushed into the bell; in the piano the soft (left) pedal is pressed; and with drums, cloths are placed over the skins, or sponge-headed drumsticks are used.

mv abbreviation for *mezzo voce*.

mystery play *or* **miracle play** a form of dramatic entertainment based on sacred subjects and given under church auspices which was used before the development of either OPERA or ORATORIO.

Nabis a group of painters working in France in the 1890s. Influenced by the works and ideas of

◊Gauguin and by oriental art, they worked in flat areas of strong colour, avoiding direct representation in favour of a symbolic approach of mystical revelation.

nachdruck (*German*) 'ACCENT' or 'emphasis'.

Nachschlag (*German*) 'after beat', a grace or ORNAMENT, like a short APPOGGIATURA, but occurring at the end instead of at the beginning of a NOTE.

nachspiel (*German*) a POSTLUDE.

Nachtmusik (*German*) 'night music', that is, music suitable for performing in the evening or suggestive of night.

naive art works by untrained artists whose style is noted for its innocence and simplicity. Scenes are often depicted literally, with little attention to formal perspective and with an intuitive rather than studied use of pictorial space, composition and colour. Naive painters work independently of contemporary trends or movements in art, and their works are often fresh and invigorating by comparison.

naker the medieval English name for a small KETTLEDRUM (often with snares, *see* SIDE DRUM) of Arabic origin, from which TIMPANI developed. Nakers were always used in pairs.

national anthem a SONG or HYMN that is formally adopted by a country and sung or played at official occasions.

nationalism a late 19th-century and early 20th-century music movement in which a number of composers set out to write work which would express their national identity, often by reference to FOLK music and by evocation of landscape. It was in part a reaction to the dominance of German music.

natural a NOTE that is neither sharpened nor flattened.

Naturalism a term deriving from the late-19th French literary movement of the same name, denoting fiction characterized by close observation and documentation of everyday life, with a strong emphasis on the influence of the material world on individual behaviour. Naturalistic NOVELS therefore tended to adopt a very deterministic approach to life and fiction, and most practitioners of the form would have described themselves as socialists or social Darwinists. The influence of the school on writers in English was patchy, however, there are strong elements of naturalism in the work of Arnold ◊Bennett.

natural key KEY of C major.

Nazarenes an art movement based on the Brotherhood of St Luke formed in Vienna in 1809. It involved painters of German and Austrian origin, who worked mainly in Italy. Inspired by the medieval guild system, they worked cooperatively with a common goal of reviving Christian art. Influences included German medieval art and Italian RENAISSANCE painting.

neck the narrow projecting part of a stringed instrument that supports the FINGERBOARD; at the end of the neck lies the PEG-BOX which secures the strings and enables them to be tuned.

nel battere (*Italian*) on the BEAT or down stroke (DOWN-BEAT).

Neoclassicism a term denoting any movement in the Arts emphasizing the virtues of imitating the style and precepts of the great classical writers and artists (Neoclassicist principles in literature derive mostly from the writings of ◊Aristotle – principally the *Poetics* – and some observations by the Roman poets, ◊Virgil and ◊Horace). The hallmarks of Neoclassicism are traditionally defined as balance, moderation, attention to formal rules – such as the dramatic *Unities*, in which time and space is strictly ordered around a sequential plot with a beginning, middle and end – avoidance of emotional display and distrust of enthusiasm, and the assumption that human nature has changed little since CLASSICAL times. Neoclassicism is frequently contrasted with ROMAN- TICISM, and, interestingly, the strongest period of Neoclassicism in art and architecture is contemporary with the peak period of Romanticism, the late 18th and early 19th centuries.

In literature, Neoclassicism is generally held to have begun with ◊Petrarch in the mid-14th century. The greatest Neoclassical English poets are ◊Dryden and ◊Pope; the greatest Neoclassical English critic is sometimes said to be Dr ◊Johnson, but his espousal of Neoclassical virtues was heavily qualified, particularly in response to French criticism of ◊Shakespeare as virtually a 'literary savage'. Neoclassicism had been a powerful influence on French literature of the 17th century, but has never won wide acceptance among writers in English, not even in the 18th century.

In art and architecture, Neoclassicism was the dominant style in Europe in the late 18th and early 19th centuries. It followed on from, and was essentially a reaction against, BAROQUE and ROCOCO styles. Classical forms were employed to express the reasoned enlightenment of the age, and Neoclassical painters adhered to the Classical principles of order, symmetry and calm. At the same time, they felt free to embrace Romantic themes.

In music, Neoclassicism was a 20th-century movement that reacted against the overtly Roman-tic forms of the late 19th century. Composers who adhered to the philosophy (in particular ◊Stravinsky, ◊Hindemith, ◊Poulenc and ◊Prokofiev) attempted to create new works with the balance and restraint found in the work of 18th-century composers, especially J. S. ◊Bach.

Neoimpressionism *see* POINTILLISM.

Neoplatonism a synthesis of Platonic and mystical concepts that originated in the Greek-speaking Mediterranean world of the 3rd century. Plotinus (*c*.205–*c*.270) was the main figure behind the synthesis. His ideas, such as the notion of a world soul and the perception of the poet as an inspired prophet who sees the real world behind the shadowy illusions of the material world, have been influential on many English writers, notably ◊Shelley.

neume a sign used in musical NOTATION from the 7th to 14th centuries before the invention of the STAVE. It indicated PITCH.

Neue Sachlichkeit (*German*) 'new objectivity', the title of an exhibition of postwar figurative art which then came to represent any art concerned with objective representation of real life, such works being the opposition to EXPRESSIONIST subjectivity.

new comedy *see* comedy.

Nibelungenlied an anonymous medieval German epic recounting the hero Siegfried's capture of a hoard of gold belonging to a race of dwarfs (the Nibelungs). The epic forms the basis for Wagner's great opera cycle, *The Ring of the Nibelungs*, and was influential on ◊Tolkien's *Lord of the Rings*.

niente (*Italian*) 'nothing'; used in *quasi niente* 'almost nothing', indicating a very soft tone.

Nihilism a philosophical movement originating in mid-19th century Russia that rejected all established authority and values. The revolutionary Bazarov in ◊Turgenev's *Fathers and Sons* is the first significant fictional portrait of a nihilist. Much of ◊Dostoevsky's work, e.g. *The Possessed*, is concerned with exposing the essential shallowness and banality of nihilism. Nihilist characters crop up frequently in English fiction of the late 19th and early 20th century, e.g. in Joseph ◊Conrad's *The Secret Agent*.

ninth an INTERVAL of nine NOTES, in which both the first and last notes are counted.

No *see* Noh.

Nobel prize any of the annual awards endowed by the Swedish chemist Alfred Nobel (1833–96) for significant contributions in the fields of chemistry, physics, literature, medicine or physiology and peace, with economics added later. Winners of the

Nobel prize for literature include ◊Kipling (1907), ◊Yeats (1923), ◊Shaw (1925), T. S. ◊Eliot (1948), ◊Pasternak (1958), ◊Hemingway (1954), ◊Beckett (1969), ◊Golding (1983) and ◊Walcott (1992).

nobile, nobilmente (*Italian*) 'noble, nobly'.

nocturne literally a 'night piece', i.e. a piece of music, often meditative in character and suggesting the quietness of night. The form was invented by John ◊Field and later perfected by ◊Chopin.

Noh *or* **No** a form of highly stylized drama originating in 14th-century Japan. The typical Noh play is short, slow-paced, draws heavily on classical Japanese symbolism, and usually involves song, dance, mime and intricately detailed costume.

noire (*French*) literally 'black', a CROTCHET or quarter note.

nomenclature *see* NOTATION.

nonet a group of nine instruments, or a piece of music for such a group.

normal pitch standard PITCH.

nota (*Latin, Italian*) 'NOTE', so *nota bianca* means 'white note' or half-note; *nota buona* is an accented note; *nota cambita* or *cambiata* is a CHANGING NOTE; *nota caratteristica* is a LEADING NOTE; *nota cattiva* is an unaccented note.

notation *or* **nomenclature** the symbols used in written music to indicate the PITCH and RHYTHM of NOTES, the combination and duration of TONES, as well as the graces and shades of expression without which music can become mechanical.

note (1) a SOUND that has a defined PITCH and duration. (2) a symbol for such a sound. (3) the KEY of a PIANO or other keyboard instrument.

novel a sustained fictional prose narrative. Although most of the essential characteristics of the form can be found in ancient texts, such as the Greek Romances of the 3rd century, and in the works of writers of the Roman world, the novel as the term is generally understood, with complex characterization and multilayered strands of plot and character development, is essentially a creation of 18th-century writers in English, particularly ◊Defoe, ◊Richardson and ◊Fielding.

In the early years of the 19th century, the dual tradition of the novel – adventure in the great world outside, and exploration of personality – reached striking new levels in the novels of Sr Walter ◊Scott and Jane ◊Austen respectively, with the latter setting standards of characterization that have rarely been matched. In the course of the 19th century, writers throughout Europe and America, e.g. ◊Dickens, ◊Stendhal, George ◊Eliot, ◊Tolstoy, ◊Dostoyevsky, ◊Balzac, ◊Melville and Henry ◊James, developed the novel into a highly sophis-

ticated vehicle for exploring human consciousness. The 20th century has seen many adaptations of the traditional form of the novel, from the use of the STREAM OF CONSCIOUSNESS technique by Joyce and Woolf, to the MAGIC REALISM of South American writers, with such oddities as the so-called *antinovels* of writers such as the French novelist Alain Robbe-Grillet (1922–) occurring along the way.

novella a short version of the NOVEL; a tale usually leading up to some point, and often, in early versions of the form, of a satirical or scabrous nature.

nuance a subtle change of speed, TONE, etc.

number (1) an integral portion of a musical composition, particularly in opera where it can mean an ARIA, DUET, etc. (2) one of the works on a programme.

Nunc Dimittis (*Latin*) the Song of Simeon, 'Lord, now lettest Thou Thy servant depart in peace', which is sung at both Roman Catholic and Anglican evening services. It has been set to music by numerous composers.

nursery rhyme a (usually traditional) short verse or song for children. Many nursery rhymes, e.g. 'Ring a Ring a Roses' (which refers to the Great Plague) have their roots in ancient and occasionally unsavoury events.

nut (1) the part of the BOW of a stringed instrument that holds the horsehair and that incorporates a screw that tightens the tension of the hairs. (2) the hardwood ridge at the PEG-BOX end of a stringed instrument's FINGERBOARD that raises the strings above the level of the FINGERBOARD.

o when placed over a NOTE in a musical SCORE for strings, indicates that the note must be played on an open string or as a harmonic.

ob abbreviation for OBOE and OBBLIGATO.

obbligato (*Italian*) 'obligatory', a term that refers to a PART that cannot be dispensed with in a performance (some parts can be optional). However, some 19th-century composers used the word to mean the exact opposite, i.e. a part that was optional.

ober (*German*) 'over', 'upper', as OBERWERK.

objet trouvé *see* FOUND OBJECT.

oblique motion two parallel MELODY lines, or PARTS: one moves up or down the SCALE while the other stays on a consistent NOTE.

oboe a WOODWIND instrument with a conical bore and a double REED. The instrument has a history dating back to ancient Egyptian times. SHAWMS evolved from these Egyptian predecessors and became known as HAUTBOIS instruments in the 17th and 18th centuries. The modern oboe (the word is a corruption of 'hautbois') dates from the 18th century. The established variations of the instrument are: the oboe (TREBLE), the COR ANGLAIS (ALTO), or the BASSOON (tenor), and the double bassoon (BASS).

ocarina a small, egg-shaped WIND INSTRUMENT, often made of clay, which is played in a way similar to a RECORDER. It was invented in the mid-19th century and is still made, mainly as a toy.

octave an INTERVAL of eight NOTES, inclusive of the top and bottom notes, e.g. C to C.

octet a group of eight instruments or VOICES, or a piece for such a group.

octobass a huge kind of three-stringed DOUBLE BASS, some four metres in height, which incorporated hand- and pedal-operated levers to STOP the immensely thick strings. It was invented by J.B. Vuillaume in Paris in 1849, but proved impractical.

oeuvre (*French*) a 'work' (OPUS).

offertory an ANTIPHON sung (or music played on the ORGAN) while the priest prepares the bread and wine at a communion service.

ohne (*German*) 'without', hence *ohne worte*, 'without words'.

oil paint a paint made by mixing colour pigments with oil (generally linseed oil) to produce a slow-drying, malleable sticky substance. Oil paint has been the dominant medium in European art since the 15th century because of the range of effects that can be produced.

old comedy *see* comedy.

Ondes Martenot *or* **Ondes Musicales** an ELECTRONIC musical instrument patented in 1922. It was used by ◊Messiaen.

Op, op abbreviation for OPUS.

Op Art *or* **Optical Art** an abstract art that uses precise, hard-edged patterns in strong colours that dazzle the viewer and make the image appear to move.

open harmony *see* HARMONY.

open note (1) in stringed instruments, an OPEN STRING. (2) in BRASS or WOODWIND instruments, a NOTE produced without using VALVES, CROOKS, or KEYS.

open string any string on an instrument that is allowed to vibrate along its entire length without being stopped (*see* STOPPING).

Oper (*German*) 'opera'.

opera a dramatic work in which all, or most of, the text is sung to orchestral ACCOMPANIMENT. It is a formidable musical form, and has a history dating back to Italy in the 16th century. This Italian form of opera reached its culmination in the operas of

◊MONTE-VERDI, in particular his *Orfeo*, but thereafter developed into a rigidly prescribed form of art. The growth of the science of HARMONY and the development of the modern ORCHESTRA led to a revolt against Italian opera, headed by ◊Gluck, ◊Mozart and, later, by ◊Weber in Germany. ◊Wagner gave a new impetus to operatic composition. His approach assumed that music that detracted from interest in the progress of the drama was bad music, and that the purpose of music, as of architecture, of lighting, of costume, and of acting, was to enforce the dramatic interest of the text.

Opera demands a LIBRETTO, an orchestra, singers, an ample stage and, only too often, considerable funds to produce. It also requires some suspension of disbelief, for opera is the convention of unreality, but in that unreality lies its ability to work magic.

opéra-bouffe (*French*), **opera buffa** (*Italian*) similar to OPÉRA COMIQUE but with the dialogue sung throughout and spoken only occasionally in *opera buffa* modelled on the French style.

opéra comique (*French*) literally 'a comic OPERA', but in fact an opera consisting of dramatic pieces with music and dancing and instrumental ACCOMPANIMENT, often along tragic rather than comic lines. Like the German SPINGSPIEL, all or nearly all the dialogue is spoken.

opera seria (*Italian*) 'serious OPERA', as usually applied to work of the 17th and 18th centuries.

operetta a short OPERA or, more usually, a term taken to mean an opera with some spoken dialogue and a romantic plot with a happy ending.

Optical Art *see* OP ART.

opus (*Latin*) 'work'; a term used by composers (or their cataloguers) to indicate the chronological order of their works. It is usually abbreviated to Op. and is followed by the catalogued number of the work.

oratorio the musical setting of a religious or epic LIBRETTO for soloists, CHORUS and ORCHESTRA, performed without the theatrical effects of stage and costumes, etc. Oratorio had its beginnings in the MYSTERY PLAYS of the Middle Ages. At first they were performed in MADRIGAL style, and became popular throughout Italy. From Italy, where it was soon overshadowed by OPERA, the oratorio spread to the rest of Europe. The church CANTATAS of J. S. ◊Bach and his PASSIONS can be regarded as its highest expression in northern Germany. In England it was ◊Handel's recourse when opera was no longer profitable or was forbidden. His *Messiah* and ◊Haydn's *Creation* and *The Seasons* were the culmination of the form, but later composers also wrote oratorios, for example ◊Mendelssohn's *Elijah*, and ◊Elgar's *Dream of Gerontius*.

orchestra a group of instruments and their players. The word comes from Greek and means 'dancing place'. This was a space in front of the stage, in which a raised platform was built for the accommodation of the CHORUS. The early composers of OPERA applied the name to the place allotted to their musicians, and it is now used to designate the place, the musicians or the instruments. Orchestras have grown over the centuries in response to larger auditoriums. A standard, modern orchestra contains four families of instruments: STRINGS, WOODWIND, BRASS and PERCUSSION. The exact number of players within each section can vary, and extra instruments can be called for by a particular SCORE.

orchestration the art of writing and arranging music for an ORCHESTRA.

organ a keyboard wind instrument, played with the hands and feet, in which pressurized air is forced through PIPES to sound NOTES. PITCH is determined by the length of the pipe. There are essentially two types of pipe; FLUE PIPES, which are blown like a WHISTLE, and REED PIPES in which air is blown over vibrating strips of metal. Flue pipes can be 'stopped' (blocked off at one end) to produce a sound an OCTAVE lower than when open. There are a number of keyboards on an organ, one of which is operated by the feet (PEDAL board). Those operated by the hands are called MANUALS, and there are four common categories: the solo (used for playing SOLO MELODIES), the swell (on which notes can be made to sound louder or softer, *see also* SWELL ORGAN), the great (the manual that opens up all the most powerful pipes), and the choir (which operates the softer sounding pipes, *see also* CHOIR ORGAN). In addition there are a number of 'STOPS' (buttons or levers) that can alter the pitch or TONE of specific pipes. The organ dates back to before the time of Christ and has gone through many stages of evolution. ELECTRONIC organs (*see* HAMMOND ORGAN) have been invented and these tend to produce sounds rather different from those in which pumped air is actually used.

organum (1) (*Latin*) 'ORGAN'. (2) measured music, as opposed to unmeasured PLAINSONG, an early form of POLYPHONY.

ornaments and graces embellishments to the NOTES of a MELODY, indicated by symbols or small notes. They were used frequently in the 17th and 18th centuries.

Orpheus the legendary poet and musician of Greek mythology whose name has been adopted by nu-

merous musical societies, etc, and has also been used as the title of several collections of vocal music. The story of Orpheus and his search in the underworld for his wife Eurydice has been the subject of many musical forms.

Orphism *or* **Orphic Cubism** a brief but influential art movement developed out of Cubist principles. Their aim was to move away from the objectivity of CUBISM towards a more lyrical and colourful art. The artists were influenced in part by Italian FUTURISM, and typical works use juxtaposed forms and strong colours. The movement had a deep influence on some of the German EXPRESSIONISTS and on SYNCHROMISM.

oscillator an ELECTRONIC INSTRUMENT that converts electrical energy into audible sound.

ossia (*Italian*) 'or', 'otherwise', 'else'; used to indicate an alternative PASSAGE of music.

ostinato (*Italian*) 'obstinate'; a short PHRASE or other pattern that is repeated over and over again during the course of a composition.

ottava (*Italian*) OCTAVE.

overblow to increase the wind pressure in a WOODWIND instrument to force an upper PARTIAL note instead of its fundamental note, thus producing a harmonic.

overtones *see* HARMONICS.

overture a piece of music that introduces an an OPERA, ORATORIO, BALLET or other major work. Overtures may be built out of the principal THEMES of the work that is to follow, or may be quite independent of them. ◊Beethoven composed no fewer than four overtures to his only opera, *Fidelio*, and ◊Verdi's *Otello* and other operas have no overture whatever. Overtures are nearly always in the SONATA FORM, being, in fact, similar to the first MOVEMENT of a SYMPHONY, on a somewhat larger scale. The CONCERT OVERTURE is often an independent piece, written for performance in a concert hall.

p (1) abbreviation for PIANO (*Italian*), meaning 'soft'. (2) abbreviation for 'PEDAL' (ORGAN).

pandora a plucked stringed instrument of the CITTERN family. It was particularly popular in England during the 16th century.

panpipes a set of graduated PIPES, stopped at the lower end, which are bound together by thongs. Each pipe makes a single NOTE, and sound is produced by blowing across the open end. They are popular in South America and parts of Eastern Europe.

pantomime a combination of dancing and gesticulation by which a drama may be represented without words, although accompanied by music. In Britain, the term is now applied to a musical show with dialogue, traditionally based on a fairy tale and performed at Christmas time.

pantonality *see* ATONAL.

parameter a 20th-century term used to describe aspects of SOUND that can be varied but which nevertheless impose a limit. It is particularly applied to ELECTRONIC MUSIC with regard to volume, etc.

part a VOICE or instrument in a group of performers, or a piece of music for it.

parte (*Italian*) 'PART', so *colla parte* means 'with the part'.

part-song a composition for unaccompanied VOICES in which the highest part usually sings the MELODY while the lower parts sing accompanying HARMONIES.

passacaglia (*Italian*) a type of slow and stately dance originating in Spain, for which keyboard music was written in the 17th century. It has come to mean a work in which such a THEME recurs again and again.

passage a FIGURE or PHRASE of music; a RUN.

passage work a piece of music that provides an opportunity for VIRTUOSO playing.

passamezzo (*Italian*) 'half-step'; a quick Italian dance in DUPLE TIME that became popular throughout Europe in the late 16th century.

passepied (*French*) a French dance in triple TIME, like a quick MINUET, that is thought to have originated in Brittany. It was incorporated into French ballets of the mid-17th century.

passing note a NOTE that is dissonant with the prevailing HARMONY but that is nevertheless useful in making the TRANSITION from one CHORD or KEY to another.

Passion music the setting to music of the story of Christ's Passion (the story of the Crucifixion taken from the Gospels). The first dramatic representation of the Passion is said to have been made in the 4th century by St Gregory Nazianzen (329–89), bishop of Antioch. It was sung throughout. From the 13th century, the Passion was changed to PLAINSONG melodies by priests in churches during Holy Week. The most celebrated of later Passions are those of Johann Sebastian ◊Bach, the *St John Passion* and the *St Matthew Passion*.

pastel a paint medium of powdered colour mixed with gum arabic to form a hard stick. When applied to paper, the colour adheres to the surface. Pastel was used to great effect by ◊Degas.

pasticchio (*Italian*) literally 'pie', a dramatic entertainment that contains a selection of pieces from various composers' works.

pastoral any piece of literature celebrating the country way of life. The first pastoral poems of

any significance are those of Theocritus (*c*.310–*c*.250 BC), a Greek poet whose work established the standard frame of the form: shepherds and shepherdesses singing to one another of their loves in a world of peace and plenty in which the sun always shines. Death, however, is occasionally present in the form of a shepherd lamenting the death of a friend (◊Shakespeare's 'Dead shepherd' couplet from *As You Like It* is a moving, later example of this convention being used as a tribute to Christopher ◊Marlow. Theocritus' form was used by ◊Virgil in his *Eclogues*, so establishing a tradition that lasted for centuries. Later practitioners of the form include ◊Petrarch, ◊Milton and ◊Shelley.

pastorale (1) a vocal or instrumental MOVEMENT or composition in COMPOUND triple TIME, which suggests a rural subject; it usually has long BASS notes that imitate the sounds of the BAGPIPE drone. (2) a stage entertainment based on a PASTORAL.

patter song a kind of comic SONG that has a string of tongue-twisting syllables and is usually sung quickly to minimal ACCOMPANIMENT. It is often found in opera.

Pauken (*German*) KETTLEDRUMS.

pausa (*Italian*) 'REST'.

pause a symbol over a NOTE or REST to indicate that this should be held for longer than its written value.

pavan, pavane a stately court dance, normally in slow DUPLE TIME, which was occasionally incorporated into instrumental music in the 16th century.

pavillon (*French*) literally a 'tent', so, with reference to the shape, the bell of a BRASS INSTRUMENT.

peal a set of church bells or, as a verb, to ring a set of church bells.

ped abbreviation for PEDAL.

pedal the part of an instrument's mechanism that is operated by the feet, such as on a PIANO, ORGAN or HARP. The *forte*, or loud, pedal on a piano by raising the dampers enriches the TONE. The *piano*, or soft, pedal enables the player to strike only one or two strings or to reduce the volume of tone. Harp pedals sharpen, flatten or neutralize one NOTE throughout the COMPASS.

Organ pedals produce notes of the lower register independently of the MANUAL or alter the arrangement of the registers.

peg-box the part of a stringed instrument that houses the pegs that anchor and tune the strings.

pencil a mixture of graphite and clay in stick form and covered by a hard casing. The greater the clay element, the harder is the pencil. Graphite replaced lead as the principal component in the 16th century. Until the end of the 18th century, the word 'pencil' also denoted a fine brush.

penny whistle *see* TIN WHISTLE.

pentatonic scale a SCALE composed of five notes in an octave. It is found in various types of folk music from Scottish to Chinese.

per (*Italian*) 'by' or 'for'.

percussion (1) the actual striking of a DISCORD after it has been prepared and before its RESOLUTION. (2) the mechanism by which the tongue of a REED is struck with a HAMMER as air is admitted from the wind chest, thus ensuring immediate 'speaking'.

percussion instrument an instrument that produces TONE when struck, such as the PIANO or XYLOPHONE, but more especially one of the family of instruments that produce SOUND when struck or shaken, for example, MARACAS, DRUMS, TRIANGLE.

perdendosi (*Italian*) 'losing itself', i.e. dying away both in volume of TONE and in speed.

perfect interval *see* INTERVAL.

period a complete musical sentence (*see* PHRASE).

perpetuum mobile (*Latin*) 'perpetually in motion', i.e. a short piece of music with a repetitive NOTE pattern that is played quickly and without any PAUSES.

perspective in art, the representation of a three-dimensional view in a two-dimensional space by establishing a vanishing point in the distance at which parallel lines converge, the objects or figures in the distance being smaller and closer together than objects or figures nearer the viewer. Perspective is demonstrated in the works of ◊Giotto, and its rules were formulated by the Italian writer, architect, sculptor and painter Leon Battista Alberti (1404–72) in *De Pictura* (1435), but by the 20th century these were being abandoned by artists.

pesante (*Italian*) 'heavy', 'ponderous' or 'solid'.

PF abbreviation for PIANOFORTE, PIANO FORTE, and PIÙ *forte*.

Phantasie (*German*) *see* FANTASIA.

philharmonic (*Greek*) literally 'music loving'; an adjective used in the titles of many orchestras, societies, etc.

phrase a short melodic section of a composition, of no fixed length, although it is often four BARS long.

piacere (*Italian*) 'pleasure', so *a piacere* means 'at [the performer's] pleasure'.

piacevole (*Italian*) 'pleasantly'.

piangevole (*Italian*) 'sadly'.

pianissimo (*Italian*) 'very quiet'.

piano (1) (*Italian*) 'soft'. (2) the common abbreviated form of PIANOFORTE.

piano accordion *see* ACCORDION.

piano à queue. (*French*) a grand PIANO.

piano carré. (*French*) a square PIANO.

pianoforte, piano a keyboard instrument that was invented by Bartolomeo Cristofori in Florence in 1709 and for which important works were being written by the end of the 18th century. Most modern instruments usually have 88 KEYS and a COMPASS of $7^1/_3$ OCTAVES, although it is possible to find larger versions. The keys operate HAMMERS that strike STRINGS at the back of the instrument. These strings can run vertically (*upright piano*) or horizontally (*grand piano*). Most pianos have one string for the very lowest NOTES, two parallel strings for the middle REGISTER notes and three strings for the highest notes. Normally, when a NOTE is played, a damper (*see* DAMP) deadens the strings when the key returns to its normal position, but a sustaining (right) PEDAL suspends the action of the dampers and allows the note to coninue sounding. The soft (left) pedal mutes the sound produced, either by moving the hammers closer to the strings so that their action is diminished, or by moving the hammers sideways so that only one or two strings are struck. On some pianos, a third, SOSTENUTO pedal, allows selected notes to continue sounding while others are dampened.

pianola *see* PLAYER PIANO.

pibroch (*Gaelic*) a type of Scottish BAGPIPE music with the form of THEME and VARIATIONS.

picaresque novel a type of NOVEL in which the hero (very rarely, the heroine – ◊Defoe's Molly Flanders is the best-known example in English) undergoes an episodic series of adventures. The term derives from the Spanish *picaro*, a rogue or trickster. Many examples appear in 16th-century Spanish literature, when the genre first established itself, but picaresque novels have been appearing since the very earliest days of the novel. The first major English example is Thomas ◊Nashe's lurid *The Unfortunate Traveller*, which appeared in the 1590s along with other, lesser examples. Several great novels of the 18th and 19th centuries, e.g. ◊Smollett's *Roderick Random*, ◊Fielding's *Tom Jones* and ◊Dickens' *Pickwick Papers*, have picaresque elements but inhabit an entirely different emotional and moral world from that of the earlier examples of the genre. Thus, Tom Jones will have a characteristically picaresque sexual escapade with Molly Seagrim, but will not abandon her to the mob, as his scabrous fictional predecessors would cheerfully have done.

piccolo a small FLUTE with a PITCH an OCTAVE higher than a concert flute. It is used in orchestras and MILITARY BANDS.

pick a common expression for plucking the strings on a GUITAR.

pietà the Italian word for 'pity', used in art to denote a painting or sculpture of the body of the dead Christ being supported by the Virgin, often with other mourning figures.

pipe a hollow cylinder in which vibrating air produces SOUND. On many instruments, the effective length of the pipe can be altered to produce a range of NOTES by means of holes that are opened or closed by the fingers.

pipe organ an American term for a real ORGAN, as opposed to an AMERICAN ORGAN.

pistons the VALVES on BRASS instruments that allow players to sound different NOTES.

pitch the height or depth of a SOUND, which determines its position on a SCALE.

più (*Italian*) 'more', so *più* ALLEGRO means 'faster', and *più* FORTE 'more loudly'.

pizz. an abbreviation of PIZZICATO.

pizzicato (*Italian*) 'plucked' (with specific reference to using the fingers to pluck the STRINGS on a bowed instrument).

plainsong the collection of ancient MELODIES to which parts of Roman Catholic services have been sung for centuries. The best-known is GREGORIAN CHANT. Plainsong is usually unaccompanied and sung in UNISON. It is also in free RHYTHM, i.e. it does not have BARS but follows the prose rhythm of the PSALM or prayer.

Platonism the theory of forms devised by ◊Plato in which objects as we perceive them are distinguished from the idea of the objects, a theory that has had a strong influence on many writers, e.g. ◊Donne, ◊Wordsworth, and, most of all, ◊Shelley (*see also* NEOPLATONISM). Plato's speculations are contained in dialogue form in several works, e.g. the *Symposium* and *Phaedo*, and in *The Republic* (an examination of the principles of good government).

player piano *or* **pianola** a mechanical PIANO operated pneumatically by a perforated roll of paper.

plectrum a small piece of horn, plastic or wood that is used to pluck the STRINGS of a GUITAR, MANDOLIN, ZITHER, etc.

plein air (*French*) 'open air', used of paintings that have been produced out of doors and not in a studio. Plein air painting was particularly popular with the BARBIZON SCHOOL and became a central tenet of IMPRESSIONISM.

poco (*Italian*) 'little' or 'slightly', so *poco* DIMINUENDO means 'getting slightly softer' and *poco a poco* means 'little by little'.

poem an arrangement of words, especially in METER, often rhymed, in a style more imaginative than ordinary speech.

poet laureate a poet appointed to a court or other formal institution. In Britain, the post is held for life and the poet laureate is expected, although not forced, to write a poem to commemorate important events. Ben ◊Jonson held the post unofficially, being succeeded by Sir William Davenant (1606–68). The first official poet laureate was ◊Dryden, who was also the first to lose the post before his death, in the upheaval surrounding the 'Glorious Revolution'. Dryden's successor in 1689 was Thomas Shadwell (*c.*1642–92). Nahum Tate (1652–1715) held the office from 1692, and Nicholas Rowe (1674–1718) from 1715. Laurence Eusden (1688–1730) succeeded in 1718, more for political reasons than for poetic ones. He is better known for being mentioned in ◊Pope's *Dunciad* for his drinking habits than for any skill as a poet. His successor in 1730, the actor and dramatist Colley Cibber (1671–1757), also became a target of the *The Dunciad*, Pope personifying him as 'Dullness' in the final edition. The appointment in 1757 of the dramatist William Whitehead (1715–85) also attracted satirical comment. Thomas Warton (1728–90) held the post from 1785, and Henry James Pye (1745–1813) from 1790. The reputation of the post revived a little with the appointment of ◊Southey in 1813 and considerably with ◊Wordsworth (1843–50) and ◊Tennyson (1850–92) but suffered a setback with the appointment of Alfred Austin (1835–1913) in 1896. The 20th-century holders of the post were Robert Bridges (1844–1930) from 1913, ◊Masefield (1930–67), ◊Day-Lewis (1968–72), ◊Betjeman (1972–84), and Ted ◊Hughes (1984–).

poi (*Italian*) 'then', SO SCHERZO DA CAPO, *poi la* CODA means 'repeat the scherzo, then play the coda'.

point the tip of a BOW; the opposite end to the part that is held (the heel).

point d'orgue (*French*) 'ORGAN point'. It can indicate a harmonic PEDAL (a NOTE sustained under changing HARMONIES); the sign for a PAUSE; or a CADENZA in a CONCERTO.

pointillism a scientific and logical development of IMPRESSIONISM pioneered by the pointillist painters, notably ◊Seurat and ◊Pissarro. The brokenly applied brushwork of ◊Monet and ◊Renoir was extended and refined to a system of dots of pure colour, applied according to scientific principles, with the intention of creating an image of greater purity and luminosity.

In music, the term was borrowed to describe a style in which NOTES seem to be isolated as 'dots' rather than as sequential parts of a MELODY. It is applied to the works of certain 20th-century composers, such as ◊Webern.

polacca (*Italian*) *see* POLONAISE.

polka a ROUND DANCE in quick 2/4 TIME from Czechoslovakia. It became popular throughout Europe in the mid-19th century.

polonaise (*French*), **polacca** (*Italian*) a stately ballroom dance of Polish origin in moderately fast 3/4 TIME. It was used by ◊Chopin in sixteen strongly patriotic piano pieces.

polyphony (*Greek*) literally 'many sounds', i.e. a type of music in which two or more PARTS have independent melodic lines, arranged in COUNTERPOINT. The blending of several distinct MELODIES is what is aimed for, rather than the construction of a single melody with harmonized ACCOMPANIMENT.

polyptych a painting, usually an ALTARPIECE, consisting of two or more paintings within a decorative frame. *See also* DIPTYCH, TRIPTYCH.

polytonality the use of two or more KEYS at the same time.

ponticello (*Italian*) 'BRIDGE' (of a stringed instrument).

Pop Art a realistic art style that uses techniques and subjects from commercial art, comic strips, posters, etc. The most notable exponents include ◊Lichtenstein, ◊Oldenburg, and Robert Rauschenberg (1925–).

pop music short for 'popular' music, i.e. 20th-century music specifically composed to have instant appeal to young people. There are many types of pop music, with influences ranging from JAZZ and FOLK to ROCK and REGGAE.

portamento (*Italian*) literally 'carrying'; an effect used in singing or on bowed instruments in which sound is smoothly 'carried' or slid from one NOTE to the next without a break.

portraiture the art of painting, drawing or sculpting the likeness of someone, either the face, the figure to the waist, or the whole person. Portraits vary from the idealized or romanticized to the realistic.

position a term used in the playing of stringed instruments for where the left hand should be placed so that the fingers can play different sets of NOTES; e.g. first position has the hand near the end of the strings, second position is slightly further along the FINGERBOARD.

poster paint *see* GOUACHE.

posthorn a simple (valveless) BRASS instrument similar to a BUGLE, but usually coiled in a circular form.

Postimpressionism a blanket term used to describe the works of artists in the late 19th century who rejected IMPRESSIONISM. It was not a movement in itself, and most of the artists it refers to worked in widely divergent and independent styles. They include ◊Braque, ◊Picasso, ◊Cézanne, ◊Gauguin, Van ◊Gogh and ◊Matisse. The name was coined by the English art critic Roger Fry (1866–1934), an enthusiastic supporter of modern art, who organized the first London exhibition of Postimpressionist painters in 1912.

postlude the closing section of a composition.

Poststructuralism *see* STRUCTURALISM.

pot-pourri (*French*) a medley of well-known tunes played at a concert.

pp, PP an abbreviation for PIANISSIMO, 'very soft'; *ppp* means 'even softer'.

precentor the official in charge of music, or the leader of the singing, at a cathedral, monastery, etc. *See also* CANTOR.

precipitato, precipitoso (*Italian*) 'precipitately', hence also 'impetuously'.

prelude an introductory piece of music or a self-contained PIANO piece in one MOVEMENT.

Pre-Raphaelite Brotherhood a movement that was founded in 1848 by Holman ◊Hunt, ◊Millais and ◊Rossetti, who wanted to raise standards in British art. They drew their imagery from medieval legends and literature in an attempt to provide an escape from industrial materialism. They sought to recreate the innocence of Italian painting before ◊Raphael, and were influenced by the works of the NAZARENES. They had a large following, partly due to the support of the critic John ◊Ruskin, which included William ◊Morris. The movement broke up in 1853.

presto (*Italian*) 'lively'; *prestissimo* indicates the fastest speed of which a performer is capable.

prima donna (*Italian*) the 'first lady', i.e. the most important female singer in an opera.

primary colours the colours red, blue and yellow, which in painting cannot be produced by mixing other colours. Primary colours are mixed to make *secondary colours*: orange (red and yellow), purple (red and blue) and green (blue and yellow).

primo (*Italian*) 'first', as the first or top PART of a PIANO DUET (the lower part being termed *secondo*, 'second').

principal (1) the LEADER of a section of an orchestra (e.g. principal HORN). (2) a singer who regularly takes leading parts in an OPERA company, but not the main ones (e.g. a principal TENOR).

principal subject the first SUBJECT in a SONATA FORM or a RONDO.

Prix de Rome (*French*) an annual prize awarded by the French government to artists of various disciplines who had been sent to study in Rome for four years. ◊Berlioz, ◊Bizet and ◊Debussy were all winners, but ◊Ravel failed after many attempts. It was first awarded in 1803, and was discontinued in 1968.

programme music music that attempts to tell a story or evoke an image. The term was first used by ◊Liszt to describe his SYMPHONIC POEMS. ◊Beethoven made occasional incursions into the realm of programme music, notably in his symphony no. 6 in F (1808), the *Pastoral Symphony*, each of the five movements of which has an evocative title: '1. *Allegro ma non molto*, The pleasant feelings aroused in the heart on arriving in the country; 2. *Andante con moto*, Scene at the brook; 3. *Allegro*, Jovial assembly of country folk, interrupted by 4. *Allegro*, thunderstorm; 5. *Allegretto*, pleasurable feelings after the storm, mixed with gratitude to God'.

progression MOTION from NOTE to note, or from CHORD to chord.

Promenade Concerts an annual season of concerts given in London's Royal Albert Hall. The 'Proms' were instituted in 1895 by Robert Newman and were conducted until 1944 by Sir Henry ◊Wood. Cheap tickets are available for standing-room; people do not, however, walk about. Several cities around the world (such as Boston, Massachusetts) have similar concert seasons.

psalm a poem (song) from the Old Testament's Book of Psalms. Attributed to King David, psalms were inherited by the Christian churches from the earlier service of the Jews. The word is from Greek, and means 'to pluck a string', hence HARP song.

psalmody the singing of PSALMS to music or the musical setting of a psalm.

psalter a book of PSALMS and psalm tunes.

psaltery a medieval stringed instrument, similar to the DULCIMER except that the STRINGS are plucked and not struck. It is usually trapezium-shaped and is generally played horizontally.

pulse *see* BEAT.

Pult (*German*) 'desk', i.e. the music stand that two orchestral players share.

punta (*Italian*) 'POINT'; so *a punta d'arco* means 'at the point of the BOW', indicating that only the tip of the bow should be used to play the strings.

quadrille a French dance that was particularly fashionable in the early 19th century. It comprised five sections in alternating 6/8 and 2/4 time.

quadruple counterpoint four-PART COUNTERPOINT so constructed that all the parts may be transposed.

quadruplet a group of four NOTES of equal value played in the time of three.

quadruple time *or* **common time** the TIME of four CROTCHETS (quarter notes) in a BAR, indicated by the time signature 4/4 or C.

quarter note (US) *see* CROTCHET

quarter tone half a SEMITONE, which is the smallest INTERVAL traditionally used in Western music.

quartet a group of four performers; a composition for four SOLO instruments or for four VOICES.

quasi (*Italian*) literally 'as if' or 'nearly'; so *quasi niente* means 'almost nothing', or as softly as possible.

quattrocento an Italian term that refers to 15th-century Italian art, often used descriptively of the early RENAISSANCE period.

quaver a NOTE that is half the length of a CROTCHET and the eighth of a SEMIBREVE (whole note).

quickstep a MARCH in quick TIME, which also developed into the modern ballroom dance.

quintet a group of five performers; a composition for five SOLO instruments or for five VOICES.

quintuple time five BEATS, usually CROTCHETS, in a BAR, i.e. 5/4 TIME.

quodlibet (*Latin*) 'what you will'; a term used to describe a collection of tunes that are cleverly woven together to create an amusing entertainment.

R an abbreviation for RIPIENO or for 'right'.

racket *or* **rackett** *or* **ranket** a WOODWIND instrument with a double REED used between the late 16th and early 18th centuries. It came in four sizes (SOPRANO, TENOR, BASS, double bass) and created a distinctive buzzing sound.

rag a piece of RAGTIME music, notably as developed by ◊Joplin.

raga a type of Indian SCALE or a type of MELODY based on such a scale. Each raga is associated with a mood and with particular times of the day and year.

ragtime a style of syncopated (*see* SYNCOPATION) popular dance music, dating from the late 19th century, which was adopted by many composers, e.g. ◊Debussy with *Cakewalk*, and ◊Stravinsky with *Ragtime*. The combination of ragtime and BLUES led to the development of JAZZ. Scott ◊Joplin was a famous ragtime composer.

rallentando (*Italian*) 'slowing down'.

ranket *see* RACKET.

rap a term for an influential type of POP MUSIC of the late 20th century, which has a pulsating RHYTHM and in which LYRICS for SONGS are usually spoken to the BEAT and not sung.

rattle a type of PERCUSSION INSTRUMENT that traditionally consists of a hollowed-out gourd filled with seeds that rattle when shaken. An alternative type is a contraption in which a strip of wood held in a frame strikes against a cog-wheel as the frame is twirled round. It is occasionally required as a percussion instrument in orchestras.

re *or* **ray** in the TONIC SOL-FA, the second NOTE of the MAJOR SCALE.

Realism in literature, a true and faithful representation of reality in fiction. Once defined, discussion of the term usually breaks down into personal prejudices of various kinds. The process of distinguishing the term from NATURALISM is particularly fraught; any artist must select from the chaos of life in order to create, and in the process of choosing must impose some personal, even if banal, vision on the world, and in the process either convince the reader that the world portrayed is a real one, or fail. In the last analysis, there is only good writing and bad writing. ◊Balzac is regarded as the founding father of literary realism.

In art, Realism is taken to be, in general, the objective representation of scenes. The term is used particularly of the 19th-century French painters, e.g. ◊Daumier and ◊Courbet, who broke away from CLASSICISM and ROMANTICISM.

rebec *or* **rebeck** a small instrument with a pear-shaped body and, usually, three strings that were played with a BOW. It developed from the Arabian *rebab* and was used in Europe from the 16th century.

recapitulation *see* SONATA FORM.

recit. an abbreviation of RECITANDO or RECITATIVE.

recital a public concert given by just one or two people, e.g. a singer with PIANO ACCOMPANIMENT.

recitando (*Italian*) 'reciting', i.e. speaking rather than singing.

recitative a way of singing (usually on a fixed NOTE) in which the RHYTHM and lilt are taken from the words, and there is no tune as such. It is commonly used in OPERA and ORATORIO.

reciting note in PLAINSONG, the NOTE on which the first few words of each verse of a PSALM are sung.

recorder a straight, end-blown FLUTE, as opposed to a side-blown (concert) flute. Notes can be played by opening or closing eight holes in the instrument with the fingers. Recorders come in CONSORTS (families): DESCANT, TREBLE, TENOR and BASS.

Redskins and Palefaces a distinction between authors who write about outdoor, manly topics such as men at war (Redskins), and those who write about the problems and relationships of indoors, 'cultivated' people. It was formulated by an American academic, Leslie Fielder, and works easily in terms of much American literature, e.g. ◊Hemingway is as obviously a Redskin as Henry ◊James is a Paleface. It is less helpful elsewhere.

reed the small part found in many blown instruments that vibrates when air is blown across it and actually creates the sound. It is usually made of cane or metal. In single-reed instruments (e.g. CLARINET, SAXOPHONE), the reed vibrates against the instrument itself; in double-reed instruments (e.g. COR ANGLAIS, BASSOON), two reeds vibrate against each other; in free-reed instruments (e.g. HARMONIUM, CONCERTINA), a metal reed vibrates freely within a slot.

reed organ the generic term for a number of instruments that have no PIPES and use FREE REEDS to produce their NOTES. Examples are the ACCORDION and the HARMONIUM.

reed pipe an ORGAN pipe with a metal reed in the mouthpiece, which vibrates when air is passed over it.

reel a Celtic dance, usually in quick 4/4 TIME and in regular four-BAR PHRASES.

refrain the CHORUS of a BALLAD.

regal a portable REED ORGAN of the 16th and 17th centuries.

reggae a type of Jamaican POP MUSIC with a heavy and pronounced RHYTHM and strongly accented UPBEAT. Its best-known exponent was Bob ◊Marley.

register (1) a set of ORGAN PIPES that are controlled by a single STOP. (2) a part of a singer's vocal COMPASS, e.g. CHEST register, HEAD register, etc. The term is also applied to certain instruments.

related keys *see* MODULATION.

relative major, relative minor terms used to describe the connection between a MAJOR KEY and a MINOR KEY that share the same KEY SIGNATURE, e.g. A minor is the relative key of C major.

relief a sculptural form that is not freestanding. The three-dimensional shape is either carved, e.g. in stone, wood, ivory, etc, or built up, as in metal, etc. Relief sculpture can be *low relief* (*basso relievo* or *bas-relief*), where the depth of the pattern is less than half; *medium relief* (*mezzo relievo*), where the depth is roughly half; or *high relief* (*alto relievo*), where practically all the medium has been removed. The extremely low-relief technique of *stiacciato*, 'drawing in marble', was devised by ◊Donatello.

Renaissance in literature, the revival of the arts that occurred in Europe in the 14th-16th centuries, as a result of the rediscovery of the writing of the great classical writers, notably the works of ◊Plato and ◊Aristotle. In the case of the latter, the 'rediscovery' occurred in terms of reinterpretation; instead of taking the texts of Aristotle as literal, unchallengable authority, as medieval scholars had tended to do, the new thinkers, such as Francis ◊Bacon (the father of modern scientific method) approached written authority with a new, sceptical eye. The Renaissance period in England is usually given as 1500–1660, i.e. from the visit of ◊Erasmus in 1599 to the RESTORATION.

In art, the early Renaissance was established in Italy with the works of ◊Giotto, in a spectacular move away from Gothic conventions and ideals. The sculptors ◊Pisano and ◊Donatello emulated Greek and Roman sculpture in an expression of the new humanist and aesthetic values of the 'age of reason'. The movement reached a peak between 1500 and 1520 with the works of ◊Leonardo da Vinci, ◊Michelangelo and ◊Raphael. The Northern Renaissance took place as ideas spread to Germany, the Netherlands and the rest of Europe during the early 16th century.

repeat two or four DOTS in the spaces of the STAVE that indicate that the PASSAGE so marked is to be played through twice.

répétiteur (*French*) a person hired to teach musicians or singers their PARTS, particularly in OPERA.

replica (*Italian*) 'repeat', so *senza replica* means 'without repetition'.

representational art *see* FIGURATIVE ART.

reprise a musical repetition; it is often found in musical comedies when songs heard in one act are repeated in another.

reredos *see* ALTARPIECE.

Requiem a MASS for the dead in the Roman Catholic Church, so called because of the opening words at the beginning of the INTROIT, *Requiem aeternam dona eis, Domine* ('Grant them eternal rest, O Lord'). It is sung annually in commemoration of the dead on All Souls' Day and may also be sung at the funeral and on the anniversary of the death of an individual. Besides the Introit, the other chief divisions are the KYRIE; the Gradual, *Requiem aeternam* and tract, *Absolve Domine*; the Sequence, DIES IRAE; the *Offertorium, Domine Jesu Christi* ('Lord Jesus'); the SANCTUS; the BENEDICTUS; the AGNUS DEI, the *Communion, Lux aeterna* ('Light eternal'); and sometimes the *Responsorium, Libera me* ('Deliver me'); and the *Lectio, Taedet animam meam*.

Notable settings of the Requiem have been composed by ◊Palestrina, ◊Mozart, ◊Brahms, ◊Beethoven and ◊Verdi.

resolution a term for a process in HARMONY by which a piece moves from DISCORD to CONCORD.

resonance the intensification and prolongation of a SOUND or a musical NOTE produced by sympathetic VIBRATION.

responses the PLAINSONG replies of a CHOIR or congregation to SOLO CHANTS sung by a priest.

rest a sign employed in NOTATION indicating silence.

Restoration (1) the re-establishment of the British monarchy in 1660, following the return to England of Charles II in that year. (2) the period of Charles II's reign (1660–85). The characteristics of Restoration literature are wit, salaciousness, and religious and philosophical questioning.

Restoration comedy *see* COMEDY OF MANNERS.

resultant tone *see* COMBINATION TONE.

retable *see* ALTARPIECE.

retardation a SUSPENSION in which a DISCORD is resolved upwards by one step rather than downwards.

retrograde motion a term for music that is played backwards.

revenge tragedy a form of tragedy that appeared in the late Elizabethan period, heavily influenced by the bloodthirsty language and plots of ◊Seneca's plays, in which revenge, often for the death of a son or father, is the prime motive. Thomas ◊Kyd's *The Spanish Tragedy* (1588–9) is the earliest example, ◊Shakespeare's *Hamlet* (1602) the greatest. ◊Marlowe's *The Jew of Malta* (1592) is another notable example. *See also* JACOBEAN TRAGEDY.

reveille (pronounced 'revally') a BUGLE call used by the British Army to awaken soldiers.

rf, rfz abbreviations for RINFORZANDO.

RH abbreviation for 'right hand'.

rhapsody the title commonly given by 19th- and 20th-century composers to an instrumental composition in one continuous movement. Rhapsodies are often based on FOLK tunes, and are nationalistic or heroic in tone.

rhythm the regular recurrence of beat, accent or silence in the flow of sound, especially of words and music. In MUSIC NOTATION, rhythm is determined by the way in which NOTES are grouped together into BARS, the number and type of BEATS in a bar (as governed by the TIME SIGNATURE), and the type of emphasis (ACCENT) that is given to the beats. Along with MELODY and HARMONY, it is one of the essential characteristics of music.

rhythm and blues a type of POP MUSIC that combines elements of BLUES and JAZZ. It developed in the USA and was widely accepted by white audiences and pop musicians. ROCK 'N' ROLL evolved from rhythm and blues.

rhythm-names *see* TIME-NAMES.

rhythm section the name given to the PERCUSSION and DOUBLE BASS section of a JAZZ band; it provides the all-important BEAT.

ribs the sides uniting the back and belly of an instrument of the VIOLIN family.

rigaudon *or* **rigadoon** a jaunty dance from southern France that has two or four BEATS to the BAR. It was used in French ballets and operas, and it became popular in England in the late 17th century.

rigoroso (*Italian*) 'rigorously', i.e. in exact TIME.

rinforzando (*Italian*) literally 'reinforcing', i.e. a sudden strong ACCENT on a NOTE or CHORD.

ripieno (*Italian*) literally 'full'; a term used to describe PASSAGES that are to be played by the whole BAROQUE orchestra, rather than only a soloist.

risoluto (*Italian*) 'resolute' or 'in a resolute manner'.

rit. an abbreviation of RITARDANDO.

ritardando (*Italian*) 'becoming gradually slower'.

ritenuto (*Italian*) 'held back' (in TEMPO), i.e. slower.

ritmo, ritmico (*Italian*) 'RHYTHM', 'rhythmic'.

ritornello (*Italian*) literally a 'small repetition'. (1) a short PASSAGE for the whole orchestra in a BAROQUE ARIA or CONCERTO, during which the soloist is silent. (2) a short instrumental piece, played between scenes in early OPERA.

rock a type of POP MUSIC that evolved from ROCK 'N' ROLL in the USA during the 1960s. It mixes COUNTRY AND WESTERN with RHYTHM AND BLUES and is usually played loudly on electric instruments.

rock 'n' roll a type of POP MUSIC, with a strong, catchy RHYTHM, that evolved in the USA during the 1950s and is often associated with 'jiving' (fast dancing that requires nimble footwork). Elvis ◊Presley was one of its greatest early exponents.

Rocky Mountain School an American school of painters who painted landscapes of this formidable countryside in the middle of the 19th century. Its principal member was Albert Bierstadt (1830–1902), a German-born American landscape painter.

Rococo a style in art following on from BAROQUE and even more exaggerated in terms of embellishments and mannered flourishes. It became established around the beginning of the 18th century and spread throughout Europe, lasting up until the advent of NEOCLASSICISM in the 1760s. The main exponents of the style were ◊Fragonard and ◊Watteau in France and, to a lesser extent, ◊Tiepolo in Italy and ◊Hogarth in England. It continued in some areas to the end of the century, particularly in church decoration.

Music from the same period is sometimes similarly termed.

roll a TRILL on PERCUSSION INSTRUMENTS produced by sounding NOTES so rapidly that they overlap and appear to produce a continuous sound.

romance a love SONG or composition of a romantic character.

Romanticism a term denoting any movement in the arts which emphasizes feeling and content as opposed to form and order. The Romantic movement can be roughly dated from the late 18th century to the early 19th century, although the contrast between the need to express emotion and the desirability of following artistic rules dates back as far as the great Athenian dramatists. Other distinctive features of the Romantic movement are: the supremacy of individual over collective judgment; a 'progressive' faith in the reformability and essential goodness of humanity; the supremacy of 'natural' and 'organic' virtues over society's artifical construction. The extent and meaning of Romanticism in 18th-century English literature is still a matter of hot debate. It is certainly true that elements of what we call Romanticism can be found in poets such as ◊Cowper and ◊Smart, and even in Dr ◊Johnson's writings, but the first great works of Romantic literature are ◊Blake's works of the 1790s and ◊Wordsworth and ◊Coleridge's *Lyrical Ballads* (1798). Other prominent Romantic poets are ◊Byron, ◊Shelley, ◊Keats, ◊Heine and ◊Scott. *Compare* NEOCLASSICISM.

In art, the movement dates from the late 18th until the mid-19th century. It was a reaction to the balanced harmony and order of Classicism, and identified with the Romantic writers of the age. In response to increasing industrialization, Romantic painters viewed nature from a nostalgic point of view, imbuing landscapes with powerful emotions, often in a melancholic or melodramatic way. Notable Romantic artists include ◊Fuseli, ◊Goya, ◊Delacroix, ◊Géricault, ◊Friedrich, ◊Constable, ◊Turner and the visionary Blake.

In music, Romanticism lasted from *c*.1820 to *c*.1920. During this phase music tended to be more poetic, subjective and individualistic than previously. Lyricism, drama and often nationalistic feeling were characteristic of Romantic music.

Rome, Prix de *see* PRIX DE ROME.

ronde (*French*) literally 'round', as a noun a SEMIBREVE.

rondo a form of instrumental music that incorporates a recurring THEME, either in an independent piece or (more usually) as part of a MOVEMENT. It usually starts with a lively tune (the 'SUBJECT'), which is repeated at intervals throughout the movement. Intervening sections are called EPISODES, and these may or may not be in different keys from the subject. Rondo forms often occur in the final movements of SYMPHONIES, SONATAS and CONCERTOS.

root the lowest ('fundamental' or 'generating')

NOTE of a CHORD. Hence, for example, the chord C-E-G has a root of C

rosin a hard resin that is applied to the hair of BOWS used to play VIOLINS, etc. It causes increased friction between the hairs of the bow and the strings.

round a short CANON in which each PART enters at equal INTERVALS and in UNISON.

round dance a dance in which partners start opposite each other and subsequently form a ring.

roundelay a poem with certain lines repeated at intervals, or the tune to which such a poem was sung.

rubato (*Italian*) literally 'robbed', i.e. the taking of TIME from one NOTE or PASSAGE and passing it on to another note or passage.

rumba a sexually suggestive and fast Afro-Cuban dance in syncopated (*see* SYNCOPATION) 2/4 TIME.

run a SCALE or succession of NOTES rapidly played, or, if vocal, sung to one syllable.

Russian Futurism *see* FUTURISM.

S abbreviation for SEGNO, SENZA, SINISTRA, SOLO, SORDINO, SUBITO.

sackbut an instrument of the 15th century, probably originating in Spain. It is similar to the TROMBONE, but smaller.

sacra conversazione (*Italian*) 'holy conversation', in art denoting a painting in one panel of the Virgin and Child with saints.

St Cecilia *see* CECILIA, ST.

Saite (*German*) a 'string'.

salon (*French*) 'room', now also denoting an art exhibition. In the 19th century, the Salon was the annual exhibition of the ACADÉMIE FRANÇAISE, whose powerful and conventional jury increasingly refused to show the work of innovative artists. In 1863 Napoleon III ordered that there be an exhibition of artists' work rejected by the Salon, the Salon des Refusés.

saltando (*Italian*) literally 'leaping', i.e. an instruction to the string player to bounce the BOW lightly off the string.

saltarello a festive Italian FOLK dance in 3/4 or 6/8 TIME.

samba a Brazilian carnival dance in 2/4 TIME but with syncopated (*see* SYNCOPATION) rhythms.

samisen *see* SHAMISEN.

sämtlich (*German*) 'complete', as in *sämtliche Werke,* the 'complete works'.

Sanctus (*Latin*) 'Holy, holy, holy'; a part of the Ordinary of MASS in the Roman Catholic Church. It has been set to music by many composers.

sarabande (*French*) a slow dance in 3/2 or 3/4 TIME, which came to Italy from Spain.

Satz (*German*) a 'MOVEMENT' or 'piece of music'.

sautille (*French*) a SALTANDO.

Savoy Operas a name for the light operas written by ◊Gilbert and Sullivan, which were first performed at the Savoy Theatre in London by the D'OYLY CARTE company.

saxhorn a family of BUGLE-like BRASS instruments patented by the Belgian instrument-maker Adolphe Sax (1814–94) in 1845. They were innovative in that they had VALVES, as opposed to the KEYS normally associated with the bugle family.

saxophone a family of instruments patented by Adolphe Sax (*see* SAXHORN) in 1846 which, although made of brass, actually belong to the WOODWIND group because they are REED instruments. Saxophones come in many different sizes (e.g. SOPRANO, TENOR) and are commonly used in JAZZ bands as well as orchestras.

scala (*Italian*) 'staircase', from which the Teatro alla SCALA gets its name, but in music a RUN or SCALE.

Scala, La *or* **Teatro alla Scala** Milan's, and Italy's, premier opera house, which was opened in 1778.

scale an ordered sequence of NOTES that ascend or descend in PITCH. The most frequently used scales in European music are the MAJOR and MINOR scales, which use TONES (whole notes) and SEMITONES (half-notes) as steps of progression.

scat singing a type of singing used in JAZZ in which nonsense SOUNDS rather than words are sung.

scena (*Italian*) 'scene', a division of an act marked by a change of scenery. In OPERA, a solo movement of dramatic purpose, generally an extended aria.

scherzando, scherzoso (*Italian*) 'playful', or 'lively', as of a phrase or movement.

scherzetto (*Italian*) a short SCHERZO.

scherzo (*Italian*) 'joke', i.e. a cheerful, quick piece of music, either vocal or instrumental. The third MOVEMENT (of four) in many SYMPHONIES, SONATAS, etc, often takes the form of a scherzo.

schnell (*German*) 'quick'.

school in art, a group of artists who hold similar principles and work in a similar style. In art history, it also denotes that a painting has been executed by a pupil or assistant. In music, the characteristics of certain composers, whose style made a school.

Schottische (*German*) 'Scottish'; a ROUND DANCE, similar to the POLKA, that was popular in the 19th century. It is not in fact Scottish, but is so called because it is what those on the Continent thought a Scottish dance should be like.

Schrammel quartet a Viennese ENSEMBLE usually comprising two VIOLINS, a GUITAR and an ACCORDION, or the music composed for such an ensemble. It takes its name from Joseph Schrammel (1858–93), who wrote WALTZES for such a group.

scordatura (*Italian*) 'mistuning', i.e. the TUNING of stringed instruments to abnormal NOTES, so as to produce special effects.

score music written down in such a way that it indicates all the PARTS for all the performers, i.e. the whole COMPOSITION. A *full* or *orchestral score* is one with separate STAVES for each part. A *piano score* is one in which all the instrumental parts are represented on two staves. A *vocal score* is a piano score with two additional staves for the vocal parts. A *short close* or *compressed score* has more than one part to the stave.

scoring the writing of a SCORE.

Scotch snap the name for a RHYTHM that leaps from a short NOTE to a longer note. It is found in many Scottish FOLK tunes.

scraper a PERCUSSION INSTRUMENT in which SOUND is produced by scraping a stick over a series of notches cut into a piece of wood or bone.

scroll the decorative end of the PEG-BOX of a VIOLIN (or other stringed instrument), which may be carved into a curl resembling a scroll, or an animal head.

sec (*French*), **secco** (*Italian*) 'unornamented', 'plain'.

secondary colours *see* PRIMARY COLOURS.

Section d'Or a group of painters in France who associated between 1912 and 1914 and whose aim was to hold group exhibitions and encourage debate of their aesthetic ideals. They admired the work of ◊Cézanne and drew inspiration from FUTURISM.

segno (*Italian*) 'sign', used in NOTATION to mark a repeat, usually as *al segno*.

segue (*Italian*) 'follows', i.e. a direction to start playing the following MOVEMENT without a break.

seguidilla a Spanish dance in 3/8 or 3/4 TIME in the style of the BOLERO, but much faster.

semibreve a 'half of a BREVE'; the NOTE with the longest time-value normally used in modern Western NOTATION. In US notation, this is called a 'whole note'.

semidemisemiquaver an alternative name for a HEMIDEMISEMIQUAVER or sixty-fourth note.

Semiotics *see* STRUCTURALISM.

semiquaver a NOTE with half the time-value of a QUAVER, and a sixteenth the time-value of a SEMIBREVE. In US NOTATION, it is called a 'sixteenth-note'.

semitone 'half a TONE'; the smallest INTERVAL regularly used in modern Western music.

semplice (*Italian*) 'unornamented', 'in a simple manner'.

sempre (*Italian*) 'throughout', 'continually'; as *sempre* FORTE, 'loud throughout'.

sentimental comedy a form of English COMEDY that arose in the early 18th century, focusing on the problems of middle-class characters. The plays always end happily and feature strongly contrasting good and bad characters and high emotional peaks. The form was developed by Richard ◊Steele in a conscious reaction to the excesses of Restoration comedy (see COMEDY OF MANNERS). Examples include Steele's *The Tender Husband* (1705) and, notably, *The Conscious Lovers* (1722). The form led on to MELODRAMA (*see also* DOMESTIC TRAGEDY).

senza (*Italian*) 'without', so *senza* SORDINO means 'without MUTE' (in music for strings).

septet a group of seven performers or a piece of music written for such a group.

septuplet a group of seven NOTES of equal time-value to be played in the time of four or six.

sequence (1) the repetition of a short PASSAGE of music in a different PITCH. (2) a form of HYMN in Latin used in the Roman Catholic Mass, such as DIES IRAE and STABAT MATER.

serenade (1) a love SONG, traditionally sung in the evening and usually accompanied by a GUITAR or MANDOLIN. (2) a DIVERTIMENTO performed during an evening entertainment.

serenata (*Italian*) an 18th-century form of secular CANTATA or a short OPERA composed for a patron.

serialism a method of composition developed by ◊Schoenberg in which all SEMITONES are treated as equal, i.e. tonal values are eliminated. *See also* TWELVE-NOTE MUSIC.

serpent an obsolete BASS WOODWIND instrument with several curves in it (hence its name). It was used during the 16th century in church orchestras and MILITARY BANDS.

seventh an INTERVAL in which two NOTES are seven steps apart (including the first and last), for example F to E.

sevillana a Spanish FOLK DANCE originally from the city of Seville. It is similar to the SEGUIDILLA.

sextet a group of six performers or a piece of music written for such a group.

sextolet *or* **sextuplet** a group of six notes to be performed in the time of four notes.

Sezession (*German*) 'secession', adopted as a name in the 1890s by groups of painters in Austria and Germany when they broke away from official academies to work and exhibit in contemporary styles, e.g. IMPRESSIONISM. In Germany, the first German Sezession was in Munich in 1892, followed by the Berlin Sezession of 1899, which in turn in 1910 repudiated Die BRÜCKE, which resulted in the latter forming the *Neue Sezession*. In

Austria, the Vienna Sezession was organized by ◊Klimt in 1897.

sf *or* **sfz** abbreviation for SFORZANDO.

sforzando (*Italian*) 'forcing', i.e. a strong ACCENT placed on a NOTE or CHORD.

sfumato a subtle modelling of light and shade between figures and background, deployed by ◊Leonardo da Vinci when it represented a remarkable departure from the RENAISSANCE art stress on strong lighting and outline.

shake an alternative term for TRILL.

shamisen *or* **samisen** a Japanese long-necked LUTE with three strings. It has no FRETS and is plucked with a PLECTRUM.

shanai a double REED instrument from India, similar to a SHAWM.

shanty a SONG, with a pronounced RHYTHM, that was sung by sailors to help them coordinate their actions in the days of sailing ships. Shanties usually follow a format in which SOLO verses are followed by a CHORUS.

sharp the sign that raises the PITCH of the line or space on which it stands on a STAVE by a SEMITONE.

shawm *or* **shawn** a double-REED WOODWIND instrument that dates from the 13th century. It was developed from Middle Eastern instruments and produced a coarse, shrill sound. It was a forerunner of the OBOE.

sheng a sophisticated Chinese mouth organ (*see* HARMONICA), dating back some 3000 years.

shift a change of position of the hands when playing on a string instrument.

shofar *or* **shophar** an ancient Jewish WIND INSTRUMENT made from a ram's horn, which is still used in synagogues.

shop ballad *see* BALLAD.

siciliano a slow dance from Sicily in 6/8 or 12/8 TIME, with a characteristic lilting RHYTHM.

side drum *or* **snare drum** a cylindrically shaped DRUM that is the smallest usually used in an orchestra. Snares, made of gut or sprung metal, are stretched across the bottom parchment and vibrate against it when the upper membrane of parchment is struck; this gives the drum its characteristic rattling sound. The snares can be released so that a more hollow sound is produced.

signature *see* KEY SIGNATURE; TIME SIGNATURE.

signature tune a few BARS of catchy music that are associated with a performer or broadcast show.

similar motion the simultaneous PROGRESSION of two or more PARTS in the same direction.

simple interval any INTERVAL that is an octave or less. *Compare* COMPOUND INTERVAL.

simple time *see* COMPOUND TIME.

sine tone an electronically produced NOTE that is entirely 'pure'.

sinfonia (*Italian*) 'SYMPHONY', i.e. an instrumental piece. It is also a term used for a small orchestra.

sinfonietta a short SYMPHONY or a symphony for a small orchestra.

singing the act of producing musical TONE by means of the VOICE.

single chant *see* ANGLICAN CHANT.

Singspiel (*German*) literally 'sing-play', i.e. a comic opera in German with spoken dialogue replacing the sung RECITATIVE.

sinistra, sinistro (*Italian*) 'left', as in MANO *sinistra*, 'left hand'.

sistrum an ancient type of RATTLE in which loose wooden or metal discs are suspended on metal bars strung across a frame.

sitar a type of Indian LUTE, which is believed to have originated in Persia. It has movable metal FRETS and three to seven 'MELODY' strings; below these strings lie twelve or so SYMPATHETIC STRINGS, which create a droning sound. The sitar is plucked with a long wire PLECTRUM. It has a distinctive 'twangy' sound and is usually played in consort with the TABLA. Ravi ◊Shankar is perhaps the world's best player of the sitar.

Six, Les (*French*) 'The Six', the name given in 1920 to six young French composers by the poet and music critic Henri Collet who, with another of their champions, the poet Jean ◊Cocteau, was passionately anti-◊Wagner. The six were Georges Auric (1899–1983), Louis Durey (1888–1979), ◊Honegger, ◊Milhaud, ◊Poulenc and Germaine Tailleferre (1892–1983). Subsequently a number of other composers also became members of the group, but it ceased to have an effective function after 1925.

sixteenth-note (US) *see* SEMIQUAVER.

sketch in art, a preliminary drawing made by an artist to establish points of composition, scale, etc. In music, a short PIANO or instrumental piece.

skiffle a type of POP MUSIC played in England during the 1950s. Skiffle BANDS relied on American idioms (for example BLUES) and attempted to become 'authentic' by incorporating home-made instruments (such as tea-chest 'basses') into their outfits.

slancio, con (*Italian*) 'with impetus'.

sleigh bells small metal bells with steel balls inside which are mounted together in groups to produce a richly textured jingling sound. They are traditionally hung on sleighs, but are occasionally used in orchestras to create special effects.

slide (1) a passing from one to note to another without an INTERVAL. (2) a mechanism on the TRUMPET and TROMBONE that lengthens the tube to allow a new series of harmonics.

slide trombone *see* TROMBONE.

slide trumpet an early form of TRUMPET that had a slide similar to that used in the TROMBONE. It became obsolete when the VALVE trumpet was invented.

slur a curved line that is placed over or under a group of NOTES to indicate that they are to be played, or sung, smoothly, that is, with one stroke of the BOW (VIOLIN music) or in one breath (singing).

smorzando (*Italian*) 'fading' or 'dying away', i.e. the music is to become softer and slower.

snare drum *see* SIDE DRUM.

soave (*Italian*) 'soft' or 'gentle'.

soca music a type of powerful, rhythmic dance music from the English-speaking islands of the Caribbean. It evolved from soul (hence *so*) and calypso (*ca*).

socialist realism the name given to official art in the former Soviet Union, which was intended to glorify the achievements of the Communist Party.

social realism a form of realism, in which an artist's political viewpoint (usually on the left) affects the content of his work.

soft pedal *see* PIANO.

soh in the TONIC SOL-FA, the fifth NOTE (or DOMINANT) of the MAJOR SCALE.

solemnis (*Latin*) 'solemn', as in MISSA *Solemnis*, 'Solemn Mass'.

solenne (*Italian*) 'solemn'.

solennelle (*French*) 'solemn'.

sol-fa *see* TONIC SOL-FA.

solfeggio (*Italian*) a type of singing exercise in which the names of the NOTES are sung. *See* TONIC SOL-FA.

solo (*Italian*) 'alone', i.e. a piece to be performed by one person, with or without ACCOMPANIMENT.

solo organ a manual on an ORGAN with strong, distinctive STOPS, used for individual effect.

sonata originally a term for any instrumental piece to distinguish it from a sung piece or CANTATA. However, during the 17th century two distinct forms of sonata arose: the *sonata da* CAMERA (chamber sonata), in which dance movements were played by two or three stringed instruments with a keyboard ACCOMPANIMENT, and the *sonata da chiesa* (church sonata), which was similar but more serious. In the 18th century the sonata came to be a piece in several contrasting movements for keyboard only or for keyboard and one SOLO instrument.

sonata form a method of arranging and constructing music that is commonly used (since *c*.1750)

for SYMPHONIES, SONATAS, CONCERTOS, etc. There are three sections to sonata form: the EXPOSITION (in which the SUBJECT or subjects are introduced), the DEVELOPMENT (in which the subject(s) are expanded and developed), and the 'recapitulation' (in which the exposition, usually modified in some way, is repeated).

sonata-rondo form a type of RONDO, popular with such composers as BEETHOVEN, which is a combination of rondo and SONATA FORM.

sonatina a short SONATA.

song (1) a musical setting of poetry or prose. (2) a poem that can be sung. (3) a name used to designate the second SUBJECT of a SONATA.

song cycle a set of SONGS that have a common THEME or have words by a single poet. ◊Schubert, ◊Schumann and ◊Mahler wrote notable song cycles.

sopra (*Italian*) 'above', so *come sopra* means 'as above'.

soprano the highest PITCH of human VOICE, with a range of about two OCTAVES above approximately middle C. The term is also applied to some instruments, such as soprano SAXOPHONE (the highest pitched saxophone). *See also* MEZZO.

sordino (*Italian*) 'MUTE'.

sospirando (*Italian*) 'sighing'.

sospiro (*Italian*) literally 'sigh', in music meaning a CROTCHET rest

sostenuto (*Italian*) 'sustained'.

sotto (*Italian*) 'below', as in *sotto voce,* which literally means 'under the VOICE' or whispered.

soul music a type of emotionally charged music developed by black musicians in America. 'Soul', as it is usually called, derives from BLUES and GOSPEL music with the addition of ROCK RHYTHMS.

sound a term in ACOUSTICS for TONES resulting from regular VIBRATIONS, as opposed to noise.

sound hole the opening in the BELLY of a stringed instrument, e.g. the f-shaped holes in a violin or the round hole in a guitar.

soundpost a piece of wood connecting the BELLY of a stringed instrument (such as a VIOLIN) to the back. It helps to distribute VIBRATIONS through the body of the instrument.

sousaphone a giant TUBA that encircles the player's body which was designed for his band by the American bandmaster and composer John Philip Sousa (1854-1932) who formed a successful MILITARY-style BAND that toured the world giving concerts.

Spanish guitar the classic GUITAR with a narrow waist, six strings and a central SOUND HOLE.

special effects a non-specific term used of any extraordinary noises or SOUNDS that may be required of an orchestra, or part of an orchestra, to satisfy the demands of a composer, such as COW BELLS, etc.

species a discipline used in teaching strict COUNTERPOINT, developed by Johann Fux (1660–1741), who listed five rhythmic patterns ('species') in which one VOICE PART could be combined with another.

spinet a type of small HARPSICHORD.

spirito (*Italian*) 'spirit', so *con spirito* means 'with spirit'.

spiritual a type of religious FOLK SONG or HYMN that was developed by black (and white) Americans in the 18th and 19th centuries. Spirituals are characterized by strong SYNCOPATION and simple MELODIES. They were superseded by GOSPEL music.

Sprechgesang (*German*) literally 'speech-song', i.e. a type of singing that is half speech. It was used by ◊Schoenberg in his song cycle *Pierrot Lunaire*.

Stabat Mater (*Latin*) 'The Mother stood', the initial words of a hymn describing the sorrows of the Virgin at the Crucifixion. It was set to music, other than its original PLAINSONG, by, for example, ◊Palestrina; later settings include those by ◊Haydn, ◊Schubert and ◊Rossini.

stabile a non-moving MOBILE, as devised by Alexander Calder.

staccato (*Italian*) 'detached', i.e. NOTES should be shortened and played with brief INTERVALS between them.

staff *see* STAVE.

stave *or* **staff** a set of horizontal lines (usually five) on which music is written. Each line, and the gaps between them, represent a different PITCH.

steel drum a PERCUSSION INSTRUMENT ('pan') made by West Indian musicians (particularly from Trinidad) out of discarded oil drums. Each DRUM can be tuned to play a range of notes by beating and heat-treating different sections of the 'head', i.e. the top of the drum.

Steinway a firm of piano manufacturers founded by Henry Steinway [originally Heinrich Steinweg] (1797–1871) in New York in 1853. A London branch opened in 1875.

stem the line, or 'tail', attached to the head of all notes smaller than a SEMIBREVE.

stesso (*Italian*) 'same', so *lo stesso* TEMPO means 'the same speed'.

stiacciato *see* relief.

Stijl, De a group of Dutch artists, founded to spread theories on ABSTRACT ART, principally through the *De Stijl* magazine, which was published 1917–28.

The group rejected the representational in art, believing that art's object was to convey harmony and order, achieved by the use of straight lines and geometrical shapes in primary colours or black and white. Their ideas had great influence, particularly on BAUHAUS, on architecture and on commercial art.

still life a genre of painting depicting inanimate objects such as fruit, flowers, etc, begun by Dutch artists seeking secular commissions after the Reformation and the loss of Church patronage. Within the genre, the *vanitas* still life contains objects symbolic of the transcience of life, e.g. skulls, hour-glasses, etc, while others contain religious symbols, such as bread and wine. In the 18th century, new life was given to the form by the French painter Jean-Baptiste Chardin (1699–1779) and in the 19th century ◊Cézanne's use of it in his experiments with structure influenced the CUBISTS.

stomp a BLUES composition in which the BEAT is literally stamped on the floor.

stop a handle or knob on an ORGAN that admits or prevents air from reaching certain PIPES and that can therefore be used to modify the potential output of the MANUALS and PEDALS.

stopping on stringed instruments, the placing of fingers on a string to shorten its effective length and raise its PITCH.

Stradivari *or* **Stradivarius** a violin made by Antonio Stradivari (1644-1737) an Italian violin-maker whose instruments are unsurpassed for the quality of their sound.

strathspey *see* REEL.

stream of consciousness a term coined by William ◊James in his *The Varieties of Religious Experience* (1902) which has been adapted by literary critics to denote a fluxive method of narration in which characters voice their feelings with no 'obtrusive' authorial comment and with no orthodox dialogue or decription. The term has particular reference to the work of ◊Joyce and ◊Woolf.

Streichquartett (*German*) a 'STRING QUARTET'.

strepitoso (*Italian*) 'boisterously'.

stretto 'close together', i.e. a quickening of TEMPO.

stride (piano) a JAZZ piano technique characterized by the use of single bass notes on the first and third beats and chords on the second and fourth.

string a vibrating cord for the production of tone, in the piano of drawn cast steel wire, in instruments of the violin family of catgut or spun silk, and in the guitar of catgut or wire.

stringendo (*Italian*) 'tightening', i.e. increasing tension, often with accelerated TEMPO.

string quartet a group of four performers who use stringed instruments (two VIOLINS, VIOLA and

CELLO); or a piece of music written for such a group.

strings a general term for the stringed instruments of the VIOLIN family.

Stroh violin a VIOLIN made of metal (invented by Charles Stroh in 1901) that incorporates a TRUMPET bell and does not have a normal violin body.

Structuralism in literary criticism, a critical approach to literature in which the text being studied is viewed as a 'cultural product' that cannot be 'read' in isolation and in which the text is held to absorb its meaning from the interconnected web of linguistic codes and symbols of which it is but a part. The process of studying the codes, etc., and their relation to each other, is called *Semiotics*. A Structuralist approach to the novels of Fennimore ◊Cooper, for example, would include recognition of the linguistic and cultural conventions underlying the author's use of language, with particular reference to the significance of both the 'Noble Savage' myth in Western culture and the emerging frontier myth in American culture. The heroic persona of the Deerslayer, and his portrayal as a transient figure between the mythic world of primitive America and the swelling wave of modern civilization, results in a creative tension ripe for hours of happy exploration and parallel.

Figures associated with the development of Structuralist theory are the Canadian critic Marshall ◊McLuhan, whose studies of mass culture and communication include *The Gutenberg Galaxy* (1962) and *The Medium is the Message* (1967); the linguist Ferdinand de ◊Saussure; the anthropologist Claude Levi-Strauss (1908–); the critic Roland ◊Barthes; and the psychoanalyst Jacques Lacan (1901–81). Structuralism is thus an approach drawing on a wide range of disciplines, with some critics, e.g. followers of Lacan, focusing on the play between unconscious and conscious concepts, while others, e.g. followers of Saussure, will focus on the linguistic relativism that emerges between the *signifier* (the spoken word) and the *signified* (the mind's concept of the word). *Deconstruction*, a concept developed by the French philosopher Jacques Derrida (1930–), is a term for the process or 'strategy' of examining the elements (signs) of language in isolation from other elements, thus exposing the contradictions inherent within language. It is also called *Poststructuralism*.

The psychologist Jean ◊Piaget usefully defines structure as composed of wholeness, transformation and self-regulation. Thus, ◊Homer's *Iliad*, for example, (a) is a work with a unity of structure conforming to the conventions of EPIC poetic form; (b) includes recognizable 'types' of characters who appear in other such works, e.g. warriors, who may also behave in ways outside the expected form, as when the young Achilles hides himself amongst women dressed as a girl; (c) can alter its meaning according to external factors, e.g. the reader's understanding of the characters' behaviour can vary at different times, as fresh experiences alter the reader's perception.

Stück (*German*) 'piece'.

study in art, a drawing or painting of a detail for use in a larger finished work. In music, a piece written to demonstrate technique in playing a musical instrument or using the voice.

subdominant the fourth NOTE of the MAJOR or MINOR SCALE.

subito (*Italian*) 'suddenly', as in PIANO *subito,* meaning 'suddenly soft'.

subject a musical THEME on which a composition (or part of a composition) is constructed, e.g. the first and second SUBJECTS in the EXPOSITION in SONATA FORM; the subject in a FUGUE; also the leading VOICE (first PART) of a fugue.

submediant the sixth NOTE of the MAJOR or MINOR SCALE.

subsidiary theme any THEME that is less important than the main theme(s) of a composition.

suite a collection of short pieces that combine to form an effective overall composition; the BAROQUE suite was a set of (stylized) dances.

sul, sull' (*Italian*) 'on' or 'over', so *sul ponticello* means 'over the BRIDGE' (in VIOLIN bowing).

Suprematism a Russian art movement based on principles of nonobjectivity. It was begun by the painter Kasimir Malevich (1878–1935) in 1913 and evolved on a parallel with Constructivism. *White on White* by Malevich is typical of the work of the movement. The influence of Suprematism spread through the BAUHAUS to Europe and the US.

Surrealism an avant-garde art movement of the 1920s and 1930s in France, which grew out of DADA and was inspired by the dream theories of Sigmund ◊Freud and by the literature and poetry of ◊Rimbaud and ◊Baudelaire. Surrealism took from Dadaism a love for the juxtaposition of incongruous images, the purpose of which in the Surrealist view was to express the workings of the unconscious mind. The term 'Surrealism' had been coined by the poet Guillaume ◊Apollinaire, but the movement really got going with the publication by the poet André Breton (1896–1966) of his first *Surrealist Manifesto* in 1924. According to Breton (who was much influenced by Freud), the 'higher reality' could only be achieved in art by freeing the mind from the lower world of superficial rationality.

In art, its influences include the works of de ◊Chirico. There were two main trends: *automatism*, or *free association*, was explored in the works of ◊Miró, ◊Ernst and others, who sought deliberately to avoid conscious control by using techniques of spontaneity to express the subconscious. The world of dreams was the source of inspiration for the incongruously juxtaposed, often bizarre, but precisely painted imagery of ◊Dali, and ◊Magritte.

suspension a device used in HARMONY, in which a NOTE sounded in one CHORD is sustained while a subsequent chord is played (or sung), producing a DISSONANCE that is then resolved.

sustaining pedal *see* PIANO.

swell organ a MANUAL on an ORGAN. The notes played on this manual can become louder and softer by the opening and closing of the shutters on the swell box, which encloses the PIPES.

swing a type of American POPULAR MUSIC of the 1935-45 era; it was played by BIG BANDS and had an insistent RHYTHM. Glenn ◊Miller and his ENSEMBLE were influential in its development.

Symbolism in literature, a French poetry movement of the late-19th century that rejected the dictates of both Realism and naturalism by seeking to express a state of mind by a process of suggestion rather than by attempting to portray 'objective reality'. As ◊Mallarmé put it, 'not the thing, but the effect produced'. Other prominent poets associated with Symbolism include ◊Verlaine and ◊Rimbaud. The movement has strong links with the world of Impressionist composers, such as ◊Debussy. Poets outside the French-speaking world who were influenced by Symbolism include T. S. ◊Eliot, ◊Pound, ◊Rilke, and, notably, ◊Yeats, whose superb epigram 'Three Movements' serves as an epitaph both for the movement and for his own involvement with it. Several important plays of the late 19th century, e.g. ◊Chekhov's *The Seagull* (1895) and ◊Ibsen's *The Master Builder* (1896), also display the influence of Symbolism.

In art, also in late 19th-century France, Symbolism represented a response to the intrinsically visual work of the Impressionists and fell into two distinct trends: some painters were inspired by the images of Symbolist Literature, while others, including ◊Gauguin, Van ◊Gogh and the NABIS explored the symbolic use of colour and line to express emotion.

sympathetic strings strings on certain instruments, such as the SITAR, which are not actually plucked or bowed, but which are set in sympathetic VIBRATION and produce a NOTE without being touched, when the same note is played on a 'melody' string.

symphonic poem *or* **tone poem** an orchestral composition, a form of PROGRAMME MUSIC, usually in one MOVEMENT, which attempts to interpret or describe an emotion, idea, or story. The term was coined by ◊Liszt.

symphony in essence, a prolonged or extended SONATA for an orchestra. Most symphonies have four MOVEMENTS (sections) that, although interrelated, tend to have recognized forms, for example, a quick first movement, a slow second movement, a MINUET third movement, and a vibrant fourth movement (FINALE).

Synchromism an art movement originating in the US in 1913 the members of which were concerned with the balanced arrangement of pure colour, or 'colours together'. The movement influenced a number of American painters.

syncopation an alteration to the normal arrangement of accented BEATS in a BAR. This is usually done by placing ACCENTS on beats or parts of a beat that do not normally carry an accent.

synthesizer an ELECTRONIC instrument, operated by a keyboard and switches, that can generate and modify an extensive range of SOUND.

tabla a pair of Indian DRUMS, beaten with the hands, which are often used to accompany the SITAR in classical Indian music.

table an alternative name for the upper surface, or BELLY, of instruments of the VIOLIN family.

tabor an early type of SIDE DRUM.

tace (*Italian*) 'silent'.

Tafelmusik (*German*) 'table music', i.e. music sung during a banquet as an entertainment.

tail the STEM attached to the head of a MINIM (half-note) or a smaller NOTE.

tail piece the piece of wood at the base of a VIOLIN to which the strings are attached.

tail pin the metal rod at the bottom of a CELLO or DOUBLE BASS, which can be pulled out to adjust the height of the instrument above the floor.

Takt (*German*) 'TIME', so *im Takt* means 'in time'.

talon (*French*) the 'nut' or heel of a BOW.

tambourin (1) a lively 18th-century piece in the style of a FOLK DANCE from Provence, usually in 2/4 TIME. (2) a narrow DRUM, played along with a PIPE, as the ACCOMPANIMENT to dancing.

tambourine a small, shallow DRUM with a single skin fastened over a circular frame. Small metal CYMBALS (jingles) are slotted into the frame and rattle when the instrument is shaken or beaten with the hand.

tampon a drumstick that has a head at each end, held in the middle to produce a DRUM ROLL.

tango a Latin-American dance in moderately slow 2/4 TIME, originating from Argentina. It makes use of syncopated RHYTHMS (*see* SYNCOPATION) and became popular in Europe in the 1920s.

tanto (*Italian*) 'so much', as in ALLEGRO *non tanto*, meaning 'quick, but not too quick'.

Tanz (*German*) 'dance'.

tap dance a dance in which the feet are used to tap out a RHYTHM. Tap dancing was made popular by performances in films by Fred ◊Astaire during the 1930s. Special shoes with steel plates at the toe and heel are usually worn.

tarantella a very fast, wild FOLK DANCE from southern Italy in 6/8 TIME, and gradually increasing in speed. ◊Chopin used the form in a concert piece.

tasto (*Italian*) the keyboard of a PIANO or the FINGERBOARD of a stringed instrument.

tattoo originally a night DRUM beat calling soldiers to their quarters, now a military display.

Te Deum laudamus (*Latin*) 'We Praise Thee, O God'; a Christian HYMN sung at MATINS. Numerous composers (e.g ◊Purcell, ◊Handel, ◊Verdi) have set it to music.

tempera a paint medium made by mixing colour pigments with egg. It was much used until the 15th century and the development of oil paint.

temperament the way in which INTERVALS between NOTES have been 'tempered', or slightly altered, in Western music so that the slight discrepancy in seven OCTAVES is distributed evenly over the range. In EQUAL TEMPERAMENT an octave is divided into twelve SEMITONES, which means that, for example, D SHARP is also E FLAT: this is a compromise, for strictly there is a marginal difference between D sharp and E flat.

tempo (*Italian*) 'TIME'. The time taken by a composition, therefore the speed at which it is performed, hence the pace of the BEAT. *A tempo* means 'in time'. It can also mean a movement of a SONATA or SYMPHONY, e.g. *il secondo tempo*, 'the second MOVEMENT'.

ten. (*Italian*) an abbreviation for TENUTO.

tenor (1) the highest adult male VOICE with a range an OCTAVE to either side of middle C. (2) as a prefix to the name of an instrument, it indicates the size between an ALTO member of the family and a BASS, for example tenor SAXOPHONE. (3) the RECITING NOTE in PSALM singing. (4) an obsolete term for a VIOLA (tenor VIOLIN).

tenor drum a DRUM, frequently used in MILITARY BANDS, between a SIDE DRUM and BASS DRUM in size and PITCH, and without snares.

tenor violin a VIOLONCELLO.

tenuto (*Italian*) 'held'; a term that indicates that a NOTE should be held for its full value, or in some cases, even longer.

ternary form a term applied to a piece of music divided into three self-contained parts, with the first and third sections bearing strong similarities.

ternary time *see* TRIPLE TIME.

terzett, terzetto (*Italian*) *see* TRIO.

tessitura (*Italian*) 'texture'; a term that indicates whether the majority of NOTES in a piece are high up or low down in the range of a VOICE (or instrument).

tetrachord a group of four NOTES.

theme the MELODY, or other musical material, that forms the basis of a work or a MOVEMENT and which may be varied or developed. It may return in one form or another throughout a composition.

thirty-second note (US) a DEMISEMIQUAVER.

thorough bass *see* FIGURED BASS.

thunder stick *or* **bull roarer** *or* **whizzer** an instrument consisting of a flat piece of wood fastened to a piece of string. When the piece of wood is whirled around the head, it creates a roaring sound.

tie a curved line that joins two NOTES of the same PITCH together, indicating that they should be played as one long note.

timbre (*French*) the quality of TONE, or the characteristic SOUND of an instrument.

time the rhythmic pattern (number of BEATS in a BAR) of a piece of music, as indicated by the TIME SIGNATURE. DUPLE TIME has two beats in a bar, triple time has three beats in a bar, and so on.

time-names *or* **rhythm-names** a French method of teaching TIME and RHYTHM, in which beats are given names, such as 'ta', 'ta-te', etc.

time signature a sign placed at the beginning of a piece of music that indicates the number and value of BEATS in a BAR. A time signature usually consists of two numbers, one placed above the other. The lower number defines the unit of measurement in relation to the SEMIBREVE (whole note); the top figure indicates the number of those units in a bar, for example, 3/4 indicates that there are three CROTCHETS (quarter notes) in a bar.

timpani *or* **kettledrums** the main orchestral PERCUSSION INSTRUMENTS, consisting of bowl-shaped shells over which the membrane is stretched. The shell is supported on a frame at the base of which, in 'pedal timpani', is the foot PEDAL that can alter the PITCH of the drum as it is played. The drum can also be tuned (*see* TUNING) by screws, which alter the tension of the membrane.

Tin Pan Alley the nickname given to West 28th

Street in New York, where the popular-song publishing business used to be situated. It consequently became a slang expression for the POPULAR MUSIC industry.

tin whistle *or* **penny whistle** a metal whistle-FLUTE, similar to a RECORDER but with six finger holes. It produces high-pitched sounds and is commonly used to play FOLK music.

toccata (*Italian*) 'touched'; a type of music for a keyboard instrument that is intended to show off a player's 'touch' or ability.

tonality the use of a KEY in a composition.

tondo the Italian word for 'round', used in art to denote a circular picture or sculpture.

tone (1) an INTERVAL comprising two SEMITONES, for example the interval between C and D. (2) (US) a musical NOTE. (3) the quality of SOUND, for example good TONE, SHARP tone, etc. (4) in PLAINSONG, a MELODY.

tone poem *see* SYMPHONIC POEM.

tonguing in the playing of a WIND INSTRUMENT, this means interrupting the flow of breath with the tongue so that detached NOTES are played, or the first note of a PHRASE is distinguished.

tonic the first NOTE of a MAJOR or MINOR SCALE.

tonic sol-fa a system of NOTATION and sight-singing used in training, in which NOTES are sung to syllables. The notes of the major scale are: DOH, RE, ME, FAH, SOH, LA, TE, DOH (doh is always the TONIC, whatever the KEY). The system was pioneered in England by John Curwen (1816–80) in the mid-19th century.

tosto (*Italian*) 'rapid', as in PIÙ *tosto*, 'quicker'.

trad jazz literally 'traditional JAZZ'; a term referring to the type of comparatively simple jazz, with a strong MELODY, as played in New Orleans, which preceded the development of BEBOP.

tragedy a form of drama in which a hero or heroine comes to a bad end. The cause of the protagonist's failure can be either a personal flaw or a circumstance beyond his or her control, or both. The earliest tragedies known, those by ◊Aeschylus, ◊Sophocles and ◊Euripides, are still among the greatest. The first critical study of the form is in ◊Aristotle's *Poetics*, where Aristotle defines tragedy as an imitation (*mimesis*) of a serious, complete action on a grand scale, 'grand' meaning a momentous action involving highly placed characters in society. The protagonist will make an 'error of judgment' (*hamartia*, often been rendered, a bit misleadingly, as 'tragic flaw') but the protagonist's main fault is as often the result of having to undertake a certain action at a certain time, as in any character defect; thus, in Sophocles'

Antigone, the heroine is in the position of having to choose between divine and human law. The protagonist is usually a good person, but not perfect, and progresses from happiness to misery. Another important concept in the *Poetics* is that of catharsis, the 'purging' (or purification, cleansing) of the emotions of pity and fear aroused in the spectators by the play.

The great tradition of the Greek tragedians was filtered through the plays of ◊Seneca to the dramatists of the Renaissance, although Renaissance tragedies, such as those of ◊Shakespeare, differ significantly from those of the past. The interplay in Shakespeare's tragedies between the heroic and the ironic or comic commonplace worlds, e.g. the banter between the rustic clown and Cleopatra at the end of *Antony and Cleopatra*, is profoundly foreign to the world of the *Poetics*, just as the pagan and fate-haunted world of the Greeks was ultimately alien to Renaissance dramatists brought up in the Christian tradition.

Several 20th-century dramatists, e.g. Arthur ◊Miller and ◊O'Neill, have tried, with debatable results, to adapt the form of Athenian tragedy to the modern stage. The most that can be said for these is that they may result in effects similar to those of the originals. The fact that such dramas invariably have as subtext the notion that this state of affairs is reformable, and the equally invariable ironic presentation of the protagonists as 'losers of history', rather than as victims of forces as permanent as they are merciless, puts such plays as Miller's *Death of a Salesman* at a further remove from the Athenian drama than, say, Shakespeare's *King Lear*, with its bleak vision of a fallen world in which bloody tyranny is an everpresent threat. *See also* DOMESTIC TRAGEDY, HEROIC TRAGEDY, JACOBEAN TRAGEDY, REVENGE TRAGEDY.

tragicomedy *see* COMEDY.

tranquillo (*Italian*) 'calm'.

transcription *see* ARRANGEMENT.

transition (1) the changing from one KEY to another during the course of a composition. (2) a PASSAGE linking two sections of a piece, which often involves a change of key.

transposing instruments instruments that sound NOTES different from those actually written down, e.g. a piece of music in E flat for the B flat CLARINET would actually be written in F.

transposition the changing of the PITCH of a composition. Singers sometimes ask accompanists to transpose a SONG higher or lower so that it is better suited to their voice range.

treble the highest boy's VOICE.

treble clef G CLEF on the second line of the STAVE, used for TREBLE VOICES and instruments of medium or high PITCH, such as VIOLINS, FLUTES, OBOES, CLARINETS, HORNS and TRUMPETS.

trecento the Italian term for the 14th century.

tremolando (*Italian*) 'trembling'.

tremolo (*Italian*) the rapid repetition of a single NOTE, or the rapid alternation between two or more notes.

triad a CHORD of three NOTES that includes a third and a FIFTH.

triangle a PERCUSSION INSTRUMENT comprising a thin steel bar bent into a triangle but with one corner left unjoined. It is normally struck with a thin metal bar.

trill *or* **shake** an ORNAMENT in which a NOTE is rapidly alternated with the note above. It is used in both vocal and instrumental pieces.

trio (1) a group of three performers, or a piece of music written for such a group. (2) the middle section of a MINUET, as found in SONATAS, SYMPHONIES, etc. It was originally a section scored for three PARTS.

triplet a group of three NOTES played in the time of two notes.

triptych a painting, usually an altarpiece, consisting of three hinged parts, the outer two folding over the middle section. *See also* diptych, polyptych.

tritone an INTERVAL consisting of three WHOLE TONES.

tromba marina a long, stringed instrument of the 15th century. It consisted of a long, tapered box with one string, mounted on top, which was played with a BOW; inside the box were some twenty SYMPATHETIC STRINGS.

trombone a BRASS instrument that has changed little for 500 years. The body of the instrument has a cylindrical bore with a bell at one end and a MOUTH-PIECE at the other. A U-shaped SLIDE is used for lengthening or shortening the tubing and therefore for sounding different NOTES. TENOR and BASS trombones are often used in orchestras.

trope (*Latin*) an addition of music or words to traditional PLAINSONG LITURGY.

troppo (*Italian*) 'too much', as in ALLEGRO *non troppo*, meaning 'fast but not too fast'.

troubadour a poet-musician of the early Middle Ages who originally came from the South of France and sang in the Provençal language.

trumpet a BRASS instrument that has a cylindrical bore with a funnel-shaped MOUTHPIECE at one end and a bell (flared opening) at the other. The modern trumpet has three valves (operated by PISTONS) which bring into play extra lengths of tubing and are therefore used to change the pitch of the instru-

ment. Trumpets are used in orchestras, JAZZ BANDS and MILITARY BANDS. 'Trumpet' is also a generic term used to describe any number of very different types of instrument that are found all over the world.

tuba a large BRASS instrument with a wide conical bore, a large cup-shaped MOUTHPIECE, and a large bell that faces upwards. It can have between three and five valves and comes in three common sizes: TENOR (EUPHONIUM), BASS and double bass. Tubas are found in orchestras and military bands.

tubular bells *see* BELLS.

tune a MELODY or AIR.

tuning the adjusting of the PITCH of an instrument so that it corresponds to an agreed note, for example, an orchestra will usually have all its instruments tuned to the note of A.

tuning fork a two-pronged steel device that, when tapped, will sound a single, 'pure' note. It was invented by John Shore in 1711 and is used to tune instruments, etc.

tutti (*Italian*) 'all'; in orchestral music, a *tutti* passage is one to be played by the whole orchestra.

twelve-note music *or* **twelve-tone system** a method of composition formulated and advanced by ◊Schoenberg. In the system, the twelve CHROMATIC NOTES of an OCTAVE can only be used in specific orders, called 'note rows'; no note can be repeated twice within a note row, and the rows must be used complete. In all, there are forty-eight ways in which a note row can be arranged (using INVERSION, RETROGRADE MOTION and inverted retrograde motion), and it is with note rows that compositions are constructed.

ukelele a small, four-stringed GUITAR that was developed in Hawaii during the 19th century. It was a popular music-hall instrument during the 1920s.

unison the sounding of the same NOTE or its OCTAVE by two or more VOICES or instruments.

Unities *see* NEOCLASSICISM.

up beat the upward movement of a conductor's BATON or hand, indicating the unstressed (usually the last) BEAT in a BAR.

upright *see* PIANO.

Urtext (*German*) 'original text'.

Utopian novel a form of NOVEL developed from *Utopia* (1516) a fantasy of a supposedly ideally organized state written by Sir Thomas ◊More. The work spawned a host of imitations throughout the centuries but the impact of 20th-century totalitarianism has lessened enthusiams for the form. Aldous ◊Huxley's *Island* (1962) is a modern example of a Utopian novel; the same author's *Brave New World* (1932) is an example of its opposite, a 'dystopia' or 'bad place'.

Utrecht School a movement in Dutch art begun by Gerrit van Honthorst (1590–1656), Hendrick Terbrugghen (1588–1629) and Dirck van Baburen (c.1595–1624), who were in Rome between 1610 and 1620 and were strongly influenced by ◊Caravaggio, whose style they took back to the Netherlands, thus influencing in turn such northern masters as ◊Vermeer and ◊Rembrandt.

V an abbreviation for VIOLINO, VOCE, VOLTA.

VA an abbreviation of VIOLA.

valse (*French*) *see* WALTZ.

valve a device attached to horns, trumpets and other brass instruments to lengthen or reduce the extend of tubing, hence lowering or raising the pitch respectively, to complete the scale.

vamp to improvise an ACCOMPANIMENT.

variation the modification or DEVELOPMENT of a THEME.

vaudeville (*French*) originally a type of popular, satirical SONG sung by Parisian street musicians. In the 18th century these songs (with new words) were incorporated into plays, and the word came to mean the last song in an opera in which each character sang a verse. In the 19th century stage performances with songs and dances were called 'vaudevilles', and the Americans used the term to describe music-hall shows.

veloce (*Italian*) 'fast'.

Venite (*Latin*) the first word of PSALM 95, '*Venite, exultemus Domino*' ('O come let us sing unto the Lord'), which is sung as a prelude to psalms at Anglican MATINS.

verismo (*Italian*) 'realism'; the term is used to describe a type of opera that was concerned with representing contemporary life of ordinary people in an honest and realistic way, e.g. *Cavalleria Rusticana* by ◊Mascagni.

verse (1) a line of poetry or a stanza of a poem. (2) a composition in METER, especially of a light nature. (3) a short section of a chapter in the Bible.

versification (1) the art of making verses. (2) the METER or verses of a poem.

Vespers the seventh of the Canonical Hours (services of the day) in the Roman Catholic Church. Many composers (such as ◊Mozart) have written musical settings for the service.

vibraphone *or* **vibes** an American instrument, similar to the GLOCKENSPIEL, which consists of a series of metal bars that are struck with mallets. Underneath the bars hang tubular resonators, containing small discs that can be made to spin by means of an electric motor. When the NOTES are sustained the spinning discs give the sound a pulsating quality.

vibration a term in ACOUSTICS for the wave-like MOTION by which a musical TONE is produced. Sound vibrations are mechanical; radio vibrations are electro-magnetic and inaudible.

vibrato (*Italian*) literally 'shaking', i.e. a small but rapid variation in the PITCH of a NOTE.

vierhändig (*German*) 'four-handed', i.e. a piano duet.

villanella (*Italian*) literally a 'rustic SONG', a popular PART-SONG of the 17th century.

Vingt, Les a group of 20 Belgian painters who exhibited together in Brussels for ten years from 1884. Their exhibitions also included works by innovative French painters, e.g. ◊Seurat, ◊Gauguin, ◊Cézanne and Van ◊Gogh.

viol a family of stringed instruments played with a BOW, which were widely used in the 16th and 17th centuries. The instruments came in several sizes and designs, but they all usually had six strings and FRETS. Although they were similar in appearance to members of the VIOLIN family, they were constructed differently and gave a much softer sound.

viola originally a general term for a bowed stringed instrument. However, it is now the name of the ALTO member of the VIOLIN family. It has four strings.

viola da braccio (*Italian*) literally an 'arm VIOL'; a generic term for any stringed instrument played on the arm. It came to mean a VIOLIN or VIOLA.

viola da gamba (*Italian*) literally a 'leg VIOL', a term originally used of those members of the viol family played vertically between the legs or on the lap, but it came to be used exclusively for the BASS viol.

viola d'amore (*Italian*) literally a 'love VIOL', i.e. a tenor VIOL with seven strings (instead of six) and seven or fourteen SYMPATHETIC STRINGS, so called because it had a particularly sweet TONE.

violin a stringed instrument, played with a BOW, which was introduced in the 16th century. It was developed independently of the VIOL from the medieval FIDDLE. It has no FRETS and just four strings. The violin family includes the violin itself (TREBLE), VIOLA (ALTO) and VIOLONCELLO or 'cello' (TENOR). The DOUBLE BASS developed from the double BASS VIOL, but it is now included in the violin family.

violoncello, cello the tenor of the VIOLIN family, dating from the 16th century. It is held vertically between the legs of the seated player, and the TAIL PIN rests on the ground. It has four strings, which are played with a BOW.

virginal a keyboard instrument dating from the 16th century in which the strings are plucked by quills. It was similar to the HARPSICHORD, except that it had an oblong body with strings running parallel to the keyboard. The word has also been used to describe any member of the harpsichord family.

virtuoso (*Italian*) a skilled performer on the VIOLIN or some other instrument. The word was formerly synonymous with 'amateur'.

vivace (*Italian*) 'lively'.

vivamente (*Italian*) 'in a lively way'.

vivo (*Italian*) 'lively'.

vocalization control of the VOICE and vocal sounds, and the method of producing and phrasing NOTES with the voice.

vocal score *see* SCORE.

voce (*Italian*) 'VOICE', as in *voce di petto*, 'CHEST VOICE'.

voice (1) the SOUND produced by human beings by the rush of air over the vocal chords, which are made to vibrate. There are three categories of adult male voice (BASS, BARITONE and TENOR); three female categories (CONTRALTO, MEZZO-SOPRANO and SOPRANO); and two boy categories (TREBLE and ALTO). (2) Parts in contrapuntal (*see* COUNTERPOINT) compositions are traditionally termed 'voices'.

volti subito (*Italian*) 'turn over quickly' (of a page).

voluntary (1) an improvised piece of instrumental music (16th century). (2) an ORGAN SOLO (sometimes improvised) played before and after an Anglican service.

Vorticism a short-lived English Cubist art movement devised by Wyndham ◊Lewis, who also edited the two issues of its magazine *Blast* (1914, 1915). *See* FUTURISM.

waltz a dance in triple TIME. Waltzes evolved in Germany and Austria during the late-18th century and became particularly popular in Vienna.

watercolour a paint medium of colour pigments mixed with water-soluble gum arabic. When moistened, a watercolour paint produces a transparent colour that is applied to paper, usually white, the paper showing through the paint.

wedding march a tune played at the start or end of a wedding service. The two most famous wedding marches are ◊Men-delssohn's 'Wedding March' from his incidental music to *A Midsummer Night's Dream*, and ◊Wagner's 'Bridal Chorus' (better known as 'Here Comes the Bride') from the opera *Lohengrin*.

whistle (1) a toy FLUTE. (2) the making of a musical sound with the lips and breath without using the vocal cords, the hollow of the mouth forming a RESONANCE BOX. Whistling PITCH is an OCTAVE higher than is generally supposed.

whizzer *see* THUNDER STICK.

whole note (US) a SEMIBREVE.

whole-tone scale a SCALE in which all the INTERVALS are whole-tones, i.e. two SEMITONES.

wind instrument a musical INSTRUMENT the SOUND of which is produced by the breath of the player or by means of bellows.

wit 'What oft was thought but ne'er so well expressed' (◊Pope, *An Essay on Criticism*). The term has had many meanings and shades of meanings through the years, and can denote either the thing itself or a notable practitioner of it. The main shifts of meaning to bear in mind are: (a) Elizabethan usage, meaning intelligence or wisdom; (b) early 17th-century usage, meaning ingenious thought, 'fancy', and original figures of speech, as in ◊Donne's verse and METAPHYSICAL POETRY; (c) the period relevant to Pope's definition, roughly from the mid-17th century to the last half of the 18th, which is discussed below; (d) 19th-century to modern times usage, meaning an amusing, perhaps surprising observation, usually involving paradox.

Wit as defined by Pope must be distinguished from what we now regard as humour: Wit can be malicious, humour is benevolent. ◊Swift is witty, ◊Addison and ◊Steele, who set the pattern for the future, are both witty and humorous. Wits tended to congregate in groups of like-minded intellectuals sharing similar political, social and religious views.

By the late half of the 18th century, the term largely ceased to be used to describe intellectual groups, implying as it did either heartless frivolity or political faction. Intellectual groupings became looser, although certainly no less formidable: a literary gathering in 1780 London, for example, could include figures as diverse as ◊Sheridan, Dr ◊Johnson, ◊Burke and Edward ◊Gibbon, all of whom could be described as sharp-witted heavyweight thinkers and controversialists, although not 'wits' in the previous sense of the term.

woodwind a term for a group of blown instruments that were traditionally made of wood (some of which are now made of metal, e.g. FLUTES, OBOES, CLARINETS and BASSOONS, etc.

xylophone a PERCUSSION INSTRUMENT made up of hardwood bars arranged like a keyboard on a frame. It is played by striking the bars with mallets. Xylophones used in orchestras have steel resonators suspended beneath each bar.

zarzuela a type of Spanish comic opera that has a satirical theme and includes dialogue. It usually comprises just one act.

zither the generic term for a range of stringed instruments. The European zither consists of a flat box which is strung with a variety of different kinds of string (up to 40). The player uses a PLECTRUM to play MELODIES on one set of strings while the fingers on the other hand pluck a series of open strings to form a DRONE ACCOMPANIMENT.

FAMOUS PEOPLE

This section of the *Family Encyclopedia* provides essential details of the lives of important people in history and the modern world. The entries are in A to Z order, and there are cross-references, indicated in SMALL CAPITALS, to enable you to relate a person to other major people, and therefore to events.

Achebe, Chinua (1930–) Nigerian novelist and poet whose work focuses on Ibo society and the legacy of colonialism. He won the Nobel Prize for literature in 1989.

Acheson, Dean [Gooderham] (1893–1971) American lawyer and statesman who was responsible for formulating and developing several important strands of American foreign policy, notably the MARSHALL Plan and the establishment of NATO.

Adams, Ansel [Easton] (1902–84) American photographer, noted for his detailed, deep-focus studies of American landscape.

Addison, Joseph (1672–1719) English essayist and poet. With his friend Richard STEELE he founded the influential magazine *The Spectator* in 1711.

Adenauer, Konrad (1876–1967) German statesman who was imprisoned twice by the Nazi regime (1934, 1944). As chancellor of West Germany (1949–63) he had a major role in world politics.

Adler, Alfred (1870–1937) Austrian psychiatrist. He was an associate of FREUD, whose emphasis on sexuality he rejected, founding a school of psychoanalysis based on the individual's quest to overcome feelings of inadequacy (the 'inferiority complex').

Aeschylus (524–456 BC) Greek dramatist regarded as the founder of Greek tragedy. Seven of his plays survive, including *Prometheus Bound* and the *Oresteia* trilogy.

Akihito *see* **Hirohito**.

Albee, Edward [Franklin] (1928–) American dramatist. His plays include *Who's Afraid of Virginia Woolf* (1962) and *A Delicate Balance* (1966).

Alexander VI, Pope *see* BORGIA, RODRIGO.

Alexander the Great (356–323 BC) Macedonian king. The pupil of ARISTOTLE, he inherited the kingdom of Macedon from his father **Philip II** (382–336 BC). He conquered Greece in 336, Egypt in 331, and the Persian Empire by 328. He extended his conquests to the east and defeated an Indian army in 326. He died in Babylon, and was buried in the city he founded, Alexandria.

Ali, Muhammad [Cassius Clay] (1942–) American boxer and world heavyweight champion (1964–67, 1974–78, 1978).

Allen, Woody [Allen Stewart Konigsberg] (1935–)

Alexander the Great

American film director, actor and writer noted for his satirical films about the neuroses of New York intellectuals.

Allende [Gossens], Salvador (1908–73) Chilean politician. He was elected president of his country in 1970, thus becoming the first freely elected Marxist president in Latin America. He was overthrown and killed in a coup that brought PINOCHET to power.

Amin [Dada], Idi (1925–) Ugandan dictator. He ruled Uganda from 1971 to 1979 (appointing himself president for life in 1976), but was overthrown.

Amis, Sir Kingsley (1922–95) English novelist and poet whose first novel *Lucky Jim* (1954), a satire on academic life, is a comic masterpiece. His later novels are darker in tone. He was the father of **Martin Amis** (1949–), who is also a novelist.

Amundsen, Roald (1872–1928) Norwegian explorer and navigator, leader of the first expedition to reach the South Pole in 1911.

Andersen, Hans Christian (1805–75) Danish writer, best known now for his fairy tales, e.g. 'The Emperor's New Clothes'.

Anderson, Carl David *see* HESS, VICTOR FRANCIS.

Andropov, Yuri [Vladimirovich] (1914–84) Soviet statesman. Former head of the KGB and president of the USSR (1983–84).

Angelou, Maya (1928–) American dramatist, poet and short-story writer. One of the leading black writers of the 20th century.

Antony, Mark [Marcus Antonius] (*c*.83–30 BC) Roman soldier who fought with Julius CAESAR in the Gallic wars, and after Caesar's assassination defeated Brutus and Cassius at the battle of Philipi (42). He deserted his wife for the Egyptian queen,

Cleopatra (69–30 BC), their forces being defeated by Augustus at Actium. He and Cleopatra committed suicide.

Apollinaire, Guillaume [Apollinaris Kostrowitzky] (1880–1918) French art critic and writer who had great influence among avant-garde artists and poets at the beginning of the 20th century. A friend of Picasso and champion of Cubism, he also supported Orphism and Futurism and originated the term Surrealism.

Aquinas, Thomas *see* Thomas Aquinas.

Aquino, [Maria] Corazon (1933–) Filipino politician. She was elected President of the Philippines in 1986, following the assassination of her husband Benigno Aquino, the most prominent opponent of Marcos. Aquino's administration survived three military coups and she was succeeded as president in 1992 by **Fidel Ramos** (1928–).

Arafat, Yasser (1929–) Palestinian leader, who helped found the anti-Israeli guerrilla force *Al Fatah*, and became chairman of the Palestine Liberation Organization in 1968. In 1993 he signed the Israeli-Palestinian Peace Accord with Israeli Prime Minister, Yitzhak Rabin, which conferred Palestinian autonomy in certain of the long-disputed occupied territories.

Aragon, Louis (1897–1982) French poet, essayist and novelist. One of the founders of both Dadaism and Surrealism.

Arden, John (1930–) English dramatist, regarded as one of the leading left-wing playwrights of his generation.

Aristophanes (*c*.448–380 BC) Greek comic dramatist, eleven of whose comedies survive. The objects of his satire ranged from politicians to his fellow dramatists (Euripides being one) and his plays are a valuable record of the intellectual debates of the day. The most popular of his plays now is *Lysistrata*.

Aristotle (384–322 BC) Greek philosopher. He taught at Plato's Academy for 20 years, and became tutor to Alexander the Great, and formed his own school (the Lyceum) in Athens in 335. His works, including *Nicomachean Ethics, Poetics* and *Politics*, were re-introduced to the Western world in the Middle Ages via Arabian scholarship and had a profound influence on almost every field of intellectual inquiry until the Renaissance.

Armstrong, [Daniel] Louis 'Satchmo' (1900–1971) American jazz trumpeter, singer and leader of many popular jazz bands. He had a genius for improvisation and became one of the best-loved entertainers of the 20th century. He also appeared in several films, including *High Society* (1956).

Louis Armstrong

Armstrong, Neil [Alden] (1930–) American astronaut. He commanded the Apollo 11 moon landing mission, in which he became the first man to walk on the moon.

Arp, Jean *or* **Hans** (1887–1966) German-born French sculptor and painter. A founder of Dadaism, his work was abstract in form.

Arthur (*fl.* 6th century AD) a possibly mythical Celtic warrior-king of post-Roman Britain, who may have organized resistance against the Saxon invaders. He is credited in legend with having won a battle over the Saxons at 'Mount Badon', and is supposed to have been buried at Glastonbury.

Ashcroft, Dame Peggy (1907–91) English actress. One of the most popular stage and film actresses of her generation, she won an Oscar for her role in *A Passage to India* (1984).

Ashdown, Paddy [Jeremy John Dunham Ashdown] (1941–) English Liberal politician. He was elected leader of the Liberal and Social Democratic Party in 1988.

Ashe, Arthur (1943–93) American tennis player. The first black tennis player to win the US open (1968), the Australian open (1970), and the Wimbledon men's competition (1975).

Ashton, Sir Frederick [William Mallandaine] (1906–88) Ecuadorian-born British choreographer and co-founder of the Royal Ballet. His ballets include *A Month in the Country* (1976).

Asquith, Herbert Henry [1st Earl of Oxford and Asquith] (1852–1928) British statesman. He was leader of the Liberal Party (1908–26) and prime minister (1908–16).

Astaire, Fred [Frederick Austerlitz] (1899–1987) American dancer, singer and actor. His partnership with **Ginger Rogers** [Virginia McMath] (1911–95) resulted in a series of classic song-and-dance films, e.g. *Top Hat* (1935).

Astor, Nancy Witcher [Langhorne] [Viscountess Astor] (1879–1964) American-born British politician. She was elected to parliament as a Conservative MP, becoming the first woman to take her seat in Parliament.

Atatürk, Kemal [Mustafa Kemal Atatürk] (1881–1938) Turkish general and statesman, regarded as the creator of the modern Turkish state.

Attenborough, Sir David [Frederick] (1926–) English naturalist and broadcaster. He is Richard ATTENBOROUGH's brother.

Attenborough, Sir Richard [Samuel] (1923–) English film director, producer and actor. He was created a life peer in 1993.

Attlee, Clement [Richard] [1st Earl Attlee] (1883–1967) British statesman. Leader of the Labour Party (1939–55) and prime minister (1945–51), his 1945 administration introduced widespread nationalization and a programme of social security reforms.

Auden, W[ystan] H[ugh] (1907–73) English-born American poet. The leading left-wing poet of his generation, he later drifted away from Marxism towards a Christian and socially conservative position.

Augustine, Saint [Augustine of Hippo] (354–430) Latin Church Father. Born in what is now Tunisia, his father was a pagan and he was brought up a Christian by his mother. Reacting against the licentious life he led in Carthage, he converted to Manichaeanism for a while before returning to the Church. He was Bishop of Hippo (396–430) and wrote a spiritual autobiography, *Confessions*, and *City of God*, a major work of Christian apologetics.

Augustine, Saint (d. 604) Italian monk. He was dispatched to Britain in 597 by Pope GREGORY I to convert the Anglo-Saxons to Christianity and impose the authority of Rome on the Celtic Church. He became the first Archbishop of Canterbury in 601.

Augustus [originally Gaius Octavianus] (63 BC–14 AD) Roman emperor. After adoption by his uncle Julius CAESAR in 44 BC, he took the name Gaius Julius Caesar Octavianus. He became the first emperor of Rome in 31 BC after defeating Mark ANTONY.

Aurelius, Marcus (121–80) Roman emperor and philosopher. Renowned by his contemporaries for his nobility and learning, he spent much of his reign in war against the incoming 'barbarians' in the eastern part of the Empire.

Austen, Jane (1775–1817) English novelist. Her six great novels, *Sense and Sensibility* (1811), *Pride and Prejudice* (1813), *Mansfield Park* (1814), *Emma* (1816) and *Northanger Abbey* and *Persuasion* (1818) are set within the confines of the society in which she lived, the well-bred es-

sentially rural middle class of Regency England. She is renowned for her masterly dialogue, finely tuned satire and moral sense.

Ayckbourn, Alan (1939–) English dramatist noted for satirical comedies including *The Norman Conquests* (1974).

Ayer, Sir A[lfred] J[ules] (1910–89) English philosopher, whose work is based on 'logical positivism' and the rejection of metaphysics and has been highly influential on British 'common-sense' philosophy.

Baade, Wilhelm Heinrich Walter (1893–1960) German-born American astronomer who made a valuable contribution to the understanding of gallactic and stellar evolution.

Baader, Andreas *see* MEINHOFF, ULRIKE.

Babbage, Charles (1792–1871) English mathematician. His primitive 'calculating machines' are regarded as the precursors of the modern computer.

Bacall, Lauren *see* BOGART, HUMPHREY.

Bach, Johann Sebastian (1685–1750) German composer. His works include some of the greatest music in several forms, e.g. his six cello suites, choral masterpieces such as the *St Matthew Passion*, and works for harpsichord, clavichord and the organ, such as *The Well-Tempered Clavier* and *The Art of Fugue*. Four of his sons were also composers: **Wilhelm Friedemann** (1710–84), **Karl Philipp Emanuel** (1714–88), **Johann Christoph Friedrich** (1732–95), and **Johann** [John] **Christian** (1735–82), who became a court musician in London, where he was known as 'the English Bach', and influenced MOZART.

Bacon, Francis (1561–1626) English philosopher and statesman. He served both Elizabeth I and her successor James VI and I in various offices until his conviction and disgrace for bribery in 1621. His writings on philosophy and the need for rational scientific method are landmarks in the history of human thought.

Bacon, Francis (1909–92) Irish-born British painter. His controversial works feature twisted and contorted human shapes, often in weird landscapes or spaces, reflecting a personal view of repulsion at human condition.

Baez, Joan (1941–) American folksinger, renowned for her 'protest songs' on civil rights and the Vietnam war in the 1960s. She published her autobiography, *Daybreak*, in 1968 and has since published other writings.

Bailey, David (1938–) British photographer who came to prominence in the 1960s and is famous for his images of that period.

Baird, John Logie (1888–1946) Scottish engineer

who invented a mechanically scanned system of television in the mid-1920s.

John Logie Baird

Baker, Dame Janet [Abbott] (1933–) English mezzosoprano who became one of Britain's most popular opera singers in the 1960s and had parts created for her by composers such as BRITTEN. She was appointed a DBE in 1976.

Baldwin, James [Arthur] (1924–87) American novelist, dramatist and essayist. His main concern was with the role of blacks in American society, and, to a lesser extent, that of homosexuals. His novels include *Go Tell it on the Mountain* (1953).

Baldwin, Stanley [1st Earl Baldwin of Bewdley] (1867–1947) British statesman and Conservative prime minister (1923–24, 1924–29, 1935–37) Notable aspects of his premierships include the passing of a state of emergency during the 1926 General Strike, his refusal to accept Wallis Simpson as Edward VIII's wife (*see* WINDSOR) and his perceived failure to deal with the rise of European totalitarianism.

Balzac, Honoré de (1799–1860) French novelist. His great collection of novels and short stories describe the lives of French men and women of every class. His works include *La Comédie humaine* ('the Human Comedy').

Bancroft, Anne *see* BROOKS, MEL.

Banda, Hastings [Kamuzu] (1905–) Malawi statesman, first president of Nyasaland from 1963 and president of Malawi (formerly Nyasaland) from 1966. One of the longest-ruling leaders in the world, he was appointed president for life in 1971.

Banks, Sir Joseph (1743–1820) English botanist and explorer. He sailed with COOK on his 1768–71 voyage round the world as a representative of the Royal Society and discovered many species of animals and plants.

Bannister, Sir Roger (1929–) British athlete and doctor who became the first man run a mile in less than four minutes in May 1954, when he recorded a time of 3 minutes and 59.4 seconds at Oxford.

Banting, Sir Frederick [Grant] (1891–1941) Canadian physician whose research into diabetes with the American physiologist **Charles Herbert Best** (1899–1978) resulted in the isolation of the hormone insulin in a form suitable for treating diabetes.

Barber, Samuel (1910–81) American composer who worked with recognizably 19th-century harmonies and forms. His works include the popular *Adagio for Strings* (1936).

Bardeen, John (1908–91) American physicist and electrical engineer. He won two Nobel prizes, the first in 1956 for research that led to the invention of the transistor, the second in 1972 for research into the theory of superconductivity.

Bardot, Brigitte (1934–) French actress. The leading 'sex symbol' of the 1950s. Since the 1970s she has devoted herself to animal rights and welfare.

Barenboim, Daniel (1942–) Argentinian-born Israeli concert pianist and conductor. He married the cellist Jacqueline DU PRÉ in 1967 and was awarded the Legion of Honour in 1987.

Barnard, Christian [Neethling] (1922–) South African surgeon, who performed the first heart transplant in 1967.

Barrie, Sir J[ames] M[atthew] (1860–1937) Scottish dramatist and novelist, remembered principally for *Peter Pan* (1904).

Barth, Karl (1886–1968) Swiss Protestant theologian whose theology was based on an orthodox 'theocentric' conception of divine grace and is seen as a reaction against the simplifications of 19th-century liberal theology. He was a committed and courageous opponent of Nazism.

Barthes, Roland (1915–80) French literary and cultural critic, whose semiological studies were regarded as necessary reading in the 1960s and 70s. His works include *Elements of Sociology*.

Bartók, Béla (1881–1945) Hungarian composer and pianist. He was a noted collector of folk songs, upon which many of his works are based. His works include the ballet *The Miraculous Mandarin*, string quartets and *Concerto for Orchestra*.

Basie, Count [William] (1904–84) American jazz composer and bandleader. He was a jazz pianist of great ability, and his big band featured singers such as Ella FITZGERALD and Frank SINATRA.

Bates, H[erbert] E[rnest] (1905–74) English novelist and short-story writer. His works include the comic novel *The Darling Buds of May* (1958) featuring the Larkins, an unruly farming family.

Baudelaire, Charles (1821–67) French poet noted for his fascination with the macabre and alleged Satanism. He became a leading Symbolist poet and his works include *Les Fleurs du mal*.

Beadle, George Wells *see* TATUM, EDWARD LAWRIE.

Beaumont, Sir Francis *see* FLETCHER, JOHN.

Beauvoir, Simone de (1908–86) French novelist and essayist whose works explore the female predicament from the standpoint of existential feminism, e.g. *The Second Sex* (1949)

Simone de Beauvoir

Beaverbrook, Max [William Maxwell Aitken, 1st Baron Beaverbrook] (1879–1964) Canadian-born British newspaper proprietor and Conservative politician. In World War I he served as minister of information (1918) and in World War II as minister of aircraft production (1940–41).

Bechet, Sidney (1897–1959) American jazz saxophonist and clarinettist. He never learned to read music but became one of the greatest soprano saxophone virtuosos of the 20th century.

Becket, Thomas à (1118–70) English saint. Of Norman descent, he became Chancellor of England in 1155 and Archbishop of Canterbury in 1162. Relations between Becket and Henry II deteriorated due to the former's strong allegiance to Church rather than king, and Becket was murdered by four of the king's knights. Becket was canonized in 1173.

Beckett, Samuel (1906–89) Irish dramatist and novelist. His works, generally bleak and existentialist in philosophy, include the play *Waiting for Godot* (1952) and the short novel *Malone Dies* (1951).

Becquerel, Antoine Henri *see* CURIE, MARIE.

Bede, the Venerable, Saint (*c.*673–735) Anglo-Saxon monk and historian. Prodigiously learned, he settled for life in the monastery at Jarrow in 682.

Beecham, Sir Thomas (1879–1961) English conductor, noted for his interpretations of the works of DELIUS, STRAUSS and SIBELIUS, and for his sharply witty ripostes.

Beerbohm, Sir [Henry] Max[imilian] (1872–1956) English parodist, caricaturist and essayist. The parodies of authors such as Henry JAMES and

KIPLING established him as one of the greatest of all parodists. His only novel is *Zuleika Dobson* (1912).

Beethoven, Ludwig van (1770–1827) German composer. Regarded as the greatest Romantic composer, he became famous throughout Europe in the 1790s as a brilliant pianist with a special gift for improvisation. His works include a violin concerto, nine symphonies, piano sonatas, string quartets, masses, and one of the greatest operas, *Fidelio* (1805).

Begin, Menachem (1913–92) Polish-born Israeli statesman. He was commander of the Irgun militant Zionist group (1943–48) and prime minister of Israel (1977–84). He and SADAT were awarded the Nobel Peace Prize in 1978, after Egypt and Israel signed a peace treaty.

Behan, Brendan (1923–64) Irish dramatist and poet. He was an Irish Republican Army supporter from an early age, and two of his works were directly based on his imprisonment for IRA activity, *The Quare Fellow* (1954) and *Borstal Boy* (1958).

Beiderbecke, [Leon] Bix (1903–31) American jazz cornetist, pianist and composer, who is regarded as one of the few white jazz musicians to have had any significant influence on the development of jazz.

Bell, Alexander Graham (1847–1922) Scottish-born American inventor and scientist. He succeeded in producing a device for transmitting the voice in 1875, and patented the telephone the following year. He founded the Bell Telephone Company in 1877, and patented the gramophone in 1887.

Alexander Graham Bell

Bellini, Giovanni (*c.*1430–1516) Italian painter. The most pro-minent of a family of noted artists, including his father **Jacopo** (*c.*1400–*c.*1470) and his brother **Gentile** (*c.*1429–*c.*1507).

Belloc, Hilaire (1870–1953) French-born English poet, essayist, historian and Liberal MP noted for his prolific output of all kinds of books, and for his robust Roman Catholicism, anti-imperialism and nationalism.

Bellow, Saul (1915–) Canadian-born American novelist, widely regarded as one of the greatest living writers. His novels include *Dangling Man* (1944) and *More Die of Heartbreak* (1987). He was awarded the Nobel prize for literature in 1976.

Ben Bella, [Mohammed] Ahmed (1916–) Algerian statesman. A leading figure of his country's independence movement in the late 1940s and 1950s, he became prime minister in 1962 shortly after independence. He was deposed in 1965 following a military coup and was imprisoned until 1980.

Ben-Gurion, David [David Gruen] (1886–1973) Polish-born Israeli statesman. He settled in Palestine in 1906, where he was active in the socialist wing of the Zionist movement. He was the first prime minister of Israel (1948–53) and was prime minister again, 1955–63.

Benn, Tony [Anthony Neil Wedgwood Benn, formerly Viscount Stansgate] (1925–) British politician. Since the late 1970s he has been regarded as one of the leading figures of the radical left in the Labour Party.

Bennett, [Enoch] Arnold (1867–1931) English novelist, dramatist and essayist. His most popular novels centred on industrial life in the Black Country Potteries and include the Clayhanger trilogy, *Clayhanger*, *Hilda Lessways* and *These Twain*.

Bentham, Jeremy (1748–1832) English philosopher. He is famous for his proposition that the prime aim of political and philosophical inquiry should be the 'greatest happiness of the greatest number', expounded in his *Introduction to the Principles of Morals and Legislation* (1789).

Bentine, Michael *see* MILLIGAN, SPIKE.

Bentley, Edmund Clerihew (1875–1956) English journalist, noted for his classic detective novel, *Trent's Last Case* (1913), and his invention of the clerihew.

Berenson, Bernard (1865–1959) Lithuanian-born American art critic. Highly influential in the early development of art history.

Berg, Alban (1885–1935) Austrian composer. He studied under SCHOENBERG and his atonal twelve-tone technique. His works include songs, chamber works and the operas *Wozzeck* and *Lulu*.

Bergman, Ingmar (1918–) Swedish film and stage director. His films, which include *The Seventh Seal* (1956), *Wild Strawberries* (1957) and *Cries and Whispers* (1972), are claustrophobic psycho-

logical studies which have been very influential. *Fanny and Alexander* (1982) won an Oscar for Best Foreign Language Film.

Bergman, Ingrid (1915–82) Swedish actress regarded as one of the most talented and beautiful actresses of her generation. Her films include *Casablanca* (1942).

Ingrid Bergman

Bergson, Henri Louis (1859–1941) French philosopher whose writings expound his theory of a 'vital spirit' moving in the world, bridging the apparent chasm between metaphysics and science. He was awarded the Nobel prize for literature in 1927.

Beria, Lavrenti Pavlovich (1899–1953) Georgian-born Soviet politician. He rose to power in the 1930s under STALIN, and became head of the secret police (1938–53). After Stalin's death, he was tried for treason and executed.

Berio, Luciano (1925–) Italian composer whose works are based on a system of serialism and often feature electronic components.

Berkeley, Busby [William Busby Enos] (1895–1976) American film director, noted especially for his elaborate, often surreal, dance choreography.

Berlin, Irving [Israel Baline] (1888–1989) Russian-born American songwriter. He began his career as a street singer, and eventually wrote around a thousand songs, many of which featured in highly successful shows and in several film musicals.

Berlin, Sir Isaiah (1909–) Latvian-born British philosopher and historian. His works focus on the history of ideas, with particular reference to historical determinism.

Berlioz, Hector (1803–69) French composer. Regarded as a founder of modern orchestral techniques, his works include *Symphonie Fantastique* and the opera *The Trojans*.

Bernstein, Leonard (1918–90) American composer and conductor. He was musical director of the New York Philharmonic (1958–70), and a tireless popularizer of classical music. His works include chamber and choral music and several very popular musicals, e.g. *West Side Story* (1957).

Bertolucci, Bernardo (1940–) Italian film director. His films,which include *The Conformist* and *The Last Emperor*, are among the most influential in modern cinema.

Best, Charles Herbert *see* BANTING SIR FREDERICK.

Best, George (1946–) Northern Irish soccer player. One of the world's finest and most entertaining wingers, his carer slowly folded in a haze of alcohol abuse and general dissipation.

Betjeman, Sir John (1906–84) English poet and essayist whose work was popular with critics and public alike. He was appointed poet laureate in 1972 and also wrote widely on architecture.

Bevan, Aneurin (1897–1960) Welsh statesman. He was Labour MP for Ebbw Vale for 30 years (1929–60), and one of the main spokespeople for the radical socialist opposition during World War II. As minister of health (1945–51), he oversaw the formation of the welfare state (*see* BEVERIDGE).

Beveridge, William Henry [1st Baron Beveridge] (1879–1963) Indian-born English economist. His 'Beveridge Report', the *Report on Social Insurance and Allied Services* (1942), became the basis for the welfare state introduced by ATTLEE's administration.

Bevin, Ernest (1881–1951) English trade unionist and statesman. He helped found the Transport and General Workers Union (1922) and was minister of labour (1940–45) in the coalition war government and Labour's foreign secretary (1945–51).

Bhutto, Zulfikar Ali (1928–79) Pakistani statesman. He was the first civilian president of Pakistan (1971–73), then prime minister (1973–77). He was deposed in a coup and executed. His daughter, **Benazir Bhutto** (1953–), was prime minister of Pakistan (1988–90), was re-elected in 1993 before being defeated in the 1997 Pakistani election.

Zulfikar Ali Bhutto

Biko, Steve [Bantu Stephen Biko] (1947–77) South African black radical leader who helped found the Black People's Convention in order to build confidence in South African blacks that they could defeat apartheid. His death in police custody while awaiting trial was universally regarded as murder.

Birtwistle, Sir Harrison (1934–) English composer regarded with Maxwell DAVIES and others as a leading postwar composer. His works include *Gawain and the Green Knight* (1991)).

Bismarck, Prince Otto von (1815–98) German statesman who was prime minister of Prussia (1862–90) and defeated first Austria during the 'Seven Weeks' War' (1866) and then France (1870–71). He became the first chancellor of united Germany and was dubbed the 'Iron Chancellor'.

Bizet, Georges (1838–75) French composer who is best known for his operas, e.g. *The Pearl Fishers*, *Carmen*, but who also wrote a symphony, several songs and the *L'Arlésienne* suits.

Blake, William (1757–1827) English poet and artist. His main poetic and artistic theme was innocence crippled by cynical experience. He is one of the greatest of English Romantic poets.

Blériot, Louis (1872–1936) French aviator and aeronautical engineer. A pioneer in aircraft design, he made the first flight across the English Channel in one of his monoplanes in 1909.

Bliss, Sir Arthur [Edward Drummond] (1891–1975) English composer whose work includes a choral symphony, ballets and film music. He was Master of the Queen's Music (1953–75).

Blum, Léon (1872–1950) French statesman, the first socialist and Jewish prime minister of France (1936–37, 1938, 1946–47).

Blunt, Anthony [Frederick] (1907–83) English art historian, who was knighted in 1956, and was appointed Surveyor of the Queen's Pictures in 1945, a post he held until 1972. He was stripped of his knighthood in 1979, following the public revelation that he had been a Soviet spy.

Blyton, Enid [Mary] (1897–1968) English children's writer whose most well-known books have featured such characters as Noddy and Big Ears and, for older children, *The Famous Five*.

Boas, Franz (1858–1942) German-born American anthropologist whose emphasis on linguistic structure and scientific methodology has been very influential on anthropology.

Boccioni, Umberto (1882–1916) Italian painter and sculptor who became the leading Futurist artist of the early 20th century and one of the movement's principal theorists.

Bogarde, Sir Dirk [Derek Jules Gaspard Ulric Niven van den Bogaerde] (1920–) English actor and author whose films include *Death in Venice* (1970). Since the early 80s he has concentrated on writing, producing several volumes of autobiography and several novels.

Bogart, Humphrey [De Forest] (1899–1957) American actor. He formed one of the best-known screen partnerships with his (fourth) wife **Lauren Bacall** [Betty Joan Perske] (1924–).

Bohr, Niels [Henrik David] (1885–1962) Danish physicist. He was the first to apply quantum theory to explain the stability of the nuclear model of the atom. He was awarded the 1922 Nobel prize for physics.

Bolivar, Simon (1783–1830) Venezeulan-born revolutionary. He overthrew Spanish rule in Venezuela, Ecuador, Colombia and Peru. Upper Peru was renamed Bolivia in his honour.

Bond, Edward (1934–) English dramatist and screenwriter. His work was often controversial and public debate over one his play *Saved* (1965) led to the abolition of stage censorship in Britain.

Bonhoeffer, Dietrich (1906–45) German Lutheran pastor and theologian who was active in the anti-Nazi Resistance during World War II and hanged by the Gestapo in 1945. His writings are among the key spiritual works of the 20th century.

Bonnard, Pierre (1867–1947) French painter and lithographer. His work is notable for being intensely colourful, especially in his interior work.

Booth, William (1829–1912) English religious leader who established a Christian mission in London's East End in the 1860s and founded the Salvation Army in 1878.

Borg, Bjorn (1956–) Swedish tennis player who won five consecutive Wimbledon championship titles (1976–80).

Borges, Jorge Luis (1899–1986) Argentinian short-story writer, poet and critic who had a remarkable gift for creating short fictions with a beguiling metaphysical content.

Borgia, Cesare (1476–1507) Italian soldier and politician. The son of **Rodrigo Borgia** (1431–1503), he became a cardinal in 1493 after his father became pope (as Alexander VI) in 1492. He attempted to bring Italian affairs under his control in an atmosphere of intrigue, war and assassination, and was the model for MACHIAVELLI's *Prince*. His sister **Lucrezia** (1480–1519) was a patron of the arts and also acquired a reputation for conspiracy.

Bosch, Hieronymous [Jerome van Aeken *or* Aken] (*c*.1450–1516) Dutch painter, known for his fantastic and often grotesque allegorical paintings which use imagery drawn from folk tales and religious symbolism.

Bose, Sir Jagadis Chandra (1858–1937) Indian physicist and plant physiologist. He invented the crescograph, a device that automatically records plant movements.

Bose, Subhas Chandra (1897–1945) Indian nationalist leader. He was president of the Indian National Congress (1938–39) and, in collaboration with the Japanese during World War II, organized the Indian National Army to combat British rule in India.

Boswell, James *see* JOHNSON, SAMUEL.

Botha, Louis (1862–1919) South African general and statesman. As general of the Transvaal army, he led the Boer forces against the British during the Boer War. He supported the Allies during World War I, and became first prime minister of South Africa (1910–19).

Botha, P[ieter] W[illem] (1916–) South African politician. As prime minister (1978–84) and then president (1984–89), he introduced limited reforms of apartheid.

Botham, Ian (1955–) English cricketer. A talented all-rounder, he captained England (1980–81) and scored 5,057 runs in Test matches (including 14 centuries).

Botticelli, Sandro (1444–1510) Florentine painter, best known for his graceful and serene religious works.

Boulez, Pierre (1925–) French composer and conductor who developed a composition style based on total serialism and electronic instruments.

Boycott, Geoffrey (1940–) English cricketer. He is regarded as one of England's greatest modern batsmen. He captained Yorkshire (1970–78) and played for England (1964–74, 1977–81).

Bradman, [Sir] Don[ald George] (1908–) Australian cricketer. A brilliant batsman, he scored 117 centuries during the 1930s and 1940s. He was Australian captain (1936–48).

Don Bradman

Bragg, Melvyn (1939–) English novelist and broadcaster. He is best known as a television presenter of arts programmes.

Brahms, Johannes (1833–97) German composer. He regarded himself as firmly in the Classical (as opposed to Romantic) tradition. His works include the great choral *German Requiem* (1869), four symphonies and chamber music.

Branagh, Kenneth (1961–) Irish-born British actor and director. He founded the Renaissance Theatre Company in 1986 and has appeared regularly in films with his wife, the actress **Emma Thompson** (1959–).

Brando, Marlon (1924–) American actor. Brando's many celebrated screen performances display the highly influential method acting style.

Brandt, Willy [Herbert Ernst Karl Frahm] (1913–92) German statesman. He was active in the German Resistance during World War II and became mayor of Berlin (1957–66) and chancellor of West Germany (1969–74). He was awarded the Nobel Peace Prize in 1971.

Braque, Georges (1882–1963) French painter. The term Cubism was coined in 1909 to describe his works. He also pioneered the use of collage in modern painting.

Brattain, Walter *see* SHOCKLEY, WILLIAM BRADFORD.

Braudel, Fernand (1902–85) French historian. His influential works focused on socio-economic trends and the changing relationship between man and the environment rather than on politics or military events.

Brecht, Bertolt (1898–1956) German dramatist who devised the theatre of alienation which he used in *The Threepenny Opera* (1928), for which WEILL wrote the music, and developed in plays such as *Galileo* (1938) and *Mother Courage* (1941). He settled in the US during Nazi rule in Germany, returning to East Berlin in 1949 where he founded the Berliner Ensemble theatre.

Breughel *see* BRUEGHEL.

Brezhnev, Leonid Ilyich (1906–82) Soviet statesman. He helped organize KHRUSHCHEV's downfall in 1964, and became general secretary of the Communist Party (1977–82) and Soviet president (1977–82). The period of his rule is now described as the 'period of stagnation' in the USSR.

Brian, [William] Havergal (1876–1972) English composer, frequently described as 'post-Romantic'. His more than 30 works include the huge *Gothic Symphony* (1919–27) and five operas.

Britten, [Edward] Benjamin [1st Baron Britten] (1913–76) English composer and pianist. His

works include the operas *Peter Grimes* (1945) and *A Midsummer Night's Dream* (1960, chamber music, orchestral works and song cycles. His works are noted for their romantic lyricism. He was created a peer in 1976.

Brontë, Anne (1820–49), **Charlotte** (1816–55) and **Emily** (1818–48) English novelists and poets. Charlotte's *Jane Eyre*, based on her experiences as a teacher and governess, was published in 1847, and Anne's *Agnes Grey* and Emily's *Wuthering Heights* followed in 1848. Anne's *Tenant of Wildfell Hall* was published in 1848, in which year Emily died of consumption. Charlotte wrote two more novels: *Shirley* (1849) and *Villette* (1853).

Brook, Peter [Stephen Paul] (1925–) English stage and film director based in Paris. He is regarded as one of the finest experimental directors of the modern era.

Brooke, Rupert [Chawner] (1887–1915) English poet. His war poems were very popular with the public for their idealized vision of the nobility of war.

Brooks, Mel [Melvin Kaminsky] (1926–) American comedian, film writer and director, best known for his fast-moving, irreverent comedy films. His wife is the actress **Anne Bancroft** [Anna Maria Italiano] (1931–) who is noted for her serious acting.

Brown, George MacKay *see* DAVIES, SIR PETER MAXWELL.

Browning, Robert (1812–89) English poet renowned for his innovative experiments in form and narrative skill. His wife, **Elizabeth Barrett Browning** (1806–61), whom he married in 1846, was also a major poet.

Brubeck, Dave (1920–) American jazz composer and pianist. He studied musical composition with SCHOENBERG and MILHAUD, forming his 'Dave Brubeck Quartet' in 1951.

Bruce, Lenny (1925–66) American comedian who developed an influential style of satirical and often scabrous comedy which frequently brought him into conflict with the authorities.

Bruegel *or* **Brueghel, Pieter (the Elder)** (*c*.1525–69) Flemish painter who painted peasant scenes and allegories. He is best known for his magnificent landscape painting *The Hunters in the Snow* (1565). His sons **Jan Brueghel** (1568–1625) and **Pieter Brueghel (the Younger)** (*c*.1564–1637) were also painters.

Brunel, Isambard Kingdom (1806–59) English engineer. He designed steamships and in the late 1820s planned and designed the Clifton Suspension Bridge. His father, **Sir Marc Isambard**

Brunel (1769–1849), was a French engineer who designed a tunnel under the Thames in London.

Isambard Kingdom Brunel

Buber, Martin (1878–1965) Austrian-born Jewish theologian and existentialist philosopher. His philosophy, which centres on the relationship between man and God, has had a large impact on both Jewish and Christian theology.

Buchan, John [1st Baron Tweedsmuir] (1875–1940) Scottish novelist, statesman and historian. He wrote several best-selling adventure novels, was created a peer in 1935, and was appointed governor-general of Canada (1935–40).

Buchman, Frank [Nathan Daniel] (1878–1961) American evangelist who founded the Oxford Group and the longer-lived Moral Rearmament, which were intended to provide ideological alternatives to both capitalism and communism.

Bunche, Ralph Johnson (1904–71) American diplomat and UN official. The grandson of a slave, he became the first Black to be awarded the Nobel Peace Prize, in 1950, for his attempt at reconciling Israel and the Arab states (1948–49).

Bunyan, John (1628–88) English author. Writer of several devotional works, the most famous being the remarkable allegory *Pilgrim's Progress* (1678–84).

Buñuel, Luis (1900–1983) Spanish-born film director. His early films were made in collaboration with Salvador Dali, and he is regarded as a master of Surrealist cinema.

Burgess, Anthony [John Anthony Burgess Wilson] (1917–93) English novelist, critic and composer. His novels include the controversial futuristic fantasy of juvenile crime *A Clockwork Orange* (1962, filmed by Kubrick in 1971).

Burgess, Guy (1911–63) English diplomat and spy. Recruited by Soviet Intelligence in the 1930s, he worked for MI5 during World War II and served with Philby at the British Embassy in Washington DC after the war. With his fellow agent, **Donald Maclean** (1913–83), he fled to the USSR in 1951.

Burke, Edmund (1729–97) Anglo-Irish statesman and philosopher. He entered parliament for the Whig party in 1765 and was soon established as the dominant political thinker of the day. His works include the cornerstone of conservative political thought, *Reflections on the Revolution in France* (1790).

Burnet, Sir Frank Macfarlane *see* Medawar, Sir Peter Brian.

Burns, Robert (1759–96) Scottish poet, renowned as both a lyric poet and a satirist. The son of a farmer, his identification with folk tradition and his rebellious lifestyle also contribute to the unwavering popularity of his work in Scotland.

Burroughs, William S[eward] (1914–) American novelist. A friend of Ginsberg and Kerouac, Burroughs became a heroin addict in the 1940s. His luridly obscene fiction features the squalid, nightmarish underworld of drug addiction.

Burton, Richard [Richard Jenkins] (1925–84) Welsh actor, regarded by his peers and critics as one of the most talented actors of his generation. He formed a screen partnership with Elizabeth Taylor, to whom he was married twice (1964–70, 1975–76).

Busby, Sir Matt[hew] (1909–94) Scottish footballer and manager of Manchester United (1946–69). Many members of his highly regarded team of 1958 died in a plane crash at Munich. His rebuilt team of 'Busby Babes' became the first English team to win the European Cup (1968).

Bush, George [Herbert Walker] (1924–) American Republican politician and 41st president of the US. The son of a wealthy senator, he served in the US Navy (1942–45), became US ambassador to the UN (1971–73), special envoy to China (1974–75) and CIA director (1976). He served under Reagan as vice-president (1980–88) and was elected president in 1988.

Buthelezi, Chief Gatsha (1928–) South African Zulu chief and politician. He helped found the paramilitary organization Inkatha.

Butler, R[ichard] A[usten], [Baron] (1902–82) English Conservative politician. As minister of education (1941–45), he introduced the Education Act of 1944. He was also chancellor of the exchequer (1951–55), home secretary (1957–62) and foreign secretary (1963–64). He was created a life peer in 1965.

Byron, Lord George Gordon Noel [6th Baron Byron of Rochdale] (1788–1824) English poet. His works include many superb lyrics and the long satirical poem *Don Juan* (1819–21).

Cadbury, George (1839–1922) English businessman, social reformer and philanthropist. With his brother **Michael Cadbury** (1835–99) he established the model village of Bournville, near Birmingham, for the Cadbury work force.

Caesar, [Gaius] Julius (100–44 BC) Roman soldier and historian. He negotiated and formed the 'First Triumvirate' with the politician **Marcus Licinius Crassus** (*c*.114–53 BC) and the statesman and general **Pompey** [Gnaeus Pompeius Magnus] (106–48 BC) in 60, after which he fought in Gaul for nine years, and invaded Britain in 55 and 54. Appointed dictator by the Senate in 49, he defeated Pompey at Pharsalia in 48, and was himself assassinated by a largely aristocratic group of conspirators.

Julius Caesar

Cage, John (1912–92) American composer. His experimental music, e.g. *4 minutes 33 seconds* (1952), in which the performers remain silent, has been derided and admired in equal proportions.

Cagney, James (1899–1986) American film actor, originally a dancer but best remembered for his many portrayals of gangsters, e.g. in *The Public Enemy* (1931), *Angels with Dirty Faces* (1938).

Caligula [Gaius Caesar Augustus Germanicus] (12–41) Roman emperor (37–41) who became tyrannical and was assassinated by a conspiracy and succeeded by his uncle, CLAUDIUS.

Callaghan, [Leonard] James [Baron Callaghan of Cardiff] (1912–) British Labour statesman prime minister (1976–79). After a vote of no confidence in his premiership in the House of Commons, he called a general election, which Labour lost.

Callas, Maria (1923–77) American-born Greek operatic soprano, renowned both for her voice and acting skills, which made her one of the most revered opera singers of the 20th century.

Calvin, John (1509–64) French religious reformer who had to flee from France to Switzerland, where, in 1536, he published his *Institutes of the Christian Religion*, a summation of his Protestant faith and the founding text of Calvinism. He set-

Maria Callas

tled in Geneva, where he established the first Presbyterian government.

Calvino, Italo (1923–85) Cuban-Italian novelist, essayist and critic. His early novels belonged in the Italian realist tradition, while his later, highly complex explorations of fantasy and myth have been compared to Latin American 'magic realism'.

Campbell, Sir Malcolm (1885–1948) English racing driver. He was awarded a knighthood in 1931, the year he set a land speed record of 246 mph, for his achievements in setting land and water speed records. His son, **Donald [Malcolm] Campbell**, (1921–67) held the water speed record, at a speed of 276 mph, but he died on Lake Coniston while trying to break it.

Campbell, Mrs Patrick [Beatrice Stella Tanner] (1865–1940) English actress. She was regarded as one of the finest (and wittiest) actresses of her generation.

Campbell-Bannerman, Sir Henry (1836–1908) British statesman. He was Liberal prime minister (1905–08) and played a major part in healing rifts in the Liberal party after the Boer War.

Camus, Albert (1913–60) French novelist, essayist and dramatist, and a leading Existentialist writer. He joined the French Resistance during the war, and was awarded the Nobel prize for literature in 1957. His major works include the novel *The Outsider* and *The Rebel*, a study of 20th-century totalitarianism.

Canaletto, Giovanni Antonio Canal (1697–1768) Venetian painter. An unrivalled architectural painter with an excellent sense of composition, his work includes many views of Venice.

Capa, Robert [André Friedmann] (1913–54) Hungarian photographer who became one of the best-known war photographers of the century. He was killed by a mine in Vietnam.

Capek, Karel (1890–1938) Czech dramatist, novelist and essayist. With his brother **Josef Capek**

(1887–1945), he wrote *The Insect Play* (1921), a prophetic satire on totalitarianism.

Capone, Al[phonse] (1899–1947) Italian-born American gangster. Nicknamed 'Scarface', he established his powerful criminal empire, specializing in bootleg liquor, prostitution and extortion, in Chicago during the prohibition era. He was eventually jailed for tax evasion and died of syphilis.

Capote, Truman (1924–84) American novelist and socialite. His varied works include light romances, e.g. *Breakfast at Tiffany's* (1958), as well as realist explorations of murder.

Capra, Frank (1897–1991) Italian-born American film director. His comedies, usually portraying an ultimately successful struggle by a decent, everyday American against the flawed political system, were enormously popular in the 1930s.

Caravaggio, Michelangelo Merisi da (1573–1610) Italian painter. Noted for his bold, expressive use of chiaroscuro, his religious paintings caused controversy by using everyday people as the models for his Biblical characters.

Carlyle, Thomas (1795–1881) Scottish historian and essayist. Hailed by many of his contemporaries as a great social critic and philosopher, he frequently attacked the materialism of the Industrial Age.

Carnap, Rudolf (1891–1970) German-born American philosopher. Regarded as a leading logical positivist, he attempted to develop a formal language that would remove ambiguity from scientific language.

Carné, Marcel (1909–) French film director. His films include the highly acclaimed theatrical epic *Les Enfants du paradis* (1944), filmed during the German occupation of France.

Carnegie, Andrew (1835–1919) Scottish-born American industrialist and philanthropist who believed that personal wealth should be used for the benefit of all members of society.

Carreras, José [Maria] (1946–) Spanish lyric tenor, one of the finest tenors (with DOMINGO, PAVAROTTI) of the late 20th century.

Carroll, Lewis (pseud. of Charles Lutwidge Dodgson) (1832–98) English author, clergyman and mathematician. His most famous works are the two remarkable 'Alice' books, *Alice's Adventures in Wonderland* and *Through the Looking Glass*.

Carter, Jimmy [James Earl Carter] (1924–) American Democratic statesman and 39th president of the US (1977–81). A successful peanut farmer, he became governor of Georgia (1974–77) and defeated Gerald FORD in the 1976 presidential campaign. His administration made significant attempts at linking overseas trade with human rights issues.

Cartier-Bresson, Henri (1908–) French photographer and film director. His documentary black-and-white photographs were taken without prior composition and an uncropped frame.

Caruso, Enrico (1873–1921) Italian tenor. Born in Naples, he sang most of the great tenor roles in Italian and French opera and is regarded as perhaps the most outstanding operatic tenor of all time. He was one of the first great singers to make recordings.

Casals, Pablo (1876–1973) Spanish cellist, pianist and composer. His recordings of BACH's cello suites and of the DVORAK cello concerto are particularly highly regarded.

Casement, Sir Roger [David] (1864–1916) British consular official and Irish nationalist. While working for the British colonial service, he exposed the repression of the people of the Congo by its Belgian rulers in 1904. Knighted in 1911, he adopted Irish nationalism shortly afterwards and was hanged for treason.

Castro [Ruz], Fidel (1927–) Cuban statesman, prime minister (1959–76) and president (1976–). He led a coup in 1959, and shortly afterwards announced his conversion to communism. He survived several attempts at overthrow by exiled opponents and the CIA, e.g. the Bay of Pigs invasion in 1961, but survived with Soviet subsidies which began to decrease in the late 1980s.

Fidel Castro

Catherine II ('the Great') (1729–96) Russian empress. She became Empress on the death of her husband, **Peter III** (1728–62), who was murdered by one of her lovers. She consolidated and expanded the Russian Empire by conquest, and scandalized European opinion by having a supposedly legion number of paramours and heading an intrigue-raddled court, while patronizing Enlightenment philosophers such as VOLTAIRE and DIDEROT.

Cavell, Edith [Louisa] (1865–1915) English nurse. She treated both German and Allied casualties in Brussels during the German occupation, and was executed by the German authorities, who accused her of helping British soldiers to escape to Holland.

Caxton, William (c.1422–91) English printer and translator. His *Recuyell of the Historyes of Troy* (printed at Bruges, 1475), is the first book to be printed in English.

Ceausescu, Nicolae (1918–89) Romanian dictator. Secretary general of the Romanian Communist Party from 1969 and president of Romania from 1974, his regime was overthrown by dissident Communists in 1989, and he and his wife were executed.

Cecil, William, 1st Baron Burghley, (1520–98) English statesman who served, with skill and dexterity, both HENRY VIII and MARY I (converting to Roman Catholicism under Mary), and was one of the prime architects of ELIZABETH I's succession.

Cervantes [Saavedra], Miguel de (1547–1616) Spanish novelist, dramatist and poet. His most famous work was the satirical masterpiece *Don Quixote de la Mancha*.

Cézanne, Paul (1839–1906) French painter whose works include landscapes and still lifes. He was very influential on succeeding generations of painters, notably the Cubists.

Chadwick, Sir James (1891–1974) English physicist. He discovered the neutron in 1932, and was awarded the Nobel prize for physics in 1935.

Chagall, Marc (1887–1985) Russian-born French painter. His vividly coloured work features unusual compositions drawing on symbolism from Russian and Jewish folk art.

Chain, Sir Ernst Boris (1906–79) German-born British biochemist. He prepared penicillin for clinical use, and with FLOREY and Alexander FLEMING, shared the 1945 Nobel prize for physiology or medicine.

Chamberlain, [Arthur] Neville (1869–1940) British statesman and Conservative prime minister (1937–40). He pursued a policy of appeasement towards the totalitarian powers of Germany, Italy and Japan in the 1930s. He died shortly after illness forced his resignation from CHURCHILL's war cabinet.

Chandler, Raymond [Thornton] (1888–1959) American novelist and screenwriter. His detective novels, e.g. *Farewell, My Lovely* (1940) are classics of the genre.

Chanel, Coco [Gabrielle Bonheur Chanel] (1883–1971) French couturière and perfumer who originated the thin, low-waist style for women's dresses.

Chaplin, Sir Charlie [Spencer] (1889–1977) English comedian and film director. His gentleman-tramp character with a beguiling shuffle, bowler hat and cane, a familiar figure in many films including *City Lights*, became perhaps the most famous comic creation of the 20th century. With the advent of sound, Chaplin was unable to repeat his success from the era of silent films.

Charlemagne (c.742–814) king of the Franks and Holy Roman emperor. In 771 he became sole ruler of the Frankish kingdom and spent the early part of his reign conquering (and converting to Christianity) neighbouring kingdoms. He led an army into Spain to fight the Moors in 778. In 800 he was crowned emperor after crushing a Roman revolt against the Pope.

Charles, Prince [Charles Philip Arthur George, Prince of Wales] (1948–) heir apparent to ELIZABETH II of the United Kingdom. He married **Lady Diana Spencer** (1961–) in 1981, and they had two children. They divorced in 1992.

Charles, Ray [Ray Charles Robinson] (1930–) American singer, pianist and songwriter. Originally a blues/jazz singer, he became one of the most popular singers in the world.

Charlton, Bobby [Robert Charlton] (1937–) English footballer, capped over 100 times for England. His brother **Jack Charlton** [John Charlton] (1935–), also an England player, became the manager of the Irish international team and led them to the World Cup finals in 1990 and 1994.

Chatham, William Pitt, 1st Earl of *see* **Pitt, William**.

Chaucer, Geoffrey (c.1340–70) English poet. His great narrative skill is displayed at its finest in *The Canterbury Tales* (c.1387), a masterpiece of wit and humour in which various pilgrims tell each other stories.

Geoffrey Chaucer

Chekhov, Anton Pavlovich (1860–1904) Russian dramatist and short-story writer. A physician, he became one of the greatest writers of his age, his works, notable for their wit and dramatic power, including the plays *Uncle Vanya* and *Three Sisters* and the short story 'The Lady with the Little Dog'.

Cherenkov, Pavel Alekseievich (1904–1990) Soviet physicist. In the mid–1930s, he discovered the form of radiation known as Cherenkhov radiation, and was awarded the 1958 Nobel prize for physics.

Cherubini, Luigi (1760–1842) Italian composer who settled in Paris. He wrote operas, e.g. *Médée*, and other works and was much admired by BEETHOVEN, whose opera, *Fidelio*, followed the story of *Médée* closely in its study of feminine psychology and stress on democratic values.

Chesterton, G[ilbert] K[eith] (1874–1936) English essayist, novelist, critic and poet. With his friend BELLOC, he became known as a gifted disputant for what they saw as the glory of old, rural Roman Catholic England.

Chiang Ch'ing *or* **Jiang Qing** (1913–) Chinese Communist politician and actress. She married MAO as his third wife in 1939 and was the main force behind the savage purges of the Cultural Revolution in the late 1960s. After Mao's death, her power waned, and she was arrested in 1976 with three confederates (the 'Gang of Four') and charged with murder and subversion. She was sentenced to death in 1981, the sentence later being suspended.

Chiang Kai-shek *or* **Jiang Jie Shi** (1887–1975) Chinese general and statesman. He was president of China (1928–38, 1943–49), then, after losing the civil war to MAO TSE-TUNG and his forces, fled the mainland to establish the nationalist republic of China in Formosa, of which he was president (1950–57).

Chiang Kai-shek

Chirico, Giorgio de (1888–1978) Greek-born Italian painter whose dreamlike pictures of open, de-

serted squares were hailed by the Surrealists as precursors of their own works in the early 1920s.

Chomsky, Noam [Avram] (1928–) American linguist, philosopher and political activist. His innovative work in linguistics is based on the principles that humans are born with an innate capacity for learning grammatical structures. He was a notable opponent of the Vietnam war.

Chopin, Frédéric [François] (1810–49) Polish pianist and composer. His emotional, melancholy works, often regarded as quintessentially Polish in mood, include over 50 mazurkas, two piano concertos and 25 preludes.

Chou En-Lai *or* **Zhou En Lai** (1898–1976) Chinese Communist statesman. He was foreign minister (1949–58) and prime minister (1949–76) of the People's Republic of China and was regarded as a moderate during the chaos of China's Cultural Revolution in the late 1960s.

Christie, [Dame] Agatha [Clarissa Mary] (1890–1976) English detective story writer whose ingeniously plotted novels, e.g. The Murder of Roger Ackroyd, established her as one of the great writers in the genre.

Churchill, Sir Winston [Leonard Spencer] (1874–1965) British Conservative statesman and writer. After an adventurous early life which included escape from imprisonment by Louis BOTHA during the Boer War, he held several posts under both Liberal and Conservative governments. He opposed CHAMBERLAIN's policy of appeasement in the 1930s and served as prime minister (1940–45) during World War II. His works include *History of the English-Speaking Peoples*. He was awarded the Nobel prize for literature (1953).

Winston Churchill

Clapton, Eric (1945–) English guitarist. Recognized as one of the most influential rock guitarists, he played with the Yardbirds (1963–65) and Cream (1966–68).

Clark, Jim [James Clark] (1936–68) Scottish racing driver who was World Champion in 1963 and

1965 and winner of 25 Grand Prix events. He was killed in a crash in West Germany.

Claudius (10 BC–54) Roman emperor who extended the Empire, initiated the conquest of Britain in 43 and extended Roman citizenship. His fourth wife (and niece), Agrippina, was the mother of his successor NERO.

Cleese, John [Marwood] (1939–) English comedy actor and writer. He was one of the main talents involved in the highly influential TV comedy series, *Monty Python's Flying Circus* (1969–74).

Clemenceau, Georges [Eugène Benjamin] (1841–1929) French statesman. A leading left-winger, he was an outspoken critic of the French government's war policy in the early days of World War I. He was prime minister (1906–19, 1917–20) and his forceful negotiation of the Versailles Treaty is believed to have led directly to World War II.

Cleopatra *see* ANTONY, MARK.

Clinton, Bill [William Jefferson Davis Clinton] (1946–) American politician and 42nd President of the United States. A lawyer, he became Arkansas attorney general (1974–79), then state governor (1979–81, 1983–92). He defeated George BUSH in the 1992 presidential election and was reelected for a second term in 1996. *See also* GORE.

Clive, Kitty *see* **Garrick, David**.

Clive, Sir Robert [1st Baron Clive of Plassey] (1725–74) English general and administrator in India. He worked for the East India Company (1743–46) before joining the Indian army. An MP in England (1760–62) and governor of Bengal (1764–67).

Cockcroft, Sir John Douglas (1897–1967) English nuclear physicist. With the Irish physicist **Sir Ernest [Thomas Sinton] Walton** (1903–), he produced the first laboratory splitting of an atomic nucleus, for which they shared the 1951 Nobel prize for physics.

Cockerell, Sir Christopher Sydney (1910–) English engineer. He invented the hovercraft, the prototype of which first crossed the English Channel in 1959.

Cocteau, Jean (1889–1963) French film director, novelist, dramatist, poet and critic. His experimental, surreal films, including *Orphee* and *La Belle et la bete*, were highly influential on modern film-makers. The best known of his novels is *Les Enfants Terribles*.

Coleridge, Samuel Taylor (1772–1834) English poet and critic. With WORDSWORTH he published *Lyrical Ballads* in 1798, a landmark in English poetry in its rejection of a special 'poetic' language and advocacy of clear everyday language.

Colette, [Sidonie Gabrielle] (1873–1954) French novelist. Her novels, e.g. *Chéri* and *Gigi*, are often erotic and display a strong sympathy for animals and the natural world.

Collins, Michael (1890–1922) Irish Republican politician. A Sinn Fein leader, he negotiated the 1922 peace treaty with Britain that resulted in the establishment of the Irish Free State. He was killed in an ambush during the civil war that followed.

Coltrane, John [William] (1926–67) American jazz saxophonist. A virtuoso on the tenor and soprano saxophones, he became an influential and popular jazz musician.

Columba, St (521–97) Irish missionary. Accused of being involved in one of Ulster's many bloody civil conflicts, as penance he fled to the Western Isles of Scotland to proselytise for Christianity. He established a monastic settlement on the island of Iona.

Columbus, Christopher (1451–1506) Italian navigator. Under the patronage of Spain, he led an expedition to seek a western route to the Far East. He discovered the New World in 1492, making landfall in the West Indies and made two subsequent voyages in 1493 and 1498, reaching South America on the third.

Compton, Arthur Holly (1892–1962) American physicist. He was a prominent researcher into X-rays, gamma rays and nuclear energy, and discovered the Compton effect. He was awarded the 1927 Nobel prize for physics.

Compton, Denis [Charles Scott] (1918–) English cricketer, who played for Middlesex and England (1937–57) and who was regarded as one of the best all-rounders ever in cricket.

Compton-Burnett, Dame Ivy (1892–1969) English novelist. Her novels were mostly in dialogue and featured the traumas of upper-middle-class Edwardian family life.

Connery, Sean [Thomas] (1930–) Scottish film actor. One of the most charismatic film actors of his generation, he initially achieved worldwide fame as Ian FLEMING's character James Bond. He went to achieve notable success in *The Name of the Rose* (1986) and *The Untouchables* (1987) for which he was awarded a Best Supporting Actor Oscar.

Connors, Jimmy (1952–) American tennis player. He achieved great success in the mid-70s to early 80s, winning the US Open championships on five occassions and Wimbledon twice.

Conrad, Joseph [Teodor Josef Konrad Korzeniowsky] (1857–1924) Polish-born English writer.

He qualified as a master mariner in 1886 and his many works often featured isolated characters in exotic locations. His novels include *Heart of Darkness*.

Joseph Conrad

Constable, John (1776–1837) English painter. Drawing inspiration from nature, he produced works such as *View on the Stour*, which were initially more influential in France than England. He and Turner are considered to be the most important English landscape painters.

Constantine I (*c*.274–337) Roman emperor. He became the first Christian emperor in 312, when, before a battle, he reportedly saw a cross in the sky inscribed 'In this sign conquer'. He moved the capital to Byzantium (which he renamed Constantinople) in 330. Christianity became the Empire's official religion in 324.

Cook, (Captain) James (1728–79) English explorer who charted and claimed the east coast of Australia for Britain and discovered New Caledonia. On his third and last voyage to the Pacific (1776–79), he was killed by islanders in Hawaii.

Cooper, Gary [Frank James Cooper] (1901–61) American film actor who specialized as the quiet, courageous hero in many westerns and adventure films, e.g. *High Noon*.

Cooper, James Fenimore (1789–1851) American novelist. His adventure novels, such as *The Last of the Mohicans*, established the enduring 'frontier myth' of America.

Copernicus, Nicolaus (1473–1543) Polish astronomer. His great work *De Revolutionibus* (1543) sets out his theory that the earth and planets revolve around the sun.

Copland, Aaron (1900–1990) American composer, pianist and conductor. He was influential in American music in incorporating elements from traditional American folk songs in his compositions. His works include the ballet score *Rodeo*.

Coppola, Francis Ford (1939–) American film director and screenwriter. His most successful films include the modern classics, *The Godfather* (1972) and *Apocalypse Now* (1979).

Corday, Charlotte *see* MARAT, JEAN PAUL.

Corman, Roger (1926–) American film director and producer. Known primarily in the 1950s and 1960s as a creator of cheap B movies he also fostered the careers of many prominent American directors and actors, including COPPOLA, DE NIRO and SCORSESE.

Cosgrave, W[illiam] T[homas] (1880–1965) Irish nationalist politician. He became first president of the Irish Free State (1922–32). His son **Liam Cosgrave** (1920–) became Fine Gael prime minister of the Republic of Ireland (1973–77).

Courbet, Gustave (1819–77) French painter. Considered to be the founder of Realism, his work was frequently condemned as 'socialistic' because he scorned the established classical outlook.

Cousteau, Jacques [Yves] (1910–) French oceanographer. He invented the aqualung (1943) and developed techniques of underwater cinematography that were influential in raising awareness of the world's oceans.

Coward, Sir Noel [Pierce] (1899–1973) English dramatist, actor and composer. His witty, sophisticated comedies and amusing songs were regarded as mildly shocking in their day and include *Private Lives* and *Blithe Spirit*.

Cowper, William (1731–1800) English poet. Best known in his own day as an engaging satirist and nature poet, his darker religious poems are now seen as of more lasting importance.

Cranmer, Thomas (1489–1556) English prelate. Appointed Archbishop of Canterbury in 1533 (while secretly married), he pronounced the annulment of the marriage of HENRY VIII to Catherine of Aragon in 1533. A moderate Protestant reformer, he was executed by MARY I.

Crassus *see* CAESAR, JULIUS.

Crawford, Joan [Lucille le Sueur] (1908–77) American film actress. One of the first leading women in Hollywood although many of her films were formulatic melodramas, such as *Mildred Pierce* (1945).

Crick, Francis [Harry Compton] (1916–) English molecular biologist. With James Dewey WATSON, he discovered the structure of DNA, and was awarded the 1962 Nobel prize for physiology or medicine.

Crippen, Hawley Harvey (1862–1910) American doctor who poisoned his wife in London in 1910. His dramatic capture on board ship involved the first use of radio for police purposes.

Cripps, Sir [Richard] Stafford (1889–1952) British Labour statesman. A leading left-winger, he became chancellor of the exchequer (1947-50), and introduced a programme of high taxation and wage restraint to deal with Britain's economic problems.

Cromwell, Oliver (1599–1658) English soldier and statesman. A Puritan country squire and MP, he was a noted critic of CHARLES I during the 1628–29 Parliament. He displayed a strong grasp of military skill during the opening year of the Civil War (1642), forming his 'Ironsides' regiment the following year. He led Parliament's New Model Army to victory at Naseby (1645) and crushed Welsh and Scottish rebellions before signing Charles I's death warrant in 1649. After a murderous conquest of Ireland in 1649, his victory over a second Scottish rebellion in 1651 ended the Civil War. He was nominated 'Lord Protector' of the Commonwealth in 1653, and established an authoritarian rule, dissolving parliament when it displeased him. He was succeeded as Protector (1658–59) by his son, **Richard Cromwell** (1626–72).

Cromwell, Thomas (c.1485–1540) English statesman. Of humble origin, he rose to power through the patronage of Cardinal WOLSEY, and became HENRY VIII's chief adviser. He fostered the passing of Reformation legislation, established the king's legal status as head of the Church in England, and oversaw the dissolution of the monasteries.

Cronin, A(rchibald) J(oseph) (1896–1981) Scottish novelist, dramatist and physician. He gave up medicine after completing his first novel in 1931. However, his knowledge of medicine and his experience of his native Scotland provided the inspiration for the enormously successful television series, *Dr Finlay's Casebook*.

Crosby, Bing [Harry Lillis Crosby] (1904–77) American singer and actor. His relaxed, jazz-influenced style of 'crooning' made him one of the most popular and imitated singers of the century. He appeared in over sixty films and formed a notable comedy partnership with BOB HOPE in the 'Road to ...' series.

Cummings, E[dward] E[stlin] (1894–1962) American poet, novelist and artist. His experimental free verse and distinctive use of typography influenced many other poets.

Curie, Marie (1867–1934) Polish-born French chemist. With her husband **Pierre Curie** (1859–1906), also a chemist, and the physicist **Antoine Henri Becquerel** (1852–1908), she was awarded the 1903 Nobel prize for physics for work on ra-

dioactivity, the first woman to win a Nobel prize. She subsequently became the first person to win two Nobel prizes when her discovery of radium and polonium led to her being awarded the 1911 prize for chemistry.

Marie Curie

Cyrano de Bergerac, Savinien (1619–1655) French soldier, poet and dramatist. Most famous in the popular imagination for having an enormous nose and for having (reputedly) fought around a thousand duels. His works include several satires, and his life was dramatized by Edmond ROSTAND in a verse drama.

Dahl, Roald (1916–90) English author (of Norwegian parentage) known primarily for his entertaining children's stories, e.g. *Charlie and the Chocolate Factory*, *Mathilda*, and collections of humorous poems. Stories for adults include the collection *Kiss, Kiss*.

Daladier, Edouard (1884–1970) French socialist statesman. He was prime minister (1933, 1934, 1938–40) and signed the Munich Pact of 1938. He denounced the Vichy government in 1943, and was then imprisoned for the duration of the war.

Dalai Lama [Tenzin Gyatso] (1935–) Tibetan spiritual and temporal leader. He became the 14th Dalai Lama in 1940, and fled Tibet in 1959 following the Chinese invasion of his country. He was awarded the 1989 Nobel Peace Prize in recognition for his commitment to the non-violent liberation of his homeland.

Dale, Sir Henry Hallett (1875–1968) English physiologist. He and **Otto Lowei** (1873–1961) were awarded the 1936 Nobel prize for physiology or medicine for their work on the chemical basis of nerve impulse transmission.

Dali, Salvador (1904–89) Spanish surrealist painter. His finely executed paintings, or 'dream photographs', did much to popularize the surrealist movement. He also collaborated with the film director Luis BUÑUEL on the films *Un Cien Andalou* and *L'Age d'or*. Dali was a memorably eccentric individual, often given to paranoia and acts of exhibitionism.

Salvador Dali

D'Annunzio, Gabriele (1863–1938) Italian poet, novelist, dramatist and political adventurer. As a writer, the sensuous imagery of much of his work has been widely admired and his oratory was credited with Italy's joining the allies in World War I. He seized the city of Fiume in 1919, which he ruled until 1920 and became a supporter of MUSSOLINI.

Dante (Alighieri) (1265-1321) Italian poet. Expelled from Florence in 1309 for political reasons, he spent 20 years in wandering exile, during which he wrote his masterpiece, the *Divine Comedy*. His literary influence was enormous, and resulted in Tuscan becoming the language of literary Italy.

Danton, Georges Jacques (1759-94) French revolutionary. After the fall of the monarchy in 1792, he became minister of justice and voted for the death of King LOUIS XVI. His efforts to moderate the Revolutionary Terror failed, and he was out-manoeuvred by ROBESPIERRE and executed.

Darwin, Charles [Robert] (1809-82) English naturalist. The grandson of the physician and poet **Erasmus Darwin** (1731-1802), he sailed as a naturalist to South America and the Pacific (1831-36) where his studies among the rich animal and plant life of the area formed the basis for his revolutionary theory of evolution by natural selection. Darwin's theory rapidly found acceptance not only amongst scientists but among society at large.

Charles Darwin

Daumier, Honoré (1808-79) French cartoonist, painter and sculptor. One of the most proficient satirists of all time, his caricatures of the king resulted in a term in jail. He was also an innovative lithographer, painter and sculptor.

David, Jacques-Louis (1748-1825) French painter. The leading artist of the French Revolution, he was imprisoned after the death of ROBESPIERRE but survived to become painter to NAPOLEON.

Davies, Sir Peter Maxwell (1934–) English composer. With BIRTWISTLE, he founded the Pierrot Players (later called the Fires of London). Since 1970 he has been based in Orkney and has frequently collaborated with the Orcadian poet, novelist and short-story writer **George Mackay Brown** (1921–1996).

Davis, Bette [Ruth Elizabeth Davis] (1908–89) American actress, whose electrifying and commanding screen presence made her a highly rated film actresses. Her films include *All About Eve*.

Davis, Sir Colin [Rex] (1927–) English conductor. Noted particularly for his interpretations of BERLIOZ, he was conductor of the BBC Symphony Orchestra (1967–71).

Davis, Miles [Dewey] (1926–91) American jazz trumpeter, composer and bandleader. The leading exponent of the influential 'cool jazz' school.

Davisson, Clinton Joseph *see* THOMSON, SIR GEORGE PAGET.

Davy, Sir Humphry (1778-1829) English chemist. An ingenious experimenter, he discovered many new metals, e.g. sodium and potassium, and developed ground-breaking studies in electrochemistry. He was knighted in 1812, and invented the 'Davy lamp' for miners in 1815.

Dawes, Charles G[ates] (1865–1951) American banker. He devised the Dawes Plan of 1924 for German reparation payments after World War I. He was US vice-president (1925–29) and was awarded the 1925 Nobel Peace Prize.

Day, Doris [Doris Kappelhoff] (1924–) American film actress, famous for her light-hearted, girl-next-door image.

Dayan, Moshe (1915–81) Israeli general and statesman. He commanded the Israeli forces during the Sinai invasion (1956) and was minister of defence during the Six Day War of 1967. He played an important part in the talks leading to the Israel-Egypt peace treaty of 1979.

Day Lewis, Cecil (1904–72) Irish-born English poet. In the 1930s he was regarded as part of the 'AUDEN generation' of left-wing poets. He became poet laureate in 1968.

Dean, Christopher *see* TORVILL, JAYNE.

Dean, James [Byron] (1931–55) American film actor. He became a cult figure in the 1950s for his portrayal of troubled, disaffected adolescence, e.g. in *East of Eden* and *Rebel Without a Cause*. He died in a car crash.

Debussy, [Achille] Claude (1862–1918) French composer. Regarded as the founder of impressionism in music and one of the strongest influences on modern music, his works include orchestral pieces, e.g. *La Mer*, and the opera *Pelléas et Mélisande*.

Defoe, Daniel (1600-1731) English novelist and pamphleteer. His works include two remarkable novels, *Robinson Crusoe* (1719) and *Moll Flanders* (1722) although he was better known in his lifetime as a skilled and prolific propagandist.

Degas, [Hilaire Germain] Edgar (1834-1917) French painter and sculptor. He met Monet in the 1860s, after which he began exhibiting with the Impressionists. He is especially noted for his paintings and pastel drawings of racehorses and ballet dancers.

de Gaulle, Charles [André Joseph Marie] (1890–1970) French general, statesman and first president (1958–69) of the Fifth Republic. An opponent of the Vichy regime led by PÉTAIN, he fled to Britain in 1940, where he led the Free French forces. Elected president of the provisional government in 1945, he resigned in 1946 after disagreement over his executive powers. He was asked to form a government in 1958, during the Algerian crisis. He granted independence to France's colonies in Africa (1959–60), oversaw increased economic prosperity, fostered France's independent nuclear deterrent policy and strongly opposed the UK's entry into the Common Market. His party won a large majority in the election following the student riots of 1968. He resigned in 1969, after being defeated on constitutional reform.

Charles de Gaulle

de Klerk, F[rederik] W[illem] (1936–) South African statesman who succeeded P. W. BOTHA as leader of the ruling National party and president (1989) and continued the policy of dismantling apartheid, in 1990 legalizing the African National Congress and organizing MANDELA's release from prison. He presided over South Africa's first free elections in 1994 but was defeated by Mandela, since when he has served as Second Deputy President.

Delacroix, Eugène (1798-1863) French painter. His early work, while Romantic in subject matter, owes much to classical composition. He studied CONSTABLE's work and in turn influenced other artists, particularly those of the Barbizon School.

de la Mare, Walter [John] (1873–1956) English poet and novelist. Much of his work was written for children and the loss of childhood innocence is a major theme in his work. His works include the poetry collection *The Listeners*.

Delaunay, Robert (1885–1941) French painter who founded the movement called 'Orphism', the name given by APOLLINAIRE to Delaunay's introduction of colour abstraction into his Cubist-style paintings. He influenced many other artists, notably KLEE.

Delius, Frederick (1862–1934) English composer. Unconnected to any traditional school, he is noted for his six operas, including *A Village Romeo and Juliet*, and large orchestral pieces.

de Mille, Cecil B[lount] (1881–1959) American film producer and director. With GOLDWYN, he is credited with creating the mass movie industry of Hollywood. His films were extravagantly produced epics that achieved enormous success throughout the world, e.g. *The Ten Commandments* (1923).

Dempsey, Jack [William Harrison Dempsey] (1895–1983) American boxer. An ex-miner who became one of the most popular boxers of his day, he was world heavyweight champion (1919–26). Known as the 'Manassa Mauler', he retired from fighting to become a boxing referee and, for a number of years, a restaurateur.

Deng Xiaoping *or* **Teng Hsiao-p'ing** (1904–1997) Chinese Communist statesman. He took part in the 'long march' of 1934–5, and was elected to the central committee of the Communist Party in 1949, beconning General Secretary in 1956. Denounced in the Cultural Revolution of the late 1960s as a 'capitalist roader', he re-emerged as a powerful figure in the late 1970s. He introduced economic reforms and developed friendly relations with the West but also sanctioned the

Tienanmen Square massacre of dissident students in June 1989. At the time of his death he held none of the major state or party offices but was still regarded as the dominant political power in the country.

Deng Xiaoping

De Niro, Robert (1943–) American actor, who is regarded as one of the finest modern screen actors, with a remarkable facility for submerging himself in a wide variety of roles. His films include *Taxi Driver*, *Midnight Run* and *The Godfather II*, for which he won an Oscar.

Depardieu, Gerard (1948–) French actor. Established in the 1970s as a leading man in French films, his strong performances and independent character have made him popular worldwide, notably in *Danton*, *Cyrano de Bergerac* and *Green Card*.

Derain, André (1880–1954) French painter. He was influenced by PICASSO and BRAQUE, and became one of the leading Fauvist painters.

Desai, Morarji [Ranchhodji] (1896–1995) Indian statesman. He held several posts under NEHRU, founded the Janata party in opposition to Indira GANDHI's Congress Party, which he defeated in the 1977 general election, and was prime minister (1977–79).

Descartes, René (1596-1650) French philosopher and mathematician. He proposed a dualistic philosophy based on the separation of soul and body, mind and matter, and sought to establish his system on mathematics and pure reason, his most famous dictum being '*Cogito ergo sum*', 'I think, therefore I am'.

De Sica, Vittorio (1902–74) Italian film director and actor. His early films, e.g. *Bicycle Thieves* (1948), are regarded as among the finest Italian neo-realist films for their compassionate insight into the lives of the poor.

De Valera, Eamon (1882–1975) American-born Irish statesman. He was sentenced to death by the British government for his part in the 1916 Easter Rising, but was reprieved after US intervention. He became president of Sinn Féin (1917–1926). He opposed the Anglo-Irish Treaty (1921), gave largely symbolic leadership to the anti-Treaty forces during the civil war (1922–23), and was imprisoned (1923–24). He became prime minister (1932–48, 1951–54, 1957–59) and president (1959–73).

Devine, George [Alexander Cassady] (1910–65) English stage director and administrator. He became one of the most prominent influences on the British stage as artistic director of the English Stage Company at the Royal Court.

De Vries, Hugo [Marie] (1848–1935) Dutch botanist and geneticist. He rediscovered the genetic principles first put forward by MENDEL, and developed the theory of evolution through the mutation of genes.

Diaghilev, Sergei [Pavlovich] (1872–1929) Russian ballet impresario. Founder of the Ballet Russe de Diaghilev in 1911, he became a very influential ballet impresario, drawing on the talents of composers such as STRAVINSKY and artists such as PICASSO.

Dickens, Charles [John Huffam] (1812-70) English novelist. His prolific output included plays, pamphlets and lectures, as well as novels and short stories. Immensely popular with both the American and British reading public, he continues to be one of the most widely read writers in the English language. His novels include *David Copperfield*, *Bleak House* and *Great Expectations*.

Charles Dickens

Dickinson, Emily (1830-86) American poet. Although only seven of her *c*.2000 poems were printed in her lifetime she became recognized as a uniquely gifted poet following the wider publication of her work in the 1890s.

Diderot, Denis (1713-84) French philosopher. With others, he edited the great *Encyclopédie*, 17 volumes of which appeared under Diderot's overall direction between 1751 and 1772.

Dietrich, Marlene [Maria Magdelene von Losch] (1902–92) German-born American singer and film actress, notable for her strong sexual presence and husky, alluring voice.

Dior, Christian (1905–57) French couturier who created the 'New Look' of the late 1940s, with a narrow waist and full pleated skirt, which was very popular in the austerity of postwar Europe.

Dirac, Paul Adrien Maurice (1902–84) English physicist. He devised a complete mathematical formulation of EINSTEIN's theory of relativity and predicted the existence of antimatter. In 1933 he shared the Nobel prize for physics with SCHRÖDINGER.

Disney, Walt[er Elias] (1901–66) American cartoonist and film producer. His cartoon films of the 1930s and 40s achieved high critical and popular acclaim, Mickey Mouse and Donald Duck being two of his famous creations. He built Disneyland amusement park in California (1955), and planned Disney World in Florida (1971).

Disraeli, Benjamin [1st Earl of Beaconsfield] (1804–81) British statesman and novelist. He became a Tory member of parliament in 1837 and prime minister (1868, 1874-80). He promoted protectionism and a romantic imperialism which appealed to Queen VICTORIA and most of the British public, much to the ire of his great rival, GLADSTONE.

Benjamin Disraeli

Dobzhansky, Theodosius (1900–75) Russian-born American geneticist. His seminal studies of genetic variation linked Darwin's evolutionary theory with MENDEL's heredity laws.

Doenitz, Karl *see* DÖNITZ, KARL.

Dolci, Danilo (1924–) Italian social reformer. Described as the 'GANDHI of Italy', he built schools and community centres in poverty-stricken Sicily in the face of fierce opposition from an unholy alliance of church, state and Mafia.

Dollfus, Engelbert (1892–1934) Austrian statesman. A devout Roman Catholic, he became leader of the Christian Socialist Party and was elected chancellor (1932–34). He opposed the German Anschluss and was assassinated by Austrian Nazis.

Domingo, Placido (1941–) Spanish tenor who studied in Mexico City. He is regarded as one of the finest modern operatic tenors for his sophisticated vocal technique and considerable acting ability.

Dominic, St (*c*.1170-1221) Spanish monk. Noted for his asceticism, he founded the Dominican Order of monks and helped the forces of the Inquisition in their barbarous treatment of the Albigensians in Southern France. He was canonized in 1234.

Donatello [Donato di Niccolò di Betto Bardi] (*c*.1386-1466) Florentine sculptor. One of the leading sculptors of the early Renaissance, his most famous work is the huge bronze statue of *David* (1430s).

Dönitz *or* **Doenitz, Karl** (1891–1980) German admiral. He was commander of the German navy (1943–45). As head of the Nazi state following HITLER's suicide, he surrendered unconditionally to the Allies, and was sentenced at Nuremberg to ten years imprisonment for war crimes.

Donleavy, J[ames] P[atrick] (1926–) American-born Irish novelist. His most popular works are comic adventures set in Dublin's undergraduate community, e.g. *The Ginger Man*.

Donne, John (1573-1631) English poet and divine. After occasional hazardous adventures, such as accompanying Sir Francis DRAKE on his raid on Cadiz, he became Dean of St Paul's in 1621 and was regarded as one of the greatest preachers of his day. His poetry is among the finest metaphysical verse.

Doré, Gustave (1832-83) French sculptor, painter and illustrator. Trained as a caricaturist, he became well known for his book illustrations and realistic drawings of London slums.

Dostoyevsky, Fyodor Mikhailovich (1821-81) Russian novelist. His novels, e.g. *Crime and Punishment*, are profound explorations of sin and redemption through suffering. With TOLSTOY he was profoundly influential on modern literature.

Douglas-Home, Sir Alec [Baron Home of the Hirsel] (1903–95) Scottish Conservative politician. He became the 14th Earl of Home in 1951, renouncing his title in 1963 to contest (and win) the seat of Kinross after succeeding Harold MACMILLAN as prime minister. The furore over his unexpected emergence as party leader resulted in reform of the Tory leadership election process.

Doyle, Sir Arthur Conan (1859–1930) Scottish novelist, short-story writer and physician. His most famous creation is the amateur detective Sherlock Holmes.

Drake, Sir Francis (*c.*1540-1596) English navigator and pirate. He became a highly popular hero to the English following his successful depradations upon Spanish ships and settlements in the Caribbean in the early 1570s. He circumnavigated the world (1577-80) and was one of the leading lights in the victory over the Spanish Armada (1588).

Dreyfus, Alfred (1859-1935) French army officer. Imprisoned in 1894 on Devil's Island on a false charge of espionage, the 'Dreyfus affair' scandalized much of Europe for the anti-semitism of the prosecution case. ZOLA's magnificent pamphlet *J'accuse* (1898) was written in his defence, and Dreyfus was released in 1906.

Dryden, John (1631-1700) English poet, dramatist and critic. One of the most important literary figures of his time, he was also a highly significant contributor to the religious and political controversies of the day. He was also the first great English critic, and an outstanding poet. His works include the social comedy *Marriage à la Mode* and the verse tragedy *All for Love*. He became poet laureate in 1668 but lost the post in 1688 in the political upheaval surrounding the replacement of King JAMES II by WILLIAM III and Mary.

Dubcek, Alexander (1921–92) Czech statesman. As first secretary of the Communist Party (1968–69), he introduced political reforms which ended with the Russian invasion of 1968. Following Czechoslovakian independence, he was appointed chairman of the federal assembly (1989).

Alexander Dubcek

Dubuffet, Jean (1901–85) French painter. He devised the concept of 'Art Brut' in reaction against 'museum art', and made paintings assembled from bits of rubbish, broken glass, etc.

Duchamp, Marcel (1887–1968) French-born American painter and sculptor. One of the early pioneers of Dadaism, he introduced the concept of the 'found object'.

Dulles, John Foster (1888–1959) American Republican statesman and lawyer. He was secretary of state (1953–59) under EISENHOWER, and developed the confrontational foreign policy of 'brinkmanship' in the Cold War against the USSR.

Dumas, Alexandre [*Dumas père*] (1802-70) French novelist and dramatist, whose entertaining Romantic novels, e.g. *The Three Musketeers*, achieved instant and lasting popularity. His illegitimate son, also called **Alexandre Dumas** [*Dumas fils*] (1824-95) also wrote novels and plays, e.g. *Camille*.

Du Maurier, Dame Daphne (1907–89) English novelist and short-story writer. Several of her works have been made into successful films, e.g. the novels *Rebecca* and *Jamaica Inn*, and the stort story 'Don't Look Now'.

Duncan, Isadora (1878–1927) American dancer and choreographer. She developed a free, interpretative style of dancing that was very influential on the development of modern dance (e.g. on DIAGHILEV). Her ardent feminism and unconventional lifestyle alienated many of her contemporaries. She died in a terrible accident when her the scarf she was wearing became caught in the rear wheel of the car in which she was travelling.

du Pré, Jacqueline (1945–87) English cellist. She became recognized as one of the world's finest cellists in the 1960s and married BARENBOIM in 1967, with whom she frequently performed. Her performing career came to an end in 1973, after she developed multiple sclerosis. Although confined to a wheelchair, she pursued an active teaching career until her death.

Dürer, Albrecht (1471-1528) German engraver and painter. A leading figure of the Northern Renaissance, his work is outstanding in its attention to detail and its emotional content. A superb draughtsman, his albums of engravings were highly influential on other artists.

Durrell, Lawrence [George] (1912–90) English poet, novelist and travel writer. His masterpiece is the series of sexual and linguistically elaborate novels comprising the 'Alexandria Quartet'.

Dvorak, Antonin (1841-1904) Czech composer. Strongly influenced by Slavonic folk music, his work was widely praised and he was made a director of the New York Conservatory in 1891. His most famous work is *Symphony No. 9 from the New World*.

Dyck, Sir Anthony van (1599-1641) Flemish painter. Renowned for his unique and influential style of portraiture, investing his sitters with character and refinement of detail, he became court painter to CHARLES I in 1632.

Dylan, Bob [Robert Allen Zimmerman] (1941–) American folk/rock singer and songwriter. He became the most prominent 'protest' folksinger in the 1960s and his lyrics are very highly regarded by some critics. *See also* BAEZ.

Earhart, Amelia (1898–1937) American aviator. She was the first woman to make a solo flight across the Atlantic (1932), after which she became a celebrity. She disappeared on a flight across the Pacific while attempting a round-the-world flight.

Eastman, George (1854–1932) American inventor of photographic equipment and philanthropist. His invention of the Kodak roll-film camera revolutionized the photographic industry, as did his development of colour photography in the late 1920s.

Eco, Umberto (1932–) Italian critic, novelist and semiologist. A leading literary critic, he is also highly regarded as a writer of fiction. His best-known work is the medieval philosophical whodunit, *The Name of the Rose.*

Eddy, Mary Baker (1821–1910) American religious leader. A faith healer, she devised a system of healing based on the Bible which she called 'Christian Science', and founded the Church of Christ, Scientist, in Boston (1879).

Eden, Sir [Robert] Anthony [1st Earl of Avon] (1897–1977) British Conservative statesman. He served several terms as foreign minister and was prime minister (1955–57). He resigned following the Suez Crisis, when British and French occupation of Egypt after NASSER's nationalization of the Suez Canal received worldwide condemnation.

Edison, Thomas [Alva] (1847–1931) American inventor. One of the most prolific and successful inventors of all time, he patented over a thousand inventions, including the gramophone, the incandescent electric light bulb, and the microphone.

Edward, Prince *see* ELIZABETH II.

Ehrlich, Paul (1854–1915) German bacteriologist. He did significant research into immunology and chemotherapy, and developed a cure for syphilis (1910). He was awarded the 1908 Nobel prize for physiology or medicine.

Eichmann, [Karl] Adolf (1906–62) Austrian Nazi leader and war criminal. He oversaw the 'Final Solution' (deportation of Jews to death camps) and escaped to Argentina at the end of the war but was captured and tried for crimes against humanity by the Israelis in 1960 and executed.

Eijkman, Christiaan (1858–1930) Dutch physi-

cian who discovered that beriberi is caused by nutritional deficiency. His research led to the discovery of 'essential food factors', i.e. vitamins. He shared the 1929 Nobel prize for physiology or medicine with Sir Frederick HOPKINS.

Einstein, Albert (1879–1955) German-born physicist and mathematician. His formulations of the special theory of relativity (1906) and general theory of relativity (1916), and research into quantum theory, mark him as one of the greatest of all thinkers. He was awarded the 1921 Nobel prize for physics. Being Jewish and a pacifist, he was forced to flee Nazi Germany in 1933, and became a US citizen in 1940.

Eisenhower, Dwight D[avid] (1890–1969) American general and Republican statesman, known as 'Ike'. He became supreme commander of the Allied forces in 1943 and 34th president of the US (1953–60).

Eisenstein, Sergei Mikhailovich (1898–1948) Soviet film director. He served with the Red Army (1918–20) during the Civil War and became one of the most influential directors of all time with films such as *Battleship Potemkin* (1925), in which he deployed his theory of film montage.

Albert Einstein

Elgar, Sir Edward [William] (1857–1934) English composer. A master of many styles, he became recognized as the leading British composer with works such as the *Enigma Variations*, the oratorio *The Dream of Gerontius* and his famous *Cello Concerto.*

El Greco [Domenikos Theotocopolous] (1541–1614) Cretan-born Spanish painter, sculptor and architect. He studied in Italy before settling in Toledo where he worked in an emotional and spiritually evocative style, using a palette of cold blues and greys at a time when the vogue was for warmer colours. His works include *The Burial of Count Orgaz.*

Eliot, George [pseud. of Mary Ann Evans] (1819–80) English novelist. Her novels, e.g. *The Mill on* *the Floss* and *Middlemarch*, deal with the problems of ethical choice in the rapidly changing rural environment of 19th-century England. She lived, unmarried, with her partner, the English writer **George Henry Lewes** (1817–78).

George Eliot

Eliot, T[homas] S[tearns] (1888–1965) American-born English poet and critic. His early poetry, e.g. *The Waste Land*, is concerned with the breakdown of civilized values in the postwar 'Jazz Era'. He also wrote verse dramas and published critical works.

Elizabeth II (1926–) queen of the United Kingdom from 1952. The daughter of GEORGE VI. She married Prince PHILIP in 1947, and has four children: Prince CHARLES, Princess Anne (1950–), Prince Andrew (1960–) and Prince Edward (1964–). She is regarded as the most formal of all modern European monarchs.

Ellington, Duke [Edward Kennedy Ellington] (1899–1974) American jazz composer, pianist and band-leader. Regarded as one of the finest jazz composers, his many works include 'Mood Indigo' and 'Sophisticated Lady'.

Elton, Charles Sutherland (1900–) English ecologist. His field studies of animal communities in their environments raised awareness of the ability of animals to adapt to changing habitats.

Emerson, Ralph Waldo (1803–82) American essayist, philosopher and poet who developed a philosophy of 'transcendentalism', based upon the authenticity of the individual conscience against both church and state.

Engels, Friedrich *see* MARX, KARL.

Epicurus (341–271 BC) Greek philosopher. He founded the Epicurean school of philosophy, which teaches that the highest good and proper study of mankind is pleasure, and that this can be attained through a life of simplicity and moderation.

Erasmus, Desiderius (*c*.1466–1536) Dutch humanist. One of the leading scholars of the Renais-

sance, he was a strong advocate of tolerance in an intolerant age. His works include *Praise of Folly*, written partly in tribute to his friend Thomas MORE.

Ernst, Max (1891–1976) German-born French painter. He was a leading member of both the Dada and Surrealist movements and pioneered the use of collage and photomontage.

Euripides (480–406 BC) Greek dramatist. He was the youngest of the three great Greek tragedians, the others being AESCHYLUS and SOPHOCLES. Nineteen of his plays are extant, the most notable including *Medea* and *The Trojan Women*.

Evans, Sir Arthur [John] (1851–1941) English archaeologist. His excavations of the palace of Knossos in Crete resulted in the rediscovery of Minoan civilization.

Evans, Dame Edith [Mary Booth] (1888–1976) English actress, notable for her command of a wide variety of roles. She created the role of Lady Utterwood in SHAW's *Heartbreak House* and gave a definitive performance as Lady Bracknell in Wilde's *The Importance of Being Earnest*.

Eysenck, Hans [Jürgen] (1916–) German-born British psychologist. He has been a notable critic of FREUD's theory of psychoanalysis and holder of controversial views on the role of genetic factors in determining intelligence.

Fairbanks, Douglas [Douglas Elton Ullman] (1883–1939) American film actor and producer, who became one of the leading stars of silent films. His son, **Douglas Fairbanks Jr** (1909–), was also an actor. His first wife was Joan CRAWFORD.

Faraday, Michael (1791–1867) English chemist and physicist. He was an assistant to DAVY and discovered electromagnetic induction and investigated electrolysis.

Farouk, King *see* NASSER, GAMAL ABDEL.

Farquhar, George (1678–1707) Irish dramatist. His lightly satirical plays, e.g. *The Beaux' Stratagem*, mark an important transitional stage between the bawdy world of Restoration comedy and the more decorous 18th-century stage.

Fassbinder, Rainer Werner (1946–82) German film director. His films, e.g. *The Bitter Tears of Petra Von Kant*, are noted for their social comment, particularly on post-war Germany.

Faulkner, William [Harrison] (1897–1962) American novelist. Considered a master of the modern novel, his best-known work, *The Sound and the Fury*, deals with social tensions in the Old South. He was awarded the 1949 Nobel prize for literature.

Fawcett, Dame Millicent [Millicent Garrett]

(1847–1929) English feminist. She became first president of the National Union of Women Suffrage Societies (1897–1919), and opposed the more militant tactics of PANKHURST.

Fellini, Federico (1920–93) Italian film director. The best known of his highly individual films is *La Dolce Vita* (1960), a cynical portrayal of Roman high society.

Fermat, Pierre de (1601–65) French mathematician. The founder of number theory, he initiated, with PASCAL, the study of probability theory.

Fermi, Enrico (1901–54) Italian-born American physicist. Awarded the Nobel prize for physics in 1938 for his work on radioactive substances and nuclear bombardment, he fled to the US. He built the first nuclear reactor at Chicago in 1942.

Ferrier, Kathleen (1912–53) English contralto. A highly regarded singer whose tragically short career ended with her death from cancer, she created the title role in BRITTEN's *Rape of Lucretia* and sang regularly with Bruno WALTER.

Feydeau, Georges (1862–1921) French dramatist, noted for his many bedroom farces, e.g. *Hotel Paradise*.

Feynman, Richard (1918–88) American physicist. He shared the 1965 Nobel prize for physics for his work in quantum electrodynamics.

Field, John (1782–1837) Irish composer and pianist who settled in Russia (1804–32) then in London. He is particularly noted for his 19 *Nocturnes*, and was an influence on CHOPIN.

Fielding, Henry (1707–54) English novelist and dramatist. His early satirical plays, e.g. *Pasquin*, provoked the British government into passing strict censorship laws and his unconventional novels which followed caused similar controversy. His greatest work, *Tom Jones*, surveys the whole of English society with masterly insight and compassion. He also wrote important tracts on social problems and worked tirelessly against legal corruption.

Fields, W. C. [William Claude Dukenfield] (1880–1946) American comedian, noted for his hard drinking, red nose, gravel voice and antipathy to children and animals.

Fischer, Bobby [Robert James Fischer] (1943–) American chess player. A grandmaster at 15 he became the first US player to win the world championship (1972) when he won against SPASSKY.

Fitzgerald, Ella (1918–1996) American jazz singer, whose highly praised vocal range, rhythmic subtlety and clarity of tone made her one of the most popular singers of her day.

Fitzgerald, F[rancis] Scott [Key] (1896–1940) American novelist and short-story writer. His works are moralistic fables set in 1920s 'Jazz Age' High Society and include *The Great Gatsby* and *Tender is the Night*.

F. Scott Fitzgerald

Flagstad, Kirsten (1895–1962) Norwegian soprano, noted for her roles in Wagner's operas. She is regarded as one of the finest Wagnerian singers of all time.

Flaherty, Robert [Joseph] (1884–1951) American documentary film director. His films, e.g. *Nanook of the North*, set high standards for all following documentary film makers.

Flaubert, Gustave (1821–80) French novelist. His masterpiece is his first published novel, *Madame Bovary*, a study of self-deception, adultery and suicide in rural France. Flaubert is noted for his meticulously impersonal and objective narrative.

Fleming, Sir Alexander (1881–1955) Scottish bacteriologist. He discovered the antibacterial qualities of the enzyme lysozome, the substance he dubbed 'penicillin'. He shared the 1945 Nobel prize for physiology or medicine with CHAIN and FLOREY.

Fleming, Ian [Lancaster] (1908–64) English novelist. His series of novels featuring the British secret agent James Bond, e.g. *Goldfinger*, were enormous successes and have all been filmed.

Fletcher, John (1579–1625) English dramatist. One of the most popular dramatists of his day, he frequently collaborated with other dramatists, such as SHAKESPEARE, but most notably with **Sir Francis Beaumont** (1584–1616), from 1606–13). Their best-known works are tragicomedies such as *The Maid's Tragedy*.

Florey, Howard Walter, Baron (1898–1968) Australian pathologist. He shared the 1945 Nobel prize for physiology or medicine with Sir Alexander FLEMING and CHAIN for their work on penicillin.

Flynn, Errol [Leslie Thomas Flynn] (1909–59) Australian-born American film actor. His starring roles in the swashbuckling tradition, e.g. *Captain Blood*, earned him considerable popularity.

Foch, Ferdinand (1851–1929) French general and

marshal of France (1918). He was given command of the Allied forces in March 1918, and led the Allies to victory following the arrival of US troops in July 1918.

Fokine, Michel (1880–1942) Russian-born American ballet dancer and choreographer. With Diaghilev in Paris, he created a new style of ballet, in which all the elements, dance, music, costume and *mise en scène*, formed a coherent whole.

Fonda, Henry (1905–82) American film actor, often seen as the epitome of 'decent' America, a man determined to set injustices right. His daughter, **Jane [Seymour] Fonda** (1937–) also won recognition as a fine actress, although her outspoken opposition to the Vietnam war was heavily criticised.

Fonteyn, Dame Margot [Margaret Hookham] (1919–91) English ballerina. Regarded as one of the finest classical ballerinas of the century, she partnered Nureyev at the age of 43.

Dame Margot Fonteyn

Foot, Michael [Mackintosh] (1913–) British Labour politician. A leading left-winger, pacifist and CND member, he was secretary of state of employment (1974–76) and leader of the House of Commons (1976–79), and succeeded Callaghan as leader of the Labour Party (1980–83). His many books include biographies of Swift and Bevan.

Ford, Ford Madox [Ford Hermann Hueffer] (1873–1939) English novelist, poet and critic. Writer of over 80 novels, e.g. *The Good Soldier*, he also founded the *Transatlantic Review* in 1924, and gave generous encouragement to many writers.

Ford, Gerald R[udolph] (1913–) American Republican statesman and 38th president of the US (1974–77). He replaced Agnew as Nixon's vice-president in 1973, becoming president the following year, after Nixon's impeachment and resignation.

Ford, Henry (1863–1947) American car designer and manufacturer. His Model T Ford, first introduced in 1908, was enormously successful and its production line manufacture became a role model for much of industry.

Ford, John (1586–c.1639) English dramatist, notable for such revenge tragedies as *'Tis Pity She's a Whore*. His bleak, objective vision of human suffering and his command of blank verse were highly praised by T. S. Eliot.

Ford, John [Sean Aloysius O'Fearna] (1895–1973) American film director. He is regarded as one of the greatest directors for his epic and poetic vision of history, particularly that of the American West, e.g. *Stagecoach* and *The Searchers*.

Forster, E[dward] M[organ] (1879–1970) English novelist and critic. His novels, e.g. *Howards End*, are mainly concerned with moral and ethical choices, and the personal relationships of educated, middle-class people.

Fox, Charles James (1749–1806) English Whig statesman. A formidable orator, he was strongly opposed to the wars against the American and French revolutionaries and was a vigorous opponent of the slave trade.

Fox, George (1624–91) English religious leader. Brought up a Puritan, he preached opposition to established religion and advocated peace and toleration. He founded the Society of Friends (known popularly as 'Quakers') in 1647.

Fragonard, Jean-Honoré (1732–1806) French painter. One of the greatest exponents of Rococo art, his early works were historical scenes on a grand scale, but he is known for his smaller, picturesquely pretty canvases.

Francesca, Piero della *see* Piero della Francesca.

Francis of Assisi, St [Giovanni Bernardone] (1181–1226) Italian monk. He abandoned a military career to care for the poor and founded a 'brotherhood' of friars in 1210, and, in 1212, an order for women, the Poor Clares. He preached poverty, chastity and obedience to the Church, and received the stigmata in 1224. He was canonized in 1228.

Franck, César [Auguste] (1822–90) Belgian-born French composer. His best works were written late in life, e.g. his string quartet, and his work only recieved public acclaim after his death.

Franck, James (1882–1964) German-born American physicist. With the German physicist **Gustav Ludwig Hertz** (1887–1975), he shared the 1925 Nobel prize for physics for work on the quantum theory, notably the effects of bombarding atoms with electrons.

Franco, Francisco (1892–1975) Spanish general and dictator. He led the right-wing rebellion against the Spanish Republican government during the Spanish Civil War (1936–39). He became leader of the Fascist Falange Party in 1937, and ruled Spain from 1939 until his death.

Frank, Anne (1929–45) German-born Dutch Jewish girl. Her journal describing her family's experiences while hiding from the Nazis is one of the most moving accounts of the terrible suffering of the Jewish people during the World War II.

Franklin, Benjamin (1706–90) American author, statesman and author. He helped draft the American Declaration of Independence, and played an active role in American political life for most of his long life. He published a highly entertaining *Autobiography* and also invented the lightning conductor.

Franklin, Rosalind Elsie (1920–58) British chemist. Her X-ray crystallography research into DNA contributed to the discovery of its structure by James WATSON and CRICK.

Frayn, Michael (1933–) English dramatist and novelist, noted for his dry, sardonic humour.

Frazer, Sir James George (1854–1941) Scottish scholar and anthropologist, whose study of religious customs and myth influenced FREUD and many 20th-century writers.

Joe Frazier (1944–) American boxer, born in Philadelphia, who held the world heavyweight boxing title from 1968–1973, becoming the first man to beat MUHAMMED ALI in a profession boxing match in the process. Before turning profesional, Frazier was also the 1964 Olympic heavyweight boxing gold medallist.

Joe Frazier

Frege, [Friedrich Ludwig] Gottlob (1848–1925) German mathematician and philosopher. He is regarded as having laid the foundations for both modern mathematical logic and the philosophy of language.

French, Sir John [Denton Pinkstone], [1st Earl of Ypres] (1852–1925) English field marshal. He commanded the British Expeditionary Force in France (1914–15) and became Lord Lieutenant of Ireland (1918–21) during the Anglo-Irish War.

Freud, Lucian (1922–) German-born British painter. The grandson of Sigmund FREUD, he is renowned for his nudes and portraits, often painted from odd angles in an extreme-realist style.

Freud, Sigmund (1856–1939) Austrian psychiatrist, who founded psychoanalysis. His writings have been enormously influential on 20th-century thought. The main tenet of Freudian theory is that neuroses and dreams are repressed manifestations of sexual desire. His stress on the importance of sex was rejected by ADLER and JUNG. His daughter **Anna Freud** (1895–1982) pioneered child psychology in the UK.

Friedman, Milton (1912–) American economist. His controversial monetarist theory of economics, stressing the need for minimal government intervention, became the dominant economic theory of the 1980s. He was awarded the 1976 Nobel prize for economics.

Friedrich, Casper David (1774–1840) German Romantic painter. Largely uninfluenced by other artists or trends, his work was highly controversial in its treatment of landscape.

Frisch, Otto Robert (1904–79) Austrian-born British nuclear physicist. He and his aunt, Lise MEITNER, discovered nuclear fission, and their work led directly to the invention of the atom bomb.

Frisch, Ragnar *see* TINBERGEN, JAN.

Frost, Robert [Lee] (1874–1963) American poet. His quiet, lyrical poems, e.g. 'Stopping by Woods on a Snow Evening', have been admired for their enigmatic use of symbolism.

Fry, C[harles] B[urgess] (1872–1956) English sportsman, regarded as one of the greatest all-round sportsmen ever, representing England in athletics, cricket and soccer.

Fry, Christopher (1907–) English dramatist, whose verse dramas, e.g. *The Lady's Not for Burning*, were popular with both critics and public.

Fry, Elizabeth (1780–1845) English prison reformer. A Quaker and preacher, she campaigned for prison reform and founded hostels for the homeless.

Fuchs, Klaus [Emil Julius] (1911–88) German-born British physicist. He began work on British atom-bomb research in 1941 and was jailed in 1950 for 14 years for passing details to the Soviet Union.

Fuchs, Sir Vivian Ernest (1908–) English explorer and scientist. He led the Commonwealth Trans-Antarctic Expedition (1955–58), which made the first overland crossing of Antarctica.

Fugard, Athol (1932–) South African dramatist, whose plays, e.g. *Boesman and Lena*, explore the tragedy of racial tension caused by apartheid in South Africa.

Fuller, [Richard] Buckminster (1895–1983) American architect and engineer. He invented the 'geodesic dome', a lightweight framework consisting of a set of polygons in the shape of a shell.

Furtwängler, Wilhelm (1886–1954) German conductor. He became one of the most popular conductors in Europe, particularly for his highly charged interpretations of WAGNER's music.

Fuseli, Henry [Johann Heinrich Füssli] (1741–1825) Swiss-born British painter. His paintings are mannered and Romantic with a sense of the grotesque and macabre.

Gable, [William] Clark (1901–60) American film actor. His rugged good looks, sardonic wit and easy-going charm made him one of the most popular film stars of his day.

Gabor, Dennis (1900–79) Hungarian-born British engineer. He was awarded the 1971 Nobel prize for physics for his invention (in 1947) of the hologram.

Gaddafi *or* **Qaddafi, Moammar al** (1942–) Libyan statesman and military dictator. He took power in a coup in 1969 and became president in 1977. Regarded almost universally as an unpredictable and often dangerous leader, Gaddafi has openly supported terrorist groups around the world.

Moammar al Gaddafi

Gagarin, Yuri [Alekseevich] (1934–68) Soviet cosmonaut, who became, in 1961, the first man in space, when his Vostok satellite circled the earth. He died in a plane crash.

Gainsborough, Thomas (1727–88) English painter. He worked as a portrait painter, but his keen interest in landscape painting pervades most of his work, his sitters often being portrayed out of doors. He developed a light, rapid painting style based on a delicate palette.

Gaitskell, Hugh [Todd Naylor] (1906–63) British Labour politician. He was regarded as being on the right of his party, having introduced, as chancellor of the exchequer (1950–51), national health

service charges. He was Labour Party leader (1955–63).

Galbraith, John Kenneth (1908–) Canadian-born American economist and diplomat. He has been notably critical of the wastefulness of capitalist society. He was American ambassador to India (1961–63). His works include *The Affluent Society*.

Galilei, Galileo (1564–1642) Italian astronomer, mathematician and natural philosopher. An innovative thinker and experimenter, he demonstrated the isochronism of the pendulum and showed that falling bodies of differing weight descend at the same rate. He also developed the refracting telescope and became convinced of the truth of COPERNICUS's theory that the earth revolved around the sun. He was unable to prove it and was forced to retract his support publicly.

Galileo Galilei

Gallup, George Horace (1901–84) American statistician. He developed the opinion poll into a sophisticated device, the 'Gallup Poll', for testing public opinion, most notably on elections.

Galsworthy, John (1867–1933) English novelist and dramatist. His plays, e.g. *Strife*, attacked social injustice. His novels include the Forsyte saga triology. He was awarded the 1932 Nobel prize for literature.

Galton, Sir Francis (1822–1911) English scientist and explorer. The cousin of DARWIN, he travelled widely in Africa and made significant contributions to meteorology and to heredity. He also developed the science of fingerprinting.

Gama, Vasco da (*c.*1469–1525) Portuguese navigator. He discovered the route to India round the Cape of Good Hope (1497–99), and became Portuguese viceroy in India in 1524.

Gandhi, Indira (1917–84) Indian stateswoman and prime minister. The daughter of NEHRU, she became prime minister (1966–77) of India. Her second term of office (1980–84) saw much ethnic strife and she was assassinated by her Sikh bodyguards. Her son **Rajiv** (1944–91) became prime minister in 1984, and was killed in a suicide bomb attack.

Gandhi, Mahatma [Mohandas Karamchand Gandhi] (1869–1948) Indian nationalist statesman and spiritual leader ('Mahatma' means 'Great Soul'). A passionate advocate of non-violent resistance, Gandhi's long campaign against British rule in India, using tactics of civil disobedience through passive resistance and hunger strikes, had great influence on world public opinion. He also struggled for reconciliation between Hindus and Moslems, and championed the cause of the Hindu Harijan caste of 'untouchables'. He was assassinated by a Hindu extremist, in the wake of India's independence and partition.

Mahatma Gandhi

Garbo, Greta [Greta Lovisa Gustafson] (1905–90) Swedish-born American film actress noted for her austere and remote beauty. Her films include *Queen Christina*.

Garibaldi, Giuseppe (1807–82) Italian patriot. Forced into exile in 1834, he returned during the year of revolutions, 1848, and took part in the defence of Rome against the French, and, in 1860, with a force a thousand strong, he took Naples and Sicily for the newly united Italy. He is regarded as the most significant figure in the struggle for Italian independence.

Garland, Judy [Frances Gumm] (1922–69) American film actress and singer. She became one of the most loved child stars of the cinema in *The Wizard of Oz* and later starred in such films as *Easter Parade* and *A Star is Born*.

Garrick, David (1717–79) English actor and dramatist. A pupil of Samuel Johnson at Lichfield, he accompanied him to London in 1737, soon making his mark as an actor. At home with tragedy, comedy or farce, he dominated the English stage for many years, and was actor-manager of Drury Lane Theatre 1747–76.

Gaskell, Mrs Elizabeth [Elizabeth Cleghorn Stevenson] (1810–65) English novelist. Her novels, e.g. *North and South*, are often concerned with the injustices of the 'two-nation' society of 19th-century England, although her most popular novel is *Cranford*, a gentle study of life in a small village.

Gauguin, Paul (1848–1903) French painter. One of the greatest exponents of Postimpressionism, his interest in primitive art led to him settling in the South Pacific islands, where he painted some of his most important masterpieces.

Gay, John (1685–1732) English dramatist and poet. His masterpiece is *The Beggars' Opera* (1728), the ballad opera on which BRECHT based his *Threepenny Opera*.

Gell-Mann, Murray (1929–) American physicist. He introduced the quark hypothesis into physics and was awarded the 1969 Nobel prize for physics for his research into particle physics.

Genet, Jean (1910–86) French dramatist and novelist. His works are often based on his experiences in the criminal underworld where he spent much of his life.

Genghis Khan [originally Temujin] (*c.*1162–1227) Mongol leader who united the Mongol tribes and conquered China, establishing an empire that stretched from the Black Sea to the Pacific.

Géricault, Théodore (1791–1824) French Romantic painter. The realism and baroque dynamism of his work had a huge influence on many painters.

Gershwin, George (1898–1937) American composer and pianist. He and his brother, the lyricist **Ira Gershwin** (1896–1983), created several very popular musicals now considered masterpieces of American music.

Getty, J[ean] Paul (1892–1976) American industrialist and art collector, renowned for his wealth, miserliness and acquisition of works of art.

Ghirlandaio, Domenico (1449–94) Florentine painter who ran a workshop with his brothers **Benedetto** (1458–97) and **Davide** (1452–1525), where he produced frescos and altarpieces for a number of churches in Florence. His son **Ridolfo** (1483–1561) was MICHELANGELO's tutor.

Giacometti, Alberto (1901–66) Swiss sculptor and painter. He became a Surrealist in 1930 and was influenced by SARTRE's existentialism.

Giap, Vo Nguyen (1912–) Vietnamese general. He led the Viet Minh army against the French and commanded the North Vietnamese army against the US during the Vietnam War.

Gibbon, Edward (1737–94) English historian. His masterpiece is his *History of the Decline and Fall of the Roman Empire* (1776–88), a work that remains a great historical study.

Gielgud, Sir [Arthur] John (1904–) English stage and film actor and producer. Regarded as one of the leading Shakespearian actors of the 20th century, he has also made appearances in several Hollywood films.

Gierek, Edward (1913–) Polish Communist statesman. He became leader of the Polish United Workers Party following GOMULKA's resignation in 1971. He presided over increasing industrial unrest and the rise of the union Solidarity.

Gilbert, Sir W(illiam) S(chwenck) (1836–1911) English dramatist and librettist. His collaboration with the composer **Sir Arthur Sullivan** (1842–1900) resulted in the popular 'Savoy Operas', e.g. *The Gondoliers* and *The Mikado*, of which there are 13 in all and which have retained their popularity.

Gill, [Arthur] Eric [Rowton] (1882–1940) English sculptor, engraver, typographer and writer. His work has been influential in several areas of art and design.

Gillespie, Dizzy [John Birks Gillespie] (1917–93) American jazz trumpeter and bandleader, renowned as a virtuoso trumpeter.

Ginsberg, Allen (1926–) American poet, regarded as the leading poet of the Beat Generation, who had much influence on the hippy culture of the 1960s.

Giorgione [Giorgio da Castelfranco] (*c.*1477–1510) Venetian painter. Little of his work has survived, although he is one of the most influential painters of his time, the importance of his work lying in his treatment of landscape, imbuing it with strong atmosphere and moods.

Giotto di Bondone (1267–1337) Florentine painter and architect who developed spatial perspective and fully rounded figures in a departure from the flat, decorative imagery of the Byzantine era.

Giscard d'Estaing, Valéry (1926–) French statesman. He served as minister of finance under DE GAULLE (1962–66) and POMPIDOU (1969–74) and was elected president (1974–81) following the latter's death.

Gladstone, William Ewart (1809–98) British statesman. He became the leader of the Liberals in 1867, and was subsequently prime minister four times: 1868–74, 1880–5, 1886, and 1892–4. He had a long rivalry with his Tory opponent, DISRAELI, whose imperialism he vehemently opposed.

Glashow, Sheldon *see* WEINBERG, STEVEN.

Glass, Philip (1937–) American composer. One of the leading avant-garde composers of the 1970s, he is noted for his deep interest in Eastern harmonies and use of repeated motifs.

Gluck, Christoph Wilibald von (1714–87) German composer. He is especially noted for his operas, e.g. *Orfeo and Eurydice*.

Godard, Jean-Luc (1930–) French film director who is regarded as one of the most influential New Wave French film directors of the 1950s with such films as *Breathless* and *Week-End*.

Gödel, Kurt (1906–78) Austrian-born American logician and mathematician. 'Godel's theorem', shows the existence of undecidable elements in arithmetic systems.

Godwin, William (1756–1836) English novelist and philosopher. He questioned the validity of established goverment and institutions, notably marriage, and had a strong influence on many radicals, including SHELLEY, who married his daughter Mary.

Goebbels, [Paul] Joseph (1897–1945) German Nazi politician. Head of the Nazi Party propaganda section in 1929 and minister of enlightenment and propaganda (1933–45). He committed suicide after shooting his wife and children.

Goering, Hermann *see* GÖRING, HERMANN.

Goethe, Johann Wolfgang von (1749–1832) German poet, dramatist, novelist, philosopher, scientist and statesman. He was one of the most learned and influential figures of his time. His masterpiece is the verse drama *Faust*.

Gogh, Vincent Van (1853–90) Dutch painter who studied theology before taking up painting in 1880. His art was thoroughly unacademic in its realistic subject matter and bold, expressionistic style, e.g. *The Potato Eaters*, but he was later influenced by the colours of Degas and Gauguin. He spent the last two years of his life in southern France, partly in an asylum, a period of intense creativity arising out of personal anguish, e.g. *The Cornfield*, painted at the scene where he shot himself.

Vincent Van Gogh

Gogol, Nikolai Vasilievich (1809–52) Russian short-story writer, dramatist and novelist. His two greatest works are scathing satires on Russian bureaucracy and incompetence, the play *The Government Inspector* and the novel *Dead Souls*.

Golding, William [Gerald] (1911–93) English novelist. His first novel *The Lord of the Flies* established him as a major modern novelist. He was awarded the Nobel prize for literature in 1983.

Goldoni, Carlo (1707–93) Italian dramatist. Around 150 of his 250 plays are comedies, frequently featuring satirical attacks on the aristocracy and invariably set in his native Venice.

Goldsmith, Oliver (1728–1774) Irish poet and essayist who settled in London in 1756. An entertaining essayist and one of the leading poets of his day, he did not receive acclaim until late in life. His two greatest works are the novel *The Vicar of Wakefield* and his hugely successful play *She Stoops to Conquer*.

Goldwyn, Samuel [Samuel Goldfish] (1882–1974) Polish-born American film producer. One of the founders of the Hollywood movie business, forming Metro-Goldwyn-Mayer with Louis B. MAYER in 1924, he was famous for his (mainly apocryphal) 'Goldwynisms', e.g. 'Include me out'.

Goodman, Benny [Benjamin David] (1909–86) American jazz clarinetist and bandleader. Known as the 'King of Swing', he was one of the first white jazz bandleaders to hire Black players.

Gorbachev, Mikhail Sergeevich (1931–) Soviet Communist statesman. He became general secretary of the Soviet Communist party in 1985 and soon began instituting far-reaching social and political reforms. He became 'executive president' in 1990, with wide-ranging powers, facing strong opposition from radicals such as YELTSIN and from hard-line Communists. His powers were insufficient to withstand the break-up of the USSR, and he resigned in December 1991.

Mikhail Gorbachev

Gore, Al[bert] (1948–) American politician. A former investigative reporter, tobacco and livestock farmer, and developer, he became vice-president of the USA in 1992.

Göring *or* **Goering, Hermann [Wilhelm]** (1893–1946) German Nazi politician and military leader. He served HITLER as Prussian prime minister, minister of the interior and air minister (1933–45), organizing the rebuilding of the Luftwaffe.

Gorki *or* **Gorky, Maxim** [Aleksey Maximovich Peshkov] (1868–1936) Russian novelist, dramatist and short-story writer. A firm communist, he helped formulate the doctrine of socialist realism in the USSR in the 1930s.

Gorky, Arshile [Vosdanig Manoog Adoian] (1904–48) Armenian-born American painter. Originally a surrealist, he developed an abstract approach that was influential on action painters.

Goya y Lucientes, Francisco de (1746–1828) Spanish painter and printmaker. His strong, freeflowing technique and powerful pictorial style are demonstated in early portraits of the royal family, to whom he was court painter, and in later works inspired by the behaviour of the French army in their invasion of Spain.

Grace, W[illiam] G[ilbert] (1848–1915) English cricketer and physician, one of the first English cricketers to become a national institution. He was also noted for his cunning gamesmanship.

Graham, Billy [William Franklin Graham] (1918–) American evangelist. His evangelical crusades go all over the world.

Graham, Martha (1893–1991) American dancer and choreographer who is regarded as one of the founders of modern dance.

Grahame, Kenneth (1859–1932) Scottish author. His masterpiece, *The Wind in the Willows* (1908), is a children's classic.

Grainger, Percy [Aldridge] (1882–1961) Australian-born American pianist and composer who was notable enthusiast for folk songs, on which many of his works are based.

Grant, Cary [Archibald Alexander Leach] (1904–86) English-born American film actor who became one of Hollywood's leading stars in light comedy roles and thrillers.

Cary Grant

Grant, Ulysses S[impson] (1822–85) American soldier and 18th US President. Commander of the Union forces in the Civil War, he was elected president in 1869. He established universal suffrage for all citizens, regardless of colour.

Grappelli, Stéphane (1908–) French jazz violinist. A founder of the Quintette de Hot Club de France, which became the leading European jazz group, he is still regarded as the finest jazz violinist ever.

Grass, Günter [Wilhelm] (1927–) German novelist, dramatist and poet. His works include the novel *Die Blechtrommel* (*The Tin Drum*, 1959), a grimly comic satire on the collapse of the Third Reich as seen through the eyes of a boy.

Graves, Robert [Ranke] (1895–1985) English poet, novelist and critic. His works include his classic autobiographical account of World War I soldiering, *Goodbye to all That* (1929), several

Gray, Thomas (1716–71) English poet. His best-known poem, 'Elegy Written in a Country Churchyard' (1751), is one of the most-quoted poems in the English Language.

Greene, [Henry] Graham (1904–1991) English novelist. Regarded as one of the greatest modern novelists, he converted to Roman Catholicism in 1926 and his religious beliefs play an important part in work, e.g. *The Heart of the Matter*.

Greer, Germaine (1939–) Australian feminist, writer and broadcaster who is best known for her controversial work, *The Female Eunuch* (1970).

Grieg, Edvard [Hagerup] (1843–1907) Norwegian composer of Scottish descent. Strongly influenced by Norwegian folk music, his works include music for IBSEN's *Peer Gynt*.

Grierson, John (1898–1972) Scottish documentary film director and producer, described as the 'father of British documentary'.

Griffith, Arthur (1871–1922) Irish nationalist leader. He founded Sinn Fein in 1905, and (with Michael COLLINS) signed the Anglo-Irish Treaty of 1921. He became the first president of the Irish Free State in 1922.

Griffith, D[avid] W[ark] (1875–1948) American film director and producer. His great technical skill was highly influential on other film-makers. In 1919, he founded United Artists with CHAPLIN, FAIRBANKS and PICKFORD.

Gris, Juan [José Victoriano Gonzàlez] (1887–1927) Spanish painter. He settled in Paris in 1906, where he became an associate of PICASSO and BRAQUE, and one of the leading Cubist painters.

Gromyko, Andrei Andreyevich (1909–89) Soviet statesman and diplomat. He was Soviet foreign minister (1957–1985) and a Politburo member (1973–89) and adapted effortlessly to each stage of relations with the West, from Cold War through 1970s detente to the GORBACHEV era.

Gropius, Walter (1883–1969) German-born architect who was a director of the influential Bauhaus school in Weimar and later in his life was a prefessor of architecture at Harvard. He is regarded as one of the most innovative and influential architects of the 20th century.

Grunewald, Matthias (*c*.1460–1528) German painter noted for his use of perspective, Gothic imagery, strong colour and an expressionistic style of distortion.

Guevara, Che [Ernesto Guevara] (1928–67) Argentinian-born Communist revolutionary. He joined Fidel CASTRO's forces in the Cuban revolution (1956–59). He subsequently led a guerrilla group in Bolivia, where he was killed by government troops.

Che Guevara

Guinness, Sir Alec (1914–) English stage and film actor. Regarded as one of the most versatile stage actors of his generation, he became a household name through his films, e.g. *Kind Hearts and Coronets* and *Star Wars*.

Gulbenkian, Calouste Sarkis (1869–1955) Turkish Armenian-born British financier, industrialist, diplomat and philanthropist. He endowed the Gulbenkian Foundation for the arts and sciences. His son, **Nubar Sarkis Gulbenkian** (1896–1972), was an Iranian diplomat and a philanthropist.

Guthrie, Sir [William] Tyrone (1900–71) English actor and theatrical producer. His productions included a controversial Hamlet in modern dress and several important Shakespeare productions.

Guthrie, Woody [Woodrow Wilson Guthrie] (1912–67) American folksinger and writer. His songs, which attack racial bigotry and the economic exploitation of the poor and immigrants, were a strong influence on 1960s 'protest' singers.

Hadrian [Publius Aelius Hadrianus] (76–138) Roman soldier and emperor. He spent much of his reign travelling through the empire, the boundaries of which he was concerned to make firm. He built Hadrian's Wall in the north of England.

Hahn, Otto (1879–1968) German physical chemist. With MEITNER and others, he undertook significant research which led to the discovery of nuclear fission. He was awarded the 1944 Nobel prize for chemistry.

Haig, Douglas, 1st Earl (1861–1928) British field marshal. In World War I he was appointed commander in chief of the British forces on the western front (1915–18). The terrible losses of soldiers under his command led to fierce criticism of his tactics. He founded the British Legion.

Haile Selassie [title of Ras Tafari Makonnen] (1892–1975) emperor of Ethiopia (1930–36, 1941–74). He lived in Britain during the occupation of his country by Italy (1936–41). In the early 1960s he helped establish the Organization of African Unity. The famine of 1973 created unrest which led to his deposition in a military coup. He is worshipped as a god by the Rastafarian cult.

Haile Selassie

Haitink, Bernard [Johann Herman] (1929–) Dutch conductor, renowned as an interpreter of Bruckner and MAHLER.

Hall, Sir Peter [Reginald Frederick] (1930–) English stage director and theatre manager. He was director of the Royal Shakespeare Company (1960–68), assistant director and then director of the National Theatre (1973–88), and artistic director of the Old Vic from January 1997.

Halley, Edmund (1656–1742) English astronomer and mathematician. In 1583, he calculated the orbit of the comet now named after him, and correctly predicted its return in following years.

Hals, Frans (c.1581–1666) Dutch painter. Noted for his lively and innovative group portraiture that moved away from formal trends. His most famous work is *The Laughing Cavalier*.

Hamilton, Alexander (1757–1804) American statesman. He founded the Federalist party in 1787 and, as first secretary of the Treasury (1789–95), founded the US federal bank.

Hammarskjöld, Dag [Hjalmar Agne Carl] (1905–61) Swedish secretary general of the the United Nations (1953–61). His period of office was a turbulent one, and he died in a plane crash during the Congo crisis. He was posthumously awarded the 1961 Nobel Peace Prize.

Hammerstein II, Oscar (1895–60) American songwriter and librettist, best known for his musicals written with Richard RODGERS, e.g. *South Pacific* and *The Sound of Music*.

Hammett, [Samuel] Dashiell (1894–1961) American novelist. He wrote realistic crime novels based on his own experiences as a Pinkerton detective. His best known are *The Maltese Falcon* and *The Thin Man*.

Hampden, John (1594–1643) English parliamentarian. Prosecuted in 1637 for refusing to pay CHARLES I's unpopular ship-money tax. He raised a regiment for Parliament during the Civil War and died of wounds received in battle.

Hancock, Tony [Anthony John Hancock] (1924–68) English comedian. His popular BBC radio and TV series, *Hancock's Half Hour*, established his well-known comic persona of the belligerent misfit. He committed suicide.

Handel, George Frederick [Georg Friedrich Händel] (1685–1759) German-born English composer. He became court composer to the Hanover court in 1710 and wrote over 40 operas, e.g. *Semele*, and many concertos and oratorios, e.g. *The Messiah*, as well as chamber and orchestral music, e.g. *Water Music*.

Hannibal, (247–182 BC) Carthaginian general. During the Second Punic War with Rome, he invaded Italy and crossed the Alps in 218. For 15 years he campaigned in Italy but was finally defeated in 204.

Hardie, [James] Keir (1856–1915) Scottish Labour politician. He was the first leader of the parliamentary Labour Party (1906–7). A committed pacifist, he withdrew from politics following the failure of parties of the Left in Europe to oppose World War I.

Hardy, Godfrey Harold (1877–1947) English mathematician, noted for his work on analytic number theory.

Hardy, Oliver *see* LAUREL, STAN.

Hardy, Thomas (1840–1928) English novelist, short-story writer and poet. As well known for his influential novels, e.g. *Far from the Madding Crowd*, as for his poetry, he is now ranked, with

ELIOT and YEATS, as one of the three great modern poets in English.

Thomas Hardy

Harlow, Jean [Harlean Carpentier] (1911–37) American film actress who became one of the screen's main sex symbols of the 1930s, with her tough, wise-cracking 'platinum blonde' image.

Harriman, W[illiam] Averell (1891–1986) American diplomat. He was the main negotiator of the nuclear test-ban treaty of 1963, between the US, UK and USSR. He was also governor of New York (1955–58).

Harris, Sir Arthur Travers (1892–1984) English air force officer, nicknamed 'Bomber Harris' for his advocacy of heavy bombing raids on German cities during World War II. The policy lasted from 1942 to the firebombing of Dresden in 1944.

Harrison, George (1943–) English singer-song-writer. He played lead guitar for the Beatles (1962–70).

Hart, Lorenz [Milton] (1895–1943) American lyricist who is best known for his collaborations with the composer Richard RODGERS, e.g. *The Boys from Syracuse* and *Pal Joey*.

Hartley, L[eslie] P[oles] (1895–1972) English novelist, whose best-known works are subtle portrayals of social and sexual intrigue in Edwardian England, e.g. *The Go-Between*.

Harvey, William (1578–1657) English physician. Court physician to both JAMES VI and CHARLES I, he published his discovery of the circulation of the blood in 1628.

Hasek, Jaroslav (1883–1923) Czech novelist and short-story writer. His masterpiece is *The Good Soldier Svejk* (1925), based on his own experiences in the Austro-Hungarian army.

Hastings, Warren (1732–1818) British administrator in India, the first governor-general of Bengal (1773–85). He established the East India Company as one of the most powerful forces in India. He resigned office in 1784 and was impeached before the House of Commons in 1788 for corruption.

Haughey, Charles [James] (1925–) Irish Fianna Fáil politician. He was prime minister of the Republic of Ireland (1979–81, 1982, 1988–92) but was forced to resign after several scandals.

Havel, Vàclav (1936–) Czech dramatist and statesman. His plays satirized the brutality and corruption of Czech communism and he was imprisoned for several years after the Soviet invasion of 1968. He was elected his country's president in 1989 but resigned in 1992. In 1993 he became president of the newly-formed Czech Republic.

Vàclav Havel

Hawke, Robert [James Lee] (1929–) Australian trades unionist and Labor statesman. He was prime minister (1983-92).

Hawking, Stephen William (1942–) English physicist. Widely regarded as perhaps the greatest physicist since EINSTEIN, his research into the theory of black holes has been highly acclaimed. He has suffered from a rare crippling nervous disease since the early 1960s and is confined to a wheelchair.

Hawks, Howard (1896–1977) American film director and producer. His films include several classics starring BOGART and Bacall, e.g. *The Big Sleep*, John WAYNE, and Marilyn MONROE.

Hawthorne, Nathaniel (1804–64) American novelist and short-story writer. New England Puritanism profoundly shaped his life and work, as in his masterpiece, *The Scarlet Letter*.

Haydn, [Franz] Joseph (1732–1809) Austrian composer. An innovative composer, he established the form of both the symphony and the string quartet. His huge oeuvre includes over 100 symphonies, 84 string quartets and the oratorio *The Creation*.

Hayek, Friedrich August von (1899–1992) Austrian-born British economist. A supporter of free-market policies and against government economic management, he shared the 1974 Nobel prize for economics with MYRDAL.

Hazlitt, William (1778–1830) English essayist and

critic. Highly influential in his own day, he remains one of the most important literary critics, especially for his essays on his contemporaries.

Healey, Denis [Winston] (1917–) English Labour politician. Chancellor of the exchequer (1974–79) and deputy leader of his party (1980–83), he is widely regarded as one of the most impressive modern British politicians.

Heaney, Seamus [Justin] (1939–) Irish poet and critic who is regarded by many critics as the finest Irish poet since YEATS. In 1995 he was awarded the Nobel Prize for Literature.

Hearst, William Randolph (1863–1951) American newspaper publisher and politician. In the late 1920s he owned more than 25 daily newspapers and built a spectacular castle at San Simeon in California. He was congressman for New York (1903–7). WELLES' film, *Citizen Kane*, is a thinly disguised account of his life.

Heath, Edward [Richard George] (1916–) British Conservative statesman and prime minister (1970–74). A fervent pro-European, he negotiated Britain's entry into the Common Market in 1973. He has also been active and influential in world politics.

Hegel, Georg Wilhelm Friedrich (1770–1831) German philosopher. His highly influential works, which describe how the Absolute is being reached by man's evolving powers of consciousness, influenced Karl MARX.

Heidegger, Martin (1889–1976) German philosopher. He is usually described as an existentialist, despite his disclaimer of the label, and his concepts, such as 'angst', had a great deal of influence on existentialists such as SARTRE.

Heifetz, Jascha (1901–87) Lithuanian-born American violinist. His flamboyant and expressive interpretation of music from BACH to WALTON has been widely acclaimed.

Heine, Heinrich (1797–1856) German poet and critic. His masterpiece is his *Book of Songs* (1827), which includes some of the finest lyric poems ever written.

Heisenberg, Werner Karl (1901–76) German theoretical physicist. He was awarded the 1932 Nobel prize for physics for his work on quantum theory.

Heller, Joseph (1923–) American novelist. His most popular novel, *Catch–22* (1961), is a grim, surrealist satire on military life and logic.

Hemingway, Ernest [Millar] (1899–1961) American novelist and short-story writer, whose laconic narrative style made a big impression on his contemporaries. Major novels include *A Farewell to Arms* and *For Whom the Bell Tolls*. He was awarded the Nobel prize for literature in 1954. He committed suicide.

Hendrix, Jimi [James Marshall Hendrix] (1942–70) American rock guitarist, singer and songwriter. With his trio, the Jimi Hendrix Experience, he became perhaps the most influential of all rock guitarists. He died of alcohol and drug abuse.

Jimi Hendrix

Henson, Jim [James Maury Henson] (1936–90) American puppeteer and film producer who created the engaging cast of 'muppets', including Kermit the Frog and Miss Piggy.

Henze, Hans Werner (1926–) German composer. His works, which often reflect his enthusiasm for left-wing causes, include the opera *Elegy for Young Lovers* with a libretto by AUDEN.

Hepburn, Katharine (1909–93) American film and stage actress, noted for her wit and versatility. She had a long personal and acting relationship with Spencer TRACY.

Hepworth, Dame [Jocelyn] Barbara (1903–75) English sculptor. She became one of Britain's leading abstract sculptors in the 1930s, noted for her strong, often monumental carving.

Herbert, George (1593–1633) English Anglican priest and poet. His poems are among the greatest devotional poems in the language, and are characteristic of metaphysical poetry in their subtle, paradoxical exploration of spiritual themes.

Herrick, Robert (1591–1674) English Anglican priest and poet. Many of the poems, which are often delicately sensual, are surprisingly direct in their sympathy for the traditional (virtually pagan) customs of English country life.

Hertz, Gustav Ludwig *see* FRANCK, JAMES.

Hertzog, James Barry Munnik (1866–1942) South African statesman. He founded the Nationalist Party (1913) and advocated non-cooperation with Britain during World War I. He became prime minister (1924–39) and founded the Afrikaner Party in 1941.

Herzog, Werner (1942–) German film director. Bizarre enterprises, e.g. the building of an opera house up the Amazon in *Fitzcarraldo*, are a notable feature of his films.

Heseltine, Michael *see* THATCHER, MARGARET HILDA.

Hess, Dame Myra (1890–1965) English pianist. She was also a much acclaimed concert pianist and an influential teacher.

Hess, [Walter Richard] Rudolf (1894–1987) German Nazi politician. Deputy leader of Nazi party (1934–41). On the eve of Hitler's invasion of Russia, he flew to Scotland, apparently in the hope of negotiating peace terms with Britain. He spent the rest of his life imprisoned.

Hess, Victor Francis (1883–1964) Austrian-born American physicist. He shared the 1936 Nobel prize for physics with the American physicist **Carl David Anderson** (1905–) for his research into cosmic rays.

Hesse, Hermann (1877–1962) German-born Swiss novelist, short-story writer and poet. His fiction reflects his fascination with oriental mysticism, spiritual alienation and worldly detachment. He was awarded the Nobel prize for literature in 1946.

Heston, Charlton [John Charlton Carter] (1923–) American film and stage actor, renowned principally for his physique, noble profile and commanding presence in religious epics.

Heyerdahl, Thor (1914–) Norwegian anthropologist. His practical demonstration of his theory that South Americans emigrated to Polynesia on rafts of balsa wood caught the public imagination and he subsequently launched similar expeditions.

Hillary, Sir Edmund [Percival] (1919–) New Zealand explorer and mountaineer. He and the Tibetan sherpa, **Tenzing Norgay** (1914–86), made the first ascent of Mount Everest in 1953. His other exploits include an overland trek to the South Pole in 1958.

Himmler, Heinrich (1900–45) German Nazi leader. He was chosen by HITLER to head the SS in 1929, and by 1936 was in command of the German police structure. Through his secret police, the Gestapo, he organized repression first in Germany then in occupied Europe, and oversaw the construction of the Nazi concentration and death camp system and the genocide of the Jews.

Hindemith, Paul (1895–1963) German composer and violist. The Nazis banned his works for their 'impropriety', and he settled in the US in 1939. Highly prolific, he wrote operas, symphonies, song cycles, ballet and chamber music.

Hindenburg, Paul von Beneckendorff und von (1847–1934) German field marshal and statesman. He shared command of the German forces in World War I (1916–18), and became president of Germany (1925–34). He defeated HITLER in the presidential election of 1932 but was persuaded to appoint Hitler chancellor in 1933.

Hines, Earl [Kenneth] 'Fatha' (1903–83) American jazz pianist, bandleader and songwriter. He became one of the most influential jazz pianists of the 1930s and 1940s.

Hirohito (1901–89) Japanese emperor (1926–89). A direct descendant of Japan's first emperor, Jimmu, he ruled Japan as a divinity until her defeat in 1945 by Allied forces in the World War II. After this he became a constitutional monarch, known primarily for his marine biology research. He was succeeded by his son **Akihito** (1933–).

Hirohito

Hiss, Alger (1904–) American state department official. A highly respected public servant, he was jailed (1950–54) for spying for the USSR.

Hitchcock, Sir Alfred (1899–1980) English film director, based in Hollywood from 1940, whose suspenseful thrillers, e.g. *The Thirty-Nine Steps*, have long been regarded as masterpieces. Known as the 'Master of Suspense', Hitchcock also appeared in every one of his films, in a crowd scene or in a minute cameo role.

Hitler, Adolf (1889–1945) Austrian-born German dictator. He co-founded the National Socialist Workers' Party in 1919, and was jailed for nine months following his part in the failed Munich coup of 1923, during which time he wrote *Mein Kampf* ('my struggle'), an anti-semitic 'testament' of his belief in the superiority of the Aryan race. He was appointed chancellor by HINDENBURG in 1933 and consolidated his brutal regime through HIMMLER'S Gestapo. He allied himself temporarily with STALIN in 1939, in which year he invaded Poland, beginning World War II. He invaded Russia in 1941 but his troops suffered great losses at Stalingrad in 1943, forcing them to retreat. As the Allies began to win the war, Hitler faced opposition from within Germany. Having survived an assassination attempt in 1944, he committed suicide in Berlin in the final days of the war. Hitler's war resulted in *c.*40 million dead.

Adolf Hitler

Ho Chi Minh [Nguyen That Tan] (1890–1969) Vietnamese statesman. A Marxist nationalist, he led the Viet Minh forces, with US help, against the occupying Japanese during World War II and became president of Vietnam (1945–54), during which time he led his forces to victory against French colonial rule. He became president of North Vietnam (1954–69) after the country's partition at the 1954 Geneva conference.

Hockney, David (1937–) English painter and etcher. Associated with the Pop Art movement in his early work, he is now regarded as one of the world's leading representational painters.

Hodgkin, Sir Alan Lloyd (1914–) English physiologist. With **Sir Andrew Fielding Huxley** (1917–) and Sir John Carew ECCLES, he shared the 1963 Nobel prize for physiology or medicine for research into nerve impulses.

Hodgkin, Dorothy [Dorothy Mary Crowfoot] (1910–) English chemist. She was awarded the 1964 Nobel prize for chemistry for her work on the molecular structures of penicillin, insulin and vitamin B12.

Hofstadter, Robert *see* MOSSBAUER, RUDOLF LUDWIG.

Hogarth, William (1697–1764) English artist. Trained as an engraver in the rococo tradition, by 1720 he had established his own illustration business. He then began his series of 'conversation pieces' and was executing some fine portraits. He also produced a remarkable series of paintings following a sequential narrative, the best known of which is *Marriage à la Mode*. He wrote a treatise on aesthetic principles entitled *The Analysis of Beauty*.

Holbein, Hans (the Younger) (*c.*1479–1543) German painter. He painted mainly portraits and reli-

gious paintings, the most memorable of the latter being the *The Death of Christ* (1521). He became court painter to HENRY VIII.

Holiday, Billie 'Lady Day' [Eleanora] (1915–59) American jazz singer. She became one of the most influential jazz singers of her time, with her sad, elegiac and subtle interpretations of popular songs.

Holly, Buddy [Charles Hardin Holley] (1936–59) American rock singer, songwriter and guitarist. He was one of the first rock singers to use the back-up of lead, rhythm and bass guitars, with drums. He died in a plane crash.

Holst, Gustav [Theodore] (1874–1934) English composer of Swedish descent. His best-known composition is *The Planets* (1917). Much of his music was inspired by the English landscape and by Thomas HARDY.

Homer (*c.*800 BC) Greek poet, author of the two great epic poems *The Iliad*, the story of the Greek war against Troy, and *The Odyssey,* which describes the adventures of the Greek hero Odysseus (known to the Romans as Ulysses) on his voyage home from the war. The characters and events of the poems have had a profound influence upon western literature.

Homer

Honecker, Erich (1912–94) East German Communist politician. Appointed head of state in 1976, he fell from power in 1989 following the wide social unrest that followed GORBACHEV's statement that the USSR would no longer intervene in East German affairs. He was charged in 1990 with treason and corruption following the re-unification of Germany.

Honegger, Arthur (1892–1955) French composer. One of the group of Parisian composers dubbed 'Les Six' his works include ballet music, symphonies, film scorces and *Pacific 231*, a musical portrait of a train.

Hoover, Herbert [Clark] (1874–1964) American Republican statesman and 31st president of the US (1929–33). He succeeded COOLIDGE as president in

1929, and was widely perceived as failing to cope with the crisis of the Great Depression.

Hoover, J[ohn] Edgar (1895–1972) American public servant and founder of the Federal Bureau of Investigation (1924–1972). He made the FBI into a highly effective federal crime-fighting force in the 1930s, but also used his organization's considerable powers against anyone perceived as 'radical' in politics.

Hope, Bob [Leslie Townes Hope] (1903–) English-born American comedian and film actor, known for his snappy wisecracks.

Hopkins, Sir Frederick Gowland (1861–1947) English biochemist. He shared the 1929 Nobel prize for physiology or medicine with EIJKMAN for his discovery of 'accessory food factors', which came to be called vitamins.

Hopkins, Gerard Manley (1844–89) English Jesuit priest and poet. A convert, he frequently expressed in his poems the keen conflict he felt between his desire to serve God as both priest and poet. None of his work was published in his lifetime.

Hopper, Edward (1882–1967) American artist. Regarded as the foremost realist American painter, his paintings have a still, introspective and often mysterious quality.

Horace [Quintus Horatius Flaccus] (65–8 BC) Roman poet and satirist. He looked to the literature of Greece for inspiration, but his sardonic, realistic and tightly controlled language is wholly Roman. His *Odes*, *Satires* and *Epistles* have been much imitated.

Housman, A[lfred] E[dward] (1859–1936) English poet and scholar. A distinguished classical scholar, his few works of poetry were published posthumously, e.g. *A Shropshire Lad*.

Howe, Sir Geoffrey *see* MAJOR, JOHN.

Hoyle, Sir Fred (1915–) English astronomer, mathematician, broadcaster and writer. He became the main proponent of the theory of the universe which holds that the universe is basically unchanging (as opposed to the big-bang theory).

Hubble, Edwin Powell (1889–1953) American astronomer. His discovery of galactic 'red shift' and other research established the theory of the expanding universe.

Hughes, Howard [Robard] (1905–76) American industrialist, aviator and film producer. He greatly extended his inherited oil wealth and made several epic flights, including a record round-the-world trip. He became increasingly eccentric and went into seclusion in 1966.

Hughes, Ted [Edward James Hughes] (1930–) English poet, noted for his violent poetic imagery

drawn from the natural world. He was married (1956–63) to Sylvia PLATH. He was appointed poet laureate in 1984.

Hugo, Victor (1802–85) French novelist, dramatist and poet. His socially challenging dramas established Hugo as the leader of the French literary Romantics. His novels include *The Hunchback of Notre Dame* and *Les Misérables*.

Hume, David (1711–76) Scottish philosopher, economist and historian. An empiricist and sceptic, Hume disallowed human speculation much beyond what could be perceived by the senses. His works include *A Treatise of Human Nature*. He has been claimed by some modern economics to have been a pro-monetarist for his discussion of the 'hidden hand' guiding market forces.

Hunt, Leigh *see* KEATS, JOHN.

Hunt, William Holman (1827–1910) English painter. A founder of the Pre-Raphaelite movement, he sought inspiration in direct study from nature and natural composition.

Hussein, [Ibn Talal] (1935–) king of Jordan. He lost the West Bank of his country to Israel after the Six Day War of 1967, and has trod an uneasy diplomatic line between friendship with the West and his efforts on behalf of the Palestinians.

Hussein, Saddam (1937–) Iraqi dictator. He became president of Iraq in 1979, and established a reputation for ruthlessness in the suppression of his opponents. After his invasion of Kuwait in 1990, UN forces forced his withdrawal in the Gulf War of 1991.

Huston, John [Marcellus] (1906–87) American film director. His films include several classics, e.g. *The Maltese Falcon*, from a story by HAMMET. His last film, *The Dead*, from a short story by JOYCE, starred his daughter, the actress **Anjelica Huston** (1951–).

Hutton, Sir Leonard ('Len') (1916–90) English cricketer. A Yorkshire player throughout his long career, he became the first professional player to captain England regularly (1952–54).

Huxley, Sir Andrew Fielding *see* HODGKIN, SIR ALAN LLOYD

Huxley, Sir Julian [Sorell] (1887–1975) English biologist. He became one of Britain's best-known scientists and humanists and was the first director-general of UNESCO (1946–48). His brother, **Aldous [Leonard] Huxley** (1894–1963) was a novelist, short-story writer and essayist. His early work depicted the brittle word of 1920s English intellectual life, but his masterpiece is *Brave New World*, a chilling fable of a future totalitarian state.

Huxley, Thomas Henry (1825–95) English biologist. He became the most prominent scientific defender of Darwin's theory of evolution. He gradually lost his belief in a deity and coined the term 'agnostic'.

Ibn Saud, Abdul Aziz (1880–1953) king of Saudi Arabia. He became the first king of Saudi Arabia (1932–53), and negotiated terms with American oil companies after the discovery of oil in his country (1938).

Ibsen, Henrik (1828–1906) Norwegian dramatist. His early verse dramas (e.g. *Peer Gynt*), plays of social realism (e.g. *Ghosts*), and later symbolic plays were all hugely influential on later dramatists, e.g. Shaw.

Ignatius Loyola, St (1491–1556) Spanish saint. A former solder who was severely wounded in action, he had a spiritual conversion and founded the Society of Jesus (the Jesuits) in 1534.

Ingres, Jean Auguste Dominique (1780–1867) French painter. One of the greatest exponents of neoclassical art, his excellent draughtsmanship influenced Degas, Matisse and Picasso.

Ionesco, Eugène (1912–94) Romanian-born French dramatist. His plays, regarded as masterpieces of the Theatre of the Absurd, include *The Bald Prima Donna* and *The Lesson*.

Irving, Washington (1783–1859) American essayist and historian. His best-known stories are 'Rip Van Winkle' and 'The Legend of Sleepy Hollow'. He wrote a biography of Washington.

Isherwood, Christopher [William Bradshaw] (1904–86) English-born American novelist and dramatist. His best-known works are set in pre-World War II Berlin.

Ives, Charles Edward (1874–1954) American composer. His works are frequently experimental but based firmly within the American tradition.

Jackson, Glenda (1936–) English actress. Highly regarded on film and stage, she became a Labour MP in 1992.

Jackson, Jesse (1941–) American Democrat politician. A Baptist minister and one of Martin Luther King's aides, he campaigned twice for the Democratic presidential nomination.

Jackson, Michael [Joe] (1958–) American pop singer. The youngest of five brothers who as children formed the Jackson 5, he became a solo performer in the late 1970s.

Jagger, Mick [Michael Philip Jagger] (1943–) English singer and songwriter, and lead singer with the Rolling Stones rock group, the original members of which, with Jagger, were the guitarist and co-writer with Jagger of many of their songs, **Keith Richard** (1943–), bass guitarist **Bill**

Wyman (1936–), drummer **Charlie Watts** (1941–) and guitarist **Brian Jones** (1944–69).

James, Henry (1843–1916) American-born British novelist, short-story writer and critic. Much of his work is concerned with the contrast between American innocence and the older, wiser European culture, e.g. *Daisy Miller*.

Henry James

James, M[ontague] R[hodes] (1862–1936) English scholar and ghost-story writer. His stories are a mix of dry, scholarly wit with a horrifically reticent undertone of supernatural terror.

James, Dame P. D. [Phyllis Dorothy White] (1920–) English novelist. Her crime novels have been much admired for their wit.

James, William (1842–1910) American philosopher and psychologist. His works include *The Varieties of Religious Experience*, in which he coined the term 'stream of consciousness'.

Janáček, Leos (1854–1928) Czech composer. His works, heavily influenced by Czech folk music and culture, include the operas *The Cunning Little Vixen* and *The House of the Dead*, and two highly regarded string quartets.

Jefferson, Thomas (1743–1826) American statesman. He was the main creator of the Declaration of Independence in 1776 and became secretary of state (1790–93) under Washington. He was elected the 3rd president of the United States (1801–9).

Jenkins, Roy [Baron Jenkins of Hillhead] (1920–) Welsh Labour and Social Democrat politician. He resigned from the Labour Party in 1981 to co-found the Social Democratic Party. He was SDP MP for Glasgow Hillhead (1982–87).

Jenner, Edward (1749–1823) English physician. He investigated the traditional belief that catching cowpox gave protection against smallpox and discovered that vaccination was efficacious in preventing smallpox.

Jesus Christ (c.6 BC–c.30 AD) founder of Christianity. *The New Testament* records that he was born in

Bethlehem, the son of Joseph and Mary, and Christians have traditionally believed that he is the Son of God. The Book of Acts describes how the Christian gospel was spread through the Mediterranean world by his disciples, notably **Paul** and **Peter**, the latter being recognized as the founder of the Roman Catholic church.

Jiang Jie Shi *see* CHIANG KAI-SHEK.

Jiang Qing *see* CHIANG CH'ING.

Jinnah, Mohammed Ali (1876–1948) Pakistani statesman. An early member of the Indian Muslim League, he became convinced of the need for Indian partition into Hindu and Muslim states and was the first governor-general of Pakistan (1947–48).

Joan of Arc (*c*.1412–31) French patriot. From a peasant family, she had a vision when she was 13, urging her to free France from the invading English. She helped raise the siege of Orléans in 1429, and brought Charles VII to Rheims to be crowned king of France. Captured by the English in 1430, she was condemned for witchcraft and burned at the stake. She was canonized in 1920.

John, Augustus [Edwin] (1878–1961) Welsh painter who was a superb draughtsman and portraitist. His sister, **Gwen John** (1876–1939), was also a painter.

John Paul II [Karel Jozef Wojtyla] (1920–) Polish pope (1978–) The first Polish pope and first non–Italian pope for 450 years. His primacy has been notable for his conservatism in matters such as abortion and the celibacy of the priesthood.

Pope John Paul II

Johns, Jasper (1930–) American painter, sculptor and printmaker. His work, especially his use of everyday images such as the stars and stripes, was very influential on later Pop Artists.

Johnson, Amy (1903–41) English aviator. She was the first woman to fly solo from England to Australia (1930). Her other records include a solo flight from London to Cape Town (1936). She was presumed drowned, after baling out over the Thames Estuary while serving as a transport pilot in World War II.

Johnson, Lyndon B[aines] (1908–73) American Democrat statesman. Following John F. KENNEDY's assassination in 1963, he became the 36th president of the US (1963–69). With Civil Rights agitation and the increasing unpopularity of the Vietnam War his time in office was a troubled one.

Johnson, Dr Samuel (1709–84) English critic, lexicographer and poet. One of the greatest literary figures of the 18th century who wrote on many subjects and was considered the moralist of the age. His works include the *Dictionary of the English Language* (1755), an important edition of SHAKESPEARE, several essays and one novel. In 1763 he met **James Boswell** (1740–95) with whom he toured the Western Isles. Boswell's biography of Johnson is considered to be the finest in the language.

Jones, Brian *see* JAGGER, MICK.

Jonson, Benjamin (1572–1637) English dramatist. After a turbulent early life, during which he served as a soldier and killed a fellow soldier in a duel, he began writing comedies which are particularly noted for their satirical dialogue and use of the 'Theory of Humours' and which established him as a great dramatist. His plays include *Bartholomew Fair* (1614), *Volpone* (1616) and *The Alchemist* (1616). He became the first poet laureate in 1616.

Joplin, Scott (1868–1917) American pianist and composer. His ragtime compositions, e.g. *Maple Leaf Rag*, were enormously popular in the USA, selling over a million copies of sheet music.

Joyce, James [Augustine Aloysius] (1882–1941) Irish novelist and short-story writer. He left Ireland in 1902, returning briefly twice. His works include athe short-story collection *Dubliners* (1914), and two great novels, *Portrait of the Artist as a Young Man* (1914–15) and *Ulysses* (1922), the latter one of the key novels of the 20th century.

Joyce, William (1906–46) American-born British traitor (of Anglo-Irish descent). Dubbed 'Lord Haw-Haw' by the British public, he broadcast rabid Nazi propaganda to Britain during World War II and was executed for treason in 1946.

Juan Carlos (1938–) king of Spain (1975–). Nominated by FRANCO in 1969 as his successor, Juan Carlos carefully steered his country towards democracy after Franco's death in 1975.

Jung, Carl Gustav (1875–1961) Swiss psychiatrist. He began his career as a follower of Freud, but split with him after challenging his concentration on sex. His theory of the 'collective unconscious', and his use of the term 'archetype' to denote an image or symbol drawn from this store, has been highly influential.

Kafka, Franz (1883–1924) Czech-born German novelist and short-story writer. His novels, *The Trial* and *The Castle*, and several of his short stories, notably 'Metamorphosis', are established classics of 20th-century literature.

Kandinsky, Wassily (1866–1944) Russian-born French painter. He co-founded (with KLEE and MARC) the *Blaue Reiter* group in 1912, and is regarded as the first major abstract artist.

Kant, Immanuel (1724–1804) German philosopher. His works include *The Critique of Pure Reason* (1781), in which he adopts (in response to HUME's empiricism) an idealist position, arguing that our knowledge is limited by our capacity for perception, and *The Critique of Practical Reason* (1788), in which he expounds his theory of ethics based upon 'categorical imperatives'.

Karajan, Herbert von (1908–89) Austrian conductor. His recordings, notably of BEETHOVEN's symphonies, are held by some critics to be definitive.

Kauffmann, Angelica (1741–1807) Swiss painter. Influenced by neoclassicism, she settled in London in 1776 where she was noted for her portraits (of REYNOLDS and GOETHE among others) and history paintings. She was a founder of the Royal Academy.

Kaunda, Kenneth [David] (1924–) Zambian politician. He became president of Zambia when his country became independent in 1964.

Keaton, Buster [Joseph Francis Keaton] (1895–1966) American film comedian and director. Widely regarded as one of the all-time great comedians of the cinema, with his 'deadpan' expression and remarkable acrobatic skill, his silent comedy films include *The Navigator* (1924) and *The General* (1926).

Keats, John (1795–1821) English poet. His first sonnets were published by **Leigh Hunt** (1784–1859) in *The Examiner*, the weekly paper Hunt founded in 1808 with his brother **James Hunt** (1774–1848). Although savagely criticised early in his career, Keats went on to write some of the greatest works of English romantic literature.

Keller, Helen [Adams] (1880–1968) American writer. She became deaf and blind when 19 months old, and was taught to read and write by the partially sighted Anne Sullivan.

Kelly, Gene [Eugene Curran Kelly] (1912–96) American dancer, choreographer and film direc-

tor, who was noted for his athleticism and witty dancing style.

Kelly, Grace [Patricia] (1929–82) American film actress. She married **Prince Rainier III** (1923–) of Monaco in 1956, and gave up her career.

Kennedy, John Fitzgerald (1917–63) American Democratic politician, who became 35th president of the US (1961–63). He was the first Roman Catholic and the youngest man elected to the presidency. His period of office, cut short by his assassination in Dallas, was subsequently seen by many as a period of hope and social reform, with most of his 'New Frontier' legislation being implemented by Lyndon JOHNSON. His brother **Robert ('Bobby') [Francis] Kennedy** (1925–68), who became attorney general (1961–64) and senator for New York (1965–68) and furthered civil rights legislation, was assassinated. Another brother, **Edward [Moore] Kennedy** (1932–), became a senator in 1962, and was widely regarded as a future president until the 'Chappaquidick' incident of 1969, in which a girl passenger in his car was drowned.

John F. Kennedy

Kenyatta, Jomo (*c.*1893–1978) Kenyan politician. He was jailed for six years (1952–58) for his leadership of the Mau-Mau rebellion. He became prime minister of Kenya on independence in 1963 and president (1964–78).

Kerenski, Alexsandr Feodorovich (1881–1970) Russian revolutionary leader. A member of the Social Democratic Party's liberal wing, he became prime minister of the Russian provisional government of 1917, but was deposed by LENIN's Bolsheviks.

Kern, Jerome [David] (1885–1945) American composer and songwriter. A highly prolific writer of music and songs, he had a huge influence on the American musical tradition.

Kerouac, Jack [Jean-Louis Lebris de Kérouac] (1922–69) American novelist and poet who was a much imitated central figure of the Beat Generation.

Keynes, John Maynard, 1st Baron (1883–1946) English economist. He argued that unemployment was curable through macroeconomic management of monetary and fiscal policies and advocated the creation of employment through government schemes.

Khomeini, Ayatollah [Ruholla] (1900–89) Iranian religious leader who established a theocratic dictatorship that crushed all dissent and declared his intention of 'exporting' the Shiite revolution to other Islamic countries. He aroused Western anger by proclaiming a death sentence against Salman RUSHDIE in 1989.

Khrushchev, Nikita Sergeyevich (1894–1971) Soviet politician. He was first secretary of the Communist Party (1953–64) and prime minister (1958–64). He promoted peaceful co-existence with the West, but entered into various disputes with China over economic and border issues. His standing in the USSR was somewhat compromised after the Cuban Missile Crisis of 1962, when his climbdown against JOHN F. KENNEDY prevented a war with the USA. He was deposed in 1964 in the Kremlin coup that brought BREZHNEV to power.

Kierkegaard, Søren Aabye (1813–55) Danish theologian and philosopher. Regarded as the founder of existentialism, he rejected the spiritual authority of organized religion and emphasized the centrality of individual choice.

Kim Il Sung

Kim Il Sung (1912–94) North Korean marshal and Communist politician. He became prime minister (1948–72) and president (1972–94) of the Democratic People's Republic of Korea (popularly known as North Korea), establishing a Stalinist dictatorship based on a personality cult of himself as the 'great leader'. In 1950 he ordered the invasion of South Korea, having never accepted the partition of 1945. This precipitated the Korean War which ended three years later with the parti-

tion kept largely intact along the 38th parallel which was established in 1948. Kim Il Sung's domestic policy centred on the nationalization of industry and the collectivization of agriculture. His son, **Kim Jong Il** (1942–), succeeded him as president and, like his father, has continued to reject political reform despite the detrimental effect this has had on the economy since the collapse of Soviet communism.

King, Billie Jean (1943–) American tennis player. Regarded as one of the finest women players ever, she won twenty Wimbledon titles between 1965 and 1980.

King, Martin Luther, Jr (1929–68) American civil rights leader and Baptist minister. Influenced by GANDHI's policy of nonviolent resistance, he organized opposition to segregationist policies in the Southern US. He was awarded the 1964 Nobel Peace Prize and was assassinated in 1968.

Martin Luther King

Kinnock, Neil [Gordon] (1942–) Welsh Labour politician. Elected leader of the Labour Party in 1983 in succession to MICHAEL FOOT, he gradually moderated the Party's policies and had great success in marginalizing the hard left of the party. He resigned as Party leader in 1992, following Labour's fourth successive general election defeat.

Kipling, [Joseph] Rudyard (1865–1936) Indian-born English short-story writer, poet and novelist. Although best known for his stories for children (e.g. *The Jungle Book*), he was also a caustic observer of Anglo-Indian society and critical of many aspects of colonialism in his writing.

Kissinger, Henry [Alfred] (1923–) German-born American statesman. Shared the 1973 Nobel Peace Prize with the North Vietnamese negotiator **Le Duc Tho** (1911–) for the treaty ending US involvement in Vietnam. As secretary of state (1973–76) he fostered détente with the Soviet Union and China, and helped negotiate peace between Israel and Egypt in 1973.

Kitchener of Khartoum, [Horatio] Herbert, 1st

Earl (1850–1916) Anglo-Irish field marshal. Commander in chief of the British forces during the Boer War of 1901–2, and of the British forces in India (1902–9), he was appointed secretary for war in 1914 and had mobilized Britain's largest-yet army by the time of his death by drowning when his ship hit a mine.

Klee, Paul (1879–1940) Swiss painter and etcher who developed a style of mainly abstract work characterized by doodle-like drawings. He was a member of the *Blaue Reiter* group.

Paul Klee

Klemperer, Otto (1885–1973) German-born conductor. A great interpreter of both classical and contemporary works, he became director of the Los Angeles Symphony Orchestra in 1936 and director of the Budapest Opera (1947–50).

Klimt, Gustav (1862–1918) Austrian painter. An excellent draughtsman, his early works were influenced by Impressionism, Symbolism and Art Nouveau. He was a founder of the Vienna Sezession and had a great influence on younger artists.

Knox, John (*c*.1513-1572) Scottish Protestant reformer. He was noted for his antagonistic relationship with Mary Queen of Scots and for his single-minded determination in the pursuit of religious reformation.

Kodály, Zoltán (1882–1967) Hungarian composer. Like his friend BARTÓK, he was much influenced by the traditional music of his country. His works include the comic opera *Hary Janos*.

Kohl, Helmut (1930–) German Christian Democrat statesman. He became chancellor of West Germany (1982–90) and the first chancellor of reunited Germany in 1990.

Kokoschka, Oskar (1886–1980) Austrian-born painter and dramatist. A leading Expressionist painter, noted particularly for his landscapes and portraits, he fled to Britain in 1938.

Korda, Sir Alexander [Sandor Kellner] (1893–1956) Hungarian-born British film director and producer. His films include *The Private Life of*

Henry VIII and *The Third Man*. He was knighted in 1942.

Kreisler, Fritz (1875–1962) Austrian-born American violinist and composer. ELGAR's violin concerto was dedicated to him, and he became one of the most popular violinists of his day.

Kubrick, Stanley (1928–) American film director and producer. His films include the anti-war classic *Paths of Glory* (1957), the black nuclear war comedy *Dr Strangelove* (1963), the innovative science fiction classic *2001: A Space Odyssey* (1968), and the still highly controversial *A Clockwork Orange* (1971).

Kundera, Milan (1929–) Czech novelist. His masterpiece is *The Unbearable Lightness of Being*, a love story set against the background of repression following the Russian invasion of Czechoslovakia in 1968.

Kurosawa, Akira (1910–) Japanese film director. His films include the samurai classics *The Seven Samurai* and *Yojimbo* (remade in the west as *The Magnificent Seven* and *A Fistful of Dollars*), and samurai versions of *Macbeth*, titled *Throne of Blood*, and *King Lear*, titled *Ran*. Kurosawa was happiest with the epic form, and also had a 'family' of actors he used regularly.

Kyd, Thomas (1558–94) English dramatist. His most important work is his revenge tragedy, *The Spanish Tragedy*, which served as a model for SHAKESPEARE's *Titus Andronicus*. A close associate of Christopher MARLOWE, he died in poverty after being accused of denying the divinity of Christ.

Laing, R[onald] D[avid] (1927–89) Scottish psychiatrist. He became a counterculture guru in the 1960s for his revolutionary ideas about mental disorders.

Lamarck, Jean [Baptiste Pierre Antoine de Monet, Chevalier de] (1744–1829) French naturalist. His theory of the evolution of species through the acquisition of inherited characteristics prepared the ground for DARWIN's theory of evolution.

Lamb, Charles (1775–1834) English essayist and critic. A friend of HAZLITT, WORDSWORTH and COLERIDGE, his writings display the great charm his friends describe. With his sister, **Mary Anne Lamb** (1764–1847), he adapted several of SHAKESPEARE's plays to prose form. In 1796, in a fit of insanity, Mary killed their mother, and Charles looked after her until his death.

Landseer, Sir Edwin Henry (1802–73) English painter. Highly regarded for his animal studies, his notable works include *The Monarch of the Glen* (1850) and the lions modelled for Trafalgar Square, London, in 1867.

Lansbury, George (1859–1940) English Labour politician. Noted for his support for women's suffrage and pacifism, he became leader of his party (1931–35) when MacDonald joined the National Government. His daughter, **Angela Lansbury** (1925–) became a popular film actress in the 1940s.

Lardner, Ring [Ringgold Wilmer Lardner] (1885–1933) American journalist and short-story writer, whose stories of American low life are noted for their cynical wit.

Larkin, Philip [Arthur] (1922–85) English poet. Known for his dark, sardonic lyricism, he is regarded as one of the greatest of all modern English poets. He also wrote two novels and a collection of essays on jazz.

Philip Larkin

Larwood, Harold (1904–95) English cricketer. His use of 'bodyline' tactics in the 1932–33 tour of Australia created great controversy, causing diplomatic tension in relations between Australia and the UK.

Lasdun, Sir Denys Louis (1914–) English architect. Influenced by Le Corbusier, his buildings include the University of East Anglia and the National Theatre in London.

Lasker, Emanuel (1868–1941) German chess player. His reign as world champion (1894–1921), is still a record.

Laski, Harold [Joseph] (1893–1950) English political scientist and socialist propagandist. He was a highly influential spokesman for Marxism through his position as teacher, writer and Labour Party power-broker (he was party chairman 1945–46), and through his friendships with politicians such as Roosevelt.

Lauda, Niki [Nikolas Andreas Lauda] (1949–) Austrian racing driver. World champion in 1975, 1977 and 1984, he suffered dreadful injuries in the 1976 German Grand Prix. He retired in 1985.

Lauder, Sir Harry [Hugh MacLennan] (1870–1950) Scottish music-hall comedian and singer, who made an international career out of his Scottish comedy routines and songs.

Laughton, Charles (1899–1962) English-born American stage and film actor, renowned for his larger-than-life performances in many memorable films, e.g. *The Private Life of Henry VIII*.

Laurel, Stan [Arthur Stanley Jefferson] (1890–1965), English-born American comedian, and **Oliver Hardy** (1892–1957) American comedian. Laurel began his career on the English music-hall stage (understudying Chaplin at one point), and Hardy performed with a minstrel troupe before going into films. They formed their Laurel (thin, vacant and bemused one) and Hardy (fat, blustering one) partnership in 1929, and made some very funny films, e.g. *Another Fine Mess* (1930).

Oliver Hardy and Stan Laurel

Laval, Pierre (1883–1945) French statesman. Prime minister (1931–32, 1935–36, 1942–44), he sided openly during the occupation with the Germans and was executed for treason in 1945 by the victorious Free French.

Lavoisier, Antoine Laurent (1743–94) French chemist. Regarded as the founder of modern chemistry, he discovered oxygen and established its role in combustion and respiration.

Lawrence, D[avid] H[erbert] (1885–1930) English novelist, poet and short-story writer. His novels caused much controversy for their frank treatment of sex. *Lady Chatterley's Lover* (1928), was not published in its unexpurgated form until 1960.

Lawrence, Gertrude [Gertrud Alexandra Dagmar Lawrence-Klasen] (1898–1952) English actress, noted for her professional relationship with Noel Coward, many of whose plays had parts written for her.

Lawrence, Sir Thomas (1769–1830) English painter. The leading portraitist of his time, he was made court painter in 1792.

Lawrence, T[homas] E[dward] (1888–1935) Welsh-born Anglo-Irish soldier and author, known as 'Lawrence of Arabia'. In World War I, he helped the Arab revolt against the Turks and was instrumental in the conquest of Palestine (1918).

Lawson, Nigel *see* MAJOR, JOHN.

Leach, Bernard Howell (1887–1979) English potter, who revolutionized the production of pottery by creating reasonably priced, attractively designed studio pottery.

Leadbelly [Huddie Ledbetter] (1888–1949) American blues singer. Discovered in a Louisiana prison in 1933, he later recorded several songs that soon became recognized as blues/folk classics, e.g. 'Rock Island Line' and 'Goodnight, Irene'.

Leakey, Louis Seymour Bazett (1903–72) Kenyan-born British archaeologist and anthropologist, and **Mary Douglas Leakey** (1913–) English archaeologist. Married in 1936, the Leakeys made several important discoveries about humanity's origins in East Africa. Their son, **Richard Erskine Frere Leakey** (1944–), is also a prominent (Kenyan) archaeologist.

Lean, Sir David (1908–91) English film director. Highly regarded for his compositional skill and craftsmanship, his films include many classics, e.g. *Brief Encounter* (1946), *Great Expectations* (1946) and *Dr Zhivago* (1965).

Leavis, F[rank] R[aymond] (1895–1978) English literary critic. With his wife, **Queenie Dorothy Leavis** (1906–81), he made a major impact on literary criticism from the 1930s attacking the modern age of mass culture and advertising.

Le Carré, John [David John Moore Cornwell] (1931–) English novelist. His popular novels are sombre anti-romantic narratives of Cold War espionage, e.g. *The Spy Who Came in from the Cold*.

Le Corbusier [Charles Edouard Jeanneret] (1887–1965) Swiss-born French architect and town planner. One of the most influential (and most praised and reviled) architects and planners of the century, his work is characterized by use of reinforced concrete and modular, standardized units of construction (based upon the proportions of the human figure), with the house famously defined as a 'machine for living in'.

Lederberg, Joshua (1925–) American geneticist. He shared the 1958 Nobel prize for physiology or medicine (with George Beadle and Edward TATUM) for his bacterial research.

Le Duc Tho *see* KISSINGER, HENRY.

Lee Kuan Yew (1923–) Singaporean politician. He became Singapore's first prime minister (1965–90), establishing a strict regime noted for its economic achievements and authoritarianism.

Léger, Fernand (1881–1955) French painter. A leading Cubist, he was much influenced by industrial imagery and machinery.

Lehár, Franz (1870–1948) Hungarian composer and conductor, noted for his operettas, e.g. *The Merry Widow*.

Leibnitz *or* **Leibniz, Gottfried Wilhelm** (1646–1716) German philosopher and mathematician. Renowned for the range and depth of his intellect, he made important contributions to many different scientific fields.

Leigh, Vivien [Vivien Mary Hartley] (1913–67) Indian-born English stage and film actress. She became an international star with *Gone With the Wind* (1939), in which she co-starred with Clark GABLE.

Lenin, Vladimir Ilyich [Vladimir Ilyich Ulyanov] (1870–1924) Russian revolutionary leader and Marxist philosopher. He was instigator of the Bolshevik October Revolution which overthrew KERENSKI's government and leader of the Bolsheviks in the Civil War (1918–21). The failure of his economic policy after the war led to the institution of the New Economic Policy of 1921, which fostered limited private enterprise. He was a brilliant demagogue and an influential philosopher.

Lennon, John [Winston] (1940–80) English guitarist, singer and songwriter. With Paul McCARTNEY, George HARRISON and Ringo STARR, he formed the Beatles, the most popular rock group ever. The band's success was based on the Lennon/ McCartney songwriting partnership; their songs achieved phenomenal popularity. He married **Yoko Ono** (1933–) and pursued a solo career after the Beatles split in 1969. He was assassinated in New York.

John Lennon

Leonardo da Vinci (1452–1519) Florentine painter, draughtsman, engineer, musician and thinker. The outstanding genius of his time, his greatest paintings include *The Last Supper* (1489) and his portrait of *Mona Lisa* (1504). His later years were devoted to scientific studies, and his work in mechanics, aeronautics, physiology and anatomy displays an understanding far beyond his times.

Le Pen, Jean-Marie (1928–) French politician. He founded the right-wing National Front in 1972, a crypto–fascist party with anti-immigrant policies.

Lermontov, Mikhail Yurievich (1814–41) Russian novelist and poet. His masterpiece is the novel *A Hero of our Time* (1840), a brilliant study of a disaffected, Byronic young aristocrat.

Lessing, Doris [May] (1919–) Iranian-born English novelist and short-story writer. Her novels include the 'Children of Violence' quintet series which explores the social and political undercurrents of modern society.

Lewes, George Henry *see* ELIOT, GEORGE.

Lewis, C[live] S[taples] (1898–1963) English novelist and critic. His works include studies of medieval literature (e.g. *The Allegory of Love*, works of Christian apologetics (e.g. *The Problem of Pain*), and science fiction novels (e.g. *Out of the Silent Planet*). He is best remembered for his enchanting Narnia stories for children (e.g. *The Lion, the Witch and the Wardrobe*).

Lewis, Jerry Lee (1935-) American rock singer and pianist noted for his flamboyant playing style and primitive rock 'n' roll lyrics.

Lewis, [Harry] Sinclair (1885–1951) American novelist. His work is particularly noted for its satirical view of small-town American life. He was the first American to win the Nobel prize for literature, in 1930.

Lewis, [Percy] Wyndham (1884–1957) English painter, novelist and critic. An influential writer with unfashionable right-wing views, he was also a leading member of the Vorticist group of artists. His best-known fictional work is the novel *The Apes of God*.

Lichtenstein, Roy (1923–) American painter and sculptor. He became the leading Pop Art painter of the 1960s with his highly coloured reproductions of sections of advertisements and cartoon strips.

Liddell, Eric Henry (1902–45) Scottish athlete, nicknamed the 'Flying Scot', who refused to compromise his sabbatarian principles by running on a Sunday during the 1924 Olympics (the subject of the award-winning film *Chariots of Fire*). He became a missionary and died in a Japanese prisoner-of-war camp.

Liddell Hart, Sir Basil Henry (1895–1970) English soldier and military historian, noted for his persistent advocacy of mechanized warfare and the development of air power after World War I.

Liebknecht, Karl *see* LUXEMBOURG, ROSA.

Ligeti, György Sándor (1923–) Hungarian composer. He fled to Vienna in 1956, where he soon became established as one of Europe's leading avant-garde composers.

Limbourg *or* **Limburg, Jean, Paul** and **Herman de** (all *fl.* 1400–16) Dutch illuminators. Their masterpiece, the unfinished *Les Tres Riches Heures*, is one of the greatest illuminated manuscripts of all time.

Lincoln, Abraham (1809–65) American statesman. From a poor background, he trained as a lawyer and became an Illinois congressman in 1846. He became the 16th president of the United States in 1861 and led the Union to victory in 1865. Firmly opposed to slavery, he finally declared emancipation 1863. He was assassinated while attending the theatre.

Abraham Lincoln

Lindbergh, Charles Augustus (1902–74) American aviator. He became the first man to fly the Atlantic solo and nonstop with his 1927 flight in the monoplane *Spirit of St Louis*. The kidnap and murder of his infant son in 1932 made world headlines.

Lippi, Fra Filippo (*c.*1406–69) Florentine painter. He took up painting while a monk and later forsook his vows to marry the mother of his son, **Filippino Lippi** (1457–1504), who also became a painter. His lyrical and fluid style invest his paintings with a wistful melancholy. An innovative painter, he was one of the first artists to explore and develop the *Madonna and Child* theme.

Liszt, Franz *or* **Ferencz** (1811–86) Hungarian pianist and composer. Recognized as one of the greatest pianists of his day, he made several important contributions to musical form and was influential in his experimentation.

Littlewood, Joan (1914–) English theatre director. Her theatre company, Theatre Workshop, formed in 1945, became one of the major left-wing companies.

Livingstone, David (1813–73) Scottish missionary and explorer. His discoveries during his African expeditions include Lake Ngami (1849) and the Victoria Falls (1855). He was also a vigorous campaigner against the slave trade. His last expedition was a search for the source of the Nile, in the course of which he himself was 'discovered' by the Welsh-American adventurer **Henry Morton Stanley** (1841–1904).

Lloyd, Clive Hubert (1944–) Guyanian-born West Indian cricketer. A fine batsman and fielder, he captained the West Indies team (1974–78, 1979–85).

Lloyd, Harold [Clayton] (1893–1971) American film comedian. He made hundreds of short silent films and is noted for his dangerous stunts.

Lloyd George, David, 1st Earl Lloyd George of Dwyfor (1863–1945) Welsh Liberal statesman. As chancellor of the exchequer (1908–15), he introduced far-reaching reforms in British society, notably the introduction of old-age pensions (1908) and the National Insurance Act (1911). Formerly a pacifist, he became minister of munitions (1915–16) and prime minister (1916–22) of coalition governments.

Lloyd Webber, Sir Andrew (1948–) English composer. With the librettist **Tim Rice** (1944–), he composed several highly successful musicals, notably *Joseph and the Amazing Technicolour Dreamcoat, Jesus Christ Superstar* and *Evita*. Other successes were *Cats*, adapted from T. S. ELIOT's *Old Possum's Book of Practical Cats, Starlight Express, Phantom of the Opera* and *Sunset Bouevard*.

Locke, John (1632–1704) English philosopher. His influential theory of the political nature of man saw the social contract as resting on a 'natural law', which, if ignored by rulers, allowed them to be overthrown.

Lodge, David (1935–) English novelist and critic. His best-known novels are entertaining satires on academic life, e.g. *Small World*, and his novel *Nice Work* is a hybrid of the campus novel with the 19th-century 'condition of England' tradition.

Longfellow, Henry Wadsworth (1807–82) American poet. His narrative poems based on American legends and folk tales were among the most popular of the 19th century, e.g. *The Song of Hiawatha*, the hypnotic unrhymed rhythms of which were much parodied by later writers, e.g. CARROLL.

Lonsdale, Dame Kathleen (1903–71) Irish physicist, noted for her innovative work in X-ray crystallography. She was the first woman to be elected as a fellow of the Royal Society, in 1945.

Lorca, Federigo García (1899–1936) Spanish poet and dramatist. His dramatic masterpiece is his trilogy of tragedies on the plight of oppressed Spanish women, *Blood Wedding, Yarma* and *The House of Bernarda Alba*. He was killed by Fascist forces near the beginning of the Spanish Civil War.

Lorenz, Konrad [Zacharias] (1903–89) Austrian ethologist and zoologist. He shared the 1973 Nobel prize for physiology or medicine (with Niko TINBERGEN and Karl von FRISCH) for his work on animal behaviour.

Lorre, Peter [Laszlo Lowenstein] (1904–64) Hungarian stage and film actor. His first major film part, as the pathetic child murderer, in LANG's *M* (1931) established him as a star. Other films include *Casablanca* and *The Maltese Falcon*.

Losey, Joseph (1909–84) American film director. Blacklisted during the MCCARTHY era, he came to work in England and had a great influence on the British film industry with films such as *The Servant, Accident* and *The Go-Between*.

Louis, Joe [Joseph Louis Barrow] (1914–81) American boxer, nicknamed the 'Brown Bomber'. He was world heavyweight champion for a record 12 years.

Lowei, Otto *see* DALE, SIR HENRY HALLETT.

Lowell, James Russell (1819–91) American poet, essayist and diplomat. His best-known verse was written in 'Yankee' dialect and inspired by his fervent abolitionism.

Lowell, Robert [Traill Spence] (1917–77) American poet. His verse is intensely personal and marked by private symbolism.

Lowry, L[aurence] S[tephen] (1887–1976) English painter. His paintings, which depict thin, dark 'matchstick' figures against a background of Northern Industrial life, became very popular in the mid–1960s.

Lowry, [Clarence] Malcolm (1909–57) English novelist and poet. His novels, e.g. *Under the Volcano*, often feature thinly veiled accounts of incidents from his own adventurous life.

Loyola, St Ignatius *see* IGNATIUS LOYOLA, ST.

Lully, Jean-Baptiste [Giambattista Lulli] (1632–87) French composer of Italian origin. He worked in the French court, where he composed many operas and comedy ballets. He died from an abscess after striking his foot with his conductor's baton.

Lumière, Auguste Marie Louis Nicolas (1862–1954) and **Louis Jean Lumière** (1864–1948) French chemists and cinematographers. They invented the first operational cine camera and projector and a colour photography process.

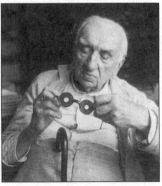

Louis Jean Lumière

Luther, Martin (1483–1546) German religious reformer. An Augustinian monk, who suffered a crisis of faith that led to his proclaiming a break with Rome following the nailing of his '95 theses' on the church door of Wittenberg. The Lutheran Reformation spread rapidly throughout Germany. CALVIN, ZWINGLI and others subsequently preached their variants of the new religion.

Luthuli or **Lutuli, Chief Albert John** (1898–1967) South African nationalist. He became president of the African National Congress (1952–60), and was awarded the 1961 Nobel Peace Prize for his advocacy of nonviolent resistance to apartheid.

Lutoslawski, Witold (1913–) Polish composer and teacher. He has written extensively, including chamber, piano and vocal music, but is best known for his orchestral works, e.g. *Concerto for Orchestra*.

Luxemburg, Rosa (1871–1919) Polish-born German revolutionary and socialist theorist. With **Karl Liebknecht** (1871–1919) she founded the revolutionary Spartacus League in Berlin on the outbreak of World War I and later the German Communist Party. She and Liebknecht were killed after the failed revolt of 1919.

MacArthur, Douglas (1880–1964) American general. He was commander of the US Far East forces in 1941. When the Japanese forced him to withdraw from the Philippines in 1942, he pledged 'I shall return'. He was appointed supreme Allied commander in the southwest Pacific in 1942 and gradually rolled back the Japanese forces, accepting their surrender in 1945. He also commanded the UN forces at the beginning of the Korean War (1950–51), being dismissed his command by TRUMAN.

McCarthy, Joseph R[aymond] (1908–57) American politician He became a Republican senator in 1946 and embarked upon a crusade against supposed communist sympathizers in public life (1950–54). His wide and increasingly bizarre accusations against innocent people came to an end shortly after he was accused, during a televised hearing, of having no shame.

McCartney, Sir Paul (1942–) English rock guitarist, singer and songwriter. He was a member of the Beatles (1961–70) with John LENNON, George HARRISON and Ringo STARR. With Lennon, he formed one of the most successful songwriting partnerships of the century. After the band's break-up, he formed the group Wings.

McCullers, Carson [Smith] (1917–67) American novelist and short-story writer. Her works, many of them filmed, usually centre on loners and mis-fits and include *The Heart is a Lonely Hunter* and *The Ballad of the Sad Café*.

MacDiarmid, Hugh [Christopher Murray Grieve] (1892–1978) Scottish poet and critic. Noted for his Communist and Nationalist sympathies, he influenced many Scottish writers, particularly with his masterpiece, *A Drunk Man Looks at the Thistle*.

MacDonald, [James] Ramsay (1866–1937) Scottish statesman. He became the first British Labour prime minister (1924, 1929–31) and was prime minister of the (mostly Conservative) coalition government of 1931–35.

MacGonagall, William (*c*.1830–*c*.1902) Scottish poet, renowned for his memorably awful doggerel verse.

Machiavelli, Niccolo (1469–1527) Italian statesman and political theorist. His treatise on the art of ruling, *The Prince* (1513), takes a dim view of human nature, seeing humanity as essentially corrupt and therefore best ruled by whatever method ensures the stability of the state, even if the method entails merciless cruelty.

McIndoe, Sir Archibald [Hector] (1900–1960) New Zealand plastic surgeon. One of the world's leading plastic surgeons, he pioneered facial surgery on burns victims.

Mackenzie, Sir [Edward Montague] Compton (1883–1972) English novelist, best known for his series of very popular comic novels set in the Scottish Western Isles, e.g. *Whisky Galore*.

Maclean, Donald *see* BURGESS, GUY.

McLuhan, [Herbert] Marshall (1911–80) Canadian critic and educator. His studies of mass culture and communication include the influential *The Medium is the Message*.

Macmillan, Sir [Maurice] Harold, 1st Earl of Stockton (1894–1986) English Conservative statesman and prime minister (1957–63) in succession to EDEN. Christened 'Supermac' by the cartoonist VICKY, he won the General Election of 1959 on the slogan 'You've never had it so good' and gained much international respect for his 'wind of change' speech in South Africa in 1958. His later years as premier were darkened by the PROFUMO scandal, and his manoeuvres against Rab BUTLER cost the latter the Tory leadership, which went to DOUGLAS-HOME. At the end of his life, he made a notably cutting speech in the House of Lords against THATCHER's privatization policies, 'selling the family silver'.

MacMillan, Sir Kenneth (1929–92) Scottish choreographer. He became the Royal Ballet's principal choreographer in 1977. He was also director of the Royal Ballet (1970–77).

MacNeice, [Frederick] Louis (1907–63) Irish poet and scholar. He was one of the leading AUDEN generation of poets, and his collections include *Letters from Iceland* (1937, written with Auden).

Madonna (Madonna Louise Veronica Ciccone; 1958–) American singer and film actress. After studying performing arts and dance in Michigan and New York, Madonna began performing with New York rock bawnds before making her first recording as a singer in 1982. Her phenomenal success has been due in part to her ability to exploit her talent through video and stage performance, often causing controversy. Her career in films has been less successful, although the film *Evita* finally brought great acclaim for her acting ability.

Maeterlinck, [Count] Maurice (1862–1949) Belgian poet, writer and playwright. Trained as a lawyer, he turned to writing poetry under the influence of the Symbolist poets. His masterpiece is *Pelléas et Mélisande* (1892), the basis for the opera by DEBUSSY. He was awarded the Nobel prize for literature in 1911.

Magritte, René (1898–1967) Belgian painter. He became a major Surrealist painter in Paris in the 1930s, devising a style dubbed 'magic realism' for its incongruous, dreamlike juxtaposition of carefully detailed everyday objects in dreamlike situations, e.g. men in bowler hats raining from the sky.

Mahler, Gustav (1860–1911) Austrian composer and conductor. Of Jewish birth, he became a Roman Catholic but remained subject to anti-semitic gibes while conductor of the Vienna State Opera (1897–1907). Regarded as both the last of the great Romantic composers of the 19th century and the first great composer of the modern era, his works include nine symphonies, song cycles, and the great symphonic song cycle, *The Song of the Earth* (1908).

Major, John (1943–) English Conservative politician. He became an MP in 1979 and was appointed a junior minister by THATCHER in 1981. His rise in the late 1980s was spectacular: he replaced **Sir Geoffrey Howe** (1926–) as foreign secretary in 19mao Tse-tung89, and later that year replaced **Nigel Lawson** (1932–) as chancellor. After Thatcher's resignation in late 1990, he was selected as Tory leader and prime minister.

Makarios III [Mikhail Khristodoulou Mouskos] (1913–77) Cypriot archbishop and statesman. Archbishop of the Orthodox Church in Cyprus, he became first president of Cyprus (1959–77) after independence.

Malan, Daniel F[rançois] (1874–1959) South African politician. A fervent believer in a racially divided society, he was prime minister (1948–54) and was responsible for the apartheid legislation.

Malcolm X [Malcolm Little] (1925–65) American black nationalist leader. A convert to Islam, he became an advocate of violence in response to racism only if used in self-defence.

Mallarmé, Stéphane (1842–98) French Symbolist poet. His impressionistic free-verse works and literary theorizing had a strong influence on the development of the Symbolist movement.

Malory, Sir Thomas (*fl.* 15th century) translator, largely from French sources, of a collection of Arthurian legends. The work includes several episodes, e.g. the quest for the Holy Grail, that have been recycled by generations of writers.

Mandela, Nelson [Rolihlahla] (1918–) South African lawyer and nationalist leader. Leader of the banned African National Congress, he was imprisoned in 1964 for life by the South African government. Upon his release in 1990, he helped to dismantle apartheid and was elected president in the first free elections in 1994. He seperated from his second wife, **Winnie Mandela** (1934–) in 1992.

Nelson Mandela (on left)

Mandelstam, Osip (1891–1938) Russian poet. Denounced for reading a satirical poem about STALIN, he and his wife, **Nadezhda Mandelstam** (1899–1980), were sent into exile in Siberia, where he died. Nadezhda later wrote accounts of their life together.

Manet, Edouard (1832–83) French painter. His direct approach and fresh, painterly style was influenced by the Impressionists, although he never exhibited with them.

Manley, Michael [Norman] (1923–) Jamaican statesman. He became leader of the socialist People's National Party in 1969, and prime minister (1972–80). He lost two subsequent elections, but won the 1989 election with a much less radical policy programme. He is regarded as a spokesman for the Third World.

Mann, Thomas (1875–1955) German novelist and

critic, primarily concerned with the role of the artist and the purpose of artistic creation in modern society. His works include *Death in Venice* (1912), *The Magic Mountain* (1930), and *The Confessions of the Confidence Trickster Felix Krull* (1954), a comedy. He was awarded the Nobel prize for literature in 1929 and fled Nazi Germany in 1933.

Mao Tse-tung *or* **Mao Ze Dong** (1893–1976) Chinese Communist statesman and Marxist philosopher. He was a founder of the Chinese Communist Party (1922). Following the Japanese occupation (1937–45), during which Nationalists and Communists collaborated against the Japanese, the Communists won the resumed civil war and Mao established his People's Republic (1949). His dictatorship became murderous as he sought to break traditional patterns of Chinese family life and launched his 'Cultural Revolution' (1966–69).

Mao Tse-tung

Marat, Jean Paul (1743–93) French revolutionary and journalist. A prominent supporter of DANTON and ROBESPIERRE, he made repeated calls for increased executions during the establishment of the Revolution and was stabbed to death in his bath by the Girondist aristocrat **Charlotte Corday** (1768–93).

Marc, Franz (1880–1916) German painter. With KANDINSKY, he founded the *Blaue Reiter* group of expressionist artists.

Marceau, Marcel (1923–) French mime artist. Regarded as the world's leading mime artist.

Marciano, Rocky [Rocco Francis Marchegiano] (1923–69) American boxer. He became world heavyweight champion (1952–56), and never lost a professional fight.

Marconi, Guglielmo, Marchese (1874–1937) Italian physicist and electrical engineer. He shared the 1909 Nobel prize for physics for his development of wireless telegraphy and later developed shortwave radio transmissions.

Marcos, Ferdinand [Edralin] (1917–89) Filipino politician. He was president of the Philippines

(1965–86). An autocratic ruler, he declared martial law in 1972, after which he ruled by oppressive and idiosyncratic decree. He was deposed in 1986 after the popular unrest that brought AQUINO to power, and lived in exile in Hawaii with his wife, **Imelda**.

Markiewicz, [Constance Georgine] Countess (1868–1927) Irish nationalist. A member of Sinn Fein involved in the Easter Rising of 1916, she became the first woman to be elected to the British Parliament in 1918, but refused to take her seat.

Markova, Dame Alicia [Lilian Alicia Marks] (1910–) English ballerina. She was a member of DIAGHILEV's Ballet Russe (1924–29) and then of the Vic-Wells Ballet, where she became prima ballerina (1933–35).

Marks, Simon, [1st Baron Marks of Broughton] (1888–1964) English businessman. He inherited the Marks and Spencer chain of shops helped build it into a respected retail empire.

Marley, Bob [Robert Nesta Marley] (1945–81) Jamaican singer and songwriter. With his group, the Wailers, he became the world's leading reggae singer.

Bob Marley

Marlowe, Christopher (1564–93) English dramatist and poet. He was one of the first English dramatists to use blank verse to great dramatic and poetic effect in his plays, the most famous of which are *Tamburlaine the Great* and his masterpiece, *Doctor Faustus*. He was probably a secret agent in the employ of the Elizabethan government and was killed in a tavern brawl.

Marshall, Alfred (1842–1924) English economist. His works have been of great influence on modern economics. He devised concepts such as 'elasticity', 'consumer surplus' and 'time analysis'.

Marshall, George C[atlett] (1880–1959) American general and statesman. He was chief of staff of the US army during World War II, and, as US secretary of state, oversaw the Marshall Aid Plan, for which he was awarded the 1953 Nobel Peace Prize.

Marvell, Andrew (1621–78) English poet. A (passive) supporter of Parliament during the English Civil War, he became member of parliament for Hull in 1659, a position he held until his death. His verse satires were much enjoyed by the wits of the day, even by those whose vices were attacked. His strange, metaphysical poems, e.g. 'The Garden' and 'Upon Appleton House', display a talent for symbolism and metaphor. His poem celebrating Oliver CROMWELL's suppression of the Irish rebellion, 'An Horatian Ode upon Cromwell's Return from Nature', is a great political poem, with its cool, restrained appreciation of Cromwell's stature.

Marx, Karl (1818–83) German philosopher. His theories on class struggle dominated 20th-century political thought from the Bolshevik Revolution to the collapse of the communist regimes of eastern Europe in 1989–91. *Das Kapital*, his study of the economics of capitalism, appeared in 1867; subsequent volumes, edited by **Friedrich Engels** (1820–95), appeared in 1885 and 1895. He also wrote *The Communist Manifesto* with Engels in 1848, and was one of the founders of the 'First International' in 1864.

Karl Marx

Marx Brothers An American comedy group of brothers consisting of **Arthur Marx (Harpo)** (1893–1964), **Milton (Gummo)** (1894–1977), **Herbert Marx (Zeppo)** (1901–79), **Julius Marx (Groucho)** (1895–1977) and **Leonard Marx (Chico)** (1891–1961). The anarchic humour of the Marx Brothers' films was enormously popular with both critics and public, with Groucho in particular enjoying a cult status among intellectuals.

Masaccio [Tommaso di Ser Giovanni di Mone] (1401–*c*.1428) Florentine painter. A key figure of the early Renaissance and in the development of perspective.

Mascagni, Pietro (1863–1945) Italian composer. His works include the perennial favourite, the one-act *Cavalleria Rusticana*.

Masefield, John [Edward] (1878–1967) English poet, whose best-known poem, from *Salt-Water Ballads* (1902) is 'I must go down to the sea again'. Many later poems, e.g. *The Everlasting Mercy* (1911), caused scandal with their frank treatment of rural themes. He was appointed poet laureate in 1930.

Mata Hari [Margarethe Geertruida Zelle] (1876–1917) Dutch spy. A dancer in Paris with many lovers, she became a German spy and was shot for treason.

Matisse, Henri (1869–1954) French painter and sculptor. In the period before World War I he became a leading Fauvist. A superb draughtsman, he also designed ballet sets for DIAGHILEV. He continued to paint more abstract and decorative works, and made use of cut-outs and collages in simple compositions.

Matthews, Sir Stanley (1915–) English footballer. Regarded as one of the greatest wingers of all time (the 'Wizard of Dribble'), he won 54 international caps in a career that spanned 22 years.

Maugham, W[illiam] Somerset (1874–1965) English novelist and dramatist. Trained as a doctor, he used his experiences working in the London slums for his first novel, *Liza of Lambeth* (1897). His best-known novels are *Of Human Bondage* (1915) and *The Moon and Sixpence* (1919), the latter based on the life of the painter Paul GAUGUIN. He was a British secret agent during World War I, and his experiences then form the basis of his spy novel, *Ashenden* (1928).

Maxwell, [Ian] Robert [Robert Hoch] (1923–91) Czech-born British newspaper proprietor, publisher and politician. His mysterious death by drowning off the Canary Islands was followed by revelations of his mishandling of his companies' assets.

Mayer, Louis B[urt] [Eliezer Mayer] (1885–1957) Russian-born American film producer. He joined with GOLDWYN to form Metro-Goldwyn-Mayer in 1924, and became one of the most powerful of the Hollywood moguls.

Mead, Margaret (1901–78) American anthropologist. Her works, which include *Coming of Age in Samoa*, argue that cultural conditioning shapes personality, rather than heredity.

Medawar, Sir Peter Brian (1915–87) Brazilian-born British zoologist. He shared the 1960 Nobel prize for physiology or medicine with the Australian virologist **Sir Frank Macfarlane Burnet** (1899–1985) for his work on immunological tolerance.

Medici, Lorenzo de' (1449–92) Florentine aristocrat and statesman. Styled 'The Magnificent', he was a poet and a noted patron of the arts. His tomb in Florence was designed by MICHELANGELO.

Meinhoff, Ulrike (1934–76) German terrorist. With **Andreas Baader** (1943–77) and others, she founded the 'Red Army Faction' in 1970, an ultra-leftist terrorist organization dedicated to using violence to bring about the collapse of West German 'capitalist tyranny'.

Meir, Golda (1898–1978) Russian-born Israeli stateswoman. Active in the fight for a Jewish state, she was minister of labour (1949–56) and of foreign affairs (1956–66) before becoming Israel's first female prime minister (1969–74).

Golda Meir

Meitner, Lise (1878–1968) Austrian-born Swedish physicist. She and Otto HAHN discovered the radioactive element protactinium (1918). With her nephew, Otto FRISCH, and others, she discovered the process of nuclear fission in the late 1930s.

Melba, Dame Nellie [Helen Porter Mitchell] (1861–1931) Australian soprano. Renowned for her light, pure voice, she became one of the world's leading prima donnas in the late 1880s.

Melville, Herman (1819–91) American novelist, short-story writer and poet. His masterpiece is the novel *Moby Dick* (1851), a complex and symbolic narrative featuring the revengeful Captain Ahab. His short novel *Billy Budd, Foretopman* was made into an opera by BRITTEN.

Memling *or* **Memlinc, Hans** (*c*.1430–1494) German-born Dutch painter. He was a prolific and popular artist, and was a successful portraitist.

Mendel, Gregor Johann (1822–84) Austrian monk who was also a biologist and botanist. He discovered that traits such as colour or height had two factors (hereditary units) and that these factors do not blend but can be either dominant or recessive.

Mendelssohn, Felix [Jakob Ludwig Felix Mendelssohn-Bartholdy] (1809–47) German composer. The grandson of the philosopher **Moses Mendelssohn** (1729–86), he became one of the leading Romantic composers. His works include

five symphonies, the opera *Elijah* (1846), songs and the overtures *A Midsummer Night's Dream* (1826) and *Fingal's Cave* (1832). His performance of BACH's *St Matthew Passion* resulted in a resurgence of interest in the composer.

Mengistu, Mariam Haile (1937–) Ethiopian dictator. He participated in the 1974 coup that toppled HAILE SELASSIE, and established a brutal dictatorship after a further coup in 1977.

Menotti, Gian Carlo (1911–) Italian-born American composer. His operas, for which he also wrote the librettos, employ a number of musical styles.

Menuhin, Sir Yehudi (1916–) American-born British violinist. An infant prodigy, he became one of the world's leading virtuosos and founded a school (in 1962) for musically gifted children.

Messerschmitt, Willy [Wilhelm Messerschmitt] (1898–1978) German aircraft designer and manufacturer. His planes include the first jet combat aircraft.

Messiaen, Olivier (1908–92) French composer and organist. His rhythmically complex works were often heavily influenced by religious mysticism.

Messmer, Otto (1894–1985) American cartoonist. His 'Felix the Cat' became the first cartoon superstar.

Meyerbeer, Giacomo (1791–1864) a German-born composer who visited Italy and wrote operas in the style of ROSSINI. His best-known work is *L'Africaine*.

Michelangelo Buonarotti (1475–1564) Florentine painter, sculptor, draughtsman, architect and poet, an outstanding figure of the Renaissance. His masterpiece was the ceiling paintings for the Sistine Chapel (1508–12). He also worked on the tombs of Lorenzo and Giuliano de Medici, and on the rebuilding of St Peter's. He was an accomplished poet, and wrote fine sonnets.

Middleton, Thomas (*c*.1570–1627) English dramatist. His two powerful tragedies, *The Changeling* and *Women Beware Women*, are now highly regarded. His other works include the satirical comedy *A Trick to Catch the Old One* and a political satire, *A Game at Chesse*, which almost resulted in his imprisonment. He collaborated with many other dramatists.

Milhaud, Darius (1892–1974) French composer. A member of 'Les Six', he was a highly prolific composer. His works, were mostly polytonal and often influenced by jazz.

Mill, John Stuart (1806–73) English philosopher and economist. A follower of BENTHAM, he elaborated the philosophy of the 'greater good' in his philosophy of utilitarianism. His most popular work is the defence of personal freedom *On Liberty* (1859).

Millais, Sir John Everett (1829–96) English painter. Along with Holman HUNT and ROSSETTI, he founded the Pre-Raphaelite Brotherhood and was known for his posed, studied tableaux in clashing colours.

Miller, Arthur (1915–) American dramatist. His tragedies include three classics of the American stage: *Death of a Salesman, The Crucible*, a comment on McCarthyism in the USA, and *A View from the Bridge*, inspired by Greek drama. He was married to Marilyn MONROE (1955–61) for whom he wrote the screenplay for her last film, *The Misfits* (1961).

Arthur Miller

Miller, [Alton] Glenn (1904–44) American composer, band-leader and trombonist. His dance band became one of the most popular in the world.

Millet, Jean-François (1814–75) French painter. He earned his living as a portraitist, and exhibited his first major genre painting, *The Winnower*, in 1848. He was labelled a social-realist although his work had no direct political import.

Millett, Kate (1934–) American feminist. Her works are cornerstones of feminist fundamentalism.

Milligan, Spike [Terence Allan Milligan] (1918–) Anglo-Irish comedian and writer. With Peter SELLERS, the Welsh comedian and singer **Harry Secombe** (1921–) and the Anglo-Peruvian comedian **Michael Bentine** (1921–96), he co-wrote and performed in the radio comedy series *The Goon Show* (1951–59), which became a highly influential comedy series, with its manic wit and surreal invention.

Millikan, Robert Andrews (1868–1953) American physicist. He was awarded the 1923 Nobel prize for physics for his determination of the charge on the electron.

Milne, A[lan] A[lexander] (1882–1956) English writer and dramatist. His children's books featur-

ing Winnie the Pooh are much loved classics of children's literature.

Milton, John (1608–74) English poet. One of the most formidably learned of all English poets, he had a European-wide reputation by his late twenties. His early poems, e.g. the elegy *Lycidas* (1637), are steeped in the humanist tradition, which looked to classical literature for ethical principles and modes of expression, and to Scripture and the Christian tradition for faith. He supported Parliament during the Civil War and wrote tracts attacking royalty and episcopacy. His most famous prose work is the tract *Aeropagitica* (1644), a rousing defence of the liberty of free speech. His masterpiece is the great epic poem on the Fall of Man, *Paradise Lost* (1667–74). Other notable works include the verse drama *Samson Agonistes* (1671), in which the blind hero represents Milton himself in Restoration England.

Miró, Joan (1893–1983) Spanish painter. Influenced by PICASSO, his work became increasingly abstract over the years and was influential on the abstract expressionist painters.

Mitchell, R[eginald] J[oseph] (1895–1937) English aircraft designer. He designed the Supermarine Spitfire (1934–36).

Mitterrand, François [Maurice Marie] (1916–96) French statesman. He became leader of the Socialist Party in 1971 and the first socialist president of France (1981–95).

Mobuto, Sese Seko Kuku Ngbendu Wa Za Banga [Joseph Désiré Mobuto] (1930–) Zairean dictator. He assumed complete power over the Congo in 1965, changing the country's name to Zaire in 1971.

Modigliani, Amedeo (1884–1920) Italian painter and sculptor. His best-known works are his African-influenced sculptures of elongated figures.

Mohammed *or* **Muhammad** (*c*.570–*c*.632) Arab prophet and founder of Islam. Born in Mecca, the son of a merchant, he began having revelations, sometime after 600, that he was the last prophet of Allah and His channel of communication with the world. He gathered together a band of followers and established himself at Medina in 622, from where, after several battles, his forces conquered Mecca in 629, and shortly after all Arabia.

Molière [pseud. of Jean-Baptiste Poquelin] (1622–73) French dramatist. His great comedies are as popular now as when they were first performed; only SHAKESPEARE's plays have been more widely performed. The plays include *Tartuffe*, a satire on religious hypocrisy, *The Misanthrope*, a study of a cynic in love, and *The Imaginary Invalid*, a hilarious depiction of hypochondria and quack medicine.

Molotov, Vyacheslav Mikhailovich [Vyacheslav Mikhailovich Scriabin] (1890–1986) Russian statesman. He negotiated the non-aggression pact with Nazi Germany and became minister for foreign affairs (1953–56).

Mondrian, Piet [Pieter Cornelis Mondriaan] (1872–1944) Dutch painter. He developed a style of painting based on grids of lines against strong colours and co-founded the De Stijl group.

Monet, Claude Oscar (1840–1926) French Impressionist painter. His *Impression: Sunrise* gave its name to the movement. With RENOIR and SISLEY, he began the direct studies of nature and changing light that were to characterize their works. His works include the *Haystacks* (1891) and *Rouen Cathedral* (1894) series.

Monk, Thelonius [Sphere] (1920–82) American jazz pianist and composer. His compositions include the classic 'Round Midnight'.

Monroe, Marilyn [Norma Jean Baker *or* Mortenson] (1926–62) American film actress. She became the leading "dumb blonde" sex symbol in the movies with such films as *Gentleman Prefer Blondes* (1953). Her other films include *Bus Stop* (1956) and the classic comedy *Some Like It Hot* (1959). Her last film, *The Misfits* (1961), was written by her third husband, Arthur MILLER. Her death was apparently due to an overdose of sleeping pills.

Marilyn Monroe

Montaigne, Michel Eyquem de (1533–92) French essayist. The dominant theme in his work was antidogmatic scepticism and he did much to establish the essay as a literary form .

Montessori, Maria (1870–1952) Italian educationalist. Her method of encouraging the child to learn at her or his own pace without restraint, was very influential on modern pedagogy.

Monteverdi, Claudio Giovanni Antonio (1567–1643) Italian composer. He introduced many new elements to the opera form and is regarded as the first major opera composer. His works include *Orfeo* and *The Coronation of Poppea*.

Montgomery of Alamein, Bernard Law, 1st Viscount (1887–1976) English soldier. In World War II he was given command of the 8th Army in Egypt in 1942, and won the Battle of Alamein, a victory recognized by CHURCHILL as a turning point in the war. He later commanded the Allied land forces on D-Day.

Moore, Henry [Spencer] (1898–1986) English sculptor. His monumental sculptures, often semiabstract in style but always based on organic form, resulted in him becoming the best known of modern sculptors.

More, Sir Thomas (1478–1535) English statesman and Roman Catholic saint. He was HENRY VIII's Lord Chancellor, and his refusal to recognize the annulment of Henry's marriage to Catherine of Aragon and declaration of supremacy over the Church in England led to his execution for treason. He was widely recognized as an honourable man, and his execution revulsed moderate opinion throughout Europe. His greatest work is his fantasy of a supposedly ideally organized state, *Utopia*. More was canonized in 1835, and has always been admired for his firm principles. His involvement in heresy trials, however, and his dispute with **William Tyndale** (*c.*1495–1536), English translator of the Bible, show a less attractive side of his character.

Morgan, Thomas Hunt (1866–1945) American geneticist and biologist. He was awarded the 1933 Nobel prize for physiology or medicine for his research into chromosomes and heredity.

Morisot, Berthe (1841–95) French painter, who exhibited in all but one of the Impressionist shows.

Moro, Aldo (1916–78) Italian Christian democrat statesman. He was prime minister (1963–68, 1974–76) and brought the Communist Party into close cooperation with his centre-left coalition shortly before his abduction and murder by the Red Brigade.

Morris, Desmond [John] (1928–) English zoologist. His studies of animal and human behaviour have been bestsellers.

Morris, William (1834–96) English poet, romance writer and artist. His influence on the arts and crafts movement was immense, as was his influence on the development and character of British socialism.

Morrison, Jim (1943–71) American rock singer and songwriter. His band, The Doors, became a huge cult after his death (from alcohol and drug abuse).

Morton, Jelly Roll [Ferdinand Joseph Lemott] (1885–1941) American jazz pianist, composer and bandleader who is regarded as one of the founders of New Orleans jazz.

Mosley, Sir Oswald [Ernald] (1896–1980) English Fascist leader. First elected to Parliament as a Conservative (1918–22), he became an Independent (1922–24), then a Labour MP (1924, 1929–31), and finally founder and leader of the British Union of Fascists (1932–36). The thuggery and demagoguery of his movement failed to attract much support, and he was interned during World War II.

Mossbauer, Rudolf Ludwig (1929–) German physicist. He shared the 1961 Nobel prize for physics with the American physicist **Robert Hofstadter** (1915–) for his discovery of the 'Mossbauer effect', involving gamma radiation in crystals.

Mountbatten, Louis [Francis Victor Albert Nicholas], [1st Earl Mountbatten of Burma] (1900–79) British naval commander and statesman. Supreme Allied Commander in South-East Asia (1943–45) and viceroy of India (1947), he oversaw the transfer of power to the independent governments of India and Pakistan. He was assasinated by the IRA.

Moussorgsky, Modest *see* **Mussorgsky, Modest**.

Mozart, Wolfgang Amadeus (1756–91) Austrian musician and composer. A child prodigy, he began composing at the age of 5. One of the most lyrical of all composers, his works include the operas *The Marriage of Figaro, Don Giovanni, Cosi fan tutte* and *The Magic Flute*, over 40 symphonies, concertos, string quartets, sonatas, 18 masses, and the unfinished *Requiem*.

Mugabe, Robert [Gabriel] (1924–) Zimbabwean statesman. He became leader of the Zimbabwe African National Union and prime minister (1980–) following the end of white minority rule. He merged his ruling party with the Zimbabwe African People's Union in 1988 to form a one-party state.

Mulliken, Robert Sanderson (1896–1986) American chemist and physicist. He was awarded the 1986 Nobel prize for chemistry for his work on molecular structure and on chemical bonding.

Munch, Edvard (1863–1944) Norwegian painter. An Expressionist, his works, e.g. *The Scream*, are noted for their strong use of primary colours and emotions.

Munthe, Axel [Martin Frederik] (1857–1949) Swedish physicist and psychiatrist. His autobiographal book, *The Story of San Michele*, describing his experiences while practising medicine, became a world bestseller.

Murdoch, [Keith] Rupert (1931–) Australian-born American newspaper tycoon. He inherited an Australian newspaper group from his father and expanded his media empire in Britain and America. His expansion into the US market necessitated his acquisition of US citizenship in 1985.

Musgrave, Thea (1928–) Scottish composer. Her early works were often on Scottish themes. Later compositions, often in serial form, include choral works and concertos.

Mussolini, Benito [Amilcare Andrea] (1883–1945) Italian dictator. Originally a socialist, he founded his fascist 'Blackshirt' party in 1919, and was elected to parliament in 1921, establishing himself as dictator ('Il Duce') in 1922. He formed the Axis with HITLER in 1937 and declared war on the Allies in 1940. He was deposed in 1943 and later executed by partisans.

Benito Mussolini

Mussorgsky *or* **Moussorgsky, Modest Petrovich** (1839–81) Russian composer. His best-known works include the opera *Boris Godunov* and the piano piece 'Pictures at an Exhibition'.

Muzorewa, Bishop Abel *see* SMITH, IAN.

Myrdal, [Karl] Gunnar (1898–1987) Swedish economist. He shared the 1974 Nobel prize for economics with HAYEK, largely for his work on the application of economic theory to the economies of the Third World.

Nabokov, Vladimir (1899–1977) Russian-born American novelist, who wrote in both Russian and English. His most famous novel is *Lolita* (1955).

Nader, Ralph (1934–) American lawyer and consumer protectionist. Nader and his followers ('Nader's Raiders') publicized many such cases of consumer abuse in the late 1960s and 70s.

Nagy, Imre (1896–1958) Hungarian statesman. He was appointed prime minister (1953–55) and forced to resign after attempting to liberalize communist policies. He became premier again in 1956, but was replaced after the Soviet invasion of that year.

Nansen, Fridtjof (1861–1930) Norwegian explorer, scientist and statesman. He traversed Greenland (1888–89) and almost reached the North Pole in 1895, achieving a record latitude. He was appointed commissioner for refugees (1920–22) by the League of Nations and awarded the 1922 Nobel Peace Prize.

Napoleon I [Napoleon Bonaparte] (1769–1821) emperor of France (1804–15). A brilliant and ruthless military leader, he established an empire throughout Europe, defeating coalitions of the other major powers. His invasion of Russia in 1812, and the murderous campaign in the Pyrenees against WELLINGTON's forces, led to the defeat of his armies at Leipzig in 1813 and Allied victory in 1814. Napoleon retired to Elba, from whence, in 1815, he came back to France, beginning the 'Hundred Days' campaign which resulted in his defeat at Waterloo, and subsequent banishment to St Helena, where he died.

Napoleon I

Nashe, Thomas (1567–1601) English writer of pamphlets and tracts on various subjects, which were usually satirical and often contain barbs directed against his many literary and religious enemies. He was particularly virulent against the Puritans.

Nasser, Gamal Abdel (1918–70) Egyptian soldier and statesman. He took a leading part in the coup that deposed **King Farouk** (1920–65) in 1952, and became prime minister in 1954. He became president (1956–70) and precipitated the Suez Crisis by nationalizing the Suez Canal (1956).

Navratilova, Martina (1956–) Czech-born American tennis player. Regarded as the one of the world's greatest tennis players, she defected to the US in 1975.

Nehru, Jawaharlal (1889–1964) Indian nationalist leader and statesman. The son of the nationalist lawyer, **Motilal Nehru ('Pandit' Nehru)** (1861–1931), he joined the Indian National Congress in

1919 and was imprisoned many times in the 1930s and 40s for his nationalist views. He became the first prime minister of India (1947–64) following independence and the partition of the subcontinent into India and Pakistan. His daughter Indira GANDHI became prime minister in 1966.

Jawaharlal Nehru

Nelson, Horatio, [Viscount Nelson] (1758–1805) English naval commander. Renowned for his tactics, he became rear-admiral in 1797 after defeating the Spanish fleet at the battle of Cape St Vincent. The following year, he won a striking victory over the French at the battle of the Nile, and was killed by a sniper during his defeat of the French at Trafalgar in 1805.

Nero (37–68) Roman emperor. He succeeded CLAUDIUS in 54, and soon became infamous for his debauchery, vanity and paranoia. He had many people put to death or forced to kill themselves.

Nervi, Pier Luigi (1891–1979) Italian architect and engineer. An exponent of the virtues of reinforced concrete, his designs include the Pirellie skyscraper in Milan.

Newman, Cardinal John Henry (1801–90) English theologian. His spirited defence of his faith, *Apologia pro Vita Sua* (1864), was much admired by believers and non-believers alike.

Newman, Paul (1925–) American film actor. His films include *Hud* (1963), *Cool Hand Luke* (1967), *Butch Cassidy and the Sundance Kid* (1969) and *The Color of Money* (1986), the last earning him an Oscar. A political activist of the moderate left, he has also raised considerable sums of money for charity through sales of his own-name salad dressing.

Newton, Sir Isaac (1642–1727) English scientist, philosopher and mathematician. According to legend, observing the fall of an apple inspired him to discover the law of gravity. He also discovered (independently of LEIBNITZ) the differential calculus, the reflecting telescope, and devised the three laws of motion.

Nicholas II (1868–1918) Russian tsar (1895–1917). A weak ruler, alternating between bursts of liberalization and repression, his authority was seriously weakened by Russia's defeat in the war with Japan (1904–5). He was deposed by the Bolsheviks in 1917, who later murdered him and his family.

Nicklaus, Jack [William] (1940–) American golfer. One of the greatest golfers of all time, he won more major tournaments than any other player in history.

Nielsen, Carl [August] (1865–1931) Danish composer. The first prominent polytonal Danish composer, his works include six symphonies, two operas and concertos.

Niemöller, Martin (1892–1984) German Lutheran pastor. An outspoken opponent of HITLER and Nazi ideology, he was imprisoned in concentration camps (1937–45). He was president of the World Council of Churches (1961–68) and a prominent pacifist.

Nietzsche, Friedrich Wilhelm (1844–1900) German philosopher and poet, whose works were highly critical of traditional morality and Christianity and proclaimed the advent of the superman. He has been very influential on many 20th-century writers and was claimed by HITLER to be a spiritual forebear of Nazism, but Nietzsche, who despised anti-Semitism, would have rejected this.

Nijinsky, Vaslav (1890–1950) Russian ballet dancer and choreographer. He became a protégé of DIAGHILEV, and is regarded as one of the greatest ballet dancers of all time.

Niven, David [James David Graham Nevins] (1909–83) Scottish film actor who established himself as the model urbane Englishman in many Hollywood productions.

Nixon, Richard Milhous (1913–94) American Republican politician. The 37th president of the US (1969-74) he became the first president to resign from office, in 1974, following the 'Watergate' scandal. He was pardoned in 1974. While in office, he ended the Vietnam war and established rapprochement with China.

Nkrumah, Kwame (1909–72) Ghanaian statesman. He was the first president of Ghana (1957–66) after independence.

Nolan, Sir Sidney [Robert] (1917–92) Australian painter. His paintings draw heavily upon Australian history and folklore.

North, Frederick, 8th Lord North [2nd Earl of Guilford] (1732–92) English statesman. As prime minister (1770–82) during the reign of George III, he implemented the king's policy that led to the loss of the American colony.

Nostradamus [Michel de Notredame] (1503–66) French astrologer and physician. He published two books of cryptic prophecies in rhymed quatrains which enjoyed a huge vogue.

Novello, Ivor [Ivor Novello Davies] (1893–1951) Welsh songwriter, composer and actor. His songs include 'Keep the Home Fires Burning', which was hugely popular with British soldiers during World War I, and 'We'll Gather Lilacs'.

Nuffield, William Richard Morris, 1st Viscount (1877–1963) English car manufacturer and philanthropist. He developed a Henry FORD-like system of mass production of cars, notably the Morris Oxford and the Morris Minor.

Nureyev, Rudolf (1939–93) Russian ballet dancer and choreographer. Regarded as the successor to NIJINSKY, he formed a famous partnership with FONTEYN in 1962.

Nyerere, Julius [Kambarage] (c.1922–) Tanzanian statesman. He became president (1962–85) and negotiated the union of Tanganyika and Zanzibar (1964), which formed Tanzania. Widely regarded as Africa's leading statesman. His invasion of Uganda in 1978 brought AMIN's dictatorship to an end.

Obote, [Apollo] Milton (1924–) Ugandan politician. He became Uganda's first prime minister (1962–66) after independence, and became president (1966–71) after deposing King Mutesa II. He was in turn deposed by AMIN and became president again (1980–85) after Amin's overthrow. He was deposed again in 1985.

O'Casey, Sean (1880–1964) Irish dramatist. His early plays, e.g. *Juno and Paycock*, reflect the patriotism that followed the Easter Rising of 1916.

Oistrakh, David Feodorovitch (1908–74) Russian violinist. A widely admired virtuoso whose son, **Igor Davidovitch Oistrakh** (1931–), also has an international reputation as a violinist.

Olivier, Laurence [Kerr], [Baron Olivier of Brighton] (1907–89) English stage and film actor and director. Regarded as the leading British actor of the modern era, he played all the major SHAKESPEARE roles and became an international film star. He was director of the National Theatre (1962–73). His second wife (of three) was VIVIEN LEIGH.

O'Neill, Eugene [Gladstone] (1888–1953) American dramatist. His greatest play *Long Day's Journey into Night* (1940–41), a study of family breakdown, was not performed until three years after his death. He was awarded the Nobel prize for literature in 1936 and won the Pullitzer prize on three occassions.

Ono, Yoko *see* LENNON, JOHN.

Oppenheimer, J[ulius] Robert (1904–67) American nuclear physicist. He resigned from the Los Alamos atom bomb project after the dropping of the bombs on Hiroshima and Nagasaki, and argued for cooperation with the USSR on the control of nuclear weapons.

Robert Oppenheimer

Orff, Carl (1895–1982) German composer. His best-known work is the popular *Carmina Burana* (1937), a 'secular oratorio' based on medieval poems.

Ortega, Daniel (1945–) Nicaraguan politician. A leader of the Sandinista resistance movement that overthrew the dictatorship in 1979, he became president (1985–90). The REAGAN administration gave backing to the right-wing 'Contra' forces in their guerrilla war against the Sandinistas, and Ortega was defeated in the 1989 election.

Ortega y Gasset, José (1883–1955) Spanish philosopher who argued that democracy in the modern era could easily lead to tyrannies of either the left or right.

Orwell, George [Eric Arthur Blair] (1903–50) Indian-born English novelist and essayist. His two greatest novels have become classics: *Animal Farm* (1945), a grim allegory of the history of the Soviet Union, and *Nineteen Eighty-Four* (1949), an even grimmer picture of a totalitarian world.

Osborne, John [James] (1929–94) English dramatist. His first play, *Look Back in Anger* (1956), gave its name to the 'Angry Young Men', a group of young playwrights who replaced the drawing-room comedies of 1950s British theatre with realistic dramas of working-class life.

Oswald, Lee Harvey (1939–63) American alleged assassin of President KENNEDY. He was arrested shortly after Kennedy's murder in Dallas in 1963, and was himself shot dead before he could come to trial by **Jack Ruby** (1911–64).

Ovid [Publius Ovidius Naso] (43 BC–*c*.17 AD) Roman poet. His sensual, witty love poems have always been admired, but his long narrative poem *Metamorphoses*, which describes myths in which characters change their forms, is of greater significance. It was used as a source book by many, e.g. SHAKESPEARE.

Owen, David [Anthony Llewellyn] (1938–) English politician. Founder of the Social Democratic Party with JENKINS and others. He refused to accept the merger of the SDP with the Liberal Party in 1987, but dissolved the SDP in 1990. He was made a life peer and worked as a UN peace negotiator in the Bosnian conflict of the early 1990s.

Owens, Jesse [James Cleveland Owens] (1913–80) American athlete. One of the finest athletes of his generation, he won four gold medals in the 1936 Berlin Olympics. Adolf HITLER left the stadium to avoid congratulating the black, non-Aryan athlete.

Jesse Owens

Padarewski, Ignace Jan (1860–1941) Polish pianist, composer and statesman. Widely regarded as the greatest pianist of his day, he served as prime minister for ten months in 1919.

Pahlavi, Mohammed Reza (1919–80) shah of Iran. He succeeded his father in 1941 and gradually established a dictatorship which was undermined by religious fundamentalists led by KHOMEINI and forced to flee his country in 1979.

Paine, Thomas (1737–1809) English-born American political theorist and pamphleteer. His highly influential pamphlet *Common Sense* (1776), which argued for American independence, was recognized by WASHINGTON as being a significant contribution to the Revolution. Paine returned to England in 1787 and published a defence of democratic principles, *The Rights of Man* (1791–92), in reply to Edmund BURKE's *Reflections on the Revolution in France*. In danger of arrest he moved to France where he was elected to the National Convention. He sided with the moderates, was imprisoned by ROBESPIERRE's faction, and released after 11 months, having narrowly escaped execution.

Paisley, Ian [Richard Kyle] (1926–) Northern Ireland Protestant clergyman and Unionist politician.

A highly vocal opponent of Irish nationalism and Roman Catholicism.

Palestrina, Giovanni Pierluigi da (*c.*1525–1594) Italian composer. One of the greatest Renaissance composers, his compositions are practically all choral church works, including more than 90 masses, hymns, motets and madrigals.

Palmer, Samuel (1805–81) English painter and engraver, noted for his pastoral landscapes. He was a follower of Blake, who deeply influenced the visionary mysticism of his work.

Pankhurst, Emmeline (1858–1928) English suffragette and feminist. She and her daughter **Dame Christabel Harriette Pankhurst** (1880–1958) founded the Women's Social and Political Union in 1903, a campaigning organization for women's suffrage. Her daughter **Sylvia Pankhurst** (1882–1960) was also a suffragette as well as a pacifist.

Papandreou, Andreas George (1919–96) Greek socialist politician and Greece's first socialist prime minister (1981–89).

Parker, Charlie *or* **Bird** [Charles Christopher Parker] (1920–55) American jazz alto saxophonist. He became the leading exponent of 'bop' jazz in the 1940s and worked with Dizzy GILLESPIE.

Parker, Dorothy [Rothschild] (1893–1967) American journalist, poet and short-story writer, noted for her dry wit and sharply ironic epigrams and satires.

Dorothy Parker

Parnell, Charles Stewart (1846–91) Irish politician. An ardent Home Ruler, he became MP for Cork in 1880 and organized in parliament a masterly campaign of obstruction with the aim of disrupting Parliament and ultimately gaining Home Rule for Ireland. GLADSTONE became a convert to the cause, but in 1890, his career began to crumble after he was cited in a divorce case.

Pascal, Blaise (1623–62) French theologian, mathematician and physicist. He made important discoveries in hydraulics and invented a calculating machine.

Pasternak, Boris [Leonidovich] (1890–1960) Russian poet and novelist. A highly original and passionate lyric poet in a regime that demanded safe verse praising its achievements forced him to turn to translation for a living; his translations of SHAKESPEARE's plays are still highly valued. His great novel *Dr Zhivago* was first published in Italy in 1958. He was awarded the Nobel prize for literature but was forced to decline it by the Soviet authorities.

Pasteur, Louis (1822–95) French chemist who discovered that fermentation is due to the presence of microorganisms and developed the process of pasteurization to destroy them. He also developed immunization processes against the diseases rabies and anthrax.

Patrick, Saint (*fl.* 5th century) British missionary. He was sold into slavery in Ireland as a youth, escaped to France and became a monk. He returned to Ireland as a missionary and converted many to his faith, and became the patron saint of Ireland.

Patton, George S[mith] (1885–1945) American general. In World War II he commanded the Allied invasion of North Africa (1942–43), and led the 3rd US army across France and Germany to the Czech border (1944–45).

Paul, Saint (d. *c.*67 AD) Christian apostle and missionary to the Gentiles. A pharisee, he was a notable persecutor of Christians before his 'Damascus Road' conversion. Many of the Epistles in the New Testament are his. According to tradition, he was executed during the reign of Nero.

Pauling, Linus Carl (1901–94) American chemist. He was awarded the 1954 Nobel prize for physics for his research into chemical bonding and molecular structure, and the 1962 Nobel Peace Prize for his criticisms of nuclear testing and his campaign for a multilateral test ban.

Pavarotti, Luciano (1935–) Italian tenor. He is regarded as one of the most powerful tenor singers of the modern era.

Pavlov, Ivan Petrovich (1849–1936) Russian physiologist. He was awarded the 1904 Nobel prize for physiology or medicine for his work on the physiology of digestion, and conducted experiments on the conditioning of reflexes.

Pavlova, Anna (1885–1931) Russian ballerina. FOKINE choreographed *The Dying Swan* for her. She also worked with DIAGHILEV, and became one of the most famous ballerinas in the world.

Pears, Sir Peter (1910–86) English tenor. A close associate of Benjamin BRITTEN, several of whose tenor opera roles were written for him.

Peary, Robert Edwin (1856–1920) American naval commander and Arctic explorer. He is credited with being the first man to reach the North Pole (1909).

Peel, Sir Robert (1788–1850) British statesman. He became Home Secretary in 1828 and founded the Metropolitan police. He was prime minister twice (1834–35, 1841–6). By his last year of office he had accepted and promoted the free-trade principles that disrupted the Tory party.

Peierls, Sir Rudolf Ernst (1907–) German-born British physicist. With Otto FRISCH, he demonstrated the feasibility of an atom bomb during World War II.

Pelé [Edson Arantes do Nascimento] (1940–) Brazilian footballer, universally recognized as one of the most skilful and entertaining players of all time.

Pepys, Samuel (1633–1703) English diarist and Admiralty official. His diary (written in code) was first published in 1825. The full uncensored version was published in 11 volumes (1970–83), and includes fascinating detail of life in 17th-century London.

Samuel Pepys

Perón, Juan Domingo (1895–1974) Argentinian dictator. He was president (1946–55), was deposed by the army and lived in exile until re-elected president (1973–74). His success was based to a large extent on his first wife, **Eva [Duarte] Perón** (1919–52), an ex-actress nicknamed 'Evita'.

Eva Perón

Perry, Fred[erick John] (1909–95) English-born American tennis and table-tennis player. He was one of the most successful lawn tennis players of the 1930s, winning every major tournament.

Pétain, Henri Philippe Omer (1856–1951) French soldier and statesman. Appointed marshal of France in 1918 in recognition of his generalship during World War I, he headed the collaborationist Vichy government (1940–44) and was sentenced to death at the end of World War II (later commuted to life imprisonment).

Peter, Saint (died *c*.67 AD) Disciple of JESUS CHRIST and Christian apostle. A fisherman, he became one of Jesus's leading disciples and played an equally prominent role in establishing Christianity after the crucifixion and is regarded by Roman Catholics as the first pope. He is believed to have been martyred in Rome.

Peterson, Oscar [Emmanuel] (1925–) Canadian jazz pianist and composer. His Oscar Peterson Trio became one of the best-known small jazz groups of the 1950s.

Petrarch [Francesco Petrarca] (1304–74) Italian lyric poet and humanist. His work popularized the sonnet form, and he is recognized as the first major poet of the Renaissance.

Petronius [Gaius Petronius Arbiter] (d. AD *c*.66) Roman courtier and satirist. His great satirical novel, the *Satyricon*, is an important landmark in Western literature.

Philby, Kim [Harold Adrian Russell Philby] (1911–88) English diplomat, journalist and secret-service double agent. He became a Soviet agent in 1933 and was recruited to the British Secret Service in 1940. He became head of anti-communist espionage (1944–46) and worked in the British embassy in Washington DC (1949–51). He worked as a foreign correspondent (1956–63) before fleeing to the USSR.

Piaf, Edith [Edith Giovanna Gassion] (1915–63) French singer and songwriter. Nicknamed 'Little Sparrow' for her small size and frail appearance, her songs include the anthemic '*Non, je ne regrette rien*'.

Piaget, Jean (1896–1980) Swiss psychologist. His studies of children's intelligence and perception have been highly influential on modern educationalists.

Picasso, Pablo (1881–1973) Spanish painter and sculptor. Regarded as the most influential artist of the modern era, with BRAQUE, he was the founder of Cubism. His 'blue period' (1901–4) works include *The Blue Room* (1901); Cubist works include *Les Demoiselles d'Avignon* (1906–7). He designed costumes and scenery for DIAGHILEV from 1917, exhibited with the Surrealists in the mid–1920s, and created his strongest and perhaps best-known image *Guernica* (1937) in response to the fascist bombing of that Basque town during the Spanish Civil War.

Pablo Picasso

Pickford, Mary [Gladys Mary Smith] (1893–1979) Canadian-born American film star. Known for her silent roles and as co-founder of the United Artists film studio (1919) with Charlie CHAPLIN and D. W. GRIFFITH.

Piero della Francesca (*c*.1416–92) Italian early Renaissance painter. While working in Florence he was deeply influenced by MASACCIO, who inspired the monumental grandeur of his subsequent works. From *c*.1460 he was working at the Urbino court, where he painted some of his finest works.

Pindar (*c*.518–*c*.438 BC) Greek lyric poet, noted for his odes and carefully constructed, elaborate poems which became influential in late-17th century England.

Pinochet [Ugarte], Augusto (1915–) Chilean general and dictator. He led the 1973 coup that deposed ALLENDE and became president (1974–90). He ruthlessly suppressed dissent and instituted monetarist economic polities.

Pinter, Harold (1930–) English dramatist, known for his halting, menacing dialogue and sinister pauses, e.g. *The Caretaker*.

Pirandello, Luigi (1867–1936) Italian dramatist and novelist. His two best-known plays are *Six Characters in Search of an Author* and *Henry IV*, both of which question theatrical conventions. He was awarded the Nobel prize for literature in 1934.

Pisano, Giovanni (*c*.1245–*c*.1314) Pisan sculptor. A leading sculptor of his time, his works are expressive and elegant in the Gothic tradition. His father **Nicola Pisano** (*c*.1220–*c*.1284) was also a notable sculptor and instrumental in the development of Christian art towards Roman rather than Gothic influences.

Pissarro, Camille (1830–1903) West Indian-born French Impressionist painter. Influenced by both CONSTABLE and TURNER, he exhibited in all eight Impressionist exhibitions.

Pitt, William [Pitt the Elder, 1st Earl of Chatham] (1707–88) English statesman. He led Britain to victory in the Seven Years War (1756–63) with France. He resigned in 1761, and served again 1766–68. His son **William Pitt** (the Younger) (1759–1806) became prime minister in 1783.

Planck, Max [Karl Ernst Ludwig] (1858–1947) German physicist. He formulated the quantum theory and was awarded the 1918 Nobel prize for physics.

Plath, Sylvia (1932–63) American poet and novelist. Her intense and highly expressive style has been very influential. She was married to TED HUGHES.

Plato (*c*.427–*c*.347 BC) Greek philosopher, regarded as the main founder of Western philosophy. Taught by SOCRATES, he was in turn ARISTOTLE's tutor. His many works, which take the form of dialogues, notably the *Symposium, Phaedo* and *The Republic*, have influenced almost every subsequent age and tradition.

Poe, Edgar Allan (1809–49) American short-story writer, poet and critic. His macabre, highly Gothic horror stories are studies in pathological obsession, and his detective stories, e.g. 'The Murders in the Rue Morgue', have been highly influential.

Pollock, [Paul] Jackson (1912–56) American painter. He became the leading exponent of action painting, a development of Abstract Expressionism, in the late 1940s.

Pol Pot *or* **Saloth Sar** [Kompong Thom] (1929–) Cambodian Communist politician. The establishment of his Maoist dictatorship cost the lives of up to three million people. The Khmer Rouge regime was overthrown by the Russian-backed Vietnamese invasion of 1979 but he remains a powerful figure.

Pompey *see* CAESAR, JULIUS.

Pompidou, Georges [Jean Raymond] (1911–74) French statesman. He was prime minister (1962–68), but was dismissed in 1968 by de Gaulle following the May student riots in Paris but elected president (1969–74).

Pope, Alexander (1688–1744) English poet. His mastery of the rhymed couplet, his deadly satire and gift for sustaining metaphor place him as one of the greatest English poets.

Popper, Sir Karl [Raimund] (1902–94) Austrian-born British philosopher. He established the concept of 'falsifiability' as the criterion by which to judge whether or not a particular proposition can be said to be scientific.

Porter, Cole [Albert] (1893–1964) American song-writer and composer. His highly popular songs, admired for their wit and lyricism, include 'Begin the Beguine' and 'Night and Day'.

Poulenc, Francis (1899–1963) French composer. A member of 'Les Six', he is particularly noted for his his settings of verses from poetry.

Pound, Ezra [Weston Loomis] (1885–1972) American poet and critic. A generous supporter of younger writers, e.g. T. S. ELIOT, HEMINGWAY, he lived in Italy from 1925 and broadcast propaganda against the Allies during World War II. He was committed to a US asylum after the war until 1958, when he returned to Italy.

Poussin, Nicolas (1594–1665) French painter who is noted for his carefully composed pictures in a classical style.

Powell, [John] Enoch (1912–) English Conservative politician. An outspoken opponent of immigration into Britain and of the Common Market. He was an Ulster Unionist MP (1974–87).

Powell, Michael (1905–90) English film producer and director. With the screenwriter **Emeric Pressburger** (1902–88), he made several films that have subsequently been hailed as important works, e.g. *The Life and Death of Colonel Blimp*.

Presley, Elvis [Aaron] (1935–77) American rock singer. He became one of the most popular singers in the world in the mid-1950s and was an outstanding interpreter of ballads.

Elvis Presley

Priestley, J[ohn] B[oynton] (1894–1984) English novelist and dramatist. His astute social commentaries made him one of the best-known literary figures of the day. His huge output includes the play *An Inspector Calls*.

Profumo, John Dennis (1915–) English Conservative politician. Secretary of state for war (1960–63), he resigned after admitting misleading the House of Commons about a sexual affair.

Prokofiev, Sergei Sergeyevich (1891–1953) Russian composer and pianist. His works include seven symphonies, ballets, piano and violin concertos, and the well-known orchestral 'fairy tale', *Peter and the Wolf* (1936).

Proust, Marcel (1871-1922) French novelist, essayist and critic, known for his long semi-autogiographic novel, *A la recherche du temps perdu*. By subjecting a mass of detail to an analytical eye and by using a circular form, Proust broke new ground in conveying the complexity of life and time and the importance ofmemory

Puccini, Giacomo (1858–1924) Italian composer. His operas, e.g. *La Bohème, Tosca, Madama Butterfly*, are regarded as the last great lyrical and dramatic works in the Italian tradition.

Purcell, Henry (*c.*1659–95) English composer. His works include incidental music for the theatre, songs, church music and six operas, notably *Dido and Aeneas* and *The Fairy Queen*.

Pushkin, Aleksandr Sergeyevich (1799–1837) Russian poet, novelist and dramatist. Widely regarded as Russia's greatest poet, the best known of his works are the verse novel *Eugene Onegin* and the historical tragedy *Boris Gudonov*. He died in a duel.

Qaddafi, Moammar al- *see* GADDAFI, MOAMMAR AL-.

Quant, Mary (1934–) English fashion designer. Her most famous design was the miniskirt, which became the symbol of 'swinging sixties' London.

Mary Quant

Quayle, Sir [John] Anthony (1913–89) English actor and director. He appeared in several films, but was principally a stage actor, founding his own classical touring company, Compass.

Quisling, Vidkun (1887–1945) Norwegian Fascist leader. He was installed as prime minister (1942-45) by the Nazis, and was executed for treason after the war.

Rabelais, François (*c.*1494–*c.*1553) French monk, physician and satirist, noted for his huge, rambling and often licentious prose fantasy *Gargantua and Pantagruel*. The adjective 'Rabelaisian' is used to denote language that is robustly bawdy.

Rabin, Yitzhak (1922–1995) Israeli politician and Prime Minister (1974–77; 1992–1995). A veteran of the 1967 Arab-Israeli War, in his second term of office, Rabin did much to convince Israelis that their security would not be compromised by making concessions to Palestinians, in order to secure peace. In 1994 he was a signatory of the Israeli-Palestinian Peace Accord, along with YASSER ARAFAT. He was assassinated by a Jewish extremist while speaking at a peace rally in Tel Aviv. He was succeeded as Prime Minister by **Shimon Peres** (1923–) who served until 1996, when his Labour Party was defeated in elections by the Likud Party and **Binyamin Netanyahu** (1949–) became Prime Minister.

Rachmaninov, Sergei (1873–1943) Russian composer and pianist. Influenced by Tchaikovsky, his music was very much in the 19th-century romantic tradition. His works include three symphonies, four piano concertos, the *Rhapsody on a Theme by Paganini* (1934) and many songs.

Raleigh, Sir Walter (1552–1618) English courtier, poet and explorer. He became a favourite of ELIZABETH I after returning from a punitive expedition to punish Irish rebels in 1580, and organized unsuccessful attempts to colonize Virginia with English settlers in the 1580s. In 1595, he travelled to the Orinoco and participated in the English raid on Cadiz in 1596. He was imprisoned after Elizabeth's death, was released to search for treasure on the Orinoco in 1616, and was executed on his return.

Sir Walter Raleigh

Rambert, Dame Marie [Cyvia Rambam] (1888–1982) Polish-born British ballet dancer, teacher and producer. After working with DIAGHILEV and NIJINSKY, she settled in Britain in 1917. She formed the Ballet Club in 1931, which, renamed Ballet Rambert in 1935, became the most influential ballet company in Britain.

Rameau, Jean Philippe (1683–1764) French composer, organist and harpsichordist. He published an influential textbook on harmony and composed motets and canatas. When he was 50, he started to write operas, with VOLTAIRE contributing several libretti.

Raphael [Raffaello Sanzio] (1483–1520) Italian painter. A leading figure of the High Renaissance, his portrayals of the Madonna and Holy Family combined Christian ideals with the grace and grandeur of classical antiquity.

Rasputin, Grigori Efimovich (*c.*1871–1916) Russian monk. Claiming to having healing powers, he became a cult figure among the Russian aristocracy, and an influential member of the royal household. He was assassinated in 1916.

Ravel, Maurice (1875–1937) French composer. He was one of the leading impressionist composers of his time. His works include the small orchestral piece *Boléro*.

Ravi Shankar (1920–) Indian sitar player and composer. Regarded as one of India's greatest modern musicians, he became world-famous after teaching George HARRISON to play the sitar.

Ray, Man (1890–1976) American photographer and painter, he was a leading exponent of Surrealist photography.

Ray, Satyajit (1921–92) Indian film director. His films, popular in art houses the world over, include the *Apu* trilogy of life in rural India, i.e. *Pather Panchali* (1955), *Aparajito* (1956) and *Apu Sansar* (1959), and *The Chess Players* (1977).

Reagan, Ronald [Wilson] (1911–) American film actor, Republican statesman and 40th president of the US (1981–89). He appeared in around fifty films and was president of the Screen Actors Guild (1947–52, 1959–60). As US president he pursued strong monetarist economic policies and a strong anti-communist foreign policy.

Ronald Reagan

Redgrave, Sir Michael [Scudamore] (1908–85) English stage and film actor. One of the finest actors of his generation, with a distinctively intellectual approach to his craft. He was married to the actress **Rachel Roberts** (1910–), and their daughter, **Vanessa Redgrave** (1937–), became a highly successful stage and film actress.

Reed, Sir Carol (1906–76) English film director. His films include *The Third Man* (1949), written by Graham GREENE and starring Orson WELLES, a bleak thriller set in postwar Vienna that is one of the most highly praised films ever made.

Reinhardt, Django [Jean Baptiste Reinhardt] (1910–53) Belgian guitarist, he formed the influential Quintette de Hot Club de Paris with GRAPPELLI.

Rembrandt Harmensz, van Rijn (1606–69) Dutch painter, draughtsman and etcher. His remarkable series of self-portraits, painted over 40 years, reveal the depth and spiritual development of his work.

Renoir, Pierre Auguste (1841–1919) French Impressionist painter. His form of Impressionism developed the use of perspective, solidity of form and preliminary sketches. His son, **Jean Renoir** (1894–1979) was a film director. His memorable films, e.g. *La Grande Illusion* and *La Règle du Jeu*, are often described as 'humanist' for their compassion and sense of human unity.

Resnais, Alain (1922–) French film director. One of the best known of the French 'New Wave' directors, his films include the romance *Hiroshima mon amour* (1959) and the experimental, "Surrealist" *Last Year in Marienbad* (1961).

Reynolds, Albert (1933–) Irish politician. Elected in 1977 as a Fianna Fáil member, he held several posts under HAUGHEY but was sacked by him (1991). He eventually succeeded as *taoiseach* (prime minister) (1992).

Reynolds, Sir Joshua (1723–92) English painter and art theorist. A portraitist, influenced by his studies of Renaissance and Baroque painting and classical sculpture, he became first president of the Royal Academy.

Rice, Tim *see* LLOYD WEBBER, ANDREW.

Richard, Cliff [Harry Roger Webb] (1940–) Indian-born English singer and film actor, and an institution in British popular music.

Richard, Keith *see* JAGGER, MICK.

Richards, Viv (1952–) Antiguan-born West Indian cricketer. One of the best batsmen and fielders in modern cricket, he was captain of the West Indies (1985–91).

Richardson, Sir Ralph [David] (1902–83) English stage and film actor. Ranked with GIELGUD and OLIVIER as among the finest British actors of the 20th century, he was equally at home with the classics and modern roles. Notable film roles include Buckingham in Olivier's film of *Richard III*,

the head of the secret service in *Our Man in Havana* and God in *Time Bandits* (1981).

Richardson, Samuel (1689–1761) English novelist. All his novels were written in epistolary form, and all were hugely popular. The first was *Pamela; or Virtue Rewarded*, in which a servant girl achieves an upwardly mobile marriage by resisting seduction (the work was attacked by FIELDING in his parody *Shamela* for its dubious morality). Richardson subsequently published *Clarissa Harlowe* and *Sir Charles Grandison*.

Richthofen, Manfred, Baron von (1882–1918) German fighter pilot, nicknamed the 'Red Baron'. He was credited with shooting down 80 allied aircraft.

Rimbaud, Arthur (1854–91) French poet. An early Symbolist, he stopped writing poetry at the age of nineteen. Some of the pieces in his collection of hallucinatory, vivid prose poems, *Les Illuminations*, were set to music by BRITTEN.

Rimsky-Korsakov, Nikolay Andreyevich (1844–1908) Russian composer. His music is typically Russian and he freely used local history, folk tunes, legends and myths as sources of inspiration. His works include 16 operas (e.g. *The Golden Cockerel*), three symphonies and numerous orchestral pieces.

Robeson, Paul [Le Roy] (1898–1976) American bass singer and actor. He qualified as a lawyer before becoming a highly popular stage actor in the 1920s. Notable performances include *Showboat* (1927) and *Othello* (1940). His warm, sensitive recordings of spirituals and folk songs were also very popular. A noted advocate of civil rights for Blacks, he came under strident attack in the US for supposed Communist sympathies and spent much of his life from the early 1960s in seclusion.

Paul Robeson

Robespierre, Maximilien Marie Isidore de (1758–94) French lawyer and revolutionary. He was elected to the National Assembly (1789) at the beginning of the French Revolution and became leader of the Jacobin group. He launched the infamous Reign of Terror, but was eventually guillotined himself.

Robinson, Mary (1944–) Irish barrister, politician and president (1990–) of the Republic of Ireland. Notably liberal in her policies, she won wide support from parties opposed to her opponent.

Rochester, John Wilmot, 2nd Earl of (1647–80) English poet and courtier. Renowned for his savage wit and depravity, his verse is among the most sexually explicit in English.

Rodgers, Richard [Charles] (1902–79) American composer. With the librettist Lorenz HART, he created musicals such as *The Pal Joey*). After Hart died, Rodgers collaborated with HAMMERSTEIN on several more successful musicals, e.g. *Oklahoma*.

Rodin, Auguste (1840–1917) French sculptor. He was responsible for reviving sculpture as an independent form rather than as an embellishment or decoration with works such as *The Thinker*.

Rogers, Ginger *see* ASTAIRE, FRED.

Rogers, Richard (1933–) Italian-born English architect. His designs include the Lloyds building (1979) in London, a steel and glass confection that typifies the controversial nature of his work.

Rolls, Charles Stewart (1877–1910) English motor car manufacturer and aviator. He joined Henry ROYCE in car manufacture in 1906.

Rommel, Erwin (1891–1944) German soldier. During World War II, he commanded the Afrika Korps in North Africa, earning the nickname the 'Desert Fox' for his brilliant tactics. He committed suicide after the discovery of his complicity in an assassination attempt on HITLER.

Franklin D. Roosevelt

Roosevelt, Franklin D[elano] (1882–1945) American Democratic statesman. He became 32nd president of the US (1933–45) and, in order to deal with the crisis of economic collapse, instituted far-reaching 'New Deal' reforms. He was a popular and highly effective leader during World War II, dying shortly after the Yalta summit meeting with Churchill and Stalin. His wife, **[Anna] Eleanor Roosevelt** (1884–1962), was an active and popular First Lady, supporting her husband during his illness with polio. After his death she worked with the UN as US representative to the General Assembly (1946–52).

Roosevelt, Theodore (1858–1919) American Republican statesman. 26th president of the US (1901–9). He legislated against big business monopolies, intervened forcefully during the Panama civil war to protect the construction of the Panama Canal and won the 1906 Nobel Peace Prize for mediating the end of the Russo-Japanese war.

Rossetti, Christina Georgina (1830–94) English poet, noted for her reflective, occasionally melancholic religious poems. She also wrote for children. Her brother, **Dante Gabriel Rossetti** (1828–82), was both a poet and an artist, and a founder of the Pre-Raphaelite school of painting.

Rossini, Gioacchino Antonio (1792–1868) Italian composer, noted especially for his light operas, e.g. *The Barber of Seville*, *The Thieving Magpie*. His other works include a *Stabat Mater*.

Rostand, Edmond (1868–1918) French dramatist and poet. His best-known work is the verse drama *Cyrano de Bergerac* (1897).

Rostropovich, Mstislav (1927–) Russian cellist. One of the outstanding cellists of modern times, he has also given many recitals as a pianist, often accompanying his wife, the soprano **Galina Vishnevskaya** (1926–), in song recitals.

Rothko, Mark [Marcus Rothkovitch] (1903–70) Russian-born American painter. Having passed through Expressionism and Surrealism, he adopted the abstract expressionist style of painting, creating large canvases with almost luminous rectangles of colour.

Rousseau, Henri Julien ['Le Douanier'] (1844–1910) French painter. His naive style remained unaffected by all trends, and he defied conventions of colour and perspective in his exotic imaginary landscapes and painted dreams.

Rousseau, Jean-Jacques (1712–78) Swiss-born French philosopher. His most notable fictional works are the novels *Julie, or the New Héloïse* (1761) and *Emile* (1762), the former describing a highly improbable menage à trois, the latter a didactic work on how to educate children (whom he saw as naturally good). These works and others, notably the political tract *The Social Contract* (1762), which begins with the famous statement 'Man is born free, and is everywhere in chains', were profoundly influential on the intellectual ferment that led to the French Revolution. His very frank autobiography, *Confessions*, was published posthumously and set a fashion for this style of reminiscence.

Royce, Sir [Frederick] Henry (1863–1933) English engineer. In partnership with Charles ROLLS, he founded the car firm Rolls-Royce in 1906.

Rubens, Sir Peter Paul (1577–1640) Flemish painter and diplomat. He entered the service of the Duke of Mantua in Italy and became court painter to the Spanish viceroys in Antwerp. His masterpiece is the triptych *Descent from the Cross*. In 1629 he was sent to England to negotiate peace with Charles I, who knighted him.

Rubinstein, Artur (1888–1982) Polish-born American pianist. An outstanding concert pianist, he was particularly noted for his CHOPIN recitals.

Ruby, Jack *see* OSWALD, LEE HARVEY.

Rushdie, [Ahmed] Salman (1947–) Indian-born British novelist. After the publication of *Satanic Verses* (1988), Ayatollah KHOMEINI pronounced a death sentence for blasphemy on him.

Ruskin, John (1819–1900) English writer, artist and influential art critic. His works, which include *Modern Painters*, dictated Victorian taste for over half a century. He was an enthusiast for Gothic art, the Pre-Raphaelite movement and TURNER, and was a strong critic of the values and ugliness of Industrial England.

Russell, Bertrand [Arthur William] 3rd Earl Russell (1872–1970) British philosopher. He made notable contributions to mathematical and philosophical theory and, with the help of his student, WITTGENSTEIN, logical positivism. He was awarded the 1950 Nobel prize for literature.

Russell, Ken (1927–) English film director. He is especially noted for his film biographies of musicians, e.g. *Lisztomania*.

Rutherford, Ernest [1st Baron Rutherford of Nelson] (1871–1937) New Zealand physicist. He was awarded the 1908 Nobel prize for chemistry. In 1911 he deduced the existence of the atom's structure and was the first scientist to split the atom.

Sabin, Albert Bruce (1906–) Polish-born American micro-biologist. He developed the Sabin polio vaccine in the mid–1950s.

Sadat, [Mohammed] Anwar El (1918–81) Egyptian statesman. He succeeded NASSER as president in 1970. After the Arab-Israeli Yom Kippur War of 1973, he came under pressure to work towards peace in the Middle East. He signed a peace treaty with BEGIN, for which they were awarded the 1978 Nobel Peace Prize. He was assassinated by Islamic fundamentalist soldiers during a military parade.

Anwar El Sadat

Sade, Donatien Alphonse François, Marquis de (1740–1814) French soldier and novelist. His highly licentious works include several novels, e.g. *Justine*. The term 'sadism' derives from the dominant theme in his life and work.

Sakharov, Andrei Dimitrievich (1921–89) Russian physicist and dissident. He developed the Russian hydrogen bomb in the 1950s and subsequently campaigned for international control of nuclear weapons. He was awarded the 1975 Nobel Peace Prize.

Saladin [Salah al-Din al-Ayyubi] (1137–93) Sultan of Egypt and Syria. The leader of the Arab world during the Crusades.

Salam, Abdus *see* WEINBERG, STEVEN.

Salk, Jonas Edward (1914–95) American physician and micro-biologist. He developed the Salk vaccine against polio.

Saloth Sar *see* POL POT.

Sappho (b. *c*.650 BC) Greek poet. The Greeks regarded her as one of the greatest of all lyric poets, but only short fragments of her work survivies.

Sappho

Sartre, Jean-Paul (1905–80) French philosopher, novelist and dramatist. His attempts at reconciling Existentialist philosophy with Marxism are now of historical interest. His novels, however, e.g. *Nausea* (1938), are highly readable. Several of his plays, e.g. *Huis clos* (1944), are frequently performed.

Satie, Erik [Alfred Leslie] (1866–1925) French composer (his mother was a Scottish composer). His simple, classically inspired compositions were influential on DEBUSSY and RAVEL.

Saussure, Ferdinand de (1857–1913) Swiss linguist. Regarded as one of the founders of modern linguistics, he established the 'structuralist' approach to language as a social phenomenon, focusing on the arbitrary relationship between the word as 'linguistic sign' and the thing it signifies.

Schoenberg *or* **Schönberg, Arnold [Franz Walter]** (1874–1951) Austrian composer (US citizen from 1941). His early works, e.g. *Gurrelieder* (1900), are lush chromatic compositions in the late Romantic tradition. He then began composing atonal works and eventually developed his serial or 'twelve-tone' method.

Schopenhauer, Arthur (1788–1860) German philosopher. Renowned for his pessimistic outlook on life, he emphasized the active role of the will as the creative force in human thought.

Schubert, Franz [Peter] (1797–1828) Austrian composer. His works include nine symphonies, the eighth (in B minor) being the 'Unfinished', string quartets, and other chamber music. His songs, as in the song cycles *Die schöne Müllerin* and *Die Winterreise*, are regarded as some of the finest ever written, include settings of lyrics by HEINE and GOETHE, and others.

Schumacher, Ernst Friedrich (1911–77) German-born British economist. His book, *Small is Beautiful* (1973), became a founding text of the conservationist movement.

Schumann, Robert Alexander (1810–56) German composer. Noted for his espousal of Romantic values in music, his works include four symphonies, songs, and much fine piano music. His wife **Clara Schumann** (1819–96) was also a pianist and composer.

Schuschnigg, Kurt von (1897–1977) Austrian statesman. A staunch opponent of HITLER, he became chancellor in 1934 and was imprisoned by the Nazis (1938–45).

Scorsese, Martin (1942–) American film director. His films include *Taxi Driver* (1976), *Raging Bull* (1980) and *Goodfellas* (1990), these three films starring the actor most associated with Scorsese's work, Robert DE NIRO.

Scott, Robert Falcon (1868–1912) English explorer. He led two Antarctic expeditions (1901–4, 1910–12). He died with four companions on his last expedition, returning from the South Pole after having reached it a month after AMUNDSEN. His son, **Sir Peter [Markham] Scott** (1909–89), was

a naturalist and artist whose television documentaries and many books were notably influential in promoting conservation.

Scott, Sir Walter (1771–1832) Scottish novelist and poet. His early, highly Romantic narrative poems, set in the Scottish past, e.g. *The Lady of the Lake*, established his popularity with both the reading public and the literary world, BYRON being particularly generous in his praise. His historical novels (a genre he refined and made into an art form), particularly *Waverley* (1814), *The Heart of Midlothian* (1818) and *Ivanhoe* (1819), were enormously influential and spawned a host of imitators.

Scriabin *or* **Skryabin, Alexander Nikolayevich** (1872–1915) Russian composer and pianist. His compositions often involved extra-musical effects, e.g. *Prometheus*, a piece for piano accompanied by coloured light projected on a screen. He envisaged all the arts coming together in one great performance.

Searle, Ronald [William Fordham] (1920–) English cartoonist and writer. Known primarily as the creator of the monstrous St Trinian's schoolgirls, who feature in several of his works, he is regarded, particularly in the US and France, as one of the finest graphic artists of the 20th century. His haunting book *To the Kwai—and Back: War Drawings 1939–45* is a record of his experiences as a Japanese prisoner of war.

Secombe, Harry *see* MILLIGAN, SPIKE.

Segovia, Andrés [Marquis of Salobreña] (1894–1987) Spanish guitarist. An internationally recognized virtuoso, he initiated a revival of interest in the classical guitar.

Sellers, Peter (1925–80) English actor and comedian. One of the founders of the Goon Show (*see* Spike MILLIGAN). He achieved further popularity as Inspector Clouseau in such films as *The Pink Panther* (1963).

Seneca [Lucius Annaeus Seneca] (*c.*4 BC–65 AD) Roman dramatist and Stoic philosopher. His verse tragedies were very influential on Elizabethan dramatists such as SHAKESPEARE.

Sennett, Mack [Michael Sinnott] (1880–1960) Canadian-born American film director and producer. He produced the manic 'Keystone Cop' comedies, which achieved international success. He also produced Charlie CHAPLIN's first films.

Seurat, Georges (1859–91) French painter, a leading neo-Impressionist. He developed the system of pointillism in which the painting is built up from tiny areas of pure colour.

Shackleton, Sir Ernest Henry (1874–1922) Anglo-Irish explorer. He served in Robert SCOTT's

Antarctic expedition, and commanded two further expeditions (1908–09, 1914–16).

Shakespeare, William (1564–1616) English dramatist and poet. His status as the greatest of all poets and dramatists has rarely been challenged. His plays are generally divided into three groups. The first group (late 1580s–*c*.1594) consists of histories, e.g. the *Henry VI* trilogy, early comedies such as *The Two Gentlemen of Verona*, and the tragedy of *Romeo and Juliet*. The second group (*c*.1595–*c*.1599) includes histories such as *King John, Henry IV Parts I and II*, the comedies *A Midsummer's Night's Dream* and *As You Like It*, and the tragedy *Julius Caesar*. The third group (*c*.1600–*c*.1612) includes the great tragedies *Hamlet, Othello, King Lear, Macbeth, Antony and Cleopatra, Coriolanus* and *Timon of Athens*, the so-called "dark comedies," *Troilus and Cressida, All's Well That Ends Well*, and *Measure for Measure*, and tragicomedies such as *The Winter's Tale* and *The Tempest*. Shakespeare's other major works are the narrative poems *Venus and Adonis* (1593) and *The Rape of Lucrece* (1594), and the magnificent *Sonnets*, which feature a romantic triangle between the poet (Shakespeare) a dark lady (identity unknown) and a beautiful young nobleman (possibly the Earl of Southampton).

William Shakespeare

Shankar, Ravi *see* RAVI SHANKAR.

Shankly, Bill [William Shankly] (1913–81) Scottish footballer and manager. Regarded as one of the outstanding football managers of the century, he transformed Liverpool into one of the most successful clubs of modern times. A renowned football fanatic, he once notoriously observed that football was more important than life or death.

Shaw, George Bernard (1856–1950) Anglo-Irish dramatist and critic. He began his literary career as a drama, literary and music critic in the 1880s, and after a false start in novel-writing, began writing plays in the 1890s. The plays, e.g. *Man and Superman* (1903), *Major Barbara* (1905) and *Pygmalion*

(1913), have been very successful thanks to Shaw's mastery of witty dialogue. He was awarded the Nobel prize for literature in 1925.

George Bernard Shaw

Shelley, Mary Wollstonecraft (1797–1851) English novelist. Her masterpiece is *Frankenstein, or the Modern Prometheus* (1818), a Gothic fantasy that has been hailed as the first science fiction novel.

Shelley, Percy Bysshe (1792–1822) English poet. His talent for public scandal emerged at Oxford University, where he was expelled for co-writing a tract entitled *The Necessity of Atheism* (1811), in which year he eloped with Harriet Westbrook. Two years later, he published his poem *Queen Mab*, which celebrates a future republican millenium of free love and vegetarianism. In 1814, he eloped with Mary Godwin (*see* Mary Wollstonecraft SHELLEY) and her step-sister, Jane 'Claire' Clairmont. Harriet committed suicide by drowning in 1816, in which year he married Mary. Shelley and his entourage moved to Italy in 1818, where he drowned in a sailing accident. His poems are among the greatest of English Romantic poetry. The highlights are: *Prometheus Unbound* (1820); *Adonais* (1821), his elegy on the death of KEATS; and several of the finest poems in the English language, notably *Ode to the West Wind* and *To a Skylark* (both 1820).

Sheridan, Richard Brinsley (1751–1816) Irish dramatist and politician, noted for his superb comedies of manners, *The Rivals* (1775) and *School for Scandal* (1777), both of which are firm repertory favourites. His other major play is *The Critic* (1779), a burlesque satirizing the conventions of tragedy. He was also highly regarded for his oratory in parliament.

Shockley, William Bradford (1910–89) American physicist. He shared, with BARDEEN and the Chinese-born American physicist **Walter Brattain** (1902–87), the 1956 Nobel prize for physics for his development of the junction transistor.

Shostakovich, Dimitri Dimitriyevich (1906–75) Russian composer. Many of his works, e.g. the opera *Lady Macbeth of Mtensk*, were attacked for their disregard of socialist realism. His works include 15 symphonies, 15 string quartets and song cycles.

Sibelius, Jean (1865–1957) Finnish composer. His works reflect his strong Finnish nationalism and often draw on the Finnish traditional epic, *Kalevala*.

Sidney, Sir Philip (1554–86) English poet, soldier and courtier. His works include *Arcadia* (1590), the first major English pastoral poem; the sonnet sequence *Astrophel and Stella* (1591), which inspired a host of imitations; and *A Defence of Poetry* (1595), a spirited defence of English as a medium for writing great poetry. His died in action against the Spaniards in the Netherlands.

Sihanouk, Prince Norodom (1922–) Cambodian statesman, formerly (elected) king of Cambodia (1941–55). He abdicated (in favour of his father) to become prime minister (1955–60) after independence from France in 1955, becoming head of state in 1960. He was deposed by a military coup in 1970 and fled to China, forming an alliance with POL POT's Khmer Rouge, who seized Cambodia in 1975. He again became head of state in 1975 and was deposed by Pol Pot the following year. After the Vietnamese invasion of 1979, Sihanouk formed a government in exile, in an uneasy alliance with the Khmer Rouge.

Sikorski, Wladyslaw (1881–1943) Polish general and statesman. Premier of the Polish government in exile during World War II and commander in chief of the Free Polish armed forces.

Sikorsky, Igor Ivan (1889–1972) Russian-born American aeronautical engineer. He built the first four-engined aircraft in 1913 and the first successful helicopter in 1939.

Simpson, Wallis *see* WINDSOR, DUKE OF.

Sinatra, Frank [Francis Albert Sinatra] (1915–) American singer and film actor. Regarded as one of the finest modern popular singers, with a finely tuned jazz-like sense of phrasing.

Frank Sinatra

Singer, Isaac Bashevis (1904–91) Polish-born American Yiddish writer. Much of his fiction deals with the now vanished world of Polish Judaism, e.g. *The Magician of Lublin*. He was awarded the Nobel prize for literature in 1978.

Sisley, Alfred (1839–99) French painter of English extraction. Influenced by the Impressionists Renoir and Monet, he painted mainly carefully composed and sensitively coloured landscapes.

Skryabin *see* SCRIABIN.

Smith, Adam (1723–90) Scottish economist and philosopher. His book *Inquiry into the Nature and Causes of the Wealth of Nations* (1776), with its advocacy of free trade was of huge influence in the development of modern capitalist societies.

Smith, Bessie [Elizabeth Smith] (1895–1937) American blues singer, nicknamed the 'Empress of the Blues'. She became very popular with jazz audiences in the 1920s.

Smith, Ian [Douglas] (1919–) Zimbabwean politician. He was prime minister of Rhodesia (1964–79), and declared UDI (unilateral declaration of independence) from Britain in 1965 in order to maintain white minority rule. Majority rule came in 1979, with **Bishop Abel Muzorewa** (1925–) serving as caretaker premier (1979–80). MUGABE's Zanu party won the 1980 election. Smith resigned his leadership of his party in 1987.

Smith, Stevie [Florence Margaret Smith] (1902–71) English poet and novelist. Her graceful, melancholic, and occasionally fiercely funny verse has been much admired.

Smollett, Tobias [George] (1721–71) Scottish surgeon and novelist. He served in the Royal Navy as a ship's surgeon and took part in an attack upon a Spanish port in the West Indies. In the early 1740s, he set up a surgical practice in London. His picaresque novels, of which the most important are *The Adventures of Roderick Random*, *The Adventures of Peregrine Pickle*, and his masterpiece, *The Expedition of Humphrey Clinker*, are cleverly plotted satirical works rich in characterization, which achieved lasting popularity. His works influenced DICKENS.

Smuts, Jan Christian (1870–1950) South African statesman and philosopher. He commanded Boer forces during the Boer War, and became prime minister (1919–24, 1939–48).

Sobers, Gary [Sir Garfield St Auburn] (1936–) West Indian cricketer. Regarded as one of the finest all-rounders of all time.

Socrates (470–399 BC) Greek philosopher, the tutor of PLATO. The sources of his teachings are many and widely varied, but it is Plato's Socrates, with a gift for answering questions with another question, who has come down to posterity in Plato's 'Socratic' dialogues. The central theme in Socrates' thinking is a quest for truth through rigorous self-examination. He was forced to commit suicide by the Athenians for supposedly corrupting youths through teaching them 'impiety'.

Solti, Sir Georg (1912–) Hungarian-born British conductor. His recording of WAGNER's Ring cycle was particularly renowned.

Solzhenitsyn, Aleksandr Isayevich (1918–) Russian novelist and historian. He exposed the corruption and cruelty of Russian society in works such as *One Day in the Life of Ivan Denisovich* (based on his experiences in a Soviet labour camp and published in the USSR during a brief thaw in cultural restrictions), and *The First Circle* and *The Gulag Archipelago*, which had to be published abroad. He was awarded the Nobel prize for literature in 1970, and was deported from the USSR in 1974. He settled in Vermont in the USA, and returned to Russia in 1994.

Sondheim, Stephen [Joshua] (1930–) American songwriter and composer. He studied with HAMMERSTEIN and wrote the lyrics for BERNSTEIN's *West Side Story*, before writing the music and lyrics for several musicals, e.g. *A Funny Thing Happened on the Way to the Forum* and *Into the Woods*.

Sophocles (*c*.496–406 BC) Greek dramatist. Seven of his *c*.120 plays are extant. He was the most popular of the three great Athenian dramatists (the others being AESCHYLUS and EURIPIDES). The plays include *Oedipus Rex*, *Oedipus at Colonus*, and *Antigone*.

Sopwith, Sir Thomas [Octave Murdoch] (1888–1989) English aeronautical engineer. He designed and built the Sopwith Camel, one of the most successful fighter planes of World War I.

Southey, Robert (1774–1843) English poet. Closely associated with WORDSWORTH and COLERIDGE, he was poet laureate from 1813.

Spassky, Boris Vasilyevich (1937–) Russian chess player. He was world champion (1969–72).

Spence, Sir Basil (1907–76) Indian-born Scottish architect. He is famous for a few highly prestigious buildings, such as Coventry Cathedral, and notorious for some disastrous council housing.

Spencer, Lady Diana *see* CHARLES, PRINCE.

Spencer, Sir Stanley (1891–1959) English painter. An isolated figure in modern art, he is best known for his series of religious paintings, and, in his capacity as a war artist, his *Shipbuilding on the Clyde* series of panels.

Spenser, Edmund (*c*.1552–99) English poet, noted

particularly for his long allegorical poem *The Faerie Queene*, which describes the adventures of 12 knights (who represent 12 virtues). Many of the adventures begin at the court of Gloriana, the Faerie Queen (an idealized version of ELIZABETH I).

Spielberg, Steven (1947–) American film director and producer. His many films include some of the most successful ever made, e.g. *Jaws*, *E.T.* and *Raiders of the Lost Ark*.

Spock, Dr [Benjamin McLane] (1903–) American pediatrician. He advocated a 'permissive', non-authoritarian approach to the raising of infants, which was popular in the 1960's.

Stalin, Joseph [Josef Vissarionovich Dzhugashvili] (1879–1953) Soviet dictator. Born in Georgia, he was expelled in 1899 from an orthodox seminary in Tiflis for expounding Marxism. After the Bolshevik Revolution, he manoeuvred his way into absolute power, shrewdly playing off his 'rightist' allies against TROTSKY and other 'leftists'. He forcibly collectivized Soviet agriculture in the 1930s and developed the Soviet Union's industrial base, using (and killing) many millions of prisoners as slave labour. His purges of the 1930s destroyed most of the surviving old Bolsheviks, as well as the army leadership. He signed a peace treaty with HITLER in 1939, and seized Poland's eastern territories after Hitler's September invasion. Forceful resistance from the Red Army, notably in the defence of Leningrad, the battle of Stalingrad, and in massive infantry and tank battles, led directly to the defeat of Hitler's regime and the occupation of eastern Europe by Stalin's forces.

Joseph Stalin

Stanislavsky, Konstantin (1863–1938) Russian director and actor, who was co-founder of the Moscow Art Theatre in 1897. The influence of his theory of acting (later dubbed 'method') has been immense. His theory is contained in such works as *An Actor Prepares* (1929) and *Building a Character* (1950).

Starr, Ringo [Richard Starkey] (1940–) English

rock drummer and singer. He was the Beatles' drummer (1962–70).

Steel, Sir David [Martin Scott] (1938–) Scottish Liberal politician. He became leader of the Liberal Party in 1976 and, with David OWEN, the SDP (1981–88).

Steele, Sir Richard (1672–1729) Anglo-Irish essayist and dramatist. With ADDISON, he contributed many notable essays to the *The Tatler* and *The Spectator*. His plays were moral, Christian responses to the excess of Restoration drama and had a strong influence on 18th-century drama.

Steiner, Rudolf (1861–1925) Austrian philosopher. Influenced by theosophy, he formed his own movement of 'anthroposophy' in 1912, dedicated to developing the innate human capacity for spiritual perception, through activities such as art and dance.

Sterne, Laurence (1713–68) Irish-born English novelist. His novel *The Life and Opinions of Tristram Shandy* (1759–67) is noted for its eccentric style and humour.

Stevenson, Robert Louis [Balfour] (1850–94) Scottish novelist, poet and essayist. He trained as an advocate in Edinburgh, but decided in his twenties to be a writer. By the time his first important fictional work, *Treasure Island* (1883), had been published he had established himself as an author of note with essays, poems and two travel books, *Travels with a Donkey in the Cevennes* (1879) and *The Silverado Squatters* (1883). His masterpiece is *The Strange Case of Dr Jekyll and Mr Hyde* (1886), a disturbing story of dual personality. Other works include *Kidnapped* (1886), *The Master of Ballantrae* (1889), and the unfinished *Weir of Hermiston* (1896). He settled in Samoa, where he died.

Stewart, Prince Charles Edward [Louis Philip] (1720–88) also known as 'Bonnie Prince Charlie' or the 'Young Pretender', he led the Jacobite revolt against the Hanoverian King George III in 1745, fleeing Scotland when his Highland soldiers were defeated at Culloden in 1746. He died in Rome.

Stewart, Jackie [John Young Stewart] (1939–) Scottish racing driver. World champion 1969, 1971, 1973, he retired in 1973.

Stockhausen, Karlheinz (1928–) German composer. Regarded as the leading exponent of twelve-tone music, he also used electronic sounds in his work.

Stokowski, Leopold (1882–1977) British-born American conductor. A popularizer of classical music, he is best known for his collaboration with Walt DISNEY on the film *Fantasia*.

Stopes, Marie [Charlotte Carmichael] (1880–1958) Scottish scientist and birth-control pioneer. She began her career as a palaeobotanist, but the breakdown of her marriage led her to the study of sex education, in which field she soon became a world authority. She established a birth control clinic in Holloway in London (1920), which gave free contraceptive advice to the poor.

Marie Stopes

Stoppard, Tom (1937–) Czech-born British dramatist. His plays, e.g. *Rosencrantz and Guildenstern are Dead*, are noted for their sharp, witty wordplay and fast, cleverly plotted action.

Stowe, Mrs Harriet [Elizabeth] Beecher (1811–96) American novelist. Her great anti-slavery novel *Uncle Tom's Cabin, or, Life Among the Lowly* (1852) has been described as a factor leading to the American Civil War.

Strauss, Richard (1864–1949) German composer and conductor. His works include a series of richly orchestrated tone poems, e.g. *Also Sprach Zarathustra*, several operas, e.g. *Elektra* and *Der Rosenkavalier*, and *Four Last Songs*.

Strauss, Johann (the Younger) (1825–99) Austrian violinist, conductor and composer. One of a musical family, he wrote light music, especially Viennese waltzes, e.g. the 'Blue Danube' waltz, and 16 operettas, e.g. *Die Fledermaus*.

Stravinsky, Igor Fyodorovich (1882–1971) Russian composer. He composed ballet scores for DIAGHILEV, e.g. *Petrushka* (1911) and *The Rite of Spring* (1913), the first performance of the latter provoking a riot but now regarded as a milestone in modernist music. He later composed several austerely neoclassical works, such as the opera-oratorio *Oedipus Rex* (1927), which also displayed the influence of SCHOENBERG's serial techniques.

Strindberg, Johan August (1849–1912) Swedish dramatist and novelist. His highly innovative works, e.g. the play *Miss Julie*, have influenced many 20th-century dramatists.

Stubbs, George (1724–1806) English painter and engraver, best known for his paintings of horses.

Sun Yat-sen *or* **Sun Zhong Shan** (1866–1925) Chinese nationalist leader and statesman. He played a leading role in the overthrow of the Manchu dynasty and became the first president of the Republic of China in 1911–12.

Sun Yat-sen

Sutherland, Dame Joan (1926–) Australian soprano. She became one of the world's leading bel canto operatic sopranos.

Sutherland, Graham [Vivian] (1903–80) English painter. He was an official war artist (1941–45) and subsequently became a portrait painter of note.

Swift, Jonathan (1667–1745) Anglo-Irish divine, poet and satirist. His first important satirical works, published in 1704, were *The Battle of the Books*, a defence of the merits of classical literature against the claims of the moderns, and *A Tale of a Tub*, an attack on religious extremism. In politics, Swift began as a Whig but soon became a staunch Tory. He became Dean of St Patrick's Cathedral in Dublin in 1713 and published several tracts defending the Irish poor against their overlords, including the savage *A Modest Proposal* (1729). His masterpiece is *Gulliver's Travels* (1726), which culminates in Gulliver's voyage to the Houyhnhnms ('whinims'), intelligent horses whose nobility is contrasted with the brutality of humanity.

Swinburne, Algernon [Charles] (1837–1909) English poet and critic, noted for his sensuous verse which frequently created scandal, not just for their sexuality but for the author's clear dislike of Christianity.

Synge, [Edmund] John Millington (1871–1909) Irish dramatist, noted for his poetic rendering of Irish peasant speech. His masterpiece is *The Playboy of the Western World* (1907), a highly controversial comedy.

Tagore, Rabindranath (1861–1941) Indian poet and philosopher. Regarded by many Bengalis as their greatest writer, he was awarded the 1913 Nobel prize for literature.

Tati, Jacques [Jacques Tatischeff] (1908–82) French actor and film director. An ex-rugby player, he became an international comedy star with his Monsieur Hulot creation, an engagingly incompetent character at odds with the modern world. Five Hulot films were made, including *Mr Hulot's Holiday* and *Mon Oncle*.

Tatum, Art[hur] (1910–56) American jazz pianist. He was an acclaimed virtuoso of jazz piano music in the 'swing' mode.

Tatum, Edward Lawrie (1909–75) American biochemist. With the American geneticist **George Wells Beadle** (1903–89), he demonstrated that biochemical reactions in cells are controlled by particular genes. With LEDERBERG, he discovered the phenomenon of genetic recombination in bacteria. All three shared the 1958 Nobel prize for physiology or medicine.

Tavener, John Kenneth (1944–) English composer. He studied under Lennox BERKELEY, and became noted particularly for his religious compositions, e.g. the cantata *Cain and Abel*.

Taverner, John (*c.*1490–1545) English composer. He taught at Oxford University and is best known for his religious works.

Taylor, A[lan] J[ohn] P[ercivale] (1906–90) English historian. Often a controversial figure (he argued that World War II was produced by accident as much as by HITLER's design), he was admired by his peers for his research and insight into modern European history, and became the historian best known to the British public through his live lectures to television audiences.

Taylor, Elizabeth (1932–) English-born (of American parents) American film actress. Her films as a child include *National Velvet* and *Little Women*. Regarded as one of the most beautiful film stars of her generation, her films include *Cat on a Hot Tin Roof* and *Who's Afraid of Virginia Woolf?* The latter film co-starred her 5th husband, Richard BURTON, whom she married twice.

Elizabeth Taylor

Tchaikovsky, Peter Ilyich (1840–93) Russian composer. Notable for his strong melodic sense and rejection of an overtly nationalistic and 'folk' approach to composition, his works include tone poems, a violin concerto, operas, e.g. *Eugene Onegin*, the ballets *Swan Lake* and *The Sleeping Beauty*, and the so-called 'Pathétique Sixth Symphony.

Teilhard de Chardin, Pierre (1881–1955) French Jesuit theologian, philosopher and palaeontologist. He developed a theory of evolution, claiming that it was compatible with Roman Catholic teaching.

Te Kanawa, Dame Kiri (1944–) New Zealand soprano. She is regarded as one of the world's leading operatic sopranos.

Telemann, George Philipp (1681–1767) German composer. A highly prolific composer of the Baroque era, his works include over 40 operas and over 40 Passions.

Teller, Edward (1908–) Hungarian-born American physicist. Known as the 'father of the hydrogen bomb' for his ground-breaking work in that field.

Temple, Shirley (1928–) American film actress and Republican politician. She became the world's leading child film star and later developed a career in politics.

Teng Hsiao-p'ing *see* DENG XIAO PING.

Tennyson, Alfred Lord [1st Baron Tennyson] (1809–92) English poet. He first came to public notice with *Poems, Chiefly Lyrical* (1830). Subsequent volumes established him as a highly popular poet, and he was appointed poet laureate in 1850, the year in which he published his great elegy for his dead friend A. H. Hallam, *In Memoriam*. He became a much respected public figure, with several of his poems, e.g. 'Locksley Hall' (1842), being regarded as oracular statements on the spirit of the age.

Tenzing Norgay *see* HILLARY, EDMUND.

Thackeray, William Makepeace (1811–63) Indian-born English novelist and essayist, noted particularly for the witty social satire of both his novels and his non-fiction works, e.g. *The Book of Snobs* (1846–47), a funny description of the varieties of snobbery. His masterpiece is *Vanity Fair* (1847–48), a decidedly non-moralistic tale of the opportunistic 'anti-heroine' Becky Sharp, set during the Napoleonic wars. His other works include the novels *Pendennis* (1850) and *Henry Esmond* (1852).

Thatcher, Margaret Hilda (1925–) English Conservative stateswoman. She became MP for Finchley in 1959 and, as secretary of state for education and science (1970–74) ended provision of free school milk. She defeated HEATH in the Tory leadership campaign of 1975, becoming the first woman to lead a major British political party. As prime minister (1979–1990), she launched an ideological crusade (dubbed 'Thatcherism') against what she perceived as the entrenchment of socialism in Britain, the principal elements of her attack being free-market policies and the privatization of nationalized industries. Her period of office was marked by rising unemployment and by the Falklands War of 1982, the success of which many believed to be the decisive factor in the Conservative's huge election victory of 1983 over FOOT's Labour party. Her policies were widely disliked by her political opponents and by moderate Tories, the latter forming small resistance groups around figures such as Heath and **Michael Heseltine** (1933–). Increasing dissension within her cabinet, over such issues as the highly controversial Community Charge (or Poll Tax) and the disarray of the Health Service, led to her resignation in 1990 and the election of MAJOR as prime minister.

Margaret Thatcher

Theresa of Calcutta, Mother [Agnes Gonxha Bojaxhiu] (1910–) Yugoslavian-born Roman Catholic nun and missionary. She founded the Order of the Missionaries of Charity in 1950. Venerated by many people as a living saint, her work in Calcutta with orphans and the dying led to her being awarded the 1979 Nobel Peace Prize.

Thomas, Dylan [Marlais] (1914–53) Welsh poet. His best-known single work is *Under Milk Wood* (1954), a radio drama in poetic prose.

Dylan Thomas

Thomas à Becket *see* BECKET, THOMAS À.

Thomas Aquinas, Saint (*c*.1225–74) Italian theologian and philosopher. His writings established the need for both reason and faith in Christianity, and have become a cornerstone in the teachings of the Roman Catholic Church.

Thomson, Sir George Paget (1892–1975) English physicist. He shared the 1937 Nobel prize for physics with the American physicist **Clinton Joseph Davisson** (1881–1958) for their (independent) discovery of the diffraction of electrons by crystals.

Thomson, Sir Joseph John (1856–1940) English physicist. He was awarded the 1906 Nobel prize for physics for his discovery (1906) of the electron, one of the most significant discoveries in physics. Seven of his assistants went on to win Nobel Prizes.

Thoreau, Henry David (1817–62) American philosopher whose advocacy of self-sufficiency and passive resistance to tyranny has been very influential, GANDHI being his most notable admirer.

Thurber, James [Grover] (1894–1961) American humorist, cartoonist and essayist, much of whose work first appeared in *New Yorker* magazine, including his most famous story, 'The Secret Life of Walter Mitty'.

Tiepolo, Giambattista (1696–1770) Italian artist, the greatest decorative fresco painter of the Rococo period.

Tillich, Paul [Johannes] (1886–1965) German-born American theologian and philosopher. His highly influential writings address the problems of matching traditional Christianity with an increasingly secular society.

Tinbergen, Jan (1903–) Dutch economist (brother of Niko TINBERGEN). He shared the 1969 Nobel prize for economics with the Norwegian economist **Ragnar Frisch** (1895–1973) for their work in the field of econometrics.

Tinbergen, Niko[laas] (1907–88) Dutch ethologist (brother of Jan TINBERGEN). He shared the 1973 Nobel prize for physiology or medicine (with LORENZ and Karl von FRISCH) for his groundbreaking studies of animal behaviour.

Tintoretto, Jacopo (1518–94) Venetian painter noted for his dynamic, highly imaginative use of lighting and highlighting.

Tippett, Sir Michael (1905–) English composer. His works include several operas, e.g. *The Midsummer Marriage*, the oratorio *A Child of our Time*, symphonies, song cycles and chamber music.

Titian [Tiziano Vecelli] (c.1490–1576) Venetian painter, one of the great figures of world art. He studied with Giovanni BELLINI and worked with GIORGIONE, whom he succeeded as the master of Venetian painters for some 60 years.

Tito, Marshal [Josip Broz] (1892–1980) Yugoslav statesman. He fought with the Bolsheviks during the Russian civil war and became secretary-general of the Yugoslav Communist Party in 1937. In 1941, after the German invasion of Yugoslavia, he organized a partisan force to fight the occupiers and succeeded in diverting British aid from other guerrilla forces to his own. After the war, he established a Communist government and broke with STALIN. He succeeded in preserving a fragile Yugoslav unity, but 11 years after his death the break-up of the Yugoslav state began with Slovenia, then Croatia, declaring independence from the Serbian-dominated state in 1991.

Marshal Tito

Tojo, Hideki (1885–1948) Japanese soldier. He became minister of war (1940–44) and prime minister (1941–44). He resigned in 1944, and was executed as a war criminal.

Tolkien, J[ohn] R[onald] R[euel] (1892–1973) South African-born British fantasy writer and scholar. Probably the most influential (and best-selling) fantasy writer, the works on which his fame rests are *The Hobbit* and *Lord of the Rings*.

J. R. R. Tolkien

Tolstoy, Count Leo [Nikolayevich] (1828–1910) Russian novelist, dramatist, short-story writer and philosopher. His spiritual self-questioning resulted in some of the world's greatest works of fiction, notably *War and Peace*, a panoramic epic of the Napoleonic invasions of Russia, and *Anna Karenina*, a tragic tale of adulterous love which raises profound questions about personal social morality.

Torvill, Jayne (1957–) and **Dean, Christopher** (1958–) English ice-dance skaters. They became world champions (1981–83) and Olympic champions (1984).

Toscanini, Arturo (1867–1957) Italian conductor. Regarded as one of the most authoritarian conductors of all time, he was renowned for his devotion to authenticity and for his remarkable musical memory.

Toulouse-Lautrec, Henri [Marie Raymond] de (1864–1901) French painter and lithographer. Influenced by VAN GOGH and DEGAS, his subjects were café clientele, prostitutes and cabaret performers in and around Montmartre.

Tracy, Spencer (1900–1967) American film actor. Noted for his straightforward, yet commanding, performances, he had a long personal and professional relationship with Katharine HEPBURN.

Trevelyan, George Macaulay (1876–1962) English historian. His highly readable works include *History of England* (1926) and *English Social History* (1944) and several biographies.

Trollope, Anthony (1815–82) English novelist, whose more than 50 books are dominated by two main novel sequences: the 'Barsetshire' novels, which focus on the provincial lives of the gentry, clergy and middle classes and the 'Palliser' novels of political life. His mother, **Mrs Frances Trollope** (1780–1863), was also a prolific author.

Trotsky, Leon [Lev Davidovich Bronstein] (1879–1940) Russian revolutionary. An advocate of 'permanent revolution', he believed that socialism could not be built in one country alone, and supported the Mensheviks against Lenin's Bolsheviks. Forced into exile by STALIN in 1929, he was later assassinated by a Russian agent.

Trueman, Freddy [Frederick Sewards Trueman] (1931–) English cricketer. He was a notable fast bowler.

Truffaut, François (1932–84) French film director, critic and actor. One of the *Cahiers du Cinema* group of film critics, his first film, the semi-autobiographical *The Four Hundred Blows*, was widely praised. His other films include *Jules et Jim* and *The Last Metro*. He has acted in several films, e.g. SPIELBERG's *Close Encounters of the Third Kind*.

Truman, Harry S (1884–1972) American Democratic statesman. He became 33rd president of the US (1945–52) after Franklin D. ROOSEVELT's death, and authorized the dropping of the atom bombs on Hiroshima and Nagasaki. He initiated the change in US foreign policy towards the Soviet Union expressed as the 'Truman doctrine', a policy of containment of communism and aid towards groups or nations resisting communism, and approved the MARSHALL Plan of aid for Britain and Western Europe.

Turgenev, Ivan Sergeyevich (1818–83) Russian novelist, short-story writer and dramatist. His novels, e.g. his masterpiece *Fathers and Sons* (1862), explore such major issues of Russian life as serfdom and revolutionary change. The best known of his plays is *A Month in the Country* (1850).

Turing, Alan Mathison (1912–54) English mathematician. Regarded as one of the most important computer theoreticians, he developed the concept of an idealized computer called the 'Universal Automaton' (later known as the 'Turing Machine'). The computer, he posited, would be able to modify its own program through a sequence on paper tape of 1s and 0s. Turing also took an important part in the vitally important code-breaking project at Bletchley Park in World War II, which deciphered the German 'Enigma' codes. He committed suicide.

Turner, Joseph Mallord William (1775–1851) English painter. Precociously talented, he exhibited his first work aged 15. After a trip to Italy in 1819 he became interested in gradations of shifting light and atmosphere. The works of the next two decades represent his finest period, CONSTABLE describing paintings such as *The Fighting Téméraire* (1839) as 'airy visions painted with tinted steam'. He found an influential champion in RUSKIN.

Mark Twain

Twain, Mark [pseud. of Samuel Langhorne Clemens] (1835–1910) American novelist, short-story writer and humorist. His two most famous novels,

The Adventures of Tom Sawyer (1876) and *The Adventures of Huckleberry Finn* (1884), have become classics of children's literature.

Tyndale, William *see* MORE, SIR THOMAS.

Ustinov, Sir Peter [Alexander] (1921–) British actor, director, dramatist and raconteur (of Russian-French parentage). His plays include *The Love of Four Oranges* (1951) and *Romanoff and Juliet* (1956). Other works include an autobiography, *Dear Me* (1977). He is best known as an engaging raconteur.

Valentino, Rudolph [Rodolfo Guglielmi di Valentina d'Anton-guolla] (1895–1926) Italian-born American film actor. He became the leading screen personification of the romantic hero in many films, e.g. *The Son of the Sheik*. He died of peritonitis.

Rudolph Valentino

Van Allen, James Alfred (1914–) American physicist. He discovered, through detectors on the US satellite *Explorer I*, the Van Allen radiation belts outside the Earth's atmosphere.

Vanbrugh, Sir John (1664–1726) English dramatist and architect, noted for his witty comedies, e.g. *The Relapse* and *The Provok'd Wife*. He was also one of the finest architects of his day, the most famous of his buildings being Blenheim Palace.

Van der Post, Sir Laurens [Jan] (1906–1997) South African novelist, travel writer and mystic. His works are strongly influenced by JUNG and display a strong sympathy for the 'primitive' peoples of the world. His novels include *The Seed and the Sower* (1963), based on his experiences a a prisoner of war of the Japanese in World War II and filmed as *Merry Christmas, Mr Lawrence*. His travel books include two African classics, *Venture to the Interior* and *The Lost World of the Kalahari*.

van Eyck, Jan (*c*.1385–1441) Dutch painter. A master in the medium of oil painting, his paintings are both realistic and charged with a serene, spiritual atmosphere.

Van Gogh, Vincent *see* GOGH, VINCENT VAN.

Varèse, Edgard (1883–1965) French-born American composer and conductor. His compositions are noted for their combination of the extreme registers of instruments with taped and electronic sounds.

Vasari, Giorgio (1511–74) Italian painter, writer and architect, noted particularly for his *Lives of the Most Eminent Painters, Sculptors and Architects*, an invaluable source book for the lives of Renaissance artists.

Vaughan Williams, Ralph (1872–1958) English composer. Like HOLST, he was heavily influenced by traditional English music, particularly English folk song. His works include the choral *Sea Symphony*, *Fantasia on a Theme by Thomas Tallis*, operas, ballet music and song cycles.

Velàzquez *or* **Velàsquez, Diego Roderiguez de Silva y** (1599–1660) Spanish painter. His earliest paintings were *bodegones*, a type of genre painting peculiar to Spain, consisting largely of domestic scenes, e.g. *An Old Woman Cooking Eggs* (1618). On RUBENS' advice he travelled to Italy in 1628, where he was influenced by TITIAN and TINTORETTO, developing a lighter palette and finer brushwork. His works include the portrait of *Pope Innocent X* (1650) and *Surrender of Breda* (1634–35).

Verdi, Giuseppe (1813–1901) Italian composer. His post-1850 operas, which include *Il Trovatore* (1853), *La Traviata* (1853), *La Forza del Destino* (1862), *Aida* (1871), *Otello* (1887) and *Falstaff* (1893), were hugely popular and remain constant favourites within the repertory of most opera companies. He became a deputy in the first Italian parliament of 1860.

Verlaine, Paul (1844–96) French poet, regarded with his friend and lover RIMBAUD as an early Symbolist. Collections of his verse include *Romances Without Words*.

Vermeer, Jan *or* **Johannes** (1632–75) Dutch painter. He is best remembered for his small-scale intimate interior scenes, carefully composed and lit, usually by daylight through a window.

Verne, Jules (1828–1905) French novelist, whose innovative fantasy novels, e.g. *Voyage to the Centre of the Earth* (1864) and *20,000 Leagues Under the Sea* (1969), are regarded as the earliest great science fiction novels.

Verwoerd, Hendrik Frensch (1901–66) South African politician and prime minister (1958–66). He fostered apartheid, banned the African National Congress (1960) and took South Africa out of the Commonwealth in 1961.

Vicky [Victor Weisz] (1913–66) German-born British cartoonist. He was one of the leading left-wing political cartoonists and caricaturists of his day.

Vidal, [Eugene Luther] Gore (1925–) American novelist, dramatist and critic. His American historical fiction provides an unofficial and waspishly entertaining alternative history of the US and its leaders, e.g. *Burr* (1973) and *1876* (1976). His work includes several important essay collections, e.g. *The Second American Revolution* (1982) and *Armageddon* (1987).

Villa-Lobos, Heitor (1887–1959) Brazilian composer and conductor. His popular works combined elements of traditional Brazilian music with the European classical tradition.

Virgil [Publius Vergilius Maro] (70–19 BC) Roman poet. One of the most influential poets of all time, his masterpiece is the epic poem the *Aeneid,* which charts the progrss of the Trojan hero Aeneas from the fall of Troy to the founding of the Roman state.

Visconti [di Modrone], Count Luchino (1906–76) Italian film director. He began his career as a stage designer, then worked as an assistant to RENOIR. His films include *The Damned* (1969) and *Death in Venice* (1971), both starring BOGARDE.

Vishnevskaya, Galina *see* ROSTROPOVICH, MSTISLAV.

Voltaire [pseud. of François Marie Arouet] (1694– 1778) French philosopher, poet, historian, essayist, dramatist and essayist. Regarded as one of the most important of the French philosophers of the Enlightenment, his most influential single work is the *Philosophical Letters* (1734), a collection of witty, acerbic attacks on the tyranny of the *ancien régime*. His writings, with those of ROUSSEAU (with whom he disputed bitterly), are often described as the main intellectual roots of the French Revolution. His other works include the remarkable novel *Candide* (1759), which takes a markedly pessimistic view of human endeavour.

von Braun, Wernher (1912–77) German-born American rocket engineer, he designed the V–1 and V–2 rocket bombs for Germany and later the Saturn moon rockets in the US.

Wagner, [Wilhelm] Richard (1813–83) German composer. He achieved great success with his third opera, *Rienzi* (1842). The operas which followed, *The Flying Dutchman* (1843) and *Tannhäuser* (1845) were not so popular, and, in trouble with the authorities for his radical sympathies, he fled to Paris. He established his own theatre in Bayreuth in 1876, where he staged his *Ring of the Niebelung* cycle. His strongly Romantic works revolutionized opera, with their use of leitmotif and dramatic power.

Walcott, Derek [Anton] (1930–) St Lucian-born West Indian poet and dramatist who has lived most of his life in Trinidad. His poetry includes *Omeros*, a reworking in a Caribbean setting of themes from HOMER's *Odyssey* and *Iliad* and DANTE's *Divine Comedy*. His plays draw on Creole traditions and imagery. He was awarded the Nobel prize for literature in 1992.

Waldheim, Kurt (1918–) Austrian diplomat. After service as a Nazi intelligence officer in World War II, he entered the Austrian diplomatic service and became secretary-general of the United Nations (1972–82) and president of Austria (1986–). Revelations about his role in the Nazi genocide machine in Yugoslavia during the war surfaced in the late 1980s.

Walesa, Lech (1943–) Polish trade union leader and statesman. He became the leader of the free trade union, Solidarity, in 1980, which forced substantial concessions from the Polish government. After the imposition of martial law in 1981, Solidarity was banned and he was imprisoned (1981– 82). After his release, he was awarded the 1983 Nobel Peace Prize. A skilled negotiator, Walesa succeeded in getting Solidarity re-legalized, and, in 1989, in the first free elections in eastern Europe since the 1940s, a Solidarity government was formed with Walesa as president. He was defeated in the presidential elections of 1995.

Lech Walesa

Wallace, Alfred Russel (1823–1913) Welsh naturalist. Independently of DARWIN, he devised a theory of evolution by natural selection, which Darwin acknowledged.

Waller, Fats [Thomas Wright Waller] (1904–43) American jazz pianist and composer, he was an exponent of the 'stride' school of jazz piano and noted for his humorous lyrics.

Wallis, Sir Barnes [Neville] (1887–1979) English aeronautical engineer. He designed the 'bouncing bombs' used in the famous 'Dambuster' bombing raids of 1943.

Walpole, Sir Robert, [1st Earl of Orford] (1676–1745) English statesman. As Chancellor and First Lord of the Treasury he was, effectively, Britain's first prime minister from 1721–42, George I having little interest in the government of Britain. His son **Horace Walpole** [4th Earl of Orford] (1717–97) was an author, noted for his vast correspondence and for his fascination with the 'Gothic', which expressed itself in his conversion of his house, Strawberry Hill, into a pseudo-medieval castle, and the Gopthic novel *The Castle of Otranto*, which spawned a host of imitations.

Walter, Bruno [Bruno Walter Schlesinger] (1876–1962) German-born American conductor. Noted for his concerts and recordings of the great German Romantic composers, he is particularly associated with the works of his friend MAHLER.

Walton, Sir Earnest *see* COCKCROFT, SIR JOHN DOUGLAS.

Walton, Izaak (1593–1683) English author, best known for *The Compleat Angler*, a treatise on the art of angling.

Walton, Sir William [Turner] (1902–83) English composer. His works include a setting of Edith Sitwell's poem *Façade* for voice and instruments. His other works include the oratorio *Belshazaar's Feast* (1930–31), and several film scores, e.g. for OLIVIER's *Henry V* (1944).

Warhol, Andy [Andrew Warhola] (1930–87) American pop artist and film-maker. He became the prime exponent of Pop Art in the early 1960s with his deliberately mundane works such as his reproductions of Campbell's soup cans and his repetitive portraits of contemporary icons, such as PRESLEY, MAO and Marilyn MONROE. His films include the three-hour *Sleep*.

Andy Warhol

Washington, George (1732–99) American general and 1st president of the United States (1789–97). He became commander of the American armed forces during the War of Independence in 1785 and led them to victory. After the Philadelphia Convention of 1787, he was president of the new country (1789–97).

Watson, James Dewey (1928–) American biologist. He and Francis CRICK discovered the 'double helix' structure of DNA, for which they shared (with Maurice WILKINS) the 1962 Nobel prize for physiology or medicine.

Watson-Watt, Sir Robert Alexander (1892–1973) Scottish physicist. He played a major role in the development of radar.

Watt, James (1736–1819) Scottish engineer. His improvements to the steam engine led directly to the rapid expansion of the Industrial revolution.

Watteau, Jean-Antoine (1684–1721) French painter, an outstanding exponent of Rococo art. His work is noted for a delicacy of colour and sensitivity of composition not achieved by his imitators.

Watts, Charlie *see* JAGGER, MICK.

Waugh, Evelyn [Arthur St John] (1903–66) English novelist known for his brilliant satires, e.g. *Vile Bodies* (1930) on the brittle postwar world of upper-class England, *Scoop* (1938) on war reporting, and *The Loved One* (1948) on Californian burial practices. His work also had a deeper tone, resulting from his conversion to Catholicism in 1930. His best-known novel, *Brideshead Revisited* (1945), although still a satire displays a growing spiritual concern. His masterpiece is his *Sword of Honour* trilogy, based on his own experiences with the Communist partisans in Yugoslavia in World War II.

Evelyn Waugh

Wayne, John [Marion Michael Morrison] (1907–79) American film actor. A screen actor of outstanding presence, he is best known as the star of many classic westerns.

Webb, Beatrice [Potter] (1858–1943) and **Webb, Sidney James,** [Baron Passfield] (1859–1947) English social reformers and economists. Married in 1892, the Webbs were, with George Bernard SHAW and H. G. WELLS, the leading propagandists of Fabian socialism. They co-founded the London School of Economics (1895), founded the *New Statesman* (1913) and produced many of pamphlets and articles.

Weber, Carl Maria Ernst von (1786–1826) German composer, conductor and pianist. He is considered to be the creator of German romantic opera, using French opera as a framework and introducing German themes. He had a colossal influence on subsequent composers up to, and including, WAGNER. His works include the operas *Der Freischütz*, *Euryanthe* and *Oberon*, two symphonies, concertos, piano sonatas and many songs.

Weber, Max (1864–1920) German sociologist. Regarded as one of the founders of sociology, he devised the concept of 'ideal types' of real situations for comparative purposes.

Webern, Anton von (1883–1945) Austrian composer, and one of the leading exponents of the serial form of composition.

Webster, John (*c.*1578–*c.*1632) English dramatist, noted for two very powerful tragedies, *The White Devil* and *The Duchess of Malfi*. His bleak and chilling dialogue has rarely been matched.

Weil, Simone (1909–43) French philosopher. From an intellectual Jewish family, she chose to live as a farm and industrial labourer during the 1930s, and worked for the Republican forces during the Spanish Civil War. She later developed a strong interest in Roman Catholic mysticism. She worked for the French Resistance in London, where she starved herself to death in sympathy with the inmates of the Nazi camps.

Weill, Kurt (1900–1950) German composer, noted especially for his collaborations with BRECHT, e.g. *The Threepenny Opera* (1928). He fled from Germany in 1935, settling in the US.

Weinberg, Steven (1933–) American physicist. He devised a theory of the unity of the forces operating on elementary particles that was independently arrived at by the Pakistani physicist **Abdus Salam** (1926–), and later developed by the American physicist **Sheldon Glashow** (1932–). All three shared the 1979 Nobel prize for physics.

Weizmann, Chaim [Azriel] (1874–1952) Russian-born chemist and Israeli statesman. A distinguished scientist, he participated in the negotiations for a Jewish homeland and became first president of Israel (1949-52).

Welles, [George] Orson (1915–85)American stage and film director and actor. He achieved notoriety with his radio production of WELLS's *War of the Worlds* in 1938, which sparked off mass panic in the US. He co-wrote, produced and directed one of

the greatest films of all time, *Citizen Kane*, based on the life of HEARST. His other films include *The Magnificent Ambersons*, *Macbeth* and *Othello*. His other acting roles include, most notably, Harry Lime in REED's masterpiece, *The Third Man*.

Orson Welles

Wellington, 1st Duke of [Arthur Wellesley] (1769–1852) Anglo-Irish soldier and statesman, nicknamed the 'Iron Duke'. He distinguished himself as a commander in India, and was appointed commander of the British forces during the Peninsular War of 1808–14 and led the Allied forces to victory against NAPOLEON at Trafalgar in 1815. Respected for defeating Napoleon, he became Tory prime minister (1828–30) and opposed reform.

Wells, H[erbert] G[eorge] (1866–1946) English novelist and short-story writer. His science fiction works include several classics, e.g. *The Time Machine* (1895), *The War of the Worlds* (1898) and *The Shape of Things to Come* (1933). He was a propagandist with the WEBBS and others for Fabian socialism, and his novels on contemporary themes generally address their subject matter from a 'progressive' viewpoint, e.g. *Ann Veronica*, and sometimes with a comic or satirical element, e.g. *Love and Mr Lewisham*and *The History of Mr Polly*.

Wesley, John (1703–91) English evangelist. The 15th son of the poet and clergyman, **Samuel Wesley** (1662–1735), he joined a small group of devout Anglicans, formed by his brother **Charles Wesley** (1707–88), who subsequently became known as 'Methodists', which name was used to describe the expanding movement, and which remained within the Church of England in Wesley's lifetime.

West, Mae (1892–1980) American vaudeville artist, dramatist and film actress. Several of her plays were banned for obscenity, notably *Sex*, for which she was briefly imprisoned. She became a major star, renowned for her sardonic wit and powerful sexuality, with such films as *She Done Him Wrong*.

Whistler, James Abbott McNeill (1834–1903) American painter. He settled in London in 1859, where he was influenced by the Pre-Raphaelites and by Japanese art. He became famous as a portraitist, with works such as *Arrangement in Grey and Black*, a portrait of the his mother. He became involved in a court case with RUSKIN, who had sneered at Whistler's work in a review of an 1877 exhibition (Whistler was awarded a penny damages).

Whitman, Walt[er] (1819–92) American poet. His collection, *Leaves of Grass*, was first published in 1855 and the book is regarded as the most important single volume of poems in American literature.

Walt Whitman

Whittle, Sir Frank (1907–) English aeronautical engineer. He designed the first operational jet engine for aircraft and the first successful flight was made in a Gloster in 1941.

Wilberforce, William (1759–1833) English philanthropist and politician. His long campaign to end the British slave trade led to its abolition in 1807.

Wilde, Oscar [Fingal O'Flahertie Wills] (1854–1900) Irish dramatist, poet, essayist and wit. He first came to notice in the late 1870s, and lived up to the 'Bunthorne' image of an aesthete presented in GILBERT and Sullivan's comic opera *Patience*. After the publication of *Poems* (1881), he made a highly successful tour of America, published his children's stories *The Happy Prince and Other Tales* in 1888, with his only novel *The Picture of Dorian Gray* following in 1890. The first of his great plays, *Lady Windermere's Fan* appeared in 1892. The succeeding plays, *A Woman of No Importance* (1893), *An Ideal Husband* (1895) and *The Importance of Being Earnest* (1895) established him as the most important dramatist of the age, with their superbly witty dialogue and biting satire. He was jailed for homosexuality (1895–57), and during his imprisonment wrote *De Profundis*. After his release, Wilde fled to France, where he wrote *The Ballad of Reading Gaol* (1898) and died in poverty.

Oscar Wilde

Wilder, Billy [Samuel Wilder] (1906–) Austrian-born American film director and screenwriter. He emigrated to the US in the 1930s, winning Oscars for *The Lost Weekend*, *Sunset Boulevard* and *The Apartment*. Other films include *Double Indemnity*, *The Seven Year Itch* and *Some Like it Hot*.

Wilkins, Maurice Hugh Frederick (1916–) New Zealand physicist and biologist. His research into DNA structure resulted in CRICK and WATSON's discovery of the 'double helix' structure of DNA, for which Wilkins, Crick and Watson shared the 1962 Nobel prize for physics.

Williamson, Malcolm (1931–) Australian-born British composer. Master of the Queen's Music since 1975, his works include several operas and music for film and television.

Wilmot, John *see* ROCHESTER, 2ND EARL OF.

Wilson, Sir [James] Harold [Baron of Rievaulx] (1916–95) English Labour statesman. He served in World War II as a civil servant and became an MP in 1945. He held various ministerial posts before succeeding GAITSKELL as Labour leader in 1963. He was prime minister (1964–70, 1974–76). Originally on the soft left of his party, he became a defender of US policy on Vietnam and imposed a statutory incomes policy to deal with the a balance of payments crisis in the mid–1960s. He unexpectedly resigned in 1976, with CALLAGHAN succeeding him as prime minister.

Wilson, [Thomas] Woodrow (1856–1924) American Democratic statesman. He was 28th president of the USA (1913–21). Re-elected in 1916 on a policy of neutrality, he declared war on Germany following the sinking of US vessels. His 'fourteen points' speech of January 1918 set out US conditions for ending the war, including the disbandment of the German, Austro-Hungarian and Ottoman empires, and imposed the armistice with Germany on Britain and France.

Windsor, Duke of [formerly Edward VIII] (1894–1972) English monarch, (1936). Highly popular with the British public for his apparent concern at the lot of the unemployed, he abdicated to marry the American divorcée, **Wallis Simpson** (1896–1986), after BALDWIN had made plain his opposition to the notion of Mrs Simpson becoming queen. He married her in 1937, after which they lived in exile, the Duke becoming governor of the Bahamas during World War II.

Wittgenstein, Ludwig [Josef Johann] (1889–1951) Austrian-born British philosopher. He studied under Bertrand RUSSELL (1912–13), who observed that he was soon learning as much from his pupil as he had taught him. On the outbreak of World War I, he returned to Austria to serve as an artillery officer and was captured. While a POW, he wrote and sent to Russell his *Tractatus Logico-Philosophicus*, a series of aphoristic propositions on the boundaries of language and philosophy in relation to the world. During the 1920s, influenced by TOLSTOY's asceticism, he gave his considerable inherited wealth away and worked as a schoolteacher. His posthumous *Philosophical Investigations* retracts the confident assertions of the *Tractatus*, focusing instead on the concept of language as a series of games in which 'The meaning of the word is its use in language.'

Wodehouse, P[elham] G[renville] (1881–1975) English novelist and short-story writer. His most famous literary creations are Bertie Wooster and his butler Jeeves.

Wollstonecraft, Mary *see* SHELLEY, MARY WOLLSTONECRAFT.

Wolsey, Cardinal Thomas (c.1475–1530) English cleric and statesman. He was made a privy councillor (1511) by HENRY VIII, who then appointed him archbishop of York (1514–30) and lord chancellor (1515–29). Instructed by the king in 1527 to negotiate with the pope the annulment of the king's marriage, his failure after two years of pressurizing Rome led to his dismissal as lord chancellor (being succeeded by Sir Thomas MORE). He died on the journey from York to London to face charges of treason.

Wood, Sir Henry [Joseph] (1869–1940) English conductor. He founded the London Promenade Concerts (the 'Proms') in 1895, which he conducted until his death.

Woolf, [Adeline] Virginia (1882–1941) English novelist and critic. Her novels, including *To the Lighthouse* and *Orlando*, are written in a fluid, poetic style using stream of consciousness narration and are recognized as being among the most innovative of the 20th century. She suffered from bouts of manic depression throughout her life, particu-

larly after finishing a novel, and finally committed suicide by drowning herself.

Virginia Woolf

Wordsworth, William (1770–1850) English poet. The main figure of the English Romantic movement, his work successfully blended the personal with the natural and social worlds into a coherent whole. He enjoyed a close artistic relationship with COLERIDGE. He became poet laureate in 1843.

Wright, Orville (1871–1948) and **Wright, Wilbur** (1867–1912) American aviators and brothers. Cycle manufacturers, they designed and built the first heavier-than-air flying machine.

Wycherley, William (1641–1715) English dramatist noted for the witty dialogue of his popular Restoration comedies.

Wyman, Bill *see* JAGGER, MICK.

Yeats, W[illiam] B[utler] (1865–1939) Anglo-Irish poet and dramatist. His early works reflect his concern with Irish myth and legend. His play, *Cathleen ni Houlihan* (1902) demonstrated his support of Irish patriotism, and he later feared it had sent men to their deaths against the British. Following Irish independence, he became a member of the Irish senate. He was awarded the Nobel prize for literature in 1923. His brother, **Jack Yeats** [John Butler Yeats] (1870–1957) was an illustrator, particularly of comic strips and children's books, before turning to painting and writing.

W. B. Yeats

Yeltsin, Boris Nikolayevich (1931–) Russian politician. A member of the Communist party since 1960, he was brought into the Soviet Politburo by GORBACHEV in 1985, in which year he was appointed head of the Moscow party organization. His subsequent assault on the ingrained inefficiency and corruption of that body resulted in his demotion from the post and from the Politburo. In the free elections of 1989, he was elected to the Congress of People's Deputies, and won an overwhelming majority of votes in the Russian presidential election of 1990.

Zeffirelli, Franco (1923–) Italian stage and film director and designer. His films include *Romeo and Juliet* and the TV film *Jesus of Nazareth*.

Zhivkov, Todor (1911–) Bulgarian Communist statesman. Prime minister (1962–71) and president (1971–89). Under his rule, Bulgaria became the most servile of the satellites of the Soviet Union.

Zhou En Lai *see* CHOU EN-LAI.

Zola, Emile (1840–1902) French novelist. Regarded as the most prominent exponent of Naturalism in the novel, and a highly able propagandist for socialism and for social justice.

Zwingli, Huldreich or **Ulrich** (1484–1531) Swiss religious reformer. After LUTHER and CALVIN, he became the most influential of the Protestant reformers.

BRITISH KINGS AND QUEENS

This section of the *Family Encyclopedia* provides a concise account of the monarchs of the British Isles, from tribal times to the present day, giving an historical overview and details of the signicant events of the reigns. The entries are arranged in an A–Z format and the cross-references, given in SMALL CAPITALS, will enable you to gain a complete historical perspective of the monarchs of Britain.

Introduction

Roman Britain

The British islands first became known to the Romans through Caesar's two expeditions in 55 BC and 54 BC. The main island was generally known by them as *Britannia*, but it was not until the time of Claudius, nearly a hundred years later, that the Romans made a serious attempt to convert Britain into a Roman province. There is evidence to suggest that there were tribal rulers in Britain as early as 4500 years ago, but the contemporary records of the Romans give the first clear proof that there were Iron Age tribal kings and queens. The tribes were migrants from mainland Europe who spoke Celtic tongues akin to Gaelic, Irish and Welsh and used coins similar to those used in Greece and Rome. Among these tribes were the Atrebates (based around Sussex), Brigantes (who founded York as their capital), Catuvellauni (based around Hertfordshire), Regni (based in Sussex) and the Iceni (based around Norfolk and Suffolk). After the Roman conquests some of the leaders of these tribes were made 'client' rulers under Rome and retained their power; others such as Boadicia of the Iceni tribe rebelled and were crushed.

The heart of Roman rule was in the south-east of England, but Roman armies also came into contact with the tribes of the north. The battle at 'Mons Graupius', thought to have been in Aberdeenshire, gave the Romans a temporary foothold, but the rugged terrain, as much as the resistance of the tribes, meant they had to retire behind the wall built in AD 120 by Hadrian, between the Solway and the Tyne. Thus the southern part of the island alone remained Roman and became specially known as Britannia while the northern portion was distinctively called Caledonia.

Eight Roman Emperors were proclaimed in Britannia, and it became an important part of the Empire, but by 400 AD attacks from northern Europe, Scotland and Ireland had almost exhausted the resources of the province. Hadrian's Wall was abandoned and troops withdrawn from important outposts until, around 415 AD, the formal rule of the Romans ended.

Early English Kings

After the Romans withdrew, Romano-British rulers had to defend their territories against attack from fierce northern tribesmen. They also had to contend with invasions by Germanic tribes eager to drive them out and settle new lands. From what are now called the Netherlands, Germany, Denmark and France came the Angles, Saxons and Jutes to set up hundreds of petty states. The rulers of these states claimed the war god Woden as their ancestor and, with the consent of a council of advisers known as the *Witan*, bequeathed their crowns to their sons.

In time the larger of these states swallowed the smaller until, around 700, there were about forty established statelets. Eventually a group of the seven most prominent states emerged, generally referred to as the *Heptarchy*, which dominated most of the southern part of Britain. This consisted of the kingdoms of Kent, the South Saxons (in Surrey and Sussex), the East Angles (East Anglia), the East Saxons (Essex and Middlesex), the West Saxons or Wessex (Devon, Dorset, Somerset and part of Cornwall), Northumbria (Northumberland, York and Durham and also south- east Scotland) and the Kingdom of Mercia (which included Gloucester, Leicester and Chester). To the north were the Picts and Scots, to the west were the Celts of Wales. The Britons fleeing the European invaders had settled in Strathclyde (the south-west of Scotland) and Cornwall. The battle for supremacy between these seven kingdoms persisted until at length Wessex overran Mercia, Sussex and all the lands south of the Humber, and the whole came to be known as England, or Angleland.

These were uncertain times, however, and the Angle kings were frequently defeated by the ambitions of Danish invaders. Their advance was checked for a time by Alfred the Great (who also ensured that those he defeated were converted to Christianity), and many were diverted to northern France where their settlements became Normandy. Alfred was the first to bring about some form of national unity among the Anglo-Saxons, but his descendants could not claim to rule all England, and fought against the Danes, with varying degrees of success, to maintain power. By 1016 England had again been made to submit to foreign rule as the Danish king, Canute, made England part of a greater northern European empire. When this empire finally collapsed England was left disunited, a state of affairs that was finally exploited by a Norman, Duke William of Normandy, who began his conquest by defeating the English forces at Hastings in 1066.

Picts and Scots

The inhabitants of the northern lands, known to the Romans as Caledonians and to the Britons as the people of Albion, could be distinguished into Goidels (or Gaels) and Picts. The Gaels were of a similar stock to the British, and, since the eighteenth century, British and Gaels have alike been termed 'Celtic'. This usage originated in the discovery of an affinity between the language of Brittany and those of the Welsh and of the Scottish Highlanders. It is, however, important to make a distinction between the Brythonic Celts of Wales, Cumbria and Strathclyde, and the Goidelic Celts or Gaels of the Northern Highlands. By the sixth century these northern tribes were more commonly known as Picts, which means literally 'painted people'.

As well as the Picts, a body of Celts from Ireland, known as Scots, had, by the middle of the sixth century, settled in what is the present-day county of Argyll and founded the kingdom of Dalriada. This kingdom consisted of Argyll, part of Northern Ireland and the islands of the Inner Hebrides.

The amalgamation of the Scots, Picts, Britons and Angles into one kingdom was a slow and lengthy process. The foundation for the ultimate union of the Scots and the Picts was laid by St Columba, who travelled to the already christianized kingdom of Dalriada in 563. The actual occasion of the union of Scots and Picts, however, was to come via a combination of factors. Dynastic intermarriage, religious unity and Scandanavian aggression all played their part, but it was the strong leadership of Kenneth mac-Alpin, king of Scots, which finally brought the kingdoms together. Under his guidance the Scots became the dominant force in the land, and the idea of the Picts as an independent people largely died out.

The amalgamation of the Lothians and of Strathclyde with Alba (the kingdom of Picts and Scots) was delayed for two centuries. Strathclyde was finally merged through dynastic inheritance, but it took the collapse of Nothumbrian power and the consolidation of England under the successors of Alfred the Great to unite the Scots and Angles. Successive monarchs of Alba tried to add Lothian to their dominions, but it was not until 1018 that it was finally annexed and the whole of the historical kingdom of Scotland was ruled by one king, who maintained the title of 'King of Scots'.

Early Welsh Kings

Previous to the Roman occupation, Wales appears to have been chiefly inhabited by three British tribes, called the Silures, Dimaetae, and Ordovices.

During the latter period of the Roman occupation the subjected part of the island was divided into four provinces, of which one, including the country from the Dee to the Severn, was called *Britannia Secunda*. It was after the invasion of the Saxons that the country aquired a distinctive national character as the refuge of the vanquished Britons who were gradually driven to the west. From this period until the final conquest of the country by Edward I, there is little but a succession of petty wars between the rival chiefs or kings into which both countries were divided during a great part of the Saxon period, or the more systematic efforts of the larger monarchy to absorb the smaller. By war and marriage three main territories were established: Gwynedd in North Wales; Powys in the central region; and Deheubarth in South Wales. Of the three, Gwynedd spread its influence the farthest, and sometimes dominated all Wales. By 1200, however, Welsh kings had been reduced to lords and princes owing homage to the King of England. The last of the Welsh princes, Llewllyn, who revolted against Edward I, was defeated by the Earl of Mortimer in 1284. Since that time the principality has been incorporated with England, and Wales has given the title of Prince of Wales to the heir apparent of the British Crown.

Early Irish Kings

As in Western Europe generally, the earliest inhabitants of Ireland are believed to have been of Iberian race and therefore akin to the modern Basques. They were followed by the Celts, different tribes of whom probably arrived at different times. Among these the Scots were the latest, and subsequently got the upper hand, so that their name became generally applied to all the inhabitants. The land was divided into five principal kingdoms—Ulster, Munster, Leinster, Connaught and Meath—which were each divided in turn into lesser divisions or chiefdoms. Each king or chieftain desired to become the *Ard-Ri* (high king) of the whole island, but for centuries no one achieved any real degree of supremacy.

Following the introduction of Christianity and Christian literature by St Patrick, Ireland became an important seat of learning. Its internal condition, however, was far from satisfactory. Divided among a number of hostile kings or chiefs, it had been long torn by internal wars and for nearly two centuries ravaged by the Danes, numbers of whom settled in the country, when, in the beginning of the eleventh century, Brian Boru united the greater part of the island under his sceptre and subdued the northern invaders. After the death of Brian the island relapsed into its former state of division and anarchy. In this state of matters Henry II of England obtained a Papal Bull giving him the right to subdue it. In 1172 Henry entered Ireland himself, and partly through the favour of the clergy, the great princes did homage to him and acknowledged his supremacy. Many Norman barons and their followers now settled in the country, but the English power was far from being established over it. For long only a part was recognized as English territory (generally known as 'the Pale'), and this was governed by various nobles subject to a viceroy. The greater part of the island remained unconquered, and English supremacy was threatened frequently by rebellions of Irish chiefs and barons. It was not until the reign of Henry VIII that the title of 'king', instead of 'lord', of Ireland was properly established. The chieftains of Ulster remained fiercely independent, however, and continued to seek allies on the continent to rid Ireland of the English. During the reign of Elizabeth I they were eventually forced to flee, and this 'Flight of the Earls' became a turning point in the history of England's involvement in Ireland for it was then that the 'Plantation'—the settlement of English and Scottish Protestants in Ulster—began.

Ireland remained under English, and then British, kings and queens until the time of Edward VIII's abdication in 1936. Independence was finally achieved in that year when the Irish Free State abolished the monarchy and formally became a republic. British monarchs continue to reign over Northern Ireland, comprising most of what had been the ancient kingdom of Ulster.

*

Aed (d.878) King of Scots. Reigned in 878. The son of KENNETH MAC-ALPIN, he succeeded his brother CONSTANTINE I as king but reigned for less than a year. He was reputedly murdered by his first cousin, GIRIC, the son of DONALD I, a rival for the throne.

Aed Find (d.c.778) King of Scots. Reigned from 768 to 778. He invaded the Pictish heartland early in his reign but the results of his efforts are not known.

Aed Finnliath (d.c.879) High king of Ireland. Reigned from 862 to 868. In 861 he joined with a Norse force against the ruling high king, MAEL SECHNAILL. On his death a year later Aed became high king and turned against his former allies. Over the next five years of his reign he was greatly troubled by inroads made by the Vikings but won an important battle against them at Killineery in 868.

Aelfwald (d.749) King of East Anglia. Reigned from 713 to 749. An under-king of Mercia under the powerful ETHELBALD from 740.

Aelle (d.588) King of Deira. Reigned from c.560 to 588. A son of IDA, he was the first king of this small kingdom (roughly Humberside) to establish independence from Bernicia (roughly Northumberland). On his death the kingdom was taken by his brother, ETHELRIC (who ruled Bernicia after his father), and his two sons fled the kingdom. Ethelric became the first king of Northumbria.

Aelle (d.867) King of Northumbria. Reigned from 866 to 867. He was chosen to be king following the removal of the unpopular OSBERT. Soon after a large Norse army, which had invaded in 865, took advantage of the royal in-fighting to seize York, the chief city of Northumbria. After resolving their dispute, Osbert and Aelle mounted a joint attack in 867. They were at first successful, but at length lost their army in a climactic battle. Osbert was killed in the fighting but Aelle was sacrificed to the Norse war god Odin in a gruesome ritual. The leader of the Norsemen, Ivar the Boneless, justified the killing by falsely claiming that Aelle had slain his father by throwing him into a pit of vipers.

Aelle (d.c.514) King of Sussex. According to the *Anglo-Saxon Chronicle* he established the South Saxon kingdom after landing around 477 and defeating a force of Britons at a place called Cymenesora. With his third son CISSA he beseiged an old Roman fort at Pevensey held by Britons in 491 and massacred the occupants. He came to be regarded as *Bretwalda*, or overlord among the Anglo-Saxon kings of southern England although at this time the success of the invaders was not certain and the title was little more than honorary.

Aesc (d.512) King of Kent. Reigned from 488 to 512. Son of HENGEST and nephew of HORSA, the first Jutish kings of Kent. In 488 his father died and Aesc succeeded to the kingdom. He was regarded as a powerful commander and many of his battles against the Welsh and the men of Kent are recorded in the *Anglo-Saxon Chronicle*. Although his father and uncle are regarded as the first kings, Aesc, also given as 'Oisc', provided the dynastic name of 'Oiscingas'.

Aescwine King of Essex. Reigned from c.527 to c.587. Thought to have been the founder of the small kingdom of East Saxons which, with London as its capital, was later absorbed by Wessex.

Aescwine King of Wessex. Reigned from 674 to 676. The son of CENFUS and a descendant of King COEL.

Aiden (Mac Gabrain) (d.606) King of Scots. Reigned from c.575 to 606. Ordained at Dunadd by his cousin Colum Cille (St Columba) in what was probably the first Christian coronation in Britain, Aiden was a descendant of FERGUS MOR and did much to consolidate the kingdom of Dalriada. Influenced by St Columba, he sought an alliance with the king of the O'Neill dynasty in Ulster to protect the monastic community founded by St Columba on the small island of Iona. This was acheived at Druim Cett soon after Aiden's coronation. Sometime between 580 and 585 he captured the Orkneys and Isle of Man respectively, thus extending his kingdom. During his reign he fought many battles against the Picts, Britons and Angles but was finally defeated in 603 by ETHELFRITH at Degsastan, near Liddesdale. As Aiden lost most of his army and a son in the battle the Northumbrian ruler was free to advance further north and west. At his death he was over seventy years old and was succeeded by his son EOCHAID.

Alchred King of Northumbria. Reigned from 765 to 774. He is known to have been a supporter of missions to Europe and was in contact with the Frankish king Charlemagne. Deposed in favour of ETHELRED, he sought shelter among the Picts.

Aldfrith (d.705) King of Northumbria. Reigned from 685 to 705. The son of Oswy's mistress, Aldfrith combined statesman-like qualities with a reputation as a man of letters. In his pursuit of learning it is said he exchanged an estate in return for a book which had been brought by a monk from Rome. He was succeeded by his eight-year-old son, OSRED.

Aldwulf King of East Anglia. Reigned from 663 to 713. Son of ETHELHERE and a Northumbrian princess, Hereswith of Deira.

Aldwulf King of Sussex. He reigned briefly *c.*765 and is known to have made a land grant to the Church.

Alexander I (the Fierce) (1077-1124) King of Scots. Reigned from 1107 to 1124. The fifth son of MALCOLM III (Canmore), the first king of all Scotland, and Margaret of England, he was born at Dumbarton and succeeded his elder brother EDGAR as king. He ruled only that part of the kingdom to the north of the Forth and Clyde as the south (Cumbria, Strathclyde and southern Lothian) had been bequeathed to David, Edgar's younger brother, for him to govern as Earl. As a devout patron of the church, he did much to strengthen the religious independence of his kingdom. He revived the see at Dunkeld, established the Augustinian priory at Scone and the abbey on the island of Inchcolm. During his reign many new castles were also built, including a new royal castle at Stirling.

He maintained good relations with HENRY I and encouraged Norman settlement in lowland Scotland where they began to introduce feudalism as a system for holding land and built castles in the Norman style. He also took an active part in Henry's campaign against the rebellious Welsh kings in 1114 and married Sybilla, his illegitimate daughter. The marriage produced no legitimate offspring, although descendents of his natural son, Malcolm Mac-Heth, later made unsuccessful claims to parts of Caithness. He was succeeded by his younger brother DAVID I upon his death at Stirling. He was interred at Dunfermline's Abbey Church.

Significant events of Alexander I's reign
- 1112 – Berwick-upon-Tweed becomes the first royal burgh in Scotland.
- 1114 – With Alexander's help Henry I receives oaths of obedience from Welsh kings.

Alexander II (1198-1249) King of Scots. Reigned from 1214 to 1249. The son of WILLIAM I, 'the Lion', and Ermengarde de Beaumont, he was born at Haddington and succeeded his father at the age of sixteen. His accession was immediately disputed and an army was gathered by Donald macWilliam, a rival to the throne. They soon landed from Ireland but were defeated by the Earl of Mar who ordered the execution of the leaders.

During Alexander's reign there was no serious conflict with England and he was married to King JOHN's eldest daughter, Princess Joan in 1221 but the relationship was far from stable. As the young Scottish king had red hair, John boasted of how he would 'hunt the red fox cub from his lair', but John was being forced by his barons to sign the Magna Carta and Alexander stood firmly with them. Following a successful campaign in the north of England the barons paid homage to Alexander, but following King John's death the barons broke the agreements made with Alexander and he raised an army against them. Alexander was unsuccessful and, constrained to come to terms with England, had to do homage to the new king, HENRY III. He was an able politician, however, and gave up his claims to Northumberland, Westmoreland and Cumberland in order to agree the peace. This led to the Treaty of York of 1237, which fixed the borders between the two countries roughly where they are today and Alexander received estates in northern England worth £200 a year. It was an uneasy peace, however, and it was seven years until the relationship improved with the betrothal of Henry's three-year-old daughter, Margaret, to Alexander's son and heir

He was an intelligent and energetic monarch; as well as being a good soldier he governed wisely, introducing measures to modernize the administration of his kingdom. As a devoted patron of the church he was responsible for the founding of several important monasteries and abbeys. Despite these progressive achievements he was still very much a king of his times; this was demonstrated by his punishment of those present at the brutal murder of Bishop Adam of Caithness when he ordered that the hands and feet of eighty men be cut off.

He died after falling ill with a fever at Kerrera, an island opposite Oban, whilst on route to join his naval fleet for an expedition in which he hoped to wrest the Hebrides from King Haakon IV of Norway. Princess Joan died childless in 1238; Alexander's second wife, Marie de Coucy, daughter of Baron de Coucy of Picardy, provided his successor, ALEXANDER III. He is buried at Melrose Abbey.

Significant events of Alexander II's reign
- 1215 – Alexander receives homage of the northern English barons.

- 1216 – King John dies of dysentery. Henry III ascends his father's throne.
- 1221 – Alexander marries Henry's sister, Joan.
- 1237 – Treaty of York fixes Scottish border with England.

Alexander III (1241-1286) King of Scots. Reigned from 1249 to 1286. The only son of ALEXANDER II and Marie de Coucy, he succeeded his father when still only eight years of age. He married Princess Margaret, eldest daughter of HENRY III of England, to whom he had been betrothed, two years later. Rival court factions jostled for positions during his minority and rebel lords, led by the Comyns of Monteith and Buchan, attempted to take control of the government when Alexander was sixteen by holding the king hostage. This angered Henry III of England who offered to keep an army in northern England until the difficulty was resolved. Negotiations at Jedburgh finally settled the matter and the Comyns were reconciled to their rivals in government and to the English king. Later the same year a council of ten people were chosen to care for Alexander until he reached his majority in 1262 and the dispute between Henry, the Comyns and the Scottish government was formally ended.

It was not long after assuming full powers that Alexander began to assert his position as king. Like his father, he was eager to bring the Hebrides under his sway, and this he was able to accomplish in a few years after defeating an invasion force led by the Norse king, Haakon, at Largs. He bought the Isle of Man and the Hebrides from Norway's Magnus VI for £2666 and in 1266 signed the Treaty of Perth which established his control over all the Western Isles. This brought to an end nearly four centuries of Norse domination of the Hebrides. For the first time the mainland and islands of Scotland had been brought together under one sovereign, although Orkney and Shetland still belonged to Norway.

Alexander was strenuous in asserting the independence both of his kingdom and of the Scottish Church, but his relationship with England remained peaceful. At his wedding, when he was only ten years old, he is said to have nearly been tricked into doing homage to Henry but had the presence of mind to evade Henry's approaches. The king of England never gave up, however, and continued to appeal to Alexander. Following Henry's death he attended EDWARD I's coronation only after receiving written confirmation that the independence of the kingdom of Scotland would not be compromised. For the rest of his reign he managed to keep on good terms with his neighbour, despite refusing to do homage except for those lands he held in the north of England.

Alexander's main difficulty was in securing a successor. His wife's death in 1275 was followed by that of his two sons, David and Alexander, both childless. In order to secure the succession he married Yolande of Dreaux; but less than six months later, while riding in the dark along the cliffs between Edinburgh and Kinghorn in Fife to meet his wife, his horse stumbled and threw him. The body of the last Celtic king was found at the foot of the cliffs the next morning. The marriage of his daughter Margaret to Eric II of Norway had brought a daughter, however, and this infant, MARGARET, the Maid of Norway, had been recognized as heir to the throne by Parliament two years before. She was proclaimed as Queen of Scotland, despite the unpopularity of female succession, but died on the sea-crossing from Norway and never put a foot on Scottish soil.

Significant events of Alexander III's reign
- 1263 – Alexander defeats Norse invaders at the Battle of Largs.
- 1266 – Alexander gains control of the Hebrides under the Treaty of Perth.
- 1274 – Alexander attends coronation of Edward I of England.
- 1285 – Margaret, the Maid of Norway, is recognized as heir presumptive.

Alfred the Great (849-899) King of Wessex. Reigned from 871 to 899. Alfred was born at Wantage in Berkshire, his father being ETHELWULF, son of EGBERT, king of the West Saxons. He succeeded his brother ETHELRED in 871, at a time when the Danes, or Norsemen, had extended their conquests widely over the country; they had completely overrun the Kingdom of the West Saxons (or Wessex) by 878. Alfred, the king of Wessex, was obliged to flee in disguise. At length he gathered a small force and having fortified himself on the Isle of Athelney, formed by the confluence of the Rivers Parret and Tone amid the marshes of Somerset, he was able to make frequent raids on the enemy. It was during his time here that he, according to legend, disguised himself as a harper and entered into the camp of King GUTHRUM. Having ascertained that the Danes felt themselves secure, he returned to his troops and led them against the enemy. He gained such a decided victory that fourteen days afterwards the Danes begged for peace. This battle took place in May 878 near Edington in Wiltshire. Later he took London and repulsed a

Danish seaborne invasion. Alfred allowed the Danes who were already in the country to remain, on condition that they gave hostages, took a solemn oath to quit Essex, and embraced Christianity. Under these terms, known as the Peace of Wedmore, their king, Guthrum, was baptized, together with thirty of his followers. In return they received that portion of the east of England now occupied by the counties of Norfolk, Suffolk, and Cambridge, as a place of residence. By this time all outside of Danish-ruled England also recognized Alfred as king of all England.

The few years of tranquillity (886–93) which followed were employed by Alfred in rebuilding the towns that had suffered most during the war, particularly London. He consolidated his power by raising around thirty forts, including Oxford and Hastings. He also set about training his people in arms and in agriculture; in improving the navy; in systematizing the laws and internal administration; and in literary labours and the advancement of learning. He re-established the monasteries (as a youth he had twice visited the Pope in Rome and had been devoutly Christian since) and schools sacked by the Danes and caused many manuscripts to be translated from Latin; he himself translated several works into Anglo-Saxon, such as the Psalms, *Æsop's Fables*, Boethius on the *Consolation of Philosophy*, the *History of Orosius* and Bede's *Ecclesiastical History*. He also drew up several original works in Anglo-Saxon and initiated the translation of Latin records to form the *Anglo-Saxon Chronicle*.

These labours were interrupted, about 894, by an invasion of the Norsemen, who, after a struggle of three years in which they tried to settle lands in the south of England, were finally driven out. Alfred died in 901 and is buried at Winchester. He had married, in 868, Alswith, the daughter of a Mercian nobleman, and left three daughters and two sons, Edward, who succeeded him, and Ethelward, who died in 922. Their descendants ruled England until 1066.

Alpin (d.c.840) King of Scots. Reigned from c.837 to c.840. This semi-legendary king ruled Dalriada for only a short period before being killed by an unknown assailant. He was the 34th king of Dalriada and is thought to be the son of Eochaid 'the Venomous', perhaps by a Pictish mother, and the father of Kenneth mac-Alpin and Donald I.

Alric King of Kent. Reigned from 747 to 762. Succeeded his brother Eadbert I as joint ruler with his other brother, Ethelbert II.

Anarawd King of Gwynedd. Reigned from 878 to 916. The son of Rhodi Mawr, he shared with his five brothers the rule of lands won by his father. In 885 he made an alliance with the Norse king of Dublin against the other kings of Wales who were themselves allied with King Alfred of Wessex. This made him the most powerful ruler in Wales and the alliance held for eight years before he finally made peace with Alfred.

Androco (*fl*.AD 20) High king of the British tribes. Son of Caswallon, he ruled the Catuvellauni tribe based around Hertfordshire and followed his father as the high king of the various tribes he had united against the Romans.

Anna (d.654) King of East Anglia. Reigned from c.633 to 654. A devout Christian son of Ine who produced four daughters known for their piety. He was killed in battle fighting Penda of Mercia.

Anne (1665-1714) Queen of England and later Great Britain and Ireland. Reigned from 1702 to 1714. The second daughter of James II, then Duke of York, and Anne Hyde, daughter of the Earl of Clarendon. With her father's permisson she was educated according to the principles of the English Church. In 1683 she was married to Prince George, son of King Ferdinand III of Denmark and brother of King Christian V of Denmark. On the arrival of the Prince of Orange in 1688, Anne wished to remain with her father; but she was prevailed upon by Lord Churchill (afterwards Duke of Marlborough) and his wife to join the triumphant party. After the death of William III she ascended the English throne and was coronated at Westminster Abbey.

Her character was essentially weak, and she remained distant from the new political parties called the Whigs and the Tories, although she presided over meetings of the Cabinet. Most of the principal events of her reign are connected with the War of the Spanish Succession. The eighteenth century had opened with a series of events in Europe which made war inevitable as it was essential for England to have an Austrian, instead of a French prince, ascend the Spanish throne. The commander of the English army, the Duke of Marlborough was a brilliant soldier, arguably the best ever produced by England, and he soon routed a combined French and Bavarian army at Blenheim. The Queen rewarded her commander with the estate on which was built Blenheim Palace. Two years later Marlborough drove the French from the Netherlands following the victory at Ramillies and King Louis sued for peace. Instead of negotiating a Treaty with the French, however, England negotiated an Act of Union with

Scotland. In May 1707 the two Parliaments of Scotland and England, not without some difficulty, were finally united. The united kingdoms came to be called Great Britain which had as its symbolic flag the Union Jack.

Meanwhile the war with the French was vigorously prosecuted; the British forces defeated the French at Oudenarde and continued to seize valuable French colonies elsewhere. At home, however, there were disputes within the Government as the Tories, who had never been in favour of the government funding wars which served the financial interests of the Whigs, took control of the Commons. Anne then fell out with her old friend, the Duchess of Marlborough, and came under the influence of the Tories. Marlborough was soon recalled from Europe and the Treaty of Utrecht negotiated with the French. Austria was to have the Spanish Netherlands, the crowns of Spain and France were to be united, and Britain was to retain the valuable colonies of Gibraltar, Nova Scotia, Newfoundland and Minorca.

England was still divided, however, as Anne wanted to secure the succession to her brother James against the wishes of the cabinet. The Act of Settlement eventually assigned the crown to JAMES I's Protestant descendants of the House of Hanover if Anne were to die childless. Grieved at the disappointment of her wishes, she fell into a state of weakness and lethargy, and died. She was to be the last of the Stuart monarchs.

The reign of Anne was distinguished not only by the brilliant successes of the British army, but also on account of the number of admirable and excellent writers who flourished at this time, among whom were Pope, Swift, and Addison. Anne bore her husband many children, all of whom died in infancy, except one son, the Duke of Gloucester, who died at the age of twelve.

Significant events of Anne's reign
- 1702 – The War of the Spanish Succession commences. The first daily newspaper, *The Daily Courant*, is printed in London.
- 1704 – The Battle of Blenheim.
- 1704 – Gibraltar is taken from Spain by Sir George Rook.
- 1705 – Barcelona is taken by the Earl of Peterborough.
- 1706 – The Battle of Ramillies.
- 1707 – The Act of Union creates Great Britain.
- 1708 – The Battle of Oudenarde. Prince George of Denmark dies. Minorca falls to the English troops.

- 1709 – The Battle of Malplaquet.
- 1710 – Tory administration takes power.
- 1711 – Marborough is relieved of his post.
- 1713 – The War of the Spanish Succession is ended by the Treaty of Utrecht. For the last time in England a death sentence for witchcraft is carried out.
- 1714 – Sophia of Hanover dies, her son George becomes heir to England's throne.

Artgal (d.871) King of Strathclyde. His kingdom was subject to raids from Norse invaders and in 870 they sacked Dumbarton, an important stronghold on the Dublin to York trade route. He escaped capture but was later put to death on the orders of CONSTANTINE I.

Arthur (*fl*.500) British tribal king. A figure best known from popular legend who may have been based on the tribal king who led the Britons to victory at Mount Badon (518) as chronicled by Nennius in *Historia Britonium* (9th century). Another writer, Gildan, a contemporary of Nennius, dates this battle at around 500 and does not connect it with the figure of Arthur. The legends of King Arthur appear in Welsh literature of the 10th century and also in Geoffrey Monmouth's *History of the Kings of Britain* in the 12th century. Later writers contributed to the tales; most famously Sir Thomas Malory, who translated and adapted from the French Arthurian romances, in *Morte d'Arthur*.

Athelstan *see* **Ethelstan**.

Balliol, Edward (1287-1364) King of Scots. Reigned from 1332 to 1341. The son of JOHN BALLIOL, he was held prisoner in Normandy and England after his father's abdication. When ROBERT the Bruce of Scotland died, leaving his young son David as heir, Balliol became EDWARD III's candidate for the throne, although he was also acknowledged by many Scots as the heir. Soon after the coronation of DAVID II he landed a considerable army at Kinghorn in Fife and with the support of dispossesed English and Scottish nobles and many mercenary soldiers, he defeated the Earl of Mar's army at Dupplin Moor. David II fled to France where he was to remain until he was seventeen. Balliol was eventually crowned at Scone in 1332 and, as he accepted Edward III as overlord of his kingdom, put Scotland in the control of England. This proved unacceptable to a great many Scots lords and he had to flee to Carlisle where he survived as a powerless king during the minority of David II. His enthronement had nonetheless entailed the surrender of much of lowland Scotland to England and the country was thrown into civil war with rival factions of nobles contesting estates. An army led by Scots lords attempting to end the siege of Berwick was subsequently defeated by an English army at Halidon Hill in 1333. Edward's army advanced again in 1334 and after taking Roxburgh, forced the Scots lords to accept Balliol as their king. He had no real power but remained a puppet of the English king until David II returned from France. Balliol was eventually dismissed with a pension by Edward III in 1356 and died on his family estate (Bailleul) in Picardy, France, in 1364.

Significant events of Edward Balliol's reign
  • 1332 – Battle of Dupplin Moor. Balliol is crowned at Scone.
  • 1333 – Edward III invades and defeats Scots at Halidon Hill.
  • 1334 – Balliol accepted as king by Scots lords.
  • 1337 – The Hundred Years' War begins.
  • 1341 – Edinburgh Castle regained from the English.
  • 1342 – David II regains the throne of Scotland.

Balliol, John (1249-1313) King of Scots. Reigned from 1292 to 1296. The son of Devorguilla Balliol, John married the daughter of the Earl of Surrey and owned vast estates in both England and Scotland. His claim to the throne in 1291 was contested by a dozen rivals (known as the Competitors), including Robert Bruce of Annadale, the grandfather of ROBERT I. The Guardians of Scotland, fearing a civil war, asked EDWARD I of England to intervene. Edward I took advantage of the situation and demanded allegiance and custody of several important castles before deciding on Balliol. Edward continued to make demands of the Scots and his interventions soon became intolerable. After demanding Scots troops for a war with France the Community of the Realm persuaded Balliol to renounce his allegiance. The Auld Alliance with France was renewed in 1295 and Edward invaded Scotland in retaliation. Following his defeat at Dunbar, John sued for peace and surrendered at Stracathro in July 1296. He was forced to relinquish the crown and the English forces which overran Scotland removed or burned many records and removed the Stone of Destiny (the Coronation Stone of Scone) to Westminster Abbey.

Imprisoned in the Tower of London, Balliol was later given Bailleul, his ancestral home in France, where he died in 1313. His son, EDWARD BALLIOL, later renewed the claim to Scotland's throne with the support of England.

Significant events of John Balliol's reign
  • 1295 – Franco-Scottish treaty begins the Auld Alliance
  • 1296 – Balliol is deposed.
  • 1296 – Edward I invades Scotland.

Bealdred King of Kent. Reigned from 807 to 825. Ruled as an under-king of Mercia under the control of Cuthred's brother, King COENWULF. He was driven out in 825 by EGBERT of Wessex who soon after became overlord of the East Angles as well as Kent, Surrey, Sussex and Essex.

Beli (d.722) King of Strathclyde. He successfully defended his kingdom from attacks by OENGUS, king of the Picts. His son and successor, TEWDWR, later won an important battle against the Picts at Mugdock, near Glasgow in 750.

Beonna King of East Anglia. Reigned *c.*760. Known to have issued a coinage during his reign.

Beonred King of Mercia. Reigned in 757. The successor to the throne of King ETHELBALD of Mercia, he was challenged by OFFA, a cousin of the former king. Disorder followed and the confederacy of kingdoms which had been built by Ethelbald was weakened as the two rivals fought. It was only a matter of months before Offa deposed him.

Beorhtwulf (d.853) King of Mercia. Reigned from 840 to 853. He succeeded WIGLAF as king and endured a period of sustained Viking attacks. He was finally routed in 852 when over 350 ships stormed London and Canterbury and he died a year later. He was succeeded by his son BURGRED.

Beornwulf (d.827) King of Mercia. Reigned from 823 to 827. He deposed CEOLWULF in 823 and soon after led an army into Wessex. After some initial success he was defeated by EGBERT and later killed near Swindon. He was succeeded by LUDECA, previously an ealdorman.

Beortric (d.802) King of Wessex. Reigned from 757 to 802. He married Eadburga of Mercia, a daughter of OFFA, and was succeeded by EGBERT.

Berhtun (d.686) King of Sussex. Reigned 686. Ruling for less than a year, he was killed during an invasion of Kent.

Boadicea (or **Boudicca**) (d.61) Queen of the Iceni tribe. Wife of PRASUTAGUS, a client of Rome, whose lands in Norfolk and Suffolk were taken by the Romans after his death. Homes and estates were burned and plundered, Boudicca was flogged and her daughters raped. In retaliation she gathered a large army and led a violent rebellion which took London, St Albans and Colchester while the Roman governor Paulinus was in Wales. When Paulinus returned he violently suppressed the rebellion and the defeated Boadicea killed herself with poison.

Bred (d.842) King of Picts. Reigned in 842. He ascended the throne on the death of his father, UURAD, but his reign did not last the year as he was deposed by his brother, KINETH.

Brian Boru (c.941-1014) high king of Ireland. Reigned from 1002 to 1014. The king of Munster from 976. He took the title of High king after defeating his great rival MAEL SECHNAILL II, the last of the O'Neill kings, at Athlone. After extending his kingdom to take in much of southern Ireland, he put an end to Norse ambitions in Ireland by his victory at the Battle of Clontarf (1014) near Dublin. The losses suffered in this conflict were great and in the confusion Brian Boru was killed in his tent by defeated and fleeing Danes. Mael Sechnaill II quickly reclaimed the title of high king but after his death in 1024 no king was able to secure the position for any notable period of time for over 150 years. This effectively ended royal rule in Ireland.

As well as a soldier Brian Boru was a patron of education and religion. He established monasteries and encouraged the preservation of the old Irish sagas as well as the writing of poetry and prose. Three O'Brien descendents remained kings of Munster until the next century (1119). Their main kingdom disappeared in 1194, but the O'Briens remained kings of north Munster until 1543. The kingdom then became an earldom under HENRY VII. It remained an earldom until 1741 when the 8th earl died childless.

Bridei I (d.586) King of Picts. Reigned from 556 to 586. His royal court at Inverness was visited by St Columba but he did not convert to Christianity.

Bridei II (d.641) King of Picts. Reigned from 635 to 641. A brother of GARNARD.

Bridei III (d.692) King of Picts. Reigned from 671 to 692. The son of DREST I's sister. He was the first of the Pictish kings to be recognized overlord of all Pictland after defeating King EGFRITH of Northumbria in the battle of Nectansmere in 685. Egfrith had installed his cousin Drest as ruler of the Picts in order to consolidate his power in the north.

Bridei IV (d.706) King of Picts. Reigned from 696 to 706. The son of DREST's sister.

Bridei V (d.763) King of Picts. Reigned from 761 to 763. The brother of OENGUS.

Brude (d.845) King of Picts. Reigned from 843 to 845. The son of UURAD's sister, he ascended the throne after killing his uncle KINETH. He was converted to Christianity by St Columba.

Bruce, Edward (1276-1318) High king of Ireland. Reigned from 1315 to 1318. One of the five sons of Robert, the Earl of Carrick and Marjorie, Countess of Carrick, and the younger brother of ROBERT I of Scotland (his three other brothers were executed following Robert's defeats at Methven and Dalry). A fearless commander, he aided his brother during his guerilla-style campaign of raids on English-held castles. By mid-1308 Edward had overrun Galloway, while his brother had control of much of the north. He failed in his attempt to take Stirling in 1313 and made a truce with Philip de Mowbray which made Bannockburn inevitable.

Invited to Ireland, where he had grown up, in 1315 by the king of Tyrone he led an army in a difficult campaign for a year before being crowned high king of Ireland. His tactic of destroying everthing in his path as he progressed through Ireland made him a widely unpopular figure. He failed to take Dublin despite assistance from his brother and was killed in battle at Dundalk in 1318 along with many of his supporters. He married Isabell of Atholl and had two sons.

Burgred King of Mercia. Reigned from 853 to 874. The son of BEORHTWULF, he married Ethelswith, daughter of ETHELWULF of Wessex, and with the assistance of the latter frequently raided north Wales. He made two treaties with the Viking Great Army (868 and 872) only to be driven from it from Repton to exile in Rome, where he died. He was buried in the Church of St Mary in the English quarter.

Cadwalla (c.658-689) King of Wessex. Reigned from 685 to 688. A member of the Wessex royal family who was forced into exile early in life. He returned in 684 and after establishing himself in Wessex attacked and subjugated the kingdom of Sussex. He also annexed the Isle of Wight and made advances as far as Kent. He was converted to Christianity and after he abdicated made a pilgrimage to Rome where he was baptised in the presence of the pope.

Cadwallon (d.633) King of Gwynedd. Reigned from c.625 to 633. An ally of King PENDA of Mercia, he helped defeat King EDWIN of Northumbria in a battle at Hatfield Chase, near Doncaster in 633. He was killed at Hexham by Edwin's nephew, OSWALD, who had returned from exile among the Scots.

Canute (c.994-1035) King of England, Denmark and Norway. Reigned (England) from 1014 to 1035. The son of SWEYN FORKBEARD. Following the death of ETHELRED, the kingdom of England fell into confusion and Canute renewed Danish attacks. He began by devastating the eastern coast, and extended his ravages in the south, where, however, he failed to establish himself until after the battle of Ashingdon and the assassination of the Saxon choice for king, EDMUND. He was accepted as king of the whole of England in 1016. At Harold's death in 1018 he gained Denmark and in 1031 he conquered Norway, thus becoming ruler of a great Danish empire. MALCOLM III of Scotland also admitted Canute's superiority and Sweden also was vassal to him.

Canute, who began his reign with barbarity and crime, afterwards became a humane and wise monarch. He restored the English customs at a general assembly, and ensured to the Danes and English equal rights and equal protection of person and property, and even preferred English subjects to the most important posts. His power was confirmed by his marriage with Emma of Normandy, Ethelred's widow. He died at Shaftesbury, leaving Norway to his eldest son, Sweyn, England to HAROLD I and Denmark to the third, HARDICANUTE.

Significant events of Canute's reign
- 1016 – Edmund Ironside is chosen as king by the Saxons. Edmund Ironside killed after the Battle of Ashingdon. Canute becomes king of all England.
- 1017 – The marriage of Canute and Emma of Normandy. Canute creates four earldoms: Wessex, Mercia, Northumbria and East Anglia.
- 1027 – Canute makes a pilgrimage to Rome.

Caradoc (d.c.54) King of the Catuvellauni tribe. A son of CUNOBELINUS, he resisted the Romans (by whom he was known as **Caratacus**) from AD 43 to 47 but was eventually captured and taken to Rome in chains. His brother TOGODUMMUS was probably killed at the Battle of the Medway in AD 43.

Cartimandua (fl.70) Queen of the Brigantes tribe. A leader of one of the largest British tribes which had York as its centre.

Caswallon (d.c.60) High king of the British tribes. Leader of the Catuvellauni tribe which settled in what is now Hertfordshire. He fought rival tribes to become high king and united them against the first Roman invasions.

Cathal O'Connor (d.1224) King of Connaught. Reigned from 1202 to 1224. The last provincial monarch in Ireland. He resisted the advances of HENRY III until his death.

Ceawlin (d.c.593) King of Wessex. Reigned from 577 to 591. The son of CYNRIC, he succeeded his father to the leadership of the West Saxons and be-

gan advancing north of the Thames. In 577 he defeated ETHELBERT of Kent at Wibbandum, killing two of his sons, and was recognized as *Bretwalda*, overlord of all the Anglo-Saxon kings. His turbulent reign came to an end following a power struggle in Wessex and he died in exile.

Cenfus King of Wessex. Reigned in 674. A grandson of CEOLWULF.

Cenred King of Mercia. Reigned from 716 to 718. He succeeded the murdered King OSRED.

Centwine King of Wessex. He reigned from 676 to 685. A brother of CENWAHL.

Cenwahl King of Wessex. Reigned from 643 to 672. The son of CYNEGLIS, Cenwahl married a sister of PENDA but had to flee from the Mercian king after abandoning her. He extended his territory after Penda's death and won an important victory in 658 at Peonnan, near Exeter. Advancing further into Devon he is said to have 'put the Britons to flight as far as the sea'. He then lost Oxfordshire to Mercia, which also seized the Isle of Wight and lands in Hampshire, and accepted baptism while in exile at the court of King ANNA of East Anglia. He ended his turbulent reign by going on pilgrimage in Rome; his successor INE followed him thirty-seven years later. In Wessex he built the Old Minster in Winchester, (648) and was buried there.

Ceorl King of Mercia. Reigned from c.606 to 626. Related to PYBBA, his daughter married the dominant king of Northumbria, EDWIN.

Ceol King of Wessex. Reigned from 591 to 597. Came to the throne after the abdication of CEAWLIN.

Ceolred King of Mercia. Reigned from 709 to 716. The son of ETHELRED, he succeeded COENRED and earned a reputation as a spoiler of monasteries. He also fought King INE of Wessex at Woodborough in Wiltshire in 715.

Ceolwulf (d.760) King of Northumbria. Reigned from 729 to 737. A brother of CENRED, he was temporarily deposed in 731 when a rival faction seized him and had him tonsured as a monk. Bede's *Ecclesiastical History of the English People* was dedicated to him.

Ceolwulf King of Wessex. Reigned from 597 to 611. He succeeded his brother, CEOL, and in 607 defeated the South Saxons who had gained lands in the kingdom following the abdication of CEAWLIN.

Ceolwulf I King of Mercia. Reigned from 821 to 823. Succeeded his brother, COENWULF, reigning for only two years before being deposed by BEORNWULF.

Ceolwulf II King of Mercia. Reigned from 874 to

c.880. The last Mercian king was chosen by the Danish overlords to rule the subordinate western half of the kingdom. He was deposed.

Cerdic King of Wessex. Reigned from 519 to 534. Recorded as the first king of the West Saxons, he landed near Southampton in 494 and fought the Britons for control of the area. He is known to have killed a British king in 508 and had a further victory at Charford on the Avon in 519. By 530 he also had control of the Isle of Wight. His son, CYNRIC, succeeded him.

Charles I (1600-1649) King of Great Britain and Ireland. Reigned from 1625 to 1649. The son of James I (JAMES VI of Scotland) and Anne, daughter of King Ferdinand II of Denmark. Charles was born at Dunfermline Palace (the last sovereign to be born in Scotland), and was never expected to become king as he was second in line to the throne after his brother, Henry, Prince of Wales. He became heir apparent after his brother died and was created Prince of Wales at the age of twelve. In early 1625, he succeeded to the throne and was crowned at Westminster Abbey. In the same year he was married by proxy to Henrietta Maria, daughter of Henry IV of France.

Charles proved himself to have inherited his father's inflexible convictions about the role of the monarchy in relation to Parliament. As he had Catholic sympathies he favoured the new High Church party of William Laud, soon to be made Archbisop of Canterbury, against the dominant parliamentarians of the time. The first Parliament which he summoned, being more disposed to state grievances than grant supplies, was dissolved after financially crippling Charles. The next year (1626) a new Parliament was summoned; but the House proved no more tractable than before, and it too was soon dissolved. In 1628 a series of military and naval disasters compelled the king to call a new Parliament, which showed itself as much opposed to arbitrary measures as its predecessor, and after voting the supplies prepared the Petition of Rights. This was essentially a reminder to Charles of what the Parliament took to be the traditional liberties of the people and asserted that any loan or tax forced by the king was illegal without the permission of Parliament. Charles was constrained to pass the Petition into law. But the assassination of the Duke of Buckingham, Charles's supporter, and the determined spirit with which the Parliament resisted the king's claim to levy tonnage and poundage on his own authority led to a rupture, and Charles again dissolved the Parliament, resolving to try and reign without one.

In this endeavour he was supported by Strafford and Laud as his chief counsellors. With their help Charles continued eleven years without summoning a Parliament, using the arbitrary courts of High Commission and Star Chamber as a kind of cover for pure absolutism, and raising money by unconstitutional or doubtful means. He made various attempts to get estates into his possession on the pretext of invalid titles, and in May 1635, the city of London estates were sequestered. In 1637 John Hampden began the career of resistance to the king's arbitrary measures by refusing to pay ship-money, the right to levy which, without authority of Parliament, he was determined to bring before a court of law. His cause was argued for twelve days in the Court of Exchequer; and although he lost it by the decision of eight of the judges out of twelve, the discussion of the question produced a very powerful impression on the public mind.

During this period Laud began to enforce his High Church discipline and Puritans, forced into exile, founded colonies in America. It was in Scotland, however, that Charles's insensitivity and arrogance came to the fore. Charles was crowned in Edinburgh with full Anglican ceremonial in 1633, and this lost him the goodwill of a number of his Scottish subjects. In 1636 the new Book of Canons was issued by the king's authority, and this attempt of Charles to introduce an Anglican liturgy into Scotland produced great opposition. The Presbyterian nobility in Scotland attempted to obstruct this imposition in Parliament but soon found that the influence of the bishops was too great. The following year saw the introduction of the Book of Common Prayer and it became apparent that Charles was acting without consultation. After repeated petitions to the king frustration led to the drawing up and subscription of the National Covenant in 1638. Thousands gathered in Edinburgh to sign this declaration of faith and obedience to the reformed religion of Scotland. Soon the Covenanters were buying arms and preparing to defend their beliefs on the battlefield and Charles was forced to seek a settlement. Suspension of the Code and Prayer Book and the calling of a General Assembly in Glasgow did not calm the Covenanters, however, and they proceeded to abolish episcopacy and defy Charles by refusing to disarm Charles attempted to reassert himself by force of arms but lack of money and troops frustrated his efforts. He was also forestalled by the effective deployment of men and arms by the Covenant army and hostilities did not immediately begin. Charles sought instead, through an agreement reached at

Berwick, to begin consultations with a view to negotiating a peaceful settlement.

The Covenanters would not disband, however, and were adamant over the abolition of the bishops. In 1640 a Parliament was summoned and Charles tried again to get military support. Again he was disappointed and an army of Covenanters moved south, taking Newcastle and occupying the northern English counties, remaining there until Charles paid an indemnity to secure their return north. Again Parliament was summoned and the stormy sessions which followed resulted in Charles agreeing to the Act of Attainder under which Charles's chief minister, the Earl of Strafford, was executed. In early 1642 Charles decided to resist and attempted to arrest the Parliamentary leaders. He failed, the City of London gave refuge to the Parliamentarians and Charles left his capital to raise the Royal standard in Nottingham in August. The king had on his side the great bulk of the gentry, while nearly all the Puritans and the inhabitants of the great trading towns sided with the Parliament.

The early successes of the king resulted in Parliament making a Solemn League and Covenant with the Scots, in return for whose help they promised to impose Presbyterianism on England. The first action, the battle of Edgehill, gave the king a slight advantage; but nothing very decisive happened until the battle of Marston Moor, in 1644. The Parliamentary army by this stage was composed of Scots, Roundheads and Sir Thomas Fairfax and Oliver Cromwell's New Model Army. The latter army fought with great vigour, representing as it did the fervour of the extreme Puritan sects, and went on to route the Royalists at the battle of Naseby. This completed the ruin of the king's cause and Charles at length gave himself up to the Scottish army at Newark. The Parliamentarians who favoured moderation in the face of mounting religious extremism had been displaced by Cromwell, however, and when the Scots decided to surrender Charles they had little influence on his fate. When the moderate Covenanters realized that their hopes for a Presbyterian settlement in England would be better served by a having a legitimate sovereign than by the turbulent Cromwell, they rallied in support of Charles. The 'Engagement' was a failure, however, and the defeat of the Royalist army at Preston (1648) sealed the King's fate. Cromwell was soon able to coerce Parliament and the more hesitating of the Presbyterians into bringing Charles to trial for high treason against the people.

Although Charles repeatedly refused to recognise the court he had the sentence of death pronounced against him. All interposition being in vain, he was beheaded before the Banqueting House in Whitehall, meeting his fate with an admirable obstinacy that seemed dignified and courageous. The execution of the king produced a feeling of revulsion throughout the country and Cromwell had to maintain his minority rule by force. Charles had nine children, notably CHARLES II and JAMES VII. He was buried at St George's Chapel, Windsor Castle.

Significant events of Charles I's reign
- 1626 – Parliament dismissed by Charles.
- 1627 – England declares war on France.
- 1628 – Assassination of Buckingham.
- 1628 – Charles reluctantly agreed the Petition of Rights.
- 1629 – Charles dismisses Parliament again, this time for eleven years.
- 1632 – Van Dyck becomes Court painter.
- 1637 – A new Book of Common Prayer is published.
- 1638 – National Covenant pledged in Scotland.
- 1640 – The Short Parliament (3 weeks) is called. The Long Parliament (until 1660) is called.
- 1641 – Abolition of the Star Chamber.
- 1642 – Civil War begins.
- 1644 – Solemn League and Covenant signed. Battle of Marston Moor.
- 1645 – Battle of Naseby.
- 1646 – Scots surrender Charles to English Parliament.
- 1648 – Royalist Scots invade England.
- 1649 – England is declared a republic. Charles is tried by Parliament and executed

Charles II (1630-1685) King of Great Britain and Ireland. Reigned from 1660 to 1685. The eldest surviving son of CHARLES I and Henrietta Maria of France and the brother of James, Duke of York, later JAMES VII. He was sent into exile in France after the battle of Edgehill (1645) and was there tutored by Thomas Hobbes among others. After his father's execution he immediately assumed the royal title. At the time, however, Cromwell was all-powerful in England and he accepted an invitation from the Scots, who had proclaimed him their king. He sailed from Holland and was crowned at Scone (1651), making him the last king to be crowned in Scotland. It would still be ten years before the restoration of the monarchy in England, but in Scotland at least the Stuart succession remained unbroken, although Charles never revisited Scotland after 1660.

An early attempt to take back the throne was frustrated by divisions among the Royalist parties and Charles reluctantly had to sign the Covenant to appease the strongest faction. Eventually he took to the field with the English Royalists, who, having gathered an army, encountered Cromwell at Worcester and were totally defeated. With great difficulty Charles escaped to France. Richard Cromwell succeeded his father in 1658 but almost immediately abdicated in the face of growing discontent with the Puritan austerities of the time. The popular Restoration, effected without a struggle by General Monk, set Charles on the throne after the Declaration of Breda, his entry into the capital being made amidst universal acclamations.

He was a witty and stylish man, and seemed at first to characterize the anti-Puritan mood of the country, becoming known by many as the 'Merry Monarch'. Within two years he married the Infanta of Portugal, Catherine of Braganza, and for a time his measures, mainly counselled by the Chancellor, Lord Clarendon, were prudent and conciliatory. In this favourable climate Charles re-established much of the royal prerogative. The privileges of the Anglican Church were restored and a pro-Catholic policy began to appear, excluding Nonconformists from holding municipal office and forcing Puritans to accept the doctrines of the Church of England. But the extravagance and licentious habits of the king soon involved the nation as well as himself in difficulties. Dunkirk was sold to the French to relieve his pecuniary embarrassment, and war, caused by commercial rivalry, broke out with Holland. A Dutch fleet entered the Thames, and burned and destroyed ships as far up as Chatham. The great plague in 1665 and the great fire of London, which burned 13,000 houses to the ground the year following, added to the disasters of the period. Amid such calamities there were mutterings of idolatry having taken root in a licentious court.

A triple alliance between England, Holland and Sweden, for the purpose of checking the ambition of Louis XIV, followed, but the extravagance of the king made him willing to become a mere pensioner of Louis XIV with whom he arranged a private treaty in 1670. This was the Secret Treaty of Dover by which he declared himself a Catholic and agreed to restore Catholicism in return for secret subsidies from Louis XIV of France. After this Charles issued a declaration attempting to free both Protestant dissenters and Roman Catholics from some of their disabilities. Parliament countered this with the Test Act which was designed to keep Roman Catholics out of public office. In 1674 Parliament also reversed Charles's foreign policy by breaking off relations with France and making peace with the Dutch.

Alarm over the 'Popish Plot' of 1678, in which an Anglican parson, Titus Oates, disclosed a Catholic plot to murder Charles and restore Catholicism to England led to further difficulties between Charles and Parliament. The plot was soon revealed to be a fabrication but the resulting furore led Parliament to attempt to exclude Charles's Catholic brother from the throne. The question of the succession was only settled after the passing of the Habeas Corpus Act in 1679 which established important measures for protecting individual rights. A new Parliament which assembled in 1680 had to be dissolved following further difficulties with the king, and yet another which met the following year at Oxford. Finally Charles, like his father, determined to govern without a Parliament, and after the discovery of an assassination conspiracy (the Rye House plot of 1683) and the execution of Lord Russell and Algernon Sidney, Charles became as absolute as any sovereign in Europe although political stability was maintained until his death in 1685. England, however, was deeply divided and uncertain of the future and thus unable to take a full part in European affairs. Advances were nonetheless made during his reign in the fields of science and architecture, areas in which Charles took great interest, encouraging and supporting the work of men like Newton, Boyle and Wren.

Charles II died from the consequences of an apoplectic fit in early 1685 after converting to Catholicism on his deathbed and having received the sacrament according to the rites of the Roman Church. He had no legitimate children but was well known for his love of women; his many mistresses included the famous Nell Gwynn and several others were raised to the highest ranks of nobility. Six of the thirteen illegitimate sons he had by them were made dukes; Monmouth (by Lucy Walters), St Albans (by Nell Gwynn), Richmond (by Louise de Querouaille), and Cleveland, Grafton and Northumberland (by Barbara Villiers).

Significant events of Charles II's reign

- 1660 – The Restoration of Charles to the throne. Samuel Pepys diary is begun.
- 1661 – Parliament meets at Westminster.
- 1662 – Act of Uniformity compels Puritans to accept the Church of England doctrines. The Royal Society receives its charter.

- 1665 – Two year long war begins with Holland. London struck by plague.
- 1666 – The Great Fire of London.
- 1670 – The Secret Treaty of Dover with France.
- 1672 – War resumes with Holland.
- 1673 – Test Act introduced to keep Catholics from office.
- 1675 – Creation of the Royal Observatory. The building of St Paul's Cathedral is begun.
- 1678 – The Popish Plot results in the persecution of the Catholics.
- 1679 – The Habeus Corpus Act introduced. Whig and Tory first used as names for political parties.
- 1681 – Exclusion Bill attempts to exclude James from the succession.
- 1683 – Rye House Plot to murder the king is uncovered.
- 1685 – Charles converts to Catholicism on his deathbed.

Cinioch (d.631) King of Picts. Son of GARTNART'S sister, the dates of his reign are not known.

Ciniod King of Picts. He reigned from 763 to 775. Son of OENGUS's sister.

Cissa King of Sussex. Reigned from c.514. Participated in the siege of Pevensey in 491 and succeeded his father AELLE. He gave his name to Chichester (Cissa's-ceaster), the royal capital. Nothing more is known of him as the *Anglo-Saxon Chronicle* does not mention the South Saxons again until 661.

Coel (Old Coel the Splendid) (fl.420) British tribal king. The 'Old King Cole' of nursery-rhyme fame was overlord of several British tribes and ruled much of lowland Scotland. His descendants ruled the kingdom of Strathclyde.

Coenred King of Mercia. Reigned from 704 to 709. The eldest son of WULFHERE. On his father's death the throne was taken by ETHELRED as he was then too young to rule. In 697 there were rebellions south of the Humber and he was declared king there in 702. Two years later Ethelred died and Coenred succeeded him as king of Mercia. He was unsuited to the royal life, however, and abdicated in 709 in favour of Ethelred's son, CEOLRED. Soon after he left England to become a monk in Rome and dedicated the rest of his life to spiritual works.

Coenwulf King of Mercia. Reigned from 796 to 821. A descendant of PENDA's youngest brother. Worcestershire is first mentioned in a land grant he made to Bishop Deneberht sometime between 814 and 820.

Cogidumnus (fl.75) King of the Regni tribe. A client king of the Romans, he called himself the leg-

ate of the Emperor in Britain and built himself a palace at Fishbourne, near Chichester.

Commius (fl.50) King of the Atrebates tribe. Leader of the tribe which settled in what are now Hampshire and Sussex after fleeing Gaul.

Conall King of Picts. Reigned from 787 to 789. The son of ALPIN II's sister, he succeeded TALORGEN but was deposed within two years by CONSTANTINE.

Constantine (d.820) King of Picts. Reigned from 789 to 820. The first Constantine to rule in Scotland was the son of ALPIN II's sister. He asserted his authority over the Scots of Dalriada sometime after 811, a task which was made easier by the frequent Viking attacks of the period.

Constantine I (d.878) King of Scots. Reigned from 862 to 877. Thought to have been the son of KENNETH MAC-ALPIN, he succeeded Kenneth's brother, DONALD, to the throne and bore the title of Constantine I although he was not the first king of that name in Scotland. During his reign his kingdom was frequently attacked by Viking forces sailing from Ireland. Following a landing in Fife in 879 his forces were routed at Dollar and in a further engagement at Forgan he lost his life. He was succeeded by his brother, AED, who soon after was murdered by a rival to the throne. Constantine's sister had married RUN, the British king of Strathclyde and their son, EOCHA, became king after Aed.

Constantine II (d.952) King of Scots. Reigned from 900 to 942. After defeating the Danes, who had killed his predecessor, DONALD II, he held an Ecclesiastical Court at Scone for the settlement of the rule and discipline of the Celtic Church. In diplomatic affairs he was the first Scottish king to acknowledge an English king as overlord (EDWARD THE ELDER, son of ALFRED THE GREAT, in 924) but this may only have been for expediency, as an ally against Norse aggression. An invasion by ETHELSTAN, son of Edward, led to a counter invasion by the Scots which resulted in the Battle of Brunanburgh, near the Humber, in 937. The Northumbrians were the victors despite Constantine having received the assistance of a Norse force from Dublin. He abdicated in 942, leaving the throne to his cousin MALCOLM I, and spent the final years of his life as Abbot at St Andrews.

Constantine III (d.997) King of Scots. Reigned from 995 to 997. The grandson of CONSTANTINE II and son of CUILEAN, his reign was short and turbulent. It is thought he may have been murdered by KENNETH III in 997 after having had KENNETH II murdered two years previously.

Cormac mac-Art (d.*c.*360) High king of Ireland. A semi-legendary warrior king who reigned from Tara, where he is said to have quarrelled with the local Druids after converting to Christianity. He choked to death on a fishbone, and folklore has it that a Druid curse was to blame.

Creoda (d.593) King of Mercia. He reigned from *c.*585 to 593. The first named king of the Mercians, he is thought to have been the son of Icel, the first continental Angle king to settle in Britain.

Cuilean (d.971) King of Scots. Ruled from 966 to 971. He was killed in battle fighting the Britons of Strathclyde.

Cunedda (*fl.*390) Welsh tribal king. A chieftain settled by the Romans in north Wales where he defended the country against attacks from Ireland. Kings of Gwynedd claim descent from him.

Cunobelinus (*fl.*43) High king of the British tribes. The Shakespearean Cymbeline. He was recognized by Augustus as the leader of the Catuvellauni tribe and high king of many others. He became an ally of the Romans and ruled from Camulodonum (Colchester). His sons later resisted the Romans.

Cuthred King of Wessex (d.756). Reigned from 740 to 756. He defeated ETHELBALD of Mercia at Burford in Oxfordshire in 752.

Cyneglis (d.641) King of Wessex. Reigned from 611 to 643. He failed in a plot to murder EDWIN of Northumbria in 626 and in 628 he was defeated in a territorial dispute with PENDA of Mercia. Towards the end of his reign he was converted to Christianity.

Cynewulf (d.786) King of Wessex. Reigned from 757 to 786. A client of OFFA of Mercia, he was murdered while visiting his mistress by Cyneheard, a brother of SIGEBERHT of Wessex, who believed that he was plotting to exile him. Cyneheard was killed by Cynewulf's bodyguards.

Cynric King of Wessex. Reigned from 534 to 560. He extended his kingdom by fighting the Britons at Salisbury in 552 and at Badbury, near Swindon in 556. He was succeeded by his son, CEAWLIN.

Dafydd ap Llywelyn (d.1246) King of Gwynedd. Reigned from 1240 to 1246. The second son and successor of LLYWELYN AP IORWERTH. He attempted to take back lands lost to the English but was forced to do homage to HENRY III and give up all the territories won by his father since 1215. He titled himself 'Prince of Wales' in 1244 and died after uniting other Welsh rulers in a second attempt to restore the kingdom. The Baron's War in England enabled him to marry Simon de Montfort's daughter but he died without an heir, and his principality was divided among the sons of his elder brother.

Dafydd ap Opwain (d.1194) King of Gwynedd. Reigned from 1170 to 1194. He married HENRY II's illegitimate half-sister, Emma.

David I (the Saint) (*c.*1081-1153) King of Scots. Reigned from 1124 to 1153. The sixth son of MALCOLM III Canmore's second marriage to Margaret, sister of Edgar the Aetheling. His early years were spent at the English court of HENRY I, and in 1100 his sister married the king; their daughter became Queen MATILDA. On the death of his elder brother EDGAR, David inherited that part of Scotland below the Forth-Clyde line. However, another brother, ALEXANDER I, succeeded Edgar, and he disputed the right of David to this territory until David strengthened his position with the support of Henry. At Alexander's death in 1124 he quickly established himself throughout his kingdom by initiating a simple form of centralized government. He was the first to introduce feudal institutions to his native land and was the first Scottish king to strike his own coinage. He also vigorously promoted education and agriculture and regularly gave informal audience to the poor in all the languages of the realm. During David's reign around a dozen royal burghs were created including Perth and Aberdeen.

Amidst baronial revolts in England David twice took an army south to support his niece Matilda against Stephen, her rival claimant for the English crown; during one of his incursions he was defeated at the Battle of the Standard near Northallerton in Yorkshire (1138).

David also acquired a considerable reputation for sanctity. While Prince of Cumbria he had begun the re-establishment or restoration of the Glasgow bishopric, and after he became king founded the bishoprics of Aberdeen, Ross, Caithness, Brechin, and Dunblane. Among the religious houses which date from his reign are Holyrood, Melrose, Jedburgh, Kelso, Dryburgh, and Newbattle. His services to the Church procured for him the popular title of saint, but the endowments so taxed the royal domains and possessions that James VI famously characterized him as 'ane sair sanct for the crown'.

In old age he spent his time gardening and establishing apple orchards. He died at Carlisle in 1153, and was succeeded by his eldest grandson, who, as MALCOLM IV, inherited a peaceful and flourishing kingdom.

David II (1324-1371) King of Scots. Reigned with

interruptions from 1329 to 1371. David was born at Dunfermline, the son of ROBERT I (the Bruce) by his second wife, Elizabeth de Burgh. At the age of four he was married to Joan, sister of EDWARD III of England, then only three years older. He succeeded to the throne on the death of his father and was acknowledged as king by the greater part of the nation.

During his minority he was troubled by those his father had disinherited and who supported the claim of EDWARD BALLIOL, the son of JOHN BALLIOL, to the throne. Balliol was backed by Edward III of England, and at first was successful at the Battle of Dupplin Moor, being coronated at Scone soon after. David fled to France following a later defeat at Halidon Hill but eventually returned to Scotland at the age of seventeen and succeeded in driving Balliol from Scotland. Still, however, the war was carried on with England with increasing rancour, till at length David was wounded and made prisoner at the Battle of Neville's Cross, near Durham (1346). After being detained in captivity for eleven years, he was ransomed for 100,000 merks, to be paid in annual instalments, but in place of making the payments he made the offer of leaving his kingdom to an English heir. Following opposition from his nobles, this plan was disallowed by the Scottish Parliament who disliked the idea of a formal union of the crowns.

David married Margaret Drummond in 1363 following the death of Joan, but died at Edinburgh Castle seven years later without having produced an heir.

Significant events of David II's reign
- 1332 – David deposed by Edward Balliol.
- 1333 – Edward III invades Scotland.
- 1341 – David II returns to claim the throne.
- 1346 – Battle of Neville's Cross. David captured and held in the Tower of London.
- 1348 – Scotland afflicted by the Black Death.
- 1357 – The Treaty of Berwick. David returns to Scotland.
- 1363 – David II offers the Scottish throne to Edward III.

Donald I (d.862) King of Scots. Reigned from 858 to 862. The son of ALPIN and brother of KENNETH I. He reputedly applied the laws of Dalriada to Pictland. Donald died at Scone leaving the throne to Kenneth I's son, CONSTANTINE I.

Donald II (d.900) King of Scots. Reigned from 889 to 900. The son of CONSTANTINE I, he was the first king of both the Scots and Picts to be referred to as *ri alban* or King of Alba. His kingdom was repeat-

edly ravaged by the Norse and he was killed in battle near Dunnottar. He was succeeded by CONSTANTINE II.

Donald III Bane (1031-1100) King of Scots. Reigned from 1093 to 1097. Donald, whose sobriquet 'Bane' means 'fair', retreated to the Hebrides on the death of his father, DUNCAN I, at the hands of his rival, MACBETH; Donald's brother, later MALCOLM III, took refuge in England. Macbeth was himself overthrown (1054) and killed by Malcolm who then became King of Scots. During his exile in the Hebrides Donald Bane had been exposed to Celtic culture and, under Malcolm's rule, had nursed a hatred of the increasing English influence he saw in the Scottish court. On Malcolm's death Donald seized the throne, at the age of 62, and attempted to reverse the anglicization of the court. His position was soon threatened, however, by Malcolm's son, DUNCAN II, who had been trained as a Norman knight during his period of detention as a hostage in England. An invasion led by Duncan, with the backing of an English and French army, dethroned Donald in 1094 but Donald regained the throne when Edmund, another of Malcolm's sons, killed Duncan. Malcolm's other surviving son, EDGAR, then had Donald blinded and imprisoned in 1097. He died three years later and became the last Scottish king to be buried on Iona.

Donald Breac (d.642) King of Scots. Reigned c.635 to 642. The tenth king of Dalriada, he invaded Ireland in 636 and was soundly beaten at the battle of Magh Rath (Moira). This was said to have activated the curse of St Columba on Scots kings who fought their own kinsmen. After returning to Scotland he was killed in a battle with the Strathclyde Britons at Strathcarron in 642. These defeats led to a sharp decline in the influence of Dalriada in Scotland.

Drest I King of Picts. Reigned from 663 to 671. The brother of GARTNAIT, he was ousted by a faction of the Picts who resented the influence and expansionist policies of his uncle, EGFRITH, the Northumbrian king. He was succeeded by BRIDEI III who crushed Egfrith's army at Nechtansmere in 685.

Drest II (d.729) King of Picts. Reigned from 724 to 729. The son of NECHTON's sister, he was one of four who claimed the title of king in 724 following the decision of Nechtan to abdicate. He was killed by his cousin OENGUS in 729.

Drest III (d.780) King of Picts. Reigned in 780. The son of ALPIN II's sister.

Drest IV (d.837) King of Picts. Reigned from 834 to 837. He was the son of UEN.

Drust (d.848) King of Picts. Reigned from 845 to 848. One of the sons of UURAD.

Dubh (d.966) King of Scots. Reigned from 962 to 966. The son of MALCOLM I, he was killed in battle.

Duncan I (1010-1040) King of Scots. Reigned from 1034 to 1040. The son of Crinan, Abbot of Dunkeld, and Bethoc, daughter of MALCOLM II, Duncan succeeded his grandfather and founded the Dunkeld dynasty. When he became king of Scots he was already the king of Strathclyde and therefore inherited a kingdom larger than any held by his predecessors. A rash and hot-headed king, he was not particularly successful in battle and in 1039 fruitlessly besieged Durham. He was also twice defeated in battle by his cousin Thorfinn, Earl of Orkney, before being killed by a rival for the throne, MACBETH, *mormaer*, or steward, of Moray at Forres, near Elgin, in 1040. He married a cousin of the Earl of Northumberland and had two sons who both became kings: MALCOLM III CANMORE and DONALD BANE. After their father's murder the two brothers fled the kingdom.

Duncan II (1060-1094) King of Scots. Reigned in 1094. The eldest son of MALCOLM III from his first marriage. His father had given him as a hostage to WILLIAM II in 1072 and he grew up in Normandy before being set free and knighted in 1087. Soon after, he married Octreda of Northumberland. On the death of his father he was seen as the true heir, and with support from the English king, William II, under whose banner he was then serving, he led an Anglo-Norman army north. He defeated his uncle DONALD III BANE to become king, but his reign was short and difficult as there was much resentment of his English supporters. Within a matter of months he was slain in battle by his half-brother, Edmund, at Mondynes near Dunnotthr and Donald Bane regained power. Donald was soon after deposed by another of Malcolm's sons, EDGAR.

Dyfnwal (d.934) King of Strathclyde. Reigned c.920 to 934. He recognized the Wessex king, EDWARD THE ELDER, as overlord in 925.

Dyfnwal (d.975) King of Strathclyde. Reigned from 934 to c.973. He killed the king of Scots, CUILEAN, in battle in 971 and died whilst on a pilgrimage to Rome.

Eadbert (d.768) King of Northumbria. Reigned from 737 to 758. Came to the throne after the abdication of his cousin CEOLWULF. His brother, Egbert, had been made Archbishop of York, and they governed Church and state in union. In 740 he led a campaign against the Picts, and in his absence ETHELBERT, King of Mercia, ravaged parts of his kingdom. He soon recovered his lands and

went on to add parts of Strathclyde to his kingdom. In 756 he was defeated by the Strathclyde Britons, and two years later he resigned his crown in favour of his son, OSWULF. Soon after, he entered the monastery of St Peters at York and remained there until his death.

Eadbert I (d.748) King of Kent. Reigned from 725 to 748. Joint ruler of the kingdom with his brother, ETHELBERT II.

Eadbert II Praen (d.c.810) King of Kent. Reigned from 796 to 798. Despite having had received the tonsure, Eadbert strongly contested the Mercian domination of Kent, and on the death of OFFA he became king. He was supported by his nobles and condemned by the Church which favoured Mercia. In 798 COENWULF invaded Kent, captured the king, and caused him to have his hands cut off and his eyes burned out. Coenwulf then imposed CUTHRED as an under-king, and the independent existence of Kent was brought to an end.

Eadric (d.688) King of Kent. Reigned from c.685 to 687. Joint ruler with the East Saxon, SUABHARD, under the overlordship of the South Saxons.

Eadwig *see* **Edwy**.

Eanfrith King of Bernicia. Reigned from 633 to 634. The son of Ethelfrith, he married a Pictish princess, and their son, TALORCEN, became a king of the Picts. He was killed by the Welsh king, CADWALLON.

Eanred (d.850) King of Northumbria. Reigned from 809 to 841. A son of EARDWULF, he did homage to EGBERT of Wessex in 827.

Eardwulf (d.762) King of Kent. Reigned from 747 to 762. The son of EARDBERT, he was a joint ruler of the kingdom with ETHELBERT II.

Eardwulf King of Northumbria. Reigned from 796 to 809. He deposed OSBALD and was succeeded by his son, ENRED.

Edbald (d.640) King of Kent. Reigned 616 to 640. On succeeding his father, ETHELBERT, he promptly renounced Christianity, which had been introduced to the kingdom by St Augustine in his father's reign. He married his stepmother before taking a daughter of the Frankish King Theudebert II of Austrasia, as his queen.

Edgar (943-975) King of England. Reigned from 957 to 975. The second son of EDMUND I and Elgifu, he replaced EDWY as ruler of Northumbria and Mercia after nobles who were discontented with his brother transferred their allegiance. Edwy still held the lands south of the Thames, but these came into Edgar's kingdom following Edwy's early death. In 973 he was created 'Emperor of Britain' at a ceremony in Bath conducted by the

Archbishop of Canterbury. The same year he was supposedly rowed on the River Dee by six or eight kings as an act of subservience. These kings included Malcolm of Strathclyde, Kenneth II of Scots, Maccus of the Isle of Man and up to five Welsh kings. It is probable that this was a conference to discuss borders in which the Scots recognized English control over Bernicia in Northumberland in return for Edgar's acknowledgment of their rule of the Lothians.

In contrast to his brother he was known for his piety and reinstated Dunstan as Archbishop of Canterbury as well as making him his chief adviser. Acting under Edgar's patronage Dunstan led a revival of Benedictine monasticism and reformed the church. In order to secure the loyalty of his clerics forty abbeys were founded in his reign and laws introduced which punished non-payment of taxes due to the Church. He also reformed the administration of the country, codified the laws of the land and clarified the boundaries between shires. He also struck a new coinage and licensed towns to mint the new silver pennies. He married twice to daughters of his ealdormen and kept a mistress. His legitimate sons were EDWARD (the Marytr), Edmund (the Aetheling) and ETHELRED II (the Unready).

Edgar (1074-1107) King of Scots. Reigned from 1097 to 1107. The fourth son of MALCOLM III's second marriage. On his father's death he went to the English court where he was sheltered by WILLIAM II. He returned to Scotland in 1096 and with the help of English troops he defeated his uncle DONALD BANE at Roscobie in Fife in early 1097. He became king in October of the same year and afterwards was practically a dependant of William II and HENRY I of England. Shortly after he took the throne his kingdom was threatened by the King of Norway, Magnus Barelegs, who brought a considerable fleet into western waters and forced Edgar to cede 'all the isles around which a ship could sail', including Kintyre. Pursuing a pro-English policy, he settled the first English knight in Lothian and married his sister Matilda to Henry I in 1100. He died unmarried, and the kingdom passed to his brother ALEXANDER I. The 'king with the Saxon name' was buried in Dunfermline.

Edmund (St Edmund) (*c.*840-870) King of East Anglia. Reigned from *c.*855 to 870. East Anglia's last Anglo-Saxon king was defeated and martyred by Danish invaders at Haegelisdum (Hellesdon in Norwich or Hoxne in Suffolk) in 870. Seeking to avert war he had tried to negotiate with the Danes but his stipulation that they convert to Christianity before making peace offended them. As a result, on his capture he was subjected to the 'blood eagle' rite and beheaded. He became protector of sailors and for a time was England's patron saint.

Edmund I (921-946) King of England. Reigned from 939 to 946. The eldest son of EDWARD the Elder and Edgifu, he succeeded his half-brother ETHELSTAN as king. Having commanded well at Brunanburgh, Edmund reclaimed those parts of the kingdom lost after his brother's death to Norse King OLAF II. He also subdued Strathclyde, which he bestowed on MALCOLM I, King of Scots, on the condition of him doing homage for it. This secured the Anglo-Scottish border. He was slain by Leofa, an exiled thief, while keeping the feast of St Augustine of Canterbury (26th May) at Pucklechurch in 946 and was buried at Glastonbury. He married twice and had two sons who became kings: EDGAR and EDWY.

Edmund II Ironside (993-1016) King of England. Reigned for seven months in 1016. The eldest surviving son of ETHELRED II and Elfled. The warrior prince, nicknamed 'Ironside' for his bravery against the Danes, attempted to oppose CANUTE's invasion of Wessex in late 1015 and could not hold Northumbria when Canute moved against it early the following year. On his father's death Edmund was chosen king by a council of Anglo-Saxon kings and ealdormen (the *Witan*) in London and proclaimed in early 1016. Canute, however, had already been elected king by a majority of Witan members gathered at Southampton. Edmund marched into Wessex and won three of the four battles there, relieving a besieged London, but support from the Mercian king EDRIC did not arrive and his cause was lost. He was finally defeated at Assandun in Essex (1016) and forced to surrender the midland and northern counties to Canute after a meeting at Olney where the two rivals agreed to partition England. He died at London of natural causes (although later sources claim he was murdered) after a reign of only seven months and was buried at Glastonbury. His infant sons fled Canute's invasion and settled in Hungary.

Edred (923-955) King of England. Reigned from 946 to 955. The youngest son of EDWARD the Elder and Edgifu, succeeded to the throne on the murder of his brother, EDMUND I, in 946 as Edmund's two sons were too young too reign. Because of ill-health the government of the kingdom appears to have been carried out by his mother and his chief minister, the Abbot Dunstan. Despite his health,

Edred was able to quell a rebellion of the Northumbrian Danes under ERIK Bloodaxe (948) but it was not until 954 that Edred was able to secure Northumbria again as part of his kingdom. This he achieved through a bloody invasion which resulted in the death of the usurper king. Edred committed Northumbria to Oswulf as an earldom. He died while still in his early twenties and was buried at Winchester.

Edward the Confessor (St Edward) (1003-1066) King of England. Reigned from 1042 to 1066. The eldest son of ETHELRED II and Emma of Normandy, he was born at Islip, in Oxfordshire, and lived in exile in Normandy from 1013 to 1042 while the Danes ruled England. On the death of his maternal brother, HARDICANUTE the Dane, he was called to the throne and thus renewed the Saxon line.

He restored the Norman influence in England as, not unnaturally, he had returned more French than English, and brought with him his Norman clergy and supporters. He moved the royal residence from the walled city of London to the Palace of Westminster, and this, more than anything else, ensured the return of a Dane to the throne after his death. London had been the wealthiest city in the land and became the centre for discontent with his rule among the powerful anti-Norman party. The real power devolved to his father-in-law, Godwin, the Earl of Wessex, and Edward was forced to remove him from government and place his own Norman supporters in high office. After a dispute in Dover, Godwin left for France in 1051 and returned a year later to take London by force. After a period of confrontation Godwin was reconciled with the king, but died shortly after, leaving his son, HAROLD, to became the Earl of Wessex. Harold returned from exile in Ireland and had little difficulty in assuming his father's position of influence. During Godwin's time in exile, however, Edward had appealed for Norman help to reassert his authority and had promised the throne of England to William, Duke of Normandy (*see* WILLIAM (I) THE CONQUEROR), in return. There was little support for such a succession in England and on Edward's death in 1066 Harold took the throne, claiming that his succession had been accepted by the dying king some time beforehand. This was contested by William, who asserted that Edward had confirmed his earlier promise some two years before his death. The succession dispute led directly to the Norman conquest of England.

Edward was a weak and superstitious but well-intentioned king, who, despite his Norman upbringing, acquired the respect of his subjects by his monkish sanctity and care in the administration of justice. His legacy was the Abbey at Westminster, and he was canonized by Pope Alexander III in 1161.

Edward the Elder (870-924) King of Wessex. Reigned from 899 to 924. The son of ALFRED the Great, he inherited Wessex from his father and defeated a Danish-backed claimant for the throne. His reign was distinguished by a series of successes over the Danes as he took control of the Danish-held Five Boroughs (Nottingham, Derby, Lincoln, Leicester and Stamford). He fortified many inland towns (including Manchester in 919) and acquired dominion over Mercia, which had been allied to his father's kingdom. With Mercian support he extended his authority to run from the English Channel to the Humber. In the north he subdued the Strathclyde king, DYFNWAL, and several Welsh tribes who later sought his protection from the Norse. Among his many sons and daughters by three wives were ELFWARD, ETHELSTAN, EDMUND I and EDRED, who all became kings. He was buried at Winchester New Minster.

Edward the Martyr (St Edward) (963-978) King of England. Reigned from 975 to 978. The only son of EDGAR and his first wife, Ethelfled, he succeeded his father at the age of twelve. His succession was disputed by supporters of Edgar's second son by his second wife, ETHELRED, and he began his reign amid a power struggle. The opportunity was taken by some members of the royal court to regain the power lost when Edgar increased the land-holding and authority of the Church. Attacks on monasteries and Church property by secular landowners increased, especially in the north of the country where it was compounded by opposition to southern rule. Edward was guided by the powerful Archbishop Dunstan but seemed powerless to stop the seizures of monastic estates and other church lands.

He was treacherously slain after only three years on the throne by a servant of his stepmother at her residence, Corfe Castle in Dorset. Travelling alone to the castle, he was seized from behind and stabbed while waiting at the gate. It is generally held that his stepmother ordered the assassination in order to make Ethelred king. The pity caused by his innocence and misfortune induced the people to regard him as a martyr, and miracles supposed to occur at his tomb later led him to be venerated as a saint

Edward I (Hammer of the Scots) (1239-1307) King of England, Wales, Scotland and Ireland. Reigned from 1272 to 1307. The eldest surviving son of HENRY III and Eleanor of Provence. The contests between his father and the barons called him early into active life. By 1265 Simon de Montfort had become leader of the opposition to King Henry and formed a Parliament which represented not only the knights of the shires but the burgesses of the towns that supported him as well. Prince Edward restored the royal authority within months by defeating and killing de Montfort at the Battle of Evesham. He then proceeded to Palestine, where he showed great valour, although no conquest of any importance was achieved. He returned on his father's death after further campaigns in Italy and France with a reputation as an excellent soldier and was crowned amid much public rejoicing at Westminster Abbey in 1274. The new king was immediately popular among the people as he identified himself with the growing tide of nationalism that was sweeping the country. The other side to this popular nationalism was displayed later in his persecution and banishment of the Jews which was the culmination of many years of anti-semitism in England. The spirit of nationalism also led to England looking to its borders. The mountainous land to the west had never been completely subdued, and following an uprising against English influence Edward commenced a war with LLEWELLYN, Prince of Wales, which ended in the annexation of that Principality to the English Crown in 1283. He secured the new Principality by building nine castles along the border and created his eldest son, Edward, Prince of Wales in 1301.

From the earliest days of his reign Edward showed great vigour as well as a degree of severity in his administration, especially in his policy of limiting the encroachment of the barons and the Church. This was achieved by restricting baronial privileges and prohibiting gifts of land to the Church. His harsh treatment of those in power whom he found to be corrupt also gave him the support of the common people, and his reforms of the administration were brought about only with their backing. Under his guidance the great common law courts consisting of the King's Bench, Exchequer and Common pleas took shape. He also called the Model Parliament which, with nobles, Churchmen and commoners, foreshadowed representative government.

Edward's great ambition, however, was to gain possession of Scotland, but the death of MARGARET, the Maid of Norway, who was to have been married to Edward's son, for a time frustrated the king's designs. The contested succession

soon gave him the opportunity to intervene, however, and he was invited by the Scots to choose between the thirteen competitors for the throne. His choice, John BALLIOL, was induced to do homage for his crown to Edward at Newcastle but was forced by the indignation of the Scottish people to throw off Edward's overlordship. An alliance between the French and the Scots followed, and Edward, then at war with the French king over possession of Gascony, marched his army north. He entered Scotland in 1296, devastated it with fire and sword, which earned him the sobriquet 'Hammer of the Scots', and removed the symbolic Stone of Destiny from Scone.

Edward assumed the administation of the country, but the following summer a new rising took place under William Wallace, the son of a knight. His successes, notably at Stirling Bridge, recalled Edward to Scotland with an army of 100,000 men. He defeated Wallace's army at Falkirk, and the supporters of Scottish independence went into hiding, but their leader was at length betrayed and executed in London as a traitor. The unjust and barbaric execution of Wallace made him a national hero in his homeland, and resistance to England became paramount among the people. All Edward's efforts to reduce the country to obedience were unavailing, and with the crowning of Robert Bruce, Earl of Carrick, as ROBERT I of Scotland, the banner of independence was again unfurled. In 1306 an enraged Edward assembled another army and marched against Bruce but only reached Burgh-on-Sands, a village near Carlisle, where he died.

He was married twice: to Eleanor of Castile, by whom he had sixteen children, and Margaret of France by whom he had three. Twelve memorials to his first wife stand between Nottingham and London to mark the journey taken by her funeral cortege. He was buried at Westminster Abbey.

Significant events of Edward I's reign

- 1277 – Edward mounts an invasion of Wales.
- 1282 – Llywelyn, the last independent Prince of Wales, is killed at Builth.
- 1284 – The Statute of Rhuddlan brings Wales under English rule.
- 1285 – The first Justices of the Peace installed.
- 1290 – England banishes the Jews. Margaret, Maid of Norway, dies before reaching Scotland.
- 1292 – John Balliol is chosen by Edward to become Scotland's king.
- 1295 – The Model Parliament is assembled. Edward mounts an invasion of France.

- 1296 – The Scots are vanquished by Edward's invading army and Balliol is deposed.
- 1297 – William Wallace gains victory over Edward at Stirling Bridge.
- 1298 – Wallace defeated at Battle of Falkirk.
- 1301 – Edward makes his son, Edward, the Prince of Wales.
- 1305 – Wallace is executed in London.
- 1306 – Robert Bruce becomes King of Scots.

Edward II (1284-1327) King of England and Wales. Reigned from 1307 to 1327. The only surviving son of EDWARD I and Eleanor of Castile, he was born at Caernarvon Castle and became the first English Prince of Wales. He succeeded his father in 1307, and was crowned at Westminster Abbey the same year. He was of a mild disposition, but indolent and fond of pleasure. With little contact with his father and surrounded by sisters, he became very reliant on his friends and fiercely loyal to them, regardless of court or public opinion. As king, this weakness for his personal favourites, Piers Gaveston and, later, the two Hugh Despensers, father and son, infuriated the powerful nobles of the land. By the Ordinances of 1311 the barons forced Edward to banish Gaveston and executed him as a public enemy when he disobeyed and returned for the second time.

Two years after this Edward assembled an immense army to check the progress of ROBERT I in Scotland who had been threatening an invasion of England. With superior tactical awareness, Bruce routed the assembled feudal forces of Edward at Bannockburn, near Stirling Castle. Barely escaping the field with his life, it soon became apparent to Edward that his father's plans for a united kingdom stood little chance of success.

Over the next few years his problems extended from France, where his Duchy of Aquataine was overrun by French soldiers, to Ireland, where Bruce had devastated the countryside and threatened the prosperity of the English of the Pale. By 1320 the king also had new favourites, the two Hugh Despensers. Their influence over the king provoked a rebellion by the barons, led by the powerful Earl of Lancaster, which was defeated at Boroughbridge in Yorkshire. Finally, in 1326 the exiled baron Roger Mortimer invaded England with his mistress, Edward's estranged wife, Isabella, to seize power. Their army was completely successful, and Edward was deposed. The Despensers, father and son, were captured and executed. Edward was imprisoned in Kenilworth, and then Berkeley Castle in Gloucestershire, and eventually mur-

dered on the orders of Isabella and Mortimer. His death was particularly gruesome as the order for his execution stipulated that no external marks should be left which would betray violence. The only way to do this was by disembowelment with a red hot iron inserted into the rectum, a conventional form of death for homosexuals at the time.

He had four children by Isabella, and his son, EDWARD, became king after him. He was buried at Gloucester Cathedral.

Significant events of Edward II's reign
- 1308 – Piers Gaveston is exiled for the first time.
- 1312 – Gaveston is put to death.
- 1314 – Edward is defeated by Robert the Bruce at Bannockburn.
- 1320 – Hugh Despenser and his son receive the favour of the king.
- 1322 – The Barons' revolt is defeated at Boroughbridge.
- 1327 – Isabella and Mortimer depose the king. The Despensers are executed.

Edward III (1312-1377) King of England and Wales. Reigned from 1327 to 1377. The eldest son of EDWARD II and Isabella of France, he was born at Windsor Castle and succeeded to the throne at the age of fourteen. During his minority, a council was elected to govern, but his mother's lover, Roger Mortimer, possessed the principal power in the state. The pride and oppression of Mortimer led to a general confederacy against him, and to his seizure and execution in 1330. Isabella received a yearly pension and quietly retired from public life.

After many years of domestic squabbles, Edward was finally in a position to improve England's international standing. First he turned his attention to Scotland. His claimant for the Scottish throne, Edward BALLIOL, the son of John BALLIOL, defeated DAVID II's army and seized the throne, forcing the Scots king into exile. A Scots army then took Balliol by surprise at Annan in Dumfriesshire and expelled him over the border. Edward, having levied a well-appointed army, invaded Scotland and defeated David II's regent, Donald, at Halidon Hill. This victory produced the restoration of Edward Balliol, who was, however, again expelled.

The ambition of the English king was diverted from Scotland by the prospect of succeeding to the throne of France. To this end Edward initiated the Hundred Years' War in 1337 which was to last intermittently until 1453. Collecting an army and accompanied by his son, the Black Prince, he crossed over to France. There he devastated the northern and eastern territories and declared himself King of France. Memorable victories followed in the Battle of Sluys, in the Battle of Crécy and at the siege of Calais. In the meantime David II, having recovered the throne of Scotland in 1346, invaded England with a large army. The campaign was a disaster, however, and he was defeated and taken prisoner at Neville's Cross, near Durham, by a much inferior force.

In 1348 a truce was concluded with France; but on the death of King Philip, in 1350, Edward again invaded France, plundering and devastating. Recalled home by a Scottish inroad, he retaliated by carrying fire and sword from Berwick to Edinburgh. In the meantime the Black Prince had penetrated from Guienne to the heart of France, fought the famous battle of Poitiers, and taken King John II prisoner. A truce was then made, the Treaty of Brétigny, which gave Edward possession of Calais, Guienne, Gascony and Poitou in return for giving up his claim to the throne. When King John died, however, and Charles V became king of France, the two countries resumed hostilities. Edward again crossed over to France and laid waste the provinces of Picardy and Champagne, but at length consented to a peace. This confirmed him in the possession of several provinces and districts of France which were entrusted to the Prince of Wales (the Black Prince), but gradually all the English possessions in France, with the exception of Bordeaux, Bayonne, and Calais, were lost.

The Black Prince died in 1376 and Edward, suffering in his later years from senile dementia, died the following year. He had 13 children by Philippa of Hainault.

Significant events of Edward III's reign

- 1330 – Mortimer is put to death.
- 1332 – Parliament is divided into the two Houses of Lords and Commons for the first time.
- 1333 – David II defeated at Halidon Hill.
- 1337 – The Hundred Years' War commences.
- 1340 – French navy is defeated at Battle of Sluys.
- 1346 – David II is captured at Neville's Cross.
- 1346 – The French are routed at the Battle of Crécy.
- 1347 – Calais is taken by England.
- 1349 – The Black Death reaches England.
- 1356 – The French are defeated at Poitiers.
- 1357 – David II is released from captivity.
- 1362 – English replaces French as the official language of government and the courts.
- 1366 – The Statute of Kilkenny imposes English law in Ireland.
- 1376 – The Black Prince dies.

Edward IV (1442-1483) King of England and Wales. Reigned from 1461 to 1483. The son of Richard Plantagenet, Duke of York, and grandson of Edmund, Earl of Cambridge and Duke of York, fourth son of EDWARD III, he became the first Yorkist king after ousting HENRY VI in the dynastic civil wars, later called the Wars of the Roses. The rival line of Lancaster descended from John of Gaunt, the third son of Edward III. The York line had intermarried with the female descendants of Lionel, the second son, which gave it the preferable right to the Crown. Before reaching his twenties Edward led an army against the Lancastrian supporters of Henry VI and defeated them at the Battle of Mortimer's Cross. This avenged the death of his father, 'the Protector', Richard, Duke of York, a claimant to the throne. He was then proclaimed king by his cousin Warwick 'the Kingmaker', and drove Henry north. Edward owed Warwick his crown, however, and hostility soon began to develop between them; Edward had allowed Warwick to govern the kingdom, but his marriage to Elizabeth Woodville, the widow of a commoner and daughter of Sir Richard Neville, caused a rift as Warwick felt his position to be threatened. Warwick rebelled against his king, defeated him at Edgecote and left England after a brief period of reconciliation. On his return he allied with the wife of Henry VI, Margaret, to restore the deposed king. Their army caused Edward to flee the realm but he returned the following year to defeat and kill Warwick at the Battle of Barnet. Margaret was also soon defeated at the Battle of Tewkesbury, and her son, Edward, was captured and killed. Henry VI was held in the Tower of London, where he was later murdered on Edward's orders.

Once he was restored, Edward secured his throne from any further Lancastrian attack by quashing rebellions in the north and began to prove himself as an able ruler in his own right. He set about improving the royal finances, which had suffered greatly under Henry VI (he was the first monarch to be solvent at his death for over three hundred years), and establishing good trading relationships abroad. He endeavoured to keep his country out of foreign entanglements and after a short campaign in France withdrew with payments. Recognizing the value of wool and cloth, he worked hard to improve trade with German cities, and England enjoyed a period of much greater prosperity in the second half of his reign than it had in the first. Law enforcement was similarly improved, and he won the respect of his commoners by establishing the

Court of Requests, which heard complaints against greedy landlords.

The dynastic disputes were not completely forgotten, however, and as the Lancastrians were no longer a threat, York turned against York. Edward's two younger brothers, the Dukes of Gloucester and Clarence, both had their eyes set on the throne, and Clarence, who had at one time allied himself with the Lancastrians, was accused of treason. He was sent to the Tower, where he was later found murdered. After Edward's death Gloucester claimed the throne and had himself crowned as RICHARD III.

Edward had ten children by Elizabeth Woodville and was succeeded by EDWARD V, who, along with his brother Richard, was declared to be illegitimate and deprived of the throne.

Significant events of Edward IV's reign
- 1461 – Edward defeats the Lancastrians at Mortimer's Cross.
- 1464 – Edward is married to Elizabeth Woodville.
- 1469 – Warwick deposes Edward.
- 1470 – Henry VI regains the crown.
- 1471 – Edward restored.
- 1476 – William Caxton brings the printing trade to England.
- 1471 – Duke of Clarence murdered in the Tower.

Edward V (1470-1483) King of England and Wales. Reigned for 77 days in 1483. The eldest son of EDWARD IV and Elizabeth Woodville, he became Prince of Wales in 1471 and was in his thirteenth year when he succeeded his father. Within weeks of becoming king he fell victim to an uncle's ambitions. Richard of Gloucester had been appointed by his dying brother, Edward IV, as Protector of the kingdom during his heir's minority. He had resented the king's marriage to Elizabeth Woodville, however, and sought to become king himself. The young prince had been brought up with his brother, Richard, under the power of the Woodville family at Ludlow Castle on the Welsh border. Suspecting that the Woodvilles would remove him from his office as Protector, Gloucester ordered that the senior members be arrested; Edward's grandfather, Earl Rivers, and an uncle were killed and his mother forced to seek shelter in Westminster Abbey. Gloucester then placed Edward and his younger brother in the Tower of London, which at that time was a royal residence as well as a prison.

Before the coronation could take place Gloucester declared that the two princes were illegitimate.

He had been informed by the Bishop of Bath and Wells that when Edward IV had married their mother he had been betrothed to Lady Eleanor Butler. A betrothal constituted the same commitment as marriage, and consequently Edward IV's marriage was declared void and his sons illegitimate. Parliament had little choice but to agree, and soon after Edward should have been crowned, Gloucester was proclaimed king and took the title RICHARD III.

Edward and Richard vanished, and although no evidence has ever been found, they were undoubtedly murdered in the Tower. Two skeletons were found in 1674 and given a full forensic examination in 1933, but their identity has never been properly established. They were most probably killed by Richard's close associate, Henry Stafford, the Duke of Buckingham, or by Henry Tudor, 2nd Earl of Richmond.

Edward VI (1537-1553) King of England and Ireland. Reigned from 1547 to 1553. The only son of HENRY VIII by Jane Seymour, was born at Hampton Court Palace and on his father's death was only nine years of age. He was too young to shape government and during his minority this was done by two Protectors: firstly Edward Seymour, Duke of Somerset, and then John Dudley, Duke of Northumberland. He grew up with a rooted zeal for the doctrines of the Reformation and made the Catholic mass illegal by the Act of Uniformity in 1549. He also ordered that icons and statues of saints be removed and destroyed and that church walls be whitewashed to cover up paintings. Objections to his reforms led to widespread disquiet and in Devon and Cornwall to a revolt which was put down with severity. His reign also produced the first Book of Common Prayer written by Thomas Cranmer, the Archbishop of Canterbury, and the later named Thirty-Nine Articles of Religion.

His reign was, on the whole, tumultuous and unsettled. In an attempt to secure Scotland an English army invaded and defeated the Scots at the Battle of Pinkie. Despite this victory, the 'Rough Wooing', as the attempt to marry Mary Queen of Scots was later called, failed. In 1551 the Protector Somerset, who had hitherto governed the kingdom with energy and ability, was deposed by the intrigues of Dudley, Duke of Northumberland, who became all-powerful. Somerset was executed two years later. At the end of 1552 Edward contracted tuberculosis of which he would die the following year. Dudley induced the dying Edward to set aside the succession of his sisters, MARY and ELIZABETH, and settle the crown upon Lady Jane

GREY, to whom he had married his son Lord Guildford Dudley. The king died, aged 15, at Greenwich Palace in 1553, and Lady Jane was proclaimed queen, though her reign was to be very brief.

Significant events of Edward VI's reign
- 1547 – The Duke of Somerset is appointed Protector of England. Scots defeated at the Battle of Pinkie.
- 1549 – The First Act of Uniformity passed.
- 1550 – John Dudley deposes the Duke of Somerset and becomes Protector.
- 1553 – Lady Jane Grey named as successor to the throne.

Edward VII (1841-1910) King of the United Kingdom of Great Britain and Ireland and British Dominions overseas; Emperor of India. Reigned from 1901 to 1910. The eldest son of Queen VICTORIA and Prince Albert of Saxe-Coburg-Gotha, he was born at Buckingham Palace and created Prince of Wales in 1841. He was educated under private tutors and at Edinburgh, Oxford and Cambridge. He visited Canada and the United States in 1860 and underwent military training at the Curragh camp in 1861. Promoted to the rank of general in 1862, he visited Palestine and the East, and next year took his seat in the House of Lords. In 1863 he was married in St George's Chapel, Windsor Castle, to Princess Alexandra, eldest daughter of Christian IX of Denmark, and from this time onwards he discharged many public ceremonial functions. Many of these duties he undertook in the place of Victoria who felt she could not trust him with domestic political affairs. This strained relationship meant that he was not given access to any state papers, cabinet reports or diplomatic correspondence. Excluded from his mother's circle of advisers, he became more frequently seen at society events, and, although he appeared to remain happily married, it was known that he had many affairs with actresses and society beauties (he had 13 known mistresses). As well as a socialite, he was a keen sportsman and enjoyed gambling, shooting and yachting. He also took a keen interest in horse-racing (he owned one Grand National and three Derby winners), motor cars (all his cars displayed the royal coat-of-arms on the sides) and the theatre, which benefited greatly from his patronage. He also had a very strong attachment to France, which, while he was Prince of Wales, very much annoyed his mother who preferred Prussia. By the end of Victoria's reign the diplomatic relationship with France was strained by territorial disputes. Soon after becoming king,

however, Edward met the French President, Emile Loubert, in an effort to improve relations. This friendly meeting laid the foundation of the *Entente Cordiale* of the following year, which settled many of the old disputes.

He succeeded to the throne on the death of Victoria in 1901 and was crowned the following year at the age of 59. A well-loved king, Edward did much to popularize the monarchy in his nine-year reign. His love of foreign travel and public ceremonial established a more ambassadorial style of monarchy, which came to replace the traditional political role of the head of state. His death from bronchitis in 1910 was felt keenly by all strata of society.

Significant events of Edward VII's reign
- 1901 – Australia is granted Dominion status.
- 1902 – The Order of Merit is created by Edward.
- 1903 – The first flight is made by Wilbur and Orville Wright. The Women's Social and Political Union formed by Emmeline Pankhurst.
- 1904 – *Entente Cordiale* is reached between Britain and France.
- 1907 – *Entente Cordial* reached between Russia and Britain.
- 1908 – The 4th Olympic Games are held in London. *Triple Entente* is reached between Britain, France and Russia.
- 1909 – Old age pensions introduced. Parliament Bill introduced to curb the power of the House of Lords.

Edward VIII (1894-1972) King of Great Britain and Ireland and British Dominions overseas; Emperor of India. Reigned in 1936 for 325 days (uncrowned). The son of George, Duke of York (GEORGE V) and Princess Mary of Teck. He was educated in England and France and joined the army on the outbreak of World War I. When it was over he made several extended tours of Europe. He was popular with the public for his great charm and concern for the plight of the unemployed during the recession. Ceremonial duties bored him, however, and he was notoriously bad at keeping appointments. His private life met with the disapproval of his father, and before his coronation he made it clear he intended to marry Mrs Wallis Simpson, an American who in 1935 was embarking on her second divorce. This was opposed by the Archbishop of Canterbury and the Prime Minister, Baldwin, who held a similar opinion of Edward's private life as Edward's father had. Rather than force a constitutional crisis, the King decided to abdicate. Edward and Mrs Simpson were created Duke and Duchess of Windsor when they married at Chateau Conde, near Tours, in 1937 and, apart from the war years when he served as Governor of the Bahamas, they lived in France. He remained on good terms with his royal relatives, but Mrs Simpson was not accepted until 1967 when she met Queen Elizabeth II for the first time in public. The Duke died in 1972; the Duchess in 1986. They had no children.

Edwin (d.633) King of Northumberland. Reigned from 616 to 633. The heir to the kingdom of Bernicia, he formed an alliance with King REDWALD of East Anglia on his return from exile and defeated his rival, King ETHELFRITH, in battle on the River Idle. He became *Bretwalda*, overlord of all Anglo-Saxon kings. The Northumbrian king married Ethelburga, daughter of Ethelfrith in 625 and embraced Christianity, being baptized at York in 627. He was finally defeated in 633 at Hatfield Chase by an alliance of the Welsh and the Mercians, under King PENDA. The city of Edinburgh derives its name from being Edwin's northern outpost.

Edwy or **Eadwig** (940-959) King of England. Reigned from 955 to 959. The eldest son of EDMUND I and Elfgifu, he succeeded his uncle EDRED to the throne in 955 at the age of fifteen. He promptly earned himself a reputation as a corrupt and incompetent ruler by leaving the coronation feast with two women. It is said that he was dragged from his bedchamber by Dunstan, the Abbot of Glastonbury, and forced to return to the table. Edwy later exiled Dunstan for his pains. His unpopularity led the Northumbrians and Mercians to renounce their allegiance to Edwy in favour of his brother, EDGAR, in 957. From then on he ruled only over the area south of the Thames.

Egbert I (d.873) King of Bernicia. Installed as ruler of this kingdom in Northumbria by the Vikings in 867.

Egbert II (d.878) King of Bernicia. Reigned from 876 to 878. The last of the Viking-installed rulers.

Egbert I King of Kent. Reigned from 664 to 673. He extended his kingdom to include Surrey.

Egbert II King of Kent. Reigned from 765 to 780. Failed to win independence from the Mercian overlords at the Battle of Otford in 766.

Egbert (Egbert III of Kent) (775-839) King of Wessex. Reigned from 802 to 839. His early years were spent in exile at the court of Charlemagne. As king of Wessex he defeated BEORNWULF of Mercia and was recognized by Northumbria in 829 as *Bretwalda*, or overall ruler of all Anglo-Saxon kings in England. He made Wessex the leading kingdom in the land and laid the basis for future unification. Under King WIGLAF, however, Mercia re-established independence in 830, and thereafter Egbert was effective ruler of Wessex only and its dependent kingdoms of Surrey, Sussex, Kent and Essex.

Egfrith (d.796) King of Mercia. Reigned from 787 to 796. The son of the powerful OFFA, he survived his father by only 141 days, having ruled jointly with him since 787.

Egfrith (d.685) King of Northumbria. Reigned from 670 to 685. The son of Oswy and Eanfled of Deira. On succeeding to the throne he consolidated his kingdom by driving the Mercians back across the Humber. In the north his expansionist policies were strongly resisted by the Picts, and in 685 his army was destroyed at Nechtansmere, near Forfar in Angus. He was killed in the battle.

Egric (d.637) King of East Anglia. Reigned from 634 to 637. A kinsman who took the throne when SIGEBERHT became a monk and entered a monastery. He was killed fighting PENDA during a Mercian invasion.

Elfward (d.924) King of Wessex. Reigned in 924. An illegitimate son of EDWARD THE ELDER, his reign lasted only a few months. His half-brother ETHELSTAN succeeded him.

Elfwold (d.788) King of Northumbria. Reigned from 779 to 788. The grandson of King EADBERT, he deposed ETHELRED to take the throne. He was murdered by a supporter of the rival dynasty and buried in Hexham church.

Elhred King of Northumbria. Reigned from 765 to 774. Claiming to belong to the Bernician royal house he deposed ETHELWALD in 765. After suffering the same fate and fleeing the kingdom he was succeeded by ETHELRED, the son of Ethelwald.

Elizabeth I (1533-1603) Queen of England and Ireland. Reigned from 1558 to 1603. The only child of HENRY VIII and Anne Boleyn, she was born at Greenwich Palace and almost immediately declared heiress to the crown. After her mother had been beheaded (1536) both she and her half-sister MARY, LATER MARY I were declared illegitimate, and she was finally placed after her half-brother, Edward, and the Lady Mary in the order of succession. On the accession of Edward as Edward VI Elizabeth was committed to the care of the Queen-Dowager Catherine; and after the death of Catherine and execution of her consort, Thomas Seymour, she was closely watched at Hatfield, where she received a classical education under William Grindal and Roger Ascham.

At the death of Edward, Elizabeth vigorously supported the title of Mary against the pretensions of Lady Jane GREY, but continued throughout the whole reign to be an object of suspicion and surveillance. In self-defence she made every demonstration of zealous adherence to the Roman Catholic faith, but her inclinations were well known. When Mary died, Elizabeth was immediately recognized as queen by Parliament. The accuracy of her judgement showed itself in her choice of advisers, Parker, a moderate divine (Archbishop of Canterbury 1559), aiding her in ecclesiastical policy; while William Cecil assisted her in foreign affairs. The first great object of her reign was the settlement of religion, to effect which a Parliament was called on 25th January and dissolved on 8th May, its object having been accomplished. The nation was prepared for a return to the Reformed faith, and Parliament was at the bidding of the Court. The ecclesiastical system devised in her father's reign was re-established, the royal supremacy asserted, and the revised Prayer Book enforced by the Act of Uniformity. While, however, the formal establishment of the reformed religion was easily completed, the security and defence of the settlement was the main object of the policy and the chief source of all the struggles and contentions of her reign. Freed from the tyranny of Mary's reign, the Puritans began to claim predominance for their own dogmas, while the supporters of the Established Church were unwilling to grant them even liberty of worship. The Puritans, therefore, like the Catholics, became irreconcilable enemies of the existing order, and increasingly stringent measures were adopted against them. But the struggle against the Catholics was the more severe, chiefly because they were supported by foreign powers so that while their religion was wholly prohibited even exile was forbidden them in order to prevent their intrigues abroad. Many Catholics, particularly priests, suffered death during her reign; but simple nonconformity, from whatever cause, was pursued with the severest penalties, and many more clergymen were driven out of the Church, by differences about the position of altars, the wearing of caps, and such like matters, than were forced to resign by the change from Rome to Reformation.

Elizabeth's first Parliament approached her on a subject which, next to religion, was the chief trouble of her reign, the succession to the Crown. They requested her to marry, but she declared her intention to live and die a virgin, and she consistently declined in the course of her life such suitors as the Duke of Alençon, Prince Erik of Sweden, the Archduke Charles of Austria, and Philip of Spain.

While, however, she felt that she could best maintain her power by remaining unmarried, she knew how to temporize with suitors for political ends and showed the greatest jealousy of all pretenders to the English succession.

With the unfortunate MARY, QUEEN OF SCOTS, were connected many of the political events of Elizabeth's reign. On her accession the country was at war with France. Peace was easily concluded (1559); but the assumption by Francis and Mary of the royal arms and titles of England led to an immediate interferene on the part of Elizabeth in the affairs of Scotland. She entered into a league with the Lords of the Congregation, or leaders of the Reformed party, and throughout her reign this party was frequently serviceable in furthering her policy. She also gave early support to the Huguenot party in France, and to the Protestants in the Netherlands, so that throughout Europe she was looked on as the head of the Protestant party. This policy aroused the implacable resentment of Philip, who strove in turn to excite the Catholics against her both in her own dominions and in Scotland. The detention of Mary in England (1568-87), where she fled to the protection of Elizabeth, led to a series of conspiracies, beginning with that under the Earls of Northumberland and Westmoreland, and ending with the plot of Babington, which finally forced the reluctant Elizabeth to order her execution. Mary's death (1587), though it has stained Elizabeth's name to posterity, tended to confirm her power among her contemporaries. The state of France consequent to the accession of Henry IV, who was assisted by Elizabeth, obviated any danger from the indignation which the execution had caused in that country; and the awe in which King James VI stood of Elizabeth and his dread of interfering with his own right of succession to England made him powerless.

Philip of Spain was not to be so appeased, the execution of Mary lending edge to other grievances. The fleets of Elizabeth had galled him in the West Indies and her arms and subsidies had helped to deprive him of the Netherlands. Soon his Armada was prepared, ready to sail for England. Accordingly he called the Queen of England a murderess and refused to be satisfied even with the sacrifice she seemed prepared to make of her Dutch allies. The Armada sailed in 1588, but the great naval force was broken up by English fireships and sent in retreat up the east coast and round the rocky shoreline of northern Scotland. It is not known how many of the ships returned to Spain, but at least one-third of the crews, 11,000 men, were lost at sea. The war with Spain dragged on until the close of Elizabeth's long reign.

During her reign the splendour of her government at home and abroad was sustained by such men as Cecil, Bacon, Walshingham and Throgmorton, but she had personal favourites of less merit who were often more brilliantly rewarded. Chief of these were Dudley, whom she created Earl of Leicester, and whom she was disposed to marry, and Essex, whose violent passions brought about his ruin when he rebelled against the government. He was beheaded in 1601, but Elizabeth never forgave herself his death. She died two years later of blood poisoning from a tonsillar abscess, having named JAMES VI of Scotland as her successor.

Significant events of Elizabeth I's reign

- 1558 – Cecil appointed Chief Secretary of State.
- 1559 – Matthew Parker appointed Archbishop of Canterbury. Elizabeth becomes head of the English Church via the Act of Supremacy.
- 1560 – Treaty of Berwick promises Scottish Protestants English aid against French. William Cecil appointed Chief Secretary of State.
- 1563 – 15,000 die in the Plague of London.
- 1568 – Mary, Queen of Scots, is imprisoned by Elizabeth.
- 1580 – Francis Drake returns to England having circumnavigated the world.
- 1586 – Mary, Queen of Scots, stands trial for treason
- 1587 – Mary, Queen of Scots, is executed.
- 1588 – The Spanish Armada is destroyed.
- 1601 – Earl of Essex is executed.

Elizabeth II (1926-) Queen of the United Kingdom of Great Britain and Northern Ireland and other realms and territories; Head of the Commonwealth and Head of State for sixteen of its members. The present queen was born at 17 Brunton Street, London, the eldest daughter of King GEORGE VI and Lady Elizabeth Bowes-Lyon. She was privately educated and at the outbreak of World War Two, under the threat of bombing, the princess and her sister Margaret were moved from London to Windsor Castle (Buckingham Palace was bombed on 12 September 1940). As a girl she had no expectations of becoming queen, but the abdication of EDWARD VIII meant her father reluctantly became George VI. In 1944 at the age of 18 she trained with the Auxiliary Transport Service, becoming a capable driver. She married in 1947 Philip Mountbatten (son of Prince Andrew of Greece and made Duke of Edinburgh on his marriage) and has four children; Charles, born 14th

November 1948 (made Prince of Wales in 1958), Anne, born 15th August 1950 (made Princess Royal in 1987), Andrew, born 19th February 1960 (later to become Duke of York), and Edward, born 10th March 1964. Elizabeth acceded to the throne on the death of her father on 6 February 1952 (whilst on tour in Africa) aged 25 and was crowned at Westminster Abbey on 2nd June 1953.

The Queen's full title in the United Kingdom is: 'Elizabeth the Second, by the Grace of God, of the United Kingdom of Great Britain and Northern Ireland and of Her other Realms and Territories Queen, Head of the Commonwealth, Defender of the Faith'. Her right to this title derives from the common-law rules of heredity and from legislation such as the Act of Settlement made in 1700 which states that only Protestant descendants of Electress Sophia of Hanover, granddaughter of JAMES VI, may succeed to the throne. The succession can only be changed if all the members of the Commonwealth that recognize the Queen as sovereign consent to the change. Only if this happens can someone other than the eldest son of the sovereign succeed to the throne.

As head of state of one of the last remaining constitutional monarchies, Queen Elizabeth's duties include the opening of Parliament and, as commander of the British armed forces, the inspection of the Trooping of the Colour. She also has the authority to pardon criminals, appoint government ministers and judges, and as Head of the Church of England it is she who appoints bishops. All these decisions are made, however, only with the advice of her government. In her long reign she has seen several Prime Ministers come and go, and all cabinet papers pass before her, as does all important diplomatic correspondence. She also sees the Prime Minister once a week and gives advice on government affairs which is rarely ignored.

A hard-working ambassador for the United Kingdom, the Queen has made state visits worldwide, a feature of these being her informal 'walkabouts', which enable her to have direct contact with her public. During her reign she has done much to bring the monarchy closer to the people, but as a result she and her family have been subjected to almost constant invasions of privacy by the world's media. The strain of being the most photographed family in the world has taken its toll on the relationships and marriages of the younger Windsors.

Significant events of the reign

- 1955 – Churchill resigns as Prime Minister and

Anthony Eden takes his place. Nationalization of the Suez Canal.

- 1957 – Macmillan becomes Prime Minister.
- 1960 – Nigeria and Cyprus gain independence.
- 1963 – Macmillan government collapses.
- 1964 – Harold Wilson becomes Prime Minister.
- 1968 – Cunard liner *Queen Elizabeth II* launched.
- 1969 – Charles invested as Prince of Wales. British troops deployed in Ulster to control sectarian disturbances.
- 1970 – Edward Heath becomes Prime Minister.
- 1973 – Britain joins the European Community.
- 1974 – Harold Wilson returns as Prime Minister.
- 1979 – Margaret Thatcher becomes first British woman Prime Minister.
- 1980 – Rhodesia gains independence as Zimbabwe.
- 1981 – Prince Charles marries Lady Diana Spencer.
- 1982 – Britain goes to war with Argentina over control of the Falkland Islands.
- 1990 – Rioters in London protest against the Poll Tax.
- 1991 – Gulf War threatens Middle East relations.
- 1992 – Windsor Castle damaged by fire. Charles and Diana announce separation.
- 1995 – The Queen agrees to pay tax on private income.
- 1994 – The Queen visits Russia.
- 1995 – The Queen visits South Africa.
- 1996 – Charles and Diana divorce.

Enfrith (d.633) King of Bernicia. Reigned in 633. The eldest son of ETHELFRITH, he married a Pictish princess but was killed after reigning for less than a year by the Welsh king, CADWALLON, at a battle near Doncaster.

Enred (d.*c*.841) King of Northumbria. Reigned from 809 to *c*.841. A son of EARDWULF, he succeeded his father in 809 and in 829 made a formal submission to EGBERT of Wessex. He maintained his kingdom without recorded incident for more than thirty years.

Eocha (d.889) King of Scots. Reigned from 878 to 889. The son of RUN, King of Strathclyde, and grandson of KENNETH mac-Alpin. He ruled jointly with his cousin, GIRIC I, the son of DONALD I, before being deposed.

Eochaid I (the Yellow-Haired) (d.*c*.629) King of Scots. The son and successor of AIDEN.

Eochaid II (d.*c*.679) King of Scots. Also known as 'Eochaid the Crook-Nose'.

Eochaid III (d.*c*.733) King of Scots. The last to rule in Irish Dalriada.

Eochaid IV (the Venomous) (d.*c*.737) King of Scots. Reigned between 733 and 737. He married a Pictish princess and his son, ALPIN, was the father of KENNETH MAC-ALPIN who became the first king of the Dalriadan Scots and the Picts.

Eormenric (d.*c*.560) King of Kent. Reigned from *c*.540 to 560. A son of AESC.

Eorpwold (d.627) King of East Anglia. Reigned from *c*.617 to 627. A son of REDWALD, he was converted to Christianity by EDWIN of Northumbria and murdered by a rival to the throne.

Eppilus (*c*.AD 15) King of the Atrebates tribe. One of the three sons of COMMIUS who divided his kingdom and used the Roman title *Rex* meaning 'king'.

Erconbert (d.664) King of Kent. Reigned from *c*.660 to 664. He married Sexburga, one of King ANNA of East Anglia's four daughters.

Eric (d.918) King of East Anglia. Reigned from 900 to 902. He succeeded his father, GUTHRUM, and was killed fighting EDWARD THE ELDER's army. He was the last Dane to rule the kingdom of East Anglia.

Erik Bloodaxe (d.954) King of York. Reigned from 947 to 954. After being deposed as King of Norway in 934 he fled to England and seized Northumbria from EDRED in 947. He was killed along with his brother and a son in an ambush at Stainmore by Edred's army.

Ethelbald (d.757) King of Mercia. Reigned from 716 to 757. By 731 Ethelbald had established himself as *Bretwalda*, or overlord, of all the kingdoms south of the Humber and styled himself *Rex Britanniae*, King of Britain. He was nonetheless troubled by frequent Welsh raids and had the fortifications of Wat's Dyke built as a bulwark against aggression. He was murdered by his bodyguard in 757 and is buried at Repton, Derbyshire.

Ethelbald (834-860) King of Wessex. Reigned from 858 to 860. He ascended the throne on the death of his father ETHELWULF, but reigned for only two years. He married his stepmother Judith. He is buried at Sherborne Abbey, Dorset.

Ethelbert I (d.616) King of Kent. Reigned from 560 to 616. Ruler of all England south of the Humber, he married Bertha, daughter of the Frankish king Charibert, *c*.589. He became the first baptized Anglo-Saxon king after receiving St Augustine's mission from Rome in 597 which landed at Ebbsfleet, near Ramsgate. He made Canterbury the centre of Christianity in southern England and is buried there in the Monastery of St Peter and St Paul. He was succeeded by EDBALD who temporarily renounced Christianity and married his stepmother.

Ethelbert II (d.762) King of Kent. Reigned from 725 to 762. The son of WIHTRED and Cynegyth, he reigned jointly to 748 with his brother, EADBERT, and then with his half-brothers, ALRIC and EARDWULF.

Ethelbert (St Ethelbert) (d.792) King of East Anglia. Reigned in 792. He was executed by his father-in-law, King OFFA of Mercia, and is the patron saint of Hereford Cathedral.

Ethelbert (836-865) King of Wessex. Reigned from 857 to 865. The third son of ETHELWULF, succeeded his elder brother ETHELSTAN in the Eastern side of the kingdom in 857, and in 860, on the death of his brother, ETHELBALD, became sole king. He was troubled by the inroads of the Danes who sacked Kent and crushed Winchester during his reign. He was buried at Sherborne Abbey in Dorset.

Ethelfrith (d.617) King of Northumbria. Reigned from 604 to 617. The third of IDA's six sons, he reigned in Bernicia from 592 and seized Deira in 604. In his efforts to expand his realm he defeated King AIDEN of the Dalradian Scots at Degastan, near Liddesdale, in 603. He also took Chester from the Welsh in 613. He married three times and had seven sons and three daughters.

Ethelheard (d.740) King of Wessex. Reigned from 726 to c.740. Succeeded INE as king in 726 although his connection to him is unknown. A land charter he made to Bishop Fortherne in 739 first mentions Devon.

Ethelhere (d.654) King of East Anglia. Reigned in 654. A younger brother of ANNA, he reigned for only a few months before being killed in the Battle of Winwaed fighting the South Saxons with PENDA of Mercia.

Ethelred King of Mercia. Reigned from 675 to 704. A brother of WULFHERE, he abdicated to become a monk.

Ethelred I (d.796) King of Northumbria. Reigned from 774 to 796. The son of ETHELWOLD MOLL, he deposed ELHRED, who had deposed his father, in 774. He earned a reputation as a tyrant after having several of his nobles executed for treachery and was briefly deposed by ELFWOLD, the grandson of King EDBERT. He returned to the throne in 790 after imprisoning OSRED, nephew of Elfwold. Osred later escaped and Ethelred was murdered at Corbridge at Hadrian's Wall. OSBALD, one of those who conspired against him, took the throne.

Ethelred II King of Northumbria. Reigned from 841 to 850. The son of EANRED.

Ethelred I (St Elthelred) (840-871) King of Wessex. Reigned from 865 to 871. The son of ETHELWULF, he succeeded his brother, ETHELBERT, at a time when the Danes were threatening the conquest of the whole kingdom. He died in consequence of a wound received in action with the Danes at the Battle of Merton in 871, and was succeeded by his brother, ALFRED. Ethelred's devout Christianity was recognised in his popular title of St Ethelred.

Ethelred (II) the Unready (968-1016) King of England. Reigned from 978 to 1016. The son of EDGAR and Emma of Norway, he succeeded his brother, EDWARD THE MARTYR, and, for his lack of vigour and capacity, earned the name of 'the Unready'. In his reign he began the practice of buying off the Danes by presents of money. After repeated payments of tribute (known as Danegeld), he effected, in 1002, a massacre of the Danes, but this led to SWEYN gathering a large force together and ravaging the country. They were again bribed to depart, but upon a new invasion, Sweyn obliged the nobles to swear allegiance to him as king of England. Ethelred fled to Normandy but returned after the death of Sweyn in 1014 when he was invited to resume the government. He died in London in the midst of a struggle with CANUTE, his Danish rival for the throne. He married twice, and among his children were EDMUND II and EDWARD THE CONFESSOR. His second marriage to the daughter of the 2nd Duke of Normandy formed an Anglo-Norman connection which provided a basis for the Norman invasion of 1066.

Ethelric King of Bernicia. Reigned from 568 to 572. A son of the Saxon king IDA, Ethelric ruled Bernicia, which, with Deira, ruled by his brother AELLE, later formed the kingdom of Northumbria. Bernicia supplied most of the kings in the merged kingdom.

Ethelstan *or* **Athelstan** (895-939) King of England. Reigned from 924 to 939. The eldest son of EDWARD THE ELDER and Egwina, he was crowned at Kingston-on Thames and became the first Saxon king with effective control of all England (with the exception of Cumbria). He was an able ruler and with the combined forces of Mercia and Wessex was victorious in his wars with the Danes of Northumberland, and with the Scots and Irish by whom they were assisted, at Brunanburh, an unidentified site in Cumbria, in 937. In this battle a son of the king of Scots and five Irish kings were killed, shattering the Viking-Scots coalition. He also summoned a number of Welsh princes to Hereford and imposed a tribute on them, as well as fixing the border between his kingdom and that of the princes at Wye. He became, therefore, in name at least, the overlord of Celtic kingdoms in Cornwall, Scotland and Wales.

Ethelstan sought to ease the plight of his poorer subjects through some of the many laws he introduced, which punished theft and corruption. He established a corps of clerks that is thought to have foreshadowed the civil service, as well as introducing a national coinage. He also strengthened relations with continental rulers through the marriages of four of his sisters to dukes in France and the Holy Roman Emperor Otto I the Great. He was buried in Malmesbury Abbey in Wiltshire.

Ethelwalh King of Sussex. Reigned before 685. He received the Isle of Wight from WULFERHERE of Mercia in 661 and in turn gave WILFRED the bishopric of Selsey.

Ethelweard King of Mercia. Reigned c.837 to 850.

Ethelwold King of Mercia. Reigned from 654 to 663. The youngest brother of ANNA.

Ethelwold (Moll) King of Northumbria. Reigned from 759 to 765. Defeated his rival for the throne, OSWIN, in 761 but was deposed by ELHRED (who claimed to be a descendant of IDA, the founder of the Bernician dynasty) four years later.

Ethelwulf (800-858) King of Wessex. Reigned from 839 to 858. He succeeded his father, EGBERT, and was chiefly occupied in repelling Danish incursions. He is best remembered for his donation to the clergy, which is often quoted as the origin of the system of tithes. ALFRED THE GREAT was the youngest of his five children.

Fergus Mor (d.501) King of Scots. The son of Erc, he was the ruler of the kingdom of Dalriada in Argyll, which also included the Inner Hebrides and part of northern Ireland. He led a group of his people, the 'Scots', from Antrim in northern Ireland to settle in western Scotland around 500 and is credited with introducing the Gaelic language to Scotland, as well as the term 'Scot' from which the country was later to take its name. All kings of Dalriada for the following 343 years claimed descent from either Fergus or Loarn, another son of Erc.

Garnard (d.635) King of Picts. Reigned from 631 to 635. The son of CINOICH's sister.

Gartnait (d.663) King of Picts. Reigned from 657 to 663. The son of TALORCEN's sister.

Gartnart (d.597) King of Picts. He reigned c.586 to 597. The son of BRIDEI I's sister,

George I (1660-1727) King of Great Britain and Ireland. Reigned from 1714 to 1727. second Elector of Hanover from 1698. The son of the 1st Elector of Hanover, Ernest Augustus, by Sophia, daughter of Frederick, Elector Palatine, and granddaughter of JAMES I, he inherited the throne through his mother following the Act of Succession of 1701, which conferred the succession on her heirs.

In 1682 he was married to Sophia Dorothea of Zell, whom, in 1694, on account of a suspected intrigue with Count Königsmark, he caused to be imprisoned and kept in confinement for the rest of her life. In 1698 he succeeded his father as the 2nd Elector of Hanover (Electors were princes of the Holy Roman Empire who elected the Emperor). He commanded the imperial army in 1707 during the War of the Spanish Succession and ascended the throne of Great Britain on the death of Queen ANNE in 1714. He arrived in Britain at the age of 54 and, as he spoke only German, had a limited knowledge of the kingdom. It was a difficult succession as there were many who wanted the Stuart dynasty to continue and within a year he was faced with a Scottish-led Jacobite rebellion. The attempt to place the 'Old Pretender', JAMES II's son, James Edward Stuart, on the throne failed, however, as a series of tactical errors led to the Jacobite army being defeated at the Battle of Sherrifmuir.

More than most monarchs, George needed good advisers and men he could trust in government. Sadly the leading men of the day were mostly corrupt and sought to take advantage of a political system that was almost entirely devoid of integrity. The poor character of the nation at large was exposed by the disaster involving the South Sea Company. Thousands invested in this trading company, but in 1720 the South Sea Bubble, as it came to be known, burst. Most of the investors lost their money and the government was engulfed in scandal. A radical change was required in the financial administration of the country and it was only with the appointment of Sir Robert Walpole as First Lord of the Treasury that confidence was restored. For twenty-one years Walpole oversaw the restructuring of government and became, in effect, the first Prime Minister. The king came to depend on him and his Whig ministry, although, because he could not speak English, he could not preside over meetings of Walpole's cabinet.

The private character of George I was bad, but he showed much good sense and prudence in government, especially of his German dominions, and was an able military leader. Other than his patronage of Handel, he had little time for the arts and was widely disliked for his treatment of his wife and for having many mistresses. He died of a stroke while visiting his German possessions. By Sophia Dorothea he had a son, George, afterwards

GEORGE II of England, and a daughter, Sophia, the mother of Frederick the Great.

Significant events of George I's reign
- 1715 – Jacobites are defeated at Sherrifmuir.
- 1716 – The Septennial Act allows for General Elections to be held every seven years.
- 1720 – The collapse of the South Sea Company.
- 1721 – Sir Robert Walpole becomes the first Prime Minister.
- 1726 – The first British circulating library opens in Edinburgh. Jonathon Swift's *Gulliver's Travels* is published.
- 1727 – Sir Isaac Newton dies.

George II (1683-1760) King of Great Britain and Ireland. Reigned from 1727 to 1760. The only son of GEORGE I and Sophia Dorothea of Zell. In 1708, then only electoral prince of Hanover, he distinguished himself at Oudenarde under Marlborough. He succeeded his father to the English throne, and inherited to the full the predilection of George I for Hanover. For the first twelve years, however, he was well served by Prime Minister Robert Walpole who kept England out of foreign entanglements and their relationship can be said to have laid the foundations of constitutional monarchy.

After 1739 Britain was involved in almost continuous warfare: first with Spain, then with France during the War of the Austrian Succession and then again with France during the Seven Years' War. The events in Europe led to the resignation of Walpole in 1742, for he was no war minister, and his place was taken for the next twenty years by Henry Pelham and his brother the Duke of Newcastle. With the British army engaged in Flanders the Scottish Jacobites took the opportunity to attempt again the restoration of a Stuart to the thone. This time it was the 'Young Pretender', Charles Edward Stuart, the son of James Edward Stuart. He landed from France in the Highlands of Scotland and raised an army among the clan chiefs loyal to the Stuarts. The Jacobite army was at first successful, taking Edinburgh with little difficulty and routing an English army at Prestonpans, before advancing south. Support hoped for in the north of England never materialized, however, and following a long and tiring retreat, the Jacobites were crushed by the British army, now returned from Flanders, under the command of William, Duke of Cumberland, son of George II, at Culloden. In the aftermath, 'Bonnie Prince Charlie' escaped to France, leaving his Highland supporters to face deportation or execution. Many clan chiefs lost their estates and possessions, and

the Highlands of Scotland, which had for so long remained distant from royal authority, where at last brought under control.

There was further success for George in the victories of the Seven Years' War. The year of 1759 saw France lose important territories in North America, and its stronghold, Quebec, was taken by General Wolfe. In India Clive defeated the French at Plassey and Madras, giving the East India Company control of the vast province of Bengal. By 1763 France had ceded all Canada to Britain and retained only two small trading posts in India. The last years of George's reign saw Britain well on its way to becoming a truly world power.

The reign was also notable for the number of great men in art, letters, war and diplomacy who then adorned Britain. George II had a keen interest in music and continued his father's patronage of Handel, but he was a king of very moderate abilities and ignorant of science or literature. He nevertheless won respect for his military abilities. At Dettingen (1743) he became the last British king to lead his troops into battle. In matters of state he was guided by his wife, Caroline of Ansbach, who was far more cultured and intelligent than her husband. By Caroline he had three sons and four daughters, notably Frederick, Prince of Wales, who predeceased George ans was the father of GEORGE III, Anne, William, Duke of Cumberland, Mary and Louisa.

Significant events of George II's reign
- 1732 – A royal charter founds Georgia.
- 1738 – The Methodist movement is founded by John and Charles Wesley.
- 1740 – War of Austrian Sucession begins.
- 1742 – Handel's *Messiah* first performed.
- 1743 – George leads troops at Dettingen.
- 1745 – The Jacobites gain victory at Prestonpans.
- 1746 – The Battle of Culloden crushes the Jacobite rising.
- 1753 – The British Museum is founded.
- 1757 – Robert Clive secures Bengal for Britain. William Pitt becomes Prime Minister.
- 1759 – Quebec is taken by James Wolfe.

George III (1738-1820) King of Great Britain and Ireland. Reigned from 1760 to 1820. From 1801 he became King of the United Kingdom of Great Britain and Ireland. Fourth Elector of Hanover (*de facto* until 1803). The eldest son of Frederick Louis, Prince of Wales, by the Princess Augusta of Saxe-Coburg-Gotha, he succeeded his grandfather, GEORGE II, in 1760. In the following year he married the Princess Charlotte Sophia of

Mecklenburg-Strelitz. The sixty years of his reign are filled with great events, amongst which are the acceleration of the Industrial Revolution, the Wilkes controversy, the American War of Independence 1775-83, the result of which the king felt acutely, the French Revolution in 1789, and the Napoleonic Wars which followed, comprising the long struggle that ended at Waterloo; and the Irish Rebellion of 1798.

He was the first Hanovarian monarch to be raised in England and took a great interest in government, although he was also thought by many to interfere too much. He had a succession of ministers, most of whom met with the king's disapproval. In 1770 he appointed Lord North and established a good working relationship which lasted to 1782. After the losses Britain suffered in the American War of Independence, however, North was held responsible, and George also had to shoulder some of the blame. William Pitt the Younger replaced North and guided Britain through the troubled times following the French Revolution. The victories of Waterloo and Trafalgar in the Napoleonic Wars restored some British pride, but the death of Nelson, Britain's greatest naval commander, was keenly felt as were the social consequences of twenty-two years of continous war with France.

In Ireland a rebellion in 1798 was followed by Pitt's attempt to solve the Irish problem by passing an Act of Union, similar to that of Scotland one hundred years earlier, whereby Ireland returned members to the British Parliament. But these members were Protestant, as Irish Catholics, although they had the vote, could not sit in Parliament. George denied Pitt's attempts to give them this right, and, after the bribing of the Irish Protestants, the assembly in Ireland dissolved itself and the country became governed by a Protestant Parliament of the newly formed United Kingdom at Westminster.

By this time, however, George had lost the ability to rule through a disease that had the appearance of a slowly worsening mental illness. It is now thought that he had a rare and incurable ailment called porphyria, in which the victim suffers delusions and displays symptoms of delirium. In the last nine years of his life the attacks became more frequent, and he died deaf, blind and in a state of permanent derangement.

George III was a man of conscientious principles and of a plain, sound understanding, though his narrow patriotism and his obstinate prejudices were often hurtful to British interests. His tastes and amusements were plain and practical, and he enjoyed touring the country. His special interest in agriculture earned him the nickname 'Farmer George'. With Queen Caroline he had fifteen children, nine being sons. His son, GEORGE, had been made Prince Regent in 1811 and succeeded to the throne on his father's death.

Significant events of George III's reign
- 1768 – James Cook begins his first voyage around the world.
- 1773 – The Boston Tea Party sparks protest against unfair British taxation.
- 1774 – The first Continental Congress meets in Philadelphia to protest at the repressive British legislation.
- 1775 – The War of American Independence commences. The first commercial steam engines are produced by Watt and Boulton.
- 1783 – Britain acknowledges the independence of the American Colonies. William Pitt the Younger becomes Prime Minister.
- 1789 – The French Revolution.
- 1791 – First publication of Thomas Paine's *Rights of Man*.
- 1793 – Britain and France go to war.
- 1798 – Irish Rebellion.
- 1800 – Act of Union with Ireland.
- 1803 – Napoleonic Wars commence.
- 1805 – Nelson dies in the Battle of Trafalgar.
- 1811 – Prince George, George III's son, becomes Prince Regent.
- 1815 – The Battle of Waterloo ends the Napoleonic Wars. The Corn Laws are passed.
- 1819 – The Peterloo Massacre: reform campaigners in Manchester are killed.

George IV (1762-1830) King of the United Kingdom of Great Britain and Ireland. Reigned from 1820 to 1830. The eldest son of GEORGE III and Charlotte Sophia. In 1811 George became Regent and, on the death of his father in 1820, king. He distinguished himself while Regent as a great patron of the arts—the first George about which that could be said—and his intelligent patronage fostered painting, literature and Regency architecture. His extravagant tastes in food and wine were well known, but in the early part of his life these weaknesses did not harm his popularity.

His secret marriage in 1875 to Maria FitzHerbert, a Catholic, caused problems, however, as the 1701 Act of Settlement prohibited the succession of a Catholic to the throne. Forced by Parliament to choose an official wife, he married his cousin, Caroline of Brunswick, in 1795 and had his huge debts paid off in return. The couple soon parted amid accusations of adultery from both sides. George lost his good standing with the people, however, when he openly accused his wife of infidelity at a public trial and forbade her from attending his coronation.

The most significant event of his reign was the Catholic Relief Act of 1829 which finally allowed Catholics access to Parliament. There were also important advances made in criminal law and labour relations. George had little interest in politics, however, and spent the majority of his time as king indulging himself with drink, food and his many mistresses. His dissipated lifestyle and his extravagance alienated him from the affection of the nation, and the image of the monarchy as a moral influence was greatly tarnished. He died in a state of obesity from internal bleeding and liver damage in 1830. As his only daughter, Princess Charlotte, wife of Leopold of Saxe-Coburg (afterwards King of the Belgians), died childless in 1817, he was succeeded by his brother, WILLIAM IV.

Significant events of George IV's reign
- 1820 – The Cato Street Conspiracy to assassinate the Cabinet is discovered. Queen Caroline is tried and George sues for divorce on grounds of adultery.
- 1824 – The National Gallery founded in London.
- 1825 – The world's first railway service, the Stockton and Darlington Railway, opens. Legalization of trade unions.
- 1828 – Wellington becomes Prime Minister.
- 1829 – The establishment of the Metropolitan Police Force by Robert Peel takes place. The Catholic Relief Act passed.

George V (1865-1936) King of Great Britain and Ireland and of the overseas British dominions, Emperor of India. Reigned from 1910 to 1936. The second son of EDWARD VII and Queen Alexandra of Denmark. He was born at Marlborough House, and after being educated by a private tutor, he and his elder brother, Prince Albert Victor, became naval cadets, and as midshipmen visited many parts of the world. Prince George attained the rank of commander in 1891, but his brother's death in 1892, which placed him in direct succession to the crown, led to his practical withdrawal from a naval career. Created Duke of York in 1892, the following year he married Princess Victoria Mary, daughter of the Duke of Teck. On the death of Queen Victoria and accession of Edward VII he became Duke of Cornwall, and later in the year was created Prince of Wales.

On the death of Edward VII the prince became king as George V.

He ascended the throne in the middle of a constitutional crisis caused by the House of Commons attempting to limit the powers of the House of Lords and within four years he was also leading the nation in the First World War. Although he was related to Kaiser Wilhelm II he had no objections to his government's decision to engage Germany in war. He quickly gained the admiration of the public by visiting the troops on the Western Front and openly disapproving of Wilhelm's military gesturing. In 1917 he changed the British royal family's name from the German Saxe-Coburg-Gotha to the English Windsor. As well as carrying out his royal duties with great conscientiousness, he made important contributions to the handing of political problems of the day. Shortly before the outbreak of the war, he summoned a conference of party leaders at Buckingham Palace for the purpose of solving the Irish question. This was an important step towards the post-war creation of the Irish Free State. Later, during the 1931 financial crisis, he intervened to persuade the leading political parties to the formation of a national coalition government.

On Christmas Day 1932 George broadcast a message to the nation (which had been written by Rudyard Kipling) and established a tradition which has been maintained every year since. His hobbies were shooting and stamp-collecting: his collection is well-known to philatelists. In later life he suffered ill-health and almost died from septicaemia, or blood-poisoning, in 1928. He never fully regained his former vigour and died of bronchitis eight years later. He was succeeded by EDWARD VIII, who abdicated before being crowned, in favour of his brother, GEORGE VI.

Significant events of George V's reign
- 1911 – Parliament Act ensures the sovereignty of the House of Commons. National Insurance Act passed.
- 1912 – The sinking of the *SS Titanic*.
- 1914 – The First World War breaks out.
- 1916 – The Easter Rising in support of Irish independence takes place in Dublin. David Lloyd George becomes Prime Minister.
- 1917 – The Russian Revolution.
- 1918 – Women over thirty gain the vote. Irish Parliament formed in Dublin.
- 1919 – Lady Astor becomes the first woman MP to take her seat in Parliament.
- 1921 – Ireland partitioned.

- 1924 – First Labour government takes power under Ramsay MacDonald.
- 1926 – A General Strike is called by Trade Unions.
- 1928 – Women over twenty-one get the vote.

George VI (1895-1952) King of the United Kingdom of Great Britain and Northern Ireland and British dominions overseas; the last Emperor of India; Head of the Commonwealth from 1949. The second son of GEORGE V and Queen Mary. Prince Albert, as he was then known, served in the navy and air force until 1919 and attended Trinity College, Cambridge, until 1920. He was called to the throne in 1936 on the abdication of his brother, EDWARD VIII.

Unprepared for the role of king, George nevertheless carried out his duties with great conscientiousness and became a popular figurehead for the nation during the Second World War. He struggled with a speech impediment in his early years but worked extremely hard to overcome it in order to perform his duties with the sense of authority that he knew was expected of him. He lived in London for the duration of the war despite frequent bombing raids and visited the troops in North Africa and France.

In 1939 he became the first British monarch to visit North America and restored much of the reputation lost by the monarchy following his brother's abdication. The post-war years saw a Labour government transforming Britain into a welfare state, and George became Head of the Commonwealth of Nations, following the fragmentation of the Empire.

He opened the 1951 Festival of Britain but died the next year after an operation for lung cancer. He had two daughters by Lady Elizabeth-Bowes Lyon (later the Queen Mother) and was succeeded by the elder, Queen ELIZABETH II.

Significant events of George VI's reign

- 1939 – World War Two breaks out.
- 1940 – The Dunkirk evacuation takes place. Winston Churchill becomes Prime Minister. The Battle of Britain prevents German invasion.
- 1941 – Pearl Harbor bombing brings the USA into the war.
- 1944 – D-Day: Allied forces land at Normandy and force German retreat.
- 1945 – Germany is defeated and war ends in Europe. Japan surrenders and World War Two ends. United Nations founded. A Labour Government is elected.
- 1947 – India and Pakistan gain independence.

- 1948 – Establishment of the National Health Service.
- 1951 – Festival of Britain opens.

Giric I (d.889) King of Scots. Reigned from 878 to 889. The cousin of AED, whom he is thought to have murdered to take the throne. He ruled jointly with EOCHA who was King of the Britons of Strathclyde and KENNETH I's grandson. He was defeated and killed in battle at Dundurn by DONALD II.

Giric II (d.1005) King of Scots. Reigned from 997 to 1005. The son of KENNETH III with whom he shared the throne. Both were killed by MALCOLM II in battle at Monzievaird in order to secure his succession to the throne.

Godfred (Crovan) (d.1095) King of the Isle of Man. Reigned from 1079 to 1095. The island kingdom had been held by Orkney rulers since *c*.990 when Godfred landed in 1079. He founded a dynasty of Norse kings which ruled this island until 1265. The legacy of Godfred's kingdom is reflected in the island's present self-governing status.

Grey, Lady Jane (1537-1554) Queen of England and Ireland (only recognized by King's Lynn and Berwick). Reigned for nine days in 1553. The daughter of Henry Grey, 3rd Marquis of Dorset, and Lady Frances Brandon, the daughter of HENRY VIII's younger sister, Mary. Lady Jane Grey was the 'The Nine Days Queen', the unfortunate victim of a scheme designed by John Dudley, the Duke of Northumberland, he father-in-law, to give his family the succession. Dudley persuaded the dying King EDWARD VI to settle the succession on his daughter-in-law and her male heirs in order to stop the throne being taken by either MARY, QUEEN OF SCOTS or MARY Tudor, both of whom were Catholics. When Edward died in 1553 the country would have nothing to do with Northumberland, and Edward's sister, Mary Tudor, was proclaimed queen by the Lord Mayor of London. Lady Jane Grey was never crowned and had only been recognized by King's Lynn and Berwick before being imprisoned in the Tower of London. She was later beheaded for treason along with her husband.

Gruffydd ap Cynan (b.1055) King of Gwynedd. Reigned from 1081 to 1137. Although born in Ireland, he invaded his ancestral kingdom three times from 1075. He secured it briefly in 1081 before he was captured and imprisoned in Chester. He escaped, reconquered Gwyned, and resisted two attempts by WILLIAM II to capture him again. He finally rendered homage to HENRY I.

Gruffydd ap Llywelyn (d.1062) King of Gwynedd. Reigned from 1039 to 1063. He briefly ruled all Wales when he annexed Deheubarth in 1055. After several years spent raiding the English border he was captured by a Wessex army at Rhuddlan in 1062 and beheaded. His descendants ruled the kingdom of Powys until 1269.

Guthrum (d.890) King of East Anglia. Reigned from 880 to 890. A Danish army commander who first attacked Wessex in 878. He was defeated in battle by ALFRED near Edington in Wiltshire and made the Treaty of Wedmore. The treaty required Guthrum and his men to embrace Christianity and accept baptism. With Alfred acting as sponsor, they honoured their oaths and were baptized at the River Aller in Somerset. Guthrum then became King of East Anglia and settled at Cirencester.

Halfran (Ragnarson) (d.895) King of York. Reigned from 875 to 883. In 875 he founded the kingdom of York which had thirteen Norse rulers in eighty years.

Hardicanute (1018-1042) King of England and Denmark. Reigned (England) from 1040 to 1042. The only son of CANUTE and Emma of Normandy. The rightful successor to Canute, he was consolidating his dominion over Denmark when his half-brother HAROLD I usurped his English throne in 1035. He came to the throne on Harold's death in 1040. Hardicanute's short and unpopular reign is noteworthy for its violence: he had Worcester burned for killing royal tax collectors, he murdered the Earl of Northumbria, and abused the body of his dead brother Harold I by having it flung into a bog. He collapsed at a drunken wedding banquet and died shortly after. He was succeeded by EDWARD THE CONFESSOR.

Harold I (Harefoot) (1016-1040) King of England. Reigned from 1035 to 1040. The second son of CANUTE and Elgifu, he succeeded his father CANUTE to the throne of England. Despite being his illegitimate son, he proclaimed himself king and had a rival claimant, Aetheling, son of ETHELRED, blinded and killed. His countrymen, the Danes, maintained him upon the throne while his half-brother was in Denmark. He exiled Emma, mother of HARDICANUTE, and defended his kingdom against vigorous attacks from the Welsh and Scots.

Harold II (1020-1066) King of England. Reigned in 1066. The second son of Godwin, Earl of Kent, and Gytha, sister of CANUTE's Danish brother-in-law. Made powerful by the inheritance of his father's lands in Wessex and Kent, he was a rival to EDWARD THE CONFESSOR for the whole kingdom of England. He successfully defeated Welsh incur-

sions and took control of Hereford before he was shipwrecked in 1064. He was then held by Duke William of Normandy (*see* WILLIAM THE CONQUEROR) and only gained his release by promising to help secure the crown for him. He himself had been named as successor by the dying Edward, however, and in early 1066 he stepped into the vacant throne. Claims of a bequest of Edward in favour of William led the latter to call upon Harold to resign the crown. Harold refused, and William prepared for invasion. William also instigated Harold's hostile brother, Tostig, to land on the northern coasts of England in conjunction with the King of Norway. The united fleet of these chiefs sailed up the Humber and landed a numerous body of men, but at Stamford Bridge, in Yorkshire, were totally routed by Harold. Tostig fell in the battle. Immediately after the battle Harold heard of the landing of the William at Pevensey in Sussex and went there with all the troops he could muster. It was a forced march of over 250 miles completed in nine days, but in the engagement which followed at Senlac, near Hastings, they were narrowly beaten.

Harold died on the field, supposedly killed by an arrow, with two of his brothers. With his death there also ended England's 600 years of rule by Anglo-Saxon kings.

Hengest *or* **Hengist** King of Kent. Reigned from 455 to 488. The first of the Jutish kings of Kent, he ruled jointly with his brother HORSA. After being invited by the British king VORTIGERN to help force back the northern Picts, Hengest, along with his brother, turned on the Britons and founded what is referred to in the *Anglo-Saxon Chronicles* as the first Saxon kingdom. His son, AESC, succeeded him.

Henry I (1068-1135) King of England. Reigned from 1100 to 1135. Often surnamed 'Beauclerk' (fine scholar). The youngest son of WILLIAM THE CONQUEROR and Matilda of Flanders. Henry was hunting with WILLIAM II when the king was accidentally killed. He immediately rode to London and claimed the throne before he could be challenged by his elder brother, Robert Curthose of Normandy, then absent as a crusader. He soon re-established by charter the laws of EDWARD THE CONFESSOR and did away with the legal abuses William II had let go unchecked. He then introduced measures to stop the seizure of Church lands, and married Matilda (Edith), a daughter of MALCOLM III of Scotland, thus conciliating in turn his people, the Church and the Scots. Robert landed an army, but was pacified with a pension and the promise of succession in the event of his brother's decease. Soon after, however, Henry invaded Normandy, captured Robert and imprisoned him in Cardiff Castle. The last years of his reign were very troubled. In 1120 his only son William was drowned returning from Normandy, where, three years later, a revolt occurred in favour of Robert's son. The Welsh also were a source of disturbance, but he was a capable commander and was never seriously threatened.

In the later years of his reign he was able to strengthen the Norman system of government and administration of justice. He also won from the Church the agreement that the bishops should acknowledge the king as overlord of their extensive secular holdings. Henry appointed as his heir his daughter, MATILDA, whom he had married first to the Emperor Henry V and then to Geoffrey Plantagenet of Anjou. This laid the basis for a much enlarged kingdom, but STEPHEN took the throne from the rightful heir, Matilda, when Henry died of fever in France.

Significant events of Henry I's reign
- 1100 – Charter of Liberties proclaimed.
  Henry marries Matilda of Scotland.
- 1101 – Robert of Normandy recognizes Henry I as king in the Treaty of Alton.
- 1104 – Crusaders capture Acre.
- 1106 – War breaks out with Normandy.
- 1120 – Henry's heir, William, is drowned.
- 1128 – Matilda marries Geoffrey Plantagenet, Count of Anjou.

Henry II (1133-1189) King of England. Reigned from 1154 to 1189. The first of the Plantagenet line was born in Normandy the son of Geoffrey, Count of Anjou, and MATILDA, daughter of HENRY I. Invested with the Duchy of Normandy, by the consent of his mother in 1150, he succeeded to Anjou and in 1152, by marriage with Eleanor of Aquitane gained Guienne and Poitou. In 1152 he invaded England to make his claim to the throne, but a compromise was effected, by which STEPHEN was to retain the crown, and Henry to succeed at his death.

He began his reign by destroying the castles, or 'dens of thieves' as he called them, built by rebellious barons in Stephen's time. A man of immense energy, he soon stamped his character on the vast kingdom which stretched from Scotland to the Pyrenees. He intended to reform the powers of the Church and began by installing Thomas à Becket as Archbishop of Canterbury. The Constitutions of Clarendon, which placed limitations on the Church's jurisdiction over crimes commited by the clergy, were contested by Becket, however, and he fled the kingdom after quarrelling with the king. They were later reconciled but quarrelled again, and Becket was murdered in Canterbury Cathedral by four knights who took Henry's request 'Will someone not rid me of this turbulent priest?' a little too literally. Although sufficiently submissive after Becket's death in the way of penance, Henry gave up only the article in the Constitutions of Clarendon which forbade appeals to the court of Rome in ecclesiastical cases.

Henry began the settlement of Ireland in 1166 after responding to a request by the king of Leinster to resolve a dynastic dispute. His support was commanded by Richard de Clare, Earl of Pembroke, commonly known as Strongbow, who successfully established England's claim to rule Ireland and forced the subservience of all Ireland's regional kings to Henry, who created himself First Lord of Ireland. An earlier papal bull, *Laudabiliter*, made by Pope Adrian IV in 1155, had given the approval of the Roman Church for an invasion to bring the Irish Church under its control.

Henry's last years were embittered by his sons, to whom he had assigned various territories. The eldest son, Henry, who had been not only declared heir to England, Normandy, Anjou, Maine and Touraine, but actually crowned in his father's lifetime, was induced by the French monarch to demand of his father the immediate resignation either of the kingdom of England or of the dukedom of Normandy. Queen Eleanor excited her other sons, RICHARD and Geoffrey, to make similar claims. WILLIAM THE LION of Scotland gave them support. A general invasion of Henry's dominions was begun in 1173 by an attack on the frontiers of Normandy and an invasion of England by the Scots, attended by considerable disturbance in England. Henry took prompt action. William the Lion was captured and forced to acknowledge the English king as overlord. Henry's sons, however, once more became turbulent, and though the deaths of Henry and Geoffrey reduced the number of centres of disturbance, the king was forced to accept humiliating terms from Richard and Philip of France.

Henry II ranks among the greatest English kings both in soldiership and statecraft. He partitioned England into four judiciary districts and appointed itinerant justices to make regular excursions through them. He revived trial by jury and established it as a right, but by the time of his death he was a defeated king, worn out by family revolts.

Significant events of Henry II's reign
- 1155 – Thomas à Becket appointed Chancellor of England.
  Pope Adrian IV's papal bull gives Henry the right to invade Ireland.
- 1162 – Becket appointed Archbishop of Canterbury.
- 1164 – The Constitutions of Clarendon issued.
- 1166 – Trial by jury is established at the Assize of Clarendon.
- 1168 – Oxford University founded.
- 1170 – Becket is murdered.
- 1171 – Henry is acknowledged as Lord of Ireland.
- 1173 – Becket is canonized.

Henry III (1207-1272) King of England. Reigned from 1216 to 1272. The eldest son of JOHN by Isabel of Angoulême, was born at Winchester and succeeded his father at the age of nine. At the time of his accession the Dauphin of France, Louis, at the head of a foreign army, supported by a faction of English nobles, had assumed the reins of government but was compelled to quit the country by the Earl of Pembroke, who was guardian of the young king until 1219. The Treaty of Lambeth followed, which established peace between France, the English barons and supporters of Henry. But as Henry approached manhood he displayed a character wholly unfit for his station. He discarded his most able minister, Hubert de Burgh, and falsely accused him of treason. After 1230, when he received homage in Poitou and Gascony, he began to bestow his chief favours upon foreigners and installed many of them in government.

His marriage in 1236 with Eleanor of Provence increased the dislike of him felt by his subjects, and although he received frequent grants of money from Parliament, on condition of confirming the Great Charter, his conduct after each ratification was as arbitrary as before. At length the nobles rose in rebellion under Simon de Montfort, Earl of Leicester and husband of the king's sister, and in 1258, a Parliament held at Oxford, known in history as the Mad Parliament, obliged the king to sign the body of resolutions known as the Provisions of Oxford. A feud arose, however, between Montfort and Gloucester, and Henry recovered some of his power. War again broke out, and Louis was called in as arbitrator, but his award being favourable to the king, Leicester refused to submit to it. A battle was fought near Lewes in which Henry was taken prisoner. A convention, called the Mise of Lewes, provided for the future settlement of the kingdom, and in 1265 the first genuine House of Commons was summoned. Leicester, however, was defeated and slain in the battle of Evesham (1265), and Henry was replaced upon the throne.

He was a selfish and petulant king, with few of the personal qualities required to command respect or obedience. In some respects, however, he redeemed himself as a patron of the arts. He established the first three colleges of Oxford and initiated the improvements to Westminster Abbey and the construction of Salisbury Cathedral. His son, EDWARD I, succeeded him.

Significant events of Henry III's reign
- 1227 – Henry takes control of the government; Hubert de Burgh remains as adviser.
- 1232 – Hubert de Burgh is dismissed; Peter des Riveaux becomes Treasurer of England.
- 1234 – A revolt led by Richard Marshal, Earl, of Pembroke, is defeated.
- 1258 – The Provisions of Oxford are prompted by a rebellion led by Simon de Montfort. Mad Parliament called.
- 1261 – Henry renounces the Provisions.
- 1264 – The Barons' War begins.
- 1265 – Montfort killed at the Battle of Evesham.

Henry IV (1366-1413) King of England and Wales. Reigned from 1399 to 1413. The eldest son of John of Gaunt, Duke of Lancaster, fourth son of EDWARD III, by Blanche of Lancaster, the daughter of Henry, Duke of Lancaster, great-grandson of HENRY III. In the reign of RICHARD II he was made Earl of Derby and Duke of Hereford, but having in 1398 preferred a charge of treason against Mowbray, Duke of Norfolk, he was banished with his adversary. In the 1390s Henry took part in crusades in Lithuania and Prussia. On the death of John of Gaunt, Richard withheld Henry's inheritance, and Henry, landing in England, deposed the king and had him imprisoned at Pontefract Castle. The recognition of Henry IV as king by Parliament was followed by Richard's death in prison the following year by self-inflicted starvation.

Henry was king by conquest and election by Parliament, however, and not by heredity. For this reason he had to accept a degree of subservience to his peers, and to conciliate the Church he accepted the *De Heretico Comburendo*, a statute that persecuted heretics, notably the followers of John Wycliffe, known as the Lollards. His position was precarious, and there were several plots to depose him, which led to the executions of several noblemen. An insurrection in Wales, however, under Owen Glyndwr proved more formidable. Glyndwr was a descendant of the last independent Prince of Wales and sought full independence for his principality. He launched a guerilla campaign against the English in 1401 and made a treaty with France by which an army was sent to help him. For the next ten years all attempts to subdue him failed.

Henry was also troubled by the Scots and the Percy family of Northumberland. In 1402 the Scots were decisively defeated by the Percys at Homildon Hill and their leader, the Earl of Douglas, was captured. An order from Henry not to permit the ransom of Douglas and other Scottish prisoners was regarded as an indignity by the Percies, who let Douglas free, made an alliance with him, and joined Glyndwr. The king met the insurgents at Shrewsbury, and the battle ended in defeat for the Percys. The Earl of Northumberland was pardoned, but a few of the insurgents were executed. A new insurrection, headed by the Earl of Nottingham and Richard le Scrope, Archbishop of York, broke out in 1405 but was suppressed. The same year, James (*see* JAMES I), the son and heir of King ROBERT III of Scotland, was captured at sea on his way to France and imprisoned in England. The rest of Henry's reign was comparatively untroubled, and he eventually died after contracting a leprosy-like illness. He was succeeded by his son, HENRY V.

Significant events of Henry IV's reign
- 1400 – Richard II dies in prison. Geoffrey Chaucer dies.
- 1401 – Statute of *De Heretico Comburendo* leads to many being burned at the stake.
- 1403 – The Percy family rebellion is defeated at Shrewsbury.
- 1404 – Glyndwr sets up Welsh Parliament.
- 1405 – French troops land in Wales.
- 1411 – Construction of the Guildhall in London begins.

Henry V (1387-1422) King of England. Reigned from 1413 to 1422. The only surviving son of HENRY IV and Mary de Bohun. He showed a wisdom in kingship in marked contrast to a somewhat reckless youth. As Prince of Wales he had fought against Welsh rebels and prided himself on his abilities as a soldier. On becoming king he acted quickly to thwart an attempt by a group of nobles to place his cousin, Edmund Mortimer, Earl of March, on the throne. He also carried on the persecution of the Lollards and sent many to their deaths. Like his father, his claim to the throne was doubtful, and he busied himself with foreign affairs in order to divert attention from domestic difficulties.

The struggle in France between the factions of the Dukes of Orleans and Burgundy afforded Henry a tempting opportunity for reviving the claims of his predecessors to the French crown. He accordingly landed near Harfleur in 1415, and though its capture cost him more than half his army, he decided to return to England by way of Calais. A large French army endeavoured to intercept him at the plain of Agincourt but was completely routed. It is thought that as many as 6,000 Frenchmen died while fewer than 400 English lost their lives. A year later the French were defeated at sea by the Duke of Bedford.

In 1417 the liberal grants of the Commons enabled Henry once more to invade Normandy with 25,000 men. The assassination of the Duke of Burgundy, which induced his son and successor to join Henry, greatly added to his power, and the alliance was soon followed by the famous Treaty of Troyes, by which Henry engaged to marry the Princess Catherine and to leave Charles VI in possession of the crown, on condition that it should go to Henry and his heirs at his death. He returned in triumph to England, but on the defeat of his brother, the Duke of Clarence, in Normandy by the Earl of Buchan, he again set out for France, drove back the army of the dauphin, and entered Paris. All his great projects seemed about to be realized when he died of fever at Vincennes, at the age of thirty-five, having reigned for ten years.

An adventurous, headstrong leader, Henry pursued his policies with great zeal and proved himself to be a shrewd military tactician as well as an able politician. His campaign in France had the effect of uniting his nobles in a common cause, thus diverting attention from domestic plots to unthrone him. His ten-month-old son, HENRY VI, succeeded him as king of both England and France.

Significant events of Henry V's reign
- 1415 – The Cambridge plot is thwarted. Battle of Agincourt.
- 1416 – Welsh leader Glyndwr dies.
- 1420 – The Treaty of Troyes makes Henry heir to the French throne

Henry VI (1421-1471) King of England and Wales. Reigned from 1422 to 1461 and from 1470 to 1471. The only son of HENRY V and Catherine of Valois, he succeeded to the throne on the death of his father when he was less than one year old. In his minority, the government of the kingdom was placed in the hands of his uncle Humphrey, Duke of Gloucester, who was made Protector of the Realm of England. A few weeks after Henry's succession Charles VI of France died, when, in accordance with the Treaty of Troyes, Henry was proclaimed King of France. His uncle John, Duke of Bedford, was appointed Regent of France. The war which followed at first proved favourable to the English, but by the heroism of Joan of Arc, who claimed to have been inspired by a vision telling her to drive the English from France, the confidence of the French people was restored. Joan of Arc was captured and burnt at the stake in Rouen, but Henry eventually lost all his possessions in France with the exception of Calais.

When Henry assumed personal rule at the age of fifteen, the government of England was being conducted by rival ministers of the Houses of York and Lancaster. The fact that he also suffered from bouts of mental illness gave these houses greater power, and their rivalry increased. In 1453 his wife, Margaret of Anjou, bore him a son, but within the year his mind had failed him to such a degree that he had to submit to the rule of a Protector, Richard, Duke of York. Fighting soon broke out between the Houses of York and Lancaster, the rival factions in government, and the appointment of York was annulled the following year, the king having recovered his faculties. York retired to the north, and being joined by his adherents, marched upon London. He encountered and defeated the king's Lancastrian army at St Albans, the first battle of the thirty years' War of the Roses. The king again becoming deranged, York was once more made Protector. Four years of peace followed, but the struggle was soon renewed. A Yorkist army led by Richard Neville, Earl of Warwick, defeated the Lancastrian forces at Bloreheath, but they recovered to win over the Yorkists at Ludford. The Lancastrians then declared York a traitor at a session of Parliament held in Coventry. Neville's army in return defeated the Lancastrians at Northampton and captured Henry.

Following this victory York was restored as Protector, and Henry's wife, Margaret, fled to Scotland. She returned with an army which defeated the Yorkists at Wakefield and killed their leader, Richard of York, who was replaced by his son, EDWARD, as Duke of York. His army then defeated the Lancastrians at Mortimer's Cross, and Warwick engaged Margaret's forces at the second Battle of St Albans. Warwick was unsuccessful at first but finally defeated Margaret's army at the Battle of Towton and declared Edward, Duke of York, as king.

Further revolts by the Lanacastrians were suppressed, and Henry was captured and imprisoned in the Tower of London following Warwick's victory at the Battle of Hexham. Edward owed his crown to Warwick, however, and it was inevitable that the next stage of the struggle would be between the kingmaker and the king. After gaining the upper hand at Edgecote, Warwick was at length banished by the king, only to return, allied with Margaret, to attempt a restoration of Henry. This they achieved in 1470, but it was to be a brief affair; Edward soon returned and defeated Warwick at the Battle of Barnet and Margaret at the Battle of Tewkesbury. Warwick and Margaret's son, Edward, were both killed in the fighting, and shortly afterwards Henry was murdered in the Tower of London. The Wars of the Roses did not end until 1485, when the Lancastrian heir and claimant to the throne, Henry Tudor (see HENRY VII), defeated RICHARD III, the brother of EDWARD IV.

Henry VI had been a pious and well-intentioned but hopelessly incompetent ruler. Throughout his life he had been deeply religious and had a passion for education and building. His principal claim to remembrance is that he founded Eton College and King's College, Cambridge.

Significant events of Henry VI's reign
- 1422 – Henry becomes king of France on the death of Charles VI.
- 1429 – Joan of Arc begins the rout of the English.
- 1431 – Joan of Arc is burned at the stake.
- 1437 – Henry VI assumes control of government.
- 1453 – The English are expelled from France, ending of the Hundred Years' War.
- 1454 – Richard, Duke of York, becomes Protector.
- 1455 – The Wars of the Roses begin when Richard is dismissed as Protector and rebels against King Henry.
- 1461 – Richard's son Edward deposes Henry.
- 1470 – Henry briefly regains the crown.

Henry VII (1457-1509) King of England. Reigned from 1485 to 1509. The first of the Tudor kings, Henry was the son of Edmund, Earl of Richmond, son of Owen Tudor and Catherine of France, widow of HENRY V. His mother, Margaret, was the only child of John, Duke of Somerset, grandson of John of Gaunt. After the battle of Tewkesbury he was taken by his uncle, the Earl of Pembroke, to Brittany, and on the usurpation of RICHARD III was naturally turned to as the representative of the House of Lancaster. In 1485 he assembled a small body of troops in Brittany, and, having landed at Milford Haven, defeated Richard III at Bosworth. Henry was proclaimed king on the field of battle, his right being subsequently recognized by Parliament. In 1486 he married Elizabeth, daughter of EDWARD IV and heiress of the House of York, and thus united the claims of the rival Houses of York and Lancaster.

His reign was troubled by repeated insurrections, of which the chief were those headed by Lord Lovell and the Staffords (1486), and the impostures of Lambert Simnel (1487) and Perkin Warbeck (1496–99). In order to strengthen England's prestige he made important marriage alliances. He brought about a match between the Infanta Catherine, daughter of Ferdinand of Aragon and Isabella of Castile, and his eldest son Arthur. On the death of Arthur, in order to retain the dowry of this princess, he caused his remaining son HENRY to marry the widow by Papal dispensation, an event which, in the sequel, led to a separation from Rome. He married his eldest daughter to JAMES IV, King of Scots, from which marriage there ultimately resulted the union of the two crowns.

The problem of the English barons was still present, however, and he set about breaking their power by reviving the Court of Star Chamber. This prevented the barons from raising private armies and allowed them to be tried if they broke the law. He also did much to strengthen England's commercial activities and took an interest in the development of trading in North America. His fiscal policies were often criticized as being largely for the benefit of the royal exchequer, and in his latter years this avarice became increasingly marked. Two exchequer judges, Empson and Dudley, were being employed in all sorts of extortion and chicanery in order to gratify the royal purse. His reign, however, was in the main beneficent. Its freedom from wars permitted the development of the internal resources of the country. His policy of depressing the feudal nobility, which proportionably exalted the middle ranks, was highly salutary. For a time, however, the power lost by the aristocracy gave an undue preponderance to that of the crown.

A cultured man, Henry also brought European scholars to England, patronized the printer William Caxton, and initiated what came to be called the 'Revival of Learning'. In his later years he suffered from arthritis and gout and died at the age of 52. His eight children by Elizabeth of York included his successor, HENRY VIII.

Significant events of Henry VII's reign
- 1485 – Henry defeats Richard III at Bosworth and is declared King of England. The Yeomen of the Guard is formed.
- 1486 – The Houses of York and Lancaster are joined with the marriage of Henry to Elizabeth of York.
- 1487 – A rebellion on behalf of the pretender Lambert Simnel is thwarted.
- 1492 – Henry defeats another attempt to dethrone him led by Perkin Warbeck.
  America discovered by Columbus.
- 1497 – John Cabot discovers Newfoundland.
- 1502 – Henry's daughter is married to James IV of Scotland.

Henry VIII (1491-1547) King of England and Ireland (from 1542). Reigned from 1509 to 1547. The second son of HENRY VII and Elizabeth of York, he succeeded his father as king at the age of eighteen. Although well educated and opinionated, the young king at first had no enthusiasm for politics or personal rule. He chose instead to leave the business of government to the very capable Cardinal Wolsey who administered with great skill and increased England's trade and standing abroad.

Henry strained the finances of his kingdom, however, through a series of costly wars. The success of the English at the Battle of the Spurs (1513) was succeeded by no adequate result, the taking of Tournay being the only fruit of this expensive expedition. In the meantime, success attended the English army at home, with JAMES IV of Scotland being completely defeated and slain at Flodden Field in 1513. Henry, however, granted peace to the Queen of Scotland, his sister, and established an influence that rendered his kingdom long secure on that side. He soon afterwards made peace with France, retaining Tournay and receiving a large sum of money. After the election of Charles V to the German Empire, both Charles and the French king, Francis I, sought the alliance of England. A friendly meeting took place between Henry and Francis at the Field of the Cloth of Gold (1520), but the interest of Charles preponderated, and Henry soon after again declared war against France.

In 1529 came the determination of the king to divorce his wife, Catherine of Aragon, who was older than he, had borne him no male heir and had, moreover, been in the first place the wife of his elder brother. The last of these points was the alleged ground for seeking divorce, though Henry was probably influenced largely by his attachment to Anne Boleyn, one of the queen's maids of honour. Wolsey, for his own ends, had at first been active in promoting the divorce but drew back and procrastinated when it became apparent that Anne Boleyn would be Catherine's successor. This delay cost Wolsey his power and the papacy its authority in England. Wolsey was accused of treason but died before he could be brought to trial. Henry eagerly caught at the advice of Thomas Cranmer, afterwards Archbishop of Canterbury, to refer the case to the Universities, from which he soon got the decision that he desired. In 1533 his marriage with Catherine was declared null and an anticipatory private marriage with Anne Boleyn declared lawful. As these decisions were not recognized by the Pope, two Acts of Parliament were obtained, one in 1534 setting aside the authority of the chief pontiff in England, the other in 1535 declaring Henry the supreme head of the Church. But although Henry discarded the authority of the Roman Church, he adhered to its theological tenets, and while, on the one hand, he executed Bishop Fisher and Sir Thomas More for refusing the oath of supremacy, he brought many of the reformers to the stake. Finding that the monks and friars in England were the most direct advocates of the Papal authority and a constant source of disaffection, he suppressed the monasteries by Act of Parliament and thereby inflicted an incurable wound upon the Catholic religion in England. The fall of Anne Boleyn, was, however, unfavourable for a time to the reformers.

Henry then married Jane Seymour, and the birth of a son in 1537 fulfilled his wish for a male heir. The death of the queen was followed in 1540 by Henry's marriage with Anne of Cleves, the negotiations of which were conducted by Thomas Cromwell, Sir Thomas More's successor as Lord Chancellor. The king's dislike of his wife, which resulted in another divorce, became extended to the minister who had proposed the union, and Cromwell's disgrace and death soon followed. A marriage with Catherine Howard in 1541 proved no happier, and in 1542 she was executed on a charge of infidelity. In 1543 he married his sixth wife, Catherine Parr, a lady secretly inclined to the Reformation, who survived the king.

In the meantime Scotland and France had renewed their alliance, and England became again involved in war. JAMES V ravaged the borders, but was defeated at Solway Moss in 1542, and in 1544 Boulogne was captured, Henry having again allied himself with Charles V. Charles, however, soon withdrew, and Henry maintained the war alone until 1546. War and his sense of isolation now so much aggravated the natural violence of Henry that his oldest friends fell victims to his tyranny. The Duke of Norfolk was committed to the Tower, and his son, the Earl of Surrey, was executed.

During his reign it is estimated that Henry had at least seventy thousand people executed for various offences. As well as the brutality shown to his subjects, he also frequently proved himself to be disloyal to his wives and advisers. His driving ambition, however, was to secure a male heir for the throne, and he cared little about his public image so long as this goal was acheived. His only son, by Jane Seymour, EDWARD VI, succeeded him.

Significant events of Henry VIII's reign
- 1509 – Henry marries Catherine of Aragon.
- 1513 – James IV of Scotland is killed at the Battle of Flodden.
- 1515- Thomas Wolsey is made Chancellor of England.
- 1517 – Martin Luther protests against the indulgences of the Roman Catholic church at Wittenberg.
- 1520 – Francis I of France meets Henry at the Field of the Cloth of Gold.
- 1529 – Cardinal Wolsey is accused of treason. Sir Thomas More becomes Chancellor of England.
- 1532 – Sir Thomas More resigns.
- 1533 – Archbishop Thomas Cranmer annuls Henry and Catherine's marriage. Henry marries Anne Boleyn. The Pope excommunicates Henry.
- 1534 – Act of Supremacy makes Henry the head of the Church in England.
- 1535 – Sir Thomas More does not accept the Act of Supremacy and is put to death.
- 1536 – Anne Boleyn is executed; Henry marries Jane Seymour. The Act of Union unites England and Wales. Dissolution of the monasteries begins.
- 1537 – Jane Seymour gives birth to a son and dies seven days later.
- 1540 – Henry marries Anne of Cleves and divorces her six months later. Henry marries Catherine Howard.
- 1542 – Catherine Howard charged with treason and put to death.
- 1543 – Henry marries Catherine Parr.

Hlothere (d.685) King of Kent. Reigned from 673 to 685. The younger brother of EGBERT I, he was sole ruler, and then from 676, joint ruler with SUAEBHARD of Essex. Early in his reign he faced an invasion from Mercia and died in battle during a later South Saxon conquest.

Horsa (d.455) King of Kent. Reigned in 455. With his brother HENGEST he became joint ruler after being invited by VORTIGERN to help fight off raids from the north. The first Jutish kings soon established themselves despite Horsa's early death in battle at Aegelsthrep (Aylesford, near Maidstone).

Hywel ab Idwal (the Bad) King of Gwynedd. Reigned from 979 to 985. A descendant of RHODI MWAR, he deposed his father IAGO IAP IDWAL.

Hywel Dda (the Good) (d.950) King of Gwynedd. Reigned from 904 to 950. A grandson of RHODI MWAR, he briefly united north and south Wales under his governorship. By marriage to Princess Elen, daughter of the king of Dyfed, he secured that kingdom (c.904) and soon extended his realm into the area of south Wales known as Deheubarth. He also absorbed Powys but had to acknowledge EDWARD THE ELDER and later ETHELSTAAN as overlords in light of threatened invasions by the Danes. The law code which he is credited with initiating (the Laws of Hywel) still survives in a 13th-century manuscript, and he was the only Welsh ruler to issue a coinage. He went on a pilgrimage to Rome in 928.

Iago I ab Idwal (d.c.980) King of Gwynedd. He reigned from 950 to 979. Deposed by his son, HYWEL AB IDWAL.

Iago II ab Idwal (d.c.1040) King of Gwynedd. He reigned from 1023 to 1039. The grandson of the first IAGO and father of GRUFFYDD AP CYNAN.

Ida (d.c.568) King of Bernicia. Reigned from 547 to 568. According to the *Anglo-Saxon Chronicle* he captured the Bernician stronghold of Banburgh and, with his son AELLE as king of Deira, effectively ruled most of Northumbria. The name for this kingdom comes from *Northanhymbre*, the Old English for 'people north of the River Humber'.

Idwal Foel (Idwal the Bald) King of Gwynedd. Reigned from 916 to 942. The son of ANARAWD, he was killed rebelling against EDMUND II of England. HYWEL DDA succeeded him.

Indulf *or* **Indulph** King of Scots. Reigned from 954 to 962. Succeeded his uncle MALCOLM I to the throne and abdicated in 962 in favour of DUBF, Malcolm's son. He died at St Andrews and was buried at Iona.

Ine (d.c.728) King of Wessex. Reigned from 688 to 726. One of the most powerful Wessex rulers, he defeated the South Saxons in battle in 722 and 725 and the Cornish Britons in 710. He set up a port at Southampton and founded the monastery at Glastonbury (his sister, Cuthburh, founded a monastery at Wimborne in Dorset). His greatest achievement, however, was the important law code he

compiled between 690 and 693, which reveals a growing sophistication in the consideration of the concepts of kingship and royal authority. He abdicated and retired to Rome.

James I (1394-1437) King of Scots. Reigned from 1406 to 1437. The son of ROBERT III by Annabella Drummond. Following the death of James's brother, Robert wished him to be conveyed to France in order to escape the intrigues of his uncle, the first Duke of Albany. The ship in which he was being conveyed was captured by an English squadron off Flamborough Head, and the prince was taken prisoner to London where he received an education from HENRY IV. To relieve the tedium of captivity, he applied himself to later poetic and literary pursuits in which he distinguished himself. Robert III died in 1406, but James was not allowed to return to his kingdom until a ransom had been paid and hostages handed over to act as security. After the Treaty of London he was freed and crowned at Scone. Before his departure he married Joan Beaufort, daughter of the Earl of Somerset. On his return to Scotland he had the second Duke of Albany and his son Murdoch executed as traitors and proceeded to carry on vigorous reforms and, above all, to improve his revenue and curb the ambition and lawlessness of the nobles. The nobility, headed by the Earl of Atholl, exasperated by the decline of their power, formed a plot against his life and assassinated him at Perth in 1437, where he was buried. His poem *The King's Quair* (or King's Book) entitles him to high rank among the followers of Chaucer. He was succeeded by his son JAMES II.

Significant events of James I's reign
- 1406 – James is captured by the English en route to France and is held in the Tower.
- 1423 – Treaty of London agreed.
- 1424 – James is freed on a ransom.

James I of England. *See* **James VI** of Scots.

James II (1430-1460) King of Scots. Reigned from 1437 to 1460. The surviving twin son of JAMES I and Queen Joan, he was only seven years old when his father was assassinated. He was the first king to be crowned at Kelso Abbey rather than Scone. During his minority the kingdom was distracted by struggles for power between his tutors Livingston and Crichton and the great House of Douglas. Crichton had the Earl Douglas murdered at what became to be known as the 'Black Dinner' at Edinburgh Castle with the young king in attendance. After assuming his full powers as king, James still found his position menaced by the great family of Douglas, and he invited the 8th Earl of Douglas to Stirling Castle to persuade him to abandon a league of nobles that had been formed in opposition to the Crown. The interview ended in the king stabbing his guest and his bodyguard killing him. The civil war that followed was won by James and, three years later Parliament announced the forfeiture of the Douglas territories.

Having finally brought the nobles under control, James, who due to a birthmark was nicknamed 'Fiery Face', consolidated his kingdom to the point where even the Lords of the Isles were involved in his attempt to take back Roxburgh from the English. He was killed by the bursting of a cannon at this siege. By his wife, Mary of Gueldres, he had four sons and was succeeded by the eldest, JAMES III.

Significant events of James II's reign
- 1451 – Glasgow University is founded.
- 1455 – James II defeats the 'Black' Douglas family.

James II (1633-1701). King of Great Britain and Ireland. Reigned from 1685 to 1688. The second son of CHARLES I and Henrietta Maria of France, he was immediately created Duke of York. During the Civil War he escaped from England and served with distinction in the French army under Turenne and in the Spanish army under Condé. At the Restoration in 1660 he got the command of the fleet as Lord High-Admiral. He had previously married Anne, daughter of Chancellor Hyde, afterward Lord Clarendon. In 1671 she died, leaving two daughters, Mary (*see* MARY II) and Anne, both of whom were subsequently sovereigns of England and Scotland. Having openly avowed the Roman Catholic faith, on the Test Act being passed to prevent Roman Catholics from holding public office he was obliged to resign his command. He was afterwards sent to Scotland as Lord High Commissioner, where he persecuted the Covenanters.

He succeeded his brother as king in 1685, and at once set himself to attain absolute power. His conversion to Catholicism had made him unpopular, however, and a rebellion was initiated by the Duke of Monmouth, the illegitimate son of Charles II. This was easily put down (the Battle of Sedgmoor in 1685 was the last to be fought on English soil) and encouraged the king in his arbitrary measures. After a series of trials, known as the Bloody Assizes, 320 rebels were executed and 800 transported as slaves.

He then accepted a pension from Louis XIV that he might more readily effect his purposes, especially that of restoring the Roman Catholic religion. The Declaration of Indulgence followed, which led to the imprisonment of seven bishops who opposed the suspension of penal laws against Roman Catholics. The bishops were later found not guilty of the charge laid at them, that of sedition, and were freed. Things came to a head in the Revolution of 1688 which immediately followed the birth of a male heir by the king's second wife, Mary of Modena (the future 'Old Pretender', recognized by Jacobites as James III of England and James VIII of Scotland). Fearing a Catholic tyranny, the king's opponents invited his son-in-law, William of Orange (later William III), husband of Mary, to claim the throne. He landed in November 1688. James found himself completely deserted and fled to France, where he was received with great kindness and hospitality by Louis XIV. Soon afterwards Parliament declared James to have abdicated, soon after and William accepted the throne. James attempted the recovery of Ireland, but the battle of the Boyne, fought in 1690, compelled him to return to France. All succeeding projects for his restoration proved equally abortive, and he spent the last years of his life in ascetic devotion. He died of a stroke at St Germain in 1701.

Significant events of James II's reign
- 1685 – Monmouth fails to depose the king. The Bloody Assizes follow.
- 1686 – James disregards the Test Act and Catholics are appointed to public office.
- 1688 – William III lands in England and James flees to France. James is deemed by Parliament to have abdicated.

James III (1451-1488) King of Scots 1460–1488. Reigned 1460-1488. The son of JAMES II and Mary of Guelders. Succeeding his father at the age of nine, the kingdom during his minority was governed in turn by Bishop Kennedy and the Boyd family. James throughout his reign was much under the influence of favourites, and he quarrelled with his brothers. One, the Earl of Mar, was reputedly murdered, and another, the Duke of Albany, forced to flee to France. Albany obtained English aid and later invaded Scotland with hopes of being crowned as Alexander IV. When James marched to meet him, the nobles seized and hanged some of his favourites, including Cochrane, an architect, who was specially unpopular. Albany was proclaimed 'Lieutenant General of the Realm' but was soon afterwards expelled.

James continued to be on bad terms with his noble, and was eventually defeated by a rebellion led by his son, later JAMES IV, in 1469. After a military defeat at Sauchieburn, near Stirling, the king was murdered, allegedly by a soldier disguised as a priest. By his marriage with Margaret, daughter of Christian I of Denmark and Norway, he brought Orkney and Shetland into the Kingdom of Scotland. They had three sons.

Significant events of James III's reign
- 1470 – Work begins on the Great Hall at Stirling Castle.
- 1472 – Scotland gains Orkney and Shetland from Norway as a royal wedding dowry.
- 1482 – Scotland loses Berwick to the English.
- 1488 – The future James IV leads the rebel noblemen at the Battle of Sauchieburn.

James IV (1473-1513) King of Scots. Reigned from 1488 to 1513. The son of JAMES III, he was in his sixteenth year when he succeeded to the throne and was, either voluntarily or by compulsion, on the side of the nobles who rebelled against his father. He was not judged to require a regent and, feeling great remorse for the manner in which he became king, carried out his duties admirably. During his reign the ancient enmity between the king and the nobility seems to have ceased. His frankness and bravery won him the people's love, and he ruled with vigour, administered justice with impartiality and passed important laws. HENRY VII, then king of England, tried to obtain a union with Scotland by political measures, and in 1503 James married his daughter, Margaret. This was later to become the basis for Stuart rule in England. A period of peace and prosperity followed. French influence, however, and the discourtesy of HENRY VIII in retaining the jewels of his sister and in encouraging the border chieftains hostile to Scotland, led to angry negotiations which ended in war.

Siding with France in 1513, James invaded England with a large force and, despite papal excommunication and his adviser's pleas for caution, engaged in battle. Together with many of his nobles he perished at Flodden Field, having fought at the heart of the battle. He was the last British king to die in battle, and it was later said of him that he was 'more courageous than a king should be'.

He is credited with sponsoring Renaissance values in Scotland and did much to broaden education in his kingdom. But by leaving an heir to the throne, JAMES V, who was barely more than a year old, he had put Scotland's independence in jeopardy.

Significant events of James IV's reign
- 1493 – James subdues the last Lord of the Isles and assumes the title himself.
- 1495 – Aberdeen University founded.
- 1496 – The Scottish Parliament passes education legislation.
- 1503 – James marries the daughter of Henry VII, Margaret Tudor.
- 1507 – Scotland's first printing press is set up in Edinburgh by Andrew Myllar.
- 1513 – James invades England.

James V (1512-1542) King of Scots. Reigned from 1513 to 1542. Succeeded his father, JAMES IV, who had fallen at Flodden when he was only eighteen months old. His cousin, the Duke of Albany, a Frenchman by birth and education, was the regent during his childhood. Owing to Albany's incompetence and the intrigues of the queen mother, Margaret of England, the period of his long minority was one of lawlessness and gross misgovernment. James assumed the reins of government in his seventeenth year. James V was culturally literate (he renovated several palaces, including Linlithgow) but was morally wanting, having at least six illegitimate children and reputedly keeping low company. In order to increase his wealth, he married Madeleine, daughter of Francis I of France, in 1537, but she died just seven months later. On her death James married Mary of Lorraine, daughter of the Duke of Guise, and obtained a large dowry.

James was able to exploit the fears of the Pope that Scotland would follow England in making the king head of the church. In return for his commitment to Rome, James received the right to appoint bishops and benefited from payments from the Church. HENRY VIII, having broken with Rome and eager to gain his nephew over to his views, proposed an interview at York, but James never came, and it is known that Henry had hoped to kidnap him. A rupture took place between the two kingdoms, and war was declared. James was ill supported by his nobles, and after some initial success in holding off an English invasion, his army was crushed at Solway Moss. The defeat destroyed him and he died a broken king only seven days after the birth of his daughter, Mary, who became MARY, QUEEN OF SCOTS.

Significant events of James V's reign
- 1532 – The Court of Session, central court of civil justice, is established.
- 1537 – James marries Madeleine de Valois, but she dies shortly afterwards.

- 1538 – James marries Mary of Guise.
- 1542 – Mary of Guise gives birth to Mary, later Mary, Queen of Scots.
- 1542 – The Scots invade England and are routed at the Battle of Solway Moss.

James VI of Scots and **I of England** (1566-1625). Reigned (as James VI) from 1567 to 1625 and (as James I) from 1603 to 1625. The only son of MARY, QUEEN OF SCOTS, and Henry Stuart, Lord Darnley. When James succeeded ELIZABETH I as King of England he had already been on the throne in Scotland for 36 years. He was first crowned at Stirling, aged 13 months, and at his coronation endured a lengthy sermon from John Knox. His childhood was passed under the direction of the Earl of Mar and the tuition of George Buchanan. He had much trouble with his nobles, a party of whom made him captive at Ruthven Castle in 1582; but a counter party soon set him free. These disputes were connected with the ecclesiastical controversies of this period, James, from his youth onwards, being determined to destroy the power of the Presbyterian clergy. When his mother's life was in danger, he did not exert himself to any great extent on her behalf, and when her execution took place he did not venture upon war. In 1589 he married Princess Anne of Denmark. James took an active interest in the North Berwick witch trials of 1591 in which several women were accused of provoking a storm in the Firth of Forth as the king was returning from Denmark with his bride. It was probably the Earl of Bothwell, an enemy of the king, who was behind these events. He was imprisoned but later escaped. James had little time for the nobility of Scotland, preferring instead to be a king of the commoners. He was also a firm believer in the Divine Right of Kings, the doctrine that holds that kings are appointed by God and are therefore unanswerable to other men.

In 1603 James succeeded to the crown of England on the death of Elizabeth and proceeded to London bringing with him his favourites from the Scottish court, which somewhat alienated the English courtiers. One of the early events of his reign was the Gunpowder Plot, in which a group of fanatical Catholics hoped to blow up the king and all his ministers. He soon allowed his lofty notions of Divine Right to become known, got into trouble with Parliament and afterwards endeavoured to rule as an absolute monarch, levying taxes and demanding loans in an arbitrary manner. In matters of religion he succeeded in establishing Episcopacy in Scotland and forced English Puritans to

conform to the Anglican Church. He also began the plantation of Scottish and English settlers in Ireland (1611) and curbed the powers of Catholic nobles there who objected to their country being treated like a colony.

James was a man of peace, however, and sought at all costs to keep his people out of war. This led to a decline of the navy and a loss of influence overseas as the government did little to support the new colonies. In 1621 the Thirty Years War of Religion, which involved almost all of Europe, began, and one of the Protestant leaders was the German prince who had married James's daughter, Elizabeth (this alliance also ultimately brought the Hanoverian family to the throne). He wished to marry his son, Charles, Prince of Wales, to a Spanish princess, but this was blocked by Parliament, and war was declared against Spain.

James, though possessed of some good abilities, had many defects as a ruler, prominent among them being subservience to unworthy favourites. He was also vain, pedantic and gross in his tastes and habits. He was well educated and enjoyed being called 'the British Solomon'. Henry IV of France is thought to have coined his more enduring nickname, 'the wisest fool in Christendom'. In his reign the authorized translation of the Bible was executed. He died at Hertfordshire of kidney failure, leaving seven children including his successor, CHARLES I.

Significant events of James's reign
- 1591 – The North Berwick witch trials.
- 1603 – James VI of Scotland ascends the English throne to become James I of England.
- 1605 – The Gunpowder Plot is thwarted. Shakespeare writes *King Lear*.
- 1607 – The English Parliament rejects proposals to unite Scotland and England. The English colony of Virginia is founded.
- 1611 – Publication of the Authorized Version of the Bible.
- 1614 – James dissolves the 'Addled Parliament' which has failed to pass any legislation.
- 1616 – Shakespeare dies.
- 1618 – Accused of treason, Sir Walter Raleigh is put to death.
- 1620 – The Pilgrim Fathers reach Cape Cod in the *Mayflower* and found New Plymouth.

John (1167-1216) King of England. Reigned from 1199 to 1216. The youngest son of HENRY II, by Eleanor of Aquitaine. Being left without any lands, he got the name of 'Sans Terre' or 'Lackland', but his brother, RICHARD I, on his accession conferred large possessions on him. He obtained the crown on the death of Richard in 1199 although the French provinces of Anjou, Touraine and Maine declared for his nephew, Arthur of Brittany, who was linearly the rightful heir, then with the king of France. A war ensued, in which John recovered the rebellious provinces and received homage from Arthur. In 1201 some disturbances again broke out in France, and the young Arthur, who had joined the malcontents, was captured and confined in the castle of Falaise and afterwards in that of Rouen, where he died. John was universally suspected of his nephew's death, and the states of Brittany summoned him before his liege lord Philip to answer the charge of murder, and in the war that followed John lost Normandy, Anjou, Maine and Touraine.

In 1205 his great quarrel with the Pope began regarding the election to the see of Canterbury, to which the Pope had nominated Stephen Langton. The result was that Innocent III laid the whole kingdom under an interdict, and in 1211 issued a Bull deposing John. Philip of France was commissioned to execute the decree and was already preparing an expedition when John made abject submission to the Pope, even agreeing to hold his kingdom as a vassal of the Pope (1213). John's arbitrary proceedings led to the rising of his nobles, and he was compelled to sign the Magna Carta, or Great Charter, in 1215. This charter set out to curtail abuses of royal power in matters of taxation, religion, justice and foreign policy. But John did not mean to keep the agreement, and obtaining a Bull from the Pope annulling the charter, he raised an army of mercenaries and commenced war. The barons, in despair, offered the crown of England to Prince Louis of France, who accordingly landed at Sandwich in 1216, and, after capturing the Tower of London, was received as the lawful sovereign. However, the issue was still doubtful when John was taken ill and died of dysentery at Newark later that year.

Significant events of John's reign

- 1202 – Wars with the French king, Philip II, commence.
- 1206 – Pope Innocent III nominates Stephen Langton as Archbishop of Canterbury.
- 1208 – The Pope's interdict prohibits almost all church services in England.
- 1209 – John is excommunicated.
- 1212 – Pope proclaims that John is not the rightful King of England.
- 1213 – John gives way to the Pope's demands.

- 1214 – French defeat the English at Bouvines. English barons meet at Bury St Edmunds.
- 1215 – John reluctantly signs the Magna Carta. Civil war breaks out when the Pope declares that John need not heed the terms of the Magna Carta.
- 1216 – The French join the fray at the invitation of the barons.

Kenneth (I) mac-Alpin (d.858) King of Scots. Reigned from 841 (Scots) and *c*.844 (Picts) to 858. The son of ALPIN, the 34th King of Dalriadan Scots. He united the kingdoms of Scots and Picts by exploiting a period of Pictish weakness due in part to devastating Scandinavian attacks on Pictland although he also had a claim to the Pictish kingship by maternal descent. An ambitious and warlike ruler, Kenneth had completely conquered the Picts by 846. In order to eliminate any opposition to his rule it is thought that he asked the heads of important Pict families to dine with him around this time and had them killed when they fell into a concealed pit he had had dug behind the benches on which they were invited to sit.

He moved the centre of his kingdom as part of the general Dalriadic migration into the lands of the Picts, possibly installing the new centre at Forteviot, the old Pictish centre. In 849 he moved the relics of St Columba to Dunkeld and either founded or enlarged a religious centre there. Kenneth's reign was not a peaceful one, and he made frequent raids on Lothian and on the Saxons and was raided by Britons, Danes and Vikings.

Kenneth also established Scone as a royal and holy centre and made it the place for the inauguration of kings of Alba, as the kingdom was then known. Although it cannot be said with any certainty it has been suggested that it was he who brought the symbolic Stone of Destiny to Scone. On his death through illness he was succeeded by his brother, DONALD I.

Kenneth II (d.995) King of Scots. Reigned from 971 to 995. The son of DUBF, he succeeded CUILEAN in 971. During his reign he was able to secure Lothian as part of his kingdom by recognizing EDGAR, king of England, as his overlord. He was murdered in 995 under mysterious circumstances following a dispute regarding the succession. He was succeeded by CONSTANTINE III, but his son later took the throne as MALCOLM II.

Kenneth III (d.1005) King of Scots. Reigned from 997 to 1005. He ruled jointly with his son, GIRIC II, and both were killed in battle at Monzievaird by MALCOLM II who then ascended the throne.

Kineth (d.843) King of the Picts. Reigned from 842

to 843. The son of UURAD, he is thought to have killed his brother BRED to take the throne. He in turn was usurped and murdered by BRUDE, his nephew.

Llywelyn ap Gruffydd (Llywelyn the Last) (d.1282) Prince of Wales. Reigned from *c*.1260 to 1282. The eldest son of Gruffydd, he styled himself 'Prince of Wales' and received recognition from HENRY III of England in 1267. His kingdom embraced Gwynedd, Powys and Deheubarth, but after refusing to do homage to EDWARD I in 1276 he lost all his lands except the western part of Gwynedd. His brother provoked another war in 1282, and Llywelyn died in a skirmish at Builth. He was buried at the monastery of Cyn Hir.

Llywelyn ap Iorwerth (Llwelyn the Great) (d.1240) King of Gwynedd. Reigned from 1202 to 1240. He reunited the formerly divided kingdom of Gwynedd in 1202 and came to dominate all other Welsh princes. He successfully evaded the attempts of King JOHN to subdue him and exploited a civil war in England to take control of Powys. He was recognized as Wales's strongest ruler in 1218 and gave himself the title of Lord of Snowdon. He was succeeded by his son DAFFYDD AP LLYWELYN.

Loeguire *see* **Niall of the Nine Hostages**.

Ludeca King of Mercia. Reigned in 827. An ealdorman who succeeded BEORNWULF, he reigned briefly before being killed in battle along with five of his earls.

Lulach (1032-1058) King of Scots. Reigned from 1057 to 1058. Installed as king on the death of MACBETH, his stepfather. Within seven months he was ambushed and killed by MALCOLM III at Strathbogie. His death was the result of an ongoing dispute as to which branch of the royal line should legitimately hold the throne of Scotland. Lulach's descendents, including his son Malsnechtai, continued to challenge the mac-Malcolm dynastys' legitimacy without success for a century.

Macbeth (1005-1057) King of Scots. Reigned from 1040 to 1057. A nephew of MALCOLM II, he was one of three kings who came to dominate 11th-century Scotland. He was heriditary *mormaer* (ruler) of Moray, and slew his cousin, King DUNCAN, at Bathgowan, near Elgin, in 1040, and proclaimed himself king. In 1050 he is said to have gone on a pilgrimage to Rome. At the death of their father, the sons of Duncan had taken refuge: Malcolm (later MALCOLM III) with his uncle Siward, Earl of Northumbria, and DONALD BANE in the Hebrides. With Siward's aid, Malcolm invaded

Scotland in 1054. A battle was fought at Dunsinane, but it was not until 1057 that Macbeth was finally defeated and slain at Lumphanan in Aberdeenshire. He was married to Gruach, granddaughter of KENNETH III, and his stepson, LULACH, reigned briefly after his death before being killed by Malcolm, who then claimed the throne.

The legends which gradually gathered round the name of Macbeth were collected by John of Fordun and Hector Boece, and reproduced by Holinshed in his *Chronicle*, where they were found by Shakespeare.

Mael Sechnaill I (d.862) High king of Ireland. His reign was threatened in 861 when his rival, AED FINNLIATH, joined with the Norse kings. He survived their attacks but died the following year and Aed became high king.

Mael Sechnaill II (d.1023) High king of Ireland. Reigned from 1002 to 1023 (interrupted). He abdicated in favour of BRIAN BORU in 1014 following his defeat at the Battle of Clontarf. Boru was killed in his tent shortly after the battle, however, and he was able to regain the title. Following Mael Sechnaill's death a civil war broke out over the succession and his sons were unable to secure the high kingship. This indirectly ended royal rule in Ireland, as for over 150 years no one could unite the various kingdoms.

Malcolm I (d.954) King of Scots. Reigned from 943 to 954. The son of DONALD II, he succeeded to the throne on the abdication of his cousin CONSTANTINE I. His kingdom was constantly under threat from hostile Norwegian forces both to the north (in Caithness and the Northern Isles under ERIK BLOODAXE) and to the south. He was granted Cumbria by EDMUND in return for recognition of Edmund's sovereignty. He attempted to stamp his authority on the northern lands but without success and was killed in battle by the men of Moray. He was succeeded by his nephew, INDULF.

Malcolm II (*c*.954-1034) King of Scots. Reigned from 1005 to 1034. The son of KENNETH II, he ascended the throne after killing his cousin, KENNETH III, who contested the inheritance, in battle at Monzieviard. On Earl Sigurd of Orkney's death in 1014 at the hands of BRIAN BORU at Clontarf in Ireland, his son Thorfinn became a vassal of Scotland and his lands in Sutherland and Caithness came under Malcolm's control. In the early years of his reign he set about attempting to annexe Bernicia and mounted raids on Northumbria. A victory over the Angles with the assistance of OWEN, king of Strathclyde, at Carham on the Tweed in 1018 secured Lothian as part of Scotland.

With Lothian and Strathclyde (probably made a sub-kingdom during Owen's reign) now under his control, Malcolm II had extended his kingdom to include the old lands of Alba and the English-speaking lands to the south. When Owen died childless in 1016, Malcolm's grandson, DUNCAN, succeeded him as king of Strathclyde. Malcolm II had no sons, and on his death DUNCAN I became king of all Scotland.

Significant events of Malcolm II's reign

- 1005 – Kenneth III killed at Monzievaird.
- 1014 – Sutherland and Caithness secured as part of Scotland.
- 1016 – Duncan installed as King of Strathclyde.
- 1018 – Battle of Carham secures the Lothians as part of Scotland.

Malcolm III Canmore (1031-1093) King of Scots. Reigned from 1058 to 1093. The 'Canmore' of his title means 'big head' or 'chief'. During the reign of MACBETH, young Malcolm was under the protection of his uncle Siward and spent his early years in exile in Northumberland. He then visited the court of EDWARD THE CONFESSOR and with the English King's aid Malcolm took the Scottish crown with the defeat of Macbeth at Dunsinane Hill in 1054 and the subsequent killing of both Macbeth (at Lumphanan 1057) and Macbeth's stepson, LULACH, who had assumed the crown on his stepfather's death. Malcolm married twice: his first wife, Ingibiorg, widow of Earl Thorfinn II of Orkney, died in 1069, leaving a son who would later become DUNCAN II; Malcolm then married Margaret, sister of the Anglo-Saxon Prince Edgar the Aetheling, who had fled the Norman invasion with her brother. Margaret had six sons by Malcolm, three of whom became kings: EDGAR, ALEXANDER I and DAVID I. Under Margaret's influence the Scottish court accepted English language and customs as the norm. The queen, an educated woman and devout Christian, encouraged religious reform, and her piety led to her canonization in 1249.

The king and queen welcomed, and indeed encouraged the influx of refugees from WILLIAM I's regime in England, which was a dangerous policy as the Norman king could see the potential menace of the pretender, Edgar the Aetheling, to the English throne residing in a hostile nation whose monarch had married his sister. Malcolm had already made incursions into Northumbria and Cumbria (in 1069 and in 1070) when, in 1072, William invaded Scotland. William forced the Scots king to accept the Treaty of Abernethy, whereby Malcolm

was obliged to acknowledge the English king as overlord. Malcolm's son Duncan was taken to England as a hostage. However, in spite of the treaty, Malcolm once again marched into England, only to be soundly defeated (1079). In 1091 Malcolm was again forced to submit to an English king, William I's successor WILLIAM II. Malcolm met his death in an ambush in 1093 on yet another expedition into England and was not long survived by his wife, Margaret, who, already ill, died four days after hearing of her husband's death. Malcolm, buried with his wife at Dunfermline, was succeeded by his brother DONALD III. His reign began more than two centuries' almost unbroken rule by the House of Dunkeld.

Malcolm IV (the Maiden) (1141-1165) King of Scots. Reigned from 1153 to 1165. The grandson of DAVID I and eldest of the three sons of Henry, the Earl of Northumberland. He ascended the throne on David I's death aged only twelve but had been proclaimed as heir before David's death and had toured the country to ensure that his succession was acceptable. At first the kingdom he inherited was peaceful but there was resentment of David's Normanizing policies and this carried over into Malcolm's reign. In 1157 HENRY II took back the territories of northern England (Northumberland, Cumberland and Westmoreland) which had been granted to David I. Malcolm nevertheless went to France the following year to fight for Henry (for which he received a knighthood) and this was taken by many as a sign of unacceptable subordination. In the west the Lord of the Isles, SOMERLED, founder of the Macdonald clan, attempted to extend the boundaries of his territory and sailed up the Clyde, but Malcolm was able to defeat his aggression. To the north the rebellious men of Moray and Galloway were also contained.

Malcolm, surnamed 'the Maiden' because he did not marry, was probably the last Gaelic-speaking monarch. Weakened by the exigences of kingship, he died in his early twenties at Jedburgh Abbey and was succeeded by his younger brother, WILLIAM I (the Lion).

Margaret (Maid of Norway) (1283-90) Queen of Scotland. Reigned from 1286 to 1290. The daughter of Erik II of Norway and Margaret, daughter of ALEXANDER III. Margaret was declared heiress to the Scottish throne in 1284 whilst still an infant, all of her grandfather's other children having died. When Alexander died accidentally two years later Margaret, aged only three, being the sole surviving descendant of the mac-Malcolm line, became queen. The Treaty of Brigham (1290) arranged the marriage of Margaret and Edward of Caernarvon, son and heir of EDWARD I of England, and guaranteed Scotland's separate existence from England, although the two nations were to be jointly ruled. Margaret set sail from Norway in September 1290 but died a shot time after reaching Orkney, never having set foot on the Scottish mainland. The Treaty of Brigham was naturally negated by Margaret's death and, with no legitimate successor to the past three generations of Scottish kings, bitter disputes over the succession ensued. During the interregnum the claim to the throne was contested by many 'competitors' the main rivals being John BALLIOL and Robert Bruce, Earl of Annandale. After conferences at Norham and Berwick in 1291, EDWARD I, the English king, found in favour of John Balliol who was crowned king of Scots at Scone in 1292. Margaret was buried at Bergen in the Orkneys.

Mary I (1516-1558) Queen of England and Ireland. Reigned from 1553 to 1558. The only surviving daughter of HENRY VIII by Catherine of Aragon, Mary was declared illegitimate when she was born but was restored to her rights when the succession was finally settled in 1544. The first undisputed female sovereign of England, she ascended the throne in early 1553, after an abortive attempt to set her aside in favour of Lady Jane GREY. One of her first measures was the reinstatement of the Roman Catholic prelates who had been superseded in the late reign. Her marriage to Philip II of Spain, united as it was with a complete restoration of the Catholic worship, produced much discontent. Insurrections broke out under Cave in Devonshire, and Wyatt in Kent, which, although suppressed, formed sufficient excuses for the imprisonment of the Princess ELIZABETH in the Tower and the execution of Lady Jane Grey and her husband, Lord Guildford Dudley. England was declared to be reconciled to the Pope and the act *De Heretico Comurendo* against heretics was revived. Nearly 300 perished at the stake, including the bishops Cranmer, Latimer and Ridley. The daily burnings repulsed the people, however, and the martyrs who perished in the fires of Smithfield in London, where most of the burnings were held, secured the triumph of Protestantism in England.

The people were also angry that England, once a powerful independent power, appeared to have become little more than a province of Spain. Philip II dragged England into a war with France which ended in the humiliating loss of Calais in 1558 after it had been held by England for over 200 years;

and Spain, with its detested Inquisition, replaced France as the enemy of the English people. This disgrace told acutely upon Mary's disordered health, and she died shortly afterwards. Feasting and dancing in the streets followed the death of the hated 'Bloody Mary' and the announcement of the succession of her Protestant half-sister, Elizabeth.

Significant events of Mary's reign
- 1554 – The announcement of Mary's intention to marry Philip II of Spain provokes an unsuccessful rebellion. Lady Jane Grey and her husband are put to death. Mary marries Philip II of Spain. Mary repeals Edward VI's religious laws and the persecution of Protestants begins.
- 1556 – Thomas Cranmer, the Archbishop of Canterbury, is burned as a heretic.
- 1557 – England enters into war with France.
- 1558 – England loses Calais.

Mary II (1662-1694) Queen of England, Scotland and Ireland. Reigned from 1689 to 1694. The elder daughter of James, Duke of York, afterwards JAMES II, by his wife, Anne Hyde, daughter of Lord Clarendon. Married in 1677 to her cousin William, Prince of Orange (later WILLIAM III). She was a popular princess in Holland and when the Revolution dethroned her father, she was declared joint-possessor of the throne with William, on whom all the administration of the government devolved. This unique arrangement lasted until her death, after which William ruled by himself.

During the absence of William in Ireland in 1690, and during his various visits to the Continent, Mary managed at home with extreme prudence and, unlike her husband, was well liked by the people. She died childless and is buried in Westminster Abbey.

Significant events of Mary II's reign
- 1689 – The Toleration Act grants freedom of worship to Protestant dissenters. A Scottish revolt is crushed. A Bill of Rights limits regal power. In Ireland Catholic forces loyal to James II besiege Londonderry.
- 1690 – James is defeated at the Battle of the Boyne.
- 1691 – The Treaty of Limerick allows freedom of worship for Catholics.
- 1692 – The Massacre of Glencoe.
- 1694 – The Bank of England is established.

Mary, Queen of Scots (1542-1587) Queen of Scotland. Reigned from 1542 to 1567. The ill-fated Mary Stuart was born at Linlithgow Palace, the daughter of JAMES V by Mary of Lorraine, a princess of the family of Guise. Her father dying when she was only seven days old, Mary was crowned at Stirling and the regency was, after some dispute, vested in the Earl of Arran (from 1554 in her mother, Mary of Guise). In 1543 the infant was betrothed to the six-year-old EDWARD, son and heir of HENRY VIII but that agreement was soon repudiated by the Scots. In retaliation Henry invaded Scotland (1544 and 1545) in what became known as the 'Rough Wooing', and following the defeat of the Scots at Pinkie his armies occupied large parts of south-eastern Scotland. Mary was sent to the island priory of Inchmahome for safety, and the Scots asked the French for help. It was duly given on the condition that Mary be sent to France, and in 1558 she was married to the dauphin, afterwards Francis II.

Mary had made a secret agreement before the marriage, however, that should she die without issue her kingdom should fall to the French Crown. Her husband died seventeen months after his succession to the crown, in December 1560, and in the minority of his brother, Charles IX, power rested with Catherine de Medici. She was not on very good terms with Mary, and three years after the accession of ELIZABETH to the English throne the widowed queen returned to Scotland.

Mary was heir presumptive to the English Crown, and Roman Catholics who did not accept the legality of Henry VIII's marriage to Elizabeth's mother, Anne Boleyn, thought that Mary had a better claim to the throne. But when she returned to Scotland she found that the influence of the Presbyterians was paramount in her kingdom. Though inclined to have Roman Catholicism again set up in Scotland, after a vain attempt to influence the leader of the Scottish Reformation, John Knox, she resigned herself to circumstances, quietly allowed her half-brother, the Protestant Earl of Moray, to assume the position of first minister, surrounded herself with a number of other Protestant advisers and dismissed the greater part of her French courtiers. She even gave these ministers her active support in various measures that had the effect of strengthening the Presbyterian party, but she still continued to have the mass performed in her own private chapel at Holyrood. At first her subjects were quiet, she herself was popular, and her court was one of the most brilliant in Europe.

The calamities of Mary began with her marriage to her cousin, Lord Darnley. He was a Roman Catholic, and immediately after the marriage the Earl of Moray and others of the Protestant lords combined against the new order of things. They were compelled to take refuge in England, and the popularity of Mary began to decline. In addition to this, Darnley proved a weak and worthless profligate, and almost entirely alienated the queen by his complicity in the brutal murder of Mary's secretary, Rizzio, though a reconciliation seemed to be effected between them about the time of the birth of their son, afterwards JAMES VI of Scotland and I of England, in 1556.

About the close of the same year, however, Darnley withdrew from the court, and in the meantime the Earl of Bothwell had risen high in the queen's favour. When the young Prince James was baptized at Stirling Castle in December 1566, Bothwell did the honours of the occasion, and Darnley, the father of the prince, was not even present. Once more, however, an apparent reconciliation took place between the king and queen. Darnley had fallen ill, and was lying at Glasgow in the care of his father. Mary visited him and took measures for his removal to Edinburgh, where he was lodged in a house called Kirk o' Field, close to the city wall. He was there tended by the queen herself, but during the absence of Mary at Holyrood the house in which Darnley lay was blown up by gunpowder.

The circumstances attending this crime were very imperfectly investigated, but popular suspicion unequivocally pointed to Bothwell as the ringleader in the outrage, and the queen herself was suspected, suspicion becoming still stronger when she was carried off by Bothwell, with little show of resistance, to his castle of Dunbar and secretly married to him. A number of the nobles now banded together against Bothwell, who succeeded in collecting a force, but on Carberry Hill, where the armies met, his army melted away. The queen was forced to surrender herself to her insurgent nobles, Bothwell making his escape to Dunbar, then to the Orkney Islands and finally to Denmark.

The confederates first conveyed the queen to Edinburgh and thence to Loch Leven Castle, where she was placed in the custody of Lady Douglas, mother of the Earl of Moray. A few days later, a casket containing eight letters and some poetry, all said to be in the handwriting of the Queen, fell into the hands of the confederates. The letters, which have come down to us only in the form of a translation, show, if they are genuine, that the writer was herself a party to the murder of Darnley. They were held to afford unmistakable evidence of the queen's guilt, and she was forced to sign a document renouncing the crown of Scotland in favour of her infant son and appointing the Earl of Moray regent during her son's minority. After remaining nearly a year in captivity Mary succeeded in making her escape from Loch Leven and, assisted by the few friends who remained loyal to her, made an effort for the recovery of her power.

Defeated by the Regent's forces at the battle of Langside, she fled to England four days later and wrote to Elizabeth entreating protection and a personal interview. This the latter refused to grant until Mary should have cleared herself from the charges laid against her by her subjects. At the end of 1568 commissioners of Elizabeth at York and Westminster heard representatives of Mary and her opponents, with a view as to whether or not she should be restored. No decision was formally made, and for one reason or another Elizabeth never granted Mary an interview, preferring instead to keep her in more or less close captivity in England, where her life was passed in a succession of intrigues for accomplishing her escape.

For more than eighteen years she continued to be the prisoner of Elizabeth, and in that time the place of her imprisonment was frequently changed, her final prison being Fotheringhay Castle, Northamptonshire. During this time there was a series of allegations that she was involved in pro-Catholic plots to depose Elizabeth. She was at last accused of being implicated in the plot by Babington against Elizabeth's life, and having been tried by a court of Elizabeth's appointing was, in late 1586, condemned to be executed. There was a long delay before Elizabeth signed the warrant, but her hand was at last forced and this was done in February the following year. Mary received the news with great serenity and was beheaded a week later in the castle of Fotheringhay.

Authorities are more agreed as to the attractions, talents and accomplishments of Mary Stuart than as to her character. Contemporary writers who saw her unite in testifying to the beauty of her person and the fascination of her manners and address. She was witty in conversation and ready in argument. In her trial for alleged complicity in Babington's plot she held her ground against the ablest statesman and lawyers of England. She was buried at Peterborough Cathedral until transferred in 1612 by her son to Henry VII's Chapel in Westminster Abbey.

Significant events of Mary's reign

- 1548 – Mary is sent to France.
- 1554 – Mary of Guise becomes Regent.
- 1558 – Mary marries the Dauphin Francis
- 1559 – The Scottish Reformation begins with John Knox's return from exile. Mary becomes Queen of France on the accession of her husband (now Francis II of France).
- 1560 – Francis II dies.
- 1561 – Mary returns to Scotland.
- 1565 – Mary marries Lord Darnley, her cousin.
- 1566 – Murder of David Rizzio, Mary's secretary.
- 1567 – Lord Darnley is murdered. Mary marries the Earl of Bothwell, James Hepburn. Rebellion by Scottish lords forces Mary's abdication.

Matilda (Empress Maud) (1102-1167) Queen of England. Reigned in 1141 (uncrowned). The daughter of HENRY I of England and Matilda (Edith), the daughter of MALCOLM III of Scotland. She was married to Henry V, the Holy Roman Emperor, at the age of twelve and ruled Germany as empress until the death of her husband in 1125. She married Geoffrey Plantagenet of Anjou in 1128 and gave birth to three children including Henry Plantagenet, who later became HENRY II of England.

She was named as heir to the English throne when Henry I's only son, William, met his death on the *White Ship* which was smashed onto rocks when returning to England from Normandy in 1120. Henry I had much persuading to do to make his barons accept his daughter as heir, twice summoning his nobility together to obtain their oath to stand by Matilda. However, the existence of another potential claimant to the throne was to produce complications for the succession. STEPHEN, whose mother, Adela of Flanders, was the daughter of William the Conqueror and whose father was Stephen of Blois, the leader of the Norman barons with substantial estates in England, believed that he was the rightful male heir, his elder brother having set aside any claim he may have had. On the death of Henry I, Stephen, who had sworn fealty to Matilda along with the rest of the nobility of England, hurried from Blois to claim the throne and, with the aid of his brother Henry, Bishop of Winchester, was installed as king. However, the early support which Stephen had enjoyed soon evaporated when it became clear that he was not as effective at government as Henry I had been. Many barons had wished to increase their power and influence and began to use Stephen's weakness as king to do so, some remembering their sworn allegiance to Matilda. Stephen also began to lose the support of the Church. In 1138 DAVID I of Scotland invaded northern England on

Matilda's behalf but was defeated at Northallerton at the Battle of the Standard. Civil war broke out in England soon afterwards. In 1139, while her husband invaded Normandy, Matilda invaded England and was welcomed by a grouping of barons who had switched their support from Stephen, including Robert of Gloucester, an illegitimate son of Henry I, who was to prove a useful ally. Matilda set up a base in the west country and at Bristol, Robert of Gloucester's stronghold, and from there further fermented the general rebellion against Stephen's rule. By 1141 the Bishop of Winchester had joined with her cause, and in that same year Matilda defeated the king's forces at the Battle of Lincoln and imprisoned Stephen in Bristol Castle. Henry of Winchester proclaimed at a council that divine intervention had indicated God's will that Matilda be queen of England and Normandy. The Queen then travelled to London to claim her crown. However, her behaviour when she arrived there did nothing to endear her to the populace of that city. She displayed the fiercely arrogant side of her nature to the Londoners and the nobility, ally or otherwise, alike and imposed a heavy tax on the city. Her falling out with Henry of Winchester led him to abandon her cause and retreat to his palace. Matilda, having come so close to gaining the crown, was chased out, the Londoners demanding Stephen's release. Matilda laid siege to the Bishop of Winchester at Wolvesey Palace, hoping to regain his support by force but was herself besieged by Stephen's Queen, Matilda of Boulogne. Matilda escaped, but Robert of Gloucester was captured. This was a major blow to Matilda's ambitions as Robert was her military leader. King Stephen was released in exchange for Robert, but with Stephen restored as king, Matilda's best chance of claiming the throne had passed. The fighting continued but was now more sporadic in nature and gradually diminished in ferocity over the next few years. Matilda was almost captured at her stronghold at Oxford in December 1142 but managed to escape by stealing away into the winter's night, travelling over six miles on foot, in her nightgown, to Wallingford where she obtained a horse. Stephen defeated her forces at the Battle of Farringdon in 1145, but she still fought on, even although her support was growing ever smaller.

In 1148 Matilda finally gave up the fight and left England for Normandy, which her husband had conquered in her absence. Matilda had failed in her attempt on the English throne, but her son, Henry Plantagenet, was more successful. After invading England in 1153 Henry and King Stephen came to an agreement under which Henry would succeed to the throne on the king's death. Stephen died the following year, and Henry II was duly crowned.

Nechton (d.*c*.724) King of Picts. Ruled from 706 to 724. The brother of BRIDEI IV, he embraced Christianity and abdicated, leaving four rivals to contest the succession.

Niall of the Nine Hostages (d.405) High king of Ireland. Reigned from 379 to 405. A semi-legendary figure who gained his name by holding nine members of different ruling dynasties hostage at the same time to secure his position as high king. He was succeeded by a son, Loeguire, who controlled much of northern and central Ireland in the year of St Patrick's mission (432). Two of Niall's other sons founded the kingdom of Aileach in 400 which had 52 known sovereigns until 1170.

Nunna King of Sussex. Reigned *c*.710 to *c*.725. He was an under-king of INE of Wessex and participated in his campaign of 710 to bring Kent under his control. He eventually freed his kingdom from subordination and offered refuge to exiles from Wessex. He made land grants to the Bishop of Selsey between 714 and 720, and one of his charters is the first to mention Sussex formally.

Octa (d.*c*.540) King of Kent. Reigned from *c*.512 to *c*.540. The grandson of HENGEST, he succeeded his father, AESC.

Oengus (d.761) King of Picts. Reigned from 728 to 761. He claimed the throne during the civil war which erupted after NECHTAN abdicated in 724. He killed his cousin, DREST, one of the four claimants, in 729 and took control of the kingdom of the Dalriadan Scots in 736. He attempted to overrun Strathclyde in 750 but was defeated at Mugdock, near Glasgow. His troubled reign lasted for over thirty years.

Offa (d.*c*.720) King of Essex. Reigned in 709. The son of SIGEHERD, he ruled for less than a year before making a pilgrimage to Rome with Cenred of Mercia and Swafred. He was tonsured and died there after founding a hostel for English pilgrims just outside Rome.

Offa (d.796) King of Mercia. Reigned from 757 to 796. A descendant of PENDA's younger brother, he defeated his rival claimants to the throne of Mercia in the civil war that followed the death of ETHELBALD. After an initial period of unrest which saw the Welsh gain territory he established Mercian superiority in all England south of the Humber. Rebellions in Kent led to a ten year period (775-785) in which the independence of that kingdom was re-established. He eventually succeeded in conquering Kent, however, and thereafter it was no more than a province of Mercia. By conquest and marriage he reduced Wessex and East Anglia to almost the same status. He issued the first major royal coinage and built a 120-mile-long defensive wall to protect Mercia from Welsh attacks, which became known as Offa's Dyke. He was also the first king of Mercia to be recognized as a significant power in Europe, and in 789 Charlemagne asked for one of his daughters as a wife for his son. He died at the height of his powers and was succeeded by his son. He had been concerned that the succession would be disputed and had his son, EGFRITH, consecrated as king before his death. Unfortunately his last wish to secure the dynasty was not fulfilled as Egfrith outlived his father by only a few months.

Olaf the White (d.c.854) King of Dublin. Reigned from 853 to c.854. He styled himself 'King of the Northmen of all Ireland and Britain' after uniting all the Viking leaders in 853.

Olaf the Red (b.c.920) King of Dublin. Reigned from c.945 to 980. Following ETHELSTAN's death in 939 he invaded Northumbria and compelled his successor, EDMUND, to cede him an area of land in the north-east Midlands, known as the Five Boroughs. He also gained land in Bernicia, and Edmund had to appeal to the Scots to assist him in keeping Olaf's advances in check. He disputed control of York with his cousin, ERIK BLOODAXE, who was finally driven out by EDRED in 952. He was the longest reigning king of Dublin.

Olaf Guthfrithson (d.941) King of Dublin. He was an ally of the Scots under CONSTANTINE II at the Battle of Brunanburgh in 937. He died in an obscure battle whilst raiding lowland Scotland.

Osbald (d.796) King of Northumbria. Reigned in 796. He was one of the conspirators who killed ETHELRED in 796 but only reigned for a few weeks before being ousted by EARDWULF.

Osbert (d.867) King of Northumbria. Reigned from 850 to 865. Expelled in favour of his brother, AELLE. Both were killed in a joint attack on York which had been seized by the Norse while they were fighting over the throne.

Osmund King of Sussex. Reigned from c.765 to 770. An under-king of the powerful OFFA of Mercia who annexed the kingdom in 772.

Osred I (d.716) King of Northumbria. Reigned from 705 to 716. The young successor of ALDFRITH in 705, he earned a reputation as a tyrant and was murdered.

Osred II King of Northumbria. Reigned from 788 to 790. A nephew of ELFWOLD, he was imprisoned by ETHELRED I and later escaped to the Isle of Man.

Osric (d.729) King of Northumbria. Reigned from 718 to 729. He was succeeded by his nephew, CEOLWULF.

Oswald (St Oswald) (605-642) King of Northumbria. Reigned from 634 to 642. The brother of ENFRITH, he lived in the Hebrides during EDWIN's reign. On his return he established himself as *Bretwalda*, or overlord, of all the Anglo-Saxon kings, with his victory over CADWALLON of Gwynedd. He then tried to check the advance of Mercia under King PENDA and died fighting him on the Welsh Marches. He was a popular figure and gained the nickname 'Bright Blade' for his abilities in battle. He was also a devout Christian and gave Bishop Aiden the island of Lindisfarne. He was the first Anglo-Saxon king to be canonized; the 5th of August is his feast day.

Oswin (d.651) King of Deira. Reigned from 642 to 651. The son of OSWALD, he contested the succession with his uncle, OSWY, and took Deira (southern Northumberland) as his kingdom. He was assassinated by his brother at Gilling in 651, and the two kingdoms of Deira and Bernicia were united.

Oswini (d.690) King of Kent. Ruled jointly from 688 to 690 with SUAEBHARD.

Oswulf (d.759) King of Northumbria. Reigned in 759. The son and successor of EADBERT, he was killed by his own bodyguard within a year of coming to power.

Oswy (602-670) King of Northumbria. Reigned from 651 to 670. The brother of OSWALD, his time as ruler saw the peak of Northumbrian power. After thirteen years ruling Bernicia in MERCIA's shadow he united his kingdom with Deira by assassinating the ruler, his nephew OSWIN, at Gilling. He went on to defeat and kill the powerful King PENDA of Mercia and many of his under-kings in 655 on the flooded River Winwaed (near Leeds) to become *Bretwalda*, or overlord, of all Anglo-Saxon kings. He also contributed to the development of religion in England by presiding over the Synod at Whitby in 664, which resolved many of the disputes between the Celtic and Roman Churches. His daughter Alchfled married PEADA of Mercia only after he was persuaded to convert to Christianity.

Owain Gwynedd (1100-1170) King of Gwynedd. Reigned from 1137 to 1170. He united the kings in the south to resist HENRY II's advances into Wales in 1165. His son married Henry's illegitimate half-sister. He is buried in Bangor Cathedral.

Owen (the Bald) (d.c.1018) King of Strathclyde. Probably the last king of Strathclyde. An ally of

King MALCOLM II of Scotland, he helped defeat Earl Uhtred of Northumbria at the Battle of Carham on the river Tweed (1018). It is thought that by this time Strathclyde was a sub-kingdom of Scotland and Owen only a vassal king.

Peada (d.656) King of Middle Anglia. Reigned from 653 to 656. The youngest son of PENDA, he was made an under-king of Middle Anglia, a kingdom created by his father in 653. He married Alchfled, the daughter of Oswy of Bernicia, who slew his father Penda in 655. Oswy gave Peada his daughter on the condition that he be converted to Christianity. In Bede's *Historia Ecclesiastica* it is claimed that Peada was murdered through the treachery of his wife.

Penda (577-655) King of Mercia. Reigned from 626 to 655. The son of PYBBA, this pagan king established Mercian supremacy by defeating the West Saxons at Cirencester in 628 and the then all-powerful EDWIN of Northumbria in 633 at Heathfield in Yorkshire with the help of CADWALLON of Gwynedd. It was under Penda that the Mercians evolved from being a tribe to being a powerful people and a formidable enemy. Penda's policy as king was to maintain the independence of his kingdom from Northumbrian domination. This led him to make two very destructive attacks on Northumbria as OSWY of Bernicia attempted to reunite the kingdom. He was finally defeated and killed during his last invasion at a battle on the flooded river at Winwaed by the much smaller army of Oswy. He features heavily in Bede's *Ecclesiastical History,* where he is portrayed as an anti-hero pagan warrior-king.

Prasutagus (d.60) King of the Iceni. He ruled the tribal kingdom as a client of the Romans but when he died his lands were seized, his daughters raped and his wife, BOADICEA, flogged. Boadicea led a rebellion agianst the Romans which was eventually quashed.

Pybba (d.c.606) King of Mercia. Reigned from c.593 to c.606. He is thought to have been the son of CREODA, the first king of the Mercians. His family of three sons and two daughters founded the Mercian dynasty.

Redwald King of East Anglia. He reigned from c.593 to c.617 and was considered to be *Bretwalda*, overlord, of all Anglo-Saxon kings in England. He helped EDWIN win the Northumbrian throne in 617 by defeating and killing ETHELFRITH at a battle on the River Idle on the Deiran frontier. He converted to Christianity but lapsed back to paganism, supposedly influenced by his wife. The pagan ship burial site at Sutton Hoo near Ipswich

is thought to commemorate his death.

Rhodi Mawr (the Great) King of Gwynedd. Reigned from 844 to 878. He resisted several attacks by the Vikings and came to dominate Powys and Deheubarth in south Wales. He was eventually forced into exile in Ireland by the Vikings and was killed in battle by a Mercian army on his return.

Richard I (the Lionheart) (1157-1199) King of England. Reigned from 1189 to 1199. The third son of HENRY II by Eleanor of Aquitaine, he was born at Beaumont Palace, Oxford. In his youth, as Duke of Aquitaine, Richard rebelled against his father, at length fighting alongside Philip II of France. On Henry's death, Richard sailed to England where he was crowned at Westminster Abbey. Very much a warrior king, Richard spent only six months of his reign in England and the principal events of his reign are connected with the third Crusade against Muslim rule in the Holy Land in which he took part, uniting his forces with those of Philip of France. In the course of this Crusade he conquered Cyprus, retook Acre and Jaffa and married the Princess Berengaria of Navarre whilst in Cyprus. He failed to take Jerusalem from Saladin but secured access for Christians to the Holy places.

Richard left Palestine in 1192 and sailed for the Adriatic but was wrecked near Aguileia. On his way home through Germany he was seized by the Duke of Austria (in spite of his disguise as a woodsman), whom he had offended in Palestine, and was given up as prisoner to Emperor Henry VI. During his seventeen-month captivity his brother, John, headed an insurrection, which was suppressed by Richard when he returned to England in 1194 after the ransom securing his release was paid (an unlikely legend tells of Richard being found by his favourite minstrel, Blondel, who sang under the walls of each castle he passed on his way back from the Crusades to England).

Richard spent the rest of his life in Normandy fighting Philip II of France. He died of a shoulder wound received whilst besieging the castle of Châlus. Legend has it that he pardoned the archer who fired the arrow that killed him. Richard, although utterly neglectful of his duties as a king, was a popular figure who owed his fame chiefly to his abilities as a military leader and to his personal bravery.

Significant events of Richard I's reign
- 1189 – The Third Crusade is launched.
- 1191 – Richard conquers Cyprus but fails to capture Jerusalem.

- 1192 – Richard captures Jaffa and makes peace with Saladin. Richard is captured by Duke Leopold of Austria.
- 1194 – A ransom is paid and Richard travels to England. Richard leaves England once more to fight Philip II of France.
- 1199 – Richard lays siege to Châlus Castle.

Richard II (1367-1400) King of England and Wales. Reigned from 1377 to 1399. The second and only surviving son of Edward the Black Prince and Joan of Kent and the grandson of EDWARD III. He was born at Bordeaux, succeeded his grandfather at the age of ten, and was crowned at Westminster Abbey. In the years of Richard's minority the government was in the hands of his uncles, firstly John of Gaunt and later Thomas of Gloucester.

These were troubled times: the continuing cost of the Hundred Years' War (1337-1453), the aftermath of the Black Death and John of Gaunt's misrule culminated in the Peasants' Revolt of 1381. The insurrection, led by Wat Tyler at the head of upwards of 10,000 men, was a reaction to the imposition of a poll tax which weighed heavily on the poor whose wages were being held down by legislation (the Statute of Labourers, 1351, held wages and prices at 1340's levels). First introduced in 1377 and again in 1379, a higher tax was imposed in 1380 with officials being sent into the country to collect arrears; violence against the tax collectors became common. On 13th June 1381 Tyler and his men had reached London, leaving in their wake a trail of destruction, and made their demands of the king, which included the abolishment of serfdom and a pardon for all who had taken part in the rebellion. The young Richard agreed to these demands but, after the danger had passed, his government reneged on the agreement, and almost two hundred peasants were killed in the reprisals which followed. However, once the unpopular poll tax was abolished Richard found the revolts of his nobles more difficult to contain. The Lords Appellant, under the Duke of Gloucester, rebelled against the unpopular government of John of Gaunt and seized control in 1387; many of Richard's advisers and personal friends were killed or banished. The Lords Appellant ruled until 1389 when the king, having reached his majority, took over the reins of power. Richard's revenge was exacted in 1397 with killing or banishment of the Lords Appellant, including the exile of Henry of Bolingbroke. John of Gaunt, the Duke of Lancaster, died in 1399 and his estates were confis-

cated by the king. These acts of despotism infuriated many of the nobles and in that same year, whilst Richard was in Ireland attempting to subdue the western part of the country, Henry of Bolingbroke (now Duke of Lancaster on John of Gaunt's death) landed in England and claimed the throne. Parliament accused Richard of having violated his coronation oaths, and he was deposed in favour of Bolingbroke. Thus the Commons became little more than a pawn in the hands of rival court factions.

The deposed king was imprisoned at Pontefract Castle, where he died the following year, either being murdered or starving himself to death. Richard was married twice, firstly to Anne of Bohemia, daughter of Emperor Charles IV, in 1382 (died 1394) and then to Isabella of France, daughter of King Charles VI of France, in 1396 (opposed by the Duke of Gloucester), but neither of the marriages produced children. Something of a tyrant, Richard was nonetheless a feeble ruler who was unable to stamp his royal authority on the kingdom. A lover of the arts, Richard took a particular interest in literature and patronized Geoffrey Chaucer. He has also been credited with conceiving the handkerchief. He was buried at Kings Langley, Hertfordshire, and reburied at Westminster Abbey in 1413.

Significant events of Richard II's reign
- 1381 – The Peasants' revolt.
- 1387 – The Lords Appellant seize power.
- 1389 – Richard assumes control of government.
- 1394 – Richard attempts to conquer the west of Ireland.
- 1397 – Reprisals are taken against the Lords Appellant and Bolingbroke is expelled. Chaucer writes *The Canterbury Tales*.
- 1398 – Richard assumes absolute rule.
- 1399 – Duke of Lancaster (later Henry IV) deposes Richard with the approval of Parliament.

Richard III (1452-1485) King of England and Wales. Reigned from 1483 to 1485. The last of the Yorkist kings was the youngest son of Richard Plantagenet, Duke of York, and Lady Cecily Neville. On the accession of his brother, EDWARD IV, he was created Duke of Gloucester and during the early part of Edward's reign served him with great courage and fidelity, taking part in the battles of 1471 against the Lancastrian supporters of HENRY VI, the rival to Edward's throne. Marriage to Anne Neville, joint-heiress of the Earl of Warwick, brought him great wealth, although disputes over the inherited estates also caused friction between himself and his younger brother, the Duke of Clarence, who married Anne's sister.

From 1480 to 1482 Richard was Lieutenant General in the North, where he won acclaim for his successes over the Scots at Edinburgh and Berwick. On the death of Edward he was appointed as Protector of the Kingdom, and his nephew, the young EDWARD V, was declared king. Richard swore fealty to the king but soon began to pursue his own ambitious schemes. The young king had grown up under the guidance of the powerful Woodville family and Richard had never been on good terms with them since he had objected to the marriage of Edward IV to Elizabeth. Before the coronation could take place, therefore, he moved against the leading members of the family. Earl Rivers, the queen's brother, and Sir Robert Grey, a son by her first husband, were arrested and beheaded at Pomfret. Lord Hastings, who was faithful to his young sovereign, was executed without trial in the Tower. With the support of the Bishop of Bath and Wells he then declared that the king and his brother were illegitimate, as their father had been betrothed to another before he married their mother, and that he, as a result, had a legal title to the crown. The Duke of Buckingham supported Richard, and Parliament had little choice but to offer him the crown. The deposed king and his brother were, according to general belief, smothered in the Tower of London by order of their uncle. The Duke of Buckingham later revolted against Richard, but this came to nothing as the rebellion was crushed and Buckingham beheaded.

Richard governed with vigour and ability and set about making financial and legal reforms, but he was not generally popular and faced increasing opposition to his slight claim to kingship. In 1485 HENRY Tudor, Earl of Richmond, head of the House of Lancaster and rival claimant to the throne, landed with a small army at Milford Haven in west Wales and soon gathered support from the disaffected nobility. Richard met him with an army of 8,000 men at Bosworth in Leicestershire. Richmond's force was initially smaller, but Lord Stanley and Sir William Stanley joined with the Lancastrians and enabled him to win a decisive victory. Richard wore his crown on the field and is said to have come within a sword's length of Henry before being cut down. His body was subjected to indignities and afterwards buried in Leicester. Henry was crowned on the field. The Wars of the Roses were at last over, and the two warring Houses were united by the marriage of

Henry VII to Edward IV's daughter, Elizabeth. The reconciliation was symbolized by the red and white rose of the House of Tudor.

Richard possessed courage as well as capacity, but his conduct showed cruelty, treachery and ambition. His personal defects were no doubt magnified by the character assassinations of historians loyal to the House of Tudor, but he remains one of the most maligned of English kings. Contrary to Shakespeare's portrayal, there is no evidence that he was a hunchback. He is buried at the Abbey of the Grey Friars in Leicester.

Significant events of Richard III's reign
- 1483 – Edward V and his brother are murdered in the Tower.
  Buckingham's rebellion is crushed.
  The College of Arms, which regulates the issue of coats of arms, is established.
- 1485 – William Caxton prints *Morte D'Arthur*.
  The Council of the North is established to govern the north of England.
- 1484 – Bail for defendants in legal courts is introduced.
  English is used for the first time for Parliamentary statutes.
- 1485 – Richard III killed in the Battle of Bosworth.

Robert (I) the Bruce (1274-1329) King of Scots (1306–29). Reigned from 1306-1329. Considered to be the greatest of the Scottish kings, Robert de Bruce VIII was born in Essex and, as the second Earl of Carrick, swore fealty to EDWARD I. In 1297 he fought with the English against William Wallace before joining the Scots in their fight for independence. He briefly returned to his allegiance with Edward until 1298, when he again joined the national party and became in 1299 one of the four regents of the kingdom.

In the three final campaigns, however, he resumed fidelity to Edward and resided for some time at his Court; but, learning that the king meditated putting him to death on information given by the traitor Comyn (a rival for the Scottish throne), he fled to Scotland, stabbed Comyn in a quarrel at Dumfries, assembled his vassals at Lochmaben Castle, and claimed the crown, which he received at Scone. Being twice defeated by the English, he dismissed his troops, retired to Rathlin Island, and was supposed to be dead, when, in the spring of 1307, he landed on the Carrick coast, defeated the Earl of Pembroke at Loudon Hill, and in two years had wrested nearly the whole country from the English. He then in successive years advanced into England, laying waste the country, and, in 1314, defeated at Bannockburn the English forces advancing under EDWARD II to the relief of the garrison at Stirling. (The Monymusk Reliquary, used by KENNETH MAC-ALPIN to carry St Columba's relics from Iona to Dunkeld, legend has it, was carried into battle at Bannockburn). He then went to Ireland to the aid of his brother Edward BRUCE, and on his return in 1318, in retaliation for inroads made during his absence, took Berwick and harried Northumberland and Yorkshire.

In the face of continuing English aggression, 31 lords and earls met at Arbroath Abbey and wrote to Pope John XXII seeking recognition of Scotland as a sovereign state independent of England (the Declaration of Arbroath, or sometimes called the Declaration of Independence). Hostilities continued until the defeat of Edward near Byland Abbey in 1323, and though in that year a truce was concluded for thirteen years, it was speedily broken. Not until 1328 was the treaty concluded by which the independence of Scotland was fully recognized (the Treaty of Northampton).

Bruce did not long survive the completion of his work, dying the following year at Cardross Castle. His heart was buried at Melrose Abbey and his other remains at Dunfermline. He was twice married: first to a daughter of the Earl of Mar, Isabella, by whom he had a daughter, Marjory, mother of ROBERT II; and then to a daughter of Aymer de Burgh, Earl of Ulster, Elizabeth, by whom he had a son, DAVID, who succeeded him.

Significant events of Robert I's reign

- 1306 – Bruce forced to flee Edward I's army.
- 1307 – Edward I dies on his way to Scotland.
- 1314 – The Battle of Bannockburn.
- 1315 – Robert's brother, Edward Bruce, is crowned as High king of Ireland.
- 1320 – The Declaration of Arbroath is drawn up and dispatched to the Pope.
- 1323 – Bruce enters into a truce with the English.
- 1327 – Edward III becomes King of England.
- 1328 – The Treaty of Northampton acknowedges the independence of Scotland.

Robert II (1316-1390) King of Scots. Reigned from 1371 to 1390. The son of Marjory, daughter of ROBERT (I) the Bruce, and of Walter, Steward of Scotland, Robert II was the first of the Steward (later changed to Stuart) kings. During DAVID II's period of imprisonment in England, 'Auld Blearie', as he was known on account of his bloodshot eyes, had acted as regent, having been recognized by Parliament in 1318 as heir to the

throne. On David II's death he was crowned at Scone. He was married twice, firstly to Elizabeth Mure (1348) who bore him nine children before marriage, and secondly, on Elizabeth's death, to Euphemia Ross who bore him four children. An Act of Parliament in 1375 settled the crown on his sons by his first wife, Elizabeth, illegitimate by ecclesiastical law. A feeble king, Robert effectively handed over power to his eldest son, John, Earl of Carrick (later ROBERT III). His reign was comparatively a peaceful one, one of the chief events being the defeat of the English at the Battle of Otterburn in 1388.

Robert III (1337-1406) King of Scots. Reigned from 1390 to 1406. The eldest son of ROBERT II and Elizabeth Mure. He was originally called John, but changed his name on his coronation in 1390. Having been crippled by being kicked by a horse, he was unable to engage in military pursuits, and he trusted the management of government affairs almost entirely to his brother, whom he created Duke of Albany. In 1398 Albany was compelled to resign his office by a party who wished to confer it on the king's eldest son, David, Duke of Rothesay. War was renewed with England in 1402 and an extended raid reached as far as Newcastle, but the Battle of Homildon Hill resulted in a crushing defeat for the Scots. In this year the Duke of Rothesay died in Falkland Castle, where he had been imprisoned, commonly believed to have been starved to death at the instigation of Albany. Dread of Albany, who had recovered the regency, induced the king to send his second son, James, to France in 1406; but the ship which carried him was captured by the English, and HENRY IV detained him as a prisoner for the next 18 years. Shortly after hearing of his son's capture Robert III died heartbroken. He was succeeded by his son JAMES I.

Rory O'Connel *or* **Roderic O'Connor** (d.1198) High king of Ireland. Reigned from 1116 to 1186. He sought Anglo-Norman assistance in his fight to take control of Leinster in 1169. An invasion by Richard de Clare, the Earl of Pembroke, followed in which Wexford, Waterford and Dublin were captured for England. In 1171 HENRY II landed near Waterford to assert his crown rights and receive homage from the native kings; O'Connel recognized Henry as his overlord in 1175.

Run (d.c.878) King of Strathclyde. He married a daughter of KENNETH MAC-ALPIN, the Scots king, and their son, EOCHA, followed CONSTANTINE and AED to the throne, thus strengthening ties between the Scots and Britons of Strathclyde.

Saebert (d.606) King of Essex. Reigned from 605 to 616. An early convert to Christianity, he established a bishopric in London in 605. He was succeeded by his sons, SEXRED and SAEWARD.

Saelred (d.746) King of Essex. Reigned from 709 to 746. He was descended from SIGEBERHT the Good. Little is known of him other than that he died a violent death.

Saeward (d.616) King of Essex. Reigned in 616. The son of SAEBERT, he reigned briefly with his brother, SEXRED. After reverting to paganism the brothers expelled Bishop Mellitus from London and were both killed by the West Saxons as a result.

Sebbi King of Essex. Reigned from 665 to 695. A joint ruler with his nephew SIGHERE, he abdicated to take monastic vows in London and is buried at Old St Paul's.

Sexred (d.616) King of Essex. Reigned in 616. The son of SAEBERT, he reigned jointly with his brother, SAEWARD. Both were killed by the West Saxons after expelling Bishop Mellitus from London.

Sigeberht (d.c.634) King of East Anglia. Reigned from 631 to 634. A half-brother of EORPWALD, he founded the bishopric of Dunwich for St Felix. He also built the monastery at Burgh Castle on the site of a Roman fort.

Sigeberht (d.759) King of Wessex. Reigned from 756 to 757. He was deposed by the council of kings and ealdormen (the *Witan*) and exiled by CYNEWULF who was elected in his place. He was murdered in revenge for the killing of one of Cynewulf's supporters.

Sigeberht I (the Little) (d.653) King of Essex. Reigned from 617 to 653. Succeeded by his son, also SIGEBERHT.

Sigeberht II (the Good) King of Essex. Reigned from 653 to 660. After his baptism he restored Christianity to the kingdom after a generation of paganism.

Sigered (d.825) King of Essex. Reigned from 798 to 825. The last king of Essex before the kingdom became absorbed by Wessex.

Sigeric King of Essex. Reigned from 758 to 798. Ruled as an under-king of Mercia before abdicating.

Sigeherd (d.c.709) King of Essex. Reigned from 695 to 709. Succeeded his father SEBBI and became joint ruler with his brother, SWALFRED.

Sighere (d.c.695) King of Essex. Reigned from 665 to 695. Son of SWITHELM, he shared the throne with his uncle, SEBBI. He married a Mercian princess.

Sledda (d.c.605) King of Essex. Reigned from 587 to 605. He married a sister of Ethelbert I of Kent.

Somerled (d.1164) King of (the Isle of) Man. Reigned from 1158 to 1164. He was the seventh king of Man and became the first Lord of the Isles after expelling the Norsemen in 1140. His nine male descendants claimed the Hebrides until 1493, and he is considered to be the founder of the powerful MacDonald clan.

Stephen (1096-1154) King of England. Reigned from 1135 to 1154. The son of Stephen, Count of Blois, and Adela, daughter of WILLIAM THE CONQUEROR. He went to the court of HENRY I (his uncle) in 1114 and received the countship of Mortain in Normandy. Despite having sworn fealty to the Empress Maud (MATILDA) when she was named as heir, he was persuaded to claim the throne on Henry's death. After convincing the Archbishop of Canterbury that he was the legitimate heir and that he had had King Henry's approval he was duly crowned king of England in 1135. However, there was some rancour from Maud's supporters and several years of unrest ensued. Many barons had sworn fealty to the Empress and some had become disillusioned with Stephen because his leadership was neither as effective nor as strong as Henry I's had been. An invasion of the north of England was undertaken by King DAVID of Scotland on Maud's behalf (1138) but the Scots were crushed near Northallerton at the Battle of the Standard (Scotland, however, retained Cumberland). The rebelliousness of his barons was more difficult to deal with: Robert, Earl of Gloucester, an illegitimate son of Henry I, had attended Stephen's court in early 1136 but now turned against him, and several nobles, such as the Earl of Chester, were only too willing to exploit the weakness of the king and the divisions in the country to enhance their own power and standing.

Stephen had not shown himself able to deal decisively with insurrection thus far, and this did not change with the coming of his rival for the throne. In September 1139 Maud arrived in England, landing at Arundel, and was welcomed by a group of barons, including the Earl of Gloucester, and with their aid she secured a base in the west country and rallied all those disillusioned with Stephen's reign. Stephen was defeated at the battle of Lincoln (1141) and imprisoned. The Empress proceeded to London but, when the Londoners called for Stephen's release and rose against her, she was forced to flee. She fled to Winchester but was besieged there by an army raised by Stephen's queen, Matilda of Boulogne. Maud escaped but Robert of Gloucester, who had become central to the campaign, was captured. Stephen regained his liberty, being exchanged in return for Robert, and he returned to the throne.

The following years saw sporadic outbreaks of warfare between the rival factions, with Stephen almost capturing Maud in her stronghold at Oxford (1142), but the fighting gradually decreased in intensity until 1148 when Maud returned to Normandy. She had, however, laid the foundations for a successful claim by her son and his descendants. The dispute over the kingship of England continued, and HENRY, eldest son of Maud and one day to become HENRY II of England, made his third and most successful invasion of England in 1153 backed by a sizeable army. Stephen fought against him but, with the intervention of the church who wished to see peace and stability restored, a compromise was reached. It was arranged that Stephen should remain king until his death and thereafter Henry would ascend the throne. Stephen died the following year (1154), never having been able to stamp his authority on the realm or bring his wayward barons to heel. Henry II was duly crowned King of England, the first of the Plantagenet kings.

Significant events of Stephen's reign
- 1135 – Stephen becomes King of England.
- 1136 – Stephen subdues baronial revolts.
- 1138 – The Earl of Gloucester defects to Matilda's camp. David I of Scotland invades northern England in support of Matilda. David loses the Battle of the Standard.
- 1139 – Matilda lands in England.
- 1141 – Stephen is captured at the Battle of Lincoln and is imprisoned. Matilda claims the throne. Robert of Gloucester is captured and exchanged for the King.
- 1142 – Matilda is besieged at Oxford but escapes.
- 1147 – Robert of Gloucester dies. Henry (later Henry II) unsuccessfully claims the throne.
- 1148 – Matilda leaves for Normandy.
- 1149 – Henry attempts to take the throne for the second time.
- 1151 – Henry becomes Count of Anjou on the death of his father.
- 1153 – Treaty of Winchester agrees that Henry will become king on Stephen's death.

Suaebhard (d.692) King of Kent. Reigned from 690 to 692. Joint ruler with WIHTRED.

Swalfred (d.c.712) King of Essex. Reigned from 695 to 709. Succeeded his father, SEBBI, and ruled jointly with his brother, SIGEHERD. With his nephew, OFFA, he visited Rome in 709.

Sweyn Forkbeard (d.1014) King of England, Norway and Demark. Reigned (England) from 1013 to 1014. The son of Harold Bluetooth of Denmark and Queen Gunild, he built his North Sea empire through conquest and marriage. In 978 he seized his father's kingdom of Denmark and began making raids on England, often demanding protection payments. In 1000 he attacked Norway and became ruler. Two years later ETHELRED, the king of England, fearing that Sweyn's empire would overrun his kingdom, ordered that the Danes settled in England be massacred. This order was impracticable in many areas where the Danes had strongholds but where the policy was carried out the consequences were terrible. Among the Danes massacred at Oxford was Sweyn's sister, Gunnhild, and Sweyn's resolve to rule England was hardened. The Massacre of St Brice's Day turned support from Ethelred and the king executed many who expressed pro-Danish sympathies. Sweyn enjoyed several early successes at Oxford and Winchester but was unable to seize London despite frequent attempts in 994. By late 1013, however, he had devastated fifteen counties and driven Ethelred from England. He was accepted soon after as king of England but died early the next year following a fall from his horse. He had two sons; Harold IV of Denmark and CANUTE, later to be king of England.

Swithhelm (d.665) King of Essex. Reigned from 660 to 665. He was baptised by St Cedd although his kingdom lapsed back to paganism following the arrival of a plague in 664. Succeeded by his son, SEBBI and his brother, SIGHERE, who ruled jointly.

Swithred King of Essex. Reigned from 746 to 758. He made Colchester the capital of his kingdom.

Talorcen (d.657) King of Picts. The son of EANFRITH of Bernicia.

Talorgen (d.787) King of Picts. He reigned from 785 to 787.

Tincommius (*fl*.15) King of the Atrebates tribe. One of COMMIUS's three sons who divided their father's kingdom and used the title of *Rex*, meaning 'king'. He was recognized by Augustus around 15 BC.

Togodummus (*fl*.50) High king of British tribes. Son of CUNOBELINUS and brother of CARADOC, he resisted the Romans and was probably killed at the Battle of the Medway.

Tytila (d.c.593) King of East Anglia. Reigned from c.578 to c.593. The successor of WUFFA.

Uen (d.c839) King of Picts. Reigned from 837 to 839. Frequent attacks by the Norse, which claimed

Uen, indirectly led to the unification of Pictland and Dalriada. Uen's brothers were also killed, as were most of the members of the major families of the Pictish kingdom, leaving a power vacuum which KENNETH MAC-ALPIN was later able to exploit to become king of both Picts and Scots.

Unuist King of Picts. The brother of DREST IV, reigned from 820 to 834.

Urien King of Rheged. A British king, descended from King COEL, who ruled this kingdom around the Solway Firth in the late 5th century. One of the last rulers, he fought the Bernicians who sought to move north and west from the Humber. The poem *Gododdin* describes one such Bernician attack at Catterick. There is evidence that Urien also allied with Strathclyde. He was assassinated by a rival tribal chief.

Uurad (d.842) King of Picts. Reigned from 839 to 842. Four of his sons claimed the throne after him.

Victoria (1819-1901) Queen of the United Kingdom of Great Britain and Ireland; from 1876 Empress of India. Reigned from 1837 to 1901. The only child of Edward, Duke of Kent, fourth son of GEORGE III, by his wife, Mary Louisa Victoria, daughter of Francis, Duke of Saxe-Coburg, and widow of Ernest, Prince of Leiningen, was born at Kensington Palace, London. Her prospect of the succession to the crown was somewhat remote, for her father might reasonably hope for a male heir; his three elder brothers were alive, and one of them, the Duke of Clarence, afterwards WILLIAM IV, had recently married. The deaths of the princess's father in 1820 and of her cousins, two daughters of the Duke of Clarence, in 1819–20, placed her next in the succession to her two elderly uncles, the Dukes of York and Clarence. The Duke of York died in 1827, and on the accession of William IV in 1830 Victoria became heiress-presumptive to the throne.

She had been brought up very quietly, but from 1830 she began to make public appearances, to the annoyance of King William, who was on very bad terms with the Duchess of Kent. Her coming-of-age, on her eighteenth birthday, was the occasion of some public rejoicings, and when she succeeded her uncle later that year, created a most favourable impression by the tact and composure which she displayed in difficult circumstances. Her accession involved the separation of the crowns of Great Britain and Hanover, the latter passing to the nearest male heir, her uncle, the Duke of Cumberland. In the first years of her reign, Queen Victoria was under the guidance of her Whig Prime Minister, Lord Melbourne, who

devoted himself to the task of training a young girl for her high responsibilities and encouraging her to involve herself in official business. Lord Melbourne and Victoria developed an affectionate relationship, so much so that there were fears that the young queen would become to closely associated with the Whig party. The General Election, which by the then existing law followed her accession, gave the Whigs a reduced but adequate majority; but in the summer of 1839 Melbourne's position in the Commons became so weak that he resigned, and the young queen encountered her first political difficulties in a controversy with Sir Robert Peel who, in taking office, proposed to replace some of the Whig Ladies of the Bedchamber by Conservatives. On her refusal, Peel declined to take office, and Melbourne returned to power for two more years; but the queen, who was taken by surprise and given insufficient time for consideration, afterwards admitted that she had been 'foolish', and no similar difficulty arose again.

The influence of Melbourne diminished after the queen's marriage in 1840 to her cousin, Prince Albert of Saxe-Coburg, whose wide interests and sagacious counsel had an important effect upon the development of the queen's character, though he never acquired popularity in his adopted country. Their first child, Victoria, afterwards the Empress Frederick, the Queen's favourite child, with whom Queen Victoria would correspond on an almost daily basis for more than forty years following 'Vicky's' marriage to the Crown Prince of Prussia in 1858, was born later the same year, and the Prince of Wales (EDWARD VII) the following year; seven other children followed between 1843 and 1857. Queen Victoria reigned during a period of tremendous change both at home and abroad. Conditions for the poor in the industrial north of the country were growing worse due to economic depression, and calls for political change were voiced by the Chartists whose demands included universal male suffrage and secret ballots, and opposition to the Corn Laws of 1815, which kept the price of bread high by banning cheap imports of corn, rumbled on. The Corn Laws were finally repealed in 1846 but the Chartist demonstration of 1848, the year of revolutions all over Europe, achieved little. The industrialization of Britain moved on apace, and the Great Exhibition of 1851, conceived by Prince Albert, displayed over 100,000 industrial products by more than 13,000 exhibitors, over half of which were British, at the purpose-built Crystal Palace in London.

Britain's Empire grew considerably during Victoria's reign. At the height of its expansion, the Union Jack flew over one quarter of the world's land surface, taking in Australia, New Zealand, Canada, many colonies in Africa and the far east and the Indian sub-continent. In India, which had been under the administration of the East India Company with limited supervision from the British government since 1600, a native uprising was to have a profound effect on the development of Britain as an Imperial Power. The East India Company's army in Bengal was an undisciplined unit; grievances over pay, disputes between officers of different castes and rumours that the Government wished to convert India to Christianity by force were among the causes of general unrest, and the court-martial of Indian troopers at Meerut in the spring of 1857 (they had refused to touch munitions which they believed had been greased with the fat of pigs and cows) and their subsequently being stripped of their uniforms turned into a mutiny later that year. Three regiments of sepoys (Indian soldiers) freed the prisoners after murdering their guards and marched on Delhi and, once there, killed every European in sight. The mutiny spread like wildfire. The worst excesses were perpetrated at Cawnpore where over 900 British and loyal Indians, men, women and children, were slaughtered. At length the revolt was overcome and Delhi re-taken. The government of India was passed to the British Crown, who promised equality and freedom of worship for all. The Government, which replaced the East India Company, was detached from those it ruled with an efficiency and impartiality that would become the model for the rest of the Empire's colonies.

Victoria came to be regarded as the figurehead for all of Britain's possessions overseas. She played a considerable part in foreign policy, and on several occasions her personal intervention improved foreign relations, especially with France. No sovereign of this country had left the island since GEORGE II (except for the brief visit of GEORGE IV to Hanover), but Queen Victoria made royal visits a part of the peace-loving diplomacy of her governments, and she paid special attention both to Louis Philippe and Napoleon III, who owed to her his reception into the royal circles of Europe. Her relations with her ministers during this period were generally cordial, though she had grave disagreements with Palmerston, whose foreign policy she distrusted and whose blunt and bullishly worded despatches to British Ambassadors overseas, sent without her consultation, she resented. Victoria considered the possibility of dismissing Palmerston but settled instead on the assurance that any such despatches should first be approved by her. Her desire for the maintenance of peace and her frequent correspondence with the Czar, which was of great help to her government, nevertheless led, just before the outbreak of the Crimean War (1854-56), to many misrepresentations of her attitude, which was supposed to be too friendly to Russia. The Queen also displayed a great fondness for Scotland and made two visits there, in 1842 and again in 1844, and oversaw the rebuilding of Balmoral Castle in Speyside, which was her favourite home.

The death of her husband (who in 1857 had been created Prince Consort) from typhoid fever in 1861 changed the whole tenor of the queen's life. Victoria wore the black of mourning for the rest of her days, and during many years of her widowhood, she lived in almost complete seclusion. She had been the first sovereign to reside in Buckingham Palace, but after 1861 she was rarely in London and preferred Balmoral and Osborne (her home in the Isle of Wight) to Windsor, and her disinclination to appear in public was the subject of numerous complaints.

Her devotion to other duties was, however, undiminished. She continued to exercise some influence on foreign policy and advocated neutrality in the Danish War of 1864. She was on terms of intimate friendship with one of the Prime Ministers of this period of her reign, Disraeli, and his Royal Titles Act of 1876, which conferred on her the dignity of Empress of India, gave her special pleasure. She did not like William Gladstone, and did not conceal her reluctance to ask him to form a government in 1880, and her distrust of his policy was increased by the course of events in Egypt and the Sudan. In the last years of her reign she welcomed the Unionist administrations of 1886 and 1890.

Queen Victoria suffered many family griefs, for two of her children and several of her grandchildren predeceased her. She felt deeply the loss of her son-in-law, the Emperor Frederick, in 1888. She had watched with anxiety the aggressive policy of Germany and Bismarck, and had warned her daughter that this country 'cannot and will not stand' the attempt of the German Empire 'to dictate to Europe', and she trusted that her son-in-law's succession would produce a change in German policy. However, by the end of the century Germany had begun arming rapidly, and ultimately this was to end in war. Her sorrow was increased by the Emperor William's treatment of his mother, but she remained on cordial terms with her grandson to the end of her life.

Her own domestic griefs rendered her sympathetic with the sorrows of her people, and as she grew older she more than recovered her early popularity and was regarded with affection by the whole Empire. This affection was illustrated by the enthusiasm for her person which was shown on the occasions of her jubilees in 1887 and 1897. Her last years were clouded by the outbreak of the Boer War in South Africa (1899-1902) and by the disasters of the opening campaign, but she lived long enough to welcome Lord Roberts on his return in January 1901 after the relief of Mafeking and the annexing of the Transvaal and the Orange Free State. She died less than three weeks later and was buried at Frogmore, near Windsor Castle.

She had reigned for almost 64 years, and few of her subjects could remember when she had not been their monarch. A great sense of loss was felt by the nation at her passing. Queen Victoria was a woman of robust physique, remarkable powers of memory, great force of character, deep sympathy and sincere religious feeling. She was very tenacious of her own opinions, but understood thoroughly the position of a constitutional sovereign, and her strong common sense kept her prejudices in check.

Significant events of Victoria's reign
- 1838 – The Anti-Corn Law League is formed.
- 1840 – The Penny Post is introduced. Victoria marries Albert of Saxe-Coburg-Gotha.
- 1841 – Robert Peel becomes Prime Minister.
- 1842 – China cedes Hong Kong to Britain by the Treaty of Nanking.
- 1845 – The Great Famine takes hold in Ireland.
- 1846 – The Corn Laws are repealed.
- 1848 – The Chartists campaign for political reforms. The year of revolutions in Europe. The *Communist Manifesto* is written by Karl Marx and Friedrich Engels.
- 1851 – The Great Exhibition is held in Hyde Park, London.
- 1852 – The Duke of Wellington dies.
- 1853 – Livingstone discovers Victoria Falls.
- 1854 – The Crimean War commences.
- 1857 – The Indian Mutiny.
- 1858 – Government of India is taken over by the British Crown.
- 1859 – *Origin of the Species* is written by Charles Darwin.
- 1861 – The American Civil War commences. Prince Albert dies.

- 1865 – The foundation of the Salvation Army. The American Civil War ends.
- 1867 – The Second Reform Act doubles the electoral franchise to over two million.
- 1868 – Gladstone becomes Prime Minister.
- 1869 – The Irish Church will cease to exist with the passing of the Disestablishment Act.
- 1870 – The Education Act makes primary education compulsory.
- 1871 – Trade Unions are legalized.
- 1872 – The secret ballot is introduced for elections.
- 1876 – Victoria is created Empress of India.
- 1884 – Electoral franchise is further extended by the Third Reform Act.
- 1886 – The Irish Home Rule Bill is defeated in Parliament.
- 1887 – Victoria's Golden Jubilee year. The Independent Labour Party is founded.
- 1893 – The Second Irish Home Rule Bill is defeated by the House of Lords.
- 1897 – Victoria's Diamond Jubilee year.
- 1899 – The Boer War begins.

Vortigern British tribal king of Kent. Reigned c.450. According to the *Anglo-Saxon Chronicle* Kent was the first of the Anglo-Saxon kingdoms, founded around 450 by Jutes from Denmark and the Rhineland. 'Vortigern', the title used for an overlord, also came to be used as the name of the tribal leader who is thought to have asked mercenaries from Jutland to help him fight off attacks from northern Picts. Led by the brothers HENGEST and HORSA, the Jutes landed at Ebbsfleet, near Ramsgate, and, after driving back the Picts, turned on Vortigern and settled the area themselves.

Wiglaf King of Mercia. Reigned from 827 to 840. He was expelled by the powerful EGBERT of Wessex in 827 but regained the throne within a year. He is buried at Repton Monastery in Derbyshire.

Wihtred King of Kent. Reigned from 690 to 725. Ruled jointly with SUAEBHARD to 692 and married three times. A strong ruler, he successfully resisted repeated Mercian attempts to control his kingdom.

Willam (I) the Conqueror (1027-1087) King of England. Reigned from 1066 to 1087. Born in Normandy, the illegitimate son of Robert, Duke of Normandy, by Arlotta, the daughter of a tanner of Falaise. His father having no legitimate son, William became the heir at his death, Robert of Normandy having made a pilgrimage to Jerusalem from which he did not return. William ruled Normandy with great vigour and displayed tremendous military ability. The opportunity of gaining a wider dominion presented itself on the death of his second cousin EDWARD THE CONFESSOR, king of England, whose crown he claimed by virtue of Edward's promise, made in 1051, that William would succeed him. HAROLD had himself sworn fealty to William after falling into the hands of his rival in 1064. However, on his deathbed Edward named Harold as his successor, and Harold duly accepted the crown. To enforce his claim to the throne William invaded England with a fleet of many hundreds of ships. The decisive victory at Hastings in 1066, in which Harold was killed, ensured his success.

After being crowned on Christmas Day he began establishing the administration of law and justice on a firm basis throughout England, conferred numerous grants of land on his own followers, and introduced the feudal constitution of Normandy in regard to tenure and services. At least 78 castles were built in this period. He also expelled numbers of the English Church dignitaries and replaced them with Normans. In the early years of his reign, however, William had to deal with rebellion against his rule from the Anglo-Saxons, many of whom had been dispossessed of their estates in favour of their Norman conquerors. Uprisings in the southwest (1067) and the north (1069-70) of the country were ruthlessly crushed, and the defeat of the rebellion of Hereward the Wake (1070-1072) effectively ended Saxon resistance. William also invaded Scotland and forced the Scots king, MALCOLM III, to recognize him as overlord at Abernethy, taking his son as a hostage. Towards the end of his reign he instituted in 1085 a general survey of the landed property of the kingdom, the record of which still exists under the title of *Doomsday Book*.

In 1087 he went to war with France, where his son had encouraged a rebellion of Norman nobles. He entered the French territory and destroyed much of the countryside, but when he burnt Mantes, his horse trod on a hot cinder and stumbled, and he was thrown forward in his saddle and received an internal injury that caused his death at the abbey of St Gervais, near Rouen. He left Normandy and Maine to his eldest son, Robert, and England to his second son, William.

Significant events of William I's reign
- 1067 – A revolt in the south-west is subdued. Work begins on the building of the Tower of London.
- 1069 – William subdues the north of England.

- 1070 – Hereward the Wake's Saxon rebellion erupts in eastern England.
- 1072 – William invades Scotland.
- 1079 – Work begins on Winchester cathedral. William is victorious against his son, Robert, at Gerberoi, Normandy. The New Forest is made a royal hunting ground.
- 1086 – The *Domesday Book* is completed.

William (I) the Lion (1143-1214) King of Scots. Reigned from 1165 to 1214. The brother of MALCOLM IV and the grandson of DAVID I, he became king at the age of 22. His first act as king was to attempt to reclaim Northumbria, which had been taken from Scotland by HENRY II in 1157. He mounted an expedition to this end in 1174, timing his move to exploit the strife which the English king was suffering after the murder of Thomas à Becket. However, whilst besieging Alnwick Castle the Scots were taken by surprise by an English force led by Geoffrey of Lincoln and Randulph of Glanville. William was captured and was taken to Henry II at Northampton, with his feet shackled beneath the belly of a horse. Henry II, who had that very day finished his public penance for Becket's murder, must have been soothed by the capture of the Scots king. William was imprisoned at Falaise Castle in Normandy where the following December, he was obliged to accept the terms of the Treaty of Falaise under which he was to acknowledge the sovereignty of England over Scotland including himself, his kingdom and the Scottish church.

Scotland had become a vassal kingdom. The chief Scottish castles were placed under English control and William's younger brother David, along with more than twenty of the Scottish nobility, were taken hostage to England. William was released a few months later but found on his return that in his absence unrest had broken out. The Celtic chiefs took advantage of William's imprisonment and the resentment felt by them towards the friendships of Scottish kings with their Norman neighbours turned into rebellion. The first uprisings took place in Galloway, where several nobles who had been loyal to William and marched with him into England now wished for Galloway to be independent from the rest of William's kingdom. The nobles seized the royal castles, expelled the king's men and requested of Henry II that he take Galloway from William and become its overlord. Only too willing to do this, Henry sent envoys to Galloway, but by the time they arrived the nobles had had a falling out. Order was not restored to Galloway for over ten tears.

The Celts of the north also rose during William's imprisonment. The men of Ross attacked Norman settlers who had been granted lands. William, after his release, travelled north to subdue his unruly subjects in 1179 and established two castles in Ross to keep order. Two years later the king was obliged to go north again, this time to subdue a rival for the throne, Donald MacWilliam, who claimed to be the great-grandson of MALCOLM III and Ingibjorg. MacWilliam, who had become a powerful chief in the North, was killed at Badenoch, but the rebellion took almost seven years to subdue and temporarily took Ross out of the king's control. With the death of Henry II in 1189 relations with England took a turn for the better. William was able to buy back Scotland's sovereignty for 10,000 marks, raised by taxation, on the accession of RICHARD I as the new English king as in dire need of money to fund the Third Crusade. Also, the two remaining castles under English control, Berwick and Roxburgh, were returned to Scotland. With the proclamation of Pope Celestine III in 1192 that the Kirk should be independent under the jurisdiction of Rome the independence of the Scottish Church from Canterbury was restored. The final clause of the Treaty of Falaise was done away with on the agreement of Richard I that William should do homage to the English king only for William's English lands. The independence of Scotland had been regained but William still had designs on Northumbria which remained in English hands. He made his claim to Richard I on the English king's return from the Crusade but to no avail. William tried to buy Northumbria, Westmorland and Cumberland from Richard and made the offer of his daughter Margaret as a bride for Richard's nephew but these attempts were all in vain.

When Richard I died and was succeeded by King JOHN, William renewed his claims to the new king, but John replied by beginning a fortress at the mouth of the River Tweed and shortly after invaded Scotland. William waited for him at Roxburgh Castle, but negotiation took the place of battle and it was agreed that, in return for 15,000 marks and two of William's heiresses, John would not build a castle at the Tweed. William's lifelong ambition to gain Northumbria was never achieved.

Even although William was unable to realize dominion over Northumbria, his reign can be seen as successful. He regained the independence of Scotland and of the Scottish Church and, through his encouragement of the growth of the towns and the creation of many Royal Burghs, was able to improve the lives of his subjects. William is also credited with being responsible for the incorporation of the lion rampant into the royal coat of arms. William married Ermengarde de Beaumont in 1186 who bore him three daughters and one son, who was to succeed him as ALEXANDER II. He died at Stirling at the age of 72 and was buried at Arbroath Abbey which he had founded in 1178.

Significant events of William I's reign
- 1174 – William invades England but is captured and imprisoned in Normandy.
  Under the Treaty of Falaise Scotland loses its independence from England. Revolts break out in Galloway and Ross.
- 1178 – Arbroath Abbey is founded.
- 1179 – William subdues the rebels in Ross.
- 1186 – Order is restored to Galloway.
- 1189 – Henry II dies and is succeeded by Richard I. William buys Scotland's sovereignty back from England for 10,000 marks.
- 1192 – Pope Celestine III declares the Scottish Church to be independent under Rome.
- 1199 – Richard I of England is succeeded by King John.

William II (Rufus) (1056-1100) King of England. Reigned from 1087 to 1100. The third son of WILLIAM I and Matilda of Flanders was born in Normandy and gained the nickname 'Rufus' on account of his florid complexion. He was nominated by his father to the English succession in preference to his elder brother Robert. The Norman barons supported Robert, however, and in 1088 attempted to depose William. The rebellion in Normandy was defeated by William, who secured the aid of Lanfranc, Archbishop of Canterbury, and the English nobles, and Robert was given the Duchy of Normandy in place of the English crown. This was not the last revolt against his rule William would have to deal with as, in 1090, a further rising by his brother necessitated an invasion of Normandy to subdue him. Robert's departure to join the First Crusade in 1096 ensured no further trouble from that quarter.

The Scots, however, proved to be difficult neighbours. An invasion of northern England by MALCOLM III of Scotland in 1091 was defeated, and the Scots king was compelled to accept William as overlord. Nonetheless a further invasion was undertaken by Malcolm two years later. Again the Scots were defeated and Malcolm III was ambushed and killed, together with his eldest son, Edward, at the Battle of Alnwick. Malcolm's son from his marriage to Ingibjorg, DUNCAN, who had resided in England since his being taken hostage by WILLIAM I in 1072, was dispatched north with William II's support at the head of an English army to wrest the throne from Malcolm's successor, DONALD III, in 1094. William's intention was for Duncan to rule Scotland as his vassal, but he only spent a few months on the throne before being killed. In 1097 William sent a second army north, this time with EDGAR, a son of Malcolm III and Margaret, who had sworn fealty to the English king, at its head. This expedition met with greater success, and Edgar was installed as king.

Further warfare was conducted, this time against the Welsh whose risings against the Norman barons in the border lands resulted in an invasion of Wales being undertaken in 1098. William also encountered trouble in ecclesiastical matters, with which he dealt in a somewhat unscrupulous manner. A characteristic incident was his contention with Anselm, Archbishop of Canterbury, in 1097 regarding Church property and the sovereignty of the Pope. Church property was regarded by William as the property of the King and as such he would not allow the election of abbots to vacant abbeys. Anselm asked for leave to receive the Pope's decision on the matter, but this too was contentious as William had not officially recognised the sovereignty of any pope. At length William did recognize Pope Urban II, but the quarrels between them escalated and resulted in Anselm's exile in France and the loss of all his lands.

William II met his death while chasing deer in the New Forest, killed by an arrow shot accidentally or otherwise from the bow of a French gentleman named Walter Tyrrel. The crown then passed to William's brother, HENRY I.

Significant events of William II's reign
- 1088 – Supporters of Robert, William II's brother, rebel in Normandy.
- 1090 – William invades Normandy.
- 1093 – Malcolm III invades England.
- 1095 – Durham Cathedral is founded. A revolt by William's northern barons is put down.
- 1096 – Robert, William's brother, joins the First Crusade.
- 1097 – Archbishop Anselm is exiled.
- 1098 – William enters Wales to subdue a rebellion.
- 1099 – Jerusalem is captured by the Crusaders.

William III (William of Orange) (1650-1702) King of England, Scotland and Ireland (respectively as William III, II and I). Reigned from 1689 to 1702.

The son of William II of Nassau, Prince of Orange, and Henrietta Mary Stuart, daughter of CHARLES I of England. During his early life in Holland all power was in the hands of the grand pensionary John De Witt, but when France and England in 1672 declared war against the Netherlands, there was a popular revolt, in which Cornelius and John De Witt were murdered, while William was declared Captain-General, Grand-Admiral, and Stadtholder of the United Provinces. In the campaign which followed he opened the sluices in the dykes and flooded the country around Amsterdam forcing the French to retreat, while peace was soon made with England. In subsequent campaigns he lost the battles of Seneffe (1674) and St Omer (1677), but was still able to keep the enemy in check.

In 1677 he was married to MARY, daughter of the Duke of York, later JAMES II of England, and the Peace of Nijmegen followed in 1678. For some years subsequent to this the policy of William was directed to curbing the power of Louis XIV, and to this end he brought about the League of Augsburg in 1686. As his wife was heir presumptive to the English throne, he had kept close watch upon the policy of his father-in-law, James II, and in 1688 he issued a declaration recapitulating the unconstitutional acts of the English king and promising to secure a free Parliament to the people. He was invited over to England by seven of the leading statesmen, who feared that James II's newly born son would be brought up as a Catholic. When he arrived suddenly at Torbay (with a fleet of 500 ships, and with 14,000 troops) the greater part of the nobility declared in his favour. James fled with his family to France, and William made his entry into London. The throne was now declared vacant and, upon William and Mary's acceptance of the Declaration of Rights, which defined the limits of regal power and fixed the succession, barring Catholics from the throne, William and Mary were proclaimed joint monarchs of England and Scotland.

The 'Glorious Revolution' was virtually complete in England, but the situation in Scotland was to take a little longer to resolve. A few months after James II's flight, John Graham of Claverhouse, the Viscount Dundee, raised the royalist standard and defeated an English army at Killiecrankie. Dundee was killed in the battle, and without his leadership the rebellion's momentum was lost. A further battle took place in August the same year at Dunkeld, but the outcome was indecisive although both claimed victory. The rising in Scotland rumbled on for a further ten months until the remaining rebels were overcome at Cromdale in early 1690. However, an incident two years later was to prove a boon for Jacobite propagandists. John Dalrymple, Master of Stair and William III's principal minister in Scotland, was placed in charge of effecting the royal decree that each Highland chief must abandon his loyalty to the deposed king and swear an oath of loyalty to his new king. The deadline for taking this oath was set for 1st January 1692, and most of the clan chiefs did as asked. MacIan of the Clan MacDonald was late in giving his oath, having first travelled mistakenly to Fort William. The oath was eventually given, but a troop of soldiers, acting on the orders of Dalrymple, nevertheless set about the slaughter of the 200 families of the Clan MacDonald, who were encamped at Glencoe. The Government inquiry that followed in 1695 was indecisive, but the Scottish Parliament decided, at length, that Dalrymple was responsible and he was subsequently dismissed. However, by now attention in Scotland had shifted to consideration of the ill-fated Darien Scheme. Scotland, excluded from any part of the wealth generated in England from its trade with its far-flung colonies, decided to establish a colony of its own. An Act passed in 1693 for the purpose of encouraging foreign trade, together with the setting up of the Company of Scotland, with powers granted by Parliament to found new colonies, laid the foundations of the scheme. The site for the proposed colony lay in Spanish territory, and Spain did not approve of the founding of a 'New Caledonia' on its property. In addition, the area became a swamp in the summer infested with fever. The two expeditions to Darien were catastrophic. Fever claimed the lives of many of the colonists, and the English possessions in the West Indies refused supplies to the stricken colony, which was then closed down by Spanish troops. It was a financial disaster that was widely felt and certainly did nothing to improve Anglo-Scottish relations.

In Ireland the accession of William III was no less troubled than it had been in Scotland. In 1689 the Catholics rose up against William's rule in support of James II, and some 30,000 or more Protestants fled to take refuge in Londonderry. James II landed with a small number of French troops sent by Louis XIV and laid siege to the city. Relief came more than 100 days later from an English fleet. James and his mostly Irish army moved south, having failed to take Londonderry, to the River Boyne, near Drogheda. In the battle with William III's forces that ensued James II's army was crushed. James escaped to France and was not to return. William, having defeated the rebellions against his reign, attempted to allow Catholics in Ireland freedom of worship in the Treaty of Limerick (1691). However, in this he was thwarted for the Protestant-dominated Irish Parliament subsequently passed harsh laws that effectively made the Catholics second-class citizens.

In the war with France William was less successful than he had been against his enemies at home, but although he was defeated at Steinkirk (1692) and Neerwinden (1693), Louis XIV was finally compelled to acknowledge him as king of England at the Peace of Ryswick in 1697. Friction between William and Louis XIV continued, however, the final straw coming on the death of James II in 1701. Louis XIV acknowledged James's son as JAMES III of England; this showed a blatant disregard for the treaty of 1697. In addition to this, Louis banned the import of English goods to France and advised Philip V of Spain to do likewise.

England, Holland and the Austrian Empire had already combined against Louis XIV in the Grand Alliance of 1701, and the War of the Spanish Succession, to prevent the union of the Spanish and French crowns, was just on the point of commencing when William died from the effects of a fall from his horse. A hard-working and able monarch, William III was nevertheless unpopular, probably due to his reserved nature, his poor understanding of the English language and his undisguised preference for his beloved Holland. He was succeeded by his sister-in-law, ANNE.

Significant events of William III's reign
- 1689 – The Bill of Rights is passed. Jacobites defeat Government forces at Killiecrankie.
- 1689 – James II lays siege to Londonderry.
- 1690 – The Jacobites are defeated at Cromdale. James II is defeated at the Battle of the Boyne.
- 1691 – The Treaty of Limerick allows freedom of worship for Irish Catholics. War with France breaks out.
- 1692 – The Massacre of Glencoe.
- 1694 – Queen Mary dies.
- 1697 – The French war is ended by the Peace of Ryswick.
- 1698 – The Darien Scheme is launched.
- 1701 – The Act of Settlement establishes the Protestant Hanoverian succession. The exiled James II dies. The War of the Spanish Succession begins.

William IV (1765-1837) King of the United Kingdom of Great Britain and Ireland (1830-1837). The third son of George III and Charlotte of Mecklenburg-Strelitz. He served in the navy, rising successively to all the grades of naval command until in 1801 he was made Admiral of the Fleet. In 1789 he had received the title of Duke of Clarence and, after retiring from the Navy in 1790, settled down with the actress Dorothea Jordan. He lived a happily domestic life with his mistress, who bore him ten children, though financial needs necessitated her frequent returns to the stage. However, in 1811, with the worsening condition of George III and uncertainty over the succession, William left Mrs Jordan and searched for a wife, at length marrying Adelaide of Saxe-Meiningen in 1818.

At the age of 64 he succeeded his brother, George IV, as king, amid much concern over his fitness for the crown. He was given to strong language, was well known for his forthright opinions, and his lack of tact had earned him the nickname 'Silly Billy'. However, his blunt speech and lack of pretence soon won him the affection of the public. He was king during a period of some considerable political upheaval, a few of the great events which render his reign memorable being the passage of the Reform Act, the abolition of slavery in the colonies and the reform of the Poor Laws. He was the last sovereign to try to chose his prime minister regardless of parliamentary support; replacing Melbourne with Peel in 1834. He died at Windsor Castle after a reign of only seven years. His two daughters by Adelaide of Saxe-Meningen had died in infancy, and, leaving no other legitimate heir, he was succeeded by his niece, Victoria.

Significant events of William IV's reign
- 1830 – The first passenger steam railways open.
- 1831 – Old London Bridge is demolished.
- 1832 – The First Reform Act greatly increases the electoral franchise.
- 1833 – The Factory Act forbids the employment of children below the age of nine. Slavery in British colonies is abolished.
- 1834 – The Tolpuddle Martyrs are transported. Workhouses are introduced under the Poor Law.
- 1837 – Charles Dickens writes *Oliver Twist*.

Wuffa (d.c.578) King of East Anglia. Reigned from 571 to c.578. Considered to be the first king of the East Angles. The term 'Wuffings' was applied to all following kings up to the time when the kingdom was merged with Mercia around 800. He was succeeded by his son, Tytila.

Wulfhere (d.675) King of Mercia. Reigned from 657 to 675. A younger brother of Peada, he led the Mercian campaign to overthrow Northumbrian hegemony and invaded Wessex in 674. He gave the Isle of Wight to King Ethelwalh of Sussex.

Reignal Table

Saxons and Danes	Reign
Egbert (d. 839) 802–39	
Aethelwulf (d. 858)	839–58
Aethelbald (d. 860), son of Aethelwulf	858–60
Aethelbert (d. 865), second son of Aethelwulf	860–65
Aethelred (d. 871), third son of Aethelwulf	865–71
Alfred (849–99), fourth son of Aethelwulf	871–99
Edward (the Elder) (d. 924), son of Alfred	899–924
Athelstan (895–939), son of Edward	924–39
Edmund (921–46), third son of Edward	939–46
Edred (d. 955), fourth son of Edward	946–55
Edwy (d. 959), son of Edmund	955–57
Edgar (c.944–975), younger son of Edmund	957–75
Edward (the Martyr) (c.963–78), son of Edgar	975–78
Aethelred (the Unready) (c.968–1016), younger son of Edgar	978–1016
Edmund (Ironside) (c.993–1016) son of Aethelred	1016
Canute (c.994–1035), king of Denmark (1014–28) and England	1016–35
Harold I (Harefoot) (d. 1040), illegitimate son of Canute	1035–40
Harthacanute (c.1019–42), son of Canute	1040–42
Edward (the Confessor) (c.1003–66), younger son of Aethelred	1042–66
Harold II (1020–66), brother-in-law of Edward	1066

House of Normandy

William I (the Conqueror) (1028–87), by conquest	1066–87
Wiliam II (Rufus) (1060–1100) third son of William I	1087–1100
Henry I (1068–1135), youngest son of William I	1100–35

House of Blois

Stephen (c.1097–1154), grandson of William I	1135–54

House of Plantagenet

Henry II (1133–89), grandson of Henry I	1154–89
Richard I (Coeur de Lion) (1157–99), third son of Henry II	1189–99
John (1166–1216), youngest son of Henry II	1199–1216
Henry III (1207–72), son of John,	1216–72
Edward I (1239–1307), son of Henry III	1272–1307
Edward II (1284–1327), son of Edward I	1307–27
Edward III (1312–77), son of Edward II	1327–77
Richard II (1367–1400), grandson of Edward III	1377–99

House of Lancaster

Henry IV (1367–1413), grandson of Edward III	1399–1413
Henry V (1387–1422), son of Henry IV	1413–22
Henry VI (1421–71), son of Henry V	1422–61, 1470–71

House of York

Edward IV (1442–83), great-grandson of Edward III	1461–70, 1471–83
Edward V (1470–1483), son of Edward IV	1483
Richard III (1452–85), brother of Edward IV	1483–85

House of Tudor

Henry VII (1457–1509), descendant of Edward III	1485–1509
Henry VIII (1491–1547), son of Henry VII	1509–47
Edward VI (1482–1553), son of Henry VIII	1547–53
Mary I (1516–58), daughter of Henry VIII	1553–58
Elizabeth I (1533–1603), younger daughter of Henry VIII	1558–1603

House of Stuart

James I (1566–1625) (James VI of Scotland), son of Mary Queen of Scots	1603–25
Charles I (1600–49), son of James I	1625–49

Commonwealth and Protectorate

Council of State	1649–53
Oliver Cromwell (1599–1658), lord protector	1653–58
Richard Cromwell (1626–1712), lord protector	1658–59
House of Stuart *(restored)*	
Charles II (1630–85), son of Charles I	1660–85
James II (1633–1701), younger son of Charles I	1685–88
William III (1650–1702), grandson of Charles I (ruled jointly with Mary II ,1689–94)	1689–1702
Mary II (1662–94), daughter of James II, (ruled jointly with William III)	1689–94
Anne (1665–1714), younger daughter of James II	1702–14
House of Hanover	
George I (1660–1727), great-grandson of James I	1714–27
George II (1683–1760), son of George I	1727–60
George III (1738–1820), grandson of George II	1760–1820
George IV (1762–1830), son of George III	1820–30
William IV (1765–1837), third son of George III	1830–37
Victoria (1819–1901), granddaughter of George III	1837–1901
House of Saxe-Coburg/Windsor *(the family name changed from Saxe-Coburg to Windsor during World War I)*	
Edward VII (1897–1972), son of Victoria,	1901–10
George V (1865–1936), son of Edward VII	1910–36
Edward VIII (1894–1972), son of George V	1936
George VI (1894–1952), second son of George V	1936–52
Elizabeth II (1926–), daughter of George VI	1952–

WORLD HISTORY

In contrast to the previous sections of the *Family Encyclopedia*, this section is laid out chronologically and thematically in order to include a précis of the enormous panorama that the history of the world represents, from the earliest beginnings of human civilization to the upheavals of the 1990s.

Before History

Origins – Scientists estimate that the earth was formed some 4600 million years ago. Fossils of the simplest animals and plants have been found in rocks dating from 1000 million years later. The early development of life within those ancient seas was inconceivably slow. The first land plants and animals evolved in the Silurian age, over 400 million years ago. The great dinosaurs ruled the earth for the 160 million years of the Mesozoic Era, which ended some 65 million years BC. The extinction of these giants provided the opportunity for the family of mammals to begin their colonization of the planet.

Some two million years ago several groups of primates, living around the forest edge in Africa, began to show characteristics which might be called 'human'. These creatures began to plan their hunting expeditions and their use of weapons. Other animals use tools – no other animals make tools for something they plan to do tomorrow!

Still, the development of man into the species *homo sapiens* remained immensely slow. Some evolutionary pathways proved to be dead ends. But the spread of the family of man was relentless. For hundreds of thousands of years, small bands of these evolving people moved into new environments, hunting and gathering their food as they went. The animal, man, proved remarkably adaptable, surviving the cold of the ice ages and the heat of the tropics.

The First Agricultural Revolution – The last ice age rolled back some 12,000 years ago, leaving the world with much the same climate that it has retained until today. Comparatively shortly afterwards some people began to introduce major changes into the timeless pattern of life.

Wild wheat and barley live naturally in the area between eastern Turkey and the Caspian Sea. At some time people – probably the women – learnt that it was possible to plant the seeds and so reduce the work of gathering. Soon these new farmers began to select which seeds produced the best crop, and so improved the quality of the crops.

The introduction of cereal farming had radical effects on human life. Tribal groups lost their mobility as they had to settle in one place to tend the crops. When, in time, one group began to produce a surplus, it had to defend its goods against attack. Settlements then needed to be fortified and a military class grew up within the community. Once a community was producing a surplus of food, some people could undertake specialised roles within the community.

The domestication of animals was, no doubt, a long process. There was no sharp dividing line between the time when the people followed herds of wild animals as hunters, and the time when they drove the animals as herders. During the same years after the last ice age, people of southern Asia and Europe domesticated sheep, cattle and pigs. In the millennia that followed, tribesmen from the mountains of northern Iran and the steppes of Central Asia tamed the horse and camel.

Scholars differ about the pattern of development of settled agriculture. The traditional view was that all innovation happened in the Fertile Crescent of the Middle East, and skills spread outward, like ripples on a pond. Others hold that settled agriculture was discovered in many different places as conditions favoured it. Certainly the new methods appeared across Europe, as well as in India and Africa in the millennia which followed. Developments in the Far East and the Americas, at least, were independent of those in the Fertile Crescent. Millet and rice were cultivated in China and South East Asia from about 6000 BC. Here chicken, water buffalo and, again, pigs were domesticated. Change came later in the Americas, where maize, the potato and other important crops were added to the world's store.

The Growth of Cities

As the agricultural age continued, so people began to gather into yet larger communities. The earliest discovered is Jericho, which grew up before 8000 BC. Two thousand years later, Catal Hüyük, in Anatolia, covered 32 acres. These cities provided protection and allowed for greater specialization of role for the inhabitants.

New skills were, indeed, needed. Copper was smelted in Anatolia in about 7000 BC, introducing the age of metals. The earliest known pottery and evidence of the first woollen textiles have both been found in Catal Hüyük.

City life also provided a centre for religious worship. A temple lay at the heart of the community, and religion and government were always closely allied to each other. The change in lifestyle brought with it a change in religious practice. Cave paintings, such as those of southern France, give a glimpse of the cults of the hunter gatherers, which focused on animals and sacred places. These have much in common with the practices of people, like some North American Indians, who lived similar lives within historical times. Settled agriculture brought with it a new emphasis on birth and fertility, symbolized by the mother goddess figures found from widely dispersed areas of this early civilized world.

Sumerian Civilization

Irrigation – As would be expected, the earliest developments in city life happened in regions which had adequate natural rainfall. Some time after 5000 BC, however, groups from the north began to settle in the dry land of Mesopotamia. Here they drained the marshes and used the water from the twin rivers Tigris and Euphrates to irrigate the fertile land.

The Rise of the Sumerian Cities – It appears as though two of the most vital inventions in the history of man – the wheel and the plough – were made in Mesopotamia in around 3500 BC. These enabled farmers to cultivate the irrigated land in a more concentrated way, so increasing the surplus production, leading to a spectacular flourishing of cities.

The most famous city, Ur of the Chaldes, was only one; also prominent were Eridu, Uruk, Bad-tibira, Nippur and Kish. Each city had its own special deity, and it served as the centre for a surrounding region of villages and farm land.

The Invention of Writing – In about 3100 BC the people of these Sumerian cities learnt how to represent their spoken language by the use of writing. The earliest characters were pictographic, and remain largely undeciphered. The Sumerians later developed the more flexible cuneiform script. The invention of writing marks the beginning of history, but the earliest documents were unremarkable. Written on tablets of clay are lists showing the ownership of jars of oil and bundles of reeds. They do show, however, that some of the inhabitants were gathering serious wealth, which could be measured in hundreds and thousands of units.

Life in Sumeria – The cities were walled, but it appears that, in the early centuries, this was not a world of warring cities. Disputes were controlled by the exchange of embassies and by dynastic marriages, rather than by conflict. The laws which governed behaviour were not particularly strict.

The area was short of both wood and stone, and the Sumerian people depended heavily on clay for building and many other functions. The skills of the artisans became ever more refined. Gold, silver, bronze and polished stones were made into fine objects for the decoration of people, homes and temples. Weavers, leather workers and potters followed their specialised crafts. The scribes of later centuries wrote down a fine oral tradition of myths, epics and hymns. The world in which small family groups of hunters lived in co-operation had now been left far behind. Everyday life was controlled by a highly developed bureaucracy, which, for good or ill, was to become a hallmark of civilization. Kings were now divine beings, who were buried with treasure, and also with their whole retinue to see them into the next life.

Egypt

In about 3200 BC King Menes united the whole of the land of the lower Nile. The deserts which stretched on both sides of the river largely protected the Egyptians against the invasions which plagued Mesopotamia. Egyptian rulers had to face the armies of Assyria and 'The People of the Sea' from the Mediterranean, but the remarkable endurance of Egyptian civilization owes much to its isolation. Despite this, the Egyptians owed much to the Sumerians. In particular, they borrowed the early Sumerian system of writing, and adapted this into their own pictorial script. *Hieroglyphics* means 'the writing of the priests' and the art remained a closely guarded secret within the priestly caste.

For more than 2000 years dynasties followed one another; the country experienced bad times as well as good, but a continuity was maintained, unparalleled in the history of the world. Even when the land later fell under foreign rulers, Egyptian culture retained its remarkable integrity.

The Nile Waters – Egypt depended on the Nile. This was a kindlier river than the Tigris and Euphrates because each year it flooded the land on either side, providing natural irrigation for the fertile soil. The whole of Egyptian life was attuned to the rise and fall of the great river. The ruler – or pharaoh, as he would later be called – was the owner of the land and the giver of its life, and the ceremonials of kingship centred on the fertility of the land. The Book of Exodus describes how the rulers of Egypt were able to organize the storage of surpluses from good years to guard against crop failures in bad years.

The Calendar – The Egyptians studied the movements of the sun and stars, and they were the first to work out the year, consisting of $364\frac{1}{4}$ days. For the farmer, this year was divided into three parts, each of four months – one of flooding, one of planting and one of harvesting.

The Capital Cities – Menes set up his capital in Memphis. Later pharaohs moved it to Thebes, but neither were true cities, like those of Sumeria. Their role was more as a centre of religion than a focus for daily life. The wide deserts provided more protection from enemies than any city walls. Because of this physical isolation, Egyptian life could remain focused on the villages, rather than on larger centres of population.

Monuments and Art – The massive monuments of ancient Egypt remain objects of wonder. Imhotep, builder of the Step Pyramid at Saqqara, has left his name as the first architect known to history. Many thousands were marshalled to build these tombs, working without winches, pulleys, blocks or tackles.

A modern visitor will look with awe at the pyramids and other great stone monuments, but it is the more modest paintings which give insight into the daily lives of the people. They show scenes of busy rural life, where peasants gather crops and hunt wild fowl by the Nile. They are happily free from the scenes of carnage and inhumanity, which is all too common in much of the art of the period. It was a world in which women had a high status and beauty was admired. No doubt the peasants had to work hard to keep not only themselves but the whole apparatus of royal and priestly rule, but the river was kind and the land was fertile, and there was usually enough for all.

Migration and Trade

Semites and Indo Europeans – The Semites were herders of sheep who originated in the Arabian peninsula. They were a warrior people, reared in the stern disciplines of life at the desert edge. The most powerful group in those early years them were a people called the Amorites. They founded cities to the north of Sumeria – Babylon, Nineveh and Damascus. The Indo-Europeans were mainly cattle herders, who made their way into Mesopotamia from the north. Their gods emerge in the Pantheon of Greece and in the Vedic deities of India.

The Indo-Europeans had learned how to tame horses from their Asian neighbours. Most importantly, they brought iron. Iron weapons and chariots gave them a technological advantage over the earlier inhabitants of Mesopotamia. Control of iron therefore became an essential precondition of political power. The slow spread of iron technology had other important effects. An iron plough could break in land which had hitherto been too hard for

agriculture. This created a rise in production, and hence an increase in population.

The Growth of Trade – Newcomers from both north and south were drawn into Mesopotamia by the rich lifestyle of the cities, but it happened that the area had no significant iron deposits and was poor in other metals. The need for raw materials was to be the driving force for the development of trade in the ancient world.

Money – It is remarkable how much trade was carried on before the development of currency as a method of exchange. Merchants from the civilized Fertile Crescent were able to take a range of manufactured goods to exchange for metals and other raw materials. Goods were also moved around the world as tribute, taxes and offerings to temples. The first coins date from about 700 BC, but their use spread slowly. Egypt, for instance, did not introduce a currency until about 400 BC.

Land Transport – The wheel was of no value in a world without roads. Columns of pack animals began to spread out into the highlands of Iran and through the Balkans into metal-rich Europe, opening up trade routes which would be trampled for centuries.

Sea Transport – Improvements in the design of ships followed. Oars and sails were developed and rigging improved; decks were made watertight. The Red Sea and Persian Gulf became navigable all the year round, and the Mediterranean at least in summer. The growth in sea transport would ultimately change the centre of gravity of early civilizations away from inland rivers towards coastal regions. Ideas and empires could now spread along sea as well as land routes.

Against this background, the empires of the ancient Near East rose and fell.

Babylon, Assyria and the Hittites

Babylon – In 1792 BC a ruler called Hammurabi came to power in the Semite city of Babylon. He can be looked upon as the first great emperor in the history of the world. Hammurabi's armies carried Babylonian power across most of the Fertile Crescent, from the Persian Gulf and the old Sumerian cities in the east, to the edge of mountains of Asia Minor and the borders of Syria in the west. Conquest was undertaken to secure essential supplies by the control of trade routes, and the exaction of tribute. Carvings show endless lines of conquered people bearing products to swell the stores of the great king, and the riches of Babylon became famous throughout the region.

Hammurabi was an absolute ruler, but he was anxious that his subject should know the laws under which they had to order their lives. He set up pillars in the temples on which were engraved all the laws which governed his kingdom, so that his subjects would be able to come and refer to them. This Code of Hammurabi was the first statement of the principle of 'An eye for an eye'.

Astrology played a vital role in all decision taking, and this led the Babylonians to study the stars closely. By 1000 BC their astrologers had plotted the paths of the sun and the planets with great accuracy, and they were able to predict eclipses. They instituted the system under which the circle is divided into 360 degrees and the hour into 60 minutes.

The first great period of Babylonian power ended when the city was destroyed by the Hittites in 1600 BC After that, Babylon remained a centre of trade and culture, but a thousand years would pass before the city would achieve a late flowering of political power, under the great king Nebuchadnezzar.

The Hittites, who destroyed the first Babylon, were an Indo-European people who had come into the area from the north, probably through the Balkans. After defeating Babylon, they dominated an even larger empire than that of Hammurabi, across the sweep of the fertile crescent, from their homeland in Anatolia, Asia Minor to the Persian Gulf and the borders of Egypt. The power of the Hittites was based on skill with iron. It was they who carried iron technology across the region.

Hittite power collapsed in its turn under pressure from the 'People of the Sea', who were also harassing Egypt. These People of the Sea, however, did not follow up their successes by founding an empire. Rather, they left a vacuum which was to be filled by the most terrible of the empires of the Ancient Near East.

Assyria – The centre of power now moved to the city of Nineveh on the middle reaches of the Tigris. Monuments of the great kings of Assyria, like Tiglath-Pileser I and Ashurbanipal show an empire based on brute military force and the use of terror to control conquered people. Whole populations, like the lost ten tribes of Israel, were moved from their homeland and resettled in other parts of the empire. In this way they lost the identity on which national resistance could be built.

Assyrian armies dominated the region from the 12th to the 7th centuries BC They marched north into the highlands of modern Turkey and Iran, looking for metals and other necessary supplies. They conquered Syria and Palestine, and, under Ashurbanipal in the mid 7th century, they even drove the Pharaohs of Egypt out of the Nile delta.

The Hebrews

Among the Semite invaders into the Near East was a group known as the Hebrews. The Bible record tells how Abraham, the father of the people, left the city of Ur to return to a purer nomadic way of life. His descendants experienced a period of bondage in Egypt, from which they emerged in about 1300 BC.

The Hebrews made their home in Palestine, and they had set up a monarchy by about 1000 BC. Hebrew power reached its peak under King Solomon, who died in 935 BC The kingdom then split; Israel, the northern kingdom, was destroyed by Assyria in 722 BC and Judah, the southern, by Babylon in 587 BC.

The Hebrews do not feature in world history by virtue of their political success, but because of their religious faith. They proclaimed a single deity whom they called Yahweh. Their sacred writings have been one of the major influences on the subsequent history of the world, so some themes need to be identified.

Monotheism – Initially Yahweh was seen as the God of the Hebrews, who was set over the gods of other peoples of the area. In time, however, Yahweh began to develop a uniqueness which challenged the existence of other gods. A writer from the period of the Babylonian exile pronounced Yahweh to be the god of the non-Hebrew, as well as the Hebrew people.

Divine Law – The rulers of Babylon and Assyria were absolute monarchs, whose word was law, and whose actions therefore could not be judged by any superior authority. The Hebrew prophetic tradition, in contrast, made it clear that a king, no less than any other person, operated under a divine law. Here, a ruler, who has unjustly taken a common man's vineyard, can be challenged by a prophet with the words 'Thou art the man!'

Man and Nature – The Hebrew creation myth, which was handed down verbally for many centuries before being written into the Book of Genesis, clearly sets man apart from the rest of creation. He is made in the image of God and given dominion over the beasts. The Bible has been the vehicle which has transmitted this perspective into Western culture.

Male-centred Religion – The Old Testament narrative describes the fierce rejection of female fertility gods of the Fertile Crescent, which were described as The Abomination of Desolation. For the Hebrews divinity was uncompromisingly male, and woman is depicted as a secondary creation, born out of man's side. This rejection of the female strand of religion would later be modified in the Catholic

Christian cult of the Virgin, but it has been influential in defining western attitudes on the relationship of the sexes.

Persia

In about the year 1000 Aryan people moved south into the land which is now Iran (the land of the Aryans). There were two dominant tribes: the Medes occupied the north of the country, while the Persians occupied the south.

In the early centuries the Medish tribes were subject to the Assyrians, but they rebelled against their masters, and in 612 BC Nineveh was sacked and the Assyrian Empire was destroyed by the army of the Medes. The success of the Medes, as of the Persians after them, was based on their successful harnessing of the horse as an instrument of war.

The power centre shifted south when the Persian King Cyrus united Medes and Persians to form what was to become the greatest empire of the Near East. At its height, it extended from Greece and North Africa in the west to the Indus valley and the edge of the Central Asian steppe in the east. Darius the Great had problems at either edge of the empire – with Greeks in the west and Scythians in the east, but the bulk of the empire held together well until 330 BC.

The official Persian religion was Zoroastrianism. This emphasised the struggle of good and evil, and was to give the Semites the concept of angels and hell fire. It did not, however, seek converts, and the people of the empire were left in peace with their own gods. Cyrus was greeted by the Jews as the instrument of Jahweh, and he even rebuilt King Solomon's temple.

Darius was not as successful a conqueror as Cyrus, but he was an administrator of genius. Once a region had been brought within the empire, the royal satrap worked to win the trust and loyalty of the conquered people. Regional traditions were respected and local people were given responsibility in managing their own affairs. The country was bound together by roads which could be used for trade and even postal services as well as for armies.

India

At its peak, the Persian Empire reached as far as the Indus valley. This was the home of another, distinct Asian civilization.

The Harappa Culture – Remains have been found in the Indus Valley of cities, dating from about 2550 BC. The pictogram writing of these early Harappa people has not been deciphered, but archaeologists have discovered houses, with bathrooms, built of burnt brick. There are remains of canals and docks, and Indian products from this period have been excavated in Mesopotamia. Rice was grown, which may indicate that the cities had contact with the Far East. Here is the first evidence of cultivated cotton.

The Harappan cities had houses, granaries and temples, but no palaces. This suggests that the civilization was centred on the priests, rather than on warrior kings, and therefore ill equipped to challenge invaders.

The Aryans – In about 1750 BC Indo-European Aryans began to penetrate into the land from the north. They herded cattle, which were to become sacred creatures. Their religion is enshrined in the oldest holy books of the world, the Vedas. From these it is possible to get an image of nomadic people, standing round a camp fire, chanting hymns to the sun and other forces of nature.

The Aryans overran the northern part of the continent, but they did not completely destroy the people who had been there before them. They slowly spread from the Indus, clearing the dense forest of the Ganges Valley and founding cities, such as Benares.

Hindu Castes – The racial structure of Aryan and non-Aryan people became enshrined in the caste system of India. There were three 'twice-born' castes, which are assumed to originate from the Aryan invaders: *Brahmins* were priests, *Ksahiyas* were warriors and the *Varsyas* were farmers and merchants. Only members of these castes were permitted to take part in the Vedic rituals.

The *Sudras*, who came below the lowest member of the twice-born castes, accommodated the conquered people. Below them were the unclean *outcasts*, who did not enjoy any caste status.

The Cults – As time passed, people looked for religious expressions which could engage the emotions more fully than the Vedic hymns. The cults surrounding the gods *Vishnu* and *Shiva*, with their consorts, fulfilled their needs. It appears that Shiva, at least, was drawn from older pre-Aryan India. The cult of Shiva, who represented the great cycle of birth and death, life and destruction, was to express the Hindu world view most completely.

Buddhism – In the early 6th century BC a prince of the warrior caste, called Siddartha Buddha, left his home to seek enlightenment. He followed strict Hindu practices of fasting but did not achieve his objective. He found that true enlightenment could only be discovered by 'letting go' of his own self and accepting that, in life, all things are change. The Buddha rejected the caste system and his teachings took his followers out of Hinduism.

Although Hinduism and Buddhism separated, any contest for supremacy lay in the mind, for there were no wars of religion, like those which were to mark the West. The two religions share the same root. Both see man as an integral part of the natural world, not as a creature set apart from, and above it, as in the Hebrew tradition.

Buddhism received a great impetus with the conversion of the north Indian king Ashoka in 260 BC. He abandoned his career of conquest and administered his kingdom in the light of the teaching, providing the people with social works and good laws. In the end, Hinduism was to retain its hold on the sub-continent, apart from Ceylon in the south and the mountains of Tibet in the north, while Buddhism made its impact farther east.

Central Asia

Across the Himalayas from India lay the great land mass of central Asia. This can be divided into three bands. Farthest north was the great wall of the forests of Siberia. The centre consists of the Asian grasslands. In the south are the deserts and mountains. The last two are influential in world history from the earliest times until about 1500 AD.

In the grasslands of the steppe lived a selection of nomadic tribes. They survived in marginal land, much as, in later times, the Plains Indians would survive on the American prairie. The nomadic life could take peoples right across the grasslands, and they often fought each other for the control of land. Because the plain could only support a small population, drought, war or other impulses could set whole peoples on the move. This would produce a knock-on effect. Ripples could grow to waves. These would then break onto the boundaries of the lands which bordered onto the steppes.

These were illiterate people, so their names and history is confused, but they appear in history as the Hsiung-nu or Huns, the Avars, the Scythians, the Turks and the Mongols. They were terrible foes, who won their battles by great mobility and superb mastery of the horse.

Farther south, in the desert region, lay the trade routes. From very early times Bactrian camels and horses carried goods along these routes, creating a link between Europe and the Near East to the west and China to the east. Most of the goods moved from east to west. At an early date, the Chinese learnt to make fine fabric from the web of the silk worm. Pepper and other spices also made light and high-value loads. It was an immense and dangerous journey, but the profits were incentive enough to keep the caravans moving.

China

Isolation and Contact – the people of China have long known of their nation as *Chung-hua,* the 'central nation'. Educated people knew well of the existence of other cultures, but they were looked on as subordinate, and, indeed, tributaries of the great nation. Although the Chinese did maintain contact with the outside world, they were little influenced by it. Chinese culture was therefore able to establish a structure in the early centuries which remained little altered throughout history.

The immediate concern of Chinese rulers, again from very early times, was to defend the northern borders against the steppe nomads. This border, which would be marked by the world's greatest building work, The Great Wall, lay along the line where the decline in rainfall made settled agriculture impracticable.

Culture and Language – The huge country centred on three rivers, the Hwang-Ho, the Yangtse and the Hsi. They were divided by great mountain ridges. A wide range of climates could be found within the nation. China has been politically divided for long periods, but she has maintained a unity of culture, beyond that achieved by any other people. An important reason for this is that, while the people of the west came to use to a phonic script, China retained the use of pictograms. The difference is fundamental. A phonic script is easily learned, but needs to reflect the sounds of a language. Therefore, people of different languages cannot communicate without first learning each other's language. A pictogram script, in contrast, is hard to learn, but it is not linked to the sound of language. It can therefore be used to bind people who speak differently. China thus developed a power to absorb and civilize the conquerors who, from time to time, spilled over her frontiers. Literacy was the property of a cultured elite, whose education had, of necessity, been centred on diligence rather than creativity. This gave Chinese culture the twin characteristics of breadth and stability.

The State – Around 1700 BC the first historical dynasty, the Shang, gained control of the northern Hwang-Ho river valley. Even a this early stage, the court had archivists and scribes. Like their successors of later dynasties, the kings saw themselves as the bringers of civilization to barbarian peoples.

About 1100 BC the Shang were overthrown by the Chou who carried royal power to the central Yangtse river valley. Then, around 700 BC, the Chou in their turn were overthrown by pastoralists from the north. This brought in the time graphically known as the Period of the Warring States.

Confucianism – During this period there lived K'ung-fu-tsu, who became known to the world as Confucius. He looked back from that period of unrest to an earlier time when the world was at peace and believed that the problems of his times arose from the fact that people had forgotten their proper duty. In an ordered world, everyone had a place in society. Some – rulers, parents, husbands – were 'higher'; others – subjects, children, wives – were 'lower'. Everyone, high and low, was bound together in ties of mutual duty and respect. The high had no more right to oppress the low than the low had to be disrespectful of the high. When these bonds were broken, the times became out of joint.

Confucianism, thus, placed emphasis on 'conservative' institutions – the state, the civil service, scholarship, and, above all, the family. It was not a religion in the sense of teaching about God, but it did bring a religious dimension to the worship of the ancestors.

Social Structure – K'ung-fu-tsu accepted the most fundamental division in Chinese society. The common peasants were not allowed to belong to a clan, and they therefore had no ancestors to worship. Their lives consisted of an endless round of toil.

For those who were, more fortunately, born into a clan, China would become a land of opportunity. Even boys from poor homes could study to pass the examinations that would open up the coveted civil service jobs. For those with more modest aspirations, growing cities offered opportunities in trade and the crafts.

The fortunate lived in an assurance that Chinese customs and the Chinese way offered the model of excellence, and all other people had to be judged according to the way in which they measured up to this standard.

MEDITERRANEAN CIVILIZATION

Early Sea-going People

Conquering the Oceans – The earliest civilizations centred around major river valleys. The rivers provided water and arteries of communication. Then the technology of sails and ship building improved to a level which enabled men to venture onto the oceans. From early times, the Red Sea and the Persian Gulf provided important communication routes, which were orientated towards the east. The Mediterranean, particularly in winter, is subject to violent storms. Further advances in marine engineering, such as the construction of watertight holds and improvements in sails and rigging, were needed before sailors could master this environment.

By about 500 BC ships were able to move freely in the Mediterranean, at least in summer, so providing easier communication than was possible on land. There was then no distinction between a fertile north and an arid north shore. The whole region was fertile. Traders and rulers therefore saw the Mediterranean basin as a single unit, bound together by its ocean highway.

Minoa and Mycenae – In about 1900 BC a civilization grew up on Crete, which has been named after its king, Minos. Its earliest writing has not been deciphered, but excavations reveal fine palaces and developed communities. Their cities stood beside the sea, and the builders were already confident enough in the control of their ships over the eastern Mediterranean to dispense with fortifications.

Objects found in Crete and Egypt show there was a lively trade between the two cultures. The Minoan sailors probably traded in timber, wood, olive oil and grapes over the whole of the Mediterranean area.

Inhabitants of the Minoan cities were the first people to enjoy the benefits of piped drains and sewers, and wall paintings show them dancing and playing sports.

The Minoans set up colonies on the mainland, of which the most important seems to have been at Mycenae. This is the name which is given to Minoan civilization as it is found on the mainland. The culture spread across the Aegean to the coast of Asia Minor and to the city of Troy at the mouth of the Bosphorus.

The first Minoan civilization was destroyed by Indo-European people who poured into the region from the north. Some of these invaders settled in the Ionian peninsula to become Greeks. A later resurgence of Minoan civilization is thought to have been under Greek influence.

The stories which were written down centuries later by the poet Homer tell of the struggles between the Mycenaens of Troy and the less advanced Indo-European invaders.

The Phoenicians – Semite people, in general, liked to keep their feet on dry land. The exception were the people who lived in the area known as Phoenicia, which is now Lebanon and southern Syria. They developed remarkable skills as sailors and for centuries their ships dominated the trade routes. Phoenician sailors reached the Atlantic Ocean and traded with tin miners in distant Cornwall. The Greek historian Heroditus even reports that one expedition rounded the southern cape of Africa.

The Phoenicians planted colonies to protect their trade routes. Most important, about 800 BC they founded the city of Carthage. The colony was strategically placed to protect the ships which brought metal from Western Europe.

Phoenicia was never a power on land, and when, in 868 BC the Assyrian king 'washed his weapons in the Mediterranean' Phoenicia lost its independence. But the rulers of the great empires needed these fine sailors, and the Phoenicians therefore exercised influence beyond their military power.

Phoenicia is best known for its sailors. It did, however, make another major contribution to western culture, by creating a phonic alphabet. The words Alpha, Beta and Gama are derived from the Phoenician words for an ox, a house and a camel.

The Greeks

In Mycenaean times, an iron-working Aryan people were moving south into the Greek peninsula. Myths of early battles with Mycenaean Troy are preserved in the works of the story teller – or tellers – given the name of Homer.

The early culture was oral, but, in around 750 BC the Greeks adopted and modified the Phoenician alphabet and committed the ancient legends to writing. These were to provide the starting point for the world's first great literary culture.

The beginning of Greek civilization was dated from the first Olympian Games, held in 776 BC. This event, held once every four years, drew together people who shared the Greek language and culture. The participants did not, however, come under one unified government.

Government – The Greek political structure was dictated by the geography of the region in which they settled. This was a land of mountain ranges, with small coastal plains, which faced outwards to the sea. Each of these plains was settled by a self-governing community, which initially contained only as many people as the land would support. This was the basis of the *polis*, or city state.

The *Iliad* provides a picture of an early feudal society of kings, nobles and common fighting men. Each city then followed its own course in working out the structure of government. The first struggle lay between the kings (monarchy) and the nobles (aristocracy). Then pressure came from other influential citizens (oligarchy) and from the mass of free male citizens (democracy). When a state plunged into chaos, a strong man (tyranny), often benevolent and public-spirited, would emerge to bring order.

The Greek concept of democracy was specific to the confined structure of the city state. It did not operate through representative institutions but through the direct participation of citizens in the decision-taking process. The meeting place, or *agora*, not the temple or the royal palace, was now the centre of city life. The citizens who met here provided the city with its law courts and its political assembly. Debate and persuasion became vital skills. People could on occasion be swept away by the power of a demagogue, but within this forum they learned to listen and to analyse argument.

The fractured nature of Greek society did not provide peace and stability. The city states might join together in games, but they were as often at war with each other. The people remained fiercely independent, more ready than any other people before, and perhaps even since, to question the structure of the society within which they lived.

Colonization – Since geography prevented expansion inland, the Greeks had an impetus to expand outwards, along the sea routes. Greek communities were established along the west and south coasts of Asia Minor, on the islands of the Aegean and as far east as Cyprus and westwards to Sicily and southern Italy, and even farther into North Africa, France and Spain. These colonies were self-governing, but they often had links with powerful city states, such as Corinth or Athens. They served as an overspill for excess population and as trading bases across the Mediterranean Sea.

The Persian Wars – the conflict between Greece and Persia has been depicted as a struggle between an oppressive empire and a freedom-loving people. Reality is more complex. Close links had long existed between the Greeks and the Persians, and many Greeks served in the Persian army. The trouble started when Greek city states in Asia Minor rebelled against Persian rule and Darius moved to put down the insurrection. The Asians were supported by the European Greeks, and this brought the Persian Empire into conflict with an alliance of Greek cities, led by Athens and Sparta. The army of Darius was defeated at Marathon in 490 BC and the navy led by his successor Xerses failed ten years later at Salamis. This war drew the boundary of the Persian Empire to the east of the area of Greek settlement.

Athens and Sparta – The alliance which had defeated Persia did not survive the victory. Athens was much the largest of the city states, with a larger population than its farm land could support. Prosperity was based on the control of silver mines worked by thousands of slaves. The city's survival depended on a structure of trade and colonies. Whatever freedom may have been enjoyed by Athenian citizens within their city, their rule of others was often oppressive. The Athenians demanded heavy tribute from client states and put down rebellion as violently as any Persian army.

Other trading states, like Corinth, felt themselves continually threatened by Athenian power. They found allies in the conservative, agricultural state of Sparta. The Peloponnesian War lasted for 27 years, and ended with the defeat of Athens in 404 BC. This led to a reaction against an over-mighty Sparta, and the destructive sequence of wars continued into the 4th century. The Greeks may have provided the world with a vocabulary of politics and an ideal of democracy, but its outstanding achievement lies, not in politics, but in broader fields of culture.

Religion – Greek myth is drawn from the common Indo-European root, which created the Vedas in India. It has provided a fertile source of inspiration for western art and literature for more than 2000 years, but it is harder to look back through the twin filters of Semitic religion and rationalism, which have shaped modern attitudes, to understand what the world of gods meant to the Greeks themselves. On the one side, there was a piety of the common man, which condemned Socrates for blaspheming against the gods; on the other side was a free-thinking strain expressed by the philosopher of 7th-century Miletus, who declared 'If an ox could paint a picture, its god would look like an ox'. The Greek religious tradition was real, but it was not an all-demanding way of life like that of the Hebrews.

Philosophy and Science – The Greeks invented organized abstract thought and took it to a level which would dominate the philosophy of the Near East and Europe until very recent times. In the Greek perspective, there was no distinction between the arts, the sciences, and, indeed, religion. All were a part of the search for truth. In the 6th century, Pythagoras did not distinguish mathematics from philosophy and religion. The two greatest Greek thinkers defined the twin, often opposing, channels through which all philosophy, and later, all theology, would flow.

Plato, a pupil of Socrates, was 23 years old when his home city of Athens was defeated by Sparta. His attempt to achieve a mental order was therefore born of the political disorder of the post war years. Plato is the apostle of the *ideal* – the abstract of perfection, whether it be for the state, the individual, or in a mathematical equation. In his philosophy, all life is a striving towards an ideal of the good, containing truth, justice and beauty, which was the only reality in an imperfect world. Plato's Academy can lay claim to being the world's first university.

Aristotle came to Plato's Academy at the age of 17 and remained his master's devoted disciple. His interests, however, took him in the opposite direction, as he came to emphasize enquiry and experiment as the source of knowledge. While Plato stressed the *ideal*, Aristotle stressed the *real*; while Plato was drawn into the abstractions of mathematics, Aristotle found himself fascinated by the complexities of biology and literary criticism. For him, truth lay not in a distant abstract, but in a 'happy medium'. Aristotle is therefore seen as the father of the scientific method.

The Arts – 5th-century Athens provided the most fertile environment for classical Greek culture. The architecture of the Parthenon, the sculptures of Praxiteles and Pheidias provide an illustration in stone of the Platonic ideal. They provided generations of architects and artists, particularly from Europe, with a standard of perfection. Literature also flourished. Aeschylus, Sophocles and Euripides used the ancient myths to explore depths of the human experience and create tragic drama, while the irreverent Aristophanes, pioneered the tradition of comedy. The disasters of the Peloponnesian Wars also inspired Thucydides to become the world's first scientific and literary historian.

The contribution of Greece to the world's cultural store is a fundamental theme of history. By the middle of the 4th century, however, the advances were largely confined to the Greek speaking world. The diffusion of Greek culture into a wider world would be the work of a young and brilliant student of Aristotle.

The Hellenistic World

Alexander the Great – The state of Macedon lay to the north of Greece, crossing the boundary which divided the civilized world from the barbarians. Philip II of Macedon developed his army into an efficient fighting machine and conquered the Greek city states. Philip died in 336 BC and was succeeded by Aristotle's pupil, his son, Alexander.

Alexander inherited his father's army and the Greek power base. The problem he faced was how he could pay the soldiers who had served Macedon so well. This search for money took Alexander on spectacular campaigns. There was ample booty to be won across the Aegean in the Persian Empire. In 334 BC the Macedonian army defeated the Persians at Issus then marched south into Egypt, where Alexander founded the city which was to carry his name. He returned north, defeated the Persians once more and sacked the capital, Persepolis. Not content, he took his army eastward into Afghanistan and the Punjab. He would have gone further, but his soldiers insisted that the time had come to turn back.

The young man was one of the great soldiers of all time. The importance of his conquests, however, was that they were the catalyst which brought together the old civilizations of the Near East and the newer Greek culture. Alexander was Greek, but he was drawn to Eastern ways. He himself married the Persian emperor's daughter, and, in a great symbolic gesture, he married 9000 of his soldiers to eastern women.

The Division of the Empire – Alexander died in Babylon in 323 BC at the age of 32, leaving no heir to succeed to his enormous empire. The land was divided between his generals. The Ptolemies based their power in Egypt, the Seleucids in the region of Syria and the Attalids around Pergamum. Parthia later became independent of the Seleuchids. These were centralized states, under absolute monarchs. The age of debate and democracy was certainly past. Over most of the Hellenistic world, this was a time of economic growth, but the Greek cities themselves declined.

Hellenistic Culture – Greek was now the official and commercial language of the whole area. The learning of the scholars became widely known and great libraries were set up at Alexandria and Pergamum. Among the books preserved were many of the writings of Plato and Aristotle. Scholars in the Greek tradition worked in different parts of the Hellenistic world. Science flourished, as it would not do again for over 1500 years. In Alexandria Euclid laid the foundation of geometry. Aristarchus correctly deduced the structure of the solar system 1800 years before Copernicus and Eratosthenes measured the circumference of the earth. Archimedes of Syracuse had the widest ranging genius of all.

Philosophers, such as the stoics, could no longer question the ways of government, so they turned their thoughts towards the inner life of man. They led a quest for virtue and true contentment. Classical Greek styles provided powerful models for painters and sculptors, but Hellenistic artists retreated from the Platonic search for an ideal and worked instead to project the humanity of their subjects.

Religion – Greek religion was too restricted a vehicle for this new, expansive world. Mystery cults began to spread which demanded a more active devotion from their followers. Two of these became increasingly dominant. From Egypt came the myth of Isis and Osiris. This told of a dying and a rising god. From Zoroastrianism came the mystery of Mithras, with its powerful image of redemption through blood.

The End of the Hellenistic World – The Hellenistic empires in their turn fell to a new power from further west in the Mediterranean. The Roman victory at Actium in 31 BC marked the end of the era. No battle, however, could put an end to Greek culture. The Roman poet Horace summed it up by saying that, although Greece was defeated, it took its conquerors prisoner.

Republican Rome

The Etruscans – In the years before 509 BC central Italy was dominated by a people called the Etruscans. They can be seen in lifelike tomb sculptures, but little is still known about their culture. They appear to have been an Indo-European people, who achieved dominance over other people by bringing iron working to a high level of perfection. Etruscan kings, the Tarquins, ruled in Rome until they were expelled, according to tradition in 509 BC. The expulsion of the kings remained a powerful myth within the Roman state. Men looked back to the days of the Tarquins as the time when the rights of the citizen were subjected to the will of a single individual.

The Structure of the State – The new Roman state was based on agriculture. Indeed, *pecunia*, the name for a flock or herd of animals, became the Latin word for money.

There were different groups within society. The old families, who took pride in their status as *patricians*, assumed power in place of the deposed kings. The remaining free people were known as the *plebs*. At first they were poor farmers with little say in affairs of state. As Rome grew, many plebeians became more wealthy and they began to look for a share in the running of affairs.

Romans, be they patricians or plebs, took immense pride in their status as citizens. Roman citizenship became a unique badge of belonging to a pure and strong society, free from the softness and corruption of the Hellenistic world around them. Every man was liable to military service which could be for as long as 16 years in the infantry or 10 years in the cavalry. Warlike virtues were admired by society and inculcated in boys through the home and education. At best, this could breed a self-sacrifice to the common good; at worst it could bring a lust for battle and bloodshed.

The organization of the Roman Republic was not unlike that of a Greek polis. The Roman forum took the place of the Greek agora. The senate, which was an assembly of patricians, wielded the real power. Two consuls, elected from its ranks, commanded the army in war and were responsible for government in time of peace. The demand of the plebs to be represented was met by the appointment of two tribunes. It therefore became possible for an unusually talented man, from a low family, to rise in the state. This structure lasted for 450 years. It carried Rome from being a small city state to dominance in the Mediterranean basin.

Early Expansion – In the early centuries, Roman armies were occupied with winning control over the Italian peninsula. If there was a ruthless character to Roman expansion, there could also be generosity in the terms given on surrender. Conquered people were given Roman citizenship and allowed a large measure of self-government. Once within Roman rule, they too were expected to provide troops for the army.

The Punic Wars – As Rome expanded, she had only one serious rival. Carthage had expanded beyond North Africa. Her ships controlled the sea and Carthaginian colonies were established in Sicily, Southern Italy and Spain. The two powers were bound to clash for supremacy in the western Mediterranean. The Romans built up a navy, and in the First Punic War (264 – 241 BC) they defeated the Carthaginians at sea and won Sicily.

The Second Punic War (218 – 201 BC) marked the decisive struggle between the two powers. When Hannibal crossed the Alps and defeated Roman armies at Lake Trasimene and Cannae, it seemed as though Roman power would be broken. In 202, however, Hannibal was in turn defeated at Zama and Carthaginian power was destroyed. In 149 BC Rome took an excuse to fight a third Punic War. This time Carthage was flattened and the ground on which the city had stood was ploughed over.

The Rise of the Generals – Victory over Carthage had been bought at a high cost, and that cost was paid by the poor. Many peasant farmers, who were citizens sold their land to the rich and so lost their means of support. This led to internal unrest.

Wars were now being fought far from home, in the Hellenistic east and to the north in the land they called Gaul. Roads were built across the empire, which enabled the army legions to move swiftly from one trouble spot to another. These distant armies could no longer be commanded by consuls, with a term of office of two years.

There therefore arose a new breed of professional generals. These men often became fabulously rich on the booty of war, and, with a loyal army at their back, they could pose a threat to the traditional institutions of the Republic. Marius made his name in Africa and Gaul, and then Sulla in the eastern Mediterranean. Julius Caesar was the most successful in this line of successful generals. In 49 BC he took an irrevocable step when he crossed the river Rubicon, which marked the boundary of Italy, and marched on Rome at the head of his army. By this action he started the chain of events, which led to his murder, and the founding of imperial Rome.

Christianity

Origins – The early years of the Roman Empire were to see the beginnings of another of the great religions of the world. The Jewish people had maintained a stubborn refusal to dilute their religion to meet the demands of Hellenistic rulers. At the time of Jesus of Nazareth, sects like the Essenes and the Zealots maintained a resistance to Roman rule.

Jesus was a Jew, but he appears to have rejected the path of political resistance and taught instead a message of the relationship of the individual to God and other men, closer to the teaching of some later rabbis. The content of Jesus' teaching was indeed to be influential, but his significance lay not in what he said but in what his disciples declared him to be.

The share of responsibility for his execution can not be determined from the documents preserved, so it is unclear whether he was executed as a danger to the Roman state or as a critic of Jewish practice. Whichever it was, his disciples declared that they had witnessed his resurrection, and proclaimed that he was the Son of God. They picked up the words of the writer from the Babylonian exile and announced him as the saviour of the world, and not just of a chosen people. The holy books of the new religion were written down in the Greek language of the Hellenistic world, rather than in the more restricted Aramaic language which Jesus himself spoke.

Christianity and the Mysteries – Paul of Tarsus carried the message in a series of missionary journeys through the Greek speaking world. There he spoke the language of the popular mysteries – of redemption through blood and of a dying and rising god. With Christianity, however, it was different, he declared. While the mysteries were based in mere images, Christianity was rooted in historical fact.

Paul and other missionaries always sought to found a Christian cell, which they called an *ecclesia*, the word used for the meeting of the Greek polis. Hellenistic culture provided a language for the new religion; Rome provided a structure which enabled it to spread. Missionaries could make use of the Roman roads, and they were not likely to be molested by bandits on the way.

There was no doubting the enthusiasm of the converts, but, for a long time, an outsider would not have readily recognized a fundamental difference between this religion and the mysteries. Heresies, like Gnosticism and Manichaeism were pulling Christianity away from its Semitic roots into the maelstrom of Hellenistic religion. The Roman army generally favoured Mithras. A long path of persecution lay ahead before Christianity would emerge as the dominant religion of the region.

Imperial Rome

The Emperor – At the battle of Actium in 31 BC, Julius Caesar's great-nephew, Octavian, brought Egypt into the empire and ended the years of civil war. Four years later he was given the title of Augustus and made consul for life. He was careful to preserve the honoured republican institutions, but the senate lapsed into impotence and all power now lay with him.

No rule of inheritance was ever established for the position of emperor. In the centuries which were to follow, incompetents would be matched by administrators and generals of ability, imbeciles by philosophers. Most emperors died violently. Succession first passed through the house of Caesar. During one century of good government, it became the practice for an emperor to adopt his successor. For long periods, however, power fell to the general who could command the largest army. But the mass of people would never see the emperor in person. For them, success or failure had to be judged on whether he was strong enough to prevent the huge empire from breaking into civil strife.

The practice of emperor worship was imported from the old Persian tradition. The act of reverence due to the god-ruler was the symbol which bound together the hugely diverse people who now lay under Roman rule. Pious Jews refused to perform this ritual, but this was recognized to be a part of their ancient tradition and it was generally overlooked. The refusal of Christians, who came from all parts and races, was looked upon as a serious threat to the unity of the empire.

Buildings – The great monuments of Rome date from the Imperial age. Augustus himself restored 82 temples, and boasted, 'I found Rome of brick and left it of marble'. Aqueducts, arches and the huge Colosseum still stand as monuments to imperial glory. The Romans were content to copy Greek styles, to which they added impressive engineering skills.

The more prosperous built homes, such as have been preserved at Pompeii and excavated across the empire. Here they built for comfort, and artists, working in paint and mosaic, expressed a less pretentious view of life with humour and grace.

Natural frontiers – The Roman armies had now carried the empire across Europe, Asia and Africa, until it had ten thousand miles of land frontier. Beyond lay barbarians, ever willing to invade and plunder. The task of defence was made easier by natural boundaries – the African and Arabian deserts and the great rivers Rhine and Danube. This line of defence had two weak points, lying on either side of the Black Sea. In Asia the entrance to the steppes lay open across the land of the Parthians. In Europe generals were tempted to go beyond the Danube, across what is now Rumania to the Carpathian Mountains. Roman armies suffered heavy defeats in both of these sectors. Claudius also carried the empire across the natural frontier of the North Sea to Britain. The expedition was designed to bring the glory of conquest and to win control of fabled metal mines of the wild island.

The City of Rome – By imperial times, Rome had grown to be a huge city. Since most of the work was done by slaves, much of the population was unemployed, and the citizens had become accustomed to a life style supported by tribute from conquered peoples. No emperor could contemplate unrest in Rome, so the citizens had to be fed and kept amused on the famous diet of bread and circuses. Entertainments were on a massive scale. The Circus Maximus alone seated 190,000 people. Claudius built the huge harbour at Ostia, where grain, wild beasts and slaves were constantly being unloaded to feed the stomachs and the jaded palates of the people. The city gave nothing back to its empire.

East and West – Gradually a distinction began to emerge between the eastern and the western parts of the empire. The West, centred on Rome itself, covered western Europe and the old Carthaginian lands of North Africa. The east included the old Hellenistic world of Greece, Asia Minor, the Near East and Egypt.

The eastern side of the empire had a better balance to life. It contained ancient cities, but none dominated the region. It was self-sufficient in grain, wood, oil, wine and other essentials, with a surplus to buy in metal from the west and luxuries from the east.

The western part of the empire was not an area of ancient civilization. Since Carthage had been flattened, it had no cities to balance metropolis of Rome, which constantly sucked in products, so upsetting the economic balance of the region.

In 285 the Emperor Diocletian appointed a co-Emperor to rule the western sector. There was now an Empire of the East and an Empire of the West. In 324 Constantine accepted the dominance of the east by taking his capital to his new city of Constantinople.

The Triumph of Christianity – By this time, Christianity had established itself as a growing force. Diocletian tried to stem the tide, but Constantine accepted the new faith. Emperor worship may now have ceased, but even a Christian emperor could not shed the concept that he was the fountain of religion. He declared himself to be the thirteenth apostle and sat as chairman of the Council of Nicea, which established Christian doctrine. This set a precedent for the control of the church by the state.

At about the same time a group of hermits came together in Egypt to form the first monastery in the Christian tradition. This was destined to grow into an influential movement, capable of confronting the ambitions of Christian rulers.

The Barbarian Invasions – The century after Constantine saw increasing pressure on the European frontier of the Western Empire. Far away in the east, the Huns were on the move, and this created pressure on western tribes. The Huns themselves erupted into Europe under Attila in 440, to be defeated at Troyes in 451, but ahead of them, as if a prow wave, came Goths, Ostrogoths, Visigoths, Franks and Vandals.

The Romans found it difficult to defend the long land frontier and they recruited barbarians to strengthen the army. In 376 about 40,000 armed Visigoths were allowed across the frontier. Then in 410 the Goths sacked Rome. Vandals, who left their name for mindless destruction, crossed through Spain into North Africa and then returned for an even more destructive assault on the great city.

In northern Europe, Angles, Saxons and Jutes crossed the North Sea, first to ravage and then to settle in the British Isles. The Celtic inhabitants, no longer protected by Roman legions, were driven back to the highland area of the West, and into Ireland, where the Christian faith survived and flourished.

Byzantium

Her Frontiers – With the ancient capital in barbarian hands, the Roman Empire can be said to have fallen. Those who lived in the Empire of the East, however, recognized no such catastrophe. In 483 Justinian succeeded in Constantinople, and he set about the task of winning back the lost western

lands. His armies recovered North Africa, Italy and Southern Spain. It appeared for a time as though the Roman Empire was still a reality. His conquests, however, were ephemeral. From his time onwards, the Empire of the East was under continual pressure.

In the East, Persia was a power of consequence once again, and behind her the steppe nomads were ever menacing. In the south the empire faced growing Arab power. In the north, Slav people were pressing into the Balkans. The Emperor Heraclius led the Imperial armies in more successful campaigns, but the pressure was ever inwards towards what was to be the Byzantine heartland of Asia Minor, Greece, the Balkans and southern Italy.

Cultural Life – The people saw themselves as being direct inheritors of the old empire. Citizens of Constantinople still visited the bath houses; they still followed the chariot races with the passion of a modern football supporter. Justinian completed the work of centuries of Roman jurists by compiling the authoritative digest of Roman law.

Byzantium, however, soon developed a distinctive character which set it apart from the old empire. This drew both from the Greco-Roman and from Eastern traditions. Constantinople remained a home of classical scholarship. Plato was particularly popular, but his thinking became overlaid by layers of mysticism. Classical features were used in buildings, but the great dome which rose over Justinian's Church of the Holy Wisdom demonstrated new skills and a new aesthetic. Secular artists still worked within Hellenistic traditions, but religious artists, in paintings and mosaics, were beginning to express a particularly eastern Christian piety.

Religion – The eastern church early developed a distinction between secular (living in the world) and religious (living out of the world) clergy. Secular clergy worked at the parish level and were allowed to marry. The ideal was set by the many hermits and monks who expressed their piety in extreme self sacrifice. Religious icons became the focus of devotion for ordinary people.

The emperor maintained Constantine's position at the head of the church. Patriarchs, bishops and priests lay under his power. Emperors decided doctrine and mercilessly persecuted many of their subjects who held 'heretical' beliefs.

In the centuries after Justinian, the eastern and the western churches drew gradually further apart. In the west, the Bishop of Rome claimed primacy and began to build a centralized structure. The church finally divided into western and eastern parts in the Great Schism of 1054. This was partly about authority, partly about abstruse issues of theology, but it mainly stemmed lack of understanding of each other's piety.

The Arabs

Mecca – The desert land of the Arabian peninsula was inhabited by fierce and independent-minded Semitic tribes people known as Arabs. They led a nomadic life of great hardship. A trade route between the Mediterranean and the Indian Ocean crossed this land, passing through Mecca. The city was also a centre for pilgrimage to the sacred stone or *kaba*. The citizens of Mecca jealously guarded the revenues of both the trade and the pilgrimage.

Early in the 7th century a merchant, called Mohammed, had a vision and started preaching the message 'There is no god but Allah'. He came into conflict with the citizens of Mecca, and in 622 he left the city to live in Medina. This is the date from which the Arab world numbers its calendar.

Islam – The prophet Mohammed had met Christians and Jews and read many of their books and the religion which he founded lies within the Semitic tradition. He preached one god, which for him, ruled out the Christian concept of the Trinity. The word 'Islam' means 'submission', for the duty of the Moslem is to submit to the will of the one god. He gave his followers the five duties – daily prayers, alms, fasting, the keeping of Friday as holy day, and the pilgrimage – but the message was one of great simplicity. Very quickly, the feuding tribes of the peninsula were given that sense of community, which has ever since been the distinguishing feature of Islam.

Mohammed taught his followers that Christians and Jews were 'people of the book'. They and their religion had, therefore to be treated with respect. Once they accepted Moslem rule, they might be taxed, but they should not be persecuted or converted by force.

The Arab Conquests – Once the Arabs were united, they started raiding towards the north in search of booty, into the lands controlled by Byzantium and Persia. Their invasions had startling and unexpected success. This was partly because the two empires had weakened one another by endless warfare. More important, however, was that taxation and religious persecution had made their governments deeply unpopular with ordinary people. To 'heretical' Christians, the tolerant Moslem invaders seemed greatly preferable to either emperor.

The Persian Empire collapsed and Byzantium was pressed ever further backwards. Jerusalem fell in

638. It seemed as though Constantinople itself would fall, but in 717 the Arab armies were driven back from the city walls. By this time the Arabs not only controlled the Near East, but also North Africa and the whole of Spain. Their armies were even crossing the Pyrenees into the plains of Europe. Here, however, they found themselves in an alien environment of cold weather and barbarous people, so they turned back towards the south. The Arab armies carried Islam over this wide empire. Many conquered people converted; indeed Christianity disappeared from its old stronghold in North Africa.

The ultimate authority within the Islamic world lay with the caliphs. In 750 the ruling Umayyad house was overthrown and the new Abbasid rulers moved the capital to Baghdad.

Arab Culture – The Arabs possessed a powerful poetic tradition before the time of Mohammed and Islamic culture was founded in literature. The Koran, with its religious message, and its classical language provided a powerful unifying bond for one of history's more stable empires. Since the depiction of the human form was forbidden, art developed as elaborate geometric pattern. As the centre of empire moved out of the Arabian peninsula to Baghdad, so eastern influences became increasingly powerful. The Arabic language remained, however, the cement of the Islamic world. Although local dialects might vary, scholars from all parts continued to use the pure language of the Koran.

The Moslems did not come as the destroyers of civilization. The men from the desert quickly absorbed the cultures which they conquered. Their scholars read the Greek philosophers and united them with the astronomy, mathematics and medicine of the east, so serving as the main channel for the ancient learning in a troubled world. Moslem civilization reached one of its peaks in Spain, where the university of Cordova was a major centre of learning.

The eastern Mediterranean remained the centre of thriving trade. War might bring temporary disruption, but trading links with the East were never long severed. From India came spices, pepper and sugar; from China came porcelain and silks. The wealthy of Byzantium had an insatiable taste for luxury goods and the Arabs soon came to share these sophisticated tastes. Byzantium controlled the overland routes to China, which ended at the Black Sea ports. The Arabs controlled the sea routes by the Persian Gulf and the Red Sea to India, with links beyond to China and the Spice Isles.

Threats to Arab Civilization – In time, Byzantium

ceased to be a threat to the Islamic Empire; indeed it seemed only a matter of time before Constantinople must fall. From the 11th century, for some 300 years, Arab civilization would be subjected to assaults by Christian crusaders from Europe and successive waves of nomadic invaders from the steppes of Asia. The latter were by far the more threatening of the two, and it was they who finally brought the great days of Near Eastern civilization to an end.

THE FORMATION OF EUROPE

Church and State

The Papacy – In the year 590 a new Bishop of Rome was elected who would later be known as Gregory the Great. He was a Roman from a senatorial family, but, in the chaos of his day, he had made the choice to become a monk. For a devout Christian the monastic life seemed the only safe course to heaven in a violent and turbulent world. But Gregory saw that it was pointless to live with regrets for past glories of Rome or hopes for help from Byzantium. The church now had a mission to the restless and threatening barbarian world. Gregory selected monks as missionaries and sent them to bring Christianity to the barbarian tribes. Best known of these was Augustine of Canterbury. At the same time, missionaries from Ireland were moving south from Scotland into England and northern Europe. The missionaries from Rome, however, succeeded in linking the growing church back to Rome.

For Gregory's successors the first priority was to establish the primacy of the bishopric of Rome, or papacy. Popes claimed that, since they stood in a direct line from St Peter, they had inherited his 'power to bind and loose'. A pope could therefore control men's eternal destiny by the weapon of excommunication. In an extreme situation, he could even place an interdict, which forbade the performance of any sacraments, on a whole country. In an age of faith this was a formidable sanction.

The Popes had first to bring the Christian clergy under their control. Ordinary parish priests were generally illiterate peasants; bishops were temporal lords who used the church as a means of expanding family lands. Most were married men, who expected to pass their lands and livings on to their children. Their prime allegiance was therefore to the king or chief, rather than to a distant pope.

Monasticism – Only the monks were free from these temporal ties. The Rule of St Benedict, which imposed poverty, chastity and obedience, was now widely accepted. The monks were also almost alone in being literate in an uncultured world. This meant that they could reach positions of influence in church and state.

The popes used monks as their representatives, and, wherever possible, promoted them to high positions within the church. In time the popes worked to extend their control over the secular clergy by forbidding clerical marriage altogether.

A Time of Turbulence – In the early centuries after the fall of Rome, the pope and his monks were able to establish respect and authority because they provided the only apparent stability in a troubled society. Groups of barbarians roamed through Europe, bound to their leaders in simple tribal ties. When they settled down and adopted Christianity, much of the old way of life continued. Society still had no recognizable political structure, in the modern sense. Disputes were still settled by traditional 'rough justice', such as the ordeal and trial by battle.

Change continued slow in the dark forests of Germany. In time, however, the Franks and other groups in the western part of the European mainland adopted a form of Latin as their language, and paid some respect to the Roman legal system. The tribes who were more cut off in the British Isles, continued to speak their own German language, and developed law, based on past rulings, as preserved in the minds of the elders. So developed the divisions between the romance and Anglo-Saxon languages, and between Roman and common law which were to become important in later western civilization.

Political and social order was beginning to emerge by the end of the 8th century, but then Viking ships brought new danger to European coasts. It is never easy to say why a people go on the move, but it appears as though population growth and weather problems disturbed the balance of marginal Scandinavian farming. Certainly the feared Norsemen set off on 'land takings' and voyages of plunder. Their ships spread out across the North Atlantic to Iceland, Greenland and North America; they emerged into the Mediterranean; they sailed down the great rivers of central Asia, setting up the Russian state, and reaching Constantinople; they won control of northern Britain and Normandy. In 1066 a family of Norse descent won the crown of England.

The Norsemen were not the only raiders. Men from the steppes, this time the Magyars, were pillaging from the east and Moslem Saracen raiders came from the south. Hardly any part of Europe escaped. The unfortunate monks of Luxeil had their monastery burned by Norsemen, Hungarians and Saracens.

The Empire – For a brief period a new power arose in Europe. In 771 the ruthless and talented Charlemagne succeeded to the whole of the Frankish kingdom. For the next 40 years he led his armies to victories on all his borders, even mounting the first counter attack against Islam in Spain. Charlemagne was more than a conqueror. He was a devout Christian, and did much to spread the faith – by the sword if necessary – across Europe. He also respected learning, and he could read himself, although writing defeated him. He encouraged the clergy to respect books and learning, founded schools and brought the best minds of the day to his court.

On Christmas Day, 800, he was crowned by Pope Leo III in the church of St Peter in Rome. The people cried, 'to Charles Augustus, crowned by God, great and peaceful Emperor of the Romans, Life and Victory. ' A new Roman Empire had been proclaimed.

The empire was based on one man's will, and, like Alexander's, it fell apart on Charlemagne's death. It was divided into three parts. The central kingdom did not survive, but the two other halves would ultimately become France and Germany. Charlemagne's eastern successor retained the title of Holy Roman Emperor, but his lands remained a loose confederacy. In 940 the Comte de Paris, Hughes Capet, won the French crown and established a monarchy which survived until the French Revolution. His family was the first European dynasty to establish the concept of a hereditary monarchy.

Powers Temporal and Spiritual – After Pope Leo III had placed the crown on Charlemagne's head, he stretched himself on the ground as a sign of honour to the Emperor. Laster popes would regret this gesture. The first objective of the Popes was to win control within the church. This involved taking the right to appoint bishops away from the temporal rulers.

In the 11th century, Pope Gregory VII and Emperor Henry IV came into conflict in the Investiture Controversy. Gregory was victorious, forcing Henry to stand barefoot in the winter snow as a sign of submission. Gregory then formulated the extreme claim that all power came from the pope, and he therefore had the right to appoint and depose kings and emperors. The Investiture Controversy

was the first of a series of disputes between church and state. They involved not only the Holy Roman Emperor but also kings of France and England.

King, Lord and Parliament

The Feudal System – In those troubled times, people were prepared to sacrifice liberty in the interest of security. Kings and emperors were remote figures, so free men bound themselves to their local lord, who could give assistance when danger was near. When a man took an oath of loyalty he gave his lands to the lord, and then received them back as the lord's vassal. He had the obligation to follow the lord to war, but as a mounted knight, to set him apart from the common serfs. The lord, in his turn, bound himself to a higher lord, and the king stood at the apex of the pyramid. Only the serfs were nobody's vassal, because they had nothing to give in exchange for protection. These common people were not allowed to leave their villages, to go to school or to get married without their lord's permission.

By the end of the 9th century, this feudal system had spread to all but the most remote areas of Europe. Kings, like other lords, were concerned to extend their lands wherever they could by war and dynastic marriage. The two-way nature of the feudal compact served as a check on royal power. In France, the Capetian kings stood at the apex of the pyramid, but for long periods their actual power did not extend beyond their own lands around Paris. So, when the King of England married a French heiress, he did homage to the French king for his lands in Aquitaine, but he did not permit any interference within his territory.

The Hundred Years War between France and England was fought sporadically from 1337 to 1453. The English king may have laid claim to the crown of France, but it remained in essence a struggle between a dynastic monarch, determined to establish direct control over feudal lands on the one side, and an over-mighty subject on the other. It was one of the catalysts which defined the meaning of the modern nation state. Some 200 years later, Shakespeare put words of nationalistic fervour into the mouths of John of Gaunt and Henry V. Such words would have been incomprehensible in the time of Charlemagne or of William the Conqueror, but they were beginning to have some meaning to their supposed speakers.

King and Parliament – William the Conqueror gave English kings more direct authority within their own realms. The feudal system was constructed to ensure that no lord could become 'over mighty'. Vassals, for their part were concerned that the king should not achieve unlimited power. In 1215 the lords forced King John to sign Magna Carta, which laid down two basic rights – that no free man could be imprisoned without a trial and the king could not raise taxes without the consent of a Great Council. In 1295, King Edward I called what became known as the Model Parliament, because it set the pattern for future parliaments. Representation was by estates – the Lords temporal, the Lords Spiritual and the Third Estate, with the first two sitting together in an upper house. It was also established that parliament had the responsibility to act as the highest court in the land, to give advice to the king, to make laws and to vote taxes.

The Rise of the Towns – The inclusion of the third estate in Edward's Model Parliament was testament to the growing importance of trade in the European economy. Wealth was no longer the preserve of landowners and the church, so, to achieve maximum income, it was now necessary to consult with the representatives of the growing towns.

As towns grew in importance, kings gave them charters, which assured them freedom from interference by local landowners. Their walls were the symbol of their independence, and magnificent churches the evidence of their wealth. Trade provided a means by which low born men could rise to positions of power within their own community, and even within the state. Different occupations were organized into guilds, which controlled terms of entry, quality standards and gave members a social structure.

The cities were often natural allies to kings who wanted to centralize power. Overmighty nobles might flourish in conditions of civil war, but merchants needed the peace, which only a strong government could provide. Kings, for their part, recognized that the growing wealth, which was the basis for national strength as well as royal revenue, was generated, not on noble estates, but inside the town walls.

The Cloth Trade – The Lord Chancellor of England, still sits in the House of Lords of a Woolsack. This was a reminder to parliament that the nation's wealth rested on the woollen trade. England, however, stood in the lowly position of a primary producer; the business in finished cloth centred round Flanders. Flemish weavers jealously guarded the trade secrets which made their cloth the most sought after in Europe. From the 13th century, the economy of northern Europe became increasingly sensitive to fluctuations in the fortunes of the cloth trade.

The Crusades

The First Crusade – In 1095 the Byzantine emperor appealed to the pope for assistance against the Turks. The pope answered the call by preaching a Holy War. The motives of those, both noble and common folk, who took the cross, were very mixed. Many of the Norman lords who took the lead saw the opportunity of a new land taking, like those of their Viking ancestors. But there was also a real devotion. The two were not incompatible. When the army arrived first in Byzantium and then in the Arab lands, they appeared like barbarians, with nothing to recommend them but their brute courage. Jerusalem fell to the crusaders, who waded through blood to give thanks for the victory.

Outremer – The crusaders established states in the conquered land. The Moslems resented these Christian enclaves in their territory and they were under constant pressure. In 1187 Saladin reconquered Jerusalem and the crusaders were unable to win it back. In 1291 the last Christian outpost fell to the army of Islam. The crusades gave the West two centuries of contact with a higher culture. Knights returned home with a taste for oriental luxury goods; some picked up an interest in learning and mathematics; methods of castle construction and siege warfare were modernized on Arab models.

The Later Crusades – Eight campaigns between the 11th and the 13th centuries are known as crusades, and include the tragic 'children's crusade' of 1212. The movement turned inwards against European heretics. The simple crusaders were not always able to distinguish which enemy they should fight. The Venetians encouraged the fourth crusade to turn on Byzantium. In 1204 Constantinople was captured by the crusaders and the city remained in Christian hands until 1261. Although the rump of empire would survive into the 15th century, Byzantium never recovered from the disaster.

Spain – At the same time, Christian forces were counter-attacking against the Moslem Moors in Spain. In 1212 the Moors were defeated and driven back to Grenada. The reconquest of the peninsula was completed in 1492. A great culture was replaced by a fanatical Christian state, in which the Inquisition was used as a tool of persecution against Moors, Jews and many Christians whose views did not please the authorities.

Learning, Art and Society

Scholarship and Authority – As long as there was no nation state, there were no sharply defined national boundaries. Latin provided a lingua franca and the church a broadly based structure within which the educated of their day could communicate.

Through the troubled times any learning remained behind monastic walls. The books of early Christian fathers were copied and became the intellectual authorities of the new world. Men had lost confidence in their own ability to reach conclusions, either through logic or through experiment, and all argument therefore referred back to authority. Even quite trivial issues of dispute would be decided by the weight of authority which could be mustered on the one side or the other.

The authors of antiquity were largely unknown until around the 12th century. Then translations began to be made into Latin from copies preserved in Islamic Spain and Sicily. The ancient dichotomy between Plato and Aristotle began to be reflected in arguments between nominalist and realist theologians. Thomas Aquinas, in particular, baptised Aristotle. This did not, however, lead to an increase in experiment; classical authors joined the Christian fathers as valid sources of authority. In southern Europe, men still lived amidst the ruins of classical civilization. The classical and the Christian came together until, as in Dante's Inferno and Michelangelo's Sistine Chapel, they became indistinguishable from each other.

By the 12th century, learning was coming out from monastic walls into the more open atmosphere of universities. The first to be opened was at Salerno, where Islamic and Byzantine influence was strong. Then came Bologna, Paris, Oxford and many others. Crowds would follow teachers, like Peter Abelard, who spoke a new and more restless language. Students at university were in religious orders of some sort, but the educational impetus continued outwards into the wider population. By later medieval times, an increasing number of lay people, particularly in the towns, were acquiring literacy.

Architecture and Painting – In early medieval times most stone buildings were either castles or monasteries. Many were fine buildings, but they added little to the techniques of antiquity. By the 12th century architects were developing their own signature. The Romanesque style of Southern Italy was based closely on a study of classical models. In northern France and England there rose magnificent cathedrals in the Gothic style. Here the pointed arch and the flying buttress enabled them to give their structures both height and light.

Most painting, likewise, remained dedicated to the church. Altar pieces showed the Virgin and child, with patrons and saints; frescoes and stained glass windows reminded illiterate worshippers of Bible stories; monks, copying psalters and books of hours, under no pressure of deadlines, painted exquisite decoration. In Italy, the school of Sienna worked under direct influence from Byzantium. But painters, like architects and stone masons, were craftsmen who were happy to work for any patron, and, as the centuries passed, an increasing number of commissions were available for secular work.

Literature – Medieval literature, like that of any other period, was made up of different strands – folk tales and myths, national histories, historical chronicles, love poems and works of devotion. Early vernacular literature, like Norse sagas and the Anglo-Saxon Beowolf, helped to create national language areas. In the early centuries, however, Latin was the language of both secular and religious writing.

By the 14th century a new vernacular literature was emerging. The supreme example of in English is *The Canterbury Tales*, written by soldier and customs officer, Geoffrey Chaucer, for an audience of other lay people.

Times of Change

The Black Death – A sickness new to the human race was first reported in the Yangtze Valley in 1334 and, according to one estimate, some 13 million Chinese died in the following years. Relentlessly it spread from east to west, leaving devastation in India and across Asia. In 1347 the plague spread across Northern Italy, and in the following years it is estimated that between a quarter and a third of the population of Europe perished. This first outbreak was the worst, but the disease returned periodically until the second half of the 17th century. Although it was a world wide phenomenon, its effects have been most closely studied in Europe.

Initially it brought economic collapse; prices of all goods fell sharply and much farm land returned to nature. As life recovered, employers were faced with an acute labour shortage. This created strains within the social structure of both town and country. Guild regulations were flouted and the feudal structure began to crumble. There were major peasant uprisings, in France in 1358, in Florence in 1378 and in England in 1381.

The Late Medieval Church – Pious Christians could not understand why God had created such destruction. The plague accentuated a Christian piety which identified God as judge and destroyer and the saints, particularly the Virgin Mary, as protector.

The church, like all institutions, moves through cycles of corruption and reformation. In early medieval times, reformation came from within. The last of these reforming movements was the founding of the friars by Francis of Assisi in 1209. Now the impetus for reform had grown weak. In the 14th century popes taxed the faithful heavily to maintain a lavish lifestyle. After the Black Death these taxes fell all the more heavily on a smaller population. One fund raising technique was the offer of indulgences, by which punishment in the next world was remitted in exchange for payment in this world. During the 14th century the church lost its independence when it moved under the protection of the King of France in Avignon. The Great Schism, when rival popes competed from Rome and Avignon, further undermined spiritual authority. Some Franciscans denounced papal luxury and were burned as heretics but their message was heard by the people.

Before the end of the 14th century John Wyclif in England and Jan Hus in Bohemia were preaching that man did not need the apparatus of the church to make contact with God. Hus founded what was to be the first protestant church, and Wyclif translated the Bible into English.

The Reformation – In the early 16th century the pope set out to raise money for the building of St Peter's by selling indulgences. In 1517, the university lecturer Martin Luther challenged the papal representative to a debate by nailing 95 theses to the door of Wittenberg cathedral. The church authorities sought to have him condemned, like heretics of old, but he found protection from German princes. The protestant movement soon won followers, particularly in the trading towns and in Northern Europe.

Luther did not initially see himself as the leader of a movement which would split the western church, but, as he preached the supremacy of the Bible and faith over the sacraments and traditional authority, the division quickly became irreconcilable. He found himself leading a mass movement, based on individual piety. Luther was the catalyst for another round in the ancient struggle between lay and secular powers. He survived to preach only because he was adopted by German princes, who saw his movement as a useful weapon against the power of the church.

Eastern Europe

Russia – It appears as though the Norsemen who settled the rivers of Russia brought no women with them, so the process of assimilation was rapid. In 980 Vladimir established the kingdom of Kiev and he married a sister of the Byzantine emperor. It is said that Russian envoys visited Constantinople and the West to decide which form of Christianity should be adopted. They were overwhelmed by the splendour of Constantinople, and the eastern link was forged. Kiev was destroyed in 1169 and the centre of power was driven northwards to Moscow. Trading and cultural ties with Byzantium were largely lost, and Russia was increasingly isolated until it was overrun by the Mongols in the 13th century. Russian independence can then be dated to the victory over the Tartars in 1380. The Grand Princes of Moscow emerged as rulers and Ivan III (1462-1505) adopted the title of Tsar (Caesar) and the double headed eagle, to substantiate the claim that, with Constantinople in Ottoman hands, the Russian monarchy had now inherited the imperial tradition.

Poland – When Vladimir made his choice of the eastern church, the Poles on his western frontier had just turned in the other direction. The missionaries who brought western Christianity to Poland also acted as forerunners for waves of land-hungry German invaders, led by the fearsome Teutonic Knights. In late medieval times, a Polish state lay across the central European plain, with its prosperity based on grain exports to the west through the port of Danzig. It was, however, already showing signs of the damage that would be caused by its geographical location as a buffer between Western Europe, the Scandinavian north, Russia and the East and the disturbed cauldron of Slavs and Magyars in the Balkan south.

In both Russia and Poland the serfs lived in great poverty, under the control of a wealthy landed class. Rulers were faced with a perpetual challenge from overmighty subjects without being able to look for the support of any considerable middle class.

THE WIDER WORLD

Asia Before the Mongols

India – For centuries after Ashoka, the Indian subcontinent was divided into warring states. The south, behind its mountain barrier, remained the home of non-Aryan people. They maintained contact with the Mediterranean civilizations through the Red Sea and Persian Gulf trading routes. In the north-west, the frontier and Indus valley remained open to Asian invaders. Invading Hunas, probably

Huns, devastated this region, as they did lands both to east and west.

In about 320 AD the Guptas united the whole of northern India. This marked the great age of Hindu culture. In the 5th century, the decimal system was invented, so opening new areas of mathematics. Sculpture and literature flourished, both achieving a broad unity of style, characterized by a warm sensuality. Buddhist culture declined as Hinduism spread across Asia and into the islands of the Pacific. The island of Bali remains today a marker of this great expansion.

The Hindu Empire was in time challenged by the rise of Islam. Moslem traders – always effective missionaries for the Prophet – would have visited the western ports in the 7th century. By the 8th century invaders were crossing the open north western frontier. By the 12th century, they controlled the Punjab and a century later they dominated the Ganges valley. The fateful religious divide was now established.

China – China was united more effectively by language and culture than it was by its political structure. The two most powerful dynasties were Han (c. 205 BC – 220 AD) and T'ang (618 – 907). Their empires were comparable to that of Rome at its most powerful. During the T'ang Dynasty trade flourished, and China became a major sea power, with trade reaching from the Persian Gulf in the west to Indonesia in the east. The great Chinese dynasties had an expectation of life of about 300 years. They were founded by a great individual who combined military and administrative skills. In later years, as the succession passed to lesser men, the state would come under pressure from nomads to the north and rebellion at home. Imperial authority was upheld by officials, who preserved the traditions of K'ung-fu-tsu. Their main tasks were to take the census and keep the land register up to date. Beyond that, they maintained only a broad supervision over local lords, who raised taxes and performed the day to day tasks of administration themselves.

Great civil works were undertaken. The country was now bound together by canals, most important of which was the Great Canal, which linked Peking with Hang-chou. Huge irrigation projects were undertaken to provide food for the ever growing population. The casting of iron, printing, the magnetic compass, the use of paper money and explosives were all pioneered in China, but the conservative structure of society militated against the fullest exploitation of her inventions.

In the periods between the great dynasties, the country relapsed into warring states. There were times of disaster, when armies ravaged large areas, but in general, conditions changed little for the mass of peasants, whose life was always more closely governed by local lords than by distant emperors. But, while Europe remained divided into her warring states, China could always be drawn together once again by a dynamic new dynasty.

The Sung Dynasty (960 – 1279) was never as powerful as its great predecessors and in 1127 it lost the northern part of the country to Chin invaders. In the following century and a half the Southern Sung Dynasty lacked military power, but the period is viewed by many as the high point of Chinese culture. The Imperial capital of Hang-chou was a centre of wealth, culture and leisured living, far beyond any other city in the world.

Sung art was influenced by the Zen school of Buddhism. Painters, such as Ma Yuan, worked with an economy of line and colour to make a visual statement about man's position within the world order. Potters made dishes which looked 'like ivory, but were as delicate as thin layers of ice'.

The Seljuk Turks – The name 'Turk' is given to widely dispersed people, originating on the Asian steppes, who spoke a common language. In the 10th century a chief called Seljuk settled with his people near Samarkand and was converted to Islam. The tribe organized an army based on slaves, mainly recruited from southern Russia and the Caucusus, who were known as mameluks. Backed by these fearless warriors, Seljuk's grandsons built an empire, from Azerbaijan and Armenia, into the ancient lands of Middle East. They overran Persia, captured Baghdad, Jerusalem and Egypt and invaded Byzantine lands in Asia Minor.

Later Seljuks rulers found it difficult to hold this vast empire together. While their efforts were largely directed against European crusades, they faced trouble in other parts of the empire. In Egypt, for instance, mameluke soldiers established a virtually independant government. The Seljuk Empire therefore became vulnerable to another and greater threat from the Asian steppes.

The Mongols

Genghis Khan and his Successors – In 1206 a chief called Temujin, better known to the world as Genghis Khan, united the Mongol tribes who lived in the area today called Mongolia. These were wild, nomadic peoples in the tradition of the horsemen who had come from the steppes throughout recorded history. He then established dominance over the more numerous Turkish peoples from the land to the north of the Himalayas. Genghis Khan came to believe that he was destined to rule the world, and he embarked on the greatest programme of conquest in history. His followers were magnificent horsemen. As a nomad people, they could survive on dried milk and the blood of their horses. Released from the constraints of supply, they were therefore uniquely mobile. They were also utterly ruthless. Cities which accepted them were often treated with leniency; those that resisted were liable to be levelled to the ground, and the population massacred. As the reputation of the Mongol horde was carried ahead, rulers capitulated to avoid the dreadful destruction.

By the time that Genghis died in 1227, the Chin Empire of northern China had fallen, and Mongol armies had swept across the open grasslands of Asia as far as Russia and the Caucasus. Still the advance continued. In 1237-8 the Russian state was overwhelmed by horsemen who rode down the frozen rivers, achieving the winter conquest that later eluded both Napoleon and Hitler. When a Great Khan died, the armies returned to their homeland to debate the issue of succession. Europe might have been overrun had Genghis Khan's successor not died in 1241. The armies did not again threaten Europe; to the Mongols, it seemed a poor land, hardly worth conquering. They did, however, return to the Middle East, capturing Baghdad and destroying the caliphate in 1258. The tide of conquest finally turned here too when, in 1260, the mamelukes of Egypt organized the armies of Islam to defeat a Mongol army at Ain Jalut, near the town of Nazareth.

Mongol China – The Mongol Empire was now divided into four, with the eastern section the portion of Genghis' grandson, Kublai. He led a Mongol assault on the Southern Sung Empire, which fell in 1279. Further expeditions were launched into South East Asia and even, unsuccessfully, against Japan. While his grandfather Genghis had devastated the north, Kublai respected the civilization which he conquered. Although he spoke little Chinese, he was a patron of literature and, like conquerors before him, he adopted Chinese ways. Mongol rule had now united the whole territory between Europe and China under a single authority and the ancient overland routes were opened once again. In 1275 members of the Polo family from far away Venice reached the court of Kublai Khan. When the young Marco Polo finally returned to Venice in 1299 he gave the west its first information about the civilization of the east. Readers in the more primitive Europe found it hard to believe that such a land of riches could exist far to the east, but some two centuries later a Genoan sailor called Christopher

Columbus would own and make notes on a copy of the Venetian's narrative.

For the Chinese, however, the Mongols remained a dynasty of foreigners. Prosperity declined sharply and there was a wave of unrest, and the Mongols were overthrown in 1367 by a new Ming Dynasty. This survived its allotted three centuries until in 1644 it was in turn overthrown by new invaders from Manchuria, who established the Manchu Dynasty.

Later Mongol Conquests – In about 1370 Timur the Lame, a chief from the region of Samarkand, proclaimed that he was the man to revive the Mongol Empire. In the next 30 years, he ravaged the Middle East, Asia Minor, Southern Russia and Northern India with a brutality only matched by his distant kinsman Genghis Khan. The ancient lands of the Fertile Crescent, so long the focus of world civilization, never recovered from his invasion. The Mongol Khanate of the Golden Hoarde in Russia was fatally weakened. He died in 1405, when on his way to carry his conquests into China.

In time the Mongol people of Central Asia and the Middle East came to accept the religion of Islam. The weakness in Mongol power lay in the fact that there was no established law of succession. Timur's successors, like other Mongols, were concerned with domestic issues as they contested succession. Fifth in line from Timur was the more attractive Babur. The kingdoms of northern India were at that time in a state of permanent warfare. In 1526 Babur won a series of victories, and by his death in 1530 he had established Mongol – or Mogul – rule in northern India.

Mogul India

Akbar – In 1556 Babur's 14-year-old grandson, called Akbar, inherited a weak and divided empire as a boy of 14. He also inherited the ancestral belief that no empire can survive unless it is continually expanding, and throughout his reign he kept his armies constantly on the offensive. He continued old Mongol tactics. When a city, like Chittor, resisted it could be utterly destroyed and its people massacred; when people accepted his authority, they found him a generous ruler. By 1600 his Mogul Empire controlled the whole of the subcontinent, except for Ceylon and Vijayanagar in the south. Akbar built a huge capital at Fathpur-Siki, which was to be the model for Mogul public buildings of incomparable grandeur, culminating in Shah Jahan's Taj Mahal.

The country was divided into provinces, but all authority sprang directly from the Emperor himself. Although a ruthless conqueror, Akbar was anxious to bind his people together effectively, and he was concerned at the religious division which existed between his Hindu and Moslem subjects. He was suspicious of all dogmatism, and devout Moslems accused him of backsliding when he abolished the poll tax payable by Hindus, and worked to find a compromise between the two religions.

The Decline of the Mogul Empire – Akbar was an outstanding ruler, but his empire suffered from weaknesses inherited from his Mongol tradition. In Europe, structures of government were coming into existence which transcended the personality of the ruler. In Mogul India, however, authority continued to be over-dependant on the ability of one man. In his last years, even Akbar was plagued by rebellious sons. The instability of the empire is illustrated by events at the end of the reign of Shah Jehan. In 1657 he fell ill, triggering a ferocious civil war between his sons. The victorious Aurangzeb was a devout and intolerant Moslem and under his rule the united empire created by Akbar began to fall apart.

The Ottomans

The Foundation of Empire – Mongol successes in central Asia created more movement of nomad tribes out of the grasslands. In the late 13th century. one Ertughrul led a band of followers, who were equally devoted to Islam and to plunder, into the Seljuk lands of the Middle East. Ertughrul's son, Othman overthrew the Seljuk sultans, and founded the great empire which was to bear his name.

Othman's successors defeated the Byzantine army. They captured Asia Minor, and, in 1361, crossed into Europe to establish Ottoman power in the Balkans. Constantinople was now an isolated fortress in Ottoman lands, and in 1453 it fell to the Sultan Mehmet II.

The Spread of Empire – Ottoman power reached its peak in the century after the fall of Constantinople. In the early 16th century Selim I marched southwards, defeating the Mamelukes of Egypt and capturing Mecca, where he was proclaimed Caliph of the Islamic world. His successor Suleiman I, the Magnificent, turned north. In 1526 he defeated the Hungarian army at the great battle of Mohacs. Three years later his armies laid siege to Vienna.

Africa

Trans-Saharan Trade – Historians of early sub-Saharan Africa are restricted by the lack of written records and the destructive capacity of termites, working on wood and mud brick. The continent, however, was far from isolated. A thriving trade existed across the Sahara trade routes between North Africa and the grassland region which lies across the continent from near the Atlantic to the Nile.

The staple product being carried southward was salt – an essential commodity for people living in a hot climate. The Moslem traders who crossed the desert carried various luxury goods, and also brought their religion and literacy in Arabic script. On the return journey they carried gold, slaves and leather goods. Before the time of Columbus, Europe was heavily dependant on African gold and 'Morocco leather' has always originated south of the Sahara. A key focal point of this trade was Timbuktu, on the Niger, which became famous as the meeting point of the camel and the canoe. The town was already well enough known to be marked on a Spanish map in the late 14th century.

The gold, and probably most of the slaves, came from the forest region still farther south, so trade reached out in both directions. Among the most active traders were the Hausa people, who were based on city states, such as Kano and Zaria. They would be late recruits to Islam and never organized into larger political units.

African Empires – Broadly based political structures did, however, come into existence in the Southern Sudan to control the two-way trade. The Empire of Ghana (8th – 11th centuries), was succeeded by Mali (12th – 14th centuries) and Songhai (14th – 16th centuries). Kings like Musa Mensa, who ruled Mali in the early 14th century, were well known for their wealth and learning across the Islamic world, and even beyond. The trade in gold appears also to have stimulated the growth of forest kingdoms, such as Benin and Oyo. These would grow in importance with the arrival of European ships on the coast in the 15th century. Far to the east, the kingdom of Ethiopia maintained its isolated Christian tradition, again with power based on trade with the north by way of the Nile.

There was also traffic in gold and slaves down the coast of East Africa. The unique stone ruins of Zimbabwe provide evidence to support the reports of inland states in this region.

America

America was the last continent to be settled by man and it remained the most isolated. Traditional hunter/gatherer lifestyles were successfully followed by people of widely differing culture across wide areas of North America and within the many forest regions of North America until they suffered under the impact of European invaders. The cultivation of maize and then of other crops, however, made possible the development of more complex civilizations.

Central America – The earliest civilization was that of the Olmecs, which flourished on the coast of the Gulf of Mexico in the 7th century BC Many of the characteristics of later civilizations of the region can already be recognized in these people. In their capital of Teotihuacan they built huge pyramids, apparently dedicated to the same gods which would be worshipped by later people of the region.

The most accomplished civilization of the region was the Maya, centred on the Yukatan peninsula, which reached its peak in the 9th century BC The Mayans used a pictogram form of writing. Like the Babylonians, they laid emphasis on the calendar and the heavenly bodies and they developed great skill in mathematics and astronomy, working out the duration of the year and learning how to predict eclipses. They were the first people in the world's history to achieve a sense of the vast span of time. Mayan sites, like those of ancient Egypt, are not cities, but vast complexes of temples and other ceremonial buildings.

The Maya were succeeded by the Toltecs, and they were overthrown in their turn by the Aztecs, who dominated the region from the 13th century. They appear to have been the first to introduce mass human sacrifice. This practice came to dominate the whole of Aztec strategy for the region. As victims were best found in warfare, they had no motivation to create conditions of peace, but rather encouraged a general unrest among subject people.

The Aztecs had a tradition that the white skinned and bearded god Quetzalcoatl would one day return from the east. When the invading Spaniards appeared to fulfil this prophesy, there were many subject people who were prepared to take their side against their feared Aztec masters.

The Andes – The long spine of the Andes is perhaps the most improbable setting for any of the world's civilizations. Between 600 and 1000 AD a people called the Huari brought some political unity to this area. In the 12th century, the Inca, based on Cuzco in modern Peru, were only one of many smaller groupings. They then conquered an empire which by the 15th century stretched 2000 miles from Quito in modern Equador to the deserts of Chile.

The Incas were a non-literate people. Instructions were carried to distant parts of the empire by messengers. Again, lacking the wheel, these messengers travelled on foot over a road network, built with great engineering skill. Inca power was centred on heavily fortified cities, where invading Spaniards were to find a wealth of beautiful objects made of gold and stone.

The Incas were not as oppressive to their subject people as the Aztecs, but there were still many who were prepared to support the small force of Spaniards who arrived in 1531 to conquer and loot the empire.

THE TRIUMPH OF EUROPE

The Background to Conquest

New Perspectives – The Mappa Mundi, in Hereford Cathedral, illustrates the medieval perspective of the world. Jerusalem lies in the centre of the world, with the three known continents – Asia, Africa and Europe – arranged around the Mediterranean Sea. Phoenician and Viking ships may have sailed the wider oceans, but these lay at the edge of the known universe.

By the 15th century, changes were taking place. The reports of Marco Polo's travels in the East were becoming widely known. No profit orientated merchant could ignore his descriptions of markets loaded with silks, velvets and damasks. He had travelled beyond China to the islands of the Pacific and described how cheaply spices could be obtained. It was still impossible to keep meat animals alive through the European winter, so all except the breeding stock was slaughtered and salted down at Michaelmas. By spring it was barely edible without

pepper, cinnamon and nutmeg to disguise the taste.

In 1400 also a copy of the Hellenistic Ptolemy's *Geography* was brought from Constantinople and published in the west. It contained many errors, but did show that the world was round and not a flat dish. During the century, this became the accepted view of scholars.

The Ottoman conquests helped stimulate interest in alternative routes to East Asia. Thorough medieval times the majority of luxury goods had been brought by the Asian overland routes. These were now threatened by a hostile power. The Genoese, traditional allies of Byzantium, were particularly threatened by the new developments. Ottomans and Venetians alike combined to shut them out from the profitable business.

Logic demanded that traders should turn their attention to the oceans that lay beyond the enclosed Mediterranean world. Luxury goods were high value and low bulk cargo. Projected returns on investment on one cargo reaching Europe were astronomical.

Technical advance – During the 15th century major technical advances were also made in Europe, which brought such a project within the bounds of the possible. Before that time European ships had been square rigged on a single main mast. Such a ship could be manned by a small crew, but could not sail efficiently into the wind. Arab ships used a lateen sail. This could sail into the wind, but such a large crew was needed that it could never go far from land where food could be obtained. Ship builders now constructed multi-masted vessels, with both square and lateen sails, which could both be handled by a small crew and sail into the wind.

If ships were to sail far out from land, then navigational techniques needed to improve. By 1500 European sailors were skilled in the use of the magnetic compass, either re-invented or brought from China, and in measuring latitude. Almost 200 years more years would pass before similar advances were made in calculating longitude. Great advances were also made in cartographical techniques, with the Dutch leading the way.

European craftsmen also developed gunnery to new levels. King John II of Portugal took particular interest in the problems of mounting modern guns on board ship. Success in these experiments meant that European ships could command the seas. In previous centuries ships came together with grappling hooks to allow soldiers to fight a conventional battle. Now the European ship could sink an enemy ship without allowing it to come close enough to bring the soldiers into action.

Population and Prices – The intellectual climate was favourable, commercial incentives were strong, and the required technology was available. As with Norsemen and Mongols, however, a further 'push factor' was needed to trigger off a major movement of European people. Demographers have shown that Western Europe had recovered from the Black Death and a cyclical population increase was in progress. Pauperism was on the increase, and also, in populations organised on the basis of primogeniture, landless younger sons of gentry families were looking for any way of making a fortune.

Historians now link the population rise with an inflationary trend which persisted through the 16th century. On average, prices quadrupled between 1500 and 1600. Since wages and savings did not always keep up with the rising prices, this created conditions of hardship which could make emigration attractive.

Religion – Christians of the period generally held that unbelievers possessed no rights. The Pope declared that Christian kings had a right to conquer heathen lands. Some Catholic friars and, later, Jesuits did identify with the cause of the native people, but even their mission stations were instruments of colonial control. The Protestant record was, if anything, worse. 300 years would pass before protestant Christians made any serious attempt to protect the rights of and to share their faith with non European people.

Asia

The Portuguese – By 1400 Portugal was free from Moslem rule and had established itself as a separate country from Spain. Its geographical position made it a natural Atlantic pioneer. In the first half of the 15th century, the king's brother, Henry 'The Navigator' established a school for sailors at Sagres, by Cape St Vincent, and sent out expeditions to explore ever further south into the Atlantic. Slowly they pushed the boundaries of exploration beyond the Azores and to Senegal.

In 1487, 27 years after Henry's death, Bartholomew Diaz rounded the Cape of Good Hope and established that the way to India lay clear. In 1498 Vasco da Gama took his ship to Calicut in south India. Indian merchants were happy to sell to the newcomers as they offered higher prices than the Arabs. He returned with a cargo of pepper, cinnamon, ginger, cloves and tin. It was reported that the King of Portugal and Vasco da Gama's other backers made a 6000% return on their investment. A century of human and financial investment had finally paid it dividend. In 1503 the Portuguese established a permanent base in India, at Cochin, followed in 1510 by Goa, and later by Seurat.

In 1509 the Arabs sent a fleet, manned by 15,000 men, to drive the Portuguese from their seas, but the European superiority in ships and gunnery proved decisive in a battle off Diu. From that time European fleets exercised control over the world's oceans. Arab and oriental sailors could no longer confront them in battle, but could only operate as pirates.

In 1517 the first Portuguese ship arrived off the coast of China and according to European custom fired their guns in salute. The Chinese found these barbaric Europeans 'crafty and cruel', but had to respect their guns which 'shook the earth'. In 1521 Portuguese ships had reached the Spice Islands. In 1557 they established their trading base at Macao, off the Chinese mainland.

The Dutch – The Portuguese successfully protected

their Africa route against encroachment by other European nations until the last years of the 16th century. Then in 1594 a group of Dutch business-men fitted out four ships to sail to the Far East. They carried the products of Europe – woollen and linen fabrics, glassware, ornaments and different kinds of ironware, including armour. The ships reached East Asia and found that the people welcomed their qual-ity goods. In 1602, the Dutch parliament, the Estates, set up the Dutch East India Company to fol-low up this initiative. The Malacca Strait, in modern Indonesia, became the focus of the empire, with headquarters on the island of Java. The Dutch then set about driving the Portuguese out of their Asian empire. Only a few Portuguese outposts, such a Goa and Macao survived the assault.

Once in control, the Dutch traders ruthlessly set about eliminating all competition. In 1623 ten Eng-lish merchants were tortured and killed at Amboyna. But they were not content to exclude other European competition from their market. Chi-nese junks were shut out of their traditional markets as even local trade was channelled into Dutch ships. By now Europe was becoming glutted with spices, so the Dutch governor, Jan Coen set about control-ling production to keep prices high. On one occa-sion he destroyed all the nutmeg trees on the Banda Islands and either killed or sold into slavery the en-tire population of 15,000 people. He burnt villages along the coast of China in an attempt to control the whole region, but complained that China, like India, was 'too extensive for discipline'.

The English – The English East India Company was founded two years earlier than the Dutch, but it lost the race to control the Spice Islands. After Amboyna, the Dutch and the English were bitter commercial rivals. The English had to accept that the prize of the Pacific trade was closed to them and had to make do as a second best with establishing themselves in India. The trade in coffee, tea and cot-ton goods was of lower value than that from further east, but the English trading stations at Madras, Bombay and Calcutta grew in importance tea gained status as a fashionable drink. When Dutch power waned in the later years of the 17th century, English ships were able to use their Indian bases for trading with China and the Pacific Islands.

The French – The French East India Company had now replaced the Dutch as the main competition. French merchants, however, operated under diffi-culties. They came from a nation whose power was centred on its land army. While naval and commer-cial interests were influential in London, they carried little weight in Paris. In time of war, French ships were exposed to the powerful English navy, and French overseas outposts were at all times starved of resources.

It long seemed impossible that any European nation would establish political control in the sub-conti-nent. Then the death of Aurangzeb in 1707 marked the end of the Mogul Empire as an effective force, and the sub-continent split into warring states. From that time, the trading companies became increas-ingly involved in politics.

America

The Spanish – On January 2nd 1492 the troops of the 'Catholic Monarchs' Isabella of Castile and her husband, Ferdinand of Aragon finally drove the Moors out of Grenada. In the cheering crowd was a Genoese sailor, Christopher Columbus. Like an-other Genoese, John Cabot, he had decided that the Indes could best be reached by sailing westwards. Both turned to western European monarchs, with a natural interest in Atlantic trade. Columbus won the support of Isabella and in August his three ships sailed from Palos, to reach San Salvador on 12 Oc-tober.

Columbus was bitterly disappointed that he did not find the eastern markets described by Marco Polo. In later voyages he explored the Caribbean Is-lands and reached the mainland. He died in 1506, still convinced that he had reached the Indes. Before then, however, another Italian, Amerigo Vespucci, this time in Portuguese pay, had established that this was indeed the continent which subsequently car-ried his name. In the year that Columbus sailed, the Spanish Pope, Alexander VI issued a Bull, award-ing to Spain and Portugal all lands already discov-ered or to be discovered 'in the West, towards the Indes or the ocean seas, with the dividing line be-tween the two on the line of longitude 45^0 West. This ruling gave Brazil to Portugal and the rest of the continent to Spain. The Spanish, however, never established effective control to the north of a line from modern Georgia in the east to California in the west.

In 1519 Magellan led an expedition to explore this new world. When the remains of his expedition re-turned in 1522, having circumnavigated the globe, the basic facts of world geography were finally es-tablished.

Meanwhile the Spaniards were establishing their power in the New World. In 1513 Balboa crossed the Isthmus of Panama and reached the Pacific Ocean. In the east, Portuguese guns could win naval battles, but they could never bring down great em-pires. In the west, however, the Spaniards found that civilizations crumbled before them. There was too large a gap between the technology of the 'New World' on one side and the firearms, horses and armour of the 'Old World' on the other. Perhaps most important, the American 'Indian' people were psychologically ill equipped to confront the brutal European soldiers. Many were killed by the newcomers; many lost the will to live when forced to work in unfamiliar ways; even more died of the plague and other diseased for which they lacked immunity. According to one estimate, 25 million people lived in what was to become New Spain when Columbus landed, but only 1½ million survived a century later.

The Spaniards may not have found silks and spices, but they found gold. What to the native Americans was a decorative metal was, to the Spaniards, the basic unit of exchange and measurement of wealth. For gold Cortes and Pizarro destroyed the Aztec and Inca civilizations. Unsuccessful searches for gold established Spanish rule in what is now the south of the United States, from Florida to the Great Plains, and the Californian coast. All kinds of gold objects were melted down and shipped back to Spain, where the new riches funded the emergence of Spain as a major power.

The gold was soon plundered and no significant mines were discovered. A sustainable flow of wealth was, however, established by the opening of silver mines in Peru. Spain now controlled both sides of the Isthmus of Panama and a merchant fleet was built on the westward, Pacific side. A trading base was established at Manila in the Philippines, and galleons carried trading goods across the wide Pacific. These luxury goods from the Orient, along with silver from Peru, were then carried across the Isthmus of Panama and loaded onto the Atlantic treasure fleet for Spain.

In the early years of colonization few women left Spain for the New World, and settlers took Indian wives. The culture, and even the religion of New Spain therefore developed a syncretism between Spanish and Indian traditions. In time the importation of black slaves from Africa further complicated the ethnic mix. It has, however, remained generally true into modern times that the social position and wealth of any individual could be gauged by skin colour.

The English – John Cabot was convinced that Columbus had got his sums wrong. He believed correctly that China was far out of range of any ship following a southerly route. By his calculation, the journey could be made at a more northerly latitude. Sailors from Bristol, England were already fishing

the Newfoundland banks and knew the North Atlantic well. Cabot therefore won support from King Henry VII of England and in 1496 reached the coast of North America. It was not obvious that the north of the continent was embedded in the Arctic ice and Cabot's son, Sebastian led a long line of English sailors in search of the North West Passage. The English sea dogs, Drake and Hawkins preferred the warm waters of New Spain to the cold northern seas. They first operated as traders, and then, after being attacked by Spaniards, as privateers.

The gold of New Spain and the luxury trade of the Orient offered instant riches. Returns on investment in North America were likely to be less spectacular. By the 1580s, however, Sir Walter Raleigh and others were advocating colonization of the land of Virginia, which was now claimed by England. Attempts were made to establish colonies in 1585 and 1589, but both failed. The first successful colony was made at Jamestown in 1607. In 1620 a group of 'Pilgrim' refugees set up a colony at New Plymouth, Massachusetts, and later moved to the better site of Boston.

The English settlements were based on a farming economy. Disease, spreading from New Spain had recently ravaged the native American tribes, leaving much of the land vacant. The surviving people practised a mixed hunting and farming economy, based on shifting cultivation, so to newcomers much of the land appeared to be empty. As land hungry settlers kept on arriving and pushing inland towards the Appalachian Mountains conflict with the Indian people was inevitable.

The Dutch – In 1614 the United New Netherlands Company established a colony at the mouth of the Hudson River. The Dutch recognized the potential of the trade in beaver fur and used the Hudson to make contact with Indian people of the interior. This settlement divided the English colonies of Virginia and New England and hostility between the two Protestant countries, aroused in the far Spice Islands, spilled over in the New World. In 1664 the English drove the Dutch from North America.

At the height of their powers, the Dutch carried their assault on the Portuguese Empire into the New World by annexing Brazil in 1637. The Portuguese settlers rebelled against them and they were driven out in 1654, leaving Brazil as the western outpost of a once great Portuguese Empire.

The French – In 1603 the French explorer de Champlain sailed into the St Lawrence River. He too was still searching for the elusive route to the east. He established settlements which were to become Montreal and Quebec and pressed on to explore the inland waterways of the interior. The French settlers were comparatively few in number and they received little support from their home government. Champlain and those who came after him exploited Indian rivalries to establish a flourishing trading empire, based on the fashion trade in beaver fur. As the animals were hunted to near extinction in the east, the 'beaver frontier' moved west, taking the hardy French after them.

French explorers followed the Great Lakes waterway into the interior and then the Mississippi to the Gulf of Mexico. Here they established the French outpost of New Orleans. The North American empire, named Louisiana, after Louis XIV, now followed the waterways in a huge, but lightly populated arc. At first the French and English colonists only came into contact with each other in the Hudson Valley. The risk of conflict grew, however, when the French tightened the noose around the English colonies by taking control of the Ohio River. At the time, however, colonial wars, which decided the fate of India and North America, were seen as little more than a sideshow beside the main European conflicts.

The Old Colonial System – The Dutch can be credited with the development of mercantilism, which became known as the old colonial system. This was not developed specifically for North America, but, when applied by the English in their American possessions, it became a root cause of later conflict between the colonists and the mother country. It was assumed that overseas colonies existed to promote the interests of the mother country, by extending its economic base. Colonists were expected to produce cash crops. Some, like rice from the Carolinas or tobacco from Virginia, could not be produced in northern Europe. Softwood timber from New England was also of vital strategic importance for ship building at a time when European forests were finally disappearing. Buying these goods from a national source saved the mother country the foreign exchange, which would be required to purchase them from abroad. By selling these crops, the colonists earned money, which would be spent on the manufactured goods. This in turn assisted the manufacturing industries and strengthened the merchant marine of the mother country. Any business between the colony and a third country had to transacted through the mother country. This had the further benefit of boosting customs revenue.

The trade-off was that the mother country was responsible for providing the colonists with protection, be it from local populations or from hostile

Europeans. This involved the Westminster government in the expense of funding wars against the French and their Indian allies. The system came under pressure when the colonies began to develop out of their original role as providers of raw materials to develop their own manufactures.

Africa

The Atlantic Slave Trade – The Portuguese were the first to discover that West Africa had human resources, which were to be exploited in a slave trade, which continued for some 350 years. A base was established on the coast as early as 1448, from which relatively small numbers of slaves were shipped back to Portugal.

An acute labour problem then began to develop in the new American plantations. The obvious solution was to recruit American Indians. Heavy field work, however, proved alien to them. Many died, often by suicide, when forced to work on European plantations. European labour was also brought in, both by the forcible transportation of convicts, and by indentured labour schemes, under which immigrants were bound to their masters for a given number of years. Again, however, expectation of life was short, and the labour problem remained unresolved.

Portuguese ships then began to take slaves direct to their colony of Brazil. In 1562 John Hawkins began the English slave trade between West Africa and the Caribbean. Dutch, French, Danes and later sailors from both North and South America joined in the business. European nations established forts on the West African coast to protect the interests of their slave traders.

It is estimated that some eight to ten million slaves were carried across the Middle Passage to America. The economies of European cities, such as Nantes, Bristol and later Liverpool were based on slaving, and the business was accepted as a part of the national commercial interest.

The individual suffering of slaves would ultimately receive wide publicity; the impact the trade had on African society is harder to quantify. European sailors rarely penetrated inland to find their own captives. Domestic slavery already existed on the continent, and Africans initially sold their own slaves to purchase European goods. In time, however, demand outstripped this source of supply. Military confederacies, such as Dahomey and Ashanti, grew up to fulfil the double function of protecting their own members, and feeding slaves to the European forts. When Europeans later penetrated the continent, they discovered that these states often acted with a savagery untypical of African society further inland. The demand for slaves created an endemic state of war which penetrated inland, far beyond any direct European contact. The resulting depopulation appears, however, to have been largely balanced by improvements in the African diet as a result of the importation of American crops, such as the yam and cassava.

Colonization – The first African colonies had the prime function of protecting and providing staging posts for national ships on the eastern trade routes. The Portuguese early established the outposts in Mozambique and Angola which would achieve the distinction of being the longest-lasting European overseas colonies. In 1652 Jan van Riesbeck set up the Dutch colony at the Cape of Good Hope, to serve the eastern convoys as a 'tavern of the seas'.

In the 18th century, the French established an interest in the Indian Ocean island of Madagascar, along with Mauritius, and Reunion. The slave coast of West Africa remained unattractive for colonization. European slavers and soldiers suffered a high mortality rate from tropical diseases, particularly yellow fever and malaria.

East Africa – At this time, East Africa lay off the main trading routes, and the region offered little to attract European merchants. Arab dhows still sailed undisturbed to Zanzibar and their caravans penetrated deep inland. Here again, slaves featured prominently as a trading commodity alongside gold and ivory. The area remained an Arab area of influence until European missionaries and traders penetrated the area in the 19th century.

THE NATIONS OF EUROPE

Italy and the European Powers

The City States and the Papacy – In the 15th century, the northern half of Italy was the most advanced part of Europe. The great trading cities of Genoa and Venice brought in wealth and broad contact was maintained, both through trade and cultural exchange with Arab and Byzantine civilizations. The country probably benefited from the fact that it was never brought under unitary political control.

The broken terrain of Tuscany and Umbria suited the development of independent city states, not unlike those of ancient Greece. Florence and Sienna, like Athens and Sparta of old, built up confederacies to counterbalance the power of the other. In the late 14th century, the banking family of Medici took power in Florence. Times were not always easy, but they led the city to its unique flowering of culture.

In the north another ring of states, with Milan as the most powerful, controlled the trading routes across the Alps. In the centre, the pope ruled the Papal States as any other temporal monarch, and involved himself in the politics of the peninsula, attempting always to extend the patrimony of St Peter. During this period the lifestyle of the popes was little different from that of any other monarch. They led troops into battle, promoted family interests, including those of their children, and built themselves enormous monuments. Julius II's decision to build himself a tomb set off the chain of events which triggered the Reformation in distant Germany; the tomb would be too large for St Peter's, so the church had to be rebuilt; this involved raising money by the granting of indulgences.

The Theory of Kingship – Within this turbulent world of Italian politics, only the fittest survived. Nicolo Machiavelli worked for the Florentine state, travelling widely as a diplomat. He wrote a book, called *The Prince*, which was based on these experiences, which contained advice for the Medici family on the theory and practice of government. Political decisions, he argued, could only be taken on a cool, indeed callous, assessment of the security needs of the state and of its ruler. Medieval concepts of the mutual duties of ruler and subjects were cut away in this first exposition of what would later come to be called 'real politik'.

Medieval monarchy was based on a feudal alliance between king and his tenants in chief. In the 16th century power was being drawn to the centre, at the expense of both the magnates and of representational institutions. For Machiavelli's prince, power was its own justification. The theory of centralization was later taken further with the formulation of the concept of the divine right of kings. Rulers, it was said, held power directly from God. Rebellion was a sin and criticism of the royal will was tantamount to treason.

Foreign Invasion – In 1494 Charles VIII of France crossed the Alps at the head of an army of 30,000 men. He laid claim to the Kingdom of Naples and on his way south, through Rome itself, his army left a trail of destruction. Other foreign armies followed. Artists still worked on, producing some of the greatest works known to man, but the days of the city state were over and Italy would henceforth be a pawn in the real politik of the great powers. In 1527 the ragged, unpaid and hungry army of the great emperor, Charles V, ran wild in the streets of Rome, and the city was sacked for the first time since the barbarian invasions.

The Empire of Charles V – Throughout medieval times, kingship was fundamentally a matter of family inheritance. Charles was the ultimate beneficiary of this dynastic system. From his mother, the mad Joanna, he inherited his grandparents' crowns of Castile and Aragon. On the paternal side, he inherited from his grandfather the title of Holy Roman Emperor, and from his grandmother the lands of the Duchy of Burgundy. As king of Castile he controlled Spanish land in the New World; as Emperor he ruled Austria, Hungary, Bohemia and much of Germany; as Duke of Burgundy he possessed the Netherlands, which was the richest part of all Europe. His empire was larger than that of Charlemagne.

France was now shut in on all sides, and its king was determined not to let Italy fall to Charles' empire. The crusading spirit was finally laid to rest as Pope and King of France allied with the Ottoman Turks against Charles.

This great empire, like that of Charlemagne, carried the seeds of its own destruction. Charles was unable to function adequately as ruler of such dispersed lands, and resentment grew, particularly in the Netherlands, at the taxes raised to support Italian wars. Charles was also depressed at his inability to control the spread of protestantism within his own lands. He abdicated in 1556 and the empire was divided. The title of Emperor passed to his brother Ferdinand I, while the more valuable western share, consisting of Spain and the old Burgundian lands, went to his son, Philip II. There were now two Hapsburg dynasties in Europe.

Protestantism and the Counter Reformation

The Spread of Protestantism – Luther's new beliefs found most followers in northern Europe, particularly in Germany and Scandinavia. The impetus behind the further spread of protestantism came not from Germany but from Geneva. John Calvin was French, but he achieved prominence in the Swiss canton. He preached a harsh form of protestantism; since God was all powerful, he had predestined a minority of people – the elect – to salvation and the rest to damnation. The elect had to show their status by a strict adherence to a way of life. In Geneva moral sins, like adultery and even disobedience by a child, were severely punished. Calvinism proved to be a more militant faith than

Lutheranism. It appealed in the Netherlands, England and the west coast of France, in Scotland and later in the Lutheran heartland of Germany.

All Protestantism stressed the direct communion of the individual with God, and it is not surprising that it early showed a capacity to fragment. In 1532 an extreme group, the Anabaptists, took control of the German city of Münster, preaching not only rejection of infant baptism but polygamy and a radical social gospel. In the extreme Protestant sects, authority lay not in any higher political or ecclesiastical power, but in the local 'gathered church'. These separatist churches were persecuted in Protestant and Catholic countries alike, but it was this tradition that would ultimately implant itself in the New England colonies of North America, and profoundly influence the development of American society.

Toleration was not a cherished ideal in 16th century Europe, but by 1530 it had become clear that protestantism was too powerful a movement to be readily suppressed. In that year the Peace of Augsburg laid down the principle of 'cuius regio, eius religio' – the country would follow the religion of the ruler. This left rulers free to persecute within their own dominions.

Sweden and England Break with Rome – Two European monarchs took their nations out of communion with the Roman church. Both were motivated by national and financial, rather than by religious reasons. In 1523 the young Swedish nobleman, Gustavus Vasa, succeeded in his struggle to make Sweden independent from Denmark, and was proclaimed king. Lutheranism had already made progress among his people. In 1527 he broke with Rome as a symbol of the new national independence, and he enriched his hard-pressed government with church lands.

Henry VIII of England had showed no personal inclination towards the reformed religion; indeed, he had written a pamphlet attacking Luther and had persecuted Protestants. In 1530, however, he became involved in a dispute with the pope over his divorce from Catherine of Aragon. Using selective intimidation, he won the support of parliament for a breach with Rome and then for the plundering of monastic lands. The Church of England, reformed in doctrine, but conservative in practice, was the creation of Henry's Archbishop of Canterbury, Thomas Cranmer. After a return to catholicism under Mary I, Elizabeth, declared that the English church should be a home for all men of goodwill. Separatist Protestants and politically active Catholics were still persecuted, but England did escape the worst violence of these years.

The Counter Reformation – The Roman church had been on the defensive against an aggressive protestantism for 25 years when Pope Paul III called his bishops together for the Council of Trent. Paul represented a new generation of Pope, anxious to clear away the scandals of the past, and re-establish the western church on a firm footing. The discussions were dominated by bishops from Spain and Italy, where protestantism had found no foothold. The Council brought in reforms, but it made no concessions to Protestant faith. By the time that the Council had finished its debates in 1563, the lines of division were clearly drawn.

Catholicism was now on the counter-offensive. As in the past, monasticism provided the papacy with its front line troops. In 1540, Ignatius Loyola, who had been a fellow student in Paris with John Calvin, established the Society of Jesus, or Jesuits. Members were bound to total loyalty to the Pope, and this provided the reforming papacy with a means of circumventing special interests within the church. Jesuits became particularly prominent in education and in missionary work.

Spain and the Netherlands

The Expulsions – Even without Charles' eastern lands, Philip II's Spain remained the dominant power in Europe. He controlled southern Italy and Sicily and succeeded in conquering Portugal. Spain's European power was now underpinned by the revenues of a two huge overseas empires.

The nation's weakness was not clearly evident at the time. When the Moors were finally defeated, Moslems and Jews had been promised security within the Christian state. The presence of infidels, however, proved too much for Catholic rulers still driven by the intolerance of the Inquisition. Moors, Jews and converted Moors, the Moriscos, were all driven out of the country. These, however, were the very tradespeople and skilled craftsmen on whom the economy rested. As a result, Spain became heavily dependent on imported goods, particularly from the prosperous Netherlands. On occasions the Panama fleet had to be diverted and sailed direct to unload its treasure in the Netherlands.

The Spanish Netherlands – The old Burgundian lands covered both the modern states of Belgium and Holland. The greatest centres of prosperity, Antwerp outstanding, lay in the south. The northern part, mostly consisting of land drained from the Rhine delta, contained the finest farmland in Europe, but, even with intensive agriculture, it could not feed the growing towns. Calvinist protestantism had won adherents in the north and in the south.

Charles V was born in the Netherlands and during his reign the two religions co-existed with reasonable tolerance. The accession of Spanish-born Philip II, however, brought change. As king, he was determined to bring the old Burgundian noble families under his control, and, as a faithful son of the church, he meant to stamp out heresy in his land. The Spanish Duke of Alva was sent with an army to bring the area under control.

Dutch Independence – In 1572 William Prince of Orange led the People of the Netherlands in revolt. As Spanish armies established control of the south, many Protestants moved north behind the protection of the dykes, and the religious division between the Catholic south and the Protestant north was established. In 1581 the followers of William of Orange declared their independence from Spain. No matter how bitter the fighting, the trade between Spain and her rebellious provinces never ceased. Philip was in no position to cut off this channel of supplies and the Dutch were happy to drain the enemy of wealth. William was murdered on Philip's orders in 1584, but the struggle continued until Spain made a truce in 1609. Almost forty years would pass before Spain finally recognized the independence of the Dutch people, but in practice Holland had established its independence from its traditional ruling house.

The Dutch Republic – The new nation was unique in that power was based on trade rather than on inherited land. A successful Dutchman did not plan for the day when he would put aside the cares of trade and live as a gentleman; his objective was to hand a thriving business to his heirs. People lived by a strict work ethic and made the most of the limited resources of their small land.

National wealth was founded on north-south trade, carrying products such as grain, timber and iron from the Baltic to the overpopulated Mediterranean lands. Dutch flyboats, little more than floating holds, plied the oceans. 'Norway was their forest, the banks of the Rhine and the Dordogne their vineyard; Spain and Ireland grazed their sheep; India and Arabia were their gardens and the sea their highway.' Scholars also provided information for sailors and laid the foundation of modern geography.

The Decline of Spain – The loss of the Netherlands was the clearest marker of Spain's fall from its position as Europe's dominant power. In 1588 a Spanish naval armada was also defeated by the English fleet. In 1640 Portugal re-established her independence under the house of Breganza. The nation could have overcome military reverse; the basic problem was that Philip II and his successors concentrated on military and colonial affairs at the expense of the economy, shattered by the mass expulsions.

The French Wars of Religion

The French Monarchy – In the middle of the 16th century, French royal power stood at a low ebb. Financial stringency led to offices being sold to the highest bidder, and, partly as a result, the size and independence of the aristocracy was ever increasing. Calvinism was strong in Brittany and Normandy, and growing in power further south on the Atlantic coast, and in Languedoc. Its strength was based on craftsmen and some poorer nobles, followed by a growing number of peasants. By 1562 there were over 1500 'Huguenot' congregations, many led by Geneva-trained pastors. The Catholics themselves were divided: the moderates, led by the Regent, Catherine de Medici, who planned to keep the peace by giving a measure of toleration to the Protestants, and an extreme Catholic party, who wanted to see heresy stamped out.

The Wars – Fighting broke out after extremist Catholics massacred a Huguenot congregation at Vassy in 1562. The ensuing wars were fought with great ferocity on both sides. In 1572, 3,000 Huguenots were massacred in Paris on St Bartholomew's Day, and in 1588 the king was ejected from his own capital by extreme Catholics. In 1584 the Huguenot Henry of Navarre became heir to the throne. He succeeded in bringing the war to an end by turning Catholic and reaching agreement with his former Protestant followers in the Edict of Nantes. This left the Huguenots with freedom of worship in large areas of the country as well as certain fortified cities, now effectively outside royal control.

Germany

The Empire after Charles V – Charles V's brother Ferdinand saw himself as a faithful Catholic and soldier of the Counter Reformation. His own lands and the south of Germany remained Catholic. The Protestant forces set against him were divided. In the north were the Lutheran powers of Denmark, Saxony and Brandenburg. The Calvinist stronghold lay to the west around the Rhine. Ferdinand dreamed of winning back the whole of Germany to catholicism, while at the same time bringing it once again under imperial rule, but he was unable to achieve his ambition because his empire was exposed on its eastern flank. In the south, the Ottoman Empire reached the peak of its power under Suleiman I, and even threatened Vienna itself. In the north, Sweden was establishing control of the Baltic

Sea while Poland and Russia both pressed on German land.

The Thirty Years War – In the early 17th century the religious divisions became more sharply fixed. In 1608-9 the Catholic League and the Calvinist Union were set up as rival military blocks. The first of a series of wars broke out in 1618, when the Calvinist Elector Palatine was elected King of Bohemia. The Catholic armies, led by virtually independent war lords, won early successes, but this rallied the Lutheran armies to the Protestant cause. The Protestant champion turned out to be Gustavus Adolphus, King of Sweden, who won a series of battles before he was killed at Lützen in 1632.

By now the religious battle lines were becoming blurred. Catholic France, under Cardinal Richelieu, was prepared to fund Protestant armies and even to intervene directly to prolong the war and so prevent and re-emergence of imperial power in Germany. This brought in Catholic and Hapsburg Spain on the imperial side.

The war was a disaster for the people of Germany. Roaming armies stripped the countryside of food; the devastation caused by the imperial sack of Magdeburg in 1629 rivalled that of a Mongol army. When the war limped to a close in 1648, the countryside was impoverished and depopulated. Ferdinand's ideal of a Catholic Germany, united under the empire was destroyed. Pro-testantism was unassailable in the north and the effective power of the emperor in the German-speaking lands was henceforth limited to his Austrian heartland. In the Treaties of Westphalia the Emperor had to accept the independence of Switzerland – a reality since the end of the 15th century – and the King of Spain that of the Netherlands. France and Spain both made achieved territorial gains in German lands. Most significant for the future, the new power of Brandenburg had emerged in the north.

Brandenburg – Prussia – In 1640 'The Great Elector' Frederick William, of the House of Hohenzollern, inherited Brandenburg and the eastern territory of Prussia. A man of great energy, he set about creating a well-run, modern state, with an efficient civil service and a highly disciplined army, which served as a model for later German armies. His work was consolidated a century later by Frederick II 'the Great'. He had no vision of a united Germany, but he ruthlessly expanded his family lands at the expense of the Empire.

The Hegemony of France

Richelieu – Henry IV was assassinated in 1610, leaving a country at peace, but with many problems.

The Huguenots were a state within a state; the nobles were over-powerful and contributed little to the nation; the peasants were desperately poor and over-taxed.

In 1624 his successor, Louis XIII, appointed Cardinal Richelieu as head of the royal council. For 18 years, he worked single-mindedly to establish royal power within the nation. He had no wish to persecute the Protestants, but he destroyed the independent Huguenot fortresses. He made examples at high level to bring the nobles under his control. Regional government was delegated to directly appointed intendants, who exercised the complete range of royal power.

Richlelieu's foreign policy was directed at limiting the power of Spain and improving national security by achieving 'natural frontiers' at the Rhine and Alps. For this, he was prepared to ally with Protestants and to prolong the misery of the Thirty Years War.

Richelieu represented the apotheosis of the Machiavellian ideal; his policy was driven by a cold analysis of *raison d'état*. He did not recognize that an improvement in the lot of the poor was essential if the state was to be securely based. Shortly after Richelieu and his royal master died in 1642-3, there was a series of popular uprisings, known as the Frondes, across the country.

Louis XIV – The young king who succeeded ruled the country until 1715. His domestic and foreign policy was a continuation of that laid down by Richelieu. All real power lay in the hands of non-noble ministers and intendants, while the nobles were emasculated by being drawn into the glittering court of Versailles.

Unlike Richelieu, however, Louis determined that he would not rule over heretics. He revoked the Edict of Nantes, facing protestants with the choice of conversion or expulsion. Like Isabella of Castille, he was hereby driving a productive group out of the nation. Economic conditions did improve, but the poor continued to suffer harshly enforced penal taxation.

Much tax revenue was spent on foreign wars. As France had organized leagues to limit the power of Spain, so now others united to contain France. The driving force in the anti-French Grand Alliance was William of Orange, Stadholder of the Netherlands. His power strengthened when, in 1688, he also became king of England as William III. The War of the Grand Alliance (1689-97) was followed by the War of the Spanish Succession (1701-14), which sought to prevent Louis from unifying the crowns of Fence and Spain by dynastic succession.

18th-century France – In 1715 France was clearly the leading power in Europe. Major losses of overseas territory to England in India and North America during the Seven Years War (1756-73) did not appear as significant at the time as they were later to become. Financial weakness, however, underlay the pageantry of the monarchy. The huge noble class – estimated at up to 250,000 strong – had lost political power but not financial and legal privilege. The state sank ever more deeply in debt but had no means of tapping the huge reserves of noble wealth. Here lay the seeds of revolution.

England

Sea Power – When Roman soldiers were posted to Britain, they considered that they were being consigned to the edge of the civilized world. Through medieval times, the British Isles remained on the periphery of the known world. The discovery of America moved the centre of gravity from the Mediterranean towards the Atlantic Ocean. Geography therefore now favoured England.

As an island nation, the English were perforce a seafaring people. By 1500, however, this seafaring tradition had not been converted into naval power. The defeat of the Spanish Armada in 1588 proved to be a turning point. The battle was won by strategy rather than by fighting force, but the Elizabethan sea-dogs created a national myth which would survive into modern times. Governments, reluctant to involve troops in European land battles, laid the greatest stress on building up naval power and securing naval supplies. The navy provided protection for the island, maintained links with overseas colonies, and secured trade routes against competition.

Monarchy and Parliament – The Tudor monarchs, Henry VIII and his daughter Elizabeth, dominated 16th-century English politics. The nobility were few in number, and were generally content to concentrate their efforts on field sports and the efficient management of their estates. While parliamentary government was withering on the continent, in England the old institutions remained robust. Henry found it convenient to use the House of Commons as his ally against the Church, and Elizabeth was able to manage parliament, even if sometimes with difficulty, both as a source of revenue, and as a channel of government.

When Elizabeth died in 1603, the succession passed to the Scots House of Stuart, which, through family and cultural ties, was more influenced by the French model. Very early, James I became involved in disputes with both the legal and parliamentary establishment. James proclaimed the divine right of kings, which, he claimed, gave the king the power to appoint and dismiss judges and to raise taxes. Jurists recovered documents such as Magna Carta from obscurity to defend ancient privileges against the new royal pretentions. Implicit in their arguments lay the notion that royal power was derived from the consent of the people – however the people might be defined. The conflict was made more acute by the fact that personality did not match pretention. The Tudors had maintained authority through the force of their personalities rather than through modern concepts of kingship. James was intelligent but personally unimpressive; his son Charles I was an inadequate recluse.

The Civil War – Charles I soon found himself in direct confrontation with parliament. In 1628 parliament presented a Petition of Right against the use of arbitrary royal power; in 1629 Charles dissolved parliament and began 11 years of direct rule. Many aspects of royal government were unpopular to influential subjects. An attempt was made to impose 'high church' worship, not only on England but also on Calvinist Scotland. An increasing number of cases were heard in royal prerogative courts rather than in the courts of common law. Direct taxes were levied without parliamentary approval. It seemed to many as though Charles would soon follow Richelieu's example and centralize all government.

The outbreak of war in Scotland brought financial disaster and Charles was forced to recall parliament in 1640. A struggle for power led to the outbreak of war in 1642. Historians have long argued the economic, religious and social issues which lay behind the conflict; certainly it was very different in nature from the violent upheavals which would later shake France and Russia. Parliamentary power was based in the rich southeast, while the king's was centred in the poorer north and west. The parliamentary victory was due both to this difference in resources and to the leadership of Oliver Cromwell, who emerged as the outstanding general in the conflict. He kept his New Model Army under such firm control that it could march across countryside and leave fields and property as they had been before the army passed.

The Commonwealth – The parliamentary broke into factions after the defeat of the king. In 1648 one faction seized power, with army support and staged the trial of Charles I. The execution of the king in 1649 provoked a shocked response across Europe. No action could have expressed the rejection of divine kingship more vividly. In 1653, Cromwell staged a military coup and assumed power as Lord Protector. Cromwell died in 1658, and a brief attempt was made to continue the protec-

torate under his son. This failed, however, and in 1660 the army again was responsible for bringing Charles II back to London.

The Glorious Revolution – The saga of the conflict between the Stuarts and their parliaments was not, however, over. Charles was mistrusted, both for his French sympathies and for his leaning towards catholicism, but he still depended on parliament for revenue. In 1685 he was succeeded by his Catholic brother, James II. Three years later James was forced to leave the country, to be replaced by his Protestant daughter, Mary and her husband, the Dutch William of Orange, who exercised the practical power. William was more interested in securing the English alliance against France than he was in pursuing power struggles with the English parliamentso he accepted laws which established that the king would henceforth require parliamentary consent to raise money and keep a standing army in peace time. It was also agreed that he could not alter or suspend any act of parliament.

In 1714 the English throne passed to the German house of Hanover. Since the new king could not speak English, day-to-day government passed to a prime minister and a cabinet, drawn from the majority party in the House of Commons. Political power had now finally passed from the monarchy to the property owning classes, who were represented in parliament. In the century which followed parliament largely used its power to improve the position of the landowning class, often at the expense of the poor. English politics had, however, run against the European tide, which favoured greater centralization in the hands of the monarch.

The Act of Union – Throughout history there had been strife between England and her smaller, poorer northern neighbour, Scotland. The union of the crowns in 1603 did not end this. In 1707, however, the two countries became formally united in the Act of Union. Two clan uprisings followed in 1715 and 1745 in favour of the exiled Stuarts but were suppressed. Scots engineers, doctors and scholars were shortly after to make a major contribution to the great surge in national prosperity of the united Great Britain.

Russia

Boyars and Serfs – Across the continent, the Russian state was following a very different pattern of development. The noble boyars held their land from the Tsar in return for defined services. Since Russia had no law of primogeniture, this class was getting ever larger, and most of its members poorer, as estates were split one generation after another. The mass of the people remained in the medieval condition of serfdom. Families were owned by their masters, had no right to move of their own free will and had no redress except in their masters' court.

The relationship of Tsar and boyars was often marked by bloody conflict. Ivan IV 'The Terrible' allied himself with the merchant class and the common people in an attempt to break noble power. He achieved many real reforms before mental disorder led him, in the latter years of his reign, to behaviour which anticipated that of Joseph Stalin in the 20th century.

National Objectives – Russian development was hindered by the lack of a warm-weather outlet to the ocean. The port of Archangel was ice-bound in winter, and all year the journey round northern Scandinavia was long and dangerous. The port of Rostov in the south was of little use as long as the Ottomans controlled the mouth of the Sea of Azov and the Dardanelles. National policy therefore became directed at winning a port on the Baltic Sea. This brought Russia into conflict with the advanced military state of Sweden, which guarded the Baltic as a Swedish lake.

The Russian Tsar could mobilize huge armies, by raising levies, but there was no adequate support structure. Forces were sent to war with the vague hope that they would be able to live off the land. Often countless thousands starved, and those who survived were in no condition to fight the world's most efficient army.

Peter the Great – Peter succeeded to the throne in 1682 at the age of 10, and suffered huge indignities from guards and boyars while still a child. Once a man, he announced his intention to bring his nation up to date and orientate it towards the West. A man of little education but enormous energy, he immersed himself in every detail of western science and technology. In a famous visit to the west he was equally at home working in disguise as a dock worker in Holland and meeting with scientists in England. His methods of enforcement were effective, if sometimes eccentric.

The vindication of Peter's work came in 1709 when his army won a decisive victory over the Swedish army under Charles XII at Poltava. Russia had won its outlet to the Baltic Sea, and here Peter decided to build his capital of St Petersburg.

Peter's great failure was that, like Louis XIV in France, he did nothing to improve the lot of the Russian poor. Someone had to pay for wars against Turkey and Sweden, for the modern weapons, for new ships and for the fine capital city. The poor were taxed and taxed again until they were left with the barest minimum necessary to stay alive. It is a measure of the depth of the misery and the capacity of the Russian people to absorb suffering that revolution did not erupt in violence for another 200 years.

THE WESTERN MIND

The Renaissance

Italy – The word *renaissance* was coined in the 19th century to describe the rebirth in Italy of the classical ideal in art, architecture and letters. The Middle Ages was looked upon as a dark period before the great transformation of the 15th and early 16th centuries. Recent study has shown that the picture was more complicated; classicism remained strong throughout the Middle Ages, and there was more cross-fertilization between Europe north and south of the Alps than had been assumed.

Any gallery visitor can, however, see the astonishing change which happened in visual perception within a comparatively few years. Across northern Italy artists experimented with new forms. In the words of Vasari, Giotto 'restored the art of design'. In Umbria, Piero della Francesca used mathematics to work out laws of perspective, well beyond any classical achievement. In Florence, Michelangelo combined an analytical eye with his huge talent to create a new vision of the human form. Even when painters and sculptors continued to work on church commissions, they now used live models to give a new sense of naturalism.

There was a keen awareness among the artistic community that they were living in an exciting new age. The Medici and other patrons commissioned works with secular themes, often drawn from Greek mythology. Artists were no longer faceless craftsmen who had produced so many medieval treasures. Art had found a new self-consciousness.

The same secular, driven innovation was reflected in music and literature. There was a passionate interest in all aspects of antiquity. Some sculptors even buried their own work and dug it up again, claiming it as a classical discovery. Old manuscripts were found in monastic libraries or brought from the east and studied with a new intensity. Enthusiasm for antiquity did not preclude Christian belief; rather the classical tradition was seen simply as one element in divine revelation, so producing a syncretism which alarmed conservative churchmen.

By the mid 16th century the Italian renaissance was losing its impetus. The first unique burst of innovation could not be maintained. The Counter Reformation church now demanded a more orthodox treatment of subject matter, both in literature and in painting. Much great work continued to be done, particularly in the Veneto. Paladio used Roman models in creating the architectural style which would bear his name, while Titian and his contemporaries were laying the foundations of what would become the baroque style. Generations of artists and patrons continued to travel to Italy to absorb the culture both of its classical past and of the present.

The Northern Renaissance – Some of the great painters of the Flemish school crossed the Alps and were much admired by Renaissance artists. Perhaps because they were not surrounded by antiquities in their home environment, however, they never made the sharp break with the gothic. Italian styles took many years to become established north of the Alps.

Northern Europe's unique contribution came in the field of scholarship and literature. Here writers were free from the restrictions of the Counter Reformation and fear of the Inquisition. Protestants wanted to make the Bible available to all. The translations of the scriptures by Martin Luther and William Tyndale were immensely influential in formalizing the written forms of German and English respectively. Traditional interpretations of the Bible were challenged when Erasmus of Rotterdam produced a version of the New Testament in the original Greek. Latin was now ceasing to be the universal language of scholarship. While a return to the vernacular liberated learning from the cloisters of the church, it also fractured the international culture, which had reached its peak in the 12th century.

A strong secular tradition now flourished in England. Chaucer had already written for the newly educated merchant class. In 1510 John Colet, Dean of St Paul's Cathedral and close friend of Erasmus, made a gesture to the secularisation of learning when he closed his cathedral school and re-founded it under the control of a trading guild. The combination of Tyndale's language and renaissance scholarship had created a uniquely favourable environment when in 1585 an actor called William Shakespeare left his native Stratford to try his luck in London.

The great flowering of French literature came in the 17th century. Corneille and Racine were still in essence renaissance writers, handling classical themes with a paladian sense of form and style.

Printing – Most importantly, the re-invention of printing by movable type provided the means of dissemination of both religious and secular literature. Whether the innovation be credited to Guttenberg of Maintz or Coster of Haarlem, the technique provided the means of dissemination of the works of any author. Books became cheaper as print runs grew longer. In the following centuries print was used to promote colonies, to circulate scurrilous pamphlets to produce works on magic as well as to disseminate works of scholarship, religion and literature. In 1702 *The Courant*, the world's first daily newspaper, was published in London. Soon after works of popular fiction began to appear. Print had become an integral part of Western life.

The Advancement of Science

The Copernican Revolution – In Hellenistic times the idea had been posited that the earth rotated round the sun, but this had not won general acceptance and in 16th century it was still generally accepted that the heavenly bodies rotated around a stationary earth. In 1543, however, the Polish scholar Copernicus published a book arguing the theory of heliocentric astronomy.

Copernicus' theory received little attention. During this time, however, Dutch craftsmen were experimenting with glass lenses. They made spectacles and also telescopes for use at sea. One of these telescopes fell into the hands of the Italian teacher, Galileo Galilei, and he turned the instrument towards the skies. By studying sun spots, the phases of Venus and the rings of Jupiter, he provided clear proof that Copernicus had been correct.

Galileo delayed publishing his findings because he recognized that they must arouse a storm of controversy. Authority, both of the Bible and of ancient authors clearly supported a geocentric universe, and the church still held to authority as the arbiter of truth. He published his findings in 1632, but, faced with the terror of the Inquisition, he recanted in 1633.

Descartes – The Turning Point – Tradition states that, after formally accepting that the earth remained stationary, Galileo muttered 'it goes on moving'. Certainly the scientific impetus continued. In 1637 the French philosopher Descartes published *Discours de la Method*, which laid down what has become known as the Cartesian method. He argued that the experimental scientific process was the arbiter of truth. The pursuit of truth now involved breaking down knowledge into ever smaller areas of study. Medicine, for instance, became concerned with analysing the symptoms of disease in minute detail – arguably at the expense of a more integrated approach to the healing process.

In his dictum, *cogito ergo sum*, Descartes proclaimed the individualism which was to be the hallmark of modern European society. Western man had at last emerged from the shadow of past authorities, religious or classical. Personally a devout Catholic, Descartes rejected authority as an arbiter of faith and proclaimed that it had to be discovered through the human intellect. This was recognized as a fundamental challenge to the church, and Louis XIV personally ensured that Descartes was denied Christian burial.

Northern Europe – The condemnation of both Galileo and Descartes, and continued activities of the Inquisition, placed scientists who lived in Catholic countries in an invidious position. In Protestant countries scientists might meet hostility from those who defended religious authority, but not persecution. The impetus for scientific innovation therefore passed to Northern Europe.

The first protestant scientist was the German, Kepler, who provided information on the movements of planets. Dutch scientists, continuing their work with lenses, developed the microscope. This opened up whole new areas of study in such areas as the biological sciences. In England the cause of experimental science was argued by the Lord Chancellor, Francis Bacon, who had early visions of its potential. In 1619, William Harvey demonstrated the mechanism of the circulation of the blood.

The advances in navigation, first in Holland, and then in France and, above all, in England, drove forward skills in cartography and geographical study. Progress in astronomy and in the construction of clocks were spin-offs from this. Landsmen could now own clocks and watches which told the time with great accuracy. People began to organize their lives around them, and to treat the hours and minutes of the day with a new respect.

The revolution started by Copernicus was completed by Isaac Newton, who published his *Principia Mathematica* in 1687. While Galileo argued the structure of the universe, Newton demonstrated the gravitational mechanism by which it worked.

Until Newton's time, man was an uncomprehending plaything of fate or divine providence. Now he began to understand that the everyday events of life were driven by a structure of causation. Later generations of scientists have maintained the process. Mendel worked out the structure of genetic inheritance, Darwin illustrated the mechanism evolution, Pasteur demonstrated the causes of disease; Crick and Watson unravelled the DNA code. These and many other insights make up the intellectual baggage of Western man.

Enlightenment to Romanticism

The Philosophes – The new enlightenment was to find its home in France, but the pattern of thinking owed much to 17th century British writers, notably John Locke, who published his *Essay Concerning Human Understanding* in 1690. Locke said that many religious issues were beyond human knowledge, and he argued for tolerance and reliance on reason and reasonableness. His work reflected a wide change of mood, singnalling the end of two centuries of religious strife; never again would the battle lines of Europe's terrible wars be drawn along religious lines.

The scientific advances of the 17th century encouraged philosophers of the following century to see the world as an ordered machine, much like one of the new clocks. There was an optimistic view that the universe was driven by a well-oiled logic, and, if people could only behave in a reasonable manner, the world's problems could be readily overcome. Past religious passions now appeared irrelevant. Many thinkers no longer saw God as an imminent cause of good or evil, but as a great watchmaker, an ultimate mover, who no longer had immediate relevance to life.

The dominant personality among the *philosophes* was the Frenchman, Voltaire. He was a satirist, rather than an original thinker, and he turned his barbed pen on anything which he saw as repressive or pretentious. Voltaire had problems with the French authorities, but he and his circle sewed seeds of scepticism about the old order which would have immense repercussions in the later years of the century.

Evangelicalism – The first reaction against the intellectual emphasis of the age came with a religious revival, which developed in parallel in England and her North American colonies, both within and outside the Church of England. John Wesley set the emotional tone of the movement, sharply in contrast with the language of the *philosophes*, when he described how he 'felt his heart strangely warmed'. This new protestantism appealed primarily not to authority, but to the conversion experience. Until then, the Protestant churches had left missionary work almost entirely to the Catholic orders, but, by the end of the century, the worldwide tide of Protestant missions was beginning to flow, with incalculable effects on non-European cultures.

Romanticism – If John Wesley represented the religious, Swiss-born philosopher, Jean Jacques Rousseau, led the secular reaction against Cartesian intellectualism. He proclaimed that man was pure when in the simple state, be that the uncorrupted form of a noble savage, or a new born child. The quest for goodness therefore involved a return to nature. Rousseau, more than any other person, taught people to look on their environment as a place of beauty; since the time of Hannibal travellers had crossed the Alps, without pausing to recognize them as anything other than a barrier on the road to Italy. Now, as if overnight, Rousseau's Swiss mountains were discovered as majestic things of beauty.

As romanticism emphasised the emotions above the intellect, so it elevated the creative artist, as the person most able to express those emotions. The great milestones of the movement, such as Wordsworth and Coleridge's *Lyrical Balads*, Beethoven's *Eroica Symphony*, the late paintings of Turner, explored new forms and emotions. This could lead to excess, but it also opened the way to the achievements, such as those of French impressionist painters and the great romantic composers.

Social Reform – At about the same time, first clearly surfacing in the 1770s, a transformation began to occur in attitudes to social issues. For centuries, Europeans had been shipping Africans to slavery with no apparent compunction. Now powerful anti-slavery movements made themselves heard in France, Denmark, England and other countries. Movements for the reform of vicious penal systems, the abolition of the 'hanging codes' and for the humane treatment of the insane can be dated to the same time. Educational reform also became a cause for the future.

Credit for this new mood of social reform has been given to the pen of Voltaire, the preaching of Wesley and the ideals of Rousseau. All played their part. In education, for instance, evangelical passion to bring truth to the poor led directly to the opening of ragged schools, while Rousseau was laying the foundations for the quite separate development of child centred learning, which was carried forward by the Swiss educator, Pestalozzi. The cause of reform was uniquely in the air and the traditional political structures were ill equipped to contain it. Europe was ready for the cataclysm of the French Revolution.

REVOLUTION

The French Revolution

The Estates General – In 1776 the British government was faced with a major revolution in its American colonies. King Louis XVI of France, recognizing this as an opportunity of regaining some of the ground lost in the Seven Years War, involved France in the conflict. In military terms the intervention was successful; in financial terms it was a disaster. The French government, always in financial straits, was now unable to function. The shortfall could no longer be met by the time honoured device of increasing taxes on the poor, but those able to pay could only be taxed with their own consent. Members of the aristocracy recognized an opportunity of winning concessions from the monarchy in return for money, and they insisted that Louis should recall the French parliament, the *Estates General*, which had not met for 150 years.

The body met in three separate houses – aristocracy, clergy and the third estate. This last house represented the property owning middle classes and was largely made up of professional men. They had no vision of themselves as revolutionaries, but they were influenced by the ideas of the *philosophes* and of the American Declaration of Independence. Louis anticipated doing his business with the other two houses before disbanding the body, but the Third Estate had equal representation with the other two, and could count on considerable support in the House of Clergy. In the summer of 1789 the Third Estate declared that it constituted a National Assembly. Louis gave way before its demands and the body set about a huge programme of constitutional, administrative and social reform.

Popular Unrest – Since the time of the Frondes, French kings had been acutely aware of the dangers of uprisings among the poor, who remained unrepresented in the National Assembly. There was unrest in many parts of the countryside, where chateaux were attacked and hated rent books burned. The most immediate danger, however, came from the poor of Paris, who found themselves caught in a spiral of inflation, most crucially in the price of bread. By the 14th July, it was estimated that only three days supply remained in the storehouses of the capital.

The mob possessed armaments, but little ammunition. This lay under close guard in the royal castle of the Bastille. On the 14th July the mob stormed the Bastille, leaving Louis quite helpless. He could not use his army because the loyalty of rank and file soldiers was in doubt. Many of the aristocracy were now fleeing France and in June 1791 Louis and his family made their bid to escape. They were captured and brought back as prisoners to the capital.

The Assembly maintained the King as a figurehead until he and signed the new Constitution in September. The body disbanded itself to make way for the new Legislative Assembly.

War and the Terror – Since the National Assembly had barred any of their number from seeking re-election, the new body was made up of inexperienced men. The dominant figures were Danton, who surrounded himself with members from the Gironde, in southern France, and the little lawyer Robespierre, whose power was based on the Jacobin Club. Protagonists of the new order now felt under siege. The King could still serve as a focal point for a royalist counter-revolution, and both Austria and Prussia were issuing threats. Robespierre argued that peace should be preserved, but Danton believed that the nation could only be united by war. He urged his fellow countrymen to 'dare and dare and dare again' and Frenchmen responded to his cry that the *patrie* was in danger. In April 1792, France declared war on Austria, and Prussia came in on the side of Austria. Early news from the war was disastrous and the capital was gripped in a fever. On 30th July a contingent marched into the capital, singing the song which would become the national anthem. They demanded that Louis should be dethroned and a republic proclaimed. The men of Marseilles were soon joined by a huge citizen army which, chanting the Marseillaise, routed the mercenary enemy soldiers at Valmy on 20th September. Two days later, France was declared a republic. Louis was placed on trial in December and executed on 21st January 1793.

The citizen's army swept across the Netherlands and at last achieved the 'natural frontiers' which had been beyond the reach of the armies of Louis XIV. The victors proclaimed liberty and equality for the poor of all lands, but in practice all too often they laid new tax burdens on those same poor to pay the cost of war.

In February 1793 France faced a coalition of Britain, Austria, Prussia, Holland, Spain, Sardinia and Italian states. Action taken against the Catholic church also provoked civil war in the conservative regions of the Vendée and Brittany. Effective power now passed from the Assembly to a Committee of Public Safety. In June, Danton's Gironde fell to Robespierre's Jacobins and the period, known as the Reign of Terror, began. Among the victims were successive waves of politicians, including both Danton and his Girondins and Robespierre and the Jacobins. As a result, power passed to a new generation of second rate men, who could not command the respect of the nation.

Meanwhile, the French citizen army, now rein-forced by the first use of conscription in modern times, was more than holding its own in the war. Britain, formidable at sea, was poorly equipped for a land war and the old enemies, Prussia and Austria, failed to co-ordinate their effort. The French armies, now led by a new generation of generals, remained firmly entrenched on the Rhine.

The Empire

The rise of Napoleon – In May 1798 a French army, led by the Corsican Napoleon Bonaparte, was sent to invade Egypt in an attempt to cut British trade routes. Land victories were made worthless when the British fleet, commanded by Admiral Nelson, destroyed the French supply fleet and so cut the army off from Europe. In August 1899 Napoleon abandoned his army and returned to France to challenge the discredited leaders of the nation. His gambler's throw succeeded, and on November 9th he staged a coup d'etat and assumed the title of First Consul. He set about centralizing power in his own hands; in 1802 he became consul for life, and in 1804, he followed the example of Charlemagne by crowning himself Emperor as Napoleon I. Any dismay at this negation of the ideals of the revolution was overwhelmed by the relief of ordinary people at the return of ordered and firm government.

Imperial Government – Napoleon had a genius for administration. After an initial purge of remaining Jacobins, he set about healing old divisions and re-uniting the country. He recognized that the continuing civil war in the Vendée could not be brought to an end unless the state came to terms with the Catholic church, so the old religion was restored to its position as the national faith. He set about recruiting the ablest men into government, regardless of whether they held republican or royalist sympathies. Most enduringly, he personally supervised a detailed revision of the whole of the French legal system into the *Code Napoleon*. Had Napoleon been content to hold the Rhine frontier and bring sound administration to France, his rule could have been outstandingly successful. But he was by instinct a general and the symbolic identification with Charlemagne at his coronation illustrated his determination to build the greatest empire that the world had seen. 'I am destined to change the face of the world,' he declared. But Napoleon, like Louis XVI, discovered that wars could only be fought at a financial cost, which had to be passed on in taxes to the people of France and the conquered countries.

The Napoleonic Wars – The great struggles of previous centuries had achieved little more than change the line of a frontier here and there. In the three years from 1805 Napoleon completely redrew the map of Europe. He owed his success to the army which he had inherited from the revolution. Opposing generals recognized that the citizens' armies of France were carried forward on a tide of national energy, which had been released by the revolution. Napoleon added to this a military professionalism, identified with the magnificent Imperial Guard. The surge of victories carried the army across Europe as far as Bohemia, north into Scandinavia and south into Italy and Spain. Ancient rulers were replaced by members of the Bonaparte family or army generals. Even then, however, Britain, Spain and Russia remained as weak points remained in the French Continental System.

Giving priority to the invasion of Britain, Napoleon gathered barges at the channel ports. Any hope of carrying out this operation, however, ended in 1805 when the French fleet was destroyed at the Battle of Trafalgar. Britain therefore remained an implacable foe across the Channel.

The victorious French army in Spain found itself unable to overcome a fierce guerilla resistance which made full use of the broken terrain. The British despatched a force under the general who would later become the Duke of Wellington. In August 1812, after a relentless campaign, Wellington led his army into Madrid.

As Madrid fell, Napoleon was on the other side of Europe, leading 450,000 men on his disastrous campaign against Russia. He had already heavily defeated the Russian army and he believed that serfs would flock to join him once they heard that he had proclaimed their emancipation. He defeated the Russian army again at Borodino and marched on across the scorched countryside to occupy Moscow. But, when the Russians burned their own capital city around his army, and winter began to set in, he was forced to order the terrible retreat. In the end, only a tenth of his great army survived the ordeal. The Imperial Guard was reduced to some 400 men; 80,000 horses had died, leaving the emperor with no effective cavalry to put in the field.

Defeats in Spain and Russia shattered the myth of invincibility and by 1814 Napoleon had lost everything. Paris fell on the 30th March and he abdicated his imperial crown on 11th April. In France, however, loyalty to the deposed emperor remained strong, and when he escaped from exile in 1815, men flocked to join his army. The hundred days adventure ended when he was defeated by the combined British and Prussian armies at Waterloo on 18th June.

Napoleon passed the remainder of his life in well-guarded exile, but the Napoleonic legend lived on. As French power declined, people remembered that it was the Little Corporal who had led them to glory.

Reaction and Revolution

The Return of the Old Order – After the defeat of Napoleon, members of the old ruling houses moved back into their palaces. The statesmen met in Vienna to reorganize the continent. The treaty took little account of nationalist aspirations. Poland was awarded to Russia; Venice and Lombardy to Austria; the Rhineland was taken from France and given to Prussia; the southern Netherlands were incorporated into Holland; Norway was made a part of Sweden.

The Austrian Prince Metternich was the main architect of this restoration of the old order. He fully recognized the huge changes in political consciousness brought about by the French Revolution, but he believed that these had to be suppressed, and that structures should return to return to their dynastic roots. He opposed all representative institutions, and established the Holy League as a coalition of powers dedicated to suppress ideas of liberty and nationalism, wherever they might show themselves. Of the major powers, only Britain -itself, however imperfectly, a representative government – stood apart to uphold a more liberal tradition.

The policy of intervention was successfully invoked when the Spanish people rose in rebellion in 1820. Austria also put down rebellion in her Italian possessions. Metternich was wise enough to see himself as the defender of a dying way of life. In 1821 the people of Greece rose against their Turkish masters. True to his principles, Metternich gave Austrian support to the Ottoman Turks, but the rebels won backing from Russia and Britain and achieved their independence in 1830.

Also in 1830 the people of Paris rose again and replaced the conservative king with his more liberal cousin, Louis Philippe, and revolutions broke out in Poland and across Germany. In the same year, the Catholic Belgians rebelled against their Dutch masters. The conservative powers threatened to intervene, but Britain, in a gesture which would be called in 84 years later, guaranteed Belgian independence.

The Year of Revolutions – The unrest of 1830 was a prelude to much greater upheavals of 1848. In January, rebellion broke out in Sicily. In February the people of Paris drove Louis Philippe, now a figure of fun, into exile. In March, Venice, Parma, Milan and Sardinia all rose against Austria. As the year progressed, there was revolution in Poland and

Hungary. Smaller German princes fell, most never to return. Uprisings in Berlin and Vienna even brought the powerful Prussian and Austrian states to the point of collapse, and the elderly Metternich had to follow Louis Philippe into exile.

The Hughes Capet French monarchy was finished for ever, and, after a period of civil war, the French people turned again to the magic name of Napoleon in the person of his nephew, Louis Napoleon. He followed family tradition by staging a coup d'état and assuming the title of Emperor Napoleon III. Across the rest of Europe, ruling houses re-established control over their dominions.

New Nations

The Unification of Italy – A decade later, Camillo Cavour, a statesman in the Italian kingdom of Savoy, set about achieving by political means what had been beyond the powers of the revolutionaries. In 1858 he met Napoleon III to discuss how Austrian rule might be ended and Italy unified under his king, Victor Emmanuel of Savoy. In 1859 French armies inflicted heavy defeats on the Austrians at Magenta and Solferino. In 1860, the popular soldier Garibaldi led 'the thousand' against the rulers of Sicily and Naples. He handed these territories over to Victor Emmanuel. For a time Austria held on to Venice, but the city fell in 1866. Finally the Papal States were brought into a united Italy in 1870. The political task was complete, but the new country faced formidable problems of poverty, and large numbers, particularly from the south, emigrated to find a better life.

The Unification of Germany – In 1862 Otto von Bismark, a nobleman of Junker descent, became Prime Minister of Prussia. His first speech was ominous for the future of European peace. 'The great issues of our day cannot be solved by speeches and majority votes, but by blood and iron.' The German-speaking people were already showing a formidable potential, but to achieve all they had to be united into a nation. Only Austria or Prussia could be the focus of that state; Bismark determined that it should be Prussia.

In 1864 the two powers collaborated to annex the German-speaking lands of Schleswig and Holstein from Denmark. Then two years later Prussia went to war with Austria. On 3rd July 1866 the Hapsburg army was devastatingly defeated at Sadowa. The Hapsburg monarchy retained Austria, but Germany was now effectively united. In 1870 Germany went to war with France and the Napoleonic legend was laid for ever on the field of Sedan.

As Bismark's army occupied Paris, there could no longer be any doubt that Germany was the dominant power in continental Europe. The violent methods by which this had been achieved were no innovation in European politics. The new state was based on admirable organization. German cities were models of organization and sanitary efficiency; a state school place was provided for every child, and illiteracy rates became the lowest in the world; the poor, who until that time had emigrated in large numbers, now showed their confidence in the government of their country by staying at home, and by playing their part in constructing the impressive industrial base of the new nation.

A CHANGING WORLD

The Infrastructure of Change

Population – The population of Europe had been growing relentlessly since the time of the Black Death. Demographers have argued why, for instance, the increase was particularly pronounced in the 16th century. It appears as though women started marrying younger, so having a longer child bearing life, but this leaves unanswered why such a social change should have occurred. The 18th century again saw a steady increase across western Europe, which pre-dated major medical advances of the following century. A modest alleviation of the harsh conditions of rural life, the improvement of the housing stock, and some advances in public health may all have contributed to a reduction in the death rate.

Rulers generally welcomed a rising population; it provided an larger man-power pool for the military, and increased the tax base of the nation. In 1798, however, an English clergyman, called Thomas Malthus, published his *Essay on Population*. The world, he argued, possesses limited resources. As population grows, so the most vulnerable – the poor – must inevitably experience disaster and hunger. Malthus' work was influential, but his warnings were not, in the short term, authenticated by events. The reason for this was that, at the same time that the population was increasing, Europe was experiencing a green revolution, which greatly increased the amount of food available to meet the growing demand.

The Second Agricultural Revolution – The first

great change in farming practice came with the introduction of settled agriculture at the beginning of historical times. Even in Babylonian cities, farming families had to produce a surplus to feed craftsmen, priests and warriors. By the beginning of the 18th century, little had changed. It is estimated that in England eight out of ten people still lived in the countryside and that, on average, one farming family had to keep one other family from the produce of its land. People still ate bread baked from their own wheat and drank beer brewed from their own barley. Animals, except for breeding stock were still slaughtered and salted down for the winter.

Once again, change originated in small, highly urbanized, 17th century Holland. Dutch farms had to be more efficient than those of their larger neighbours, and major improvements were pioneered, particularly in the development of root vegetables, largely for animal winter feed, and in high yield artificial grasses, such as alfalfa and lucerne.

In the 18th century, the English gentry, unlike their neighbours in France, lived on their estates and it became fashionable to take an interest in farming. George III set the tone by contributing articles to a farming journal. Some began to introduce the Dutch innovations on their estates. New crops and methods of rotation were introduced and selective breeding produced remarkable improvements in the quality of livestock.

These improvements could not be introduced without radical changes in the organization of the countryside. Improved agriculture could not be successfully introduced in the old communal fields, so enclosures, which had been taking place for two centuries, were given a new impetus. In the change from peasant holdings to larger farms, worked by landless labourers, many lost land and ancient rights. The production of food, however, became a much more efficient process. By the late 18th century British farmers were in a position to support a huge increase in the nation's urban population.

Financial and Human Resources – Any major economic expansion needs to be built on a sound financial base. Britain's growing international trade brought prosperity, and her island position meant that wealth did not have to be dissipated on the maintenance of a large standing army. By the standards of the time, she also had a sophisticated and well-capitalized banking industry.

It is harder to establish a link between the skills required for technological advance and the social and educational structure of the day. Few of the innovators of the new age came form the conventional academic background, which had produced Isaac

Newton; they were more typically self taught, or the products of Scottish or dissenting education.

Economic Theory – In 1776 Adam Smith published *The Wealth of Nations*, which laid the basis of modern economics. He argued the benefits of competition in a free economy, against both state control and the abuses of monopoly powers. His arguments were influential both in government and business circles, initially in Britain and later in the United States and elsewhere.

The First Industrial Revolution

Iron and Coal – Since the time of the Hittites, iron working had been centred on the great forests. The charcoal used in the smelting process consumed large quantities of timber, which was also vital for building and naval supplies. Over the centuries, the forests receded to the more remote areas. By the end of the 17th century, Britain faced something of a crisis. In the 14th century, German craftsmen of the Rhineland had learnt how to make cast iron, so that the metal could now be used to make a wider range of products, but there was an acute shortage of wood to drive the blast furnaces.

Early in the 18th century the Darby family of Shropshire finally solved the problem of how iron could be smelted from coal. As this technique became widely known, industry moved from the forests to the great coal fields, which lie across Europe in a band from mid Russia to Wales. Surface coal was soon exhausted and deep mines were sunk to exploit the seams. Iron goods could now be produced in bulk.

Steam Power – As early as Hellenistic times, it had been recognized in theory that steam could be used to drive an engine, but the technological basis was lacking. The need for pumps to drain the new deep mines made progress all the more essential. The Scottish engineer, James Watt made the essential break-through when he separated the cylinder from the condenser. As a result, industry could now be liberated, not only from the forests, but also from the banks of fast flowing rivers.

Water Transport – Before the 18th century, land transport was rudimentary over most of Europe. Once again, the Dutch had pioneered the use of the canals which drained their country for transporting loads. The French government also constructed the magnificent Canal du Midi, designed to prevent goods having to be carried around the coast of Spain. France also had a high quality road network, built by forced labour, but these, like those of the Romans, were built for military use.

In 1861 the Duke of Bridgewater opened a canal

which linked his coal mine at Worsley with the growing town of Manchester. The potential for improved communications to lubricate economic growth was illustrated when the price of coal in Manchester fell immediately by a half. The great revolution occurred, however, when steam power was applied to locomotion. Here the initiative was taken in the United States, where inland waterways provided the essential communication links for the new nation. In 1807, Robert Fulton sailed *The Steamboat* from New York to Albany in 32 hours – a journey which had previously taken four days. Two years later, steam power was applied to ocean navigation.

Railways – The world's first commercial railway was opened in Britain in 1825, between Stockton and Darlington. Huge sums of money were invested in railway building in many nations, but, despite massive construction programmes, especially in the United States and Germany, Britain retained her initial advantage. The age of cheap and rapid communication brought important social as well as economic change, as the structure of society began to reflect the new mobility.

Cotton and the Factory System – There had long been a market for fine fabrics in western Europe. By the early 18th century a substantial silk industry had grown up in France, which reduced dependence on imports. At this time, cotton was still a luxury fabric, and ready woven cloth was imported from India. Entrepreneurs then began to import raw cotton, which was put out for manufacture to domestic workers. Whole families worked immensely long hours at carding, spinning and weaving to earn a modest subsistence. The early machines were invented by enterprising craftsmen to help boost domestic output.

The first large spinning factory was built by Richard Arkwright at Cromford, near Derby in 1771. By the early 19th century all the stages of cotton cloth production had been brought within the factory system. Britain, backed by her huge merchant marine, had established a dominant position in the world supply of textiles. One machine, tended by a woman or a girl, could now do the work of many domestic workers, and traditional producers, from Britain itself to India, lost their livelihood.

Despite being a closely guarded secret, the new technology was bound to become known. The United States was already showing itself a fertile ground for industrial development and a substantial industry grew up in New England.

Urban Growth – The population of Europe continued to grow rapidly throughout the 19th century, but the increase was now concentrated in the urban centres. In the 19th century, the population of London increased from about 900,000 to some 4.7 million, that of Paris from 600,000 to 3.6 million; small country towns turned into conurbations. This growth in Europe was matched by comparable expansion of New York and the mid-western cities of the United States.

The urban centres grew faster than their service infrastructure, and so the industrial revolution became identified with slum housing, malnutrition and cholera, on a scale which remains common in the burgeoning cities of modern developing countries. For most of the workers, however, the change from rural to urban poverty was not the disaster that has often been painted. The poor had always lived on the edge of subsistence; there were indeed reverses, as during the 'hungry forties', but the overall tendency was towards an improvement in living standards. Pasteur's discovery of the germ causation of disease stimulated major sewage and other sanitation projects in the second half of the century.

The Second Industrial Revolution

The Decline of Britain – Visitors to the Great Exhibition, which was held in London in 1851, would not have readily recognized that the age of British industrial supremacy was already nearing its end. Britain possessed half the world's mileage of railway lines, and half its merchant marine. Five years later, the British inventor, Henry Bessemer, would present his Convertor, which made possible the mass production of steel. The nation's lead still appeared unassailable.

In hindsight it is possible to recognize the signs of decay. Too much investment lay in the industries of the first industrial revolution, which were vulnerable to competition from low cost countries; the educational system, both for the rich and poor was ill equipped to train in the more technical skills needed to meet the ever growing complexities of industry; British industry was already at times failing to capitalize on the skills of the inventors.

Germany and the United States – In the second half of the century two powers demonstrated great economic potential. German military expenditure funded the expansion of the mighty firm of Alfred Krupp, which was soon competing with British companies for the supply of railway and shipyard equipment. The electric dynamo was invented simultaneously in Britain and Germany, but the German firm of Siemens reaped the benefits. In 1885 Carl Benz produced the first working automobile using the internal combustion engine, so initiating the greatest transport revolution in the world's history.

The United States was also showing both creativity and economic power. Inventive geniuses, like Bell and Edison, found that the young nation, with its growing market base, provided an ideal environment for the exploitation of new technology. The telegraph, the telephone, the domestic sewing machine, mechanised agricultural machines, the safety lift, air conditioning, the electric light, the phonograph, the cine camera and the aeroplane were all American contributions to the more sophisticated second phase of industrialisation. Andrew Carnegie and Henry Ford also showed the American capacity to build great operations on the inventions of others.

Capital and Labour

Trades Unions – Throughout history, there had always been a sharp divide between the rich and the poor, but the working people within the new factory system became acutely aware of the polarization between those who owned the means of production and their employees. By gathering workers together in large units of production, the owners made it practicable for them to organize in defence of their living conditions.

Robert Owen, a working man turned successful cotton master, attempted to establish a model industrial society at New Lanark, Scotland; he introduced schools and all kinds of leisure activities for the working people, and still was able to show a profit for the mill. He became dissatisfied with this paternalistic approach and set up a co-operative venture in New Harmony in America. This proved less successful, and he returned to Britain, where he founded the ambitious Grand National Consolidated Trades Union. This was a bid to harness the power of the working people, so that they could control the industries in which they worked. Owen's union failed, as did most of the early attempts to organize labour. Unskilled workers, faced by organized management, lacked credible bargaining power. Over much of Europe, they were further weakened by being divided between opposing Christian and socialist unions.

The battle between capital and labour could be seen in its rawest state in the United States. The owners mobilized city and state authorities and hired private armies to break strikes. Also there were as yet no anti-trust laws to prevent employers from combining to achieve their objectives. The workers responded by organizing themselves into violent secret societies, like the Molly Macguires of the Pennsylvania coal mines. At Andrew Carnegie's Homestead works at Pittsburgh in 1892, the two sides confronted each other in pitched battle.

At Homestead, as elsewhere, management emerged victorious because there was always unskilled 'blackleg' labour on hand to fill the jobs of those who went on strike. At the end of the century, however, there emerged a new generation of union leaders who recognized that progress could best be made by organizing the skilled labour, which was now vital for the more sophisticated industries. In The United States, in 1886, Samuel Gompers organized these skilled trades into the more successful American Federation of Labour. In the years which followed, the rights of organized labour were increasingly recognized by the legal systems of the industrialized nations.

Socialism – During the time that Robert Owen was experimenting with new structures, continental thinkers were beginning to challenge the laissez-faire theories of Adam Smith. Most influential was the French nobleman, Claude de Saint-Simon, often looked on as the founder of socialism, who published his critique of the new industrial age in the 1820s. He argued for the replacement of the existing ruling elite by a 'meritocracy', which would manage the economy for the general good of the population, rather than for individual gain.

Saint-Simon's ideas gained ground in France. In 1848 Paris experienced two distinct revolutions. The first unseated the king; the second was a bloody confrontation between workers, proclaiming the new socialists ideas, and the bourgeoisie, who defended traditional property rights.

Communism – Karl Marx watched the destruction of the Paris workers in 1848 with distress but without surprise. He had been associated with the revolutionary movement in his native Germany before being forced into exile. He believed that history showed two struggles. The first, as in all earlier revolutions, had been between the feudal authorities and the bourgeoisie. The second, in his own day, lay between the bourgeoisie and the proletariat.

Marx held that the value of goods lay in the labour which had been expended in its production, and the interests of the proletariat lay in winning a fair return for that labour. That was in conflict with the interests of the owners, or bourgeoisie, who were dedicated to achieving a profit on the product. Within a capitalist society the proletariat was therefore alienated from the production process, and both sides were inevitably locked in class war. The objective for the proletariat was to win control of the machinery of government by revolution, and then to use the new communist state to control the 'commanding heights of the economy' – land, transport, factories and banks.

Marx and his friends tried to gather the revolutionary movement into the unity without which he believed it could never be effective. In 1864 the First International of the Communist party was held in London, with delegations from France, Germany, Italy, Switzerland and Poland, as well as Britain. The party, however, early showed its capacity for splitting into factions. When, by the beginning of the next century, none of the great industrial nations had fallen, many thought that the communist challenge had passed away. It was not anticipated that the revolution would come in Russia which, under Marx's definition, still lay within the feudal stage of development.

Change and Society

Education – The movement for educational reform can be traced to the late 18th century, but another century had to pass before change affected the lives of working people. In Germany and the United States, and later in Japan, politicians recognized that, if a nation were to remain competitive in the new world, it needed an educated labour force. All across the industrial world there was a huge increase, not only in basic education but in the provision of higher education. Literacy and numeracy were at last seen as functional skills, rather than as the prerogative of a privileged elite.

The Women's Movement – Before the middle of the 19th century individual voices had been raised to protest against the subjugation of women in western society, but the origin of a formal movement can be placed in 1848, the year of revolutions. In that year a group of women, with men supporters, met at Seneca Falls in New York State and laid out a programme which was to be the blueprint for the women's movement. The resolutions demanded voting rights, equality before the law, the right to hold property, justice in marriage, equal opportunity in education, free access to jobs and an end to the pervasive double standard in morality. The political struggle became identified with the names of Susan Anthony in the USA and later with the Pankhurst family in Britain. Many advances were made, particularly in educational provision, but the radical change came with World War One. Women who undertook a wide range of men's work could no longer be denied basic rights.

Leisure – Towards the end of the century working hours began to be reduced and, perhaps for the first time in history, the less privileged found themselves with time for leisure activities. Virtually all the major sports, which are popular across the world today, were codified during these decades, and this happened mainly in Britain, where the industrial achievements brought the earliest benefits. By the beginning of the 20th century, the bicycle and the railway excursion were giving urban dwellers a new sense of freedom. Many problems remained, but those who lived in the industrial societies experienced a genuine improvement in the quality of life. This improvement made the programmes of the revolutionaries less attractive than Karl Marx and his followers had anticipated.

AMERICA

The Birth of the United States

The Causes of Conflict – When the Seven Years War ended in 1763 it appeared that Britain had achieved her aims in the New World. The French colonies in Canada had fallen under British rule and the stranglehold on the Thirteen Colonies by French forts on the Ohio River had been broken. Very shortly, however, it became clear that strains were building up in the relationships between the American colonists and the mother country.

Under the mercantilist system, it was taken for granted that the colonies existed for the benefit of the other country. As American economies strengthened, however, they began to generate their own momentum. Slaving ships from New England, for instance, now competed directly with those from Bristol on the Guinea Coast.

The American colonists had already developed the westward momentum, which remains a feature of the nation today. Pioneers were penetrating into the rich lands to the west of the Appalachians. Britain, as the colonizing power, was responsible for security, and the London government therefore had to decide whether to expand budgets to provide protection to these pioneers. To the annoyance of many colonists, a decision was taken that a limit should be drawn along the ridge of the Appalachians. The government further decided that the American colonies should be taxed to help pay security costs. When the traditional colonial assemblies refused to vote the funds, the British government decided to establish the principle of its right to impose direct taxation. The Stamp Act, the Sugar Act and the duty on tea were all stages in the deteriorating relationships. None were in themselves onerous, but they created genuine anxiety. Sugar molasses, for instance, was turned into rum which was the staple of the slave trade. Any tax could be used to make American ships uncompetitive with their British rivals.

Independence – Tension centred on the largest city and trading port of Boston, where fighting started in 1775. In the following year representatives from the Thirteen colonies, now to become states, met in Philadelphia to declare themselves independent. The famous and highly influential Declaration of Independence, drafted by Thomas Jefferson, justified the act of rebellion in terms which drew from Locke and the *philosophes*. It declared that government derives from the consent of the governed and the misgovernment, listed in detail, broke that tie of consent.

The colonists faced serious problems in organizing themselves to fight a major European power; there was little natural unity, and money to fund the conflict proved as hard to raise as it had been under British rule. American success was largely the result of the outstanding leadership qualities of George Washington and the ability of the colonists to adapt to a guerilla style of warfare, well suited to the heavily forested terrain. In 1781, the British army surrendered at Yorktown and two years later, Britain accepted defeat.

The Constitution – It was not immediately clear, however, whether one or thirteen new nations had emerged from the conflict. Many of Washington's army remained unpaid and no mechanism existed for a central government to raise money from the states. The Constitutional Convention of 1787 was faced with serious division between the interests of large and of small states, and between those who wanted to see a strong central government and those who preferred to see real power continue to lie with the individual states. The final document, which was ratified in 1788, steered a compromise course between the interests. The Executive, Legislature and Judiciary all had their own spheres of responsibility, and acted as a control on one another by a complex structure of checks and balances.

Canada

The successful rebellion by the American colonies left the British government reluctant to expend further effort and resources on colonization. Many loyalists from the south had moved north, and the division between the French and English population remained deep. The colonies covered much sparsely populated territory and communications remained poor. The people were united only in a common hostility to any threat of annexation by the more powerful neighbour to the south.

In the first half of the 19th century, progress was made towards the establishment of a confederacy. In 1867 The British North America Act brought together four provinces into a federal Dominion of Canada. To protect minority French interests, language and education remained provincial concerns and other provinces joined the federation in subsequent years. In 1885 the last rivet was driven into the Canadian Pacific Line, bringing together east and west and opening up the prairies for agricultural development.

Latin America

To the south, the countries of Latin America remained under the colonial control of Spain and Portugal. The successful rebellion of the British colonies was shortly followed by the collapse of the old monarchies in the face of Napoleon's army. Links with the old countries were cut during the European wars, and this generated an outburst of nationalist fervour.

Spain fought a series of devastating wars to recover control of her American empire. In 1810, the Mexican priest Manuel Hidalgo y Costilla led the poor in a rising, but independence was not finally won until 1821. Power then passed, not to the poor, but to the wealthy classes of Spanish descent. Rulers like Santa Anna treated the country as a personal *hacienda*, and the situation of the poor became, if anything, worse than it had been in colonial days.

In 1811 Venezuela declared its independence under the 'Great Liberator', Simon Bolivar. He had travelled in Europe and was particularly influenced by the writings of Voltaire, and he now saw himself as the George Washington who would bring unity to the Spanish speaking countries of South America. He won a series of victories against Spain and independence seemed assured, provided the conservative European powers did not follow Metternich's plan and intervene to uphold the old order. This was prevented by American President Monroe, who warned off any intervention by proclaiming his doctrine of 'hands off America'.

Bolivar seemed on the verge of creating a United Republic of Columbia, which could be a comparable power to the USA. He was, however, unable to hold the new country together, and one part of the country after another broke off to form new nations. As in Mexico, the privileged classes preserved power to themselves. The European nations competed to invest, particularly in Argentina, and there was a steady stream of immigration from the Old World, but the old inequalities remained, and the economies of many of the new Latin American countries became dangerously dependent on single primary products.

The path to independence was smoother in Brazil. The Portuguese royal family decided not to defend its rights, and in 1822 the country was declared an independent empire by consent. Here too old social inequalities remained and, as late as 1888, Brazil was the last American country to abolish the Atlantic slave trade.

Slavery and the Civil War

King Cotton – Dr Samuel Johnson spoke for many when he poured scorn on American ideals of liberty, which were denied to the black slave population. The continuance of the institution was one of the issues discussed in the Constitutional Convention. There were three broad points of view. Opponents of slavery wished to see the institution outlawed in the new nation; representatives of the southern states would not contemplate joining a union which deprived them of their property; moderates, like Washington himself, opposed slavery, but they believed that they could let history take its course. Slavery, they argued, was outdated and it would wither away of its own accord. Events proved them wrong.

The new English cotton mills created an insatiable demand for raw cotton. The native short staple cotton was an uneconomic crop until in 1793 Eli Whitney invented a gin, which enabled it to be cleaned in large quantities. In the decades which followed huge areas of the south was given over to cotton cultivation. This created a demand for slaves. The Atlantic slave trade was declared illegal, but many were smuggled into the country; others were 'sold down the river' by plantation owners from the more northerly slave states.

This resurgence of slavery led to widespread unrest. Slave risings broke out and an increasing number of slaves used the freedom road to escape north. White and black activists combined in a highly organized anti-slavery movement. Anger rose when in 1850 Congress passed the Fugitive Slave Law, which gave southern owners the right to pursue their property into the free northern states.

Slave and Free States – The House of Representatives, elected by population, was dominated by the free states. The Senate was more finely balanced. As new states were added to the Union, the balance was maintained. California and Oregon tilted the balance towards the free states. 'Bleeding' Kansas, a fierce bone of contention, fell to the slave party. When, in 1860, a republican from Illinois called Abraham Lincoln was elected president, the slave states felt that the political balance had swung irretrievably against them. In March 1861 eleven southern states declared their secession from the Union and the following month they attacked the federal Fort Sumter. *The Civil War.* Over 600,000 men died in the four years of war which followed. The southern armies were highly motivated and generally well commanded, but they were bound to lose a long war of attrition. The industrial north had a larger population, more industrial production and more miles of railway. This was the first major war in history fought with armaments which were the products of the industrial revolution, and great battles, like Antietam and Gettysburg presaged the terrible loss of life at Verdun and on the Somme half a century later. When the war ended in April 1865, the south lay devastated. Lincoln was assassinated five days later.

Civil Rights – The war had been fought over the right of the south to secede from the Union. Slavery was abolished in the process. Lincoln's emancipation decree was given the force of law by the Thirteenth Amendment of 1865 and further amendments wrote civil rights into the constitution. In the years of reconstruction black legislators took their seats and it appeared as though political and social equality might be close. Gradually, however, by a process of manipulation and terrorisation, the white supremacists regained control of the southern states. The liberal fervour of the anti-slavery years was now spent, and the Supreme Court proved unwilling to uphold even most clearly defined constitutional rights. Disillusioned, many blacks migrated to the booming industrial cities of the north, where they encountered new forms of discrimination.

The situation only began to improve with the great Civil Rights movement of the 1960s, when Dr Martin Luther King provided a rallying point for his people's aspirations and liberal white sympathizers were again mobilized, as they had been a century before in the anti-slavery campaign.

The Westward Movement

Thomas Jefferson – Of all the founding fathers, Jefferson had the clearest vision that the new nation could become a great power, and that this had to be based on an exploitation of the great potential of the continent. He was the architect of the system whereby new states could be added to the Union. By 1803 Napoleon had decided that the Mississippi lands of Louisiana, which remained French, were of no value and Jefferson, now president, negotiated to buy them for $15,000,000 – and so double the land area of the United States. In 1804 he sent out an expedition led by Lewis and Clark to cross the continent and report back on its potential.

Jefferson's vision of the west as the land of opportunity gradually captured the American imagination. It was argued that the American people – by which was meant the white American people – had a 'manifest destiny' to possess the continent from the Atlantic to the Pacific Oceans.

The Dispossession of the Indian People – During the early years of the 19th century, Americans of European origin were pushing into traditional Indian territory beyond the Appalachians. Every expedient was used, from purchase to forced expulsion, to drive the Indian people back into the western grasslands, which remained unattractive to white settlement.

The nomadic buffalo culture of the plains Indians was based on horses originally acquired form the southern Spanish settlements, and it was therefore a comparatively recent development. In the middle of the century, migrants were attracted, not to the featureless plains with their extremes of climate, but to the far west. For a brief period, wagon trains and nomadic Indians were able to co-exist. By the 1870s, however, the white men began to move into these last hunting grounds. The buffalo were hunted, depriving the Indian people of their livlihood and providing food for railway construction workers; then the railway link with the eastern markets made the grasslands attractive for cattle farming. Finally new agricultural machinery and irrigation techniques made large scale wheat farming economic. With buffalo herds destroyed and their whole way of life undermined, the surviving Indian people were driven back into ever more arid and infertile reservations.

Oregon Country – Lewis and Clark reported on fertile land on the Pacific coast around the mouth of the Columbia River. Many Americans were prepared to go to war with Britain over British Columbia, but agreement was reached on the 49th parallel boundary. This left ample scope for colonization in the north west. The wagon trains which followed the Oregon trail brought farming families into this attractive region.

The South West – The new Mexican state claimed the whole of the south west of the country, from Texas to California and as far north as Utah and southern Wyoming. Spanish settlement had been based on missions, which were often widely dis-

persed, and the non-Indian population of the region remained low. Between 1836 and 1847 the United States and Mexico were in an intermittent state of war, which ended with the capture of Mexico City and defeat for Mexico. Under the treaty the United States won the whole of the southwest. Existing property rights of the Spanish speaking people were, in theory, protected, but, in practice, they had no means of protecting them against the newly arrived 'Anglos', who controlled the courts.

Shortly before the treaty was signed, gold was discovered at Sutter's Mill in Northern California. This set off the Gold Rush, which brought fortune hunters flocking to California from the east, and, indeed, from many parts of the world. The influx of population in turn created a farming boom and California was rapidly converted from a thinly populated region, largely consisting of mountains and desert, to the world's most rapidly expanding economy.

Immigration

From Europe – Any measurement of the population rise of Europe from the middle of the 18th to the end of the 19th centuries should properly include, not only statistics on those countries themselves, but also of the millions who emigrated to destinations in many parts of the world – as well as their descendants. Figures can not be collated, but people of European stock took over great areas of the world, often at the expense of the indigenous population.

In order to overcome human reluctance to disrupt living patterns, there needs to be a 'push factor' propelling people from their homes, and a 'pull factor' drawing them to a new environment. As in the early years of colonization, the growing wealth of the United States drew economic, political and religious refugees from Europe. The British still came. Many of the Mormons who pulled their handcarts across the plains to Utah originated from among the cotton mills of Lancashire. The depopulation of the Scottish glens provided a new stream, although most preferred to go to Canada. The Catholic Irish, angered by English protestant rule and by unjust land laws had long been ready recruits; the disastrous potato famine of 1845-6, which is estimated to have claimed the lives of a million people, turned the stream into a flood.

People now came from new countries of origin. Norwegian families, long accustomed to extremes of climate, left their marginal fiord farms to farm in the harsh environment of Wisconsin and the Dakotas. Germans fled from the political and social upheavals of their country. Peasants from southern

Italy, condemned to live on the brink of subsistence under rapacious landlords, took the boat to America. Towards the end of the century, people were coming from further east. Russian Jews fled the pogroms; Poles fled Russian oppression. All funnelled through Ellis Island to emerge, often penniless, and speaking no English, onto the streets of New York. They worked as they could, in clothing sweatshops, on construction, in domestic service. Each new national group faced discrimination as those who were settled in jobs and homes tried to protect their position from work-hungry newcomers.

The New Immigration – The capture of the south west brought a significant Spanish speaking population within the United States. Civil war in Mexico and an increasing divergency of the standards of living brought an increasing number of immigrants across the border. Most came as migrant workers, following the crops into California and far beyond. In good times, they were welcomed as cheap labour, but in times of depression they proved easy targets for discrimination. In the 20th century immigrants from Puerto Rico and other Caribbean islands have also increased the Hispanic population of the eastern side of the country.

Asian immigration began when Chinese labourers were recruited to work in the 1849 Gold Rush. Distinctive in those early days in their 'queues' and national clothing, they found themselves at the bottom of the immigrant 'heap', increasingly shut out from desirable employment and property ownership by Chinese Exclusion Acts. They, like subsequent Asian immigrants, preserved a respect for education, which enabled them to improve their status

rapidly when the legal discrimination was brought to an end.

The United States Abroad

The Continent – The Monroe Doctrine was originally proclaimed to protect emerging Latin nation seeking to establish independence from European colonial powers. In the later years of the century it was used to promote the continent as a sphere of U.S. interest.

One major thrust lay through the Caribbean towards South America. In 1903 effective control over the Isthmus of Panama was wrested from Columbia and the Panama Canal linking the two oceans was opened in 1914. War with Spain in 1898 also ended with the acquisition of Puerto Rico and Cuba.

American interests also led expansion across the Pacific. Alaska was purchased from Russia in 1867, providing the westward bridge of the Aleutian Islands. Midway Island was won in the same year, followed by Samoa and the Hawaiian group. The war with Spain finally brought Guam and the Philippines within the American empire.

Although the United States was no longer a new country, the need to absorb waves of immigrants fostered an introversion and at times an aggressive nationalism. Many American statesmen, wishing to distance their country from what they saw as the destructive quarrels of the old world, proclaimed a policy of isolationism. The history of the 20th century was to show that the world's greatest power could not successfully stand back from international events.

THE AGE OF IMPERIALISM

India

The East India Company – The battles in the 18th century between rival trading companies were fought, not to win territory, but to establish trading advantage. In 1757, however, the British East India Company's army in Bengal first captured the French trading station and then defeated the Nawab's army at Plassey. The company then found itself, by default, the inheritor of Mogul power. Now irretrievably involved in politics, it gradually extended its control over large areas of the sub-continent.

Company officials never lost sight of the fact that their objective was to turn in a profit. As the company extended its control over all internal as well as external trade, the standard of living of many Indi-

ans declined. Company officials took the opportunity of amassing private fortunes, often by corrupt means. In the days before steam ships and the opening of the Suez Canal, India was far distant from home. Men travelled out as bachelors and many took local women and lived much as Indian princes.

After the American revolution, the British government was reluctant to become involved in further colonial expansion. To bring the Company under control, however, it assumed dual control of the Indian possessions in 1784. The writings of Adam Smith had discredited the old mer-cantilist ideas, which had been the justification for early colonization. In line with prevailing doctrines of free trade, the company therefore lost its monopoly trading rights and was reduced to an administrative organization.

Modernization – The evangelical fervour of the age brought protestant missionaries of many denominations to the sub-continent. Most had a simple desire to replace the traditional religions of Hinduism and Islam. They started schools which offered western education and encouraged converts to adopt western dress and habits. These missionaries looked to the Christian rulers for protection and active encouragement.

The new generation of administrators was less directly motivated by the profit motive and more by a desire to bring the benefits of modern life to the people of India. Many had a genuine, albeit paternalistic, respect for Indian culture, and they resisted the missionaries' attempt to overturn traditional ways. These administrators did, however, believe in reform. Laws, based on western practice, were introduced to stamp out traditional practices, such as the burning of widows and the killing of infant girls. The products of the industrial revolution, such as the electric telegraph and railways were also enthusiastically introduced.

The Mutiny – The modernization programme inevitably created tension. Railways, for instance, were looked upon as a threat to the caste system. There was also powerful resentment against British acquisition of new land, particularly in the northern province of Oudh. The introduction of a new form of greased cartridge was the immediate cause of the Indian Mutiny of 1857. This was as much a traditionalist reaction against modernization as it was a rebellion against the ever expanding foreign rule. Many educated Indians, like the operators of the Delhi telegraph, died at the hands of the mutineers. The mutiny was put down with as much ferocity as it had been waged. The British parliament at last accepted direct responsibility for government. In 1877 Queen Victoria was proclaimed Empress of India and rule over the sub-continent became the symbol of British power. The true age of imperialism had begun.

The Raj – The new rulers determined that mistakes which had led to the mutiny should not be repeated. They therefore took care to respect the rights of the traditional ruling class. When early representative institutions were introduced, this ruling class was called upon to represent the Indian people. The aspirations of the rising intelligentsia were therefore overlooked. Indeed, the contempt for the educated 'westernized native' which was to be characteristic of British imperialism, was first shown in India. The first meeting of the Indian National Congress was held in Bombay in 1885, but a further 20 years would pass before independence appeared on the Congress agenda.

Throughout history, India, like China, had shown a capacity to absorb its conquerors; the British alone resisted assimilation. The new rulers of the Indian Civil Service were drawn from the elite and many acquired a knowledge of Indian language and customs, but, in the wake of the Mutiny, a barrier existed between the two races which could not be crossed. Fast and comfortable steam ships now linked Europe and India, and the journey time was much reduced when the Suez Canal opened in 1869. Administrators and traders increasingly kept their roots in Britain, while serving tours of duty overseas. Also men were now joined in India by their womenfolk. Few of these *memsahibs* had work which brought them into contact with Indian people, so their cultural values were never seriously challenged.

In the second half of the century new concepts of racial superiority were fashionable, particularly in northern Europe. Europeans had long treated other races as inferior, but they had not theorised about it. Now concepts of racial superiority were becoming fashionable, partly based on popular Darwinianism. The European rulers of India, as of other colonized people, were therefore ill equipped to understand the nationalist aspirations when they did come to the surface.

China

The Manchu Empire – In the middle of the 17th century, invaders from Manchuria overthrew the Ming emperor and established the foreign Manchu (Ch'ing) Dynasty. Following the ancient pattern, the early rulers were able men who established a working relationship with the mandarin administrators, and for more than a century the land experienced one of its more prosperous periods.

By the end of the 18th century, however, problems were growing. It is estimated that the population trebled, from 100 to 300 million between 1650 and 1800, and it would reach 420 million by 1850. In China, Malthus' forecasts on the effects of population growth proved accurate. All available land was already under cultivation, so production could not match increased demand. The situation became disastrous in the terrible northern famine of 1887-9, when some ten million people starved to death. As social problems became worse, so the quality of imperial government deteriorated into corruption and mismanagement. Resentment boiled and people remembered that the Manchu were a foreign race.

In 1786 rebellion broke out in Shantung, and this was followed in 1795 by the White Lotus Uprising on the borders of Szechwan and Shensi. These were the preludes of a century of peasant unrest on a scale far beyond anything experiences in human history before that time. It is estimated, for instance, that more people died in the T'ai-p'ing Rebellion of 1850-64 than World War One, while huge Islamic risings of the north and south west left wide areas of the country devastated. The Manchu Dynasty, however, managed to cling to power through all these upheavals.

China and the West – Since the earliest times, China had always had a favourable balance of trade with Europe. There was a demand in the West of porcelain and silks, but, apart from a few clocks and toys, Europe had little to offer in return. Towards the end of the 18th century the balance took a turn for the worse. There was a fashion in Europe for Chinoiserie, reflected in some of the art of the period. More important, tea became the staple drink of many Europeans. This could only be bought by a steady drain of bullion.

Western merchants were convinced a huge Chinese market, was waiting to be opened up, but contact was strictly controlled through a few merchants in Canton. Attempts to open the market ended in frustration. In 1793 George III of Britain sent an emissary to the Manchu court with gifts. The Emperor thanked King George for his "submissive loyalty in sending this tribute mission" from "the lonely remoteness of your island, cut off from the world by intervening wastes of sea", but the mission achieved nothing of substance.

The Opium Wars – In the early years of the 19th century British traders found that the drug opium could right the adverse balance of trade. A great deal of Indian farmland was placed under the crop and the flow of bullion into China was quickly reversed. Apart from the direct damage done by the opium, the Chinese government found that the drain of wealth quickly created financial crisis. The opium trade was a breach of Chinese law, and in 1839 a large quantity was destroyed. In the following First Opium War the Chinese forces proved ill equipped to fight a modern war, and they were defeated. In 1842, China was forcedly opened up to foreign trade and missions, and Britain won control of the trading outpost of Hong Kong.

Thirteen years later, the British Prime Minister, Lord Palmerston, decided to assert British authority once again. His declared policy that half civilized governments such as those of China "all need a dressing every eight or ten years to keep them in order" made him popular at home. He was prepared to defend British citizens against the valid operation of foreign law and in 1856 he defied parliament to take Britain, with French help, to war with China again. The imperial army was weakened by the T'ai-p'ing Rebellion and in 1860 the allied army marched into Peking and burned the Imperial palace.

European Influence and Reaction – Although China herself remained nominally independent, her influence in Asia was much reduced. Russia used the British and French invasion as a cover for occupying the northern Amur river, so winning the Pacific outlet of Valdivostok; Britain won Burma, against fierce local opposition; France defeated Chinese armies to win Indo-China; Korea won its independence, later to fall to Japan; Japan conquered Taiwan; even the United States closed in by conquering the Philippines, again in the face of fierce nationalist resistance. Within China itself, the European powers jockeyed for privileges. Even more threateningly, the country was now open to western missionaries who, along with the Christian gospel, brought cultural assumptions profoundly at odds with traditional Confucian values.

The End of the Manchu Empire – By the last years of the century the ancient civilization was in collapse. In 1898 the young emperor and his advisers decided that China must follow the Japanese example and adopt western ways. The experiment was short lived as the dowager Empress led the faction of reaction. She imprisoned the emperor and gave support to the xenophobic Boxers, who were attacking mission stations and other western interests across the country, and the embassy area of Peking was besieged. The western powers replied by sending a combined army to relieve the city. Still the Manchu rulers clung to power, but they were threatened from two directions. The army war lords were now unreliable, and outside the country, young foreign educated men plotted to overthrow the dynasty. In 1911-2 the two combined to bring down the Manchu Dynasty. The foreign educated Sun Yat-sen became president, but one year later he gave way to one of the military commanders.

Japan

The Shogunate – In 1603, at a time when the imperial family had lost effective power, the military leader, or Shogun, Tokugawa Ieyasu established power over the whole of Japan. In the centuries which followed, the Tokugawa shogunate closed Japan off from the outside world. The Japanese were prohibited from travelling abroad; Jesuit missionaries were expelled and their converts persecuted; only a few Dutch traders were allowed to operate from the city of Nagasaki. For Japanese urban entrepreneurs, however, the cost was a small price to pay for the peace and prosperity brought by the powerful shoguns. Educational reforms created a high level of literacy and a vigorous free enterprise economy was permitted to flourish in the growing towns. The growing prosperity of the towns was not matched in the countryside, where both the traditional lords, and the peasants tended to become poorer.

The Opening of Japan – In 1854 the navy officer, Matthew Perry, was commissioned by the President of the U. S. A. to open up the Japanese market. His Treaty of Kanagawa brought the years of isolation to an end. The shogunate did not long survive it, and the Emperor resumed direct power in 1868. There was now a fierce debate within Japan as to whether the country should adopt western ways wholeheartedly, or follow the example of China and remain separate. The reformers were able to point to the disastrous results of conservative policies, as applied in China. The largest feudal families voluntarily surrendered their rights and the government systematically set about the modernization of their country. The changes were based on the solid structure, bequeathed by the Tokugawa shogunate, but there remains no example in history of a comparable change in social life within a single generation. By 1900 an advanced system of state education had been constructed, western experts were imported to train the people in engineering, and young Japanese students were sent to study overseas. At first the new industries, like textiles and shipbuilding, were faithful copies of western prototypes, but they gained an increasing share in world markets. By the 1920s Japan was a formidable industrial competitor to the European nations.

Japanese Expansionism – The era of peace had left the samurai caste deprived of employment. They bequeathed an aggressive nationalism to the new state. The Japanese also recognized that European world domination had been based the use of force. The now popular motto 'Asia for the Asians' was intended as a Munroe Doctrine for a Japanese sphere of influence.

Russian power was particularly menacing. The transcontinental railway had now reached Vladivostok, and the Russians were showing interest in the newly independent Korea. In 1902 Japan concluded a treaty with Britain, which provided security against intervention and two years later she attacked Russian shipping in the Manchurian Port Arthur. The city fell in January of the next year and the Russian Baltic fleet was destroyed in the Tsushima Straits in May.

This victory of an Asian over an essentially European power in the Russo-Japanese War marked the end of the European military domination of the world, which had survived since the 15th century. In the years which followed, Japan continued to build an empire, first by the annexation of Korea and then by the acquisition of wide Chinese lands. These conquests were accepted as a fait accompli by the European powers at the Treaty of Versailles in 1919.

The Pacific

The Aborigines of Australia – Some time before 50,000 years ago great ice caps in the polar regions made the world's seas lower than they are today. The Indonesian islands formed a great peninsula, and Australia was joined to Tasmania and New Guinea in a single land mass. At this time the ancestors of the Australian aborigines arrived in their isolated home. Despite the lower waters, they had still crossed a wide stretch of ocean, making them possibly the world's first seafaring people. These early settlers brought their dogs, but the other animals would have provided a strange sight to people accustomed to the fauna of Asia. In their new home, they adopted a hunter-gatherer lifestyle, delicately in balance with the unique environment.

The Polynesians – Much later, some 4-3,000 years ago, a different race of people began to spread out across the islands of the Pacific Ocean. The methods by which they navigated their great canoes are little understood; they probably followed the paths of migrating birds, and it is suggested that they could feel the current off distant land masses on the surface of the water with their hands. Certainly they made successful voyages of up to 2000 miles to colonize unknown islands. It appears that they went via Fiji and Samoa to the remote Marquesas Islands, from where they fanned out, north to Hawaii, south east to Easter Island and south west to New Zealand, which they called Ao-te-roa, or Long White Cloud. The earliest settlers were probably fleeing from war, but in time warfare followed them to their new homes.

The Arrival of the Europeans – Australian aborigines and Polynesian islanders alike were for long protected from European ships by unfavourable trade winds. In the 17th century several Dutch sailors, operating from the East Indes, made voyages in the area, but they were not attracted by what they saw as the region offered no prospect of profitable trade. In 1770 the English Captain James Cook sailed along the east coast of Australia and landed at Botany Bay. He later commanded two more voyages through the Pacific Islands. The sailors found the Pacific island societies to be living examples of the 'noble savage' existence, extolled by writers such as Rousseau – and bequeathed the devastations of syphilis to the islanders.

Australia

The Penal Colony – The war with the American colonies shut Britain off from the penal colonies of Georgia and the Carolinas, and so gave the British government the problem of how to dispose of its surplus criminal population. As long as the war was in progress, convicts were kept in hulks, moored in river estuaries.

One of those who had landed off Cook's ship in Botany Bay was a geographer and scientist, called Joseph Banks. In the years which followed he had become the driving force behind an exploration movement, intended to open new areas of the world to British trade. Banks argued the case for establishing a penal colony in Botany Bay. The land was good, he argued, the climate mild and the natives few in number.

The first convoy sailed in 1787, under the command of Captain Arthur Phillip, carrying 571 male and 159 female convicts, supervised by over 200 marines. Most of those transported were hard core criminals, but there were also a significant number of political prisoners, particularly from rebellious Ireland. Large numbers of enforced immigrants suffered dreadfully on the long journey and in the penal settlements, and many continued to nurse resentment against the 'old country' and the forces of law and order.

Exploration – By the end of the century it was established that New South Wales in the east was linked to New Holland in the West in a single continent. In 1813 pioneers crossed the Great Dividing Range, which hemmed in the eastern coastal plain to discover the broad grasslands of the interior. By 1859 the landmass of the continent had been divided into six colonies.

It is said that, when these first white men appeared on the central plain, an aborigine scrambled into a tree and let out a long, high-pitched shriek. The establishment of European civilization in Australia was an immense achievement, but, yet again, the heaviest price would be paid by indigenous people in the age-old clash of interests between nomadic hunter-gatherers and settled agriculturalists.

Economic growth – The earliest settlers did not readily find cash crops to make the colony self sufficient. Lieutenant John Macarthur is credited with recognizing the immense potential of the interior for sheep farming. He developed new breeds which would flourish in the New South Wales grasslands

and was able to live in the style of an English country gentleman. His example was followed by emancipated convicts and a new generation of free settlers. The influx of cheap Australian wool to the home country stimulated the Yorkshire woollen industry at a time when woollen fabrics were gaining popularity in world markets. Towards the end of the 19th century the development of refrigerated ships boosted the meat trade. Then, in the early years of the 20th century, strains of wheat were developed to suit the dry climate.

It was also early established that the continent possessed great mineral wealth. Gold and copper mines were in operation by the middle of the 19th century, and the great Broken Hills complex was opened up in 1883. Despite such development in the interior, the cities proved to be the main beneficiaries. By 1901, 65 per cent of the population lived in the six capital cities.

Political Development – Progress with self-government in Canada encouraged the British government to devolve increasing political responsibility in its colonies of European settlement. In 1901 the six colonies became states to form the Commonwealth of Australia. Old links proved decisive when Australian soldiers fought with the British army in the wars of the 20th century.

During World War Two, however, the nation's leaders, recognizing that Britain could contribute little against an expansionist Japan, turned to the United States for support. Since the war, extensive non-British immigration into Australia, and the increasing orientation of Britain towards Europe has further weakened traditional ties.

New Zealand

European Settlement – In 1814 a group of missionaries arrived in the land which had been described by Captain Cook. It is estimated that, at that time there was a population of about a quarter of a million Maori people, who had lived in complete isolation for many centuries. In 1839 Edmund Gibbon Wakefield established the New Zealand Company, with a view to buying land off the tribes and organizing settlements. The British government, hoping to control the movement, formally annexed the country in the following year.

The British proclaimed equality between Maori and European people, but practice never matched theory, and the colonists' land hunger provoked the Land War of 1845-8. After further settlers arrived, many from Scotland, war broke out again in the 1860s. By 1870 the Maori people had effectively lost control of their land.

Constitution and Economy – The country was granted a constitution in 1852 and in 1907 it became a self governing dominion within the British Empire. Its dependence on agriculture, however, left it heavily dependant on the British economy. Early prosperity was based on wool and gold, but the introduction of refrigerated shipping in the last years of the 19th century favoured low cost New Zealand farmers, at the expense of their British competitors. This brought a period of prosperity, and ties with Britain were reinforced by disproportionate contribution made in two world wars. As with Australia, these have weakened in the second half of the 20th century, during which time the country has suffered from its heavy dependence on primary products.

Japan and The United States

By 1914 Britain was withdrawing from direct involvement in the Pacific region but neither the emerging Australia or New Zealand were showing potential as a regional power. The Dutch still controlled the East Indes, modern Indonesia, but only at the cost of a series of major struggles against an emerging nationalism on Bali, Sumatra, and Java. Two powers now faced each other; the United States, with forward bases in Samoa, Guam and the Philippines, and the emerging power of Japan. The foundations of a major regional conflict were already laid.

North Africa

Egypt – In classical times, North Africa had been an integral part of Mediterranean civilization. After the early flowering of Islamic civilization, however, it became increasingly cut off from the countries on the northern, Christian shore of the inland sea. Egypt was for centuries isolated under mameluke rule. When Napoleon led an army into Egypt in 1798, he took with him not only fighting men, but also scholars, who would be able to interpret the remains of the country's fabled ancient civilization.

French interest in the area survived the fall of Napoleon, and, when Mehemet Ali broke the power of the mamelukes and established effective independence from Ottoman rule, French influence remained powerful. Under Mehemet Ali and his grandson Ishmael, Egyptian power was taken south into the Sudan and along the Red Sea. Ishmael contracted with France for the construction of the Suez Canal, which was opened in 1869. He staved off financial collapse, however, by selling a controlling interest in the canal to Britain, who now controlled this lifeline to India. In 1881 the Egyptian government was threatened by a nationalist rising in Egypt and by

the Mahdi in the Sudan, and Britain responded by sending a force to protect the canal. It was not intended as an army of occupation, but Britain became involved in a protracted war in the Sudan and in administering a protectorate over Egypt.

Algeria – The coast of Algeria to the west was had long been the home of Barbary pirates. In the 1830s, France began a major advance into the area. The pirates were driven from their harbours, and the French moved south to the Atlas mountains. Here they met fierce resistance, led by Abd el Kadir, but they won control of the mountain passes. Military success was followed by an influx of French settlers into the coastal region.

Sub-Saharan Africa

An Unknown Continent – For centuries the interior of Africa had been viewed by the outside world as little more than a source of human merchandise. European merchants had shipped slaves by the million out of the west coast; Bedu and Tuareg tribesmen had driven them across the Sahara to the markets of North Africa; they had been carried across the Red Sea in dhows by Arab traders; they had been beaten into submission by Dutch settlers in the south. The slave trade had had a profoundly brutalising effect on African life, far beyond the boundaries of foreign exploration.

Yet in the early 19th century, Africa had its own political movements. In the grasslands of West Africa, an aggressive Islam was expanding in Hausaland under the Fulani Uthman dan Fodio and in Futa Jallon, under Al-hajj Umar. Far away, in the south west, the Bantu people were experiencing a period of unrest. Shaka founded the Zulu kingdom in 1818, setting neighbouring people on the move.

European Explorers and Missionaries – In the early days of the industrial revolution, there was a general view, vigorously fostered by Joseph Banks, that Africa was a land of unbounded wealth, which offered untapped opportunities for trade. Since Britain had most to sell, British interests funded the earliest explorers. Early explorers, such as Mungo Park, acted as commercial travellers, carrying samples of Lancashire textiles and other manufactures. Results were disappointing but some solid business was established on the coast in products such as palm oil.

By the middle of the century European interest was increasingly focused on 'the dark continent', and explorers, such as David Livingstone and H. M. Stanley became major celebrities. Livingstone maintained an interest in 'legitimate trade', which he hoped would displace the continuing traffic in slaves in Central Africa, but he travelled as a missionary. European civilization had now achieved an unassailable self-confidence. Romantic concepts of the noble savage were forgotten and it was readily assumed that Africa was in need of Christianity, Western customs, and the post-industrial working practices which alone could provide the basis for economic advance.

Explorers, mainly following the routes of the great rivers, penetrated deep into the continent. They were followed by missionaries from a wide range of denominations from Europe and America, who were at times almost as much at competition with each other as they were with traditional practice. Expectation of life for explorers, traders and missionaries in malarial West Africa could be measured in months until quinine was introduced as a prophylactic in the 1840s and even afterwards the coast was still considered unfit for European settlement.

The Scramble for Africa – In 1880 active European political interest in Africa was limited to the French colony in Algeria in the north and the British Cape Colony in the south. The old Portuguese colonies, various ex-slaving trading outposts and settlements of freed slaves retained only tenuous links with Europe. By 1914 in the whole continent, only Ethiopia was a truly independent nation. In the intervening decades the continent was divided up between colonizing powers. Lines were drawn on maps in European capitals; boundaries sometimes followed rivers, often placing a village in one country and its farm land in another. The colonizing movement was in places the focus of national policy; elsewhere it was the product of adventurers or commercial companies working on their own initiative.

The British assumption of control in Egypt provoked the jealousy of other European countries. In particular, the French army was suffering from the humiliation of the defeat at Sedan in 1870. Africa offered a forum for the recovery of lost military prestige.

France and Britain were the main protagonists in the northern half of the continent. French colonization followed two thrusts. The first came south across the Sahara from Algeria into the grasslands of West Africa. The second went east from Senegal along the upper Niger to Lake Chad towards the Nile. The British also had a dual thrust, south from Egypt and north from South Africa. The imperialist Cecil Rhodes dreamed of establishing an unbroken chain of British possessions from the Cape to Cairo. French and British forces met where the thrusts intersected at Fashoda in the southern Sudan in 1898, when for a time it seemed likely that the two countries would be involved in a colonial war.

Meanwhile the British had also established West African colonies, based on their old slaving stations. The Germans were active in West, South West and in East Africa. King Leopold of the Belgians gained control of the Congo as a private venture. This was taken over by the Belgian state in 1908. Spain won control of much of the north western Sahara and shared influence in Morocco with France. Italy belatedly joined the scramble by invading Libya in 1911.

In the early years colonization was largely bloodless, but the process became increasingly violent. Britain faced African revolts as far apart as the Gold Coast and Rhodesia and France the Niger and Madagascar. The brutality of King Leopold's exploitation of the Congo was exposed in 1904. In the same year a major rebellion broke out against the Germans in South West Africa, which ended when they drove the Herero people to virtual extinction in the desert. The Italian invasion of Libya was also conducted with widespread brutality. By the beginning of 1914 the re-drawn map of Africa could be seen as a symbol of a dangerously aggressive and expansionist mood within Europe.

South Africa

The Great Trek – During the Napoleonic Wars Britain occupied the Cape of Good Hope, and the territory was retained, as the Cape Colony, in 1814 for its strategic value in controlling the sea routes to the east. At that time, however, there were no British settlers, the land being shared between nomadic Bushmen and Hottentots and Afrikaner speaking Boers of Dutch and French Huguenot origin. Soon the new government began to bring in thousands of British settlers. The Boers were angered when their black slaves were freed and laws were introduced which they considered to be unduly favourable to the previously subject black people. In 1835 some 10,000 Boers left their homes at the Cape and settled on land of the Vaal and Orange Rivers. More Boers followed when the British annexed their republic of Natal. The settlers set up the new republics of Transvaal and the Orange Free State where they could live free from British interference.

The movement of the Boers from the south coincided with migrations of Bantu people, displaced by the Zulu kingdom. The two people clashed, but the main losers were the native nomadic people, who were driven to a precarious existence in the desert.

The Boer War – Resentment between Boers and the British continued to grow. The British briefly annexed the Transvaal, and, although they withdrew, this left the Boers feeling that they would never be left in peace. Then, in 1886, gold was discovered in the Transvaal, and, within a few years, a new city of Johannesburg had grown to a population of 100,000. Most of the newcomers were British, but they were excluded from the running of the republic. Angry that the world's great empire, the modern Rome, could be frustrated by a small number of intransigent Boer farmers, Cecil Rhodes provoked a confrontation. President Kruger of the Transvaal, an implacable opponent of British rule, responded with an ultimatum, and war broke out in 1899.

Liberal opinion in Britain and Europe saw the

Boers as an oppressed minority, and, when they were finally defeated in 1902, there was pressure for a generous settlement. A few voices were raised in the British parliament to defend the rights of the black peoples, but these found no support. In 1909 the four territories were brought together into the self governing Union of South Africa, which lost little time in passing laws which discriminated against the non-white peoples. In 1948 the Afrikaner speaking people won power within the country and put in place the formal structure of apartheid.

THE NATION STATE IN CRISIS

The Eastern Question

The Decline of Ottoman Power – In 1683 armies of the Ottoman empire laid siege to the city of Vienna for the second time and Europe was threatened once more from the East. The armies withdrew, but the Emperor at Istanbul still controlled almost three quarters of the Mediterranean coastline – North Africa, the Arab lands of the Middle East, the homeland of Turkey, Greece and the Balkans.

By the 18th Century statesmen recognized that the Ottoman Empire, like other empires before it, was in decline. Administration was clumsy, and the sultan had to rely on local rulers, whose loyalty was often in doubt. Also, the social military structure of the empire was becoming increasingly out of date.

Russian Objectives – Russian statesmen took the closest interest in the Ottoman decline. Peter the Great had won a warm water port, but, for both trade and strategic reasons, the country still badly needed an outlet into the Mediterranean Sea. In 1768-74 Catherine the Great fought a successful war against the Turks and won the Crimea and other territory on the north bank of the Black Sea along with rights of navigation into the Mediterranean. She also established that Russia had the right to act as protector of eastern Christians within the Turkish dominions.

Russia continued to make advances after the defeat of Napoleon. She won control of the ancestral Ottoman homeland in the grasslands east of the Caspian Sea, taking her empire as far as the mountain passes of the Himalayas. Still further east, she won the Pacific port of Vladivostok from the Chinese. Russian territorial ambitions were backed by huge military forces, and other European powers perceived her as an aggressive imperial power.

Concern focused on the fate of the Turkish European territories. In a private conversation with the

British ambassador, Tsar Nicholas I described Turkey as 'the sick man of Europe'. He implied that it would be better for the powers to consider how to share the sick man's possessions, rather than to wait and fight over them when he died. Other powers, however, preferred to support Turkey so that it could continue to act as a check on Russian ambitions in Eastern Europe.

In 1841 the European powers came together in the Convention of the Straits to guarantee Turkish independence. It was agreed then that the Bosphorus should be closed to all ships of war. This shut Russia out of the Mediterranean, and meant that she could not protect her merchant ships, now carrying increasing grain exports by the Black Sea.

The Crimean War – In 1851 Russia invaded Turkey's Danube lands, and in 1853 her navy sank the Turkish fleet, so winning back her outlet to the Mediterranean. Excitement ran high in Paris and London. Napoleon III was looking for a way of rebuilding the family's military prestige. Britain was concerned for her links with India, for, although the Suez Canal was not yet built, traffic was already following the Mediterranean route. In March 1954 the two powers declared war on Russia in support of Turkey. Combined forces were despatched to capture the Russian naval base at Sevastopol in the Crimea. The huge Russian army was unable to dislodge the invading force and in 1856 she was forced to accept peace on the terms that she would keep no fleet in the Black Sea and build no bases on its shores.

The battles of the Crimean War were made famous because the armies were followed by a journalist, who published detailed reports in the London *Times*. For the first time in history, the public was able to read first hand reports of the sufferings of the soldiers. The modern profession of nursing dates itself from the work done by Florence Nightingale and her staff in this campaign.

Disintegration – Victory over Russia in the Crimean War could not long delay the final disintegration of the Ottoman Empire. France and Britain, who had fought as allies of the Turks, were happy to help themselves to territory in North Africa. Britain also occupied Cyprus and extended her influence in the Middle East. Russia continued her forward movement in the less sensitive territory to the east of the Caspian Sea. In Eastern Europe, Greece was already independent and in the half century after the end of the Crimean War, Serbia, Rumania and Bulgaria would also break free. Russia, always ready to stand as protector of the oppressed Slav peoples, went to war with Turkey again in 1877. For a time Europe stood on the brink of another war as the powers prepared to shore up the tottering empire once again. In 1878, however, Russia, faced by the combination of Prussia, Austria and Britain, was forced to accept terms at the Congress of Berlin.

In 1907 rebellion broke out in Turkey itself. A group, who called themselves the Young Turks, demanded constitutional reforms, along European lines. In 1909 the long reigning Abdul-Hamid was deposed. The new rulers dressed their government as a constitutional monarchy, but it was effectively a dictatorship, dedicated to reviving Turkish power, at home and in the remaining Ottoman lands of the Middle East.

Nationalism

Western Europe – The Napoleonic conquests and the reactions against them had aroused fierce emotions of nationalism, which were to influence European politics. Germany and Italy began their discovery of a national identity, but the mood also affected smaller peoples, such as Belgians and Norwegians. Britain had her own problems in Ireland. The situation was complicated by the fact that Westminster politicians had to reconcile two vocal nationalist groups. The majority Catholics considered themselves to be under a foreign power, discriminated against in their religion and insecure in their land holding. The minority Protestants of the north, mostly of Scots descent, used their political connections with English conservatives to defend accustomed privileges. After 1848, however, concern on issues of nationality centred on eastern Europe.

Poland – Russia might stand as the liberator of oppressed Slav peoples in the Balkans, but, on her own western frontier, she was the oppressor. The

decline of Poland began with long wars against Sweden, which ended in 1709. Depopulated and weakened, with no natural frontiers, she stood between aggressive powers to east and west. In the last decades of the 18th century, she was partitioned between Russia, Austria and Prussia. A supposedly free Poland, created in 1815, was effectively a Russian colony. A series of nationalist rebellions were a failure and Russian administrators tried to eliminate all traces of Polish nationalism, insisting that even primary school children should be taught in the Russian language.

The Austro-Hungarian Empire – Metternich recognized very clearly that the new nationalism could undermine the whole structure of the Austrian empire. The house of Hapsburg ruled over different nationalities, speaking a wide range of languages. Defeat by Prussia and then the loss of Italy pushed the western boundaries of the once great empire back to the Austrian heartland. Alone of all the major European powers, land-locked Austria was not in a position to participate in the scramble for colonial possessions. Any expansion had to be towards the east, and foreign policy now focused on the Danube and the Balkans.

The new nationalists of the region, however, saw Austria, as much as Turkey as a threat to their aspirations. The Hungarians exploited the weakness of the Empire after the Prussian victory at Sadowa to negotiate a new Covenant with Vienna. The Hapsburgs now ruled a dual Austro-Hungarian empire, in which military and foreign policy was coordinated, but in other ways the eastern part had virtual self government. The new Hungarian section of the Empire contained a number of national minorities, and trouble was never far from the surface.

In 1878, after the war between Russia and Turkey, Bosnia and Herzegovina were placed under Austrian administration, and in 1908 they were annexed by Austria. The independent Serbia, with Russian support, now stood as the focus of pan-Slavic aspirations, and so as protector of the nationalist movements in the two territories. The Austrian government, angry at this subversion, looked for an opportunity of crushing Serbia.

Russia

Despotism – In the decades before the Crimean War, Russia was ruled by the autocratic Tsar, Nicholas I. He tried to keep all western ideas of liberalism and socialism at bay by a suffocating censorship. At the same time the administration became ever more corrupt and inefficient. The repression of this period was primarily directed against the intelligentsia, who were traditionally close to developments in the west. Nicholas did recognize, however, that the position of the serfs had become such an anomaly that it endangered the Russian state. As in France before the Revolution, these poorest people had to carry by far the bulk of the load of taxation. Nicholas declared a desire to make changes and he did make progress in codifying peasants' rights and bringing them within the legal system. He was unable, however, to tackle the medieval structure of serfdom, which tied the mass of the people to their villages, and left them as the virtual possessions of their masters.

Emancipation and Reform – Nicholas was succeeded by his son, Alexander II during the Crimean War. The failure of the superior Russian armies and the humiliating nature of the peace, left no doubt that the state needed radical overhaul. Although conservative by nature, the new Tsar supervised a major overhaul of the army, the law and the administrative system.

Most difficult, he put in train the process of emancipation for the serfs. "Better," he said, "to abolish serfdom from above, than to wait till it begins to abolish itself from below." Emancipation was pronounced in 1861, but problems still remained to be solved. Landlords needed to be compensated and a system had established whereby the peasants could buy their own land. This took the form of a tax, which left many, in practice, worse off than they had been before emancipation.

The Prelude to Revolution – The first shot was fired at Alexander only five years after his emancipation decree; he was assassinated in 1881. The reforms of his reign were matched with a continued autocracy, which aroused profound frustration, particularly amongst the intelligentsia. The education system, in particular, was subject to the tightest control by a reactionary bureaucracy. During these years also individuals within government gave support to pogroms against the Jews. After the murder of Alexander in 1881, government fell increasingly into the hands of the opponents of reform.

Opposition was divided between liberals, socialists and groups of nihilists, all of whose leaders were drawn from the intelligentsia. They appealed first to the suffering peasants, demanding a programme of land reform. Towards the end of the century, however, large numbers of peasants were leaving the land to make up the industrial proletariat of the long delayed industrial revolution. Revolutionary activists now found it productive to work in the growing slums of the cities, building up revolutionary cells of workers.

Success is the ultimate justification of autocratic government, and defeat by Japan in 1905 brought the imperial government to the brink of collapse. The battleship *Potemkin* mutinied and terrorised the Black Sea. Massive strikes, particularly by railway workers, crippled the economy. In October 1905 the Socialist groups organized themselves into the First Soviet, based on the principle of the cells which had been established in the factories. Nicholas II, like Louis XVI before him, was forced to attempt to rally national unity by calling a national parliament, or *duma*. Experiments in representative democracy were, however, half-hearted and failed. In the years which followed, the weak Tsar shut himself increasingly within his family circle, now increasingly dominated by the eccentric Rasputin. When the European war broke out in 1914, Russia was ill prepared for such a disaster.

The Armed Peace

The Alliances – In the years after 1871, there were two fixed points in European diplomacy. Austria and Russia faced each other over the control of the liberated Turkish lands in the Balkans. Fighting could break out at any time within the region, leading to the risk of 'superpower involvement'. France also, smarting from defeat at Sadowa and the occupation of Paris, was chronically hostile to Germany. She was, however, militarily weak and the autocratic powers of Austria and Russia looked on her as a threat, and so she remained isolated and impotent.

In previous centuries, alliances had been formed under the immediate threat of war, and they had disintegrated immediately after the threat was over. During these decades, however, the European powers began to form themselves into permanent alliances, committed to help each other in the event of war. By the beginning of the 20th century, a new alignment of powers had become established. Germany allied with Austria, and they were later to be joined by Italy to form the Triple Alliance. To meet this threat, France and Russia joined to form the Dual Alliance. Britain was not a significant continental power, and as late as 1898 the two countries narrowly avoided a colonial war. In 1904, however, policy changed dramatically as Britain concluded a non-binding *entente cordiale* with France, which was followed by a similar agreement with Russia. Hostility towards Germany increased as the German government set about a major naval construction programme, which was interpreted as a direct threat to Britain. The British government responded with its own programme, and a major arms race was under way.

Military Strategy – With Europe organized into armed camps, the generals considered strategy in the event of conflict. Failing to take account of the bloody attrition of the American Civil War, they assumed that events would be settled, as in Bismark's wars, by one swift, decisive campaign. Germany was faced with the prospect of fighting on two fronts. Strategists decided that, while the Russian war machine was massive, the bureaucratic inefficiency would prevent rapid deployment of forces. They therefore developed a plan that, in the event of impending war, the German army would make a first strike to knock out France, so that it could then give full attention to the eastern front.

In the early years of the century, there was a mood of militarism throughout Europe, fed by accounts of colonial wars and victories against non European people. It was most evident in Germany, where theorists declared that war was the natural state of man, but it spread much wider. In Britain, for instance, metaphors of war and sport were subtly mingled in the public school education of the nation's elite.

The Outbreak of War

Austria and Serbia – By the summer of 1918 Serbian support for rebels in Bosnia and Herzegovina had brought relationships with Austria to a low state. On the 28th June the heir to the throne, the Archduke Franz Ferdinand and his wife were murdered in the Bosnian capital of Sarajevo. Encouraged by her ally Germany, Austria used this as a pretext for invading Serbia. On the 30th July Tsar Nicholas II ordered mobilization, not only in the Balkans, but along the whole border.

First strike – It appears that, at the last moment, Kaiser Wilhelm II of Germany may have had doubts about plunging Europe into war. The British foreign minister tried to gather support for a conference to localize the conflict, but the German war machine was now moving under its own impetus. On the 3rd August Germany attacked France through undefended neutral Belgium. Italy declared that the conflict was none of her concern, so the central powers of Germany and Austria faced France and Russia. Britain had no treaty obligation to enter the war on behalf of France, but did consider herself bound by guarantee the guarantee made to Belgium after the 1830 uprising. Her formal position for taking up arms was therefore as defender of the rights of small nations. Italy later entered the war on the side of the Allies, as they were now called, while Turkey and Bulgaria aligned themselves with the Central Powers. Military enthusiasts forecast a short war, to be decided by Christmas, but the British Foreign Secretary, Sir Edward Grey warned, 'The lamps are going out all over Europe. We shall not see them lit again in our lifetime.'

War and Revolution

Stalemate – The German first strike strategy involved high risk. Russia mobilized more rapidly than anticipated the and German army was defeated at Grumbinnen on the 20th August. In early September the western offensive became bogged down on the Marne. Germany was fighting the war on two fronts, which her generals had feared. On the western front, the opposing armies dug in for their long years of attrition. The new German navy remained in port as the British fleet set about sapping German resistance by blockade. Allied attempts to break the stalemate by offensives in the Dardanelles and Salonika, were unsuccessful.

The Russian Revolution – The huge open spaces of the eastern front kept war more mobile. Early Russian success was undermined by the failure of the political structure. In March 1917 a wave of unrest swept the Tsar from power. The opposition was divided between liberal politicians, who now set up a provisional government and the socialist Soviet – itself divided between the moderates and a radical Bolshevik wing. The moderate provisional government pledged itself to continue the war, but in April the Bolshevik, Vladimir Ilyich Lenin, returned from exile and announced the arrival of world revolution. Russian workers, he claimed, should not be dying in a bosses' war. The provisional government staked every-thing on a last great offensive, but this failed, and on the 7th November, Lenin staged a Bolshevik coup d'état. In March 1918 he concluded peace between his newly born Soviet Union and the Central Powers at Brest Litovsk.

American Intervention – Germany now had to fight on only one front, but, during this period, another, even more formidable enemy had been drawn into the war. Desperate at the success of the British naval blockade, the German navy mounted its own submarine blockade of Britain. To be successful it had to attack American ships which were carrying supplies to Britain. This brought the United States into the conflict in April 1917. The German High Command recognized that the intervention of American troops would tilt the battle against the Central Powers, but it staked everything on defeating Britain and France before the Americans arrived. 1917 and 1918 saw huge and costly offensives from both sides on the western front. In the Autumn of 1918 Germany's allies, Austria and Bulgaria began to crumble, the German fleet mutinied and there was increasing unrest in the German cities. Finally the Kaiser abdicated and the generals sued for peace.

The World of Versailles

The Cost of War – All the major continental nations emerged weakened from the war. Russia, involved in civil war, was no longer a factor in international politics. France, though victorious, had suffered grievously. Austria was now not a power of significance. Germany, although defeated, was no longer surrounded by serious rivals. Loss of life had been severe in all the combatant nations, but wealth had also drained away. The United States was the main beneficiary of the war at a cost of fewer casualties than had been suffered by the Dominion of Australia. In the past she had been a major debtor nation, but now she moved into a period of being the world's main creditor.

A New World – The war was also an emotional and intellectual landmark. It was as if the great optimism, which had buoyed up a successful and expansionist Europe was suddenly pierced. The belief an inevitable tide of progress, prevalent since the time of Descartes, no longer seemed tenable in the face of sustained barbarity on European soil. Liberal thinkers, in disciplines such as theology as well as in politics, found themselves on the defensive. New absolutisms, of both left and right, emerged in confrontation, both threatening to overwhelm traditions of representative government.

The conflict had also brought permanent changes in the structure of society. Women, who had been mobilized to fill men's jobs, could no longer be denied political and a growing economic emancipation. The war brought technological advances in areas such as aeronautics and the development of motor vehicles. Output had increased to meet the demands of a technological war, and, in the process, labour unions had established a stronger position for themselves. Many felt threatened by the rapid social change, evident in almost every field of life.

In the years before the war, artists had already been working in strange and disturbing new forms. Picasso's *Demoiselles d'Avignon* and Stravinsky's *Rite of Spring* created scandal in their fields. In 1922, Joyce's *Ulysses* dispensed with the convention of the English novel. It seemed as though all recognisable values were now fractured as creative artists abandoned both classical and romantic forms to explore abstraction and an inner life, now provided with a whole new vocabulary by the works of Sigmund Freud. The arrival of jazz from America, exploiting the interaction between African and European popular music, only served to heighten the alarm of traditionalists.

The Treaties – The Treaty of Versailles, ratified by Germany in July 1919, was the first of a series of treaties imposed on the defeated Central Powers. The leading architects of the new order were President Wilson of the U. S. A., Clemenceau of France and Lloyd George of Britain. Clemenceau, recognizing the continuing potential of Germany, pressed for financial reparations, intended to retard industrial recovery, the return of Alsace-Lorainne to France, and the demilitarisation of the Rhineland. The map of eastern Europe was re-drawn, with Poland, Hungary, Czechoslovakia and Yugoslavia created as new nations. In the north, Finland, which had won independence from Russia in 1917, was joined by the three newly independent Baltic States. The pattern of nationalities was, however, more complex than could be accommodated within national boundaries, and all of these nations had substantial minorities. Of greatest significance for the future, substantial numbers of German speaking people found themselves within Czechoslovakia and Poland. The treaties attempted to protect minority rights, but there was considerable movement of peoples across national boundaries. The largest movement came at the end a war between Greece and Turkey from 1920 to 1922. In particular Greek people left the coast of Asia Minor, where they had lived since ancient times. The two communities continued in uneasy co-existence in Cyprus.

The treaties also changed the wider world. Germany's East and West African possessions were divided between Britain and France, while South West Africa, the future Namibia, was placed under the trusteeship of South Africa. The concept of trusteeship was also used to extend western influence over the old Ottoman territories of the Middle East.

The League of Nations – President Wilson hoped that his country, with its democratic tradition could take the lead in creating a new atmosphere of goodwill. He therefore proposed a League of Nations, which would serve as guardian of world peace. Wilson was to be bitterly disappointed when his own Congress refused to let the United States join the new body. Unhappy about the way in which their country had ben plunged into European affairs, the majority of Americans were anxious to return to a traditional isolationism. It became evident that the new body lacked credibility as early as 1920, when Poland successfully seized Vilna from Lithuania. Later incidents reinforced the fact that successful international collaboration to repel aggression could not be organized through the League. The Italian government took full advantage when, in Europe's final African venture, it launched an attack on Ethiopia in 1935.

Ireland – Attempts by pre-war liberal administrations to give home rule to Ireland had been frustrated by the collaboration of Ulster protestants and conservative politicians. Prime Minister Lloyd George now faced destructive guerilla warfare from nationalists. In 1921 the moderate nationalists accepted partition of the island, which left a significant Catholic minority within the Protestant dominated northern provinces. This led to civil war within the new Irish Free State, and laid the foundations of continuing strife in the north.

The World Economy

The Post-War Boom – During the 1920s world trade appeared to be returning to its pre-war vigour, but, even during these boom years, there were signs of problems ahead. The war had created an increased potential for production, but demand was stagnant. The Soviet Union was in no position to import goods from abroad, and new nations raised tariff barriers to protect fledgling industries. The United States now produced over half of all the world's manufactures, but American consumers, like the Japanese 70 years later, showed little desire to buy goods from abroad and domestic industry was protected by import duties.

As industry boomed, so the price of raw materials, including agricultural products, declined, creating problems for primary producers. At the same time, the fact that workers did not share the profits of their industries, brought outbursts of industrial unrest, such as the British General Strike of 1926.

The Great Depression – By 1928 world trade had become heavily dependent on American finance. In that year Wall Street experienced The Great Bull Market as the price of shares rose to unrealistic heights. Then, on October 28 1929, the stock market crashed. American capital for investment dried up, leading to a rapid world wide collapse of industrial confidence. Governments took what action they could to protect their own industries against imports, so further inhibiting world trade. It is estimated that at the depth of the recession in 1932 industrial production in the United States and Germany was only half of what it had been three years earlier. Unemployment reached record levels in all the industrial countries, bringing times of great hardship.

The New Deal – In America, the parties divided over the political response to the problems. The Republicans, favouring a traditional *laissez-faire* approach, were defeated in the 1932 elections by a Democratic party, led by Franklin Roosevelt. He instituted a New Deal, based on substantial public investment. The showpiece was the publicly owned Tennessee Valley Authority. Designed to provide an industrial infrastructure for one of the country's poorest regions. Roosevelt won great popularity, going on to win an unprecedented four presidential elections, but the improvement brought by the New Deal was as much psychological as practical, and real recovery had to await the stimulus of a second world war.

The Rise of the Dictators

Italy – In the elections of 1921 a new party won just 36 seats in the Italian parliament. Its leader, Benito Mussolini had a background as a socialist, but he now proclaimed that he would save Italy from the menace of communism. The party appealed to ancient Rome in its extended arm salute and the symbol of the *fasces*, which gave the movement its name. The black shirted fascists used intimidation, first to come to power and then to eliminate all political opposition. Mussolini's rule achieved some legitimacy when, in 1929, he negotiated a treaty with the highly conservative papacy.

Germany – The Austrian-born Adolf Hitler became leader of the German National Socialist, or Nazi, party in 1921. Having failed in an early attempt to take control of the Bavarian government, he set about reorganizing his party as a military movement, not hesitating to purge his own followers. He directed his appeal to a German people, who were frustrated by military defeat, humiliated by the loss of empire and European territory, and, in many cases, impoverished by hyper-inflation. Hitler's philosophy was laid out in his early book *Mein Kampf*. This described both his military ambitions for Germany, and his obsessive hatred of the Jewish people.

The struggle appeared to lie between Hitler's new right, and the parties of the left. But the left was divided. The communists, taking their orders from Moscow, attempted, and sometimes succeeded, in fomenting revolution. The social democrats were therefore forced into alliance with conservative military leaders. Capitalizing on these divisions and on the economic problems brought by the Depression, Hitler took his Nazi party to power in 1933. He then quickly set up a reign of terror. While the Jews were the prime target, political opponents, gypsies, the handicapped, and anybody not considered to be of true Aryan descent also suffered. Despite this, his popularity remained high among most Germans. His armaments and other public works programme appeared to be bringing a return of prosperity, while military success retrieved national pride.

The Soviet Union – The Bolshevik revolution of 1917 was followed by three years of civil war, during which White Russian armies, supported by foreign troops, tried to overthrow the new communist state. Lenin and his followers emerged successful, but at huge cost. It is estimated that some 13 million died in the war and the famine it caused; economic life was at a standstill. In 1921, as a emergency measure, Lenin largely freed the economy and recovery followed rapidly.

Lenin died in January 1924, leaving two men contending the succession. Trotsky proclaimed that the new society could only flourish within a communist world, and the prime task was therefore to export the revolution. His opponent, Stalin, argued that the priority was to rebuild the Soviet Union, by creating 'communism within one state'. When Stalin emerged victorious, it appeared as though the forces of moderation had prevailed.

Stalin assumed autocratic power and created a personality cult, not dissimilar to those constructed around the fascist dictators. He set himself the objective of changing the Soviet Union from a largely medieval economy, to a major modern state within a few decades. This involved the conversion of agriculture from its peasant structure by wholesale collectivization, and the rapid development of heavy industry. The programme was forced through at huge human cost. Industrially the results were dramatic. Production of coal, iron and steel and other basics increased many times over. The expansion of heavy industry was, however, bought at the expense of consumer goods, and the people, were constantly disappointed in the promised general improvement in living standards. Peasants on the collective farms, also resentful at being expected to produce low cost food for the growing cities for little return, remained obstinately unproductive.

Stalin's increasingly paranoiac behaviour was now demonstrated in a series of show trials and purges. Virtually all the old political leaders and a high proportion of military officers were executed to ensure that nobody would be able to challenge for power. Millions more suffered and died in labour camps. The new administrators of the country were tied to Stalin by a common guilt, and by an increasing web of petty corruption.

Spain – By the 1930s the days in which Spain had been a great European power were long past, and she had therefore avoided involvement in the First World War. In 1933 a right wing government came to power, which provoked rebellion by national minorities. In early 1936 a left wing government was elected with a large majority. General Franco, modelling himself on the fascist dictators, led a mutiny of the army in Morocco and invaded the mainland. The army, the political right and the Roman Catholic church aligned with Franco, while left wing groups and the national minorities aligned with the elected government. Franco received assistance from the fascist states, while the government was supported by the Soviet Union and a variety of international volunteers. The bitter war lasted until 1939, when Franco achieved the position of dictator, which he held until his death in 1975.

The Second World War in the West

German Expansion – From the beginning Hitler, followed a programme for the creation of a German empire in central Europe. His first objective was to win back land lost at Versailles; he then planned to conquer the whole of mainland Europe, including European Russia, and create an empire in which 'lower' races, such as the Slavs, would be reduced to a servile status. He exploited the weakness of the League of Nations, American isolationism, and lack of unity among other European powers in a series of successes – the recovery of the Saarland by plebiscite, the remilitarisation of the Rhineland, unification with Austria, and finally the dismemberment of Czechoslovakia. When Britain and France acquiesced to the last of these at Munich, it appeared as though no other power had the will to frustrate his ambitions.

In 1939 Hitler and Stalin concluded the Nazi-Soviet Pact to preserve Russian neutrality. The Soviet Union was awarded eastern Poland and took the opportunity to advance further into the Baltic States and Finland, where Russian armies were halted by fierce national resistance. Unlike Czechoslovakia, Poland was protected by treaty links with Britain and France, and the German invasion provoked a joint ultimatum and war. Mussolini took the opportunity of entering the war in support of Germany and invading Greece.

German Successes – After defeating Poland, in 1940, the German army repeated the 1914 tactic of invading France across Belgium. This time Paris fell and a puppet government was installed in southern France. Successful campaigns to the north and south reduced Denmark, Norway, Yugoslavia and Greece. In early 1941 the Afrika Corps landed in Libya and within two months was threatening Cairo and the Suez Canal. Britain, now rallied by the charismatic leader, Winston Churchill, held off an air offensive, intended to prepare the way for invasion, in the Battle of Britain.

On 22 June 1941 Hitler launched Operation Barbarossa against an unprepared Soviet Union. The invasion followed the logic of Hitler's master plan, but it dangerously overstretched German resources. The imbalance was made greater when the Japanese attack on Pearl Harbour brought the United States into the war in December of the same year. Stalin demonstrated his character as a national leader in rallying his people for a massively costly defence. The war turned in November 1942, when the Russians broke the German front at Stalingrad and a British army defeated the Afrika Corps at El Alamein. Once the Allies had re-established a western front with the Normandy landings of June 1944, the final defeat in 1945 was inevitable.

The years from 1939-45 gave a new and terrible meaning to warfare. The Germans mobilised conquered people for slave labour, and perpetrated mass genocide on European Jewry; the Russians deported whole national populations for alleged collaboration; residential areas of cities were targeted in indiscriminate bombing by both sides. Among some 50 million dead were an estimated 27 million Russians, 6 million Jews and $4^{1}/_{2}$ million Poles.

Europe Divided

The Yalta Settlement – The future political shape of Europe was negotiated in February 1945 at a conference at Yalta in the Crimea, attended by Stalin, Roosevelt and Churchill. Germany was to be partitioned and the countries of Eastern Europe were to form a zone of Russian influence. In the event, Austria and Greece – the latter after civil war – remained within the western sphere.

The Recovery of Western Europe – At the end of the war, western Europe was in a state of serious economic collapse. Once again, there were large movements of displaced people, and food shortages continued for years after the war. In June 1947, the US Secretary of State, George Marshall, announced a major aid programme, directed, "not against any country or doctrine, but against hunger, poverty, desperation and chaos". The Soviet Union was offered the chance of participating but turned it down. The Marshall Plan provided much needed capital for reconstruction.

The United Nations – During the last years of the war, thought was given to the reasons why the League of Nations had failed to preserve world peace. In 1945 representatives of the nations met in San Francisco to set up the new United Nations. In its constitution, great influence was given to the Security Council, which had five 'great powers' as permanent members and representatives of other nations. The right of veto was given only to the great powers, but at least all the major powers were now involved in the organization and debates were subject to the scrutiny of the world media.

The European Community – Some European leaders now argued that the nation state was no longer capable of providing a secure structure for world peace. In particular, the long standing enmity between France and Germany was no longer tolerable. In 1952, the Federal Republic of Germany, France, Italy, Holland, Belgium and Luxembourg, formed the European Economic Community. This was designed to be both a trading group, capable of competing with the new super-powers, and also a stabilizing influence on the volatile European political scene.

The Cold War in Europe – It soon became clear that European nationalism had run its destructive course, and the danger to world peace now lay in the confrontation of the United States and the Soviet Union. In the words of Winston Churchill, an 'iron curtain' had descended across Europe.

The Soviet Union emerged from the Second World War in control of a vast empire. It had inherited imperial conquests, and had further added the Baltic Republics. It also now controlled puppet regimes in Eastern Europe, bound together in the Warsaw Pact, which maintained huge land forces on its western front. Stalin's policy was still primarily directed at preserving national security, which had been so devastatingly violated by Hitler's army. He and later Soviet leaders therefore felt threatened by American superiority in nuclear weapons. A crash nuclear programme was put in hand and advances in rocketry were clearly illustrated when, in April 1961, Yuri Gregarin became the first man to be launched into space.

America and her allies in the North Atlantic Treaty Organization (NATO) relied heavily on nuclear superiority. The Americans responded to the Russian space programme and, in July 1969, with a wondering world watching on television, men were placed on the moon.

Berlin, divided between the four occupying powers, lay exposed within the Russian area of influence and in 1948-9 conflict loomed as the Russians shut off western communications with the city. A later crisis ended with the building of the Berlin Wall in 1961. This stood for the next 28 years as a potent symbol of the Cold War and the division of Europe into two hostile camps.

AFTER EMPIRE

The Expansion of Japan

The Beginnings of Aggression – The 1914-18 war brought prosperity to the rising Japanese economy. European competition in Asian markets was reduced and Japanese factories were able to export to the combatants. During this period, heavy industries, such as shipbuilding, were able to build up a firm base. Competition returned in the post-war boom years, but Japan continued to export successfully. During the boom years, companies re-invested profits in preparation for more difficult times. Japanese industry, none the less, suffered badly in the Depression. With foreign markets closed and the home market as yet undeveloped, industry worked at only a fraction of capacity. The weakness was exacerbated by the country's lack of raw materials. Foreign policy therefore became directed at the winning control of the export markets and natural resources of East Asia. China was the first target for expansion.

China's Weakness – After the fall of the Manchu Empire in 1912, China plunged back into chaos. Sun Yat-Sen's Nationalist (Kuomintang) party struggled for power with independent war lords. During this time communist cells were coming into existence. Following Marxist orthodoxy, they initially concentrated on the cities, but later, under the influence of the rising Mao Tse-tung, they worked increasingly among the mass of the peasants, who had suffered greatly during the upheavals. Under his influence, the communists built up communes in scattered and remote areas. Sun Yat-Sen died in 1925, to be succeeded as Kuomintang leader by the more conservative Chiang Kai-shek. After a period of collaboration, Chiang attempted to exterminate the communist opposition. Driven from their southern bases, the communists only survived by coming together in the Long March of 1934-5 and establishing a new northern headquarters, based on Yenan.

The War in the East

The Attack on China – Japan exploited the weakness and growing corruption of the Kuomintang government by strengthening its control over Manchuria and areas of the north in the early 1930s. In 1936 the Kuomintang and the communists made common cause against the foreigners, but in the following year, Japan launched a major assault on China. In December 1937 the Japanese army captured the capital at Nanking and the Chinese government had to retreat to remote Szechwan, leaving the Japanese in control of the north, and

most of the Pacific coast, including the major industrial cities.

Victory brought Japan into conflict with the Pacific colonial powers, and their concessionary ports were blockaded. The Americans and British responded by supplying Chiang Kai-shek along the Burma Road, and the United States renounced its commercial agreement with Japan.

Control of the Pacific – Japanese foreign policy was now set on winning control over the whole of the Pacific rim. With the outbreak of war in Europe, she allied herself with Germany. Then in 1941, as German troops were sweeping into Russia, she launched her first attack on French Indo-China. On the 7th December 1941 her air force suddenly attacked the American navy in Pearl Harbour, Hawaii, and, at the same time, she launched assaults on the Dutch in Indonesia and the British in Mayalsia and Burma. The campaigns were brilliantly successful and by mid 1942 both India and Australia were under threat.

Defeat and the Atom Bomb – The attack on Pearl Harbour put and end to isolationism and united Americans behind President Roosevelt. As the world's greatest industrial power became geared for war, the tide turned against Japan. In June 1942 the Japanese fleet suffered a reverse at the battle of Midway Island, and thereafter a relentless American offensive drove them from their Pacific conquests, while the British also fought back through Burma. From November 1944 the Japanese cities came under direct air attack. The war ended with the use of the new atomic weapon on the cities of Hiroshima and Nagasaki in August 1945.

Decolonization – The Japanese victories, and, in particular, the fall of Singapore on the 15th February 1942, involved a profound loss of face for the colonizing powers. The invading armies were seen by many Asians as liberators from western regimes. Many of those who had assumed control, under Japanese direction, now became prominent in independence movements. The United States handed over political control of the Philippines in 1946 after negotiating a continued miliary presence. The Dutch, themselves newly liberated, at first fought to preserve their possessions but in 1948 they accepted the independence of the Republic of Indonesia. The British fought a communist rebellion in Malaya before handing over to a more acceptable national government in 1957. The French became involved in a long war for Indo-China before being defeated in 1954.

China and her Neighbours

Communist China – The fall of Japan left the two forces of the Kuomintang and the communists vying for the control of China. China's miseries continued when civil war broke out in 1947. In one battle, half a million men were engaged on each side. By 1949 the communists were gaining the upper hand and in May 1950 Chiang Kai-shek retreated to Taiwan with his government.

The new communist government was faced by a huge task of reconstruction. According to Mao's estimate, some 800,000 'enemies of the people' were executed, largely from the old village landlord class. The communists had long experience with the collectivization of agriculture within their own territories, and they did not follow Stalin's example of imposing it from above. Peasants were organized to control their own operations and, despite set-backs, the conditions of life for the mass of people improved.

Mao Tse-tung capitalized on the age-old Chinese respect for authority to provide a strong central government, which had for so long been lacking. He adapted western Marxist ideology to traditional thought patterns, and showed a strong hostility to western culture, which was given full rein in the Cultural Revolution, which he launched in 1966.

To western eyes, China appeared now to be a part of a united Communist bloc, intent on achieving world dominance. In practice, however, Mao had largely rejected the Russian brand of communism. By the 1960s, acute strains were appearing in the relationship between the two countries. When China developed its own nuclear capability in 1964, it was primarily as a deterrent against potential Russian aggression. The Chinese reconquest of the old province of Tibet also led to a successful war with India in 1962.

The Korean War – In 1945 the Japanese colony of Korea was occupied by Russian troops from the north and Americans from the south. This led to partition, with both governments claiming the whole country. In 1950 the northern armies invaded the south. The United States and other western nations, with the backing of a United Nations resolution, responded by sending forces to support the south.

For a time it appeared as though the north would be defeated, but China, concerned at her own security, sent an army across the border. The American President Truman refused to become involved in a war on Chinese soil, and the war was concluded in 1953 by an armistice, which perpetuated partition.

Conflict of Ideologies – American analysts saw the communist strategy in South East Asia and being a process of 'slicing the salami'. Territories were to fall to communism, not in one major conflict, but one by one. The communist uprisings, which faced almost every nation of South East Asia in the coming decades, were in fact little co-ordinated and variously owned allegiance to Moscow, Peking or neither. That in the new nation of Indonesia was put down with great violence.

American policy became dedicated to holding the line against communism in the region. This involved providing support to non-communist regimes, including Chinese nationalist government in Taiwan. The United States was therefore deeply involved in the politics of the region.

The French defeat in Indo-China left the new country of Vietnam divided, with a communist regime under the old nationalist Ho Chi Minh established in the north. The United States became increasingly involved in the struggle, supporting unstable non-communist administrations, based in the southern capital of Saigon. In 1965, faced with the possibility of the defeat of the client regime by the northern backed Viet Cong guerillas, President Johnson authorized massive involvement in the conflict. The weight of American firepower proved ineffective against a highly motivated enemy. In 1973 the American government, confronted by a mounting anti-war campaign at home, withdrew from the conflict. In the next two years the three countries of Indo-China fell to the communists. The people of Cambodia, having experienced American bombing, now suffered from the worst aberrations of Marxism, as interpreted by the Pol Pot regime.

The Pacific Rim

Japanese Reconstruction – United States troops occupied Japan, in an enlightened manner, from 1945 until 1952. The first objective was to ensure that the expansionist phase was over. A new democratic constitution was established, the Emperor renounced his divinity, and expenditure on defence and armaments was radically curtailed. As with Germany, industrial reconstruction followed fast. After the humiliation of defeat, both nations needed to experience success. Also, the imposed limitation of defence expenditure proved a powerful boost to the civilian economy.

The Technological Revolution – Soon Japan was no longer a low cost economy and her heavy industries began to suffer some of the problems experienced in the West. By this time, however, the nation had developed skills, which enabled it to take the lead in the third, technological phase of world industrialization. Automobile production boomed, winning markets in Europe and North America, and Japanese labour and management skills proved highly suitable to the detailed work involved in the production of hi-tech goods. Supported by a huge balance of payments surplus, she has established a position of dominance in world markets in a wide range of product areas.

An Area of Growth – In the last decade of the 20th century it is clear that the region of the Pacific rim is established as a formidable competitor to the established industrialized regions of western Europe and North America. It remains, however, a region of wide diversity. South Korea, Taiwan, Singapore and – at least until re-unification with China in 1997 – Hong Kong have participated in the economic prosperity pioneered by Japan. At the other extreme, peoples of many of the nations of South East Asia continue to survive on low per capita incomes. China herself emerged from the isolation of the Cultural Revolution to re-build international links, but, unlike communist regimes to the west, it has successfully repressed those who wished to liberalize the political structure of the nation.

The Indian Sub-Continent

The Independence Movement – Indian troops made a significant contribution to the allied victory in the World War One, and in 1918 nationalist politicians looked to see their country start its progress towards the self government which had already been given to the white dominions. In 1919, however, a British general ordered troops to fire on a demonstration in Amristar, killing some 400 and injuring many more. Although the government disavowed the act, many British residents were loud in support, fuelling bitterness between the two communities. One of those radicalised by the Amristar massacre was Mohandas Gandhi, known as Mahatma (Great Soul). During the next decades, he led a civil disobedience movement, based, if not always successfully, on non-violent principles.

In the face of opposition at home, as well as from residents in India, the British government slowly moved towards accepting the principle of granting dominion status to India, and the Government of India Act of 1935 gave substantial power to elected representatives. The movement for complete independence, however, continued to grow. With the outbreak of the World War Two, some Indians sided with Japan in the hope of bringing down the colonial power.

Partition – The independence movement still faced the problem of reconciling the two major religious groupings of the sub-continent. Mohammed Ali Jinnah emerged as leader of the Muslim League, which now demanded that an independent state of Pakistan should be established at independence for the Islamic community. In 1945 a Labour government was returned in Britain and in March 1946 it made an offer of full independence. As disputes continued, it announced that Britain would withdraw not later than June 1948. Faced with this ultimatum, the Hindu leaders accepted partition – for which decision Gandhi was assassinated by an extremist Hindu.

The new state of Pakistan was established in two blocks in the north west and north east. The rulers of princely states on the boarder of the two nations were permitted to decide their allegiance, leaving Kashmir as disputed territory. Independence was marked by communal rioting, which left some half a million dead, and the mass movement of peoples in both directions across the frontier.

Independence – Jinnah died in 1948 and a decade later the army took control of Pakistan. Leadership of India fell to Jawaharlal Nehru and later passed to his daughter, Indira and grandson Rajiv Gandhi. The new country faced formidable problems. Independence was quickly followed by famine in 1951, and, with rising population it appeared as though Malthusian disaster was imminent. A combination of a reduction in the rate of population growth and an agricultural 'green' revolution has, however, improved the supply of food. The fragile ecology of eastern Pakistan, however, continued to bring disasters and a cyclone in 1970 led to rebellion, which, with Indian help, brought into being the separate Islamic state of Bangladesh.

Partition did not bring the end of India's communal problems. Indira Gandhi was assassinated by discontented Sikh nationalists, and her son Rajiv by Tamils of the south. Mahatma Gandhi allied himself with outcasts and hoped to see the end of the caste system, but this has been frustrated by a resurgence of Hindu fundamentalism. For all its problems, however, India remains the world's largest democracy.

Sub-Saharan Africa

Decolonization – British governments of both parties continued the policy of giving independence to colonies, which had begun with India. The sheer size of the British Empire had meant that expatriate manpower was spread thinly. The second layer of administration was already staffed by African personnel and a machinery of local government was in place. Riots in the Gold Coast in 1948 gave notice of a growing nationalist movement. In 1957 the Gold Coast became the first independent country, within the British commonwealth, under its new name, Ghana. Three years later the much larger Nigeria became a sovereign state.

Across the continent, in Kenya and Rhodesia the problem was complicated by the presence of a white settler population, bitterly opposed to any move towards majority rule. The most serious challenge was posed by the Mau Mau disturbances in Kenya of 1952-6. Many of the white minority left the country when it received its independence in 1963. Two years later the white minority government of Southern Rhodesia declared unilateral independence and seceded from the British commonwealth. The British government failed to take effective action against this colonial rebellion and war continued between the white government and black nationalist groups, until the latter won to set up the state of Zimbabwe in 1976.

French governments, disillusioned by prolonged war in Indo-China and Algeria, gave independence at an even faster pace. In 1960 the Sub-Saharan colonies were offered either complete separation, in which case, they would receive no continued assistance, or association with France, within a French Community.

The two largest African empires were therefore dismantled within a few years with comparatively little strife. Independence for the remaining colonies proved a more painful process. In 1960, the Belgians withdrew from the Congo, which became the state of Zaire. Until that time, Africans held no positions of responsibility, and there was little preparation for the event. When the mineral-rich area of Katanga attempted to secede, the Cold War superpowers became involved in the ensuing civil war.

The last African empire was also the oldest. The Portuguese colonies of Guinea Bissau, Mozambique and Angola only achieved independence after prolonged struggle.

After Independence – The emergent nations faced formidable problems. Some new nations spent unwisely on military prestige projects, but even where this was avoided, as, for instance, in Tanzania, falling world commodity prices led to a serious reduction in government revenue. Industrialization has proved unattainable, both through lack of capital, but also because it has proved hard for products from new nations to break into the controlled markets of the developed world.

Independent African nations found themselves caught in the Malthusian nutcracker of increasing population and falling revenue. This led to a decline in already low living standards and a failure by governments to deliver the public health and education programmes expected within a newly liberated nation. This exacerbated traditional communal rivalries, which in turn frequently erupted into civil war, like the Nigerian Biafra War of 1967-70 and later struggles in the Sudan, Ethiopia and the Horn of Africa. Political instability led to the emergence of authoritarian, often military, regimes. An already difficult situation has been made worse for nations immediately south of the Sahara by climatic change and desertification, which have destroyed large areas of productive land.

South Africa – The violently imposed apartheid system led to South Africa being increasingly ostracized from the world community. She withdrew from the British Commonwealth in 1961, and was later expelled from the United Nations. Economic sanctions imposed by the U. S. A. and a world sporting boycott had an effect and in the early 1990s the legal apparatus of apartheid was dismantled, but, with the white community continuing to control the police and armed forces, the political structure of a future multiracial nation remained unresolved. In 1990, however president F.W. de Klerk, instigated the release from prison of Nelson Mandela, the leader of the banned African National Congress. In 1994, the first free elections were held and Nelson Mandela was elected president. The economic sanctions against South Africa have since been lifted and relations with the international community restored.

Latin America

Capital and Industrialization – In the years before World War One there was heavy European involvement in the economy of Latin America. The war then led to a drying up of European capital and the United States became the main investor in the region.

The world depression of the 1930s hit the region hard. The price of primary products, the mainstay of the economies, collapsed. After the war many of the larger nations instituted industrialization programmes, at times with a degree of success, but this was achieved only by borrowing the required capital, which left the nation with a heavy burden of debt and vulnerable to currency and interest rate fluctuations on the international market.

Economic problems created political instability. The rural poor had always lived in conditions of poverty, but they did not pose the same immediate problem to political stability as the growing and highly volatile urban populations.

Political Structures – The economic problems of the region meant that reforming governments did not have the revenue to deliver the social programmes needed to combat deprivation. When reforms were attempted they created inflation which weakened the economic base of society. Reforming democracies have therefore been under constant pressure from more authoritarian sys-tems of government. These took three broad forms – popularist leaders, military regimes and revolutionary governments.

The archetype polularist leader was Getulio Vargas, who came to power in Brazil in 1930, and the best known Juan Peron, who ruled the Argentine from 1943-55 and then returned briefly in 1973. Both drew comparison with European dictators, but they had wide support among the urban poor, who believed that they alone could take on powerful vested interests on behalf of the people. They depended, however, on army support, and both were vulnerable when this was withdrawn.

Cuba and Revolution – The revolutionary movement had early roots in Mexico, but it became focused on Cuba with the success of Fidel Castro's revolution in 1959. An attempt by the United States to undermine the revolution came to disaster at the Bay of Pigs in 1961. In the following years Cuba, now aligned with Russia, exported revolution into Latin America. Che Guevara, a symbol to the new left across the world, was killed fighting with Bolivian guerillas in 1967. The United States became involved, supporting anti-communist regimes within the region, even when these had a poor human rights record. The democratic left wing government of Salvadore Allende in Chile, for instance, was overthrown by the military in 1973 with American support. Contra rebels against the Cuban inspired government of Nicaragua were funded from Washington, and the government of the island of Grenada was overthrown by American invasion in 1983.

The Missile Crisis – In 1962 Cuba was the focus of the most dangerous crisis of the Cold War. In October intelligence reports showed that sites were being built on Cuba from which missiles would be able to reach any city in the United States. President Kennedy demanded that all missiles in Cuba should be withdrawn and announced that ships bringing more would be intercepted. The super-powers stood poised for nuclear confrontation, but the Russian President Khrushchev broke the crisis by agreeing to withdraw the missiles. President Kennedy had successfully reasserted the Munroe Doctrine that the American continent would remain an area of United States influence, and the powers would not again come so close to open war.

The Middle East and North Africa

The New Turkey – In 1918, a proposal was put forward that Turkey itself should be divided into French, British and Italian spheres of influence. The successful general, Mustafa Kemal, led resistance against Greek and French forces, and established independence for the new, smaller nation. He set about a process of modernization of the nation, which went as far as westernizing its script and converting the country into a secular state. His people gave him the name of Ataturk – father of the Turks.

The Mandates – The old Ottoman lands of the Islamic Middle East, now finally separated from the Ottoman Empire, had acquired new strategic importance with the early development of oil reserves – although the scale and future importance of these were not as yet recognized. National boundaries were drawn up and the region was divided between France and Britain under the mandate of the League of Nations. This implied that the newly defined countries were destined to move towards self governing status. France was awarded Lebanon and Syria, although she had to take possession of the latter by force, and continued to rule it with considerable oppression. Britain received Palestine, Iraq, and Trans-Jordan, and she also controlled the emirates of the Persian Gulf. In 1932, Britain largely withdrew from Iraq, but the Palestinian mandate turned out to be something of a poisoned chalice.

The Founding of Israel – The objective of founding a Jewish national home in Palestine was first put forward in a Zionism Congress as early as 1897. It was to be a refuge for Jewish people who were persecuted in the pogroms of eastern Europe, and it also attracted many from minority Jewish communities within the Arab world. In 1917, the British government gave support to the project, with the contradictory provision that it should not interfere with the rights of the indigenous people. The movement was given further impetus by German persecution of the Jews under Hitler. In the post war years, large numbers of European Jews sought entry, and the British authorities had the impossible task of reconciling the opposing interests. In 1947 the United Nations voted for the partition of Palestine in the face of opposition from the Arab states and in 1948 the British withdrew. In the ensuing war, large numbers of Arabs left their homes for refugee camps in the neighbouring countries. The Arab states refused to accept the existence of a Jewish state in the Islamic heartland. The refugees remained, unsettled, waiting to return to their homeland as Israel and her neighbours continued in a state of war.

North African Independence – After World War Two the British presence in Egypt was restricted to a defensive force in the canal zone and, by 1956, Libya, Tunisia and Morocco had shaken off foreign ties. Armed conflict centred on Algeria, where over one million French settlers resisted any move towards independence. The country was declared an integral part of metropolitan France and a bitterly fought dispute continued from 1954-62. When General de Gaulle finally decided to give independence, colonists allied with army generals and France itself was taken to the brink of civil war.

Nasser and Pan-Arabism – In 1952 a group of Egyptian army officers overthrew the monarchy. Two years later, Gamal Abdel Nasser became president of the country. His objective was to establish Egypt as the unquestioned leader of a new and more coherent Arab people. Lacking oil resources, however, Egypt remained a poor country and Nasser planned a development programme based on the construction of the Aswan High Dam on the Nile. When the Americans and British withdrew offers of funding, Nasser turned to the communist bloc for support, so introducing cold war politics into the Middle East.

In 1956 he nationalized the company which administered the Suez Canal. In October the Israelis invaded Egyptian territory, ostensibly to destroy guerilla bases and this was followed by a joint attack by the British and French on the Suez Canal zone. World opinion was outraged, and the American government applied pressure which forced the invaders to withdraw.

The Suez fiasco left Nasser as the leading figure within the Arab world, but his attempts to take this towards political union were unsuccessful. In 1967 he closed the Straits of Tiran to Israeli shipping and the Israeli army launched a 'first strike' in what has become known as the Six Day War. After a successful campaign, Israel controlled new territory, including, from Jordan, the whole West Bank of the River Jordan, and, from Syria, the tactically important Golan Heights. Jerusalem, a city of great symbolic importance to all three Semitic religions, now passed under full Israeli control. Successive Israeli governments, in time reinforced by the possession of nuclear weapons, have failed to comply with United Nations resolutions demanding withdrawal from the occupied territories. Indeed, increasing numbers of Jewish immigrants have been estab-

lished in West Bank settlements. As Israel's neighbour, the Lebanon, collapsed into civil war, many Arabs resorted to international terrorism.

The Oil Crisis – In 1961 Britain withdrew from her interests 'east of Suez'. Much of the, now increasingly vital, oil production of the region, however, remained under the control of western companies. A further outbreak of hostilities between Israel and her neighbours in 1973 led the Arab countries to 'play the oilcard' by taking more direct control over their own reserves and withholding supplies from Israel's allies in the developed world. This led to an increase in price, which had a sharp effect on the world economy. The Arab nations, and other oil-producing nations, led by Saudi Arabia now organized themselves into O.P.E.C., with a view to controlling world prices. This was less successful than had been anticipated because the depression caused by the price rise restricted world demand, and Britain and Norway, opening new North Sea reserves, stood outside the cartel.

In 1978 Nasser's successor, President Sadat, made peace with Israel under American sponsorship at Camp David. This did not end the conflict within the region, but rather took Egypt out of the mainstream of Arab politics.

Iran and Islamic Fundamentalism – With Egypt returned to the American sphere of influence after Camp David, the Soviet Union turned increasingly to the radical, though mutually hostile, governments of Syria and Iraq. The United States, looking for a buffer between the Soviet Union and the oil-rich Middle East, put heavy backing behind the conservative and corrupt administration of the Shah of Iran. In 1979 discontent erupted into revolution, and the Shah was replaced by a fundamentalist regime, dominated by the Ayatollah Khomeni. This sparked a wave of Islamic fundamentalism which gave expression to pent up Arab anger at the imposition of alien values by aggressive western societies. Equally hostile to capitalist and to communist ideologies, Islamic fundamentalism has threatened governments of different complexion from Afghanistan to Algeria. Indeed, the failure of the 1979 Soviet invasion of Afghanistan in support of a crumbling Marxist regime, demonstrated militant Islam to be a highly effective barrier against further Russian expansion in the region.

Iraq – In 1979 Iraq came under the full control of a determined and ruthless leader, Sadaam Hussein. He had ambitions to revive Nasser's pan-Arab vision, this time based on Iraqi military power. He received wide western and Arab backing when he took his country to war with Iran, but he failed to achieve any of his war objectives. In 1990, he attacked and occupied Kuwait, provoking an international response in 1991, which left his country damaged, but his own power intact.

The Collapse of the Russian Empire

Cracks in the Structure – As early as 1953, the year that Stalin died, there were signs of unrest among subject people of the Russian Empire. Yugoslavia, while remaining communist, had already loosened her ties with the Soviet bloc. Anti-Soviet riots in East Germany in 1953 and in Poland in 1955 were followed by rebellion in Hungary in 1956. The last was suppressed by military force, launched under the cover of the Anglo-French attack on Suez. The profound unpopularity of Russian domination and of the repressive puppet regimes continued to be demonstrated by a haemorrhage of refugees crossing from East to West Germany. In 1961 the East German authorities responded by building that ultimate symbol of the Cold War – the Berlin Wall. In 1968 a reforming communist government in Czechoslovakia was again overthrown by Soviet tanks. By this time large Russian forces were also tied down on the eastern frontier to check an increasingly hostile China.

Collapse – Meanwhile the government was coming increasingly under strain within the Soviet Union. Khrushchev's denunciation of Stalin at the Twentieth Congress of the Soviet Communist Party and the termination of the worst excesses of the secret police enabled citizens to express dissatisfaction. Industrialization had been bought at the expense of the production of consumer goods, The corrupt and petty bureaucracy was increasingly exposed, and agriculture remained in the disastrous condition bequeathed by Stalin's collectivization.

In 1985 President Gorbachev inherited a collapsing empire. Constricted by domestic pressures, he chose not to intervene when, in a few dramatic months of late 1989 and early 1990, communist governments of Eastern Europe collapsed under popular pressure and new regimes declared themselves independent of Soviet control. The tearing down of the Berlin Wall, and subsequent reunification of Germany was the most powerful symbol of change. The situation was little better in the republics which constituted the Soviet Union. The people were increasingly dissolutioned by falling living standards and inefficient government. Powerful nationalist forces, from the southern republics of Armenia to Azerbaijan to the old Baltic States in the north now threatened to break up the Soviet Union from within. In August 1991 an attempt by communist 'hard-liners' to restore the old system in a coup d'état failed, leaving the central Soviet government stripped of any real power. As one republic after another announced secession it was quickly clear that the world possessed another 'sick man' – with all the attendant dangers.

The collapse of the Russian Empire at least signalled the end of super power confrontation. Faced with mounting problems at home, Gorbachev looked for support from America and other western nations and the Strategic Arms Limitation Treaty of July 1991 began the long process of disarmament. The benefits for world peace were illustrated when the Soviet Union refrained from backing Iraq in the Gulf War, so preventing a regional conflict being inflated into a confrontation of super-powers. At the time of writing, the formidable problems of the former Soviet Union itself, and of the emerging democratic states of Eastern Europe remain unresolved, but the fear of the human race being destroyed by its own weapons has – rightly or wrongly – been overtaken by new environmental concerns, which centre on the rate in which post-industrial man is consuming the natural resources of the planet.

WORLD FACTS

This section of the Family Encyclopedia provides essential information on the geography of the world. Arranged in an A–Z format, it gives a political, physical, economic and climactic picture of over 200 countries, together with helpful maps and fact boxes. It also covers major and historical regions, cities and towns and the earth's physical features, from oceans to deserts and from mountains to lakes. The entries are cross-referenced, indicated by the SMALL CAPITALS, to enable you to gain a complete overview of the world's features.

✳

Aachen (Aix-la-Chapelle) a historic university city and spa town in western GERMANY. (Pop. 247000)

Aarhus *see* **Århus**.

Aba an industrial town in southern NIGERIA. (Pop. 271,000)

Abadan a major oil-refining port on an island in the SHATT AL ARAB waterway, southern IRAN. (Pop. 308,000)

Abeokuta an industrial town in western NIGERIA. (Pop. 387,000)

Aberdeen a major NORTH SEA oil city and fishing port in north-east SCOTLAND. (Pop. 218,200)

Aberdeenshire a county in north-east SCOTLAND, formerly part of Grampian region. (6318 sq km/2439 sq miles; pop. 223,600)

Abidjan a major port and the chief city of CÔTE D'IVOIRE. (Pop. 1,850,000)

Åbo *see* **Turku**.

Abruzzi a region of southern central ITALY; its capital is Aquila. (Pop. 1,578,000)

Abu Dhabi the largest sheikhdom of the UNITED ARAB EMIRATES, of which the city of Abu Dhabi is the capital. (67,350 sq km/26,000 sq miles; pop. emirate 535,700/city 244,000)

Abuja the new capital of NIGERIA, in the centre of the country, still under construction but inaugurated in 1992. (Pop. 379,000)

Abu Simbel the celebrated site of temples of ancient EGYPT, by Lake NASSER.

Acapulco a large port and beach resort on the PACIFIC coast of MEXICO. (Pop. 592,000)

Accra the capital and main port of GHANA. (Pop. urban area 1,781,000)

Aconcagua the highest mountain of the ANDES, in ARGENTINA. (6960 m/22,835 ft)

Adan *see* **Aden**.

Adana a city and province in southern TURKEY. (Pop. city 916,000)

Ad Dawhah *see* **Doha**.

Addis Ababa (Adis Abeba) the capital of ETHIOPIA, in the centre of the country. (Pop. 1,739,000)

Adelaide the state capital of SOUTH AUSTRALIA. (Pop. 1,050,000)

Aden (Adan) a major port in southern YEMEN, formerly the capital of South Yemen. (Pop. 417,000)

Adirondack Mountains a mountain range in NEW YORK State, USA. The highest peak is Mount Marcy (1629 m/5344 ft).

Adis Abeba *see* **Addis Ababa**.

Adriatic Sea a branch of the MEDITERRANEAN SEA, between ITALY, SLOVENIA and CROATIA.

Aegean Sea a branch of the MEDITERRANEAN SEA between GREECE and TURKEY.

Afghanistan is a landlocked country in southern ASIA. The greater part of the country is mountainous with several peaks over 6000 m (19,686 ft) in the central region. The climate is generally arid with great extremes of temperature. There is considerable snowfall in winter which may remain on the mountain summits the year round. The main economic activity is agriculture and although predominantly pastoral, successful cultivation takes place in the fertile plains and valleys. Natural gas is produced in northern Afghanistan and over 90% of this is piped across the border to the former USSR. Other mineral resources are scattered and so far underdeveloped. The main exports are Karakuls (Persian lambskins), raw cotton and foodstuffs. Since the Russian withdrawal from Afghanistan in 1989, the country has still been troubled by, mainly ethnic, conflict.

Quick facts:
Area : 652,225 sq km (251,773 sq miles)
Population : 18,052,000
Capital : Kabul
Other major cities : Herat, Kandahar, Mazar-i-Sharif
Form of government : Republic
Religions : Sunni Islam, Shia Islam
Currency : Afghani

Africa the second largest continent in the world, with the MEDITERRANEAN SEA to the north, the ATLANTIC OCEAN to the west and the INDIAN OCEAN to the east. There are 52 nations within Africa, excluding WESTERN SAHARA. (30,300,000 sq km/11,700,000 sq miles; pop. 642,000,000)

Agadir a port and popular tourist resort in MOROCCO. (Pop. 439,000)

Agartala the state capital of TRIPURA in north-eastern INDIA. (Pop. 158,000)

Agra a city in central INDIA, and site of the Taj Mahal. (Pop. 956,000)

Ahmadabad (Ahmedabad) an industrial city in western INDIA. (Pop. 3,298,000)

Ahvaz a port on the Karun River in southern IRAN. (Pop. 725,000)

Aix-en-Provence a university city in southern FRANCE. (Pop. 127,000)

Aix-la-Chapelle *see* **Aachen**.

Ajman the smallest emirate of the UNITED ARAB EMIRATES. (65 sq km/25 sq miles; pop. emirate 42,000/ town 27,000)

Akron a city in the north-east of the state of OHIO, USA (Pop. city 223,600/metropolitan area 660,000)

Alabama a state in southern USA The state capital is MONTGOMERY. (133,667 sq km/51,606 sq miles; pop. 4,137,500)

Alamein *see* **El Alamein**.

Alamo, The an old Spanish mission near SAN ANTONIO, TEXAS, USA Davy Crockett and 186 other Texans died defending the fort here against a Mexican army in 1836.

Alaska the largest and most northerly state of the USA The state capital is JUNEAU. (1,518,800 sq km/ 586,400 sq miles; pop. 587,800)

Albacete a town and province of south-eastern SPAIN. (Pop. town 141,100)

Albania is a small mountainous country in the eastern MEDITERRANEAN. Its immediate neighbours are GREECE, SERBIA and THE FORMER YUGOSLAV REPUBLIC OF MACEDONIA, and it is bounded to the west by the ADRIATIC SEA. The climate is typically Mediterranean and although most rain falls in winter, severe thunderstorms frequently occur on the plains in summer. Winters are severe in the highland areas and heavy snowfalls are common. All land is state owned, with the main agricultural areas lying along the Adriatic coast and in the Korce Basin. Industry is also nationalized and output is small. The principal industries are agricultural product processing, textiles, oil products and cement. Most trade is with neighbouring Serbia and The Former Yugoslav Republic of Macedonia.

Albania has been afflicted by severe economic problems and in late 1996 public dissatisfaction with the government erupted into civil unrest, leading to a major revolt by citizen militias during which the government forces lost control, particularly in the south of the country. By March 1997 the country was on the brink of collapse and large numbers of refugees were leaving.

Quick facts:
Area : 28,748 sq km (11,100 sq miles)
Population : 3,422,000 (estimate prior to 1997, when significant numbers of refugees were leaving the country)
Capital : Tirana (Tiranë)
Other major cities : Durrës, Shkodër, Elbasan
Form of government : Socialist Republic
Religion : Constitutionally atheist but
    mainly Sunni Islam
Currency : Lek

Albany the capital city of NEW YORK State, USA (Pop. 99,700)

Al Basrah *see* **Basra**.

Albert, Lake a lake in the GREAT RIFT VALLEY in East AFRICA, shared between UGANDA and ZAÏRE. (Also known as Lake Mobuto Sese Seko.) (5180 sq km/ 2000 sq miles)

Alberta a province of western CANADA; EDMONTON is its capital. (661,190 sq km/255,285 sq miles; pop. 2,628,000)

Ålborg a city and port in northern DENMARK. (Pop. 157,000)

Albuquerque a university city on the RIO GRANDE in NEW MEXICO, USA (Pop. 398,500)

Alcalá de Henares a town in central SPAIN, birthplace of Miguel de Cervantes (1547-1616), author of *Don Quixote*. (Pop. 166,000)

Alderney one of the CHANNEL ISLANDS. (8 sq km/3 sq miles; pop. 2,300)

Aleppo (Halab) an industrial city of ancient origins in SYRIA. (Pop. 1,445,000)

Alexandria the main port of EGYPT, on the NILE delta. (Pop. 3,380,000)

Al Fujayrah the second smallest emirate in the UNITED ARAB EMIRATES and also the name of a small town in the emirate. (117 sq km/45 sq miles; pop. emirate 38,000/town 760)

Al Furat *see* **Euphrates, River.**

Algarve the southern province of PORTUGAL. (4,945 sq km/1909 sq miles; population 341,000

Algeria is a huge country in northern AFRICA, which fringes the MEDITERRANEAN SEA in the north. Over four-fifths of Algeria is covered by the SAHARA DESERT to the south. Near the north coastal area the

ATLAS MOUNTAINS run east-west in parallel ranges. The climate in the coastal areas is warm and temperate with most of the rain falling in winter. The summers are dry and hot with temperatures rising to over 32°C. Inland beyond the Atlas Mountains conditions become more arid and temperatures range from 49°C during the day to 10°C at night. Most of Algeria is unproductive agriculturally, but it does possess one of the largest reserves of natural gas and oil in the world. Algeria's main exports are oil-based products, and some fruit and vegetables.

Quick facts:
Area : 2,381,741 sq km (919,590 sq miles)
Population : 27,940,000
Capital : Algiers (Alger)
Other major cities : Oran, Constantine,
  Annaba
Form of government : Republic
Religion : Sunni Islam
Currency : Algerian dinar

Algiers (El Djazair, Alger) the capital of ALGERIA, on the MEDITERRANEAN coast. (Pop. 1,722,000)

Aliákmon, River the longest river in GREECE. (Length 297 km/184 miles)

Alicante a port and popular beach resort, and also the name of the surrounding province, on the MEDITERRANEAN coast of SPAIN. (Pop. town 275,000)

Alice Springs a desert settlement in the NORTHERN TERRITORY of AUSTRALIA. (Pop. 18,400)

Aligarh a university town in north INDIA. (Pop. 481,000)

Allahabad a holy city in INDIA on the confluence of the rivers GANGES and YAMUNA. (Pop. 858,000)

Alma-Ata (Almaty) a trading and industrial city and capital of KAZAKHSTAN. (Pop. 1,500,000)

Al Madinah *see* **Medina.**

Al Manamah the capital and main port of BAHRAIN. (Pop. 143,000)

Al Mawsil *see* **Mosul.**

Alps, The a mountain range in southern central EUROPE that spans the borders of SWITZERLAND, FRANCE, GERMANY, AUSTRIA, SLOVENIA and ITALY.

Alsace a region in the north-east of FRANCE.

Altai an area of high mountain ranges in central Asia on the borders of CHINA and the RUSSIAN FEDERATION where they meet at the western end of MONGOLIA.

Amager a fertile island to the south of COPENHAGEN, in DENMARK.

Amalfi a small, picturesque town on a spectacular part of the west coast of ITALY. (Pop. 7000)

Amarillo an industrial city in north-west TEXAS. (Pop. 161,100)

Amazon, River (Amazonas) with the River NILE, one of the world's two longest rivers. It rises in the ANDES of PERU and flows east through BRAZIL to the ATLANTIC OCEAN. (Length 6440 km/4000 miles)

Amboina *see* **Ambon.**

Ambon (Amboina) an island and the capital of the so-called Spice Islands in the MALUKU group in eastern central INDONESIA. (813 sq km/314 sq miles; pop. 73,000)

America the continent lying between the ATLANTIC and the PACIFIC OCEANS. For convenience it is divided into three zones: North America (USA, CANADA, MEXICO and GREENLAND: 23,500,000 sq km/9,000,000 sq miles; pop. 354,000,000), Central America (the area between the southern MEXICO border and the PANAMA-COLOMBIA border together with the CARIBBEAN: 1,849,000 sq km/714,000 sq miles; pop. 63,000,000), and South America (the area to the south of the PANAMA-COLOMBIA border: 17,600,000 sq km/6,800,000 sq miles; pop. 284,000,000).

American Samoa *see* **Samoa, American.**

Amiens an industrial city and capital of the SOMME department of northern FRANCE. (Pop. 156,000)

Amindivi Islands *see* **Lakshadweep.**

Amman the capital of JORDAN, in the north-east of the country. (Pop. 1,272,000)

Amritsar an industrial city in northern INDIA and home of the Golden Temple, the most sacred shrine of the Sikhs. (Pop. 709,000)

Amsterdam the capital and commercial centre of the NETHERLANDS, a historic port set on the IJSSELMEER. (Pop. 1,091,000)

Amudar'ya, River a central Asian river forming much of the border between TAJIKISTAN and AFGHANISTAN before flowing through UZBEKISTAN into the ARAL SEA. Its ancient name was OXUS. (Length 2620 km/1630 miles)

Amundsen Sea an arm of the South PACIFIC in ANTARCTICA.

Amur, River (Heilong Jiang) a river which runs along the border between CHINA and the RUSSIAN FEDERATION, flowing east into the PACIFIC OCEAN. (Length 4510 km/2800 miles)

Anatolia the historical name for the Asian part of Turkey.

Anchorage the largest city and port in ALASKA, USA, on its southern coast. (Pop. 246,000)

Andalusia (Andalucía) a region of south-western Spain, with a coast on the MEDITERRANEAN and ATLANTIC. (Pop. 6,441,800)

Andaman and Nicobar Islands two groups of islands in the Bay of BENGAL, administered by INDIA. (Pop. 280,000)

Andaman Sea a branch of the Bay of BENGAL, lying between the ANDAMAN ISLANDS and MYANMAR.

Andes a high mountain range that runs down the entire length of the western coast of South AMERICA. The highest peak is Mount ACONCAGUA, in ARGENTINA (6960 m/22,835 ft).

Andhra Pradesh a state in south-east INDIA. The capital is HYDERABAD. (275,068 sq km/106,203 sq miles; pop. 66,508,008)

Andorra is a tiny state, situated high in the eastern PYRÉNÉES, between FRANCE and SPAIN. The state consists of deep valleys and high mountain peaks which reach heights of 3000 m/9843 ft. Although only 20 km/12 miles wide and 30 km/19 miles long, the spectacular scenery and climate attract many tourists. About 10 million visitors arrive each year, during the cold weather when heavy snowfalls makes for ideal skiing, or in summer when the weather is mild and sunny and the mountains are used for walking. Tourism and the duty-free trade are now Andorra's chief sources of income. Natives who are not involved in the tourist industry may raise sheep and cattle on the high pastures.

> *Quick facts:*
> Area : 457 sq km (170 sq miles)
> Population : 65,000
> Capital : Andorra-la-Vella
> Form of government : Co-principality
> Religion : RC
> Currency : Franc, Peseta
>
>

Andorra-la-Vella the capital of ANDORRA. (Pop. 22,000)

Andros the largest of the islands of the BAHAMAS (4144 sq km/1600 sq miles; pop. 8900)

Angara, River a river in the RUSSIAN FEDERATION flowing from Lake BAIKAL into the YENISEY River. (Length 1825 km/1135 miles)

Angel Falls a narrow band of water falling 979 m (3212 ft) from a high plateau in south-eastern VENEZUELA to form the world's highest waterfall.

Angkor the ruined ancient capital of the Khmer empire in CAMBODIA.

Anglesey an island off the north-western tip of WALES. (715 sq km/276 sq miles; pop. 68,000)

Angola is situated on the ATLANTIC coast of west central AFRICA, Angola lies about 10°S of the equator. It shares borders with CONGO, ZAÏRE, ZAMBIA and NAMIBIA. Its climate is tropical with temperatures constantly between 20°C and 25°C. The rainfall is heaviest in inland areas where there are vast equatorial forests. The country is also rich in minerals, however deposits of manganese, copper and phosphate are as yet unexploited. Diamonds are mined in the north-east and oil is produced near LUANDA. Oil production is the most important aspect of the economy, making up about 80% of export revenue. However, the Angolan economy has been severely damaged by the civil war of the 80s and early 90s.

> *Quick facts:*
> Area: 1,246,700 sq km (481,351 sq miles)
> Population : 10,844,000
> Capital : Luanda
> Other major cities : Huambo, Lobito, Benguela
> Form of government : People's Republic
> Religions : RC, Animism
> Currency : Kwanza
>
>

Anguilla an island in the LEEWARD ISLANDS group of the CARIBBEAN, now a self-governing British dependency. (91 sq km/35 sq miles; pop. 6500)

Angus a county in the east of Scotland, part of Tayside region until 1996. The administrative centre is Forfar. (2181 sq km/842 sq miles; pop. 218,150)

Anhui (Anhwei) a province of eastern CHINA. Its capital is HEFEI. (130,000 sq km/50,000 sq miles; pop. 56,180,000).

Anjou a former province of western FRANCE, in the valley of the River LOIRE.

Ankara the capital of TURKEY, in the eastern central part of Asian Turkey. (Pop. 2,559,000).

Annaba (Bone) a historic town and seaport on the

MEDITERRANEAN coast of ALGERIA. (Pop. 348,000)

Annam the old name used by the French for the central MIEN TRUNG region of VIETNAM.

Annapolis the capital of the state of MARYLAND, USA (Pop. 34,070)

Annapurna a mountain of the HIMALAYAS, situated in NEPAL. (8172 m/26,810 ft)

Anshan a steel-manufacturing city in LIAONING province, northern CHINA. (Pop. 1,370,000)

Antakya (Antioch) a city of ancient origins in southern TURKEY. (Pop. 124,000)

Antalya a port and resort on the MEDITERRANEAN coast of TURKEY. (Pop. 378,000)

Antananarivo the capital of MADAGASCAR, in the centre of the island. (Pop. 1,250,000)

Antarctic Circle Latitude 66° 32' south. At the southern winter solstice, the sun does not rise, nor does it set at the summer solstice, at this line, or in higher latitudes.

Antarctic Ocean (Southern Ocean) the waters that surround ANTARCTICA made up of the southern waters of the ATLANTIC, INDIAN and PACIFIC OCEANS.

Antarctica an ice-covered continent around the SOUTH POLE consisting of a plateau and mountain ranges reaching a height of 4500 m (15,000 ft). It is uninhabited apart from temporary staff at research stations. (14,000,000 sq km/5,100,000 sq miles)

Antigua and Barbuda is located on the eastern side of the Leeward Islands, a tiny state comprising three islands—Antigua, Barbuda and the uninhabited Redonda. Antigua's strategic position was recognized by the British in the 18th century when it was an important naval base, and later by the USA who built the island's airport during World War II to defend the CARIBBEAN and the PANAMA CANAL. The climate is tropical although its average rainfall of 100 mm (4 inches) makes it drier than most of the other islands of the WEST INDIES. On Antigua, many sandy beaches make it an ideal tourist destination, and tourism is the main industry. Barbuda is surrounded by coral reefs and the island is home to a wide range of wildlife.

Quick facts:
Area : 442 sq km (170 sq miles)
Population : 66,000
Capital : St John's
Form of government : Constitutional Monarchy
Religion : Christianity
 (mainly Anglicanism)
Currency : East
 Caribbean dollar

Antioch *see* **Antakya**

Antilles the major chain of islands in the CARIBBEAN SEA, divided into two groups: the Greater Antilles (which includes CUBA and PUERTO RICO) to the west; and the Lesser Antilles (including e.g. MARTINIQUE and BARBADOS) to the east.

Antrim a county and town in NORTHERN IRELAND. (2831 sq km/1093 sq miles; pop. county 642,267/ town 22,242)

Antwerp (Antwerpen, Anvers) the capital of the province of Antwerp and the main port of BELGIUM. (Pop. 465,000)

Anvers *see* **Antwerp**.

Anyang a city of ancient origins in the HENAN province of eastern CHINA. (Pop. 420,000)

Aomori a port on HONSHU Island, JAPAN. (Pop. 288,000)

Apennines (Appennino) the mountain range which forms the 'backbone' of ITALY. The highest peak is Monte Corno (2912 m/9554 ft).

Apia the capital of WESTERN SAMOA. (Pop. 37,000)

Appalachian Mountains a chain of mountains which stretches 2570 km (1600 miles) down eastern North AMERICA from CANADA to ALABAMA in the USA. The highest peak is Mount MITCHELL (2037 m/6684 ft).

Appennino *see* **Apennines**.

Apulia *see* **Puglia**.

Aqaba the only port in JORDAN, situated on the Gulf of Aqaba in the RED SEA. (Pop. 35,000)

Aquitaine a region and former kingdom of southwestern FRANCE.

Arabian Gulf *see* **Gulf, The.**

Arabian Sea a branch of the INDIAN OCEAN between INDIA and the Arabian Peninsula.

Arafura Sea a stretch of the PACIFIC OCEAN between NEW GUINEA and AUSTRALIA.

Aragon a region and former kingdom of north-east SPAIN.

Aral Sea a large, salty lake, to the east of the CASPIAN SEA, which lies on the border between UZBEKISTAN and KAZAKHSTAN. (64,750 sq km/ 25,000 sq miles)

Aran Islands (Oileáin Arann) three small islands Inishmore, Inishmaan and Inisheer off County GALWAY in the Republic of IRELAND. (44 sq km/18 sq miles; Pop. 4600)

Ararat, Mount (Büjük Agri Dagi) the mountain peak in eastern TURKEY where Noah's Ark is said to have come to rest after the Great Flood. (5165 m/17,000 ft)

Arauca, River a major tributary of the ORINOCO River which forms part of the border between COLOMBIA and VENEZUELA. (Length 1000 km/620 miles)

Archangel (Arkhangel'sk) a port on the DVINA Delta on the WHITE SEA in the RUSSIAN FEDERATION. (Pop. 421,000)

Arctic the regions that lie to the north of the ARCTIC CIRCLE.

Arctic Circle latitude 66° 32' north. The sun does not set above this line at the northern summer solstice, nor does it rise above this line at the winter solstice.

Arctic Ocean the ice-laden sea to the north of the ARCTIC CIRCLE. (14,100,000 sq km/5,440,000 sq miles)

Ardabil a town in IRAN, famous for its knotted carpets. (Pop. 311,000)

Ardennes a hilly and forested region straddling the borders of BELGIUM, LUXEMBOURG and FRANCE.

Arequipa a city and department of PERU. (Pop. city 620,000)

Argentina, the world's 8th largest country, stretches from the Tropic of Capricorn to Cape HORN on the southern tip of the South American continent. To the west a massive mountain chain, the ANDES, forms the border with CHILE. The climate ranges from warm temperate over the PAMPAS in the central region, to a more arid climate in the north and west, while in the extreme south conditions although also dry are much cooler. The vast fertile plains of the Pampas once provided Argentina with its main source of wealth, but as manufacturing industries were established in the early 20th century agriculture suffered badly and food exports were greatly reduced. A series of military regimes has resulted in an unstable economy which fails to provide reasonable living standards for the population.

Quick facts:
Area : 2,766,889 sq km (1,302,296 sq miles)
Population : 34,663,000
Capital : Buenos Aires
Other major cities : Cordoba, Rosaria,
  Mendoza, La Plata
Form of government : Federal Republic
Religion : RC
Currency : Peso

Århus a port and second largest city in DENMARK. (Pop. 271,000)

Arizona a state in the south-west of the USA. The capital is PHOENIX. (295,024 sq km/113,902 sq miles; pop. 3,832,000)

Arkansas a state in the south of the USA. The state capital is LITTLE ROCK. (137,539 sq km/53,104 sq miles; pop. 2,394,000)

Arkansas, River a tributary of the River MISSISSIPPI in the USA, flowing from the ROCKY MOUNTAINS through the states of KANSAS, OKLAHOMA and ARKANSAS. (Length 2335 km/1450 miles)

Arkhangel'sk *see* **Archangel**.

Armagh a county and city in NORTHERN IRELAND. (1254 sq km/484 sq miles; pop. county 118,820/ city 12,700)

Armenia (1) the former independent kingdom that straddled the borders of modern TURKEY, IRAN, GEORGIA, and AZERBAIJAN. **(2)** is the smallest republic of the former USSR and part of the former kingdom of Armenia which was divided between Turkey, Iran and the former USSR. It declared independence from the USSR in 1991. It is a landlocked Transcaucasian republic, and its neighbours are Turkey, Iran, Georgia and Azerbaijan. The country is very mountainous with many peaks over 3000 m (9900 ft). Agriculture is mixed in the lowland areas. The main crops grown are grain, sugar beet and potatoes, and livestock reared include cattle, pigs and sheep. Mining of copper, zinc and lead is important, and industrial development is increasing. Hydro-electricity is produced from stations on the river Razdan as it falls 1000 m (3281 ft) from Lake Sevan to its confluence with the River Araks.

Quick facts:
Area : 29,800 sq km (11,500 sq miles)
Population : 3,603,000
Capital : Yerevan
Other major city : Kumayri (Leninakan)
Form of government : Republic
Religion : Armenian Orthodox
Currency : Dram

Arnhem a town in the NETHERLANDS, scene of a battle in 1944 between British (and Polish) paratroops and the German army. (Pop. 308,600)

Arnhem Land an Aboriginal reserve in the NORTHERN TERRITORY of AUSTRALIA.

Arno, River the main river of TUSCANY in ITALY,

flowing westward through FLORENCE to PISA on the coast. (Length 245 km/152 miles)

Aruba a CARIBBEAN island off the coast of VENEZUELA, formerly one of the NETHERLANDS ANTILLES. The capital is ORANJESTAD. (193 sq km/75 sq miles; pop. 67,000; cur. Aruba guilder)

Arunachal Pradesh a state of northern INDIA, bordering TIBET. The capital is Itanagar. (Pop. 864,558)

Ascension Island a tiny volcanic island in the South ATLANTIC OCEAN, forming part of the ST HELENA DEPENDENCIES. (Pop. 1625)

Ashkhabad the capital of TURKMENISTAN. (Pop. 407,000)

Asia the largest continent, bounded by the ARCTIC, PACIFIC and INDIAN OCEANS, plus the MEDITERRANEAN and RED SEAS (43,600,000 sq km/16,800,000 sq miles; pop. 3,075,000,000). East Asia is taken to mean those countries to the north-east of BANGLADESH (e.g. CHINA); South Asia refers to the countries on the Indian subcontinent (e.g. INDIA); and South-East Asia includes those countries to the south-east of China, including the islands to the west of NEW GUINEA (e.g. INDONESIA).

Asmara (Asmera) the capital of ERITREA. (Pop. 430,000)

Assam a state in north-eastern INDIA. (78,438 sq km/30,284 sq miles; pop. 24,295,000)

Assisi a small town in UMBRIA, central ITALY, and birthplace of St Francis (1182-1226). (Pop. 25,000)

Assyria an empire of ancient MESOPOTAMIA founded around 3000 BC and reaching the height of its power in the 7th century BC. Its main cities were Assur and NINEVEH on the River TIGRIS, now in modern IRAQ.

Astrakhan a port near the CASPIAN SEA, situated on the delta of the River VOLGA in the RUSSIAN FEDERATION. (Pop. 512,000)

Asturias a region of northern SPAIN. The capital is OVIEDO. (Pop. 1,227,000)

Asunción the capital and the only major city of PARAGUAY. (Pop. 729,300)

Aswan a city in southern EGYPT by the River NILE. The Aswan High Dam, completed 1971, is 13 km (8 miles) to the south. (Pop. 220,000)

Atacama Desert an extremely dry desert lying mainly in northern CHILE.

Athabasca, River a river in CANADA which flows north from the ROCKY MOUNTAINS to Lake Athabasca. (Length 1231 km/765 miles)

Athens (Athinai) the historic capital, and the principal city, of GREECE. (Pop. city 885,700/metropolitan area 3,097,000)

Athos a group of monasteries on Mount Athos (2033 m/6670 ft) on a peninsula in northern GREECE.

Atlanta the capital and largest city of the state of GEORGIA in the USA (Pop. 394,850)

Atlantic Ocean the second largest ocean, lying between North and South AMERICA, EUROPE and AFRICA. (82,200,000 sq km/31,700,000 sq miles)

Atlas Mountains a series of mountain chains stretching across North AFRICA from MOROCCO to TUNISIA.

Auckland the largest city and chief port of NEW ZEALAND, on North Island. (Pop. 896,600)

Augsburg a historic city in BAVARIA, GERMANY. (Pop. 265,000)

Augusta (1) a city and river port on the Savannah River in GEORGIA, USA (Pop. city 46,000/metropolitan area 368,300). **(2)** the state capital of MAINE, USA (Pop. 22,000)

Auschwitz a small town in POLAND and site of the biggest of the Nazi concentration camps 1940-45. (Pop. 35,600)

Austin the capital city of the state of TEXAS, USA (Pop. city 492,320/metropolitan area 645,400)

Australasia a general term for AUSTRALIA, NEW ZEALAND and neighbouring islands.

Australia, the world's smallest continental landmass, is a vast and sparsely populated island state in the southern hemisphere. The most mountainous region is the GREAT DIVIDING RANGE which runs down the entire east coast. Because of its great size, Australia's climates range from tropical monsoon to cool temperate and also large areas of desert. Central and south QUEENSLAND are subtropical while north and central NEW SOUTH WALES are warm temperate. Much of Australia's wealth comes from agriculture, with huge sheep and cattle stations extending over large parts of the interior. These have helped maintain Australia's position as the world's leading producer of wool. Cereal growing is dominated by wheat. Mineral extraction is also very important.

Quick facts:
Area : 7,300,848 sq km (2,966,150 sq miles)
Population : 18,114,000
Capital : Canberra
Other major cities : Adelaide, Brisbane,
  Melbourne, Perth, Sydney
Form of government : Federal Parliamentary State
Religion : Christianity
Currency : Australian dollar

Australian Capital Territory the small region which surrounds CANBERRA, the capital of AUSTRALIA. (2432 sq km/939 sq miles; pop. 240,000)

Austria is a landlocked country in central EUROPE and is surrounded by seven nations. The wall of mountains which runs across the centre of the country dominates the scenery. In the warm summers tourists come to walk in the forests and mountains and in the cold winters skiers come to the mountains which now boast over 50 ski resorts. Agriculture in Austria is based on small farms, many of which are run by single families. Dairy products, beef and lamb from the hill farms contribute to exports. More then 37% of Austria is covered in forest, resulting in the paper-making industry near GRAZ. Unemployment is very low in Austria and its low strike record has attracted multinational companies in recent years. Attachment to local customs is still strong and in rural areas men still wear lederhosen and women the traditional dirndl skirt on feast days and holidays.

Quick facts:
Area : 83,855 sq km (32,367 sq miles)
Population : 8,015,000
Capital : Vienna (Wien)
Other major cities : Graz, Linz, Salzburg
Form of government : Federal Republic
Religion : RC
Currency : Schilling

Auvergne a mountainous region of central FRANCE.

Avignon a historic city on the River RHONE in southern FRANCE, the seat of the Pope, 1309–77. (Pop. 181,000)

Avila a town and province in the mountainous central region of SPAIN, famous as the birthplace of St Teresa (1515-82). (Pop. Town 41,800)

Avon a former county in the west of ENGLAND which was formed in 1974 from parts of Somerset and Gloucestershire and the city of Bristol. It ceased to exist in 1996 following local government restructuring.

Axios, River a river flowing through the BALKANS to GREECE and the AEGEAN SEA. (Length: 388km/241miles)

Axum an important historic town in ETHIOPIA, once a royal capital where the Queen of Sheba is said to have ruled. (Pop. 20,000)

Ayers Rock a huge rock, sacred to the Aborigines, rising sharply out of the plains in the NORTHERN TERRITORY of AUSTRALIA. (348 m/1142 ft)

Ayrshire a county in south-west SCOTLAND, formerly part of Strathclyde region. (.

Ayutthaya a town with the extensive ruins of the city that was the capital of THAILAND from 1350 to 1767. (Pop. 113,300)

Azarbaijan a region of northern IRAN. Its population shares the same language as the people of neighbouring republic of AZERBAIJAN. (Pop. 4,613,000)

Azerbaijan, a republic of the former USSR, declared itself independent in 1991. It is situated on the south-west coast of the CASPIAN SEA and shares borders with IRAN, ARMENIA, GEORGIA and the RUSSIAN FEDERATION. The Araks river separates Azerbaijan from the region known as AZARBAIJAN in northern Iran. The country is semi-arid, and 70% of the land is irrigated for the production of cotton, wheat, maize, potatoes, tobacco, tea and citrus fruits. It has rich mineral deposits of oil, natural gas, iron and aluminium. The most important mineral is oil, which is found in the BAKU area from where it is piped to Batumi on the Black Sea. There are steel, synthetic rubber and aluminium works at Sumgait just north of the capital Baku.

Quick facts:
Area : 86,600 sq km (33,400 sq miles)
Population : 7,559,000
Capital : Baku
Other major cities : Kirovabad, Sumgait
Form of government : Republic
Religions : Shia Islam, Sunni Islam, Russian
  Orthodox
Currency : Manat

Azores three groups of small islands in the North ATLANTIC OCEAN, belonging to PORTUGAL. The capital is PONTA DELGADA. (2335 sq km/901 sq miles; pop. 336,100)

Baalbek a ruined city dating back to Phoenician and Roman times in the BEQA'A VALLEY in LEBANON.

Babylon one of the great cities of ancient MESOPOTAMIA, situated in modern IRAQ.

Baden-Baden a famous spa town in south-west GERMANY dating from Roman times. (Pop. 50,000)

Baden-Württemburg the southern state of GERMANY bordering FRANCE and SWITZERLAND. (35,751 sq km/13,803 sq miles; pop. 9,911,000)

Baffin Bay a huge bay within the ARCTIC CIRCLE between BAFFIN ISLAND in CANADA and GREENLAND.

Baffin Island a large, mainly ice-bound island in north-east CANADA. (507,451 sq km/195,927 sq miles)

Baghdad the capital of IRAQ, in the centre of the country, on the River TIGRIS. (Pop. 3,841,280)

Bahamas, The consist of an archipelago of 700 islands located in the ATLANTIC OCEAN off the south-east coast of FLORIDA. The largest island is ANDROS (4144 sq km/1600 sq miles), and the two most populated are Grand Bahama and New Providence where the capital NASSAU lies. Winters in the Bahamas are mild and summers warm. Most rain falls in May, June, September and October, and thunderstorms are frequent in summer. The islands have few natural resources, and for many years fishing and small-scale farming was the only way to make a living. Now, however, tourism, which employs over two-thirds of the workforce, is the most important industry and has been developed on a vast scale. About three million tourists, mainly from North AMERICA, visit the Bahamas each year.

Quick facts:
Area : 13,939 sq km (5382 sq miles)
Population : 277,000
Capital : Nassau
Other important city : Freeport
Form of government : Constitutional Monarchy
Religion : Christianity
Currency : Bahamian dollar

Bahrain is a Gulf State comprising 33 low-lying islands situated between the QATAR peninsula and the mainland of SAUDI ARABIA. Bahrain Island is the largest, and a causeway linking it to Saudi Arabia was opened in 1986. The highest point in the state is only 122.4 m (402 ft) above sea level. The climate is pleasantly warm between December and March, but very hot from June to November.

Most of Bahrain is sandy and too saline to support crops but drainage schemes are now used to reduce salinity and fertile soil is imported from other islands. Oil was discovered in 1931 and revenues from oil now account for about 75% of the country's total revenue. Bahrain is being developed as a major manufacturing state, the first important enterprise being aluminum smelting. Traditional industries include pearl fishing, boat building, weaving and pottery.

Quick facts:
Area : 691 sq km (267 sq miles)
Population : 539,000
Capital : Manama
Form of government : Monarchy (Emirate)
Religions : Shia Islam, Sunni Islam
Currency : Bahraini dollar

Baikal, Lake the world's deepest freshwater lake, and the largest by volume, situated in south-east SIBERIA in the RUSSIAN FEDERATION. (31,500 sq km/ 12,150 sq miles)

Baile Atha Cliath *see* **Dublin**.

Baja California a huge 1300 km (800 mile) long peninsula belonging to MEXICO which stretches south from CALIFORNIA in the USA into the PACIFIC OCEAN. (Pop. 1,736,000)

Bakhtaran formerly called Kermanshah, a large city in IRAN on the old trading routes between TEHRAN and BAGHDAD. (Pop. 624,000)

Baku (Baky) a port on the CASPIAN SEA and the capital of the republic of AZERBAIJAN. (Pop. 1,149,000)

Balaklava *see* **Sevastopol**.

Balaton, Lake a lake in western HUNGARY, famous as a tourist resort. (601 sq km/232 sq miles)

Bâle *see* **Basle**.

Balearic Islands a group of islands in the western MEDITERRANEAN SEA belonging to SPAIN and famous as tourist resorts. The main islands are MAJORCA (Mallorca), MINORCA (Menorca), IBIZA (Iviza), Formentera and Cabrera. (5000 sq km/ 1930 sq miles; pop. 755,000)

Bali a small island lying off the eastern tip of JAVA, distinguished by being the only island in INDONESIA to have preserved a predominantly Hindu culture intact. The main town and capital is DENPASAR. (5591 sq km/2159 sq miles; pop. 2,787,000)

Balkans the south-eastern corner of EUROPE, a broad, mountainous peninsula bordered by the ADRIATIC SEA, the IONIAN SEA, the AEGEAN SEA and the BLACK SEA. ALBANIA, BULGARIA, GREECE, ROMANIA, SLOVENIA, BOSNIA & HERZEGOVINA, CROATIA, other territories of the former YUGOSLAVIA and the European part of TURKEY are all in the Balkans.

Balkhash, Lake a massive lake in KAZAKHSTAN, near the border with CHINA. (22,000 sq km/8500 sq miles)

Ballarat a historic gold-mining town in VICTORIA, AUSTRALIA, the scene of the 1854 rebellion known as the Eureka Stockade. (Pop. 75,200)

Baltic Sea a shallow sea in northern EUROPE, completely surrounded by land masses except for the narrow straits that connect it to the NORTH SEA. It's coastline is 8000 km/5000 miles in length and stretches from SCANDINAVIA, to POLAND, the BALTIC STATES and RUSSIA.

Baltic States a collective name given to ESTONIA, LATVIA and LITHUANIA.

Baltimore the largest city in the state of MARYLAND, USA (Pop. city 726,100/metropolitan area 2,434,000)

Baluchistan a province of south-western PAKISTAN, bordering IRAN and AFGHANISTAN. (Pop. 4,908,000)

Bamako the capital of MALI. (Pop. 658,300)

Bandar Abbas a port in southern IRAN on the Strait of HORMUZ, at the neck of THE GULF. (Pop. 250,000)

Bandar Seri Begawan the capital of BRUNEI. (Pop. 56,300)

Banda Sea a part of the PACIFIC OCEAN, in eastern INDONESIA.

Bandung a large inland city in western JAVA, INDONESIA. (Pop. 1,462,700)

Banffshire a former county in north-east Scotland.

Bangalore a large industrial city in central southern INDIA. (Pop. 4,087,000)

Bangkok (**Krung Thep**) the capital of THAILAND, on the River CHAO PHRAYA. (Pop. 5,876,000)

Bangladesh was formerly the Eastern Province of PAKISTAN. It is bounded almost entirely by INDIA and to the south by the Bay of BENGAL. The country is extremely flat and is virtually a huge delta formed by the GANGES, BRAHMAPUTRA and Meghna rivers. The country is subject to devastating floods and cyclones which sweep in from the Bay of Bengal. Most villages are built on mud platforms to keep them above water. The climate is tropical monsoon with heat, extreme humidity and heavy

rainfall in the monsoon season. The short winter season is mild and dry. The combination of rainfall, sun and silt from the rivers makes the land productive, and it is often possible to grow three crops a year. Bangladesh produces about 70% of the world's jute and the production of jute-related products is a principal industry.

Quick facts:
Area : 143,998 sq km (55,598 sq miles)
Population : 118,342,000
Capital : Dacca (Dhaka)
Other major cities : Chittagong, Khulna
Form of government : Republic
Religion : Sunni Islam
Currency : Taka

Bangui the capital of the CENTRAL AFRICAN REPUBLIC, in the south of the country. (Pop. 597,000)

Bangweulu, Lake a large lake in northern ZAMBIA. (9800 sq km/3784 sq miles)

Banja Luka a city of ancient origins on the Vrbas River in north-west BOSNIA & HERZEGOVINA. (Pop. 195,000)

Banjarmasin a port on the southern coast of KALIMANTAN, INDONESIA. (Pop. 444,000)

Banjul the capital of the GAMBIA. Formerly called Bathurst. (Pop. urban area 150,000)

Barbados is the most easterly island of the WEST INDIES and lies well outside the group of islands which makes up the Lesser ANTILLES. Most of the island is low-lying and only in the north does it rise to over 340 m/1116 ft at Mount Hillaby. The climate is tropical, but the cooling effect of the North-east Trade winds prevents the temperatures rising above 30°C (86°F). There are only two seasons, the dry and the wet, when rainfall is very heavy. At one time the economy depended almost exclusively on the production of sugar and its by-products molasses and rum, and although the industry is now declining, sugar is still the principal export. Tourism has now taken over as the main industry and it employs approximately 40% of the island's labour force. The island is surrounded by pink and white sandy beaches and coral reefs which are visited by almost 400 000 tourists each year.

Quick facts:
Area : 430 sq km (166 sq miles)
Population : 264,000
Capital : Bridgetown
Form of government : Constitutional
  Monarchy
Religions : Anglicanism, Methodism
Currency : Barbados dollar

Barbuda *see* **Antigua and Barbuda**.

Barcelona the second largest city in SPAIN, and the name of the surrounding province. It is a major port on the MEDITERRANEAN SEA. (Pop. city 1,653,200)

Barents Sea a part of the ARCTIC OCEAN to the north of NORWAY.

Bari a major port on the ADRIATIC coast of ITALY. (Pop. 342,000)

Baroda *see* **Vadodara**.

Barossa Valley a wine-producing region in SOUTH AUSTRALIA, 50 km (30 miles) north of ADELAIDE.

Barquisimeto an industrial city in western VENEZUELA. (Pop. 745,000)

Barranquilla the largest port on the CARIBBEAN coast of COLOMBIA. (Pop. 1,049,000)

Basel *see* **Basle**.

Bashkiria (Baskir Republic) a republic of the RUSSIAN FEDERATION, in the southern URALS. The capital is UFA. (143,500 sq km/55,400 sq miles; pop. 3,860,000)

Basle (Basel, Bâle) a city in northern SWITZERLAND and the name of the surrounding canton. (Pop. city 200,000)

Basque Region an area straddling the border of SPAIN and FRANCE on the Atlantic coast. The Spanish Basque Region was created an autonomous region of northern Spain in 1979 (7300 sq km/2818 sq miles; pop. 2,176,800). The French Basque Region comprises the department of Pyrenees-Atlantique (7633 sq km/2947 sq miles; pop. 556,000).

Basra the second city of IRAQ, and its main port. (Pop. 850,000)

Bassein a trading city on the delta of the IRRAWADDY River in MYANMAR (Burma). (Pop. 355,600)

Basseterre the capital of ST KITTS AND NEVIS. (Pop. 15,000)

Basse Terre the capital of the French island of GUADELOUPE, situated on the island called Basse Terre. (Pop. town 14,000/island 141,000)

Bass Strait the stretch of water spanning the 290 km (180 miles) which separate the mainland of AUSTRALIA from TASMANIA.

Bath a beautifully preserved spa town in the south-west ENGLAND. (Pop.79,900)

Bathurst *see* **Banjul**.

Baton Rouge the state capital of LOUISIANA, USA, situated on the MISSISSIPPI River. (Pop. city 224,700/metropolitan area 538,000)

Bavaria (Bayern) the largest state in GERMANY. (70,553 sq km/27,241 sq miles; pop. 11,552,000)

Bayern *see* **Bavaria**.

Bayeux a market town in NORMANDY, FRANCE, and the home of the huge 11th-century Bayeux tapestry depicting the Norman conquest of ENGLAND. (Pop. 15,300)

Bayonne the capital of the French BASQUE region. (Pop. 129,730)

Bayreuth a town in BAVARIA, GERMANY, famous for the theater built by the composer Richard Wagner (1813-83), where his operas are still staged every summer. (Pop. 71,800)

Beaufort Sea a part of the ARCTIC OCEAN to the north of North AMERICA.

Beaujolais a famous wine-producing region of FRANCE situated on the River SAONE between LYONS and Macon.

Bechuanaland the former name of BOTSWANA (until 1966).

Bedfordshire a county in central southern ENGLAND; the county town is Bedford. (1235 sq km/ 477 sq miles; pop. 524,000)

Beijing (Peking) the capital of CHINA, situated in the north-east of the country. (Pop. city 6,690,000; urban area 10,819,000)

Beirut (Beyrouth) the capital and main port of LEBANON. (Pop. 1,500,000)

Belarus (Belorussia, Byelorussia), a republic of the former USSR, declared itself independent in 1991. It borders POLAND to the west, UKRAINE to the south, LATVIA and LITHUANIA to the north, and the RUSSIAN FEDERATION to the east. The country consists mainly of a low-lying plain, and forests cover approximately one third of the country. The climate is continental with long severe winters and short warm summers. Although the economy is overwhelmingly based on industry, including oil refining, food processing, woodworking, chemicals, textiles and machinery, output has gradually declined since 1991 and problems persist in the supply of raw materials from other republics that previously formed parts of the USSR. Agriculture, although seriously affected by contamination from the CHERNOBYL nuclear accident of 1986, accounts for approximately 20% of employment, the main crops being flax, potatoes and hemp. The main livestock raised are cattle and pigs. Extensive forest areas also contribute in the supply raw materials for woodwork and paper-making.

Quick facts:
Area : 207,600 sq km (80,150 sq miles)
Population : 10,355,000
Capital : Minsk
Other major cities : Gomel, Mogilev, Vitebsk
Form of government : Republic
Religions : Russian Orthodox, RC
Currency : Rouble

Belau a republic consisting of a group of islands in the western PACIFIC formerly known as Palau. It has an agreement of free association with the USA. Copper is the chief export and fishing rights are sold to foreign fleets. The main language is English. The capital is KOROR. (494 sq km/191 sq miles; pop. 15,100; cur. US dollar)

Belém a major port of BRAZIL situated to the north of the mouth of the River AMAZON. (Pop. 1,246,000)

Belfast the capital and by far the largest city of NORTHERN IRELAND. (Pop. 297,000)

Belgium is a relatively small country in north-west EUROPE with a short coastline on the NORTH SEA. The MEUSE river divides Belgium into two distinct geographical regions. To the north of the river the land slopes continuously for 150 km/93 miles until it reaches the North Sea where the coastlands are flat and grassy. To the south of the river is the forested plateau area of the ARDENNES. Between these two regions lies the Meuse valley. Belgium is a densely populated industrial country with few natural resources. Agriculture is based on livestock production but employs only 3% of the workforce. The metal-working industry, originally based on the small mineral deposits in the Ardennes, is the most important industry, and in the northern cities new textile industries are producing carpets and clothing. Nearly all raw materials are now imported through the main port of ANTWERP.

Quick facts:
Area : 30,519 sq km (11,783 sq miles)
Population : 10,100,630
Capital : Brussels
Other major cities : Antwerp, Charleroi, Ghent, Liege
Form of government : Constitutional Monarchy
Religion : RC
Currency : Belgian franc

Belgrade (Beograd) the capital of SERBIA, on the confluence of the Rivers DANUBE and Sava. (Pop. 1,137,100)

Belize is a small Central American country on the south-east of the YUCATAN Peninsula in the CARIBBEAN SEA. Its coastline on the Gulf of HONDURAS is approached through some 550 km/342 miles of coral reefs and keys (cayo). The coastal area and north of the country are low-lying and swampy with dense forests inland. In the south the Maya Mountains rise to 1100 m/3609 ft. The subtropical climate is warm and humid and the trade winds bring cooling sea breezes. Rainfall is heavy, particularly in the south, and hurricanes may occur in summer. The dense forests which cover most of the country provide valuable hardwoods such as mahogany. Most of the population make a living from forestry, fishing or agriculture. The main crops grown for export are sugar cane, citrus fruits (mainly grapefruit), bananas and coconuts. Despite this approximately only 5% of Belize's total land area is cultivated and industry is very underdeveloped, causing many people to emigrate to find work.

Quick facts:
Area : 22,965 sq km (8867 sq miles)
Population : 205,000
Capital : Belmopan
Other major city : Belize City
Form of government : Constitutional Monarchy
Religion : RC
Currency : Belize dollar

Bellinghausen Sea a part of the PACIFIC OCEAN off ANTARCTICA, due south of South AMERICA.

Belmopan the capital of BELIZE. (Pop. 3740)

Belo Horizonte an industrial city, and the third largest city of BRAZIL, in the south-east of the country. (Pop. 2,049,000)

Belorussia *see* **Belarus**.

Belostock *see* **Bialystok**.

Benares *see* **Varanasi**.

Bengal a former Indian state which was divided at the partition of INDIA in 1947 into two parts: WEST BENGAL in India, and East Pakistan (now BANGLADESH).

Bengal, Bay of the massive bay occupying the broad sweep of the INDIAN OCEAN between INDIA and MYANMAR, to the south of BANGLADESH.

Benghazi a major port at the eastern end of the Gulf of SIRTE in LIBYA. (Pop. 485,000)

Benidorm one of the most popular MEDITERRANEAN seaside resorts of SPAIN. (Pop. 74,900)

Benin on the southern coast of West AFRICA is an ice cream cone-shaped country with a very short coastline on the Bight of Benin. The coastal area has white sandy beaches backed by lagoons and low-lying fertile lands. In the north-west the Atakora Mountains are grassy plateaux which are deeply cut into steep forested valleys. The climate in the north is tropical and in the south equatorial. There are nine rainy months each year so crops rarely fail. Farming is predominantly subsistence, with yams, cassava, maize, rice, groundnuts and vegetables forming most of the produce. The country is very poor, although since the late '80s economic reforms have been towards a market economy and Western financial aid has been sought. The main exports are palm oil, palm kernels, and cotton. Tourism is now being developed but as yet facilities for this are few except in some coastal towns.

Quick facts:
Area : 112,622 sq km (43,483 sq miles)
Population : 5,160,000
Capital : Porto-Novo
Other major city : Cotonou
Form of government : Republic
Religions : Animism, RC, Sunni Islam, Christian
Currency : CFA Franc

Ben Nevis *see* **Grampian Mountains**.

Benue, River a river which flows through CAMEROON and NIGERIA to the Gulf of GUINEA. (Length 1390 km/865 miles)

Benxi an industrial city in LIAONING province in northern CHINA. (Pop. 767,000)

Beograd *see* **Belgrade**.

Beqa'a a long, fertile valley running north to south in LEBANON, between the Lebanon and Anti-Lebanon Mountains.

Bergamo a historic and industrial city in northern ITALY. (Pop. 115,000)

Bergen (1) an old port in south-west NORWAY, and now that country's second largest city. (Pop. 195,000). **(2)** *See* **Mons**.

Bering Sea a part of the PACIFIC OCEAN between ALASKA and eastern RUSSIAN FEDERATION.

Bering Strait the stretch of sea, 88 km (55 miles) wide, that separates the RUSSIAN FEDERATION from ALASKA in the USA.

Berkeley a city in north CALIFORNIA in the USA situated on SAN FRANCISCO Bay. (Pop. 101,100)

Berkshire a county of central southern ENGLAND; the county town is Reading. (1256 sq km/485 sq miles; pop. 796,000)

Berlin the capital of GERMANY, situated in the north of the country on the River Spree. Until 1990 it was divided in two by the infamous Berlin Wall. (Pop. 3,454,000)

Bermuda consists of a group of 150 small islands in the western ATLANTIC OCEAN. It lies about 920 km/572 miles east of Cape HATTERAS on the coast of the USA. The hilly limestone islands are the caps of ancient volcanoes rising from the sea-bed. The main island, Great Bermuda, is linked to the other islands by bridges and causeways. The climate is warm and humid with rain spread evenly throughout the year. Bermuda's chief agricultural products are fresh vegetables, bananas and citrus fruit. Many foreign banks and financial institutions operate from the island, taking advantage of the lenient tax laws. Its proximity to the USA and the pleasant climate have led to a flourishing tourist industry.

Quick facts:
Area : 54 sq km (21 sq miles)
Population : 61,000
Capital : Hamilton
Form of government : Colony under British administration
Religion : Protestantism
Currency : Bermuda dollar

Berne (Bern) the historic capital of SWITZERLAND, and also the name of the surrounding canton. (Pop. city 135,600)

Berwickshire *see* **Borders**.

Besançon a town of ancient origins in the JURA region of eastern FRANCE. (Pop. 123,000)

Bethlehem a town in the WEST BANK area of ISRAEL, celebrated by Christians as the birthplace of Jesus Christ. (Pop. 30,000)

Béthune an industrial town in north-eastern FRANCE. (Pop. 259,700)

Beuten *see* **Bytom**.

Beyrouth *see* **Beirut**.

Bhopal an industrial city in central INDIA. It was the scene of a massive industrial accident in 1984 in which some 3000 people were killed by gas leaking from a major chemical plant. (Pop. 1,063,000)

Bhutan is surrounded by INDIA to the south and CHINA to the north. It rises from foothills overlooking the BRAHMAPUTRA river to the southern slopes of the HIMALAYAS. The Himalayas, which rise to over 7500 m/24,608 ft in Bhutan, make up most of the country. The climate is hot and wet on the plains but temperatures drop progressively with altitude, resulting in glaciers and permanent snow cover in the north. The valleys in the centre of the country are wide and fertile and about 95% of the workforce are farmers. Yaks reared on the high pasture land provide milk, cheese and meat. Rice is grown on the lowest ground. Vast areas of the country still remain forested as there is little demand for new farmland. Bhutan is one of the world's poorest and least developed countries; it has little contact with the rest of the world and the number of visitors is limited to 1500 each year.

Quick facts:
Area : 46,500 sq km (17,954 sq miles)
Population : 1,442,000
Capital : Thimpu
Form of government : Constitutional Monarchy
Religion : Buddhism, Hinduism
Currency : Ngultrum

Biafra *see* **Iboland**.

Bialystok (Belostock) an industrial city in north-east POLAND, producing primarily textiles. (Pop. 272,000)

Bianco, Monte *see* **Blanc, Mont**.

Bielefeld an industrial city in western GERMANY. (Pop. 325,000)

Bielsko-Biala (Bielitz) an industrial city in southern POLAND. (Pop. 184,000)

Bihar a state in north-east INDIA. The capital is PATNA. (Pop. 86m,339,000)

Bikini an atoll in the MARSHALL ISLANDS famous as the site of US nuclear weapons tests between 1946 and 1962.

Bilbao a port and industrial city in the BASQUE region of northern SPAIN. (Pop. 372,200)

Bioko an island in the Gulf of GUINEA, formerly called Fernando Póo, governed by EQUATORIAL GUINEA. (2017 sq km/780 sq miles; pop. 57,000)

Birmingham (1) the main city of the industrial WEST MIDLANDS and the second largest city in the UK (Pop. city 1,008,400). **(2)** the largest city in the state of ALABAMA, USA. (Pop. city 264,984/ metropolitan area 895,200)

Biscay, Bay of the broad bay, notorious for its rough weather, formed by the ATLANTIC OCEAN between northern SPAIN and BRITTANY in north-west FRANCE.

Bishkek, formerly Frunze, the capital of KYRGYZSTAN. (Pop. 616,000)

Bismarck the state capital of NORTH DAKOTA, USA. (Pop. city 51,300/metropolitan area 86,100)

Bismarck Sea a branch of the PACIFIC OCEAN to the north of PAPUA NEW GUINEA.

Bissau (Bissão) a port and the capital of GUINEA-BISSAU. (Pop. 109,500)

Black Country the industrial area of the British MIDLANDS around BIRMINGHAM. The name was derived from effects of the heavy pollution caused by heavy industry, however, anti-pollution legislation has considerably improved the area.

Black Forest (Schwarzwald) an extensive area of mountainous pine forests in south-west GERMANY (4660 sq km/1800 sq miles).

Black Hills a range of hills rising to 2207 m (7242 ft) on the border between the states of SOUTH DAKOTA and WYOMING in the USA.

Blackpool the largest seaside holiday resort in the UK, in LANCASHIRE. (Pop. 156,000)

Black Sea a sea lying between south-east EUROPE and western ASIA; it is surrounded by land except for the BOSPHORUS channel, leading to the MEDITERRANEAN SEA.

Blanc, Mont (Monte Bianco) the highest mountain in Western EUROPE, on the border between FRANCE and ITALY, just to the south of CHAMONIX. (4808 m/ 15,770 ft)

Blantyre the largest city in MALAWI. (Pop. 399,000)

Bloemfontein the judicial capital of SOUTH AFRICA, and the capital of the ORANGE FREE STATE. (Pop. 300,000)

Blue Mountains (1) a range of mountains rising to 1100 m (3609 ft) in NEW SOUTH WALES in AUSTRALIA, some 65 km (40 miles) from SYDNEY. **(2)** the mountains in eastern JAMAICA rising to 2256 m (7402 ft) at Blue Mountain Peak. The region has given its name to the high quality coffee produced there.

Bochum an industrial city in the RUHR region of western GERMANY. (Pop. 401,000)

Bodensee *see* **Constance, Lake**.

Bodh Gaya a small town in eastern INDIA, which is the site of Buddhist religion's most revered shrine. Gautama, the Lord Buddha, achieved enlightenment here in about 500BC. (Pop. 15,700)

Bodrum a port on the south-eastern MEDITERRANEAN coast of TURKEY. Known in the ancient world as Halicarnassus, its Mausoleum (since destroyed) was one of the Seven Wonders of the Ancient World. (Pop. 13,090)

Bogotá the capital of COLOMBIA, set on a plateau of the eastern ANDES in the centre of the country. (Pop. 5,789,000)

Bohemia formerly an independent kingdom (9th to 13th centuries), now a western region of the CZECH REPUBLIC which includes the capital PRAGUE. The name is derived from the earliest known inhabitants of the region, the Celtic Boii. In the mistaken belief that Gypsies came from Bohemia, the term Bohemian came to be applied to artists and writers who led rather unconventional lifestyles.

Bohol one of the VISAYAN ISLANDS in the central area of the PHILIPPINES. (3862 sq km/1491 sq miles; pop. 759,370)

Boise the state capital of IDAHO, USA. (Pop.135,500)

Bolivia is a landlocked republic of Central South AMERICA through which the great mountain range of the ANDES runs. It is in the Andes that the highest navigable lake in the world, Lake TITICACA, is found. On the undulating depression south of the lake, the Altiplano, is the highest capital city in the world, LA PAZ. To the east and north-east of the mountains is a huge area of lowland containing tropical rainforests (the Llanos) and wooded savanna (the Chaco). The north-east has a heavy rainfall while in the south-west it is negligible. Temperatures vary with altitude from extremely cold on the summits to cool on the Altiplano, where at least half of the population lives. Although rich in natural resources, e.g. oil, tin, Bolivia remains a poor country because of lack of

funds for their extraction, lack of investment and political instability. Agriculture produces foodstuffs, sugar cane and cotton for export, and increased production of coca, from which cocaine is derived, has resulted in an illicit economy.

Quick facts:
Area : 1,098,581 sq km (424,164 sq miles)
Population : 7,715,000
Capital : La Paz (administrative capital),
 Sucre (legal capital)
Other major city : Cochabamba
Form of government : Republic
Religion : RC
Currency : Boliviano

Bologna the capital of the EMILIA ROMAGNA region in north-eastern ITALY. (Pop. 401,000)

Bolton a textile manufacturing town in the county of LANCASHIRE, ENGLAND. (Pop. 210,000)

Bombay the largest city in INDIA, a major port and the capital of MAHARASHTRA state in central western INDIA. It is now India's most important industrial city. (Pop. 12,572,000)

Bonaire a Caribbean island off the coast of VENEZUELA formerly administered by the Dutch and still a part of the NETHERLAND ANTILLES (288 sq km/111 sq miles; pop. 9700)

Bondi Beach a famous surfing beach in the suburbs of SYDNEY, AUSTRALIA.

Bone *see* **Annaba**.

Bonin Islands a group of small volcanic islands in the PACIFIC OCEAN belonging to JAPAN. (Pop. 2430)

Bonn the capital of former West Germany, which will remain the administrative centre of GERMANY until the government moves to BERLIN. (Pop. 300,000)

Bophuthatswana was a former 'homeland' OF SOUTH AFRICA that consisted of seven separate territories. Under the new South African Constitution implemented in April 1994, 'homelands' ceased to exist and were incorporated into the nine newly delineated provinces. (40,330 sq km/15,571 sq miles)

Bordeaux a major port on the GIRONDE estuary in south-western FRANCE. The region is famous for its wines. (Pop. 650,125)

Borders an administrative region of southern SCOTLAND, created in 1975 out of the former counties of

Berwickshire, Peeblesshire, Roxburghshire, Selkirkshire and part of Midlothian. (4662 sq km/ 1800 sq miles; pop. 101,000)

Borkum *see* **Friesian Islands**.

Borneo one of the largest islands in the world, now divided between three countries. Most of the island is known as KALIMANTAN, a part of INDONESIA. The northern coast is divided into the two states of SARAWAK and SABAH, which are part of MALAYSIA, and the small independent Sultanate of BRUNEI. (751,900 sq km/290,320 sq miles)

Borobodur the great Buddhist temple near YOGYAKARTA in southern central JAVA, INDONESIA, dating from about AD750.

Bosnia & Herzegovina, a republic of former YUGO-SLAVIA, was formally recognized as an independent state in March 1992. It is a very mountainous country and includes part of the Dinaric Alps, which are densely forested and deeply cut by rivers flowing northwards to join the Sava river. Half the country is forested, and timber is an important product of the northern areas. One quarter of the land is cultivated, and corn, wheat and flax are the principal products of the north. In the south, tobacco, cotton, fruits and grapes are the main products. Bosnia & Herzegovina has large deposits of lignite, iron ore and bauxite, but there is little industrialization. Despite the natural resources the economy has been devastated by civil war which began in 1991 following the secession of CROATIA and SLOVENIA from the former Yugoslavia. Dispute over control of Bosnia and Herzegovina continued, leading to UN intervention in an attempt to devise a territorial plan acceptable to all factions. A peace agreement signed in late 1995 has resulted in the division of the country into two self-governing provinces. The population of the state was significantly diminished when refugees from the civil war fled between 1992 and 1993.

Quick facts:
Area : 51,129 sq km (19,741 sq miles)
Population : 3,500,000
Capital : Sarajevo
Other major cities : Banja Luka, Mostar, Tuzla
Form of government : Republic
Religions : Eastern Orthodox, Sunni Islam, RC
Currency : Dinar

Bosphorus the narrow strip of water, some 29 km (18 miles) long and no more than 4 km (2.5 miles) wide, which provides the navigable link between the MEDITERRANEAN SEA and the BLACK SEA by way of the Sea of MARMARA. The Bosphorus separates the European part of TURKEY from its Asian part.

Boston an ATLANTIC port and the state capital of MASSACHUSETTS, USA. (Pop. city 551,700/metropolitan area 5,439,000)

Botany Bay a bay now in the suburbs of SYDNEY, AUSTRALIA, discovered by Captain James Cook in 1770.

Bothnia, Gulf of the most northerly arm of the BALTIC SEA, bordered by FINLAND and SWEDEN.

Botswana is a landlocked republic in southern AFRICA which straddles the Tropic of Capricorn. Much of the west and south-west of the country forms part of the KALAHARI Desert. In the north the land is marshy around the Okavango Delta, which is home for a wide variety of wildlife. With the exception of the desert area, most of the country has a subtropical climate. In winter, days are warm and nights cold while summer is hot with sporadic rainfall. The people are mainly farmers and cattle rearing is the main activity. After independence in 1966 the exploitation of minerals started. Diamonds became an important revenue earner and the copper from the nickel/copper complex at Selebi-Pikwe was also exported. Mineral resources in the north-east are now being investigated and the exploitation of these resources has facilitated a high rate of economic growth withihn the country. About 17% of the land is set aside for wildlife preservation in National Parks, Game Reserves, Game Sanctuaries and controlled hunting areas.

Quick facts:
Area : 581,730 sq km (224,606 sq miles)
Population : 1,326,800
Capital : Gaborone
Other major cities : Francistown, Molepolole, Selibe-Pikwe
Form of government : Republic
Religions : Animism, Christian
Currency : Pula

Bouaké the second largest city of the CÔTE D'IVOIRE. (Pop. 362,000)

Bougainville the easternmost island belonging to Papua New Guinea, and a part of, though politically separate from, the chain of islands forming the Solomon Islands. (10,620 sq km/4100 sq miles; pop. 128,000)

Bournemouth a coastal resort in the county of Dorset in southern England. (Pop. 160,000)

Boyne, River a river flowing into the Irish Sea on the east coast of the Republic of Ireland. Famous for its prehistoric remains, it was the site of the battle (1690) in which Protestant William of Orange defeated Catholic James II. (Length 115 km/ 70 miles)

Boyoma Falls a series of seven cataracts over 90 km (56 miles) where the Lualaba River beomes the Zaïre River. They were formerly called Stanley Falls after the British explorer Sir Henry Morton Stanley.

Brabant the central province of Belgium around the capital, Brussels. (3358 sq km/1297 sq miles; pop. 2,254,000)

Brac *see* **Dalmatia**.

Bradford a city in the county of West Yorkshire, England, which came to prominence as the centre of the woollen industry in the 19th century. (Pop. 480,000)

Bragança a small inland town of medieval origins in Portugal, and the original home of the family which ruled Portugal from 1640 to 1910. (pop. 33,050)

Brahmaputra a major river of South Asia, flowing from the Himalayas in Tibet through Assam in northern India to join the River Ganges in Bangladesh. (Length 2900 km/1802 miles)

Braila a port in Romania on the River Danube, 140 km (87 miles) inland from the Black Sea. (Pop. 236,000)

Brasília the capital, since 1960, of Brazil. (Pop. 1,596,000)

Brazov an industrial city in central Romania. (Pop. 324,000)

Bratislava (Pressburg) the second largest city in the former Czechoslovakia, and the capital of the newly independent Slovakia. (Pop. 448,800)

Braunschweig *see* **Brunswick**.

Brazil is a huge South American country bounded to the north, south and east by the Atlantic ocean. It is the fifth largest country in the world and covers nearly half of South America. The climate is mainly tropical, but altitude, distance from the sea and prevailing winds cause many variations. In the Amazonia area it is constantly warm and humid, but in the tropical areas winters are dry and summers wet. Droughts may occur in the north-east,

where it is hot and arid. About 14% of the population is employed in agriculture and the main products exported are coffee, soya beans and cocoa. Brazil is rich in minerals and is the only source of high grade quartz crystal in commercial quantities. It is also a major producer of chrome ore and it is now developing what is thought to be the richest iron ore deposits in the world.

Quick facts:
Area : 8,511,965 sq km (3,285,488 sq miles)
Population : 156,500,000
Capital : Brasília
Other major cities : Belo Horizonte, Porto Alegre, Recife, Rio de Janeiro, Salvador, São Paulo
Form of government : Federal Republic
Religion : RC
Currency : Cruzeiro

Brazzaville the capital of the Congo, on the River Zaïre. (Pop. 938,000)

Breconshire *see* **Powys**.

Breda a historic and manufacturing city in the Netherlands. (Pop. city 125,000/Greater Breda 165,000)

Bremen a major port on the River Weser, near to the North Sea coast of Germany, and also the name of the surrounding state. (Pop. 552,000)

Bremerhaven a port on the North Sea coast of Germany, 55 km (34 miles) to the north of Bremen. (Pop. 131,000)

Brescia a city in northern Italy. (Pop. 193,000)

Breslau *see* **Wroclaw**.

Brest (1) a important naval port situated on an inlet on the tip of Finistere in north-western France. (Pop. 202,000). **(2) (Brzesc)** an inland port in Belarus situated on the River Bug on the border between Belarus and Poland. (Pop. 269,000)

Brezhnev *see* **Naberezhnyye Chelny**.

Bridgeport a manufacturing city on the coast of the state of Connecticut, USA. (Pop. city 137,000/ metropolitan area 441,500)

Bridgetown the capital of Barbados. (Pop. city 8000/urban area 108,000)

Brighton a famous seaside resort on the south coast of England in the county of East Sussex. (Pop. 155,000)

Brindisi a port on the east coast of Italy at the southern end of the Adriatic Sea. (Pop. 92,000)

Brisbane a port on the east coast of Australia, and

the state capital of QUEENSLAND. (Pop. city 777,000/urban area 1,327,000)

Bristol a major city and port in south-west ENGLAND. (Pop. 402,000)

Britain *see* **Great Britain**.

British Columbia the western seaboard province of CANADA. The capital is VICTORIA. (929,730 sq km/ 358,968 sq miles; pop. 3,448,000)

British Indian Ocean Territory the Chagos Archipelago, a group of five coral atolls in the middle of the INDIAN OCEAN. (52 sq km/20 sq miles)

British Isles the name given to the group of islands in north-western EUROPE formed by GREAT BRITAIN and IRELAND, and the surrounding islands.

British Virgin Islands *see* **Virgin Islands, British**.

Brittany (Bretagne) the region of FRANCE occupying the extreme north-western peninsula, overlooking the ATLANTIC. (27,200 sq km/10,449 sq miles)

Brno (Brünn) an industrial city in the south-east of the CZECH REPUBLIC. (Pop. 390,000)

Bromberg *see* **Bydgoszcz**.

Bruges (Brugge) a historic town and capital of the province of West FLANDERS, BELGIUM. (Pop. 116,000)

Brunei is a sultanate located on the north-west coast of BORNEO in South-East ASIA. It is bounded on all sides by the SARAWAK territory of MALAYSIA, which splits the sultanate into two separate parts. Broad tidal swamplands cover the coastal plains and inland Brunei is hilly and covered with tropical forest. The climate is tropical marine, hot and moist, with cool nights. Rainfall is heavy (2500 mm/98 inches) at the coast but even heavier (5000 mm/197 inches) inland. The main crops grown are rice, vegetables and fruit, but economically the country depends on its oil industry, which employs 7% of the working population. Oil production began in the 1920s and now oil and natural gas account for almost all exports. Other minor products are rubber, pepper, sawn timber, gravel and animal hides.

Quick facts:
Area : 5,765 sq km (2,226 sq miles)
Population : 276,000
Capital : Bandar Seri Begawan
Other major cities : Kuala Belait, Seria
Form of government : Monarchy
  (Sultanate)
Religion : Sunni Islam
Currency : Brunei dollar

Brünn *see* **Brno**.

Brunswick (Braunschweig) a historic town in northern GERMANY, and the capital of the Dukes of Saxony. (Pop. 255,000)

Brussels (Brussel, Bruxelles) a historic city and the capital of BELGIUM. It plays a central role in EUROPE as the administrative headquarters for the European Community. (Pop. 950,000)

Bryansk an industrial city in the west of the RUSSIAN FEDERATION. (Pop. 482,000)

Bubiyan Island a large island belonging to KUWAIT situated in the very north of The GULF and close to the border with IRAQ.

Bucaramanga a city in the north of COLOMBIA, close to the border with VENEZUELA. (Pop. 351,000)

Bucharest (Bucuresti) the capital of ROMANIA, in the south-east of the country. (Pop. 2,065,000)

Buckinghamshire a county in central southern ENGLAND; the county town is Aylesbury. (1883 sq km/727 sq miles; pop. 658,000)

Budapest the capital of HUNGARY, comprising Buda and Pest, which lie on opposite sides of the River DANUBE. (Pop. 2,009,400)

Budweiss *see* **Ceské Budejovice**.

Buenos Aires the capital of ARGENTINA. (Pop. city 2,961,000/metropolitan area 12,600,000)

Buffalo a city and port in NEW YORK state situated at the eastern end of Lake ERIE. (Pop. city 323,280/metropolitan area 1,194,000)

Bug, River a river which flows north-west from the UKRAINE, forming the border with POLAND before turning west into Poland and joining the Narew and VISTULA rivers. (Length: 813 km/480 miles)

Büjük Agri Dagi see **Ararat, Mount**

Bujumbura the capital of BURUNDI, situated at the northern end of Lake TANGANYIKA. (Pop. 235,400)

Bukhara an old trading city in UZBEKISTAN. (Pop. 228,000)

Bulawayo the second city of ZIMBABWE, in the south-west of the country. (Pop. 622,000)

Bulgaria is a south-east European republic located on the east Balkan peninsula and has a coast on the BLACK SEA. It is bounded to the north by ROMANIA, west by SERBIA and THE FORMER YUGOSLAV REPUBLIC OF MACEDONIA and south by GREECE and TURKEY. The centre of Bulgaria is crossed from west to east by the Balkan Mountains. The south of the country has a Mediterranean climate with hot dry summers and mild winters. Further north the temperatures become more extreme and rainfall is higher in summer. Traditionally Bulgaria is an agricultural country and a revolution in farming during the 1950s has led to great increases in output.

This was due to the collectivization of farms and the use of more machinery, fertilizers and irrigation. Each agricultural region now has its own specialized type of farming. Increased mechanization led to more of the workforce being available to work in mines and industry. However, following the break up of the former Soviet Union, with whom Bulgaria had particularly close trade links, the country has suffered very high rates of inflation and unemployment in the early 90s. The tourist trade has flourished though, with over 10,000,000 people visiting the Black Sea resorts annually.

> *Quick facts:*
> Area : 110,912 sq km (42,823 sq miles)
> Population : 8,473,000
> Capital : Sofia (Sofiya)
> Other major cities : Burgas, Plovdiv,
>   Ruse, Varna
> Form of government : Republic
> Religion : Eastern Orthodox
> Currency : Lev
>
>

Burgas a major port on the BLACK SEA coast of BULGARIA. (Pop. 226,000)

Burgos an industrial town in northern SPAIN, and the name of the surrounding province. (Pop. town 166,000)

Burgundy (Bourgogne) a region of central FRANCE, famous for its wine. (31,600 sq km/12,198 sq miles)

Burkina Faso (Burkina) a landlocked state in West AFRICA, Burkina lies on the fringe of the SAHARA, to the north. The country is made up of vast monotonous plains and low hills which rise to 700 m (2297 ft) in the south-west. Precipitation is generally low, the heaviest rain falling in the south-west, while the rest of the country is semi-desert. In the last two decades the country has been stricken by drought. The dusty gray plains in the north and west have infertile soils which have been further impoverished by overgrazing and overcultivation. About 90% of the people live by farming, and food crops include sorghum, beans and maize. Some cotton, livestock and oil seeds are exported. There is a great poverty and shortage of work and many of the younger population go to

GHANA and CÔTE D'IVOIRE for employment. The only main industries are textiles and metal products.

> *Quick facts:*
> Area : 274,200 sq km (105,869 sq miles)
> Population : 9,780,000
> Capital : Ouagadougou
> Form of government : Republic
> Religions : Animist, Sunni Islam
> Currency : Franc CFA
>
>

Burma *see* **Myanmar.**

Bursa a city in north-western TURKEY, and also the name of the surrounding province, of which it is the capital. (Pop. 835,100)

Burundi is a small densely populated country in central east AFRICA, bounded by RWANDA to the north, TANZANIA to the east and south, and ZAÏRE to the west. It has a mountainous terrain, with much of the country above 1500 m (4921 ft). The climate is equatorial but modified by altitude. The savanna in the east is several degrees hotter than the plateau and there are two wet seasons. The soils are not rich but there is enough rain to grow crops in most areas. The main food crops are bananas, sweet potatoes, peas, lentils and beans. Cassava is grown near the shores of Lake TANGANYIKA. The main cash crop is coffee, accounting for 90% of Burundi's export earnings. There is a little commercial fishing on Lake Tanganyika, otherwise industry is very basic. Since 1994 Burundi has been afflicted by ethnic conflict between the majority Hutu and minority Tutsi. Between 1994 and 1995 it is estimated that 150,000 were killed as a result of ethnic violence and the political situation remains highly volatile.

> *Quick facts:*
> Area : 27,834 sq km (10,747 sq miles)
> Population : 5,958,000
> Capital : Bujumbura
> Form of government : Republic
> Religion : RC
> Currency : Burundi franc
>
>

Buryat Republic an autonomous republic of the RUSSIAN FEDERATION, situated in the south-east, between Lake BAIKAL and MONGOLIA. (351,300 sq km/135,600 sq miles; pop. 1,014,000)

Bute an island in the Firth of CLYDE in SCOTLAND. The main town is Rothesay. (120 sq km/46 sq miles)

Bydgoszcz (Bromberg) a historic and industrial city in central POLAND. (Pop. 382,00)

Byelorussia *see* **Belarus**.

Bytom (Beuthen) an industrial city in south-west POLAND. (Pop. 232,000)

Byzantium *see* **Istanbul**.

Cadiz a port of Phoenician origins on the ATLANTIC coast of southern SPAIN; also the name of the surrounding province. (Pop. town 155,000)

Caen a city in the NORMANDY region of northern FRANCE. (Pop. 192,000)

Caerdydd *see* **Cardiff**.

Caernarfonshire a former county in north WALES which was merged as part of GWYNEDD IN 1975.

Cagliari the capital of the Italian island of SARDINIA. (Pop. 180,00)

Cairngorm Mountains a range forming part of the GRAMPIAN MOUNTAINS in SCOTLAND.

Cairns a port on the north-east coast of QUEENSLAND, AUSTRALIA and a tourist resort catering for visitors to the GREAT BARRIER REEF. (Pop. 48,000)

Cairo (El Qahira) the capital of EGYPT, in the north of the country on the River NILE; it is the largest city in AFRICA. (Pop. city 6,800,000/urban area 13,000,000)

Caithness *see* **Highland Region**.

Calabria the region which occupies the southern 'toe' of Italy. The main town is REGGIO DI CALABRIA. (15,100 sq km/5,829 sq miles; pop. 2,153,700)

Calais an old port in northern FRANCE situated on the narrowest part of the ENGLISH CHANNEL, opposite DOVER in ENGLAND. (Pop. 102,000)

Calcutta one of the largest cities in INDIA, a major port and industrial centre situated in the north-east of the country, on the HUGLI River. (Pop. 10,916,000)

Calgary the second largest city in the province of ALBERTA, CANADA. (Pop. 754,000)

Cali an industrial city in southern COLOMBIA. (Pop. 1,678,000)

Calicut (Kozhikode) a port on the west coast of southern INDIA. (Pop. 801,100)

California the most populous state of the USA on the PACIFIC coast. The state capital is SACRAMENTO, but LOS ANGELES is the biggest city. (411,015 sq km/158,693 sq km; pop. 30,895,000)

California, Gulf of the narrow inlet which separates the mainland part of MEXICO from the peninsula of BAJA CALIFORNIA. It is also known as the Sea of Cortes.

Callao the port serving LIMA, the capital of PERU. (Pop. 638,000)

Calvados a department of northern FRANCE, a part of the region of NORMANDY. It is famous for its apple-based liqueur called Calvados. (Pop. 590,000)

Camargue the broad, flat area of sea marshes in the delta of the River RHONE in the centre of the MEDITERRANEAN coast of FRANCE.

Cambodia is a South-East Asian state bounded by THAILAND, LAOS and VIETNAM and its southern coast lies on the Gulf of Thailand. The heart of the country is saucer-shaped, and gently rolling alluvial plains are drained by the MEKONG river. The Dangrek Mountains form the frontier with Thailand in the north-west. In general Cambodia has a tropical Monsoon climate and about half of the land is tropical forest. During the rainy season the Mekong swells and backs into the TONLE SAP (Great Lake), increasing its size threefold. Almost 162,000 hectares of land are flooded by this seasonal rise of the Mekong and this area is left with rich silt when the river recedes. Crop production depends entirely on the rainfall and floods but production was badly disrupted during the civil war and yields still remain low. Despite the gradual rebuilding of the infrastructure in the early 1990s, Cambodia remains one of the world's poorest nations.

Quick facts:
Area : 181,035 sq km (69,898 sq miles)
Population : 9,280,000
Capital : Phnom-Penh
Other major cities : Battambang, Kampong Cham,
Form of government : People's Republic
Religion : Buddhism
Currency : Riel

Cambrian Mountains a range of mountains which forms the 'backbone' of WALES.

Cambridge (1) a famous university city in eastern ENGLAND. (Pop. 113,000) **(2)** a city in MASSACHUSETTS, USA, home of Harvard University and the Massachusetts Institute of Technology. (Pop. 93,550)

Cambridgeshire a county in eastern ENGLAND; the county town is CAMBRIDGE. (3409 sq km/1316 sq miles; pop. 645,000)

Cameron Highlands an upland area of MALAYSIA where tea and and vegetables are grown.

Cameroon is a triangular-shaped country of diverse landscapes in west central AFRICA. It stretches from Lake CHAD at its apex to the northern borders of EQUATORIAL GUINEA, GABON and the CONGO in the south. The landscape ranges from low-lying lands, through the semi-desert SAHEL, to dramatic mountain peaks and then to the grassy savanna, rolling uplands, steaming tropical forests and hardwood plantations. Further south are the volcanoes, including Mount CAMEROON, and the palm beaches at Kribi and Limbe. The climate is equatorial with high temperatures and plentiful rain. The majority of the population lives in the south where they grow maize and vegetables. In the drier north where drought and hunger are well known, life is harder. Bananas, coffee and cocoa are the major exports although oil, gas and aluminum are becoming increasingly important.

> *Quick facts:*
> Area : 475,442 sq km (183,568 sq miles)
> Population : 12,800,000
> Capital : Yaoundé
> Other major city : Douala
> Form of government : Republic
> Religions : Animism, RC, Sunni Islam
> Currency : Franc CFA

Cameroon, Mount an active volcano in west CAMEROON. (4095 m/13,435 ft)

Campania a region of central southern ITALY, on the west coast around NAPLES. (13,595 sq km/5249 sq miles; pop. 5,624,000)

Campinas a modern industrial town 75 km(47 miles) north of SÃO PAULO in BRAZIL. (Pop. 846,000)

Cam Ranh Bay a naval base in VIETNAM developed by the US Navy which subsequently played an important part in the Soviet military presence in South-East ASIA.

Canada is the second largest country in the world, and the largest in North AMERICA. Canada is a land of great climatic and geographical extremes. It lies to the north of the USA and has PACIFIC, ATLANTIC and ARCTIC coasts. The ROCKY MOUNTAINS and Coast Mountains run down the west side, and the highest point, Mount Logan (5951 m/19,524 ft), is in the YUKON. Climates range from polar conditions in the north, to cool temperate in the south with considerable differences from west to east. More than 80% of its farmland is in the prairies that stretch from ALBERTA to MANITOBA. Wheat and grain crops cover three-quarters of the arable land. Canada is rich in forest reserves which cover more than half the total land area. The most valuable mineral deposits (oil, gas, coal and iron ore) are found in Alberta. Most industry in Canada is associated with processing its natural resources.

> *Quick facts:*
> Area : 9 ,970,610 sq km (3,849,674 sq miles)
> Population : 28,150,000
> Capital : Ottawa
> Other major cities : Toronto, Montréal,
>   Vancouver, Québec City
> Form of government : Federal
>   Parliamentary State
> Religions : RC, United Church of Canada,
>   Anglicanism
> Currency : Canadian dollar

Canary Islands a group of islands belonging to SPAIN, some 95 km (60 miles) off the coast of WESTERN SAHARA. The main islands are Tenerife, Gran Canaria, Fuertaventura, Gomera, Lanzarote, La Palma. (7273 sq km/2808 sq miles; pop. 1,493,000)

Canaveral, Cape a long spit of land on the east coast of the state of FLORIDA, USA. It is the USA's main launch site for space missions and the home of the John F. Kennedy Space Center. For a time it was known as Cape Kennedy.

Canberra the capital of AUSTRALIA, lying about halfway between SYDNEY and MELBOURNE in the south-east of the country. (Pop. 310,000)

Cancún a tiny island just off the YUCATAN coast of MEXICO, connected to the mainland by a causeway, and now a popular holiday resort. (Pop. 70,000)

Cannes a famous beach resort on the Côte d'Azur in southern France. (Pop. 72,800)

Cantabria a province on the Atlantic coast of northern Spain. (Pop. 510,800)

Canterbury an historic cathedral city in the county of Kent in southern England. (Pop. 133,000)

Canton *see* **Guangzhou**.

Cape Breton Island an island off the eastern coast of Canada which forms part of the province of Nova Scotia. (10,349 sq km/3970 sq miles; pop. 170,000)

Cape Town a major port on the south-western tip of South Africa, and the country's legislative capital. (Pop. metropolitan1,912,000)

Cape Verde, one of the world's smallest nations, is situated in the Atlantic Ocean, about 640 km (400 miles) north-west of Senegal. It consists of 10 islands and 5 islets. The islands are divided into the Windward group and the Leeward group. Over 50% of the population live on São Tiago on which is Praia, the capital. The climate is arid with a cool dry season from December to June and warm dry conditions for the rest of the year. Rainfall is sparse and the islands suffer from periods of severe drought. Agriculture is mostly confined to irrigated inland valleys and the chief crops are coconuts, sugar cane, potatoes and cassava. Bananas and some coffee are grown for export. Fishing for tuna and lobsters is an important industry but in general the economy is shaky and Cape Verde relies heavily on foreign aid.

Quick facts:
Area : 4033 sq km (1575 sq miles)
Population : 370,000
Capital : Praia
Form of government : Republic
Religion : RC
Currency : Cape Verde escudo

Cappadocia an area of eastern Turkey to the south-east of Ankara, noted for the extraordinary sugarloaf shapes of its volcanic rock formations, into which cave houses have been carved.

Capri a rocky island at the southern end of the Bay of Naples on the west coast of Italy, famous as a fashionable holiday retreat. (10.4 sq km/4 sq miles; pop. 16,500)

Caprivi Strip a narrow corridor of land, 450 km(280 miles) long, which belongs to Namibia and gives it access to the Zambezi River along the border between Botswana to the south and Angola and Zambia to the north.

Caracas the capital of Venezuela, in the north-east of the country. (Pop. 2,784,000)

Cardamom Mountains a range of mountains rising to 1813 m (5948 ft) which line the coast of Cambodia and separate the interior from the Gulf of Thailand.

Cardiff (Caerdydd) the capital of Wales, situated in the south-east of the principality. (Pop. 307,000)

Cardiganshire a former county in west Wales.

Cardigan Bay the long, curving bay which, as part of the Irish Sea, forms much of the west coast of Wales.

Caribbean, The a term that refers to the islands lying within the compass of the Caribbean Sea.

Caribbean Sea a part of the western Atlantic Ocean, bounded by the east coast of Central America, the north coast of South America and the West Indies.

Carinthia (Kärnten) the southern state of Austria, which borders Italy and Slovenia. (9533 sq km/3681 sq miles; pop 552,000)

Carlow a landlocked county in the south-east of the Republic of Ireland. The county town is also called Carlow. (900 sq km/347 sq miles; pop. county 41,000)

Carlsbad *see* **Karlovy Vary**.

Carmarthenshire a county in south Wales; the county town is Carmarthen. (Pop. 169,000)

Carmel, Mount a ridge of land rising to 528 m(1746 ft) in northern Israel, mentioned in the Bible, and the place where the Carmelite Order of mendicant friars originated in the 12th century.

Caroline Islands a scattered group of islands in the western Pacific Ocean which now make up the Federated States of Micronesia and the separate state of Belau.

Carpathian Mountains a broad sweep of mountains stretching for nearly 1000 km (625 miles) down the border between Slovakia and Poland and into central Romania. They rise to 2663 m (8737 ft) at their highest point.

Carpentaria, Gulf of the broad gulf of shallow sea between the two hornlike peninsulas of northern Australia.

Carrara a town 50 km (31 miles) north of Pisa in Italy, famous for centuries for its marble quarries. (Pop. 68,500)

Carson City the state capital of Nevada, USA. (Pop. 42,840)

Cartagena (1) a major port on the CARIBBEAN coast of COLOMBIA. (Pop. 688,000). **(2)** a port of ancient origins on the MEDITERRANEAN coast of SPAIN. (Pop. 180,000)

Carthage the site of a great trading port of the ancient MEDITERRANEAN world, now in modern TUNISIA.

Casablanca (Dar el Beida) the main port and largest city of MOROCCO. (Pop. 3,079,000)

Cascade Range a range of mountains stretching some 1125 km (700 miles) parallel to the coast of northern CALIFORNIA in the USA and into southern CANADA. The highest point is at Mount Rainier (4392 m/14,410 ft) in WASHINGTON State.

Caspian Sea the largest inland (salt) sea in the world, supplied mainly by the River VOLGA. It lies to the north of IRAN, which shares its coasts with AZERBAIJAN, GEORGIA, KAZAKHSTAN and TURKMENISTAN.

Cassai *see* **Kasai**.

Cassel *see* **Kassel**.

Castile (Castilla) a former kingdom of SPAIN, occupying most of the central area, now divided into two regions, Castilla-La Mancha and Castilla-León.

Castries the capital of ST LUCIA. (Pop. 56,000)

Catalonia (Cataluña) an autonomous region of SPAIN, in the north-east, centring on BARCELONA. (Pop. 5,958,000)

Catania a major port and the second largest city in SICILY. (Pop. 378,500)

Catskill Mountains a range of mountains in NEW YORK State, USA, famed for their scenic beauty. The highest peak is Slide Mountain (1281 m/4204 ft).

Caucasus (Kavkaz) the mountainous region between the BLACK and CASPIAN SEAS, bounded by the RUSSIAN FEDERATION, GEORGIA, ARMENIA and AZERBAIJAN. It contains EUROPE's highest point, Mount ELBRUS. (5642m/18,510 ft)

Cauvery *see* **Kaveri**.

Cavan a county in the north of the Republic of IRELAND, part of the ancient province of ULSTER; Cavan is also the name of the county town. (1890 sq km/730 sq miles; pop. county 52,800)

Caveri *see* **Kaveri**.

Cawnpore *see* **Kanpur**.

Cayenne the capital of GUIANA (FRENCH). (Pop. 42,000)

Cayman Islands a group of three islands in the CARIBBEAN SEA 240 km (150 miles) north-west of Jamaica which form a British Crown colony. The capital is Georgetown, on Grand Cayman. (260 sq km/100 sq miles; pop. 22,000)

Cebu one of islands in the central PHILIPPINES, forming part of the VISAYAN group; also the name of its capital city. (5088 sq km/1964 sq miles; pop. island 2,092,000/city 490,231)

Celebes *see* **Sulawesi**.

Celebes Sea a sea between the islands of eastern INDONESIA and the PHILIPPINES.

Central African Republic is a landlocked country in central AFRICA bordered by CHAD in the north, CAMEROON in the west, SUDAN in the east and the CONGO and ZAÏRE in the south. The terrain consists of a 610–915-m (2000–3000-ft) high undulating plateau with dense tropical forest in the south and a semi-desert area in the east. The climate is tropical with little variation in temperature throughout the year. The wet months are May, June, October and November. Most of the population live in the west and in the hot, humid south and south-west. Over 86% of the working population are subsistence farmers and the main crops grown are cassava, groundnuts, bananas, plantains, millet and maize. Livestock rearing is small-scale because of the prevalence of the tsetse fly. Gems and industrial diamonds are mined and vast deposits of uranium have been discovered.

Quick facts:
Area : 622,984 sq km (240,535 sq miles)
Population : 3,173,000
Capital : Bangui
Form of government : Republic
Religions : Animism, RC
Currency : Franc CFA

Central Region a former local government area of SCOTLAND formed in 1975 out of the old counties of Clackmannanshire and parts of Perthshire and Stirlingshire.

Cephalonia *see* **Ionian Islands**.

Ceram *see* **Seram**.

Ceredigion a county in west Wales (1800 sq km/694 sq miles; pop. 70,000)

Ceské Budejovice (Budweiss) a historic town in the south of BOHEMIA, CZECH REPUBLIC, famous for its Budvar beer. (Pop. 175,450)

Ceuta a Spanish enclave in northern MOROCCO administered by SPAIN. (Pop. 80,000)

Cévennes the name given to the southern part of the Massif Central in France.

Ceylon *see* **Sri Lanka**.

Chad, a landlocked country in the centre of northern Africa, extends from the edge of the equatorial forests in the south to the middle of the Sahara Desert in the north. It lies more than 1600 km (944 miles) from the nearest coast. The climate is tropical with adequate rainfall in the south but the north experiences semi-desert conditions. In the far north of the country the Tibesti Mountains rise from the desert sand more than 3000 m (9843 ft). The southern part of Chad is the most densely populated and its relatively well-watered savanna has always been the country's most arable region. Recently, however, even here the rains have failed. Normally this area is farmed for cotton (the main cash crop), millet, sorghum, groundnuts, rice and vegetables. Fishing is carried out in the rivers and in Lake Chad. Cotton ginning is the principal industry. Chad remains one of the poorest countries in the world, a result of drought and the civil war, which lasted from 1960 to 1988. Some unrest ontinues in the country.

Quick facts:
Area : 1,284,000 sq km (495,750 sq miles)
Population : 6,100,000
Capital : N'Djamena
Other major cities : Sarh, Moundou
Form of government : Republic
Religions : Sunni Islam, Animism
Currency : Franc CFA

Chad, Lake a large lake in western Chad, on the border with Niger and Nigeria. (26,000 sq km/ 10,000 sq miles)

Chamonix a winter sports resort in France, just to the north of Mont Blanc. (Pop. 11,000)

Champagne a region of north-eastern France famous for the sparking wine called champagne. It now forms part of the administrative region called Champagne-Ardennes.

Chandigarh a modern city 230 km (143 miles) north of Delhi, India, planned by the French architect Le Corbusier (1887–1965). (Pop. 576,000)

Changchun the capital of Jilin province, China. (Pop. 2,070,000)

Changhua a historic city situated near the west coast of Taiwan. (Pop. 1,206,400)

Chang Jiang (Yangtze) the world's third longest river. It rises in Tibet and flows across central China into the East China Sea. (Length 6380 km/ 3965 miles)

Changsha the capital of Hunan province, China. (Pop. 1,510,000)

Channel Islands a group of islands in the English Channel, close to the coast of France, which are British Crown dependencies. The main islands are Jersey and Guernsey, but the group also includes the smaller inhabited islands of Alderney, Sark and Herm. (Pop. 145,700)

Chao Phrya, River a river running from north to south down the west side of Thailand and through its capital, Bangkok. (Length 100 km/62 miles)

Chapala, Lake the largest lake in Mexico, near Guadalajara. (2460 sq km/950 sq miles)

Charleroi an industrial city in central Belgium. (Pop. 206,200)

Charleston (1) the state capital of West Virginia, USA. (Pop. city 57,100/metropolitan area 267,000). **(2)** An old port on the Atlantic coast of South Carolina, USA. (Pop. 81,300)

Charlotte Amalie the capital of the US Virgin Islands, on St Thomas. (Pop. 11,800)

Charlottetown a port and the provincial capital of Prince Edward Island, Canada. (Pop. 45,000)

Chartres a market town, capital of the department of Eure-et-Loire, in northern France, 80 km (50 miles) west of Paris. It is famous for its early 13th-century cathedral, with original stained-glass windows. (Pop. 80,340)

Chattanooga an industrial city and railway town in Tennessee, USA. (Pop. city 152,890)

Chechen Republic one of the autonomous republics of the Russian Federation. The republic declared its independence in 1991 and concern that this might undermine the solidity of the Russian Federation resulted in an invasion of the region by Russia. This initial invasion was unsuccessful and after negotiation failed to reach a resolve, Russian troops again invaded in late 1994. After nearly two years of fighting and an estimated loss of more than 20,000 lives, the Russian forces agreed to withdraw, after negotiating an agreement that the republic would delay pressing for sovereignty until 2001. (19,300 sq km/7450 sq miles; pop. 1,204,000)

Cheju Do an island belonging to South Korea, some 90 km (56 miles) off its southern tip, and dominated by the sacred volcano, Mount Halla (1950 m/6398 ft). (1828 sq km/706 sq miles; pop. 463,000)

Cheltenham a well-preserved spa town in the

county of GLOUCESTERSHIRE, famous for the music and literature festivals that are held there and for its racecourse, home to the Cheltenham Gold Cup horse race. (Pop. 107,000)

Chelyabinsk an industrial city in the RUSSIAN FEDERATION. (Pop. 1,170,000)

Chemnitz an industrial city in south-east GERMANY, named Karl-Marx-Stadt during the period of Communist rule in former East Germany (until 1990). (Pop. 310,000)

Chengdu the capital of SICHUAN province, CHINA. (Pop. 2,810,000)

Chenstokhov *see* **Czestochowa**.

Chernobyl a city about 90 km (55 miles) north of KIEV, in the UKRAINE. In April 1986 one of the reactors in its nuclear power station exploded, releasing a radioactive cloud that caused fallout in many parts of the world. It is the world's worst nuclear accident.

Chesapeake Bay an inlet, 314 km (195 miles) long, on the east coast of the USA, shared by the states of VIRGINIA and MARYLAND.

Cheshire a county in north-west ENGLAND; the county town is Chester. (2322 sq km/897 sq miles; pop. 956,000)

Cheviot Hills a range of hills, 60 km (37 miles) long, which line the border between SCOTLAND and the county of NORTHUMBERLAND in ENGLAND.

Cheyenne the state capital of WYOMING, USA. (Pop. 51,900)

Chiang Mai *see* **Chiengmai**.

Chianti the winemaking region of central TUSCANY, ITALY.

Chiba a large industrial city on HONSHU Island, JAPAN. (Pop. 851,000)

Chicago the largest city in the state of ILLINOIS, and the third largest city in the USA (after NEW YORK and LOS ANGELES). (Pop. city 2,768,500/metropolitan area 8,410,000)

Chichén Itzá a village in YUCATAN, MEXICO, which is the site of a complex of major ruins of the Mayans and Toltecs.

Chiengmai (Chiang Mai) a city in THAILAND, in the north-west of the country, famous for its temples and the crafts produced in the surrounding villages. (Pop. 167,000)

Chihuahua a city in northern central MEXICO, and the name of the surrounding province, of which it is the capital. (Pop. 530,000)

Chile lies like a backbone down the PACIFIC coast of the South American continent. Its Pacific coastline is 4200 km (2610 miles) long. Because of its enormous range in latitude it has almost every kind of climate from desert conditions to icy wastes. The

north, in which lies the ATACAMA DESERT, is extremely arid. The climate of the central region is Mediterranean and that of the south cool temperate. 60% of the population live in the central valley where the climate is similar to southern CALIFORNIA. The land here is fertile and the principal crops grown are wheat, sugar beet, maize and potatoes. It is also in the central valley that the vast copper mine of El Teniente is located. This is one of the largest copper mines in the world and accounts for Chile's most important source of foreign exchange.

Quick facts:
Area : 756,945 sq km (292,258 sq miles)
Population : 13,813,000
Capital : Santiago
Other major cities : Arica, Talcahuano, Viña del Mar
Form of government : Republic
Religion : RC
Currency : Chilean peso

Chiltern Hills a range of hills to the north-west of LONDON, ENGLAND, rising to 260 m (850 ft).

China, the third largest country in the world, covers a large area of East ASIA. In western China most of the terrain is very inhospitable—in the north-west there are deserts which extend into MONGOLIA and the RUSSIAN FEDERATION, and much of the south-west consists of the ice-capped peaks of TIBET. The south-east has a green and well watered landscape comprising terraced hillsides and paddy fields. Most of China has a temperate climate but in such a large country wide ranges of latitude and altitudes produce local variations. China is an agricultural country, and intensive cultivation and horticulture is necessary to feed its population of over one billion. Since the death of Mao in 1976, and under the leadership of Deng Xiao Ping, China has experienced a huge modernization of agriculture and industry due to the supply of expertize, capital and technology from Japan and the West. The country has been opened up to tourists and to a degree has adopted the philosophy of free enterprise, resulting in a dramatic improvement in living standards for a significant proportion of the population. However, the change towards a market economy has created internal political problems within the last decade. Pro-democracy demonstrations in 1989 resulted in the Tianmen Square massacre, which was condemned throughout the world, and raised questions regarding China's human rights approach. Deng Xiao Ping had been a *de facto* leader since announcing his retirement in 1989, but his influence on the country was profound until his death in February 1997.

Quick Facts:
Area : 9,571,300 sq km (3,695,500 sq miles)
Population : 1,200,000,000
Capital : Beijing (Peking)
Other major cities: Chengdu, Guangzhou, Shanghai, Tianjin, Wuhan
Form of government: People's Republic
Religions : Buddhism, Confucianism, Taoism
Currency : Yuan

China Sea a part of the PACIFIC OCEAN, off the east coast of CHINA.

Chindwin, River a river in MYANMAR, flowing parallel to the north-west border before joining the IRRAWADDY River in the centre of the country. (Length 1130 km/700 miles)

Chios (Khios) an island in the AEGEAN SEA, belonging to GREECE but lying only 8 km (5 miles) from the coast of TURKEY. It is said to have been the home of the poet Homer (*c*.800BC). (904 sq km/ 349 sq miles; pop. 49,900)

Chisinau *see* **Kishinev**.

Chittagong the main port of BANGLADESH and its second largest city. (Pop. 2,040,000)

Cholula a town in south-east MEXICO famous for its extensive Toltec ruins dating from 500BC. (Pop. 160,000)

Chongqing (Chungking) an industrial city on the CHANG JIANG river, CHINA, and the largest city in SICHUAN province. (Pop. 2,980,000)

Chonju a historic city in the south-west of South KOREA. (Pop. 517,000)

Chonnam *see* **Kwangju**.

Christchurch (1) the largest city on South Island, NEW ZEALAND. (Pop. 308,000) **(2)** a town in the county of DORSET in south ENGLAND. (Pop. 41,000)

Christmas Island (1) an island in the eastern INDIAN OCEAN, 400 km (250 miles) to the south of

JAVA, administered by AUSTRALIA since 1958. (142 sq km/55 sq miles; pop. 3500). **(2) (Kiritimati)** is the PACIFIC OCEAN's largest coral atoll, situated at the north-eastern end of the KIRIBATI group. (432 sq km/167 sq miles; pop. 1300)

Chubu Sangaku a national park in central HONSHU Island which contains two of the highest mountains in JAPAN, Mount Hotaka (3190 m/10,466 ft) and Mount Yari (3180 m/10,434 ft)

Chungking *see* **Chongqing**.

Churchill, River a river which flows into the HUDSON BAY at the port of Churchill after a journey through SASKATCHEWAN and MANITOBA. (Length 1600 km/1000 miles)

Chuvash Republic one of the autonomous republics of the RUSSIAN FEDERATION. (18,300 sq km/ 7050 sq miles; pop. 1,314,000)

Cincinnati a city in the south-west of the state of OHIO, USA, on the OHIO River. (Pop. city 370,500/ metropolitan area 1,673,500)

CIS *see* **Commonwealth of Independent States**.

Ciskei was a former 'homeland' of SOUTH AFRICA. Under the new South African Constitution implemented in April 1994, 'homelands' ceased to exist. The area is now part of Eastern Cape, South Africa.

Citaltépetl a volcanic peak to the south-east of MEXICO CITY, and at 5747 m (18,855 ft) the highest point in MEXICO.

Clare a county on the west coast of IRELAND; the county town is Ennis. (3188 sq km/1230 sq miles; pop. 91,000)

Clermont-Ferrand a city in the AUVERGNE region of central FRANCE. (Pop. 254,000)

Cleveland (1) a county of north-east ENGLAND created in 1974 out of Durham and Yorkshire to administer the industrial region along the River Tees, known as Teeside (583 sq km/225 sq miles; pop.). **(2)** a port and industrial city on the southern side of Lake ERIE, in OHIO, USA. (Pop. city 502,540/ metropolitan area 2,890,000)

Cluj-Napoca a city of ancient origins in central ROMANIA. (Pop. 322,000)

Clwyd a former county in north-east WALES created in 1974 out of the county of FLINTSHIRE and parts of Merionethshire and DENBIGHSHIRE. In 1996 the structure was again changed and Clwyd ceased to exist as a county.

Clyde, River a river in south-west SCOTLAND which flows north-west to form an estuary 100 km (60 miles) long, called the Firth of Clyde, with GLASGOW at its head. (Length 170 km/105 miles)

Coast Ranges a range of mountains lining the western coast of the USA, stetching 1600 km(1000

miles) from the borders with CANADA to LOS ANGE-LES. The highest point is in the San Jacinto Mountains (3301 m/10,831 ft)

Cobh a town and port in Cork Harbor on the south coast of IRELAND, some 10 km (6 miles) from the city of CORK. It was formerly a port of call for transatlantic steamships. (Pop. 6200)

Cochin a port on the south-western tip of INDIA. (Pop. 564,000)

Cochin China the name given to the region around the MEKONG delta during the French occupation of VIETNAM.

Cockburn Town the capital of TURKS AND CAICOS. (Pop. 4000)

Cocos Islands (Keeling Islands) a cluster of 27 small coral islands in the eastern INDIAN OCEAN, equidistant from SUMATRA and AUSTRALIA, and administered by Australia since 1955. (14 sq km/6 sq miles; pop. 700)

Cod, Cape a narrow, low-lying peninsula of sand dunes and marshland on the coast of MASSACHU-SETTS, USA, where the Pilgrim Fathers landed in 1620.

Cologne (Köln) a city and industrial centre on the River RHINE, GERMANY. (Pop. 693,000)

Colombia is situated in the north of South AMERICA and most of the country lies between the equator and 10° north. The ANDES, which split into three ranges (the Cordilleras) in Colombia, run north along the west coast and gradually disappear toward the CARIBBEAN SEA. Half of Colombia lies east of the Andes and much of this region is covered in tropical grassland. Toward the AMAZON Basin the vegetation changes to tropical forest. The climates in Colombia include equatorial and tropical according to altitude. Very little of the country is under cultivation although much of the soil is fertile. The range of climates result in an extraordinary variety of crops of which coffee is the most important. Colombia is rich in minerals and produces about half of the world's emeralds. It is South America's leading producer of coal, and oil has recently been discovered.

Quick facts:
Area : 1,141,748 sq km (439,735 sq miles)
Population : 34,900,000
Capital : Bogotá
Other major cities : Barranquilla, Cali, Cartagena, Medellin
Form of government : Republic
Religion : RC
Currency : Peso

Colombo a major port and the capital of SRI LANKA. (Pop. urban area 1, 860,000)

Colorado an inland state of central western USA; the state capital is DENVER. (270,000 sq km/ 104,247 sq miles; pop. 3,464,700)

Colorado, River a river which rises in the ROCKY MOUNTAINS in the state of COLORADO, USA, and flows south-west to the Gulf of CALIFORNIA, forming the GRAND CANYON on its way. (Length 2330 km/1450 miles)

Colorado Springs a spa and resort city in the state of COLORADO, USA. (Pop. city 296,000)

Columbia (1) the state capital of SOUTH CAROLINA, USA. (Pop. city 98,830) **(2)** a city in Missouri, USA. (Pop. 73,100)

Columbia, District of *see* **Washington D.C.**

Columbia, River a river which flows northwards from its source in BRITISH COLUMBIA, CANADA, before turning south into WASHINGTON STATE, USA and entering the PACIFIC OCEAN at PORTLAND, OREGON. (Length 1950 km/1210 miles)

Columbus the state capital of OHIO, USA. (Pop. city 643,000/metropolitan area 1,394,000)

Commonwealth of Independent States (CIS) an organization created in 1991 to represent the common monetary, commercial and trade interests of eleven independent states of the former Soviet Union. The name is a translation of *Sodruzhestvo Nezavisimikh Gosudarstv*. The eleven member states are: ARMENIA, AZERBAIJAN, BELARUS, KAZAKHSTAN, KYRGYZSTAN, MOLDOVA, RUSSIAN FEDERATION, TAJIKISTAN, TURKMENISTAN, UKRAINE and UZBEKISTAN. The former Soviet states of ESTONIA, LATVIA, LITHUANIA and GEORGIA did not join the CIS on gaining independence.

Comorin, Cape the southern tip of INDIA, where the INDIAN OCEAN, the BAY OF BENGAL and the ARABIAN SEA MEET.

Comoros, The consist of three volcanic islands in the INDIAN OCEAN situated between mainland AFRICA and MADAGASCAR. Physically four islands make up the group but the island of MAYOTTE remained a French dependency when the three western islands became a federal Islamic republic in 1975. The islands are mostly forested and the tropical climate is affected by Indian monsoon winds from the north. There is a wet season from November to April. Only small areas of the islands are cultivated and most of this land belongs to foreign plantation owners. The chief product was formerly sugar cane, but now vanilla, copra, maize, cloves and essential oils are the most important products. The forests provide timber for building and there is a small fishing industry.

Quick facts:
Area : 1862 sq km (719 sq miles)
  excluding Mayotte
Population : 510,000
Capital : Moroni
Form of government : Federal Islamic Republic
Religion : Sunni Islam
Currency : Comorian franc

Conakry the capital of GUINEA, a port partly located on the island of Tumbo. (Pop. 810,000)

Concord the state capital of NEW HAMPSHIRE, USA. (Pop. 35,637)

Congo formerly a French colony, the Republic of the Congo is situated in west central AFRICA where it straddles the equator. The climate is equatorial, with a moderate rainfall and a small range of temperature. The Bateke Plateau has a long dry season but the Congo Basin is more humid and rainfall approaches 2500 mm (98 inches) each year. About 62% of the total land area is covered with equatorial forest from which timbers such as okoume and sapele are produced. Valuable hardwoods such as mahogany are exported. Cash crops such as coffee and cocoa are mainly grown on large plantations but food crops are grown on small farms usually worked by the women. A manufacturing industry is now growing and oil discovered offshore accounts for much of the Congo's revenues.

Quick facts:
Area : 342,000 sq km (132,046 sq miles)
Population : 2,700,000
Capital : Brazzaville
Other major city : Pointe-Noire
Form of government : Republic
Religion : RC
Currency : Franc CFA

Congo, River *see* **Zaïre, River**.

Connaught (Connacht) one of the four old provinces into which IRELAND was divided (17,130 sq km/6612 sq miles; pop. 423,000)

Connecticut a state on the north-eastern seaboard of the USA, in NEW ENGLAND; the capital is HARTFORD. (12,973 sq km/5009 sq miles; pop. 3,279,160)

Connemara a famously beautiful part of County GALWAY on the west coast of IRELAND centring upon the distinctive peaks of the Twelve Bens.

Constance, Lake (Bodensee) a lake in Europe surrounded by GERMANY to the north, SWITZERLAND to the south and AUSTRIA to the east. (536 sq km/207 sq miles)

Constanta a major port on the BLACK SEA coast of ROMANIA. (Pop. 349,000)

Constantine (Qacentina) an ancient walled city in the north-eastern corner of ALGERIA. (Pop. 449,000)

Constantinople *see* **Istanbul**.

Cook, Mount the highest mountain in NEW ZEALAND, on South Island. (3753 m/12,316 ft)

Cook Islands a group of 15 islands in the South PACIFIC, independent since 1965 but associated with NEW ZEALAND. The capital is Avarua. (240 sq km/93 sq miles; pop. 18,000; cur. Cook Islands dollar/ New Zealand dollar = 100cents)

Cook Strait the strait that separates North Island and South Island of NEW ZEALAND, 26 km (16 miles) across at its widest point.

Cooper Creek a river flowing into Lake EYRE in SOUTH AUSTRALIA from its source in central QUEENSLAND. The upper stretch is known as the Barcoo River. (Length 1420 km/800 miles)

Copacabana a famous beachside suburb of RIO DE JANEIRO, BRAZIL.

Copán, a village in HONDURAS, is the site of a great Mayan city which flourished between AD450 and 800 before being abandoned.

Copenhagen (København) a port and the capital of DENMARK, located on the islands of ZEALAND and AMAGER. (Pop. urban area 1,337,00)

Coral Sea a part of the PACIFIC OCEAN, off the north-east coast of AUSTRALIA.

Córdoba (Cordova) (1) a city in southern Spain, famous for its cathedral which was built originally as a mosque; also the name of the surrounding province. (Pop. city 316,000) **(2)** the second city of ARGENTINA, and the name of the surrounding province. (Pop. city 1,179,000)

Corfu (Kérkira) the most northerly of the IONIAN ISLANDS, in western GREECE; the capital is also called Corfu. (592 sq km/229 sq miles; pop. 97,100)

Corinth (Korinthos) a town of ancient origin in the PELOPONNESE in western GREECE, built near the Corinth Ship Canal. It is the site of the ruined temple of Apollo

Cork the second largest city in the Republic of IRE-

LAND, at the head of a large natural harbor which cuts into the southern coast. Also the name of the county of which it is the county town. (County 7459 sq km/2880 sq miles; pop. county 402,300; pop. city 174,000)

Cornwall the county occupying the south-western tip of ENGLAND; the county town is Truro. Administratively, Cornwall incorporates the ISLES OF SCILLY. (3546 sq km/1369 sq miles; pop. 479,000)

Coromandel Coast the coast of south-eastern INDIA around MADRAS.

Coromandel Peninsula the central peninsula reaching northwards from North Island, NEW ZEALAND.

Corpus Christi a port in TEXAS on the Gulf of MEXICO. (Pop. 266,412)

Corsica (Corse) an island in the MEDITERRANEAN SEA lying to the north of SARDINIA, governed by FRANCE. (8680 sq km/3350 sq miles; pop. 249,000)

Corunna (La Coruña) a port and manufacturing town in north-west SPAIN, and also the name of the surrounding province. (Pop. town 251,300)

Costa Brava a strip of coastline to the north-east of BARCELONA in SPAIN, famous for its beaches and its popular resorts.

Costa Rica with the PACIFIC OCEAN to the south and west and the CARIBBEAN SEA to the east, Costa Rica is sandwiched between the central American countries of NICARAGUA and PANAMA. Much of the country consists of volcanic mountain chains which run north-west to south-east. The climate is tropical with a small temperature range and abundant rain. The dry season is from December to April. The most populated area is the Valle Central which was first settled by the Spanish in the 16th century. The upland areas have rich volcanic soils which are good for coffee growing and the slopes provide lush pastures for cattle. Coffee and bananas are grown commercially and are the major agricultural exports. Costa Rica's mountainous terrain provides hydro electric power, which makes it almost self-sufficient in electricity, and attractive scenery for its growing tourist industry.

Quick facts:
Area : 51,100 sq km (19,730 sq miles)
Population : 3,323,000
Capital : San José
Other major city : Límon
Form of government : Republic
Religion : RC
Currency : Colon

Costa Smeralda the 'emerald coast' on the northeast side of the MEDITERRANEAN island of SARDINIA, famed for its watersports and its upmarket resorts.

Côte d'Azur the coast of south-east FRANCE, famous for its beaches and resorts such as ST TROPEZ, CANNES and NICE.

Côte d'Ivoire a former French colony in west AFRICA, Côte d'Ivoire is located on the Gulf of GUINEA with GHANA to the east and LIBERIA to the west. The south-west coast has rocky clif fs but further east there are coastal plains which are the country's most prosperous region. The climate is tropical and affected by distance from the sea. The coastal area has two wet seasons but in the north, there is only the one. Côte d'Ivoire is basically an agricultural country which produces cocoa, coffee, rubber, bananas and pineapples. It is the world's largest producer of cocoa and the fourth largest producer of coffee. These two crops bring in half the country's export revenue. Since independence industrialization has developed rapidly, particularly food processing, textiles and sawmills.

Quick facts:
Area : 322,463 sq km (124,503 sq miles)
Population : 13,316,000
Capital : Yamoussoukro
Other major cities : Abidjan, Bouaké, Daloa
Form of government : Republic
Religions : Animism, Sunni Islam, RC
Currency : CFA Franc

Cotonou a port and the main business centre of BENIN. (Pop. 537,000)

Cotswold Hills a range of hills in western central ENGLAND, lying to the east of the River SEVERN.

Coventry an industrial city in the WEST MIDLANDS of ENGLAND. (Pop. city 305,000/urban area 348)

Cracow (Krakow) the third largest city in POLAND, and the capital during medieval times. (Pop. 751,000)

Craiova an industrial city in south-west ROMANIA. (Pop. 303,000)

Cremona a town on the River Po in central northern ITALY, famous for the manufacture of violins, especially those of Antonio Stradivari (?1644-1737). (Pop. 80,800)

Crete (Krití) the largest and most southerly of the islands of GREECE, with important ruins of the Minoan civilization at KNOSSOS. The capital is HERAKLION. (8366 sq km/3229 sq miles; pop. 537,000)

Crimea (Krym) a diamond-shaped peninsula jutting out into the northern part of the BLACK SEA and an autonomous region of the UKRAINE. (25,900 sq km/10,000 sq miles; pop. 2,500,000)

Croatia (Hrvatska), a republic of former YUGOSLAVIA, made a unilateral declaration of independence on June 25, 1991. Sovereignty was not formally recognized by the international community until early in 1992. Located in south-east EUROPE, it is bounded to the west by the ADRIATIC SEA, to the north by SLOVENIA and ROMANIA, and to the south by BOSNIA & HERZEGOVINA. Western Croatia lies in the Dinaric Alps. The eastern region, drained by the rivers Sava and DRAVA, is low-lying and agricultural. The chief farming region is the Pannonian Plain. Over one third of the country is forested and timber is a major export. Deposits of coal, bauxite, copper, oil and iron ore are substantial, and most of the republic's industry is based on the processing of these. In Istria in the north-west and on the Dalmatian coast tourism was a major industry until the Croatia became embroiled in the Serbo-Croat war prior to its secession in 1992. Following the formal recognition of Croatia's independence by the international community, the fighting abruptly ceased, however, the tourism industry continued to suffer from the effects of the on-going hostilities in other parts of the former Yugoslavia.

Quick facts:
Area : 56,538 sq km (21,824 sq miles)
Population : 4850,500
Capital : Zagreb
Other major cities : Rijeka, Split
Form of government : Republic
Religions : RC, Eastern Orthodox
Currency : Kuna

Crozet Islands a group of some 20 islands and islets in the SOUTHERN OCEAN, forming part of the FRENCH SOUTHERN AND ANTARCTIC TERRITORIES. (300 sq km/116 sq miles)

Cuango, River *see* **Kwango, River.**

Cuba is the largest and most westerly of the Greater ANTILLES group of islands in the WEST INDIES. It is strategically positioned at the entrance to the Gulf of MEXICO and lies about 140 km (87 miles) south of the tip of FLORIDA. Cuba is as big as all other CARIBBEAN islands put together and is home to a third of the whole West Indian population. The climate is warm and generally rainy and hurricanes are liable to occur between June and November. The island consists mainly of extensive plains and the soil is fertile. The most important agricultural product is sugar and its by-products, and the processing of these is the most important industry. Most of Cuba's trade was with other communist countries, particularly the former USSR, and the country's economy has suffered as a result of a US trade embargo.

Quick facts:
Area : 110,861 sq km (42,803 sq miles)
Population : 10,905,000
Capital : Havana (La Habana)
Other major cities : Camaguey, Holguin, Santiago de Cuba
Form of government : Socialist Republic
Religion : RC
Currency : Cuban peso

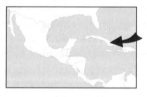

Cubango, River *see* **Kavango, River.**

Cúcuta a city in northern COLOMBIA on the border with VENEZUELA. (Pop. 450,000)

Cuenca a city in southern ECUADOR, founded by the Spanish in 1557, but also the site of a number of important Inca ruins. (Pop. 195,000)

Cuernavaca an old resort town in the mountains 80 km (50 miles) to the south of MEXICO CITY. (Pop. 557,000)

Culloden an area of moorland about 8 km (5 miles) to the east of INVERNESS in SCOTLAND, the scene of a bloody battle in 1746 in which the forces of the Young Pretender, Bonnie Prince Charlie, were defeated by the English under the Duke of Cumberland.

Cumberland *see* **Cumbria**.

Cumbria a county in north-west ENGLAND, created in 1974 from the old counties of Cumberland, Westmorland and a part of LANCASHIRE. (6809 sq km/2629 sq miles; pop. 483,000)

Curaçao an island in the CARIBBEAN lying just off the coast of VENEZUELA but a part of the NETHERLANDS ANTILLES. (444 sq km/171 sq miles; pop. 170,000)

Curitaba an industrial city in southern BRAZIL. (Pop. city 1,250,000)

Cuzco a city set in the ANDES mountains in PERU, and the name of the surrounding province. It was a centre of the Inca empire, and there are numerous Inca remains in the region, including MACHU PICCHU. (Pop. 258,000)

Cyclades (Kikládhes) a group of some 220 islands in the middle of the AEGEAN SEA and belonging to GREECE. The best known are Tínos, Andros, Mílos, Míkonos, DELOS, Náxos, Paros, Kithnos, Serifos, Ios and Síros. (Pop. 88,400)

Cyprus is an island which lies in the eastern MEDITERRANEAN about 85 km (53 miles) south of TURKEY. It has a long thin panhandle and is divided from west to east by two parallel ranges of mountains which are separated by a wide central plain open to the sea at either end. The highest point is Mount OLYMPUS (1951 m/6401 ft) in the south-west. The climate is Mediterranean with very hot dry summers and warm damp winters. This contributes towards the great variety of crops grown e.g. early potatoes, vegetables, cereals, tobacco, olives, bananas, and grapes. The grapes are used for the strong wines and sherries for which Cyprus is famous. Fishing is a significant industry, but above all the island depends on visitors and it is the tourist industry which has led to a recovery in the economy since 1974.

Quick facts:
Area : 9251 sq km (3572 sq miles)
Population : 740,000
Capital : Nicosia
Other major cities : Limassol, Larnaca
Form of government : Republic
Religions : Greek Orthodox, Sunni Islam
Currency : Cyprus pound

Czechoslovakia *see* **Czech Republic; Slovakia.**

Czech Republic, The, was newly constituted on January 1,1993, with the dissolution of the 74-year-old federal republic of Czechoslovakia. It is landlocked at the heart of central EUROPE, bounded by SLOVAKIA, GERMANY, POLAND and AUSTRIA. Natural boundaries are formed by the Sudeten Mountains in the north, the Erzgebirge, or Ore Mountains, to the north-west, and the Bohemian Forest in the south-west. The climate is humid continental with warm summers and cold winters. Most rain falls in summer and thunderstorms are frequent. Agriculture, although accounting for only a small percentage of the national income, is highly developed and efficient. Major crops are sugar beet, wheat and potatoes. Over a third of the labour force is employed in industry, the most important being iron and steel, coal, machinery, cement and paper. Recently investment has gone into electronic factories and research establishments.

Quick facts:
Area : 78,864 sq km (30,449 sq miles)
Population : 10,325,700
Capital : Prague (Praha)
Other major cities : Brno, Ostrava, Plzen
Form of government : Republic
Religions : RC, Protestantism
Currency : Koruna

Czestochowa (Chenstokhov) an industrial city in southern POLAND. Its Jasna Gora Monastery is a national shrine and pilgrimage centre. (Pop. 258,000)

Dacca *see* **Dhaka.**

Dachau a market town in BAVARIA, GERMANY, and the site of the first concentration camp to be built by the Nazis, in 1935. (Pop. 35,000)

Dagestan an autonomous republic of the RUSSIAN FEDERATION lying to the west of the CASPIAN SEA. The capital is MAKHACHKALA. (50,300 sq km/ 19,400 sq miles; pop. 1,890,000)

Dakar the main port and capital of SENEGAL. (Pop. 1,730,000).

Dalian *see* **Lüda.**

Dakota *see* **North Dakota; South Dakota.**

Dal, Lake the most famous of the lakes of KASHMIR, India, by SRINAGAR.

Dallas a city in north-east TEXAS, USA. (Pop. city 1,022,500/metropolitan area 4,215,000)

Dalmatia (Dalmacija) the coast of CROATIA, on the

ADRIATIC SEA. The main islands of the coast are Krk, Rab, Losinj, Brac, Hvar, Korcula and Mljet. The principal tourist centre is DUBROVNIK.

Damascus (Dimashq) the capital of SYRIA, an oasis town. (Pop. 1,451,000)

Damavand, Mount an extinct volcano, and the highest peak in the ELBURZ MOUNTAINS, IRAN. (5670 m/18,600 ft)

Danube, River (Donau) the longest river in Western EUROPE, rising in the BLACK FOREST in GERMANY, and passing through AUSTRIA, SLOVAKIA, HUNGARY and SERBIA of former Yugoslavia, forming much of the border between BULGARIA and ROMANIA before turning north and forming a delta on the BLACK SEA. (Length 2850 km/1770 miles)

Danzig *see* **Gdansk**.

Dardanelles the narrow ribbon of water, some 80 km(50 miles) long, in TURKEY which connects the AEGEAN SEA to the Sea of MARMARA (and from thence the BLACK SEA). GALLIPOLI is on the peninsula to the north. The Dardanelles were known as the Hellespont to the ancient Greeks.

Dar el Beida *see* **Casablanca**.

Dar es Salaam the largest town and main port of TANZANIA. It was the national capital until 1974 when the seat of government was transferred to DODOMA. (Pop. 1,371,000)

Darién the eastern province of PANAMA, a narrow neck of land on the border with COLOMBIA, and the only gap in the Pan-American Highway, which otherwise runs from ALASKA to CHILE.

Darjiling (Darjeeling) a town in WEST BENGAL, INDIA near the border with NEPAL, famous for its tea. (Pop. 282,200)

Darling, River a river flowing from southern QUEENSLAND through NEW SOUTH WALES in AUSTRALIA before converging with the MURRAY RIVER. (Length 3057 km/1900 miles)

Dartmoor an area of remote moorland in the county of DEVON, ENGLAND. (945 sq km/365 sq miles)

Darwin the capital of the NORTHERN TERRITORY, AUSTRALIA. (Pop. 73,000)

Datong (Tatung) an industrial city in SHANXI province, CHINA. (Pop. 798,000)

Davao a city in the southern part of the island of MINDANAO, PHILIPPINES, and now that country's second largest city. It is also the name of the surrounding region. (Pop 868,000)

Davis Strait the broad strait, some 290 km (180 miles) across at its narrowest, separating BAFFIN ISLAND in CANADA and GREENLAND.

Dead Sea a small sea on the border between ISRAEL and JORDAN into which the River JORDAN flows and

does not exit. It is one of the lowest places on Earth (396 m/1299 ft below normal sea level) and the body of water with the world's highest salt content. (1049 sq km/395 sq miles)

Death Valley a low-lying area of desert and salt beds in eastern CALIFORNIA, USA. At 80 metres/250 feet below sea level it is the lowest point in the USA. It is extremely arid and hot, with temperatures rising to 51°C (125°F).

Debrecen an agricultural and industrial centre in eastern HUNGARY, which has grown up around the original medieval town. (Pop. 217,000)

Deccan the broad, triangular plateau which forms much of the southern part of INDIA.

Dehra Dun a town in northern INDIA, in the foothills of the HIMALAYAS. It is famous for a military academy established by the British in the 1930s. (Pop. 367,000)

Delaware a state on the east coast of the USA, and the second smallest in the USA after RHODE ISLAND. The capital is DOVER. (5328 sq km/2057 sq miles; pop. 690,880)

Delft a small city in central western NETHERLANDS, famous since the 16th century for its distinctive blue and white pottery. (Pop. 89,300)

Delhi, including NEW DELHI, the capital of INDIA, in the north of the country, on the YAMUNA River. (Pop. 8,375,190)

Delos the smallest of the islands in the CYCLADES group, Greece, said to be the birthplace of Apollo.

Delphi the ruins of the Temple of Apollo on Mount Parnassos, 166 km (102 miles) north-west of ATHENS, GREECE. It was the seat of the most important oracle of ancient Greece

Demerara, River a river in central GUYANA which flows through the capital, GEORGETOWN. It has given its name to the type of brown sugar which is grown in the region. (Length 320 km/200 miles)

Denbighshire a county in the north-west of Wales. In 1974 it was merged, together with parts of FLINTSHIRE and Merionethshire, into the regions of GWYNEDD and Clwyd. In 1996 the county was re-established as part of the reorganization of local authorities. (840 sq km/324 sq miles; pop. 91,000)

Den Haag *see* **Hague, The**.

Denmark is a small European state lying between the NORTH SEA and the entrance to the BALTIC. It consists of a western peninsula and an eastern archipelago of 406 islands only 89 of which are populated. The country is very low lying and the proximity of the sea combined with the effect of the Gulf Stream result in warm sunny summers and cold cloudy winters. The scenery is very flat and monotonous but the soils are good and a wide variety of crops can be grown. It is an agricultural country and three-quarters of the land is cultivated mostly by the rotation of grass, barley, oats and sugar beet. Animal husbandry is however the most important activity, its produce including the famous bacon and butter. Despite Denmark's limited range of raw materials it produces a wide range of manufactured goods and is famous for its imaginative design of furniture, silverware and porcelain.

Quick facts:
Area : 43,077 sq km (16,632 sq miles)
Population : 5,215,710 (excluding the Faeroe Islands)
Capital : Copenhagen (København)
Other major cities : Ålborg, Århus, Odense
Form of government : Constitutional Monarchy
Religion : Lutheranism
Currency : Danish krone

Denmark Strait the arm of the North ATLANTIC OCEAN which separates ICELAND from GREENLAND, some 290 km (180 miles) apart.

Denpasar the capital of the island of BALI, INDONESIA. (Pop. 82,140)

Denver the state capital of COLORADO, USA. (Pop. city 483,850/metropolitan area 2,089,000)

Derby a city of Saxon and Danish origins in the county of DERBYSHIRE, ENGLAND. (Pop. 230,500)

Derbyshire a county in north central ENGLAND; the county town is MATLOCK. (2631 sq km/1016 sq miles; pop. 954,000)

Derry *see* **Londonderry**.

Des Moines the state capital of IOWA, USA. (Pop. city 195,540)

Detroit a major industrial city and GREAT LAKES port in the state of MICHIGAN, USA. (Pop. city 1,012,110/metropolitan area 5,246,000)

Devon a county in south-west ENGLAND; the county town is EXETER. (6715 sq km/2593 sq miles; pop. 1,053,000)

Dhahran a commercial centre with an important international airport in eastern SAUDI ARABIA. It is also a centre for petroleum extraction business. It was used as a military base during the 1991 Gulf War

Dhaka (Dacca) the capital of Bangladesh, on the delta of the Rivers GANGES and BRAHMAPUTRA. (Pop. urban area 6,537,000)

Dhanbad a city in north-east INDIA and a centre for the coal mining industry of the Damodar Valley. (Pop. 818,000)

Dhaulagiri, Mount a mountain peak of the Great HIMALAYAS in NEPAL. (8172 m/26,810 ft)

Dhodhekanisos *see* **Dodecanese**.

Dien Bien Phu a small town in north-west VIETNAM where the French army was decisively defeated by the Communist and Nationalist Vietminh forces in 1954, effectively forcing the French to leave Vietnam.

Dijon the historic capital of the Bourgogne region (BURGUNDY) in western central FRANCE, famous in particular for its mustard. (Pop. 231,000)

Dimashq *see* **Damascus**.

Diyarbakir a city on the River TIGRIS in south-eastern TURKEY, and the name of the province of which it is the capital. (Pop. city 381,000)

Djerba *see* **Jerba**.

Djibouti is situated in north-east AFRICA and is bounded almost entirely by ETHIOPIA except in the south-east where it shares a border with SOMALIA and in the north-west where it shares a border with ERITREA. Its coastline is on the Gulf of ADEN. The land which is mainly basalt plains has some mountains rising to over 1500 m (4922 ft). The climate is hot, among the world's hottest, and extremely dry. Only a tenth of the land can be farmed even for grazing so it has great difficulty supporting its modest population. The native population are mostly nomadic, moving from oasis to oasis or across the border to Ethiopia in search of grazing land. Most foodstuffs for the urban population in Djibouti city are imported. Cattle, hides and skins are the main exports.

Quick facts:
Area : 23,200 sq km (8958 sq miles)
Population : 520,000
Capital : Djibouti (population 340,700)
Form of government : Republic
Religion : Sunni Islam
Currency : Djibouti franc

Dnepr, River *see* **Dnieper, River**.

Dnepropetrovsk an industrial and agricultural city on the River DNIEPER in the UKRAINE. It was formerly (1787-96 and 1802-1920) known as Ekaterinoslav. (Pop. 1,190,000)

Dnestr, River *see* **Dniester, River**.

Dnieper (Dnepr), River the third longest river in EUROPE after the VOLGA and the DANUBE, flowing south through the RUSSIAN FEDERATION and the UKRAINE to the BLACK SEA via KIEV. (Length 2285 km/1420 miles)

Dniester (Dnestr), River a river flowing through the UKRAINE and MOLDOVA to the BLACK SEA. (Length 1411 km/877 miles)

Dodecanese (Dhodhekanisos) a group of twelve islands belonging to GREECE in the eastern AEGEAN SEA near the coast of TURKEY. They are scattered between Samos in the north and Karpathos in the south and include Patmos, Kalimnos, KOS and RHODES, the largest in the group. They are also called the Southern Sporades. (2663 sq km/1028 sq miles

Dodoma the capital (since 1974) of TANZANIA, in the centre of the country. (Pop. 88,475)

Doha (Ad Dawhah) the capital of QATAR. (Pop. 220,000)

Dolomites a range of mountains in north-eastern ITALY, near the border with AUSTRIA. The highest point is Mount Marmolada (3342 m/10,964 ft)

Dominica discovered by Columbus, Dominica is the most northerly of the WINDWARD ISLANDS in the WEST INDIES. It is situated between the islands of MARTINIQUE and GUADELOUPE. The island is very rugged and with the exception of 225 sq km (87 sq miles) of flat land, it consists of three inactive volcanoes, the highest of which is 1447 m (4747 ft). The climate is tropical and even on the leeward coast it rains two days out of three. The wettest season is from June to October when hurricanes often occur. The steep slopes are difficult to farm but agriculture provides almost all Dominica's exports. Bananas are the main agricultural export but copra, citrus fruits, cocoa, bay leaves and vanilla are also revenue earners. Industry is mostly based on the processing of the agricultural products.

Quick facts:
Area : 751 sq km (290 sq miles)
Population : 871,200
Capital : Roseau
Form of govt : Republic
Religion : RC
Currency : Franc

Dominican Republic forms the eastern portion of the island of HISPANIOLA in the WEST INDIES. It covers two-thirds of the island, the smaller portion consisting of HAITI. The west of the country is made up of four almost parallel mountain ranges and between the two most northerly is the fertile Cibao valley. The south-east is made up of fertile plains. Although well endowed with fertile land, only about 30% is cultivated. Sugar is the main crop and mainstay of the country's economy. It is grown mainly on plantations in the south-east plains. Other crops grown are coffee, cocoa and tobacco. Some mining of gold, silver, platinum, nickel and aluminium is carried out but the main industries are food processing and making consumer goods. The island has fine beaches and the tourism industry is now very important to the economy.

Quick facts:
Area : 48,734 sq km (18,816 sq miles)
Population : 7,680,000
Capital : Santo Domingo
Other major city : Santiago de los Caballeros
Form of government : Republic
Currency : Dominican peso

Don, River a river flowing southwards into the SEA of AZOV from its source to the south of MOSCOW. (Length 1870 km/1165 miles)

Donau *see* **Danube, River**.

Donbass *see* **Donets Basin**.

Donegal the northern-most county of the Republic of IRELAND, on the west coast. The county town is also called Donegal. (Pop. county 127,900)

Donets Basin (Donbass) a coal mining region and major industrial area in the eastern UKRAINE.

Donetsk the main industrial centre of the DONETS BASIN. (Pop. 1,120,000)

Dongbei (Manchuria) the north-eastern region of CHINA, covering part of the NEI MONGOL AUTONOMOUS REGION and the three provinces, HEILONGJIANG, JILIN and LIAONING.

Dordogne, River a river of south-western France which rises in the MASSIF CENTRAL and flows west to the GIRONDE estuary. (Length 475 km/295 miles)

Dordrecht a river port and industrial city of medieval origin 19 km (12 miles) south-east of ROTTER-DAM in the NETHERLANDS. (Pop. 212,200)

Dorset a county of south-west ENGLAND; the county town is Dorchester. (2654 sq km/1025 sq miles; pop. 618,000)

Dortmund a major city in the industrial RUHR region of western GERMANY. (Pop. 602,000)

Douala the main port of CAMEROON, on the Gulf of GUINEA. (Pop. 884,000)

Douro (Duero), River a river flowing west from its source in northern central SPAIN and across northern PORTUGAL to the ATLANTIC OCEAN near OPORTO. (Length 895 km/555 miles)

Dover (1) a port in the county of KENT, ENGLAND, overlooking the ENGLISH CHANNEL at its narrowest point, opposite CALAIS, FRANCE. (Pop. 33,000). **(2)** The state capital of DELAWARE, USA. (Pop. 28,230)

Dover, Strait of the stretch of water separating ENGLAND and FRANCE, where the ENGLISH CHANNEL meets the NORTH SEA. The ports of DOVER and CALAIS are situated on either side of its narrowest point, 34 km (21 miles) across.

Down a county of NORTHERN IRELAND, on the east coast; the county town is Downpatrick. (2448 sq km/945 sq miles; pop. 339,200)

Drakensberg Mountains a range of mountains which stretch 1125 km (700 miles) across LESOTHO and neighbouring regions of SOUTH AFRICA. The highest point is Thabana Ntlenyana (3482 m/11,424 ft).

Drake Passage the broad strait, some 640 km (400 miles) wide, which separates Cape HORN on the southern tip of South AMERICA and ANTARCTICA.

Drava (Drau), River a river flowing from eastern AUSTRIA to CROATIA and SERBIA, where it forms much of the border with HUNGARY before joining the DANUBE. (Length 718 km/447 miles)

Dresden a historic city on the River ELBE in the south of eastern GERMANY. Formerly the capital of SAXONY, it was noted particularly for its fine porcelain. (Pop. 480,000)

Duarte, Pico a mountain peak in central DOMINICAN REPUBLIC which is the highest point in the WEST INDIES. (3175 m/10,417 ft)

Dubai (Dubayy) the second largest of the UNITED ARAB EMIRATES, at the eastern end of The GULF. Most of the population lives in the capital, also called Dubai. (3900 sq km/1506 sq miles; pop. emirate 296,000/city 266,000)

Dublin (Baile Atha Cliath) the capital of the Republic of IRELAND, on the River LIFFEY, and also the name of the surrounding county. Its main port area is at Dun Laoghaire. (Pop. county 1,024,000/city 481,000)

Dubrovnik (Ragusa) a pretty medieval port on the ADRIATIC coast of CROATIA, for long a popular tourist destination. (Pop. 31,200)

Duero, River see **Douro, River**.

Duisburg a major inland port situated at the confluence of the Rivers RHINE and RUHR in GERMANY. (Pop. 537,000)

Duluth a port and industrial centre on Lake SUPERIOR, in the state of MINNESOTA, USA. (Pop. city 85,430)

Dumfries and Galloway a region of south-west SCOTLAND created out of the old counties of Dumfriesshire, Kirkudbrightshire and Wigtownshire. The regional capital is Dumfries. (6370 sq km/2459 sq miles; pop. 145,200)

Dunbartonshire see **Strathclyde**.

Dundee a port on the east coast of SCOTLAND, on the north side of the Firth of Tay, and the administrative centre of TAYSIDE region. (Pop. 172,000)

Dunfermline see **Fife**.

Dunkirk (Dunkerque) a port and industrial town in north-eastern FRANCE, close to the border with BELGIUM. It was virtually destroyed in 1940 when British, French and Belgian forces were trapped by the advancing German army, but were successfully evacuated to Britain in a fleet of small boats. (Pop. 191,000)

Dun Laoghaire see **Dublin**.

Durango a mineral-rich state in northern MEXICO, with a capital called (Victoria de) Durango. (Pop. state 1,200,000/city 348,000)

Durban a port on the east coast of SOUTH AFRICA, and the largest city of the province of NATAL. (Pop. 1,137,000)

Durham a city of north-east ENGLAND, and the name of the county of which it is the county town. (County 2436 sq km/940 sq miles; pop. county 607,800/city 89,000)

Dushanbe an industrial city and the capital of TAJIKISTAN. (Pop. 595,000)

Düsseldorf a major commercial and industrial centre in the RUHR region of western GERMANY, situated on the River RHINE 34 km (21 miles) north of COLOGNE. (Pop. 575,000)

Dvina, River the name of two quite separate rivers. The West (Zapadnaya) Dvina flows from its source to the west of MOSCOW into the BALTIC SEA at RIGA in LATVIA. The North (Severnaya) Dvina flows through the north-west of the RUSSIAN FEDERATION to the WHITE SEA at ARCHANGEL. (Length West Dvina 1020 km/635 miles; North Dvina 1320 km/820 miles)

Dwarka one of the seven holy cities of the Hindus, situated on the tip of the peninsula to the south of the Gulf of KUTCH, in north-west INDIA. It is said to be the capital of the god Krishna. (Pop. 21,400)

Dyfed a former county in south-west WALES, created in 1974 out of the old counties of Pembrokeshire, Cardiganshire and Carmarthenshire. It ceased to exist with local authority reorganization of 1996.

Dzungaria *see* **Xinjiang Uygur Autonomous Region.**

East Anglia an old Anglo-Saxon kingdom occupying the bulge of the east coast of ENGLAND between the THAMES estuary and The WASH, and now covered by the counties of NORFOLK, SUFFOLK, and parts of CAMBRIDGESHIRE and ESSEX.

Easter Island (Isla de Pascua) a remote and tiny island in the South PACIFIC OCEAN annexed by CHILE in 1888. About 1000 years ago it was settled by Polynesians who set up over 600 huge stone statues of heads on the island. (120 sq km/46 sq miles; pop. 2000)

East Timor *see* **Timor.**

Ebro, River a river flowing across north-eastern SPAIN, from its source near the north coast to the MEDITERRANEAN SEA south of TARRAGONA. (Length 909 km/565 miles)

Ecuador is an Andean country situated in the northwest of the South American continent. It is bounded to the north by COLOMBIA and to the east and south by PERU. The country contains over thirty active volcanos. Running down the middle of Ecuador are two ranges of the Andes which are divided by a central plateau. The coastal area consists of plains and the eastern area is made up of tropical jungles. The climate varies from equatorial through warm temperate to mountain conditions according to altitude. It is in the coastal plains that plantations of bananas, cocoa, coffee and sugar cane are found. In contrast to this the highland areas are adapted to grazing, dairying and cereal growing. The fishing industry is important on the PACIFIC Coast and processed fish is one of the main exports. Oil is produced in the eastern region and crude oil is Ecuador's most important export.

Quick facts:
Area : 283,561 sq km (109,484 sq miles)
Population : 10,981,000
Capital : Quito
Other major cities : Guayaquil,
　Cuenca
Form of government : Republic
Religion : RC
Currency : Sucre

Edinburgh the capital of SCOTLAND, a university city and commercial centre, on the Firth of Forth (the estuary of the River Forth). (Pop. 447,550)

Edmonton the capital of ALBERTA, CANADA. (Pop. 840,000; city 574,000)

Edo *see* **Tokyo.**

Edward (Rutanzige), Lake a lake in the GREAT RIFT VALLEY, on the border between UGANDA and ZAÏRE. (2135 sq km/820 sq miles)

Egypt is situated in north-east AFRICA, acting as the doorway between Africa and ASIA. Its outstanding physical feature is the river NILE, the valley and delta of which cover about 35,580 sq km (13,737 sq miles). The climate is mainly dry but there are winter rains along the Mediterranean coast. The temperatures are comfortable in winter but summer temperatures are extremely high particularly in the south. The rich soils deposited by floodwaters along the banks of the Nile can support a large population and the delta is one of the world's most fertile agricultural regions. 96% of the population live in the delta and Nile valley where the main crops are rice, cotton, sugar cane, maize, tomatoes and wheat. The main industries are food processing and textiles. The economy has been boosted by the discovery of oil and although not in large quantities it is enough to supply Egypt's needs and leave surplus for export. The Suez Canal shipping and tourism connected with the ancient sites are also important revenue earners.

Quick facts:
Area : 1,001,449 sq km (386,662 sq miles)
Population : 58,980,000
Capital : Cairo (El Qahira)
Other major cities : Alexandria, El Gîza,
　Port Said
Form of government : Republic
Religions : Sunni Islam, Christianity
Currency : Egyptian pound

Eifel an upland area of western GERMANY between the MOSELLE River and the border with BELGIUM.

Eiger, The a mountain in southern central SWITZERLAND, renowned among climbers for its daunting north face. (3970 m/13,025 ft)

Eilat *see* **Elat.**

Eindhoven an industrial city in the southern central part of the NETHERLANDS. (Pop. 194,000)

Eire *see* **Ireland, Republic of.**

Ekaterinburg *see* **Yekaterinburg.**

Ekaterinoslav *see* **Dnepropetrovsk.**

El Alamein a village on the MEDITERRANEAN coast of EGYPT, to the south-west of ALEXANDRIA, which gave its name to the battle fought between Allied troops under General Montgomery and German troops under General Rommel in 1942.

Elat (Eilat) a port and tourist resort in the very south of ISRAEL at the tip of the Gulf of AQABA, an arm of the RED SEA. (Pop. 19,500)

Elba an island lying about 10 km (6 miles) off the coast of TUSCANY, ITALY. (223 sq km/86 sq miles; pop. 29,000)

Elbe, River a largely navigable river flowing northward from its source in the CZECH REPUBLIC through GERMANY to HAMBURG, and then into the NORTH SEA. (Length 1160 km/720 miles)

Elbrus, Mount the highest mountain in EUROPE, situated in the western CAUCASUS, RUSSIAN FEDERATION. (5642 m/18,510 ft)

Elburz Mountains a range of mountains in northern IRAN, between TEHRAN and the CASPIAN SEA. The highest peak is the extinct volcano, DAMAVAND (5670 m/18,600 ft).

El Djazair *see* **Algiers.**

Eleuthera *see* **Bahamas.**

El Faiyum (Fayum) a large and fertile oasis to the west of the River NILE in EGYPT. (Pop. 250,000)

El Gezira a major irrigation scheme in SUDAN between the Blue NILE and the White Nile.

El Gîza a sprawling suburb of CAIRO, EGYPT, at the edge of which stand the three most famous pyramids of the Ancient Egyptians. (Pop. 2,144,000)

Elisabethville *see* **Lubumbashi.**

El Khartum *see* **Khartoum.**

El Mansura a city in the delta of the River NILE in northern EGYPT. (Pop. 371,000)

El Paso a city in western TEXAS, USA, close to the border with MEXICO. (Pop. city 463,000/metropolitan area 544,000)

El Qahira *see* **Cairo.**

El Salvador is the smallest and most densely populated state in Central AMERICA. It is bounded north and east by HONDURAS and has a PACIFIC coast to the south. Two volcanic ranges run from east to west across the country. The Lempa river cuts the southern ranges in the centre of the country and opens as a large sandy delta to the Pacific Ocean. Although fairly near to the equator, the climate tends to be warm rather than hot and the highlands have a cooler temperate climate. The country is predominantly agricultural and 32% of the land is used for crops such as coffee, cotton, maize, beans, rice and sorghum, and a slightly smaller area is used for grazing cattle, pigs, sheep and goats. A few industries such as food processing, textiles and chemicals are found in the major towns.

> *Quick facts:*
> Area : 21,041 sq km (8124 sq miles)
> Population : 5,743,000
> Capital : San Salvador
> Other major cities : Santa Ana, San Miguel
> Form of government : Republic
> Religion : RC
> Currency : Colón

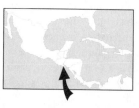

Elsinore (Helsingør) a town of medieval origins on the island of ZEALAND, DENMARK, to the north of COPENHAGEN. Kronborg Castle, which dominates the town, is the setting for Shakespeare's play *Hamlet*. (Pop. 57,000)

Emilia-Romagna a region on the east coast of northern central ITALY; the capital is BOLOGNA. (22,123 sq km/8542 sq miles; pop. 3,984,000)

Emmenthal the valley of the River Emme, in SWITZERLAND, famous for its distinctive cheese.

Empty Quarter *see* **Rub al-Khali.**

Enewetak *see* **Marshall Islands.**

Engel's an industrial town on the River VOLGA, in the RUSSIAN FEDERATION. (Pop. 216,000)

England the country occupying the greater part of the island of GREAT BRITAIN, and the largest of the countries that make up the UNITED KINGDOM. SCOTLAND lies to the north and WALES to the west. The capital is LONDON. (130,410 sq km/50,351 sq miles; pop. 48,708,000)

English Channel the arm of the eastern ATLANTIC OCEAN which separates the south coast of ENGLAND from FRANCE.

Enschede an industrial town in the eastern part of the NETHERLANDS, close to the border with GERMANY. (Pop. 147,000)

Entebbe a town with an international airport on Lake VICTORIA, UGANDA. It was the capital until 1962. (Pop. 42,000)

Enugu a coal-mining state in southern central NIGERIA, its capital is the city of Enugu . It was the capital of Biafra (IBOLAND) during the Civil War (1967-70). (Pop. state 3,161,000)

Eolian (Lipari) Islands a group of small volcanic islands which lie between the north coast of SICILY and mainland ITALY. The main islands are STROMBOLI, LIPARI, Salina, Panarea and Vulcano. (Pop. 12,500)

Ephesus the ruins of an ancient Greek city on the east coast of TURKEY, overlooking the AEGEAN SEA. Its Temple of Diana (or Artemis) was one of the Seven Wonders of the Ancient World.

Eptanisos *see* **Ionian Islands**.

Equatorial Guinea lies about 200 km (124 miles) north of the Equator on the hot humid coast of West AFRICA. The country consists of a square-shaped mainland area (Mbini) with its few small offshore islets, and the islands of Bioko and Pagalu. The climate is tropical and the wet season in Bioko and Pegalu lasts from December to February. Bioko is a very fertile volcanic island and it is here the capital Malabo is sited beside a volcanic crater flooded by the sea. It is also the centre of the country's cocoa production. The country now relies heavily on foreign aid. There is, however, much potential for a tourist industry.

Quick facts:
Area : 28,051 sq km (10,830 sq miles)
Population : 440,000
Capital : Malabo
Other major city : Bata
Form of government : Republic
Religion : RC
Currency : Franc CFA

Ercolano *see* **Herculaneum**.

Erfurt a historic town and tourist centre in central GERMANY. (Pop. 215,000)

Erie, Lake one of the five GREAT LAKES (the second smallest after Lake ONTARIO), on the border between CANADA and the USA (25,670 sq km/9910 sq miles)

Eritrea, formerly an autonomous province of ETHIOPIA, gained independence in May 1993 shortly after a landslide vote in favor of sovereignty. Bounded by DJIBOUTI, SUDAN and ETHIOPIA, Eritrea has acquired Ethiopia's entire coastline along the RED SEA. Eritrea's climate is hot and dry along its desert coast but is colder and wetter in its

central highland regions. Most of the population depend on subsistence farming. Future revenues may come from developing fishing, tourism and oil industries. Eritrea's natural resources include gold, potash, zinc, copper, salt, fish and probably oil.

Quick facts:
Area : 93,679 sq km (36,170 sq miles)
Population : 3,850,000
Capital : Asmara
Other major cities : Mitsiwa, Keren
Form of government : In transition
Religion : Sunni Islam, Christianity
Currency : Ethiopian birr

Erzurum a market town in western TURKEY, and the name of the surrounding province. (Pop. town 252,700)

Escorial, El (San Lorenzo del Escorial) a small town 40 km (25 miles) north-west of MADRID, famous for its splendid royal palace built (1563-84) for Philip II. (Pop. 9500)

Esfahan (Isfahan) a city in central IRAN noted for its magnificent blue-tiled mosques and other Islamic buildings. (Pop. 987,000)

Eskisehir a spa town in western TURKEY and the name of the surrounding province. (Pop. town 367,300)

Espiritu Santo *see* **Vanuatu**.

Esseg *see* **Osijek**.

Essen an industrial city in western GERMANY, and the largest in the RUHR region. (Pop. 627,000)

Essex a county in south-east ENGLAND; the county town is Chelmsford. (3674 sq km/1419 sq miles; pop. 1,569,900)

Estonia lies to the north-west of the RUSSIAN FEDERATION and is bounded to the north by the Gulf of FINLAND, to the west by the BALTIC SEA and to the south by LATVIA. It is the smallest of the three previous Soviet Baltic Republics. Agriculture and dairy farming are the chief occupations and there are nearly three hundred agricultural collectives and state farms. The main products are grain, potatoes, vegetables, meat, milk and eggs. Livestock includes cattle, sheep, goats and pigs. Almost 22% of Estonia is forested and this provides material for sawmills, furniture, match and pulp industries.

The country has rich, high quality shale deposits and phosphorous has been found near Tallinn. Peat deposits are substantial and supply some of the electric power stations.

Quick facts:
Area : 45,100 sq km (17,413 sq miles)
Population : 1,530,000
Capital : Tallinn
Other major cities : Tartu, Narva
Form of government : Republic
Religion : Eastern Orthodox, Lutheranism
Currency : Rouble

Ethiopia is a landlocked, east African country with borders on Sudan, Kenya, Somalia, Djibouti and Eritrea. Most of the country consists of highlands which drop sharply toward Sudan in the west. Because of the wide range of latitudes, Ethiopia has many climatic variations between the high temperate plateau and the hot humid lowlands. The country is very vulnerable to drought but in some areas thunderstorms can erode soil from the slopes reducing the area available for crop planting. Coffee is the main source of rural income and teff is the main food grain. The droughts in 1989-90 have brought much famine. Employment outside agriculture is confined to a small manufacturing sector in Addis Ababa. The country is wrecked with environmental, economic and political problems that culminated in May 1993 when one of Ethiopia's provinces, Eritrea, became independent.

Quick facts:
Area : 1,221,900 sq km (471,778 sq miles)
Population : 55,500,000
Capital : Addis Ababa (Adis Abeba)
Other major cities : Dire Dawa, Nazret
Form of government : In transition
Religion : Ethiopian Orthodox, Sunni Islam
Currency : Ethiopian birr

Etna, Mount the largest volcano in Europe, situated near the east coast of Sicily, Italy, and still highly active. (3323 m/10,902 ft)

Euboea (Evvoia) a large island in the Aegean Sea lying close to the east coast of mainland Greece and joined to the mainland by a bridge. (3655 sq km/1411 sq miles; pop. 188,400)

Euphrates, River (Al Furat) one of the great rivers of the Middle East, flowing from its source in eastern Turkey, across Syria and central Iraq to The Gulf. (Length 2720 km/1690 miles)

Europe a continent that is divided from Asia by a border that runs down the Ural Mountains to the Caspian Sea and then west to the Black Sea. For convenience it is commonly divided into two areas: Eastern Europe (the countries that have or had Communist governments since the Second World War) and Western Europe. (9,699,000 sq km/ 3,745,000 sq miles; pop. 498,000,000)

Everest, Mount the highest mountain in the world, situated on the border between Nepal and China in the eastern Himalayas. (8848 m/29,028 ft)

Everglades a vast area of subtropical swampland on the western side of southern Florida, USA.

Evvoia *see* **Euboea.**

Eyre, Lake a large salt lake in South Australia. (8900 sq km/3400 sq miles)

Faeroe (Faroe) Islands (Føroyar) a group of 18 islands in the North Atlantic Ocean belonging to Denmark, which lie approximately halfway between Iceland and Scotland. (1399 sq km/540 sq miles; pop. 47,000)

Fair Isle a small island situated between the Orkney and Shetland Islands to the north of Scotland, famous for the distinctive, patterned sweaters made there. (pop. 75)

Faisalabad (Lyallpur) an industrial city and agricultural centre in north-east Pakistan. (Pop. 1,105,000)

Faiyum *see* **El Faiyum.**

Falkland Islands (Islas Malvinas) a British Crown Colony consisting of two large islands and some 200 smaller ones lying about 650 km (410 miles) east of southern Argentina. The capital is Port Stanley. (12,173 sq km/4700 sq miles; pop. 2120, excluding military personnel)

Fao (Al Faw) a port and oil tanker terminal in Iraq, at the mouth of the Shatt al Arab waterway.

Faro the capital of the Algarve province of Portugal. (Pop. 28,200)

Faroe Islands *see* **Faeroe Islands.**

Fatehpur Sikri a magnificent deserted palace complex some 150 km (93 miles) south of Delhi, India, built as a capital by the Moghul Emperor Akbar in 1570 but abandoned in 1584.

Fayum *see* **El Faiyum**.

Fermanagh a lakeland county in the south-west of NORTHERN IRELAND; the county town is Enniskillen. (1676 sq km/647 sq miles; pop. 54,450)

Fernando Póo *see* **Bioko**.

Ferrara a historic city in north-eastern ITALY in the Po Valley. (Pop. 150,300)

Fès (Fez) a city in northern MOROCCO, the oldest of that country's four imperial cities. (Pop. 735,000)

Fife a region of eastern SCOTLAND. The administrative centre is Glenrothes. (1308 sq km/505 sq miles; pop. 351,000)

Fiji is one of the largest nations in the western PACIFIC and consists of some 320 islands and atolls, but only 150 are inhabited. It is situated around the 180° International Date Line and lies about 17° south of the Equator. Fiji has high rainfall, high temperatures and plenty of sunshine all year round. The two main islands, Viti Levu and Vanua Levu, are extinct volcanoes and most of the islands in the group are fringed with coral reefs. The south-east of the islands have tropical rain forests but a lot of timber has been felled and soil erosion is a growing problem. The main cash crop is sugar cane although copra, ginger and fish are also exported. Tourism is now a major industry.

Quick facts:
Area : 18,274 sq km (7056 sq miles)
Population : 773,000
Capital : Suva
Form of government : Republic
Religion : Christianity, Hinduism
Currency : Fiji dollar

Finistère the department of FRANCE occupying the tip of the BRITTANY peninsula. (Pop. 828,000)

Finisterre, Cape the north-west corner of SPAIN.

Finland lies at the eastern limit of western Europe with the RUSSIAN FEDERATION to the east and the Gulf of BOTHNIA to the west. Most of the country is lowlying except for the north which rises to over 1000 m (3281 ft) in LAPPLAND. It is covered with extensive forests and thousands of lakes. The climate has great extremes between summer and winter. Winter is very severe and lasts about six months. Even in the south snow covers the ground for three months in winter. Summers are short but quite warm with light rain throughout the country. Finland is largely self-sufficient in food and produces great surpluses of dairy produce. Most crops are grown in the south-west. In the north reindeer are herded and forests yield great quantities of timber for export. Major industries are timber products, wood pulp and paper, machinery and shipbuilding, which has developed due to the country's great need for an efficient fleet of ice breakers.

Quick facts:
Area : 338,000 sq km (130,500 sq miles)
Population : 5,125,000
Capital : Helsinki (Helsingfors)
Other major cities : Turku, Tampere
Form of government : Republic
Religion : Lutheranism
Currency : Markka

Finland, Gulf of the easternmost arm of the BALTIC SEA, with the southern coast of FINLAND to the north, ST PETERSBURG at its eastern end, and ESTONIA to the south.

Firenze *see* **Florence**.

Flanders (Vlaanderen, Flandre) A Flemish-speaking coastal region of northern BELGIUM, now divided into two provinces, East and West Flanders. (6115 sq km/2361 sq miles; pop. both provinces 2,452,000)

Flinders Range mountains in the eastern part of SOUTH AUSTRALIA, stretching over 400 km (250 miles). St Mary Peak is the highest point (1188 m/3898 ft).

Flintshire a county in north-east WALES, the administrative centre is Mold. (Pop. 144,000)

Florence (Firenze) one of the great Renaissance cities of ITALY, straddling the River ARNO. It is also the capital of the region of TUSCANY. (Pop. 453,300)

Flores a volcanic island in the SUNDA group in INDONESIA, lying in the chain which stretches due east of JAVA. (17,150 sq km/6622 sq miles; pop. 803,000)

Flores Sea a stretch of the PACIFIC OCEAN between FLORES and SULAWESI.

Florida a state occupying the peninsula in the south-eastern corner of the USA. The state capital is TALLAHASSEE. (151,939 sq km/58,664 sq miles; pop. 13,679,000)

Florida, Straits of the waterway which separates the southern tip of FLORIDA, USA from CUBA, some 145 km (90 miles) to the south.

Flushing (Vlissingen) a port on the south-west coast of the NETHERLANDS. (Pop. 46,400)

Fly, River a largely navigable river flowing from the central mountains in western PAPUA NEW GUINEA to its broad estuary on the Gulf of PAPUA to the south. (Length 1200 km/750 miles)

Foggia a city in the PUGLIA region of south-eastern ITALY. (Pop. 158,400)

Fontainebleau a town 55 km (35 miles) south-east of PARIS, FRANCE, with a 16th-century royal château and a famous forest. (Pop. 39,400)

Formosa *see* **Taiwan**.

Fortaleza a major port on the north-eastern coast of BRAZIL. (Pop. city 1,709,000)

Fort-de-France a port and the capital of the island of MARTINIQUE. (Pop. 102,000)

Fort Knox a military reservation in KENTUCKY, USA, 40 km (25 miles) south-west of LOUISVILLE; also the site of the principal depository of the country's gold bullion. (Pop. 38,300)

Fort Lamy *see* **Ndjamena**.

Fort Lauderdale a city and resort on the east coast of FLORIDA, USA, 40 km (25 miles) north of MIAMI. (Pop. city 149,000)

Fort Worth a city in north-east TEXAS, USA, just to the west of DALLAS and part of a Dallas-Fort Worth conurbation (the South-west Metroplex). (Pop. city 454,500)

Foshan an industrial city in GUANGDONG province, CHINA. (Pop. 500,000)

France is the largest country in western EUROPE and has a coastline on the ENGLISH CHANNEL, the MEDITERRANEAN SEA and on the ATLANTIC OCEAN. The lowest parts of the country are the great basins of the north and south-west from which it rises to the MASSIF CENTRAL and the higher ALPS, JURA and PYRÉNÉES. Climate ranges from moderate maritime in the north-west to Mediterannean in the south. Farming is possible in all parts of France. The western shores are ideal for rearing livestock, while the PARIS Basin is good arable land. It is in the south-west around BORDEAUX that the vineyards produce some of the world's best wines. The main industrial area of France is in the north and east, and the main industries are iron and steel, engineering, chemicals, textiles and electrical goods.

Quick facts:
Area : 547,026 sq km (211,208 sq miles)
Population : 58,285,000
Capital : Paris
Other major cities : Bordeaux, Lyon, Marseille, Toulouse
Form of government : Republic
Religion : RC
Currency : Franc

Frankfort the state capital of KENTUCKY, USA. (Pop. 26,800)

Frankfurt (Frankfurt am Main) a major financial, trade and communications centre in central western GERMANY, on the River MAIN. (Pop. 660,000)

Frankfurt an der Oder a town on the River ODER in eastern GERMANY, on the border with POLAND. (Pop. 84,800)

Fraser, River a river flowing through southern BRITISH COLUMBIA, CANADA, from its source in the ROCKY MOUNTAINS to the Strait of GEORGIA by VANCOUVER. (Length 1370 km/850 miles)

Fredericton the capital of NEW BRUNSWICK, CANADA. (Pop. 43,750)

Fremantle *see* **Perth** (Australia).

Freetown the main port and capital of SIERRA LEONE. (Pop. 505,000)

Freiburg (Freiburg im Breisgau) the largest city in the BLACK FOREST in south-west GERMANY, close to the border with FRANCE. (Pop. 197,000)

French Guiana (Guyane) *see* **Guiana (French)**.

French Polynesia a total of about 130 islands in the South PACIFIC OCEAN administered as overseas territories by FRANCE.

French Southern and Antarctic Territories a set of remote and widely scattered territories in ANTARCTICA and the ANTARCTIC OCEAN administered by FRANCE. They include the Crozet Islands and KERGUELEN.

Fresno a city in central eastern CALIFORNIA. (Pop. city 376,000)

Friesian (Frisian) Islands a string of sandy, low-lying islands that line the coasts in the south-eastern corner of the NORTH SEA. The West Friesians (including Terchelling and Texel) belong to the NETHERLANDS; the East Friesians (including Borkum and Norderney) belong to GERMANY; and the North Friesians are divided between Germany and DENMARK.

Frunze *see* **Bishkek.**

Fuji, Mount (Fujiyama) the highest peak in JAPAN, a distinctive volcanic cone 100 km (62 miles) to the south-west of TOKYO. (3776 m/12,389 ft)

Fujian (Fukien) a coastal province in south-east China. The capital is FUZHOU. (120,000 sq km/ 46,350 sq miles; pop. 30,048,000)

Fukuoka a port and the largest city on the island of KYUSHU, JAPAN. (Pop. 1,269,000)

Funafuti the capital of TUVALU, and the name of the atoll on which it is sited. (2.4 sq km/0.9 sq miles; pop. 3400)

Funchal the capital of MADEIRA. (Pop. 47,000)

Fundy, Bay of lies between NOVA SCOTIA and NEW BRUNSWICK, CANADA. It has the world's largest tidal range: 15 m (50 ft) between low and high tide.

Fünen (Fyn) the second largest of the islands of DENMARK, in the centre of the country. (2976 sq km/1048 sq miles; pop. 434,000)

Fushun a mining city in LIAONING province, CHINA, situated on one of the largest coalfields in the world. (Pop. 1,330,000)

Fuzhou an important port and the capital of FUJIAN province, CHINA. (Pop. 1,380,000)

Fyn *see* **Fünen**

Gabès, Gulf of a branch of the MEDITERRANEAN SEA which, with the Gulf of SIRTE to the east, makes a deep indent in the coast of north AFRICA.

Gabon is a small country in west-central AFRICA which straddles the Equator. It has a low narrow coastal plain and the rest of the country comprises a low plateau. Three quarters of Gabon is covered with dense tropical forest. The climate is hot, humid and typically equatorial with little seasonal variation. Until the 1960s timber was virtually Gabon's only resource and then oil was discovered. By the mid-1980s it was Africa's sixth largest oil producer and other minerals such as manganese, uranium and iron ore were being exploited. Much of the earnings from these resources were squandered and most of the Gabonese people remain subsistence farmers. The country has great tourist potential but because of the dense hardwood forests transport links with the interior are very difficult.

Gaborone the capital of BOTSWANA, in the south-east of the country. (Pop. 134,000)

Galapagos Islands a group of 15 islands on the Equator administered by ECUADOR, but located some 1100 km (680 miles) to the west of that country. (7428 sq km/2868 sq miles; pop. 9780)

Galati an inland port on the River DANUBE in eastern ROMANIA, close to the border with the MOLDOVA. (Pop. 326,000)

Galicia a region in the very north-west corner of SPAIN. (Pop. 2,754,000)

Galilee the most northerly region of ISRAEL, bordering LEBANON and SYRIA, with the Sea of Galilee (Lake TIBERIAS) on its eastern side.

Gallipoli (Gelibolu) the peninsula which marks the northern side of the DARDANELLES in TURKEY, and also the name of a port on the peninsula. In World War I, Allied troops (particularly Australians and New Zealanders) suffered heavy losses here in an unsuccessful attempt to take control of the Dardanelles.

Galloway *see* **Dumfries and Galloway.**

Galveston a port in TEXAS, USA, sited on an island in the Gulf of MEXICO. (Pop. city 62,400/metropolitan area 215,400)

Galway a county in the central part of the west coast of IRELAND. The county town is also called Galway, or Galway City. (5940 sq km/2293 sq miles; pop. county 180,000/city 50,800)

Gambia, the smallest country in AFRICA, pokes like a crooked finger into SENEGAL. The country is divided along its entire length by the river Gambia which can only be crossed at two main ferry crossings. Gambia has two very different seasons. In the dry season there is little rainfall, then the south-west monsoon sets in with spectacular storms producing heavy rain for four months. Most Gambians live in villages with a few animals, and grow enough millet and sorghum to feed themselves. Groundnuts are the main and only export crop of any significance. The river provides a thriving local fishing industry and the white sandy beaches on the coast are becoming increasingly popular with foreign tourists.

Quick facts:
Area : 267,667 sq km (103,346 sq miles)
Population : 1,316,000
Capital : Libreville
Other major city : Port Gentile
Form of government : Republic
Religion : RC, Animism
Currency : Franc CFA

Quick facts:
Area : 11,295 sq km (4361 sq miles)
Population : 1,144,000
Capital : Banjul
Form of government : Republic
Religion : Sunni Islam
Currency : Dalasi

Gambia, River a major river of west AFRICA, flowing into the ATLANTIC OCEAN from its source in GUINEA, through SENEGAL and then through GAMBIA, for which it provides a central and vital focus. (Length 1120 km/700 miles)

Gand *see* **Ghent**.

Ganges, River (Ganga) the holy river of the Hindus, flowing from its source in the HIMALAYAS, across northern INDIA and forming a delta in BANGLADESH as it flows into the Bay of BENGAL. (Length 2525 km/1568 miles)

Gansu a mountainous province in northern central CHINA. The capital is LANZHOU. (450,000 sq km/170,000 sq miles; pop. 19,600,000)

Garonne, River a major river of south-western FRANCE, flowing north from its source in the central PYRÉNÉES in SPAIN to BORDEAUX, where it contributes to the GIRONDE estuary. (Length 575 km/355 miles)

Gascony (Gascogne) the historic name of an area in the south-western corner of FRANCE bordering SPAIN.

Gaza Strip a finger of coastal land stretching from the Egyptian border to the MEDITERRANEAN port of Gaza. It borders with ISRAEL to its east and north. It was administered by EGYPT after the creation of Israel in 1948, and became home to numerous Palestinian refugees. It was taken over by Israel in the Six-Day War of 1967 but under an agreement with the Palestine Liberation Organization in 1993, Israel agreed to the gradual withdrawal of military forces and limited self-rule for the territory. (Pop. 650,000)

Gaziantep a town in southern central TURKEY, close to the border with SYRIA, and also the name of the surrounding province. (Pop. town 603,000)

Gdansk (Danzig) The main port of POLAND, on the BALTIC SEA. (Pop. 465,000)

Gdynia (Gdingen) a port on the BALTIC coast of POLAND 16 km (10 miles) north-west of GDANSK. (Pop. 252,000)

Geelong a port and the second largest city of VICTORIA, AUSTRALIA. (Pop. 151,000)

Gelsenkirchen an industrial and coal mining town in the RUHR region of GERMANY. (Pop. 290,000)

Geneva (Genève; Genf) a city in the extreme south-west of SWITZERLAND, at the western end of Lake Geneva, and close to the border with FRANCE. It is also the name of the surrounding canton. (Pop. city 167,200)

Genoa (Genova) the major seaport of north-west ITALY, and the capital of LIGURIA. (Pop. 786,000)

Gent *see* **Ghent**.

Georgetown (1) a port and the main city of PINANG Island, MALAYSIA. (Pop. 325,000). **(2)** the main port and capital of GUYANA. (Pop. 200,000) **(3)** the

capital and main port of the CAYMAN ISLANDS. (Pop. 12,970)

Georgia (1) a state in the south-east of the USA, named after George II by English colonists in 1733; the state capital is ATLANTA. (152,576 sq km/58,910 sq miles; pop. 6,773,000). **(2)** is a republic in the south-west of the former USSR occupying the central and western parts of the CAUCASUS. It shares borders with TURKEY, ARMENIA, AZERBAIJAN and the RUSSIAN FEDERATION. It is bounded to the west by the BLACK SEA. Almost 40% of the country is covered with forests. Agriculture, which is the main occupation of the population, includes tea cultivation and fruit growing, especially citrus fruits and viticulture. The republic is rich in minerals, especially manganese. Industries include coal, timber, machinery, chemicals, silk, food processing and furniture. Georgia declared itself independent in 1991.

Quick facts:
Area : 69,773 sq km (26,911 sq miles)
Population : 5,493,000
Capital : Tbilisi
Other major cities : Kutaisi, Rustavi, Batumi
Form of government : Republic
Religion : Russian Orthodox
Currency : Rouble

Georgia, Strait of the southern part of the stretch of water which separates VANCOUVER ISLAND from the coast of BRITISH COLUMBIA in CANADA.

Germany is a large country in northern central EUROPE which comprises the former East and West German republics, re-united in 1990. In the north is the North German Plain which merges with the North Rhinelands in the west. Further south, a plateau which stretches across the country from east to west, is divided by the river RHINE. In the south-west the BLACK FOREST separates the Rhine Valley from the fertile valleys and scarplands of Swabia. The Bohemian Uplands and Erz Mountains mark the border with the CZECH REPUBLIC. Generally the country has warm summers and cold winters. Agricultural products include wheat, rye, barley, oats, potatoes and sugar beet. The main industrial and most densely populated areas are in the RHUR Valley. Principal industries are mechanical and electrical engineering. Chemical and textile industries are found in the cities along the Rhine and motor vehicle industry in the large provincial cities. The country depends heavily on imports.

Quick facts:
Area : 365,755 sq km (137,738 sq miles)
Population : 81,075,000
Capital : Berlin, Bonn (Seat of government)
Other major cities : Cologne, Frankfurt,
  Hamburg, Leipzig, Munich, Stuttgart
Form of government : Republic
Religions : Lutheranism, RC
Currency : Deutsche Mark

Gezira *see* **El Gezira**.

Ghana is located on the southern coast of West AF-RICA between CÔTE D'IVOIRE and TOGO. In 1957, as the former British GOLD COAST, it became the first black African state to achieve independence from European colonial rule. It has palm-fringed beaches of white sand along the Gulf of GUINEA and where the great river VOLTA meets the sea there are peaceful blue lagoons. The climate on the coast is equatorial and towards the north there are steamy tropical evergreen forests which give way in the far north to tropical savanna. The landscape becomes harsh and barren near the border with BURKINA FASO. Most Ghanaians are village dwellers whose homes are made of locally available materials. The south of the country has been most exposed to European influence and it is here that cocoa, rubber, palm oil and coffee are grown. Ghana has important mineral resources such as manganese and bauxite. Most of Ghana's towns are in the south but rapid growth has turned many of them into unplanned sprawls.

Quick facts:
Area : 238,537 sq km (92,100 sq miles)
Population : 17,460,000
Capital : Accra
Other major cities : Kumasi, Tamale,
  Sekondi-Takoradi
Form of government : Republic
Religion : Protestant,
  Animism, RC
Currency : Cedi

Ghats the two ranges of mountains that line the coasts of the DECCAN peninsula in INDIA: the Eastern Ghats (rising to about 600 m/2000 ft) and the Western Ghats (1500 m/5000 ft).

Ghent (Gent; Gand) a medieval city spanning the Rivers Lys and SCHELDE and the capital of the province of East FLANDERS, BELGIUM. (Pop. city 230,000/ metropolitan area 490,000)

Gibraltar a self-governing British Crown Colony on the south-western tip of SPAIN, where a limestone hill called the Rock of Gibraltar rises to 425 m (1394 ft). Its commanding view over the Strait of Gibraltar has made the territory strategically significant. Spain lays claim to Gibraltar, but the UK is reluctant to relinquish it. English is the official language, although Spanish is also spoken. The capital is Gibraltar Town. (6.5 sq km/2.5 sq miles; pop. 32,200; cur. Gibraltar pound = 100 pence)

Gibraltar, Strait of the narrow waterway, 13 km (8 miles) at its narrowest, which connects the MEDITERRANEAN SEA to the ATLANTIC OCEAN, with SPAIN to the north and MOROCCO to the south.

Gibson Desert a desert of sand and salt marshes in central WESTERN AUSTRALIA, with the GREAT SANDY DESERT to the north and the GREAT VICTORIA DESERT to the south.

Gifu a town in central HONSHU Island, JAPAN. (Pop. 411,700)

Gijón a port and industrial town in AUSTURIAS, in the centre of the north coast of SPAIN. (Pop.260,000)

Gilbert Islands *see* **Kiribati**.

Gilgit a mountain district in the northern state of JAMMU AND KASHMIR, INDIA, noted for its great beauty. The small town of Gilgit perches startlingly beneath a dramatic rock face.

Gironde the long, thin estuary stretching some 80 km (50 miles) which connects the Rivers DORDOGNE and GARONNE to the Atlantic coast of south-west FRANCE.

Giuba, River *see* **Jubba, River**.

Giza *see* **El Gîza**.

Glamorgan, Vale of a county of south Wales, the administrative centre is Barry. (Pop. 119,200)

Glasgow a port on the River CLYDE, a major industrial centre and the largest city in SCOTLAND. (Pop. city 618,430)

Gliwice (Gleiwitz) an industrial city in southern POLAND. (Pop. 214,000)

Gloucestershire a county in western ENGLAND; the county town is Gloucester. (2638 sq km/1019 sq miles; pop. 549,500)

Goa a state on the west coast of INDIA, 400 km (250 miles) south of BOMBAY, which was captured by the Portuguese in 1510 and remained under the control of PORTUGAL until it joined India in 1961. (3702 sq km/1429 sq miles; pop. 1,169,800)

Gobi Desert a vast expanse of arid land which occupies much of MONGOLIA and central northern CHINA. Temperatures range from very hot to extremely cold over the year. (1,295,000 sq km/ 500,000 sq miles)

Godavari, River a river which runs across the middle of the DECCAN peninsula in INDIA from its source in the Western GHATS near BOMBAY to its delta on the central zeast coast. (Length 1465 km/910 miles)

Godthåb (Nuuk) The capital of GREENLAND. (Pop. 12,200)

Godwin Austen *see* **K2**.

Golan Heights an area of high ground in southwest SYRIA on the border with northern ISRAEL. The Heights were captured by Israel in the Arab-Israeli War of 1967 and annexed by Israel in 1981. (2225 m/7300 ft)

Gold Coast (1) the name given to a string of beach resorts on the east coast of QUEENSLAND, AUSTRALIA, to the south of BRISBANE. **(2)** *See* **Ghana**.

Golden Triangle the remote and mountainous region where the borders of THAILAND, MYANMAR and LAOS meet, noted in particular for its opium cultivation and as one of the world's main sources of the drug heroin.

Gomel an industrial city in south-eastern BELARUS. (Pop. 503,000)

Gomera *see* **Canary Islands**.

Good Hope, Cape of the tip of the narrow Cape Peninsula which extends from the south-western corner of SOUTH AFRICA.

Gor'kiy (Gorky) *see* **Nizhniy Novgorod**.

Gothenburg (Göteborg) a major port on the KATTEGAT and the second largest city in SWEDEN. (Pop. 444,550)

Gotland an island in the BALTIC SEA which forms a county of SWEDEN. (3140 sq km/1210 sq miles; pop. 58,200)

Göttingen a university town in central GERMANY and an important trading centre in medieval times. (Pop. 138,000)

Gouda a historic town in eastern NETHERLANDS, famous for its cheese. (Pop. 59,200)

Gozo *see* **Malta**.

Grampian formerly an administrative region of north-eastern SCOTLAND created out of the former counties of Aberdeenshire, Kincardineshire, Banffshire and part of Morayshire. In 1996 Grampian ceased to exist as an administrative region.

Grampian Mountains a range of mountains that stretch across northern SCOTLAND to the south of

Loch NESS. The mountains rise to their highest point at BEN NEVIS (1344 m/4409 ft), the highest peak in the UK.

Granada a city in the SIERRA NEVADA of central southern SPAIN. An administrative centre during the Moorish occupation of Spain, during which its famous Alhambra Palace was built (1248-1345). Granada is also the name of the surrounding province. (Pop. city 270,000)

Gran Canaria *see* **Canary Islands**.

Grand Bahama *see* **Bahamas**.

Grand Canyon the dramatic gorge of the COLORADO RIVER, in places over 1.5 km (1 mile) deep, in north-western ARIZONA.

Grand Rapids a city 40 km (14 miles) to the east of Lake Michigan in the state of MICHIGAN, USA. (Pop. city 191,000)

Graz the second largest city in AUSTRIA, in the south-east of the country. (Pop. 238,000)

Great Australian Bight the arm of the SOUTHERN OCEAN which forms the deep indentation in the centre of the southern coastline of Australia.

Great Australian Desert the collective word for the deserts that occupy much of the centre of Australia. (3,830,000 sq km/1,480,000 sq miles)

Great Barrier Reef the world's most extensive coral reef, lining the coast of QUEENSLAND, AUSTRALIA, stretching some 2000 km (1250 miles).

Great Bear Lake the fourth largest lake in North AMERICA, in the remote north-west of CANADA. It drains into the MACKENZIE RIVER. (31,153 sq km/ 12,028 sq miles)

Great Britain the island shared by ENGLAND, SCOTLAND and WALES, and which forms the principal part of the UNITED KINGDOM OF GREAT BRITAIN AND NORTHERN IRELAND.

Great Dividing Range a range of mountains which runs down the east coast of AUSTRALIA, from QUEENSLAND in the north, across NEW SOUTH WALES to VICTORIA in the south, some 3600 km (2250 miles) in all. The highest point is Mount KOSCIUSKO. (2230 m/7316 ft)

Greater Manchester *see* **Manchester**.

Greater Sunda Islands *see* **Lesser Sunda Islands**.

Great Indian Desert *see* **Thar Desert**.

Great Lakes the largest group of freshwater lakes in the world, drained by the ST LAWRENCE RIVER. There are five lakes, four of which (Lakes HURON, SUPERIOR, ERIE and ONTARIO) are on the border of CANADA and the USA; the fifth (Lake MICHIGAN) is in the USA.

Great Plains a vast area in North AMERICA of flat and undulating grassland east of the ROCKY MOUNTAINS and stretching from northern CANADA to TEXAS,

USA. It includes the PRAIRIES, most of which are now ploughed for cereal and fodder crops.

Great Rift Valley a series of geological faults which has created a depression stretching 6400 km (4000 miles) from the valley of the River JORDAN across the RED SEA and down East AFRICA to MOZAMBIQUE.

Great Salt Lake a salt lake in north-west UTAH, USA, lying just to the north-west of SALT LAKE CITY. (5200 sq km/2000 sq miles)

Great Sandy Desert the desert region in the northern part of WESTERN AUSTRALIA.

Great Slave Lake a lake drained by the MACKENZIE RIVER in the southern part of the NORTHWEST TERRITORIES of CANADA. (28,570 sq km/11,030 sq miles)

Great Smoky Mountains part of the APPALACHIAN MOUNTAINS, running along the border between TENNESSEE and NORTH CAROLINA. The highest point is Clingmans Dome (2025 m/6643 ft).

Great Victoria Desert A vast area of sand dunes straddling the border between WESTERN AUSTRALIA and SOUTH AUSTRALIA.

Greece the Greek peninsula is the most south-easterly extension of EUROPE. The Pindus Mountains divide Greece from the Albanian border in the north to the Gulf of CORINTH in the south. About 70% of the land is hilly with harsh mountain climates and poor soils. The Greek islands and coastal regions have a typical MEDITERRANEAN climate with mild rainy winters and hot dry summers. Winter in the northern mountains is severe with deep snow and heavy precipitation. Agriculture is the chief activity and large scale farming is concentrated on the east coasts. The main industries are small processing plants for tobacco, food and leather. Fishing is an important activity around the 2000 islands which lie off the mainland. Tourism is also a major industry.

Quick facts:
Area : 131,944 sq km (50,944 sq miles)
Population : 10,500,000
Capital : Athens (Athinai)
Other major cities : Patras, Piraeus, Thessaloníki
Form of government : Republic
Religion : Greek Orthodox
Currency : Drachma

Greenland a huge island to the north-east of North AMERICA, most of which lies within the ARCTIC CIRCLE. A province of DENMARK, the island was granted home rule in 1979. The economy is heavily reliant on fishing and most of the population is Eskimo.

Quick facts:
Area: 2,175,600 sq km (840,000 sq miles)
Population: 59,000
Capital: Gothåb (Nuuk)
Form of government: Self-governing part of the Danish realm
Religion: Lutheranism
Currency: Danish krone

Greenwich a borough of east LONDON, ENGLAND, on the south bank of the River THAMES. It was the site of the Royal Observatory, and since 1884 has been accepted to be on 0° meridian from which all lines of longitude are measured. Greenwich Mean Time is the time at 0° longitude, against which all world time differences are measured.

Grenada is the most southerly of the Windward Island chain in the CARIBBEAN. Its territory includes the southern GRENADINE Islands to the north. The main island consists of the remains of extinct volcanoes and has an attractive wooded landscape. In the dry season its typical climate is very pleasant with warm days and cool nights but in the wet season it is hot day and night. Agriculture is the islands main industry and the chief crops grown for export are cocoa, nutmegs, bananas and mace. Apart from the processing of its crops Grenada has little manufacturing industry although tourism is an important source of foreign revenue. It is a popular port of call for cruise ships.

Quick facts:
Area : 344 sq km (133 sq miles)
Population : 95,000
Capital : St Georges
Form of government : Constitutional Monarchy
Religion : RC, Anglicanism, Methodism
Currency : East Caribbean dollar

Grenadines a string of some 600 small islands that lie between ST VINCENT to the north and GRENADA to the south. Most of them belong to St Vincent, but the largest, Carriacou, is divided between St Vincent and Grenada. Other islands include Union, MUSTIQUE and Bequia.

Grenoble a manufacturing city in south-east FRANCE, in the foothills of the ALPS. (Pop. 154,000)

Groningen the largest city in the north-east of the NETHERLANDS, and also the name of the surrounding province. (Pop. city 209,000)

Guadalajara a major city of central western MEXICO. (Pop. 2,847,000)

Guadalcanal an island at the southern end of the archipelago where HONIARA, capital of the SOLOMON ISLANDS, is located. The bitterly contested battle in 1942–3 here between U.S. forces and the occupying Japanese marked a turning point in the U.S. PACIFIC campaign which eventually led to the defeat of the Japanese in 1945.

Guadeloupe a group of islands in the LEEWARD ISLANDS in the eastern CARIBBEAN which since 1946 has been an overseas department of FRANCE. The principal island is Guadeloupe (divided into two parts, Basse Terre and Grande Terre). The other islands include Marie Galante, La Désirade, Iles des Saintes, ST BARTHÉLÉMY and St Martin. The capital is BASSE TERRE. (1779 sq km/687 sq miles; pop. 440,00)

Guam the largest of the MARIANA ISLANDS in the western PACIFIC OCEAN. (549 sq km/212 sq miles; pop. 133,000)

Guangdong a province of south-east CHINA. The capital is GUANGZHOU (Canton). (210,000 sq km/81,000 sq miles; pop. 62,830,000)

Guangxi-Zhuang an autonomous region of southern CHINA on the border with VIETNAM. To the south of the city of Guilin, around the Gui Jiang River, is a famous landscape of towering rock hills which rise up from the watery plains. The regional capital is NANNING. (230,000 sq km/89,000 sq miles; pop. 42,245,000)

Guangzhou (Canton) a major port in south-east CHINA, the country's sixth largest city and the capital of GUANGDONG province. (Pop. 3,600,000)

Guantanamo a city in the south-east of CUBA, and also the name of the surrounding province. The USA has a naval base at nearby Guantanamo Bay. (Pop. city 203,000)

Guatemala is situated between the PACIFIC OCEAN and the CARIBBEAN SEA where North AMERICA meets Central America. It is a mountainous country with a ridge of volcanoes running parallel to the Pacific coast. It has a tropical climate with little or no variation in temperature and a distinctive wet season. The Pacific slopes of the mountains are exceptionally well watered and fertile and it is here that most of the population are settled. Coffee growing on the lower slopes dominates the economy. A small strip on the coast produces sugar, cotton and bananas. Industry is mainly restricted to the processing of the agricultural products. Guatemala is politically a very unstable country and civil conflict has practically destroyed tourism.

Quick facts:
Area : 108,889 sq km (42,042 sq miles)
Population : 10,624,000
Capital : Guatemala City
Other major cities : Puerto Barrios,
   Quezaltenango
Form of government : Republic
Religion : RC
Currency : Quetzal

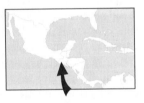

Guatemala City the capital of Guatemala, in the south-east of the country. (Pop. city 1,675,589)

Guayaquil the main port and the largest city of ECUADOR. (Pop. 1,508,000)

Guernica a small town in the BASQUE country of north-east SPAIN where the Basque parliament used to assemble. In 1937, during the Spanish Civil War, it was heavily bombed from the air by German forces. (Pop. 17,836)

Guernsey one of the CHANNEL ISLANDS, lying in the centre of the group and some 50 km (30 miles) off the coast of FRANCE. The capital is St Peter Port. (78 sq km/30 sq miles; pop. 59,000)

Guiana (French) *or* **Guyane** is situated on the north-east coast of South AMERICA and is still an overseas department of FRANCE. It is bounded to the south and east by BRAZIL and to the west by SURINAME. The climate is tropical with heavy rainfall. Guiana's economy relies almost completely on subsidies from France. It has little to export apart from shrimps and the small area of land which is cultivated produces rice, manioc and sugar cane. Recently the French have tried to develop the tourist industry and exploit the extensive reserves of hardwood in the jungle interior.

Quick facts:
Area : 91,000 sq km (35,135 sq miles)
Population : 154,000
Capital : Cayenne
Form of government : French overseas
  department
Religion : RC
Currency : Franc

Guinea, formerly a French West African territory, is located on the coast at the 'bulge' in AFRICA. It is a lush green beautiful country about the same size as the UNITED KINGDOM. It has a tropical climate with constant heat and a high rainfall near the coast. Guinea has great agricultural potential and many of the coastal swamps and forested plains have been cleared for the cultivation of rice, cassava, yams, maize and vegetables. Further inland on the plateau of Futa Jalon dwarf cattle are raised and in the valleys bananas and pineapples are grown. Coffee and kola nuts are important cash crops grown in the Guinea highlands to the south-west. Minerals such as bauxite, iron ore and diamonds are mined but development is hampered by lack of transport.

Quick facts:
Area : 245,857 sq km (94,925 sq miles)
Population : 7,200,000
Capital : Conakry
Other major cities : Kankan, Labé
Form of government : Republic
Religion : Sunni Islam
Currency : Guinea franc

Guinea, Equatorial *see* **Equatorial Guinea.**
Guinea, Gulf of the arm of the ATLANTIC OCEAN which creates the deep, right-angled indent in the west coast of AFRICA.
Guinea-Bissau formerly a Portuguese territory, Guinea-Bissau is located south of SENEGAL on the ATLANTIC coast of West AFRICA. It is a country of stunning scenery and rises from a deeply indented and island-fringed coastline to a low inland pla-

teau. The adjacent Bijagos archipelago forms part of its territory. The climate is tropical with abundant rain from June to November but hot dry conditions for the rest of the year. Years of Portuguese rule and civil war have left Guinea-Bissau impoverished, and it is one of the poorest West African states. The country's main aim is to become self-sufficient in food, and the main crops grown are groundnuts, sugar cane, plantains, coconuts and rice. Fishing is an important export industry.

Quick facts:
Area : 36,125 sq km (13,948 sq miles)
Population : 1,050,000
Capital : Bissau
Form of government : Republic
Religion : Animism, Sunni Islam
Currency : Peso

Guiyang an industrial city in central southern CHINA, and capital of GUIZHOU province. (Pop. 1,490,000)
Guizhou a province of central southern CHINA. The capital is GUIYANG. (170,000 sq km/65,600 sq miles; pop. 32,390,000)
Gujarat a state lining the north-west coast of INDIA, on the border with PAKISTAN. The capital is Gandhinagar. (196,024 sq km/75,665 sq miles; pop. 41,309,582)
Gujranwala a textile city in the province of PUNJAB, PAKISTAN, some 65 km (40 miles) north of LAHORE. (Pop. 912,000)
Gulf, The the huge inlet to the south of IRAN which is connected to the ARABIAN SEA by the Strait of HORMUZ. It is often referred to as the Persian Gulf, or the Arabian Gulf.
Guyana, the only English-speaking country in South AMERICA, is situated on the north-east coast of the continent on the ATLANTIC OCEAN. The country is intersected by many rivers and the coastal area comprises tidal marshes and mangrove swamps. It is on this coastal area that rice is grown and vast plantations produce sugar. The jungle in the south-west has potential for the production of minerals, hardwood and hydroelectric power, but 90% of the population live in the coastal area where the climate is moderated by sea breezes. The country is deeply divided politically and noth-

ing has been done to improve productivity with the result that today the country is in an economic crisis.

Quick facts:
Area : 214,969 sq km (83,000 sq miles)
Population : 830,000
Capital : Georgetown
Other major city : New Amsterdam
Form of government : Cooperative Republic
Religion : Hinduism,
 Protestantism, RC
Currency : Guyana dollar

Guyane *see* **Guiana (French)**.

Gwalior a city in central INDIA, 280 km (174 miles) south-east of DELHI. (Pop. 720,000)

Gwangju *see* **Kwangju**.

Gwent a county in south-east WALES, bordering the SEVERN estuary just to the east of CARDIFF. The county was created in 1974 and more or less coincides with the old county of Monmouthshire. The county town is Cwmbran. (1376 sq km/532 sq miles; pop. 440,000)

Gwynedd a county in north-west WALES which includes the island of ANGLESEY. It was created in 1974 out of the former county of Caernarfonshire, and parts of Denbighshire and Merionethshire. The administrative centre is Caernarfon. (3868 sq km/1493 sq miles; pop. 232,000)

Haarlem a city in central western NETHERLANDS, 18 km (11 miles) from AMSTERDAM. (Pop. 150,000)

Hagen a steel town in the industrial RUHR region of western GERMANY. (Pop. 216,000)

Hague, The (Den Haag; 's-Gravenhage) the administrative centre of the NETHERLANDS, on the west coast. (Pop. 445,000)

Haifa the main port of ISRAEL. (Pop. 246,000)

Hainan Island a large tropical island in the SOUTH CHINA SEA belonging to CHINA, and the southernmost extremity of that country. (33,670 sq km/13,000 sq miles; pop. 6,557,000)

Haiphong a port in the north of VIETNAM, 90 km (55 miles) east of the capital, HANOI. It is Vietnam's third largest city after HO CHI MINH CITY and Hanoi. (Pop. 1,448,000)

Haiti occupies the western third of the large island of HISPANIOLA in the CARIBBEAN. It is a mountainous country, the highest point reaching 2680 m (8793 ft) at La Selle. The mountain ranges are

separated by deep valleys and plains. The climate is tropical but semi-arid conditions can occur in the lee of the central mountains. Hurricanes and severe thunderstorms are a common occurrence. Only a third of the country is arable, yet agriculture is the chief occupation. Many farmers grow only enough to feed their own families, and the export crops—coffee, sugar and sisal—are grown on large estates. Severe soil erosion caused by extensive forest clearance has resulted in a decline in crop yields. Haiti is the poorest country in the Americas and has experienced many uprisings and attempted coups.

Quick Facts:
Area : 27,750 sq km (10,714 sq miles)
Population : 7,180,000
Capital : Port-au-Prince
Other major cities : Les Cayes,
 Gonaïves, Jérémie
Form of government : Republic
Religion : RC, Voodooism
Currency : Gourde

Hakodate a port at the southern tip of HOKKAIDO Island, JAPAN. (Pop. 304,000)

Halicarnassus *see* **Bodrum**.

Halifax (1) the capital of NOVA SCOTIA, CANADA. (Pop. metropolitan area 320,000) **(2)** a town in West YORKSHIRE, ENGLAND. (Pop. 115,000)

Halle an industrial town and inland port served by the Saale River in central GERMANY. (Pop. 295,000)

Halmahera *see* **Maluku**.

Hamah an industrial city in eastern SYRIA, on the River ORONTES. (Pop. 254,000)

Hamamatsu a city in southern HONSHU Island, JAPAN. (Pop. 535,000)

Hamburg the main port of GERMANY, situated on the River ELBE. (Pop. 1,670,000)

Hamelin (Hameln) a town in northern GERMANY. It is famous for its legendary Pied Piper, who in 1284 is said to have rid the town of a plague of rats by playing his pipe to them and luring them to their deaths in the River WESER. (Pop. 56,300)

Hamersley Range part of the Pilbara Range in WESTERN AUSTRALIA. The highest peak is Mount Bruce (1235 m/4052 ft).

Hamhung (Hamheung) A port and industrial city

on the east coast of Democratic People's Republic of KOREA. (Pop. 770,000)

Hamilton (1) the capital of BERMUDA. (Pop. 6000) **(2)** a port and industrial city at the western end of Lake ONTARIO, CANADA. (Pop. city 306,430/metropolitan area 542,090). **(3)** a town in the north-western part of NORTH ISLAND, NEW ZEALAND. (Pop. 101,300). **(4)** a town and the administrative centre of North Lanarkshire region of SCOTLAND, 17 km (10 miles) south-east of GLASGOW. (Pop. 52,000)

Hammerfest a town in the very north of NORWAY, and one of the world's most northerly settlements. (Pop. 6900)

Hampshire a county of central southern ENGLAND; the county town is WINCHESTER. (3773 sq km/1456 sq miles; pop. 1,605,000)

Hangzhou (Hangchow) a port and industrial city on the east coast of central CHINA, at the head of an estuary called Hangzhou Wan. Hangzhou is at the southern end of the Grand Canal, which links it to BEIJING 1100 km (690 miles) to the north. (Pop. 1,330,000)

Hankow *see* **Wuhan**.

Hannover *see* **Hanover**.

Hanoi the capital of VIETNAM, in the north of the country. (Pop. 3,050,000)

Hanover (Hannover) a historic city in central northern GERMANY. (Pop. 518,000)

Haora (Howrah) an industrial city in WEST BENGAL, INDIA, on the HUGLI River, facing CALCUTTA. (Pop. 946,000)

Harare the capital of ZIMBABWE; it was formerly called Salisbury (until 1982). (Pop. 1,184,000)

Harbin the largest city of northern CHINA, situated in central DONGBEI (Manchuria), and capital of HEILONGJIANG province. (Pop. 2,930,000)

Hari Rud a river which flows westwards from central AFGHANISTAN, through the city of HERAT before turning north to form part of the border with IRAN, ending in TURKMENISTAN. (Length 1125 km/700 miles)

Harrisburg the state capital of PENNSYLVANIA, USA. (Pop. city 53,400)

Hartford the state capital of CONNECTICUT, USA. (Pop. city 132,000/metropolitan area 1,030,400)

Haryana a state in north-west INDIA, formed in 1966. (44,212 sq km/17,066 sq miles; pop. 16,318,000)

Harz Mountains a range of mountains, noted for their forests, in central GERMANY. The highest peak is Brocken (1142 m/3747 ft).

Hastings a historic port and resort on the south coast of ENGLAND, in the county of EAST SUSSEX. The Battle of Hastings of 1066, in which the Eng-

lish were defeated by the Normans, was fought nearby. (Pop. 81,000)

Hatteras, Cape the tip of a chain of islands lining the coast of NORTH CAROLINA, USA, notorious for its violent weather.

Havana (La Habana) the capital of CUBA, a port on the north-west coast of the island. It is also the name of the surrounding province. (Pop. 2,100,400)

Hawaii a group of 122 islands just to the south of the Tropic of Cancer, some 3700 km (2300 miles) from the coast of CALIFORNIA. Since 1959 they have formed a state of the USA. The main islands are OAHU, MAUI and Hawaii Island, which at 10,488 sq km (4049 sq miles) is by far the largest. HONOLULU, the state capital, is on Oahu. (16,705 sq km/6450 sq miles; pop. 1,108,000)

Hebei a province in northern CHINA which surrounds (but does not include) BEIJING. The capital is SHIJIAZHUANG. (180,000 sq km/70,000 sq miles; pop. 61,082,000)

Hebrides some 500 islands lying off the west coast of SCOTLAND, consisting of the Inner Hebrides to the south-east, whose main islands are Tiree, Jura, Coll, Mull, Eigg and SKYE, and the Outer Hebrides to the north-west whose islands include Lewis and Harris, the Uists, Benbecula and Barra.

Hefei an industrial city in central eastern CHINA, capital of ANHUI province. (Pop. 1,110,000)

Heidelberg a university town in south-west GERMANY on the NECKAR RIVER. (Pop. 140,000)

Heilongjiang a province of DONGBEI (Manchuria) in northern CHINA; the capital is HARBIN. (464,000 sq km/179,000 sq miles; pop. 35,200,000)

Heilong Jiang, River *see* **Amur, River.**

Hejaz (Hijaz) a mountainous region which lines the RED SEA, formerly an independent kingdom but since 1932 a part of SAUDI ARABIA.

Helena the state capital of MONTANA, USA. (Pop. 24,600)

Heligoland (Helgoland) A small island and former naval base in the NORTH SEA off the coast of GERMANY. (2.1 sq km/0.5 sq miles)

Hellespont *see* **Dardanelles**.

Helsingfors *see* **Helsinki**.

Helsingør *see* **Elsinore**.

Helsinki (Helsingfors) the capital and chief industrial centre and port of FINLAND. (Pop. 515,765)

Henan a province of central CHINA; the capital is ZHENGZHOU. (160,000 sq km/62,000 sq miles; pop. 85,500,000)

Heraklion (Iraklion) the capital and main port of the island of CRETE. (Pop. 111,000)

Herat a city in western AFGHANISTAN on the HARI RUD River. (Pop. 177,000)

Hercegovina *see* **Bosnia & Herzegovina**.

Herculaneum (Ercolano) an excavated Graeco-Roman town on the Bay of NAPLES which was buried by the eruption of Mount VESUVIUS, along with POMPEII, in AD79.

Hereford and Worcester a county in the west of ENGLAND, on the border with Wales, which was created in 1974 when the old counties of Herefordshire and Worcestershire were combined. The county town is Worcester. (3927 sq km/1516 sq miles; pop. 699,900)

Herefordshire *see* **Hereford and Worcester**.

Hermon, Mount a mountain in southern LEBANON near the borders with SYRIA and ISRAEL. It is the source of the River JORDAN. (2814 m/9332 ft)

Hertfordshire a county in south-east ENGLAND, to the north of LONDON. The county town is Hertford. (1634 sq km/631 sq miles; pop. 1,005,400)

Herzegovina *see* **Bosnia & Herzegovina**.

Hessen a state in central western GERMANY. The capital is WIESBADEN. (21,112 sq km/8151 sq miles; pop. 5,500,000)

Highland Region an administrative region in northern SCOTLAND comprising the most northerly part of the mainland and many of the Inner HEBRIDES. It is the largest county in the U.K. It was created in 1975 out of the old counties of Caithness, Nairnshire, Sutherland, most of Inverness-shire, Ross and Cromarty and parts of Argyll and Morayshire. The capital is INVERNESS. (26,136 sq km/10,091 sq miles; pop. 196,000)

Highlands the rugged region of northern SCOTLAND, which includes the GRAMPIAN mountains and the North West Highlands.

Himachal Pradesh a state in northern INDIA, in mountainous country bordering TIBET. (83,743 sq km/32,333 sq miles; pop. 5,111,000)

Himalayas the massive mountain range stretching some 2400 km (1500 miles) in a broad sweep from the northern tip of INDIA, across NEPAL, BHUTAN and southern TIBET to ASSAM in north-eastern India. The average height of the mountains is some 6100 m (20,000 ft), rising to the world's tallest peak, Mount EVEREST (8848 m/29,028 ft).

Himeji an industrial city and port in southern HONSHU Island, JAPAN. (Pop. 452,900)

Hims *see* **Homs**.

Hindu Kush a range of mountains which stretches some 600 km (370 miles) at the western end of the HIMALAYAS, straddling the web of borders where AFGHANISTAN, TAJIKISTAN, CHINA, INDIA and PAKISTAN meet. The highest peak is Tirich Mir (7690 m/25,229 ft) in Pakistan.

Hiroshima an industrial city in south-western HONSHU Island, JAPAN. Three quarters of the city was destroyed on August 6, 1945 when the world's first atomic bomb was dropped here, killing 78,000 people. (Pop. 1,090,000)

Hispaniola the name of the large CARIBBEAN island that is shared by HAITI and the DOMINICAN REPUBLIC. (76,200 sq km/29,400 sq miles)

Hitachi an industrial city on the east coast of HONSHU Island, JAPAN. (Pop. 202,100)

Hobart a port and capital of the island of TASMANIA, AUSTRALIA, on the south-east coast. (Pop. 184,000)

Ho Chi Minh City (Saigon) the largest city in VIETNAM, and the capital of former independent South Vietnam. (Pop.3,294,000)

Hoggar (Ahaggar) a remote mountain range rising from the desert landscape of southern ALGERIA, and noted for the weathered shapes of its rock formations. The highest peak is Mount Tahat (2918 m/9573 ft).

Hohe Tauern a part of eastern ALPS in southern AUSTRIA, rising to the highest point at Grossglockner (3797 m/12,460 ft), Austria's highest peak.

Hohhot an industrial city and the capital of the NEI MONGOL AUTONOMOUS REGION (Inner Mongolia), CHINA. (Pop. 870,000)

Hokkaido the most northerly of the main islands of JAPAN, and the second largest after HONSHU. The capital is SAPPORO. (78,509 sq km/30,312 sq miles; pop. 5,678,000)

Holland a name generally applied to the NETHERLANDS, but in fact the term really applies to the central coastal region which comprise the two provinces of Noord Holland and Zuid Holland.

Hollywood a suburb in the northern part of LOS ANGELES in CALIFORNIA, USA. It has long served as the base for the USA's powerful film industry.

Homs (Hims) an industrial city of ancient origins on the River ORONTES in SYRIA. (Pop. 518,000)

Honduras is a fan-shaped country in Central AMERICA which spreads out toward the CARIBBEAN SEA at the Gulf of HONDURAS. Four fifths of the country is covered in mountains which are indented with river valleys running toward the very short PACIFIC coast. There is little change in temperatures throughout the year and rainfall is heavy, especially on the Caribbean coast where temperatures are also higher than inland. The country is sparsely populated and although agricultural, only about 25% of the land is cultivated. Honduras was once the world's leading banana exporter and although that fruit is still its main export, agriculture is now more diverse. Grains, coffee and sugar are important crops, and these are grown mainly on the coastal plains of the Pacific and Caribbean. The forests are not effectively exploited and industry is small-scale.

Quick Facts:
Area : 112,088 sq km (43,277 sq miles)
Population : 5,940,000
Capital : Tegucigalpa
Form of government : Republic
Religion : RC
Currency : Lempira

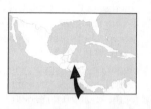

Hong Kong is a British Dependent Territory located in the SOUTH CHINA SEA and consists of Hong Kong Island (once a barren rock), the peninsula of KOWLOON and about 1000 sq km (386 sq miles) of adjacent land known as the New Territories. Hong Kong is situated at the mouth of the Pearl River about 130 km (81 miles) south-east of GUANGZHOU (Canton). The climate is warm subtropical with cool dry winters and hot humid summers. Hong Kong has no natural resources, even its water comes from reservoirs across the Chinese border. Its main assets are its magnificent natural harbor and its position close to the main trading routes of the PACIFIC. Hong Kong's economy is based on free enterprise and trade, an industrious work force and an efficient and aggressive commercial system. Hong Kong's main industry is textiles and clothing which accounts for 38% of its domestic exports.

Quick Facts:
Area : 1074 sq km (416 sq miles)
Population : 6,000,000
Form of government : Colony under British
   administration until summer 1997 when
   China will take over
Religion : Buddhism, Taoism, Christianity
Currency : Hong Kong dollar

Honiara the capital of the SOLOMON ISLANDS, situated on GUADALCANAL. (Pop. 37,000)

Honolulu the state capital of HAWAII, USA, on the south coast of the island of OAHU. (Pop. city 371,000/ metropolitan area 836,000)

Honshu the central and largest of the islands of JAPAN. (230,988 sq km/89,185 sq miles; pop. 99,250,000)

Hooghly, River *see* **Hugli, River.**

Hormuz (Ormuz), Strait of the narrow strait at the mouth of The GULF between the horn-like protrusion of the MUSANDAM peninsula of OMAN to the south, and IRAN to the north.

Horn, Cape (Cabo de Hornos) the southern tip of South AMERICA, represented by a spattering of remote islands belonging to CHILE off TIERRA DEL FUEGO.

Houston the largest city in TEXAS, USA. (Pop. city 1,690,000/metropolitan area 3,711,000)

Howrah *see* **Haora.**

Hrvatska *see* **Croatia.**

Huang He (Hwang Ho; Yellow River) the second longest river in CHINA after the CHANG JIANG (Yangtze), flowing from the QINGHAI mountains across northern central China to the YELLOW SEA, south of BEIJING. (Length 5464 km/3395 miles)

Huascaran a peak in the ANDES in central PERU, and that country's highest mountain. (6768 m/22,205 ft)

Hubei a landlocked province of central CHINA. (180,000 sq km/69,500 sq miles; pop. 53,969,000)

Hudson Bay a huge bay in north-eastern CANADA, hemmed in to the north by BAFFIN ISLAND, and connected to the ATLANTIC OCEAN by the Hudson Strait.

Hudson River a river flowing from its source in the ADIRONDACK MOUNTAINS in NEW YORK State, USA, to the ATLANTIC OCEAN at New York City. The ERIE Canal joins the Hudson River to link New York to the GREAT LAKES. (Length 492 km/306 miles)

Hué the capital and powerbase of the rulers of VIETNAM from 200BC to the 19th century, located in the central coastal region of the country. (Pop. 211,100)

Hugli (Hoogly) a major branch of the River GANGES which forms at its delta and flows through CALCUTTA and the surrounding industrial conurbations into the Bay of BENGAL. (Length 193 km/120 miles)

Hull *see* **Kingston upon Hull.**

Humber the estuary of the Rivers OUSE and TRENT which cuts deep into the east coast of ENGLAND to the north of the WASH. (Length 60 km/35 miles)

Hunan an inland province of south-east CHINA. The capital is CHANGSHA. (210,000 sq km/81,000 sq miles; pop. 60,660,000)

Hungary landlocked in the heartland of EUROPE, Hungary is dominated by the great plain to the east of the river DANUBE which runs north-south across the country. In the west lies the largest lake in Central Europe, Lake BALATON. Winters are severe, but the summers are warm and although wet in the west, summer droughts often occur in the east. Hungary experienced a modest boom in its economy in the 1970s and 1980s. The government invested money in improving agriculture by mechanizing farms, using fertilizers and bringing new land under cultivation. Yields of cereals for breadmaking and rice have since soared and large areas between the Danube and Tisza rivers are now used to grow vegetables. Industries have been carefully developed where adequate natural resources exist. New industries like electrical and electronic equipment are now being promoted and tourism is fast developing around Lake Balaton.

> *Quick Facts:*
> Area : 93,032 sq km (35,920 sq miles)
> Population : 10,546,000
> Capital : Budapest
> Other major cities : Debrecen, Miskolc, Pécs, Szeged
> Form of government : Republic
> Religion : RC, Calvinism, Lutheranism
> Currency : Forint

Hunter Valley the valley of the Hunter River, lying 100 km (60 miles) north-west of SYDNEY, AUSTRALIA. It is particularly noted for its wine.

Huntingdonshire *see* **Cambridgeshire**.

Huron, Lake one of the GREAT LAKES, lying at the centre of the group on the border between CANADA and the state of MICHIGAN in the USA. (59,570 sq km/23,000 sq miles)

Hwang Ho *see* **Huang He.**

Hyderabad (1) the capital of the state of ANDHRA PRADESH in eastern south INDIA. (Pop. 3,146,000). **(2)** a city on the INDUS delta 160 km (100 miles) north-east of KARACHI, PAKISTAN. (Pop. 1,040,000)

Hydra (Idhra) a small island in the AEGEAN SEA, off the east coast of the PELOPONNESE, GREECE, noted as a haven for tourists where motor traffic is prohibited.

Iasi a historic city in north-eastern ROMANIA, near the border with MOLDOVA. (Pop. 338,000)

Ibadan the second largest city in NIGERIA, some 120 km (75 miles) north of LAGOS. It is a busy market town, and noted for its university. (Pop. 1,295,000)

Ibiza (Iviza) *see* **Balearic Islands.**

Iboland a densely populated region of south-east-ern NIGERIA inhabited by the Ibo people. The attempt by the region to break away from Nigeria (1967-70) under the name of Biafra caused a civil war that led to a famine, killing over a million people.

Ica, River see Putumayo, River.

Içel *see* **Mersin**.

Iceland is a large island situated in a tectonically unstable part of the North ATLANTIC OCEAN, just south of the ARCTIC CIRCLE. The island has over 100 volcanoes, at least one of which erupts every five years. One ninth of the country is covered with ice and snowfields and there are about seven hundred hot springs which are an important source of central heating. The climate is cool temperate but because of the effect of the North Atlantic Drift it is mild for its latitude. The south-west corner is the most densely populated area as the coast here is generally free from ice. Only 1% of the land is cultivated mostly for fodder and root crops to feed sheep and cattle. The island's economy is based on its sea fishing industry which accounts for 70% of exports. Wool sweaters and sheepskin coats are also exported.

> *Quick facts:*
> Area : 103,000 sq km (39,768 sq miles)
> Population : 266,790
> Capital : Reykjavík
> Form of government : Republic
> Religion : Lutheranism
> Currency : Icelandic króna

Idaho a inland state in the north-west of the USA. The state capital is BOISE. (216,413 sq km/83,557 sq miles; pop. 1,066,000)

Idhra *see* **Hydra**.

Idlib a large commercial and agricultural centre in north-western SYRIA. (Pop. 428,000)

Ieper *see* **Ypres.**

IJsselmeer formerly a large inlet of the NORTH SEA known as the Zuiderzee on the north-eastern coast of the NETHERLANDS, but after the creation of the dam called the Afsluitdijk across its mouth, it has filled with water from the River IJssel and is now a freshwater lake, bordered by fertile areas of reclaimed land (polders).

Ile de France a region and former province of FRANCE with PARIS at its centre, now consisting of eight separate departments. (12,012 sq km/4638 sq miles; pop. 10,735,000)

Illinois a state in the MIDWEST of the USA, bordering Lake MICHIGAN to the north. The capital is SPRINGFIELD, but CHICAGO is its main city. (146,075 sq km/56,400 sq miles; pop. 11,613,000)

Imjin River a river which flows from its source in southern NORTH KOREA across the border into SOUTH KOREA and to the YELLOW SEA. In 1951, during the Korean War, it was the scene of a heroic stand by the British 1st Gloucester Regiment. (Length 160 km/100 miles)

Inagua *see* **Bahamas.**

Inch'on (Incheon) a port and industrial city on the western (YELLOW SEA) coast of South KOREA, 39 km (24 miles) west of SEOUL. (Pop. 1,680,000)

India is a vast country in South ASIA which is dominated in the extreme north by the world's youngest and highest mountains, the HIMALAYAS. At the foot of the Himalayas, a huge plain, drained by the INDUS and GANGES rivers, is one of the most fertile areas in the world and the most densely populated part of India. Further south the ancient DECCAN plateau extends to the southern tip of the country. India generally has four seasons, the cool, the hot, the rainy and the dry. Rainfall varies from 100 mm (3.94 inches) in the north-west desert to 10,000 mm (394 inches) in Assam. About 70% of the population depend on agriculture for their living and the lower slopes of the Himalayas represent one of the world's best tea growing areas. Rice, sugarcane and wheat are grown in the Ganges plain. India is self-sufficient in all of its major food crops.

Quick Facts:
Area : 3,287,590 sq km (1,269,346 sq miles)
Population : 897,560,000
Capital : New Delhi
Other major cities : Bangalore, Bombay, Calcutta, Delhi, Hyderabad, Madras
Form of government : Federal Republic, Secular Democracy
Religion : Hinduism, Islam, Sikkism, Christianity, Jainism, Buddhism
Currency : Rupee

Indiana a state in the MIDWEST of the USA to the south-east of Lake MICHIGAN. The state capital is INDIANAPOLIS. (93,994 sq km/36,291 sq miles; pop. 5,658,000)

Indianapolis the state capital of INDIANA. (Pop. city 746,540)

Indian Ocean the third largest ocean, bounded by ASIA to the north, AFRICA to the west and AUSTRALIA to the east. The southern waters merge with the ANTARCTIC OCEAN. (73,481,000 sq km/28,364,000 sq miles)

Indonesia is made up of 13,667 islands which are scattered across the INDIAN and PACIFIC OCEANS in a huge crescent. Its largest landmass is the province of KALIMANTAN which is part of the island of BORNEO. SUMATRA is the largest individual island. JAVA, however, is the dominant and most densely populated island. The climate is generally tropical monsoon and temperatures are high all year round. The country has one hundred volcanoes, and earthquakes are frequent in the southern islands. Overpopulation is a big problem especially in Java, where its fertile rust coloured soil is in danger of becoming exhausted. Rice, maize and cassava are the main crops grown. Indonesia has the largest reserves of tin in the world and is one of the world's leading rubber producers. Indonesia's resources are not as yet fully developed but there is great potential for economic development.

Quick Facts:
Area : 1,904,570 sq km (735,358 sq miles)
Population : 189,907,000
Capital : Jakarta
Other major cities : Badung, Medan, Semarang, Surabaya
Form of government : Republic
Religion : Sunni Islam, Christianity, Hinduism
Currency : Rupiah

Indore a textile-manufacturing city, and once the capital of the princely state of Indore, in western MADHYA PRADESH, central INDIA. (Pop. 1,104,000)

Indus, River one of the great rivers of ASIA, whose valleys supported some of the world's earliest civilizations, notably at MOHENJO DARO. It flows from its source in TIBET, across the northern tip of INDIA before turning south to run through the entire length of PAKISTAN to its estuary on the ARABIAN SEA, south of KARACHI. (Length 3059 km/1900 miles)

Inner Mongolia *see* **Nei Mongol Autonomous Region.**

Inverness a town in north-eastern SCOTLAND at the head of the MORAY FIRTH and at the eastern end of Loch NESS. (Pop. 62,000)

Iona a small island off the south-western tip of MULL in western SCOTLAND where the Irish monk St Columba founded a monastery in AD563. (8 sq km/3 sq miles)

Ionian Islands (Eptanisos) the seven largest of the islands which lie scattered along the west coast of GREECE in the IONIAN SEA. They are CORFU, PAXOÍ, Cephalonia, Levkás, ITHACA, Zákinthos and Kíthira. (Pop. 182,700)

Ionian Sea that part of the MEDITERRANEAN SEA which lies between southern ITALY and GREECE. It is named after Io, a mistress of the Ancient Greek god Zeus.

Ios (Nios) *see* **Cyclades.**

Iowa a state in the MIDWEST of the USA bounded on the east and west by the upper reaches of the MISSISSIPPI and MISSOURI rivers. The capital is DES MOINES. (145,791 sq km/56,290 sq miles; pop. 2,802,000)

Iran lies across The GULF from the Arabian peninsula and stretches from the CASPIAN SEA to the ARABIAN SEA. It is a land dominated by mountains in the north and west, with a huge expanse of desert in its centre. The climate is hot and dry, although more temperate conditions are found on the shores of the Caspian Sea. In winter, terrible dust storms sweep the deserts and almost no life can survive. Most of the population live in the north and west, where TEHRAN is situated. The only good agricultural land is on the Caspian coastal plains, where rice is grown. About 5% of the population are nomadic herdsmen who wander in the mountains. Most of Iran's oil is in the south-west, and other valuable minerals include coal, iron ore, copper and lead. Precious stones are found in the north-east. Main exports are petrochemicals, carpets and rugs, textiles, raw cotton and leather goods.

Quick Facts:
Area : 1,648,000 sq km (636,296 sq miles)
Population : 66,000,000
Capital : Tehran
Other major cities : Esfahan, Mashhad, Tabriz
Government : Islamic Republic
Religion : Shia Islam
Currency : Rial

Iraq is located in south-west ASIA, wedged between The GULF and SYRIA. It is almost landlocked except for its outlet to The Gulf at SHATT AL ARAB. Its two great rivers, the TIGRIS and the EUPHRATES, flow from the north-west into The Gulf at this point. The climate is arid with very hot summers and cold winters. The high mountains on the border with TURKEY are snow covered for six months of the year, and desert in the south-west covers nearly half the country. The only fertile land in Iraq is in the basins of the Tigris and Euphrates where wheat, barley, rice, tobacco and cotton are grown. The world's largest production of dates also comes from this area. Iraq profited from the great oil boom of the 1970s, but during the war with Iran oil terminals in The Gulf were destroyed and the Trans-Syrian Pipeline closed. Iraq is now wholly reliant on the pipeline from KIRKUK to the MEDITERRANEAN.

Quick Facts:
Area : 434,924 sq km (167,935 sq miles)
Population : 19,951,000
Capital : Baghdad
Other major cities : Al-Basrah, Al Mawsil
Form of government : Republic
Religion : Shia Islam, Sunni Islam
Currency : Iraqi dinar

Ireland an island off the west coast of GREAT BRITAIN, almost four fifths of which is the independent Republic of IRELAND, while the remainder is NORTHERN IRELAND, which is a province of the UK. (84,402sq km/32,588 sq miles)

Ireland, Republic of is one of Europe's most westerly countries, situated in the ATLANTIC OCEAN and separated from GREAT BRITAIN by the IRISH SEA. It has an equable climate, with mild south-west winds which makes temperatures uniform over most of the country. The Republic extends over four fifths of the island of Ireland and the west and south-west is mountainous, with the highest peak reaching 1041 m (3416 ft) at Carrauntoohil. The central plain is largely limestone covered in boulder clay which provides good farmland and pasture. The rural population tend to migrate to the cities, mainly Dublin, which is the main industrial centre and the focus of radio, television, publish-

ing and communications. Lack of energy resources and remoteness from major markets has
slowed industrial development, although the
economy has improved in recent years.

Quick Facts:
Area : 70,284 sq km (27,137 sq miles)
Population : 3,621,000
Capital : Dublin (Baile Atha Cliath)
Other major cities : Cork, Galway,
  Limerick, Waterford
Form of government : Republic
Religion : RC
Currency : Punt = 100 pighne

Irian Jaya the western half of the island of NEW
GUINEA, which has been part of INDONESIA since
1963. (421,9800 sq km/162,927 sq miles; pop.
1,650,000)

Irish Sea the arm of the ATLANTIC that separates the
islands of IRELAND and GREAT BRITAIN.

Irkutsk an industrial city on the Trans-Siberian
Railway lying near the southern end of Lake
BAIKAL in the RUSSIAN FEDERATION. (Pop. 626,000)

Irrawaddy, River the central focus of MYANMAR,
(Burma) flowing from its two primary sources in
the north of the country to MANDALAY and then
south to its delta in the Bay of BENGAL. (Length
2000 km/1250 miles)

Irtysh, River a largely navigable river flowing
northwards from its source near the border between north-west CHINA and MONGOLIA across the
centre of KAZAKHSTAN and through OMSK to join
the River OB' on its journey to the ARCTIC OCEAN.
(Length 4440 km/2760 miles)

Ischia a beautiful volcanic island at the northern
end of the Bay of NAPLES. (46 sq km/18 sq miles;
pop. 43,900)

Isfahan *see* **Esfahan**.

Iskenderun a port of ancient origin in southern
TURKEY, in the north-eastern corner of the MEDI-
TERRANEAN SEA. (Pop. 173,600)

Islamabad the capital of PAKISTAN since 1967, in
the north of the country. (Pop. 266,000)

Israel occupies a long narrow stretch of land in the
south-east of the MEDITERRANEAN. Its eastern

boundary is formed by the GREAT RIFT VALLEY,
through which the river JORDAN flows to the DEAD
SEA. The south of the country is made up of a triangular wedge of the NEGEV Desert which ends at
the Gulf of AQABA. The climate in summer is hot
and dry, in winter it is mild with some rain. The
south of the country is arid and barren. Most of the
population live on the coastal plain bordering the
Mediterranean where TEL AVIV-JAFFA is the main
commercial city. Israel's agriculture is based on
collective settlements known as Kibbutz. The
country is virtually self-sufficient in foodstuffs
and a major exporter of its produce. Jaffa oranges
are famous throughout EUROPE. A wide range of
products is processed or finished in the country,
and main exports include finished diamonds, textiles, fruit, vegetables, chemicals, machinery and
fertilizers.

Quick Facts:
Area : 20,770 sq km (8019 sq miles)
Population : 5,451,000
Capital : Jerusalem
Other major cities : Tel Aviv-Jaffa, Haifa
Form of government : Republic
Religion : Judaism, Sunni Islam, Christianity
Currency : Shekel

Issyk-Kul' a lake in southern central KAZAKHSTAN,
set in the high mountains that line the border with
CHINA. (6280 sq km/2424 sq miles)

Istanbul the largest city in TURKEY, built mainly on
the western bank of the BOSPHOROUS, with a commanding view of shipping entering the BLACK SEA.
It was founded by the Greeks in 660BC and was
known as Byzantium; between AD330 and 1930 it
was called Constantinople. (Pop. 6,620,000)

Italy is a republic in southern EUROPE, which comprises a large peninsula and the two main islands
of SICILY and SARDINIA. The ALPS form a natural
boundary with its northern and western European
neighbours, and the ADRIATIC SEA to the east separates it from the countries of former Yugoslavia.
The APENNINE Mountains form the backbone of
Italy and extend the full length of the peninsula.
Between the Alps and the Apennines lies the Po
valley, a great fertile lowland. Sicily and Sardinia

are largely mountainous. Much of Italy is geologically unstable and it has four active volcanoes, including ETNA and VESUVIUS. Italy enjoys warm dry summers and mild winters. The north is the main industrial centre and agriculture is well mechanized. In the south farms are small and traditional. Industries in the north include motor vehicles, textiles, clothing, leather goods, glass and ceramics. Tourism is an important source of foreign currency.

Quick Facts:
Area : 301,225 sq km (116,304 sq miles)
Population : 57,181,000
Capital : Rome (Roma)
Other major cities : Milan, Naples, Turin, Genoa,
  Palermo, Florence
Form of government : Republic
Religion : RC
Currency : Lira

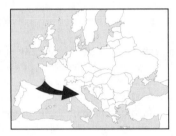

Ithaca (Ithaki) the smallest of the IONIAN ISLANDS, situated off the west coast of GREECE. Odysseus (Ulysses), the hero of Homer's *Odyssey*, was a son of the royal house of Ithaca. (93 sq km/36 sq miles; pop. 3650)

Ivanovo a textile manufacturing city in the RUSSIAN FEDERATION, 240 km (150 miles) north-east of Moscow. (Pop. 480,000)

Iviza *see* **Balearic Islands.**

Ivory Coast *see* **Côte d'Ivoire.**

Iwo Jima the largest in the group of islands called the Volcano Islands belonging to JAPAN, which lie some 1200 km (745 miles) south of TOKYO in the PACIFIC OCEAN. It was the scene of bitter fighting in 1945 at the end of World War II, when U.S. troops took the island from the Japanese. (21 sq km/8 sq miles)

Ixtacihuatl a volcanic peak south of MEXICO CITY, which is twinned with neighbouring POPOCATÉPETL. (5286 m/17,342 ft)

Izmir (Smyrna) a port of ancient Greek origin on the AEGEAN coast of TURKEY, to the south of ISTANBUL. (Pop. 1,757,500)

Izmit (Kocaeli) a port and naval base on the Sea of MARMARA, 90 km (55 miles) south-east of ISTANBUL. (Pop. 236,100)

Jackson the state capital of MISSISSIPPI. (Pop. city 196,00)

Jacksonville a port on the north-east coast of FLORIDA, USA. (Pop. city 661,000)

Jaffna a port on the tip of the northern peninsula of SRI LANKA, and the main centre for the Tamil population of the island. (Pop. 130,000)

Jaipur the capital of the state of RAJASTHAN, INDIA. (Pop. 1,450,000)

Jakarta the capital of INDONESIA, a port on the north-western tip of JAVA. (Pop. 7,885,500)

Jamaica is an island state in the CARIBBEAN SEA about 150 km (93 miles) south of CUBA. The centre of the island comprises a limestone plateau and this is surrounded by narrow coastal flatlands and palm fringed beaches. The highest mountains, the BLUE MOUNTAINS, are in the east of the island. The climate is tropical with high temperatures at the coast, with slightly cooler and less humid conditions in the highlands. The island lies right in the middle of the hurricane zone. The traditional crops grown are sugar cane, bananas, peppers, ginger, cocoa and coffee, and new crops such as winter vegetables fruit and honey are being developed for export. Despite this the decline in the principal export products, bauxite and alumina, has resulted in near economic stagnation. Tourism is a particularly important industry, as is the illegal trade in cannabis.

Quick facts:
Area : 10,990 sq km (4243 sq miles)
Population : 2,513,000
Capital : Kingston
Other major cities : Montego Bay,
  Spanish Town
Form of government : Constitutional Monarchy
Religion : Anglicanism, RC, other
  Protestantism
Currency : Jamaican dollar

James Bay the southern arm of the HUDSON BAY, CANADA, which extends 440 km (273 miles) into ONTARIO and QUEBEC.

Jammu and Kashmir the state in the very north of INDIA, bordering CHINA, PAKISTAN and a small portion of AFGHANISTAN. Some of the world's highest mountains are located in the northern portion of this Indian state, including K2 (Godwin Austen), the second highest mountain in the world. (222,236 sq km/85,783 sq miles; pop. 7,718,700)

Jamshedpur an industrial city in north-east INDIA which grew up around steel foundries set up by Jamshedji Tata in 1907–11. (Pop. city 460,000)

Jamuna, River the name given to the river formed by the BRAHMAPUTRA and the Tista as it flows through BANGLADESH to join the GANGES.

Japan is located on the eastern margin of ASIA and consists of four major islands, HONSHU, HOKKAIDO, KYUSHU and SHIKOKU, and many small islands. It is separated from the mainland of Asia by the Sea of Japan. The country is made up of six chains of steep serrated mountains, which contain about 60 active volcanoes. Earthquakes are frequent and widespread and often accompanied by giant waves (tsunami). Summers are warm and humid and winters mild, except on Hokkaido which is covered in snow in winter. Japan's agriculture is highly advanced with extensive use made of fertilizers and miniature machinery for the small fields. Fishing is important. Japan is the second largest industrial economy in the world. It is very dependent on imported raw materials, and its success is based on manufacturing industry, which employs about one third of the workforce.

Quick facts:
Area : 377,708 sq km (145,834 sq miles)
Population : 124,764,200
Capital : Tokyo
Other major cities : Osaka, Nagoya, Sapporo, Kobe, Kyoto, Yokohama
Form of government : Constitutional Monarchy
Religion : Shintoism, Buddhism, Christianity
Currency : Yen

Japan, Sea of a part of the PACIFIC OCEAN that lies between JAPAN and the Korean peninsula.

Java (Jawa) the central island in the southern chain of islands of INDONESIA. The capital is JAKARTA. (130,987 sq km/50,574 sq miles; pop. 107,514,000)

Java Sea an arm of the PACIFIC OCEAN that separates JAVA and BORNEO.

Jedda *see* **Jiddah.**

Jefferson City the state capital of MISSOURI, USA. (Pop. 35,500)

Jena a university town in southern central GER-

MANY, 80 km (50 miles) south-west of LEIPZIG. The Prussian army was defeated by the French under Napoleon here in 1806. (Pop. 110,000)

Jerba (Djerba) an island in the Gulf of GABES belonging to TUNISIA. It has become a popular tourist resort in recent years. (67 sq km/42 sq miles; pop. 92,300)

Jerez de la Frontera (Jerez) a town in south-west SPAIN, just inland from CADIZ, famous for the sweet wine to which it has given its name, sherry. (Pop. 185,000)

Jericho a town in the WEST BANK area occupied by ISRAEL since 1967, on the site of a city that dates back to about 7000BC. (Pop. 15,000)

Jersey the largest of the British CHANNEL ISLANDS. The capital is St Helier. (117 sq km/45 sq miles; pop. 84,000)

Jerusalem the capital of ISRAEL, although not internationally recognized as such, and a historic city considered holy by Muslims, Christians and Jews. (Pop. 544,000)

Jiangsu a heavily populated but highly productive province on the central east coast of CHINA. The capital is NANJING. (100,000 sq km/38,600 sq miles; pop. 67,050,000)

Jiangxi an inland province of south-eastern CHINA. Its capital is NANCHANG. (164,800 sq km/64,300 sq miles; pop. 37,700,000)

Jiddah (Jedda) a port on the RED SEA coast of SAUDI ARABIA, and one of the country's main centres of population. (Pop. 1,400,000)

Jilin (Kirin) a province of central DONGBEI (Manchuria) in northern CHINA. The capital is CHANGCHUN. (180,000 sq km/69,500 sq miles; pop. 24,660,000)

Jinan the capital of SHANDONG province, situated close to the HUANG HE River, 360 km (225 miles) to the south of BEIJING. (Pop. 2,300,000)

Jodhpur a city in central RAJASTHAN, INDIA, on the perimeter of the THAR DESERT. The city has given its name to the riding breeches that first became popular here. (Pop. 666,000)

Jogjakarta *see* **Yogyakarta.**

Johannesburg the centre of the RAND goldmining area of SOUTH AFRICA and now that country's largest town. (Pop. 1,610,000)

John o'Groats the village traditionally held to be on the most northerly point of mainland SCOTLAND and GREAT BRITAIN.

Johor Baharu (Johore) a port and growing city in MALAYSIA situated on the southern tip of the Malay Peninsula opposite SINGAPORE, to which it is connected by a causeway. It is also the capital of the state of Johor. (Pop. city 325,000)

Jordan almost landlocked except for a short coastline on the Gulf of AQABA, Jordan is bounded by
SAUDI ARABIA, SYRIA, IRAQ and ISRAEL. Almost
80% of the country is desert and the rest comprises
the East Bank Uplands and Jordan Valley. In general summers are hot and dry and winters cool and
wet, with variations related to altitude. The east
has a desert climate. Only one fifth of the country
is fertile enough to be farmed but the country is
self-sufficient in potatoes, onions and poultry
meat. The agricultural system is intensive and efficient. AMMAN is the main industrial centre of the
country and the industries include phosphates, petroleum products, cement, iron and fertilizers. The
rich Arab states such as Saudi Arabia give Jordan
substantial economic aid.

Quick facts:
Area : 97,740 sq km (37,737 sq miles)
Population : 4,095,000
Capital : Amman
Other major cities : Irbid, Zarga
Form of government : Constitutional Monarchy
Religion : Sunni Islam
Currency : Jordan dinar

Jordan, River a river flowing southwards from
Mount HERMON in southern LEBANON, through
northern ISRAEL to Lake TIBERIAS (Sea of Galilee)
and then on through Jordan into the DEAD SEA,
where it evaporates. The WEST BANK to the north
of the Dead Sea is disputed territory which has
been occupied by Israel since the Six Day War in
1967. (Length 256 km/159 miles)

Juan de Fuca Strait the channel to the south of
VANCOUVER ISLAND on the border between CANADA
and the USA, through which ships from VICTORIA,
VANCOUVER and SEATTLE can pass to reach the PA-
CIFIC OCEAN.

Juan Fernández Islands a group of three remote
islands in the PACIFIC OCEAN belonging to CHILE
and some 650 km (400 miles) due west of SAN-
TIAGO. (181 sq km/62 sq miles)

Judaea the southern part of ancient PALESTINE, oc-
cupying the area of modern ISRAEL between the
MEDITERRANEAN coast to the west and the DEAD
SEA and River JORDAN to the east.

Jumna, River *see* **Yamuna, River.**

Juneau The state capital of ALASKA. (Pop. 28,360)

Jungfrau a famous peak in the Bernese Oberland
range in the Swiss ALPS, popular with climbers but
now also ascended by cable car. (4158 m/13,642
ft)

Jura a large upland band of limestone in eastern
central FRANCE which lines the border with SWIT-
ZERLAND, giving its name to a department in France
and a canton in Switzerland. A further extension
continues across southern GERMANY as far as NU-
REMBERG (the Swabian and Franconian Jura).

Jutland (Jylland) a large peninsula stretching
some 400 km (250 miles) northwards from GER-
MANY to separate the NORTH SEA from the BALTIC
SEA. Most of it is occupied by the mainland part of
DENMARK, which calls it Jylland, while the south-
ern part belongs to the German state of SCHLESWIG-
HOLSTEIN.

Jylland *see* **Jutland.**

K2 (Godwin Austen) the second highest mountain
in the world after Mount EVEREST, situated in the
KARAKORAM mountain range in the state of JAMMU
AND KASHMIR, INDIA. (8611 m/28,250 ft)

Kabul the capital and main city of AFGHANISTAN, in
the north-east of the country on the Kabul River.
(Pop. 1,420,000)

Kachchh *see* **Kutch.**

Kagoshima a port on the south coast of KYUSHU Island, JAPAN. (Pop. 537,000)

Kaifeng a city of ancient origins in HENAN province, CHINA. (Pop. 500,000)

Kairouan a city in northern TUNISIA, to Muslims the
most holy city of the MAGHREB. (Pop. 75,000)

Kalahari a region of semi-desert occupying much
of southern BOTSWANA and straddling the border
with SOUTH AFRICA and NAMIBIA.

Kalgoorlie a town in southern WESTERN AUSTRALIA
which has grown up around its gold and nickel reserves. (Pop. 27,000)

Kalimantan the greater part of BORNEO, which is
governed by INDONESIA. (538,718 sq km/208,000
sq miles; pop. 9,100,000)

Kalimnos (Calino) *see* **Dodecanese.**

Kalinin *see* **Tver.**

Kaliningrad (Königsberg) a port and industrial
city on the BALTIC coast belonging to the RUSSIAN
FEDERATION, in an enclave between LITHUANIA and
POLAND. Founded in the 13th century, it was called
Königsberg and was the capital of East PRUSSIA,
but was ceded to the former USSR in 1945 and renamed after Mikhail Kalinin, the Soviet President
1937–46. (Pop. 408,000)

Kalmyk (Kalmuck) Republic an autonomous republic of the RUSSIAN FEDERATION, lying to the

north-west of the CASPIAN SEA. (75,900 sq km/ 29,300 sq miles; pop. 327,000)

Kamchatka a peninsula, some 1200 km (750 miles) long, which drops south from eastern SIBERIA in the RUSSIAN FEDERATION into the north PACIFIC OCEAN.

Kampala the capital and main city of UGANDA, situated on Lake VICTORIA. (Pop. 773,000)

Kampuchea *see* **Cambodia.**

Kananga a city in central southern ZAIRE, founded in 1894 as Luluabourg. (Pop. 372,000)

Kanazawa a historic port on the central northern coast of HONSHU Island, JAPAN. (Pop. 443,000)

Kandahar a city in Afghanistan, the capital of Kandahar province, situated in the south-eastern part of the country, near the border with PAKISTAN. (Pop. 226,000)

Kandy a town in the central mountains of SRI LANKA, which was once the capital of the Sinhalese kings and is sacred to Buddhists. (Pop. 104,000)

Kangchenjunga the world's third highest mountain (after Mount EVEREST and K2), situated in the eastern HIMALAYAS, on the borders between NEPAL, CHINA and the Indian state of SIKKIM. (8585 m/ 28,165 ft)

Kano a historic trading city of the Hausa people of northern NIGERIA, the third largest city in Nigeria after LAGOS and IBADAN. (Pop. city 700,000)

Kanpur (Cawnpore) an industrial city in northern central INDIA. (Pop. 2,110,000)

Kansas a state in the GREAT PLAINS of the USA. The state capital is TOPEKA. (213,064 sq km/82,264 sq miles; pop. 2,515,00)

Kansas City an industrial city on the MISSOURI RIVER which straddles the border between the states of MISSOURI and KANSAS. (Pop. metropolitan area 1,566,700)

Kaohsiung the second largest city in TAIWAN and a major port, situated in the south-west of the island. (Pop. 1,405,000)

Karachi a port and industrial centre, and the largest city in PAKISTAN. (Pop. 5,181,000)

Karaganda an industrial city in the mining region of KAZAKHSTAN. (Pop. 610,000)

Karakoram a range of mountains at the western end of the HIMALAYAS on the border between CHINA and INDIA.

Kara Kum (Karakumy) a sand desert in southern TURKMENISTAN, to the east of the CASPIAN SEA, and on the borders with IRAN and AFGHANISTAN.

Kara Sea a branch of the ARCTIC OCEAN off the central northern coast of the RUSSIAN FEDERATION.

Karbala a town in central IRAQ, 90 km (55 miles)

south of BAGHDAD. As the site of the tomb of Hussein bin Ali and his brother Abbas, grandsons of the prophet Mohammad, it is held sacred by the Shia Muslims. (Pop. 107,500)

Karelia a autonomous region of the RUSSIAN FEDERATION which straddles the border with FINLAND. (172,400 sq km/66,500 sq miles; pop. 792,000)

Kariba Dam a hydroelectric dam on the River ZAMBEZI on the border between ZAMBIA and ZIMBABWE.

Karl-Marx-Stadt *see* **Chemnitz.**

Karlovy Vary (Carlsbad; Karlsbad) a spa town in the CZECH REPUBLIC. (Pop. 122,300)

Karlsruhe (Carlsruhe) an industrial city in the valley of the River RHINE, in south-western GERMANY. (Pop. 280,000)

Karnak the site of the extensive ruins of a temple complex dating from about 1560-1090BC, on the eastern bank of the River NILE in central EGYPT.

Karnataka a state in south-west INDIA. The capital is BANGALORE. (191,791 sq km/74,031 sq miles; pop. 44,817,000)

Kärnten *see* **Carinthia.**

Karoo (Karroo) two separate regions of semi-desert, the Great Karoo and the Little Karoo, lying between the mountain ranges of southern Western Cape, SOUTH AFRICA.

Kasai (Cassai), River a major river of ZAÏRE. (Length 2150 km/1350 miles)

Kashmir formerly a princely state on the north-western border of India, Kashmir now refers to a mountainous region spanning the length of the Indian state of JAMMU AND KASHMIR. It has been subject to dispute since the partition of INDIA and PAKISTAN in 1947.

Kassel (Cassel) an industrial city in central GERMANY. (Pop. 202,000)

Kasvin *see* **Qazvin.**

Kathmandu (Katmandu) the capital and principal city of NEPAL. (Pop. 419,000)

Katowice (Kattowitz) an industrial city in central southern POLAND. (Pop. 367,300)

Kattegat (Cattegat) the strait, 34 km (21 miles) at its narrowest, at the entrance to the BALTIC SEA which separates SWEDEN from the JUTLAND peninsula of DENMARK.

Kaunas (Kovno) an industrial city and former capital of LITHUANIA. (Pop. 415,000)

Kavango (Cubango), River a river, known formerly as the Okavango, which flows south-east from central ANGOLA to form the border with NAMIBIA before petering out in the swampy inland Okavango Delta in northern BOTSWANA. (Length 1600 km/1000 miles)

Kaveri (Caveri, Cauvery), River a holy river of southern India, flowing south-east from the DECCAN plateau to the coast on the Bay of BENGAL. (Length 800 km/497 miles)

Kavkaz *see* **Caucasus.**

Kawasaki an industrial city on the east coast of HONSHU Island, JAPAN, forming part of the TOKYO-YOKOHAMA conurbation. (Pop. 1,195,000)

Kazakhstan, the second largest republic of the former USSR, extends approximately 3000 km (1864 miles) from the coast of the CASPIAN SEA to the north-west corner of MONGOLIA. The west of the country is low-lying, the east hilly, and in the south-east mountainous areas include parts of the TIAN SHAN and ALTAI ranges. The climate is continental and very dry with great extremes of temperature. Much of the country is semi-desert. Crops can only be grown in the wetter north-west regions or on land irrigated by the SYRDAR'YA river. Extensive pastoral farming is carried out in most of the country, and cattle, sheep and goats are the main livestock reared. The country is rich in minerals, particularly copper, lead, zinc, coal, tungsten, iron ore, oil and gas. Kazakhstan declared itself independent in 1991.

Quick facts:
Area : 2,717,300 sq km (1,050,000 sq miles)
Population : 17,185,000
Capital : Alma-Ata
Other major city : Karaganda
Form of government : Republic
Religion : Sunni Islam
Currency : Rouble

Kazan' an industrial city and capital of the TATARSTAN REPUBLIC in central RUSSIAN FEDERATION. (Pop. 1,107,000)

Keeling Islands *see* **Cocos Islands.**

Kefallinia *see* **Ionian Islands.**

Kells a market town in County MEATH, IRELAND. It was the site of a monastery founded in the 6th century by St Columba, which was the source of the illuminated Book of Kells.

Kelsty *see* **Kielce.**

Kemerovo an industrial city in the coal mining region of southern SIBERIA in the RUSSIAN FEDERATION. (Pop. 520,000)

Kennedy, Cape *see* **Canaveral, Cape.**

Kent a county in the extreme south-east of ENGLAND. The county town is Maidstone. (3732 sq km/1441 sq miles; pop. 1,546,000)

Kentucky a state in east central USA. The state capital is FRANKFORT. (104,623 sq km/40,395 sq miles; pop. 3,754,000)

Kenya located in east AFRICA, Kenya straddles the Equator and extends from Lake VICTORIA in the south-west, to the INDIAN OCEAN in the south-east. Highlands run north to south through central Kenya and are divided by the steep-sided GREAT RIFT VALLEY. The coastal lowlands have a hot humid climate but in the highlands it is cooler and rainfall heavier. In the east it is very arid. The south-western region is well watered with huge areas of fertile soil and this accounts for the bulk of the population and almost all its economic production. The main crops grown for domestic consumption are wheat and maize. Tea, coffee, sisal, sugar cane and cotton are grown for export. Oil refining at MOMBASA is the country's largest single industry, and other industry includes food processing and textiles. Tourism is an important source of foreign revenue.

Quick facts:
Area : 580,367 sq km (224,080 sq miles)
Population : 28,113,080,000
Capital : Nairobi
Other major cities : Mombasa, Kisumu
Form of government : Republic
Religions : RC, Protestantism, other Christianity, Animism
Currency : Kenya shilling

Kenya, Mount a towering extinct volcano in central Kenya, the second highest mountain in AFRICA after Mount KILIMANJARO. (5200 m/17,058 ft)

Kerala a state occupying the western coast of the southern tip of INDIA. The capital is TRIVANDRUM. (38,863 sq km/15,005 sq miles; pop. 29,033,000)

Kerguelen the largest in a remote group of some 300 islands in the southern INDIAN OCEAN forming part of the FRENCH SOUTHERN AND ANTARCTIC TERRITORIES, now occupied only by the staff of a scientific base. (3414 sq km/1318 sq miles)

Kérkira *see* **Corfu.**

Kermanshah *see* **Bakhtaran.**

Kerry a county in the south-west of the Republic of IRELAND, noted for the rugged beauty of its peninsulas and its green dairy pastures. The county town is Tralee. (4701 sq km/1815 sq miles; pop. 122,800)

Key West a port and resort at the southern end of Florida Keys, a chain of coral islands off the southern tip of FLORIDA, USA. (Pop. 24,900)

Khabarovsk a major industrial city in south-eastern SIBERIA, lying just 35 km(22 miles) north of the border with CHINA. (Pop. 600,000)

Khajuraho a town in northern MADHYA PRADESH, INDIA, noted in particular for its Hindu and Jain temples which are famed for their intricate and erotic sculpture.

Kharg Island a small island in the northern GULF where IRAN has constructed a major oil terminal.

Khar'kov a major industrial and commercial centre of the UKRAINE. (Pop. 1,618,000)

Khartoum (El Khartum) the capital of SUDAN, situated at the confluence of the Blue NILE and White Nile. (Pop. city 561,000)

Khios *see* **Chios.**

Khone Falls a massive set of waterfalls on the River MEKONG in southern LAOS, which effectively prevents the Mekong from being navigable beyond this point. With a maximum width of 10.8 km (6.7 miles), these are the widest falls in the world.

Khorasan the north-eastern province of IRAN, bordering AFGHANISTAN and TURKMENISTAN. The capital is MASHHAD. (Pop. 3,267,000)

Khulna a port and district in south-west BANGLADESH. (Pop. town 877,000)

Khuzestan (Khuzistan) a province in south-western IRAN, and the country's main oil-producing area. The capital is AHVAZ. (Pop. 2,702,000)

Khyber Pass a high pass (1072 m/3518 ft) over the Safed Koh mountains connecting PESHAWAR in PAKISTAN with KABUL in AFGHANISTAN. It has been of great strategic importance throughout history.

Kiel a port and shipbuilding city on the BALTIC coast of northern GERMANY. It stands at the mouth of the Kiel Ship Canal which permits ocean-going ships to cross the JUTLAND peninsula from the BALTIC to HAMBURG and the NORTH SEA. (Pop. 248,500)

Kielce (Kelsty) an industrial city in central southern POLAND. (Pop. 214,200)

Kiev (Kiyev) the capital of the UKRAINE, situated on the DNIEPER River. Founded in the 6th century, it is now a major industrial city. (Pop. 2,640,000)

Kigali the capital of RWANDA. (Pop. 156,000)

Kikládhes *see* **Cyclades.**

Kildare a county in the south-east of the Republic of IRELAND, famous for its racehorses and the racecourse, The Curragh. The county town is Naas. (1694 sq km/654 sq miles; pop. 123,100)

Kilimanjaro, Mount Africa's highest mountain, in north-eastern TANZANIA. (5895 m/19,340 ft)

Kilkenny a county in the south-east of the Republic of IRELAND, and the name of its county town. (2062 sq km/769 sq miles; pop. county 74,000/city 10,100)

Killarney a market town in county KERRY, in the Republic of IRELAND, which is at the centre of a landscape of lakes and mountains much admired and visited for its beauty. (Pop. 7960)

Kimberley a town in Northern Cape, SOUTH AFRICA, which is at the centre of South Africa's diamond mining industry. (Pop. 153,900)

Kimberleys, The a vast plateau of hills and gorges in the north of WESTERN AUSTRALIA. (420,000 sq km/162,000 sq miles)

Kingston the capital and main port of JAMAICA. (Pop. 644,000)

Kingston upon Hull (Hull) a city and port in eastern ENGLAND, situated on the north side of the HUMBER estuary. (Pop. 265,000)

Kingstown the capital of ST VINCENT and a port, famed for its botanical gardens. (Pop. 27,000)

Kinshasa the capital of ZAÏRE, on the banks of the River ZAÏRE. It is the largest city in Central AFRICA. (Pop. 3,741,000)

Kirghizia (Kirgizia) *see* **Kyrgyzstan.**

Kirin *see* **Jilin.**

Kiribati comprises three groups of coral atolls and one isolated volcanic island spread over a large expanse of the central PACIFIC. The group includes Banaba Island, the Phoenix Islands and some of the Line Islands. The climate is maritime equatorial with a high rainfall. Most islanders are involved in subsistence agriculture. The principal tree is the coconut which grows well on all the islands. Palm and breadfruit trees are also found. Soil is negligible and the only vegetable which can be grown is calladium. Tuna fishing is an important industry and Kiribati had granted licences to the former USSR to fish its waters. Phosphate sources have now been exhausted and the country is heavily dependent on overseas aid.

Quick facts:
Area : 726 sq km (280 sq miles)
Population : 77,000
Capital : Tarawa
Government : Republic
Religions : RC,
　Protestantism
Currency : Aus. dollar

Kiritimati *see* **Christmas Island.**

Kircudbrightshire *see* **Dumfries and Galloway.**

Kirkuk an industrial city and regional capital in the Kurdish north of IRAQ. (Pop. 420,000)

Kirov (Vyatka) an industrial city in east central RUSSIAN FEDERATION, founded in the 12th century. (Pop. 490,000)

Kisangani a commercial centre and regional capital in northern ZAÏRE, on the River ZAÏRE. It was originally called Stanleyville. (Pop. 374,000)

Kishinev (Chisinau) the capital of MOLDOVA. (Pop. 655,000)

Kistna, River *see* **Krishna, River.**

Kitakyushu a major industrial city situated in the north of KYUSHU Island, JAPAN. (Pop. 1,026,500)

Kitchener-Waterloo two towns in southern ONTARIO, CANADA, which have become twin cities, 100 km (62 miles) west of TORONTO. (Pop. 356,000)

Kíthira (Cerigo) *see* **Ionian Islands.**

Kivu, Lake a lake in the GREAT RIFT VALLEY on the border between RWANDA and ZAÏRE. (2850 sq km/ 1100 sq miles)

Kiyev *see* **Kiev.**

Kizil Irmak, River the longest river in TURKEY, flowing westwards from the centre of the country near SIVAS, before curling north to the BLACK SEA. (Length 1130 km/700 miles)

Klaipeda a major port and shipbuilding centre on the BALTIC coast of LITHUANIA. (Pop. 203,000)

Klondike, River a short river flowing through YUKON TERRITORY in north-western CANADA to meet the Yukon River at Dawson. Gold was discovered in the region in 1896, causing the subsequent goldrush. (Length 160 km/100 miles)

Knock a village in County MAYO, in the west of the Republic of IRELAND, where a group of villagers witnessed a vision of the Virgin Mary and other saints in 1879. It has now become a Marian shrine of world importance. (Pop. 1400)

Knossos the site of an excavated royal palace of the Minoan civilization, 5 km (3 miles) south-east of HERAKLION, the capital of CRETE. The palace was built in about 1950BC and destroyed in 1380BC.

Knoxville an industrial city in eastern TENNESSEE, USA, and a port on the TENNESSEE RIVER. (Pop. city 167,290/metropolitan area 589,400)

Kobe a major container port and shipbuilding centre at the southern end of HONSHU Island, JAPAN. (Pop. 1,457,000)

København *see* **Copenhagen.**

Koblenz (Coblenz) a city at the confluence of the Rivers RHINE and MOSELLE in western GERMANY, and a centre for the German winemaking industry. (Pop. 113,000)

Kola Peninsula a bulging peninsula in the BARENTS SEA in the extreme north-west of the RUSSIAN FEDERATION, to the east of MURMANSK.

Köln *see* **Cologne.**

Kolyma, River a river in north-eastern SIBERIA, flowing north from the gold-rich Kolyma mountains into the East Siberian Sea. (Length 2600 km/1600 miles)

Komi Republic an autonomous republic in the north of the RUSSIAN FEDERATION, which produces timber, coal, oil and natural gas. (415,900 sq km/ 160,600 sq miles; pop. 1,255,000)

Komodo a small island of Indonesia in the Lesser SUNDA group, between SUMBAWA and FLORES, noted above all as the home of the giant monitor lizard, the Komodo Dragon. (520 sq km/200 sq miles)

Königsberg *see* **Kaliningrad.**

Konya a carpet-making town and capital of the province of the same name in central southern TURKEY, 235 km (146 miles) south of ANKARA. (Pop. town 513,000)

Korcula *see* **Dalmatia.**

Korea, Democratic People's Republic of (formerly North Korea) occupies just over half of the Korean peninsula in east ASIA. The Yala and Tumen rivers form its northern border with CHINA and the RUSSIAN FEDERATION. Its southern border with South KOREA is just north of the 38th parallel. It is a mountainous country, three quarters of which is forested highland or scrubland. The climate is warm temperate, although winters can be cold in the north. Most rain falls during the summer. Nearly 90% of its arable land is farmed by cooperatives which employ over 40% of the labour force and rice is the main crop grown. North Korea is quite well endowed with fuel and minerals. Deposits of coal and hydro-electric power generate electricity, and substantial deposits of iron ore are found near P'YONGYANG and Musan. 60% of the labour force are employed in industry, the most important of which are metallurgical, building, cement and chemicals.

Quick facts:
Area : 120,538 sq km (46,540 sq miles)
Population : 23,472,000
Capital : P'yongyang
Other major cities : Chongjin, Nampo
Form of government : Socialist Republic
Religions : Chondoism, Buddhism
Currency : N. Korean won

Korea, Republic of (formerly South Korea) occupies the southern half of the Korean peninsula and stretches about 400 km (249 miles), from the Korea Strait to the demilitarized zone bordering North KOREA. It is predominantly mountainous with the highest ranges running north to south along the east coast. The west is lowland which is extremely densely populated. The extreme south has a humid warm temperate climate while farther north it is more continental. Most rain falls in summer. Cultivated land represents only 23% of the country's total area and the main crop is rice. The country has few natural resources but has a flourishing manufacturing industry and is the world's leading supplier of ships and footwear. Other important industries are electronic equipment, electrical goods, steel, petrochemicals, motor vehicles and toys. Its people enjoy a reasonably high standard of living brought about by hard work and determination.

Quick facts:
Area : 98,484 sq km (38,025 sq miles)
Population : 44,563,000
Capital : Seoul (Soul)
Other major cities : Pusan, Taegu, Inch'on
Form of government : Republic
Religions : Buddhism, Christianity
Currency : South Korean won

Korea Strait the stretch of water, 64 km (40 miles) at its narrowest, separating the southern tip of South KOREA from JAPAN. It is also sometimes known as the Tsushima Strait, after the island of that name.

Korinthos *see* **Corinth.**

Koror the capital of BELAU. (Pop. 15,500)

Kos (Cos) one of the DODECANESE ISLANDS, belonging to GREECE, in the AEGEAN SEA, noted as the birthplace (*c.*460BC) of Hippocrates, the father of medicine. (290 sq km/112 sq miles; pop. 20,300)

Kosciusko, Mount the highest mountain in AUSTRALIA, a peak in the SNOWY MOUNTAINS range in southern NEW SOUTH WALES. (2230 m/7316 ft)

Kosice a rapidly growing industrial city, and the regional capital of eastern SLOVAKIA. (Pop. 235,000)

Kosovo an autonomous province in the south-west of SERBIA in the FEDERAL REPUBLIC OF YUGOSLAVIA. About 75% of the population are ethnic Albanians

while many ethnic Serbs have left the province. The capital is PRISTINA. (10,817 sq km/4202 sq miles; pop. 1,584,000)

Kovno *see* **Kaunas.**

Kowloon a mainland territory of HONG KONG, lying opposite and to the north of Hong Kong Island. (Pop. 800,000)

Kra, Isthmus of the narrow neck of land, only some 50 km (30 miles) wide and shared by MYANMAR and THAILAND, which joins peninsular MALAYSIA to the mainland of South-East ASIA.

Krakatau (Krakatoa) a volcano which erupted out of the sea between JAVA and SUMATRA in INDONESIA in 1883 in an explosion that was heard 5000 km (3100 miles) away, and which killed 36,000 people. Today the site is marked by a more recent volcano called Anak Krakatau (Son of Krakatau).

Krakow *see* **Cracow.**

Krasnodar an agricultural centre and industrial city in the RUSSIAN FEDERATION near the BLACK SEA. (Pop. city 623,000)

Krasnoyarsk a mining city on the Trans-Siberian Railway in central southern SIBERIA in the RUSSIAN FEDERATION. (Pop. 912,000)

Krefeld a textile town specializing in silk in western GERMANY, near the border with the NETHERLANDS. (Pop. 224,000)

Krishna (Kistna), River a river that flows through southern INDIA from its source in the Western GHATS to the Bay of BENGAL. (Length 1401 km/871 miles)

Kristiania *see* **Oslo.**

Krití *see* **Crete.**

Krivoy Rog a city in the DONETS BASIN mining region of the UKRAINE. (Pop. 689,000)

Krk (Veglia) a richly fertile island belonging to CROATIA, in the northern ADRIATIC SEA. (408 sq km/ 158 sq miles; pop. 1500)

Krung Thep *see* **Bangkok.**

Krym *see* **Crimea.**

Kuala Lumpur the capital of MALAYSIA, sited on the banks of the Kelang and Gombak Rivers. (Pop. 1,145,100)

Kuanza, River *see* **Cuanza, River.**

Kumamoto a city in the west of KYUSHU Island, JAPAN, noted for its electronics industries. (Pop. 579,300)

Kumasi a town in central southern GHANA, and the capital of the Ashanti people. (Pop. 490,000)

Kunming an industrial and trading city, and capital of YUNNAN province in southern, central CHINA. (Pop. 1,500,000)

Kurashiki a city in south-western HONSHU Island, JAPAN. Although an industrial centre, it still preserves much of its medieval heritage. (Pop. 418,000)

Kurdistan a region of the MIDDLE EAST occupied by the Kurdish people spanning the borders of IRAQ, IRAN and TURKEY. Although proposals for an independent Kurdistan were agreed in 1920, this plan has never been realized. Greater autonomy was in principle granted to the Kurdish people in Iraq after the 1991 Gulf War.

Kuril (Kurile) Islands a long chain of some 56 volcanic islands stretching between the southern coast of the KAMCHATKA peninsula in eastern RUSSIAN FEDERATION and HOKKAIDO Island, northern JAPAN. The archipelago was taken from Japan by the former USSR in 1945; this remains an issue of contention between the Russian Federation and Japan. (15,600 sq km/6020 sq miles)

Kursk a major industrial city in the RUSSIAN FEDERATION, 450 km (280 miles) south of MOSCOW. It was the scene of a devastating tank battle in 1943 which left the city in ruins. (Pop. 433,000)

Kurukshetra a sacred Hindu city in northern INDIA, 140 km (87 miles) north of DELHI. (Pop. 186,100)

Kutch (Kucchh) an inhospitable coastal region on the border between PAKISTAN and INDIA, which floods in the monsoon season and then dries out into a baking, salty desert. (45,652 sq km/17,626 sq miles)

Kuwait is a tiny Arab state on The GULF, comprising the city of Kuwait at the southern entrance of Kuwait Bay, a small undulating desert wedged between IRAQ and SAUDI ARABIA and nine small offshore islands. It has a dry desert climate, cool in winter but very hot and humid in summer. There is little agriculture due to lack of water; major crops produced are melons, tomatoes, onions and dates. Shrimp fishing is becoming an important industry. Large reserves of petroleum and natural gas are the mainstay of the economy. It has about 950 oil wells, however 600 were fired during the Iraqi occupation in 1991 and are unlikely to resume production for several years. Apart from oil, industry includes boat building, food production, petrochemicals, gases and construction.

Quick facts:
Area : 18,049 sq km (6969 sq miles)
Population : 1,575,980
Capital : Kuwait city (Al Kuwayt)
Government : Constitutional
 Monarchy
Religion : Sunni Islam,
 Shia Islam
Currency : Kuwait dinar

Kuwait City (Al Kuwayt) the chief port and capital of Kuwait, it is also a financial centre. (Pop. metropolitan 400,000)

Kuybyshev *see* **Samara**.

Kuznetsk *see* **Novokuznetsk.**

Kwai, River two tributaries of the Mae Khlong River in western THAILAND, the Kwai Yai (Big Kwai) and the Kwai Noi (Little Kwai). During World War II, Allied prisoners of war were forced by their Japanese captors to build a railroad line and a bridge over the Kwai Yai at the cost of some 110,000 lives.

Kwajalein one of the largest atolls in the world, with a lagoon covering some 2800 sq km (1100 sq miles). The island forms part of the MARSHALL ISLANDS in the PACIFIC OCEAN, and is leased to the USA as a missile target.

Kwangju (Gwangju; Chonnam) an industrial city and regional capital in the south-western corner of South KOREA. (Pop. 1,145,000)

Kwango (Cuango), River a river which rises in northern ANGOLA and flows northwards to join the River KASAI in ZAÏRE. (Length 110 km/68 miles)

KwaZulu-Natal a province of South Africa in the east of the country. (91,481 sq km/39,736 sq miles;pop 8,594,000)

Kyongju an ancient city in the south-east of South KOREA which was the capital of the Silla kingdom from 57BC to AD935. (Pop. 142,000)

Kyoto situated in central southern HONSHU Island, this was the old imperial capital of JAPAN from AD794 to 1868. (Pop. 1,452,000)

Kyrgyzstan, a central Asian republic of the former USSR, independent since 1991. It is located on the border with north-west China. Much of the country is occupied by the TIAN SHAN Mountains which rise to spectacular peaks. The highest is Pik Pobedy 7439 m (24,406 ft), lying on the border with China. In the north-east of the country is Issyk-Kul', a large lake heated by volcanic action, so it never freezes. Most of the country is semi-arid or desert, but climate is greatly influenced by altitude. Soils are badly leached except in the valleys, where some grains are grown. Grazing of sheep, horses and cattle is extensive. Industries include non-ferrous metallurgy, machine building, coal mining, tobacco, food processing, textiles and gold mining, hydroelectricity and the raising of silkworms.

Quick facts:
Area : 198,501 sq km (76,642 sq miles)
Population : 4,500,000
Capital : Bishkek
 (formerly Frunze)
Government : Republic
Religion : Sunni Islam
Currency : Rouble

Kyushu the most southerly of JAPAN's main islands, and the third largest after HONSHU and HOKKAIDO. (43,065 sq km/16,627 sq miles; pop. 13,302,000)

Laatokka, Lake *see* **Ladoga, Lake**.

Labrador the mainland part of the province of NEWFOUNDLAND, on the east coast of CANADA. (295,800 sq km/112,826 sq miles)

Laccadive Islands *see* **Lakshadweep**.

Ladakh a remote and mountainous district in the north-eastern part of the state of JAMMU AND KASHMIR, INDIA, noted for its numerous monasteries which preserve the traditions of Tibetan-style Buddhism. The capital is LEH.

Ladoga (Ladozhskoye; Laatokka), Lake Europe's largest lake, in western RUSSIAN FEDERATION, to the north-east of ST PETERSBURG. (18,390 sq km/7100 sq miles)

Lagos the principal port and former capital (until 1992) of NIGERIA, situated on the Bight of BENIN. (Pop. 1,347,000)

Lahore a city in eastern central PAKISTAN, close to the border with INDIA. (Pop. 2,953,000)

Lake District a region of lakes and mountains in the county of CUMBRIA, in north-west ENGLAND. It includes England's highest peak, Scafell Pike (978 m/3208 ft), and a series of lakes famed for their beauty, notably Windermere, Coniston and Ullswater.

Lake of the Woods a lake spattered with some 17,000 islands in south-western ONTARIO, CANADA, on the border with the USA. (4390 sq km/1695 sq miles)

Lakshadweep a territory of INDIA consisting of 27 small islands (the Amindivi Islands, Laccadive Islands and Minicoy Islands) lying 300 km (186 miles) off the south-west coast of mainland India. (32 sq km/12 sq miles; pop. 51,700)

La Mancha a high, arid plateau in central SPAIN, some 160 km (100 miles) south of MADRID, the setting for *Don Quixote*, a 17th-century novel by Miguel de Cervantes.

Lambaréné a provincial capital in eastern central GABON, famous as the site of the hospital founded by Albert Schweitzer (1875–1965). (Pop. 28,000)

Lanarkshire a county in south-west Scotland, now divided into two administrative regions: North Lanarkshire (with Motherwell as its administrative headquarters) and South Lanarkshire (with HAMILTON as its administrative headquarters). (Pop. 634, 260)

Lancashire a county of north-west ENGLAND, once the heart of industrial Britain. The county town is Preston. (3043 sq km/1175 sq miles; pop. 1,424,000)

Lancaster (1) a city in Lancashire, formerly the county town of LANCASHIRE, situated on the river Lune. (Pop. 135,000) **(2)** a city in PENNSYLVANIA, USA; the state capital of Pennsylvania from 1799 until 1812, and very briefly the capital of the USA in 1777. (Pop. 57,170)

Landes a department of the AQUITAINE region on the coast of south-west FRANCE. (12,950 sq km/5000 sq miles)

Land's End the tip of the peninsula formed by CORNWALL in south-west ENGLAND, and the most westerly point of mainland England.

Languedoc-Rousillon a region of FRANCE which lines the MEDITERRANEAN coast from the River RHONE to the border with SPAIN. (27,376 sq km/10,567 sq miles; pop. 2,139,000)

Lansing the state capital of MICHIGAN, USA. (Pop. city 126,720)

Lantau the largest of the islands which form part of the New Territories of HONG KONG. (150 sq km/58 sq miles)

Lanzarote *see* **Canary Islands**.

Lanzhou a major industrial city and the capital of GANSU province, central CHINA. (Pop. 1,480,000)

Laois a county in the centre of the Republic of IRELAND. The county town is Portlaoise. (1718 sq km/664 sq miles; pop. 52,000)

Laos is a landlocked country in South-East ASIA which is ruggedly mountainous, apart from the MEKONG river plains along its border with THAILAND. The ANNAM mountains, which reach 2500 m (8203 ft), form a natural border with VIETNAM. It has a tropical monsoon climate with high temperatures throughout the year and heavy rains in summer. Laos is one of the poorest countries in the world and its development has been retarded by war, drought and floods. The principal crop is rice, grown on small peasant plots. There is some export of timber, coffee and electricity. All manufactured goods must be imported. The capital and largest city, VIENTIANE, is the country's main trade outlet via Thailand.

Area : 236,800 sq km (91,428 sq miles)
Population : 4,605,000
Capital : Vientiane
Form of government : People's Republic
Religion : Buddhism
Currency : Kip

La Paz a city set high in the ANDES of BOLIVIA, and the capital and seat of government. (Pop. 1,115,000)

Lapland *see* **Lappland**.

La Plata a port on the estuary of the River PLATE (Rio de la Plata) in north-eastern ARGENTINA, 56 km (35 miles) south-east of BUENOS AIRES. (Pop. 644,000)

Lappland (Lapland) the region of northern SCAN-DINAVIA and the adjoining territory of the RUSSIAN FEDERATION, traditionally inhabited by the no-madic Lapp people; also a province of northern FINLAND, called Lappi. The capital of the province is Rovaniemi. (93,057 sq km/35,929 sq miles; pop. 202,000.

Laptev Sea part of the ARCTIC OCEAN bordering central northern SIBERIA.

Larnaca a port, with an international airport, on the south-east coast of CYPRUS. (Pop. 63,000)

Lascaux a set of caves in the DORDOGNE department of south-west FRANCE where (in 1940) Paleolithic wall paintings dating back to about 15,000 BC were discovered.

Las Palmas de Gran Canaria the main port and largest city of the CANARY ISLANDS, on the island of Gran Canaria. (Pop. 348,000)

Las Vegas a city in the south-east of the state of NEVADA, USA. This state's liberal gaming laws have allowed Las Vegas to develop as an interna-tionally famous centre for gambling and entertain-ment. (Pop. city 295,500)

Latakia (Al Ladhiqiyah) a city on the MEDITERRA-NEAN coast of SYRIA, founded by the Romans, and now that country's main port. (Pop. 284,000)

Latium *see* **Lazio**.

Latvia is a BALTIC state that regained its independ-ence in 1991 with the break-up of the former So-viet Union. It is located in north-east EUROPE on the BALTIC SEA and is sandwiched between ESTO-NIA and LITHUANIA. It has cool summers, wet sum-mers and long, cold winters. Latvians traditionally lived by forestry, fishing and livestock rearing. The chief agricultural occupations are cattle and dairy farming and the main crops grown are oats, barley, rye, potatoes and flax. Latvia's population is now 70% urban and agriculture is no longer the mainstay of the economy. Cities such as Riga, the capital, Daugavpils, Ventspils and Liepaja now produce high quality textiles, machinery, electrical appliances, paper, chemicals, furniture and food-stuffs. Latvia has extensive deposits of peat which is used to manufacture briquettes. It also has de-posits of gypsum and in the coastal areas amber is frequently found.

Quick facts:
Area : 63,700 sq km (24,595 sq miles)
Population : 2,558,000
Capital : Riga
Other major cities : Liepaja
 Daugavpils, Jurmala
Form of government : Republic
Religion : Lutheranism
Currency : Lats

Launceston the second city and port of TASMANIA, AUSTRALIA, at the head of the Tamar river estuary. (Pop. 89,000)

Lausanne a city on the north shore of Lake GENEVA, SWITZERLAND, and capital of the French-speaking canton of Vaud. (Pop. city 123,000)

Lazio (Latium) a region occupying the central western coast of ITALY around ROME, the regional capital. (17,203 sq km/6642 sq miles; pop. 5,146,000)

Lebanon is a mountainous country in the eastern MEDITERRANEAN. A narrow coastal plain runs par-allel to its 240-km (149-mile) Mediterranean coast and gradually rises to the spectacular Lebanon Mountains, which are snow covered in winter. The Anti Lebanon Mountains form the border with SYRIA, and between the two ranges lies the BEQA'A Valley. The climate is Mediterranean with short warm winters and long hot and rainless summers. Rainfall can be torrential in winter and snow falls on high ground. Lebanon is an agricultural coun-try, the main regions of production being the Beqa'a Valley and the coastal plain. Main products in-clude olives, grapes, citrus fruits, apples, cotton, tobacco and sugar beet. Industry is small scale and manufactures include, cement, fertilizers and jew-ellery. There are oil refineries at TRIPOLI and SIDON.

Quick facts:
Area : 10,400 sq km (4015 sq miles)
Population : 2,970,000
Capital : Beirut (Beyrouth)
Other important cities : Tripoli, Zahle
Form of government : Republic
Religions : Shia Islam,
 Sunni Islam, Christianity
Currency : Lebanese pound

Lebowa was a former 'homeland' of SOUTH AFRICA. 'Homelands' ceased to exist under the South African Constitution implemented in April 1994. The area is now part of Northern Transvaal, SOUTH AFRICA.

Lecce a historic city in the PUGLIA region of ITALY. (Pop. 101,000)

Leeds an important industrial town on the River Aire in West YORKSHIRE, in northern ENGLAND. (Pop. 722,000)

Leeward and Windward Islands (1) the Lesser ANTILLES in the southern CARIBBEAN are divided into two groups. The northern islands in the chain, from the VIRGIN ISLANDS to GUADELOUPE are the Leeward Islands; the islands further south, from DOMINICA to GRENADA, form the Windward Islands. **(2)** the SOCIETY ISLANDS of FRENCH POLYNESIA are also divided into Leeward and Windward Islands.

Leghorn *see* **Livorno**.

Le Havre the largest port on the north coast of FRANCE. (Pop. city 179,900/metropolitan 254,000)

Leicester a historic cathedral city, and the county town of LEICESTERSHIRE. (Pop. 293,400)

Leicestershire a county in central ENGLAND. Since 1974 it has also incorporated the former county of Rutland. The county town is LEICESTER. (2553 sq km/986 sq miles; pop. 917,000)

Leiden (Leyden) a university city in western NETHERLANDS on the River Oude RIJN. (Pop. city 113,000/metropolitan 192,000)

Leinster one of the four ancient provinces into which IRELAND was divided, which covered the south-eastern quarter of the country. (19,630 sq km/7557 sq miles; pop. 1,860,000)

Leipzig an industrial city and important cultural centre in south-eastern GERMANY. (Pop. city 503,000)

Leitrim a county in the north-west of the Republic of IRELAND, with a small strip of coast and a northern border with FERMANAGH in NORTHERN IRELAND. The county town is Carrick-on-Shannon. (1525 sq km/589 sq miles; pop. 27,600)

Léman, Lake another name for Lake GENEVA.

Le Mans a university city in north western FRANCE, famous for the 24-hour car race held annually at a circuit nearby. (Pop. connurbation 189,000)

Lemberg *see* **L'vov**.

Lena, River a river, navigable for much of its length, which flows across eastern SIBERIA, from its source close to Lake BAIKAL to the LAPTEV SEA in the north. (Length 4270 km/2650 miles)

Leningrad *see* **St Petersburg**.

Lens a sprawling industrial city in the coal-mining region of northern FRANCE. (Pop. 323,400)

Léon (1) a major manufacturing city in central MEXICO. (Pop. 872,000) **(2)** a historic city, founded by the Romans, in north-west SPAIN, and capital of the province of the same name. (Pop. city 146,300)

Léopoldville *see* **Kinshasa**.

Leptis Magna the well-preserved ruins of an ancient Roman port on the MEDITERRANEAN coast of LIBYA.

Lesbos a large, fertile island in the eastern AEGEAN SEA, belonging to GREECE, but only 10 km (6 miles) from the coast of TURKEY. It was the birthplace of the poet Sappho (*c*.612–580BC). (1630 sq km/630 sq miles; pop. 103,700)

Lesotho is a small landlocked kingdom entirely surrounded by the Republic of SOUTH AFRICA. Snow-capped mountains and treeless uplands, cut by spectacular gorges, cover two thirds of the country. The climate is pleasant with variable rainfall. Winters are generally dry with heavy frosts in lowland areas and frequent snow in the highlands. Due to the mountainous terrain, only one eighth of the land can be cultivated and the main crop is maize. Yields are low because of soil erosion on the steep slopes and over-grazing by herds of sheep and cattle. Wool, mohair and diamonds are exported but most foreign exchange comes from money sent home by Lesotho workers in South Africa. Tourism is beginning to flourish, the main attraction to South Africans being the casinos in the capital MASERU, as gambling is prohibited in their own country.

Quick facts:
Area : 30,355 sq km (11,720 sq miles)
Population : 1,943,000
Capital : Maseru
Form of government : Constitutional monarchy
Religions : RC, other Christianity
Currency : Loti

Lesser Sunda Islands (Nusa Tenggara) a chain of islands to the east of JAVA, INDONESIA, stretching from BALI to TIMOR. (The Greater Sunda Islands comprise BORNEO, SUMATRA, JAVA and SULAWESI.)

Levkas *see* **Ionian Islands**.

Lexington a city in central KENTUCKY, USA, named after the Battle of Lexington in MASSACHUSETTS

(1775) which marked the beginning of the American Revolutionary War. (Pop. 232,562)

Leyden see **Leiden**.

Leyte an island of the VISAYAN group in the central PHILIPPINES. The main town is Tacloban. (7213 sq km/ 2785 sq miles; pop. 1,480,000)

Lhasa the capital of TIBET, an autonomous region of CHINA. It lies 3606 m (11,830 ft) above sea level. (Pop. 107,000)

Liaoning a coastal province of DONGBEI (Manchuria), north-east CHINA, bordering North KOREA. The capital is SHENYANG. (140,000 sq km/54,000 sq miles; pop. 39,460,000)

Liberia is located in West AFRICA and has a 560-km (348-mile) coast stretching from SIERRA LEONE to CÔTE D'IVOIRE. It is the only African country never to be ruled by a foreign power. It has a treacherous coast with rocky cliffs and lagoons enclosed by sand bars. Inland the land rises to a densely forested plateau dissected by deep, narrow valleys. Farther inland still, there are beautiful waterfalls and the Nimba Mountains rise to over 1700 m (5577 ft). Agriculture employs three quarters of the labour force and produces cassava and rice as subsistence crops and rubber, coffee and cocoa for export. The Nimba Mountains are rich in iron ore, which accounts for 70% of export earnings. There is potential for tourism to develop. Forest and animal reserves are magnificent and the beaches and lagoons are beautiful but so far the facilities are average.

Quick facts:
Area : 111,369 sq km (43,000 sq miles)
Population : 2,644,000
Capital : Monrovia
Form of government : Republic
Religion : Animism,
　Sunni Islam, Christianity
Currency : Liberian dollar

Libreville the capital and main port of GABON. It is so called ('Freetown') because it was originally a settlement for freed slaves. (Pop. 352,000)

Libya is a large north African country which stretches from the south coast of the MEDITERRANEAN to, and in some parts beyond, the Tropic of Cancer. The SAHARA DESERT covers much of the country extending right to the Mediterranean coast at the Gulf of SIRTE. The only green areas are the scrublands found in the north-west and the forested hills near BENGHAZI. The coastal area has

mild wet winters and hot dry summers but the interior has had some of the highest recorded temperatures of anywhere in the world. Only 14% of the people work on the land, the main agricultural region being in the north-west near TRIPOLI. Many sheep, goats and cattle are reared and there is an export trade in skins, hides and hairs. Libya is one of the world's largest producers of oil and natural gas. Other industries include food processing, textiles, cement and handicrafts.

Quick facts:
Area : 1,759,540 sq km (679,358 sq miles)
Population : 4,280,000
Capital : Tripoli (Tarabulus)
Other major cities : Benghazi, Misurata
Form of government : Socialist People's Republic
Religion : Sunni Islam
Currency : Libyan dinar

Liechtenstein the principality of Liechtenstein is a tiny central European state situated on the east bank of the River RHINE, bounded by AUSTRIA to the east and SWITZERLAND to the west. In the east of the principality the ALPS rise to 2599 m (8527 ft) at Grauspitze. The climate is mild alpine. Once an agricultural country, Liechtenstein has rapidly moved into industry in the last thirty years, with a variety of light industries such as textiles, high quality metal goods, precision instruments, pharmaceuticals and ceramics. It is a popular location for the headquarters of foreign companies, in order that they can benefit from the lenient tax laws. Tourism also thrives, beautiful scenery and good skiing being the main attractions.

Quick facts:
Area : 160 sq km (62 sq miles)
Population : 30,000
Capital : Vaduz
Form of government : Constitutional Monarchy
Religion : RC
Currency : Swiss franc

Liège (Luik) a historic city in eastern Belgium, and capital of the province of Liège, built on the confluence of the Rivers Meuse and Ourthe. (Pop. city 194,000/province 999,600)

Liffey, River the river upon which Dublin, the capital of the Republic of Ireland, is set. (Length 80 km/49 miles)

Liguria the region of north-western Italy which fronts the Gulf of Genoa; it has a border with France. (5415 sq km/2091 sq miles; pop. 1,719,000)

Ligurian Sea the northern arm of the Mediterranean Sea to the west of Italy, which includes the Gulf of Genoa.

Lille-Roubaix-Tourcoing a conurbation of industrial towns in north-eastern France. (Pop. 959,000)

Lilongwe the capital of Malawi, and the second largest city in the country after Blantyre. (Pop. 268,000)

Lima the capital of Peru, situated on the banks of the River Rimac, 13 km (8 miles) from the coast. (Pop. 6,484,000)

Limassol the main port of Cyprus in the southern part of the island. (Pop. 137,000)

Limerick a city and port on the River Shannon, and the county town of the county of Limerick, in the south-west of the Republic of Ireland. (County 2686 sq km/1037 sq miles; pop. county 161,900; pop. city 52,000)

Limoges a city in eastern central France, famous for its richly decorated porcelain. It is the capital of the Limousin region. (Pop. city 136,000)

Limousin a region of east-central France in the foothills of the Massif Central, famous in particular for its Limousin cattle. (Pop. 723,000)

Limpopo, River a river which flows northwards from its source in Northern Transvaal to form part of the border between South Africa and Botswana before crossing southern Mozambique to reach the Indian Ocean. (Length 1610 km/1000 miles)

Lincoln (1) a historic city, with a fine cathedral dating from the 11th century, and the county town of Lincolnshire, England. (Pop. 84,600). **(2)** The state capital of Nebraska, USA. (Pop. city 180,400/metropolitan area 197,500)

Lincolnshire a county on the east coast of central England. The county town is Lincoln. (5885 sq km/2272 sq miles; pop. 591,780)

Lindisfarne a small island, also known as Holy Island, just off the east coast of Northumberland in north-east England. It has an 11th-century priory built on the site of a monastery founded in the 7th century.

Linz a city and port on the River Danube in northern Austria. (Pop. 203,000)

Lion, Golfe de (Gulf of Lions) the arm of the Mediterranean Sea which forms a deep indent in the southern coast of France, stretching from the border with Spain to Toulon, and centreing on the delta of the River Rhone.

Lipari the largest of the volcanic Eolian Isalnds that lie off the north coast of Sicily, Italy. (38 sq km/15 sq miles)

Lisbon (Lisboa) the capital and principal port of Portugal, situated on the broad River Tagus, approximately 15 km (9 miles) from the Atlantic coast. (Pop. city 950,000)

Lithuania lies to the north-west of the Russian Federation and Belarus and is bounded to the north by Latvia and west by Poland. It is the largest of the three former Soviet Baltic Republics. Before 1940 Lithuania was a mainly agricultural country but has since been considerably industrialized. Most of the land is lowland covered by forest and swamp, and the main products are rye, barley, sugar beet, flax, meat, milk and potatoes. Industry includes heavy engineering, shipbuilding and building materials. Oil production has started from a small field at Kretinga in the west of the country, 16 km (10 miles) north of Klaipeda. Amber is found along the Baltic coast and used by Lithuanian craftsmen for making jewellery.

Quick facts:
Area : 65,200 sq km (25,174 sq miles)
Population : 3,724000
Capital : Vilnius
Other major cities : Kaunas, Klaipeda, Siauliai
Form of government : Republic
Religion : RC
Currency : Rouble

Little Rock the state capital of Arkansas, USA. (Pop. city 176,900)

Liverpool a major port on the estuary of the River Mersey in north-west England. (Pop. 479,000)

Livorno (Leghorn) a port and industrial city on the coast of Tuscany, northern Italy. (Pop. 175,300)

Ljubljana an industrial city on the River Sava, and the capital of Slovenia. (Pop. 305,200)

Lodz an industrial city and the second largest city in POLAND, located in the centre of the country. It was rebuilt after being destroyed in World War II. (Pop. 848,500)

Logan, Mount the highest mountain in CANADA, and the second highest in North America after Mount MCKINLEY. It is situated in south-west YUKON, on the border with ALASKA. (5951 m/19,524 ft)

Loire, River the longest river in FRANCE, flowing northwards from the south-eastern MASSIF CENTRAL and then to the west to meet the ATLANTIC OCEAN just to the west of NANTES. Its middle reaches are famous for their spectacular châteaux, and for the fertile valley which produces fine white wines. (Length: 1020 km/635 miles)

Lombardy (Lombardia) the central northern region of ITALY, which drops down from the ALPS to the plain of the River Po, one of the country's most productive areas in both agriculture and industry. MILAN is the regional capital. (23,854 sq km/9210 sq miles; pop. 898,700)

Lombok an island of the LESSER SUNDA group, east of BALI. (5435 sq km/2098 sq miles; pop. 1,300,200)

Lomé the capital and main port of TOGO, situated close to the border with GHANA. (Pop. 283,000)

London (1) the capital city of the UNITED KINGDOM, situated in the south-east of ENGLAND, which straddles both banks of the River THAMES near its estuary. It consists of 32 boroughs, including the City, an international centre for trade and commerce. (Pop. 6,961,000). **(2)** an industrial city in south-western ONTARIO, Canada. (Pop. 284,000)

Londonderry (Derry) the second largest city in NORTHERN IRELAND after BELFAST, and the county town of the county of Londonderry. (County 2076 sq km/801 sq miles; pop. county 84,000; pop. city 62,000)

Longford a county in the centre of the Republic of IRELAND, with a county town of the same name. (1044 sq km/403 sq miles; pop. 31,100)

Long Island an island off the coast of NEW YORY State, stretching some 190 km (118 miles) to the north-east away from the city of NEW YORK. Its western end forms part of the city of New York (the boroughs of Brooklyn and Queens) but the rest is a mixture of residential suburbs, farmland and resort beaches. (3685 sq km/1423 sq miles)

Lord Howe Island a small island lying some 600 km (375 miles) to the east of the coast of NEW SOUTH WALES, AUSTRALIA, now a popular resort. (16 sq km/6 sq miles; pop. 300)

Lorraine a region of north-east FRANCE, with a border shared by BELGIUM, LUXEMBOURG and GERMANY. The regional capital is METZ. (Pop. 2,320,000)

Los Angeles a vast, sprawling city on the PACIFIC OCEAN in southern CALIFORNIA, USA, the second largest city in the USA after NEW YORK. (Pop. city 3,096,700/conurbation 12,372,600)

Lothian a former local government region in south-east central SCOTLAND, with EDINBURGH as its administrative centre. It was created in 1975 out of the former counties of Midlothian, and East and West Lothian. In 1996 the region was again reorganized into East and West Lothian, Midlothian and the city of Edinburgh.

Louisiana a state in central southern USA, on the lower reaches of the MISSISSIPPI River, and with a coastline on the Gulf of MEXICO. The state capital is BATON ROUGE. (125,675 sq km/48,523 sq miles; pop. 4,481,000)

Louisville a city and commercial centre, in northern KENTUCKY, USA, on the OHIO River. (Pop. city 289,800/metropolitan area 962,600)

Lourdes one of the world's most important Marian shrines, in the foothills of the central PYRÉNÉES, FRANCE. It became a place of miraculous healing after a series of visions of the Virgin Mary witnessed by Bernadette Soubirous in 1858. (Pop. 17,600)

Lourenço Marques *see* **Maputo**.

Louth a county on the east coast of the Republic of IRELAND, bordering NORTHERN IRELAND. The county town is Dundalk. (823 sq km/318 sq miles; pop. 88,500)

Lualaba, River a river that flows northwards across the eastern part of ZAÏRE from the border with ZAMBIA before joining the River Lomami to form the River ZAÏRE. (Length 1800 km/1120 miles)

Luanda The capital of ANGOLA, and a major port on the ATLANTIC OCEAN. (Pop. 700,000)

Lübeck a BALTIC port in northern GERMANY, lying some 20 km (12 miles) from the coast on the River Trave. (Pop. 80,000)

Lublin (Lyublin) a city and agricultural centre in south-eastern POLAND. (Pop. 320,000)

Lubumbashi the principal mining town of ZAÏRE, and the capital of the Shaba region in the southeast of the country. It was founded in 1910 and known as Elisabethville until 1966. (Pop. 600,000)

Lucca a town in north-western TUSCANY, ITALY. Surrounded by impressive 16th-century fortifications, its medieval street plan incorporates the site of a Roman amphitheater. (Pop. 89,100)

Lucerne (Luzern) a city set on the beautiful Lake Lucerne in central SWITZERLAND, retaining much of its medieval past; also the name of the surrounding canton. (Pop. city 67,500)

Lucknow the capital of the state of UTTAR PRADESH in central northern INDIA. (Pop. 1,669,000)

Lüda (Dalian) an industrial city and port in LIAONING province, north-eastern CHINA. (Pop. 4,000,000)

Ludhiana a town in central PUNJAB, INDIA, home of the respected Punjab Agricultural University. (Pop. 1,043,000)

Ludwigshafen a town, industrial centre and river port on the River RHINE in south-western GERMANY. (Pop. 163,000)

Lugansk a major industrial city of the eastern UKRAINE in the DONETS BASIN. Renamed Voroshilovgrad in 1970 after Kliment Voroshilov, president of the former USSR (1953–60). Its name reverted to Lugansk in 1990. (Pop. 491,000)

Luik *see* **Liège**.

Luluabourg *see* **Kananga**.

Lumbini the birthplace of Buddha (Prince Siddhartha Gautama, *c.*563–488BC), part of the village of Rummindei in central southern NEPAL.

Luoyang a city of ancient origins, founded in about 2100BC, in HENAN province in eastern central CHINA. As a principal centre of the Shang dynasty (18th-12th centuries BC), the area is rich in archaeological remains. (Pop. 500,000)

Lusaka the capital of ZAMBIA, situated in the southeast of the country. (Pop. 538,500)

Luton an industrial town in BEDFORDSHIRE, ENGLAND, 50 km (30 miles) north of LONDON. It is a centre of the British car-manufacturing industry, and also has a busy international airport. (Pop. 165,000)

Luxembourg the Grand Duchy of Luxembourg is a small independent country bounded by BELGIUM on the west, FRANCE on the south and GERMANY on the east. In the north of the Duchy a wooded plateau, the Oesling, rises to 550 m/1804 ft and in the south a lowland area of valleys and ridges is known as the Gutland. Northern winters are cold and raw with snow covering the ground for almost a month, but in the south winters are mild and summers cool. In the south the land is fertile and crops grown include maize, roots, tubers and potatoes. Dairy farming is also important. It is in the south, also, that beds of iron ore are found and these form the basis of the country's iron and steel industry. In the east Luxembourg is bordered by the MOSELLE river in whose valley wines are produced.

Quick facts:
Area : 2586 sq km (998 sq miles)
Population : 406,500
Capital : Luxembourg (pop. 75,800)
Form of government : Constitutional Monarchy
Religion : RC
Currency : Luxembourg franc

Luxor a town that has grown up around one of the great archaeological sites of ancient EGYPT. It is situated on the east bank of the River NILE in the centre of the country, just south of the ancient capital, THEBES, and 3 km (2 miles) from KARNAK. (Pop. 78,000)

Luzern *see* **Lucerne**.

Luzon the largest island of the PHILIPPINES, in the north of the group, with the nation's capital, MANILA, at its centre. (104,688 sq km/40,420 sq miles; pop. 29,400,000)

L'vov (Lemberg) a major industial city of medieval origins in the western UKRAINE. (Pop. 688,000)

Lyallpur *see* **Faisalabad**.

Lyons (Lyon) the second largest city in FRANCE after PARIS, situated at the confluence of the Rivers RHÔNE and SAONE in the south-east of the country. (Pop. 1,236,100)

Lyublin *see* **Lublin**.

Maas, River *see* **Meuse**.

Macáu (Macao) a tiny Portuguese province on the coast of south CHINA, opposite HONG KONG. Occupied by the Portuguese since 1557, the territory will be handed back to China in 1999. The vast majority of the population is Chinese, and the dominant language is Cantonese. Tourism and the handling of trade destined for China are major sources of income. The capital is Macau City. (15.5 sq km/6 sq miles; pop. 490,000; cur. Pataca = 100 avos)

Macdonnell Ranges the parallel ranges of mountains of central AUSTRALIA, in the southern part of the NORTHERN TERRITORY, near to ALICE SPRINGS. The highest peak is Mount Ziel (1510 m/4954 ft).

Macedonia (1) the largest region of GREECE, occupying most of the northern part of the mainland, and with northern borders with ALBANIA, The FORMER YUGOSLAV REPUBLIC OF MACEDONIA and BULGARIA. (Pop. 1,936,900). **(2) Macedonia, The**

Former Yugoslav Republic of, under the name of Macedonia, declared its independence from Yugoslavia in 1991. However Greece, angered at the use of 'Macedonia'—also the name of the neighbouring Greek province—imposed a trade embargo and convinced the UN to refuse to recognize the nation's independence. In 1993, Macedonia was admitted to the UN after changing its official name to The Former Yugoslav Republic of Macedonia. In 1995 an agreement was reached with Greece whereby both countries would respect the territory, sovereignty and independence of the other, with Macedonia agreeing to adopt a new flag. A landlocked country, Macedonia shares its borders with Albania, Bulgaria, Greece and Yugoslavia. Its terrain is mountainous, covered with deep valleys and has three large lakes. Its climate consists of hot, dry summers and relatively cold winters. It is the poorest of the six former Yugoslav republics but sustains itself through agriculture and the coal industries. Some of its natural resources include chromium, lead, zinc, nickel, iron ore and timber.

Quick facts:
Area : 25,713 sq km (9929 sq miles)
Population : 2,173, 000
Capital : Skopje
Other major cities : Tetova, Prilep
Form of government : Republic
Religion : Eastern Orthodox, Muslim
Currency : Dinar

Maceió a port on the central east coast of BRAZIL. (Pop. 699,800)

Macgillicuddy's Reeks a range of mountains in the south-west of the Republic of IRELAND which includes the country's highest peak at CARRAUNTOOHIL (1040 m/3414 ft).

Machu Picchu the ruins of a great Inca city, set spectacularly on a mountain ridge high in the ANDES near CUZCO, in south central PERU. It was abandoned in the 16th century and rediscovered only in 1911.

Mackenzie, River a river flowing northwards through the western part of the NORTHWEST TERRITORIES of CANADA from the GREAT SLAVE LAKE to the ARCTIC OCEAN. (Length 4250 km/2640 miles)

McKinley, Mount the highest mountain in North AMERICA, located in the Denali National Park in southern ALASKA, USA. (6194 m/20,320 ft)

MacMurdo Sound an arm of the ROSS SEA on the International Date Line. The American MacMurdo Base on Ross Island is the largest scientific research station in ANTARCTICA.

Madagascar is an island state situated off the south-east coast of AFRICA, separated from the mainland by the MOZAMBIQUE Channel. Madagascar is the fourth largest island in the world and the centre of it is made up of high savanna-covered plateaux. In the east, forested mountains fall steeply to the coast and in the south-west, the land falls gradually through dry grassland and scrub. The staple food crop is rice and 80% of the population grow enough to feed themselves. Cassava is also grown but some 58% of the land is pasture and there are more cattle than people. The main export earners are coffee, vanilla, cloves and sugar. There is some mining for chromite and an oil refinery at Toamasina on the east coast. Upon independence in 1960, Madagascar became known as the Malagasy Republic, but was changed back by referendum in 1975.

Quick facts:
Area : 587,041 sq km (226,657 sq miles)
Population : 15,206,000
Capital : Antananarivo
Other major cities : Fianarantsoa, Mahajanga, Toamasina
Form of government : Republic
Religions : Animism, RC, Protestantism
Currency : Malagasy franc

Madeira the main island in a small group in the eastern ATLANTIC OCEAN which have belonged to PORTUGAL since the the 16th century, lying some 1000 km (620 miles) due west of CASABLANCA in MOROCCO. The capital is FUNCHAL. (813 sq km/314 sq miles; pop. 300,000)

Madhya Pradesh the largest state in INDIA, in the centre of the country. The capital is BHOPAL. (442,700 sq km/170,921 sq miles; pop. 66,135,400)

Madison the state capital of WISCONSIN, USA. (Pop. city 191,300/metropolitan area 333,000)

Madras the main port on the south-east coast of IN-DIA, and the capital of the state of TAMIL NADU. (Pop. 5,422,000)

Madrid the capital of SPAIN, situated in the middle of the country, and also the name of the surrounding province. (Pop. city 3,121,000/province 4,846,000)

Madura an island off the north-eastern coast of JAVA. (4,564 sq km/1,762 sq miles; pop. 2,447,000)

Madurai a textile city in TAMIL NADU, in the southern tip of INDIA, famous for its temples. (Pop. 904,000)

Mae Nam Khong, River *see* **Salween, River**.

Mafikeng (Mafeking) a town in North West SOUTH AFRICA, scene of the Relief of Mafeking in 1890, ending an eight-month siege of British troops by the Boers. (Pop. 29,400)

Magdalena, River a river which flows northwards through western COLOMBIA and into the CARIBBEAN at BARRANQUILA. (Length 1550 km/965 miles)

Magdeburg a city and inland port on the River ELBE in eastern GERMANY, 120 km (75 miles) south-west of BERLIN. (Pop. 289,000)

Magellan, Strait of the waterway, 3 km (2 miles) across at its narrowest, which separates the island of TIERRA DEL FUEGO from the southern tip of mainland South AMERICA. It was discovered by the Portuguese navigator Ferdinand Magellan (1480–1521) in 1520.

Maghreb (Maghrib) the name by which the countries of north-west AFRICA, MOROCCO, ALGERIA and TUNISIA are often called collectively.

Magnitogorsk an industrial city specializing in iron and steel, in the southern URAL MOUNTAINS in the RUSSIAN FEDERATION, founded in 1930. (Pop. 430,000)

Maharashtra a state in the centre of the west coast of INDIA, with BOMBAY as its capital. (307,690 sq km/118,768 sq miles; pop. 78,706,700)

Mahore *see* **Mayotte**.

Main, River a river that snakes its way westwards from its source near BAYREUTH in central GERMANY, passing through FRANKFURT AM MAIN before joining the River RHINE at MAINZ. (Length 524 km/325 miles)

Maine a state in the north-eastern corner of the USA, bordering CANADA. The state capital is AUGUSTA. (86,027 sq km/33,215 sq miles; pop. 1,228,000)

Mainz a city and inland port on the confluence of the Rivers RHINE and MAIN in western central GERMANY. (Pop. 185,000)

Majorca (Mallorca) the largest of the BALEARIC IS-LANDS, in the western MEDITERRANEAN. The capital is PALMA. (3639 sq km/1405 sq miles; pop. 561,200,000)

Majuro an atoll of three islands (Dalap, Uliga and Darrit) which together form the capital of the MARSHALL ISLANDS. (Pop. 8700)

Makassar *see* **Ujung Padang**.

Makassar Strait the broad stretch of water, 130 km (81 miles) across at its narrowest, which separates the islands of BORNEO and SULAWESI in INDONESIA.

Makeyevka an industrial city in the DONETS BASIN in the southern UKRAINE. (Pop. 426,000)

Makhachkala a port and industrial city on the west coast of the CASPIAN SEA, and the capital of the republic of DAGESTAN. (Pop. 320,000)

Makkah *see* **Mecca**.

Malabar Coast the name given to the coastal region of the state of KERALA in south-western INDIA.

Malabo a port and the capital of EQUATORIAL GUINEA, situated on the north coast of BIOKO Island. (Pop. 37,200)

Malacca *see* **Melaka**.

Malacca, Strait of the busy waterway, just 50 km (31 miles) wide at its narrowest, which separates the island of SUMATRA in INDONESIA from the southern tip of MALAYSIA, with SINGAPORE at its eastern end.

Malaga a port, manufacturing city, and tourist resort on the MEDITERRANEAN coast of the autonomous region of ANDALUCIA, southern SPAIN. Also the name of the province of which it is the capital. (Pop. town 531,000)

Malagasy Republic *see* **Madagascar**.

Malawi lies along the southern and western shores of the third largest lake in AFRICA, Lake MALAWI. To the south of the lake the Shire river flows through a valley, overlooked by wooded, towering mountains. The tropical climate has a dry season from May to October and a wet season for the remaining months. Agriculture is the predominant occupation and many Malawians live on their own crops. Plantation farming is used for export crops. Tea is grown on the terraced hillsides in the south and tobacco on the central plateau, with sugar and maize also important crops. Malawi has bauxite and coal deposits but due to the inaccessibility of their locations, mining is limited. Hydroelectricity is now being used for the manufacturing industry but imports of manufactured goods remain high, and the country remains one of the poorest in the world. Malawi was formerly the British colony of Nyasaland, a name meaning 'Land of the Lake', which was given to it by the 19th-century explorer, David Livingstone.

Quick facts:
Area : 118,484 sq km (45,747 sq miles)
Population : 9,800,000
Capital : Lilongwe
Other major cities : Blantyre, Mzuzu, Zomba
Form of government : Republic
Religions : Animism, RC,
  Presbyterianism
Currency : Kwacha

Malawi (Nyasa), Lake a long, narrow lake which runs down most of the eastern side of Malawi and forms Malawi's border with TANZANIA and MOZAMBIQUE. (23,300 sq km/9000 sq miles)

Malaysia the Federation of Malaysia lies in the SOUTH CHINA SEA in south-east ASIA, and comprises peninsular Malaysia on the Malay Peninsula and the states of SABAH and SARAWAK on the island of BORNEO. Malaysia is affected by the monsoon climate. The north-east monsoon brings rain to the east coast of peninsular Malaysia in winter, and the south-west monsoon brings rain to the west coast in summer. Throughout the country the climate is generally tropical and temperatures are uniformly hot throughout the year. Peninsular Malaysia has always had thriving rubber-growing and tin dredging industries and now oil palm growing is also important on the east coast. Sabah and Sarawak have grown rich by exploiting their natural resources, the forests. There is also some offshore oil and around the capital, KUALA LUMPUR, new industries such as electronics are expanding.

Quick facts:
Area : 330,434 sq km (127,580 sq miles)
Population : 20,174,000
Capital : Kuala Lumpur
Other major cities : Ipoh, George Town,
  Johor Baharu
Form of government : Federal
  Constitutional Monarchy
Religion : Islam
Currency : Ringgit

Maldives the Republic of Maldives lies 640 km (398 miles) south-west of SRI LANKA in the INDIAN OCEAN and comprises 1200 low-lying coral islands grouped into 12 atolls. Roughly 202 of the islands are inhabited, and the highest point is only 1.5 m (5 ft) above sea level. The climate is hot and humid and affected by monsoons from May to August. The islands are covered with coconut palms, and some millet, cassava, yams and tropical fruit are grown. Rice, the staple diet of its islanders, is imported. Fishing is an important occupation and the chief export is now canned or frozen tuna. Tourism is now developing fast and has taken over fishing as the major foreign currency earner.

Quick facts:
Area : 298 sq km (115 sq miles)
Population : 254,000
Capital : Malé
Form of govt : Republic
Religion : Sunni Islam
Currency : Rufiyaa

Malé the main atoll of the MALDIVES, and the town which is the country's capital. (2.6 sq km/1 sq mile; pop. 55,000)

Mali is a landlocked state in West AFRICA. The country mainly comprises vast and monotonous plains and plateaux. It rises to 1155 m (3790 ft) in the Adrar des Iforas range in the north-east. The SAHARA in the north of the country is encroaching southwards. Mali is one of the poorest countries in the world. In the south there is some rain and plains are covered by grassy savanna and a few scattered trees. The river NIGER runs through the south of the country and small steamboats use it for shipping between Koulikoro and Gao. Only a fifth of the land can be cultivated. Rice, cassava and millet are grown for domestic consumption and cotton for export. Droughts in the 1970s resulted in thousands of cattle dying, crop failure, famine and disease killing many of the population. Iron ore and bauxite have been discovered but have yet to be exploited.

Quick facts:
Area : 1,240,192 sq km (478,838 sq miles)
Population : 10,700,000
Capital : Bamako
Other major cities : Segou,
  Mopti
Form of government : Republic
Religions : Sunni Islam,
  Animism
Currency : Franc CFA

Mallorca *see* **Majorca**.

Malmö a port in south-west SWEDEN, on the narrow channel which separates Sweden from COPENHAGEN in DENMARK. (Pop. 489,900)

Malta, a small republic in the middle of the MEDITERRANEAN, lies just south of the island of SICILY. It comprises three islands, Malta, Gozo and Comino, which are made up of low limestone plateaux with little surface water. The climate is Mediterranean with hot, dry sunny summers and little rain. Winters are cooler and wetter. Malta is virtually self-sufficient in agricultural products and exports potatoes, vegetables, wine and cut flowers. The British military base on Malta was once the mainstay of the economy but after the British withdrew in the late 1970s, the naval dockyard was converted for commercial shipbuilding and repairs, which is now one of the leading industries. Tourism has also boomed and the island has become popular with people who wish to retire to a suunier climate.

> *Quick facts:*
> Area : 316 sq km (122 sq miles)
> Population : 367,000
> Capital : Valletta
> Form of government : Republic
> Religion : RC
> Currency : Maltese pound

Maluku (Moluccas) a group of some 1000 islands in eastern INDONESIA. They are known as the Spice Islands, for they were once the only source of cloves and nutmegs in the world. The principal islands are Halmahera, Seram and Buru. The capital is AMBON. (74,505 sq km/28,766 sq miles; pop. 1,814,000)

Man, Isle of an island of the BRITISH ISLES, in the IRISH SEA, halfway between ENGLAND and IRELAND. It is a British Crown possession, not a part of the UK, and has its own parliament, the Court of Tynwald. The capital is Douglas. (572 sq km/221 sq miles; pop. 69,800)

Managua the capital of NICARAGUA, situated on the edge of Lake Managua. (Pop. 974,000)

Manaus a major port on the River AMAZON in BRAZIL, lying 1600 km (1000 miles) from the sea. (Pop. 1,011,000)

Manchester a major industrial and commercial city in north-west ENGLAND, and the administrative centre for the metropolitan county of Greater Manchester. It is connected to the estuary of the River MERSEY by the Manchester Ship Canal. (County 1290 sq km/498 sq miles; pop. city 431,100/county 2,594,778)

Manchuria *see* **Dongbei**.

Mandalay the principal city of central MYANMAR (formerly BURMA), and a port on the River IRRAWADDY. (Pop. 533,000)

Manila the capital of the PHILIPPINES. The city is an important port and commercial centre, and is sited on LUZON island. The surrounding urban area is known as Metro Manila (Pop. city 1,559,000/Metro Manila 7,832,000)

Manipur a small state of INDIA in the far north-east, on the border with MYANMAR. The capital is Imphal. (22,327 sq km/8618 sq miles; pop. 1,826,700)

Manitoba the most easterly of the prairie provinces of CANADA. The capital is WINNIPEG. (650,087 sq km/250,998 sq miles; pop. 1,092,600)

Mannheim an inland port and industrial city on the confluence of the Rivers RHINE and NECKAR. (Pop. 300,000)

Mansura *see* **El Mansura**.

Mantua (Mantova) a city in the valley of the River Po, in the LOMBARDY region of ITALY, retaining much of its medieval heritage. (Pop. 60,900)

Maputo the capital and main port of MOZAMBIQUE. It was formerly known as Lourenço Marques. (Pop. 1,070,500)

Maracaibo the second largest city in VENEZUELA, in the north-west. (Pop. 1,364,000)

Maracaibo, Lake a shallow lake in north-west VENEZUELA, linked to the CARIBBEAN SEA by a channel. It contains one of the richest oil fields in the world. (13,280 sq km/5127 sq miles)

Marbella a popular resort on the MEDITERRANEAN coast of southern SPAIN, in the province of MALAGA. (Pop. 80,700)

Marburg *see* **Maribor**.

Marche (Marches) a region of central eastern ITALY, lining the ADRIATIC coast. The capital is ANCONA. (9694 sq km/3743 sq miles; pop. 1,435,600)

Mar del Plata a coastal city and beach resort on the north-east coast of ARGENTINA, 400 km (250 miles) south of BUENOS AIRES. (Pop. 520,000)

Margarita an island belonging to VENEZUELA, lying just off its north-eastern coast, rapidly becoming a popular tourist destination. (929 sq km/355 sq miles; pop. 38,000)

Marianske Lazne (Marienbad) a spa town in western Czech Republic. (Pop. 18,000)

Maribor (Marburg) an industrial city in SLOVENIA. (Pop. 153,000)

Marie Galante *see* **Guadeloupe**.

Marienbad *see* **Marianske Lazne**.

Marigot *see* **St Martin**.

Marmara, Sea of a small sea lying between the DARDANELLES and the BOSPHORUS, providing a vital link in the route between the MEDITERRANEAN SEA and the BLACK SEA. The surrounding coasts all belong to TURKEY.

Marmolada, Mount *see* **Dolomites**.

Marquesas Islands a group of a dozen or so fertile, volcanic islands in the north-eastern sector of FRENCH POLYNESIA, lying about 1400 km (875 miles) north-east of TAHITI. (1270 sq km/490 sq miles; pop. 7,500)

Marrakech (Marrakesh) a historic oasis city in central western MOROCCO, founded in the 11th century and formerly the country's capital. (Pop. 6565,000)

Marseilles (Marseille) the largest port in FRANCE, on the MEDITERRANEAN coast, and France's third largest city after PARIS and LYONS. (Pop. 1,087,00)

Marshall Islands Formerly part of the US administered UN territory, this self-governing republic, independent since 1991, comprises a scatter of some 1250 coral atolls and 34 main islands, arranged in two parallel chains, Ratak and Ralik, located in eastern Micronesia in the western Pacific Ocean, and lying to the north-west of KIRIBATI. The climate is tropical maritime, with little variation in temperature, and rainfall that is heaviest from July to October. The republic remains in free association with the USA and the economy is almost totally dependent on US-related payments for use of the islands as bases. Attempts are being made to diversify the economy before US aid runs out in 2001. The main export is copra.

Quick facts:
Area : 181 sq km/70 sq miles
Population : 55,000
Capital : Dalag-Uliga-Darrit (on Majuro atoll)
Form of govt : Republic in free association with the USA
Religion : Protestant
Currency : US dollar

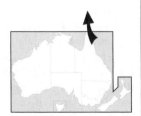

Martinique one of the larger of the islands in the WINDWARD ISLANDS group in the southern CARIBBEAN, lying between DOMINICA and ST LUCIA. It is administered as a department of FRANCE. FORT-DE-FRANCE is the capital. (1079 sq km/417 sq miles; pop. 360,000)

Maryland a state on the central east coast of the USA, virtually divided in two by CHESAPEAKE BAY. The state capital is ANNAPOLIS. (31,600 sq km/12,198 sq miles; pop. 4,781,500)

Masbate *see* **Visayan Islands**.

Maseru the capital of LESOTHO. (Pop. 289,000)

Mashhad (Meshed) a major trading centre and the capital of KHORASAN province in north-eastern IRAN. (Pop. 1,759,000)

Mason-Dixon Line the state boundary between PENNSYLVANIA and MARYLAND to its south, surveyed 1763–7 by Charles Mason and Jeremiah Dixon. It is considered to be the traditional border between the North and the South of the USA.

Masqat *see* **Muscat**.

Massachusetts one of the NEW ENGLAND states on the north-eastern coast of the USA. The capital is BOSTON. (21,386 sq km/8257 sq miles; pop. 6,016,400)

Massif Central the rugged upland region which occupies much of southern central FRANCE to the west of the River RHONE. The highest point is at Puy de Sancy (1885 m/6184 ft).

Matsuyama a port and industrial city on the north coast of SHIKOKU Island, JAPAN. (Pop. 454,000)

Matterhorn (Monte Cervino) a distinctive, pyramid-shaped peak on the border between ITALY and SWITZERLAND, 5 km (3 miles) south of ZERMATT. (4477 m/14,688 ft)

Maui the second largest island of HAWAII, USA. (1885 sq km/727 sq miles; pop. 63,000)

Mauna Kea a dormant volcano in the north of the island of HAWAII, USA. (4205 m/13,796 ft)

Mauna Loa an active volcano in the centre of the island of HAWAII. (4169 m/13,677 ft)

Mauritania, a country nearly twice the size of France, is located on the west coast of AFRICA. About 47% of the country is desert, the SAHARA covering much of the north. The only settlements found in this area are around oases, where a little millet, dates and vegetables can be grown. The main agricultural regions are in the Senegal river valley in the south. The rest of the country is made up of the drought-stricken SAHEL grasslands. The majority of the people are traditionally nomadic herdsmen, but the severe droughts since the 1970s have killed about 70% of the nation's animals, and the population has settled along the Senegal river.

As a result, vast shanty towns have sprung up around all the towns. Deposits of iron ore and copper provide the country's main exports, and development of these and the fishing industry on the coast form the only hope for a brighter future.

Quick facts:
Area : 1,025,520 sq km (395,953 sq miles)
Population : 2,268,000
Capital : Nouakchott
Form of government : Republic
Religion : Sunni Islam
Currency : Ouguiya

Mauritius is a beautiful island with tropical beaches which lies about 20° south in the INDIAN OCEAN, 800 km (497 miles) east of MADAGASCAR. The islands of Rodrigues and Agalega are part of Mauritius. Mauritius is a volcanic island with many craters surrounded by lava flows. The central plateau rises to over 800 m (2625 ft), then drops sharply to the south and west coasts. The climate is hot and humid, and south-westerly winds bring heavy rain in the uplands. The island has well-watered fertile soil, ideal for the sugar plantations that cover 45% of the island. Although the export of sugar still dominates the economy, diversification is being encouraged. The clothing and electronic equipment industries are becoming increasingly important and tourism is now the third largest source of foreign exchange.

Quick facts:
Area : 2040 sq km (788 sq miles)
Population : 1,112,669
Capital : Port Louis
Form of government : Republic
Religions : Hinduism, RC,
  Sunni Islam
Currency : Mauritius rupee

Mayo a county on the west coast of the Republic of IRELAND, noted for its rugged splendor. The county town is Castlebar. (5400 sq km/2084 sq miles; pop. 110,700)

Mayotte (Mahore) part of the COMOROS Island group, lying between MADAGASCAR and the mainland of AFRICA. Unlike the other three islands in the group, Mayotte voted to remain under the administration of France when the Comoros Islands became independent in 1974. (373 sq km/144 sq miles; pop. 73,000)

Mbabane the capital of SWAZILAND. (Pop. 42,000)

Meath a county on the east coast of the Republic of IRELAND, to the north of DUBLIN. The county town is Navan. (2336 sq km/902 sq miles; pop. 105,600)

Mecca (Makkah) a city in central western SAUDI ARABIA, 64 km (40 miles) east of the RED SEA port of JIDDAH. An important trading city on caravan routes in ancient times, it was the birthplace of the Prophet Mohammed, and as such is the holiest city of Islam. (Pop. 618,000)

Medan a major city in northern SUMATRA, INDONESIA. (Pop. 1,686,000)

Medellín the second largest city in COLOMBIA after the capital BOGOTA, situated in the centre of the country, 240 km (150 miles) north-west of the capital. (Pop. 1,608,000)

Medina (Al Madinah) the second holiest city of Islam after MECCA. The Prophet Mohammed fled from Mecca to Medina, 350 km (217 miles) to the north, to escape persecution in AD622 (year 0 in the Islamic lunar calendar). (Pop. 500,000)

Mediterranean Sea a large sea bounded by southern EUROPE, North AFRICA and south-west ASIA. It is connected to the ATLANTIC OCEAN by the Strait of GIBRALTAR.

Médoc one of the prime wine-producing regions of FRANCE, a flat, triangular-shaped piece of land between the GIRONDE estuary and the ATLANTIC OCEAN.

Meerut an industrial town of northern INDIA, 60 km (40 miles) north-east of DELHI. The Indian Mutiny began here in 1857. (Pop. 847,000)

Meghalaya a predominantly rural state in the hills of north-eastern INDIA. The capital is Shillong. (22,429 sq km/8658 sq miles; pop. 1,760,600)

Meissen a historic town on the River ELBE, 20 km (12 miles) to the north-west of DRESDEN, in southeastern GERMANY. It is famous above all for its fine porcelain, produced here since 1710. (Pop. 38,200)

Meknès a former capital, with a fine 17th-century royal palace, in northern MOROCCO. (Pop. 495,000)

Mekong, River the great river of South-East ASIA, flowing from TIBET, through southern CHINA, LAOS and CAMBODIA before forming a massive and highly fertile delta in southern VIETNAM and flowing into the SOUTH CHINA SEA. (Length 4184 km/ 2562 miles)

Melaka (Malacca) a port on the south-west coast of MALAYSIA, overlooking the Strait of MALACCA, once a key port in Far Eastern trade. (Pop. 584,000)

Melanesia the central and southern group of islands in the South PACIFIC OCEAN, including the SOLOMON ISLANDS, VANUATU, FIJI and NEW CALEDONIA.

Melbourne the second largest city in AUSTRALIA after SYDNEY and the capital of the state of VICTORIA. (Pop. 3,081,000)

Melos *see* **Cyclades**.

Memel *see* **Klaipeda**.

Memphis a city on the River MISSISSIPPI in the south-west corner of TENNESSEE, USA, on the border with and extending into ARKANSAS. (Pop. city 610,000)

Menai Strait the narrow strait, 180 m (590 ft) across at its narrowest, separating mainland WALES from the island of ANGLESEY, spanned by road and rail bridges.

Mendoza a trading, processing and wine-producing centre in the foothills of the ANDES, in western ARGENTINA. (Pop. 775,000)

Menorca *see* **Minorca**.

Merida the historic capital of the YUCATAN province of eastern MEXICO. (Pop. 557,000)

Merionethshire *see* **Gwynedd**.

Mersey, River a river in north-west ENGLAND. It forms an estuary to the south of LIVERPOOL which is deep and wide enough to permit access for ocean-going ships to Liverpool and MANCHESTER (via the Manchester Ship Canal. (Length 110 km/ 70 miles)

Merseyside a metropolitan county created in 1974 out of parts of LANCASHIRE and CHESHIRE, centring on River MERSEY, with LIVERPOOL as its administrative centre. Replaced in 1986 by a residual body covering former functions. (652 sq km/252 sq miles; pop. 1,376,800)

Mersin (İçel) the principal MEDITERRANEAN port of TURKEY, in the central south of the country, to the north of CYPRUS. (Pop. 422,000)

Meshed *see* **Mashhad**.

Mesolóngion (Missolonghi) a small town in south-western GREECE, on the north coast of the Gulf of PATRAS. It is remembered for its heroic role in the Greek struggle for independence from TURKEY, when it held out against three sieges 1824–6. (Pop. 10,200)

Mesopotamia the 'Fertile Crescent' of land lying between the Rivers TIGRIS and EUPHRATES, mainly in modern IRAQ, where some of the world's earliest civilizations arose (Sumer, ASSYRIA and BABYLON). The name means 'land between the rivers.'

Messina a historic port, founded in the 8th century BC, in north-east SICILY, overlooking the narrow Strait of Messina (6 km/4 miles wide at its narrowest) which separates Sicily from mainland ITALY. (Pop. 271,000)

Metz the capital of the industrial LORRAINE region in eastern FRANCE, situated on the River MOSELLE, close to the border with GERMANY. (Pop. 193,000)

Meuse (Maas), River a river which flows north-west from its source in the LORRAINE region of FRANCE, across central BELGIUM and into the NETHERLANDS, where it joins part of the delta of the River RHINE before entering the NORTH SEA. (Length 935 km/580 miles)

Mexico, the most southerly country in North AMERICA, has its longest border with the USA. to the north, a long coast on the PACIFIC OCEAN and a smaller coast in the west of the Gulf of MEXICO. It is a land of volcanic mountain ranges and high plateaux. The highest peak is Citlaltepetl, 5699 m (18,697 ft), which is permanently snow capped. Coastal lowlands are found in the west and east. Its wide range of latitude and relief, produce a variety of climates. In the north there are arid and semi arid conditions while in the south there is a humid tropical climate. 30% of the labour force are involved in agriculture growing maize, wheat, kidney beans and rice for subsistence and coffee, cotton, fruit and vegetables for export. Mexico is the world's largest producer of silver and has large reserves of oil and natural gas. Developing industries are petrochemicals, textiles, motor vehicles and food processing.

> *Quick facts:*
> Area : 1,958,201 sq km (756,061 sq miles)
> Population : 93,342,000
> Capital : México City
> Other major cities : Guadalajara,
>   Monterrey, Puebla de Zaragoza
> Form of government : Federal Republic
> Religion : RC
> Currency : Mexican peso
>
>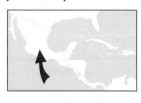

Mexico, Gulf of an arm of the ATLANTIC OCEAN, bounded by the FLORIDA peninsula in the south-east USA. and the YUCATAN peninsula in MEXICO, with the island of CUBA placed in the middle of its entrance.

México City the capital of MEXICO, and the most populous city in the world. It lies to the south of the country on a high plateau 2200 m (7350 ft) above sea level. (Pop. 15,048,000)

Miami a major city and resort on the ATLANTIC coast of south-east FLORIDA, USA. (Pop. city 3,309,000)

Michigan a state in north central USA, formed out of two peninsulas between the GREAT LAKES, with Lake MICHIGAN in the middle. The capital is LANSING. (150,780 sq km/58,216 sq miles; pop. 9,433,700)

Michigan, Lake one of the GREAT LAKES, and the only one to lie entirely within the USA. (57,750 sq km/22,300 sq miles)

Micronesia one of the three main groupings of islands of the PACIFIC OCEAN, lying to the north-west of the other two main groupings, MELANESIA and POLYNESIA. They stretch from BELAU to KIRIBATI.

Micronesia, Federated States of formerly part of the US administered UN Trust Territory of the Pacific, known as the Caroline Islands, this self-governing republic became independent in 1991. It comprises an archipelago of over 600 islands, most of which are uninhabited and are located in the western Pacidic Ocean 1600km (1000 miles) north of PAPUA NEW GUINEA. The climate is tropical maritime, with high temperatures and rainfall all year round, but a pronounced precipitation peak between July and October. Micronesia is still closely linked to the USA, with a heavy reliance on aid. Attempts are being made to diversify the economy whose exports are mainly fishing and copra.

Quick facts:
Area : 701 sq km (271 sq miles)
Population : 125,000
Capital : Palikir
Form of government :Republic
Religion : Christianity
Currency : US dollar

Middle East a non-specific term used to describe an area of south-west ASIA, which is mainly Islamic and/or Arabic-speaking. Countries included are: EGYPT, TURKEY, IRAN, IRAQ, SYRIA, JORDAN, ISRAEL, SAUDI ARABIA, LEBANON, YEMEN, OMAN, the UNITED ARAB EMIRATES, QATAR, BAHRAIN and KUWAIT.

Middlesbrough the county town of CLEVELAND, ENGLAND. (Pop. 146,000)

Middle West *see* **Midwest**.

Mid Glamorgan a former county in central southern WALES, which was formed in 1974 out of part of the former counties of Breconshire, Glamorgan and Monmouthshire. It is now part of the Rhondda, Cynon, Taff administrative authority.

Midlands, The a term used to describe the central industrial counties of England: DEBYSHIRE, NORTHAMPTONSHIRE, NOTTINGHAMSHIRE, STAFFORDSHIRE, WARWICKSHIRE, LEICESTERSHIRE, and WEST MIDLANDS.

Midlothian a region in east SCOTLAND, south of EDINBURGH city. (356 sq km/137 sq miles; pop. 79,900)

Midway Islands two atolls in the north PACIFIC OCEAN, some 2000 km (1242 miles) north-west of HAWAII. They have been possessions of the USA since 1867, and were the scene of a decisive US naval victory against the Japanese in 1942. (3 sq km/2 sq miles; pop. 2200)

Midwest (Middle West) a term used to describe the fertile north-central part of the USA. States in the Midwest include OHIO, MICHIGAN, INDIANA, ILLINOIS, WISCONSIN, MINNESOTA, IOWA and MISSOURI, but others, such as KANSAS, are also often included.

Mien Bac the northern region of VIETNAM, called Tonkin (Tongking) when a part of French Indochina.

Mien Trung the central region of VIETNAM, called Annam when a part of French Indochina.

Mikinai *see* **Mycenae**.

Míkinos *see* **Cyclades**.

Milan (Milano) the major industrial and commercial centre of northern ITALY, and the country's second largest city after ROME, situated in central LOMBARDY. (Pop. 1,359,000)

Milos *see* **Cyclades**.

Milwaukee a port on the west side of Lake MICHIGAN, and the main industrial centre of WISCONSIN, USA. (Pop. city 628,100/metropolitan area 1,629,000)

Minch, The the broad channel separating north-west SCOTLAND from the Outer HEBRIDES or WESTERN ISLES.

Mindanao the second largest island of the PHILIPPINES. (94,631 sq km/36,537 sq miles; pop. 10,100,000)

Mindoro an island in the western central PHILIPPINES. (10,347 sq km/3,995 sq miles)

Minicoy Islands *see* **Lakshadweep**.

Minneapolis-St Paul a major agricultural and commercial centre in south-east Minnesota, USA, on the River Mississippi, and adjoining St Paul. (Pop. city 368,400/ metropolitan area 2,464,100)

Minnesota a state in north central USA. The state capital is St Paul. (217,736 sq km/84,068 sq miles; pop. 4,468,100)

Minorca (Menorca) the second largest of the Balearic Islands (after Majorca). The capital is Mahon. (689 sq km/266 sq miles; pop. 55,500)

Minsk a major industrial city, and the capital of Belarus. (Pop. 1,613,000)

Miquelon *see* **St Pierre and Miquelon**.

Miskolc a city in the north-east of Hungary, and the country's second largest city after Budapest. (Pop. 191,000)

Mississippi a state in central southern USA with a small coastline on the Gulf of Mexico. The state capital is Jackson. (123,600 sq km/47,710 sq miles; pop. 2,615,200)

Mississippi, River the second longest river in the USA. It rises in Minnesota and runs south the length of the country to the Gulf of Mexico. (Length 3779 km/2348 miles)

Missolonghi *see* **Mesolóngion**.

Missouri a state in the Midwest of the USA. The state capital is Jefferson City. (180,600 sq km/ 69,712 sq miles; pop. 5,190,700)

Missouri, River the main tributary of the Mississippi with which it is the longest river in North America. It rises in Montana, flows north, east and south-east to join the Mississippi at St Louis. (Length 3969 km/2466 miles)

Mitchell, Mount *see* **Appalachian Mountains**.

Mizoram a state of India, in the hilly north-east, on the border with Mayanmar. The capital is Aizawl. (21,100 sq km/8,145 sq miles; pop. 686,200)

Mobile a port on the coast of Alabama, USA, on the Gulf of Mexico. (Pop. city 202,000)

Mobutu Sese Seko, Lake *see* **Albert, Lake**.

Modena an industrial city retaining many vestiges of its medieval past, in north-eastern Italy. (Pop. 177,000)

Mogadishu (Muqdisho) the capital and main port of of Somalia. (Pop. 1,000,000)

Mohave Desert *see* **Mojave Desert**.

Mohenjo Daro an ancient city in the valley of the River Indus, in southern Pakistan. It was inhabited for about 1000 years from 2500BC.

Mojave (Mohave) Desert a desert in southern California, stretching from Death Valley to Los Angeles. (38,850 sq km/15,000 sq miles)

Moldova (Moldavia) (1) the neighbouring region of north-east Romania. **(2)** was a Soviet socialist republic from 1940 until 1991 when it became independent of the former USSR. It is bounded to the west by Romania and to the north, east and south by Ukraine. The republic consists of a hilly plain that rises to 429 m (1408 ft) in the centre. Its main rivers are the Prut in the west and the Dniester in the north and east. Moldova's soils are fertile, and crops grown include wheat, corn, barley, tobacco, sugar beet, soybeans and sunflowers. There are also extensive fruit orchards, vineyards and walnut groves. Beekeeping and silkworm breeding are widespread throughout the country. Food processing is the main industry. Other industries include metal working, engineering and the manufacture of electrical equipment.

> *Quick facts:*
> Area : 33,700 sq km (13,000 sq miles)
> Population : 4,434,000
> Capital : Chisinau
> Other major cities : Tiraspol, Bendery
> Form of government : Republic
> Religion : Russian Orthodox
> Currency : Rouble

Molise a region of eastern Italy, on the Adriatic coast, between Abruzzi and Puglia. (4,400 sq km/ 1,698 sq miles; pop. 336,500)

Molotov *see* **Perm'**.

Moluccas *see* **Maluku**.

Mombasa the second city of Kenya and an important port on the Indian Ocean. (Pop. 457,000)

Monaco is a tiny principality on the Mediterranean. It is surrounded landwards by the Alpes Maritimes department of France. It comprises a rocky peninsula and a narrow stretch of coast. It has mild moist winters and hot dry summers. The old town of Monaco-Ville is situated on a rocky promontory and houses the royal palace and the cathedral. The Monte Carlo district has its world-famous casino and La Condamine has thriving businesses, shops, banks and attractive residential areas. Fontvieille is an area reclaimed from the sea where now marinas and light industry are located. Light industry includes chemicals, plastics, electronics, engineering and paper but it is tourism that is the main revenue earner.

Quick facts:
Area : 1.95 sq km/(0.75 sq miles)
Population : 32,000
Capital : Monaco-Ville
Form of government : Constitutional Monarchy
Religion : RC
Currency : Franc

Monaco-Ville the capital of Monaco, sited on a rocky headland that sticks out into the Mediterranean Sea. (Pop. 3,100)

Monaghan a county in the central north of the Republic of Ireland, with a county town of the same name. (1290 sq km/498 sq miles; pop. county 51,300)

Mönchengladbach an industrial city in the south-west of the Ruhr region of western Germany, 25 km (16 miles) west of Dusseldorf. (Pop. 265,000)

Mongolia is a landlocked country in north-east Asia which is bounded to the north by the Russian Federation and by China to the south, west and east. Most of Mongolia is mountainous and over 1500 m (4922 ft) above sea level. In the north-west are the Hangayn Mountains and the Altai, rising to 4362 m (14,312 ft). In the south there are grass-covered steppes and desert wastes of the Gobi. The climate is very extreme and dry. For six months the temperatures are below freezing and the summers are mild. Mongolia has had a nomadic pastoral economy for centuries and cereals, including fodder crops, are grown on a large scale on state farms. Industry is small scale and dominated by food processing. The collapse of trade with the former Soviet Union has created severe economic problems and Mongolia is increasingly looking to Japan and China for trade and economic assistance.

Quick facts:
Area : 1,566,500 sq km (604,826 sq miles)
Population : 2,363,000
Capital : Ulan Bator (Ulaanbaatar)
Other major cities : Darhan, Erdenet
Form of govt : Republic
Religion : Buddhist,
   Shamanist, Muslim
Currency : Tugrik

Monmouthshire a county in east Wales, on the borders with England. (840 sq km/308 sq miles; pop. 84,000)

Monrovia the capital and principal port of Liberia. (Pop. 490,000)

Mons (Bergen) a town in south-west Belgium. (Pop. 91,700)

Montana a state in the north-west of the USA, on the border with Canada. The state capital is Helena. (318,100 sq km/147,143 sq miles; pop. 822,300)

Monte Carlo an elegant coastal town and resort in Monaco, famed in particular for its casinos. (Pop. 13,200)

Montenegro (Crna Gora) the smallest of the republics of former Yugoslavia, in the south-west on the Adriatic Sea and bordering Albania. The capital is Titograd. (13,812 sq km/5331 sq miles; pop. 615,000)

Monterey a resort town on the Pacific coast of central California, USA, 135 km (85 miles) south-east of San Francisco. It is well known for its annual jazz festival. (Pop. 32,000)

Monterrey an industrial city in north-east Mexico, the country's third largest city after Mexico City and Guadaljara. (Pop. 2,522,000)

Montevideo the capital of Uruguay, and an important port on the River Plate estuary. (Pop. 1,384,000)

Montgomery the state capital of Alabama, USA. (Pop. city 192,000)

Montgomeryshire *see* **Powys**.

Montpelier the state capital of Vermont, USA. (Pop. 8,200)

Montpellier a university and trading city in central southern France, the capital of the Languedoc-Roussillon region. (Pop. 248,000)

Montréal the second largest city in Canada after Toronto, on the St Lawrence River, in the south of the province of Quebec. Two-thirds of the population are French-speaking Québecois. (Pop. 3,127,000)

Montserrat a British Crown colony in the Leeward Islands, in the south-eastern Caribbean. The capital is called Plymouth. (102 sq km/39 sq miles)

Monza a city in northern Italy, 12 km (8 miles) north-east of Milan. (Pop. 120,000)

Moravia a historical region of the Czech Republic, east of Bohemia, west of Slovakia, with Poland to the north and Austria to the south.

Moray Firth an inlet of the North Sea cutting some 56 km (35 miles) into the eastern coast of north-east Scotland, with Inverness at its head.

Moray a county in north Scotland (2238 sq km/864 sq miles; pop. 86,000).

Morocco, in north-west AFRICA, is strategically placed at the western entrance to the MEDITERRANEAN SEA. It is a land of great contrasts with high rugged mountains, the arid SAHARA and the green ATLANTIC and Mediterranean coasts. The country is split from south-west to north-east by the ATLAS mountains. The north has a pleasant Mediterranean climate with hot dry summers and mild moist winters. Farther south winters are warmer and summers even hotter. Snow often falls in winter on the Atlas mountains. Morocco is mainly a farming country, wheat, barley and maize are the main food crops and it is one of the world's chief exporters of citrus fruit. Morocco's main wealth comes from phosphates, reserves of which are the largest in the world. The economy is very mixed. Morocco is self sufficient in textiles, it has car assembly plants, soap and cement factories and a large sea fishing industry. Tourism is a major source of revenue.

> *Quick facts:*
> Area : 446,550 sq km (172,413 sq miles)
> Population : 26,857,000
> Capital : Rabat
> Other major cities : Casablanca, Fez,
>   Marrakech
> Form of government : Constitutional Monarchy
> Religion : Sunni Islam
> Currency : Dirham

Moroni the capital of the COMOROS islands. (Pop. 22,000)

Moscow (Moskva) the capital of the RUSSIAN FEDERATION, sited on the Moskva River. It is an ancient city with a rich heritage, and is the political, industrial and cultural focus of the country. (Pop. 8,957,000)

Moselle (Mosel), River a river which flows northwards from the south-eastern LORRAINE region of eastern FRANCE to form part of the border between LUXEMBOURG and GERMANY before flowing eastwards to meet the River RHINE at KOBLENZ. (Length 550 km/340 miles)

Mosul (Al Mawsil) a historic trading city on the banks of the River TIGRIS in north-west IRAQ, and an important centre for the surrounding oil-producing region. (Pop. city 571,000)

Mourne Mountains a mountain range of noted scenic beauty in the south of COUNTY DOWN, NORTHERN IRELAND. The highest point is Slieve Donard (852 m /2795 ft).

Mozambique is a republic located in south-east AFRICA. A coastal plain covers most of the southern and central territory, giving way to the western highlands and north to a plateau including the Nyasa Highlands. The ZAMBEZI river separates the high plateaux in the north from the lowlands in the south. The country has a humid tropical climate with highest temperatures and rainfall in the north. Normally conditions are reasonably good for agriculture but a drought in the early 1980s, followed a few years later by severe flooding, resulted in famine and more than 100,000 deaths. A lot of industry was abandoned when the Portuguese left the country and, due to lack of expertise, was not taken over by the local people. There is little incentive to produce surplus produce for cash and food rationing has now been introduced. This also has led to a black market which now accounts for a sizable part of the economy.

> *Quick facts:*
> Area : 801,590 sq km (309,494 sq miles)
> Population : 17,800,000
> Capital : Maputo
> Other major cities : Beira, Nampula
> Form of government : Republic
> Religions : Animism, RC,
>   Sunni Islam
> Currency : Metical

Mozambique Channel the broad strait, some 400 km (250 miles) across at its narrowest, which separates the island of MADAGASCAR from mainland AFRICA.

Mühlheim an der Ruhr an industrial city and port on the River RUHR, in the RUHR region of western GERMANY. (Pop. 178,000)

Mulhouse an industrial city in the ALSACE region of eastern FRANCE. (Pop. 224,000)

Mull an island just off the central western coast of SCOTLAND. (925 sq km/357 sq miles)

Multan an industrial city in PUNJAB province in eastern central PAKISTAN. (Pop. 722,000)

Munich (München) a historic and industrial city in southern GERMANY, and capital of BAVARIA. (Pop. 1,256,000)

Munster one of the four historic provinces of IRE-

LAND, covering the south-west quarter of the country.

Münster an inland port and industrial centre on the DORTMUND-EMS Canal in north-western GERMANY. (Pop. 267,000)

Muqdisho *see* **Mogadishu**.

Murcia a trading and manufacturing city in south-eastern SPAIN, and capital of the province of the same name. (Pop. city 342,000)

Murmansk the largest city north of the ARCTIC CIRCLE, a major port and industrial centre on the KOLA PENINSULA in the far north-western corner of the RUSSIAN FEDERATION. (Pop. 432,000)

Murray, River a major river of south-east AUSTRALIA, which flows westwards from its source in the SNOWY MOUNTAINS to form much of the boundary between the states of NEW SOUTH WALES and VICTORIA. It is joined by the River DARLING before flowing across the south-eastern corner of SOUTH AUSTRALIA and into the INDIAN OCEAN. (Length 2570 km/1600 miles)

Mururoa an atoll in the south-eastern sector of FRENCH POLYNESIA, used by FRANCE since 1966 as a testing ground for nuclear weapons.

Musandam a rocky, horn-shaped peninsula, jutting out into The GULF to form the southern side of the Strait of HORMUZ. It belongs to OMAN, but is separated from it by part of the UNITED ARAB EMIRATES.

Muscat (Masqat) the historic capital of OMAN. The neighbouring port of Muttrah has developed rapidly in recent decades to form the commercial centre of Muscat. (Pop. (with Muttrah) 350,000)

Muscovy a principality of RUSSIA from the 13th to 16th centuries, with MOSCOW as the capital.

Mustique a privately-owned island situated in the GRENADINES, to the south of ST VINCENT, in the south-eastern CARIBBEAN. (Pop. 200)

Muttrah *see* **Muscat**.

Myanmar the Union of Myanmar (formerly Burma) is the second largest country in South-East Asia. The heartland of the country is the valley of the IRRAWADDY. The north and west of the country are mountainous and in the east the Shan Plateau runs along the border with THAILAND. The climate is equatorial at the coast, changing to tropical monsoon over most of the interior. The Irrawaddy river flows into the ANDAMAN SEA, forming a huge delta area which is ideal land for rice cultivation. Rice is the country's staple food and accounts for half the country's export earnings. Myanmar is rich in timber and minerals but because of poor communications, lack of development and unrest among the ethnic groups, the resources have not been fully exploited.

Quick facts:
Area : 676,552 sq km (261,218 sq miles)
Population : 44,613,000
Capital : Yangon (formerly Rangoon)
Other major cities : Mandalay, Moulmein, Pegu
Form of government : Republic
Religion : Buddhism
Currency : Kyat

Mycenae (Mikinai) an ancient city in the north-east of the PELOPONNESE, GREECE, inhabited from 1580 to 1100BC.

Mysore an industrial city in the state of KARNATAKA, southern INDIA. (Pop. 652,000)

Naberezhnyye Chelny an industrial town in the RUSSIAN FEDERATION. Renamed Brezhnev in 1984 after the Soviet president Leonid Brezhnev (1906–82). Name reverted to Naberezhnyye Chelny in 1988. (Pop. 517,000)

Nagaland a primarily agricultural state in the hilly far north-eastern corner of INDIA, bordering MYANMAR. (16,721 sq km/6,456 sq miles; pop. 1,125,600)

Nagasaki a port and industrial city on the west coast of KYUSHU Island, JAPAN. Three days after the first atomic bomb destroyed HIROSHIMA, a second was dropped on Nagasaki (August 9, 1945), killing 40,000 people. (Pop. 441,000)

Nagorno Karabakh a disputed autonomous enclave in AZERBAIJAN, which is claimed by ARMENIA. Three quarters of the population are Armenian.

Nagoya a port and industrial centre on the south-eastern coast of HONSHU, JAPAN. (Pop. 2,159,000)

Nagpur a commercial centre and textile manufacturing city on the DECCAN plateau of MAHARASHTRA state, central INDIA. (Pop. 1,661,000)

Nairobi the capital of KENYA and a commercial centre, in the south-west highland region. (Pop. 1,429,000)

Nakhichevan an autonomous republic, enclaved by ARMENIA, but a part of AZERBAIJAN. The capital is also called Nakhichevan.

Namib Desert a sand desert lining the coast of NAMIBIA in south-western AFRICA.

Namibia is situated on the Atlantic coast of south-west AFRICA. There are three main regions in the country. Running down the entire Atlantic coastline is the NAMIB DESERT, east of which is the Central Plateau of mountains, rugged outcrops, sandy valleys and poor grasslands. East again and north is the KALAHARI DESERT. Namibia has a poor rainfall, the highest falling at Windhoek, the capital. Even here it only amounts to 200–250 mm (8–10 inches) per year. It is essentially a stock-rearing country with sheep and goats raised in the south and cattle in the central and northern areas. Diamonds are mined just north of the River ORANGE, and the largest open groove uranium mine in the world is located near Swakopmund. One of Africa's richest fishing grounds lies off the coast of Namibia, and mackerel, tuna and pilchards are an important export.

Quick facts:
Area : 824,292 sq km (318,259 sq miles)
Population : 1,610,000
Capital : Windhoek
Form of government : Republic
Religions : Lutheranism, RC, other Christianity
Currency : Rand

Nanchang an industrial city and commercial centre in central south-eastern CHINA, and the capital of JIANGXI province. (Pop. 1,440,000)

Nancy a manufacturing city in north-east FRANCE, and former capital of LORRAINE. (Pop. 329,000)

Nanjing (Nanking) a major industrial and trading city built on the lower reaches of the CHANG JIANG (Yangtze) river, and the capital of JIANGSU province, central eastern CHINA. (Pop. 2,490,000)

Nanning the capital of the GUANGXI-ZHUANG autonomous region in the extreme south-east of CHINA. (Pop. 960,000)

Nansei-shoto *see* **Ryukyu Islands**.

Nantes a port and commercial centre in north-western FRANCE and capital of the Loire Atlantique department. (Pop. 496,000)

Naples (Napoli) the third largest city in ITALY after ROME and MILAN, is a port situated on the spectacular Bay of Naples. (Pop. 1,072,000)

Nara a historic city in south HONSHU Island, JAPAN, the capital of Japan in the 8th century. (Pop. 356,000)

Nashville the state capital of TENNESSEE, USA, an industrial city famous as the traditional home of Country and Western music. (Pop. city 495,000)

Nassau capital of the BAHAMAS, situated on the north

side of New Providence Island. (Pop. 190,000)

Nasser, Lake a massive artificial lake on the River NILE in southern EGYPT, created when the ASWAN High Dam was completed in 1971. It was named after the former president (1956–70) of Egypt, Gamal Abdel Nasser (1918–70). (5000 sq km/ 1930 sq miles)

Natal (1) a port city on the north-east tip of BRAZIL, and capital of the state of Rio Grande do Norte. (Pop. 607,000) **(2)** a province on the eastern coast of SOUTH AFRICA. The capital is PIETERMARITZBURG, but the main city is DURBAN. (91,785 sq km/35,429 sq miles; pop. 2,145,000)

Nauru is the world's smallest republic. It is an island situated just 40 km (25 miles) south of the Equator and is halfway between AUSTRALIA and HAWAII. It is an oval-shaped coral island only 20 km (12 miles) in diameter and is surrounded by a reef. The centre of the island comprises a plateau which rises to 60 m (197 ft) above sea level. The climate is tropical with a high and irregular rainfall. The country is rich, due entirely to the deposits of high quality phosphate rock in the central plateau. This is sold for fertilizer to AUSTRALIA, NEW ZEALAND, JAPAN and South KOREA. Phosphate deposits are likely to be exhausted by 1995 but the government is investing overseas.

Quick facts:
Area : 21 sq km (8 sq miles)
Population : 12,000
Capital : Yaren
Form of government : Republic
Religions : Protestantism, RC
Currency : Australian dollar

Navarra (Navarre) a province in the mountainous north-eastern part of SPAIN. The capital is PAMPLONA. (10,420 sq km/4023 sq miles; pop. 513,000)

Naxos a fertile island in the southern AEGEAN SEA, the largest of the CYCLADES. (428 sq km/165 sq miles)

Nazareth a town in northern ISRAEL, and the childhood home of Jesus. (Pop. 64,00)

Ndjamena (N'Djamena) the capital of CHAD, in the south-east of the country. It was founded by the French in 1900 and named Fort Lamy. (Pop. 530,000)

Neagh, Lough the largest freshwater lake in the British Isles, in the east of NORTHERN IRELAND. (381 sq km/147 sq miles)

Nebraska a state in the MIDWEST of the USA, in the very centre of the country. The capital is LINCOLN. (200,018 sq km/77,227 sq miles; pop. 1,600,500)

Neckar, River a tributary of the River RHINE, rising in the BLACK FOREST in the south-west of GERMANY. (365 km/227 miles)

Negev a desert in southern ISRAEL.

Negros the fourth largest island of the PHILIPPINES. (12,704 sq km/4905 sq miles; pop. 2,800,000)

Nei Mongol Autonomous Region (Inner Mongolia) a region of north-eastern CHINA, bordering MONGOLIA. The capital is HOHHOT. (1,200,000 sq km/450,000 sq miles; pop. 21,457,000)

Neisse, River a tributary of the ODER, which flows north from its source in the CZECH REPUBLIC to form part of the border between GERMANY and POLAND. (Length 256 km/159 miles)

Nepal is a long narrow rectangular country, landlocked between CHINA and INDIA on the flanks of the eastern HIMALAYAS. Its northern border runs along the mountain tops. In this border area is EVEREST (8848 m /29,028 ft), the highest mountain in the world. The climate is subtropical in the south, and all regions are affected by the monsoon. Nepal is one of the world's poorest and least developed countries, with most of the population trying to survive as peasant farmers. It has no significant minerals, however with Indian and Chinese aid roads have been built from the north and south to KATHMANDU. The construction of hydroelectric power schemes is now underway.

Quick facts:
Area : 140,800 sq km (54,362 sq miles)
Population : 21,953,000
Capital : Kathmandu
Form of government : Constitutional
  Monarchy
Religion : Hinduism, Buddhism
Currency : Nepalese rupee

Netherlands, The situated in north-west EUROPE, the Netherlands is bounded to the north and west by the NORTH SEA. Over one-quarter of the Netherlands is below sea level and the Dutch have tack-

led some huge reclamation schemes to add some land area to the country. One such scheme is the IJSSELMEER, where four large areas (polders) reclaimed have added an extra 1650 sq km (637 sq miles) for cultivation and an overspill town for AMSTERDAM. The Netherlands has mild winters and cool summers. Agriculture and horticulture are highly mechanized, and the most notable feature is the sea of glass under which salad vegetables, fruit and flowers are grown. Manufacturing industries include chemicals, machinery, petroleum, refining, metallurgy and electrical engineering. The main port of the Netherlands, ROTTERDAM, is the largest in the world.

Quick facts:
Area : 37,330 sq km (15,770 sq miles)
Population : 15,495,000
Capital : Amsterdam
Seat of government : The Hague
  (Den Haag, 's-Gravenhage)
Other major cities : Eindhoven, Rotterdam
Form of government : Constitutional Monarchy
Religions : RC, Dutch reformed, Calvinism
Currency : Guilder

Netherlands Antilles an overseas division of the NETHERLANDS, spread over the southern CARIBBEAN. The principal islands are: CURACAO, ST MARTIN, ST EUSTATIUS, and BONAIRE. ARUBA was part of the group until 1986.

Neusatz *see* **Novi Sad**.

Neva, River the river which flows through ST PETERSBURG. (Length 74 km/45 miles)

Nevada a state in the west of the USA, consisting mostly of desert. The state capital is CARSON CITY. (286,298 sq km/110,540 sq miles; pop. 1,336,000)

Nevis *see* **St Kitts and Nevis**.

Newark a major port city in NEW JERSEY. (Pop. 275,200)

New Britain the largest offshore island belonging to PAPUA NEW GUINEA, in the BISMARCK Archipelago. (36,500 sq km/14,100 sq miles; pop. 253,000)

New Brunswick a state on the coast in south-east CANADA, bordering the USA. The state capital is FREDERICTON. (73,436 sq km/28,354 sq miles; pop. 725,600)

New Caledonia the main island of a group called by the same name in the South Pacific, which form an overseas territory of France. The capital is Noumea. (19,103 sq km/7,376 sq miles; pop. 164,000)

Newcastle a port and industrial city in New South Wales, Australia. (Pop. 429,000)

Newcastle-under-Lyme a town in the Potteries region of Staffordshire, central England. (Pop. 123,100)

Newcastle upon Tyne a historic and industrial city in the county of Tyne and Wear, north-east England. (Pop. 283,600)

New Delhi became the official capital of India in 1911. (Pop. 301,000)

New England the name given to north-eastern states of the USA: Maine, Vermont, New Hampshire, Connecticut, Massachusetts and Rhode Island.

Newfoundland the province in the extreme east of Canada. The capital is St John's. (405,700 sq km/ 156,600 sq miles; pop. 571,600)

New Guinea one of the world's largest islands, divided into two parts: independent Papua New Guinea in the east and Irian Jaya, a state of Indonesia, in the west.

New Hampshire a state of New England, in the north-west of the USA. The state capital is Concord. (24,097 sq km/9304 sq miles; pop. 1,116,000)

New Haven a port in Connecticut, USA. (Pop. city 130,500)

New Hebrides *see* **Vanuatu.**

New Jersey a state on the Atlantic coast in the north-east of the USA. The state capital is Trenton. (20,295 sq km/7836 sq miles; pop. 7,820,300)

New Mexico a state in the south-west of the USA, bordering Mexico. The state capital is Santa Fe. (315,115 sq km/121,666 sq miles; pop. 1,581,800)

New Orleans an important and historic port in southern Louisiana, on the Mississippi delta. (Pop. city 496,900/metropolitan area 1,303,000)

Newport a port and naval base in Rhode Island, USA. (Pop. 29,900)

Newport News a major eastern seaboard port in Virginia, USA. (Pop. 170,000)

New Providence *see* **Bahamas.**

New South Wales the most populous of the states of Australia, situated in the south-east of the country. The capital is Sydney. (801,430 sq km/ 309,433 sq miles; pop. 5,570,000)

New Territories *see* **Hong Kong.**

New York (City) the most populous city in the

USA, its most important port, and a major financial centre. It is sited on the mouth of the Hudson River, and comprises five boroughs: Manhattan, the Bronx, Queens, Brooklyn and Staten Island. (Pop. city 7,322,600)

New York (State) a populous state in the north-east of the USA, on the Atlantic coast. The state capital is Albany (127,200 sq km/49,099 sq miles; pop. 18,109,500)

New Zealand lies south-east of Australia in the South Pacific. It comprises two large islands, North Island and South Island, Stewart Island and the Chatham Islands. New Zealand enjoys very mild winters with regular rainfall and no extremes of heat or cold. North Island is hilly with isolated mountains and active volcanoes. On South Island the Southern Alps run north to south, and the highest point is Mount Cook (3753 m/12,313 ft). The Canterbury Plains lie to the east of the mountains. Two-thirds of New Zealand is suitable for agriculture and grazing, meat, wool and dairy goods being the main products. Forestry supports the pulp and paper industry and a considerable source of hydroelectric power produces cheap electricity for the manufacturing industry which now accounts for 30% of New Zealand's exports.

Quick facts:
Area : 270,986 sq km (104,629 sq miles)
Population : 3,567,000
Capital : Wellington
Other major cities : Auckland,
  Christchurch, Dunedin, Hamilton
Form of government : Constitutional Monarchy
Religions : Anglicanism, RC, Presbyterianism
Currency : New Zealand dollar

Ngaliema, Mount *see* **Ruwenzori.**

Niagara Falls spectacular waterfalls on the Niagara River, situated on the Canada-USA border between Lakes Erie and Ontario.

Niamey the capital of Niger. (Pop. 398,000)

Nicaragua lies between the Pacific Ocean and the Caribbean Sea, on the isthmus of Central America, and is sandwiched between Honduras to the north and Costa Rica to the south. The east coast contains forested lowland and is the wettest part of the island. Behind this is a range of volcanic mountains and the west coast is a belt of savanna lowland running parallel to the Pacific coast. The western region, which contains the two huge lakes, NICARAGUA and Managua, is where most of the population live. The whole country is subject to devastating earthquakes. Nicaragua is primarily an agricultural country and 65% of the labour force work on the land. The main export crops are coffee, cotton and sugar cane. All local industry is agriculture-related.

Quick facts:
Area : 130,000 sq km (50,193 sq miles)
Population : 4,544,000
Capital : Managua
Form of government : Republic
Religion : RC
Currency : Córdoba

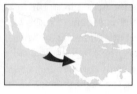

Nicaragua, Lake a large lake in the south-west of NICARAGUA. (8264 sq km/3191 sq miles)

Nice a city, harbor and famous resort town of the CÔTE D'AZUR, south-eastern FRANCE. (Pop. 516,000)

Nicosia the capital of CYPRUS, situated in the centre of the island. (Pop. 177,000)

Niger is a landlocked republic in west AFRICA, just south of the Tropic of Cancer. Over half of the country is covered by the encroaching SAHARA DESERT in the north, and the south lies in the drought-stricken SAHEL. In the extreme south-west corner, the river NIGER flows through the country, and in the extreme south-east lies LAKE CHAD, but the rest of the country is very short of water. The people in the south-west fish and farm their own food, growing rice and vegetables on land flooded by the river. Farther from the river, crops have failed as a result of successive droughts since 1968. In the north, where the population are traditionally herdsmen, drought has wiped out whole clans. Uranium mined in the Aïr mountains is Niger's main export.

Quick facts:
Area : 1,267,000 sq km (489,189 sq miles)
Population : 9,149,000
Capital : Niamey
Form of govt : Republic
Religion : Sunni Islam
Currency : Franc CFA

Niger, River a river in West AFRICA flowing through GUINEA, MALI, NIGER and NIGERIA to the Gulf of GUINEA. (Length 4170 km/2590 miles)

Nigeria is a large and populous country in west AFRICA, and from the Gulf of GUINEA it extends north to the border with NIGER. It has a variable landscape, from the swampy coastal areas and tropical forest belts of the interior, to the mountains and savanna of the north. The two main rivers are the NIGER and the BENUE, and just north of their confluence lies the Jos Plateau. The climate is hot and humid and rainfall, heavy at the coast, gradually decreases inland. The dry far north is affected by the Harmattan, a hot dry wind blowing from the SAHARA. The main agricultural products are cocoa, rubber, groundnuts and cotton. However, only cocoa is of any significance for export. The country depends on revenue from petroleum exports but fluctuations in the world oil market have left Nigeria with economic problems.

Quick facts:
Area : 923,768 sq km (356,667 sq miles)
Population : 108,448,000
Capital : Abuja (New Federal Capital)
 Lagos (Capital until 1992)
Other major cities : Ibadan, Kano, Ogbomsho
Form of government : Federal republic
Religions : Sunni Islam, Christianity
Currency : Naira

Nijmegen a city of eastern central NETHERLANDS, close to the border with GERMANY. (Pop. 247,000)

Nikolayev a port and industrial city on the north coast of the BLACK SEA, in the UKRAINE. (Pop. 501,000)

Nile, River (An Nil) a river in AFRICA and, with the AMAZON, one of the two longest rivers in the world. It rises in BURUNDI, flows into Lake VICTORIA and then flows northwards through UGANDA, SUDAN and EGYPT to its delta on the MEDITERRANEAN. The river is called the White Nile (Bahr el Abiad) until it reaches KHARTOUM, where it is joined by its main tributary, the Blue Nile (Bahr el Azraq), which rises in ETHIOPIA. (Length 6695 km/ 4160 miles)

Nîmes a city in southern FRANCE, overlooking the River RHONE. (Pop. 139,000)

Nineveh the excavated remains of the ancient capital of ASSYRIA, near MOSUL in northern IRAQ.

Ningbo a port and industrial city in ZHEJIANG province, in central eastern CHINA. (Pop. 1,100,000)

Ningxia-Hui Autonomous Region a region of central northern CHINA, south of Inner Mongolia. The capital is YINCHUAN. (170,000 sq km/65,620 sq miles; pop. 4,655,000)

Nis (Nish) a historic city in the east of SERBIA. (Pop. 179,000)

Nizhniy Novgorod an industrial city in the RUSSIAN FEDERATION on the River VOLGA, formerly known as Gor'kiy (Gorky). (Pop. 1,451,000)

Nizhniy Tagil an industrial city in the central URAL MOUNTAINS, RUSSIAN FEDERATION. (Pop. 437,000)

Nordkapp *see* **North Cape**.

Norfolk (1) a county of EAST ANGLIA, ENGLAND. The county town is NORWICH. (5355 sq km/2068 sq miles; pop. 768,500) **(2)** a port and naval base in the state of VIRGINIA, USA. (Pop. city 261,200/ metropolitan area 1,497,000)

Normandy an area of central northern FRANCE, now divided into two regions, Haute Normandie and Basse Normandie. (Pop. 3,148,000)

Northampton the county town of NORTHAMPTON-SHIRE. (Pop. 187,600)

Northamptonshire a county in central ENGLAND. The county town is NORTHAMPTON. (2367 sq km/ 914 sq miles; pop. 594,800)

North Cape (Nordkapp) one of Europe's most northerly points—500 km (310 miles) north of the ARCTIC CIRCLE in NORWAY.

North Carolina a state on the south-eastern coast of the USA. The state capital is RALEIGH. (136,198 sq km/52,586 sq miles; pop. 6,836,300)

North Dakota a state in the west of the USA. The state capital is BISMARCK. (183,022 sq km/70,665 sq miles; pop. 634,000)

Northern Ireland a province of the UK, occupying most of the northern part of the island of IRELAND. It is divided into six counties. The capital is BELFAST. (14,121 sq km/5452 sq miles; pop. 1,641,700)

Northern Marianas a group of 14 islands in the western PACIFIC which in 1978 became a commonwealth of the USA. The capital is SUSEPE, on the island of SAIPAN.

Northern Territory a territory of northern AUSTRALIA. The capital is DARWIN. (1,346,200 sq km/ 519,770 sq miles; pop. 157,000)

North Island *see* **New Zealand**.

North Korea *see* **Korea, People's Democratic Republic of**.

North Pole the northernmost point on the earth's axis.

North Sea a comparatively shallow branch of the ATLANTIC OCEAN that separates the BRITISH ISLES from the European mainland.

Northumberland a county in north-eastern ENGLAND. The county town is Morpeth. (5033 sq km/ 1943 sq miles; pop. 307,700)

Northwest Territories a vast area of northern CANADA, occupying almost a third of the country's whole land area. The capital is YELLOWKNIFE. (3,426,000 sq km/1,322,552 sq miles; pop. 54,000)

Norway occupies the western half of the Scandinavian peninsula in northern EUROPE, and is surrounded to the north, west and south by water. It shares most of its eastern border with SWEDEN. It is a country of spectacular scenery of fjords, cliffs, rugged uplands and forested valleys. Two-thirds of the country is over 600 m/1969 ft and it has some of the deepest fjords in the world. The climate is temperate as a result of the warming effect of the Gulf Stream. Summers are mild and although the winters are long and cold, the waters off the west coast remain ice-free. Agriculture is chiefly concerned with dairying and fodder crops. Fishing is an important industry and the large reserves of forest provide timber for export. Industry is now dominated by petrochemicals based on the reserves of Norwegian oil in the NORTH SEA.

> *Quick facts:*
> Area : 323,895 sq km (125,056 sq miles)
> Population : 4,361,000
> Capital : Oslo
> Other major cities : Bergen, Trondheim, Stavanger
> Form of government : Constitutional Monarchy
> Religion : Lutheranism
> Currency : Norwegian krone

Norwegian Sea a sea lying between NORWAY, GREENLAND and ICELAND; to the north it joins the ARCTIC OCEAN, and to the south, the ATLANTIC.

Norwich the county town of NORFOLK, in eastern ENGLAND. (Pop. 127,800)

Nottingham the historic county town of NOTTING-

HAMSHIRE, situated on the River TRENT. (Pop. 282,500)

Nottinghamshire a county in the MIDLANDS of ENGLAND. The county town is NOTTINGHAM,. (2164 sq km/836 sq miles; pop. 1,030,900)

Nouakchott the capital city of MAURITANIA, near the ATLANTIC coast. (Pop. 600,000)

Nouméa the capital and chief port of NEW CALEDONIA. (Pop. 65,000)

Nova Scotia a province on the eastern coast of CANADA. The capital is HALIFAX. (52,841 sq km/20,401 sq miles; pop. 897,500)

Novi Sad (Ujvidek; Neusatz) a city on the River DANUBE and the capital of VOJVODINA, an autonomous province of SERBIA. (Pop. 257,700)

Novokuznetsk an industrial city in central southern SIBERIA. (Pop. 614,000)

Novosibirsk a major industrial city in central RUSSIAN FEDERATION. (Pop. 1,472,000)

Nuku'alofa the capital and main port of TONGA. (Pop. 29,000)

Nullarbor Plain a huge, dry and treeless (the name is from the Latin for 'no trees') plain which borders the GREAT AUSTRALIAN BIGHT, in WESTERN and SOUTH AUSTRALIA.

Nuremberg (Nürnberg) a city in BAVARIA, central southern GERMANY. (Pop. 499,000)

Nürnberg *see* **Nuremberg**.

Nusa Tenggara *see* **Lesser Sunda Islands**.

Nuuk *see* **Godthåb**.

Nyasa, Lake *see* **Malawi, Lake**.

Nyasaland *see* **Malawi**.

Oahu the third largest of the islands of HAWAII, where the state capital, HONOLULU, and PEARL HARBOR are located. (1,525 sq km/589 sq miles; pop. 838,500)

Oakland a port on SAN FRANCISCO BAY in central western CALIFORNIA, USA. (Pop. city 373,000)

Ob', River a river in the RUSSIAN FEDERATION which rises near the border with MONGOLIA and flows northwards to the KARA SEA. (Length 5570 km/3460 miles)

Oberammergau a village in BAVARIA in south-west GERMANY, famed for the Passion play which it puts on every ten years.

Oceania a general term used to describe the central and southern islands of the PACIFIC OCEAN including those of AUSTRALIA and NEW ZEALAND. (8,500,000 sq km/3,300,000 sq miles; pop. 27,000,000)

Oder, River a river in central EUROPE rising in the CZECH REPUBLIC and flowing north and west to the BALTIC SEA; it forms part of the border between GERMANY and POLAND. (Length 912 km/567 miles)

Odessa a major BLACK SEA port in the UKRAINE. (Pop. 1,096,000)

Offaly a county in the centre of the Republic of IRELAND. The county town is Tullamore. (1998 sq km/771 sq miles; pop. 58,500)

Ogaden a desert region of south-eastern ETHIOPIA, claimed by SOMALIA.

Ohio a MIDWEST state of the USA, with a shoreline on Lake ERIE. The capital is COLUMBUS. (106,765 sq km/41,220 sq miles; pop. 11,021,400)

Ohio River a river in the eastern USA, formed at the confluence of the Allegheny and Monongahela Rivers. It flows west and south and joins the MISSISSIPPI at Cairo, ILLINOIS. (Length 1575 km/980 miles)

Okavango, River see **Kavango (Cubango), River**.

Okayama a commercial city in south-west HONSHU Island, JAPAN. (Pop. 605,000)

Okhotsk, Sea of a part of the north-western PACIFIC OCEAN bounded by the KAMCHATKA peninsula, the KURIL islands, and the east coast of SIBERIA.

Oklahoma a state in the south-west of the USA. The state capital is OKLAHOMA CITY. (173,320 sq km/66,919 sq miles; pop. 3,205,200)

Oklahoma City the state capital of OKLAHOMA. (Pop. city 454,000/metropolitan area 984,000)

Olympia (1) the original site of the Olympic Games, centring upon a temple to Zeus, on the PELOPONNESE in south-western GREECE. **(2)** a port and the state capital of WASHINGTON, on the west coast of the USA. (Pop. city 36,800)

Olympus, Mount a group of mountains in central mainland GREECE, the home of the gods of ancient Greek myth. The highest peak is Mytikas (2917 m/9570 ft).

Omaha a city in eastern NEBRASKA, USA. (Pop. city 339,700)

Oman situated in the south-east of the Arabian peninsula, Oman is a small country in two parts. It comprises a small mountainous area, overlooking the Strait of HORMUZ, which controls the entrance to The GULF, and the main part of the country, consisting of barren hills rising sharply behind a narrow coastal plain. Inland the hills extend into the unexplored Rub' al Khali (The Empty Quarter) in SAUDI ARABIA. Oman has a desert climate with exceptionally hot and humid conditions from April to October. As a result of the extremely arid environment, less than 1% of the country is cultivated, the main produce being dates. The economy is almost entirely dependent on oil, providing 90% of its exports. Over 15% of the resident population is made up by foreign workers. Oman has some deposits of copper and there is a smelter at Sohar.

Quick facts:
Area : 212,457 sq km (82,030 sq miles)
Population : 2,252,000
Capital : Muscat (Musqat)
Form of government : Monarchy (sultanate)
Religion : Ibadi Islam, Sunni Islam
Currency : Rial Omani

Oman, Gulf of a branch of the ARABIAN SEA leading to the Strait of HORMUZ.

Omdurman a city situated across the River NILE from KHARTOUM, the capital of SUDAN. (Pop. 526,000)

Omsk an industrial city in central western SIBERIA, on the Trans-Siberian Railway. (Pop. 1,193,000)

Ontario a province of central CANADA. The capital is TORONTO. (1,068,582 sq km/412,580 sq miles; pop. 9,114,000)

Ontario, Lake the smallest and most easterly of the GREAT LAKES; it drains into the ST LAWRENCE River. (19,550 sq km/7550 sq miles)

Oporto (Porto) a port in north-west PORTUGAL, and the country's second largest city after LISBON. (Pop. 1,174,000)

Oran (Wahran) a MEDITERRANEAN port and the second largest city of ALGERIA. (Pop. 664,000)

Orange, River the longest river in southern AFRICA, rising in LESOTHO and flowing west to the ATLANTIC. (Length 2090 km/1299 miles)

Orange Free State a landlocked province in central SOUTH AFRICA, with its capital at BLOEMFONTEIN. (127,993 sq km/49,405 sq miles; pop. 1,863,000)

Oranjestad (1) the capital of ARUBA, and an important port. (Pop. 20,000) **(2)** the capital of ST EUSTATIUS, and a port. (Pop. 1200)

Oregon a state in the north-west of the USA, on the PACIFIC. The state capital is SALEM. (251,180 sq km/96,981 sq miles; pop. 2,971,600)

Orinoco, River a river in northern South AMERICA. It rises in southern VENEZUELA and flows west, then north and finally east to its delta on the ATLANTIC. It forms part of the border between COLOMBIA and Venezuela. (Length 2200 km/1370 miles)

Orissa an eastern state of INDIA. The capital is Bhubaneshwar. (155,707 sq km/60,103 sq miles; pop. 31,512,000)

Orkney Islands a group of some 90 islands off the north-east coast of SCOTLAND. The capital is Kirkwall. (976 sq km/377 sq miles; pop. 19,800)

Orlando a city in central FLORIDA, and the focus for visitors to Disney World and Cape CANAVERAL. (Pop. city 174,000)

Orléans a city in north central FRANCE, on the River LOIRE. (Pop. 243,000)

Orontes, River the river flowing through HOMS and HAMAH in SYRIA. (Length 384 km/238 miles)

Osaka a port on south HONSHU Island, and the third largest city in JAPAN after TOKYO and YOKOHAMA. (Pop. 2,589,000)

Osijek a city in eastern CROATIA, on the DRAVA River. It was formerly called Esseg. (Pop. 165,000)

Oslo the capital of NORWAY, and its main port, in the south-east of the country. From 1624 to 1925 it was called Christiania (or Kristiania). (Pop. city 714,000)

Otranto, Strait of the waterway separating the heel of ITALY from ALBANIA.

Ottawa the capital of CANADA, in eastern ONTARIO, on the OTTAWA River. (Pop. 921,000)

Ottawa, River a river of central CANADA which flows into the ST LAWRENCE at MONTREAL. (Length 1271 km/790 miles)

Ouagadougou the capital of BURKINA, situated in the centre of the country. (Pop. 634,000)

Oviedo a steel making city in northern SPAIN, capital of the province of ASTURIAS. (Pop. 202,000)

Oxford an old university city, and county town of OXFORDSHIRE, ENGLAND. (Pop. 132,800)

Oxfordshire a county in southern central ENGLAND. The county town is OXFORD. (2611 sq km/1008 sq miles; pop. 590,400)

Oxus, River *see* **Amudar'ya, River**.

Pacific Ocean the largest and deepest ocean on Earth, situated between ASIA and AUSTRALIA to the west and the AMERICAS to the east. (165,384,000 sq km/63,838,000 sq miles)

Padang a port and the capital of West SUMATRA, INDONESIA, on the west coast. (Pop. 477,000)

Padua a historic city in VENETO, north-east ITALY. (Pop. 214,000)

Pagan the ruined 11th-century capital of MYANMAR, 160 km (100 miles) south-west of MANDALAY.

Painted Desert a desert of colorful rocks in northern ARIZONA, USA. (19,400 sq km/7500 sq miles)

Pakistan lies just north of the Tropic of Cancer and has as its southern border the ARABIAN SEA. The valley of the INDUS river splits the country into a highland region in the west, and a lowland region in the east. A weak form of tropical monsoon climate occurs over most of the country and conditions in the north and west are arid. Temperatures are high everywhere in summer but winters are cold in the mountains. Most agriculture is subsistence, with wheat and rice as the main crops. Cotton is the main cash crop, but the cultivated area is restricted because of waterlogging and saline soils. Pakistan's wide range of mineral resources have not been extensively developed and industry concentrates on food processing, textiles and consumer goods.

> *Quick facts:*
> Area : 796,095 sq km (307,372 sq miles)
> Population : 143,595,000
> Capital : Islamabad
> Other major cities : Faisalabad, Hyderabad,
>   Karachi, Lahore
> Form of government : Federal Islamic Republic
> Religion : Sunni Islam, Shia Islam
> Currency : Pakistan rupee
>
>

Palau *see* **Belau**.

Palembang a port and the capital of South SUMATRA, on the south-east coast. (Pop. 1,084,000)

Palenque the ruins of one of the great Mayan cities in southern MEXICO.

Palermo the capital of SICILY, ITALY, on the north-west coast. (Pop. 697,000)

Palestine an ancient historic region on the eastern shore of the Mediterranean Sea, also known as 'The Holy Land' because of its symbolic importance for Christians, Jews and Muslims. It was part of the Ottoman Empire from the early part of the 16th century until 1917, when Palestine was captured by the British. The Balfour Declaration of 1917 increased Jewish hopes that they may be enabled to establish a Jewish state. This was realised in 1948 with the UN creation of the state of ISRAEL. This created hostility among Israel's Arab neighbours and Palestinians indigenous to the area, many of whom left. Since this time the territory has been disputed, leading to a series of wars between the Arabs and Israelis and more recently to conflict between Israeli forces and the Palestine Liberation Organization. The disputed territories are the WEST BANK, the GAZA STRIP, the GOLAN HEIGHTS, and JERUSALEM. In 1994 limited autonomy of some of these disputed areas was granted to the appointed Palestinian National Authority, and Israeli military forces began a withdrawal of the area. However, the whole peace process has been compromised by continuing the violent conflict that has erupted since the assassination of the Israeli Prime Minister, Yitzak Rabin, by Jewish extremists in 1995.

Palma (Palma de Mallorca) The capital of MAJORCA and of the BALEARIC ISLANDS. (Pop. 308,600)

Palma, La *see* **Canary Islands**.

Palm Beach a resort on an island off the east coast of FLORIDA, USA, with the manufacturing centre of West Palm Beach on the mainland opposite. (Pop. Palm Beach 9,800)

Palmyra an ancient ruined city in SYRIA. Its Biblical name is Tadmor.

Pamir a region of high plateaux in central Asia which straddles the borders of TAJIKISTAN, AFGHANISTAN and CHINA.

Pampas the flat grasslands of central ARGENTINA.

Pamplona a city in north-eastern SPAIN, famous for its bull-running festival in July. (Pop. 182,000)

Panama is located at the narrowest point in Central AMERICA. Only 58 km (36 miles) separates the CARIBBEAN SEA from the PACIFIC OCEAN at Panama, and the PANAMA CANAL which divides the country is the main routeway from the Caribbean and ATLANTIC to the Pacific. The climate is tropical with high temperatures throughout the year and only a short dry season from January to April. The country is heavily forested and very little is cultivated. Rice is the staple food. The economy is heavily dependent on the Canal and income from it is a major foreign currency earner. The country has great timber resources, and mahogany from these is an important export. Other exports are shrimps and bananas. In 1989 the country was briefly invaded by US military forces in order to depose the corrupt dictator, General Noriega.

> *Quick facts:*
> Area : 77,082 sq km (29,761 sq miles)
> Population : 2,629,000
> Capital : Panama City
> Other major cities : San Miguelito, Colón
> Form of government : Republic
> Religion : RC
> Currency : Balboa
>
>

Panama Canal a canal 64 km (40 miles) long that runs through the centre of PANAMA, linking the CARIBBEAN to the PACIFIC. It was completed in 1914.

Panama City the capital of PANAMA, situated at the Pacific end of the PANAMA CANAL. (Pop. 584,000)

Panay *see* **Visayan Islands**.

Papeete the capital of FRENCH POLYNESIA, on the north-west coast of TAHITI. (Pop. 26,000)

Papua New Guinea in the south-west PACIFIC, comprises the eastern half of the island of NEW GUINEA, together with hundreds of islands of which NEW BRITAIN, BOUGAINVILLE and New Ireland are the largest. The country has a mountainous interior surrounded by broad swampy plains. The climate is tropical with high temperatures and heavy rainfall. Subsistence farming is the main economic activity although some coffee, cocoa and coconuts are grown for cash. Timber is cut for export and fishing and fish processing industries are developing. Minerals such as copper, gold, silver and oil form the mainstay of the economy. The country still receives valuable aid from AUSTRALIA, which governed it before independence.

Quick facts:
Area : 462,840 sq km (178,703 sq miles)
Population : 4,292,000
Capital : Port Moresby
Form of government : Constitutional Monarchy
Religion : Protestantism, RC
Currency : Kina

Paracel Islands a group of islands lying some 300 km (185 miles) east of VIETNAM, owned by CHINA but claimed by Vietnam.

Paraguay, located in central South AMERICA, is a country without a coastline and is bordered by BOLIVIA, BRAZIL and ARGENTINA. The climate is tropical with abundant rain and a short dry season. The River Paraguay splits the country into the Chaco, a flat semi-arid plain on the west, and a partly forested undulating plateau on the east. Almost 95% of the population live east of the river, where crops grown on the fertile plains include cassava, sugar cane, maize, cotton and soya beans. Immediately west of the river, on the low Chaco, are huge cattle ranches which provide meat for export. The world's largest hydro-electric dam has been built at Itaipú and cheap power from this has stimulated industry. Industry includes food processing, vegetable oil refining, textiles and cement.

Quick facts:
Area : 406,752 sq km (157,047 sq miles)
Population : 4,979,000
Capital : Asunción
Other major city : Ciudad Alfredo Stroessner
Form of government : Republic
Religion : RC
Currency : Guaraní

Paraguay, River a major river of South AMERICA. It rises in BRAZIL and flows south through PARAGUAY to join the PARANA River. (Length 1920 km/1190 miles)

Paramaribo the capital, principal city and main port of SURINAME. (Pop. 201,000)

Parana, River the second largest river in South AMERICA after the AMAZON. It rises in BRAZIL and flows south to join the River PLATE. (Length 4200 km/2610 miles)

Paris the capital of FRANCE, in the north of the country, on the River SEINE. (Pop. city 2,175,000/ Greater Paris 9,319,000)

Parma a historic city in northern ITALY, in EMILIA-ROMAGNA. (Pop. 171,000)

Páros an island in the CYCLADES group, GREECE. (194 sq km/75 sq miles)

Pasadena a city in south-west CALIFORNIA, USA. (Pop. 132,600)

Pascua, Isla de *see* **Easter Island**.

Patagonia a cold desert in southern ARGENTINA and CHILE.

Patna the capital of the state of BIHAR, in north-east INDIA, on the River GANGES. (Pop. 1,099,000)

Patras a port and the main city of the PELOPONNESE, GREECE. (Pop. 154,000)

Peace River a river in western CANADA, a tributary of the Slave/MACKENZIE River, rising in BRITISH COLUMBIA. (Length 1923 km/1195 miles)

Pearl Harbor a harbor and naval base on OAHU, HAWAII; the Japanese attack on the US fleet in 1941 drew the USA into World War II.

Pécs the main city of south-west HUNGARY. (Pop. 172,000)

Peeblesshire *see* **Borders**.

Peking *see* **Beijing**.

Pelée, Mount an active volcano on Martinique, which destroyed the town of St Pierre in 1902. (1397 m/4583 ft)

Peloponnese a broad peninsula of southern Greece, joined to the northern part of the country by the isthmus of Corinth.

Pembrokeshire a county in west Wales, the county town is Haverfordwest. (1590 sq km/614 sq miles; population 114,000)

Penang a state of west Malaysia comprising Penang Island and the mainland province of Wellesley.

Pennines a range of hills that runs down the middle of northern England from the border with Scotland to the Midlands, rising to 894 m (2087 ft) at Cross Fell.

Pennsylvania a state of the north-eastern USA situated mainly in the Appalachian Mountains. The capital is Harrisburg. (117,412 sq km/45,333 sq miles; pop. 11,995,400)

Perm' an industrial port on the Kama River in the western Urals of the Russian Federation. It was known as Molotov 1940-57. (Pop. 1,108,000)

Perpignan a cathedral town in south-western France. (Pop. 158,000)

Persepolis the capital of ancient Persia, south-east of Tehran, destroyed by Alexander the Great in 330bc.

Persia *see* **Iran**.

Persian Gulf *see* **Gulf, The**.

Perth (1) the state capital of Western Australia, which includes the port of Fremantle. (Pop. 1,193,000). **(2)** a city and former capital of Scotland, 55 km (35 miles) north of Edinburgh. (Pop. 44,000)

Perthshire a former county of Scotland, now part of Perth and Kinross.

Peru is located just south of the Equator, on the Pacific coast of South America. The country has three distinct regions from west to east: the coast, the high sierra of the Andes, and the tropical jungle. The climate on the narrow coastal belt is mainly desert, while the Andes are wet, and east of the mountains is equatorial with tropical forests. Most large-scale agriculture is in the oases and fertile, irrigated river valleys that cut across the coastal desert. Sugar and cotton are the main exports. Sheep, llamas, vicuñas and alpacas are kept for wool. The fishing industry was once the largest in the world but recently the shoals have become depleted. Peru's main source of wealth is oil, but new discoveries are needed as present reserves are near exhaustion. In general, the economy has recently been damaged due to the declining value of exports, natural disasters and guerrilla warfare.

Quick facts:
Area : 1,285,216 sq km (496,235 sq miles)
Population : 25,588,000
Capital : Lima
Other major cities : Arequipa, Callao, Cuzco, Trujillo
Form of govrnment : Republic
Religion : RC
Currency : Sol

Peshawar a historic town in north-west Pakistan at the foot of the Khyber Pass. (Pop. 556,000)

Petra an ancient ruined city, carved out of pink limestone. Founded in *c.*1000bc in southern Jordan.

Petrograd *see* **St Petersburg**.

Philadelphia a port and city in south-east Pennsylvania, the fourth largest city in the USA. (Pop. city 1,552,572)

Philippines comprise a group of islands in the western Pacific which are scattered over a great area. There are four main groups, Luzon and Mindoro to the north, the Visayan Islands in the centre, Mindanao and the Sulu Archipelago in the south, and Palawan in the south-west. Manila, the capital, is on Luzon. Most of the island group is mountainous and earthquakes are common. The climate is humid with high temperatures and high rainfall. Typhoons are frequent. Rice and maize are the main subsistence crops and coconuts, sugarcane, pineapples and bananas are grown for export. Copper is a major export and there are deposits of gold, nickel and petroleum. Major industries include textiles, food processing, chemicals and electrical engineering.

Quick facts:
Area : 300,439 sq km (116,000 sq miles)
Population : 67,167,000
Capital : Manila
Other major cities : Cebu, Davao, Quezon City
Form of government : Republic
Religions : Sunni Islam,
  RC, Animism
Currency : Philippine peso

Phnom Penh the capital of Cambodia, in the south of the country. (Pop. 900,000)

Phoenix the state capital of ARIZONA, USA. (Pop. city 1,012,230)

Piedmont (Piemonte) a region of north-west ITALY. The main town is TURIN.

Pierre the capital of SOUTH DAKOTA, USA. (Pop. 12,900)

Pietermaritzburg a city in eastern SOUTH AFRICA and capital of NATAL. (Pop. 229,000)

Pigs, Bay of (Bahia de Cochinos) a bay on the south coast of CUBA where exiled Cubans, backed by the USA, made a disastrous invasion attempt in 1961.

Pilsen *see* **Plzen**.

Piraeus the main port of GREECE, close to ATHENS, on the AEGEAN SEA. (Pop. 170,000)

Pisa a city in north-west ITALY on the River ARNO, famous for its leaning bell tower. (Pop. 104,300)

Pitcairn Island an island and British colony in the south PACIFIC, where mutineers from H.M.S. *Bounty* settled (after 1790).

Pittsburgh an industrial city in western PENNSYLVANIA, USA. (Pop. city 366,800/metropolitan area 2,406,000)

Plate, River (Rio de la Plata) the huge estuary of the PARANA and URUGUAY Rivers in south-east South AMERICA, with URUGUAY to the north and ARGENTINA to the south.

Plenty, Bay of the broad inlet on the north coast of the North Island of NEW ZEALAND.

Plovdiv a major market town in BULGARIA. (Pop. 379,000)

Plymouth (1) a port and naval base in south-west ENGLAND and the place from which the Pilgrim Fathers set sail in the *Mayflower* in 1620. (Pop. 255,800). **(2)** the capital of the island of MONTSERRAT. (Pop. 4,000) **(3)** a town in MASSACHUSETTS, which has grown from the first European settlement in NEW ENGLAND, established by the Pilgrim Fathers of the *Mayflower*. (Pop. 37,100)

Plzen (Pilsen) an industrial city in western BOHEMIA, CZECH REPUBLIC. Pilsner lager beer was first produced here in 1842. (Pop. 175,000)

Po, River the longest river in ITALY, flowing across the north of the country from the ALPS across a fertile plain to the ADRIATIC SEA. (Length 642 km/405 miles)

Pohnpei (Ponape) the island on which KOLONIA, the capital of the Federated States of MICRONESIA, stands.

Pointe-à-Pitre the main port of GUADELOUPE. (Pop. 23,000)

Poitiers a historic university city in south central FRANCE (Pop. 82,500).

Poland is situated on the North European Plain. It borders GERMANY to the west, the CZECH REPUBLIC and SLOVAKIA to the south and BELARUS and UKRAINE to the east. Poland consists mainly of lowlands and the climate is continental, marked by long severe winters and short warm summers. Over one-quarter of the labour force is involved in, predominantly small-scale, agriculture. The main crops are potatoes, wheat, barley, sugar beet and fodder crops. The industrial sector of the economy is large scale. Poland has large deposits of coal and reserves of natural gas, copper and silver. Vast forests stretching inland from the coast supply the paper and furniture industries. Other industries include food processing, engineering and chemicals.

Quick facts:
Area : 312,677 sq km (120,725 sq miles)
Population : 38,587,000
Capital : Warsaw (Warszawa)
Other major cities : Gdansk, Kraków, Lódz, Wroclow
Form of government : Republic
Religion : RC
Currency : Zloty

Polynesia the largest of the three island divisions of the PACIFIC, the others being MICRONESIA and MELANESIA. The group includes SAMOA, the COOK, SOCIETY and MARQUESAS Islands, and TONGA.

Pomerania a region of north-west POLAND, on the BALTIC coast.

Pompeii an ancient city near NAPLES which was smothered by ash from an eruption of VESUVIUS in AD79.

Ponape *see* **Pohnpei**.

Pondicherry the former capital of French INDIA, in the south-east of the country, founded in 1683 and governed by France until 1954. (Pop. 401,000)

Ponta Delgada a port and the capital of the AZORES, on SÃO MIGUEL Island. (Pop. 22,200)

Poona (Pune) a historic and industrial city east of BOMBAY, in western INDIA. (Pop. 2,485,000)

Popocatépetl a volcano, twinned with IXTACIHUATL, 65 km (40 miles) south-east of MEXICO CITY. (5452 m/17,887 ft)

Port-au-Prince the main port and capital of HAITI. (Pop. 1,402,000)

Port Elizabeth a port and industrial city in Eastern Cape, SOUTH AFRICA. (Pop. 853,000)

Port Harcourt the second port of NIGERIA after LAGOS. (Pop. 371,000)

Port Jackson the great natural harbor also called Sydney Harbour, in south-east AUSTRALIA.

Portland (1) a port on the ATLANTIC coast of the USA, in MAINE. (Pop. city 62,800). **(2)** a port on the Williamette River in OREGON, USA. (Pop. city 445,500/metropolitan area 1,897,000)

Port Louis the capital and main port of MAURITIUS, on the east coast of the island. (Pop. 144,000)

Port Moresby the capital and main port of PAPUA NEW GUINEA, in the south-east. (Pop. 174,000)

Porto *see* **Oporto**.

Porto Alegre a port and regional capital of southern BRAZIL. (Pop. city 1,263,000)

Port of Spain the capital and chief port of TRINIDAD AND TOBAGO. (Pop. city 60,000)

Porto Novo the administrative capital of BENIN. (Pop. 179,000)

Port Said the port at the MEDITERRANEAN end of the SUEZ CANAL, EGYPT. (Pop. 460,000)

Portsmouth a port and major naval base in southern ENGLAND. (Pop. 189,300)

Port Stanley the capital of the FALKLAND ISLANDS. (Pop. 2,000)

Portugal, in the south-west corner of EUROPE, makes up about 15% of the Iberian peninsula. The most mountainous areas of Portugal lie to the north of the river TAGUS. In the north-east are the steep sided mountains of Tras-os-Montes and to south of this the DOURO valley, running from the Spanish border to OPORTO, on the ATLANTIC coast. South of the Tagus river is the Alentajo, with its wheat fields and cork plantations, and this continues to the hinterland of the ALGARVE with its beautiful groves of almond, fig and olive trees. Agriculture employs one quarter of the labour force, and crops include wheat, maize, grapes and tomatoes. Manufacturing industry includes textiles and clothing, footwear, food processing and cork products. Tourism, particularly in the south, is the main foreign currency earner.

Quick facts:
Area : 92,389 sq km (35,671 sq miles)
Population : 10,600,000
Capital : Lisbon (Lisboa)
Other major cities : Braga, Coimbra, Oporto, Setúbal
Form of government : Republic
Religion : RC
Currency : Escudo

Port-Vila the capital and chief port of VANUATU. (Pop. 20,000)

Posen *see* **Poznan**.

Potsdam a city just 25 km (16 miles) south-west of BERLIN, GERMANY. (Pop. 139,000)

Powys a county in mid-WALES created in 1974 out of Breconshire, Montgomeryshire, and Radnorshire. Since recent local government reorganistion in Wales, it now includes the soem communities from the former county of Clwyd. The administrative centre is Llandrindod Wells. (5,200 sq km/2007 sq miles; pop. 122,000)

Poznan (Posen) a historic city in central western POLAND. (Pop. 590,000)

Prague (Praha) the capital and principal city of the CZECH REPUBLIC, situated on the Vltava River. (Pop. 1,216,000)

Praia the capital of CAPE VERDE. (Pop. 69,000)

Prairies *see* **Great Plains**.

Pressburg *see* **Bratislava**.

Pretoria the administrative capital of SOUTH AFRICA, 48 km (30 miles) north of JOHANNESBURG in Pretoria-Witwatersrand-Vereeniging. (Pop. 1,080,000)

Prince Edward Island the smallest of the provinces of CANADA, an island in the Gulf of ST LAWRENCE. The provincial capital is Charlottetown. (5660 sq km/2185 sq miles; pop. 129,900)

Principe *see* **São Tomé and Principe**.

Pristina the capital of the autonomous province of KOSOVO in SERBIA. (Pop. 216,000)

Provence a historical region of coastal south-east FRANCE.

Providence a port, and the state capital of RHODE ISLAND, USA. (Pop. city 155,418)

Prussia a historical state of GERMANY, centring on its capital, BERLIN.

Puebla a major city 120 km (75 miles) south-east of MEXICO CITY, and the capital of a state of the same name. (Pop. city 1,055,000)

Puerto Rico is the most easterly of the Greater ANTILLES and lies in the CARIBBEAN between the DOMINICAN REPUBLIC and the US VIRGIN ISLANDS. It is a self-governing commonwealth in association with the USA and includes the main island, Puerto Rico, the two small islands of Vieques and Culebra and a fringe of smaller uninhabited islands. The climate is tropical, modified slightly by cooling sea breezes. The main mountains on Puerto Rico are the Cordillera Central, which reach 1338 m (4390 ft) at the peak of Cerro de Punta. Dairy farming is the most important agricultural activity but the whole agricultural sector has been overtaken by industry in recent years.

Tax relief and cheap labour encourages American businesses to be based in Puerto Rico. Products include textiles, clothing, electrical and electronic goods, plastics and chemicals. Tourism is another developing industry.

Quick facts:
Area : 8897 sq km (3435 sq miles)
Population : 3,689,000
Capital : San Juan
Form of government : Self-governing
  Commonwealth (USA)
Relgion : RC, Protestantism
Currency : US dollar

and the reserves of natural gas are enormous. In order to diversify the economy, new industries such as iron and steel, cement, fertilizers, and petrochemical plants have been developed.

Quick facts:
Area : 11,000 sq km (4247 sq miles)
Population : 594,000
Capital : Doha (Ad Dawhah)
Form of government : Monarchy
Religions : Wahhabi Sunni Islam
Currency : Qatari riyal

Puglia (Apulia) a region of south-east ITALY. The regional capital is BARI. (19,250 sq km/7500 sq miles; pop. 4,081,500)

Pune *see* **Poona**.

Punjab (1) a state in north-western INDIA. The capital is CHANDIGARH. (50,362 sq km/19,440 sq miles; pop. 20,190,800) **(2)** a fertile province in the north of PAKISTAN. The capital is LAHORE.(205,344 sq km/79,283 sq miles; pop. 47,292,000)

Pusan a major port, and the second largest city in SOUTH KOREA after SEOUL. (Pop. 3,798,000)

Putumayo, River a river of north-west South AMERICA, rising in the ANDES and flowing south-east to join the AMAZON. (Length 1900 km/1180 miles)

P'yongyang (Pyeongyang) an industrial city and the capital of North KOREA. (Pop. 2,639,000)

Pyrénées a range of mountains that runs from the Bay of BISCAY to the MEDITERRANEAN, along the border between FRANCE and SPAIN. The highest point is Pico d'Aneto (3404 m/11,170 ft).

Qacentina *see* **Constantine**.

Qatar is a little emirate which lies halfway along the coast of The GULF. It consists of a low barren peninsula and a few small islands. The climate is hot and uncomfortably humid in summer and the winters are mild with rain in the north. Most fresh water comes from natural springs and wells or from desalination plants. The herding of sheep, goats and some cattle is carried out and the country is famous for its high quality camels. The discovery and exploitation of oil has resulted in a high standard of living for the people of Qatar. The Dukhan oil field has an expected life of forty years

Qazvin (Kasvin) a historic town in north-west IRAN. (Pop. 279,000)

Qingdao a city in SHANGDONG province in north-eastern CHINA. (Pop. 1,300,000)

Qinghai a province of north-western CHINA. The capital is XINING. (720,000 sq km/280,000 sq miles; pop. 4,457,000)

Qiqihar a manufacturing city in HEILONGJIANG province, CHINA. (Pop. 1,070,000)

Qom (Qum) a holy city in central northern IRAN. (Pop. 681,000)

Québec the largest province of CANADA, in the east of the country, and also the name of the capital of the province. The majority of the population are French-speaking. (1,358,000 sq km/524,300 sq miles; pop. province 6,811,800/city 646,000)

Queen Charlotte Islands a group of some 150 islands lying 160 km (100 miles) off the west coast of CANADA. (9790 sq km/3780 sq miles; pop. 2,500)

Queen Charlotte Strait a waterway, some 26 km (16 miles) wide, between the north-eastern coast of VANCOUVER ISLAND and mainland CANADA.

Queensland the north-eastern state of AUSTRALIA. The state capital is BRISBANE. (1,272,200 sq km/491,200 sq miles; pop. 2,650,000)

Quercy a former province of south-western FRANCE, around Cahors.

Quetta the capital of the province of BALUCHISTAN, PAKISTAN. (Pop. 286,000)

Quezon City a major city and university town, now a part of Metro MANILA, and the administrative capital of the PHILIPPINES from 1948 to 1976. (Pop. 1,667,000)

Quito the capital of ECUADOR, lying just south of the Equator, 2850 m (9350 ft) high in the ANDES. (Pop. 1,101,000)

Qum *see* **Qom**.

Rabat the capital of MOROCCO, in the north-west, on the ATLANTIC coast. (Pop. incl. Salé) 1,545,000)

Radnorshire *see* **Powys**.

Ragusa *see* **Dubrovnik**.

Rainier, Mount *see* **Cascade Range**.

Rajasthan a state of north-west INDIA. The state capital is JAIPUR. (342,239 sq km/132,104 sq miles; pop. 43,880,600)

Raleigh the state capital of NORTH CAROLINA, USA. (Pop. city 222,500)

Ranchi an industrial town in the state of BIHAR, INDIA. (Pop. 614,000)

Rand, The *see* **Witwatersrand**.

Rangoon (Yangon) the capital of MYANMAR, and an important port on the mouth of the Rangoon River. (Pop. 2,513,000)

Rarotonga the largest of the COOK ISLANDS, with the capital of the islands, Avarua, on its north coast. (67 sq km/26 sq miles)

Ras al Khaimah one of the UNITED ARAB EMIRATES, in the extreme north-east, on the MUSANDAM peninsula. (1700 sq km/660 sq miles; pop. 116,000)

Ravenna a city in north-eastern ITALY, noted for its Byzantine churches. (Pop. 136,000)

Rawalpindi a military town of ancient origins in northern PAKISTAN. (Pop. 795,000)

Recife a city and regional capital on the eastern tip of Brazil. (Pop. 1,290,000)

Red River (1) a river of the southern USA, rising in TEXAS and flowing east to join the MISSISSIPPI. (Length 1639km/ 1018miles). **(2)** **(Song Hong; Yuan Jiang)** a river that rises in south-west CHINA and flows south-east across the north of VIETNAM to the Gulf of TONGKING. (Length 800 km/500 miles)

Red Sea a long, narrow sea lying between the Arabian Peninsula and the coast of north-east AFRICA. In the north it is connected to the MEDITERRANEAN SEA by the SUEZ CANAL.

Reggio di Calabria a port on the toe of southern ITALY. (Pop. 178,000)

Reggio nell'Emilia a town of Roman origins in north-eastern ITALY. (Pop. 133,000)

Regina the capital of the province of SASKATCHEWAN, CANADA. (Pop. 192,000)

Reims *see* **Rheims**.

Renfrewshire a county in south-west SCOTLAND. In 1975 it became part of STRATHCLYDE region. Since the local authority reorganization of 1996, the two local authorities of Renfrewshire, and East Renfrewshire have been created. (Renfrewshire: 261 sq km/101 sq miles; pop. 177,000. East Renfrewshire: 172 sq km/67 sq miles; pop. 86,800)

Rennes an industrial city in north-eastern France. (Pop. 245,000)

Reno a gambling centre in NEVADA, USA. (Pop. city 140,000)

Réunion an island to the east of MADAGASCAR, an overseas department of France. The capital is SAINT-DENIS. (2515 sq km/970 sq miles; pop. 655,000)

Reykjavik the capital and main port of ICELAND, on the south-west coast. (Pop. 103,000)

Reynosa a town in north-eastern MEXICO, on the border with the USA. (Pop. 281,000)

Rheims (Reims) a historic city in FRANCE, and the centre of champagne production. (Pop. 206,000)

Rhine (Rhein, Rhin, Rijn), River one of the most important rivers of EUROPE. It rises in the Swiss ALPS, flows north through GERMANY and then west through the NETHERLANDS to the NORTH SEA. (Length 1320km/ 825miles)

Rhode Island the smallest state in the USA. The state capital is PROVIDENCE. (3144 sq km/1214 sq miles; pop. 1,001,300)

Rhodes (Rodhos) the largest of the DODECANESE group of islands belonging to GREECE. (1399 sq km/540 sq miles; pop. 88,500)

Rhône, River a major river of EUROPE, rising in the Swiss ALPS and flowing west into FRANCE, and then south to its delta on the Golfe de LION. (Length 812 km/505 miles)

Richmond the state capital of VIRGINIA. (Pop. city 202,300/metropolitan area 796,100)

Ridings, The *see* **Yorkshire**.

Riga a BALTIC port, and the capital of LATVIA. (Pop. 840,000)

Rijeka (Fiume) a port on the ADRIATIC, in CROATIA. (Pop. 206,000)

Rijn, River *see* **Rhine, River**.

Rimini a popular resort on the ADRIATIC SEA, north-eastern ITALY. (Pop. 130,000)

Rio Bravo *see* **Rio Grande**.

Rio de Janeiro a major port and former capital (1763–1960) of BRAZIL, situated in the south-east of the country. (Pop. 5,336,000)

Rio Grande (Rio Bravo) a river of North AMERICA, rising in the state of COLORADO, USA, and flowing south-east to the Gulf of MEXICO. For much of its length it forms the border between the USA and MEXICO. (Length 3078 km/1885 miles)

Rioja, La an autonomous area in the south of the BASQUE region of SPAIN, famous for its fine wine. (Pop. 263,000)

Riyadh the capital and commercial centre of SAUDI ARABIA, founded on an oasis. (Pop. 2,000,000)

Road Town the capital of the British VIRGIN IS-LANDS. (Pop. 6000)

Roca, Cabo da a cape sticking out into the ATLAN-TIC in central PORTUGAL, to the west of LISBON, the western-most point of mainland EUROPE.

Rockall a tiny, rocky, uninhabited island lying 400 km (250 miles) west of IRELAND, and claimed by the UK.

Rocky Mountains (Rockies) a huge mountain range in western North AMERICA, extending some 4800 km (3000 miles) from BRITISH COLUMBIA in CANADA to NEW MEXICO in the USA.

Rodhos *see* **Rhodes**.

Romania apart from a small extension towards the BLACK SEA, Romania is almost a circular country. It is located in south-east EUROPE and bordered by UKRAINE, HUNGARY, SERBIA and BULGARIA. The CARPATHIAN MOUNTAINS run through the north, east and centre of Romania and these are enclosed by a ring of rich agricultural plains which are flat in the south and west but hilly in the east. The core of Romania is Transylvania within the Carpathian arc. Romania has cold snowy winters and hot summers. Agriculture in Romania has been neglected in favor of industry but major crops include maize, sugar beet, wheat, potatoes and grapes for wine. There are now severe food shortages. Industry is state owned and includes mining, metallurgy, mechanical engineering and chemicals. Forests support timber and furniture making industries in the Carpathians.

Quick facts:
Area : 237,500 sq km (91,699 sq miles)
Population 22,863,000
Capital : Bucharest (Bucuresti)
Other major cities : Brasov, Constanta, Timisoara
Form of government : Republic
Religions : Romanian Orthodox, RC
Currency : Leu

Rome (Roma) the historic capital of ITALY, on the River TIBER, in the centre of the country near the west coast. (Pop. 2,723,000)

Rosario an industrial and commercial city on the River PARANA in ARGENTINA. (Pop. 1,096,000)

Roscommon a county in the north-west of the Republic of IRELAND, with a county town of the same name. (2462 sq km/950 sq miles; pop. county 51,900)

Roseau the capital of DOMINICA. (Pop. 21,000)

Ross and Cromarty a former county of north-west SCOTLAND which includes many islands; part of the HIGHLAND region.

Ross Sea a large branch of the ANTARCTIC OCEAN, south of NEW ZEALAND.

Rostock a major port on the BALTIC coast of GER-MANY. (Pop. 237,000)

Rostov-na-Donau (Rostov-on-Don) a major industrial city on the River DON, near the north-western extremity of the Sea of AZOV in south-eastern RUSSIAN FEDERATION. (Pop. 1,004,000)

Rotterdam the largest city in the NETHERLANDS and the busiest port in the world. (Pop. city 582,200/ Greater Rotterdam 1,069,000)

Roubaix *see* **Lille-Roubaix-Turcoing**.

Rouen a port on the River SEINE in northern FRANCE. (Pop. 380,000)

Rousillon *see* **Languedoc-Rousillon**.

Roxburghshire *see* **Borders**.

RSFSR *see* **Russian Soviet Federated Socialist Republic**.

Rub al-Khali the so-called 'Empty Quarter', a vast area of sandy desert straddling the borders of SAUDI ARABIA, OMAN and YEMEN. (650,000 sq km/ 251,000 sq miles)

Ruhr, River the river in north-western GERMANY whose valley forms the industrial heartland of western Germany. It joins the RHINE at DUISBURG. (Length 235 km/146 miles)

Rushmore, Mount a mountain in the Black Hills of SOUTH DAKOTA, USA, noted for the huge heads of four US presidents (Washington, Jefferson, Lincoln, Theodore Roosevelt) which were carved into its flank 1927-41. (170 m/5600 ft)

Russia the old name for the Russian Empire, latterly used loosely to refer to the former USSR or the RUSSIAN FEDERATION.

Russian Federation, The, which is the largest country in the world, extends from Eastern EUROPE through the URAL Mountains east to the PACIFIC OCEAN. The CAUCASUS Range forms its boundary with GEORGIA and AZERBAIJAN, and it is here that the highest peak in Europe, Mt ELBRUS, is located. In the east, SIBERIA is drained toward the ARCTIC OCEAN by the great rivers Ob', Yenisey, LENA and their tributaries. Just to the south of the Central Siberian Plateau lies Lake BAIKAL, the world's deepest freshwater lake. The environment ranges from vast frozen wastes in the north to subtropical deserts in the south. Agriculture is organized into either state or collective farms, which mainly produce sugar beet, cotton, potatoes and vegetables. The country has extensive reserves of coal, oil, gas, iron ore and manganese. Major industries include iron and steel, cement, transport equipment, engineering, armaments, electronic equipment and chemicals. The Russian Federation declared itself independent in 1991.

Quick facts:
Area : 17,075,400 sq km (6,592,800 sq miles)
Population : 148,385,000
Capital : Moscow (Moskva)
Other major cities : St Petersburg (formerly Leningrad), Nizhniy Novgorod, Novosibirsk
Form of government : Republic
Religions : Russian Orthodox, Sunni Islam, Shia Islam, RC
Currency : Rouble

Russian Soviet Federated Socialist Republic (RSFSR) the former name for the RUSSIAN FEDERATION.

Rutanzige, Lake *see* **Edward, Lake.**

Rutland once the smallest county of ENGLAND, now a part of LEICESTERSHIRE.

Ruwenzori a mountain range on the border between ZAÏRE and UGANDA, also known as the Mountains of the Moon. The highest peak is Mount Ngaliema (Mount Stanley) (5109 m/16,763 ft).

Rwanda is a small republic in the heart of central AFRICA which lies just 2° south of the Equator. It is a mountainous country with a central spine of highlands from which streams flow west to the ZAÏRE river and east to the NILE. Active volcanoes are found in the north where the land rises to about 4500 m (14,765 ft). The climate is highland tropical with temperatures decreasing with altitude. The soils are not fertile and subsistence agriculture dominates the economy. Staple food crops are sweet potatoes, cassava, dry beans, sorghum and potatoes. The main cash crops are coffee, tea and pyrethrum. There are major reserves of natural gas under Lake Kivu in the west, but these are largely unexploited. In 1994, half a million refugees fled to escape civil war in which an estimated 200,000 people were killed. In July 1994, victory in the civil war was claimed by the Tutsi-dominated rebel Rwandan Patriotic Front, who announced the formation of a coalition government, headed by Faustin Twagiramungu, a Hutu. Although the Rwandan Patriotic Frontgave its assurance that there would be no reprisal against the defeated Hutus, and the refugees returned home, the situation remains volatile.

Quick facts:
Area : 26,338 sq km (10,169 sq miles)
Population : 7,899,000
Capital : Kigali
Form of government : Republic
Religions : RC, Animism
Currency : Rwanda franc

Ryazan an industrial city 175 km (110 miles) southeast of MOSCOW, RUSSIAN FEDERATION. (Pop. 533,000)

Ryukyu Islands (Nansei-shoto) a chain of islands belonging to JAPAN stretching 1200 km (750 miles) between Japan and TAIWAN. (Pop. 1,179,000)

Saarbrücken an industrial city of western GERMANY, near the border with France. (Pop. 191,000)

Sabah the more easterly of the two states of MALAYSIA on northern coast of the island of BORNEO. (80,429 sq km/29,353 sq miles; pop. 1,470,000)

Sacramento the state capital of CALIFORNIA. (Pop. city 382,800/metropolitan area 1,563,000)

Sacramento, River the longest river in CALIFORNIA, USA. (Length: 560 km/350 miles)

Sahara Desert the world's largest desert, spanning much of northern AFRICA, from the ATLANTIC to the RED SEA, and from the MEDITERRANEAN to MALI, NIGER, CHAD and SUDAN.

Sahel a semi-arid belt crossing AFRICA from SENEGAL to SUDAN, separating the SAHARA from tropical Africa to the south.

Saigon *see* **Ho Chi Minh City.**

St Barthélémy a small island dependency of GUADELOUPE. (Pop. 5,000)

St Christopher (St Kitts) and Nevis the islands of St Christopher (popularly known as St Kitts) and Nevis lie in the LEEWARD group in the eastern CARIBBEAN. In 1983 it became a sovereign democratic federal state with Elizabeth II as head of state. St Kitts consists of three extinct volcanoes linked by a sandy isthmus to other volcanic remains in the south. Around most of the island sugar cane is grown on fertile soil covering the gentle slopes. Sugar is the chief export crop but market gardening and livestock are being expanded on the steeper slopes above the cane fields. Industry includes sugar processing, brewing, distilling and bottling. St Kitts has a major tourist development at Frigate Bay. Nevis, 3 km (2 miles) south, is an extinct volcano. Farming is declining and tourism is now the main source of income.

Quick facts:
Area : 262 sq km (101 sq miles)
Population : 45,000
Capital : Basseterre
Form of government : Constitutional
  Monarchy
Religions : Anglicanism, Methodism
Currency : East Caribbean dollar

St Croix the largest of the US Virgin Islands. The main town is Christiansted. (218 sq km/84 sq miles)

Saint-Denis the capital of Réunion island. (Pop. 123,000)

St-Etienne an industrial city 50 km (30 miles) south-west of Lyons, France. (Pop. 313,000)

St Eustatius (Statia) an island of the Netherlands Antilles, in the Leeward Islands. The capital is Oranjestad. (Pop. 1100)

St George's the capital of Grenada, and the island's main port. (Pop. city 7,000)

St Helena a remote island and British colony in the South Atlantic. Napoleon Bonaparte was exiled here by the British from 1815 until his death in 1821. (122 sq km/47 sq miles; pop. 5600)

St Helena Dependencies the islands of Ascension and Tristan da Cunha are so-called dependencies of St Helena, a British colony.

St Helens, Mount an active volcano in the Cascade Range of western Washington State, USA. It erupted in 1980, causing widespread destruction. (2549 m/8364 ft)

Saint John a port at the mouth of the Saint John River, on the Atlantic coast of New Brunswick, Canada. (Pop. 172,000)

Saint John River a river of the eastern USA, which rises in Maine and flows north-west through Canada to the Bay of Fundy. (Length 673 km/418 miles)

St John the smallest of the main islands of the US Virgin Islands. (52 sq km/20 sq miles)

St John's (1) the capital and main port of Antigua. (Pop. 38,000). **(2)** a port and the capital of Newfoundland, Canada. (Pop. 172,000)

St Kitts and Nevis *see* **St Christopher and Nevis**.

St Lawrence, Gulf of an arm of the Atlantic Ocean in north-eastern Canada, into which the St Lawrence River flows.

St Lawrence, River a commercially important river of south-east Canada, which flows north-east from Lake Ontario to the Gulf of St Lawrence, forming part of the border between Canada and the USA. (Length 1197 km/744 miles)

St Lawrence Seaway a navigable waterway that links the Great Lakes, via the St Lawrence River, to the Atlantic Ocean.

St Louis a city in eastern Missouri, USA, on the River Mississippi. (Pop. city 383,700/metropolitan area 2,519,000)

St Lucia is one of the Windward Islands in the eastern Caribbean. It lies to the south of Martinique and to the north of St. Vincent. It was controlled alternately by the French and the British for some two hundred years before becoming fully independent in 1979. St Lucia is an island of extinct volcanoes and the highest peak is 950 m (3117 ft). In the west are the peaks of Pitons which rise directly from the sea to over 750 m (2461 ft). The climate is wet tropical with a dry season from January to April. The economy depends on the production of bananas and, to a lesser extent, coconuts. Production, however, is often affected by hurricanes, drought and disease. Tourism is becoming an important industry and Castries, the capital, is a popular calling point for cruise liners.

Quick facts:
Area : 622 sq km (240 sq miles)
Population : 147,000
Capital : Castries
Form of Government : Constitutional
  Monarchy
Religion : RC
Currency : East Caribbean dollar

St Martin one of the LEEWARD ISLANDS, in the southeastern CARIBBEAN, divided politically into two, one a part of GUADELOUPE (France); the other (Sint Maarten) a part of the NETHERLANDS ANTILLES. The capital of the French side is Marigot; of the Dutch side Philipsburg. (54 sq km/21 sq miles; pop. 28,500)

St Paul the state capital of MINNESOTA, twinned with the adjoining city of MINNEAPOLIS. (Pop. city 268,300/ metropolitan area 2,618,000)

St Petersburg (Sankt Peterburg) a former capital of RUSSIA and the current RUSSIAN FEDERATION second-largest city. It is an industrial city, important cultural centre and major port on the BALTIC SEA. From 1914-24 it was known as Petrograd; then, until 1991, Leningrad. (Pop. 5,004,000)

St Pierre and Miquelon two islands to the south of NEWFOUNDLAND, CANADA, which are an overseas territory administered by FRANCE. (240 sq km/93 sq miles; pop. 6300)

St Thomas the principal tourist island of the US VIRGIN ISLANDS. The capital is CHARLOTTE AMALIE. (83 sq km/32 sq miles)

St Vincent, Cape *see* **São Vincente, Cabo de.**

St Vincent and the Grenadines St Vincent is an island of the Lesser Antilles, situated in the eastern CARIBBEAN between ST LUCIA and GRENADA. It is separated from Grenada by a chain of some 600 small islands known as the Grenadines, the northern islands of which form the other part of the country. The largest of these islands are Bequia, MUSTIQUE, Canouan, Mayreau and Union. The climate is tropical, with very heavy rain in the mountains. St Vincent Island is mountainous and a chain of volcanoes runs up the middle of the island. The volcano, Soufrière (1234 m/4049 ft), is active and it last erupted in 1979. Farming is the main occupation on the island. Bananas for the UK are the main export, and it is the world's leading producer of arrowroot starch. There is little manufacturing and the government is trying to promote tourism.

Quick facts:
Area : 388 sq km (150 sq miles)
Population : 112,000
Capital : Kingstown
Form of government : Constitutional Monarchy
Religions : Anglicanism, Methodism, RC
Currency : E. Caribbean dollar

Saipan the largest and most heavily populated of the NORTHERN MARIANAS. The island group's capital, SUSUPE, is on the western side. (122 sq km/47 sq miles; pop. 39,000)

Sakhalin a large island to the north of JAPAN, but belonging to the RUSSIAN FEDERATION. (76,400 sq km/29,500 sq miles; pop. 650,000)

Salamanca an elegant university town in western SPAIN, and the name of the surrounding province. (Pop. town 167,000)

Salem (1) a city in MASSACHUSETTS, USA. (Pop. city 38,100). **(2)** the state capital of OREGON, USA. (Pop. city 112,100)

Salonika (Thessaloníki) the second largest city in GREECE after ATHENS. (Pop. metropolitan 706,000)

Salt Lake City the state capital of UTAH. (Pop. city 165,800/metropolitan area 1,128,000)

Salvador a port on the central east coast of BRAZIL and capital of the state of Bahia. (Pop. 2,056,000)

Salvador, El *see* **El Salvador.**

Salween, River a river rising in TIBET and flowing south through MYANMAR, forming part of the border with THAILAND, to the ANDAMAN SEA. (Length 2900 km/1800 miles)

Salzburg a city in central northern AUSTRIA, and the name of the surrounding state, of which it is the capital. (Pop. city 144,000)

Samar the third largest island of the PHILIPPINES. (13,080 sq km/5050 sq miles; pop. 1,271,000)

Samara a major industrial city and port on the River Volga in the RUSSIAN FEDERATION. Founded in 1586, it was renamed Kuybyshev, in 1935, after the Revolutionary leader Valerian Kuybyshev. Its name reverted in 1991. (Pop. 1,271,000)

Samarkand an ancient city in eastern UZBEKISTAN. It was the main junction of the 'Silk Road' between CHINA and the Mediterranean, and in the 14th century was the centre of the empire built up by the great Mongol conqueror, Timur. (Pop. 370,000)

Samoa, American an American territory, comprising a group of five islands, in the central South PACIFIC. The capital is PAGO PAGO. (197 sq km/76 sq miles; pop. 46,800)

Sámos a GREEK island 2 km (1 mile) off the coast of TURKEY. (Pop. 31,600)

Samsun a port on the BLACK SEA coast of TURKEY, and the name of the surrounding province, of which it is the capital. (Pop. town 304,000)

San'a the capital of YEMEN, situated in the middle of the country. (Pop. 427,000)

San Antonio an industrial centre in southern TEXAS, USA. (Pop. city 966,400/metropolitan area 1,379,000)

San Diego a major port and industrial city in southern CALIFORNIA, USA. (Pop. city 1,148,900/metropolitan area 2,601,000)

San Francisco a PACIFIC port and commercial centre in CALIFORNIA. (Pop. city 728,900/metropolitan area 6,410,600)

San Francisco Bay an inlet of the PACIFIC OCEAN in western CALIFORNIA, USA, joined to the ocean by the Golden Gate Strait.

San José the capital of COSTA RICA, in the centre of the country. (Pop. 303,000).

San Jose a city in CALIFORNIA, USA, and the focus of 'Silicon Valley.' (Pop. city 810,300)

San Juan the capital of PUERTO RICO, and a major port. (Pop. 1,816,000)

Sankt Peterburg *see* **St Petersburg**.

San Lorenzo del Escorial *see* **Escorial, El**.

San Luis Potosi an elegant colonial city and provincial capital in north-central MEXICO. (Pop. city 526,000)

San Marino is a tiny landlocked state in central ITALY, lying in the eastern foothills of the APENNINES. It has wooded mountains and pasture land clustered around the limestone peaks of Monte Titano which rises to 739 m (2425 ft). San Marino has a mild MEDITERRANEAN climate. The majority of the population work on the land or in forestry. Wheat, barley, maize and vines are grown, and the main exports are wood machinery, chemicals, wine, textiles, tiles, varnishes and ceramics. Some 3.5 million tourists visit the country each year, and much of the country's revenue comes from the sale of stamps, postcards, souvenirs and duty-free liquor. Italian currency is in general use but San Marino issues its own coins.

Quick facts:
Area : 61 sq km (24 sq miles)
Population : 26,000
Capital : San Marino
Form of government : Republic
Religion : RC
Currency : Lira

San Miguel de Tucumán a regional capital in north-western ARGENTINA. (Pop. 622,000)

San Pedro Sula the second largest city in HONDURAS. (Pop. 461,000)

San Salvador (1) the capital and major city of EL SALVADOR. (Pop. 1,522,000) **(2)** a small island in the centre of the BAHAMAS, the first place in the New World reached by Columbus (1492).

San Sebastián a port and industrial city in north-eastern SPAIN. (Pop. 178,000)

Santa Barbara a resort and industrial centre in southern CALIFORNIA, USA. (Pop. city 85,100)

Santa Fe the state capital of NEW MEXICO, USA. (Pop. city 59,000)

Santander a port and industrial city in north-eastern SPAIN. (Pop. 195,000)

Santiago the capital and principal city of CHILE. (Pop. 5,343,000)

Santiago de Compostela a university city and centre of pilgrimage in north-western SPAIN. (Pop. 105,500)

Santiago de Cuba a port and provincial capital in southern CUBA. (Pop. 419,000)

Santo Domingo the capital and main port of the DOMINICAN REPUBLIC. (Pop. 2,200,000)

Santos the largest port in BRAZIL, 60 km (38 miles) south-east of SÃO PAULO. (Pop. 429,000)

São Francisco, River a river of eastern BRAZIL, important for its hydroelectric dams. (Length 2900 km/1800 miles)

São Miguel the largest island in the AZORES. (770 sq km/298 sq miles; pop. 132,000)

Saône, River a river of eastern FRANCE which merges with the River RHÔNE at LYONS. (Length 480 km/300 miles)

São Paulo a major industrial city in south-eastern BRAZIL, and capital of the state called São Paulo. (Pop. city 9,480,000)

São Tomé the capital of São Tomé and Principe. (Pop. 36,000)

São Tomé and Príncipe volcanic islands which lie off the west coast of AFRICA. São Tomé is covered in extinct volcanic cones, reaching 2024 m (6641 ft) at the highest peak. The coastal areas are hot and humid. Príncipe is a craggy island lying to the north-east of São Tomé. The climate is tropical with heavy rainfall from October to May. 70% of the workforce work on the land, mainly in state-owned cocoa plantations. Small manufacturing industries include food processing and timber products.

Quick facts:
Area : 964 sq km (372 sq miles)
Population : 133,000
Capital : São Tomé
Form of government : Republic
Religion : RC
Currency : Dobra

São Vincente, Cabo de (Cape St Vincent) the south-western corner of PORTUGAL.

Sapporo a modern city, founded in the late 19th century as the capital of HOKKAIDO Island, JAPAN. (Pop. 1,732,000)

Saragossa *see* **Zaragoza.**

Sarajevo the capital of BOSNIA & HERZEGOVINA. (Pop. 526,000)

Saransk an industrial town, capital of the republic of Mordovia in the RUSSIAN FEDERATION. (Pop. 316,000)

Saratov an industrial city and river port on the VOLGA, RUSSIAN FEDERATION. (Pop. 916,000)

Sarawak a state of MALAYSIA occupying much of the north-western coast of BORNEO. (125,204 sq km/48,342 sq miles; pop. 1,669,000)

Sardinia (Sardegna) the second largest island of the MEDITERRANEAN after SICILY, also belonging to ITALY, lying just south of CORSICA. The capital is CAGLIARI. (24,089 sq km/9301 sq miles; pop. 1,645,000)

Sargasso Sea an area of calm water in the ATLANTIC between the WEST INDIES and the AZORES, where seaweed floats on the surface. It is a major spawning ground for eels.

Sark *see* **Channel Islands.**

Saskatchewan a province of western CANADA, in the GREAT PLAINS. The capital is REGINA. (651,900 sq km/251,000 sq miles; pop. 995,300)

Saskatchewan, River a river of CANADA, rising in the ROCKY MOUNTAINS and flowing westwards into Lake WINNIPEG. (Length 1930 km/1200 miles)

Saskatoon a city on the SASKATCHEWAN RIVER. (Pop. 210,000)

Saudi Arabia occupies over 70% of the Arabian Peninsula. Over 95% of the country is desert and the largest expanse of sand in the world, 'Rub'al-Khali' ('The Empty Quarter'), is found in the south-east of the country. In the west, a narrow, humid coastal plain along the RED SEA is backed by steep mountains. The climate is hot with very little rain and some areas have no precipitation for years. The government has spent a considerable amount on reclamation of the desert for agriculture, and the main products are dates, tomatoes, watermelons and wheat. The country's prosperity, however, is based almost entirely on the exploitation of its vast reserves of oil and natural gas. Industries include petroleum refining, petrochemicals and fertilizers. As a result of the Gulf War in 1990-91, 460km/285mi of the Saudi coastline has been polluted by oil threatening desalination plants and damaging the wildlife of saltmarshes, mangrove forest, and mudflats.

Quick facts:
Area : 2,149,690 sq km (829,995 sq miles)
Population : 18,395,000
Capital : Riyadh (Ar Riyah)
Other major cities : Mecca, Jeddah,
  Medina, Ta'if
Form of government : Monarchy
Religions : Sunni Islam, Shia Islam
Currency : Rial

Savannah the main port of GEORGIA, USA. (Pop. city 139,000)

Savoie (Savoy) a mountainous former duchy in south-east FRANCE, which has been a part of France since 1860 and is now divided into two departments, Savoie and Haute Savoie.

Scafell Pike *see* **Lake District.**

Scandinavia the countries on, or near, the Scandinavian peninsula in north-east EUROPE, usually taken to include NORWAY, SWEDEN, DENMARK and FINLAND.

Scapa Flow an anchorage surrounded by the ORKNEY ISLANDS, famous as a wartime naval base.

Schelde, River a river of western EUROPE rising in FRANCE and then flowing through BELGIUM and the NETHERLANDS to the NORTH SEA. (Length 435 km/270 miles)

Schlesien *see* **Silesia.**

Schleswig-Holstein the northern-most state of GERMANY. The capital is KIEL. (Pop. 2,637,000)

Schwarzwald *see* **Black Forest.**

Scilly, Isles of a group of islands off the south-west tip of England. The main islands are St Mary's, St Martin's and Tresco. (Pop. 2000)

Scotland a country of the UK, occupying the northern part of GREAT BRITAIN. The capital is EDINBURGH. (78,762 sq km/30,410 sq miles; pop. 5,106,000)

Seattle a port in WASHINGTON State, USA. (Pop. city 519,600/metropolitan area 3,131,000)

Seine, River a river of northern FRANCE, flowing through PARIS to the ENGLISH CHANNEL. (Length 775 km/482 miles)

Selkirkshire *see* **Borders.**

Semarang a port and textile city on the north coast of JAVA, INDONESIA. (Pop. 1,005,000)

Sendai a city in the east of HONSHU Island, JAPAN. (Pop. 951,000)

Senegal is a former French colony in West AFRICA which extends from the most western point in Africa, Cape Verde, to the border with MALI. Senegal is mostly low-lying and covered by savanna. The Fouta Djalon mountains in the south rise to 1515 m (4971 ft). The climate is tropical with a dry season from October to June. The most densely populated region is in the southwest. Almost 80% of the labour force work in agriculture, growing groundnuts and cotton for export and millet, maize, rice and sorghum as subsistence crops. Senegal has been badly affected by the drought that has afflicted the SAHEL and relies on food imports and international aid.

> *Quick facts:*
> Area : 196,722 sq km (75,954 sq miles)
> Population : 8,308,000
> Capital : Dakar
> Other major cities : Kaolack, Thies, St Louis
> Form of government : Republic
> Religions : Sunni Islam, RC
> Currency : Franc CFA

Senegal River a West African river that flows through GUINEA, MALI, MAURITANIA, and SENEGAL to the ATLANTIC OCEAN. (Length 1790 km/1110 miles)

Seoul (Soul) the capital of South KOREA, in the north-west of the country. (Pop. 10,628,000)

Sepik River a major river of PAPUA NEW GUINEA. (Length 1200 km/750 miles)

Seram (Ceram) an island in the MALUKU group, INDONESIA. (17,148 sq km/6621 sq miles)

Serbia (Srbija) a land-locked republic, and the largest republic of former YUGOSLAVIA. The capital is BELGRADE. (102,170 sq km/39,448 sq miles; pop. 10,881,000)

Sevastopol' a BLACK SEA port of the UKRAINE. (Pop. 371,000)

Severn, River the longest river in the UK, flowing through WALES and the west of ENGLAND. (Length 350 km/220 miles)

Seville (Sevilla) a historic, now industrial city in southern SPAIN, and also the name of the surrounding province. (Pop. city 714,000)

Seychelles are a group of volcanic islands which lie in the western INDIAN OCEAN about 1200 km (746 miles) from the coast of East AFRICA. About forty of the islands are mountainous and consist of granite while just over fifty are coral islands. The climate is tropical maritime with heavy rain. About 90% of the people live on the island of Mahé which is the site of the capital, VICTORIA. The staple food is coconut, imported rice and fish. Tourism accounts for about 90% of the country's foreign exchange earnings and employs one-third of the labour force. The Seychelles were a one party socialist state until 1991, when a new constitution was introduced. The first free elections were held in 1993.

> *Quick facts:*
> Area : 455 sq km (176 sq miles)
> Population : 75,000
> Capital : Victoria
> Form of government : Republic
> Religion : RC
> Currency : Seychelles rupee

's-Gravenhage *see* **Hague, The**.

Shaanxi a province of north-western CHINA. The capital is XIAN. (195,800 sq km/75,579 sq miles; pop. 32,282,000)

Shandong a province of northern CHINA, with its capital at JINAN. (153,300 sq km/59,174 sq miles; pop. 84,393,000)

Shanghai the largest and most westernized city in CHINA. An important port, it is situated on the delta of the CHANG JIANG (Yangtze) River, in the east of the country. It is also the capital of the municipal province of the same name. (Pop. city 8,930,000/ province 13,400,000)

Shannon, River a river of the Republic of IRELAND, and the longest river in the BRITISH ISLES. It flows south-west into the ATLANTIC OCEAN near LIMERICK. (Length 386 km/240 miles)

Shanxi a province of northern CHINA, with its capital at TAIYUAN. (157,100 sq km/60,641 sq miles; pop. 28,759,000)

Sharjah one of the United Arab EMIRATES. Its capital city is also called Sharjah. (2,600 sq km/1,004 sq miles; pop. emirate 269,000/ city 125,000)

Shatt al Arab a waterway flowing into The GULF along the disputed border between IRAN and IRAQ, formed where the Rivers EUPHRATES and TIGRIS converge some 170 km (105 miles) from the coast.

Sheffield a major industrial city in South YORK-SHIRE, ENGLAND. (Pop. 530,100)

Shenandoah, River a river, of great significance during the American Civil War, that flows through northern VIRGINIA, USA. (Length 90 km/55 miles)

Shenyang the capital of LIAONING province, CHINA. (Pop. 4,500,000)

Shetland Islands a group of some 100 islands lying 160 km (100 miles) north-east of mainland SCOT-LAND. The capital is Lerwick. (1426 sq km/550 sq miles; pop. 22,900)

Shijiazhuang the capital of HEBEI province, CHINA. (Pop. 1,610,000)

Shikoku the smallest of the four main islands of JA-PAN. (Pop. 4,227,200)

Shiraz a provincial capital of IRAN, south-east of TEHRAN. (Pop. 965,000)

Shropshire a county of west central ENGLAND; the county town is Shrewsbury. (3490 sq km/1347 sq miles; pop. 416,500)

Siam *see* **Thailand.**

Siberia a huge tract of land, mostly in northern RUSSIAN FEDERATION, that extends from the URAL MOUNTAINS to the PACIFIC coast. It is renowned for its inhospitable climate, but parts of it are fertile, and it is rich in minerals.

Sichuan (Szechwan) the most heavily populated of the provinces of CHINA, in the south-west of the country. The capital is CHENGDU. (570,000 sq km/220,000 sq miles; pop. 107,218,000)

Sicily (Sicilia) an island hanging from the toe of ITALY, and the largest island in the MEDITERRANEAN. The capital is PALERMO. (25,708 sq km/9926 sq miles; pop. 5,196,800)

Siena (Sienna) a historic town of TUSCANY, in central ITALY. (Pop. 60,800)

Sierra Leone, on the ATLANTIC coast of West AFRICA, is bounded by GUINEA to the north and east and by LIBERIA to the south-east. The coastal areas consist of wide swampy forested plains and these rise to a mountainous plateau in the east. The highest parts of the mountains are just under 2000 m (6562 ft). The climate is tropical with a dry season from November to June. The main food of Sierra Leoneans is rice and this is grown in the swamplands at the coast. In the tropical forest areas, small plantations produce coffee, cocoa and oil palm. In the plateau much forest has been cleared for growing of groundnuts. Most of the country's revenue comes from mining. Diamonds are panned from the rivers and there are deposits of iron ore, bauxite, rutile and some gold.

Quick facts:
Area : 71,740 sq km (27,699 sq miles)
Population : 4,467,000
Capital : Freetown
Form of government : Republic
Religion : Animism, Sunni Islam, Christianity
Currency : Leone

Sierra Madre Occidental the mountain range of western MEXICO.

Sierra Madre Oriental the mountain range of eastern MEXICO.

Sierra Nevada (1) a mountain range in southern SPAIN. **(2)** a mountain range in eastern CALIFORNIA, USA.

Si Kiang *see* **Xi Jiang**.

Sikkim a state in north-eastern INDIA. The capital is Gangtok. (7300 sq km/2818 sq miles; pop. 403,600)

Silesia (Schlesien) a region straddling the borders of the CZECH REPUBLIC, GERMANY and POLAND.

Simbirsk a city of the eastern URALS, in the RUSSIAN FEDERATION, on the River VOLGA. It was formerly known as ULYANOVSK. (Pop. 638,000)

Simla a hill station and the capital of the state of HIMACHAL PRADESH, northern INDIA.

Simpson Desert an arid, uninhabited region in the centre of AUSTRALIA.

Sinai a mountainous peninsula in north-eastern EGYPT, bordering ISRAEL, between the Gulf of AQABA and the Gulf of SUEZ.

Sind a province of south-eastern PAKISTAN. The capital is KARACHI. (140,914 sq km/54,407 sq miles; pop. 19,029,000)

Singapore, one of the world's smallest yet most successful countries, comprises 1 main island and 59 islets which are located at the foot of the Malay peninsula in South-East ASIA. The main island of Singapore is very low-lying, and the climate is hot and wet throughout the year. Only 1.6% of the land area is used for agriculture and most food is imported. Singapore has the largest oil refining centre in Asia. The country has a flourishing manufacturing industry for which it relies heavily on imports. Products traded in Singapore include machinery and appliances, petroleum, food and beverages, chemicals, transport equipment, paper products and printing, and clothes. The Jurong Industrial Estate on the south of the island has approximately 2,300 companies and employs nearly 141,000 workers. Tourism is an important source of foreign revenue.

Quick facts:
Area : 622 sq km (240 sq miles)
Population : 2,990,000
Capital : Singapore
Form of government : Parliamentary Democracy
Religions : Buddhism, Sunni Islam, Christianity and Hinduism
Currency : Singapore dollar

Sinkiang Uygur Autonomous Region *see* **Xinjiang Uygur Autonomous Region**.

Sint Maarten *see* **St Martin**.

Siracusa *see* **Syracuse**.

Síros *see* **Cyclades**.

Sirte, Gulf of a huge indent of the MEDITERRANEAN SEA in the coastline of LIBYA.

Sivas an industrial town in central TURKEY. (Pop. 222,000)

Sjaelland *see* **Zealand.**

Skagerrak the channel, some 130 km (80 miles) wide, separating DENMARK and NORWAY. It links the NORTH SEA to the KATTEGAT and BALTIC SEA.

Skiathos the westernmost of the Greek Sporades (DODECANESE) Islands.

Skopje the capital of THE FORMER YUGOSLAV REPUBLIC OF MACEDONIA. (Pop. 563,000)

Skye an island off the north-west coast of SCOTLAND; the largest of the Inner HEBRIDES. The main town is Portree. (1417 sq km/547 sq miles; pop. 12,000)

Slavonia (Slavonija) a part of CROATIA, south-east of ZAGREB, mainly between the DRAVA and Slava Rivers.

Sligo a county on the north-west coast of the Republic of IRELAND, with a county town of the same name. (1796 sq km/693 sq miles; pop. county 54,700)

Slovakia (Slovak Republic) was constituted on January 1, 1993 as a new independent nation, following the dissolution of the 74-year old federal republic of CZECHOSLOVAKIA. Landlocked in central Europe, its neighbours are the CZECH REPUBLIC to the west, POLAND to the north, AUSTRIA and HUN-

GARY to the south, and a short border with UKRAINE in the east. The northern half of the republic is occupied by the Tatra Mountains which form the northern arm of the CARPATHIAN Mountains. This region has vast forests and pastures used for intensive sheep grazing, and is rich in high-grade minerals. The southern part of Slovakia is a plain drained by the DANUBE and its tributaries. Farms, vineyards, orchards and pastures for stock form the basis of southern Slovakia's economy.

Quick facts:
Area : 49,032 sq km (18,931 sq miles)
Population : 5,400,000
Capital : Bratislava
Other major city : Kovice
Form of government : Republic
Religion : RC
Currency : Koruna

Slovenia is a republic which made a unilateral declaration of independence from former YUGOSLAVIA on June 25, 1991. Sovereignty was not formally recognized by the European Community and the United Nations until early in 1992. It is bounded to the north by AUSTRIA, to the west by ITALY, to the east by HUNGARY, and to the south by CROATIA. Most of Slovenia is situated in the Karst Plateau and in the Julian Alps. Although farming and livestock raising are the chief occupations, Slovenia is very industrialized and urbanized. Iron, steel and aluminium are produced, and mineral resources include oil, coal and mercury. Tourism is an important industry. The Julian Alps are renowned for their scenery, and the Karst Plateau contains spectacular cave systems. The north-east of the republic is famous for its wine production.

Quick facts:
Area : 20,251 sq km (7817 sq miles)
Population : 2,000,000
Capital : Ljubljana
Other major cities : Maribor, Celje
Form of government : Republic
Religion : RC
Currency : Tolar

Smolensk an industrial city in the RUSSIAN FEDERA-TION, on the River DNIEPER. (Pop. 352,000)

Smyrna *see* **Izmir**.

Snake, River a river of the north-west USA, which flows into the COLUMBIA River in WASHINGTON State; it is used for irrigation and to create hydro-electric power. (Length 1670 km/1038 miles)

Snowdonia a mountainous region in the north of WALES. The highest peak is Mount Snowdon (1085 m/3560 ft).

Snowy Mountains a range of mountains in south-eastern AUSTRALIA, where the River Snowy has been dammed to form the complex Snowy Moun-tains Hydroelectric Scheme. The highest peak is Mount KOSCIUSKO (2230 m/7316 ft).

Society Islands a group of islands at the centre of FRENCH POLYNESIA. They are divided into the Windward Islands, which include TAHITI and Moorea; and the Leeward Islands, which include Raiatea and Bora-Bora. (Pop. 142,000)

Socotra an island in the north-western INDIAN OCEAN, belonging to YEMEN.

Sofia (Sofiya) the capital of BULGARIA, in the west of the country. (Pop. 1,221,000)

Solent, The a strait in the ENGLISH CHANNEL that separates the Isle of WIGHT from mainland ENGLAND.

Solomon Islands the Solomon Islands lie in an area between 5° and 12° south of the Equator to the east of PAPUA NEW GUINEA, in the PACIFIC OCEAN. The nation consists of six large islands and innumer-able smaller ones. The larger islands are moun-tainous and covered in forests with rivers prone to flooding. GUADAL-CANAL is the main island and the site of the capital, HONIARA. The climate is hot and wet and typhoons are frequent. The main food crops grown are coconut, cassava, sweet potatoes, yams, taros and bananas. The forests are worked commercially, and the fishing industry is develop-ing with the help of the Japanese. Other industries include palm-oil milling, fish canning and freez-ing, saw milling, food, tobacco and soft drinks.

Quick facts:
Area : 28,896 sq km (11,157 sq mi)
Population : 378,000
Capital : Honiara
Form of government : Constitutional Monarchy
Religions : Anglicanism,
  RC, other Christianity
Currency : Solomon Is.
  dollar

Somalia is shaped like a large number seven and lies on the horn of AFRICA's east coast. It is bounded north by the Gulf of Aden, south and east by the INDIAN OCEAN, and its neighbours include DJIBOUTI, ETHIOPIA, and KENYA. The country is arid and most of it is low plateaux with scrub vegeta-tion. Its two main rivers, the Juba and Shebelle, are used to irrigate crops. Most of the population live in the mountains and river valleys and there are a few towns on the coast. Main exports are live animals, meat, hides and skins. A few large-scale banana plantations are found by the rivers. Years of drought have left Somalia heavily dependent on foreign aid, and many of the younger population are emigrating to oil-rich Arab states.

Quick facts:
Area : 637,657 sq km (246,199 sq miles)
Population : 9,180,000
Capital : Mogadishu
Other major cities : Hargeisa, Baidoa, Burao,
  Kismaayo
Form of government : Republic
Religion : Sunni Islam
Currency : Somali shilling

Somerset a county in the south-west of ENGLAND; the county town is Taunton. (3458 sq km/1335 sq miles; pop. 477,500)

Somme, River a river of northern FRANCE, the scene of a devastating battle during World War I. (Length 245 km/152 miles)

Song Hong *see* **Red River**.

Soul *see* **Seoul**.

Soúnion, Cape a cape overlooking the the southern AEGEAN SEA, 60 km (37 miles) south-west of ATH-ENS, GREECE.

South Africa is a republic that lies at the southern tip of the African continent and has a huge coast-line on both the ATLANTIC and INDIAN OCEANS. The country occupies a huge saucer-shaped plateau, surrounding a belt of land which drops in steps to the sea. The rim of the saucer rises in the east, to 3482 m (11,424 ft), in the DRAKENSBERG. In gen-eral the climate is healthy with plenty of sunshine and relatively low rainfall. This varies with lati-tude, distance from the sea, and altitude. Of the to-tal land area 58% is used as natural pasture. The main crops grown are maize, sorghum, wheat, groundnuts and sugarcane. A drought-resistant variety of cotton is also now grown. It is South Africa's extraordinary mineral wealth which overshadows all its other natural resources. These include gold, coal, copper, iron ore, manganese and chrome ore, and diamonds. A system of apartheid existed in South Africa from 1948 until the early 1990s, effectively denying black South Africans civil rights and promoting racial segregation. During this time the country was subjected to international economic and political sanctions. In 1990 F. W. de Klerk, then president, lifted the ban on the outlawed African National Congress and released its leader, Nelson Mandela, who had been imprisoned since 1962. This heralded the dismantling of the apartheid regime and in the first multi-racial elections, held in 1994, the ANC triumphed, with Mandela voted in as the country's president. Since this time South Africa has once again become an active and recognized member of the international community.

Quick facts:
Area : 1,221,037 sq km (471,442 sq miles)
Population : 44,000,000
Capital : Pretoria (Administrative),
  Cape Town (Legislative)
Other major cities. Johannesburg, Durban,
  Port Elizabeth, Bloemfontein
Form of government : Republic
Religions : Dutch reformed, Independent
  African, other Christianity, Hinduism
Currency : Rand

Southampton a major port in southern ENGLAND. (Pop. 211,600)

South Australia a state in central southern Australia, on the GREAT AUSTRALIAN BIGHT. ADELAIDE is the state capital. (984,380 sq km/380,069 sq miles; pop. 1,388,000)

South Carolina a state in the south-east of the USA, with a coast on the ATLANTIC OCEAN. The state capital is COLUMBIA. (80,432 sq km/31,055 sq miles; pop. 3,602,900)

South China Sea an arm of the PACIFIC OCEAN between south-east CHINA, MALAYSIA and the PHILIPPINES.

South Dakota a state in the western USA. The state capital is PIERRE. (199,552 sq km/77,047 sq miles; pop. 708,400)

Southern Alps a range of mountains on the South Island of NEW ZEALAND.

Southern Ocean *see* **Antarctic Ocean**.

South Georgia an island in the South ATLANTIC, and a dependency of the FALKLAND ISLANDS. (3755 sq km/1450 sq km)

South Glamorgan a former county in south WALES. In 1996 the county ceased to exist, becoming part of the VALE OF GLAMORGAN and part of the district of CARDIFF.

South Island *see* **New Zealand**.

South Korea *see* **Korea, Republic of**.

South Pole the most southerly point of the Earth's axis, in ANTARCTICA.

South Sandwich Islands a group of islands in the South ATLANTIC which are dependencies of the FALKLAND ISLANDS. (340 sq km/130 sq miles)

Soweto a group of townships to the south of JOHANNESBURG, SOUTH AFRICA. (Pop. 597,000)

Spa a spa town in BELGIUM which was the origin of the general term 'spa.' (Pop. 10,100)

Spain is located in south-west EUROPE and occupies the greater part of the Iberian peninsula, which it shares with PORTUGAL. It is a mountainous country, sealed off from the rest of Europe by the PYRÉNÉES, which rise to over 3400 m (11,155 ft). Much of the country is a vast plateau, the Meseta Central, cut across by valleys and gorges. Its longest shoreline is the one that borders the MEDITERRANEAN SEA. Most of the country has a form of Mediterranean climate with mild moist winters and hot dry summers. Spain's principal agricultural products are cereals, vegetables and potatoes, and large areas are under vines for the wine industry. Industry represents 72% of the country's export value, and production includes textiles, paper, cement, steel and chemicals. Tourism is a major revenue earner, especially from the resorts on the east coast.

Quick facts:
Area : 504,782 sq km (194,896 sq miles)
Population : 39,664,000
Capital : Madrid
Other major cities : Barcelona, Seville,
  Zaragosa, Malaga, Bilbao
Form of government : Constitutional Monarchy
Religion : RC
Currency : Peseta

Spitsbergen A large island group in the SVALBARD archipelago, 580 km (360 miles) to the north of NORWAY. (39,000 sq km/15,060 sq miles)

Split the largest city on the coast of DALMATIA, CROATIA. (Pop. 207,000)

Sporades (Sporadhes) *see* **Dodecanese**.

Spratly Islands a group of islands in the SOUTH CHINA SEA between VIETNAM and BORNEO. Occupied by JAPAN during World War II, they are now claimed by almost all the surrounding countries.

Springfield (1) the state capital of ILLINOIS, USA. (Pop. city 106,400). **(2)** a manufacturing city in MASSACHUSETTS, USA. (Pop. city 153,500)

Srbija *see* **Serbia**.

Sri Lanka is a teardrop-shaped island in the INDIAN OCEAN, lying south of the Indian peninsula from which it is separated by the Palk Strait. The climate is equatorial with a low annual temperature range but it is affected by both the north-east and south-west monsoons. Rainfall is heaviest in the south-west while the north and east are relatively dry. Agriculture engages 47% of the work force and the main crops are rice, tea, rubber and coconuts. Amongst the chief minerals mined and exported are precious and semiprecious stones. Graphite is also important. The main industries are food, beverages and tobacco, textiles, clothing and leather goods, chemicals and plastics. Attempts are being made to increase the revenue from tourism. Politically, Sri Lanka has been afflicted by ethnic divisions between the Sinhalese and Tamils. In the 1980s attempts by the Tamil extremists to establish an independent homeland bought the north-east of the country to the brink of civil war and the situation remains extremely volatile.

Quick facts:
Area : 65,610 sq km (25,332 sq miles)
Population : 18,3597,000
Capital : Colombo
Other major cities : Dehiwela-Mt. Lavinia,
  Moratuwa, Jaffna
Form of government : Republic
Religions : Buddhism, Hinduism, Christianity,
  Sunni Islam
Currency : Sri Lankan rupee

Srinagar the capital of the state of JAMMU AND KASHMIR, northern INDIA. (Pop. 595,000)

Staffordshire a MIDLANDS county of ENGLAND. The county town is Stafford. (2716 sq km/1049 sq miles; pop. 1,054,400)

Stalingrad *see* **Volgograd**.

Stanley *see* **Port Stanley**.

Stanley Falls *see* **Boyoma Falls**.

Stanley, Mount *see* **Ngaliema, Mount**.

Stanleyville *see* **Kisangani**.

Statia *see* **St Eustatius**.

Stettin *see* **Szczecin**.

Stockholm the capital of SWEDEN, and an important port on the BALTIC SEA. (Pop. 1,539,000)

Stockton-on-Tees a town in the county of CLEVELAND, ENGLAND. (Pop. 178,200)

Stoke-on-Trent a major city in the 'Potteries' of the MIDLANDS of ENGLAND. (Pop. 254,200)

Strasbourg an industrial city and river port in eastern FRANCE, the capital of the ALSACE region, and the seat of the European Parliament. (Pop. 388,000)

Stratford-upon-Avon a town in WARWICKSHIRE, ENGLAND, the home of William Shakespeare (1564–1616). (Pop. 109,500)

Strathclyde an former administrative region in western SCOTLAND. It was created in 1975 out of the former counties of Ayrshire, Lanarkshire, Renfrewshire, Bute, Dunbartonshire and parts of Stirlingshire and Argyll. In 1996 local government restructuring it was broken up into 12 different authorities.

Stromboli an island with an active volcano in the EOLIAN ISLANDS, to the north of SICILY. (Pop. 400)

Stuttgart a major industrial centre and river port of the NECKAR river in south-western GERMANY. (Pop. 594,000)

Sucre the legal capital of BOLIVIA. (Pop. 131,000)

Sudan is the largest country in AFRICA, lying just south of the Tropic of Cancer in north-east Africa. The country covers much of the upper NILE basin and in the north the river winds through the Nubian and Libyan deserts, forming a palm-fringed strip of habitable land. In 1994, the country was divided into 26 states, compared to the original nine. The climate is tropical and temperatures are high throughout the year. In winter, nights are very cold. Rainfall increases in amount from north to south, the northern areas being virtually desert. Sudan is an agricultural country, subsistence farming accounting for 80% of production. Cotton is farmed commercially and accounts for about two-thirds of Sudan's exports. Sudan is the world's greatest source of gum arabic used in medicines and inks. This is the only forest produce to be exported.

> *Quick facts:*
> Area : 2,505,813 sq km (967,494 sq miles)
> Population : 29,980,000
> Capital : Khartoum (El Khartum)
> Other major cities : Omdurman,
>   Khartoum North, Port Sudan
> Form of government : Republic
> Religions : Sunni Islam, Animism, Christianity
> Currency : Sudanese pound

Sudd a vast swampland on the White NILE in SU-DAN.

Sudetenland *see* **Sudety**.

Sudety (Sudetenland) a mountainous region straddling the border between the CZECH REPUBLIC and POLAND.

Suez Canal a canal in north-east EGYPT, linking the MEDITERRANEAN to the RED SEA. It was completed in 1869.

Suez (El Suweis) a town situated at the southern end of the SUEZ CANAL. (Pop. 388,000)

Suez, Gulf of a northern arm of the RED SEA that leads to the SUEZ CANAL.

Suffolk a county in EAST ANGLIA, ENGLAND. The county town is Ipswich. (3800 sq km/1467 sq miles; pop. 649,300)

Sulawesi (Celebes) a large, hook-shaped island in the centre of INDONESIA. (179,370 sq km/69,255 sq miles; pop. 12,521,000)

Sulu Archipelago a chain of over 400 islands off the south-west PHILIPPINES, stretching between the Philippines and BORNEO.

Sulu Sea a part of the PACIFIC OCEAN which lies between the PHILIPPINES and BORNEO.

Sumatra the main island of western INDONESIA. (473,607 sq km/182,860 sq miles; pop. 36,882,000)

Sumba one of the LESSER SUNDA ISLANDS, INDONESIA, to the south of SUMBAWA and FLORES.

Sumbawa one of the Lesser Sunda Islands, InDO-NESIA, between LOMBOK and FLORES.

Sunda Islands see Lesser Sunda Islands.

Sunda Strait the strait, 26 km(16 miles) across at its narrowest, which separates JAVA and SUMATRA.

Sunderland an industrial town in the north-east of ENGLAND. (Pop. 297,200)

Superior, Lake the largest and most westerly of the GREAT LAKES. (82,400 sq km/31,800 sq miles)

Surabaya the second largest city of INDONESIA after JAKARTA, on the north-east coast of JAVA. (Pop. 2,421,000)

Surat a port on the west coast of INDIA, in western GUJARAT. (Pop. 1,517,000)

Suriname is a republic in north-east South AMERICA, bordered to the west by GUYANA, to the east by GUIANA, and to the south by BRAZIL. The country, formerly a Dutch colony, declared independence in 1975. Suriname comprises a swampy coastal plain, a forested central plateau, and southern mountains. The climate is tropical with heavy rainfall. Temperatures at PARAMARIBO average 26-27°C all year round. Rice and sugar are farmed on the coastal plains but the mining of bauxite is what the economy depends on. This makes up 80% of exports. Suriname has resources of oil and timber but these are so far underexploited. The country is politically very unstable and in need of financial aid to develop these resources.

> *Quick facts:*
> Area : 163,265 sq km (63,037 sq miles)
> Population : 421,000
> Capital : Paramaribo
> Form of government : Republic
> Religions : Hinduism, RC, Sunni Islam
> Currency : Surinam guilder

Surrey a county of central southern ENGLAND. The county town is Guildford. (1655 sq km/639 sq miles; pop. 1,043,800)

Sussex, East a county in south-east ENGLAND; the county town is Lewes. (1795 sq km/693 sq miles; pop. 726,400)

Sussex, West a county in south-east ENGLAND; the county town is Chichester. (1989 sq km/768 sq miles; pop. 722,200)

Susupe the capital of the NORTHERN MARIANAS, on the island of SAIPAN.

Suva the capital and main port of Fiji. (Pop. city 75,000/metropolitan area 145,000)

Suzhou a city in Jiangshu province, China. (Pop. 706,000)

Svalbard an archipelago in the Arctic Ocean to the north of Norway, which includes Spitsbergen. A convention (1920) granted sovereignty to Norway, but all signatories can exploit the mineral reserves. (62,049 sq km/23,958 sq miles; pop. 4000)

Sverdlovsk *see* **Yekaterinburg**.

Swansea a port in south Wales. (Pop. 231,000)

Swaziland a landlocked hilly enclave almost entirely within the borders of the Republic of SOUTH AFRICA. The mountains in the west of the country rise to almost 2000 m (6562 ft), then descend in steps of savanna toward hilly country in the east. The climate is subtropical moderated by altitude. The land between 400 m (1312 ft) and 850 m (2789 ft) is planted with orange groves and pineapple fields, while on the lower land sugar cane flourishes in irrigated areas. Asbestos is mined in the north-west of the country. Manufacturing includes fertilizers, textiles, leather and tableware. Swaziland attracts a lot of tourists from South Africa, mainly to its spas and casinos.

Quick facts:
Area : 17,360 sq km (6716 sq miles)
Population : 849,000
Capital : Mbabane
Other major cities : Big Bend, Manzini, Mhlume
Form of government : Monarchy
Religion : Christianity, Animism
Currency : emalangeni

Sweden is a large country in northern EUROPE which makes up half of the Scandinavian peninsula. It stretches from the BALTIC SEA north, to well within the ARCTIC CIRCLE. The south is generally flat, the north mountainous, and along the coast there are 20,000 or more islands and islets. Summers are warm but short and winters are long and cold. In the north snow may lie for four to seven months. Dairy farming is the predominant agricultural activity. Only 7% of Sweden is cultivated, with the emphasis on fodder crops, grain and sugar beet. About 57% of the country is covered in forest, and the sawmill, wood pulp and paper industries are all of great importance. Sweden is one of the world's leading producers of iron ore, most of which is extracted from within the Arctic Circle.

Other main industries are engineering and electrical goods, motor vehicles and furniture making.

Quick facts:
Area : 449,964 sq km (173,731 sq miles)
Population : 8,893,000
Capital : Stockholm
Other major cities : Göteborg, Malmö, Uppsala, Orebro
Form of government : Constitutional Monarchy
Religion : Lutheranism
Currency : Krona

Switzerland is a landlocked country in central EUROPE, sharing its borders with FRANCE, ITALY, AUSTRIA, LIECHTENSTEIN and GERMANY. The ALPS occupy the southern half of the country, forming two main east-west chains divided by the rivers RHINE and RHÔNE. The climate is either continental or mountain type. Summers are generally warm and winters cold, and both are affected by altitude. Northern Switzerland is the industrial part of the country and where its most important cities are located. BASLE is famous for its pharmaceuticals and ZÜRICH for electrical engineering and machinery. It is also in this region that the famous cheeses, clocks, watches and chocolates are produced. Switzerland has huge earnings from international finance and tourism.

Quick facts:
Area : 41,293 sq km (15,943 sq miles)
Population : 7,268,000
Capital : Berne (Bern)
Other major cities : Zürich, Basle, Geneva, Lausanne
Form of government : Federal republic
Religions : RC, Protestantism
Currency : Swiss franc

Sydney the largest city and port in AUSTRALIA, and the capital of NEW SOUTH WALES. (Pop. 3,657,000)

Syracuse (1) a city in the centre of NEW YORK State. (Pop. city 162,800). **(2) (Siracusa)** an ancient seaport on the east coast of SICILY, ITALY. (Pop. 127,000)

Syrdar'ya, River a river of central ASIA, flowing through KAZAKHSTAN to the ARAL SEA. (Length 2860 km/1780 miles)

Syria is a country in south-west ASIA which borders on the MEDITERRANEAN SEA in the west. Much of the country is mountainous behind the narrow fertile coastal plain. The eastern region is desert or semi-desert, a stony inhospitable land. The coast has a Mediterranean climate with hot dry summers and mild wet winters. About 50% of the workforce get their living from agriculture, sheep, goats and cattle are raised, and cotton, barley, wheat, tobacco, fruit and vegetables are grown. Reserves of oil are small compared to neighbouring IRAQ but it has enough to make the country self-sufficient and provide three quarters of the nation's export earnings. Industries such as textiles, leather, chemicals and cement have developed rapidly in the last 20 years.

Quick facts:
Area : 185,180 sq km (71,498 sq miles)
Population : 14,614,000
Capital : Damascus (Dimashq)
Other major cities : Halab, Homs,
  Latakia, Hama
Form of government : Republic
Religion : Sunni Islam
Currency : Syrian pound

Szczecin (Stettin) a port in north-west POLAND. (Pop. 414,000)

Szechwan *see* **Sichuan**.

Table Mountain a flat-topped mountain overlooking CAPE TOWN in south-west SOUTH AFRICA. (1087 m/3567 ft)

Tabriz a city in north-west IRAN. (Pop. 1,089,000)

Tadmor *see* **Palmyra**.

Tadzhikistan *see* **Tajikistan**.

Taegu the third largest city of South KOREA, in the south-east of the country. (Pop. 2,229,000)

Tagus (Tajo; Tejo), River a major river of south-west EUROPE, which rises in eastern SPAIN and flows west and south-west through PORTUGAL to the ATLANTIC OCEAN west of LISBON. (Length 1007 km/626 miles)

Tahiti the largest of the islands of FRENCH POLYNESIA in the South PACIFIC. The capital is PAPEETE. (1042 sq km/402 sq miles; pop. 116,000)

Taichung (T'ai-chung) a major commercial and agricultural centre in western TAIWAN. (Pop. 817,000)

Tainan (T'ai-nan) a city in south-west TAIWAN. (Pop. 700,000)

Taipei (T'ai-pei) the capital and largest city of TAIWAN, in the very north of the island. (Pop. 2,653,000)

Taiwan is an island which straddles the Tropic of Cancer in East ASIA. It lies about 160 km (99 miles) off the south-east coast of mainland CHINA. It is predominantly mountainous in the interior, the tallest peak rising to 3997 m (13,114 ft) at Yu Shan. The climate is warm and humid for most of the year. Winters are mild and summers rainy. The soils are fertile, and a wide range of crops, including tea, rice, sugar cane and bananas, is grown. Taiwan is a major international trading nation with some of the most successful export-processing zones in the world, accommodating domestic and overseas companies. Exports include machinery, electronics, textiles, footwear, toys and sporting goods.

Quick facts:
Area : 36,179 sq km (13,969 sq miles)
Population : 21,100,000
Capital : Taipei (T'ai-pei)
Other major cities : Kaohsiung, Taichung,
  Tainan
Form of government : Republic
Religions : Taoism, Buddhism, Christianity
Currency : New Taiwan dollar

Taiwan Strait the stretch of water that separates Taiwan from mainland CHINA.

Taiyuan the capital of SHANXI province, CHINA. (Pop. 1,720,000)

Tajikistan, a republic of southern central former USSR, declared itself independent in 1991. It is situated near the Afghani and Chinese borders.

The south is occupied by the PAMIR mountain range, whose snow-capped peaks dominate the country. More than half the country lies over 3000 m (9843 ft). Most of the country is desert or semi-desert, and pastoral farming of cattle, sheep, horses and goats is important. Some yaks are kept in the higher regions. The lowland areas in the Fergana and AMUDAR'YA valleys are irrigated so that cotton, mulberry trees, fruit, wheat and vegetables can be grown. The Amudar'ya river is also used to produce hydro-electricity for industries such as cotton and silk processing. The republic is rich in deposits of coal, lead, zinc, oil and uranium, which are now being exploited.

Quick facts:
Area : 143,100 sq km (55,250 sq miles)
Population : 6,102,000
Capital : Dushanbe
Form of Government : Republic
Religion : Shia Islam
Currency : Rouble

Tajo *see* **Tagus**.

Taklimakan Desert the largest desert in CHINA, consisting mainly of sand, in the west of the country.

Takoradi (Sekondi-Takoradi) a main port of GHANA, in the south-west of the country. (Pop. 175,000)

Tallahassee the state capital of FLORIDA, USA. (Pop. city 130,400)

Tallinn a port on the BALTIC SEA, and the capital of ESTONIA. (Pop. 435,000)

Tamil Nadu a state in south-east INDIA. The state capital is MADRAS. (130,058 sq km/50,215 sq miles; pop. 55,638,300)

Tammerfors *see* **Tampere**.

Tampa a port and resort on the west coast of FLORIDA, USA. (Pop. city 284,700/metropolitan area 2,107,000)

Tampere (Tammerfors) the second largest city in FINLAND after HELSINKI, in the south-west of the country. (Pop. 179,000)

Tana (Tsana), Lake a lake in the mountains of north-west ETHIOPIA, and the source of the Blue NILE. (3673 sq km/1418 sq miles)

Tanganyika *see* **Tanzania**.

Tanganyika, Lake the second largest lake in AFRICA after Lake VICTORIA, in the GREAT RIFT VALLEY, between TANZANIA and ZAÏRE, although BURUNDI and ZAMBIA also share the shoreline. (32,893 sq km/12,700 sq miles)

Tangier (Tanger) a port on the north coast of MOROCCO, on the Strait of GIBRALTAR. (Pop. 420,000)

Tangshan an industrial and coal-mining city in HEBEI province, CHINA. (Pop. 1,044,000)

Tanzania lies on the east coast of central AFRICA and comprises a large mainland area and the islands of Pemba and ZANZIBAR. The mainland consists mostly of plateaux broken by mountainous areas and the east African section of the GREAT RIFT VALLEY. The climate is very varied and is controlled largely by altitude and distance from the sea. The coast is hot and humid, the central plateau drier, and the mountains semi-temperate. 80% of Tanzanians make a living from the land, but productivity is low and there is no surplus from the crops, mainly maize, that they grow. Cash crops include cotton and coffee. The islands are more successful agriculturally and have important coconut and clove plantations. Tanzania's mineral resources are limited and of low grade, and there are few manufacturing industries.

Quick facts:
Area : 945,087 sq km (364,898 sq miles)
Population : 29,710,000
Capital : Dodoma
Other major cities : Dar es Salaam,
  Zanzibar, Mwanza, Tanga
Form of government : Republic
Religions : Sunni Islam, RC, Anglicanism,
  Hinduism
Currency : Tanzanian shilling

Tarabulus *see* **Tripoli**.

Taranto a port and naval base on the south coast of ITALY, on the Gulf of TARANTO. (Pop. 230,000)

Taranto, Gulf of an inlet of the MEDITERRANEAN SEA between the 'toe' and the 'heel' of ITALY.

Tarawa the main atoll and capital of the group of islands forming KIRIBATI. (Pop. 20,000)

Tarragona a port of ancient origins on the MEDI-

TERRANEAN coast of north-eastern SPAIN, and the name of the surrounding province. (Pop. city 115,000)

Tarsus an agricultural centre in south-east TURKEY, the birthplace of St Paul. (Pop. 188,000)

Tashkent the capital of UZBEKISTAN, in the north-east, near the border with KAZAKHSTAN. (Pop. 2,094,000)

Tasmania an island state to the south of AUSTRALIA, separated from the mainland by the BASS STRAIT. The capital is HOBART. (67,800 sq km/26,171 sq miles; pop. 448,000)

Tasman Sea a branch of the PACIFIC OCEAN that separates AUSTRALIA and NEW ZEALAND.

Tatar Republic (Tatarstan) an autonomous republic of the RUSSIAN FEDERATION, south-west of MOSCOW, around the River VOLGA. In 1992 it voted to become a sovereign state. The capital is KAZAN'. (68,000 sq km/26,250 sq miles; pop. 3,537,000)

Tatra Mountains a range of mountains that lines the border between POLAND and SLOVAKIA The highest peak is Gerlachovka (2663 m/8737 ft).

Tatung *see* **Datung**.

Tayside an administrative region of SCOTLAND formed in 1975 out of the former counties of Angus, Kinrossshire and part of Perthshire. The administrative centre is DUNDEE. (7511 sq km/2900 sq miles; pop. 395,000)

Tbilisi the capital of GEORGIA, in the centre of the republic. (Pop. 1,279,000)

Tegucigalpa the capital of HONDURAS, in the south of the country. (Pop. 679,000)

Tehran the capital of IRAN, in the central north of the country. (Pop. 6,476,000)

Tejo *see* **Tagus**.

Tel Aviv-Jaffa the largest city of ISRAEL, former capital and the main financial centre. It was combined with the old port of Jaffa in 1950. (Pop. 353,000)

Tenerife the largest of the CANARY ISLANDS. The capital is Santa Cruz. (2058 sq km/795 sq miles; pop. 558,000)

Tennessee a state in southern central USA. The state capital is NASHVILLE. (109,412 sq km/42,244 sq miles; pop. 5,025,300)

Tennessee, River a river which flows south-west from the APPALACHIAN MOUNTAINS of NORTH CAROLINA and then through ALABAMA, TENNESSEE and KENTUCKY to join the OHIO RIVER. It is an important source of irrigation and hydro-electric power. (Length 1049km/ 652miles)

Tevere, River *see* **Tiber**.

Texas a state in the south-west of the USA, bordering MEXICO. It is the nation's second largest state.

The capital is AUSTIN. (691,200 sq km/266,803 sq miles; pop. 17,682,600)

Thailand, a country about the same size as FRANCE located in South-East ASIA, is a tropical country of mountains and jungles, rain forests and green plains. Central Thailand is a densely populated, fertile plain and the mountainous Isthmus of Kra joins southern Thailand to MALAYSIA. Thailand has a subtropical climate with heavy monsoon rains from June to October, a cool season from October to March, and a hot season from March to June. The central plain of Thailand contains vast expanses of paddy fields which produce enough rice to rank Thailand as the world's leading exporter. The narrow southern peninsula is very wet, and it is here that rubber is produced. Thailand is the world's third largest exporter of rubber.

Quick facts:
Area : 513,115 sq km (198,114 sq miles)
Population : 58,432,00
Capital : Bangkok (Krung Thep)
Other major cities : Chiengmai, Nakhon
  Ratchasima, Songkhla
Form of government : Constitutional Monarchy
Religions : Buddhism, Sunni Islam
Currency : Baht

Thailand, Gulf of a branch of the SOUTH CHINA SEA lying between the Malay peninsula and the coasts of THAILAND, CAMBODIA and VIETNAM.

Thames, River a major river of southern ENGLAND flowing eastwards from its source in the Cotswold Hills, past LONDON to its estuary on the NORTH SEA. (Length 338 km/210 miles)

Thar Desert (Great Indian Desert) a desert in north-west INDIA, covering the border between RAJASTHAN and PAKISTAN.

Thebes the ruins of an ancient city on the River NILE in central EGYPT. It was the capital of Ancient Egypt for about 1000 years from 1600BC.

Thessaloníki *see* **Saloniki**.

Thimphu (Thimbu) the capital of BHUTAN, in the west of the country. (Pop. 30,000)

Thousand Islands a group of over 1000 islands scattered in the upper ST LAWRENCE River, between the USA and CANADA.

Tianjin (Tientsin) a major industrial city in HEBEI

province, the third largest city in CHINA after SHANGHAI and BEIJING. (Pop. 5,000,000)

Tiber (Tevere), River a river of central ITALY, rising to the east of FLORENCE and flowing south to ROME and then to the MEDITERRANEAN SEA. (Length 405 km/252 miles)

Tibet (Xizang Autonomous Region) a region of south-west CHINA, consisting of a huge plateau high in the HIMALAYAS. Formerly a Buddhist kingdom led by its spiritual leader, the Dalai Lama, it was invaded by China in 1950 and has been gradually desecrated. (1,221,600 sq km/471,660 sq miles; pop. 2,190,000)

Tientsin *see* **Tianjin**.

Tierra del Fuego the archipelago at the southern tip of South AMERICA, belonging to ARGENTINA and CHILE and separated from the mainland by the Strait of MAGELLAN.

Tigray a province of northern ETHIOPIA, bordering ERITREA, whose people have been fighting a separatist war against the central government. The capital is Mekele. (Pop. 2,400,000)

Tigris, River a major river of the MIDDLE EAST, rising in eastern TURKEY, flowing through SYRIA and IRAQ and joining the EUPHRATES to form a delta at the SHATT AL ARAB waterway as it enters The GULF. (Length 1900 km/1180 miles)

Tijuana a border city and resort in north-west MEXICO, at the northern end of the BAJA CALIFORNIA. (Pop. 743,000)

Timbuktu a town in central MALI at the edge of the SAHARA DESERT. (Pop. 20,500)

Timisoara an industrial city in south-west ROMANIA. (Pop. 325,000)

Timor an island at the eastern end of the LESSER SUNDA ISLANDS, INDONESIA. The eastern half of the island was a possession of PORTUGAL, but was annexed in 1975 by Indonesia, although sovereignty was not internationally recognized. East Timor, which includes an enclave in the west of Timor, remains a fiercely disputed territory. (30,775 sq km/11,883 sq miles)

Timor Sea the arm of the INDIAN OCEAN between the north-west coast of AUSTRALIA and the island of TIMOR.

Tipperary a county in the south of the Republic of IRELAND. It includes the town of Tipperary, but Clonmel is the county town. (4255 sq km/1643 sq miles; pop. 132,600)

Tiranè (Tirana) the largest city and capital of ALBANIA, in the centre of the country. (Pop. 251,000)

Tiruchchirappalli (Trichinopoly) an industrial city in central TAMIL NADU in southern INDIA. (Pop. 711,000)

Titicaca, Lake the largest lake in South AMERICA, in the ANDES, on the border between BOLIVIA and PERU. (8135 sq km/3141 sq miles)

Tobago an island to the north-east of TRINIDAD, forming part of the republic of TRINIDAD and TOBAGO. (Pop. 50,000)

Togo is a tiny country with a narrow coastal plain on the Gulf of GUINEA in West AFRICA. Grassy plains in the north and south are separated by the Togo Highlands, which run from south-west to north-east and rise to nearly 1000 m (3281 ft). High plateaux, mainly in the more southerly ranges, are heavily forested with teak, mahogany and bamboo. Over 80% of the population are involved in agriculture with yams and millet as the principal crops. Coffee, cocoa and cotton are grown for cash. Minerals, especially phosphates, are now the main export earners. Togo's exports are suffering from the recession in its major markets in Western EUROPE.

Quick facts:
Area : 56,785 sq km (21,925 sq miles)
Population : 4,140,000
Capital : Lomé
Form of government : Republic
Religions : Animism,
  RC, Sunni Islam
Currency : Franc CFA

Tokyo the capital of JAPAN, a port on the east coast of HONSHU Island. Its original name was Edo (until 1868). (Pop. city 11,927,700/Greater Tokyo 25,000,000)

Toledo (1) a historic city of central SPAIN, on the River TAGUS. (Pop. 62,000). **(2)** a city and GREAT LAKE port in OHIO, USA. (Pop. city 329,300)

Tonga is situated about 20° south of the Equator and just west of the International Date Line in the PACIFIC OCEAN. It comprises over 170 islands and only about one-fifth of them are inhabited. It comprises a low limestone chain of islands in the east and a higher volcanic chain in the west. The climate is warm with heavy rainfall. The government owns all the land, and males can rent an allotment for growing food. Yams, cassava and taro are grown as subsistence crops, and fish from the sea supplements their diet. Bananas and coconuts are grown for export. The main industry is coconut processing.

Quick facts:
Area : 750 sq km (290 sq miles)
Population : 107,000
Capital : Nuku'alofa
Form of government : Constitutional Monarchy
Religions : Methodism, RC
Currency : Pa'anga

Tongking (Tonkin) *see* **Mien Bac**.

Tonle Sap a lake in central CAMBODIA which swells and quadruples in size when the River MEKONG floods. (In flood 10,400 sq km/4000 sq miles)

Topeka the state capital of KANSAS, USA. (Pop. city 120,000)

Torino *see* **Turin**.

Toronto the largest city of CANADA, and the capital of ONTARIO, situated on Lake Ontario. (Pop. 3,893,000)

Torres Strait the strait which separates the north-eastern tip of AUSTRALIA from NEW GUINEA.

Tortola the main island of the British VIRGIN ISLANDS.

Toscana *see* **Tuscany**.

Toulon a major naval base and port in south-east FRANCE, on the MEDITERRANEAN. (Pop. 438,000)

Toulouse a city of south-west FRANCE, on the GARONNE River. (Pop. 650,000)

Touraine a former province of north-west FRANCE, around TOURS.

Tours a town in western FRANCE, on the River LOIRE. (Pop. 282,000)

Trablous *see* **Tripoli**.

Trabzon (Trebizond) a port on the BLACK SEA in north-eastern TURKEY. (Pop. 144,000)

Trafalgar, Cape the south-western tip of SPAIN.

Transylvania a region of central and north-western ROMANIA.

Trebizond *see* **Trabzon.**

Trent, River is the main river of the MIDLANDS of ENGLAND, flowing north-east from Staffordshire to the HUMBER. (Length 270 km/170 miles)

Trenton a city in eastern USA on the Delaware River in western NEW JERSEY, of which it is the capital. British troops were defeated here (1776) during The American War of Independence. (Pop. 87,800)

Trieste a port on the ADRIATIC SEA in north-east ITALY, close to the border with SLOVENIA. (Pop. 228,000)

Trichinopoly *see* **Tiruchiràppalli**.

Trinidad and Tobago form the third largest British Commonwealth country in the WEST INDIES and are situated off the ORINOCO Delta in north-eastern VENEZUELA. The islands are the most southerly of the Lesser ANTILLES. Trinidad consists of a mountainous Northern Range in the north and undulating plains in the south. Tobago is more mountainous. The climate is tropical with little variation in temperatures throughout the year and a rainy season from June to December. Trinidad is one of the oldest oil-producing countries in the world. Output is small but provides 90% of Trinidad's exports. Sugar, coffee and cocoa are grown for export, but food now accounts for 10% of total imports. Tobago depends mainly on tourism for revenue.

Quick facts:
Area : 5130 sq km (1981 sq miles)
Population : 1,295,000
Capital : Port-of-Spain
Form of government : Republic
Religions : RC, Hinduism, Anglicanism, Sunni Islam
Currency : Trinidad and Tobago dollar

Tripoli (Trablous, Tarabulus) (1) the capital and main port of LIBYA, in the north-west. (Pop. 990,000). **(2)** a port in northern LEBANON. (Pop. 500,000)

Tripura a tiny state in north-eastern INDIA. The capital is AGARTALA. (10,491 sq km/4051 sq miles; pop. 2,744,800)

Tristan da Cunha a group of four remote, volcanic islands in the middle of the South ATLANTIC OCEAN, which form part of the ST HELENA DEPENDENCIES. (110 sq km/42 sq miles; pop. 299)

Trivandrum a port on the southern tip of INDIA, and the state capital of KERALA. (Pop. 826,000)

Trujillo a city and provincial capital in north-west PERU. (Pop.509,000)

Tsana, Lake *see* **Tana, Lake**.

Tsushima Strait *see* **Korea Strait**.

Tucson a city in southern ARIZONA, USA. (Pop. city 415,000)

Tulsa a city in north-eastern OKLAHOMA, on the ARKANSAS River. (Pop. 375,300)

Tunis the capital and main port of TUNISIA. (Pop. 1,395,000)

Tunisia is a North African country which lies on the south coast of the MEDITERRANEAN SEA. It's bounded by ALGERIA to the west and LIBYA to the south. Northern Tunisia consists of hills, plains and valleys. Inland mountains separate the coastal zone from the central plains before the land drops down to an area of salt pans and the SAHARA DESERT. Climate ranges from warm temperate in the north, to desert in the south. 40% of the population are engaged in agriculture, producing wheat, barley, olives, tomatoes, dates and citrus fruits. The mainstay of Tunisia's modern economy, however, is oil from the Sahara, phosphates, and tourism on the Mediterranean coast.

Quick facts:
Area : 163,610 sq km (63,170 sq miles)
Population : 8,906,000
Capital : Tunis
Other major cities : Sfax, Bizerta, Djerba
Form of government : Republic
Religion : Sunni Islam
Currency : Tunisian dinar

Turin (Torino) a major industrial town on the River PO, and the capital of the PIEDMONT region, in north-west ITALY. (Pop. 953,000)

Turkey with land on the continents of EUROPE and ASIA, Turkey forms a bridge between the two. It guards the sea passage between the MEDITERRANEAN and the BLACK SEA. Only 5% of its area, Thrace, is in Europe and the much larger area, known as ANATOLIA, is in Asia. European Turkey is fertile agricultural land with a Mediterranean climate. Asiatic Turkey is bordered to the north by the Pontine Mountains and to the south by the Taurus Mountains. The climate here ranges from Mediterranean to hot summers and bitterly cold winters in the central plains. Agriculture employs over half the workforce. Major crops are wheat, rice, tobacco and cotton. Manufacturing industry includes iron and steel, textiles, motor vehicles and Turkey's famous carpets. Hydroelectric power is supplied by the TIGRIS and EUPHRATES. Tourism is a fast-developing industry.

Quick facts:
Area : 779,452 sq km (300,946 sq miles)
Population : 61,303,000
Capital : Ankara
Other major cities : Istanbul, Izmir, Adana, Bursa
Form of government : Republic
Religion : Sunni Islam
Currency : Turkish lira

Turkmenistan, a central Asian republic of the former USSR, declared itself a republic in 1991. It lies to the east of the CASPIAN SEA and borders IRAN and AFGHANISTAN to the south. Much of the west and central areas of Turkmenistan are covered by the sandy KARA KUM Desert. The east is a plateau, which is bordered by the AMUDAR'YA river. The climate is extremely dry, and most of the population live in oasis settlements near the rivers. Agriculture is intensive around the settlements and consists of growing cereals, fruit, cotton and rearing Karakul sheep. There are rich mineral deposits, especially natural gas. Silk, oil and sulphur are also produced.

Quick facts:
Area : 488,100 sq km (186,400 sq miles)
Population : 4,100,000
Capital : Ashkhabad
Form of government : Republic
Religion : Sunni Islam
Currency : Rouble

Turks and Caicos Islands a British colony in the north-eastern WEST INDIES consisting of some 14 main islands. The capital is Cockburn Town on Grand Turk. (430 sq km/166 sq miles; pop. 15,000)

Turku (Åbo) a port in south-west FINLAND, on the Gulf of BOTHNIA. (Pop. 162,000)

Tuscany (Toscana) a region of central western ITALY. The capital is FLORENCE. (Pop. 3,600,000)

Tuvalu is located just north of FIJI, in the South PA-
CIFIC, and consists of nine coral atolls. The group
was formerly known as the Ellice Islands, and the
main island and capital is Funafuti. The climate is
tropical with temperatures averaging 30°C and annual rainfall ranges from 3000–4000 mm (118–
157 inches). Coconut palms are the main crop and
fruit and vegetables are grown for local consumption. Sea fishing is extremely good and largely
unexploited, although licenses have been granted
to JAPAN, TAIWAN and South KOREA to fish the local
waters. Most export revenue comes from the sale
of elaborate postage stamps to philatelists.

> *Quick facts:*
> Area : 24 sq km (10 sq miles)
> Population : 10,000
> Capital : Funafuti (or Fongafale)
> Form of government : Constitutional
>   Monarchy
> Religion : Protestantism
> Currency : Australian dollar

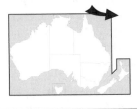

Tver an industrial city on the navigable part of the
River VOLGA, RUSSIAN FEDERATION, 160 km (100
miles) north-west of MOSCOW. Founded in 1181, it
was renamed Kalinin in 1932 after Mikhail
Kalinin, the Soviet president 1937–46. Its name
reverted to Tver in 1991. (Pop. 460,000)

Tyne and Wear a metropolitan county in north-east
ENGLAND, created in 1974 out of parts of Durham
and Northumberland. The administrative centre is
SUNDERLAND. (540 sq km/208 sq miles; pop.
1,134,000)

Tyrol a province of western AUSTRIA, in the ALPS.
The capital is Innsbruck. (Pop. 619,600)

Tyrone a county in the west of NORTHERN IRELAND.
The county town is Omagh. (3266 sq km/1260 sq
miles; pop. 160,000)

Tyrrhenian Sea a part of the MEDITERRANEAN SEA
between SICILY, SARDINIA, and mainland ITALY.

UAE *see* **United Arab Emirates**.

Udaipur a historic city in southern RAJASTHAN, IN-
DIA. (Pop. 308,000)

Ufa an industrial city and capital of BASHKIRIA, in
the RUSSIAN FEDERATION. (Pop. 1,100,000)

Uganda is a landlocked country in east central AF-
RICA. The Equator runs through the south of the

country, and for the most part it is a richly fertile
land, well watered, with a kindly climate. In the
west are the RUWENZORI Mountains, which rise to
over 5000 m (16,405 ft) and are snow-capped. The
lowlands around Lake VICTORIA, once forested,
have now mostly been cleared for cultivation. Agriculture employs over three quarters of the labour
force, and the main crops grown for subsistence
are plantains, cassava and sweet potatoes. Coffee
is the main cash crop and accounts for 90% of the
county's exports. Attempts are being made to expand the tea plantations in the west, to develop a
copper mine and to introduce new industries to
KAMPALA, the capital.

> *Quick facts:*
> Area : 235,880 sq km (91,073 sq miles)
> Population : 21,466,000
> Capital : Kampala
> Other major cities : Jinja, Masaka, Mbale
> Form of government: Republic
> Religions : RC, Protestantism, Animism,
>   Sunni Islam
> Currency : Uganda shilling

Ujung Padang a major port in the south-west of
SULAWESI, INDONESIA. It was formerly known as
Makassar. (Pop. 913,000)

Ujvidek *see* **Novi Sad**.

UK *see* **United Kingdom**.

Ukraine, formerly a Soviet socialist republic, declared itself independent of the former USSR in
1991. Its neighbours to the west are POLAND,
SLOVAKIA, HUNGARY and ROMANIA, and it is
bounded to the south by the BLACK SEA. To the east
lies the RUSSIAN FEDERATION and to the north the
republic of BELARUS. Drained by the DNEPR,
DNESTR, Southern BUG and Donets rivers, Ukraine
consists largely of fertile steppes. The climate is
continental, although this is greatly modified by
the proximity of the Black Sea. The Ukrainian
steppe is one of the chief wheat-producing regions
of Europe. Other major crops include corn, sugar
beet, flax, tobacco, soya, hops and potatoes. There
are rich reserves of coal and raw materials for industry. The central and eastern regions form one of
the world's densest industrial concentrations.

Manufacturing industries include ferrous metal-lurgy, machine building, chemicals, food process-ing, gas and oil refining.

> **Quick facts:**
> Area : 603,700 sq km (233,100 sq miles)
> Population : 52,027,000
> Capital : Kiev
> Other major cities : Dnepropetrovsk,
>   Donetsk, Kharkov, Odessa
> Form of government : Republic
> Religions : Russian Orthodox, RC
> Currency : Rouble
>
>

Ulan Bator (Ulaanbaatar) the capital of MONGO-LIA, in the central north of the country. (Pop. 601,000)

Ulan-Ude the capital of the BURYAT REPUBLIC, in the RUSSIAN FEDERATION. (Pop. 384,000).

Ulster one of the four ancient provinces into which IRELAND was divided, covering the north. It is often used to refer to NORTHERN IRELAND, but three coun-ties of Ulster are in the Republic of Ireland (DON-EGAL, MONAGHAN and CAVAN).

Ulyanovsk *see* **Simbirska**.

Umbria a land-locked region of central, eastern ITALY, bordering TUSCANY. (Pop. 822,700).

Umm al Qaywayn one of the seven UNITED ARAB EMIRATES, on The GULF. Its capital is also called Umm al Qaywaym. (750 sq km/290 sq miles; pop. emirate 29,000)

United Arab Emirates (UAE) the United Arab Emirates is a federation of seven oil-rich sheikdoms located in The GULF. As well as its main coast on the Gulf, the country has a short coast on the Gulf of OMAN. The land is mainly flat sandy desert ex-cept to the north on the peninsula where the Hajar Mountains rise to 2081 m (6828 ft). The summers are hot and humid with temperatures reaching 49°C, but from October to May the weather is warm and sunny with pleasant, cool evenings. The only fertile areas are the emirate of RAS AL KHAYMAH, the coastal plain of AL FUJAYRAH and the oases. ABU DHABI and DUBAI are the main industrial cen-tres and, using their wealth from the oil industry, they are now diversifying industry by building alu-minium smelters, cement factories and steel-rolling mills. Dubai is the richest state in the world.

> **Quick facts:**
> Area : 83,600 sq km (32,278 sq miles)
> Population : 2,800,000
> Capital : Abu Dhabi
> Other major cities : Dubai, Sharjh,
>   Ras al Khaymah
> Form of government: Monarchy (emirates)
> Religion : Sunni Islam
> Currency : Dirham
>
>

United Kingdom (UK) situated in north-west EU-ROPE, the United Kingdom of GREAT BRITAIN and NORTHERN IRELAND, comprises the island of Great Britain and the north-east of Ireland, plus many smaller islands, especially off the west coast of SCOTLAND. The south and east of Britain is low-ly-ing, and the PENNINES form a backbone running through northern ENGLAND. Scotland has the larg-est area of upland, and WALES is a highland block. Northern Ireland has a few hilly areas. The climate is cool temperate with mild conditions and an even annual rainfall. The principal crops are wheat, bar-ley, sugar beet, fodder crops and potatoes. Live-stock includes cattle, sheep, pigs and poultry. Fishing is important off the east coast. The UK is primarily an industrial country, although the re-cent recession has left high unemployment and led to the decline of some of the older industries, such as coal, textiles and heavy engineering. A growing industry is electronics, much of it defence-related.

> **Quick facts:**
> Area : 244,880 sq km (94,548 sq miles)
> Population : 58,306,000
> Capital : London
> Other major cities : Birmingham,
>   Manchester, Glasgow, Liverpool
> Form of government : Constitutional
>   Monarchy
> Religion : Anglicanism, RC,
>   Presbyterianism, Methodism
> Currency : Pound sterling
>
>

United States of America (USA) stretches across central north AMERICA, from the ATLANTIC OCEAN in the east to the PACIFIC OCEAN in the west, and from CANADA in the north to MEXICO and the Gulf of Mexico in the south. It consists of fifty states, including outlying ALASKA, north-west of Canada, and HAWAII in the Pacific Ocean. The climate varies a great deal in such a large country. In Alaska there are polar conditions, and in the Gulf coast and in Florida conditions may be subtropical. Although agricultural production is high, it employs only 1.5% of the population because primarily of its advanced technology. The USA is a world leader in oil production. The main industries are iron and steel, chemicals, motor vehicles, aircraft, telecommunications equipment, computers, electronics and textiles. The USA is the richest and most powerful nation in the world.

Quick facts:
Area : 9,809,431 sq km (3,787,421 sq miles)
Population : 263,563,000
Capital : Washington D.C.
Other major cities : New York, Chicago,
  Detroit, Houston, Los Angeles, Philadelphia,
  San Diego, San Francisco
Form of government : Federal Republic
Religion : Protestantism, RC, Judaism,
  Eastern Orthodox
Currency : US dollar

Upper Volta *see* **Burkina**.

Uppsala an old university town in eastern central SWEDEN. (Pop. 181,000)

Ural Mountains (Urals, Uralskiy Khrebet) a mountain range in western RUSSIAN FEDERATION. Running north to south from the ARCTIC to the ARAL SEA, the Urals form the traditional dividing line between EUROPE and ASIA. The highest point is Mount Narodnaya (1894 m/6214 ft).

Uruguay is one of the smallest countries in South AMERICA. It lies on the east coast of the continent, to the south of BRAZIL, and is bordered to the west by the Uruguay river, Rio de la Plata to the south, and the ATLANTIC OCEAN to the east. The country consists of low plains and plateaux. In the south-

east, hills rise to 500 m (1641 ft). About 90% of the land is suitable for agriculture but only 10% is cultivated, the remainder being used to graze vast herds of cattle and sheep. The cultivated land is made up of vineyards, rice fields and groves of olives and citrus fruits. Uruguay has only one major city in which half the population live. The country has no mineral resources, oil or gas, but has built hydroelectric power stations at Palmar and Salto Grande.

Quick facts:
Area : 177,414 sq km (68,500 sq miles)
Population : 3,186,000
Capital : Montevideo
Form of government : Republic
Religions : RC, Protestantism
Currency : Uruguayan nuevo peso

Ürümqi (Urumchi) the capital of the XINJIANG AUTONOMOUS REGION of north-west CHINA. (Pop. 1,130,000)

USA *see* **United States of America**.

US Virgin Islands *see* **Virgin Islands, US**.

Utah a state in the west of the USA. The state capital is SALT LAKE CITY. (212,628 sq km/82,096 sq miles; pop. 1,811,200)

Utrecht a historic city in the central NETHERLANDS. (Pop. city 231,000/Greater Utrecht 543,000)

Uttar Pradesh the most populous state of INDIA, in the north of the country. The state capital is LUCKNOW. (294,400 sq km/113,638 sq miles; pop. 138,760,400)

Uzbekistan, a central Asian republic of the former USSR, declared itself independent in 1991. It lies between KAZAKHSTAN and TURKMENISTAN and encompasses the southern half of the ARAL SEA. The republic has many contrasting regions. The TIAN SHAN region is mountainous, the Fergana region is irrigated and fertile, the KYZLKUM Desert is rich in oil and gas, the lower Amudar'ya river region is irrigated and has oasis settlements, and the Usturt Plateau is a stony desert. Uzbekistan is one of the world's leading cotton producers, and Karakul lambs are reared for wool and meat. Its main industrial products are agricultural machinery, textiles and chemicals. It also has significant reserves of natural gas. Economic growth has been checked by concerns about political instability and much of the economy remains based on the centralized state-owned model.

> *Quick facts:*
> Area : 447,400 sq km (172,741 sq miles)
> Population : 22,833,000
> Capital : Tashkent
> Other major city : Samarkand
> Form of government : Republic
> Religion : Sunni Islam
> Currency : Rouble

Vadodara (Baroda) an industrial city in south-east GUJARAT, INDIA. (Pop. 1,115,000)

Valencia (1) a port on the MEDITERRANEAN coast of SPAIN, and the capital of the province of the same name. (Pop. city 764,000). **(2)** An industrial city in northern VENEZUELA. (Pop. 1,032,000)

Vale of Glamorgan a county in south Wales created in the 1996 council reorganization from parts of South Glamorgan and part of the disctrict of Cardiff. (340 sq km/134 sq miles; pop. 119,000)

Valladolid an industrial city in north-west SPAIN. (Pop. city 337,000)

Valle d'Aosta a French-speaking region of northwest ITALY. The capital is Aosta. (Pop. 116,000)

Valletta the capital of MALTA. (Pop. 102,000)

Valparaíso the main port of CHILE, in the centre of the country. (Pop. 296,000)

Van, Lake a salt lake in eastern TURKEY. (3675 sq km/1419 sq miles)

Vancouver a major port and industrial centre in south-east BRITISH COLUMBIA, CANADA, on the mainland opposite VANCOUVER ISLAND with access to the Pacific Ocean. (Pop. 1,603,000)

Vancouver Island the largest island off the PACIFIC coast of North AMERICA, in south-west CANADA. The capital is VICTORIA. (32,137 sq km/12,408 sq miles)

Vanuatu, formerly known as the New Hebrides, is located in the western PACIFIC, south-east of the SOLOMON ISLANDS and about 1750 km (1087 miles) east of AUSTRALIA. About eighty islands make up the group. Some of the islands are mountainous and include active volcanoes. The largest islands are Espírtu Santo, Malekula and Efate, on which

the capital Vila is sited. Vanuatu has a tropical climate which is moderated by the south-east trade winds from May to October. The majority of the labour force are engaged in subsistence farming, and the main exports include copra, fish and cocoa. Tourism is becoming an important industry.

> *Quick facts:*
> Area : 12,189 sq km (4706 sq miles)
> Population : 167,000
> Capital : Vila
> Form of government : Republic
> Religion : Protestantism, Animism
> Currency : Vatu

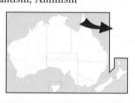

Varanasi (Benares) a holy Hindu city on the banks of the River GANGES in UTTAR PRADESH, north-eastern INDIA. (Pop.1,026,000)

Vatican City State lies in the heart of ROME on a low hill on the west bank of the river TIBER. It is the world's smallest independent state and headquarters of the Roman Catholic Church. It is a walled city made up of the Vatican Palace, the Papal Gardens, St Peter's Square and St Peter's Basilica. The state has its own police, newspaper, coinage, stamps and radio station. The radio station, 'Radio Vaticana,' broadcasts a service in thirty-four languages from transmitters within the Vatican City. Its main tourist attractions are the frescoes of the Sistine Chapel, painted by Michelangelo Buonarroti (1475–1564). The Pope exercises sovereignty and has absolute legislative, executive and judicial powers.

> *Quick facts:*
> Area : 0.44 sq km (0.17 sq miles)
> Population : 1000
> Capital : Vatican City (Citta del Vaticano)
> Form of government : Papal Commission
> Religion : RC
> Currency : Vatican City lira

Veglia *see* **Krk**.

Veneto a region of north-eastern ITALY, centring upon VENICE. (Pop. 4,398,100)

Venezia *see* **Venice**.

Venezuela forms the northernmost crest of South America. Its northern coast lies along the CARIBBEAN SEA and it is bounded to the west by COLUMBIA and to the south-east and south by GUYANA and BRAZIL. In the north-west a spur of the ANDES runs south-west to north-east. The river ORINOCO cuts the country in two, and north of the river run the undulating plains known as the Llanos. South of the river are the Guiana Highlands. The climate ranges from warm temperate to tropical. Temperatures vary little throughout the year and rainfall is plentiful. In the Llanos area cattle are herded across the plains, and this region makes the country almost self-sufficient in meat. Sugar cane and coffee are grown for export but petroleum and gas account for 95% of export earnings. The oil fields lie in the north-west near LAKE MARACAIBO, where there are over 10,000 oil derricks.

Quick facts:
Area : 912,050 sq km (352,143 sq miles)
Population : 21,810,000
Capital : Caracas
Other major cities : Maracaibo, Valencia, Barquisimeto
Form of government : Federal Republic
Religion : RC
Currency : Bolívar

Venice (Venezia) a historic port built on islands at the head of the ADRIATIC SEA in north-eastern ITALY. The principal thoroughfares are canals. (Pop. 306,000)

Vermont a state in the north-east of the USA, bordering CANADA. The state capital is MONTPELIER. (24,887 sq km/9609 sq miles; pop. 571,300)

Verona a historic and industrial city in VENETO, northern ITALY. (Pop. 255,000)

Versailles a town just to the west of PARIS, FRANCE, which grew up around the palace built there by Louis XIV in the 1660s. (Pop. 91,000)

Vesuvius an active volcano to the south-east of NAPLES, in south-west ITALY, notorious for having buried POMPEII in ash during an eruption in AD79. (1281 m/4203 ft)

Viangchan *see* **Vientiane**.

Victoria (1) a state in south-eastern AUSTRALIA. The state capital is MELBOURNE. (227,620 sq km/

87,884 sq miles; pop. 4,184,000). **(2)** a port on the south-eastern coast of VANCOUVER ISLAND, south-west CANADA, and the capital of BRITISH COLUMBIA. (Pop. 288,000). **(3)** former port and capital of HONG KONG, in the north-west of Hong Kong Island. (4) the capital of the SEYCHELLES, on the island of Mahé. (Pop. 25,000)

Victoria Falls one of the greatest waterfalls in the world, where the River ZAMBEZI tumbles some 108 m (355 ft), on the border between ZAMBIA and ZIMBABWE.

Victoria, Lake the largest lake in AFRICA, and the second largest freshwater lake in the world after Lake SUPERIOR. Its shoreline is shared by UGANDA, KENYA and TANZANIA. (69,485 sq km/26,828 sq miles)

Vienna (Wien) the capital of AUSTRIA, on the River DANUBE, in the north-east of the country. (Pop. 1,560,000)

Vientiane (Viangchan) the capital of LAOS, on the River MEKONG in the north-east of the country, near the border with THAILAND. (Pop. 449,000)

Vietnam is a long narrow country in south-east ASIA which runs down the coast of the SOUTH CHINA SEA. It has a narrow central area which links broader plains centred on the RED and MEKONG rivers. The narrow zone, now known as MIEN TRUNG, is hilly and makes communications between north and south difficult. The climate is humid with tropical conditions in the south and subtropical in the north. The far north can be very cold when polar air blows over Asia. Agriculture employs over three quarters of the labour force. The main crop is rice but cassava, maize and sweet potatoes are also grown for domestic consumption. Rubber, tea and coffee are grown for export. Major industries are food processing, textiles, cement, cotton and silk manufacture. Vietnam, however, remains underdeveloped and is still recovering from the ravages of many wars this century.

Quick facts:
Area : 331,689 sq km (128,065 sq miles)
Population : 74,580,000
Capital : Hanoi
Other major cities : Ho Chi Minh City, Haiphong
Form of government : Socialist Republic
Religion : Buddhism, Taoism, RC
Currency : Dong

Vilnius the capital of LITHUANIA. (Pop. 576,000)

Virginia a state in the east of the USA, with a coast on the ATLANTIC OCEAN. The capital is RICHMOND (105,600 sq km/40,762 sq miles; pop. 6,394,500).

Virgin Islands, British a British Crown colony in the eastern CARIBBEAN, to the east of PUERTO RICO. The British islands are in the east of the Virgin Island group. Sixteen of the islands are inhabited, including Virgin Gorda and TORTOLA, the site of the capital, ROAD TOWN. (153 sq km/59 sq miles; pop. 20,000)

Virgin Islands, US a territory of the USA in the eastern CARIBBEAN, to the east of PUERTO RICO. The US islands are in the west and south of the Virgin Island group. The main islands are ST JOHN, ST CROIX and ST THOMAS, the site of the capital, CHARLOTTE AMALIE. (344 sq km/133 sq miles; pop. 105,000)

Visayan Islands a group of islands in the centre of the PHILIPPINES, which includes NEGROS, CEBU, LEYTE, BOHOL, Panay and SAMAR.

Vistula, River a river of central and northern PO-LAND, flowing northwards through CRACOW and WARSAW to the BALTIC SEA. (Length 1090 km/677 miles)

Vlaanderen *see* **Flanders**.

Vladivostok a major port on the Pacific coast in the far east of the RUSSIAN FEDERATION, 50 km (30 miles) from the border with CHINA. (Pop. 675,000)

Vlissingen *see* **Flushing**.

Vojvodina an autonomous province in the north of SERBIA. The capital is NOVI SAD. (21,506 sq km/8301 sq miles; pop. 2,050,000)

Volga, River a largely navigable river of western RUSSIAN FEDERATION, flowing south from its source, to the north-east of MOSCOW, to the CAS-PIAN SEA. It is the longest river in EUROPE. (Length 3690 km/2293 miles)

Volgograd a port and major industrial city on the River VOLGA. It was called Stalingrad from 1925 to 1961. (Pop. 1,031,000)

Volta, Lake a major artificial lake that occupies much of eastern GHANA, formed by the damming of the Volta River. (8480 sq km/3251 sq miles)

Volta, River a river in GHANA, fed by the Black Volta and the White Volta, which flows south to the Bight of BENIN. It was dammed to form Lake VOLTA. (Length, including lake, 480 km/298 miles)

Voronezh an industrial city 450 km (280 miles) south of MOSCOW, RUSSIAN FEDERATION. (Pop. 958,000)

Voroshilovgrad *see* **Lugansk**.

Wahran *see* **Oran**.

Waikato, River the longest river in NEW ZEALAND, flowing north-west from the centre of North Island to the TASMAN SEA. (Length 350 km/220 miles)

Wales a principality in the south-west of GREAT BRITAIN, forming a part of the UK. CARDIFF is the capital. (20,768 sq km/8017 sq miles; pop. 2,913,000)

Wallis and Futuna Islands three small islands forming an overseas territory of FRANCE in the south-west PACIFIC, to the north-east of FIJI. The capital is Mata-Utu. (367 sq km/143 sq miles; pop. 13,700)

Warsaw (Warszawa) the capital of POLAND, on the River VISTULA, in the eastern central part of the country. (Pop. 1,655,000)

Warwickshire a county of central ENGLAND. The county town is Warwick. (1981 sq km/765 sq miles; pop. 496,000)

Wash, The a shallow inlet formed by the NORTH SEA in the coast of East Anglia, between the counties of LINCOLNSHIRE and NORFOLK.

Washington a state in the north-west of the USA, on the border with CANADA, and with a coast on the PACIFIC. The capital is OLYMPIA. (172,416 sq km/66,570 sq miles; pop. 4,409,000)

Washington D.C. the capital of the USA, on the Potomac River. It stands in its own territory, called the District of Columbia (D.C.), between the states of VIRGINIA and MARYLAND, close to the ATLANTIC coast. (179 sq km/69 sq miles; pop. city 585,221/metropolitan area 4,360,000)

Waterford a county in the south of the Republic of IRELAND. The county town is also called Waterford. (1838 sq km/710 sq miles; pop. county 91,600)

Weimar a historic city in southern central GER-MANY. (Pop. 80,000)

Wellington the capital of NEW ZEALAND and a port in the south-west of North Island. (Pop. 326,000)

Weser, River a river in the north-west of GERMANY, flowing through BREMEN and BREMERHAVEN to the NORTH SEA. (477 sq km/196 sq miles)

West Bank a piece of disputed territory to the west of the River JORDAN, including a part of JERUSA-LEM, which was taken by ISRAEL from Jordan in the Arab-Israeli war of 1967, and has been occupied by Israel since then. New Israeli settlements here have incited growing resentment among the Palestinian population. (5858 sq km/2262 sq miles)

West Bengal a state in eastern INDIA, bordering BANGLADESH. CALCUTTA is the capital. (87,900 sq km/33,929 sq miles; pop. 67,982,700)

Western Australia a state occupying much of the western half of AUSTRALIA. The capital is PERTH.

(2,525,500 sq km/974,843 sq miles; pop. 1,478,000)

Western Isles the regional island authority covering the Outer HEBRIDES of western SCOTLAND. The administrative centre is Stornaway, on the Isle of Lewis. (2900 sq km/1120 sq miles; pop. 29,300)

Western Sahara a disputed territory of western AFRICA, with a coastline on the ATLANTIC OCEAN. Consisting mainly of desert, it is rich in phosphates. It was a Spanish overseas province until 1975 and is now claimed by MOROCCO, against the wishes of an active separatist movement. The main town is Laâyoune (El Aaiún). (266,770 sq km/103,000 sq miles; pop. 220,000)

Western Samoa lies in the Polynesian sector of the PACIFIC OCEAN, about 720 km (447 miles) northeast of FIJI. It consists of seven small islands and two larger volcanic islands, Savai'i and Upolu. Savai'i is largely covered with volcanic peaks and lava plateaux. Upolu is home to two-thirds of the population and the capital APIA. The climate is tropical with high temperatures and very heavy rainfall. The islands have been fought over by the Dutch, British, Germans and Americans, but they now have the lifestyle of traditional Polynesians. Subsistence agriculture is the main activity, and copra, cocoa and bananas are the main exports. Many tourists visit the grave of the Scottish writer Robert Louis Stevenson (1850–94) who died here and whose home is now the official home of the king.

Quick facts:
Area : 2831 sq km (1093 sq miles)
Population : 169,000
Capital : Apia
Form of government : Constitutional Monarchy
Religion : Protestantism
Currency : Tala

West Glamorgan a former county in South WALES, created in 1974 from part of Glamorgan and the borough of SWANSEA. In 1996 it was broken up into the county boroughs of SWANSEA, Neath-Port Talbot and Bridgend. (817 sq km/315 sq miles; pop. 368,000)

West Indies a general term for the islands of the CARIBBEAN SEA.

Westmeath a county in the central north of the Republic of IRELAND. The county town is Mullingar. (1764 sq km/681 sq miles; pop. 61,900)

West Midlands a metropolitan county of central ENGLAND, created in 1974, with its administrative centre in BIRMINGHAM. (899 sq km/347 sq miles; pop. 2,627,800)

West Virginia a state of eastern USA. The capital is CHARLESTON. (62,341 sq km/24,070 sq miles; pop. 1,808,900)

Wexford a county in the south-east of the Republic of IRELAND. The county town is also called Wexford. (2352 sq km/908 sq miles; pop. county 102,293)

White Sea an arm of the BARENTS SEA off the northwest of the RUSSIAN FEDERATION, which is almost enclosed by the bulge of the KOLA peninsula.

Whitney, Mount a mountain in the Sequoia National Park in eastern CALIFORNIA, with the highest peak in the USA outside ALASKA. (4418 m/14,495 ft)

Wichita a city in southern KANSAS, USA, on the ARKANSAS River. (Pop. city 311,700)

Wicklow a county in the south-west of the Republic of IRELAND. The county town is also called Wicklow. (2025 sq km/782 sq miles; pop. county 97,300)

Wien *see* **Vienna**.

Wiesbaden an old spa town in western GERMANY, and capital of the state of HESSEN. (Pop. 271,000)

Wight, Isle of an island and county off the south coast of ENGLAND, separated from the mainland by the SOLENT. The county town is Newport. (380 sq km/147 sq miles; pop. 124,700)

Wiltshire a county in central southern ENGLAND. The county town is Trowbridge. (3476 sq km/1342 sq miles; pop. 587,500)

Winchester a historic city in southern ENGLAND, and the county town of HAMPSHIRE. (Pop. 101,800)

Windward Islands *see* **Leeward and Windward Islands**.

Winnipeg the capital of MANITOBA, CANADA, in the south of the state. (Pop. 652,000)

Winnipeg, Lake a lake in the south of MANITOBA, CANADA, which drains into HUDSON BAY via the Nelson River. (23,553 sq km/9094 sq miles)

Wisconsin a state in the north central USA, bordering Lake SUPERIOR and Lake MICHIGAN. The state capital is MADISON. (145,500 sq km/56,163 sq miles; pop. 4,992,700)

Witwatersrand (The Rand) a major gold-mining and industrial area of Pretoria-Witwatersrand-Vereeniging, SOUTH AFRICA.

Wollongong a major port and industrial centre in

NEW SOUTH WALES, AUSTRALIA, 80 km (50 miles) south of SYDNEY. (Pop. 238,000)

Wolverhampton an old industrial town in the WEST MIDLANDS of ENGLAND. (Pop. 245,100)

Worcestershire *see* **Hereford and Worcester**.

Wroclaw (Breslau) an industrial city on the River ODER in south-west POLAND. (Pop. 643,000)

Wuhan (Hankow) the capital of HUBEI province, south-east CHINA. (Pop. 3,870,000)

Wyoming a state in the west of the USA. The state capital is CHEYENNE. (253,597 sq km/97,914 sq miles; pop. 464,700)

Xiamen a port on the east coast of CHINA, in FUJIAN province. (Pop. 470,000)

Xi'an the capital of SHAANXI province, CHINA, and an industrial centre and former capital of China. (Pop. 2,410,000)

Xi Jiang (Si Kiang) the third longest river in CHINA, flowing across the south-west of the country from YUNNAN to its delta on the SOUTH CHINA SEA near GUANGZHOU (Canton). (Length 2300 km/1437 miles)

Xining an industrial city and the capital of QINGHAI province, in western CHINA. (Pop. 552,000)

Xinjiang (Sinkiang) Uygur Autonomous Region a region of north-west CHINA, bordering MONGOLIA, the RUSSIAN FEDERATION, AFGHANISTAN and INDIA. It is also known as Dzungaria. The capital is ÜRÜMQI. (1,646,799 sq km/635,829 sq miles; pop. 15,156,000)

Xizang Autonomous Region *see* **Tibet**.

Yamoussoukro the new capital of CÔTE D'IVOIRE, in the centre of the country. (Pop. 126,000)

Yamuna (Jumna), River a major river of north INDIA, a tributary of the GANGES. (Length 1376 km/855 miles)

Yangon *see* **Rangoon**.

Yangtze Kiang *see* **Chang Jiang**.

Yaoundé the capital of CAMEROON, in the south-west of the country. (Pop. 750,000)

Yaren the capital of NAURU. (Pop. district 8000)

Yekaterinburg an industrial city to the east of the URAL MOUNTAINS, RUSSIAN FEDERATION. It was called Ekaterinburg before 1924 and Sverdlosk from 1924 to 1992. (Pop. 1,413,000)

Yellowknife a city on the GREAT SLAVE LAKE, CANADA, and capital of the NORTHWEST TERRITORIES. (Pop. 11,700)

Yellow River *see* **Huang He**.

Yellow Sea a branch of the PACIFIC OCEAN between the north-east coast of CHINA and the peninsula of KOREA.

Yemen is bounded by SAUDI ARABIA in the north, OMAN in the east, the Gulf of ADEN in the south, and the RED SEA in the west. The country was formed after the unification of the previous Yemen Arab Republic and the People's Democratic Republic of Yemen (South Yemen) in 1989. However, at this point there was no active integration of the two countries and politocally the country remained divided between north and south. In 1994 a civil war, which lasted three months, broke out between the former North and South Yemen. Most of the country comprises rugged mountains and trackless desert lands. The country is almost entirely dependent on agriculture even though a very small percentage is fertile. The main crops are coffee, cotton, millet, sorghum and fruit. Fishing is an important industry. Other industry is on a very small scale. There are textile factories, and plastic, rubber and aluminium goods, paints and matches are produced. Modernization of industry is slow because of lack of funds.

Quick facts:
Area : 527,970 sq km (203,849 sq miles)
Population : 14,609,000
Capital : Sana'a, Commercial Capital : Aden
Form of government : Republic
Religion : Zaidism, Shia Islam, Sunni Islam
Currency : Riyal and dinar

Yerevan an industrial city and the capital of ARMENIA, close to the border with TURKEY. (Pop. 1,254,000)

Yinchuan the capital of NINGXIAHUI AUTONOMOUS REGION, in north central CHINA. (Pop. 430,000)

Yogyakarta (Jogjakarta) a city of south central JAVA, and a cultural centre. (Pop. 399,000)

Yokohama the main port of JAPAN, and the country's second largest city after neighbouring TOKYO, on the south-east coast of HONSHU Island. (Pop. 3,288,000)

Yorkshire an old county of north-east ENGLAND which used to be divided into the East, West and North Ridings. In 1974, however, the county was redivided into North Yorkshire (8310 sq km/3208 sq miles; pop. 727,000); West Yorkshire (2034 sq km/785 sq miles; pop. 2,104,000) and South Yorkshire (1560 sq km/602 sq miles; pop. 1,305,000).

Yucatán a state on a broad peninsula of south-east MEXICO. (Pop. 1,035,000)

Yugoslavia, which was created in 1918, became a single federal republic after World War II under the leadership of Marshal Tito (1892–1980). The six constituent republics were SERBIA, CROATIA, SLOVENIA, BOSNIA & HERZEGOVINA, MACEDONIA and MONTENEGRO. Yugoslavia today refers only to Serbia and Montenegro, which operate as two equal republics under a federal authority. However, the situation remains particularly complex, with each republic operating its own legislature. The other republics, beginning with Slovenia and Croatia in 1991, have all declared their independence from Yugoslavia. The economy was devastated by the wars in Bosnia and Croatia, then by inflation to the degree that the financial infrastructure all but collapsed in late 1993. The economy has only just begun to take the first steps to recovery. It is largely agricultural, and produce includes wheat, maize, grpaes and citrus fruit. Exports include chemicals, machinery, textiles and clothing.

> *Quick facts:*
> Area : 102,172 sq km (39,449 sq miles)
> Population : 10,881,000
> Capital : Belgrade (Beograd)
> Other Major Cities : Nis, Titograd
> Form of government : Federal Republic
> Religions : Eastern Orthodox
> Currency : Dinar
>
>

Yukon Territory a mountainous territory in north-west CANADA centring upon the River Yukon and including the River KLONDIKE. (483,572 sq km/ 186,631 sq miles; pop. 26,500)

Yünnan a province in south-western CHINA. The capital is KUNMING. (436,200 sq km/168,400 sq miles; pop. 36,973,000)

Zagreb the capital of CROATIA. (Pop. 931,000)

Zagros Mountains a mountain range in south-west IRAN, running parallel to the border with IRAQ. The highest point is Zard Kuh (4548 m/14,918 ft).

Zaïre situated in west central AFRICA, Zaïre is a vast country with a short coastline of only 40 km (25 miles) on the ATLANTIC OCEAN. Rain forests, which cover about 55% of the country, contain valuable hardwoods such as mahogany and ebony. The country is drained by the river Zaïre and its main tributaries. Mountain ranges and plateaux surround the Zaïre Basin, and in the east the

RUWENZORI Mountains overlook the lakes in the GREAT RIFT VALLEY. In the central region the climate is hot and wet all year but elsewhere there are well-marked wet and dry seasons. Agriculture employs 75% of the population yet less than 3% of the country can be cultivated. Grazing land is limited by the infestation of the tsetse fly. Cassava is the main subsistence crop, and coffee, tea, cocoa, rubber and palms are grown for export. Minerals, mainly copper, cobalt, zinc and diamonds, account for 60% of exports.

> *Quick facts:*
> Area : 2,344,860 km (905,350 miles)
> Population : 44,504,000
> Capital : Kinshasa
> Other major cities : Lubumbashi, Mbuji-Mayi,
>   Kananga,
> Form of government : Republic
> Religion : RC, Protestantism, Animism
> Currency : Zaïre
>
>

Zaïre (Congo), River a major river of central AFRICA (the second longest river in Africa after the NILE) and, with its tributaries, forming a massive basin. It rises as the Lualaba in the south of Zaïre, then flows north and north-west, and finally southwest, forming the border between Zaïre and the CONGO before entering the ATLANTIC OCEAN. (Length 4800 km/3000 miles)

Zambezi, River a river of southern AFRICA. It rises in ZAMBIA, then flows south to form the border with ZIMBABWE, and then south-east across MOZAMBIQUE to the INDIAN OCEAN. (Length 2740 km/ 1700 miles)

Zambia, situated in central AFRICA, is made up of high plateaux. Bordering it to the south is the ZAMBEZI river, and in the south-west it borders on the KALAHARI DESERT. It has some other large rivers, including the Luangwa, and lakes, the largest of which is Lake BANGWEULU. The climate is tropical, modified somewhat by altitude. The country has a wide range of wildlife, and there are large game parks on the Luangwa and Kafue rivers. Agriculture is underdeveloped and most foodstuffs have to be imported. Zambia's economy relies heavily on the mining of copper, lead, zinc and cobalt. The poor market prospects for copper, which will eventually be exhausted, make it imperative for Zambia to develop her vast agricultural potential.

Quick facts:
Area : 752,614 sq km (290,584 sq miles)
Population : 9,500,000
Capital : Lusaka
Other major cities : Kitwe, Ndola, Mufulira
Form of government : Republic
Religion : Christianity, Animism
Currency : Kwacha

Zanzibar an island lying just off the east coast of AFRICA, in the INDIAN OCEAN. Settled by Arab traders, it was a major commercial centre by the 17th century. It became a British protectorate in 1890, and joined neighbouring Tanganyika in 1964 to form the independent republic of TANZANIA. The main town is the port also called Zanzibar. (1658 sq km/640 sq miles; pop. 571,000)

Zaporozh'ye a major industrial city on the River DNIEPER in the UKRAINE. (Pop. 898,000)

Zaragoza (Saragossa) a historic and industrial city in north-eastern SPAIN, on the River EBRO, and the name of the surrounding province. (Pop. city 607,000)

Zealand (Sjaelland) the largest island of DENMARK, on which the capital, COPENHAGEN, is sited. (1790 sq km/691 sq miles; pop. 357,500)

Zermatt a popular ski resort in south-west SWITZERLAND, close to the MATTERHORN. (Pop. 3700)

Zhangjiakou a city in HEBEI province, in north-east CHINA. (Pop. 1,100,000)

Zhejiang a province of eastern CHINA, with a coast on the East CHINA SEA. The capital is HANGZHOU. (102,000 sq km/39,780 sq miles; pop. 41,446,000)

Zhengzhou the capital of HENAN province, in east central CHINA. (Pop. 1,690,000)

Zibo an industrial city in SHANGDONG province, north-eastern CHINA. (Pop. 2,400,000)

Zimbabwe is a landlocked country in southern AFRICA. It is a country with spectacular physical features and is teeming with wildlife. It is bordered in the north by the ZAMBEZI river, which flows over the mile-wide VICTORIA FALLS before entering Lake Kariba. In the south, the River LIMPOPO marks its border with SOUTH AFRICA. Most of the country is over 300 m (984 ft) above sea level, and a great plateau between 1200 m (3937 ft) and 1500 m (4922 ft) occupies the central area. Massive granite outcrops, called *kopjes*, also dot the landscape. The climate is tropical in the lowlands and subtropical in the higher land. About 75% of the labour force are employed in agriculture. Tobacco, sugar cane, cotton, wheat and maize are exported and form the basis of processing industries. Tourism is a major growth industry.

Quick facts:
Area : 390,760 sq km (150,872 sq miles)
Population : 11,453,000
Capital : Harare
Other major cities : Bulawayo, Mutare, Gweru
Form of government : Republic
Religion : Animism, Anglicanism, RC
Currency : Zimbabwe dollar

Zwolle a market town in the Netherlands, with brewing and distilling as its main industries.

Zululand *see* **KwaZulu-Natal**.

Zürich the largest city in SWITZERLAND, in the north-east of the country, and a major industrial and financial centre. (Pop. 840,000)

Illustration Acknowledgments

Illustrations on pages 289, 300, 292, 293, 294, 295, 296, 298, 299, 300, 301, 302, 304, 305, 306, 307, 308, 309, 310, 311, 312, 313, 314, 315, 316, 317, 318, 319, 320, 321, 322, 323, 324, 325, 327, 328, 329, 330, 331, 332, 333, 334, 335, 337, 338, 339, 340, 341, 342, 343, 343, 346, 347, 348, 349, 350, 351, 352, 353, 354, 356, 357, 358, 359, 360, 361, 362, 363, 364, 365, 366, 367, 368 and 367 © The Hulton Getty Picture Collection Limited.